AUG 0 4 1986

Ion Implantation
and Ion Beam Processing
of Materials

MATERIALS RESEARCH SOCIETY SYMPOSIA PROCEEDINGS   VOLUME 27

ISSN 0272 - 9172

Volume 1—Laser and Electron-Beam Solid Interactions and Materials Processing, J.F. Gibbons, L.D. Hess, T.W. Sigmon, 1981

Volume 2—Defects in Semiconductors, J. Narayan, T.Y. Tan, 1981

Volume 3—Nuclear and Electron Resonance Spectroscopies Applied to Materials Science, E.N. Kaufmann, G.K. Shenoy, 1981

Volume 4—Laser and Electron-Beam Interactions with Solids, B.R. Appleton, G.K. Celler, 1982

Volume 5—Grain Boundaries in Semiconductors, H.J. Leamy, G.E. Pike, C.H. Seager, 1982

Volume 6—Scientific Basis for Nuclear Waste Management, S.V. Topp, 1982

Volume 7—Metastable Materials Formation by Ion Implantation, S.T. Picraux, W.J. Choyke, 1982

Volume 8—Rapidly Solidified Amorphous and Crystalline Alloys, B.H. Kear, B.C. Giessen, M. Cohen, 1982

Volume 9—Materials Processing in the Reduced Gravity Environment of Space, G.E. Rindone, 1982

Volume 10—Thin Films and Interfaces, P.S. Ho, K.N. Tu, 1982

Volume 11—Scientific Basis for Nuclear Waste Management V, W. Lutze, 1982

Volume 12—In Situ Composites IV, F.D. Lemkey, H.E. Cline, M. McLean, 1982

Volume 13—Laser-Solid Interactions and Transient Thermal Processing of Materials, J. Narayan, W.L. Brown, R.A. Lemons, 1983

Volume 14—Defects in Semiconductors II, S. Mahajan, J.W. Corbett, 1983

Volume 15—Scientific Basis for Nuclear Waste Management VI, D.G. Brookins, 1983

Volume 16—Nuclear Radiation Detector Materials, E.E. Haller, H.W. Kraner, W.A. Higinbotham, 1983

Volume 17—Laser Diagnostics and Photochemical Processing for Semiconductor Devices, R.M. Osgood, S.R.J. Brueck, H.R. Schlossberg, 1983

Volume 18—Interfaces and Contacts, R. Ludeke, K. Rose, 1983

Volume 19—Alloy Phase Diagrams, L.H. Bennett, T.B. Massalski, B.C. Giessen, 1983

Volume 20—Intercalated Graphite, M.S. Dresselhaus, G. Dresselhaus, J.E. Fischer, M.J. Moran, 1983

Volume 21—Phase Transformations in Solids, T. Tsakalakos, 1984

Volume 22—High Pressure in Science and Technology, C. Homan, R.K. MacCrone, E. Whalley, 1984

Volume 23—Energy Beam-Solid Interactions and Transient Thermal Processing, J.C.C. Fan, N.M. Johnson, 1984

Volume 24—Defect Properties and Processing of High-Technology Nonmetallic Materials, J.H. Crawford, Jr., Y. Chen, W.A. Sibley, 1984

MATERIALS RESEARCH SOCIETY SYMPOSIA PROCEEDINGS   VOLUME 27

Volume 25—Thin Films and Interfaces II, J.E.E. Baglin, D.R. Campbell, W.K. Chu, 1984

Volume 26—Scientific Basis for Nuclear Waste Management VII, G.L. McVay, 1984

Volume 27—Ion Implantation and Ion Beam Processing of Materials, G.K. Hubler, O.W. Holland, C.R. Clayton, C.W. White, 1984

Volume 28—Rapidly Solidified Metastable Materials, B.H. Kear, B.C. Giessen, 1984

Volume 29—Laser-Controlled Chemical Processing of Surfaces, A.W. Johnson, D.J. Ehrlich, H.R. Schlossberg, 1984

Volume 30—Plasma Processing and Synthesis of Materials, J. Szekely, D. Apelian, 1984

Volume 31—Electron Microscopy of Materials, W. Krakow, D. Smith, L.W. Hobbs, 1984

Volume 32—Better Ceramics Through Chemistry, C.J. Brinker, D.R. Ulrich, D.E. Clark, 1984

Volume 33—Comparison of Thin Film Transistor and SOI Technologies, H.W. Lam, M.J. Thompson, 1984

Volume 34—Physical Metallurgy of Cast Iron, H. Fredriksson, 1985

MATERIALS RESEARCH SOCIETY SYMPOSIA PROCEEDINGS   VOLUME 27

Ion Implantation and Ion Beam Processing of Materials

Symposium held November 1983 in Boston, Massachusetts, U.S.A.

EDITORS:

G. K. Hubler
Naval Research Laboratory, Washington, D.C., U.S.A.

O. W. Holland
Oak Ridge National Laboratory, Oak Ridge, Tennessee, U.S.A.

C. R. Clayton
State University of New York at Stony Brook, Stony Brook, New York, U.S.A.

C. W. White
Oak Ridge National Laboratory, Oak Ridge, Tennessee, U.S.A.

NORTH-HOLLAND
NEW YORK • AMSTERDAM • OXFORD

This work was sponsored by the Division of Materials Sciences, U.S. Department of Energy under contract W-7405-eng-26 with Union Carbide Corporation and by the U.S. Army Research Office under Grant Number DAAG29-83-M-0368.

The view, opinions, and/or findings contained in this report are those of the author(s) and should not be construed as an official Department of the Army position, policy, or decision, unless so designated by other documentation. This work was sponsored by the U.S. Army Research Office under Grant Number DAAG29-83-M-0368.

©1984 by Elsevier Science Publishing Co., Inc.
All rights reserved.

This book has been registered with the Copyright Clearance Center, Inc. For further information, please contact the Copyright Clearance Center, Salem, Massachusetts.

Published by:

Elsevier Science Publishing Company, Inc.
52 Vanderbilt Avenue, New York, New York 10017

Sole distributors outside the United States and Canada:

Elsevier Science Publishers B.V.
P. O. Box 211, 1000 AE Amsterdam, The Netherlands

Library of Congress Cataloging in Publication Data

Ion Implantation and Ion Beam Processing of Materials Symposium (1983: Boston, Mass.)

    Ion Implantation and Ion Beam Processing of Materials Symposium.

    (Materials Research Society symposia proceedings, ISSN 0272-9172; v. 27)

    Includes indexes.
    1. Materials—Effect of radiation on—Congresses. 2. Ion implantation—Congresses. 3. Ion bombardment—Congresses. I. Hubler, G. K. II. Title. III. Series.
TA418.6.I59   1983   670   84-8167
ISBN 0-444-00869-1

Manufactured in the United States of America

vii

TABLE OF CONTENTS

Preface — xvii

Acknowledgments — xix

PART I. ION MIXING

ION MIXING PROCESSES* — 3
  Marc-A. Nicolet, T. C. Banwell, and B. M. Paine

ION-INDUCED SURFACE MODIFICATION OF ALLOYS* — 13
  H. Wiedersich

THE DEPENDENCE OF ION BEAM MIXING ON PROJECTILE MASS — 25
  R. S. Averback, L. J. Thompson, and L. E. Rehn

INVESTIGATION OF A THERMAL SPIKE MODEL FOR ION MIXING OF METALS WITH SI — 31
  U. Shreter, Frank C. T. So, B. M. Paine, and M-A. Nicolet

MODIFICATION OF μm THICK SURFACE LAYERS USING KEV ION ENERGIES — 37
  L. E. Rehn, N. Q. Lam, and H. Wiedersich

ANISOTROPIC TRANSPORT OF IMPURITIES IN ION MIXING — 43
  S. Matteson, J. A. Keenan, and R. F. Pinizzotto

ION BEAM MIXING EFFECTS IN Au-Fe AND Pt-Fe THIN BILAYERS — 49
  G. Battaglin, A. Carnera, G. Celotti, G. Della Mea, V. N. Kulkarni, S. Lo Russo, and P. Mazzoldi

ION BEAM IRRADIATION OF METAL FILMS ON SiO_2 — 55
  G. J. Clark, J.E.E. Baglin, F. M. d'Heurle, C. W. White, G. Farlow, and J. Narayan

RADIATION ENHANCED DIFFUSION IN ION-IMPLANTED GLASSES AND GLASS/METAL COUPLES — 61
  G. W. Arnold

ELECTRICAL RESISTIVITY OF MULTILAYERS DURING ION BEAM ANNEALING — 67
  J. Grilhe, J. P. Riviere, J. Delafond, C. Jaouen, and C. Templier

DOSE DEPENDENCE OF ION BEAM MIXING OF Au ON AMORPHOUS AND SINGLE CRYSTALLINE Si AND Ge — 73
  D. B. Poker, O. W. Holland, and B. R. Appleton

STUDY OF ION BOMBARDMENT-INDUCED SUBSURFACE COMPOSITIONAL MODIFICATIONS IN Ni-Cu ALLOYS AT ELEVATED TEMPERATURES BY ION SCATTERING SPECTROSCOPY — 79
  N. Q. Lam, H. A. Hoff, H. Wiedersich, and L. E. Rehn

*Invited Paper

ION MIXING IN THE Ni-Sn SYSTEM 85
  L. Calliari, L. M. Gratton, L. Guzman, C. Principi, and
  C. Tosello

BOLTZMANN APPROACH TO CASCADE MIXING 91
  Irwin Manning

ENERGY DEPENDENCE OF COMPOUND GROWTH IN Au-Al AND Cu-Al 97
BILAYER SYSTEMS DURING ION BEAM MIXING
  S. U. Campisano, Chu Techang, S. Cannavó, and E. Rimini

THE EFFECT OF ATOMIC MIXING ON THE DEPTH PROFILES OF METAL 103
MARKERS IN SILICON
  B. V. King, D. G. Tonn, I.S.T. Tsong, and J. A. Leavitt

DEPTH DEPENDENCE AND CHEMICAL EFFECTS IN ION MIXING OF Ni 109
ON SiO_2
  T. C. Banwell and M-A. Nicolet

PART II. METASTABLE AND AMORPHOUS MATERIALS

MICROSTRUCTURAL DEVELOPMENTS DURING IMPLANTATION OF METALS* 117
  D. I. Potter, M. Ahmed, and S. Lamond

AMORPHIZATION OF THIN MULTILAYER FILMS BY ION MIXING AND 127
SOLID STATE REACTION
  M. Van Rossum, U. Shreter, W. L. Johnson, and M-A. Nicolet

ION MIXING KINETICS OF THIN LAYERED FILMS IN THE Fe-Al SYSTEM 133
  J. Grilhe, J. P. Riviere, J. Delafond, and C. Jaouen

AMORPHIZATION OF GARNET BY ION IMPLANTATION 139
  A. M. Guzman, T. Yoshiie, C. L. Bauer, and M. H. Kryder

AMORPHOUS PHASE FORMATION AND RECRYSTALLIZATION IN ION- 145
IMPLANTED SILICIDES
  C. A. Hewett, I. Suni, S. S. Lau, L. S. Hung, and D. M. Scott

SECOND PHASE FORMATION IN ALUMINUM ANNEALED AFTER ION 151
IMPLANTATION WITH MOLYBDENUM
  J. Bentley, L. D. Stephenson, R. B. Benson, Jr., P. A.
  Parrish, and J. K. Hirvonen

MODULATED STRUCTURES IN ION-IMPLANTED Al-Fe SYSTEM 157
  K. V. Jata, D. Janoff, and E. A. Starke, Jr.

THE DEVELOPMENT OF THE AMORPHOUS PHASE IN NiTi DURING HEAVY 163
ION OR ELECTRON BOMBARDMENT
  J. L. Brimhall, H. E. Kissinger, and A. R. Pelton

HVEM AND INTERNAL OXIDATION STUDIES OF LITHIATED NICKEL 169
  K. Seshan, P. Baldo, and H. Wiedersich

AMORPHOUS SOFT MAGNETIC MATERIAL FORMATION BY ION MIXING 175
  P. Gerard, G. Suran, B. Blanchard, J. Devenyi, M. Dupuy,
  and P. Martin

*Invited Paper

HIGH-DENSITY CASCADE EFFECTS IN ION-IMPLANTED Ag-Au ALLOYS 181
  F. R. Vozzo

A NEW HAFNIUM-BERYLLIUM SYSTEM PRODUCED BY ION IMPLANTATION 187
AND ANNEALING TECHNIQUES
  J. C. Soares, A. A. Melo, M. F. da Silva, E. J. Alves,
  K. Freitag, and R. Vianden

PART III. SEMICONDUCTORS

ION INDUCED MORPHOLOGICAL INSTABILITIES IN Ge* 195
  B. R. Appleton

THE PRODUCTION OF POROUS STRUCTURES ON Si, Ge AND GaAs BY 205
HIGH DOSE ION IMPLANTATION
  J. S. Williams, D. J. Chivers, R. G. Elliman, S. T. Johnson,
  E. M. Lawson, I. V. Mitchell, K. G. Orrman-Rossiter, A. P.
  Pogany, and K. T. Short

THERMODYNAMICS AND KINETICS OF CRYSTALLIZATION OF AMORPHOUS 211
Si AND Ge PRODUCED BY ION IMPLANTATION
  E. P. Donovan, F. Spaepen, D. Turnbull, J. M. Poate, and
  D. C. Jacobson

PHYSICAL PROPERTIES OF TWO METASTABLE STATES OF AMORPHOUS 217
SILICON
  G. K. Hubler, C. N. Waddell, W. G. Spitzer, J. E.
  Fredrickson, and T. A. Kennedy

IMPLANTATION INDUCED FILAMENTARY STRUCTURES 223
  C. Jaussaud, B. Maillot, and M. Bruel

EPITAXIAL CRYSTALLISATION OF DOPED AMORPHOUS SILICON 229
  R. G. Elliman, S. T. Johnson, K. T. Short, and J. S. Williams

HIGH CURRENT DENSITY IMPLANTATION AND ION BEAM ANNEALING IN Si 235
  O. W. Holland and J. Narayan

ION-BEAM PROCESSING OF ION-IMPLANTED Si 241
  H. B. Dietrich, R. J. Corazzi, and W. F. Tseng

ION-BEAM-INDUCED DAMAGING AND DYNAMIC ANNEALING PROCESSES 247
IN SILICON
  K. T. Short, D. J. Chivers, R. G. Elliman, J. Liu, A. P.
  Pogany, H. K. Wagenfeld, and J. S. Williams

MEGAVOLT BORON AND ARSENIC IMPLANTATION INTO SILICON 253
  P. F. Byrne, N. W. Cheung, S. Tam, C. Hu, Y. C. Shih,
  J. Washburn, and M. Strathman

LATTICE SITE LOCATION OF GROUP VII IMPURITIES IN SILICON 259
  D. O. Boerma, P.J.M. Smulders, and T. S. Wierenga

A REVIEW OF SILICON-ON-INSULATOR FORMATION BY OXYGEN 265
ION IMPLANTATION*
  R. F. Pinizzotto

*Invited Paper

FORMATION OF OXIDE LAYERS BY HIGH DOSE IMPLANTATION INTO 275
SILICON
 S. S. Gill and I. H. Wilson

CHARACTERISATION OF DEVICE GRADE SOI STRUCTURES FORMED BY 281
IMPLANTATION OF HIGH DOSES OF OXYGEN
 P.L.F. Hemment, E. A. Maydell-Ondrusz, K. G. Stephens,
 R. P. Arrowsmith, A. C. Glaccum, J. A. Kilner, and J. B.
 Butcher

CHARACTERIZATION OF N-TYPE LAYERS FORMED IN Si BY ION 287
IMPLANTATION OF HYDROGEN
 S. R. Wilson, W. M. Paulson, W. F. Krolikowski, D. Fathy,
 J. D. Gressett, A. H. Hamdi, and F. D. McDaniel

POINT DEFECT SUPERSATURATION AND ENHANCED DIFFUSION IN SPE 293
REGROWN SILICON
 S. J. Pennycook, J. Narayan, and O. W. Holland

CHARACTERIZATION OF ION IMPLANTED SILICON BY SPECTROSCOPIC 299
ELLIPSOMETRY AND CROSS SECTION TRANSMISSION ELECTRON
MICROSCOPY
 P. J. McMarr, K. Vedam, and J. Narayan

SPATIAL CORRELATION INTERPRETATION OF EFFECTS OF As^+ 305
IMPLANTATION ON THE RAMAN SPECTRA OF GaAs
 D. E. Aspnes, K. K. Tiong, P. M. Amirtharaj, and F. H. Pollak

HIGH RESOLUTION TRANSMISSION ELECTRON MICROSCOPY STUDY OF 311
Se^+ IMPLANTED AND ANNEALED GaAs
 D. K. Sadana, T. Sands, and J. Washburn

ION IMPLANTATION IN GALLIUM INDIUM ARSENIDE 317
 M. Anjum, M. A. Shahid, S. S. Gill, B. J. Sealy, and
 J. H. Marsh

ION IMPLANTATION DAMAGE IN CdS CRYSTALS USING 323
RBS/CHANNELING AND TEM
 N. R. Parikh, D. A. Thompson, R. Burkova, and V. S. Raghunathan

COMPARISON OF HEAT-PULSE AND FURNACE ISOTHERMAL ANNEALS OF 329
Be IMPLANTED InP
 B. Molnar, G. Kelner, G. O. Ramseyer, G. H. Morrison, and
 S. C. Shatas

A CHANNELING STUDY ON Mg IMPLANTED InSb SINGLE CRYSTALS 335
 H. W. Alberts

STUDY OF NEAR SURFACE STRUCTURE AND COMPOSITION FOR HIGH 341
DOSE IMPLANTATION OF Cr^+ INTO Si
 F. Namavar, J. I. Budnick, H. C. Hayden, F. A. Otter, and
 V. Patarini

THE INFLUENCE OF IMPLANTATION CONDITIONS AND TARGET 347
ORIENTATION IN HIGH DOSE IMPLANTATION OF Al^+ INTO Si
 F. Namavar, J. I. Budnick, A. Fasihuddin, H. C. Hayden,
 D. A. Pease, F. A. Otter, and V. Patarini

*Invited Paper

EFFECTS OF DOUBLE-IMPLANT ON THE EPITAXIAL GROWTH OF AMORPHOUS SILICON L. J. Chen and C. W. Nieh	353
A COMPARATIVE STUDY OF NEAR-SURFACE EFFECTS DUE TO VERY HIGH FLUENCE H$^+$ IMPLANTATION IN SINGLE CRYSTAL FZ, CZ AND WEB Si W. J. Choyke, R. B. Irwin, J. N. McGruer, J. R. Townsend, N. J. Doyle, B. O. Hall, J. A. Spitznagel, and S. Wood	359
CHARACTERIZATION OF ION IMPLANTATION DAMAGE IN CAPLESS ANNEALED GaAs H. Kanber, M. Feng, and J. M. Whelan	365
STOICHIOMETRIC DISTURBANCE IN InP MEASURED DURING ION IMPLANTATION PROCESS D. Haberland, P. Harde, H. Nelkowski, and W. Schlaak	371
DAMAGE DISTRIBUTION STUDIES IN PROTON-IMPLANTED GaAs H. A. Jenkinson, M. O'Tooni, J. M. Zavada, T. J. Haar, and D. C. Larson	377

PART IV. CERAMICS, POLYMERS, AND GRAPHITE

ION BEAM MODIFICATION OF CERAMICS* C. J. McHargue, C. W. White, B. R. Appleton, G. C. Farlow, and J. M. Williams	385
BEHAVIOR OF IMPLANTED α-Al$_2$O$_3$ IN AN OXIDIZING ANNEALING ENVIRONMENT G. C. Farlow, C. W. White, C. J. McHargue, and B. R. Appleton	395
CHANGING THE INDENTATION BEHAVIOR OF MgO BY ION IMPLANTATION P. J. Burnett and T. F. Page	401
MICROSTRUCTURAL DEVELOPMENT OF TiB$_2$ ION IMPLANTED WITH 1 MEV NICKEL P. S. Sklad, P. Angelini, M. B. Lewis, and C. J. McHargue	407
ION IMPLANTATION OF POLYMERS* M. S. Dresselhaus, B. Wasserman, and G. E. Wnek	413
IMPLANTATION-INDUCED CONDUCTIVITY OF POLYMERS B. Wasserman, G. Braunstein, M. S. Dresselhaus, and G. E. Wnek	423
CHEMICAL AND PHYSICAL INTERACTIONS IN COVALENT POLYMERS IMPLANTED WITH TRANSITION METALS P. E. Pehrsson, D. C. Weber, N. Koons, J. E. Campana, and S. L. Rose	429
STRUCTURE/MAJORITY CARRIER RELATIONSHIPS IN ION-IMPLANTED POLYMER FILMS G. E. Wnek, B. Wasserman, and I.-H. Loh	435

*Invited Paper

SYNTHESIS OF HARD Si-C COMPOSITE FILMS BY ION BEAM IRRADIATION OF POLYMER FILMS T. Venkatesan, T. Wolf, D. Allara, B. J. Wilkens, G. N. Taylor and G. Foti	439
MAGNETIC PROPERTIES OF IRON IMPLANTED POLYMERS AND GRAPHITE N. C. Koon, D. Weber, P. Pehrsson, and A. I. Schindler	445
COMPARISON OF CONDUCTIVITY PRODUCED IN POLYMERS AND CARBON FILMS BY PYROLYSIS AND HIGH ENERGY ION IRRADIATION T. Venkatesan, R. C. Dynes, B. Wilkens, A. E. White, J. M. Gibson, and R. Hamm	449
DEPTH PROFILING OF HYDROGEN IN ION-IMPLANTED POLYMERS J. D. Carlson, P. P. Pronko, and D. C. Ingram	455
TWO-DIMENSIONAL ORDERING OF ION DAMAGED GRAPHITE B. S. Elman, M. S. Dresselhaus, G. Braunstein, G. Dresselhaus, T. Venkatesan, B. Wilkens, and J. M. Gibson	461
CHANNELING STUDIES OF THERMAL REGROWTH IN ION DAMAGED GRAPHITE T. Venkatesan, B. S. Elman, G. Braunstein, M. S. Dresselhaus, and G. Dresselhaus	467
HIGH TEMPERATURE IMPLANTATION IN GRAPHITE G. Braunstein, B. S. Elman, M. S. Dresselhaus, G. Dresselhaus, and T. Venkatesan	475
STOICHIOMETRIC DETERMINATION OF GRAPHITE INTERCALATION COMPOUNDS USING RUTHERFORD BACKSCATTERING SPECTROMETRY L. Salamanca-Riba, B. S. Elman, M. S. Dresselhaus, and T. Venkatesan	481
TRANSPORT PROPERTIES AND ELECTRON MICROSCOPY OF ION IMPLANTED GRAPHITE T. C. Chieu, B. S. Elman, L. Salamanca-Riba, M. Endo, and G. Dresselhaus	487
MAGNETOREFLECTION IN ION-IMPLANTED GRAPHITE L. E. McNeil, B. S. Elman, M. S. Dresselhaus, G. Dresselhaus, and T. Venkatesan	493

PART V. NOVEL PROCESSING TECHNIQUES

DEVELOPMENT OF NEW MATERIALS BY IONIZED-CLUSTER BEAM TECHNIQUE* T. Takagi	501
DYNAMIC RECOIL MIXING FOR THE PRODUCTION OF SILICON NITRIDE FILMS H. Kheyrandish, J. S. Colligen, and A. E. Hill	513
STRUCTURE OF Al-N FILMS DEPOSITED BY A QUANTITATIVE DUAL ION BEAM PROCESS H.T.G. Hentzell, J.M.E. Harper, and J. J. Cuomo	519

*Invited Paper

SURFACE MODIFICATION BY ION BEAM ENHANCED DEPOSITION R. A. Kant and B. D. Sartwell	525
MASKLESS PATTERNING OF Cr FILMS USING FOCUSED ION BEAMS K. Gamo, K. Moriizumi, T. Matsui, and S. Namba	531
CARRIER LIFETIME REDUCTION BY ION IMPLANTATION INTO SILICON A. Mogro-Campero and R. P. Love	537
MODIFYING POLYCRYSTALLINE FILMS THROUGH ION CHANNELING R. B. Iverson and R. Reif	543
THIN POLYMERIC FILMS PRODUCED BY ION IMPLANTATION FROM FROZEN ORGANIC MOLECULES L. Calcagno, K. L. Sheng, and G. Foti	549
MAGNETOREFLECTION OF ION-IMPLANTED BISMUTH E. M. Kunoff, B. S. Elman, and M. S. Dresselhaus	553
ION IRRADIATION SMOOTHING AND FILM BONDING FOR LASER MIRRORS P. P. Pronko, A. W. McCormick, D. C. Ingram, A. K. Rai, J. A. Woollam, B. R. Appleton, and D. B. Poker	559
ENHANCEMENT IN ADHESION OF Pt FILMS ON CERAMICS BY HELIUM ION AND ELECTRON IRRADIATION AND A STUDY OF THEIR ELECTROCHEMICAL BEHAVIOR D. K. Sood, P. D. Bond, and S.P.S. Badwal	565
H IMPLANTATION IMPROVES SUPERCONDUCTIVITY IN NON-TRANSITION METALS F. Ochmann and B. Stritzker	571
ION IMPLANTATION INTO Nb/NbO/PbAuIn JOSEPHSON TUNNEL JUNCTIONS G. J. Clark and S. I. Raider	577

PART VI. APPLICATIONS: MECHANICAL

TRIBOMECHANICAL PROPERTIES OF ION-IMPLANTED METALS* I. L. Singer	585
EFFECTS OF NITROGEN AND HELIUM ION IMPLANTATION ON UNIAXIAL TENSILE PROPERTIES OF 316 SS FOIL J. A. Spitznagel, B. O. Hall, N. J. Doyle, R. Jayram, R. W. Wallace, J. R. Townsend, and M. Miller	597
HARDNESS AS A MEASURE OF WEAR RESISTANCE W. C. Oliver, R. Hutchings, J. B. Pethica, I. L. Singer, and G. K. Hubler	603
MODELING OF HIGH FLUENCE TITANIUM ION IMPLANTATION AND VACUUM CARBURIZATION IN STEEL D. Farkas, I. L. Singer, and M. Rangaswamy	609
RETENTION OF IONS IMPLANTED AT NON-NORMAL INCIDENCE K. S. Grabowski, N.E.W. Hartley, C. R. Gossett, and I. Manning	615

*Invited Paper

INDUSTRIAL APPLICATIONS OF ION IMPLANTATION* J. K. Hirvonen	621
THE CHARACTERIZATION OF NITROGEN IMPLANTED WC/Co D. W. Oblas	631
FRICTION AND WEAR BEHAVIOR OF A COBALT-BASED ALLOY IMPLANTED WITH Ti OR N S. A. Dillich, R. N. Bolster, and I. L. Singer	637
THE MECHANICAL PROPERTIES OF Si^+ AND Pb^+ IMPLANTED Al P. B. Madakson	643
WEAR BEHAVIOR OF FLAT AND GRADED PROFILE BORON-IMPLANTED BERYLLIUM K. Kumar, H. Newborn, and R. A. Kant	649
MICROSTRUCTURES OF STAINLESS STEELS EXHIBITING REDUCED FRICTION AND WEAR AFTER IMPLANTATION WITH Ti AND C D. M. Follstaedt, F. G. Yost, and L. E. Pope	655
FRICTION AND WEAR REDUCTION OF 440C STAINLESS STEEL BY ION IMPLANTATION L. E. Pope, F. G. Yost, D. M. Follstaedt, S. T. Picraux, and J. A. Knapp	661
FRICTION, WEAR, AND DEFORMATION OF SOFT STEELS IMPLANTED WITH Ti AND N I. L. Singer and R. A. Jeffries	667
PROCESSING STEELS FOR TRIBOLOGICAL APPLICATIONS BY TITANIUM IMPLANTATION I. L. Singer and R. A. Jeffries	673
THE REDUCTION OF WEAR AND WEAR VARIABILITY UNDER LUBRICATED SLIDING BY ION IMPLANTATION J. J. Au and P. Sioshansi	679
EFFECTS OF ION IMPLANTATION ON THE ROLLING CONTACT FATIGUE OF 440C STAINLESS STEEL F. M. Kustas, M. S. Misra, and P. Sioshansi	685
ENHANCEMENT OF FERROUS ALLOY SURFACE MECHANICAL PROPERTIES BY NITROGEN IMPLANTATION J.T.A. Pollock, M. J. Kenny, and P.J.K. Patterson	691
WEAR IMPROVEMENT IN Ti-6Al-4V BY ION IMPLANTATION R. G. Vardiman	699
ION-IMPLANTED Ti-6Al-4V W. C. Oliver, R. Hutchings, J. B. Pethica, E. L. Paradis, and A. J. Shuskus	705
WEAR BEHAVIOR AND STRUCTURAL CHARACTERIZATION OF A NITROGEN-IMPLANTED Ti-6Al-4V ALLOY AT DIFFERENT TEMPERATURES R. Martinella, G. Chevallard, and C. Tosello	711

*Invited Paper

PART VII. APPLICATIONS: CHEMICAL

ION BEAMS AND CATALYSIS*   719
J. A. Cairns

CORROSION BEHAVIOR OF Ni^+-ION IRRADIATED NiTi ALLOYS   729
R. Wang and J. L. Brimhall

EFFECT OF N-IMPLANTATION ON THE CORROSIVE-WEAR PROPERTIES OF SURGICAL Ti-6Al-4V ALLOY   735
J. M. Williams, G. M. Beardsley, R. A. Buchanan, and R. K. Bacon

EFFECT OF Cr^+ IMPLANTATION ON THE THERMAL OXIDATION OF Ta   741
K. S. Grabowski and C. R. Gossett

ENHANCEMENT OF URANIUM OXIDATION RESISTANCE BY MOLYBDENUM IMPLANTATION   747
E. N. Kaufmann, R. G. Musket, C. A. Colmenares, and B. R. Appleton

MODIFICATION OF THE HYDRIDING OF URANIUM USING ION IMPLANTATION   753
R. G. Musket, G. Robinson-Weis, and R. G. Patterson

APPLICATION OF ION-IMPLANTED COVALENT POLYMERS   759
D. C. Weber, M. K. Bernett, and H. Ravner

RECOIL IMPLANTATION OF ITO THIN FILMS ON GLASS SUBSTRATES   765
B. H. Rabin, B. B. Harbison, and S. R. Shatynski

HIGH ENERGY ION BEAM MIXING IN Al_2O_3 771
M. B. Lewis and C. J. McHargue

Author Index   777

Subject Index   781

PREFACE

This volume contains selected papers from the "Ion Implantation and Ion Beam Processing of Materials" symposium of the Materials Research Society which was held in Boston, Massachusett, November 14 to 17, 1983. This symposium represented the third of a biennial series of international "Ion Implantation" symposia initiated at the Materials Research Society annual meeting in 1979 (Ion Implantation Metallurgy, eds., E.M. Preece and J.K. Hirvonen) and was continued at the 1981 meeting (Metastable Materials Formation by Ion Implantation, eds., S.T. Picraux and J. Choyke). The present symposium ran for four days and included eight sessions with invited and contributed oral papers and an evening poster session for contributed papers. The number of attendees has approximately doubled with each successive symposium which is a good indication that interest in the field of ion beam modification is expanding. More than 350 participants gathered to give 130 presentations, 33 of which were by authors from among 14 countries outside North America. The background of the participants were also evenly balanced between different types of organizations with 40% from government sponsored institutions, 40% from universities, and 20% from industrial research laboratories. Because a majority of the institutions currently involved in ion beam research participated in the meeting, the papers in this volume provide an accurate cross-section of the work that is currently being conducted in the field.

The major theme of this symposium was the use of ion beams in materials research and materials processing and included sessions on the fundamental processes of ion beam interactions with solids, the application of ion beams for the improvement of material properties, and novel methods and materials or emerging new techniques in surface modification using ion beams. It was encouraging to see that a common feature of the papers from all of the sessions was the attention given to microstructural changes occuring as a result of ion bombardment and the correlation of the microstructure with other properties. It is clear that the research community is now investigating fully the unique structures that may be formed by ion implantation.

Because ion beams may be applied to virtually any material, the contents of this volume span a very diverse set of materials and material properties. For example, the materials studied include metals, semiconductors, ceramics, ceramic composites, insulators, polymers and others, while the material properties investigated include microstructural changes, thermodynamics and kinetics of phase transformations, diffusion, superconductivity, mechanical strength, corrosion, wear, conductivity, optical response, adhesion, etc. The volume is organized into seven parts. Fundamentals of ion beam-solid interactions and the microstructures formed unifies Part I and Part II (Ion Mixing; Metastable Materials) while the material being investigated unifies Parts III and IV (Semiconductors; Novel Materials). This is the first conference to emphasize ceramics, polymers and graphite which are the subject of Part IV . The ion beam processing technique unifies Part V where ion beam assisted deposition was highlighted. We feel this area will be one of rapid growth in both research and applications. Mechanical properties are emphasized in Part VI and chemical properties in Part VII. Parts VI and VII are chiefly concerned with metals but other materials such as insulators and polymers will be found there as well. All of the parts have subject overlaps in materials or properties to some extent; for example, the mechanical properties of ceramics may be found in Part IV, and Parts I and II, which deal with fundamentals, contain papers on a wide variety of materials.

Applications of ion beams (excluding semiconductors) are most advanced in the area of improving the wear resistance of tools and dies for industrial manufacturing (Part VI). However, the research presently being conducted will likely produce additional applications. Areas which appear particularily promising are Ti implantation for the wear improvement of many kinds of steels, implantation of surgical implant alloys for improved corrosion and wear, and implantation of ceramics to enhance the mechanical properties.

                G.K. Hubler   O.W. Holland

                C.R. Clayton   C.W. White

                          February 1984

ACKNOWLEDGMENTS

We wish to thank the program committee and the officers of the Materials Research Society for the opportunity to organize this symposium and for their timely and expert assistance before, during and after the meeting.

It is a sincere pleasure to acknowledge the financial contributions from government agencies and private industries that were essential to the success of the symposium. Major support was provided by the Division of Materials Sciences, Office of Basic Energy Sciences, U.S. Department of Energy (Dr. M.C. Wittels) and by the Metallurgy and Materials Science Division of the Army Research Office (Dr. P.A. Parrish) through grant number DAAG29-83-M-0368.

Additional support was contributed by:

>   Spire Corp.
>   TRW
>   Eaton-Nova
>   Zymet, Inc.
>   CTC, Inc.

Without the help of the MRS program committee and the generosity of the contributors there would have been no symposium.

We would like to thank all the participants for making the symposium such a success. A special note of thanks goes to the session chairmen for directing the sessions and guiding the discussions, and to the referees whose cooperation at the meeting advanced both the quality and timeliness of these proceedings. The contributions of all these individuals are greatly appreciated.

We in particular want to acknowledge the effort of each of the invited speakers whose presentations highlighted each session:

B.R. Appleton	S.T. Picraux
J.A. Cairns	R.F. Pinizzotto
J.J. Cuomo	D.I. Potter
M.S. Dresselhaus	I.L. Singer
J.K. Hirvonen	T. Takagi
C.J. McHargue	H. Wiedersich
M-A. Nicolet	

Finally, we are deeply indebted to M. Gray, S. Thomas, and P. Green for skillful administrative and editorial assistance. It is largely through their efforts that these proceedings have been published.

Ion Implantation
and Ion Beam Processing
of Materials

PART I

ION MIXING

ION MIXING PROCESSES

MARC-A. NICOLET, T. C. BANWELL, AND B. M. PAINE
California Institute of Technology, Pasadena, CA 91125

ABSTRACT

We consider ion mixing in the low temperature regime, where it is insensitive to temperature. Mixing of most thin markers is Gaussian, independent of irradiation flux, and varies linearly with fluence. However, the mixing in some media varies widely between markers of similar mass and appears to correlate with thermal diffusion constants. In bilayer systems, the profile of long-range mixing is exponential, and the number of mixed atoms scales linearly with fluence. This can be modeled successfully with simple collisional theory. Short range mixing scales with the square root of the fluence, but again shows strong correlations with known bulk chemical properties. We conclude that chemical driving forces and low energy transport mechanisms such as interstitial migration play major roles in ion mixing, even at low temperatures.

INTRODUCTION

Experiments with bilayer, multilayer and marker samples all demonstrate that the sample temperature is a primary factor in the ion mixing process. The available evidence suggests that two distinct regimes exist whose position in the temperature scale can be indicated by a critical temperature T_c. Above T_c, ion mixing produces results that depend on the temperature; below T_c, the outcome is insensitive to temperature. The general interpretation of this fact is that above T_c, the perturbation produced in the sample is principally determined by the thermal motion of (radiation-induced) defects. In this regime, the details of the primary collisional process are largely obliterated by the subsequent thermally-induced rearrangements of atoms. The outcome becomes sensitive not only to temperature, but also to the specific chemical and structural properties of the irradiated material. Below T_c, extensive migration of defects is absent, so the observed atomic distributions reflect the primary collisional processes of the ion-solid interaction more closely than above T_c. We consider here primarily this low temperature (i.e. $T < T_c$) regime and review some of the data that are available from marker and bilayer experiments.

MARKER EXPERIMENTS

In a marker experiment, a very thin layer of an atomic species is imbedded in an otherwise uniform host material of a different species. Backscattering spectra are taken before and after ion irradiation. The increase, Ω^2, in the energy variance of the marker signal due to ion mixing is obtained from the difference $\Omega^2_{irr} - \Omega^2_{unirr}$ of the variances after and before irradiation. The variance in depth, σ^2, is then given by $(\Omega/N[\varepsilon])^2$, where N is the atom density of the host medium and $[\varepsilon]$ is the stopping cross section factor for scattering of He ions from the marker in that host medium.

The principal facts established so far, from such marker experiments are:
(i) Marker spreading is observed only when the ion beam penetrates to the marker.
(ii) The spreading is Gaussian. Notable exceptions are Pd and Pt in Si

[1,2] (above room temperature for Pt and above 77 K for Pd).
(iii) The mixing is almost independent of temperature below $\sim$ 300 K, with the exception of Pd in Si [2].
(iv) The variance increases linearly with dose, ϕ, of the irradiation (ϕ = 0.5-20 x 10^{15} ions/cm^2).
(v) The mixing is independent of the irradiation flux.

Observations have been made on systems pairing metal-metal (Sb, Ti, Fe, W, Au, Pt in Al [3,4], W in Cu [5]), metal-semiconductor (Ni, Pd, Sn, Sb, Pt, Au, W in Si [1,2,6] and Si in Ni, Pd [1]), semiconductor-semiconductor (Si in Ge [1] and Ge in Si [1,2]), and metal-oxide (SiO_2 [6], Al_2O_3 [4]). Such generally valid facts must correspond to basic processes of the ion-solid interaction.

The symmetry of the observed mixing rules out single high-energy events as a prominent mechanism. The dependence of σ on the square root of the dose suggests instead that many low energy events are responsible, whose combined effect on the marker displacement is random in nature. A number of models have been developed along these lines, with various levels of elaboration [7-12]. In essence, these models represent the displacement of an atom as a sequence of random relocations whose variance has some value $<r^2>$ and whose frequency is determined by the energy that the irradiation deposits in such displacements, per lattice atom. The prediction then is

$$\sigma^2 = 1/3(F_D \phi /N)<r^2>/E_d , \quad (1)$$

where F_D is the energy deposited in atomic displacements per incident ion, per unit depth, N is the atom density of the medium, and E_d is an average energy that an atom dissipates each time it moves the distance r. It is implied in this picture that displacements are events that take place additively and that scale linearly with energy.

This model is consistent with the facts listed above. It further predicts that σ^2 should depend linearly on F_D and on $<r^2>/E_d$. However, these quantities are difficult to determine experimentally. Techniques to calculate F_D are available [13] and comparisons have been made between experiments and such calculations in various ways. In the first such test, markers of Pt were placed at various depths in an amorphous Si host to see if the variance σ^2 changes with depth as F_D does for a fixed irradiation energy [9]. In another experiment, markers at a fixed position in an amorphous Si matrix were irradiated with various ions [2]. The results are all consistent with a linear F_D dependence, but the uncertainties in the calculated values of the energy deposited in atomic displacements preclude very accurate tests.

The dependence on $<r^2>$ and E_d is even more difficult to verify, as direct measurements of these values are lacking and theory provides rough estimates at best. Estimates for E_d have been made by relating it to the minimum energy for self-displacements [7]. For the average distance of an individual displacement, a value of the order of $\sim$ 10 Å has been suggested on the basis that at low temperatures a Frenkel pair annihilates in most metals for shorter distances. Taking $F_D \sim E/R_p$, where R_p is the projected range and E is the kinetic energy of the incident particle, one finds that $\sigma^2/\phi \sim (E/R_p)(1/N)(10^{-15} cm^2)/20$ eV. This relationship is useful for order of magnitude estimates of marker mixing.

Since quantitative results are difficult to obtain from present theories, experiments should be conducted that expose systematic trends. For example, according to the cascade mixing model, changing the marker should change only $<r^2>/E_d$. Changing the medium and the marker will vary F_D $<r^2>/E_d$. But two media with the same marker can be compared for which F_D is expected to change little while their chemical affinity to the marker changes much. Such is the case for Si and SiO_2, for which F_D calculations according to Winterbon [13] show a difference of less than 2% for 300 keV Xe [6]. Markers of W, Pt, and Au spread indistinguishably in an SiO_2 matrix, meaning that $\langle r^2 \rangle / E_d$ is probably the same, at the two temperatures tested (LN$_2$T and RT). In amorphous Si, however, the variance σ^2 of Au is six times that of W, and that of Pt is four times that of W (see Fig. 1). These differences hold throughout the tested temperature range. If the displacements of the marker atoms are viewed as random flights generated by

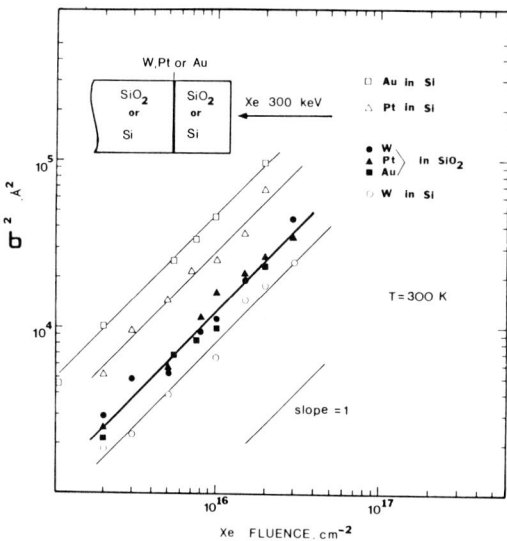

Fig. 1. Variance of mixing, σ^2, for 300 keV Xe$^+$ irradiations of thin markers of W, Pt and Au in Si and SiO$_2$ media, at room temperature (from Ref. 6). Similar results are obtained for 80 K irradiations.

low-energy collisions with the host atoms, the process is determined by collision cross sections, recoil ranges, and threshold energies for displacement. These have no temperature dependence, as is indeed observed. They depend on the density of the host atoms, the mass and the interaction potential of the atoms. To explain the results by cascade mixing requires that $\langle r^2 \rangle / E_d$ should vary considerably between different marker species in Si, but not in SiO$_2$. This cannot be ruled out, since actual values for $\langle r^2 \rangle / E_d$ are unknown. It seems unlikely, though, that the mean displacement distances and the displacement energy thresholds should vary so drastically in Si when they are so similar in SiO$_2$. Another view therefore is that the consistently similar behavior of all markers in SiO$_2$ correponds to the expected outcome for cascade mixing, because Au, Pt and W are collisionally similar heavy atoms, and that the results in Si indicate a departure from a simple collisional model.

Experiments conducted with Ti, Fe, W, Pt and Au markers in Al and Al$_2$O$_3$ support this view. In these two very different materials, F_D calculated for 300 keV Xe is quite similar (within ± 5%). From 80 K to 300 K, the variances of all markers in Al or Al$_2$O$_3$ are the same to within a factor of two. For a quantitative comparison of the mixing of one marker species in both media, caution is needed because the density of an amorphous material such as Al$_2$O$_3$ is not naturally fixed, and atom densities must be known in order to obtain σ^2 from the measured backscattering energy variances Ω^2, of a marker signal. The ambiguity is circumvented when $\sigma^2 N^2$ is compared. This quantity is derivable from a backscattering spectrum without knowing the atom density N; it expresses the variance in units of atoms per unit area, squared. Correspondingly, the displacement distance r is then given by rN.

In Fig. 2, the values of $\sigma^2 N^2$ measured for a fixed dose of 8×10^{15} ions/cm^2 at 80 K for Ne, Ar, Kr and Xe are plotted against the values of F_D/N calculated according to Winterbon [13], assuming the polyatomic media can be represented as fictitious monoatomic materials with atomic numbers and atomic masses equal to the averaged values for the real media. By the model of collision cascade mixing, these curves should have a dependence $(\sigma N)^2 = 1/3 \; \phi(F_D/N)<(rN)^2>/E_d$. Considering the experimental errors and the uncertainties in the calculated abscissa, the change with F_D/N is indeed compatible with a linear dependence, except for Au in Si. The value of $<(rN)^2>/E_d$ for Au and W in Al$_2$O$_3$ is indistinguishable and has a value of about 9×10^{30} at^2/cm^2 eV. The quantity varies by less than 30% between Al$_2$O$_3$ and Al for these two markers, and about 30% between SiO$_2$ and Al$_2$O$_3$ for Au. But the results for the Au marker in Si are clearly uniquely high. An unusually high value is also observed in Si for a Pt marker and Xe irradiation (see Fig. 1).

Some clues have emerged as to the nature of the non-collisional mixing in Si. The mixing of Au and Pt in Si is greatly enhanced relative to the mixing of such markers in SiO$_2$ or W markers in Si. It is well known that Au [14] and Pt [15] are fast diffusers in single crystalline Si [16]. This correlation can be established experimentally as well by vacuum-annealing the marker samples [4,6]. In such experiments with amorphous Si, Au and Pt markers are found to diffuse rapidly at a few hundred °C, with Au diffusing more than Pt, while W remains immobile. In Al, marker atoms (particularly Au and Pt) migrate rapidly to the boundaries of the Al film above 300°C. This has been attributed to grain boundary diffusion, and hence precludes true bulk diffusion measurements in such samples.

These facts indicate that the above model of collision cascade mixing fails when the marker is a fast diffuser. High diffusivities are explained by an interstitial diffusion mechanism that is independent of lattice

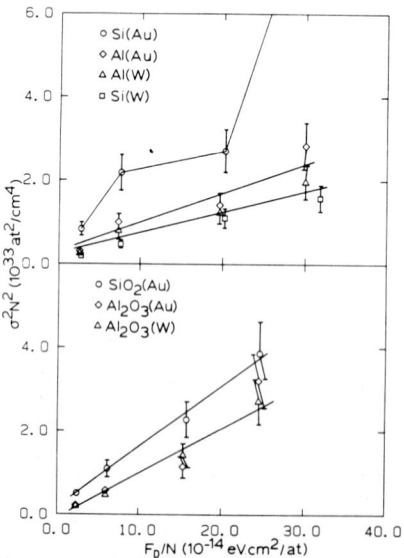

Fig. 2. Dependence of the mixing of several marker systems on F_D, the energy per incident ion, per unit depth, deposited in the matrix by atomic displacements. Irradiation was with a fixed dose of 8×10^{15} ions/cm^2 of Ne, Ar, Kr and Xe at 80 K. Marker A in medium M is denoted by M(A). The value for Si(Au) that lies beyond the range of the ordinate is $N^2\sigma^2 = 9.75 \times 10^{33}$ at^2/cm^4 for $F_D/N = 31.8 \times 10^{-14}$ eV·cm^2/at.

defects [17]. The activation energy for these processes is low ($\lesssim$ 1 eV) and diffusivities approach those of the liquid state near the melting point of the host medium ($\sim 10^{-5}$ cm^2/s). These processes are evidently activated by the impact of an ion, regardless of the ambient temperature. How this interstitial displacement process must be incorporated into the cascade mixing model, or whether different ion mixing models should be adopted in this instance is not clear at this time. It is evident, however, that additional measurements on the spreading of markers in different host media with fast and slow diffusers must be carried out to clearly establish when the simple cascade mixing model applies, and where it fails, and how. A simple quantitative model of the mixing of marker systems by interstitial migration has been proposed by Sigmund [18].

BILAYER EXPERIMENTS

The way in which atoms are rearranged at an initially abrupt interface between two media upon irradiation also provides insights into the processes of ion mixing. Experiments with bilayers are generally more complex to analyze and understand than with markers, because the ion-solid interaction is spacially inhomogeneous and can also change as the irradiation proceeds. When the two adjacent layers differ elementally, chemical effects tend to be overriding at temperatures above T_c.

Recoil Mixing

Recoil mixing refers to atomic rearrangements caused by the first collisions of incident ions (primary collisions) with target atoms. The subject has been treated theoretically by several authors, analytically [19,10,20] and numerically [21]. It is found that most primary recoils are low energy events where the target atom is displaced nearly perpendicularly to the direction of ion incidence. The high energy events are rare, but

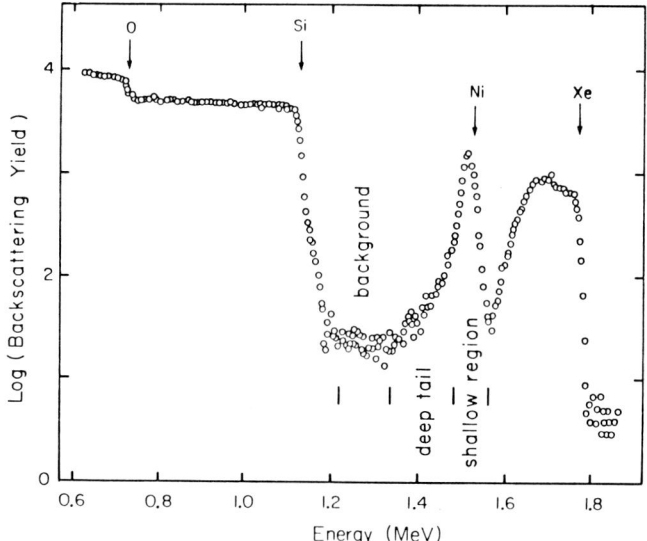

Fig. 3. After irradiation of a Ni/SiO$_2$ bilayer at room temperature (or 80 K) and removal of the metal film, the depth profile of Ni atoms in the amorphous SiO$_2$ substrate exhibits a weak exponential tail at relatively large depths and a strong peak near the interface (from Ref. 27).

produce knock-ons that are projected deeply into the substrate and produce a profile that is nearly exponential [19,21]. Because of its spatial separation, this exponential tail provides a particularly convenient means for studying recoil mixing. Experimental investigations of this process are usually performed with a thin layer of a different elemental composition than the underlying substrate. The profiling of the deep-lying atoms is done by SIMS, AUGER, RBS or nuclear reactions. Systems that have been investigated in this manner are the recoil mixing of oxygen from an SiO_2 layer into the Si substrate [22,23] of Sb into Al [24] of Cu into Al [25], and of Ni in SiO_2 [26]. In such experiments, the quality of the experimental result is much improved if the film is removed from the substrate before the analysis (see Fig. 3).

Long-range recoil mixing is proportional to the fluence [24,25,27], independent of target temperature [25,27], and agrees quantitatively with theoretical prediction [19,23]. For thin overlayers (i.e. when self-absorption in the overlayer is negligible), the number of implanted primary recoils is proportional to the overlayer thickness [20].

Cascade Mixing

The number of primary recoils increases rapidly as their energy decreases, and so does the profile of recoil-implanted atoms at shallow depths. Their total number is small, however, compared to all atoms that are intermixed in the nearest vicinity of the interface. This short-range intermixing is again attributed to numerous small displacements created by high-order, low-energy events in a collision sequence that involves atoms on either side of the interface. The magnitude of this interfacial cascade mixing will depend on the particulars of the species involved.

However, little experimental evidence exists so far which supports the simple picture of interfacial cascade mixing just described. Recent ion mixing results obtained on Ni/Hf and Ti/Hf bilayers [28] demonstrate, however, that noncollisional processes can dominate the mixing process. Nickel is a fast diffuser in Hf while Ti is not [29]. Between liquid nitrogen and room temperature, at a dose of $2 \times_o 10^{15}$ Xe/cm^2, Ni and Hf are already intermixed over a depth of several 100 Å; but the Ti/Hf interface is insignificantly altered. Collisionally, these two systems are quite similar. This difference can be explained by the different diffusivities. Additionally, the systems also differ much in their heats of formation (ΔH_f = 60 kJ/gat for Ni-Hf and ~ 0 kJ/gat for Ti-Hf). Which of these two factors dominates, if any, cannot be established from this experiment. Chemical energies in the range of 50 to 100 kJ/gat are of the same magnitude as the energy deposited into the target by the irradiating Xe ions. That chemical effects are a major factor in interfacial cascade mixing has been documented also for Ni versus Cr on Si [30] and for Au versus W on Cu [5]. Collisionally, each pair in these systems is quite similar, but the amount of intermixing observed differs much in each of these pairs. The Ni/Si bilayer intermixes more rapidly than the Cr/Si bilayer, and Ni is known to diffuse more rapidly in Si than Cr. The Cu/W bilayer intermixes significantly less rapidly than the Cu/Au bilayer, and Cu-W are mutually quite insoluable, while Au-Cu forms a solid solution.

Because interfacial cascade mixing is a low-energy process, minor interfacial alterations can affect the outcome. For example, the native oxide layer of Al initially suppresses intermixing at a Cu/Al interface [25]. At high doses such barriers break down [31].

In the limiting case where the intermixed interfacial layer remains thin enough so as not to alter the interfacial characteristics significantly, one expects the number of penetrating atoms to increase as the square root of the dose, and to be independent of temperature at low temperatures. Both facts have been established experimentally in several cases. On Si substrates, the intermixing of Nb [32], Cr [33], Pt [30] and Ni [30] has been shown to change little between 300 K and 80 K; for Ni, it is constant down to 10 K [34]. For Cu on Al, temperature-independence is established from 300 K down to 40 K [25], and for Ni on SiO$_2$, it has been verified from 80 K to above 300 K [26]. The dependence on $\phi^{1/2}$ has recently been confirmed for Cu on Al [25]. A recent detailed analysis of the whole Ni profile in SiO$_2$ has shown that for doses ranging from $\sim$ 1 to 10 x 10^{15} Xe/cm^2, the total numbers of Ni atoms per unit area in the SiO$_2$ is quantitatively described as

$$[Ni]_s = \alpha \phi + \omega \phi^{1/2} , \qquad (2)$$

where α and ω are independent of ϕ. The linear term accounts mostly for the primary recoil-implanted atoms and the parabolic term includes the remainder (see Fig. 4). Neither term depends on temperature from LN$_2$T to 300 K. Evidently, in this instance the main contribution to ion mixing comes from

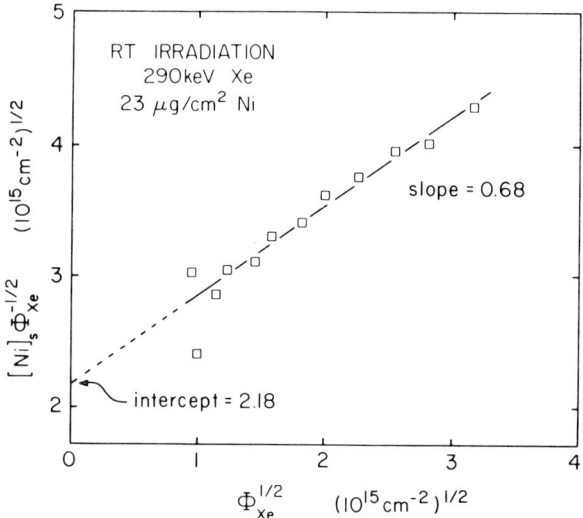

Fig. 4. The total number of Ni atoms per unit area mixed into an amorphous SiO$_2$ substrate, [Ni]$_s$, divided by (fluence)$^{1/2}$, plotted as a function of the square root of the irradiation fluence. The observed linear dependence implies that [Ni]$_s = \alpha \phi + \omega \phi^{1/2}$ (from Ref. 27).

the few primary (high energy) collisions and from the many high order (low energy) collisions. Plural (medium energy) collisions are of lesser importance. The interesting question that arises is whether this is also true in other systems.

CONCLUDING COMMENTS

There is presently no generally accepted model for interfacial cascade mixing at low temperatures. Besenbacher et al. [25] conclude that the simple model of cascade mixing (Eqn. 1) predicts too low a value to explain the intermixing observed at a Cu/Al interface. They propose that the diffusion of interstitials within the cascade volume enhances intermixing, as suggested by Sigmund [16]. Alternatively, the motion of defects may also inhibit intermixing, as suggested by Westendorp et al [5]. The marker experiments discussed initially independently provide support to the notion that the details of the diffusional processes in the irradiated medium must be included in a quantitative model of cascade mixing. But the typical temperature thresholds associated with the motion of a particular defect in isothermal experiments (e.g. interstitial) are not observed in ion mixing. That fact suggests that the transport of atoms in this regime is not governed by the ambient temperature of the medium. The picture of an isolated volume subjected to perturbations for a while and then quenched is most consistent with these observations. It is not sufficient, however, to keep track of collisional processes as if they occurred in an otherwise homogeneous dissipative medium, as some models of cascade mixing do, because transport mechanisms that are specific to the particular medium are being omitted, and chemical driving forces are ignored as well. Improved models of ion mixing will have to include these two features. On the experimental side, the challenge is to distinguish the contribution of each of these two effects. The use of marker and bilayer experiments combined to investigate particular pairs of elemental combinations offers one way toward that goal.

ACKNOWLEDGMENTS

We thank our colleagues involved in ion mixing, particularly at Caltech and at Argonne National Laboratory, whose results and thoughts helped us in shaping our present ideas on the subject. Some financial support from a Senior Fulbright Fellowship, from an IBM Grant, and from funds of the Jet Propulsion Laboratory is also acknowledged.

REFERENCES

1. S. Matteson, G. Mezey, and M-A. Nicolet in: Proceedings of the Symposium on Thin Films and Interfaces, J. E. E. Baglin and J. M. Poate, eds. (The Electrochemical Society, Princeton, 1980), Vol. 80-2, p. 242.
2. S. Matteson, B. M. Paine, M. G. Grimaldi, G. Mezey, and M-A. Nicolet, Nucl. Instr. Meth. 182/183, 43 (1981).
3. B. M. Paine, M-A. Nicolet, and T. C. Banwell in: Metastable Materials Formation by Ion Implantation, S. T. Picraux and W. J. Choyke, eds. (North-Holland, New York, 1982), MRS Symposia Proceedings Vol. 7, p. 79.
4. A. J. Barcz and M-A. Nicolet, (to be published in Appl. Phys. A).
5. H. Westendorp, Z. L. Wang, and F. W. Saris, Nucl. Instr. Meth. 194, 453 (1982).
6. A. J. Barcz, B. M. Paine, and M-A. Nicolet, (to be published in Appl. Phys. Lett.).
7. H. H. Andersen, Appl. Phys. 18, 131 (1979).
8. P. K. Haff and Z. E. Switkowski, J. Appl. Phys. 48, 3383 (1977).
9. B. Y. Tsaur, S. Matteson, G. Chapman, Z. L. Liau, and M-A. Nicolet, Appl. Phys. Lett. 35, 825 (1979).
10. P. Sigmund and A. Gras-Marti, Nucl. Instr. Meth. 182/183, 25 (1981).
11. S. Matteson, B. M. Paine, and M-A. Nicolet, Nucl. Instr. Meth. 182/183, 53 (1981).
12. U. Littmark and W. O. Hofer, Nucl. Instr. Meth. 168, 329 (1980).
13. K. B. Winterbon, Ion Implantation Range and Energy Deposition Distributions (Plenum Press, New York, 1975), Vol. 2.
14. S. D. Brotherton and J. E. Lowther, Phys. Rev. Lett. 44, 606 (1980).
15. S. D. Brotherton, P. Bradley, and J. Bicknell, J. Appl. Phys. 50, 3396 (1979).
16. J. A. Keenan and G. B. Larrabee in: VLSI Electronics, N. G. Einspruch, ed., Vol. 6, N. G. Einspruch and G. B. Larrabee, eds. (Academic Press, New York, 1983), Chap. 1, Appendix p. 48.
17. E. R. Weber, Appl. Phys. A30, 1 (1983).
18. P. Sigmund, Appl. Phys. A30, 43 (1983).
19. R. Kelly and J. B. Sanders, Surf. Sci. 57, 143 (1976), and S. Dzioba and R. Kelly, J. Nucl. Mat. 76, 175 (1978).
20. S. Matteson and M-A. Nicolet in: Metastable Materials Formation by Ion Implantation, S. T. Picraux and W. J. Choyke, eds. (North-Holland, New York, 1982), MRS Symposia Proceedings Vol. 7, p. 3.

21. L. A. Christel, J. F. Gibbons, and S. Mylroie, Nucl. Instr. Meth. 182/183, 187 (1981).
22. T. Hirao, K. Inoue, Y. Yaegashi, and S. Takayanagi, Jap. J. Appl. Phys. 18, 647 (1979).
23. B. Villepelet, F. Ferrieu, A. Grouillet, A. Golanski, J. P. Gaillard, and E. Ligean, Nucl. Instr. Meth. 182/183, 137 (1981).
24. J. Delafond, S. T. Picraux, and J. A. Knapp, Appl. Phys. Lett. 38, 237 (1981).
25. F. Besenbacher, J. Bøttiger, S. K. Nielsen, and H. J. Whitlow, Appl. Phys. A29, 141 (1982).
26. T. Banwell, B. X. Liu, I. Golecki, and M-A. Nicolet, Nucl. Instr. Meth. 209/210, 125 (1983).
27. T. Banwell and M-A. Nicolet, (This Conference).
28. M. Van Rossum, U. Shreter, W. L. Johnson, and M-A. Nicolet, (This Conference).
29. A. D. LeClaire, J. Nucl. Mat. 67&70, 70 (1978).
30. L. S. Wieluński, B. M. Paine, B. X. Liu, and M-A. Nicolet, phys. stat. sol. (a)72, 399 (1982).
31. L. S. Wieluński, C.-D. Lien, B. X. Liu, and M-A. Nicolet in: Metastable Materials Formation by Ion Implantation, S. T. Picraux and W. J. Choyke, eds. (North-Holland, New York, 1982), MRS Symposia Proceedings Vol. 7, p. 139.
32. S. Matteson, J. Roth, and M-A. Nicolet, Rad. Effects, 42, 217 (1979).
33. J. W. Mayer, B. Y. Tsaur, S. S. Lau, and L.-S. Hung, Nucl. Instr. Meth. 182/183, 1 (1981).
34. R. S. Averback, L. J. Thompson, Jr., J. Moyle, and M. Schalit, J. Appl. Phys. 53, 1342 (1982).

ION-INDUCED SURFACE MODIFICATION OF ALLOYS*

H. WIEDERSICH
Argonne National Laboratory, 9700 South Cass Avenue, Argonne, IL 60439

ABSTRACT

   In addition to the accumulation of the implanted species, a considerable number of processes can affect the composition of an alloy in the surface region during ion bombardment. Collisions of energetic ions with atoms of the alloy induce local rearrangement of atoms by displacements, replacement sequences and by spontaneous migration and recombination of defects within cascades. Point defects form clusters, voids, dislocation loops and networks. Preferential sputtering of elements changes the composition of the surface. At temperatures sufficient for thermal migration of point defects, radiation-enhanced diffusion promotes alloy component redistribution within and beyond the damage layer. Fluxes of interstitials and vacancies toward the surface and into the interior of the target induces fluxes of alloying elements leading to depth-dependent compositional changes. Moreover, Gibbsian surface segregation may affect the preferential loss of alloy components by sputtering when the kinetics of equilibration of the surface composition becomes competitive with the sputtering rate. Temperature, time, current density and ion energy can be used to influence the individual processes contributing to compositional changes and, thus, produce a rich variety of composition profiles near surfaces.

INTRODUCTION

   Ion beam modification of alloy surfaces has become an active field of study during the past few years. The field has roots in several areas: ion implantation in semiconductors, radiation effects on the structure and properties of materials, effects of ion sputtering, and recognition of the potential to improve surface properties by ion beams. Recent conferences and books addressing these aspects of ion-solid interactions are cited in references [1-7]. The motivation for a large fraction of the work on alloys is the frequently beneficial effect of ion bombardment on technologically important properties and processes such as hardness, friction, wear, corrosion, catalysis, adhesion, and reflectance. Underlying these effects are ion-induced changes in microstructure, i.e., composition, phase distribution, crystal structure and defect microstructure, in the surface and the near surface regions of the material. It has become evident that a considerable number of distinct processes contribute to the evolution of the microstructure during ion bombardment, see e.g. [8]. Energetic ions cause rearrangement of atoms of the solid in displacement cascades, remove near surface atoms by sputtering and become incorporated into the material at the end of range. Disordering of atoms may transform the crystalline structure into an amorphous phase as is frequently observed in covalently bonded elements and compounds. Whereas most pure metals appear to remain crystalline during ion implantation even at low temperatures, alloys above a critical concentration are observed to become

---

*Work supported by the U.S. Department of Energy

amorphous, e.g., Al with $\gtrsim$ 15 at.% Ni [9] and Mo with $\gtrsim$ 20 at.% P [10]. Ordered alloys become disordered at sufficiently low temperatures; however, at temperatures at which point defects are mobile, the bombardment may accelerate the ordering reaction in partially ordered alloys. Excess interstitials and vacancies agglomerate in the form of dislocation loops and voids, leading to a complex defect microstructure near the implanted surface.

With the exception of the accumulation of the implanted species, the processes mentioned thus far can occur in the surface layer of a uniform alloy target without affecting the alloy component distribution. Several processes, however, initiate and maintain non-uniform distributions of alloying elements within the surface region during bombardment. Preferential loss of certain elements by sputtering reduces the concentration of those elements in the first few layers of the target. Gibbsian adsorption, in which the free energy of the surface is reduced by an increase in concentration of surface-active elements at the surface, leads to a preferential loss of surface-active elements even in the absence of true preferential sputtering. Defect fluxes from the damage region to the surface and toward the interior of the material may transport certain alloying components preferentially and, thus, produce a non-uniform concentration distribution. This process is termed radiation-induced segregation (RIS). Displacement mixing and radiation-enhanced and thermal diffusion counteract any non-uniform concentration distribution present prior to ion bombardment or induced by RIS or sputtering.

In this paper we give a short review of the mechanisms that have been found important in the development of alloy composition and microstructure near surfaces during ion bombardment, i.e., defect production and agglomeration, displacement mixing, radiation-enhanced diffusion, preferential sputtering, Gibbsian adsorption and radiation-induced segregation.

DEFECT PRODUCTION AND AGGLOMERATION

The development of the microstructure during ion bombardment at ion energies that result in ion ranges larger than the characteristic scale of the evolving microstructure is best treated by the concepts which have evolved in the radiation damage community. The scale ranges from a few nanometers at very low temperatures to tens of micrometers at high temperatures. We will concentrate in this section on the processes that occur in the "interior" of the damage range of the incoming ions, where the loss of defects and of alloying elements to the surface and to the undamaged region beyond the range is of minor consequence. For a conceptual understanding of the microstructural evolution of a material during ion bombardment, a treatment of the production and annihilation of defects and defect clusters by chemical rate theory is useful [see e.g. 11,12]. This theory assumes that defects are produced randomly at a density corresponding to the energy density deposited in nuclear collisions [see e.g. 13,14]. The defects are classified as "immobile", e.g., collapsed vacancy loops or void embryos, and potentially "mobile", e.g., interstitial atoms, vacancies, point defect-solute complexes and small clusters of point defects.

At temperatures sufficiently low so that neither interstitials nor vacancies migrate, the defect microstructure consists of displacement cascades which can be visualized as a core region from which atoms have been ejected to appear as interstitial atoms in the surrounding region. The severe rearrangement of atoms in the cascade may result in an amorphous region or, for ordered alloys, in a disordered zone. This is illustrated in Fig. 1a for cascades produced by 100 keV Cu⁺ ions in Cu₃Au taken from the work Black et al. [15]. Disordered regions occur at every cascade site. The vacancy rich regions collapse to dislocation loops in a fraction of the cascades as shown in Fig. 1b and 1c. With increasing dose, cascade regions overlap, and vacancy and interstitital type defects are created in close proximity of existing defects and defect clusters, so that spontaneous recombination of opposite defects occurs with increasing frequency. The material approaches a steady state when essentially each new defect is created within the recombination volume of an existing defect. The corresponding defect concentrations are expected to be $\lesssim 10^{-4}$ atom fraction. This recombination process will contribute to displacement mixing once cascade overlap becomes significant.

At temperatures where one defect type is mobile, the mobile defects annihilate at stationary defects and defect clusters, or form aggregates. In metals interstitials and small interstitial clusters usually become mobile at lower temperatures than vacancies. The ultimate steady state of the irradiated material in this temperature range will not be significantly different than that at lower temperatures, except that the density of the mobile species is greatly reduced, and that of the immobile species is correspondingly increased.

The steady state density of the less mobile defect starts to decrease from its low temperature limiting value when its thermal jump frequency becomes larger than the frequency of production of new defects within the recombination volume. The slow defect then samples a number of potential annihilation sites before a new defect is created in its recombination volume. As the temperature and, therefore, the jump frequency of the slow defects increase, their concentration decreases. Losses of mobile interstitials and vacancies occur by formation of immobile defect clusters, and by annihilation at stationary or slowly moving defect sinks. Frequently, interstitial dislocation loops grow by preferential absorption of

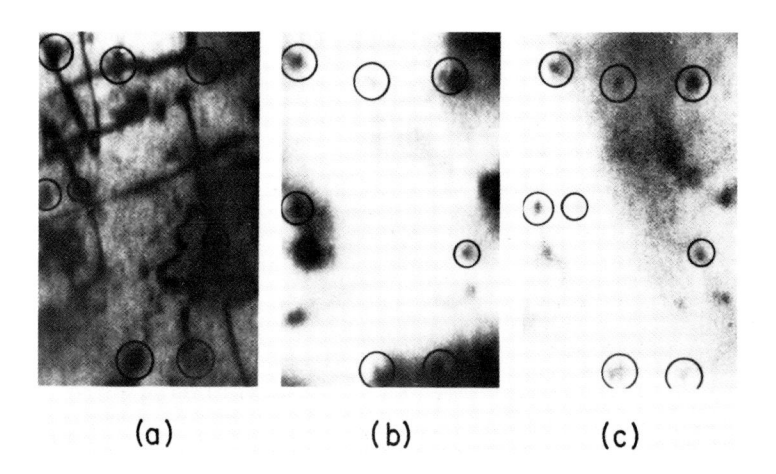

FIG. 1. Cascades produced in-situ by 100 keV Cu⁺ in Cu₃Au and observed at low temperature: (a) (110) dark field superlattice image, showing disordered zones at cascade sites, (b) (220) dark field and (c) bright-field kinematical images showing loops at some cascade sites.

interstitials until they interact and form dislocation networks; vacancies preferentially agglomerate into voids. Defect annihilation at sinks induces defect concentration gradients and, hence, defect fluxes from the interior of crystalline regions to spatially discrete sinks. These fluxes, and the excess of mobile point defects are the predominant causes of microstructural developments of alloys during irradiation at elevated temperatures: radiation-induced and radiation-enhanced phase transformations, radiation-induced dislocation structures and void swelling [5,11,16].

Finally, at high temperatures little or no microstructural changes occur during irradiation. Defect clusters quickly decompose and high vacancy concentrations promote thermal annealing processes and eliminate long range migration of interstitials to sinks.

DISPLACEMENT MIXING AND RADIATION-ENHANCED DIFFUSION

Spatial redistribution of alloying elements requires diffusion or diffusion-like processes. As pointed out above, the production and annihilation of defects result in spatial relocation of atoms. We will use here the term "radiation-enhanced diffusion" for diffusion of elements under thermodynamic driving forces by thermally activated motion of defects during irradiation. The term "displacement mixing" will be used for the collection of processes which lead to redistribution of atoms without thermally activated defect motion. Cascade, ballistic and ion-beam mixing are frequently used instead of displacement mixing.

Several processes contribute to displacement mixing even though a precise distinction is sometimes difficult. Atoms are knocked off their original site and relocated. Relocation may also occur by replacement chains. Relaxation can also contribute to displacement mixing, e.g., by collapse of vacancy rich cascade cores. Extensive defect migration and defect recombination is expected during the "cooling phase" of high-energy-density cascades. Similarly, the high kinetic energy density during the evolution of a cascade will induce diffusion of preexisting defects. Finally, spontaneous recombination between preexisting and newly formed defects contributes to atomic mixing. A number of theoretical treatments for displacement mixing exist [17-21]. However, none includes all the processes mentioned.

A simple semi-empirical approach to describe displacement mixing follows along the suggestions of Anderson [17] and Matteson et al. [20]. Neglecting any anisotropy, the mixing process is considered as a random walk of atoms in three dimensions. The mean square distance, R^2, an atom has traveled after N uncorrelated jumps of length r_n is given by

$$R^2 = \sum_{n=1}^{N} r_n^2 \equiv N \langle r^2 \rangle \tag{1}$$

where $\langle r^2 \rangle$ is the mean square length of the individual jumps. If the N jumps have taken place in a time interval t, one obtains the following relation between the diffusion coefficient D, the jump rate $N/t \equiv \nu$ and the mean square jump distance $\langle r^2 \rangle$ [22]

$$D = (1/6) \nu \langle r^2 \rangle. \tag{2}$$

In this approximation, displacement mixing can be described by the usual diffusion formalism in a way very similar to thermal and radiation-enhanced diffusion. The quantities that must be estimated to apply eq. (2) are the rate with which atoms are changing sites, ν, and the mean square distance, $\langle r^2 \rangle$, associated with the site changes.

This task is especially simple for high temperature thermal diffusion in crystalline solids, where diffusion occurs by exchange of atoms with neighboring vacancies [22]. Therefore, we will discuss the application of eq. (2) to thermal and radiation-enhanced diffusion before returning to displacement mixing. Denoting the nearest neighbor distance by b, we have $\langle r^2 \rangle = b^2$ for atom-vacancy exchange. Neglecting all complications, such as correlation effects and solute-vacancy binding, the jump-frequency of an atom is the product of the probability of having a vacancy next to it, ZC_v, and the exchange frequency of the atom with the vacancy, ν_v; here C_v is the atomic fraction of vacancies and Z the coordination number. Thus,

$$D_{thermal} = (b^2/6) \; ZC_v \; \nu_v. \qquad (3)$$

Both ν_v and C_v contain Boltzmann factors with the activation enthalpy for migration and the formation enthalpy of vacancies, respectively. The straight line denoted "thermal" in the Arrhenius plot of the diffusion coefficient, Fig. 2, represents the thermal diffusion coefficient.

As discussed previously, the vacancy concentration is increased significantly above the thermal concentration at intermediate and low temperatures during ion bombardment, and diffusion of atoms via vacancies is described by eq. (3) with C_v representing the radiation-enhanced vacancy concentration. The presence and migration of interstitials during irradiation also contributes to diffusion. Hence, we can represent the diffusion coefficient during irradiation approximately as [24-26]

$$D_{rad} \simeq (b^2/6) Z \; [C_v \; \nu_v + C_i \; \nu_i] \qquad (4)$$

where we have ignored potential differences in jump-distances and coordination numbers for intersititials and vacancies. The quantity in square brackets is calculated from rate theory [11, 12, 24]. At steady state, the contribution from the interstitials to D_{rad} equals that from the excess vacancies. The solid line in Fig. 2 is calculated for radiation conditions typical of high energy ion bombardment experiments.

The diffusion coefficient for displacement mixing is more difficult to estimate. The jump rate ν of atoms in eq. (2) should be proportional to the displacement rate, K. From molecular dynamics calculations King and Benedek found that $\simeq 40$ atoms change sites for each stable Frenkel pair produced at low primary recoil energies [27]. Similar numbers have been derived from experimentally determined disordering rates of ordered alloys at liquid He temperatures [28, 29]. The mean square jump distance $\langle r^2 \rangle$ should be a few times b^2, because a fraction of the atoms will be relocated by more than a nearest neighbor distance for larger recoil momenta in off-close-packed directions. Additional contributions to displacement mixing are expected from induced migration of preexisting defects and spontaneous defect recombination when cascade overlap becomes important. The diffusion coefficient due to displacement mixing can be written as

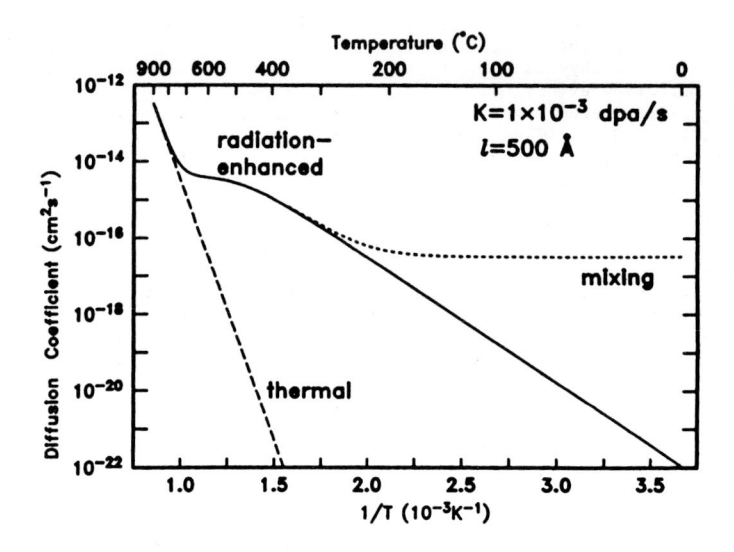

FIG. 2. The steady state diffusion coefficient in a 50 nm thick foil of Ni at a damage rate $K = 10^{-3}$ dpa/s typical of ion bombardment, calculated from the analytical solution to the rate equations due to Lam et al. [23]. A sink density of $10^{10}/cm^2$ was assumed. The thermal diffusion coefficient and the diffusion coefficient due to displacement mixing (see text) are also shown.

$$D_{mix} \simeq (b^2/6)\, \eta\, K \qquad (5)$$

where η should be on the order of 10^2 to 10^3 at steady state. Ion beam mixing of bilayer and of marker specimens at low temperature yield values for η in this range [30, 31]. The dotted line in Fig. 2 indicates the magnitude of mixing expected without thermally activated motion of defects at low temperatures. The diffusion coefficient from athermal mixing exceeds that expected from thermal motion of defects up to the temperature regime where vacancies become rather mobile.

PREFERENTIAL SPUTTERING AND GIBBSIAN ADSORPTION

Preferential loss of certain alloying elements from the surface induces compositional changes which spread into the interior by the diffusion processes discussed in the previous section. The yield or number of atoms of species i per incident ion in the flux of sputtered ions can be written as [8]

$$Y_i = \int_0^\infty p_i(x)\, C_i(x)\, dx \simeq \bar{p}_i\, C_i^s \qquad (6)$$

where $p_i(x)$ is the probability per unit depth that an i-atom at depth $x > 0$ is ejected from the surface, $x = 0$, and $C_i(x)$ is the atomic fraction of i in the alloy at depth x. Written in this form, the distinction between "true" preferential sputtering or primary effects and secondary effects in alloy sputtering is made explicit [31]. The sputter probability contains the physical variables that are directly related to the individual sputtering events, such as the type and energy of the incoming ion, the type of sputtered atom and its surface binding energy. Secondary effects, i.e., Gibbsian adsorption and radiation-induced segregation influence the yield because of their effects on the composition of the alloy in the near surface region.

The sputtered atoms come predominantly from a shallow surface layer [32]. For low energy sputtering the contributions fall off approximately exponentially with depth, with a decay length on the order of two atomic layers. Therefore, the integral in eq. (6) can be replaced to a good approximation by $\bar{p}_i C_i^s$ where $\bar{p}_i$ is the average total probability for an i-atom present in the surface layer to be sputtered off per incident ion and C_i^s is the average atomic concentration of i in the layer. The thickness of this layer is not well defined, but should be on the order of two atomic layers as recent results on high temperature sputtering of Ni-Cu alloys indicate [33].

The differences in the ejection probabilities, $\bar{p}$, for the component atoms of an alloy result primarily from differences (1) in the energy and momentum transferred to atoms of unlike masses in similar collisions with the same projectile, and (2) in the energy required by component atoms to overcome their specific surface binding energy. Anderson has recently given an extensive compilation and discussion of preferential sputtering in multicomponent metals [34].

Continuous sputtering of a semi-infinite alloy target of uniform bulk composition results eventually in a steady state in which the composition of the sputtered atom flux equals that of the bulk, i.e.,

$$Y_1 : Y_2 : Y_3 \ldots = C_1^b : C_2^b : C_3^b \ldots \qquad (7)$$

where Y_i and C_i^b are the yield and the atomic fraction in the bulk alloy of component i, respectively, of the element i. In addition, the sputtered flux contains the bombarding ion species corresponding to a yield of unity at steady state. Preferential sputtering is accommodated by an appropriate change in the near surface concentration at steady state. Combining the approximation given in eq. (6) with eq. (7) we obtain

$$(C_1^s/C_1^b) : (C_2^s/C_2^b) : (C_3^s/C_3^b) \ldots = 1/\bar{p}_1 : 1/\bar{p}_2 : 1/\bar{p}_3 \ldots \qquad (8)$$

i.e., the ratio of the surface and bulk concentrations of an element, at steady state, is inversely proportional to the sputter probability of that element. It should be emphasized that C_i^s is the concentration of i properly averaged over the depth of origin of sputtered atoms and, therefore, is heavily weighted towards the first few atom layers.

Whereas preferential sputtering tends to pin the surface concentration at a value different from the bulk value, see eq. (8), Gibbsian adsorption tends to minimize the surface free energy by increasing the concentration of surface-active elements in the outermost atom layer relative to that of deeper layers. When atom exchange by thermal or radiation-enhanced diffusion between the top and underlaying atom layers occurs rapidly enough compared to the sputtering rate, the concentration ratios

C_i^G/C_i will approach their thermal equilibrium values, e.g., in a binary alloy i, j [35]

$$C_i^G/C_i = C_j^G/C_j \exp(H_{ij}/kT) \qquad (9)$$

where C^G and C are the atom fractions of i or j in the adsorption layer (~ 1 atom layer thick [36]) and the adjacent atom layer, respectively. H_{ij} is the enthalpy change resulting from the exchange between a j-atom in the surface and an i-atom in the adjacent atom layer, k the Boltzmann constant and T the absolute temperature. Enhanced loss of i will occur for $C_i^G/C_i > 1$, because the origin of sputtered atoms is heavily weighted towards the first atom layer. As a consequence, C_i in the subsurface layer will be reduced in an attempt to approach the equilibrium condition, eq. (9). We note that the thickness of the Gibbsian adsorption layer is expected to be smaller than the sputter depth [33]. Hence, the region for the proper averaging of C_i^s may include a portion of the depleted subsurface region.

Preferential loss of elements from the surface during sputtering leads, of course, to a corresponding subsurface depletion. At low temperature, the region of depletion will be spread out by displacement mixing only. With the estimate of D_{mix} given in the previous section and noting that the surface recession rate should be on the order of one atomic plane per dpa (displacement per atom), the altered layer thickness should be less than a few tens of atom layers for high energy ions, and not exceed the damage range for low energy ions.

At elevated temperatures radiation-enhanced and thermal diffusion can lead to significantly increased altered layer thicknesses [33,37]. This is especially noteworthy for the case of low energy ions typically used for sputtering. At temperatures at which the mobility of the slower defects is significant, vacancies and interstitials escape from the damage region. Hence, an increased diffusion coefficient extends far into the target. For example, the silicon depleted layers in Ni-Si alloys are 200 atom layers thick after prolonged sputtering with 5 keV Ar^+ ions at 700°C [38].

RADIATION-INDUCED SEGREGATION

Excess point defects produced by ion bombardment migrate thermally over significant distances at elevated temperature before being eliminated by recombination or annihilated at sinks such as surfaces, grain boundaries and dislocations. The spatial separation between defect production and annihilation leads to persistent defect fluxes, e.g., towards the surface, or from the peak damage region towards regions of lower defect production in front of and behind the peak damage region. Motion of defects requires motion of atoms, i.e., a vacancy exchanges sites with a neighboring atom, an interstitial atom jumps into an adjacent interstice, or interstitialcy motion forces a substitutional atom into an interstitial site while returning a different atom to a substitutional site. In alloys, defects will frequently migrate preferentially via atoms of some of the alloying components. This, in turn, will couple fluxes of alloying elements to defect fluxes. The combination of persistent defect fluxes and the preferential coupling of certain alloying elements to the defect fluxes leads to a non-uniform distribution of elements within the microstructure of an initially uniform alloy phase. This phenomenon of radiation-induced segregation (RIS) is rather common, and has been reviewed recently in some detail [39,40].

The underlying concepts are as follows. A flux of atoms of equal magnitude and direction is associated with interstitials, $J_a^i = J_i$. The flux of atoms induced by vacancies is also of the same magnitude, but opposite in direction to that of the vacancy flux, $J_a^v = -J_v$. Each of the atom fluxes can be proportioned among the components of the alloy identified by subscripts k,

$$J_a^i = \sum J_k^i = J_i \sum \alpha_k^i \, , \quad J_a^v = \sum J_k^v = J_v \sum \alpha_k^v \tag{10}$$

where α_k^i and α_k^v are the coupling constants between element k and the interstitital and the vacancy flux, respectively. A net flux of element k occurs unless $J_k^v = -J_k^i$, i.e., the vacancy-induced flux exactly compensates the interstitital-induced flux. At steady state with respect to defects, the flux of vacancies to sinks (or out of the peak damage region) equals that of interstitials, because both defects are produced in equal numbers, and recombination eliminates defects in equal numbers. This assures that the rate of the total number of atoms arriving at sinks or leaving the peak damage region quickly approaches zero; however, local compositional changes will still occur unless the coupling constants of each element to vacancies and interstititals are equal in magnitude and opposite in sign, $\alpha_k^v = -\alpha_k^i$.

Radiation-induced segregation has been established for numerous alloys [40]. It is most pronounced at intermediate temperatures, where defect mobilities are high. At lower temperatures, the high density of defect clusters which develops during bombardment suppresses long range migration of defects and, hence, segregation over significant distances. At high temperatures, high thermal diffusivities prevent the build-up of significant local concentration differences. In solid solution alloys, RIS frequently increases the local concentration of solute in the vicinity of defect sinks sufficiently to exceed the solubility limit. This is illustrated in the micrographs of Fig. 3, which shows the precipitation of Ni_3Si on several types of defect sinks during ion bombardment of a solid solution Ni-Si alloy.

Radiation-induced segregation can also lead to spatial redistribution of phases within the microstructure of polyphase alloys. For example, enrichment of the solvent (Ni) in the vicinity of defect sinks in the two-phase Ni-12.8 at. % Al alloy leads to the dissolution of Ni_3Al precipitates near the surface, grain boundaries and dislocation loops [41]. The complex redistribution of precipitates in the near surface region of a two-phase Ni-12.7 at. % Si alloy during ion bombardment at elevated temperature is illustrated in Fig. 4 which shows a cross section of the specimen from the surface to beyond the damage range [42]. Defect fluxes ending at the sample surface have deposited sufficient Si to form a continuous film of Ni_3Si on the surface. The Si originated, at least in part, just below the surface film as evidenced by a precipitate depleted zone. At the peak damage region, ~1 μm from the surface, the Ni_3Si precipitates have been entirely dissolved and a band of increased precipitate volume fraction has formed on either side of the peak damage region. Analysis by energy-dispersive x-ray spectroscopy has shown that the alloy in the peak damage region is almost entirely depleted of silicon, whereas the Si concentration in the precipitate depleted zone near the Ni_3Si surface film is close to the solubility limit.

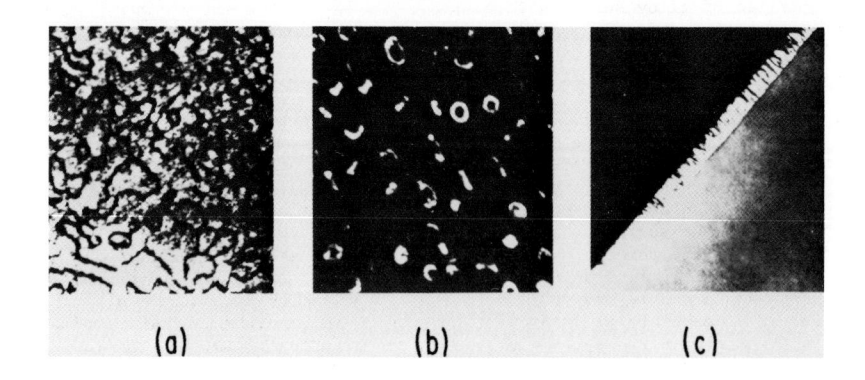

Fig. 3. Precipitation of Ni$_3$Si on defect sinks in solid solution Ni-Si alloys during irradiation; (a) the domain structure of the continuous surface film of Ni$_3$Si; (b) toroidal Ni$_3$Si precipitates that form on interstitial dislocation loops; (c) a Ni$_3$Si film that covers a grain boundary. Courtesy of P. R. Okamoto and K.-H. Robrock.

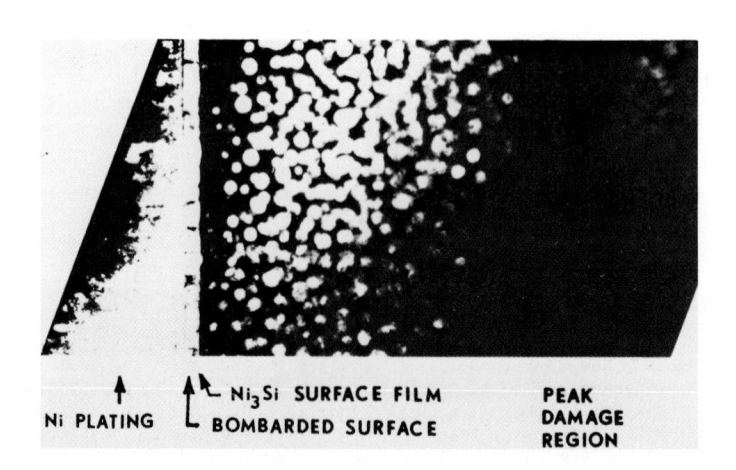

Fig. 4. Redistribution of Ni$_3$Si precipitates in a Ni-12.7 at. % Si alloy during bombardment with 250 keV protons at 500°C. Shown is a cross section prepared after plating the bombarded surface with Ni. The dark field image is produced by using a superlattice reflection of the Ni$_3$Si phase. The original uniformly distributed precipitate phase has formed a continuous film at the surface, depleting a subsurface layer of precipitates. The defect fluxes leaving the peak damage region removed most of the Si from this region. Courtesy of C. Allen, P. R. Okamoto and N. J. Zaluzec.

SUMMARY

Ion beams can modify the microstructure, composition and, hence, the properties of alloy surfaces in a variety of ways. The impact of the ions produces point defects and defect clusters. This process results in displacement mixing even at very low temperatures. It also may result in disordering of ordered alloys or amorphization. In materials that remain crystalline, a high density of defects and defect clusters develops with increasing dose at low temperatures. At intermediate temperatures, interstitial and vacancy defects become increasingly mobile, and the defect structures become less dense and well defined in the form of dislocation loops and networks, and, in many alloys, voids. With the reduced defect density, vacancies and interstitials migrate significant distances before annihilating at surfaces, dislocations and grain boundaries. The presence of excess mobile defects leads to radiation-enhanced diffusion, and the persistent defect fluxes lead to redistribution of alloy components and phases within and beyond the damage range. Selective loss of certain alloying elements by preferential sputtering and, indirectly, by Gibbsian adsorption or RIS and sputtering, leads to depletion of these elements from the near surface region. The depleted layer can spread significantly by displacement mixing, radiation-enhanced diffusion and RIS. It is evident that a wide variety of microstructures, and phase and composition distributions can be produced in near surface regions of alloys by appropriate choices of energy, type and current density of the bombarding ions, the irradiation temperature, and the initial structure and composition of the target.

Acknowledgements

Many stimulating discussions with my colleagues at ANL have contributed much to the integrated understanding of the processes contributing to the microstructural changes in ion bombarded surfaces described here. Special thanks go to L. E. Rehn for also critically commenting on the paper. M. A. O'Connor deserves acknowledgement for her able efforts to bring the manuscript into its final, pleasing appearance.

REFERENCES

1. R. E. Benenson, E. N. Kaufmann, G. L. Miller and W. W. Scholz, eds., Ion Beam Modifications of Materials (North-Holland, Amsterdam 1981); also see Nucl. Instrum. Methods 182/183, (1981).
2. B. Biasse, G. Destafanis and J. P. Gailliard, eds., Ion Beam Modifications of Materials (North-Holland, Amsterdam 1983); also see Nucl. Instrum. Methods 209/210 (1983).
3. J. M. Preece and J. K. Hirvonen, eds., Ion Implantation Metallurgy (TMS-AIME, New York 1980).
4. N. L. Peterson and S. D. Harkness, eds., Radiation Damage in Metals (American Society for Metals, Metals Park, Ohio 1976).
5. F. V. Nolfi, Jr., ed., Phase Transformations During Irradiation (Applied Science Publishers, London and New York, 1983).
6. J. M. Poate, G. Foti and D. C. Jacobson, eds., Surface Modification and Alloying by Laser, Ion and Electron Beams (Plenum Press, New York and London, 1983).
7. S. T. Picraux and W. J. Choyke, eds., Metastable Materials Formation by Ion Implantation (Elsevier Science Publishing Co., New York 1982).
8. H. Wiedersich, H. H. Anderson, N. Q. Lam, L. E. Rehn and H. W. Pickering, pp. 261-285 in Ref. 6.
9. S. T. Picraux and D. M. Follstaedt, in Ref. 6, pp. 287-321.
10. G. Linker, Nucl. Instrum. Methods 182/183, 501 (1981).

11. H. Wiedersich in Ref. 4, pp. 157-193.
12. A. D. Brailsford and R. Bullough, J. Nucl. Mater. 69 & 70, 434 (1978).
13. K. L. Merkle in Ref. 3, pp. 58-94; M. T. Robinson in Ref. 3, pp. 1-27.
14. J. A. Davies in Ref. 6, pp. 189-209.
15. T. J. Black, M. L. Jenkins and M. A. Kirk in: Proc. EMAG 83 (Electron Microscopy and Analysis Group - 1983), Guildford, England, August 1983, in press.
16. H. Wiedersich in: Advanced Techniques for Characterizing Microstructures, F. W. Wiffen and J. A. Spitznagel, eds. (AIME, New York 1982) pp. 15-30.
17. H. H. Anderson, Appl. Phys. 18, 131 (1979).
18. U. Littmark and H. O. Hofer, Nucl. Instrum. Methods 168, 329 (1980).
19. P. Sigmund and A. Gras-Marti, Nucl. Instrum. Methods 182/183, 25 (1981).
20. S. Matteson, B. M. Paine and M.-A. Nicolet, Nucl. Instrum. Methods 182/183, 53 (1981).
21. P. Sigmund, Appl. Phys. A30, 43 (1983).
22. P. G. Shewmon, Diffusion in Solids (McGraw-Hill, New York, 1963).
23. N. Q. Lam, S. J. Rothman and R. Sizmann, Radiat. Eff. 23, 53 (1974).
24. R. Sizmann, J. Nucl. Mater, 69/70, 386 (1968).
25. H. Wiedersich and N. Q. Lam, in Ref. 5, pp. 1-46.
26. S. J. Rothman, in Ref. 5, pp. 189-211.
27. W. E. King and R. Benedek, J. Nucl. Mater. 117, 26 (1983).
28. M. A. Kirk and T. H. Blewitt, Met. Trans. A9, 1729 (1978); also see J. Nucl. Mater. 108/109, 124 (1982).
29. R. H. Zee, M. W. Guinan and G. L. Kulcinski, J. Nucl. Mater. 114, 190 (1983).
30. R. S. Averback, L. J. Thompson and L. E. Rehn, these proceedings; also see S.-J. Kim, R. S. Averback, P. Baldo and M.-A. Nicolet, submitted to Appl. Phys. Lett.
31. P. Sigmund in: Sputtering by Particle Bombardment I, R. Behrish, ed., Topics in Applied Physics (Springer, Berlin 1981) pp. 11-71.
32. G. Falcone and P. Sigmund, Appl. Phys. 25, 307 (1981).
33. N. Q. Lam, H. A. Hoff, H. Wiedersich and L. E. Rehn, these proceedings.
34. H. H. Anderson in: Physics of Ionized Gases (SPIG 1980), M. Matić, ed. (Boris Kidric Institute of Nuclear Science, Beograd, Yugoslavia, 1980) p. 421.
35. P. Wynblatt and R. C. Ku in: Interfacial Segregation, W. C. Johnson and J. M. Blakely, eds. (American Society for Metals, Metals Park, Ohio 1979) p. 115.
36. Y. S. Ng, T. T. Tsong and S. B. McLane, Jr., Phys. Rev. Lett. 42, 588 (1979).
37. L. E. Rehn, N. Q. Lam and H. Wiedersich, these proceedings.
38. L. E. Rehn, V. T. Boccio and H. Wiedersich, Surface Sci. 128, 37 (1983).
39. H. Wiedersich and N. Q. Lam in Ref. 5, pp. 1-46.
40. L. E. Rehn and P. R. Okamoto in Ref. 5, pp. 237-290.
41. D. I. Potter in Ref. 5, pp. 213-245.
42. C. Allen, P. R. Okamoto and N. J. Zaluzec, to be published.

THE DEPENDENCE OF ION BEAM MIXING ON PROJECTILE MASS*

R. S. AVERBACK, L. J. THOMPSON AND L. E. REHN
Materials Science and Technology Division, Argonne National Laboratory,
9700 S. Cass Ave., Argonne, Illinois 60439, U.S.A.

ABSTRACT

Ion beam mixing in Pt-Si bilayered samples was measured during irradiation with projectiles ranging in mass from 4 amu (He) to 131 amu (Xe) at 10 K, 300 K and 373 K. Using deposited damage energy as a basis for comparing the different irradiations, it was found that the heavier ions were more efficient than the lighter ones for inducing mixing. Moreover, it was observed that the mixing was essentially independent of temperature below 373 K. These results are interpreted on the basis that the mixing is caused by the stimulated motion of defects during the cooling phase of energetic cascades.

INTRODUCTION

Theoretical attempts to explain the low-temperature diffusion of solids during ion irradiation have had little success. The collisional mixing theory of Sigmund and Gras-Marti [1], for example, greatly underestimates the magnitude of diffusion; moreover, it incorrectly predicts a shift of light impurities relative to heavy ones [2]. A second collisional theory, which was proposed by Matteson et al. [3], and which is based on a "random flights" model, is able to predict the magnitude of mixing; however, it predicts also that light marker atoms should diffuse substantially more than heavy ones and this is contrary to observation [4]. To overcome the failures of ballistic mixing theories, Sigmund has recently proposed that the large magnitude of mixing arises from the migration of defects produced in energetic displacement cascades [5]. In reference 5, it seems to be implied that the defect motion results from a thermally activated process. This would be difficult to reconcile with the observations that the magnitude of mixing does not dramatically decrease when the specimen is cooled to very low temperatures, 10 K [6,7], for which thermally-induced defect motion is thought to be suppressed. The theory itself however, only requires that defects have some mobility within the cascade region. An alternative mechanism to thermally-induced defect motion is defect motion which is stimulated by the agitation of the lattice during the cooling phase of the cascade. There is both experimental [8] and theoretical [9,10] evidence that such stimulated motion is important in cascades. In this model both the production of defects and their stimulated motion is required for low-temperature diffusion. To explore this possibility, mixing was measured for the cases that: 1) defects are produced in isolated low-energy recoil events, and 2) defects are produced in energetic displacement cascades. Specifically, the experiments consisted of measuring the interdiffusion of Pt-Si diffusion couples during irradiation with projectiles ranging in mass from 4 amu (He) to 131 amu (Xe) at temperatures between 6 K and 373 K. The He irradiation produces defects (i.e. Frenkel pairs) mostly in low-energy recoil events, whereas the Kr and Xe irradiations produce them in energetic displacement cascades. The measurements at the higher temperatures were carried out to investigate the influence of thermally-induced defect motion on mixing.

---
*Work supported by the U. S. Department of Energy.

EXPERIMENTAL

The interdiffusion of the irradiated Pt-Si bilayered specimens was measured at different temperatures using backscattering spectrometry. In the present experiments only a 300-keV He beam was available for in-situ analysis of the specimens after irradiation at 10 K. Although the spatial resolution of backscattering for a 300-keV He beam is not optimal, it was sufficient for the present purposes. The Pt and Si were sequentially vapor deposited onto sapphire substrates using electron beam heating in a cryo-pumped vacuum system which has a base pressure of $\sim 1 \times 10^{-7}$ Torr. Pt was chosen for one of the layers in the diffusion couple in order to produce cascades with high energy density. Si was used as the cover layer as it has a low sputtering yield compared to Pt; and also because the energy loss of heavy 300-keV particles in traversing the Si layer is small so that they could penetrate well beyond the Pt-Si interface. The specimens were cooled by clamping the sapphire substrate to the copper tailpiece of the cryostat. Indium was inserted between the sapphire and cryostat to prevent the sapphire from cracking under the applied load.

The irradiation doses were converted from ions/cm^2 to damage energy/A-cm^2 at the interface using a modification of the Monte Carlo computer simulation, TRIM [11]. The results of these calculations are illustrated in Fig. 1. The average damage energy on the two sides of the interface was used to normalize the different particle irradiations. This procedure introduces some ambiguity since the relative damage energies of the particles in the Pt and Si are not exactly the same; however the differences are small.

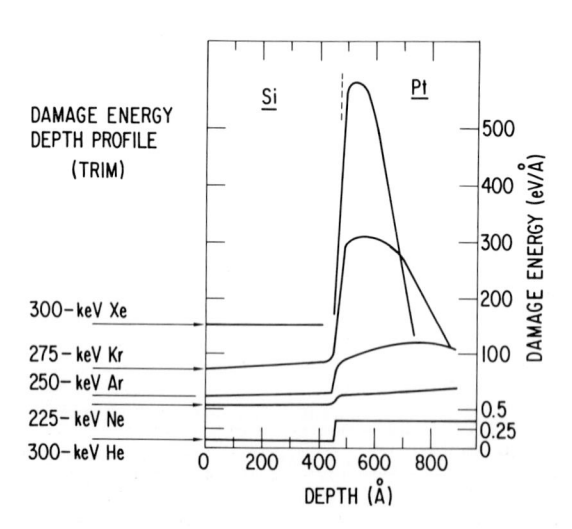

FIG. 1. Calculated damage distribution in Si-Pt diffusion couples.

RESULTS AND DISCUSSION

The mixing at the Pt-Si interface was quantified by subtracting the backscattering spectrum for each specimen prior to irradiation from the corresponding spectrum for the irradiated specimen. The magnitude of the difference in each channel was then summed over all channels. These integral yields were used as a measure of the mixing. By comparing the integral yields to corresponding yields obtained by computer simulation, the thickness of the mixed layers could be deduced. For convenience it was assumed that the mixed layers were entirely Pt_2Si. The details of this procedure are published elsewhere [6].

The results for the different irradiations are listed in Table I. Since the same deposited damage energy was not employed for all the irradiatons, the quantity t^2/ϕ, where t is the effective layer thickness of Pt_2Si formed, was used as a measure of mixing. This quantity, or mixing parameter, has been found to be independent of dose for nearly all cases of ion beam mixing, and this behavior is consistent with the measurements here. It can be observed in the table that t^2/ϕ increases as the mass of the projectile increases. This behavior is illustrated in Fig. 2. Here we have formed the ratio of the mixing parameter for each irradiation with that of Kr. This ratio is referred to here as the efficiency of mixing, i.e.,

$$\xi_i = (t^2/\phi)_i / (t^2/\phi)_{Kr}. \qquad (1)$$

TABLE I. Ion beam mixing parameter for various projectile irradiations at different temperatures.

	t^2/ϕ [a]		
	10 K	293 K	393 K
300-keV He	0.36	0.44	0.30
225-keV Ne	0.63		
250-keV Ar	0.84	0.91	
275-keV Kr	1.38	1.80	
300-keV Xe	1.37	1.40	1.20
160-keV Xe	1.50		

[a] in units of $10^3 Å^5/eV$

It was noted above that He irradiation produces defects mostly in low-energy recoil events and that Kr and Xe produce them in dense zones, and therefore plotting the efficiency as a function of projectile mass provides some indication of the dependence of mixing on cascade energy density. The overall dependence of the mixing on energy density seems consistent with the hypothesis that stimulated motion of defects in energetic cascades is an important aspect in low-temperature ion beam mixing.

Two features of the efficiency curve warrant further discussion: 1) the efficiency for 300-keV He irradiation is not zero; and 2) the efficiency appears to level off at high energy densities. It might be expected that the efficiency for irradiations which produce defects only in low-energy recoils would be nearly zero since there would be no stimulated motion. Unfortunately, 300-keV He irradiation of Pt-Si does not produce only low energy recoils; approximately 25% of the defects produced by the He irradiation result from recoils above 6-keV. Since the efficiency for the He irradiation is also ~25%, we consider the simple model that the efficiency for recoils below 6-keV is zero, and that the efficiency for recoils above 6 keV is 1. Using this simple model the efficiencies of the other irradiations can be qualitatively explained. The fraction of defects produced in recoils above 6-keV for the Ne, Ar, Kr, and Xe irradiations are 60, 75, 85, and 90% respectively. Although these percentages are not in

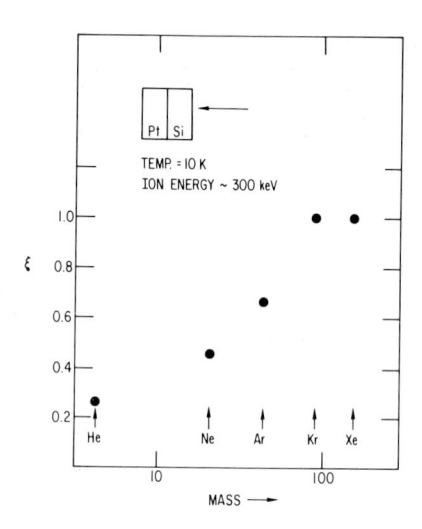

FIG. 2. Efficiency of ion beam mixing in Pt-Si as a function of projectile mass.

exact agreement with the mixing efficiencies, they are reasonable considering the crudeness of the model, the uncertainties in the data, and the fact that the details of the energy deposition along the track of the projectile were ignored (i.e. the recoils for the heavier particles are in close proximity to each other). For these reasons it would not be particularly illuminating to construct a better fitting efficiency function at this time. To examine the efficiency function in more detail, it would be desirable to obtain mixing data for irradiations which produce even lower energy recoils than 300-keV He irradiation. This would set the efficiency for low-energy recoils which was "guessed" to be zero here. One of the experimental problems inherent in such an approach, however, is attaining the sufficiently high damage levels necessary to measure the mixing, but without destroying the specimen by the high concentration of implanted ions. This was just possible with 300-keV He. Nevertheless, efforts in this direction are being made.

The efficiency behavior at high energy densities also requires additional investigation. The leveling off of the efficiency at high energy densities is indicated only by the 300-keV Xe irradiation. One problem with the Xe irradiation is that the projectiles have rather limited range in Pt and this may limit the mixing. We note however, that no large deviations in the mixing parameter were observed with dose, and the efficiency for 160-keV Xe is essentially no different than that for the 300-keV Xe. Leveling off of the mixing efficiency at high recoil energies was also observed for low-temperature Ar and Kr irradiation of Ni-Si diffusion couples [6]. In fact the general shape of the efficiency curve for Ni-Si is similar to that for Pt-Si [12]. These efficiency curves are reminiscent of efficiency curves for defect production [13], radiation induced segregation [14], and disordering [15]. These studies have shown that "cascade effects" become important at relatively low recoil energies, similar to those found here. Molecular dynamic computer simulations have also shown that stimulated motion of defects in cascades is important at quite low recoil energies [9,10].

The results for the mixing at higher temperatures are also listed in Table I. There is very little dependence of mixing on temperature in Pt-Si. If we assume that thermal defect motion is suppressed at 10 K, then we must conclude that radiation-enhanced diffusion is not important to the mixing in Pt-Si for temperatures below 373 K. This result is similar to that observed for Ni-Si [6] and Pd-Si [16] for which weak temperature dependences were observed.

CONCLUSION

The present experiments show that ion beam mixing does not scale directly with the deposited damage energy during an irradiation, but that it also depends on the energy density of the cascades produced by the irradiation particle. It was also found that the mixing in Pt-Si was essentially independent of temperature below 373 K. These results seem consistent with a model of ion beam mixing in which defect motion in cascades is stimulated by the agitation of the lattice during the cooling phase of the cascade. More complete information about the efficiency function at low energy and at very high energy would be very useful in evaluating the possible mechanisms of ion beam mixing.

REFERENCES

1. P. Sigmund and A. Gras-Marti, Nucl. Inst. & Meth., 182/183, 25 (1981).
2. B. M. Paine, J. Appl. Phys. 53, 6828 (1982).
3. S. Matteson, B. M. Paine, and M.-A. Nicolet, Nucl. Instr. & Meth., 182/183, 53 (1982).
4. S.-J. Kim, R. S. Averback, P. Baldo, and M.-A. Nicolet, submitted to Appl. Phys. Lett.
5. P. Sigmund, Appl. Phys., A30, 43 (1983).
6. R. S. Averback, L. J. Thompson, Jr., J. Moyle, and M. Shalit, J. Appl. Phys., 53, 1342 (1982).
7. S. Mantl, R. S. Averback, L. E. Rehn, and L. J. Thompson, unpublished.
8. R. S. Averback, J. Nucl. Mater. 108/109, 33 (1982).
9. M. W. Guinan and J. H. Kinney, Proc. 2nd Topical Meeting on Fusion Reactor Materials, Wa. 1981.
10. W. E. King and R. Benedek, J. Nucl. Mater. 117, 26 (1983).
11. J. P. Biersack and L. G. Haggmark, Nucl. Instr. & Meth. 174, 257 (1980).
12. K. R. Johnson, L. E. Rehn, and R. S. Averback, unpublished.
13. R. S. Averback, R. Benedek, and K. L. Merkle, Phys. Rev. B18, 4156 (1978).
14. L. E. Rehn, P. R. Okamoto, and R. S. Averback, unpublished.
15. S. P. Lamond and D. I. Potter, J. Nucl. Mater. 117, 64 (1983).
16. G. E. Chapman, S. S. Lau, S. Matteson, and J. W. Mayer, J. Appl. Phys. 50, 6321 (1979).

INVESTIGATION OF A THERMAL SPIKE MODEL FOR ION MIXING OF METALS WITH SI

U. SHRETER, FRANK C. T. SO, B. M. PAINE AND M-A. NICOLET
California Institute of Technology, Pasadena, CA 91125

ABSTRACT

A model for ion mixing of bilayer systems in a thermal spike is described. Normal thermal reactions, at a constant average temperature, are assumed for the duration of the spike. New experimental results on mixing of a Nb/Si couple with Xe ions are shown to agree with the prediction of the model that the apparent activation energy in ion mixing can depend upon the ion species. Calculations of spike mixing for Cr and Ni with Si are presented which show that a substantial part of the experimentally observed mixing rates, and their temperature dependence, can be attributed to thermal spike effects.

INTRODUCTION

Atomic mixing induced by heavy-ion irradiation has been attributed to two main mechanisms: collisional mixing, and delayed thermal diffusion of irradiation-induced defects (radiation-enhanced diffusion) [1,2]. Mixing in metal-Si bilayer systems has been interpreted as a collisional process in the temperature-independent regime (below 300 K) and radiation-enhanced diffusion in the temperature-activated regime (above 300 K).

But recent investigations of ion mixing phenomena suggest that the mixing mechanisms may be more complex than this. Radiation-enhanced diffusion would be expected to vary with the flux of the irradiation, but Averback et. al. [3] have shown that mixing of Ni/Si bilayers in the high temperature regime is independent of flux. They therefore argued that the transport mechanism cannot be normal radiation-enhanced diffusion but rather is rapid migration over distances of the order of the cascade volume, i.e. it is essentially an intra-cascade phenomenon. Also, Averback [4] has determined that mixing of Pt/Si bilayers at very low temperatures depends superlinearly on the density of energy deposited in atomic displacements. This is not consistent with a collisional mechanism based on two body collisions and suggests a collective motion or "spike". In recent work we have determined that in the formation of $CrSi_2$ by ion mixing the irradiation induces an interfacial reaction that is temperature-dependent [5] and therefore is clearly not a purely collisional effect, and also does not necessarily involve radiation-enhanced diffusion.

Additional clues as to the nature of the mixing processes in metal-Si systems can be obtained by comparing the formation of binary compounds by thermal annealing and by ion mixing. There are strong correlations in the first phases that are formed and in the reaction kinetics [6], suggesting that the ion mixing indeed involves thermal diffusion and reactions. However, the temperature dependences of the two processes are different. Ion mixing induces reactions at lower ambient temperatures and with lower apparent activation energies than thermal annealing.

These observations suggest that a thermal spike model may be appropriate for mixing of these systems: following the ion impact, the region of the collision cascade is hot for a short time during which the silicide grows by normal thermal reaction, and as the region cools by classical heat conduction the process is halted. In this paper, we present a simple version of this model, describe measurements designed to test the model qualitatively, and to reveal parameters of the model. In the light of the values of these parameters, we discuss the plausibility of the model, and explore its implications for the full range of temperatures for which data exists.

Spike models and relevant experimental data have been reviewed recently by Thompson [7]. These data covered damage production, inert gas desorption, sputtering, secondary ion emission and electron emission. To our knowlege, a spike model has not previously been applied to ion mixing.

THE MODEL

We assume that energy is deposited instantaneously in a small volume, producing an instant localized increase of temperature. To model the mixing we simply assume normal thermal reaction processes for the duration of the spike. Although the temperature must in reality change continously, we will assume a constant average temperature for a time t_o, followed by an instantanous drop to ambient temperature. The reaction in each small spike volume will have a rate proportional to $EXP(-Ea/k(T_o+T))$ were Ea is the activation energy, k is the Boltzmann constant, T the sample temperature and T_o the temperature added to T by the spike energy. Thus the thickness X of compound formed is given by the formulae that are used for conventional furnace annealing, modified to account for the repetitive nature of the ion irradiation:

$$X^2 = A \phi t_o a_o EXP(-Ea/k(T_o+T)) \qquad (1a)$$

or

$$X = B \phi t_o a_o EXP(-Ea/k(T_o+T)) \qquad (1b)$$

Here t_o is the spike duration, ϕ the ion fluence and a_o is the effective area of the spike. Ea is the activation energy, and A and B are the pre-exponential factors in the cases of parabolic and linear growth rates, respectively. We have determined Ea, A, and B from conventional furnace annealing experiments [6], assuming that they are independent of temperature up to the spike temperature.

From Eqs.1 we see that the slope in a plot of $\log(X^2)$ or $\log(X)$ vs. $1/T$ is not constant as it is in a regular Arrhenius plot, and is given by:

$$-Ea/k(1+T_o/T)^2 \qquad (2)$$

The magnitude of this apparent activation energy is lower than Ea at all temperatures T and becomes very small when $T \ll T_o$. Thus spike mixing can contribute to both the temperature-dependent and temperature-independent regimes in ion mixing. Other qualitative features of mixing in the spike model are: 1) It is a single-cascade process and should not depend on ion flux at any sample temperature. 2) The exponential dependence on T_o may make the mixing rate non-linear in energy density. 3) Kinetics and compositions of the compounds formed should be similar to those found in normal furnace annealing because spike mixing is a thermal process.

RESULTS AND DISCUSSION

The spike model predicts that the apparent activation energy of mixing may be changed in a given material by changing T_0. To check this we performed a mixing experiment in the Nb/Si system using Xe ions and compared the results to data published by Matteson et al.[1] for the mixing of the same system with 275 keV Si ions. Since estimates of the cascade volumes for these two ions indicate that the energy deposited per atom by the Xe ions in Nb will be substantially greater than that for the Si ions, the value of T_0 will be greater for the Xe implantation than for the Si implantation, making the apparent activation energy for Xe lower than that for Si (see Eq. 2.) Fig. 1 is an Arrhenius plot of the results for temperatures above 300 K. Our precise experimental conditions are described in the sketch. Indeed, we find an apparent activation energy of 0.2 ± 0.1 eV for Xe, compared to 0.9 eV found for the Si ion mixing. No such variation of the activation energy with irradiation parameters would be expected for radiation-enhanced diffusion. A quantitative comparison of the mixing with the prediction of the model is not possible, however, since reliable data on thermal formation of $NbSi_2$ are not available [8].

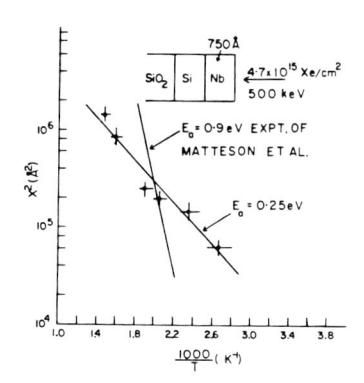

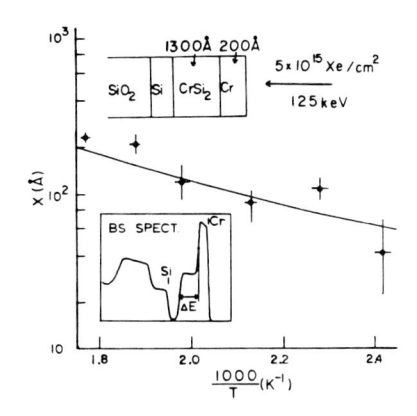

FIG. 1. Arrhenius plot of the square of the amount of $NbSi_2$ grown by ion irradiation of Nb/Si bilayers. The data points are the result of the present measurement with the system shown. A line of best fit is included. Also shown is a line with a gradient that corresponds to the activation energy found by Matteson et al. [1].

FIG. 2. Data points: measurement of the temperature-dependence of the ion mixing of Cr and Si. Initial samples were already partly reacted, and X is the irradiation-induced increase in the thickness of $CrSi_2$, deduced from E. Solid line: fit to the data with the thermal spike model.

We have not attempted to calculate the temperature and time duration of spikes for our model. However it is possible to deduce these quantities from an experimental Arrhenius plot of mixing, assuming the mixing occurs purely by the spike mechanism. But Arrhenius plots for ion mixing that have been published to date were derived from measurements of the total number of mixed atoms, and these presumably include contributions from collisional mixing, the magnitudes of which are unknown, in addition to the spike mixing component. Therefore we have conducted a new set of measurements of the formation of $CrSi_2$ by ion mixing, in which the starting samples had already been partly reacted by furnace annealing, and only the irradiation-induced increase in the well-defined thickness of fixed-composition silicide was recorded (see Eq.2). We assume that this represents only the component of the mixing from spikes, while the variable-composition mixing, which we ignored in this measurement, results from collisional mixing. The results, which are shown in Fig. 2 were fitted by a straight line, to yield a gradient of 0.2 eV at ~570 K while in thermal annealing Ea=2.0 eV [9]. Using Eq. 2 we find $T_o/T=2.2$ and a spike temperature of $T_o=1250$ K. To calculate t_o we use the following data: Thermally, $CrSi_2$ forms at a constant rate (Eq. 1b applies) of 1 Å/s at $450°C$. The pre-exponential factor B is therefore 8.6 10^5 cm/s. Irradiation with 125 keV Xe ions at $300°C$ to a fluence of 5 10^{15} /cm^2 induces growth of 200 Å of silicide. If we take the standard deviation in the distribution of energy deposited in atomic displacements as the radius of the spike area (100 Å) and substitute all the numbers in Eq. 1b we get a spike time of $t_o=5\ 10^{-11}$ s. The resulting curve for ion mixing is shown in Fig. 2. As can be seen in the figure, the calculated curve for this temperature range is almost a straight line. This justifies the use of the average slope calculated at 570 K to obtain the value of T_o. The calculated curve levels off at temperatures below 300 K to a value of only a few Å.

Mayer et al. [2] have measured the temperature dependence of the total (collisional plus spike) mixing of Cr with Si by 300 keV Xe ions. Their data are plotted in Fig. 3 as discrete points. We attribute all of the mixing observed below 300 K to collision cascades. This is reasonable since the data are essentially independent of temperature, and the backscattering spectra show essentially only error function profiles of mixed Cr and Si. We have fitted this by a constant level (dashed line). The fit of our model to the data of Fig. 2 has been scaled by fluence and added to this and plotted as a solid line. The agreement with the data in this regime is good.

Experimental formation rates in furnace annealing [6] and in ion mixing [10] are available for the Ni/Si system, and can be used to further compare the spike model with experiment. Because of the similarity in mass between Ni and Cr we take the same spike parameters (T_o, t_o and a_o) for Ni/Si mixing as found in the Cr/Si case. Using the thermal formation data for Ni_2Si (Ea=1.5eV A=0.3cm^2/s) and substituting in Eq. 1a, we calculate the temperature dependence of the thickness of silicide and plot the results in Fig 4. Unlike for Cr-Si mixing, in this case there is a substantial contribution to the low-temperature mixing from thermal spikes. The remainder of the mixing in this regime is attributed to collision cascades, which we represent as a horizontal dashed line at a sufficient level to bring the total mixing at the lowest temperature up to the experimental result. Once this is done, we find good agreement with the high temperature data.

The comparison of the spike model with the $CrSi_2$ mixing experiment gave a spike temperature of T=1250 K. If we assume that 3/2kT is equal to the total energy lost by the ion in elastic collisions, as calculated from Winterbon's tables [11], divided by the number of atoms in the spike volume, we get T_o= 1040 K, which is in reasonable agreement. Sigmund presents a more realistic formula for the effective maximum temperature at the depth Rp in a spike [12], which yields T_o= 1690 K for our conditions. Also, Sigmund gives a formula for the duration of a spike, which yields 6.1×10^{-11} s for the present system, in good agreement with our value of $t_o = 5 \times 10^{-11}$ s. Similarly, in a detailed calculation of the time evolution of thermal spikes, Sanders [13] finds temperatures and time constants of the same order of magnitude.

Thus the spike parameters used in the model, and the amounts of mixing calculated are not unreasonable. However, we point out that these parameters, as well as the annealing data used here are subject to considerable uncertainties. In particular, the extrapolation of experimental formation rates to extremely high temperatures may introduce substantial errors. Deviations in the values of the parameters in the model can change the calculated amounts of mixing by at least an order of magnitude, which may make the spike contribution to mixing small compared to experimental results. Also it is clear that refinements in the spike model, using detailed time and space evolution of temperature in the spike region [14], are not difficult to implement. However, such improvements are meaningless as long as the basic parameters are not known with reasonable accuracy. This suggests for instance, measuring reaction rates at high temperatures and short times using pulsed laser beams.

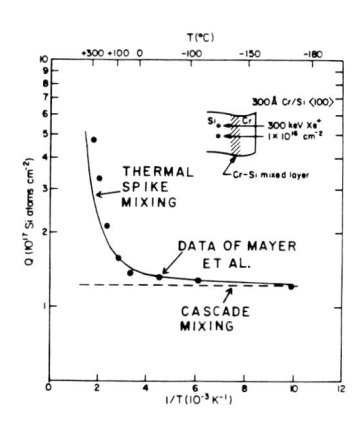

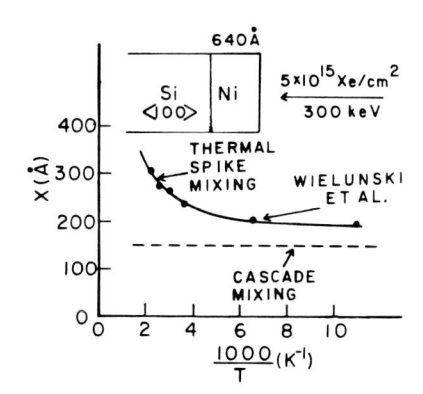

FIG. 3. Comparison of the model with the measurement by Mayer et al. [2] of the mixing of Cr and Si.

FIG. 4. Comparison of the model with the measurement by Wielunski et al. [10] of the mixing of Ni and Si.

SUMMARY

We have developed a simple thermal spike model for ion mixing of metal - Si bilayers, in which known thermal reaction data are used to calculate amounts of mixing. We find that the model can fit high temperature ion mixing results for Cr/Si with physically reasonable values for the spike parameters. The model predicts negligible mixing by spikes below 300 K in the Cr/Si system. The mixing in this regime is presumably by collision cascades.

For Ni/Si mixing down to 80 K, the model predicts significant mixing, but less than the experimentally observed values. The difference is attributed to cascade mixing. The high temperature mixing in Ni/Si is well fitted by the model.

In addition, the thermal spike concept can successfully explain the following observed characteristics of ion mixing of metal - Si bilayers:
 1. The lack of flux dependence.
 2. The non-linear dependence of mixing on the density of deposited energy.
 3. The difference in apparent activation energy for mixing with different ion species.
 4. The close correlation with results of furnace annealing.

ACKNOWLEDGMENTS

We wish to thank R. Gorris and A. Ghaffari (Caltech) for technical assistance. The Implantation part of the work was partially supported by the U.S. Department of Energy through an agreement with the National Aeronautics and Space Administration and monitored by the Jet Propulsion Laboratory, California Institute of Technology (D. Burger).

REFERENCES

1. S. Matteson, J. Roth, and M-A. Nicolet, Rad. Eff., 42, 217 (1979).
2. J. W. Mayer, B. Y. Tsaur, S. S. Lau, and L-S. Hung, Nucl. Instr. and Meth., 182/183, 1 (1981).
3. R. S. Averback, L. J. Thompson, J. Moyle, and M. Schalit, J. Appl. Phys., 53, 1342 (1982).
4. R. S. Averback in: Proceedings of the Workshop on Ion Mixing and Surface Layer Alloying, S. T. Picraux and M-A. Nicolet, Eds., (Sandia Report SAND83-1230, 1983).
5. U. Shreter, F. C. T. So, and M-A. Nicolet, submitted for publication in J. Appl. Phys.
6. M-A. Nicolet and S. S. Lau in: VLSI Electronics: Microstructure Science N. Einspruch, Series Ed., Supplement A - Materials and Process Characterization, G. Larrabee, Ed., (Academic Press, New York, 1983) p. 329.
7. D. A. Thompson, Rad. Eff., 56, 105 (1981).
8. R. J. Wagner, S. S. Lau, J. W. Mayer, and J. A. Roth, Thin Film Phenomena-Interfaces and Reactions, J. E. E. Baglin and J. M. Poate, Eds., (The Electrochemical Society, N.J., 1978), p. 59.
9. C-D. Lien, unpublished.
10. L. S. Wielunski, B. M. Paine, B. X. Liu, C.-D. Lien, and M-A. Nicolet, phys. stat. sol. (a), 72, 399 (1982).
11. K. B. Winterbon, Ion Implantation Range and Energy Deposition Distributions, Vol.2 (Plenum, New York 1975).
12. P. Sigmund, Appl. Phys. Lett., 25, 169 (1974), and Appl. Phys. Lett., 27, 53 (1975).
13. J. B. Sanders, Rad. Eff., 51, 43 (1980).
14. G. H. Vineyard, Rad. Eff., 29, 245 (1976).

MODIFICATION OF μm THICK SURFACE LAYERS USING keV ION ENERGIES*

L. E. REHN, N. Q. LAM AND H. WIEDERSICH
Materials Science and Technology Division, Argonne National Laboratory,
9700 S. Cass Ave., Argonne, Illinois 60439, U.S.A.

ABSTRACT

   Root-mean-square diffusion distances for both vacancy and interstitial defects in metals can be very large at elevated temperatures, e.g. several μm's in one second at 500°C. Consequently, defects that escape the implanted region at elevated temperature can produce compositional and microstructural changes to depths which are much larger than the ion range. Because of the high defect mobilities, and of the fact that diffusion processes must compete with the rate of surface recession, the effects of defect production (ballistic mixing), radiation-enhanced diffusion and radiation-induced segregation become spatially separated during ion bombardment at elevated temperature. Results of such experimental studies in a Cu-Ni alloy are presented, discussed and compared with predictions of a phenomenological model. Contributions to the subsurface compositional changes from radiation-enhanced diffusion and radiation-induced segregation are clearly identified.

INTRODUCTION

   Ion beam modification of materials is not restricted to depths commensurate with the projected range of the bombarding ions. At least two fundamentally different methods have been suggested for modifying layers which are much thicker than the ion penetration depth. These schemes are important, since lower ion energies and thicker modified layers are both advantageous for most technological applications.

   The simplest method for obtaining thick layers involves ion bombardment of a surface on which material is simultaneously deposited. This technique has been exploited by, for example, Greene and coworkers [1] to produce very thick layers of various semiconductor materials. An ion-beam mixing version of this simultaneous deposition and bombardment method would involve deposition of two different atom species during ion bombardment. Alternatively, a sequence of individual steps, each consisting of the deposition of a very thin layer of material and subsequent ion bombardment could be repeated until the desired layer thickness is attained.

   A second method, quite distinct from the one described above, is to induce modifications via point defects which escape from the near-surface implanted region into the specimen interior during ion bombardment. Fluxes of point defects can modify materials by enhancing equilibrium processes such as aging and interdiffusion, or by inducing nonequilibrium distributions of alloying components because of radiation-induced segregation [2]. Obviously, the mobilities and concentrations of defects in the specimen interior must be high if significant modifications are to occur over appreciable depths. In this paper it will be shown that these two conditions are apparently fulfilled in many alloy systems during low

---

*Work supported by the U. S. Department of Energy.

energy ion bombardment at moderate temperatures (300-600°C). Results from an experimental study of the effects of point defect penetration deep into the interior of a Cu-40 at.% Ni specimen during 5 keV Ar bombardment at elevated temperatures are discussed. The results are compared with predictions of a model for sputter-induced subsurface compositional changes formulated by Lam and Wiedersich [3]. This comparison reveals that two processes, namely radiation-enhanced diffusion and radiation-induced segregation, are operative at depths much greater than the ion range.

EXPERIMENTAL RESULTS

Auger electron spectroscopy (AES) and ion sputtering at room temperature were used to depth-profile subsurface changes produced in a Cu-40 at.% Ni specimen by 5 keV Ar bombardment at elevated temperatures. Details of the experimental procedure can be found elsewhere [4]. After 5 keV Ar bombardment at 600°C for two hours at a dose rate of 1.2×10^{15} ions $cm^{-2} s^{-1}$, nickel enrichment extending several μm's into the specimen was observed. (It should be noted that ~4 μm of material is removed from the surface of the specimen by sputtering during the elevated temperature bombardment.) Measurements of the additional nickel found at various depths after two hours of bombardment at 500 and at 600°C are shown in Fig. 1. Here, the difference between the nickel concentration measured after the indicated sputtering time and that measured after sputtering deep into

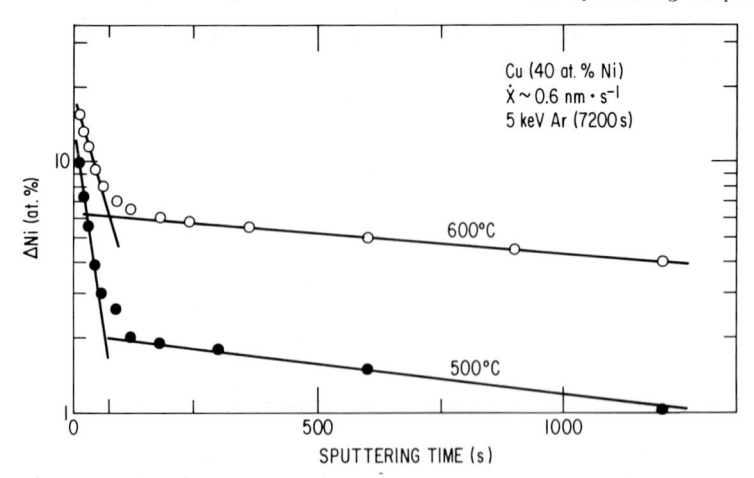

FIG. 1. Experimental measurements of subsurface nickel enrichment produced by two hours of sputtering with 5 keV Ar ions at 500 and at 600°C.

the bulk alloy is plotted on a logarithmic scale as a function of sputtering time at room temperature. The sputtering rate at room temperature is calculated to be ~0.55 nm s^{-1}. Preferential removal of copper during sputtering of Cu-Ni alloys is well known [5]. This effect increases the near-surface nickel concentration measured after sputtering to ~60 at.% Ni. However, for the conditions found in Fig. 1, i.e. net concentration changes ≤ 15% and no abrupt change in composition with depth, the differences in the nickel concentrations measured after sputtering at room temperature are essentially equal to the differences which would be obtained if no preferential sputtering occurred.

The projected range of 5 keV argon in the alloy is ⩽ 3 nm [6]. Several processes, e.g. preferential sputtering, Gibbsian segregation, ballistic mixing, argon implantation, radiation-enhanced diffusion and radiation-induced segregation, contribute to concentration changes in the near-surface, implanted layer. The relative importance of these different effects in producing the experimentally observed, nickel-enriched subsurface layer is discussed elsewhere [6,7]. The first data point given in Fig. 1, however, was obtained at a depth of ~6 nm. Hence, all the nickel enrichment depicted in Fig. 1 occurs at depths greater than the ion range, where only the two defect-flux driven processes, i.e. radiation-enhanced diffusion and radiation-induced segregation, can alter the composition.

The results in Fig. 1 indicate two different regions of nickel enrichment extending well beneath the implanted layer. The first region exhibits a relatively steep drop in nickel enrichment (from ~15 at.% to 6 at.% Ni) over depths extending from about 6 to ~50 nm (~90 s of sputtering), while the second region manifests itself as a considerably shallower concentration gradient extending several μm's into the specimen.

DISCUSSION

The mobility of both vacancy and interstitial defects in metals can be very high at elevated temperatures. This is illustrated in Table I, where the calculated root-mean-square diffusion distances, $\bar{x}$, for vacancy defects with migration enthalpies, H_m^v, of 0.6 eV (Al) and 1.1 eV (Ni) [8] are tabulated for a diffusion time of one second. We have used:

$$\bar{x} = \sqrt{4Dt} \quad \text{and} \quad D = a_o^2 \nu_o \exp[-H_m^v/kT];$$

t is the time; a_o is the lattice parameter; ν_o is the preexponential jump frequency (5 × 10^{13} s^{-1}); T is the absolute temperature.

TABLE I. Root-mean-square diffusion distances for a vacancy in one second.

H_m^v	.6 eV	1.1 eV
$\bar{x}(500°C)$	63 μm	1.3 μm
$\bar{x}(600°C)$	106 μm	3.3 μm

The diffusion distances for interstitial atoms, which typically have significantly lower migration enthalpies, are of course even larger.

The surface recedes at a rate of ~0.6 nm s^{-1} during the elevated temperature bombardment as a result of sputtering; this rate is negligible compared to the values listed in Table I. Of course, actual defect mobilities depend sensitively on the degree of clustering of like defects. However, transmission electron microscopy studies of ion bombarded pure copper [9] and pure nickel [10] specimens indicate that even large defect clusters are unstable at these temperatures. A second important consideration is that because the ion bombardment produces a continuous source of defects in the implanted layer, the defect fluxes into the specimen interior are persistent. Consequently, large numbers of both irradiation-induced vacancies and interstitials escape deep into the specimen interior at temperatures ⩾ 400°C.

We have used the model of sputter-induced compositional changes developed by Lam and Wiedersich [3] to calculate the time dependence of the expected changes in composition beneath the implanted layer of the Cu-40 at.% Ni alloy at 500°C. The parameters utilized in the calculations, including an ion current of 40 µA/cm^2, are the same as in Ref. 3. No attempt was made to fit the calculated results to the experimental data by varying parameters. The purpose here was simply to use the phenomenological model to gain an understanding of the basic mechanisms responsible for the two different regions of nickel enrichment found in Fig. 1. The results of the calculations for times of 400, 1000, 4000 and 10^6 s are displayed in Fig. 2. The ordinates of Figs. 1 and 2 are the same, i.e. the amount of subsurface nickel enrichment plotted on a logarithmic scale. The scales of the abscissas have been chosen such that the sputtering time in Fig. 1 multiplied by the calculated sputtering rate of 0.55 nm s^{-1} yields approximately the same scale as in Fig. 2. As in Ref. 3, radiation-induced segregation was included in the calculations by assuming that the copper atoms jump into vacancies faster than nickel atoms do, i.e. the migration enthalpy was set equal to .77 eV for diffusion of copper via vacancies and to .82 eV for diffusion of nickel via vacancies. In fact, the results shown for 10^6 s in Fig. 2 are identical to those shown for 500°C in Fig. 4 of Ref. 3. The calculations reveal that the experimental bombardment time of two hours is sufficient for the subsurface composition profiles to approach close to their steady state values.

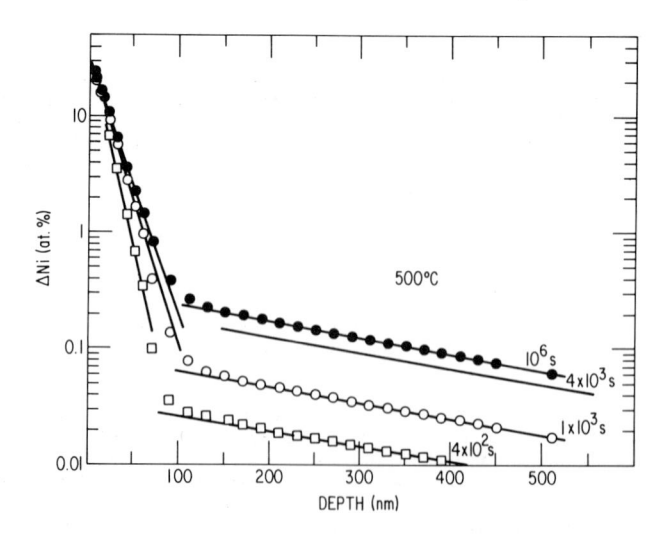

FIG. 2. Calculated time dependence of subsurface nickel enrichment during sputtering at 500°C.

Qualitatively, the experimental (Fig. 1) and calculated (Fig. 2) profiles appear very similar. Both exhibit a relatively strong and apparently exponential decrease in nickel enrichment over the first several tens of nanometers, followed by a significantly shallower, but again exponential decrease extending very deep into the specimen. The slopes characteristic of each of the two regions also appear qualitatively similar. The main discrepancy between the calculated and experimental results is in the absolute magnitude of the nickel enrichment in region II. The experimental results give values of a few percent nickel at 500°C, while the calculated enrichment is almost an order of magnitude lower.

We now turn to the question of which transport mechanisms are responsible for the nickel enrichment found in the two regions. Region I, which extends in the experimental profiles to a depth of about 50 nm is the most straightforward. As stated above, several effects combine to produce a highly nickel-enriched region over the penetration range of the argon ions. At temperatures where sufficient atomic mobility exists, diffusion processes enhanced by the presence of radiation-induced defects extend this nickel enrichment beyond the ion penetration depth. Hence, the primary mechanism responsible for Region I is radiation-enhanced diffusion. This situation has been treated by Ho [11], and subsequently by several other authors [12-14]. Assuming that the diffusion in the subsurface is enhanced uniformly, the radiation-enhanced diffusion coefficient, D, can be shown to equal the product of the altered layer thickness, δ, which is defined as the inverse of the slope in Region I, and the surface recession velocity, $\dot{x}$, resulting from sputtering. Our results yield altered layer thicknesses in Region I of 14, 20 and 40 nm for temperatures of 400 (not shown in Fig. 1), 500 and 600°C, respectively. The corresponding radiation-enhanced diffusion coefficients are 8×10^{-14}, 1.2×10^{-13} and 2.2×10^{-13} cm$^2 \cdot$s^{-1}. We note, however, that the above analysis ignores any contribution to Region I from the mechanism which produces the nickel enrichment in Region II.

Region II, to our knowledge, has never been identified previously in low energy bombardments. We have investigated the effects on Region II of varying the five mechanisms contained in the model, i.e. preferential sputtering, displacement mixing, Gibbsian adsorption, radiation-enhanced diffusion and radiation-induced segregation. Removing the preferential transport of copper atoms by vacancies causes Region II to disappear (Fig. 3). Adding preferential transport of nickel atoms by interstitials causes Region II to reappear. Preferential transport of both copper via vacancies and nickel via interstitials increases the absolute magnitude of the nickel enrichment in this region. Variations in the other four mechanisms have essentially no effect on Region II. In fact, even in the absence of all four other effects, Region II occurs if either preferential transport of copper via vacancies or nickel via interstitials is included in the calculations. We therefore conclude that radiation-induced segregation is responsible for the nickel enrichment observed in Region II.

This conclusion is consistent with previous studies of radiation-induced segregation in Cu-Ni alloys. High energy ions, which create defect fluxes from deep (~1 μm) in the specimen toward the irradiated surface produce a nickel enriched near-surface layer at elevated temperature which has been observed using Auger electron spectroscopy [15] and Rutherford backscattering [16]. That is, nickel is known to be preferentially transported in the same direction as the defect fluxes. In the present low-energy ion bombardment, the high production rate of point defects in a thin surface layer combined with the high mobilities of both the vacancy and interstitial defects generates significant defect fluxes which penetrate deep into the specimen interior. Preferential transport of nickel by these defect fluxes produces a region of nickel enrichment which extends well beyond the depths to which radiation-enhanced diffusion processes can compete with the rate of surface recession. A similar effect has been observed by Gudlaht et. al. [17] in Cu-Be marker experiments. The existence of these large and persistent defect fluxes at intermediate temperatures provides a means for modifying μm-thick layers during low-energy ion bombardment.

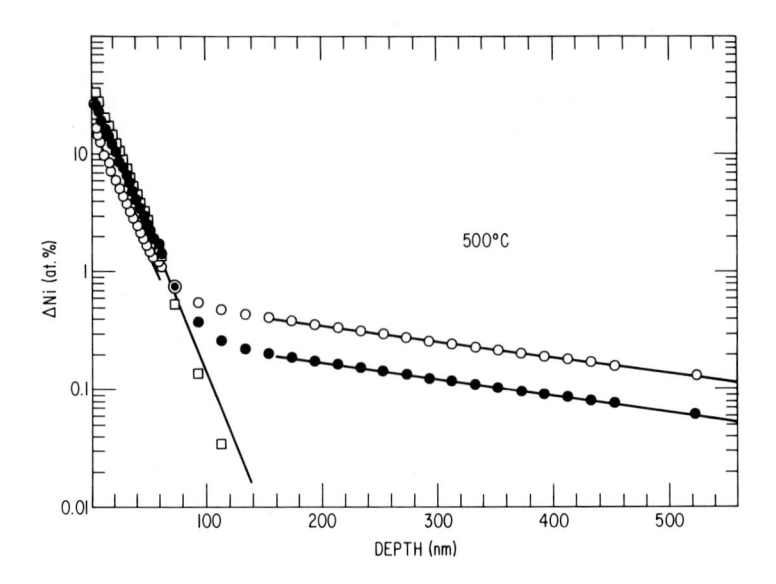

FIG. 3. Profiles calculated for no preferential transport via vacancies or interstitials (open boxes), preferential transport of copper via vacancies (filled circles), and preferential transport of copper via vacancies and nickel via interstitials (open circles).

REFERENCES

1. K. C. Cadien, M. A. Roy, S. M. Shin, J. M. Rigsbee, S. A. Barnett and J. E. Greene, Mat. Res. Soc. Symp. Proc. 7, 93 (1982).
2. L. E. Rehn, Mat. Res. Soc. Symp. Proc. 7, 17 (1982).
3. N. Q. Lam and H. Wiedersich, J. Nucl. Mater. 103 & 104, 433 (1982).
4. L. E. Rehn, V. T. Boccio and H. Wiedersich, Surf. Sci. 128, 37 (1983).
5. H. Shimizu, M. Ono and K. Nakayama, Surf. Sci. 36, 817 (1973).
6. N. Q. Lam and H. Wiedersich, Rad. Effects Letters 67(4), 107 (1982).
7. N. Q. Lam, H. A. Hoff, H. Wiedersich and L. E. Rehn, these proceedings.
8. R. W. Siegel in Positron Annihilation, P. G. Coleman et. al., eds., North Holland Publishing Co., New York, 351 (1982).
9. C. A. English, B. L. Eyre and J. Summers, Phil. Mag. 34, 603 (1976).
10. I. M. Robinson and M. L. Jenkins, Phil. Mag. A43, 999 (1981).
11. P. S. Ho, Surf. Sci. 72, 253 (1978).
12. D. G. Swartzfager, S. B. Ziemecki and M. J. Kelley, J. Vac. Sci. Technol. 19(2), 185 (1981).
13. R. Webb, G. Carter and R. Collins, Rad. Effects 39, 129 (1978).
14. H. Wiedersich in Surface Modification and Alloying, J. M. Poate and G. Foti, eds., Plenum Press, New York 261 (1983).
15. L. E. Rehn, W. Wagner and H. Wiedersich, Scripta Met. 15, 683 (1981).
16. W. Wagner, V. Naundorf, L. E. Rehn and H. Wiedersich in Effects of Radiation on Materials, H. R. Brager and J. S. Perrin, eds., ASTM, Philadelphia, PA, 895 (1982).
17. H. J. Gudladt, V. Naundorf, M.-P. Macht and H. Wollenberger, J. Nucl. Mater. 118, 73 (1983).

ANISOTROPIC TRANSPORT OF IMPURITIES IN ION MIXING

S. Matteson, J.A. Keenan, and R.F. Pinizzotto
Texas Instruments Incorporated
P.O. Box 225936 MS-147
Dallas, TX 75265

ABSTRACT

Experimental investigations of the redistribution of a thin (1nm) Sn marker layer lying between two thicker dissimilar layers (Ge and Si) during 360 keV As ion irradiation are reported. Several permutations of layer arrangements were tested, i.e. Si/Sn/Ge, Ge/Sn/Si, Si/Sn/Si and Ge/Sn/Ge. It was also found that in the dissimilar "diffusion" couples, the Sn drifts in the Ge rich direction, regardless of whether the Ge is on the surface side or the substrate side of the marker. This phenomemon of anisotropic transport is interpreted as a drift induced by a gradient in the "diffusion" coefficient. The radiation resistance of concentrated alloys is discussed in the light of this phenomenon.

INTRODUCTION

Ion Mixing or Ion Beam Mixing has received much attention in the last five years, both experimental and theoretical.[1-4] The nature of the processes at work in Ion Mixing has so far eluded precise description. It appears that, in general, thin impurity distributions are redistributed by ion mixing obeying approximately gaussian statistics, i.e. the width of the redistributed impurity profile grows as the square root of the incident ion fluence and is often independent of temperature. The rate of the increase in the width of the distribution is directly related to the energy deposited in atomic displacements.[5] However, the precise functional relationship is not well established. Moreover, small shifts in the peak position of the distributions have been reported.[6]

Attempts at describing the atomistic origins of Ion Mixing have taken three forms. Random walk models have been proposed which correctly predict the gaussian-like form of the redistribution profiles. The significant contribution of ballistic mixing in these models comes from low energy recoils of impurity atoms. The range of low energy displaced atoms is not known accurately but is essential for quantitative predictions. Some estimates suggest that simple ballistic mixing is insufficient to account for the observations. Recoil mixing, an alternative model of Ion Mixing, treats the relative displacement of the impurity atom in the host matrix by examining the high energy recoils of the various atoms. Anisotropy is an expected outcome of such processes. A prediction of this model is that the profiles would be non-gaussian.[3] This is not generally observed. More recent models have suggested that a relaxation mechanism may overshadow the prompt ballistic displacements.[7] Indeed, the displaced impurity atom may "wander" some distance from its original site before it is trapped again. This process could be expected to appear temperature independent, as well as ion dose rate independent, if the trapping sites were not mobile and if the relaxation occurred over a time which was short relative to the mean time between displacements for an atom.

There is an apparent contradiction between gaussian-like spreading and peak drifts. However, anisotropic mixing can occur even in random walk models, if a gradient in the ion mixing "diffusion" coefficient D^* exists. If the probability of displacement increases in a given direction, there will be a net flux in that direction even if the probability of displacement is isotropic, i.e. all directions of displacement are equally likely. The present experiment was conceived to test this hypothesis. Due to the depth dependence of the energy deposition occurring during ion penetration of a solid, a natural gradient in D* should occur. However, in most experimentally attainable situations large gradients are not accessible, but by the use of layered structures of widely varying atomic mass, an extreme gradient in energy deposition density can be achieved. Germanium and silicon were chosen because they differ by a factor of about 2.5 in mass but have very similar material properties. The two elements are mutually soluble at all concentrations,[8] and both can be prepared and remain as amorphous films. Tin was chosen as the marker layer since it is insoluble in both Si and Ge and is well behaved in the silicon matrix under ion mixing.[5]

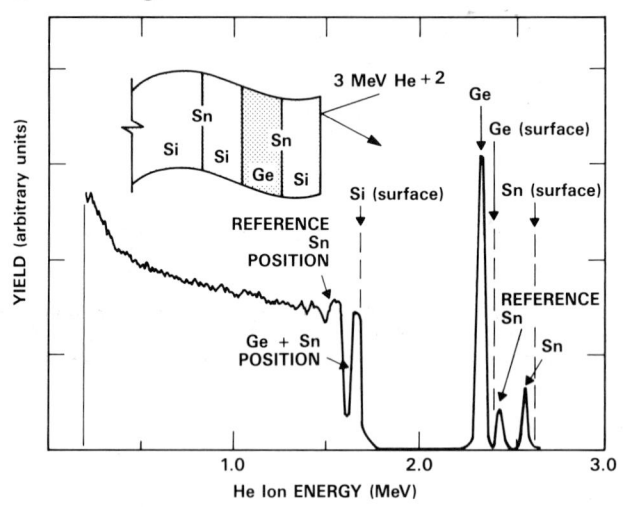

Fig. 1. Representative spectrum of Si/Sn/Ge specimen before ion mixing. Doubly charged 3 MeV He ions were used to obtain the backscattering spectrum.

EXPERIMENTAL DETAILS

Four types of layered structures were prepared by sequential electron beam evaporation of Si, Ge and Sn. The films were deposited on single crystal silicon substrates in a vacuum of better than 1×10^{-7} torr, maintained by titanium sublimation pumps and an ion pump. The four permutations of structures are Si/Sn/Ge/Si/Sn/substrate, Ge/Sn/Si/Sn/substrate, Si/Sn/Si/Sn/substrate, and Ge/Sn/Ge/Si/Sn/substrate. The most shallow Sn marker (of approximately 1 nm thickness) served as the probe of the process while the deeper Sn layer, which was near the end of the ion's range, lay between Si in all causes. The second marker was expected to be little affected by the ion; this was found to be the case. Corrections for sputtering or carbon deposition could be made by referring to the position of the deeper reference marker. The thickness of the Ge surface layers were ~ 30 nm while the thickness of the Si surface layers were ~ 60 nm. These thicknesses were chosen to place the markers at equivalent positions on the energy deposition curve in all cases.

The ion irradiation was carried out using As^{+2} ions accelerated by a potential of 180 kV. The 360 keV As ions were electrostatically scanned over portions of all four types of layered specimen. The unirradiated portions of the specimen were obtained for comparison from parts of the irradiated sample which were masked

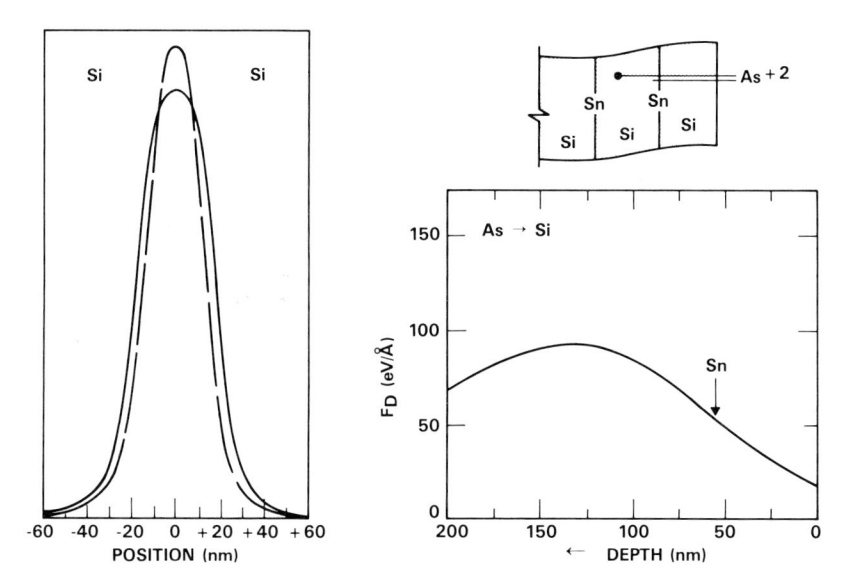

Fig. 2. Depth distribution of Sn between Si layers before (broken line) and after (solid line) irradiation with 1×10^{16} cm^{-2} As ions at 360 keV. The Sn layer spread slightly but did not shift significantly. The energy deposition function is given on the right for comparison.

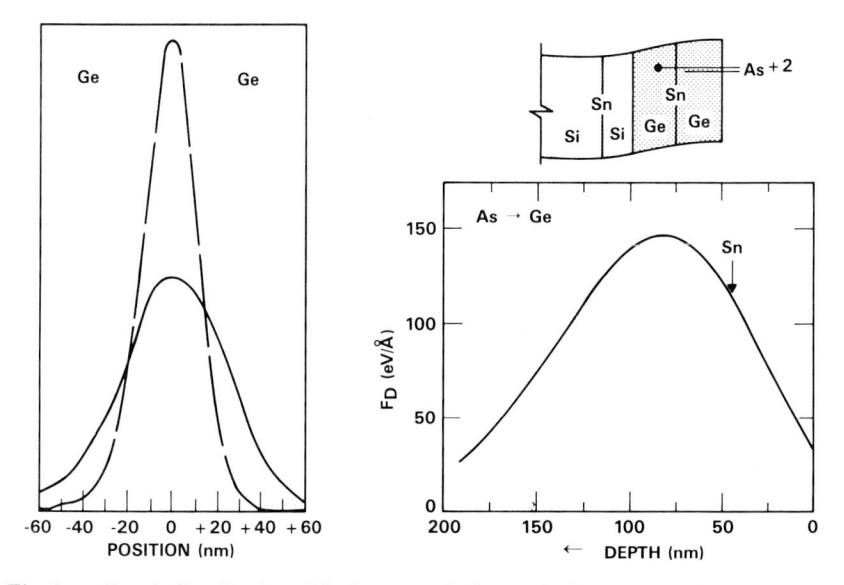

Fig. 3. Depth distribution of Sn between Ge layers before (broken line) and after (solid line) irradiation with 1×10^{16} cm^{-2} As ions at 360 keV. The Sn layer spread significantly more than in the case of Si neighboring layers, but no appreciable shift is observed. The energy deposition function is also seen to be several times larger than in the Si case.

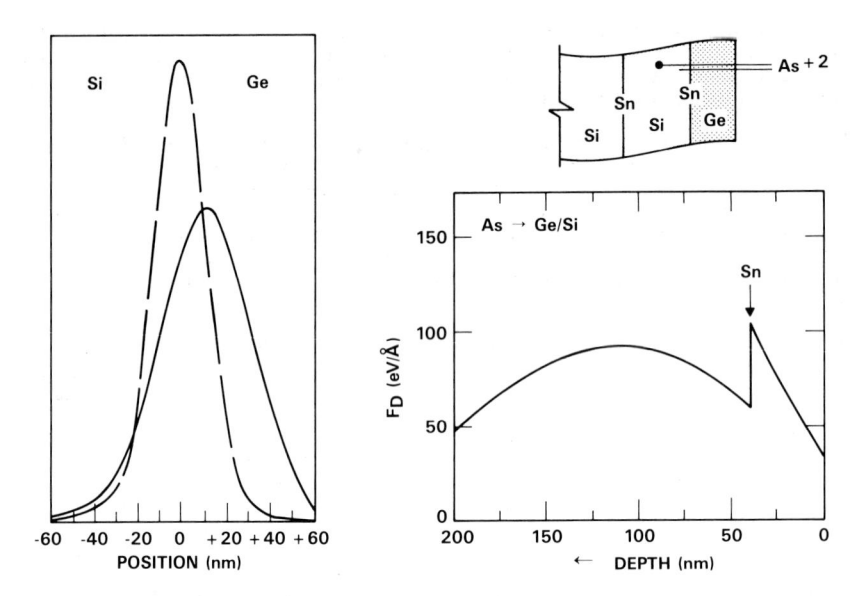

Fig. 4. Depth distribution of Sn between Si and Ge layers before (broken line) and after (solid line) irradiation with 1×10^{16} cm^{-2} As ions at 360 keV. The Sn layer spreads and drifts in the direction of increasing energy deposition (i.e. into the Ge) as seen at the right hand side of the figure.

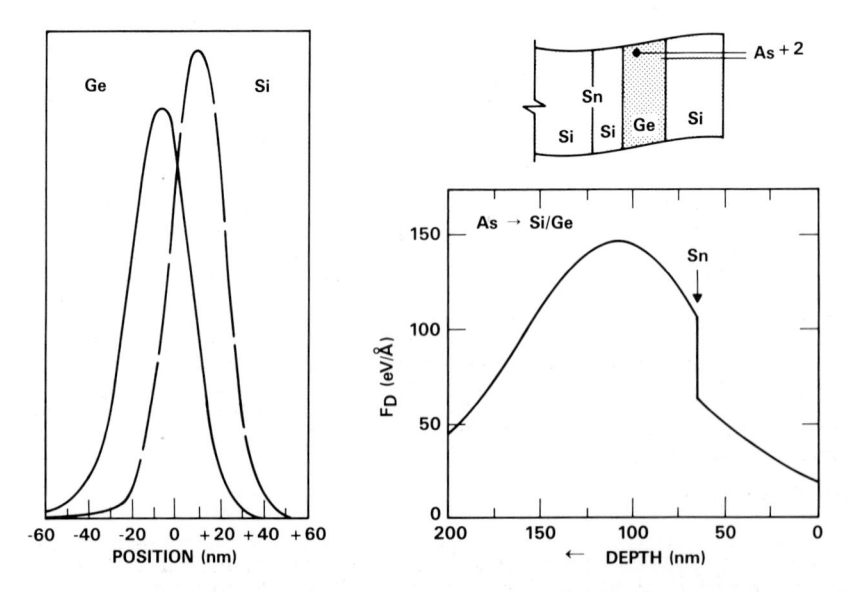

Fig. 5. Similar irradiation conditions as shown in figure 4, except the order of the layers is reversed. Once again the Sn layer drifts up the gradient in the energy deposition function (and diffusion coefficient) into the Ge rich layer.

during ion irradiation. This was done to detect spurious thermal induced effects. The temperature was maintained near room temperature by heat sinking the samples to a large aluminum block and maintaining the power density of the ion implant below about 100 milliwatts.

The redistributions produced by the ion beam were analyzed by use of Rutherford Backscattering Spectrometry with 2 to 3 MeV He ions produced in a 1 MV Pelletron. The elastically scattered He ions were energy analyzed using a silicon surface barrier detector at a laboratory scattering angle of 170°. The specimen normal was inclined to the incident ion beam by 60° to increase the depth sensitivity. A typical full spectrum is shown in figure 1. The specimen shown has Si as its outermost layer; schematically the arrangement is Si/Sn/Ge/Si/Sn/Si substrate, as illustrated in the small inset.

High resolution scanning electron micrographs were made of the cross section of cleaved samples. No large deviations from planarity were observed at the relatively low dose used in this experiment. With the use of heavier ions, roughening of sputtered single crystal Ge surfaces has been observed.[9] While unusual texture developed on the cross sectional face during the preparation of cross sectional TEM samples, this was brought about by the sputter thinning of the samples in an ion mill. In no case did the surface exhibit roughening on the face exposed to the As ion beam greater than about 10 nm, and that was of very low profile. While the non-planarity may produce some additional spreading in the marker width if irregularities were not present before irradiation, the absence of any motion of the reference marker assures one that the presence of the imperfections is insignificant in its effect on the position of the probe marker.

RESULTS AND DISCUSSIONS

The specimen in which the Sn layer had either Si alone or Ge alone as neighboring layers is shown in figures 2 and 3. One can see that the mixing of Sn in Ge is several times more efficient than in Si at room temperature. The approximate values of the mixing parameter D^*/ϕ found in this experiment are 3.4×10^{-29} cm^4 for Sn in Si and 1.7×10^{-28} cm^4 for Sn in Ge under 360 keV As irradiation, a factor of five difference in D^*. ϕ is the ion flux. The effect of sputtering of the surface Ge layer was corrected for by integrating the Ge peak signal in each spectrum and determining the shift in energy. Correcting for sputtering brought the reference peaks back into coincidence in all spectra.

In figures 2 and 3 the expanded distributions of the Sn layer before and after ion mixing show no significant change in mean position. This is to be expected since the gradient in the energy deposition function F_D is small, as is illustrated at the right side of each figure. The F_D distribution was calculated using an Edgeworth distribution having the moments obtained from interpolations of tabulated moments of the appropriate distributions.[10]. Figures 4 and 5, however, illustrate what occurs when an extreme gradient in F_D exists. In figure 4, the energy deposition increases toward the surface. The Sn distribution then drifts toward the surface. In the sample arrangement of figure 5, the increase in energy deposition and therefore the "diffusion" coefficient is toward the substrate side. Thus, the marker layer drifts inward. The Sn distribution which remains is approximately gaussian.

CONCLUSIONS

That anisotropic mixing can be brought about simply by a gradient in the probability of displacement has been demonstrated clearly in this work. The drifts produced are in the direction of increasing probability of displacement, i.e. in the direction of greater energy deposition. This observation is consistent with the experimental observations of drifts observed for thin buried marker layers in otherwise homogeneous matrices, although the anisotropy in the present work is much greater.

The implications of the present experiment are far-reaching. Consider the case of a heavy metal such as Pt on a Si substrate. In the metal rich layer Si is mixed much more efficiently than the metal is mixed in the Si rich substrate. The disparity is so great that the mixing can be thought of as taking place almost entirely in the metal rich layer. Furthermore, the mixing of Si should be greater than the heavier atom in the same matrix. This differential in the mixing rate, or diffusion coefficient, will result in sharp gradients of composition in the Si rich regions but relatively flat profiles in the metal-rich regions. With appropriate gradients at the Si-rich/Pt-rich interface, a nearly steady state condition may be established, such that the gradient in the diffusion coefficient compensates the gradient in the composition. If the product of the two gradients is a constant, then a steady state condition (constant flux) will obtain.

The large gradients which are ever present in ion mixing, especially in the mixing of concentrated binary alloys, will introduce strong anisotropic effects not often observed in standard diffusion. Attention must be given to the magnitude and character of such energy deposition gradients if a clear understanding of ion mixing is to emerge.

ACKNOWLEDGMENT

The authors gratefully acknowledge the technical assistance of R. Williams, K. Baker, and F. Clark.

REFERENCES

1. S. Matteson and M.-A. Nicolet in *Annual Reviews of Material Science*, 13 (1983) 33.

2. J.W. Mayer, B.Y. Tsaur, S.S. Lau and L-S. Hung, *Nucl. Instrum. Method* 182/183 (1981) 1.

3. P. Sigmund and A. Gras-Marti, *Nucl. Instrum. Method* 182/183 (1981) 25.

4. S.T. Picraux and W.J. Choyke-eds. *Material Research Society Symposia Proceedings*, Vol. 7, "Metastable Materials Formation by Ion Implantation," (Elesevier, New York, 1982) 1-99.

5. S. Matteson, B.M. Paine, M.G. Grimaldi, G. Mezey and M.-A. Nicolet, *Nucl. Instrum. Method* 182/183 (1981) 43.

6. B.M. Paine, *J. Appl. Phys.* 10 (1982) 6828.

7. P. Sigmund, *Appl. Physics A.* 30 (1983) 43.

8. M. Hansen, *Constitution of Binary Alloys* (McGraw-Hill, New York, 1958) 774.

9. B.R. Appleton, O.W. Holland, J. Narayan, O.E. Schow III, J.S. Williams, K.T. Short and E. Lawson, *Appl. Phys. Lett.* 41 (1982) 711.

10. K.B. Winterbon, *Ion Implantation Range and Energy Deposition Distributions*, Vol. 2 (Plenum, New York, 1975).

ION BEAM MIXING EFFECTS IN Au-Fe AND Pt-Fe THIN BILAYERS*

G. BATTAGLIN*, A. CARNERA*, G. CELOTTI**, G. DELLA MEA*,
V.N. KULKARNI*, S. LO RUSSO* AND P. MAZZOLDI*
* Unità GNSM-CNR, Dipartimento di Fisica dell'Università,
Via Marzolo 8, 35131 Padova, Italy; ** Istituto CNR-La.M.El.,
Via Castagnoli 1, 40126 Bologna, Italy

ABSTRACT

Mixing effects induced by Kr^{++} bombardment in the Au-Fe and Pt-Fe metallic systems have been studied by Rutherford backscattering and X-ray diffraction techniques. The mixed amount of Fe atoms shows a linear dependence on the square root of the Kr dose for both systems. The induced mixing appears more efficient for the Pt-Fe with respect to the Au-Fe system. In the case of Pt-Fe mixing is much more efficient when the initial bilayer structure has Pt on the top. The X-ray diffraction analysis reveals the formation of an extended solid solution of Fe in Pt, having the $Fe_{40}Pt_{60}$ composition.

INTRODUCTION

Ion mixing is a choice method in forming metastable materials with either crystalline or non-crystalline structure [1-4].
Cascade mixing and radiation enhanced diffusion are substantial in the formation of metastable phases, equilibrium compounds or solid solutions. Ion mixing appears not only a powerful technique for production of specialized alloy surfaces but is also attractive in studying the mechanisms involved in the irradiation process. In the present paper we report an investigation of the intermixing effects induced by Kr^{++} bombardment on the Pt-Fe and Au-Fe metallic systems.

EXPERIMENTAL

The Au-Fe, Pt-Fe (Fe on top in both cases) and Fe-Pt (Pt on top) bilayer structures were prepared by sequential e-beam vacuum ($\sim 10^{-6}$ torr) evaporation of the two respective metals onto Si single crystals or glass plates of 50 mm x 25 mm x 2 mm dimension.
In the second situation the element on the top (Fe) was deposited onto a tilted glass plate to produce a surface layer of varying thickness. The thicknesses for the different deposited layers ranged from 100 Å to 500 Å.
The samples were bombarded at room temperature with Kr^{1+} ions of 100 to 200 keV energy at doses ranging from 1×10^{15} to 3.5×10^{16} ions/cm^2. The beam current density was maintained at a value of 0.5 μA/cm^2 in order to avoid

----

* Work supported by Consiglio Nazionale delle Ricerche, Progetto Finalizzato Metallurgia.

heating effects.

The samples were analyzed by Rutherford backscattering spectrometry (RBS) using a 1.8 MeV ^{4}He$^+$ beam. Glancing angle X-ray diffraction (XRD) was employed to identify the phases produced by the bombardment. The surface morphology was investigated by Scanning Electron Microscopy (SEM).

RESULTS AND DISCUSSION

The scanning electron microscopic measurements revealed that the surface of the samples is flat before and after bombardment in all the cases. Therefore the backscattering spectra are free from the artifacts which would arise because of a surface structure having globules or some other inhomogeneities.

Fig.1 shows backscattering spectra from the sample regions unirradiated and irradiated with 120 KeV Kr^{++} to a dose of 3.5×10^{16} ions/cm^2 for the Au-Fe and Pt-Fe systems. The changes in the element signals from the bombarded region in comparison with those from the as-deposited one indicates the occurrence of mixing.

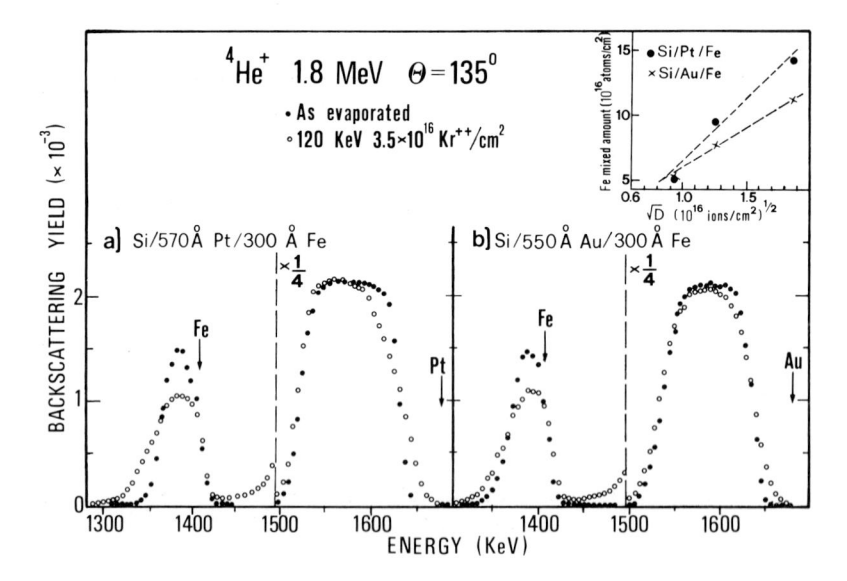

FIG. 1. Backscattering spectra of 1.8 MeV He$^+$ from the Si supported a) Pt-Fe and b) Au-Fe bilayers (Fe on top in both cases) having the indicated thicknesses, before and after Kr^{++} ion bombardment at room temperature. The RBS signals from Si are not shown in the figure. The inset shows the amount of Fe atoms in the mixed region as a function of the square root of the Kr^{++} dose (D). The error in the evaluation of the mixed Fe amount (see text) is ±15%.

The amount of Fe atoms which has been mixed is calculated by measuring the area under the Fe tail after subtraction of the contribution arising from the unreacted Fe layer. In these calculations we assume that the signals from the unreacted Fe layer are symmetric. The choice of a symmetric shape of these signals can give rise to systematic errors which have been estimated to be within ±15%. The inset in fig. 1 reports the amount of Fe atoms in the mixed region versus the square root of the Kr^{++} dose. The straight line dependence indicates that the rate limiting process [1,5,6] takes place in the mixed layer. The growth rate of the layer is proportional to the flux of atoms across the layer. Such a result excludes one-way processes, such as recoil implantation, where matter is transported in the direction of the beam only, suggesting that the mixing observed in the present cases is associated with a diffusion like transport of matter and with cascade mechanisms. It may be noted that in the case of the Pt-Fe system the mixing rate as a function of dose is higher as compared to that of the Au-Fe system. In order to better understand the mixing process mechanism in the Pt-Fe system we changed the relative position of the two metals and irradiated samples having the Pt layer on top. The backscattering spectra corresponding to the as evaporated samples and to the ones bombarded at a fluence of 2.5×10^{16} ions/cm^2 of Kr^{++} ions having 120 keV (a) and 200 keV (b) energies are reported in fig. 2.

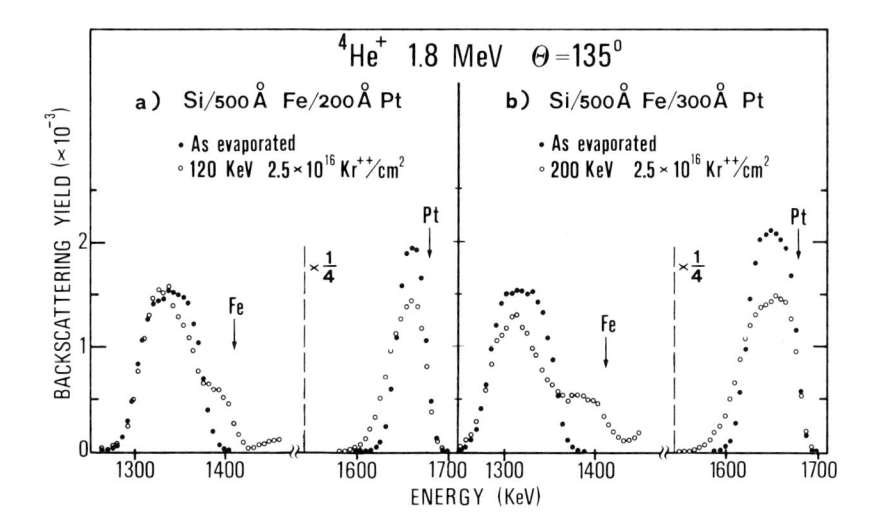

FIG. 2. Backscattering spectra of 1.8 MeV He$^+$ from the Fe-Pt system with Pt on top: a) Si substrate/500 Å Fe/200 Å Pt before and after bombardment with 120 keV Kr^{++}, and b) Si substrate/500 Å Fe/300 Å Pt before and after bombardment with 200 keV Kr^{++}. The fluence was 2.5×10^{16} ions/cm^2 and the implants were performed at room temperature. RBS signals from Si are not shown.

The mixing process appears more effective when Pt is on the top as it is evident by a comparison between fig. 1a and fig.2. For the 2.5×10^{16} Kr^{++}/cm^2 dose at the energy of 200 keV (fig. 2b) the abrupt change in the slope of the Pt back edge and the formation of a plateau in the Fe signal suggest the formation of a phase with a definite stoichiometry. Calculations based both on the height and the area measured for the Fe plateau and for the Pt spectrum indicate that the composition of the mixed region corresponds to $Fe_{40}Pt_{60}$.
The energy width of the Fe plateau and the FWHM of Pt signal match very well by considering a stopping of the $^4He^+$ ions in a Fe-Pt mixed region having the above mentioned composition. The mixed layer is about 600 Å thick. The glancing XRD measurements on this mixed region were performed using a Wallace-Ward cylindrical texture camera. The diffraction patterns are reported in fig. 3.

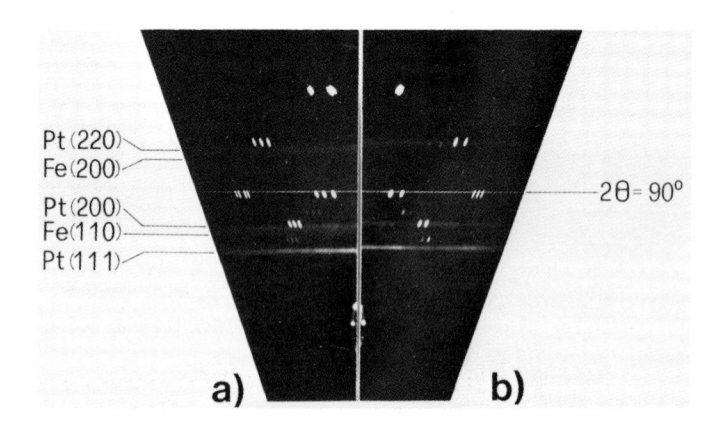

FIG. 3. Wallace-Ward cylindrical texture camera diffraction patterns of Si/500 Å Fe/300 Å Pt: a) as evaporated and b) after 200 keV Kr^{++} bombardment at room temperature to a fluence of 2.5×10^{16} ions/cm². The spots are relative to the Si single crystal substrate. The $2\theta = 90°$ is a reference line. CrK_α radiation (20 kV, 28 mA), angle of incidence of the beam = 20°.

The X-ray patterns obtained before the bombardment are shown on left side (a) and those obtained after the bombardment are shown on the right side (b) of the figure. We observe that after irradiation the X-ray lines corresponding to Pt are shifted with respect to those of pure Pt while the lines relative to Fe are not shifted but are considerably weaken because of the reduced amount of pure Fe. These results clearly show that the Fe-Pt phase produced after the bombardment has the fcc Pt structure with strained lattice parameters. The derived lattice parameter "a" is (3.827±0.007) Å for the irradiated system as opposed to (3.916±0.007) Å for pure Pt. The lattice parameters "a" [7] reported in literature for the fcc class of Pt_xFe_{1-x} alloys as a function of the composition satisfies the formula $x = 4.48 a - 16.55$ for $0.5 < x < 1$ with "a" in Å. The mean value of x, fitting with the lattice parameter of our phase, is 0.59±0.03 corresponding to $Pt_{59\pm3}Fe_{41\mp3}$ composition which is in a close agreement with the one calculated from the analysis of the RBS spectrum ($Fe_{40}Pt_{60}$, Fig. 2b).

To clarify the influence of the deposited energy densities in the mixing process we evaluated [8] the energy deposited in atomic processes in both metals, which form a layer pair. The deposited energy evaluations are obviously made only for the initial bilayer configuration of the samples. The composition at the interface changes during the bombardment and the incident particles observe different stopping values. In Table I are reported the calculated values of the energy deposited at the interface, in the two contiguous metal layers, 50 Å thick, and in the 100 Å thick layer through the interface.

TABLE I

Bilayer structure and thicknesses	Kr^{++} incident energy (keV)	Deposited energy at the interface (keV)		
		50 Å Fe	50 Å Au or Pt	Total (100 Å through the interface)
1) Si/550 Å Au/300 Å Fe	120	7.5	5.8	13.3
2) Si/570 Å Pt/300 Å Fe	120	9.0	7.5	16.5
3) Si/500 Å Fe/200 Å Pt	120	6.8	15.8	22.6
4) Si/500 Å Fe/300 Å Pt	200	5.0	13.7	18.7

Energy deposited in atomic processes by Kr^{++} ions around the interface of the Au-Fe and Pt-Fe bilayers for the configurations and energies used in the present experiment.

The influence of the relative nuclear energy deposition in the two layers is evident. For the Pt-Fe system the deposited energies in the Pt and Fe layers are comparable when the Fe film is on the top whereas the fraction of the energy deposited in the Pt layer becomes larger when the Pt film is on the top. For this last experimental situation uniform concentration of Pt and Fe has been obtained. On the basis of the energy deposition data, this fact can be understood as a consequence of enhanced diffusion of Fe in Pt because large number of defects are created in Pt due to the major part of the energy deposited in it. On the other hand for the inverse bilayer configuration, i.e. the system with Fe on top, the number of defects created in Pt is comparatively lower since the deposited energy in Pt is also relatively lower. In this case one might expect an enhanced diffusion of Pt in Fe which would give a uniform composition of Pt and Fe. The RBS spectra of fig. 1b does not indicate such a behaviour of Pt. Probably the enhanced diffusion process of Fe in Pt is much more efficient than that of Pt in Fe.

The comparison of the mixing behaviour of Au-Fe and Pt-Fe (inset of fig.1) shows that Fe mixes with Pt much more efficiently than it mixes with Au. This fact is confirmed by the results obtained after 100 keV Kr^{++} irradiation on the glass supported Au-Fe and Pt-Fe bilayer structures having top Fe thickness ranging from 100 Å to 300 Å. Of the two major processes, i.e. cascade mixing and enhanced diffusion, the former one ought to give similar effects in both systems since the masses of Au and Pt atoms are similar.

Therefore we attribute the different behaviour of the two systems to the enhanced diffusion processes which probably are more efficient in the Pt-Fe than in the Au-Fe system. Experiments with Fe-Au bilayer structure with Au on top should clarify this point.

The equilibrium phase diagram [9] for Fe-Pt binary systems shows that Fe and Pt are mutually soluble in all proportions at high temperatures ($\gtrsim 1350°C$). At lower temperatures down to 0°C the composition region extending from 60 to 80 at % Pt in Fe contains an equilibrium $FePt_3$ phase. The composition of the Fe-Pt phase which we observe in the case of 200 keV Kr irradiation (figs.2 and 3) lies in this region but we have shown above that this phase has the structure of fcc Pt with strained lattice parameter. Thus the ion mixing in this case has led to an extended solid solution of Fe in Pt.

In conclusion, the present experimental results, showing intermixing processes induced by Kr^{++} bombardment on the Pt-Fe and Au-Fe bilayers, suggest that enhanced diffusion mechanism plays a dominant role and is responsible for the different behaviour observed for the two systems under ion bombardment. Since the efficiency of enhanced diffusion depends on the density of created defects and, as a consequence, on the deposited energy, attention must be paid in the choice of the metal to be deposited on the top of a bilayer structure in order to obtain an efficient mixing between the two metallic species, taking into account beam energy, profile of deposited energy and defect mobility.

ACKNOWLEDGEMENTS

We thank Mr. G. Manente for the evaporations, Mr. E. Bolzan for the implantations and Mr. A. Rampazzo for drawings. One of us (V.N.K.) thanks the International Centre for Theoretical Physics (Trieste) for financial support in the form of a fellowship

REFERENCES

1. J.W. Mayer, B.Y. Tsaur, S.S. Lau and L.S. Hung, Nucl.Instr. Meth. 182/183, 1 (1981).
2. J. Poate, G. Foti and D. Jacobson eds., Surface Modification and Alloying, NATO Conference Series VI, Materials Science Vol.8, (Plenum, New York 1983).
3. S.S. Lau, B.X. Liu and M.A. Nicolet, Nucl.Instr.Meth. 209/210, 97 (1983).
4. B.X. Liu, W.L. Johnson and M.A. Nicolet, Nucl.Instr. Meth. 209/210, 229 (1983).
5. R.S. Averback, L.J. Thompson,Jr., J. Moyle and M. Schalit, J. Appl. Phys. 53, 1342 (1982).
6. S.U. Campisano, Chu-Te Chang, A. Lo Giudice and E. Rimini, Nucl. Instr. Meth. 209/210, 139 (1983).
7. Powder data file cards Nr. 26-1139, 29-716, 29-717, 29-718.
8. K.B. Winterbon, Ion Implantation Range and Energy Deposition Distributions, Vol. 2 (Plenum, New York 1975).
9. F.A. Shunk, Constitution of Binary Alloys (Mc Graw-Hill, New York 1969), p. 342.

ION BEAM IRRADIATION OF METAL FILMS ON SiO$_2$

G. J. CLARK,* J. E. E. BAGLIN,* F. M. d'HEURLE,* C. W. WHITE,** G. FARLOW,** and J. NARAYAN**
*IBM Thomas J. Watson Research Center, Yorktown Heights, New York 10598; **Solid State Division, Oak Ridge National Laboratory, Oak Ridge, TN 37830

ABSTRACT

Ion beam irradiation of metal film/SiO$_2$ interfaces causes reactions when the metals are those chemically capable of reducing SiO$_2$. These reactions result in the formation of metal rich silicides in the region of the interface and an increase in the adhesion of the film to the substrate. For other non-reactive metals ion irradiation causes lateral transport of metal atoms resulting in the formation of an island structure. The results obtained by ion irradiation are compared with previous studies of high temperature thermal processing of metal films on SiO$_2$.

INTRODUCTION

Metallization schemes to improve the adhesion of deposited metal films on insulating substrates are important for the fabrication of electrically conductive paths on insulating substrates for semiconductor device fabrication. In many cases, interfacial reactions between the film and the insulating substrate can lead to improved adhesion. High temperature thermal processing of metal films deposited on SiO$_2$ has been shown previously to result in the formation of metal rich silicides at the metal/SiO$_2$ interface [1-5]. Ion beam irradiation has previously been used to form silicides in the case of metal films deposited on silicon [6]. Also for some time people involved in fabricating Si devices [7] have been aware that a blanket implant of energetic ions through a metal film with a high melting point and consequently large atomic mobility would cause the formation of a silicide where the metal film was in contact with Si, whereas no reaction was observed in adjacent areas where the film is in contact with SiO$_2$. In the work reported here, we have used ion beam irradiation as a low temperature processing technique to cause interface reactions between deposited metal films and an underlying SiO$_2$ substrate. Although there has been some previous work [8] on the chemical effects in ion beam mixing of transition metals on SiO$_2$, to our knowledge this is the first reported use of ion beam irradiation to induce metal/SiO$_2$ contact reactions. In a number of cases, the result of the induced contact reactions is an increase in the adhesion of the film to the SiO$_2$ substrate.

EXPERIMENTAL

Thin metal films (~ 300 Å thick) of Nb, V, Ti, Hf, Zr, Cu and Pd were deposited in an oil free vacuum system onto thermally oxidized silicon wafers. The SiO$_2$ thickness was ~ 2000 Å. Ion beam irradiation was carried out in 10^{-7} Torr vacuum at room temperature using Xe$^+$ ions at energies of 200-350 keV. For each irradiation, the ion energy was chosen such that the peak in the deposited energy distribution would be in the vicinity of the metal/SiO$_2$ interface. A dose of 1×10^{16} ions/cm^2 was used for each irradiation. Following ion beam mixing, the films were examined by Rutherford backscattering spectrometry (RBS), secondary electron microscopy (SEM), transmission electron microscopy (TEM), glancing angle x-ray diffraction and electron diffraction.

RESULTS

The results of these experiments fall into two general categories: (a) metal/SiO$_2$ systems, which undergo interface reactions during ion beam mixing; (b) metal/SiO$_2$ systems which exhibit lateral segregation with no indication of interface reactions.

a) An example of a case in which interfacial reactions occur is shown by the RBS results for Nb (300 Å) on SiO$_2$ in Fig. 1. These RBS spectra were obtained using 2.0 MeV He$^+$ ions in a grazing exit angle geometry for enhanced depth resolution. Following ion beam mixing (Xe$^+$, 200 keV, 1×10^{16}/cm^2), there is a decrease in the scattering yield from Nb in the region of the front surface (Fig. 1). The change in scattering yield near the interface suggests an interfacial reaction between the metal and the insulating substrate. The interfacial reaction appears to have consumed about 40% of the original film thickness. In Fig. 1, Xe$^+$ irradiation results in the transport of silicon from the SiO$_2$ toward the surface, again consistent with an interfacial reaction between Nb and SiO$_2$ during ion beam mixing. The movement of Si is also through about 40% of the film. Analysis of the backscattering results shows that the amounts of reacted Nb and Si are consistent with the formation of Nb$_3$Si at the metal SiO$_2$ interface during ion beam mixing. In Fig. 1 ion beam mixing also gives rise to an increase in the oxygen content and a decrease in the scattering yield from Nb near the front surface. These results are consistent with the formation of a metal oxide at the front surface during ion beam mixing, but the resulting oxide is thin and the composition cannot be determined.

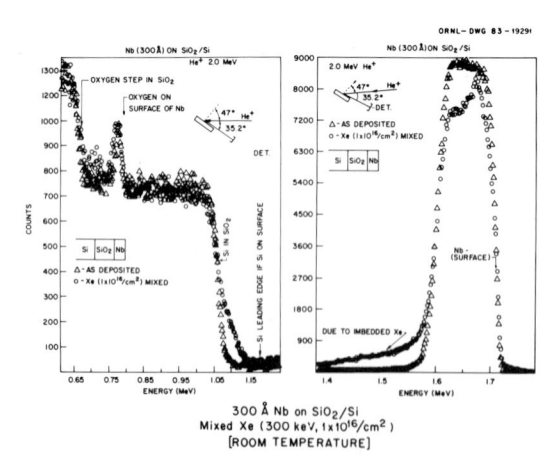

Fig. 1   Rutherford backscattering spectrum for Nb (300 Å) on SiO$_2$/Si before and after Xe$^+$ ion irradiation.

For the case of Nb on SiO$_2$, SEM images of the surface obtained before and after irradiation were smooth and featureless. Glancing angle Debye-Scherrer-Hull powder camera x-ray diffraction with a stationary target showed that a new polycrystalline phase with texture was produced by the ion beam irradiation. In Fig. 2 these powder camera results are shown for a non-irradiated (Fig. 2a) and an irradiated (Fig. 2b) part of the Nb film. These results were obtained by using Cu x-rays. In Fig. 2b the new phase is seen as a line on the irradiated sample but unfortunately the phase cannot be identified from the data. The finite length of the line indicates texture.

On the other hand, Bragg-Brentano x-ray diffraction analysis of the irradiated Nb sample (Figure 3) did not show the formation of a new texture phase. Figure 3 shows only peaks corresponding to the Nb film and the Si substrate.

new polycrystalline phase with texture

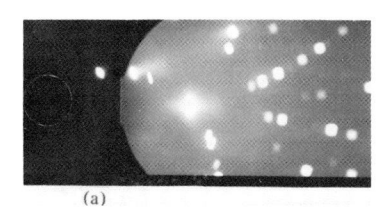

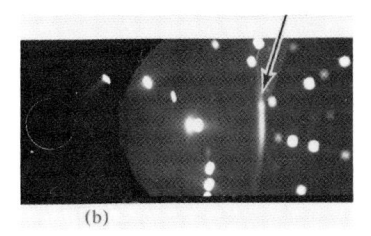

Fig. 2 X-ray diffraction pattern from a glancing angle Debye-Scherrer-Hull powder camera image of a Nb (300Å) on SiO_2/Si sample before (Fig. 2a) and after (Fig. 2b) Xe^+ irradiation (300 keV, $1.10^{16} cm^{-2}$).

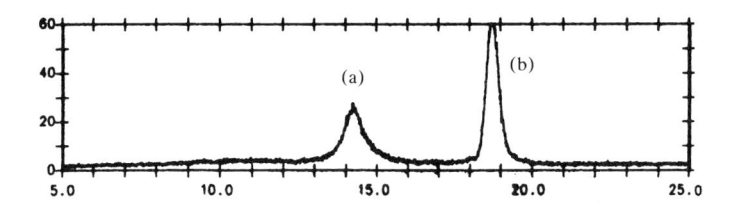

Fig. 3 Bragg-Brentano diffraction pattern from ion beam irradiated Nb on SiO_2/Si. The peaks correspond to Si(a) Nb metal (b). There is no evidence for a silicide phase in this data.

An electron diffraction pattern (Figure 4) showed a crystalline (dots) and an amorphous (ring) phase of (presumably) Nb_3Si to be present in agreement with the Rutherford backscattering analysis. The electron beam is looking at a very small spot on the sample. The Bragg-Brentano analysis possibly does not show the new Nb_3Si phase because of the small amount of the silicide present in the crystalline form.

Fig. 4 Electron diffraction pattern from ion beam irradiated Nb on SiO_2/Si. The pattern shows the presence of a crystalline (dots) and amorphous (ring) phase of Nb_3Si.

A cross-sectional TEM photograph of the Nb on SiO$_2$ sample before and after irradiation is shown in Fig. 5. The interface mixing following radiation is clearly visible in these photographs.

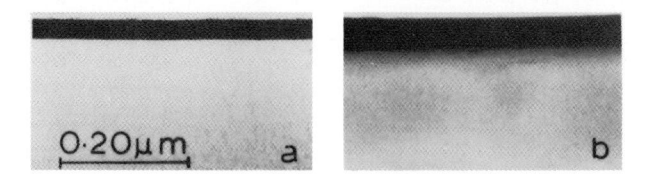

Fig. 5   Cross-sectional TEM photograph of a Nb(300Å) on SiO$_2$/Si sample before (Fig. 5a) and after (Fig. 5b) irradiation. (Xe$^+$, 300 keV, 1.10^{16}cm^{-2}).

Similar RBS and glancing angle x-ray diffraction analysis showed interfacial reactions in the V(300Å)/SiO$_2$ system, suggesting the formation of a silicide during ion irradiation (Xe$^+$, 200 keV, 1×10^{16}/cm^2). The stoichiometry of the silicide could not be determined because the extent of interfacial reaction was not as great as for the Nb/SiO$_2$ case.

b) A number of metal/SiO$_2$ systems do not appear to undergo interfacial reactions during ion irradiation. Examples of this category of results are shown by the SEM images in Fig. 6 of the surface topography of Cu (300 Å) and Pd (300 Å) on SiO$_2$ following ion irradiation. Prior to irradiation, the surfaces of the deposited films are smooth and featureless. As a result of ion irradiation, lateral transport of metal over thousands of angstroms has occurred at temperatures near room temperature. We cannot rule out the possibility that limited interfacial reaction has occurred over a very shallow depth because RBS analysis cannot be used to determine interface reactions when the surface topography has been changed so radically by the ion irradiation process.

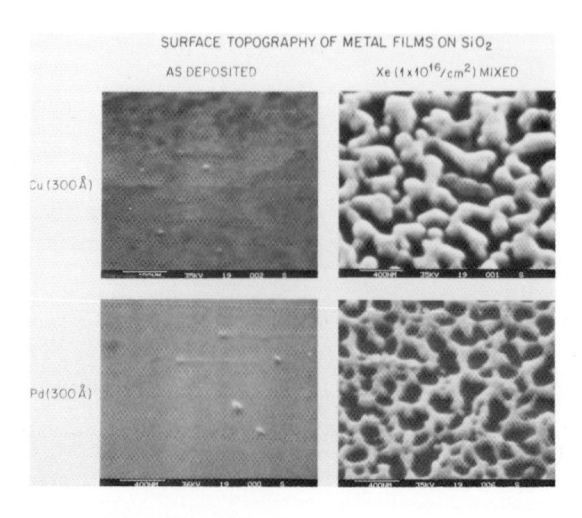

Fig. 6   Surface topography of Cu and Pd films (300Å thick) deposited on SiO$_2$ before and after Xe$^+$ irradiation (10^{16}/cm^2).

ADHESION RESULTS

Standard pull adhesion tests on the Nb film showed a factor of 4 improvement in adhesion (from 0.5 to 2.0 grams/mm^2) following ion bombardment. Improvements in adhesion (by up to a factor of 3) as a result of ion irradiation were also found for V, Ti, Hf and Zr films, again suggesting interfacial reactions in these systems. Of course, no improvement in adhesion was observed in those samples that balled up under irradiation.

DISCUSSION

The previous sections describe, for the first time, the formation of metal silicides on SiO_2 at temperatures near room temperature. We have facilitated this reaction using ion beam irradiation. Previous work by others has shown that thermal processing at elevated temperatures of similar deposited metal films on SiO_2 gives rise to interfacial reactions which result in the formation of silicides [1-5]. In that work it was found that metal rich silicides of Group IVa and Va films were formed as a result of thermal processing at temperatures greater than 700°C. For other metals (such as Cu and Pd) interfacial reactions did not occur during thermal processing and the metal was observed to coalesce to form an island structure [1].

In our experiments, the interfacial reaction is driven by the energy deposited into the collision cascade and not by simple thermal activation because the temperature at the surface during irradiation was always less than 100°C. In the cases where interfacial reactions do occur (such as Nb on SiO_2), ion beam irradiation initiates atomic motion which is driven by chemical forces until a stoichiometric metal rich silicide (believed to be Nb_3Si) is formed.

The previous thermal processing results have been explained on the basis of the heats of formation of the products compared with those of the reactants [1]. If the heat of reaction of the products is less than that of the reactants for the reaction

$$M_x + SiO_2 \rightarrow M_ySi + M_{x-y}O_2 \tag{1}$$

(where M is the metal) then the free energy change is negative and the reaction is thermodynamically possible and is often observed. This explanation favors reaction for those metal films which have a high affinity for oxygen, and it is those films which react readily.

A similar explanation to that described above is believed to operate in the ion beam mixing case also. The same metal/SiO_2 systems are observed to undergo interfacial reactions during ion irradiation and thermal processing, while other systems exhibit lateral transport or coalescence during both types of processing. It is interesting to note that for the case of Nb on SiO_2, Nb_3Si is formed in both the ion beam irradiation case and in the thermal processing case [2]. The similarity of implantation and thermal mechanisms for metal-on-silicon had been pointed out previously [9]. The present observations for metal-on-SiO_2 seems to confirm those earlier conclusions.

In those cases such as Cu and Pd on SiO_2 where island coalescence is observed during ion irradiation and during high temperature thermal processing, [1] the free energy change is positive, implying that interfacial reactions are not likely. In these cases the energy deposited during ion irradiation is transferred into atomic motion and surface diffusion takes place over thousands of angstroms. Each incident ion must mobilize and transport at lease ten metal atoms. Transport of this many atoms over these distances implies that in these systems the surface mobilities are very high. As is observed during thermal annealing the coalescence of the islands is driven by the high surface energy (both free upper surface and lower interface) of the original planar film. This also (as for the case of chemical reactions) corresponds to a decrease in free energy of the system.

SUMMARY

We have shown that ion irradiation can give rise to interfacial reactions in certain SiO_2 systems at temperatures near room temperature. The systems which react are the same as those that react during thermal processing at elevated temperatures. These are systems in which the free energy change is negative. The silicide phase formed, viz Nb_3Si in the case of Nb on SiO_2 is the same for both ion beam and thermal processing.

For systems in which the free energy change would be positive, interfacial reactions are not observed and the metal film undergoes lateral transport and coalesces into an island structure during irradiation (Pd and Cu on SiO_2). In these systems a similar coalescence has been observed during thermal processing at elevated temperatures [1].

Interfacial reactions initiated by ion irradiation give rise to increases in adhesion of the film to the SiO_2 by up to a factor of 4. This resulting increase in adhesion suggests the potential use of ion beam irradiation to increase the adherence of metal circuit lines to insulators by low temperature processing.

ACKNOWLEDGMENTS

The authors thank J. Karisinski for the Bragg-Brentano x-ray diffraction analysis.

REFERENCES

1. R. Pretorius, J. M. Harris and M. A. Nicolet, Solid State Electronics 24, 667 (1978).
2. H. Kräutle, W. K. Chu, M. A. Nicolet, J. W. Mayer and K. N. Tu, p. 193 in Applications of Ion Beams to Metals, ed. by S. T. Picraux, E. P. EerNisse and F. L. Vook, Plenum Press, New York, 1974.
3. H. Kräutle, M. A. Nicolet and J. W. Mayer, J. Appl. Phys. 45, 3304 (1974).
4. K. N. Tu, J. F. Ziegler and C. J. Kircher, Appl. Phys. Lett. 23, 493 (1973).
5. H. Kräutle, M. A. Nicolet and J. W. Mayer, Phys. Stat. Sol. 20, K33 (1973).
6. See for example, B. Y. Tsaur, S. S. Lau and L. S. Hung, Nucl. Inst. Meth. 182/183, 1 (1981), and references contained therein.
7. E. Nagasawa, M. Morimoto and H. Okabayashi in "1982 Symposium on VLSI Technology, Digest of Technical Papers," (Business Center for Academic Societies of Japan, Tokyo, 1982) p. 26.
8. T. Banwell, B. X. Liu, I. Golecki and M. A. Nicolet, Nucl. Inst. Meth. 209/210, 125 (1983).
9. F. d'Heurle, C. S. Petersson and M. Y. Tsai, J. Appl. Phys. 53, 8765 (1982).

RADIATION ENHANCED DIFFUSION IN ION-IMPLANTED GLASSES AND GLASS/METAL COUPLES*

G. W. ARNOLD
Sandia National Laboratories, Albuquerque, NM 87185

ABSTRACT

Ion implantation causes alkali migration to the surface in alkali silicate glasses. Rutherford backscattering spectrometry was used to follow this depletion. Room temperature implantations of 5×10^{16} 250 keV Xe/cm^2 in $12M_2O \cdot 88SiO_2$ (M = Li,Na,K,Rb,Cs) removes approximately equal numbers (within a factor of 2) of alkali from the glass. Low temperature (77K) implants significantly reduce the alkali loss. These results imply a radiation-enhanced diffusion mechanism in which the alkali interchanges with the products of the collision cascade, with the kinetics being limited by the radiation damage components. The results for mixed-alkali glasses $((12-x)M_2O \cdot xCs_2O \cdot 88SiO_2)$ give further evidence for this process. In glass/metal couples, radiation enhanced diffusion allows the interchange of glass network components with deposited metals. Rutherford backscattering spectrometry was used to follow the interchange of silicate and phosphate glass components with metal ions near the heavy-ion implanted interface between glass substrate and metal (Al,Zr) films.

INTRODUCTION

Alkali depletion in alkali-containing glasses subjected to electron or ion bombardment has been noted in many instances (see, e.g., [1-3]). This early work has been reviewed by Carter and Grant[4]. Charged particle irradiation resulted in movement of the alkali ions toward the surface and into the vacuum from a depth roughly corresponding to the depth of penetration of the particle. This effect makes Auger electron spectroscopy (AES) difficult in alkali-containing glasses because of the disappearance of the alkali signal with time due to the electron (0.3-3 keV) analysis beam[5-8]. Alkali mobility can also be a problem in secondary-ion-mass-spectroscopy (SIMS) which use ion-beam milling in order to depth profile elemental concentrations[9]. Rutherford backscattering spectroscopy (RBS), however, allows accurate measurement of the depth-profiles of near-surface alkali because of the higher beam energy (1-3 MeV) and the ability to simultaneously probe in depth without high beam fluence.

Ion implantation can alter the near-surface properties of glasses either through the chemical activity of the implanted species or by the implantation damage. Arnold and Borders[10-12] used RBS to demonstrate that heavy ion implantation removed alkali ions from lithia-alumina-silicate glasses in the region of the implanted-ion distribution. Arnold and Peercy[13] used inert gas implantation to bring about low-temperature crystallization of a $Li_2O \cdot 2SiO_2$ glass. Crystallization occurred because the loss of Li from the implanted region caused this region to be readily phase separable.

---

*This work performed at Sandia National Laboratories supported by the U. S. Department of Energy under contract #DE-AC04-76DP00789.

Two different models have been proposed to account for the depletion phenomenon[14-18]: a) an electric field enhanced movement for electrons and light ions (H)[14-16], and b) a model of radiation-enhanced diffusion in the implanted region in the case of heavy-ion implantation[17,18]. Radiation-enhanced migration has also been used in the interpretation of the observed interpenetration of ion-implanted semiconductor/metal film composites to form stable compounds[19].

In the present paper, RBS techniques are used to study the alkali depletion in binary alkali-silicate glasses and in mixed-alkali glasses. The results can be interpreted in terms of a radiation-enhanced diffusion mechanism. Results are also presented on preliminary work on radiation-enhanced mixing of glass and metal composites. These latter experiments could lead to useful changes in the surface properties of glass.

EXPERIMENTAL

The alkali-silicate glasses studied had the composition $12M_2O \cdot 88SiO_2$ (M = Li,Na,K,Rb,Cs). The mixed-alkali glasses were Cs-based and had the composition $(12-x)M_2O \cdot xCs_2O \cdot 88SiO_2$ (x = 3,6,9 and M = Li,Na,K,Rb). All glasses were finished either with diamond paste or alumina in non-aqueous media or on dry paper. Sample dimensions were 1.27 cm x 0.635 cm x 0.1 cm.

Implantations of Xe or N_2^+ were made using an Accelerators, Inc. ion implanter (10-250 keV) with samples at room temperature or near 77K. Rutherford backscattering (2.8 MeV ^{4}He) depth profile analyses were made using the Sandia National Laboratories tandem Van de Graaff accelerator.

Standard evaporation techniques (10^{-8} Torr base vacuum system) were used to produce the thin Al and Zr films on the samples used for studies of interpenetration of metal and glass constituents as a result of ion implant damage.

RESULTS AND DISCUSSION

Alkali Depletion

Figure 1 shows the depletion of K from a $12K_2O \cdot 88SiO_2$ glass as a function of 250 keV Xe ion fluence. These RBS spectra show only the K-edge region and have been vertically separated for clarity. The energy losses of the 2.8 MeV He beam have been converted to a depth scale as shown at the top of the figure. The projected range is $R_p \cong 1060$ Å using the empirical range relation, $R_p = (530 \times E(keV)/\rho Z_1)$, where $\rho = 2.31$ g/cm^3 and $Z_1 = 54$, established by Wu et al[20]. The K depletion profile in the implanted region at the lower fluences mirrors the expected Xe ion distribution.

Implantations were also made at 77K (beam measurements at RT). For the K, Rb, and Cs glasses, the depletion was much less (see Table I). The loss for the Li and Na glasses could not be accurately determined due to movement of these ions into the near-surface region during the 77K implantation; however, the overall loss was relatively small.

For the mixed-alkali glass system $(12-x)M_2O \cdot xCs_2O \cdot 88SiO_2$ (x = 3,6,9; M = Li,Na,K,Rb), implants of N_2^+ were used in order not to complicate the RBS profiles. Figure 2 shows the results for a room-temperature implantation of 5×10^{16} 100 keV N_2^+/cm^2 on $6K_2O \cdot 6Cs_2O \cdot 88SiO_2$ glass. The insets show the loss profiles of Cs and K. Table I shows the integrated losses for the glasses in this series as well as the RT and 77K data for the binary glasses. Figure 3 shows the amount of Cs depleted as a function of the molar concentration of Cs for the Li, Na, and K glasses. Although these ions (Li,Na,K) vary widely in mass and mobility, this plot shows that the amount of Cs depleted depends on its concentration. As seen in Table I, the total alkali depletion for these glasses is the same within a factor of two. These data imply, as do the 77K data for the binary glasses, that alkali

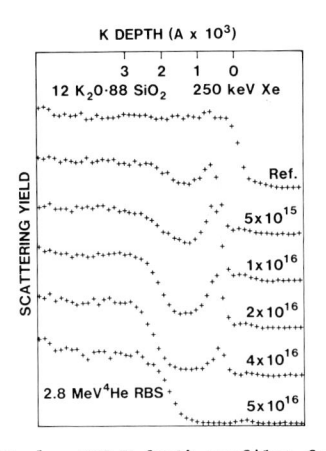

FIG. 1. RBS K depth profiles for 250 keV Xe implanted into 12K$_2$O·88SiO$_2$ glass at various fluence levels. The traces are offset for clarity.

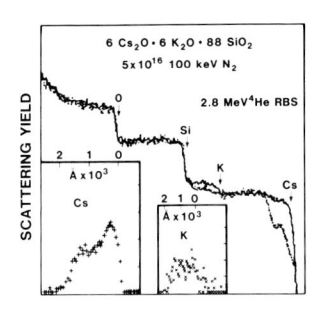

FIG. 2 RBS histogram spectra for a 6K$_2$O·6Cs$_2$O·88SiO$_2$ glass after implantation at room temperature with 5x10^{16} 100 keV N$_2$/cm^2. The inset shows the Cs and K loss profiles obtained by subtracting the implanted spectrum from the reference spectrum. The integrated loss values are given in Table I.

removal is directly related to the damage produced by implantation and is not dependent -- to first order -- on the characteristics of the individual ions. The mechanism by which alkali movement occurs is probably that of an increased local diffusion coefficient in the damaged region and the damage concentration gradient as has been discussed by Liau et al[21]. For this model of radiation enhanced migration, the rate limiting factors for the depletion are the sputtering rates for the various ions in the outermost layers of the sample. Simple partition of energy considerations would lead to the conclusion that the loss for equal energy deposition should be in the order Li>Na>K. However, this conclusion must be modified somewhat due to the fact that single M-O bond strengths[22] increase by a factor of ~ 3 in going from Li to K. In addition, incumbent H on the glass surface moves into the ion-damaged region by an exchange mechanism[13] with the alkali ions. The in-diffusion of H (measured by elastic recoil detection (ERD))

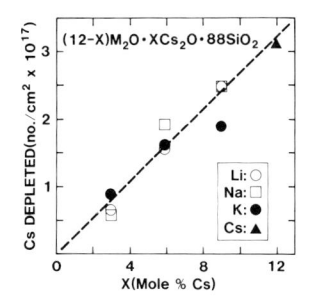

FIG. 3. Cs depleted from (12-x)M$_2$O·xCs$_2$O·88SiO$_2$ glass (M = Li,Na,K) as a function of mole % Cs for 5x10^{16} 100 keV N$_2$ ions/cm^2.

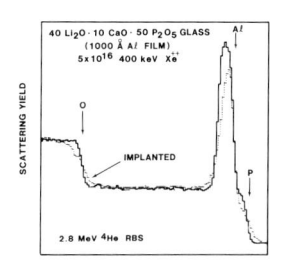

FIG. 4. RBS spectra for 40Li$_2$O·10CaO·50P$_2$O$_5$ glass with a 700Å Al film before and after 5x10^{16} 400 keV Xe^{++}/cm^2 at room temperature.

is in the order Li<K<Na. These considerations make it difficult to model the alkali-loss process precisely and accounts for some of the non-systematic losses and scatter in the data of Table I. The importance of the degree of damage in the loss process can be inferred by noting that, on the average, the loss of alkali per ion for the Xe implantation is about 2 while for the N_2 implant the loss per N ion is about 0.5, a ratio of losses of about 4. This is consistent with a ratio of energy deposition per Xe ion to that of the N ion which is about 5; i.e., the loss depends on the energy deposition or damage in the collision cascade.

Ion Interchange at Glass/Metal Interfaces

The apparent increased mobility of displaced glass constituents in ion-implanted damage regions demonstrated in the previous section has encouraged investigation of the possibility of modifying, on an atomic scale, the composition of glass surfaces through the interchange of ions at the ion-damaged interface between the glass substrate and deposited thin metal films. Figure 4 shows RBS spectra for a phosphate glass ($40Li_2O \cdot 10CaO \cdot 50P_2O_5$) with a 1000 Å Al film implanted with 400 keV Xe at a fluence of 5×10^{16} 250 keV Xe ions/cm^2. Although the extent of atomic migration is small, it is apparent that interchange of Al ions with O and P ions has occurred. The composition of the interchanged region has not been determined. Li is also distributed (not shown) throughout the film. We have previously shown results for Al on SiO_2 and for Al on a more complex phosphate glass [23]. SEM and optical microscope examinations of film texture show that films on phosphate glasses are not as likely to "ball-up" when implanted relative to films on silicate glass. Phosphate glasses have many desirable features (e.g., low melting point, high thermal expansion coefficient) but are much more susceptible to erosion in aqueous solutions or in humid environments. The present results suggest that Al and/or Zr (Zr data not shown) can be incorporated into the glass surface network. These ions are expected to increase the durability of either silicate or phosphate glasses[24]. These investigations will be extended and compared to glass modification using implanted Al and Zr instead of thin films.

CONCLUSIONS

Ion implantation into alkali-silicate glasses removes alkali by a damage-enhanced local diffusion coefficient throughout the implanted region. To first order, the number of alkali ions depleted depends on the energy deposited into displacement damage per ion and not on the properties of the ions. Ion implantation damage also makes possible the interchange of metal film atoms and glass network constituents. This latter effect may allow improvement in the adhesion and surface chemical durability of these glasses.

TABLE I. Alkali ion depletion in $xM_2O \cdot (12-x)Cs_2O \cdot 88SiO_2$·glasses for 5×10^{16} 100 keV N_2 ions/cm^2 and 5×10^{16} 250 keV Xe ions/cm^2 (Xe implant values in parentheses).

		5×10^{16} 100 keV N_2		
M	Alkali (mole %)	Depleted $M(\times 10^{16})$	Depleted $Cs(\times 10^{16})$	Depleted Total$(\times 10^{16})$
Li	3	0.78	2.51	3.29
	6	1.34	1.55	2.89
	9	2.28	0.65	2.93
	12	3.56	-	3.56
	(12)	(13.0)(250 keV Xe)		
Na	3	5.4	2.5	7.9
	6	7.6	1.92	9.52
	9	4.1	0.57	4.7
	12	7.4	-	7.4
	(12)	(11.0)(250 keV Xe)		
K	3	1.2	1.9	3.1
	6	2.9	1.6	4.5
	9	2.9	0.9	3.8
	12	6.3	-	6.3
	(12)	(12.0)(250 keV Xe:RT)		
	(12)	(4.9)(250 keV Xe:77K)		
Rb	3	2.8	0.8	3.6
	6	2.1	0.7	2.8
	9	5.0	0.7	5.7
	12	3.9	-	3.9
	(12)	(10.0)(250 keV Xe:RT)		
	(12)	(6.3)(250 keV Xe:77K)		
Cs	12	3.1		
	(12)	(6.0)(250 keV Xe:RT)		
	(12)	(1.4)(250 keV Xe:77K)		

REFERENCES

1. D. G. Bills and A. A. Evett, J. Appl. Phys. 30 (1959) 564.

2. J. L. Lineweaver, J. Appl. Phys. 34 (1963) 1786.

3. L. Holland, The Properties of Glass Surfaces (Chapman and Hall, London, 1966) p. 317.

4. G. Carter and W. A. Grant, Phys. Chem. Glasses 7 (1966) 94.

5. S. Thomas, J. Appl. Phys. 45 (1974) 161.

6. C. G. Pantano, Jr., D. B. Dove, and G. Y. Onada, Jr., J. Vac. Sci. Techn. 13 (1976) 414.

7. D. M. Usher, J. Phys. C: Solid State Phys. 14 (1981) 2039.

8. P. G. Whitkop, Surf. Sci. 10 (1981) 261.

9. B. Rauschenbach and G. Blasek, Appl. Phys. 22 (1980) 233.

10. G. W. Arnold and J. A. Borders, J. Appl. Phys. 48 (1977) 1488.

11. G. W. Arnold and J. A. Borders, in: Applications of Ion Beams to Materials, 1975, Eds. G. Carter, J. S. Collington, and W. A. Grant, Conference Series, No. 28 (The Institute of Physics, Bristol, 1976) 121.

12. J. A. Borders and G. W. Arnold, in: Ion Beam Surface Layer Analysis, Vol. 1, 1976, Eds. O. Meyer, G. Linder, and F. Kappeler (Plenum, NY, 1977) 415.

13. G. W. Arnold and P. S. Peercy, J. Non-Crystalline Solids 41 (1980) 359.

14. G. Battaglin, G. Della Mea, G. DeMarchi, and P. Mazzoldi, Rad. Eff. 64 (1982) 99.

15. G. Battaglin, G. Della Mea, G. DeMarchi, P. Mazzoldi, A. Miotello, and M. Guglielmi, J. Phys. C: Solid State Phys. 15 (1982) 5623.

16. A. Miotello and P. Mazzoldi, J. Phys. C: Solid State Phys. 15 (1982) 5615.

17. A. Miotello and P. Mazzoldi, J. Phys. C: Solid State Phys. 16 (1983) 221.

18. P. Mazzoldi, Nucl. Inst. Methods 209/210 (1983) 1089.

19. J. W. Mayer, B. Y. Tsaur, S. S. Lau, and L-S. Hung, Nucl. Inst. Methods 182/183 (1981) 1.

20. W. K. Chu, B. L. Crowder, J. W. Mayer, and J. F. Ziegler, Appl. Phys. Lett. 22 (1973) 490.

21. Z. L. Liau, J. W. Mayer, W. L. Brown, and J. M. Poate, J. Appl. Phys. 49 (1978) 5295.

22. F. V. Tooley in: The Handbook of Glass Manufacture, Vol. II, Ed. F. V. Tooley (Books for Industry, Inc., NY, 1974) p. 999.

23. A preliminary account of these investigations has been presented at the 2nd International Conference on Radiation Effects in Insulators, Albuquerque, NM, May 30-June 3, 1983. (Nucl. Inst. Methods, to be published).

24. A. Paul, J. Mats. Sci. 12 (1977) 2246.

ELECTRICAL RESISTIVITY OF MULTILAYERS DURING ION BEAM MIXING

J. GRILHE, J.P. RIVIERE, J. DELAFOND, C. JAOUEN, C. TEMPLIER
Laboratoire de Métallurgie Physique L.A. 131 du C.N.R.S.
40, avenue du Recteur Pineau 86022 Poitiers (France)

ABSTRACT

A new approach is developed, employing "in situ" electrical resistivity measurements, as a tool to study ion beam mixing of evaporated metal-metal multi or bilayers. The electrical resistivity variations measured continuously during the ion bombardment exhibit a monotonical increase and a tendency toward a saturation process allowing to detect precisely the total mixing of the film. The volume fraction of intermixed atoms can be determined within the framework of a simple conduction model. Experimental results are given in the case of Fe-Al and Al-Ag multilayers.

INTRODUCTION

The application of energetic ion beams to surface processing of materials has been recently developed in order to improve the mechanical properties of the surface on engineering components (1, 2). The possibility to produce surface alloys by a process which do not involve heating the bulk of the material represents an important advantage of ion implantation. In addition, the high defect concentration produced in the collision cascades modifies considerably the diffusion coefficients and the classical compositional limitations connected with the solid solubility are surpassed and new metastable alloys can be produced either single phase solid solution or amorphous alloys. The practical importance of these thin adherent alloys layers has been demonstrated in many cases (3) for wear and corrosion properties.

Unfortunately, high dose implantations are necessary to produce concentrated alloys, generally above 10^{17} ions/cm^2 and a physical limitation to the maximum concentration achievable is imposed by the sputtering phenomenon. Consequently, direct implantation is restricted to low concentrations and small alloyed depth.

Tsaur et al.(4, 5) have tried successfully to overcome the precedent limitation by using the ion beam mixing technique combining thin layer evaporation and ion bombardment. Successive layers are deposited on the material and bombarded with Xe ions in order to homogenize the different layers to an atomic scale. The process necessitates Xe ion doses several orders of magnitude lower than direct implantation reducing the influence of sputtering. Moreover, the average film composition can be varied easily in a wide concentration range by only adjusting the relative thickness of the individual layers. Different configurations of layered materials have been studied : bilayers, multilayers or also thin markers. However, for potential applications, the multilayer structure appears more convenient in producing alloys of desirable composition than the bilayer one because the composition is forced to a predetermined value. From the experimental stand point most of the studies have used the RBS technique (6, 7, 8) which appears powerful in very specific cases for bilayers or thin markers. But in the case of multi-layered materials, the experimental results are very scarce and their interpretation is difficult due to the RBS resolution. In order to improve our understanding of the basic mechanisms of ion beam mixing, it appears necessary to develop other experimental techniques.

In this paper, we present a new experimental approach for the determination of mixing kinetics in layered materials. The electrical resistivity is a physical property very sensitive to the presence of defects and changes in the structural state of a material, in addition it has the great advantage to be measured continuously during the particle bombardment. The possibilities of our approach are illustrated in two different multilayered structures Fe-Al and Al-Ag.

EXPERIMENTAL PROCEDURE.

Fe-Al and Al-Ag multilayers were evaporated by vacuum deposition on single crystalline quartz substrates under a clean vacuum of about 10^{-7} Torr. The thickness of the individual layers was determined during evaporation using a calibrated quartz oscillator. Three Al layers were successively evaporated on three Fe or Ag ones, giving a total thickness of 1100 Å for Fe-Al and 800 Å for Al-Ag. The individual film thickness ranged between 100 and 200 Å, they are choosen to have average film compositions Fe-60 at % - Al-40 at % and Al-75 at % - Ag 25 at %. The thickness of the outermost layer was increased of about 80 Å in order to take into account the sputtering effect. The energies of Xe ions were selected such that the ion implanted distribution is approximately the total film thickness : $e \simeq R_p + \Delta R_p$.

The electrical resistivity samples were evaporated using a masking technique with their potential and current leads attached to them; their lateral dimensions are about L = 20 mm, l = 2 mm. On the same substrate four samples connected in series are evaporated allowing four different experiments.

A 1 mA current stabilized to one part in 10^5 is passed through the samples and their electrical resistivity is deduced from the voltage drop across them with a relative accuracy of 10^{-5}. Low ion beam currents 30-300 nA/cm^2 are used in order to avoid a possible beam heating of the samples, and no temperature rise was observed using a non-implanted sample as a temperature monitor.

EXPERIMENTAL RESULTS.

Ion beam mixing experiments have been performed with Xe ions either at 77 K or 300 K, in both cases the electrical resistivity of the samples increases continuously as a function of the ion fluence ϕ. The experimental results for a Fe-Al multilayer irradiated at 300 K with 450 KeV Xe$^+$ ions are presented in Figure 1 and for an Al-Ag one irradiated at 77 K with 200 KeV Xe$^+$ ions in Figure 2.

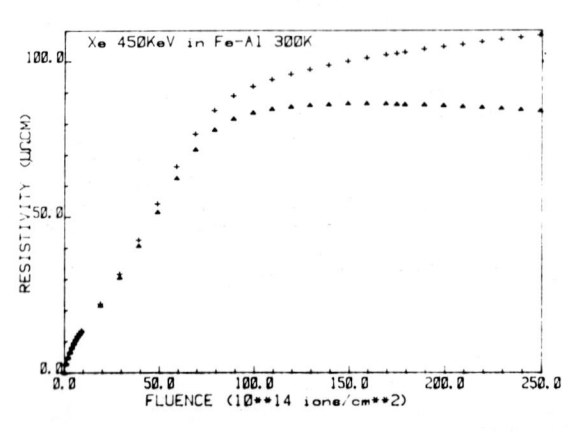

Fig. 1 : Electrical resistivity increase in a Fe-Al multilayer bombarded at 300 K with 450 KeV Xe$^+$ ions. The lower curve is corrected for the sputtering.

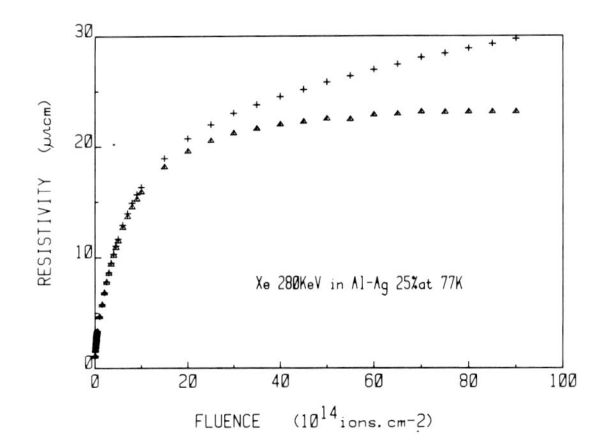

Fig. 2 : Electrical resistivity increase in an Al-Ag multilayer bombarded at 77 K with 280 KeV Xe$^+$ ions. The lower curve is corrected for the sputtering.

The upper curves in Figures 1 and 2 correspond to the direct measurements, while the lower ones are corrected for the sputtering effect. This latter one has been estimated from the final and linear part of the upper curves assuming that this regime is only a geometrical effect since the damage saturation is reached and the layers mixed. The sputtering yields S found by this treatment are 7 for Fe-Al and 9 for Al-Ag. The general evolution of the resistivity increase can be roughly divided into three main regimes :

1) the first one up to about 10^{13} ions/cm^2 corresponds to a rapid increase mainly due to the formation of defects and defect clusters in the dense collision cascades.

2) the second regime with large resistivity variations corresponds to the mixing regime itself.

3) finally, a tendancy toward a saturation process is reached for the higher doses : 10^{16} ions/cm^2 for Fe-Al and 3.10^{15} ions/cm^2 for Al-Ag. This saturation regime indicates that the total mixing of the film is obtained in a given structural state with a constant electrical resistivity value. When a metastable solid solution is formed by mixing, a smooth resistivity increase at a continuous decreasing rate is expected; though in the case where several phases or amorphization occurs, a change in the rate is expected due to the different resistivity values of the different phases. Such a behavior will be discussed in a companion paper by Grilhé (9) where many more experimental results are given. The resistivity increase $\Delta\rho(\Phi)$ reflects the growth of the volume fraction of the mixed layers $C(\Phi)$ in order to make the link between these two quantities, we consider a sample made of successive layers of two different metals A and B, Figure 3.

The intensity lines during the resistivity measurements in the sample are parallel to the different layers and the conductivity of such a layer is given by :

$$\sigma_o = \sigma_A C_A + \sigma_B C_B$$

where σ_A, σ_B are the conductivities of pure A and B; C_A and C_B are the volume fractions of elements A and B.

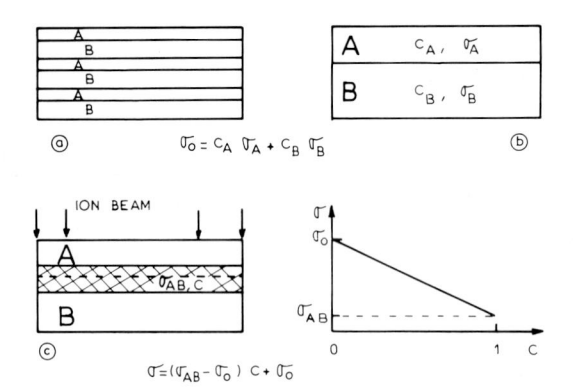

Fig. 3 : Schematic view of the ion beam mixing and its influence on the conductivity of a layered material.
a. multilayer as evaporated
b. isolated A-B interface before mixing
c. interface AB when partial intermixing has occured.

During implantation the penetration of energetic ions through the different interfaces A-B produces intermixed regions and the concentrations of the two elements A and B in the intermixed layer are close to that of a stable or metastable phase. It is important to notice that the final conductivity σ_{AB} of the stable or metastable, eventually amorphous phase formed will be always lower than the individual conductivities σ_A and σ_B of pure elements A and B. It is possible to consider that the thin intermixed layer is built up with the final composition and has the final conductivity σ_{AB} of the intermetallic compound formed at the end of the ion beam mixing. The resulting conductivity $\sigma(\Phi)$ of the multilayer at time t during the mixing process is produced by parallel layers either intermixed with a conductivity σ_{AB} or unmixed with a conductivity σ_o. Introducing the volume fraction $C(\Phi)$ of the intermixed layer at a fluence Φ, we can write the relation :

$$\sigma(\Phi) = \sigma_{AB} \, C(\Phi) + \sigma_o \, (1 - C(\Phi))$$

As the implantation proceeds, the growth of the intermixed layers occurs at the different interfaces untill the total mixing of all the couples is achieved. The kinetic behavior of the ion beam mixing process as a function of the fluence will be given by the variations of :

$$C(\Phi) = \frac{\sigma(\Phi) - \sigma_o}{\sigma_{AB} - \sigma_o}$$

On the basis of this simple conduction model, the growth of the volume fraction of intermixed atoms can be deduced from electrical resistivity measurements (10). As a representative example Figure 4 shows the application to the results of Figures 1 and 2.

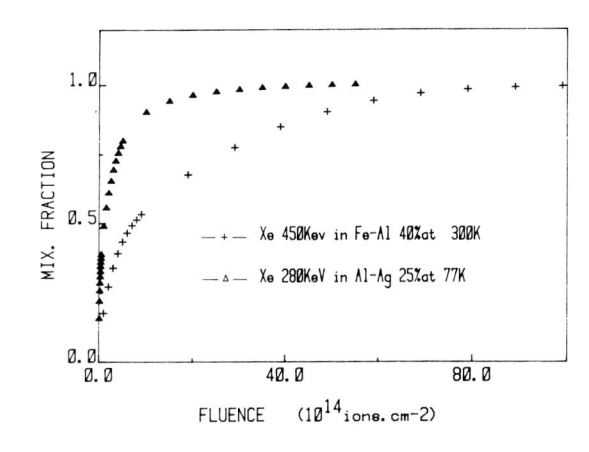

Fig. 4 : Variation of the volume fraction of intermixed atoms as a function of the fluence.

CONCLUSION.

- We have introduced a new approach to study ion beam mixing of layered materials. Instantaneous information as the mixing proceeds is given by the electrical resistivity measurements.

- It is possible to determine directly the critical dose beyond which the atomic mixing is achieved.

- A simple relation between the conductivity and the volume fraction of mixed atoms has been established in the case where a single metastable phase is produced.

- This method allows to study the mixing kinetics during the whole process and can give in principle informations on the basic mechanisms of ion beam mixing.

REFERENCES.

1. G. Dearneley, Rad. Effects 63, 1 (1982).
2. H. Herman, Nucl. Instr. Meth. 182/183, 887 (1981).
3. G. Dearneley, idem as ref. 2, p. 899.
4. B.Y. Tsaur, S.S. Lau, J.W. Mayer, Appl. Phys. Lett. 36, 823 (1980).
5. B.Y. Tsaur, S.S. Lau, L.S. Hung, J.W. Mayer, idem as ref. 2, p. 67.
6. S.T. Picraux, D.M. Follstaedt, J. Delafond, in Metastable Materials formation by ion implantation. MRS Proceeding vol. 7, p. 71, S.T. Picraux, W.J. Choyke Ed.
7. B.M. Maine, M.A. Nicolet, G.C. Banwell, idem as ref. 7, p. 79.
8. J.W. Mayer, B.Y. Tsaur, S.S. Lau, L.S. Hung, idem as ref. 2, p. 1.
9. J. Grilhé, J.P. Rivière, J. Delafond, C. Jaouen, This MRS Symposium.
10. J.P. Rivière, J. Delafond, C. Jaouen, A. Bellara, J.F. Dinhut, Applied Physics (to be published).

DOSE DEPENDENCE OF ION BEAM MIXING OF Au ON AMORPHOUS AND SINGLE CRYSTALLINE Si AND Ge

D. B. POKER, O. W. HOLLAND AND B. R. APPLETON
Solid State Division, Oak Ridge National Laboratory, Oak Ridge, TN 37831

ABSTRACT

The rate of ion induced mixing of 700 Å Au layers vapor deposited on amorphous and single crystalline Si substrates held at room temperature was measured as a function of dose using 300 keV Si, 350 keV Ar, and 525 keV Kr ion beams. Mixing profiles were measured at various fluences by Rutherford backscattering techniques and were found to be consistent with mixed layers whose thicknesses increased with ion dose. Mixing compositions, which were stoichiometric over the entire mixed region at Au-28.5 at. % Si, were found to be independent of ion species or implant fluence. For all ion species the dose dependence of mixing was closer to linear than the square root power law reported previously [1]. In addition, the mixing rate for Au on single crystalline substrates was significantly higher than Au on substrates amorphised by Si (self-ion) implantation at liquid nitrogen temperature. No difference was found between the mixing rates when the amorphous substrates were prepared by room temperature implantation. Preliminary results indicate similar behavior for the Au/Ge couple.

INTRODUCTION

The Au/Si system is considered in mixing studies as a representative metal-semiconductor system that exhibits rapid mixing, forming an amorphous alloy with constant stoichiometry over the mixed region [1,2]. It has been suggested that chemical driving forces are instrumental to the mixing process [2,3]. During ion irradiation non-equilibrium defect production and enhanced defect mobility can occur, providing an environment for the chemical potentials to equilibrate. The influence of such forces could be tested by examining the effect of mixing on single crystal and amorphous Si substrates. In this study we have measured the mixing rates of Au on these substrates to determine whether the free energy difference influences the mixing process.

EXPERIMENTAL PROCEDURE

Samples were cut from wafers of <100> Si. Half of each sample was amorphised with 150 keV ^{30}Si to a dose of 2×10^{16} cm^{-2} using either of two procedures. Initially the amorphisation was performed with the samples mounted on a liquid nitrogen cooled copper block in vacuums of 10^{-7}-10^{-8} Torr. It was suspected, however, that condensation of hydrocarbons from the vacuum were being decomposed on the sample during implantation [4,5]. In an effort to reduce the formation of carbon on the sample surface later amorphisations were performed at room temperature. These samples were then exposed to 0.254 nm UV light at an intensity of 0.15 W cm^{-2} for 30 minutes to decompose any carbon which might have formed. After either method of amorphisation the samples were organically cleaned, etched in HF, and rinsed with deionized water before mounting into an oil-free vacuum system for thin-film deposition. Films of Au, ∿700 Å thick, were evaporated onto the samples with a nominal deposition rate of 10 Å/s. The background pressure during the evaporations was less than 5×10^{-7} Torr.

Samples were ion-beam mixed with 300 keV Si, 350 keV Ar, or 525 keV Kr ions at room temperature. Typical beam densities were about 1 µA cm^{-2}. The ion energies were chosen so as to produce an ion range that significantly exceeded the Au film thickness. Analytical [6] and Monte Carlo TRIM [7] codes were used to calculate the ion ranges and damage energy profiles.

The compositional distributions after mixing were measured using Rutherford backscattering (RBS) of 2.5 MeV ^{4}He atoms. A scattering angle of 143 degrees was used with a sample tilt of -35 degrees to enhance the depth resolution capability of RBS. TRIM calculations that provided for the multiply layered samples were used to convert the implanted dose to damage energy at the interface, thereby normalizing the doses of different implant species to damage production.

The samples were alternately mixed and RBS analysed in situ without exposure to atmosphere between mixing doses. Since each sample contained regions of single crystal and amorphous Si substrate, the effects of several variables (varying Au thickness, mixing dosimetry, etc.) were eliminated.

RESULTS AND DISCUSSION

The result of ion mixing a thin film of Au on a single crystal of Si is typified in the RBS spectrum of Fig. 1, where only the yield of ions scattered from Au are shown versus depth for clarity. The virgin spectrum indicates a uniform Au film approximately 750 Å thick. After a mixing dose

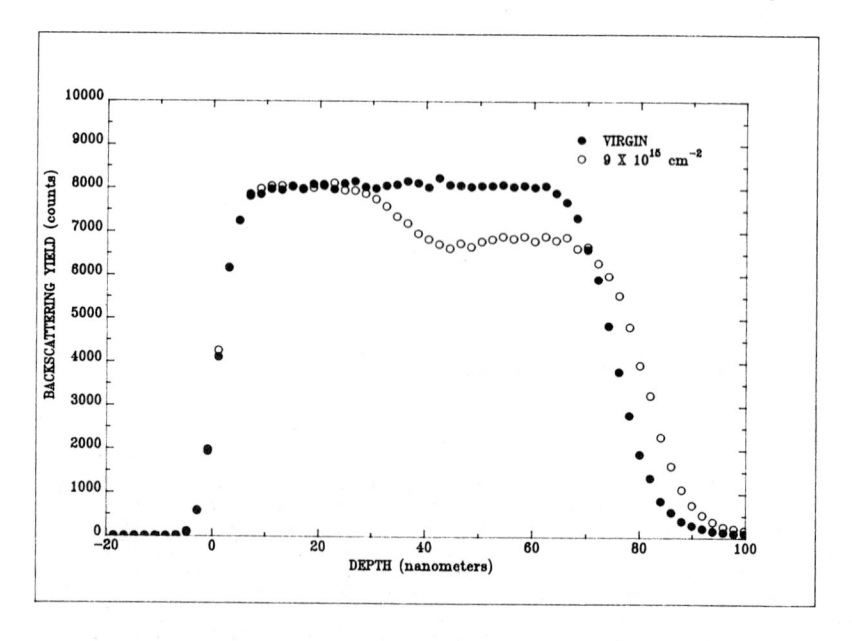

FIG. 1. Rutherford backscattering spectra of Au on single crystal Si before and after mixing by 300 keV ^{28}Si to a dose of 9×10^{15} cm^{-2}. Only the scattering from the Au is shown for clarity.

of 9×10^{15} cm^{-2} of 300 keV ^{28}Si a uniformly mixed region 375 Å wide had been formed at the Au/Si interface. The stoichiometry of the mixed region was calculated to be Au$_{72}$Si$_{28}$, similar to previous results [1]. Mixing to higher doses produced films that were shown to be amorphous by thin film X-ray diffraction measurements.

The dose dependence of mixing on single crystal and liquid nitrogen amorphised Si substrates with 300 keV ^{28}Si is shown in Fig. 2. The doses have been converted to damage energy density at the interface with the results of appropriate TRIM calculations for each case. The mixing rate on amorphous substrates was about 10% lower over the entire range of mixing fluences compared to the rate on single crystal substrates. Moreover, the mixing rates are close to linear in damage energy (dose) with a fit of the data to a power law showing the data to be linear to within a few percent. This is contrary to previous observations of mixing of comparable thickness Au films on Si where mixing rates proportional to the square root of dose were reported [1].

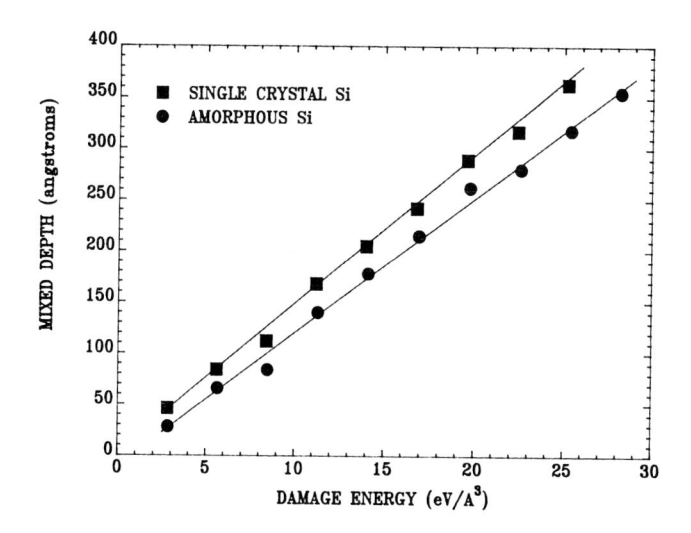

FIG. 2. Depth of the mixed Au$_{72}$Si$_{28}$ phase as a function of damage energy due to mixing by 300 keV ^{28}Si on single crystal and amorphous Si substrates. The straight lines are linear fits to the data.

The object of comparing the mixing rates on single crystal and amorphous Si substrates was to determine whether the free energy difference between the two phases affected the mixing process through differences in the chemical driving forces. The results of Fig. 2 provided support for this assumption. However, another aspect of the mixing process is whether the characteristics of the damage cascade influence the mixing rates on substrates in different phases. It has been suggested that only simple phases can be formed by ion beam mixing with light ions since the quench rates (10^{10}-10^{13} K/s) associated with cascade lifetimes are insufficient for the necessary atomic rearrangement of more complicated structures [8]. Since the amorphous Si phase is not created by an individual collision cascade during Si implantation, but rather by the overlap of displacement damage produced within many cascades [9], it was felt that the mixing rates would depend upon the phase of the substrate prior to mixing. This was consistent with the observations. However, a difference in the mixing rates between single crystal and amorphous Si substrates was observed for heavy ion irradiation as well. Since the amorphous phase is formed directly by an individual cascade for these ions, it would be difficult to explain the different rates by the cascade mixing model alone. These observations are more consistent with a model whose mixing is governed by defect production, mobility, and interactions.

A more plausible explanation is that the difference in mixing rates between single crystal and amorphous substrates results from the contamination of the Si surface during the amorphisation by condensation of hydrocarbons from the vacuum. These hydrocarbons can be reduced by the ion beam, forming a "diamond-like" carbonaceous surface layer [5] between the Si and the evaporated Au that might interfere with the mixing process. This production of carbonaceous surface layers has previously been observed with other low temperature implants [4]. To reduce the possibility of forming such layers Si samples were amorphised at room temperature and exposed to UV light as described in the experimental procedure. The mixing rate of these samples (14.6 $Å^2$/eV) was essentially identical to the single crystal results (14.5 $Å^2$/eV), as opposed to the lower value (13.0 $Å^2$/eV) for samples amorphised at liquid nitrogen temperature as shown in Fig. 2. Assuming that the Si amorphised at room temperature and treated with UV light has a less contaminated surface, these results support the hypothesis that surface (interface) impurities inhibit mixing.

Related studies were made of ion induced mixing of Au on single crystal and amorphous Ge fabricated by ^{74}Ge implantation at liquid nitrogen temperature. Results from these experiments are plotted in Fig. 3. Analagous to Si, the mixing was linear in dose with the single crystal Ge exhibiting faster mixing that the liquid nitrogen amorphous. However, in this case quantitative nuclear reaction analysis [10] showed only a few monolayers of O and C contamination formed on the Ge surface

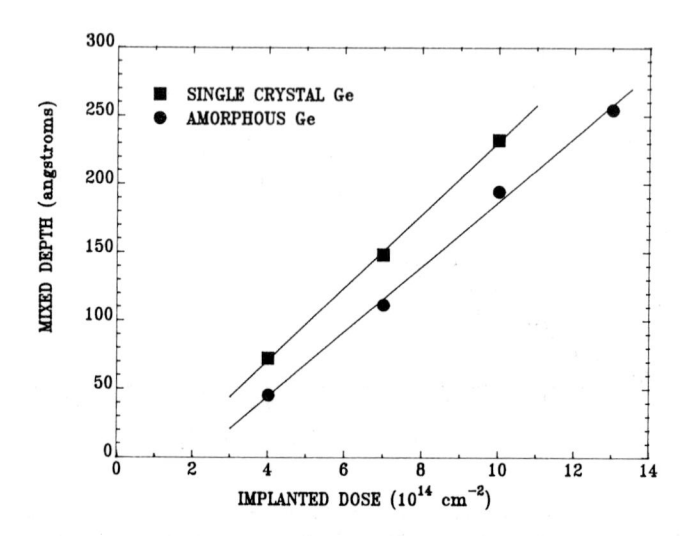

FIG. 3. Depth of the mixed phase as a function of damage energy due to mixing by 300 keV ^{74}Ge on single crystal and amorphous Ge substrates.

during implantation. If, as we suspect, the same result holds for low temperature amorphisation of Si, then the added impurities are small compared to the normal 30-50 Å of air-formed oxide and a mechanism other than the surface contamination may be responsible. Experiments to quantify the presence of carbon on the surface of Si amorphised at nitrogen temperature using enhanced elastic scattering of H from ^{12}C are planned, as well as mixing of atomically clean interfaces for the Au/Si and Au/Ge systems.

Another possible explanation is that the different mixing rates result from different amorphous phases produced by room temperature and low temperature ion implantation. Amorphous phases produced in this way have been found to have different free energies [11,12,13].

Implanting with other ion species also produced linear mixing, as shown in Fig. 4. The heavier ions were more efficient at mixing, with Ar mixing at 83% of the Kr rate and Si mixing at 72% of the Kr rate. These ratios are slightly higher than observed by Averback [14] for the mixing of Si/Pt at 10 K, where a square root dose dependence was observed. This observation of mixing that is linear in dose is contrary to most previously reported results. The repeated observations for several ion species on many samples of Si prepared under varied conditions, as well as preliminary observations on Ge, indicate that this result is not an isolated artifact. The modeling of ion induced mixing as a random walk process displaying diffusion-like behavior has been successful [15], and predicts mixing rates that follow the square root dose dependence. These models, however, are valid only for dilute solutions, usually thin markers, in a homogeneous matrix where chemical driving forces are insignificant [16]. In the system considered here there is an essentially unlimited supply of both elements available for mixing. Silicon diffuses very rapidly in Au [17]. Presumably it diffuses rapidly in the mixed phase as well. If there always exists an excess supply

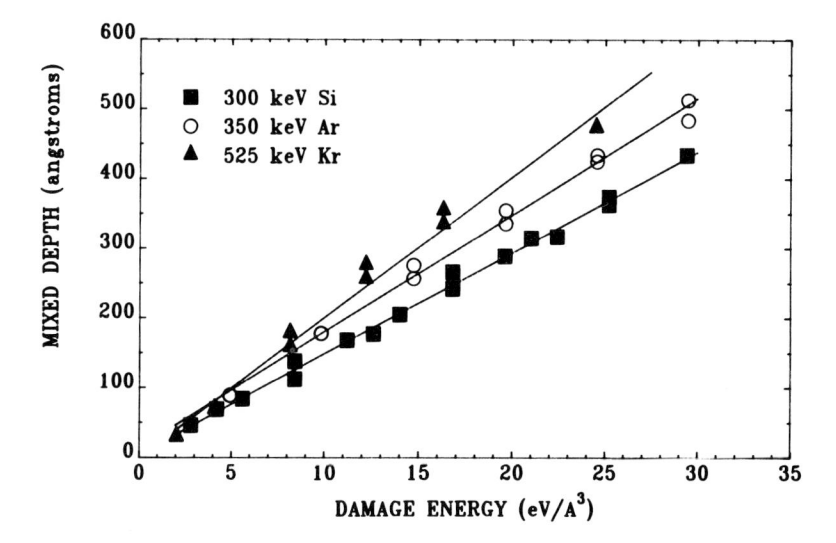

FIG. 4. Comparison of mixing rates for Au on single crystal Si by different species of mixing ions. The mixing doses been normalized to deposited damage energy at the interface. The straight lines are linear fits to the data.

of Si at the mixed phase boundaries with the Au, and if the mixed phase maintains a constant stoichiometry, then it is plausible that the mixing is not diffusion limited, but instead proceeds at a rate that is linear in the amount of damage energy deposited near the interface, as we observed. Whether this process applies here or to other systems will be the subject of continuing research.

ACKNOWLEDGEMENTS

This research was sponsored by the Division of Materials Sciences, U. S. Department of Energy under contract W-7405-eng-26 with Union Carbide Corporation. The authors would like to thank B. Larson for the thin film X-ray diffraction measurements.

REFERENCES

1. B. Y. Tsaur and J. W. Mayer, Philos. Mag. A43, 345 (1981).
2. S. S. Lau, B. Y. Tsaur, M. von Allmen, J. W. Mayer, B. Stritzker, C. W. White, and B. Appleton, Nuclear Instrum. and Methods 182/183, 97 (1981).
3. J. W. Mayer, B. Y. Tsaur, S. S. Lau, and L-S. Hung, Nuclear Instrum. and Methods 182/183, 1 (1981).
4. D. B. Poker, Radiation Effects (in press).
5. J. C. Angus, M. J. Mirtich, and E. G. Wintucky, Mat. Res. Soc. Symp. Proc. 7, 433 (1982).
6. J. P. Biersack, Nuclear Instruments and Methods 182/183, 199 (1981).
7. J. P. Biersack and L. G. Haggmark, Nuclear Instruments and Methods 174, 257 (1980).
8. M. Nastasi, L. S. Hung, and J. W. Mayer, Appl. Phys. Lett. 43, 831 (1983).
9. J. Narayan, D. Fathy, O. S. Oen, and O. W. Holland, Materials Letters (in press).
10. O. W. Holland, B. R. Appleton, and J. Narayan, J. Appl. Phys. 54, 2295 (1983).
11. John C. C. Fan and Carl H. Anderson, Jr., J. Appl. Phys. 52, 4003 (1981).
12. J. E. Fredrickson, C. N. Waddell, W. G. Spitzer, and G. K. Hubler, Appl. Phys. Lett. 40, 172 (1982).
13. J. Narayan and O. W. Holland, Appl. Phys. Lett. 41, 239 (1982).
14. R. S. Averback, Proceedings of the Workshop on Ion Beam Mixing and Surface Layer Alloying, Albuquerque, April 15-16, 1983, SAND83-1230, p.10.
15. S. Matteson, Appl. Phys. Lett. 39, 288 (1981).
16. P. Sigmund and A. Gras-Marti, Nuclear Instrum. and Methods 182/183, 25 (1981).
17. A. Hiraki, M-A. Nicolet, and J. W. Mayer, Appl. Phys. Lett. 18, 178 (1971).

STUDY OF ION BOMBARDMENT-INDUCED SUBSURFACE COMPOSITIONAL MODIFICATIONS IN
Ni-Cu ALLOYS AT ELEVATED TEMPERATURES BY ION SCATTERING SPECTROSCOPY

N. Q. LAM, H. A. HOFF, H. WIEDERSICH and L. E. REHN
Materials Science and Technology Division, Argonne National Laboratory,
Argonne, IL   60439

ABSTRACT

Changes in the subsurface composition of Ni-40 at.% Cu alloys during 3-keV Ne^+ bombardment at temperatures between 25 and 700°C were studied by means of ion scattering spectroscopy. Both the time evolution of the composition in the surface atom layer during ion bombardment and subsurface concentration profiles after rapid specimen cooling to room temperature were measured as a function of temperature. Radiation-enhanced diffusion coefficients were derived from the effective altered-layer thicknesses obtained. A comparison of the experimental measurements with theoretical calculations based on a phenomenological model enabled the identification of processes and kinetics responsible for subsurface compositional modifications.

INTRODUCTION

Subsurface compositional modification of alloys during elevated-temperature ion sputtering has been a subject of detailed investigations over the past few years. The interest in this phenomenon was motivated by recent concern about plasma contamination by sputtered particles in magnetic fusion devices and wide application of ion-beam processing in many areas of materials science.

Changes in the near-surface composition due to preferential sputtering near room temperature have been observed in many alloy systems (extensive reviews have been given recently by Betz [1], Kelly [2], and Andersen [3]). At high temperatures where additional thermally-activated processes are expected, this phenomenon is more complex. In fact, it is now established that at least five distinct processes, including preferential sputtering (PS), displacement mixing (DM), Gibbsian adsorption (GA), radiation-induced segregation (RIS) and radiation-enhanced diffusion (RED), can contribute to subsurface compositional alterations of alloys during elevated-temperature ion sputtering. Recent model calculations of Lam and Wiedersich [4-6] have provided new insights into the simultaneous effects of these processes and helped understand a number of experimental observations.

Most of the measurements of ion bombardment-induced alloy composition changes at high temperatures were made by means of Auger electron spectroscopy (AES) [7-12]. Unfortunately, owing to the finite escape depth of the Auger electrons, the AES technique cannot yield definitive information about the composition in the outermost atom layer, where the effects of PS and GA are known to be significant. In a recent work, Swartzfager et al. [13] employed the ion-scattering-spectroscopy (ISS) technique, which is extremely sensitive to the first surface atom layer. However, the time

---

*Work supported by the U.S. Department of Energy.

evolution of the alloy composition in the first atomic layer during ion bombardment was not reported. Most recently, Andersen et al. [14] used a new technique, in which the angular distribution of the sputtered-flux composition was measured, to obtain qualitative information on near-surface compositional gradients.

In the present study, both the evolution of the surface alloy composition during sputtering and subsurface concentration profiles after rapid specimen cooling to room temperature were measured as a function of temperature using ISS. Information regarding the enhancement of diffusion by irradiation was obtained from the measurements of the effective altered-layer thicknesses. The experimental observations are compared with the results of model calculations in order to identify the mechanisms and kinetics responsible for ion bombardment-induced modifications of alloy composition.

EXPERIMENTAL

The Ni-40 at.% Cu alloy was prepared by arc melting and subsequent levitation melting in an induction furnace. Rectangular specimens (~0.5 x 1.5 x 0.05 cm^3) cut from rolled material were metallographically polished to obtain an optically flat surface, and then solution annealed in a vacuum of <10^{-6} Pa at 800°C for 5 hours. The average grain size in the specimens was measured to be ~41 μm in diameter. The specimens could be resistance-heated to temperatures up to 1000°C, and rapidly cooled by interrupting the current. The temperature was measured with a chromel-alumel thermocouple spot-welded to the back side of the specimen. Before each at-temperature sputtering experiment, the specimen surface was sputter-cleaned at room temperature until a steady-state alloy surface composition was attained. The sample was then heated and maintained at temperature for ~20 min before the start of sputtering.

The ion scattering spectrometer used is a 3M (now Kratos) model 535. All experiments were performed with a normally incident beam of 3-keV Ne$^+$ ions (the static beam diameter was 125 μm). During sputtering, the ion beam was rastered over an area of 1 mm^2, and the scattered ion signal was electronically gated (to ~60%), so that only the central portion of the rastered area was probed. The ion flux was measured to be 3.75 x 10^{13} ions/cm^2.s (i.e. ~6 μA/cm^2). Taking 2.5 and 2.0 atoms/ion to be the sputtering coefficients for Cu and Ni, respectively [15], we estimated a sputtering rate $\dot{x} \simeq 9 \times 10^{-3}$ nm/s. The damage distribution for 3-keV Ne$^+$ on Ni was calculated using the TRIM code [16]. The peak damage occurs at ~1 nm and the total range is ~7 nm. With the above-mentioned ion flux, the peak-damage rate is ~3.5 x 10^{-2} dpa/s.

RESULTS AND DISCUSSION

Effect of temperature on surface and subsurface compositional modification

Figure 1a shows the ratios of Cu to Ni ISS intensities measured during ion bombardment at various temperatures. Changes of these ratios in time reflect the temporal evolution of the alloy composition in the first surface atom layer. Since the specimen was held at temperature for ~20 min before the start of sputtering, the Cu concentration at the surface, C_{Cu}^s, was very high at short sputtering times because of Gibbsian adsorption [17]. With increasing time, C_{Cu}^s decreased, owing to the dominant effects of RIS and PS, to a steady-state value. Below ~100°C, steady state was attained quickly, within ~1 min of sputtering. At higher temperatures, increasingly longer times were required to achieve steady state. In contrast to previous theoretical predictions, which were based on the

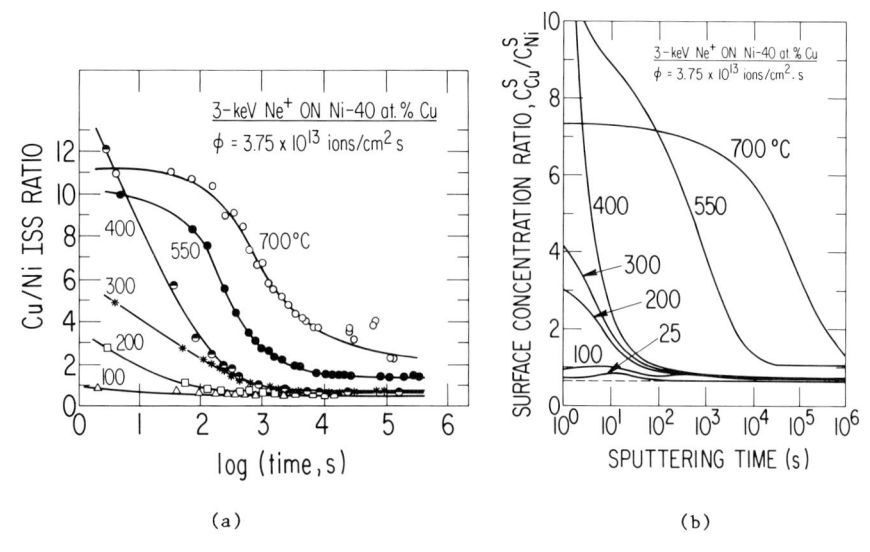

Fig. 1. Time evolution of the composition in the outermost atom layer of a Ni-40 at.% Cu alloy during sputtering at various temperatures: (a) measured by ISS; (b) calculated using the Lam-Wiedersich model, with 37% contribution of the second layer to the sputtered-atom flux.

assumption that sputtered atoms originated entirely from the first atomic layer [4-6], the steady-state C_{Cu}^S was found to be temperature dependent. This dependence can be interpreted in terms of significant contributions to the sputtered-atom flux from deeper layers. Gibbsian adsorption tends to increase the Cu concentration in the first atomic layer relative to that in the subsurface layers, as the rate of atom exchange between these layers increases with increasing temperature. The steady-state condition, i.e. the composition of the sputtered-atom flux must equal the alloy composition in the bulk, can be met with an increased C_{Cu}^S if this increase is balanced by a corresponding decrease in the deeper layers. That is, a sufficient fraction of the sputtered-atom flux must originate from the deeper layers. Previous model calculations [4-6] and experimental results [8,9] have shown a substantial Cu depletion of subsurface layers, relative to the top layer, during sputtering at high temperatures. This is confirmed by the present measurements (see Fig. 2a).

New model calculations were performed to lend support to this interpretation. Our recently-developed kinetic model, which takes into account the interplay of the five basic processes mentioned above [4-6], was slightly modified by including significant contributions of the second atom layer to the sputtered-atom flux. Details of the theoretical treatment will be given in a separate publication [17]. Some theoretical results are illustrated in Fig. 1b. The experimentally-observed temperature dependence is indeed reproduced by model calculations. However, good qualitative agreement between experiment and theoretical predictions can only be obtained, if one assumes that at least 37% of the sputtered Cu flux comes from the second atom layer [17]. This contribution seems to be somewhat

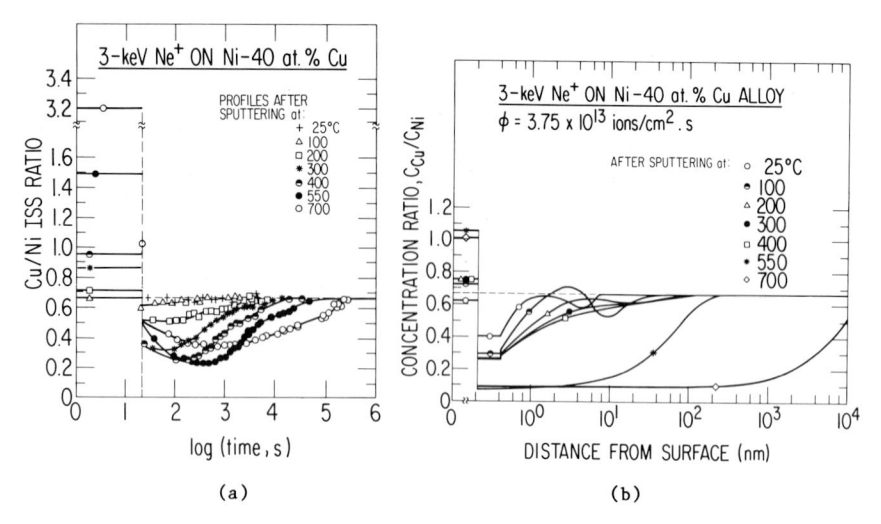

Fig. 2. Steady-state concentration profiles of the altered layers in a Ni-40 at.% Cu alloy after sputtering at various temperatures: (a) measured by ISS after rapid cooling to room temperature, $\dot{x} = 9 \times 10^{-3}$ nm/s; (b) theoretically calculated. The vertical lines indicate the boundary between the first and second atomic layers.

high, though not unreasonable, in light of the results of molecular-dynamics calculations by Harrison et al. [18].

In order to obtain information on the extent of alloy compositional alterations in the subsurface region, the ion beam and heating current were turned off simultaneously after steady-state surface composition had been achieved, and the Cu/Ni ISS intensity ratios were subsequently measured as a function of sputtering time at room temperature. The resulting Cu concentration profiles obtained are shown in Fig. 2a. Near room temperature, no subsurface compositional changes could be detected by the ISS. At 100°C, minor compositional alterations were revealed; the thickness of the region in which the alloy composition was modified was approximately equal to the damage depth. At higher temperatures, as the thermal processes (GA, RIS and RED) became more effective in modifying the alloy composition, the Cu depletion in the subsurface region was more severe and extended to much larger depths. For example, at 700°C, the modified layer was more than two orders of magnitude larger than the total damage range. The calculated Cu concentration profiles are displayed in Fig. 2b for comparison with experiment. Except for the fine structures in profiles calculated for sputtering temperatures below ~200°C, which result from the significant effect of RIS [19], the theoretical profiles are generally in good qualitative agreement with experimental measurements.

Radiation-enhanced diffusion in the altered layer

The subsurface Cu concentration profiles shown in Fig. 2a can be further analyzed in order to extract information regarding RED of Cu in the alloy. Based on the steady-state equation formulated by Ho [20] for the

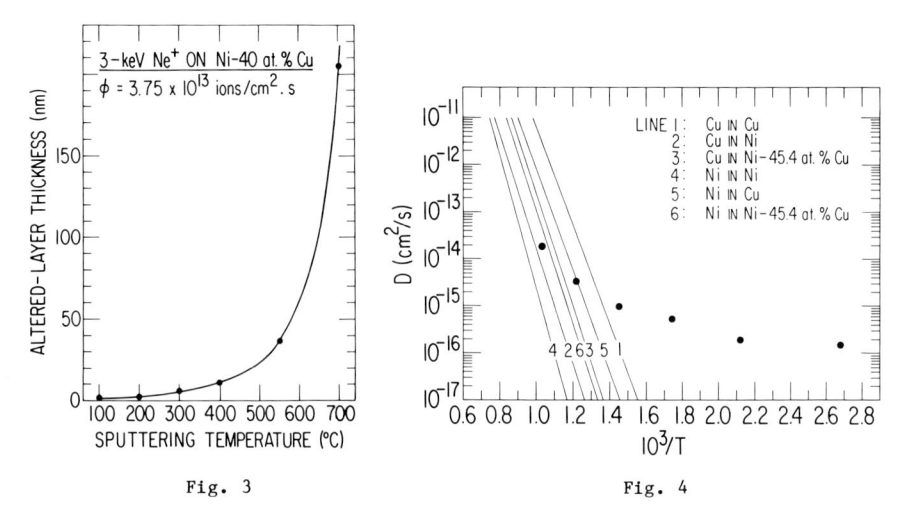

Fig. 3. Temperature dependence of the effective altered-layer thickness.

Fig. 4. Arrhenius plot of the radiation-enhanced diffusion coefficients. The numbered lines correspond to various data for thermal diffusion (lines 2, 3, and 6 were obtained by extrapolation of high-temperature data given in refs. 25 and 26).

alloy composition in the altered layer, we plot the natural log of the difference between the bulk concentration and the concentration in the altered layer versus the distance from the surface. The plot consists of two straight portions of different slopes: the first portion contains information about RED, and the second one reflects the effects of RIS. The RIS effects are discussed in more detail in ref. [21]. The first portion can be used to deduce the effective altered-layer thickness, δ, which is strongly temperature-dependent above ~400 °C, as illustrated in Fig. 3. Then, using the relationship $\delta = D/\dot{x}$ [20], we calculate the RED coefficients, D. The Arrhenius plot of the results is shown in Fig. 4. Compared with various data for thermal diffusion [22-26], the results show that diffusion in the altered region is significantly enhanced by ion bombardment-induced point defects below ~550°C. Even though defects are produced only in a 7 nm thick zone, a large number of them survive annihilation and move deeper into the region beyond the ion range, promoting diffusion. In addition, below ~400°C, RED is virtually temperature-independent, indicating that, in this temperature regime, nonequilibrium point defects annihilate mainly by diffusion to sinks. The present results are in good agreement with previous ISS measurements by Swartzfager et al. [13], and consistent with AES data of Rehn et al. [21].

CONCLUSION

1. Using ISS, we found that, during high-temperature sputtering, the steady-state surface concentration of Cu was temperature dependent, which could be explained in terms of significant contributions of the second atomic layer to the sputtered-atom flux. Results of our recent model calculations support this interpretation.

2. The coefficients for radiation-enhanced diffusion of Cu in the bombarded alloy were deduced from steady-state concentration profiles of the compositionally-altered layers whose thicknesses were strongly temperature-dependent. The results are in good agreement with previous ISS and AES measurements.

ACKNOWLEDGEMENTS

The authors have greatly benefited from useful discussions with Drs. H. H. Andersen, P. G. Régnier, and D. G. Swartzfager. Ms. Helen Mirenic's contributions in skillfully formatting and typing the paper are also gratefully acknowledged.

REFERENCES

1. G. Betz, Surface Sci. 92, 283 (1980).
2. R. Kelly, in: Symposium on Sputtering, eds. P. Varga, G. Betz and F. P. Viebőck (Institut für Allgemeine Physik, Vienna 1980) p. 390.
3. H. H. Andersen, in: Physics of Ionized Gases (SPIG 1980), ed. M. Matić (Boris Kidrič Institute of Nuclear Sciences, Beograd, Yugoslavia, 1980) p. 421.
4. N. Q. Lam and H. Wiedersich, J. Nucl. Mater. 103 & 104, 433 (1981).
5. N. Q. Lam and H. Wiedersich, Rad. Effects Lett. 67, 107 (1982).
6. N. Q. Lam and H. Wiedersich, Mat. Res. Soc. Symp. Proc. 7, 35 (1982).
7. H. Shimizu, M. Ono and K. Nakayama, J. Appl. Phys. 46, 460 (1975).
8. L. E. Rehn, S. Danyluk and H. Wiedersich, Phys. Rev. Lett. 43, 1764 (1979).
9. L. E. Rehn and H. Wiedersich, Thin Solid Films 73, 139 (1980).
10. M. Shikata and R. Shimizu, Surface Sci. 97, L363 (1980).
11. H. Shimizu, M. Ono, N. Koyama and Y. Ishida, J. Appl. Phys. 53, 3044 (1982).
12. L. E. Rehn, V. T. Boccio and H. Wiedersich, Surface Sci. 128, 37 (1983).
13. D. G. Swartzfager, S. B. Ziemecki and M. J. Kelley, J. Vac. Sci. Technol. 19, 185 (1981).
14. H. H. Andersen, B. Stenum, T. Sørensen, and H. J. Whitlow, Nucl. Instr. and Meth. (to be published).
15. H. H. Andersen and H. L. Bay, in: Sputtering by Particle Bombardment, ed. R. Behrisch (Springer, Heidelberg, 1981) p. 145.
16. J. P. Biersack and L. G. Haggmark, Nucl. Instr. and Meth. 174, 257 (1980).
17. N. Q. Lam, H. A. Hoff and H. Wiedersich, to be published.
18. D. E. Harrison, Rad. Effects 70, 1 (1983).
19. N. Q. Lam, G. K. Leaf and H. Wiedersich, J. Nucl. Mater. 88, 289 (1980).
20. P. S. Ho, Surface Sci. 72, 253 (1978).
21. L. E. Rehn, N. Q. Lam and H. Wiedersich, this volume.
22. K. Maier, C. Bassani and W. Schüle, Phys. Lett. 44A, 539 (1973).
23. K. Maier, H. Mehrer, E. Lessmann and W. Schüle, Phys. Stat. Sol. (b) 78, 689 (1976).
24. J. L. Seran, Acta Met. 24, 627 (1976).
25. H. U. Helfmeier, Z. Metallkde. 65, 238 (1974).
26. D. B. Butrymowicz, J. R. Manning and M. E. Read, in: Diffusion Rate Data and Mass Transport Phenomena for Copper Systems (National Bureau of Standards, Washington, D.C., 1977) pp. 196-199.

ION MIXING IN THE Ni-Sn SYSTEM

L. CALLIARI,[+] L.M. GRATTON,[++] L. GUZMAN,[+] G. PRINCIPI,[o+] AND C. TOSELLO[+,++]
+Istituto per la Ricerca Scientifica e Tecnologica, Povo (Trento), Italy;
++Facoltà di Scienze, Università di Trento, Italy; oIstituto di Chimica Industriale, Università di Padova, Italy

ABSTRACT

Nickel substrates were coated with a thin Sn film and implanted with 100 keV Xe^+ to fluences between 3 and $7 \cdot 10^{15}$ ions/cm^2. Surface chemistry changes were monitored using Auger and Mössbauer spectroscopies. By comparing between profiles of un-implanted and implanted specimens, it was found that ion implantation through the film caused Ni to move into the Sn film and Sn to penetrate into the Ni substrate. The Mössbauer analysis disclosed the presence of new surface phases. A substantial increase in the tarnishing resistance of Ni was found by using this treatment.

INTRODUCTION

Ion beam mixing provides a valuable alternative to high dose ion implantation for surface alloying, because the main limitation imposed by sputtering can be overcome and a whole class of stable or metastable phases may be produced.

Ion mixing studies have been reported recently in many metal systems [1], including systems in which constituents alloy readily as well as those where no alloys are known to exist under normal thermodynamic conditions. Considerable attention has been directed towards the nature of ion mixed surfaces. In some systems, terminal solubilities can be greatly extended [2,3]. In other cases, where the two constituents have different crystalline structure, it has been found as a rule for multilayered samples, that an extension of the terminal solubility limits from both sides of the phase diagram becomes structurally incompatible and an amorphous alloy is formed [4,5,6].

The formation via ion mixing of new phases with favourable properties appears to be an important mechanism for improvements in surface sensitive properties of metals, such as wear and corrosion resistance. One of the most interesting occurrences is the possible presence of an amorphous surface layer. In particular, the excellent corrosion resistance of amorphous metals prepared by other than ion implantation techniques is well known [7].

From a technological point of view, it is interesting to improve the surface properties of nickel, which is very much affected by corrosion in hydrogen sulphide environments. Tarnishing of Ni occurs basically because the sulphide-containing surface film that forms is not protective, in contrast to the protective layers that form for example on stainless steel. Conventional alloying techniques have been used in an attempt to develop tarnish-resistant nickel alloys. Among them, electrodeposited Ni-Sn alloys are known to be excellent coatings [8] from the point of view of their hardness and corrosion resistance in various atmospheres and aqueous solutions [9], including those which readily attack the pure component metals. The corrosion properties of amorphous layers from ion bombardment are for a short time under investigation [10]. Ion implantation on nickel has resulted in many cases [11, 12, 13, 14] in surface amorphization, because of the catastrophic effect induced on a lattice structure where atoms have a directional contribution to their bonding.

In this work we report on the bombardment with xenon ions of a tin deposited nickel substrate. Ion beam mixing is expected to induce both surface amorphization and new phases formation, combining in such a way the advantages of ion implantation and metallurgical treatment. Preliminary results on the corrosion behaviour of the Ni/Sn mixed surfaces are also presented.

EXPERIMENTAL

Discs of 20 mm diameter stamped from a 0.25 mm thick, 99.98% pure Ni sheet, were coated on one face with a 20 nm thick Sn film at a base pressure of 10^{-4} Pa and with a deposition rate of 0.1 nm/s. The Ni/Sn specimens were then implanted at room temperature with 100 keV Xe ions to fluences between 3 and $7 \cdot 10^{15}$ ions/cm^2, reproducible within 10%, with a current density of 7 μA/cm^2.

The as-evaporated and implanted samples were observed with a scanning electron microscope (SEM), supplied with energy-dispersive X-ray analysis (EDXS), and characterized by the following techniques:

(i) Auger Electron Spectroscopy (AES), performed with a 3 keV, 150 nA primary electron beam, the analyzed area being of the order 200 μm by 200 μm. The combination of this surface specific technique with 2 keV Ar$^+$ sputter etching gives the profiles presented below;

(ii) Conversion Electron Mössbauer Scattering (CEMS), performed at room temperature with a constant acceleration spectrometer using the ^{119}Sn resonance (CaSnO$_3$ source, nominal activity 15 mCi). The spectral profiles were analyzed with a standard minimization routine to fit a number of Lorentzian lines;

(iii) X-ray Diffraction (XRD), performed in a conventional way using the Cu Kα radiation.

Finally, corrosion tests of the samples were carried out in H$_2$S-containing environments for several hours.

TABLE I. 100 keV Xe$^+$ implants on Sn-on-Ni samples

Dose ($\cdot 10^{15}$ ions/cm^2)	Pct. retained Sn after the implant*	Remarks	Corrosion behaviour
3	70	Sn poorly mixed	fair
5	50	optimal conditions	good
7	20	excessive Sn sputter loss	bad

* as determined by integral of Auger depth profiles and/or EDXS measurements

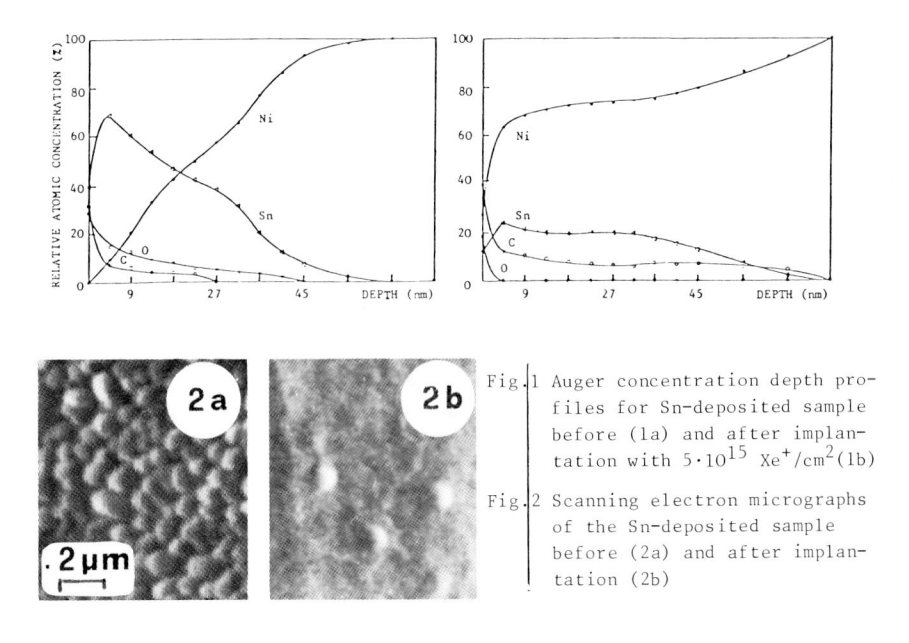

Fig. 1 Auger concentration depth profiles for Sn-deposited sample before (1a) and after implantation with $5 \cdot 10^{15}$ Xe$^+$/cm^2 (1b)

Fig. 2 Scanning electron micrographs of the Sn-deposited sample before (2a) and after implantation (2b)

RESULTS AND DISCUSSION

The ion bombardment has been carried out to three doses: 3, 5 and $7 \cdot 10^{15}$ Xe$^+$/cm^2. From Table I it comes out that the intermediate dose is the more interesting one, therefore we will consider in the following only this case. Figs. 1a and 1b show the Auger depth profiles of a sample, as-evaporated and after implantation. The relative concentrations of Sn, Ni, O and C were calculated using the elemental sensitivity factors method given in Ref.[15]. The sputtering time scale was converted into a depth scale by putting the interface between the Sn deposit and the Ni substrate at the intersection point of the depth profiles of these two elements in Fig. 1a; in such a way an approximate sputter rate of 4.5 nm/min for this system was determined. The assumption is made here that, in spite of changes in the system composition, there is roughly no change in the sputter rate.

Fig. 1a shows clearly a Sn layer on a Ni substrate even though the interface is very broad. This broadening is thought to be essentially due to the roughness of the Sn deposit, as shown in Fig. 2a, which could leave some uncovered Ni. In the implanted sample, Fig. 1b, there is no trace of the Sn layer: Sn and Ni appear together from the very beginning of the depth distribution; the sample surface looks quite smooth, Fig. 2b, except for the presence of some residual Sn grains. By comparing the area under the Sn depth profile before and after the implantation (Figs. 1a and 1b), a reduction by a factor 2 of the initial amount of Sn can be determined. This Sn loss is due to sputtering. In this experiment, the sputtering yield of tin under 100 keV Xe$^+$ bombardment could not be measured, due to the fact that quite soon the Xe ions "see" both Sn and Ni atoms. Actually, an evaluation of a sort of Sn sputtering yield under these conditions gives a value of 7 (assuming for the tin deposit a density of 7.3 g/cm^3), which is lower than the quoted value of about 11

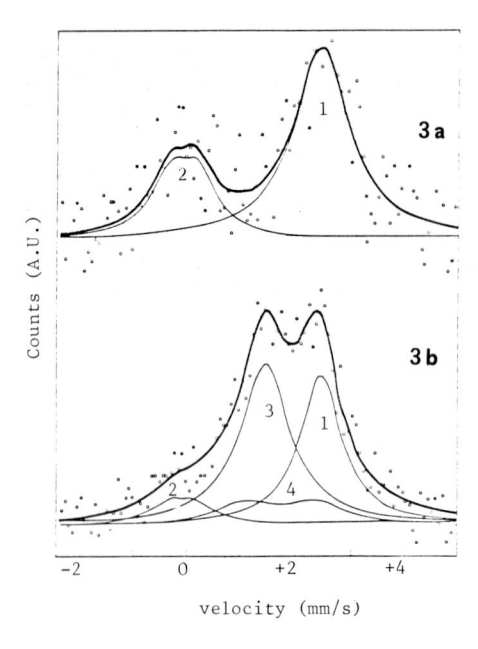

TABLE II. Computed component pct. relative areas in the Mössbauer spectra of Sn-on-Ni samples

Phase	as-evaporat.	implanted
Sn met.	70 (95)	34
SnO$_2$	30 (5)	8
Ni$_3$Sn	–	45
Ni$_3$Sn$_2$	–	13

( ) calculated mean phase percent on the basis of recoilless-free factors ratio, see text.

Fig. 3 ^{119}Sn CEMS spectra of an as-evaporated sample (a) and the same, after Xe$^+$ implantation (b). Continuous light lines are the computed subspectral components: (1) Sn$_{met.}$; (2) SnO$_2$; (3) Ni$_3$Sn; (4) Ni$_3$Sn$_2$

in Ref. [16], as expected. Noteworthy is however the fact that the residual 50% Sn atoms result, after implantation, to be mixed with Ni over a depth of 40-50 nm. The Ni to Sn atomic concentration ratio in this mixed region lies between 3 and 4.

To establish whether a structural modification has taken place in this region, CEMS analysis was performed. The obtained spectra are reported in Fig. 3. The computed subspectral relative areas of the recognized phases are reported in Table II. The values are not proportional to phase concentrations, since the recoilless-free factors of the various phases (not all reported in the literature) and the structural and compositional inhomogeneity of the investigated layers are to be taken into account. In the as-evaporated sample, Fig. 3a, the presence of tin and stannic oxide have been detected. Being the room temperature recoilless-free factors ratio f(Sn)/f(SnO$_2$) = 0.11,[17] the mean percent of SnO$_2$ in the surface layer results about 5%. No other phases are recognizable as a possible result of room temperature mutual interdiffusion of Sn and Ni. This is in agreement with the given interpretation of the broad interface in terms of inhomogeneity of the deposit. The effect of Xe$^+$ implantation, displayed in Fig. 3b, is noteworthy. It consists in the formation of a considerable amount of new phases, with Mössbauer parameters close to those of Ref. [18]: Ni$_3$Sn (singlet with isomer shift = 1.52 mm/s) and Ni$_3$Sn$_2$ (doublet with isomer shift = 1.76 mm/s and quadrupole splitting = 1.26 mm/s). The fit of this second spectrum is not fully satisfactory, for two main reasons: (i) the low statistics, due to the thin deposit of unenriched Sn, and (ii) the actual surface compounds do not necessarily correspond to equilibrium phases, so

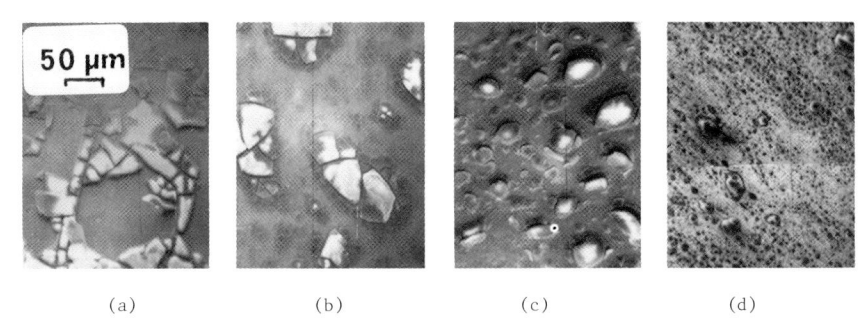

Fig. 4 SEM micrographs of (a) pure nickel, (b) xenon implanted nickel, (c) tin deposited nickel, and (d) ion mixed Sn/Ni, after 6 hour exposure to a H_2S-containing atmosphere

that their Mössbauer hyperfine parameters may differ from the used ones. It is impossible to give here the mean percent of these phases, being actually not available the relevant recoilless-free factors.

X-ray diffractograms of the as-evaporated samples reveal, together with the Ni substrate lines, the presence of the metallic Sn deposit despite its thinness (20 nm). On the other hand, the diffractograms corresponding to implanted specimens show only Ni lines. This may be an indication that the phases present in the near surface region have an amorphous structure. More specific analysis (glancing incident X-rays or TEM) is needed in order to confirm this hypothesis, even though morphologies such as the one shown in the SEM micrograph of Fig. 2b are often interpreted in terms of amorphization[19].

The effect of ion mixing on the corrosion behaviour of Ni in H_2S-containing atmosphere is quite remarkable.

As an example, tarnish films were formed on a nickel sample after several hours of exposure in H_2S-polluted air. Fig. 4 shows SEM micrographs of 4 different zones of the sample. Both the untreated nickel (a) and the Xe-implanted one (b) show severe spalling of the sulphide corrosion product. The phenomenon is less evident in the Sn-deposited sample (c): the protective action exerted by the Sn barrier seems to be incomplete due to the imperfections present in the deposit. On the contrary, the sulphur attack is drastically reduced by the more homogeneous protective layer (d) induced by ion mixing. Moreover, in the ion mixed sample there are no interface problems. Sulphur in this specimen was detected by EDXS to be 6 times lower than in an equivalent area of the only implanted one. In the formation of the protective layer, the role of tin alloying seems to be more important than those of radiation damage induced by the implantation alone.

The details about the corrosion resistance improvement of this system are developed elsewhere [20].

CONCLUSIONS

In conclusion the reported techniques agree in bringing out noticeable compositional and structural modifications in the Xe^+ implanted Ni/Sn system. Schematically: (1) A remarkable Ni/Sn interdiffusion is observed by AES; (2)

The stoichiometry of the mixed layer is close to $Ni_{75}Sn_{25}$; (3) Conventional XRD is not conclusive about surface amorphization of the specimens; (4) SEM micrographies of the ion mixed surface are indicative of an amorphous layer, and (5) CEMS characterization clearly indicates that a new structure, recognizable as a mixture of Ni_3Sn and Ni_3Sn_2, is induced.

Acknowledgements

The authors are indebted to Prof. F. Ferrari and Prof. I. Scotoni for encouragement and would like to acknowledge the collaboration of P.L. Bonora, L. Fedrizzi, A. Molinari, S. Lo Russo and P. M. Ossi, as well as valid help from I.R.S.T. technical staff. This research is partly supported by Progetto Finalizzato Metallurgia of Italian National Research Council (C.N.R.).

REFERENCES

1. B.Y. Tsaur, S.S. Lau, L.S. Hung and J.W. Mayer, Nucl. Instr. Methods 182/183, 67 (1981)
2. B.Y. Tsaur, S.S. Lau and J.W. Mayer, Appl. Phys Lett. 36, 823 (1980)
3. B.Y. Tsaur and J.W. Mayer, Appl. Phys. Lett. 37, 389 (1980)
4. S.S. Lau, B.X. Liu and M.A. Nicolet, Nucl. Instr. Methods 209/210, 97 (1983)
5. B.X. Liu, W.L. Johnson, M.A. Nicolet and S.S. Lau, Appl. Phys. Lett. 42, 45 (1983); Nucl. Instr. Methods 209/210, 229 (1983)
6. L.S. Hung, M. Nastasi, J. Gyulai and J.W. Mayer, Appl. Phys. Lett. 42, 672 (1983)
7. K. Hashimoto, K. Osada, T. Matsumoto, S. Shimodaira, Corros. Sci. 16, 71, 909, 935 (1976)
8. C.J. Evans, J. of Metals, dec. 1982, p. 43
9. M. Clarke and C.S. Britton, Corros. Sci. 3, 207 (1963)
10. W.A. Grant, Nucl. Instr. Methods 182/183, 809 (1981)
11. A. Ali, W.A. Grant and P.J. Grundy, Phil. Mag. 37 B, 353 (1978)
12. C.M. Preece, E.N. Kaufmann, A. Staudinger and L. Buene, in "Ion Implantation Metallurgy", ed. by M. Preece and J.K. Hirvonen, (AIME, 1980), p. 77
13. Q.M. Chen, H.M. Chen, X.D. Bai, J.Z. Zhang, H.H. Wang and H.D. Li, Nucl. Instr. Methods 209/210, 867 (1983)
14. C. Cohen, A.V. Drigo, H. Bernas, G. Chaumont, K. Krolas and L. Thomé, Phys. Rev. Lett. 48, 1193 (1982)
15. L.E. Davis et al, "Handbook of Auger Electron Spectroscopy", publ. by Phys. El. Div. (Perkin-Elmer, 1978)
16. H.H. Andersen and H.L. Bay, in "Sputtering by Particle Bombardment I", ed. by R. Behrisch (Springer, Berlin, 1981) p. 183
17. F. Gauzzi, A. Maddalena, G. Principi, B. Verdini, Proc. Int. Conf. Applied Mössbauer Effect, Alma-Ata (USSR), sept. 1983; ed. by Sci Acad. USSR, in press
18. J. Silver, C.A. Mackay and J.D. Donaldson, J. Mater. Sci. 11, 836 (1976)
19. P.L. Bonora, G. Cerisola, C. Tosello and S. Tosto, in "Ion Implantation into Metals", ed. V. Ashworth, W.A. Grant and R.P.M. Procter, (Pergamon, Oxford, 1982) p. 35
20. L. Fedrizzi et al, to be published.

BOLTZMANN APPROACH TO CASCADE MIXING

Irwin Manning
Naval Research Laboratory, Washington, DC 20375

ABSTRACT

The Boltzmann transport equation is used to describe a beam of ions of atomic species 1 (1-atoms) bombarding a target modelled as an amorphous mixture of 2-atoms and 3-atoms. In a manner familiar in nuclear reactor theory, the method of characteristics is used to integrate the resulting transport equations. An exact expression for the migration flux J_3 of 3-atoms is obtained in closed form. This expression can be evaluated in terms of a power series in a distance parameter s. For the case of slowly varying density N_3 of 3-atoms, Fick's law, relating J_3 to the gradient of N_3, is derived from this expression; it is given by the lowest order term of the power series. J_3 is shown to be proportional to the bombarding flux. Concomitantly, a closed expression for the mixing parameter in Fick's law is obtained, which allows a calculation of this quantity for realistic interatomic potentials. A model Kinchin-Pease displacement cascade is proposed, which is expected to allow a reasonable first approximation calculation of the mixing parameter in Fick's law. It is deduced that the mixing parameter will depend sensitively on the lattice displacement energy. This dependence constitutes a physical mechanism for chemical effect in cascade mixing, as well as for fluence and temperature dependence of cascade mixing.

INTRODUCTION

Haff and Switkowsky [1] have proposed a model for cascade mixing in which atomic migration under ion bombardment derives from a mechanism somewhat similar to that of gaseous diffusion driven by concentration gradients. Matteson [2] has examined this process from the point of view of the theory of random flights. In the present work, we propose to discuss this process on the basis of the Boltzmann transport equation.

REVIEW OF LINEAR TRANSPORT THEORY

It is assumed that the target can be modelled as being amorphous, and that all atomic interactions take place through uncorrelated binary atomic collisions. It is further assumed that, in all atomic collisions, one of the atoms is at rest in the laboratory frame. Consider the case of a beam of atomic species 1 bombarding a target of atomic species 2 containing an impurity of atomic species 3, and let $\psi_\alpha(\vec{r}, \vec{v}, t) d^3r\, d^3v$ be the number of atoms of atomic species α ($\alpha = 1, 2,$ or 3) located at time t in space d^3r about $\vec{r}$ and having velocity d^3v about $\vec{v}$. Correspondingly, $\vec{j}_\alpha(\vec{r}, \vec{v}, t) \cdot \vec{n} dS d^3v dt$ is the number of α-atoms (particles of atomic species α) with velocity d^3v around $\vec{v}$ crossing an element of area dS with unit normal $\vec{n}$ in time dt around t. Let $\sigma_{\beta\alpha}(\vec{v}' \rightarrow \vec{v}; \vec{r})v'\psi_\beta(\vec{r}, \vec{v}', t)\, d^3v' d^3v d^3r\, dt$ be the number of α-atoms emitted into the phase space region $d^3r\, d^3v$ about $\vec{r}$ and $\vec{v}$ in time dt due to collisions in which a β-atom of velocity d^3v' about $\vec{v}'$ collides with an α-atom. It follows from these definitions that [3]

$$\sum_\alpha \int \sigma_{\beta\alpha}(\vec{v}' \rightarrow \vec{v}; \vec{r}) d^3v = \sigma_\beta(\vec{v}'; \vec{r}), \qquad (1)$$

where σ_β is the total cross section for β-atoms; that is, the reciprocal mean-free path of these atoms. Let $q_\alpha(\vec{r}, \vec{v}, t) d^3r d^3v dt$ be the number of α-atoms inserted into the phase space region $d^3r d^3v$ about $\vec{r}$ and $\vec{v}$ by external sources in time dt about t, and let $Q_{\beta\alpha}(\vec{r}, \vec{v}, t) d^3r d^3v dt$ be the corresponding number of α-atoms inserted by collision in which a β-atom strikes an α-atom. From the above definitions one has

$$Q_{\beta\alpha}(\vec{r}, \vec{v}, t) = \int \sigma_{\beta\alpha}(\vec{v}' \rightarrow \vec{v}; \vec{r}) v' \psi_\beta(\vec{r}, v', t) d^3v'. \qquad (2)$$

Simple conservation of particles leads to the Boltzmann equation [3]

$$\frac{\partial \psi_\alpha}{\partial t} + \vec{v} \cdot \vec{\nabla} \psi_\alpha + v\sigma_\alpha \psi_\alpha = q_\alpha + \sum_\beta Q_{\beta\alpha}. \tag{3}$$

In a manner familiar in nuclear reactor theory, the method of characteristics of the theory of partial differential equations can be applied to integrate the above Boltzmann equation [4]. A characteristic or ray in $(\vec{r},t)$ space is determined; in our case it turns out to be the line

$$\left. \begin{array}{l} \vec{r} = \vec{r}_0 + s\vec{\Omega} \\ t = t_0 + \dfrac{s}{v} \end{array} \right\} \tag{4}$$

where $\vec{\Omega}$ is the unit vector defined by $\vec{v} = v\vec{\Omega}$, and s is a parameter determining arc length along the ray (4). Along this ray $\psi_\alpha(\vec{r},\vec{v},t)$ becomes a function $\psi_\alpha(s,\vec{v})$ of s and $\vec{v}$, with $\psi_\alpha(s,\vec{v}) = \psi_\alpha(\vec{r}_0 + s\vec{\Omega}, \vec{v}, t_0 + s/v)$. For different choices of initial conditions $(\vec{r}_0, t_0)$ the rays (4) map out all of $(\vec{r},t)$ space. Along the ray (4), the Boltzmann equation (3) becomes

$$v\frac{d\psi_\alpha}{ds} + v\sigma_\alpha \psi_\alpha = q_\alpha + \sum_\beta Q_{\beta\alpha}, \tag{5}$$

which is easily integrated to yield

$$\psi_\alpha(s,\vec{v}) = c\, e^{-F(s)} + \frac{1}{v} e^{-F(s)} \int_0^s e^{F(s')} [q_\alpha(s') + \sum_\beta Q_{\beta\alpha}(s')]\, ds' \tag{6}$$

with

$$F(s) = \int_0^s \sigma_\alpha(s')\, ds'. \tag{7}$$

We shall be concerned with the steady state, which is obtained by considering the bombarding beam to be constant in time and doing the integration in (6) for $-\infty < s' \leqslant 0$, which corresponds to times $-\infty < t \leqslant 0$. Upon making the change of variable $-s'$ to s', one obtains for the steady-state density function

$$\psi_\alpha(0) = \frac{1}{v} \int_0^\infty e^{-F(s)} [q_\alpha(-s') + \sum_\beta Q_{\beta\alpha}(-s')]\, ds' \tag{8}$$

with

$$F(s) = \int_0^s \sigma_\alpha(-s')\, ds'. \tag{9}$$

FICK'S LAW FOR CASCADE MIXING

Consider the flux of impurity atoms due to atomic collisions during ion bombardment. From the definitions above one has for this flux

$$\vec{J}_3(\vec{r},t) = \int \vec{v} \psi_3(\vec{r},\vec{v},t)\, d^3v. \tag{10}$$

The $\vec{r}$ dependence of the collision cross section derives from the spatial variation of atomic densities N_α in the target, which is explicitly displayed by writing

$$\sigma_{\beta\alpha}(\vec{v} \to \vec{v}; \vec{r}) = N_\alpha(\vec{r}) K_{\beta\alpha}(\vec{v} \to \vec{v}). \tag{11}$$

A key point of the present work is the observation that the integral form of the Boltzmann equation immediately yields the atomic migration currents in closed form. For the steady-state flux, one has from Eq. (8)

$$\vec{J}_3(\vec{r}_0) = \sum_\beta \int d^3v\, \vec{\Omega} \int_0^\infty ds'\, e^{-F(s')} \int d^3v'\, N_3(-s') K_{\beta 3}(\vec{v}' \to \vec{v}) \psi_\beta(-s', \vec{v}')\, d^3v', \tag{12}$$

with

$$F(s) = \int_0^s \sigma_3(-s')\, ds'. \tag{13}$$

The term in Eq. (12) for $\beta = 1$ describes recoil implantation effects as well as effects in which implanted atoms participate in cascades. If one describes primary knock-on production by source terms $q_\alpha(\vec{r},\vec{v},t)$ then, in the zero-fluence limit, this $\beta = 1$ term represents purely recoil implantation effects.

Subject only to the assumptions of uncorrelated binary collisions in an amorphous target, Eq. (12) is exact.

Now arrange the coordinate system so that the incoming beam is along the z axis and, in order to fix ideas, let the beam and target have plane symmetry and perpendicular incidence. One contribution to J_3 in Eq. (12) is the variation in $N_3(-s') = N_3(\vec{r}_0 - s'\vec{\Omega})$. Suppose that N_3 increases in the positive z direction. Then, at a fixed point $\vec{r}_0$ there will be more particles arriving at $\vec{r}_0$ which originate from collisions in the positive z direction (where N_3 is larger) than there are from the negative z direction, and this gives rise to a net flux. The particle density functions ψ_β in the integrand of Eq. (12) will also make a contribution to J_3 because they depend on N_β and, correspondingly, they will have a directional dependence. The parameter s' measures the distance from $\vec{r}_0$ to where these particles originate, and they are attenuated by the factor $e^{-F(s')}$, where $F(s')$ is the average number of mean-free paths these particles travel in going from the point $\vec{r}_0 - s\vec{\Omega}$, where they are created by collisions, to the point $\vec{r}_0$, where they make their contribution to J_3.

Now expand the integral of Eq. (12) in a power series in s', and keep only the lowest-order term. That term comes from the spatial dependence of the factor N_3 which, to lowest order, is

$$N_3(-s) = N_3(\vec{r}_0) - s\,\vec{\Omega}\cdot\vec{\nabla}N_3(\vec{r}_0). \tag{14}$$

The result is

$$\vec{J}_3 = -D\,\vec{\nabla}N_3(\vec{r}_0) \tag{15}$$

with

$$D = \sum_\beta \int d^3v\,\Omega_z^2\,\frac{1}{\sigma^2(\vec{v},\vec{r}_0)}\int d^3v'\,K_{\beta 3}(\vec{v}'\to\vec{v})v'\psi_\beta(\vec{r}_0,\vec{v}'). \tag{16}$$

In obtaining this result, one uses the identity $\int_0^\infty ds'\,s'\,e^{-\sigma s'} = 1/\sigma^2$, and the fact that the integrand of Eq. (12) is cylindrically symmetric about the z axis. In Eq. (16), the particle densities ψ_β are all proportional to the bombarding fluence ϕ, so the diffusion coefficient D of Eq. (16) can be written

$$D = D_0\phi, \tag{17}$$

with the mixing parameter D_0 independent of the bombarding flux.

Equation (15) is Fick's law for atomic migration, and is obtained as the lowest-order approximation to Eq. (12). This approximation is valid only when the variation of $N_3(\vec{r})$ over a mean-free path is small compared to N_3. When this is not true, one must include more terms in the power series expansion in s'. These higher terms will involve powers of N_3 and also the spatial variations of $\psi_\beta(\vec{r},\vec{v})$ due to the spatial variations of $N_\beta(\vec{r})$.

On account of the (averaging) integrations $\int d^3v$ and $\int d^3v'$ in expression (16), we can expect the result to be insensitive to the details of the density functions ψ_β for the recoil cascades; rather crude approximations will probably suffice. On the other hand, Eq. (16) indicates that D is highly sensitive to the atomic collision cross sections, and this expression allows the calculation for realistic interatomic potentials.

MODEL KINCHIN-PEASE CASCADE

In order to calculate the mixing coefficient D_0, one needs the density functions $\psi_\beta(\vec{r}_0,\vec{v}')$ for the displacement cascades. In another work, we plan to show how the integral form (6) can be used to formulate an iteration procedure for these functions which will allow the calculation of D_0 by successive approximations. Here, we formulate a model for ψ_β which allows a first approximation to D_0. We consider the case of an infinitely dilute impurity, and let us agree to ignore recoil implantation effects. In these circumstances, the terms $\beta = 1$ and $\beta = 3$ drop out in Eq. (16), leaving only the term $\beta = 2$.

The quantity needed is the density function $\psi_2(\vec{r},\vec{v},t)$ for the cascade of 2-atoms under steady-state bombardment by 1-atoms, which is obtained as follows: Let $G(\vec{r},\vec{v},t;\vec{r}_0,\vec{v}_0,t_0)$ be the density function for the displacement cascade created by a primary knock-on atom (PKA) of energy E_0, velocity $\vec{v}_0$ created at time t_0 at the location $\vec{r}_0$. We will construct G to represent a simple implementation of the Kinchin-Pease cascade [5]. Let $\mathbf{G}(\vec{r},\vec{v};\vec{r}_0,\vec{v}_0)$ be the density function for the displacement cascades in steady state, where the above PKA's have been created at a constant rate for a long time. One can show from the time-translation invariance of these Green's functions that

$$\mathbf{G}(\vec{r},\vec{v};\vec{r}_0,\vec{v}_0) = \int_0^\infty G(\vec{r},\vec{v},t;\vec{r}_0,\vec{v}_0,t_0=0)\,dt, \tag{18}$$

which will be used to obtain $\mathbf{G}$ from G. Let $q_2(\vec{r}_0,\vec{v}_0)$ be the steady-state creation rates for PKA's of velocity $\vec{v}_0$. The density function to be used in Eq. (16) will then be

$$\psi_2(\vec{r},\vec{v}) = \int \mathbf{G}(\vec{r},\vec{v};\vec{r}_0,\vec{v}_0)\,q_2(\vec{r}_0,\vec{v}_0)\,d^3v_0. \tag{19}$$

Thus, $\psi_2(\vec{r},\vec{v},t)$ can be obtained if one has the density function G for a single cascade.

For our first approximation to D_0, we will use for G a simple implementation of the Kinchin-Pease cascade [5]. The cascade is considered to consist of m steps. At the k-th step, there are 2^k recoils, each of energy $E_0/2^k$, velocity

$$v_k = v_0/2^{k/2}, \tag{20}$$

and isotropically distributed in direction. This step lasts an interval of time

$$\Delta_k = \frac{1}{\sigma(\vec{v})v_k}, \tag{21}$$

where σ is the reciprocal of mean free path, which can be obtained from the actual realistic interatomic potential. The total number m of such steps is given by the condition

$$E_d = \frac{E_0}{2^m}, \tag{22}$$

where E_d is the displacement energy. The spatial dependence of the cascade is ignored, since it is expected to make a contribution to D_0 only in higher orders. Define the points in time t_k by

$$t_k = t_0 + \sum_{k=1}^{k} \Delta_k. \tag{23}$$

Then G can be written

$$G(\vec{r},\vec{v},t) = \delta(\vec{r}-\vec{r}_0)\,2^k\frac{\delta(\vec{v}-\vec{v}_k)}{4\pi v_k^2} \tag{24}$$

for $t_k \leq t \leq t_{k+1}$, with $G = 0$ for $t > t_m$.

The steady-state Green's function is obtained from Eq. (18) as

$$\mathbf{G}(\vec{r},\vec{v}) = \frac{\delta(\vec{r}-\vec{r}_0)}{4\pi}\sum_{k=1}^{m}\frac{2^k}{\sigma(\vec{v}_k)}\frac{\delta(\vec{v}-\vec{v}_k)}{v_k^3}. \tag{25}$$

The above $\mathbf{G}$ will make a contribution to the calculation (16) of D_0 via the form $v'\psi_2(\vec{r},\vec{v}')d^3v'$. We can assess this contribution by considering the quantity

$$A \equiv \int v'\mathbf{G}(\vec{r},\vec{v}')d^3v' = \delta(\vec{r}-\vec{r}_0)\sum_{k=1}^{m}\frac{2k}{\sigma(\vec{v}_k)}. \tag{26}$$

For $\sigma(\vec{v}_k)$ a constant, 50% of the total contribution to A comes from the last step $k = m$, 75% from the last two steps, 87.5% from the last three steps, 93.75% from the last four steps, and so on. This shows that the final steps of the cascade will dominate the contributions of $\psi_2(\vec{r},\vec{v})$ to the mixing coefficient D_0. This corresponds to the low-velocity region of $\psi_2(\vec{r},\vec{v})$, where the density function constructed above can be expected to be not a bad approximation. Recall also that the expression (16) is relatively insensitive to the details of $\psi_\beta(\vec{r},\vec{v})$. These two facts together lead us to expect that the above $\psi_2(\vec{r},\vec{v})$ will provide a reasonable first approximation to D_0.

The above argument also shows that the D of Eq. (16) will be highly sensitive to the displacement energy E_d, which determines m via Eq. (22). Now, instead of the dilute solvent case $N_3 \ll N_2$, consider the general case of arbitrary N_3. If the displacement energy should then depend on the compositions N_3 and N_2 of the target, then so will the mixing coefficient D_0. This will be a chemical effect in cascade mixing. Furthermore, D_0 will have a (complex) dependence on N_1, N_2, and N_3, and therefore will vary with fluence. In addition, if E_d varies with temperature, then so will D_0.

DISCUSSION

The physical origin of cascade mixing in the present work is the same as that of Haff and Switkowski [1], and this work can be considered a refinement of their approach. Their model of cascades has one velocity and one mean free path, and they find $D = vf/3\sigma$, with f a certain fraction. We have the exact expression (12) involving the actual velocity distribution, which allows a derivation of Fick's law and the closed expression (16) for the mixing coefficient.

Emphasis should be placed on the deduction of the previous section that the mixing parameter D_0 will depend sensitively on the displacement energy E_d. If actual calculation bears out this deduction, then this constitutes a physical mechanism for chemical effect in cascade mixing. In addition, cascade mixing will depend on N_1, N_2, and N_3 (and therefore fluence) and on temperature.

ACKNOWLEDGMENT

I'm grateful to K. Grabowski, G.H. Herling, and G. K. Hubler for stimulating and helpful conversations, and to G.H. Herling for reading the manuscript.

REFERENCES

1. P.K. Haff and Z.E. Switkowski, J. Appl. Phys. 48, 3383 (1977).

2. S. Matteson, Appl. Phys. Lett. 39, 288 (1981).

3. K.M. Case and P.L. Zweifel, Linear Transport Theory (Addison-Wesley, 1977).

4. G.I. Bell and S. Glasstone, Nuclear Reactor Theory (Van Nostrand, 1970).

5. G.H. Kinchin and R.S. Pease, Rep. Prog. Phys. 18, 1 (1955).

ENERGY DEPENDENCE OF COMPOUND GROWTH IN Au-Al AND Cu-Al BILAYER SYSTEMS DURING ION BEAM MIXING

S.U.CAMPISANO, CHU TECHANG[*], S.CANNAVO' AND E.RIMINI
Istituto Dipartimentale di Fisica, Università di Catania,
Corso Italia 57, I95129, Catania, Italy

ABSTRACT

The kinetics of phase formation and growth in Au-Al and Cu-Al thin film bilayers under ion beam bombardment was investigated in detail for the dependence on the beam energy. The experimental data can be interpreted within the radiation enhanced diffusion description, taking into proper account the target configuration, the growth of a new phase and the deposited energy density distribution.

INTRODUCTION

Ion beam induced phase formation has been studied in a few thin film bilayer systems [1,2]. In the case of the Au-Al system [3] the two compounds, Au_2Al and $AuAl_2$, are formed in the sequence $Au-Au_2Al-AuAl_2-Al$. The thickness of these phases increases linearly with the square root of the fluence and the growth rate depends on the beam energy and layer thickness combination. In the Cu-Al system only the Al_4Cu_9 phase has been detected and the growth kinetics occurs with similar trend. A linear relationship between thickness and $(dose)^{\frac{1}{2}}$ suggests a diffusion like process in which atom migration can occur in both directions along the beam incidence. A "modified" radiation enhanced diffusion description [4] can explain the experimental data in both systems.

EXPERIMENTAL

Au and Al thin layers were evaporated sequentially onto glass substrates, maintained at -30°C to avoid the formation of thin layers of Au_5Al_2. The as-evaporated samples were stored at LN_2 temperature for the same reason. During the evaporation the residual pressure was less than 3×10^{-7} Torr. Typical thicknesses were 150-220 nm and 40-80 nm for Au and Al, respectively. The Cu-Al bilayers were made by the same process. The samples were irradiated at LN_2 temperature with Kr ions for doses of $10^{15}-2\times10^{16}/cm^2$ in the energy range 60-240 keV. The power density of the beam was maintained below 0.1 Watt. A thin film thermocouple [5] was used to measure the temperature during implantation and the maximum temperature was less than 100K.

2 MeV He^+ backscattering and X-ray diffraction techniques were adopted to measure the thickness and to identify the formed phases respectively. Experimental details are reported elsewhere [3].

RESULTS

The irradiation of a 200 nm thick Au and 75 nm thick Al bilayer with 140 keV Kr ions to a fluence of $2\times10^{16}/cm^2$ causes the formation of Au_2Al and

---

[*] Permanent address: Shanghai Institute of Nuclear Research. Visiting scientist on the basis of INFN - Academia Sinica cultural exchange plan.

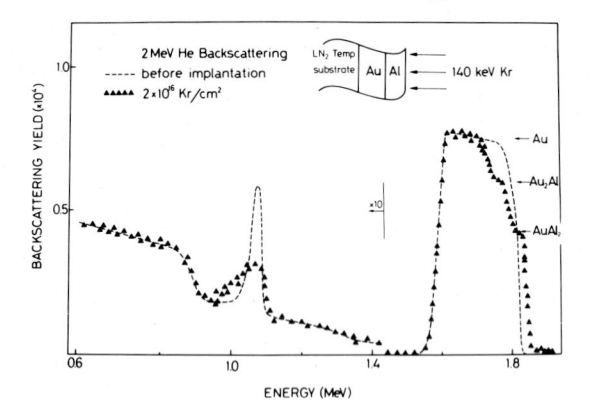

Fig.1. 2 MeV He$^+$ backscattering spectra for Au-Al sample implanted with 2x10^{16}Kr/cm^2 at 140 keV. The dashed line refers to the sample before implantation.

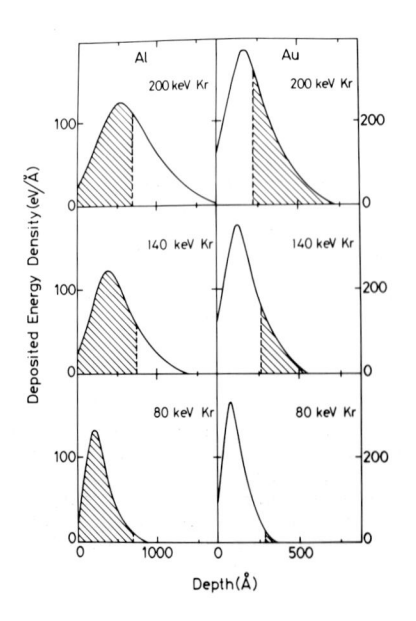

Fig.2. Calculated energy density profiles for three Kr beam energies: 200 keV, 140 keV and 80 keV, in Al (left hand side) and in Au (right hand side) bulk targets. The shadowed regions indicate the fraction of energy deposited in the Al and in Au layers respectively.

AuAl$_2$ compounds. The backscattering analysis shown in Fig.1 indicates that there are two plateaus in the Au signal corresponding to the compositions of Au$_2$Al and AuAl$_2$ respectively. The two compounds are also identified by X-ray diffraction measurements. (For Cu-Al system only Al$_4$Cu$_9$ compound is formed under irradiation). The thickness of the layers increases linearly with the square root of fluence and depends on the beam energy. The measured rates are 32 nm/(10^{16}ion/cm^2)$^{\frac{1}{2}}$ and 47 nm/(10^{16}ion/cm^2)$^{\frac{1}{2}}$ for both compounds in Au-Al system with similar thickness of the outer Al layer at 140 and 200 keV respectively. The results suggest that the distribution of energy deposited in nuclear collision plays an important role. In our experiment the target is inhomogeneous and the composition varies during bombardment. The calculation of the energy distribution will be rather complex. As a first attempt the deposited energy profile is calculated in bulk Al or Au targets using the tabulated range and energy deposition distributions |6|. The results are shown in Fig.2. The left hand side refers to bulk Al and the right hand side to bulk Au. The dashed lines indicate the interface positions and the shadowed parts the deposited energy in the considered element, the remaining part is deposited in the other element (Au). The remaining part has been converted into equivalent deposited energy in Au by the ratio $\nu(E)_{Au}/\nu(E)_{Al}$, being $\nu(E)$ the energy deposited into nuclear collision. The two shadowed areas determine the energy distribution in the Au-Al bilayer system.

As a first attempt to investigate the dependence of the reacted layer thickness on the beam energy we compare the thickness with the total energy deposited into nuclear collisions. The results for Au-Al bilayers and several doses and beam energies are shown in Fig.3. The data show no correlations between the two considered quantities. The lack of correlation can be easily attributed to the requirement that the energy must be deposited near the interface region in order to get a detectable reaction. In fact being the process limited to the near interface region only the energy deposited in this depth range must be considered. The energy density deposited at interface has been calculated as the average value between those in Al and Au, respectively. For the above three cases the deposited energy density at interface is 0.219 keV/Å, 0.112 keV/Å and 0.012 keV/Å, respectively, while total deposited energy are 110 keV, 82 keV and 50 keV respectively.

In our experimental data the growth rates for the two beam energies differ by a factor $\sqrt{2}$, while the deposited energy density at the interface differs of a factor 2 for the two cases. This suggests that the reacted layer thickness increases linearly with the square root of the energy density deposited at the interface. The parameter (fluence x $\nu(E_{int.})/\Delta x)^{\frac{1}{2}}$ may be then useful to relate the experimental data. Fig.s 4 and 5 show that the data for both Au-Al and Cu-Al systems follow nicely straight lines as a function of the above parameter.

DISCUSSION

Ion beam mixing involve several processes. Ion-impurity knock-out events is a one way process where the matter is transported in the beam direction and far from interface ($\gtrsim$ 1000 Å). Collision cascade mixing treats the matter transport as a random walk-type problem in which the atom is displaced many times in small steps in successive collision cascade and the total distance of transport is in any case $\sim 10^2$ Å. For mixing of a marker in a homogeneous target the broadening of the distribution for individual collision cascade is given |4| by

$$<\Delta t^2> = \frac{1}{3} \frac{\ell^2}{c} \phi \sigma p \qquad (1)$$

being ℓ the jump distance, ϕ the fluence, σ the cross section of a collision

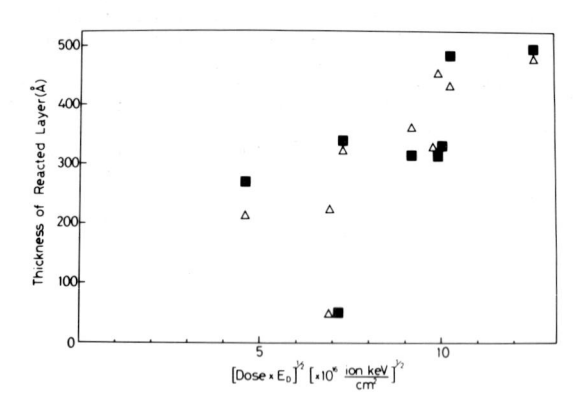

Fig.3. Thickness of the formed compounds (Au$_2$Al and AuAl$_2$) vs the parameter: (dose x total energy deposited into nuclear collisions)$^{1/2}$.
■ -Au$_2$Al   ▽ - AuAl$_2$.

Fig.4. Thickness of the formed compounds (Au$_2$Al and AuAl$_2$) as a function of the parameter: (fluence x average energy density deposited at the interface)$^{1/2}$.

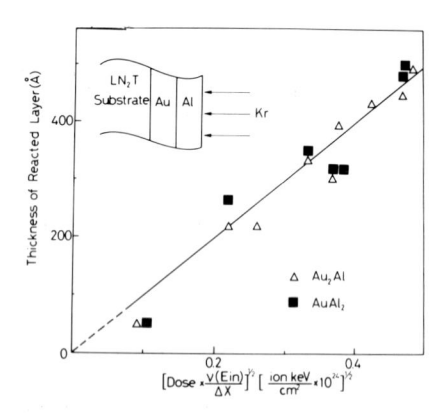

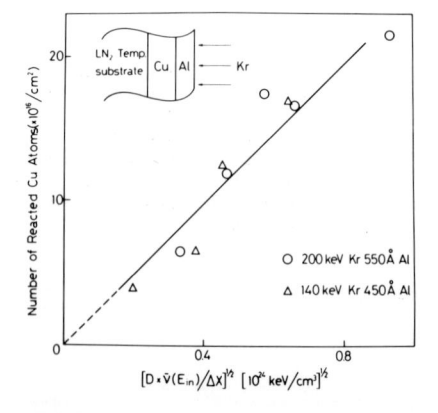

Fig.5. Amount of Cu atoms reacted in Cu-Al bilayer vs the parameter: (dose x average energy density deposited at the interface)$^{1/2}$

cascade facing the beam incidence, c the trap concentration (mean number of trap sites per lattice site) and p the probability for an atom within a damaged zone to undergo diffusional relocation. For the case of thin film bilayers and phase formation at the interface we may assume $p \simeq N_i$ i.e. equal to the interstitial concentration (atom fraction). These interstitial atoms can easily migrate over long distance before being captured. However the concentration of trap sites in our cases is determined by the amount of compound nuclei (Au_2Al, $AuAl_2$ or Al_4Cu_9 respectively). When a fast moving interstitial atom encounters the phase boundaries it can be trapped in a lattice site of this phase thus lowering the total free energy of the system. In a "marker experiment" instead the interstial atom is trapped by a vacancy, whose concentration is in turn equal to the interstitial concentration. Under these assumptions, Eq.1 becomes

$$<\Delta t^2> \simeq \frac{1}{3} \frac{\ell^2}{k} \Phi \sigma N_i \qquad (2)$$

where k is now a constant given by the trapping process. This description is valid only in a region close to the interface where the proposed trapping mechanism is effective. Moreover we can simply assume that $N_i(x) \propto \nu|E(x)|$ and then

$$<\Delta t^2> = cost. \Phi \, \nu(E_{int.}) \qquad (3)$$

as experimentally observed.

If we interpret equation (1) in terms of effective diffusion coefficient D_{eff}, we can write $D_{eff}(x) \sim \ell^2 \nu(E(x))/c$ and the depth distribution of $D_{eff}(x)$ follows that of $\nu(E(x))$. Fig.8 represents schematically the mixing process for the three beam energies used in connection with Fig.2. The rectangles are the damaged zones and the points are the defects produced in the collision cascade. In an individual damaged zone the main contribution to matter transport is due to the defects produced in $\sqrt{2 D_{eff} \tau}$ region from the interface, where τ is relaxation time (average life-time of defects). The extent of this region is few Å, meaning that only the local defects at the interface are effective. For 200 keV Kr the interface is located near the peak of the energy distribution, where most of the defects are produced by collision cascade; then a thicker reacted layer is formed. For 140 keV case the interface is at half height of the distribution, and the amount of reacted material is much less than that for 200 keV case. When ion energy is reduced to 80 keV, the interface is near the end of these distribution tails. Only few defects are effective and the reaction rate becomes negligible.

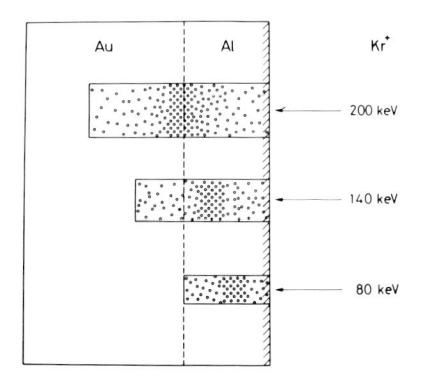

Fig.6. Schematic diagram of collision cascade for three beam energies. The rectangles are damaged zones; o represent defects produced by collision cascades. The dashed line indicates the interface.

CONCLUSION

Ion beam mixing causes the formation of Au_2Al, $AuAl_2$ and of Cu_9Al_4 in the Au-Al and Cu-Al bilayer system respectively. The growth kinetics are linear with the parameter (fluence x interfacial deposited energy density)$^{\frac{1}{2}}$. The ion mixing is characterized by diffusion-like process with effective diffusion coefficient $D_{eff}(x)$, which is proportional to interfacial deposited energy density.

The authors are greatly indebted to Mr.V.Piparo and A.Trovato for their technical assistance with the Van de Graaff accelerator and with the ion implanter. Work supported in part by Progetto Finalizzato Metallurgia-CNR, by GNSM-CNR and by MPI(40%).

REFERENCES

1. J.W.Mayer,B.Y.Tsaur,S.S.Lau and L.S.Hung, Nucl.Instr. and Meth. 182/183,1 (1981)
2. E.Besenbacker,J.Bøttiger,S.K.Nielsen,H.J.Whitlow, Appl.Phys. A29,141(1982)
3. S.U.Campisano,Chu Techang,S.Lo Giudice and E.Rimini, Nucl.Instr. and Meth. 209/210 , 139(1983)
4. P.Sigmund, Appl.Phys. A30 , 43(1983)
5. Chu Techang,S.U.Campisano and A.Trovato, Rad.Effects Letters 76,131(1983)
6. K.B.Winterbon, Ion Implantation Range and Energy Deposition Distributions, Vol.2 (Plenum Press, New York, 1975).

THE EFFECT OF ATOMIC MIXING ON THE DEPTH PROFILES OF METAL MARKERS IN SILICON

B. V. KING, D. G. TONN, I.S.T. TSONG,
Physics Department, Arizona State University, Tempe, AZ 85287

AND J. A. LEAVITT
Physics Department, University of Arizona, Tucson, AZ 85721

ABSTRACT

Atomic mixing effects of sputter depth-profiles are modeled by a diffusion theory with a depth-dependent diffusion constant D. The model is compared to SIMS depth profiles, using 5 keV Ar^+ bombardment of dilute thin-film multilayers of Al, Ag, Ti and Mo in silicon. The experimental values of D can be explained by cascade mixing and radiation enhanced diffusion within the cascade for Al, Ag and Ti markers but not for the Mo marker.

INTRODUCTION

During sputter depth-profiling of a solid, a damaged region precedes the analyzed volume. The impurity depth profile measured by SIMS or AES is then different from the true profile. The mixing effect of the impurity profile is due to a combination of prompt processes - cascade and recoil mixing and the delayed process of radiation enhanced diffusion. Isotropic cascade mixing is the dominant material transport process for incident ion energies above about 1 keV in materials in which delayed processes are absent[1].

Isotropic cascade mixing has been modeled by a diffusion equation[2] where the diffusion constant D is related to the nuclear stopping power of the incident ion F_D, the mean square target atom recoil range $<d^2>$ and effective displacement energy E_d by

$$D = \frac{0.07 F_D U}{S} \frac{4 M_I M_T}{(M_I + M_T)^2} \frac{<d^2>}{E_d} \quad (1)$$

where U is the target erosion rate (ms^{-1}), S is the target sputtering ratio (atoms/ion) and M_I, M_T are the target impurity and matrix atomic masses respectively.

In this model the ratio $<d^2>/E_d = k$ is a measure of the collisional properties of the target. For Si we have calculated a value of $4 Å^2 eV^{-1}$ from the theoretical recoil range distributions[3]. This value of k should also be found from the mixing of dilute impurity concentrations in Si. However, the values of k are found to lie between $21 Å^2 eV^{-1}$ for Pt to $45 Å^2 eV^{-1}$ for Ge derived from experiments using low fluence 300 keV Xe^+ irradiation of 10Å wide markers in Si[4]. This indicates that other processes apart from cascade mixing may dominate atomic mixing of markers in Si.

In a previous report[5] we have sought to determine a value of k from SIMS depth profiles for 7 keV Ar$^+$ irradiation of dilute Al markers in Si where F_D and D, which are in general depth-dependent, were approximated by constants over a finite range. Although the model was not entirely satisfactory we determined a value for k in the range 8-21 $Å^2$ eV^{-1}.

This model will be extended to the case of a varying D(x) by numerical solution of the diffusion equation. This requires approximating D(x) by constants D_k (k=1,...n) over n depth regions and solving the diffusion equation and (n+1) boundary conditions for the concentration [6]. We have tested this model by comparison with SIMS depth profiles. D(x) was set proportional to $F_D(x)$ [7] which has been approximated for our experimental conditions (5 keV Ar$_D^+$ irradiation at 45° incidence) by constants over 12 intervals, each 15 Å wide.

Dilute markers of Al, Ag, Mo and Ti in silicon were chosen as targets. Mo and Ti form silicides with enthalpies of formation ΔH_f = 1.36 eV and 1.40 eV respectively. This choice of metal markers allows the effects of different impurity collisional properties and compound formation on the magnitude of atomic mixing to be assessed.

EXPERIMENTAL

SIMS depth profiling is sensitive to changes in the target composition. Use of dilute multilayers allows changes in sputtering rate and ion yield to be estimated from the change in peak shape and area. The dilute thin film multilayers were made by two methods.

The Ag/Si multilayer was made by electron beam deposition of Si and simultaneous thermal evaporation of Ag. The Si was continuously deposited at 1.5Ås^{-1}. At the depths shown in Fig. 1 Ag was codeposited at varying rates to give markers with Ag concentrations between 0.4% and 6%.

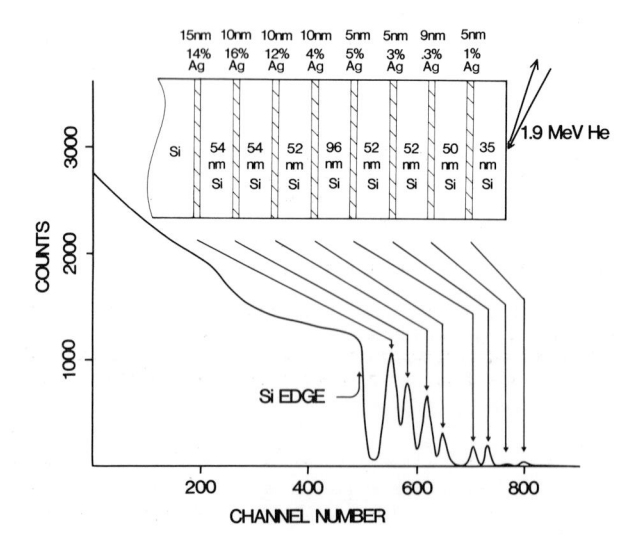

Fig. 1. Rutherford Backscattering spectrum of the Ag/Si multilayer.

The Mo/Si, Al/Si and Ti/Si multilayers were made by DC sputtering. The vacuum system was evacuated to 5x10^{-7} Torr and backfilled with argon. Silicon was continuously deposited at 20Å s^{-1}. At the depths shown in Figs. 2 and 3 the metal was codeposited at varying rates to give markers with metal concentrations between 1% and 20%.

The Ag/Si and Mo/Si multilayers were analyzed by Rutherford backscattering (Fig. 1). This verified that the marker metal concentrations were within 20% of the nominal deposition values. The impurity concentrations indicated in Figs. 1 and 3 were obtained by dividing the total number of impurity atoms per unit area, as measured by RBS, by the nominal marker widths. The concentrations indicated in Fig. 2 were the nominal deposition values, which were calibrated with a stylus profilometer. The measured impurity profile is not, however, sensitive to changes of up to 20Å in nominal marker widths of less than 50Å because the fwhm of the measured impurity peak after mixing, W_P, is approximately related to the deposited marker fwhm W_O by

$$W_M^2 = W_P^2 - W_O^2 \qquad (2)$$

where W_M is the broadening produced by the mixing process. W_M is about 100Å for 5 keV Ar$^+$ irradiation of Si and dominates W_O.

The oxygen concentration determined by RBS in the Ag/Si films was approximately 30% but less than 3% in the Mo/Si, Ti/Si and Al/Si films when analyzed by AES.

The multilayers were analyzed by SIMS using a 5 keV Ar$^+$ ion beam at 45° incidence. Secondary ions were collected normal to the target surface, energy analyzed with a pass bandwidth of 2 eV in a hemispherical electric field and mass analyzed in a RF quadrupole.

The ion beam spot, about 100 μm diameter, was rastered over a 0.5 mm^2 area and the secondary ion signal was electronically gated to accept ions from the center 6% of the irradiated area. Typical ion current densities were 40 μA cm^{-2}. The pressure during analysis was 5 x 10^{-10} Torr.

RESULTS AND DISCUSSION

Figures 2 and 3 show experimental SIMS depth profiles of metal-silicon multilayers (A) and corresponding theoretical curves (B). The samples made by DC sputtering show greater background metal concentrations in the nominally pure Si layers (0.05% for Ti, 0.1% for Mo) than the electron beam deposited sample (0.01% for Ag). The peak metal concentrations in the Ti, Mo and Ag samples were 1%, 7% and 1% respectively.

The oxygen concentration was also monitored by SIMS and varied by less than 15% for the Ti and Ag samples and by about 20% for the Mo samples. Only dilute markers are displayed in the SIMS profiles (Figures 2-3) since increasing oxygen and metal concentrations introduce distortions in SIMS measurements.

Distortions in the SIMS measurements are not evident for these dilute markers. SIMS and RBS results agree for Ag and Mo markers with areal densities up to about 5x10^{14} atoms cm^{-2}, which corresponds to a 10Å wide marker of pure metal.

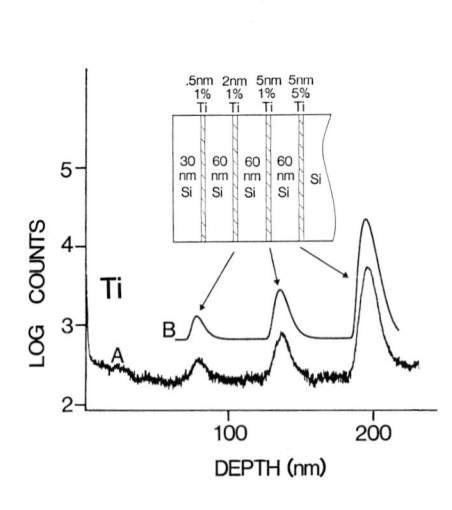

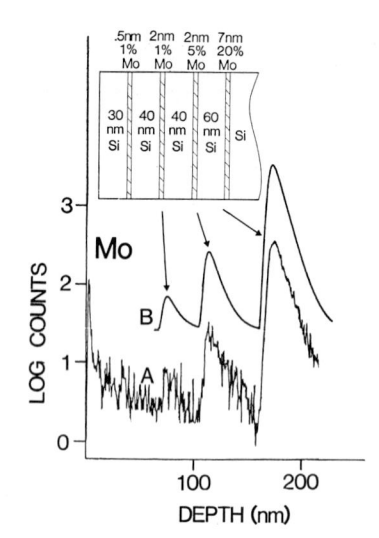

Fig. 2. Comparison of A) Experimental Ti$^+$ depth profile of the Ti/Si multilayer with B) the theoretical fit using diffusion theory and the parameters in Table 1.

Fig. 3. Comparison of A) Mo$^+$ depth profile of the Mo/Si multilayer with B) the theoretical fit. Note that in Figs. 2 and 3 the theoretical curves have been displaced vertically for clarity.

The peaks in the SIMS profiles are not gaussian which is a consequence of the sampling of only the surface impurity concentration during sputtering and is in contrast to the gaussian profiles seen in RBS spectra where the impurity concentrations at different depths are sampled concurrently. The skewness is not due to anisotropic processes since the SIMS profile (Figs.2 and 3) can be fitted by using isotropic diffusion theory.

The increase in the atomic mixing rate for Mo is confirmed by the values of k given in Table 1 which have been calculated from the value of diffusion constant required to fit the experimental peak shapes and the experimental erosion rate U. $D_{exp(m)}$ is the value of D which correponds to the maximum nuclear stopping power, F_D, of 44.7 eV Å^{-1} for 5 keV Ar irradiation of Si. The errors included in $D_{exp(m)}$ result from the difficulty in fitting the profile by eye.

The estimate of the cascade mixing contribution to the diffusion constant, $D_{c(m)}$, has been calculated from eqn. 1 using values of k = 4 Å^{2}s^{-1} and F_D = 44.7 eV Å^{-1} and is presented in Table 1. The experimental diffusion constants $D_{exp(m)}$ are larger than $D_{c(m)}$ for all the markers. The values of k in Table 1 which have been calculated from $D_{exp(m)}$ and eqn. 1, are, as a consequence, higher than the theoretical estimate of 4 Ås^{-1}. The experimental results cannot then be explained by cascade mixing alone.

Table 1 Mixing parameters calculated from theoretical fits
to the SIMS profiles. U is the experimental erosion rate, $\langle d^2 \rangle/E_d$
is the ratio of the mean square range and displacement energy of
cascade recoils. $D_{th(m)} = D_{c(m)} + D_{r(m)}$

Target	U (Å s^{-1})	$D_{exp(m)}$ ($\text{Å}^2\text{s}^{-1}$)	$D_{c(m)}$ ($\text{Å}^2\text{s}^{-1}$)	$D_{r(m)}$ ($\text{Å}^2\text{s}^{-1}$)	$D_{th(m)}$ ($\text{Å}^2\text{s}^{-1}$)	$\langle d^2 \rangle/E_d$ ($\text{Å}^2\text{eV}^{-1}$)
Al/Si	1.1	20±4	9	10	19	9±2
Ti/Si	0.36	9±3	3	3	6	13±4
Ag/Si	3.6	122±36	20	35	55	24±8
Mo/Si	2.05	350±10	12	19	31	116±35

If radiation enhanced diffusion processes occur within the cascade the variance of the gaussian profile formed by the bombardment of an initially narrow profile is given by Sigmund[8]

$$\langle \Delta x^2 \rangle \simeq \phi \pi \rho^2 f \ell^2 / 3Z \qquad (3)$$

where ρ is the r.m.s. transverse straggling, Z is the number of trap sites within the annihilation volume surrounding an interstitial, ℓ is the jump distance and $f = 1$ for interstitial diffusion, and ϕ is the fluence.

The effective diffusion constant, D_r, is related to the variance by

$$\langle \Delta x^2 \rangle = 2 D_r t \qquad (4)$$

where t is the duration of the bombardment. Then

$$D_r \simeq \frac{\pi \rho^2 \ell^2 N U}{6 Z S} \qquad (5)$$

since the current density J is given by

$$J = \frac{UN}{S} \qquad (6)$$

For our experimental conditions $\rho = 34\text{Å}$, $\ell = 2.2\text{ Å}$, $S = 1.5$ atoms/ion and $Z \simeq 23$. In Table 1 $D_{exp(m)}$ is calculated at the peak nuclear stopping power of 44.7 eVÅ^{-1} whereas D_r given in eqn. 5 is the mean value over the cascade length. To correspond to $D_{exp(m)}$ then, D_r must be multiplied by the ratio of the peak to mean nuclear stopping powers which is 2.3 for 5 keV Ar$^+$ irradiation of Si. Then

$$D_{r(m)} = 2.3 D_r = 9.5 \text{ U Å}^2\text{s}^{-1} \qquad (7)$$

This value is an underestimate for $D_{r(m)}$ since the value $Z = 23$ was taken for the case of 500 keV Xe$^+$ irradiation of Al [8] and modified for the smaller cascade volume. However the annihilation volume for metals is large [9] so Z is less than 23 for silicon.

The order of magnitude estimate for $D_{r(m)}$, added to the cascade mixing component, $D_{c(m)}$, is presented in Table 1 as $D_{th(m)}$ for each of the metals.

The values of $D_{exp(m)}$ agree with the theoretical estimates for Al and Ti but not for Mo and Ag. If allowance is made for the order of magnitude estimate for $D_{r(m)}$, Ag can also be explained by the combination of cascade mixing and radiation enhanced diffusion but Mo cannot. Since Mo has an atomic mass between Ti and Ag collisional processes can be ruled out. Compound formation is also unlikely since Ti and Mo silicides have similar formation enthalpies and the concentrations of Ti and Mo in the markers are below their concentrations in the respective metal silicides.

The presence of oxygen does not appear to change the diffusion constant for Al since the experiments on Al/Si presented here and previously [5] give similar values of k but the samples had oxygen concentrations varying by an order of magnitude. The difference in mixing parameters for Mo and Ti cannot then be explained in terms of oxygen concentration differences.

Large variations in the magnitude of mixing of the collisionally similar Cu-Au and Cu-W systems [10] have been linked to a difference in the defect mobilities in the two bilayers. Presumably a similar mechanism operates in the Mo/Si and Ti/Si systems. This can be determined by varying the temperature of the target during SIMS analysis.

CONCLUSION

A diffusion model with depth dependent diffusion constant can successfully model the atomic mixing of metal markers in silicon by 5 keV Ar. The values of diffusion constant obtained for Ag, Al and Ti agree with order of magnitude calculations of cascade mixing and radiation enhanced diffusion within the cascade. However the results for Mo markers in silicon can only be explained by an increased defect mobility.

ACKNOWLEDGEMENT

This work was supported by the U.S. Army Research Office under Contract No. DAAG29-81-K-0162. One of us (JAL) wishes to acknowledge use of the IBM/UA Ion Beam Analysis Facility.

References
1. Z.L. Ziau, B.Y. Tsaur and J.W. Mayer, J. Vac. Sci. Tech. 16, 121 (1979).
2. B.Y. Tsaur, S. Matteson, G. Chapman, Z.L. Liau and M.-A. Nicolet, Appl. Phys. Lett. 35, 825 (1979).
3. U. Littmark and W.O. Hofer, Nucl. Instr. Meth. 168, 239 (1980).
4. S. Matteson, B.M. Paine, M.G. Grimaldi, G. Mezey and M.-A. Nicolet, Nucl. Instr. Meth. 182/3, 43 (1981).
5. B.V. King and I.S.T. Tsong, (in press), Nucl. Instr. Meth.
6. B.V. King and I.S.T. Tsong, to be published.
7. K.B. Winterbon, Ion Implantation Range and Energy Deposition Distributions, Vol. 2, (Plenum, New York, 1975).
8. P. Sigmund, Appl. Phys. A30, 43 (1983).
9. A. Duesing, H. Hammerich, W. Sassin and W. Schilling, Int. Conf. on Vacancies and Interstitials in Metals, Julich, Germany, Vol. 1, p. 246 (1968).
10. M. Westendorp, Z.L. Wang and F.W. Saris, Nucl. Instr. Meth. 194, 453 (1982).

DEPTH DEPENDENCE AND CHEMICAL EFFECTS IN ION MIXING OF Ni ON SiO_2

T. C. BANWELL AND M.-A. NICOLET
California Institute of Technology, Pasadena, California 91125

ABSTRACT

We report on our studies of Ni transport induced by 290 keV Xe irradiation of thin Ni films evaporated on thermally grown SiO_2 at Xe fluences of 10^{15}-10^{16} cm^{-2} and at temperatures of 77-750 K during irradiation. A simple etching technique was used to remove the free Ni leaving the SiO_2 layer with incorporated Xe and Ni, whose profiles are directly measured using 2 MeV He^+ backscattering spectrometry.

Marker experiments are used to verify the selectivity of the etching procedure. An apparent discontinuity in the Ni concentration across the Ni-SiO_2 interface may produce the high selectivity observed within our etching process. Features associated with both cascade mixing and recoil implantation are readily discernible in the residual metal profiles. An exponential tail is evident beyond ∿ 500 Å of the SiO_2 and is insensitive to temperature. Within 500 Å of the surface, the Ni profile demonstrates a strong temperature dependence which affects both the cascade mixing and recoil implantation processes. These profiles show favorable agreement with theory for samples implanted at temperatures of ≤ 300 K, while deviations in the high temperature behavior suggest a way chemical effects may alter the collision processes.

INTRODUCTION

Atomic transport induced by heavy ion irradiation is generally attributed to two principle mechanisms; cascade mixing (and enhanced diffusion), hereafter called diffusional mixing, is associated with numerous short range displacements that may be influenced by thermally activated processes, and recoil implantation which produces long range displacements that are insensitive to temperature [1]. Diffusional mixing has a characteristic parabolic ($\phi^{1/2}$) incident ion fluence dependence [2,3]. Recoil implantation has a characteristic linear (ϕ) fluence dependence and is significant in bilayer samples where large concentration gradients exist. Theoretical investigations indicate that the profile of primary recoils from a surface layer can be described far from the interface by $\phi Be^{-\beta x}$, where x is the normal distance from the bilayer interface and B, β are constants [4,5]. Chemical effects may influence these processes, particularly when large gradients in chemical potential exist.

We study Ni transport induced by 290 keV Xe irradiation of thin Ni films on SiO_2. We are interested in the role of the Ni-SiO_2 interface and chemistry on the ion induced mixing.

EXPERIMENTAL

Substrates with a ∿ 6500 Å SiO_2 layer were prepared by 1100°C steam oxidation of polished <111> Si wafers. 250 Å Ni layers were deposited on the SiO_2 by e-beam evaporation in an oil-free vacuum system at pressures of < 10^{-7} Torr. Xenon implantations were made at sample temperatures of 77, 300 and 750-770 K with 290 keV Xe (R_p = 390 Å, ΔR_p = 140 Å in Ni) [6] to a fluence ϕ of 1-15 x 10^{15} cm^{-2}. The samples were subsequently examined by 2 MeV $^4He^+$ ion backscattering spectrometry; however, Ni-SiO_2 interfacial mixing was not discernible due to limited depth resolution and sputtering effects [7]. The free Ni was removed by etching in boiling (85°C) 12M HCl, as described elsewhere [7]. The Ni and Xe profiles within the SiO$_2$ are readily discernible by 2 MeV ^{4}He$^+$ backscatter spectrometry after removal of the interfering Ni layer. An incident angle of 60° from the sample normal and a 170° scattering angle were used.

The etching procedure circumvents the resolution limitations by transferring the uncertainty to the etching process, which can be independently ascertained. Measurements of SiO$_2$ thickness before and after Xe irradiation using O and Si backscatter signals show that an immeasurable amount (< 33 Å) of SiO$_2$ is removed during etching. Prolonged etching for 2-30 min does not alter the residual Ni or Xe profiles. SEM investigations of irradiated and etched samples did not reveal significant surface asperities.

RESULTS

A typical He backscatter spectra is shown in Fig. 1 for a RT Xe irradiated samples after etching. There are three regions of interest. Most of the Ni is present in a resolution limited peak near the SiO$_2$ surface. The Ni peak moves deeper with increasing ϕ and appears to saturate at $C_o \sim$ 4-5 $\times$ 10^{21} cm^{-3} with high Xe fluences. The first moment $\bar{x}$ of the shallow peak displays an approximately $\phi^{1/2}$ dose dependence with $d(\bar{x})^2/d\phi \sim$ 1400 Å^2/10^{15} Xe cm^{-2}. A deep exponential tail, linear in the log plot, is observable at 1.4 MeV. A flat background dominates from 1.2-1.3 MeV. Its origin is not well understood [8]. For SiO$_2$ substrates implanted with Xe, the backscatter yield is flat in the region between Si and Xe signals and varies linearly with ϕ. A similar background appears to be superimposed on the Ni tail.

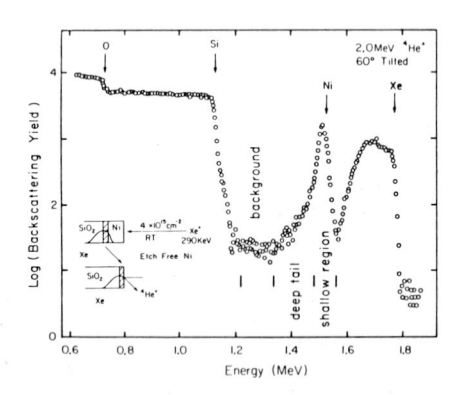

Fig. 1 He backscatter spectra for RT 4 $\times$ 10^{15} cm^{-2} Xe irradiated sample after etching.

Figure 2 shows log-log plots of the Ni areal density in the SiO$_2$ after etching; [Ni]$_s$, versus Xe fluence for the different sample temperatures during irradiation. The RT and LNT results are indistinguishable. The quantity log [Ni]$_s$ exhibits a predominantly linear dependence on log ϕ over the fluence range investigated, with d log [Ni]$_s$/d log ϕ = 0.69. A value of 1/2 would indicate purely isotropic diffusive mixing whereas a value 1 would suggest entirely recoil-like transport; the observed value is consistent with transport produced by both types of processes. Mixing is greatly suppressed in samples held at 750-770 K during Xe irradiation. A linear dose dependence is evident with these samples.

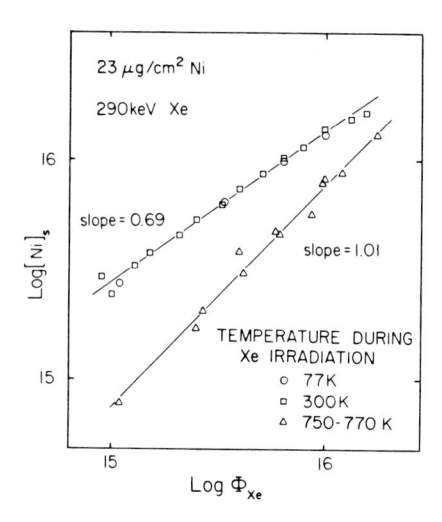

Fig. 2 Graphs of $\log[Ni]_s$ versus $\log \phi$ for different sample temperatures during irradiations.

The concentration profiles $C(x,\phi)$ in the deep exponential tails were numerically fitted to $\ln C(x,\phi) = \ln(B\phi) - \beta x$. The fitting was performed by linear regression on $\ln$(Ni counts - background) versus x for points in the apparent linear region, subject to the constraint that the area Σ(Ni counts - background) in the linear region was preserved by the fitted curve. The fitted tail was then linearly extrapolated to the known surface position for Ni to obtain $B\phi$, from which [extrapolated Ni tail]$_s$ = $B\phi/\beta$ was calculated. The results of these calculations are shown in Fig. 3. The inset figures portray the extrapolation process. A linear Xe dose dependence is observed at each irradiation temperature for the fluences considered, as anticipated for recoil implantation. The characteristic length $1/\beta$ is $\sim$ 340 Å. The significance of the difference between 750-770 K and 77, 300 K results shown in Fig. 3 is not certain. As indicated in the inset figures, the near surface peak nearly vanishes in samples irradiated at 750-770 K. There is a significant inward shift of the Ni peak. The deep tail is substantially the same as that observed at 77, 300 K for a given ϕ.

DISCUSSION

The exponential spatial profile and linear ϕ dependence of the deep Ni tail provide convincing evidence that these Ni atoms are primary recoils. Figures 2 and 3 indicate that Xe irradiation at 750-770 K produces Ni transport via primary recoil implantation alone. The deep high energy recoil implantation is relatively insensitive to localized interfacial processes. The influence of the interface on shallow primary recoils is uncertain. Theoretical calculations show that the primary recoil profile is strongly peaked at the interface [2,4,5]. At 750-770 K, the inward shift of the Ni peak therefore suggests that the low energy primary recoils are either suppressed or expelled from the SiO_2. It is most likely that the low energy primary recoils do not penetrate far into the SiO_2 at 750-770 K as a power law dependence d $\log[Ni]_s$/d log ϕ < 1 would be expected if diffusion were responsible.

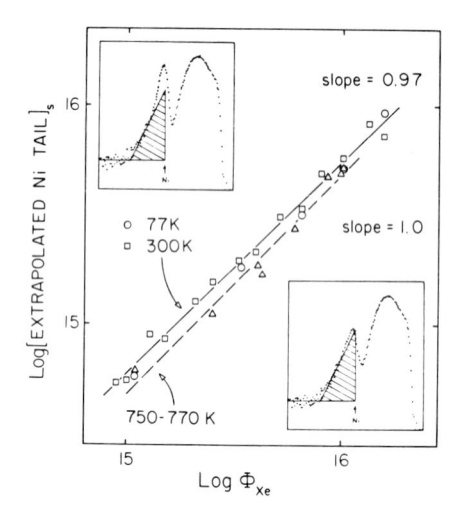

Fig. 3 Plots of extrapolated Ni tail yield versus fluence.

An apparent discontinuity in the Ni concentration of about 22:1 is maintained across the Ni-SiO_2 interface. The utility of the etching process probably arises from this discontinuity. This behavior in the 77, 300 K results is reminiscent of an interfacial equilibrium-like exchange process, which establishes a fixed Ni concentration C_o in the SiO_2 at the interface.

Both recoil implantation and cascade mixing processes contribute to Ni transport at 77, 300 K. This suggests that $[Ni]_s$ may follow

$$[Ni]_s = A\phi^{1/2} + R\phi, \qquad (1)$$

over a limited range of Xe fluence ϕ. The first term represents diffusion-like mixing and the second term is the contribution from primary recoils. Figure 4 shows a graph of $[Ni]_s \phi^{-1/2}$ versus $\phi^{1/2}$ for the 300 K results given in Fig. 3. The linear relationship is apparent. Equation (1) arises from very general considerations and provides little definite information concerning the coefficients A,R beyond dimensional arguments. One interpretation for the observed behavior is that Ni transport within the SiO_2 proceeds via an irradiation induced isotropic random walk from a conjectured surface source C_o and primary recoil implantation produces local injection of Ni atoms within the SiO_2. This model predicts A $\sim 2C_o(D/\pi)^{1/2}$, where D is an effective diffusion coefficient [2,9], and also $\bar{x}^2 \sim 1/4\pi D\phi$. The value $C_o = 4.2 \times 10^{21}$ cm^{-3} determined from this model is consistent with our previous estimate. Equation (1) is probably valid even if Ni forms clusters in the SiO_2, although the interpretation may require modification.

The transition from the low to high temperature behavior could arise from a temperature dependent C_o or D which vanishes at 750-770 K. This hypothesis was tested by performing an irradiation with 6×10^{15} cm^{-2} Xe at 300 K followed by 3.4×10^{15} cm^{-2} Xe at 770 K. The shallow peak resulting from the 300 K irradiation was not visibly altered by the second irradiation; the deep tail was that produced by the sum of the two implantation fluences. One would expect ∼ 50% reduction in the shallow peak due to diffusion if only $C_Q \to 0$ at 770 K; consequently, we must presume D → 0, possibly as a result of Ni precipitation.

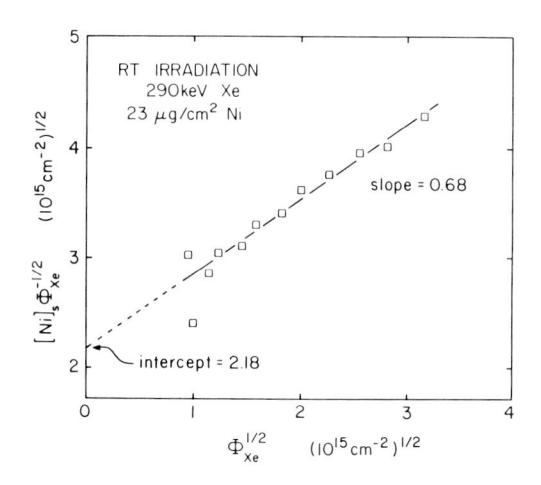

Fig. 4 Graph of $[Ni]_s \phi^{-1/2}$ versus $\phi^{1/2}$ for samples with RT Xe irradiation.

CONCLUSIONS

The etching procedure used to circumvent the energy resolution limitations of our backscatter analysis does not appear to introduce significant uncertainty to the residual Ni and Xe profiles in SiO_2. The structure of the Ni profiles suggest that both diffusional and primary recoil mixing are prevalent at 77, 300 K, whereas only the later occurs with Xe irradiations at 750-770 K. Chemistry appears to be a dominant factor at the interface from 77-770 K. The suppression of diffusional mixing at 750-770 K is also attributed to chemical effects. Clearly, further insight will require more specific studies of $Ni-SiO_2$ chemistry. Additional investigations using TEM, XPS, etc. are needed to proceed beyond the current limited phenomenological understanding.

ACKNOWLEDGMENTS

The authors gratefully acknowledge the technical assistance provided by R. Fernandez, A. Ghaffari, R. Gorris and M. Parks. SEM analysis was provided by R. E. Johnson (Rockwell International) and Dr. R. E. Davis (IBM, Tucson). The project benefitted from the interest and encouragement of Dr. T. M. Reith (IBM,Tucson). The research was financially supported in part by IBM/GDP (Tucson).

REFERENCES

1. F. Besenbacher, J. Bøttiger, S. K. Nielsen, and H. J. Whitlow, Appl. Phys. A29, 141 (1982).
2. S. Matteson and M-A. Nicolet in: Metastable Materials Formation by Ion Implantation, S. T. Picraux and W. J. Choyke, eds. (North-Holland, New

York, 1982), MRS Symposia Proceedings Vol. 7, p. 3.
3. P. Sigmund and A. Gras-Marti, Nucl. Instr. Meth. 182/183, 25 (1981).
4. S. Dzioba and R. Kelly, J. Nucl. Mat. 76, 175 (1978).
5. L. A. Christel, J. F. Gibbons, and S. Mylroie, Nucl. Instr. Meth. 182/183, 187 (1981).
6. J. P. Biersack and J. F. Ziegler in: Ion Implantation Techniques (Springer-Verlag, New York, 1982), p. 157.
7. T. Banwell, B. X. Liu, I. Golecki, and M-A. Nicolet, Nucl. Instr. Meth. 209/210, 125 (1983).
8. W. K. Chu, J. W. Mayer, and M-A. Nicolet, Backscattering Spectrometry (Academic Press, New York, 1978), pp. 209-210.
9. A. Gras-Marti, J. J. Jimenez-Rodriguez, J. Peon-Fernandez, and M. Rodriguez-Vidal, Phil. Mag. A, 45, 191 (1982).

PART II

METASTABLE AND AMORPHOUS MATERIALS

MICROSTRUCTURAL DEVELOPMENTS DURING IMPLANTATION OF METALS

D. I. POTTER, M. AHMED, AND S. LAMOND
Metallurgy Department and Institute of Materials Science
University of Connecticut, Storrs, CT 06268

ABSTRACT

The chemical and microstructural changes caused by the direct implantation of solutes into metals are examined. The particular case involving Al^+-ion implantation into nickel is treated in detail. Chemical composition profiles measured using Auger spectroscopy and Rutherford backscattering, and average near-surface chemical composition measured using an analytical electron microscope, are compared with model calculations. The microstructures that develop during implantation are investigated using transmission electron microscopy. For low fluences implanted near room temperature, these microstructures contain dislocations and dislocation loops. Dislocation loops, dislocations, and voids result from implantations at temperatures near 500°C. Higher fluences at these elevated temperatures produce precipitates when the composition of implanted solute lies in a two-phase region of the phase diagram. Implanted concentrations corresponding to intermetallic compounds produce continuous layers of these compounds. Room temperature, as compared to elevated temperature, implantation may produce the same phases at the appropriate concentrations, e.g. β'-NiAl, or different phases, depending on the relative stability of the phases involved.

INTRODUCTION

Ion implantation provides a unique method for altering the near-surface chemical composition and microstructure of materials, and thus for improving surface sensitive properties like corrosion, oxidation, and wear resistance [1,2]. These two aspects of implantation, alteration of chemical composition and microstructure, are examined in this paper, using aluminum implantation of nickel as a representative system. First, the main factors governing the concentration profiles of implanted species are presented. It is shown that the profiles calculated on the basis of these factors agree well with those measured experimentally for the system. Second, the microstructural developments in the implantation environment are described, and the phases observed in these microstructures are compared to those expected according to the equilibrium phase diagram.

Aluminum was chosen as the implanting ion because this element can be implanted into nickel to high concentrations, ≳75 at.%Al in Ni. Further, the combination of aluminum and nickel is found in many so-called "superalloys", used for high-temperature turbine materials, and thus there is considerable applied interest in the Ni-Al system. Lastly, the usefulness of these alloys has resulted in a well-established data base describing their behavior under various conditions, e.g. the Ni-Al phase diagram [3], and this simplifies the interpretation of the implanted microstructures.

EXPERIMENTAL PROCEDURES

The implantations were performed using an analyzed, 180 kV Al^+-ion beam with a flux of $\sim 1 \times 10^{14}$ ions/cm^{2}s. The specimens were contained in an ultra high vacuum chamber during implantation and the pressure was $\sim 5 \times 10^{-8}$ torr. The approximate temperatures of the specimens were measured during implantation using thermocouples attached to the blocks on which the specimens were mounted. Individual specimen temperatures were monitored using an infrared pyrometer which was responsive at temperatures above 50°C.

The concentration profiles resulting from implantation were measured using Auger electron spectroscopy (AES) and Rutherford backscattering spectroscopy (RBS). Implanted profiles measured using AES were obtained by alternately sputtering the specimen and monitoring the 1396 eV Al and 848 eV Ni Auger signals. The Al:Ni peak-to-peak heights were converted to Al:Ni concentrations by comparison with corresponding intensities measured from a series of Al-Ni standards. The RBS spectra were recorded with 1.5 MeV alpha particles, a 170° scattering angle, a detector with 15 keV resolution, and pure nickel as a reference.

The compositions in the implanted layers were also measured using energy dispersive x-ray spectroscopy. The energy dispersive spectrometer is attached to an analytical electron microscope (AEM) that also functions as a scanning and conventional transmission electron microscope. The use of the AEM for chemical analysis is demonstrated in Fig. 1, where a two phase γ/γ' alloy of Ni-17.7at% Al is examined. The microstructure, Fig. 1a, consists of ~ 1500Å particles of γ'-Ni$_3$Al in a matrix of γ-phase, the solid solution of Al in Ni. The γ' phase has the $L1_2$ crystal structure, a cubic unit cell with Al atoms at the cell corners and Ni atoms in face centered positions. The electron beam can be focused to a diameter as small as 20Å and positioned so as to excite x-rays from either phase and thus produce the spectra in Fig. 1b and Fig. 1c. The chemical compositions of the phases is found by noting that the weight fraction ratio of two elements is directly proportional to the ratio of the integrated intensities under the corresponding characteristic peaks:

$$\frac{C_{Al}}{C_{Ni}} = k \frac{I_{Al}}{I_{Ni}}$$

The proportionality constant, k, can be calculated from first principles [4] or found using standards. The latter was done here to obtain $k = 0.8 \pm 0.02$. The intensities were corrected for absorption since the foils were not sufficiently thin to neglect the correction.

RESULTS OF IMPLANTATION - CHANGES IN NEAR SURFACE CHEMICAL COMPOSITION

At low fluences the chemical composition profiles resulting from implantation can be calculated from existing physical models, e.g. ref. 5. These concentration profiles can be crudely described as gaussian, but significant deviations from this shape occur [6]. The depth at which the profile reaches a maximum, i.e. the range, varies from a few hundred to several thousand angstroms, depending on the implanting ions, their energies, and the elements in the substrate. The ranges of 180 kV Al^+-ions implanted into nickel and into aluminum are about 990 Å and 2570 Å, respectively. [5]

(a) (b) (c)

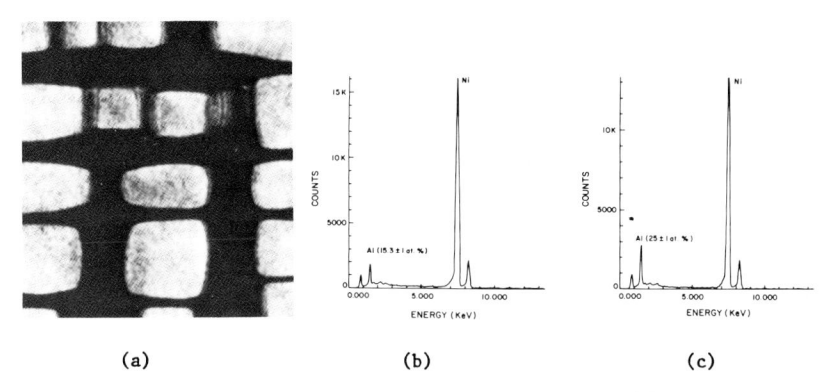

Figure 1. Use of the analytical electron microscope: (a) to produce an image of a two-phase γ/γ' alloy of Ni-17.7 at.% Al; (b) and (c) to show x-ray spectra from γ and γ', respectively, with the analyzed compositions noted.

The development of composition profiles during high fluence implantation depends on factors in addition to the physical models mentioned above. These include sputtering produced by the ion beam used for implantation, expansion of the substrate by the implanted ions, and the extent of ion beam mixing. Sputtering may limit the composition that can be achieved by direct implantation [7]. The limiting composition is reached when the rate at which the ions being implanted is equal to the rate at which they are being sputtered; the latter rate increases with increasing concentration of the ion used for implantation, and is affected by preferential sputtering as well. Recent experiments involving high fluence implantation [8] suggest that the sputtering yield is sensitive to the vacuum environment. It has been treated herein as an adjustable parameter and used to match calculated with experimental concentration profiles.

Substrate expansion due to implantation was modeled by Kräutle [9]. More recent models recognize ion beam mixing as well [2,10]. The latter process can be accounted for using a "forced" or "ballistic" diffusion coefficient to describe the spatial spreading of the the implantation profile during application of the ion beam. The magnitude of the forced diffusion coefficient, and the relative importance of this contribution in determining the final implanted profile, increases with increasing atomic number and mass of the ions used for implantation. For implantation with light ions, such as aluminum into nickel, it appears to play a minor role.

Figure 2 shows measured and calculated concentration profiles for the direct implantation of Al^+-ions into nickel substrates. These profiles are characterized by their maximum or "plateau" concentrations and the depth distribution of the concentration, i.e. the profile shape. Figure 3 shows how the experimentally measured plateau concentrations increase with increasing ion dose, reaching concentrations in excess of 75 at.% Al by the time 4×10^{18} ions/cm^2 are implanted. The calculated profiles agree well with the measured profiles, e.g. at 1.2×10^{18} ions/cm^2 in Fig. 2. Also, the calculated plateau concentrations agree well with those measured experimentally, Fig. 3. These good matches were obtained by varying the sputtering yield used in the calculations from 3.5 initially, to 1.5 at 1.2×10^{18} ions/cm^2. This assumed variation would be difficult to predict quantitatively a priori,

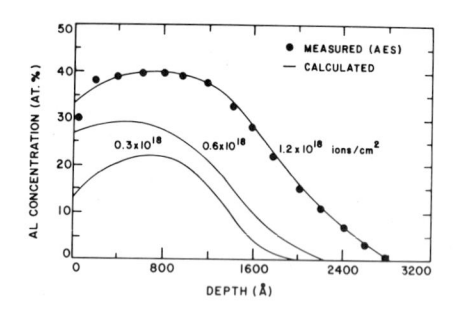

Figure 2. Experimentally determined and calculated composition profiles for nickel implanted with Al$^+$-ions to fluences indicated.

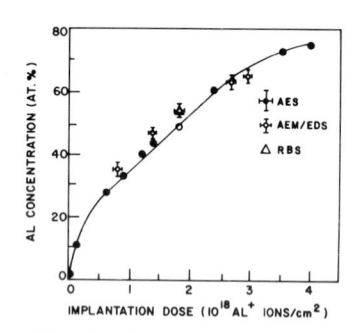

Fig. 3. Plateau aluminum concentrations implanted into nickel determined experimentally by AES, AEM and RBS, plotted versus fluence.

but can be rationalized a posteriori. Increasing aluminum concentration results in a reduced energy density deposited near the implanted surface, providing one mechanism for reduction in sputtering rates with increased fluence. Additionally, one might suggest the formation and subsequent thickening with increasing fluence of a thin layer of Al_2O_3, which is known to have a very low sputtering yield.

Figure 4 shows the implanted chemistry as determined by three methods - Rutherford backscattering, Auger spectroscopy, and analytical electron microscopy. The profile shapes determined by the first two methods agree quite well, but the RBS profile indicates higher concentrations than the AES data. Specimens implanted in different runs in the same implantation chamber were used in these two cases. The composition found by using the energy dispersive x-ray spectroscopy capability of the analytical electron microscope was intermediate between the RBS and AES compositions. The AEM analysis provides a composition that is spatially averaged through the TEM foil. The dashed line in Fig. 4 represents a hypothetical profile that is consistent with the average composition found for the specimen using AEM, and a profile that is also consistent with those measured by RBS and AES. The AEM data point in Fig. 4 was placed at a depth equal to half the measured foil thickness. Fig. 3 presents a further comparison

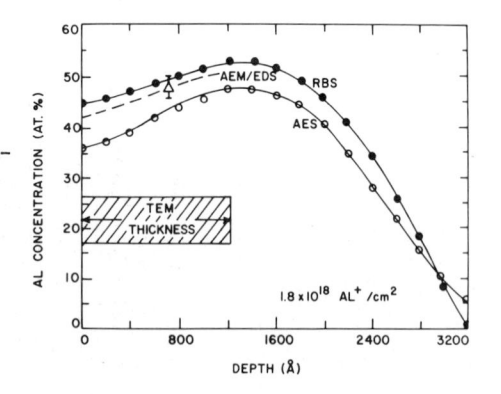

Figure 4. Comparison of implanted concentrations in nickel determined by AES, RBS and AEM.

between AEM and AES analyses. Consideration of all the data leads to the conclusion that the concentrations stated in the next section are accurate to within ± 4at.% aluminum.

RESULTS OF IMPLANTATION - MICROSTRUCTURAL DEVELOPMENTS

At low fluences, where the chemical composition has not been changed significantly, the microstructural changes due to implantation result primarily from atomic displacements. Energy deposited locally by the ions create displacement cascades with vacancy-rich centers. Collapse of the cascades to form vacancy loops represents one of the earliest effects of room temperature implantation visible in pure metals using the electron microscope. These {111} planar aggregates of vacancies have been observed in nickel, for example, by Robinson and Jenkins [11]. With increasing fluence and implanted ion concentration, solid state phase transformations may occur and result in phases not present initially. These phase transformations can be caused by the chemical changes produced by implantation, radiation damage accompanying implantation, or both. The equilibrium diagram is particularly useful in interpreting the phase changes when the chemical composition in the implanted layer is known. The microstructural developments accompanying Al^+-ion implantation of nickel are presented below. The results of elevated temperature implantation are described first, followed by results from implanting Al^+ ions into nickel at room temperature.

The electron microscope images presented in Figure 5 show results typical of implanting Al^+-ions into nickel substrates held at temperatures between 300-600°C. At very low fluences, upper portion of Fig. 5a, the major observable effect of the implantation is the production of dislocation loops. Previous analysis of loops formed at these elevated temperatures by implantation of nickel with self-ions [12] has demonstrated that these loops are aggregates of interstitial atoms on {111} planes. This interpretation is consistent with the observation in the present case that vacancies collect to form voids with further implantation, lower portion Fig. 5a. The formation of interstitial loops and voids caused by radiation damage at elevated temperatures is commonly observed in nickel and nickel-base alloys [13]. The growth of these voids has been treated by Wiedersich [14].

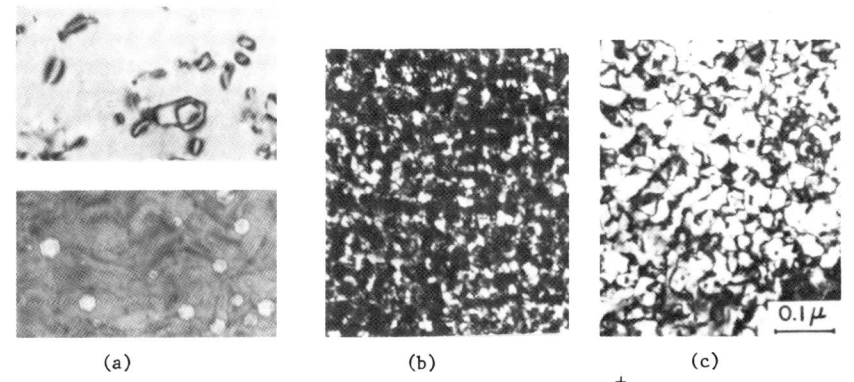

Figure 5. Microstructures of nickel implanted with Al^+-ions as follows: (a) 2×10^{15} ions/cm^2, upper, and 3×10^{17} ions/cm^2, lower, both at 500°C; (b) 3×10^{17} ions/cm^2 at 500°C; (c) 6×10^{17} ions/cm^2 at 300°C. Compositions corresponding to these doses can be read from Fig. 3.

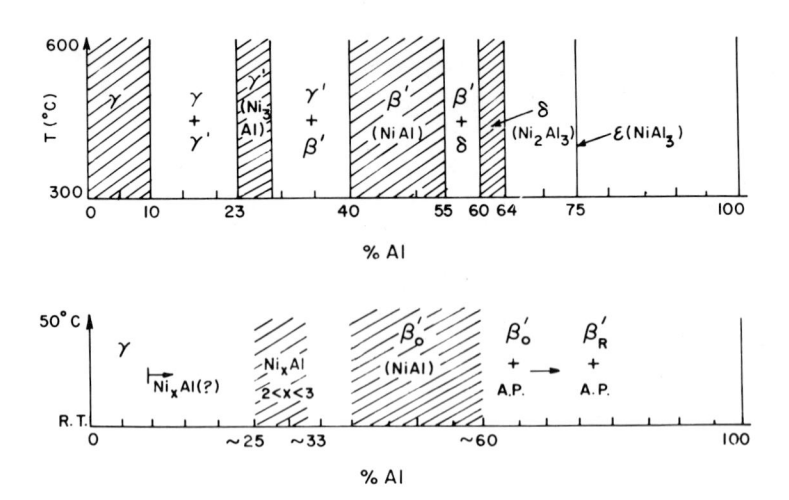

Figure 6. Phases observed as a function of composition in Ni-Al alloys: (a) upper diagram, under equilibrium conditions at elevated temperatures [3]; and (b) lower diagram, in specimens implanted near room temperature.

Further implantation of Al$^+$-ions into nickel can produce alloys whose composition thermodynamically favors separation into two phases. Radiation-enhanced diffusion provides a kinetic path to accomplish this phase separation during elevated temperature implantation. Thus, implantation to ∼14 at.% Al produced γ'-Ni$_3$Al in γ-solid solution, Fig. 5b, consistent with the equilibrium phase diagram presented in Fig. 6a. Implantation to 24 at.% Al produced a continuous layer of γ', epitaxial with the substrate, Fig. 5c, also consistent with the equilibrium diagram. Further, implantation to 45 at.% Al produced a continuous layer of β'-NiAl made up of grains or crystals ∼3500Å in size. The β' phase is an ordered structure, isotypic with CsCl.

Implantation at room temperature produced dislocations at low fluences, Fig. 7a, and these were accompanied by a high density of black spots in the background. The black spots may be dislocation loops that are too small to be resolved [11]. The density of dislocation lines increased with increasing fluence, Fig. 7b and Fig. 7c. It became difficult to distinguish dislocations from the background for fluences between 3×10^{17} and 6×10^{17} ions/cm^2, corresponding to implanted compositions from ∼14 at.%Al to ∼25 at.%Al. The developments in the background, and their role in subsequent phase transformations, remain to be investigated.

Fluences near and beyond 3×10^{17} ions/cm^2 place the implanted layer composition in the two-phase, γ + γ' region of the phase diagram, Fig. 6a. However, no indications of γ'-phase, such as superlattice electron beams, were observed even in specimens implanted to 25 at.%Al. Implantation of γ'-phase at temperature below ∼300°C causes very rapid disordering of the L1$_2$ structure [15], making it unstable in this environment. Electron beams diffracted by another phase were noted. The first indications of this phase took the form of very weak diffraction spots and these were seen even below

(a) (b) (c)

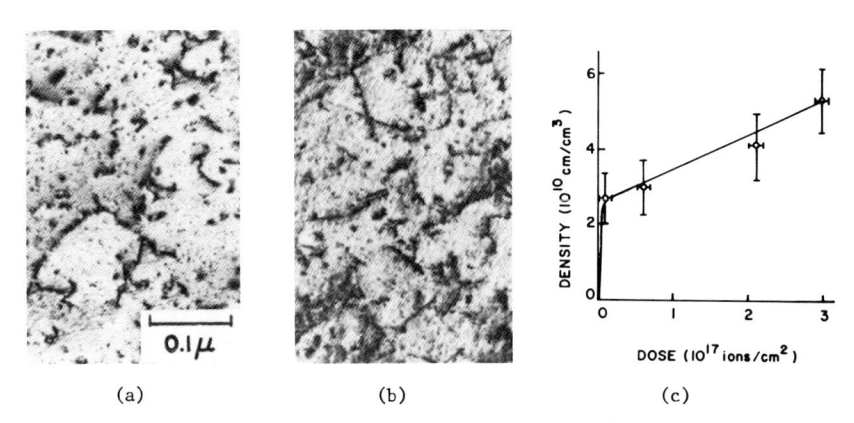

Figure 7. The development of dislocations during Al$^+$-ion implantation of nickel: (a) and (b) show images of dislocations observed after fluences of 7.5×10^{15} and 2.1×10^{17} ions/cm^2, respectively; (c) shows the dislocation density measured as a function of fluence.

10 at.%Al. The spots remained diffuse to about 30 at%Al, whereupon a sharp increase in intensity was observed, Fig. 8. The phase appeared as platelets in dark field images, Fig. 8a, with fine structure visible inside larger platelets, Fig. 8b. The fine structure was revealed by imaging with streaks found along <111> directions in the electron diffraction patterns, Fig. 8c. Intersections of these streaks with the plane of diffraction caused extra spots in the electron diffraction patterns, Fig. 8d. The platelet phase has an approximate stoichiometry of Ni$_x$Al, with 2<x<3. The uncertainty in stating the value of x results from the ±4 at.% Al uncertainty in our present ability to quantitatively measure the composition in the implanted layer, as mentioned earlier. The profuse streaking along <111> directions, and spots produced by these streaks, is consistent with Ni$_x$Al having a heavily faulted face-centered cubic or hexagonal close packed structure, with average atomic radius nearly identical to the matrix in which it forms. The diffraction patterns were not entirely consistent with phases observed in this composition range by other researchers, including hexagonal close packed structures reported by Johnson et. al.[16], a martensitic phase seen at compositions ≳ 35 at.% Al [17], or Ni$_2$Al [18].

The diffraction patterns in Fig. 9 summarize the phases observed after room temperature implantation to compositions higher than 30 at.% Al. At 44 at.% Al, Fig. 9a, the diffraction pattern contains spots along arcs and these spots come from β'-NiAl. The pattern can be indexed recognizing that β' forms with only certain orientations relative to the matrix, conforming to the the Nishiyama relation: $\{110\}_\beta,||\{111\}_\gamma$, $<1\bar{1}0>_\beta,||<\bar{2}11>_\gamma$. Diffraction patterns like that in Fig. 9a were observed from nickel specimens implanted with 44 to 67 at.% Al. A pattern of diffuse rings was superposed on the spot patterns by the time the latter concentration was achieved, Fig. 9b. The diffuse rings result from Ni$_2$Al$_3$ made amorphous by the ion beam, as described elsewhere in these proceedings [19]. Further implantation to 75 at.% Al, Fig. 9c, leads to an increase in intensity of the diffuse rings

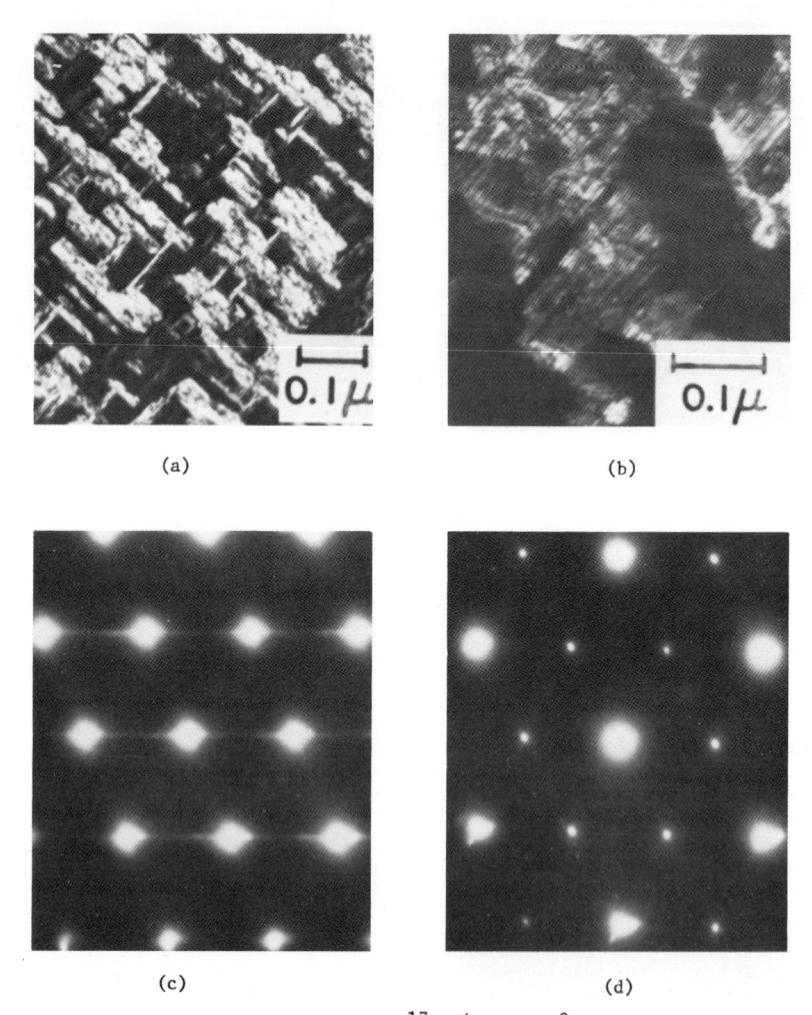

Figure 8. Nickel implanted with 8×10^{17} Al$^+$-ions/cm^2: (a) and (b) are electron microscope images showing platelets and internal structure, respectively, of Ni$_x$Al phase described in text. The 110 and 111 diffraction patterns, based on assumed FCC structure for Ni$_x$Al, are shown in (c) and (d), respectively.

relative to β' diffraction effects. The β' phase contributes sharp rings, indicating it is randomly oriented by this fluence.

The phases observed during room temperature implantation of Al$^+$-ions into nickel are summarized in Fig. 6b. This figure also permits comparison to the equilibrium diagram at elevated temperatures, Fig. 6a. The results of implantation at elevated temperatures described earlier were consistent with Fig. 6a. As noted in Fig. 6b, the Ni$_x$Al phase is first observed during room temperature implantation at compositions near 10 at.% Al. As the im-

(a) (b) (c)

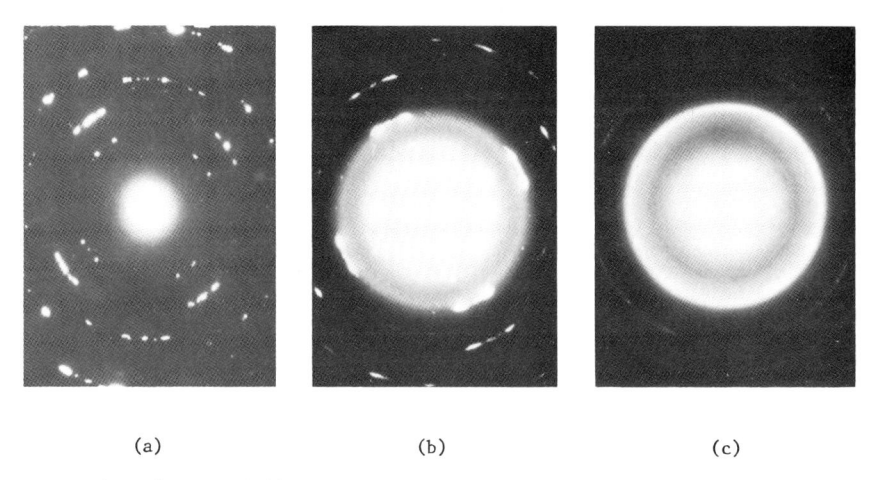

Figure 9. Electron diffraction patterns from nickel implanted to: (a) ∿44 at.% Al, (b) ∿67 at.% Al, and (c) ∿75 at.% Al.

planted composition approaches 30 at.% Al, Ni_xAl replaces the γ-phase over an implantation interval of a few atomic percent aluminum. Over a correspondingly short composition interval near 38 at.% Al, β' phase replaces Ni_xAl phase and the β' phase is oriented with respect to the substrate. This phase remains oriented and ordered ($β'_o$) until a composition of ∿60 at.% Al, whereupon it slowly gives way to an amorphous phase and loses its orientation relation with the substrate ($β'_R$).

SUMMARY

The effects of high fluence implantation on near-surface chemical composition and microstructure were described, using Al^+-ion implantation of nickel as a specific example of these effects. The development of chemical concentration profiles depends mainly on three factors: sputtering of surface atoms by the ion beam, substrate expansion, and ion beam mixing. A model based on the first two factors was used to calculate concentration profiles for the Ni-Al example system. Good agreement between calculated and experimentally measured profiles was obtained. The profiles exhibited relatively broad maxima or "plateaus" which extended to depths $\gtrsim 1200$Å. These maximum concentrations increased with increasing fluence, reaching $\gtrsim 75$ at.% Al by 4×10^{18} ions/cm^2.

Many of the microstructures produced by implantation could have been predicted from knowledge of radiation damage effects, by measuring the chemistry in the implanted layer, and referring to the equilibrium phase diagram. After low fluence implantation, the microstructures contained dislocations and dislocation loops, and, after implantation at elevated temperatures, often contained voids, as well. The phases observed after higher fluence implantation were frequently those expected from the phase diagram, particularly when the implantations were done at elevated temperature. These expectations are not always fulfilled [20] at elevated temperatures, but this was not explored in this paper. At room temperature, certain phases that appear on the phase diagram, e.g. γ'-Ni_3Al and Ni_2Al_3, were unstable due to radiation damage accompanying implantation. These phases were replaced by other phases, e.g. Ni_xAl and an amorphous phase, respectively.

ACKNOWLEDGEMENTS

We thank the following people in the University of Connecticut Physics Department who contributed to this research: J. Budnick and F. Namavar for assistance with the RBS analysis; J. Gianoupolis, H. Hayden, Q. Kessel and C. Koch for assistance with the implantations. We also thank L. McCurdy, J. Soracchi, T. Swol and D. Rock of the Institute of Materials Science. We thank V. Mayer for her helpful comments regarding the manuscript. Fruitful discussions with F. Milillo, Union College, Schenectady, New York were also appreciated. We are grateful for financial support provided by a Special Creativity Extension to our NSF grant DMR8006084, and for support of the AEM laboratory provided by the State of Connecticut and the NSF through grant DRM 8207266.

REFERENCES

1. G. Dearnaley, J. Metals 35, 18–28 (1982).
2. D. I. Potter, M. Ahmed, and S. Lamond, J. Metals 35, 17–22 (1983).
3. M. Hansen, Constitution of Binary Alloys (McGraw-Hill, New York 1958) 118.
4. J. I. Goldstein in: Intro. to Analytical Electron Microscopy, J. Hren, J. Goldstein and D. Joy, eds. (Plenum Press, New York 1979) pp. 83–120.
5. J. P. Biersack and L. G. Haggmark, Nucl. Inst. and Methods 174, 257–269 (1980).
6. A. Anttila, M. Bister and A. Forrtell, Rad. Effects 33, 13–19 (1977).
7. Z. L. Liau and J. W. Mayer in: Treatise on Materials and Technology, Vol. 18, J. K. Hirvonen, ed. (Academic Press, New York 1980) pp. 17–50.
8. F. Namavar, J. I. Budnick, H. C. Hayden, F. A. Otter and V. Patarini, these proceedings.
9. H. Kräutle, Nucl. Inst. and Methods 134, 167–172 (1976).
10. D. Farkas, I. L. Singer, and M. Rangaswamy, Abstracts Fall Meeting Metallurgical Soc. AIME, Philadelphia, PA (1983), p. 54.
11. T. M. Robinson and M. L. Jenkins, Phil. Mag. A43, 999–1015 (1981).
12. B. O. Hall and D. I. Potter, in Effects of Irradiation on Structural Materials, J. A. Sprague and D. Kramer, eds. (American Society of Testing and Materials, Philadelphia, PA 1979) pp. 32–45.
13. Radiation Effects in Breeder Reactor Structural Materials, M. L. Bleiberg and J. W. Bennett, eds. (Metallurgical Society of AIME, New York 1977).
14. H. Wiedersich, Rad. Effects 12, 111–125 (1972).
15. S. Lamond and D. I. Potter, J. Nuclear Materials 117, 64–69 (1983).
16. E. Johnson, T. Wohlenberg, and W. A. Grant, Phase Trans. 1, 23–34 (1979).
17. E. Enami, S. Nenno and K. Shimizu, Trans. Japan Inst. of Metals 14, 161–165 (1973).
18. P. F. Reynaud, J. Appl. Cryst. 9, 263–268 (1976).
19. L. A. Grunes, J. C. Barbour, L. S. Hung, J. W. Mayer and J. J. Ritsko, these proceedings.
20. Phase Transformations During Irradiation, F. Nolfi, Jr. ed. (Applied Science Pub., Essex, England 1983), also Phase Stability During Irradiation, J. R. Holland, L. K. Mansur and D. I. Potter, eds. (Metallurgical Society of AIME, Warrendale, PA 1981).

127

AMORPHIZATION OF THIN MULTILAYER FILMS BY ION MIXING AND SOLID STATE REACTION

M. VAN ROSSUM*, U. SHRETER, W. L. JOHNSON, AND M-A. NICOLET
California Institute of Technology, Pasadena, CA. 91125
*Present Address: Instituut Voor Kern- en Stralingsfysika, Leuven University, Belgium

ABSTRACT

　We have compared the formation of amorphous alloys from Ni-Hf multilayer films by ion mixing and solid state diffusion. We find that ion mixing and solid state reaction produce significant differences in the composition range of the amorphous phase inside the mixed samples. Moreover, the thermochemical parameters which are of primary importance for the solid state reaction also influence the behavior of the Ni-Hf system under ion mixing.

INTRODUCTION

　　　The formation of amorphous phases by ion mixing (IM) of thin multilayered films is now well documented [1]. To improve the understanding of this process, it would be desirable to compare amorphization by IM with other methods. A direct comparison with traditional quenching techniques is not straightforward, since the latter produce the amorphous phase from a mixed liquid or gas phase, whereas IM acts directly on the elemental constituents in the solid state. However, a novel approach has recently been developed that allows the formation of amorphous binary alloys by thermal interdiffusion of solid multilayer films [2]. Direct comparative experiments with IM can therefore be carried out on the same samples, provided that the chosen system fulfills the requirements for amorphization under both procedures. The conditions for solid state reaction (SSR) have been stated as follows [2]: (i) the binary system should exhibit a large negative heat of mixing in the solid phase in order to provide a strong chemical driving force for the reaction, (ii) one of the elements has to be an "anomalously fast diffuser" in the other, so that a temperature range can be found in which the amorphous phase formation is kinetically favored over the nucleation of crystalline compounds. On the other side, the requirements for amorphization by IM have been summarized in the "structural difference rule" [1], which states that an amorphous phase is most likely to be formed if the lattice structures of the crystalline multilayer constituents are different.
　　　The Ni-Hf system was selected for our investigation on the basis of these criteria: it fulfills the structural difference rule, the elements exhibit a large negative heat of mixing [3], and Ni is known as an anomalously fast diffuser in Hf [4]. Moreover, the large mass difference between the two elements facilitates the characterization of the mixing process by backscattering spectrometry.

EXPERIMENTAL PROCEDURE

　　　Ni-Hf multilayer films were prepared from metallic targets by electron-gun evaporation on thin pyrex or oxidized Si substrates. The thickness of the individual layers was kept below 150 Å in order to maximize the mixing efficiency. The total thickness of samples intended for IM was typically 600 Å, which matches the penetration depth (mean projected range plus one standard deviation) of 300 keV Xe^+ ions. The thickness of the multilayers intended for SSR varied between 500 and 3000 Å. Bilayer films designed for detailed analysis of the mixing process did not exceed 2000 Å in total thickness. The thermal reactions were carried out at 340°C for several hours, whereas ion mixing was achieved by Xe^+ irradiations at ambient temperature or liquid nitrogen temperature.

The depth profiles of the layers before and after IM or SSR were monitored by backscattering spectrometry, using 2 MeV He^+ ions. Structural characterization of the samples was accomplished by x-ray diffraction (variable angle diffractometer or Read-camera) and by transmission electron microscopy (TEM).

RESULTS

Multilayer films with average composition $Ni_{20}Hf_{80}$, $Ni_{55}Hf_{45}$ and $Ni_{80}Hf_{20}$ were amorphized after 12 hrs of thermal annealing in vacuo, or by ion irradiation with 10^{15} Xe^+ ions/cm^2 at room temperature. Table I gives an overview of the film structure after mixing, as observed by x-ray diffraction. It appears that IM achieves complete amorphization of the samples over a broader range of composition than SSR does. Moreover, neither the IM nor the SSR results seem to depend on the "deep eutectic" criterion, which is known to be particularly relevant for liquid quenching techniques [5].

TABLE I

Composition	X-Ray Characterization	
	After IM (10^{15} Xe/cm^2, RT)	After SSR (340°C, 12 hr)
$Ni_{80}Hf_{20}$	Amorphous	Amorphous + Ni
$Ni_{55}Hf_{45}$	Amorphous	Amorphous
$Ni_{20}Hf_{80}$	Amorphous	Amorphous + Hf

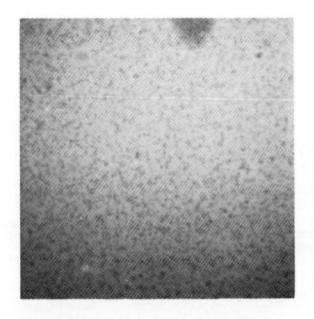

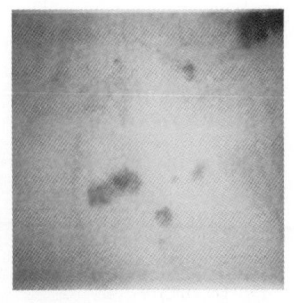

Fig. 1. Electron micrograph (1 mm = 400 Å) of a $Ni_{55}Hf_{45}$ film after ion mixing (left) and solid state diffusion (right).

The microscopic structure of the amorphous films was investigated by TEM and electron diffraction. Similar images were observed for thermally reacted and ion mixed samples. The electron micrographs (Fig. 1) show a mostly featureless background with uniformly distributed spots of size < 100 Å. The spots appeared more strongly contrasted in the ion mixed film than in the thermally reacted sample, but it is not clear whether this observation is of general significance. The electron diffraction patterns exhibit the typical broadening associated with the amorphous structure and a complete absence of Laue spots or sharp rings.

Although ion mixed and thermally reacted films appear structurally similar at the microscopic level, significant differences in the amorphization process are revealed by backscattering analysis. The experiments were performed on bilayers rather than on multilayer samples in order to optimize the depth resolution of the spectra, which is an essential requirement for a detailed study of the mixing process. Figure 2 displays the backscattering spectrum of a thermally reacted bilayer. The spectrum reflects the composition profile of the film at a particular instant during the reaction process. At the time chosen, the reaction is still incomplete and the sample contains a mixed amorphous region between the pure crystalline constituents. The composition profile of the mixed region varies approximately linearly between $Ni_{72}Hf_{28}$ and $Ni_{50}Hf_{50}$. The amorphous-crystalline interface remains sharp during the interdiffusion process, which suggests a laterally uniform growth of the amorphous phase.

These results were compared with a Ni-Hf bilayer that had been irradiated with 5×10^{15} Xe^+ ions/cm^2 at LN_2 temperature (Fig. 3). The composition profile of the ion mixed film markedly differs from the thermally reacted sample. The mixing in this case does not occur by the motion of a uniform reaction front, but rather by a gradual smearing of the Ni-Hf interface with increasing ion dose. Moreover, the mixed layer does

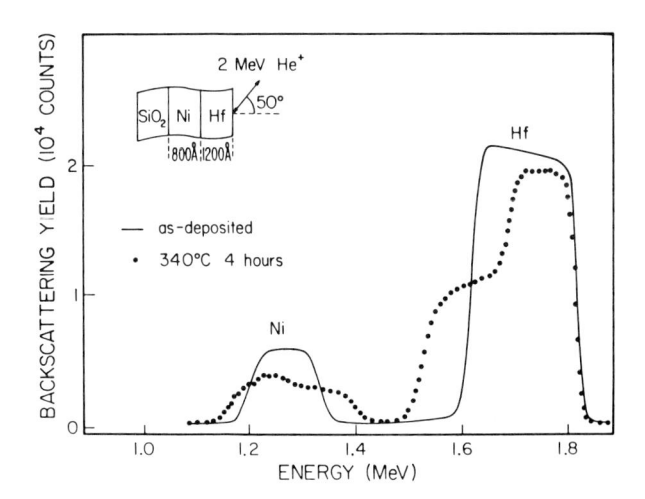

Fig. 2. Backscattering spectrum of a Hf-Ni bilayer after 4 hrs of interdiffusion at 340°C.

not appear to be subjected to any compositional limitations, since the concentration of both constituents in the mixed region changes smoothly with depth from one pure element to the other.

The details of the bilayer mixing profile help to understand the differences between multilayer mixing by IM and SSR. In particular, they explain how a single phase amorphous product can be obtained over a broader range of compositions by IM than by SSR. We believe that the origin of these differences is to be found in the kinetics of both mixing processes. In the case of SSR, a constant temperature is maintained over the whole sample during the interdiffusion process. The reaction products should therefore obey the global equilibrium conditions for the coexistence of solid phases, as determined by the value of the chemical potential at both sides of the reaction boundary. Calculation of these equilibrium conditions using the procedure described in Ref. [2] shows that a single phase amorphous Ni-Hf region should be limited to composition between $Ni_{75}Hf_{25}$ and $Ni_{40}Hf_{60}$ [6]. In contrast to this description, the ion mixing process can be depicted as the deposition of a large amount of energy in a very short time (10^{-12} s) over a small lattice volume (typically 5×10^{-18} cm^3) around the track of the incoming ions. As a consequence of this sudden local supply of energy, the cascade region finds itself in a highly excited state. No compositional limitations are expected to exist in this excited region, since Ni and Hf are completely miscible in the liquid or gaseous state. The subsequent fast quenching (10^{-11} s) of the excited region minimizes the possible relaxation of the established chemical and topological short-range order, so that atomic rearrangement processes such as segregation or recrystallization will be too slow to take place during the quenching phase. This will most probably be the case if the nucleation of a crystalline

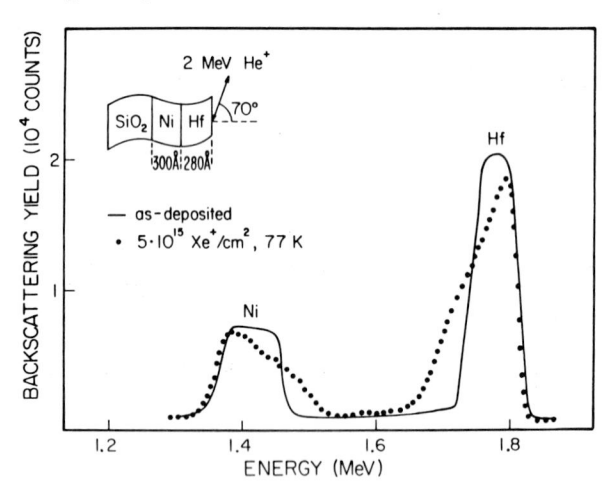

Fig. 3. Backscattering spectrum of a Hf-Ni bilayer after LN_2 irradiation with 5×10^{15} Xe$^+$ ions/cm^2 at 300 keV.

compound phase with a well-defined stoichiometry requires a large amount of atomic rearrangement to create the appropriate unit cell. Moreover, the existence of a large negative heat of mixing, such as in the Ni-Hf system, strongly disfavors any elemental phase segregation, and therefore enhances the probability of a mixed state.

During the irradiation of Ni-Hf multilayers, we observed that complete mixing could be obtained at relatively low ion doses (10^{-15} ions/cm^2) as compared to average mixing doses (listed e.g. in Ref. [1]). To investigate whether ion mixing could be influenced by some of the thermochemical characteristics of the Ni-Hf system, we compared IM of Ni-Hf and Ti-Hf bilayers, using the same beam energy and irradiation dose. Ti-Hf was chosen because this system exhibits only normal (substitutional) diffusion, has zero heat of mixing [3] and atomic masses similar to that of Ni-Hf. The thickness of both the Ni or Ti top layer was calculated to achieve maximal energy deposition at the bilayer interface. The Ti surface was protected with a thin Ni cap. Implantation of 300 keV Xe$^+$ ions was carried out a LN$_2$ temperature in order to minimize possible effects of radiation enhanced diffusion. Backscattering spectra taken after 2×10^{15} Xe$^+$ irradiation are displayed in Fig. 4. It can be observed from the slope of the backscattering signals of Ni and Hf that the intermixing at the interface is significantly higher for Ni-Hf than for Ti-Hf at the same dose. Since both systems are expected to behave similarly under collisional mixing, these results strongly suggest that other mechanisms, possibly connected with the fast diffusion or the large negative heat of mixing, play an important role in the ion mixing of the Ni-Hf system. More experiments are under way to further investigate this effect. In particular, it would be desirable to ascertain that the observed effect is not caused by impurities.

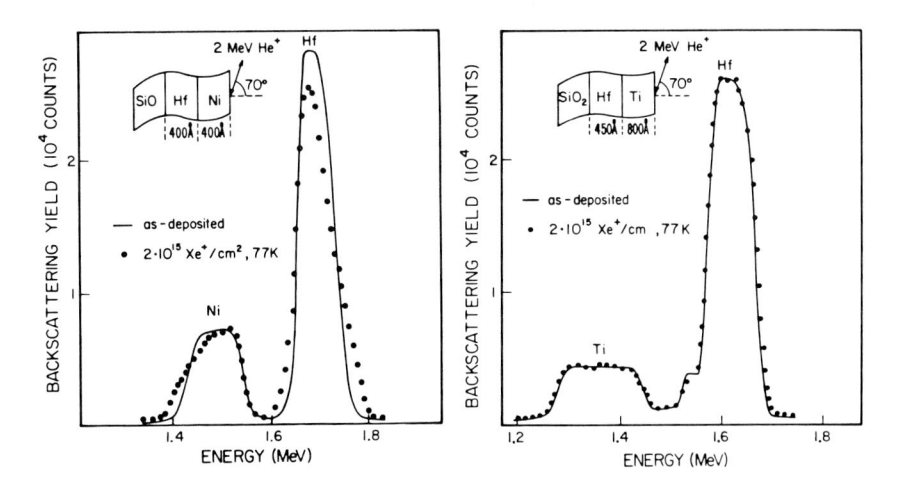

Fig. 4. Backscattering spectra of a Ni-Hf bilayer (left) and a Ti-Hf bilayer (right) after LN$_2$ irradiation with 2×10^{15} Xe$^+$ ions/cm^2 at 300 keV.

CONCLUSION

Our experiments performed on layered Ni-Hf films illustrate how an amorphous phase can be formed in the solid state by two methods involving different kinetic regimes. The solid state diffusion process takes place under constant and uniform temperature conditions and uses the heat of mixing of the alloy constituents as a chemical driving force. During ion mixing, a local nonequilibrium state is produced in the target material by atomic collisions around the track of the incoming ion, followed by a rapid quenching of the excited region. Due to different kinetics, the two methods of amorphization generate different compositions and concentration profiles across the interface of a bilayer structure, but they do not influence the microstructure of the amorphous phase formed from multilayer films. Finally, there are indications that the thermochemical characteristics of the Ni-Hf system affect its behavior under ion mixing, which seems to rule out a purely collisional description of the ion mixing process in this case.

ACKNOWLEDGMENTS

The authors wish to thank Dr. R. B. Schwartz (Argonne National Laboratory) for useful discussions and A. Ghaffari (Caltech) for sample preparation. The research was supported in part by the General Products Division of IBM, Tucson, Arizona (T. M. Reith); and the implantation part of this work was partially supported by the U.S. Department of Energy through an Agreement with the National Aeronautics and Space Administration and monitored by the Jet Propulsion Laboratory, California Institute of Technology (D. Burger). One of the authors (W.L.Johnson) acknowledges the support of the Project Agreement No. DE-AT03-81ER10870 under Contract DE-AM03-76SF00767.

REFERENCES

1. B. X. Liu, W. L. Johnson, M-A. Nicolet, and S.S. Lau, Appl. Phys. Lett. 42, 45 (1983).
2. R. B. Schwarz and W. L. Johnson, Phys. Rev. Lett. 51, 415 (1983).
3. A. R. Miedema, Philips Tech. Rev. 36, 217 (1976).
4. A. D. LeClaire, J. Nucl. Mat. 69, 70 (1978).
5. M. H. Cohen and D. Turnbull, Nature, 189, 131 (1961).
6. M. Van Rossum et al., (unpublished).

ION MIXING KINETICS OF THIN LAYERED FILMS IN THE Fe-Al SYSTEM

J. GRILHE, J.P. RIVIERE, J. DELAFOND, C. JAOUEN
Laboratoire de Métallurgie Physique, L.A. 131 du C.N.R.S.
40, avenue du Recteur Pineau 86022 Poitiers (France)

ABSTRACT

Evaporated bilayers and multilayers of Fe and Al have been studied during ion beam mixing with Xe ions using in-situ electrical resistivity measurements. Experiments have been performed in the composition range 40 - 58 at.% Al and at both temperatures 77 K and 300 K. A semi-empirical model is proposed to explain the observed kinetics. At low doses, a square root dependence of the mixed volume fraction on dose is found at 77 K but not at 300 K. The results are discussed by comparison with the different models proposed for ion beam mixing.

INTRODUCTION

Presently, surface alloys are readily produced by the technique of ion mixing (1,2,3). Fundamental studies have been carried out in order to understand the basic mechanisms of ion mixing. Up to now, most of the experimental results have been obtained using Rutherford backscattering spectrometry analysis. The models may be divided into two general classes : ballistic and thermally activated models. However, the basic mechanisms proposed for both models are not yet well understood. It appears that a precise kinetic study able to point out the fundamental mechanisms of ion mixing must be done with a technique allowing at the same time a very sensitive and continuous measure while the mixing proceeds. Electrical resistivity measurements are well suited to the preceding requirements. They are an easily measured bulk quantity capable of giving continuous measurement, and their extreme sensitivity to radiation damage has been demonstrated in irradiation studies in metals. It is interesting to apply this experimental technique to study ion mixing kinetics (4) especially in the case of multilayers, where it is not possible to measure kinetics by R.B.S.

In the present paper, we give the results of a study performed in the Fe-Al system for both bilayer and multilayer thin film structures. Since the temperature of the film is a fundamental parameter for the different mechanisms of ion mixing, the experiments have been performed at room temperature and 77 K.

EXPERIMENTAL RESULTS

The experimental procedure as well as the general behavior of the resistivity variations as a function of dose have been previously described (5). At low doses typically $<5.10^{13}$ cm^{-2} the rapid increase is mainly due to the production of point defects, in this regime the ratio of number of defects produced over the number of mixed atoms by incident ions is of the order of 100. We present only the experimental curves corrected for sputtering but not for the defect contribution at low dose. We estimate the sputtering coefficient to be 7 at the energies and target compositions used. The results of the resistivity variations during ion mixing with Xe ions are given in Fig.1 for bilayers of final composition Fe 40 at.% Al and in Fig.2 for multilayers of composition Fe 58 at.% Al.

The first point to notice on all curves of Figures 1 and 2 is the ability to detect quite precisely the critical dose which finally

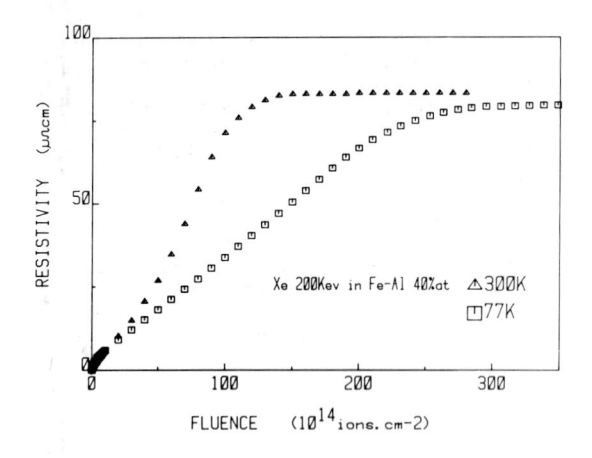

FIG.1. Electrical resistivity increase at 77K and 300K during ion beam mixing of a bilayer (final composition: Fe-40 at.% Al).

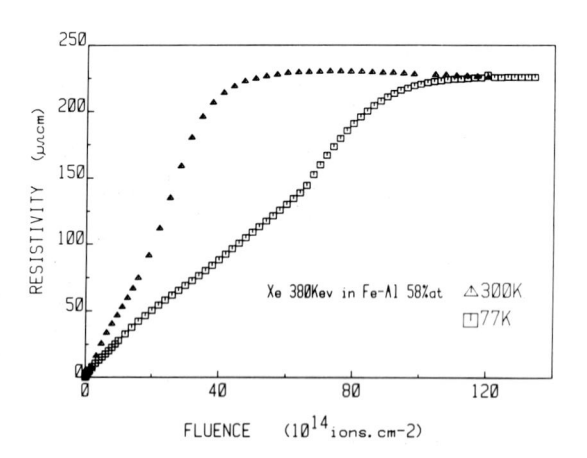

FIG.2. Electrical resistivity increase at 77K and 300K during ion mixing of a multilayer (final composition: Fe-58 at.% Al).

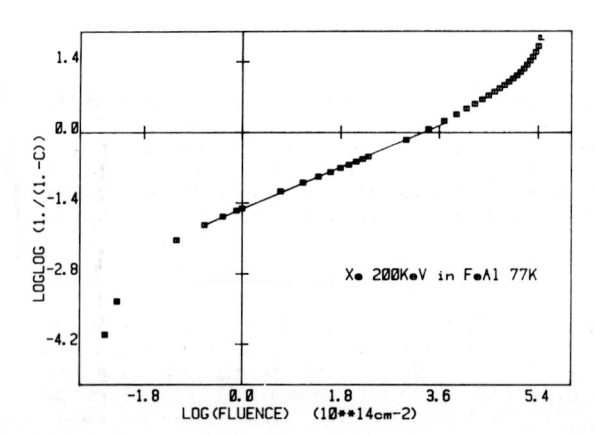

FIG.3. Variations at 77K of the mixed volume fraction C of a bilayer Fe-40 at.% Al. Mixing dominates in the fluence range $5\cdot 10^{13} - 5\cdot 10^{15}\,\text{cm}^{-2}$ and data are fitted with a straight line (slope $n = 0.47$). The deviation at low doses is due to the contribution of damage production on resistivity (not corrected in this first analysis).

corresponds to total mixing. It must be emphasized that the other experimental techniques do not allow so easily the determination of this parameter.
For both film structures, the direct comparison of the critical doses for resistivity saturation at 77 K and 300 K points out very clearly the existence of radiation enhanced diffusion at 300 K. At this temperature the mixing proceeds more rapidly and the dose necessary to realize the total mixing is about half that at 77 K. Since resistivity variations are connected with the volume fraction of mixed atoms (4,5)

$$C(\phi) = (\rho^{-1}(\phi) - \rho_0^{-1})/(\rho_{AB}^{-1} - \rho_0^{-1})$$

(ρ_0 is the initial resistivity, ρ_{AB} the final one and $\rho(\phi)$ is resistivity at a fluence ϕ), it is possible to investigate the mixing kinetics. In the following, a semi-empirical model for mixing is developed, taking into account the saturation effect observed experimentally.

ANALYSIS OF THE RESULTS

Let us consider a bi or multilayer metal-metal system being irradiated at a temperature T. The collision cascades penetrate through the different interfaces producing intermixed layers either directly or by an overlapping process. At relatively low doses, the irradiated volume consists of mixed regions of volume fraction C plus unmixed ones of volume fraction 1-C. In an increment of time dt, or dose d $\phi = \dot{\phi}$ dt, the volume fraction of the mixed regions increases by an amount dC given by :

$$dC = (1 - C) f(\phi, \dot{\phi}, T, \Theta_D) d\phi$$

The function $f(\phi, \dot{\phi}, T, \Theta_D)$ can be considered as a measure of the effectiveness with which the intermixed regions are produced. It is expected that this function is dependent on the dose ϕ, dose rate $\dot{\phi}$, temperature T, and deposited energy density Θ_D. In principle, the experimental determination of the function f ($\phi, \dot{\phi}, T, \Theta_D$) could allow a comparison with the theoretical models of ion beam mixing. When all the parameters $\dot{\phi}$, T and Θ_D are fixed, except ϕ, we are able to integrate the preceding differential expression :

$$\text{Log}(1/1-C) = \int_0^\phi f(\phi) \cdot d\phi$$

or by writing $\int_0^\phi f(\phi) \cdot d\phi = F(\phi)$, we find :

$$C(\phi) = 1 - \exp\left[-F(\phi)\right]$$

We have also to introduce the boundary conditions :
for $\phi = 0$ we have $C = 0$ and $F(\phi) = 0$.
$\phi \to \infty$ $C \to 1$ and $F(\phi) \to +\infty$.
We have considered the simplest function satisfying these conditions by choosing : $F(\phi) = A \phi^n$. This is a power law function where A is a constant, n is the power and both are likely dependent on temperature, dose rate, and deposited energy. A comparison of the proposed kinetic behavior with the experimental results is possible and leads to the determination of the two quantities A and n by writing the analytical formula into the form :

$$\text{Log Log}(1/1-C) = \text{Log } A + n \text{ Log } \phi.$$

The experimental results in Fig. 4, 5, 6 indicate clearly a good agreement with the linear variation predicted by the model in the intermediate dose regime during which most of the resistivity change occurs. For the high fluences the deviation form the linearity above 6.10^{15} ions/cm^2 can be explained, very probably, by structural changes not included in the model.

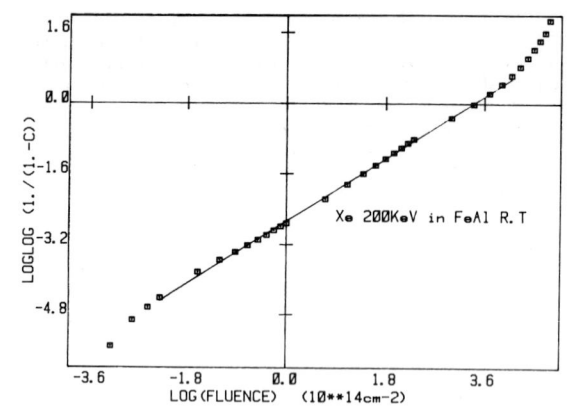

FIG.4. Case of bilayer Fe-40 at.% Al during mixing at 300K. slope n = 0.8

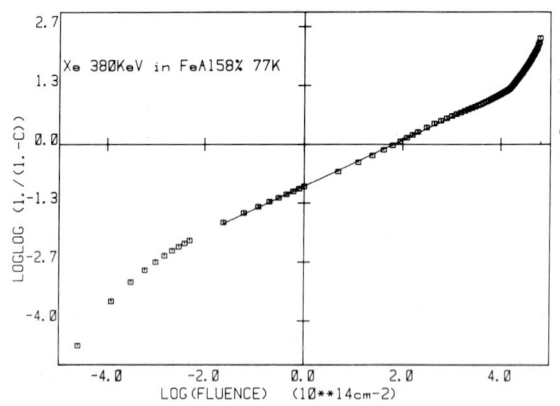

FIG.5. Case of a multilayer Fe-Al 58 at.% during mixing at 77K. slope n = 0.51.

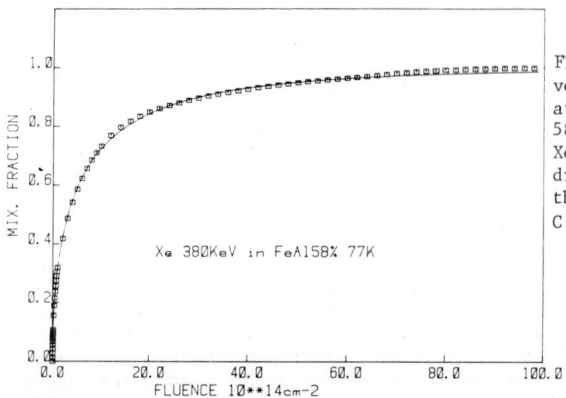

FIG.6. Variation of the volume fraction of mixed atoms : multilayer Fe-Al 58 at.% mixed with 380 KeV Xe ions at 77K. The full drawn curve correspond to the calculated one ;
$C = 1 - \exp\left[-A\Phi^{1/2}\right]$.

This result is also apparent when looking at the directly measured resistivity variation curves in Figures 1 and 2. The case of the Al-Ag system reported in another paper (5) appears more simple; the resistivity increases at a continuously decreasing rate without any irregularity suggesting that only one mechanism is occuring during ion mixing.

We give in Table I the different values found for the power n when fitting the curves in Fig. 3, 4, 5 to a linear variation.

TABLE I : Experimental values of n found for the model $C = 1-\exp(-A\phi^n)$.

	n 77 K	n 300 K
Multilayer Fe-58 at.% Al	∿ 0.5	∿ 0.8
Bilayer Fe-40 at.% Al	∿ 0.5	∿ 0.8

We have plotted in Figure 6 the experimental variations of the volume fraction of mixed atoms deduced from the resistivity curves. The corresponding kinetics given by our semi-empirical model of ion mixing : $C = 1 - \exp[-A\phi^{1/2}]$ are also given on Figure 6 and one can see that a good agreement is obtained in that case. The agreement is not very good for experiments at 300 K.

We can classify mixing in terms of long and short range mixing. Long range mixing refers to thoses processes where atoms are transported over large distances (thousand Å) while short range mixing refers to transport over short distances (tens of Å) (6). The experiments of Besenbacher et al. (7) show the existence of such a long range mixing in the Al-Cu system. That part of the process leads to a direct fluence dependence and it is less important than other contributions. So, the corresponding atomic redistribution produced by this long range mixing will give very low concentration alloys; it will not be seen by our resistivity measurements mainly sensitive to the formation of concentrated alloys of high resistivity. The main contributions to short range mixing that our technique can detect are first, the isotropic cascade mixing at low temperature (77 K) and second, mixing assisted by enhanced diffusion at higher temperatures (300 K). Let us consider the variations of Log Log (1/1-C) as a function of ϕ in the case of experiments at 77 K (Fig. 3 and Fig. 5). The variations exhibit linear portions with a slope 0.5. If Ω represents the thickness of the mixed layer (5), our result predicts at low doses a behavior $\Omega \propto \phi^{1/2}$. However, it is obvious that at high doses, the behavior predicted $\Omega \propto 1 - \exp(-A\phi^{1/2})$ by our model is more realistic.

Isotropic cascade mixing contributions has been calculated by Sigmund(8)

$$\Omega^2 = \frac{1}{2} \Gamma_0 \frac{F_0(x)}{N} \gamma_{12}^2 \frac{R_c^2}{E_d} \phi$$

where $\Gamma_0 = 0.608$, $F_0(x)$ is the deposited energy at depth x, N is the atomic density and $\gamma_{12}^2 = 4 m_1 m_2/(m_1 + m_2)^2$, E_d is the threshold displacement energy and R_c^2 is the mean square range associated with E_d. The thickness of the mixed layer deduced from the curve of Figure 6 and for a dose 10^{15} ions/cm^2 is 450 Å but it corresponds to three interfaces which implies $\Omega = 150$ Å. The deposited energy has been calculated using the TRIM program (9) leading to :

$$(R_c^2/E_d) = 32 \text{ Å}^2 \text{ eV}^{-1}$$

If we take a threshold displacement energy E_d = 30 eV we find R_c = 30 Å. Such a value appears too high for an average path length associated with an energy of 30 eV, as emphasized also by Besenbacker et al. They consider the possibility of a diffusion process inside the cascade via interstitials since they are mobile at 77 K in Al. Such a mechanism could be also efficient in the case of Fe-Al layers.

The radiation enhanced diffusion mechanism has been observed to occur at 300 K in our experiments (Figs.1 and 2). Myers (10) has deduced an analytical expression for the effective diffusion coefficient leading to $\Omega \propto \phi^{1/2}$ whatever the annihilation process may be.

In our experiments at 300 K the value found for the power is n ≃ 0.8 either for bilayers or multilayers. Such a behavior cannot be explained only by the radiation enhanced diffusion. A study including a measure of the structural state of the material obtained during mixing at 300 K is necessary to further elucidate the mechanism at 300 K.

CONCLUSION

- We have shown that the volume fraction of mixed atoms appears to be correlated with resistivity measurements.
- A semi-empirical model explaining the experimental results observed during ion mixing has been proposed in the case where a single mechanism operates. The analytic mixing kinetics are given by :

$$C(\phi) = 1 - \exp(-A\phi^n).$$

The power n has been found experimentally to be n = 1/2 at 77 K and n = 0.8 at 300 K. However, the existence of the radiation enhanced diffusion mechanism at 300 K does not appear enough to explain the observed differences in n at both temperatures. Nevertheless, the electrical resistivity technique appears to be promising tool for studying ion beam mixing.

REFERENCES

1. B.Y. Tsaur, S.S. Lau, J.W. Mayer, Appl. Phys. Lett. 36, 823 (1980).
2. B.Y. Tsaur, S.S. Lau, L.S. Hung, J.W. Mayer, Nucl. Instr. and Meth. 182 183, 67 (1981).
3. B.Y. Tsaur, S. Matteson, G. Chapman, Z.L. Liau, M.A. Nicolet, Appl. Phys. Lett. 35, 825 (1979).
4. J.P. Rivière, J. Delafond, C. Jaouen, A. Bellara, J.F. Dinhut, Applied Physics (to be published).
5. J. Grilhé, J.P. Rivière, J. Delafond, C. Jaouen, C. Templier,This M.R.S. Symposium.
6. P. Sigmund, A. Gras-Marti, Nucl. Instr. Methods, 168, 388 (1980).
7. P. Besenbacker, J. Bottiger, S.K. Nielsen, H.J. Whitlow, Appl. Phys. A 29, 141 (1982).
8. P. Sigmund, A. Gras-Marti, Nucl.Instr. Methods 182/183 , 25 (1981).
9. J.P. Biersack, L.G. Haggmark, Nucl. Instr. Methods, 174, 257 (1980).
10. S.M. Myers, Nucl. Instr. Methods, 168, 265 (1980).

AMORPHIZATION OF GARNET BY ION IMPLANTATION*

A. M. GUZMAN, T. YOSHIIE**, C. L. BAUER and M. H. KRYDER,
Carnegie-Mellon University, Pittsburgh, PA 15213, USA.

ABSTRACT

Amorphization by ion implantation has been investigated in films of $(SmYGdTm)_3Ga_{0.4}Fe_{4.6}O_{12}$ garnet by transmission electron microscopy, incorporating a special cross-sectioning technique. These films were produced by liquid phase epitaxy on (111) garnet substrates and subsequently implanted with ions of deuterium at 60 keV and doses ranging from 0.50 to 4.5×10^{16} D_2^+/cm^2 and ions of oxygen at 110 keV and doses ranging from 0.95 to 8.6×10^{14} O^+/cm^2. The amorphization process proceeds in separate stages involving the formation of isolated amorphous regions, merging of these regions into a continuous band and subsequent propagation of the amorphous band toward the implanted surface. Details of these processes are interpreted in terms of various atomic displacement mechanisms.

INTRODUCTION

Ion implantation in magnetic garnet epitaxial thin films is of special interest because of its role in altering the magnetic anisotropy of garnet crystals. The effects of ion implantation have recently been reported in various technical papers [1-3]. From the viewpoint of the garnet crystal structure, the effect of ion implantation is to generate isolated vacancies and interstitials as well as displacements of the garnet ions from their regular lattice sites. As the implantation doses increase, however, the corresponding increase in lattice defects may be sufficient to induce a crystalline to amorphous transformation, thus rendering the implanted layer magnetically inactive since amorphous garnet is believed to be paramagnetic [4].

Komenou et al.[4] have investigated the amorphization of garnet films implanted with neon ions at 100 keV using a double-crystal diffraction technique. Their results indicate that the process evolves in three distinct stages involving the formation and propagation of an amorphous band. More recently, amorphization of garnet by ion implantation with deuterium and oxygen ions has been studied by Yoshiie et al. [5] using transmission electron microscopy (TEM). Their results show that the amorphization process evolves in three slightly different stages involving the initial formation of isolated amorphous particles at low implantation dose and the subsequent merging of these particles into an amorphous band at higher doses.

We report on the investigation of the amorphization process of thin film garnets implanted with deuterium and oxygen ions by TEM, incorporating a special cross-sectioning technique. Direct observations of the structural changes and phase morphology are presented and discussed.

EXPERIMENTAL TECHNIQUES

Magnetic garnet films of $(SmYGdTm)_3Ga_{0.4}Fe_{4.6}O_{12}$, ranging from 0.6 to 1.1 μm in thickness, were grown by liquid phase epitaxy (LPE) on (111) oriented $Gd_3Ga_5O_{12}$ (GGG) substrates. These films were subsequently uniformly implanted at room temperature with deuterium ions at 60 keV and doses ranging from 0.5 to 4.5×10^{16} D_2^+/cm^2 and oxygen ions at 110 keV and doses ranging from 0.95 to 8.6×10^{14} O^+/cm^2. The ion beam was aimed at an angle about 7° off the normal to the implanted surface to avoid channeling effects. The choice of deuterium and oxygen as implant species in our investigation has been discussed elsewhere [3]. Specimens were extracted from the implanted samples for examination by TEM in the direction parallel and perpendicular to the garnet film surface. Thin foils were made by thinning the specimens by mechanical grinding to approximately 50 μm followed by ion milling with a collimated beam of argon ions accelerated by 6 kV. The thinned regions were then examined in a JEOL JEM 120CX electron microscope.

RESULTS

The experimental results are presented in a series of photomicrographs depicting bright-field images parallel (in the (111) plane) and perpendicular (in the (110) plane) to the implanted surface, the corresponding electron diffraction patterns and selected dark-field images. The perpendicular bright-field images contain information related to the implantation profiles and the LPE/GGG interface, as well as morphological features of the implanted area. The electron diffraction patterns reveal details of changes in the crystalline structure resulting from an increased implantation dose.

Figure 1 shows typical results for a specimen implanted with 0.5×10^{16} D_2^+/cm^2 and then 0.95×10^{14} O^+/cm^2. The diffraction pattern of Fig. 1(b) verifies that the implanted

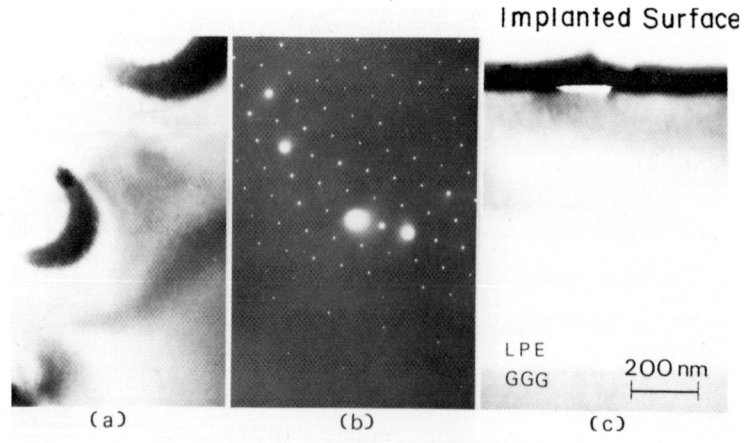

Fig. 1. Bright-field image parallel to the implanted surface (a), (b) corresponding electron diffraction pattern, and (c) bright-field image perpendicular to the implanted surface following implantation with 0.50×10^{16} D_2^+/cm^2 and then 0.95×10^{14} O^+/cm^2.

area is entirely monocrystalline with no evidence of observable lattice defects. In addition to the LPE/GGG interface visible in the cross-sectional view in Fig. 1(c), a faint band at approximately 450 nm below the implanted surface is also visible. This band corresponds to the deuterium implantation depth profile. The oxygen implantation profile, estimated to extend about 150 nm below the surface, is too faint to be easily detected.

Results for a specimen implanted with 1.5×10^{16} D_2^+/cm^2 and then 2.85×10^{14} O^+/cm^2 are shown in Figure 2. The slight graininess visible in Fig. 2(a), and the appearance of halo rings in the corresponding electron diffraction pattern in Fig. 2(b) suggest the formation of localized amorphous regions in the monocrystalline matrix. Moreover, the cross-sectional view shown in Fig. 2(c) and the inserted micro-microdiffraction patterns clearly indicate this morphology, whereas a mixture of amorphous particles, about 10 nm in diameter, in a crystalline matrix is observed in the deuterium plus oxygen implanted region extending about 150 nm below the implanted surface. The diffraction pattern of the deuterium implanted region also shows that this region is fully monocrystalline. These micro-microdiffraction patterns, each originating from an area less than 10 nm in diameter, clearly reveal the presence of amorphous particles rather than randomly oriented crystallites.

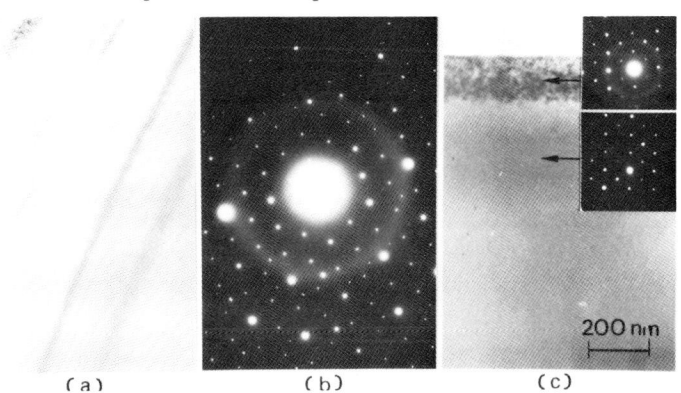

(a) (b) (c)

Fig. 2. Bright-field image parallel to the implanted surface (a), (b) corresponding electron diffraction pattern, and (c) bright-field image perpendicular to the implanted surface (with inserted micro-microdiffraction patterns) following implantation with 1.5×10^{16} D_2^+/cm^2 and then 2.9×10^{14} O^+/cm^2.

Figure 3 shows typical results for a specimen implanted with 3.0×10^{16} D_2^+/cm^2 and 5.7×10^{14} O^+/cm^2. Figures 3 (a), (b) and (c) indicate that although the implanted layer is predominantly amorphous, numerous crystallites are present. This is particularly evident in Fig. 3(b) where a fine Debye ring is superposed on a more diffuse halo ring. Moreover, the dark-field image produced from part of the Debye ring, picture in Fig. 3(c), verifies the existance of small crystallites. The cross-sectional view depicted in Fig. 3(d) and the corresponding micro-microdiffraction patterns reveal that the deuterium plus oxygen implanted region is transformed into an amorphous band, as verified by the sole presence of halo rings. The deuterium implanted region remains monocrystalline. The corresponding selected-area diffraction pattern in Fig. 3(e) shows a mixture of halo rings, sharp monocrystalline diffraction spots and weak randomly oriented diffraction spots. The halo rings and the sharp diffraction spots originate from the amorphous and monocrystalline regions. The randomly oriented diffraction spots are produced by crystallites at the implanted surface as it is revealed by the dark field image obtained from one of such spots in Fig. 3(f).

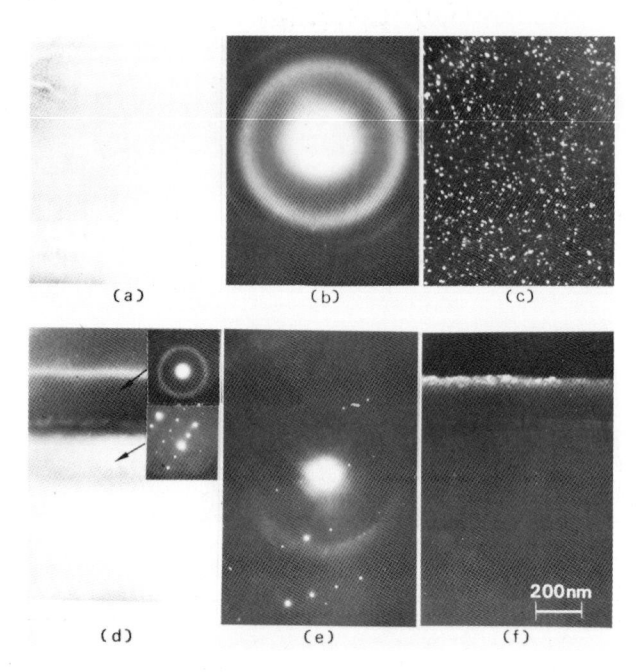

Fig. 3. Bright-field image parallel to the implanted surface (a), (b) corresponding electron diffraction pattern and (c) dark-field image, (d) bright-field image perpendicular to the implanted surface (with inserted micro-microdiffraction patterns) and (e) electron diffraction pattern and (f) dark-field image following implantation with 3.0×10^{16} D_2^+/cm^2 and then 5.7×10^{14} O^+/cm^2.

Figure 4 shows typical results for a specimen implanted with 4.5×10^{16} D_2^+/cm^2 and 8.6×10^{14} O^+/cm^2. At this implantation dose all features have disappeared from the bright-field image and the corresponding electron diffraction pattern of the deuterium plus oxygen implanted region shows only halo rings (Fig. 4a and b respectively), thus indicative of a totally amorphous state. Fig. 4(c) depicts the amorphous D_2^+ plus O^+, and the crystalline D_2^+ implanted regions.

143

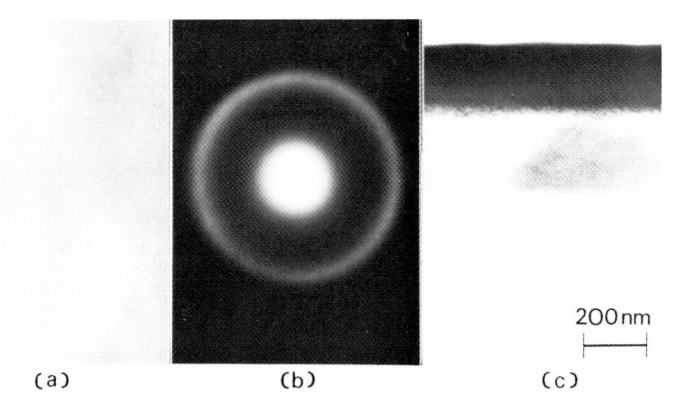

      (a)   (b)   (c)

Fig. 4. Bright-field image parallel to the implanted surface (a), (b) corresponding electron diffraction pattern, and (c) bright-field image perpendicular to the implanted surface following implantation with 4.5×10^{16} D_2^+/cm^2 and then 8.6×10^{14} O^+/cm^2.

DISCUSSION OF RESULTS

The combination of bright and dark field images and the corresponding electron diffraction patterns have shown the transformation of the implanted garnet from the crystalline to the amorphous state. These experimental results indicate that amorphization is caused by implantation with oxygen, but prior implantation with deuterium may sensitize the lattice by increasing the stress.

Table I. Parameters for deuterium and oxygen ions in garnet.[6]

Ion	Energy (keV)	Proj. range	Std. dev.	Nucl. stopp. (keV/μm)	Elect. stopp. (keV/μm)
Deuterium	60	0.492	0.102	1.09	137.1
Oxygen	110	0.136	0.081	194.4	440.1

During ion implantation the incident ions collide with the target garnet ions establishing cascades of recoil atoms. The resulting strain and damage are directly related to the energy deposited in the garnet lattice through nuclear collisions. Table I shows the values for mean projected range, projected standard deviation, and nuclear and electronic energy loss for deuterium at 60 keV and oxygen at 110 keV in garnet [6]. Implanted molecular deuterium ions, with mass 4, predominantly lose energy by electronic interactions. Their effect is to increase the stress in the implanted layer, but their contribution to damage is negligible. Thus, implantation with deuterium at a dose as high as 4.5×10^{16} ions/cm^2 does not result in amorphization. Oxygen ions lose a substantial portion of their energy by nuclear stopping as indicated by the nuclear energy loss factor in Table I. Their relatively large momentum transfer occurring during atomic collisions results in large cascades. For typical implantation doses and energies used in this work the number of atomic collisions per incident oxygen ion is several hundred [7]. At an intermediate oxygen dose the lattice ions are sufficiently displaced by these collisions to promote amorphization of isolated areas. With increasing implantation doses, the amorphization propagates toward the free surface of the garnet until the entire region becomes amorphous. The last region to be amorphized is the garnet surface not only because the damage is less at the surface but also because of stress relaxation.

* This work was supported by the Air Force Office of Scientific Research under grant number AFOSR-80-0284.

** Present address: Faculty of Engineering, Hokkaido University, Sapporo 060, Japan.

REFERENCES

1. R. Wolfe, J. C. North, W. A. Johnson, R. R. Spiwak, L. J. Varnerin and R. F. Fischer, AIP Conf. Proceedings 10, 339 (1973).
2. Y. S. Lin, G. S. Almasi and G. E. Keefe, J. Appl. Phys., 48, 5201 (1977).
3. A. M. Guzman, C. S. Krafft, X. Wang and M. H. Kryder, Nucl. Instr. and Meth., 209/210, 1121 (1983).
4. K. Komenou, I. Hirai, K Asama and M. Sakai, J. Appl. Phys., 49, 5816 (1978).
5. T. Yoshiie, C. L. Bauer and M. H. Kryder, Proceedings of 21st Intermag Conf., J. IEEE (in press).
6. J. F. Gibbons, W. S. Johnson and S. W. Mylroie, Projected Range Statistics, 2nd ed. (Halstead, New York, 1975).
7. H. Matsutera, S. Esho and Y. Hidaka, J. Appl. Phys. 53, 2504 (1982).

AMORPHOUS PHASE FORMATION AND RECRYSTALLIZATION IN ION-IMPLANTED SILICIDES

C.A. HEWETT*, I. SUNI**, S.S. LAU*, L.S. HUNG***, AND D.M. SCOTT*
* University of California at San Diego, La Jolla, CA 92093. ** Caltech, Pasadena, CA 91125.
*** Cornell University, NY 14853.

ABSTRACT

Ion implantation induced phase transformations and recrystallization during post-annealing in $CoSi_2$, $CrSi_2$, and Pd_2Si are studied. All three silicides are found to reorder at about 1/3 the melting point of the silicide. We speculate that ion-implanted silicides recrystallize by the same mechanism and that amorphous phases produced by implantation are unstable rather than metastable.

INTRODUCTION

The self-aligned gate technology is commonly used in MOSFET fabrication. The gate electrode, usually a metal or a metal silicide, is used as a mask to protect the conducting channel during the implantation of source and drain regions. The phase transformations during ion implantation and subsequent annealing are thus of fundamental interest. One qualitative measure of the ability of a material to withstand ion bombardment is its radiation hardness. Radiation soft materials are easily damaged, while radiation hard materials are much more resistant to ion beam induced damage. In this study we investigate amorphous phase formation and recrystallization of silicon ion implanted $CoSi_2$ and $CrSi_2$, believed to be radiation soft materials, and Pd_2Si, a radiation hard material. These silicides are of interest as each is capable of forming good crystalline quality epitaxial silicide layers on Si substrates; in addition, Co, Cr, and Pd silicide layers may be oxidized to form SiO_2 films on their surface without significantly altering the composition of the silicide layer [1].

EXPERIMENTAL

Thin layers of Co or Pd were e-beam deposited onto chemically cleaned <111> Si substrates. Cr layers were evaporated onto chemically cleaned SiO_2 substrates. Depositions were carried out in a vacuum chamber equipped with ion pumps. The pressure during evaporation was $\sim 10^{-7}$ Torr. The substrate temperature was held at $\sim 600°C$ for the Co evaporation; Pd and Cr were deposited onto the substrates without intentional heating. After deposition, samples were annealed in vacuum ($\sim 3 \times 10^{-7}$ Torr). Co samples were annealed at 950°C for one hour, Pd samples were annealed at 700°C for one hour, and Cr samples were annealed at 500°C for 30 minutes, forming $CoSi_2$, Pd_2Si, and $CrSi_2$, respectively. Pd_2Si was observed to be highly textured and $CrSi_2$ was observed to be polycrystalline by x-ray diffraction (Read Camera [2], CuKα radiation). $CoSi_2$ layers were observed to be epitaxial by x-ray diffraction as well as by MeV $^4He^+$ channeling measurements. The samples were then bombarded in a random direction with 60 keV to 250 keV Si^+ ions at liquid nitrogen temperatures (LNT) for all cases in this study.

For an investigation of regrowth kinetics, epitaxial $CoSi_2$ ~ 1130 Å thick with a channeling surface minimum yield of 5.5% (1.5 MeV $^4He^+$) was implanted with either 2×10^{15} Si^+/cm^2 or 5×10^{15} Si^+/cm^2 at 60 keV (Rp ~ 450 Å, $\Delta Rp \sim 200$ Å). These implantation conditions rendered the near surface region of the implanted $CoSi_2$ layer amorphous with a single crystalline $CoSi_2$ seed remaining at the $CoSi_2$-Si interface. MeV $^4He^+$ channeling techniques and electrical conductance measurements were used to monitor the regrowth kinetics.

The resistance change as a function of implanted dose for Pd_2Si and $CoSi_2$ layers was investigated by Si^+ implantation to various doses. Pd_2Si layers ~ 1860 Å thick were implanted to doses of 5×10^{14}, 1×10^{15}, 2×10^{15}, 5×10^{15}, and 2×10^{16} Si^+/cm^2 at 250 keV (Rp ~ 1400 Å, $\Delta Rp \sim 550$ Å). $CoSi_2$ layers ~ 2340 Å thick were implanted with 200 keV Si^+ ions (Rp ~ 1630 Å, $\Delta Rp \sim 400$ Å) at doses of 5×10^{14}, 8×10^{14}, 2×10^{15}, 5×10^{15}, 1×10^{16} and $2 \times 10^{16}/cm^2$. The energy of the implant was chosen to render the entire silicide layer amorphous at high doses. The amorphous-polycrystalline transformation was investigated by electrical conductance measurements as a function of temperature of these samples as well as $CrSi_2$ samples implanted with 1×10^{16} Si^+/cm^2 at 120 keV (Rp ~ 930 Å, ΔRp

~250 Å). Transmission electron microscopy (TEM) was used to examine the layers after implantation for amorphous phase formation.

EPITAXIAL REGROWTH OF CoSi$_2$

Irradiation of epitaxially grown CoSi$_2$ layers (~1130 Å, χ_{min} ~5.5%) with 60 keV ^{28}Si$^+$ ions at LNT produced an amorphous region as observed by TEM. The electron diffraction patterns taken from samples implanted to doses of 2×10^{15} and 5×10^{15} Si$^+$/cm^2 showed in both cases the presence of an amorphous halo. Channeling measurements indicated that the surface region of the CoSi$_2$ layer was damaged to the depth of 750 Å in samples implanted to 2×10^{15} Si$^+$/cm^2. The layer was almost entirely amorphized at the dose of 5×10^{15} Si$^+$/cm^2. Figure 1 shows the channeling data for a sample implanted with 5×10^{15} Si$^+$/cm^2 after annealing at 105°C for various times.

Figure 1. Channeling spectra for an 1130 Å layer of CoSi$_2$ implanted with 5×10^{15} Si$^+$/cm^2 at 60 keV and liquid nitrogen temperature as a function of annealing time at 105°C.

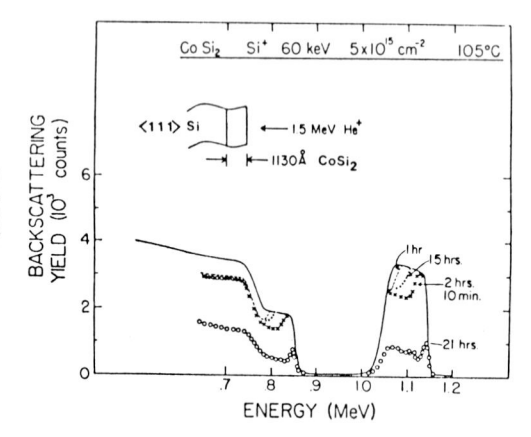

The channeling measurements show essentially layer by layer epitaxial regrowth, followed by a realignment of misoriented regions in the regrown layer after longer annealing. The regrowth rate at 105°C is about 6 Å/min. The fraction $X_T(t)$ of the volume crystallized as a function of time at a constant temperature was derived from electrical conductance measurements, and can be written in the form

$$X_T(t) = \frac{Z(t)}{d} = \frac{G(t) - G_o}{G_f - G_o} \quad (1)$$

Here $Z(t)$ is the thickness of the regrown layer and d is the thickness of the initial amorphous layer. $G(t)$ is the conductance at time t and G_o and G_f are the initial and final conductances, respectively. Figure 2 shows the normalized conductance change (regrown fraction) as a function of annealing time for four temperatures.

Figure 2. Normalized conductance change (fraction regrown) as a function of annealing time and temperature for CoSi$_2$ samples similar to those of figure 1.

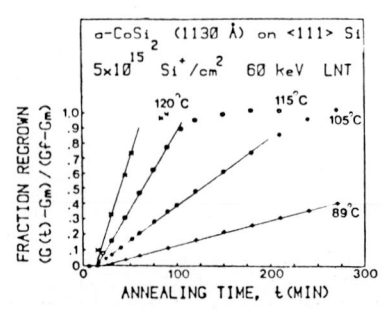

The regrowth takes place at low temperatures (< 100°C) and is linearly proportional to time. The results obtained from electrical measurements are consistent with those from channeling experiments. The Arrhenius plot for the growth rate determined in this narrow temperature range gives an activation energy $\Delta E = 0.93 \pm 0.10$ eV.

EFFECTS OF ION IRRADIATION ON THE ELECTRICAL RESISTANCE

Figure 3a shows the electrical resistance change for a 2340 Å $CoSi_2$ layer on <111> Si as a function of implantation dose (200 keV, Si^+ ions, at LNT). Figure 3b shows the electrical resistance change for an 1860 Å Pd_2Si layer on <111> Si as a function of implantation dose (250 keV, Si^+ ions, LNT).

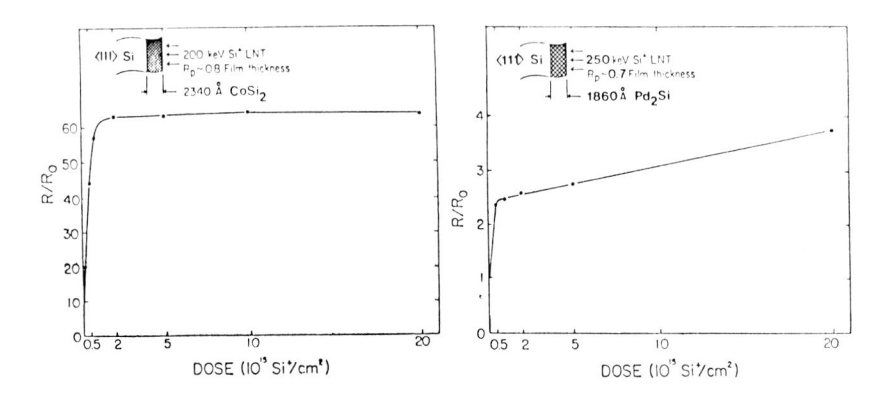

Figure 3. (a) Resistance change of a $CoSi_2$ (~ 2340 Å) layer as a function of Si^+ ion implanted dose at 200 keV and liquid nitrogen temperature (LNT). R_o is the resistance of the un-implanted sample. (b) Resistance change of a Pd_2Si (~1860 Å) layer as a function of Si^+ ion implanted dose at 250 keV and LNT.

Upon irradiation the resistance of the $CoSi_2$ layer increases rapidly and saturates at the level of $R \sim 60 R_o$ (R_o is the resistance of the as-annealed sample) after an implanted dose of 2×10^{15} Si^+/cm^2. For higher silicon doses the resistance remains unchanged up to 2×10^{16} Si^+/cm^2. In contrast, Pd_2Si layers show no saturation in resistance for doses up to 2×10^{16} Si^+/cm^2, with the resistance increasing by a factor less than four. $CrSi_2$ was implanted to only one dose, thus the dose dependence of the electrical resistance is not available. Since $CrSi_2$ is a semiconductor it is expected that ion irradiated layers have a lower resistance. Electron diffraction (TEM) showed that a diffuse halo becomes prominent in $CoSi_2$ above the dose of 2×10^{15} Si^+/cm^2, but no halo was observed in Pd_2Si for doses up to 2×10^{16} Si^+/cm^2. Both electron diffraction results and the saturation of the resistance change suggest that the $CoSi_2$ layer is rendered amorphous at a dose of 2×10^{15} Si^+/cm^2, while the Pd_2Si layer is only partially damaged.

THE AMORPHOUS-POLYCRYSTALLINE TRANSFORMATION

The recrystallization behavior of implanted samples was investigated by in-situ electrical resistance measurements. The recrystallization temperature is taken at the point where the normalized resistance of the sample is equal to one-half. Using this criterion, the recrystallization temperature was found to be ~140°C for the case of $CoSi_2$, and ~260°C for $CrSi_2$.

To analyze the recrystallization behavior we assume the relationship between the resistance R(t) and the fraction of the crystallized volume $X_T(t)$ to be

$$X_T(t) = \frac{R_o - R(t)}{R_o - R_f} \quad (2)$$

with R_o the resistance of the as implanted sample, R_f the final value, and $R(t)$ the resistance at a given time t. (It should be noted that in the case of $CrSi_2$, which is semiconducting rather than metallic in nature, equation (2) may not be correct.) We further assume that the generalized relationship for the early phase of nucleation and growth transformations holds and is given by the Avrami equation [2]:

$$X_T(t) = 1 - exp(-kt^n) \quad (3)$$

where k is a (temperature sensitive) kinetic constant and n is determined by the mode of transformation. Plotting the parameter $\ln\{1/[1-X_T(t)]\}$ vs. $\ln(t)$ for a given annealing temperature determines n. A straight line should result if equation (3) holds. Figures 4a and 4b show this linear relationship for $CoSi_2$ and $CrSi_2$ respectively.

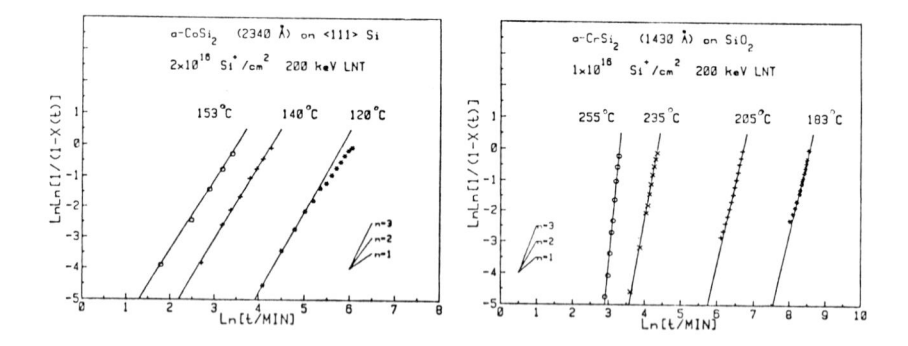

Figure 4. (a) $\ln\ln\{1/[1-X_T(t)]\}$ vs. $\ln(t/min)$ for $CoSi_2$ (~2340 Å) sample as a function of temperature for samples implanted with 2×10^{16} Si^+/cm^2 at 200 keV and liquid nitrogen temperature (LNT). The slopes of the straight lines are ~3. (b) same plot for $CrSi_2$ implanted with 1×10^{16} Si^+/cm^2 at 120 keV and LNT. Slopes are temperature dependent and range from 12.1 to 4.9.

In the case of $CoSi_2$ a value of 3 is obtained for n for the three temperatures investigated, signifying that the transformation proceeds from a fixed number of nuclei [3]. This leads us to conclude that the implanted layer still contains crystalline nuclei with the size and density of these nuclei below the TEM detection limit of approximately 30 Å. Nuclei of this size contain on the order of one hundred $CoSi_2$ unit cells and would therefore act as efficient growth centers. In the case of $CrSi_2$, n ranged from 12.1 at 255°C down to 4.9 at 183°C. A value of n greater than 4 is indicative of increasing nucleation rate [3].

DISCUSSION

The recrystallization behavior of high dose ion implanted $NiSi_2$ has been reported previously in the literature [4,5,6]. We note that the recrystallization behavior of $CoSi_2$ and $NiSi_2$ are nearly identical, the only notable difference being the amorphous appearance of ion-implanted $CoSi_2$ under TEM investigation while ion-implanted $NiSi_2$ appears to remain crystalline although highly defective. $CoSi_2$ and $NiSi_2$ both regrow epitaxially in a layer-by-layer manner just as also observed in ion-implanted Si and Ge amorphous layers [7]. Since both $NiSi_2$ and $CoSi_2$ transform with an n value of

3, we believe that ion-implanted $CoSi_2$ contains crystalline nuclei and that the structure is not completely amorphous as indicated by TEM. Ion implanted Pd_2Si is also not amorphous under TEM investigation, leading us to conclude that silicide layers which are usually metal-like in their electrical conductivity are radiation hard and very difficult to be made completely amorphous by ion implantation. That the n values for $CrSi_2$ are temperature dependent does not necessarily negate the idea of pre-existing nuclei. Also since $CrSi_2$ is a semiconductor, non-linear effects such as Schottky barrier formation between amorphous and crystalline $CrSi_2$ regions could invalidate equation (2) for estimating the fraction of crystallized volume.

Several guidelines have been proposed for the prediction of amorphous phase formation by ion mixing or ion irradiation in binary metallic alloys: (1) The structural difference rule proposed by Liu, et al [8], which states that an amorphous binary alloy will be formed by ion mixing of multilayer samples when the constituents are of different structures, regardless of the atomic sizes and electronegativities. (2) In ion mixing of thin layers of binary fcc metal systems, only the simplest intermetallics (simple structure with small number of atoms per unit cell) are found in the crystalline form after ion irradiation, i.e. intermetallics with complex crystalline structures are radiation soft. Although only three metal-metal systems (Al/Ni, Al/Pd, and Al/Pt) have been investigated thus far [10], it is believed that this rule is generally valid. (3) The heat of formation of a compound (metal-silicide in our case) is believed to correlate with radiation hardness, i.e. large heat of formation is suggestive of a high degree of covalent bonding in a silicide; therefore more sensitivity to radiation damage [11]. We tested these guidelines on six silicides for which the amorphous to crystalline transition temperatures are relatively well known. Table I lists these silicides and some of their properties.

Table I

	T_c (K)	R/R_0	ΔH/metal-atm (eV)	#atms/unit cell	T_m (K)	T_c/T_m	reference
$NiSi_2$	~ 393	~ 4	0.9	12	1266	0.31	[4]
$CoSi_2$	~ 413	~ 60	1.08	12	1599	0.27	
$PdSi_2$	~ 373	~ 2.5	0.6	8	1174	0.32	[6]
Pd_2Si	~ 423	~ 4	0.45	96	1667	0.25	
$CrSi_2$*	~ 523	~ 0.2	1.36	9	1823	0.28	
$TaSi_2$	~ 773	~ 4	1.04	9	2473	0.31	[11]

R = Saturation resistance value after ion implantation
R_0 = Resistance value before ion implantation
* = Co-sputtered

According to the structural difference rule, all six silicides should be rendered amorphous by ion irradiation, as the individual components of these systems have different crystalline structures. (In applying the structural difference rule, the diamond cubic lattice is not considered to be an fcc structure.) We note that all these silicides have complex unit cells, therefore they should be radiation soft. From a heat of formation point of view, Pd_2Si and $NiSi_2$ should be the most insensitive to ion irradiation, with $CrSi_2$ and $CoSi_2$ the most sensitive. Experimentally Pd_2Si [4] and $NiSi_2$ [5] are known to remain crystalline even after very high doses of ion irradiation. In the case of $CoSi_2$, it appears that crystalline nuclei are still present in a high-dose-implanted layer. Of particular interest is Pd_2Si: because of its complex unit cell, it should be easily amorphized. On the other hand, Pd_2Si has the smallest heat of formation, thus it should be very insensitive to ion irradiation. It appears, then, that the small heat of formation has more influence on the radiation hardness of Pd_2Si than that of unit cell complexity. The reason for this is not clear at present.

If we take the ratio of T_c (amorphous-to-crystalline transition temperature in degrees K) to T_m (melting point of the silicide), we note that all six silicides in Table I have ratios of $T_c/T_m \sim 1/3$. According to our data these silicides are not completely transformed into an amorphous phase by ion-irradiation and therefore the recrystallization occurs by the growth of pre-existing nuclei. No nucleation barrier exists for the recrystallization of these silicides, hence they are unstable rather than metastable against thermal annealing. If the recrystallization mechanism is the same for all cases, the recrystallization temperature is expected to scale with the same physical parameter, which in this case is conveniently expressed by $T_c/T_m \sim 1/3$. It is also interesting to note that amorphous binary alloys in a number of metal-metal systems, produced either by co-evaporation [12], co-sputtering [13], or ion mixing [14] have amorphous-to-crystalline transition temperatures approximately 1/3 of the average melting point of the constituent elements. In this respect the recrystalization of ion-irradiated silicides is similar to that of amorphous binary metallic alloys.

ACKNOWLEDGEMENTS

Cornell and UCSD acknowledge the financial support of NSF (Grant No. DMR-8106843, L. Toth). Caltech acknowledges the financial support of the U.S. Department of Energy through an agreement with the National Aeronautics and Space Administration and monitored by the Jet Propulsion Lab at the California Institute of Technology (D. Bickler).

We also acknowledge C. D. Lien (Caltech) for $CrSi_2$ sample preparation and the encouragement of Professor M.-A. Nicolet at Caltech and Professor J. W. Mayer at Cornell.

REFERENCES

[1] M.-A. Nicolet and S. S. Lau, "Formation and Characterization of Transition Metal Silicides", in VLSI Electronics: Microstructure Science, Norman Einspruch, Series Editor, Vol. 6, Materials and Process Characterization, Graydon Larrabee, Guest Editor, Academic Press, 1983.
[2] S. S. Lau, W. K. Chu, J. W. Mayer and K. N. Tu, Thin Solid Films, *23*, 205 (1974).
[3] J. W. Christian, The Theory of Transformations in Metals and Alloys, Part I, Equilibrium and General Kinetic Theory, 2nd ed., Pergamon Press, Oxford, 1965, Chapter 12.
[4] H. Ishiwara, K. Hikosaka and S. Furukawa, Appl. Phys. Lett. *32*, 23 (1978).
[5] M. Maenpaa, L. S. Hung, M.-A. Nicolet, D. K. Sadana and S. S. Lau, Thin Solid Films, *87*, 277, (1982).
[6] M. Maenpaa, L. S. Hung, B. Y. Tsaur, J. W. Mayer, M.-A. Nicolet, S. S. Lau, D. K. Sadana and W. F. Tseng, J. Elect. Mat. *11*, 289 (1982).
[7] S. S. Lau and W. F. van der Weg, "Solid Phase Epitaxy", in Thin Films-Interdiffusion and Reactions, J. M. Poate, K. N. Tu and J. W. Mayer, Editors. John Wiley & Sons, Inc., New York, 1978.
[8] B. X. Liu, W. L. Johnson, M.-A. Nicolet and S. S. Lau, Appl. Phys. Lett. *42*, 45 (1983).
[9] G. Goltz, R. Fernandez, M.-A. Nicolet and D. K. Sadana, in Metastable Materials Formation by Ion Implantation, S. T. Picraux and W. J. Choyke, Eds., (North-Holland, N. Y. 1982), MRS Symposia Proceedings Vol. 7, p. 227.
[10] L. S. Hung, M. Nastasi, J. Gyulai and J. W. Mayer, Appl. Phys. Lett., *42*, 672 (1983).
[11] B. Y. Tsaur and C. H. Anderson Jr., J. Appl. Phys. *53*, 940 (1982).
[12] S. Mader, J. Vac. Sci. Technol. *2*, 35 (1965).
[13] J. D. Wiley, J. H. Perepezko, J. E. Nordman and Kang-Jin Guo, IEEE Trans. Ind. Electron., *29*, 154 (1982).
[14] B. X. Liu, W. L. Johnson, M.-A. Nicolet and S. S. Lau, Nucl. Instr. and Meth., *209/210*, 229 (1983).

SECOND PHASE FORMATION IN ALUMINUM ANNEALED AFTER ION IMPLANTATION WITH MOLYBDENUM.*

J. BENTLEY,[1] L. D. STEPHENSON,[1,2] R. B. BENSON, Jr.,[2] P. A. PARRISH,[3] AND J. K. HIRVONEN.[4]
[1]Metals and Ceramics Division, Oak Ridge National Laboratory, Oak Ridge, TN 37830; [2]Department of Materials Engineering, North Carolina State University, Raleigh, NC 27607; [3]Army Research Office, Research Triangle Park, NC; [4]Zymet, Inc., Danvers, MA.

ABSTRACT

The microstructure of aluminum annealed after implantation to peak concentrations of approximately 4.4 and 11 at. % Mo was investigated by analytical electron microscopy. $Al_{12}Mo$ precipitates formed with pseudo-lamellar and continuous film microstructures. Video recordings of in-situ annealing experiments revealed the details of the phase transformations.

INTRODUCTION

The microstructures of aluminum ion-implanted with molybdenum and subjected to various heat treatments are being investigated for correlation with near-surface properties such as corrosion. Previous work by Al-Saffar et al. [1] indicated enhanced corrosion resistance, but dealt chiefly with the as-implanted condition and involved little microstructural characterization. In addition, the Al-Mo binary system is of interest because metastable phase formation was considered to be possible and the equilibrium phase diagram is poorly defined. [2,3]

EXPERIMENTAL PROCEDURES

Heavily cold-worked aluminum with a purity level of 99.999% was recrystallized by annealing at 350°C for one-half hour. Grain sizes in the range 0.3 to 0.5 mm were produced. Electropolished coupons 38 × 28 × 0.5 mm were implanted with Mo^+ ions at the Naval Research Laboratory using dual energy implant schedules. The fluences of 50 and 110 keV ions and the resultant peak concentration of molybdenum (measured by ion backscattering) at the mean projected range of ~50 nm are shown in table 1.

Disks (3 mm diam) were electrodischarge machined from as-implanted materials. TEM specimens were prepared by backthinning (electropolishing) as-implanted and annealed specimens. Carbon extraction replicas of annealed specimens were also prepared.

Post-implantation in-situ annealing was performed in a Philips EM400T/FEG analytical electron microscope with the use of a Philips single-tilt heating holder at a pressure of approximately 10^{-5} Pa (10^{-7} torr). Conventional vacuum annealing of specimens at a pressure of approximately 10^{-4} Pa (10^{-6} torr) was also performed.

---

*Research sponsored by the Division of Materials Sciences, U.S. Department of Energy under contract No. W-7405-eng-26 with Union Carbide Corporation, and by the Army Research Office.

TABLE I. Fluences of Mo⁺ ions and peak molybdenum concentrations produced in the specimens for the present study.

50 keV Mo⁺ ion fluence (ions/m^2)	110 keV Mo⁺ ion fluence (ions/m^2)	Peak concentration[a] (atomic % Mo)
4.88 x 10^{19}	6.14 x 10^{19}	4.4
1.12 x 10^{20}	1.24 x 10^{20}	11.0

[a] Measured by ion backscattering.

RESULTS AND DISCUSSION

Aluminum implanted with 4.4 at. % Mo

TEM examination of backthinned as-implanted specimens revealed displacement damage in the form of a tangled dislocation network and small dislocation loops, as shown in figure 1. The implanted layer remained crystalline and no precipitates were observed, indicating the formation of a supersaturated solid solution, in agreement with previous work. [1]

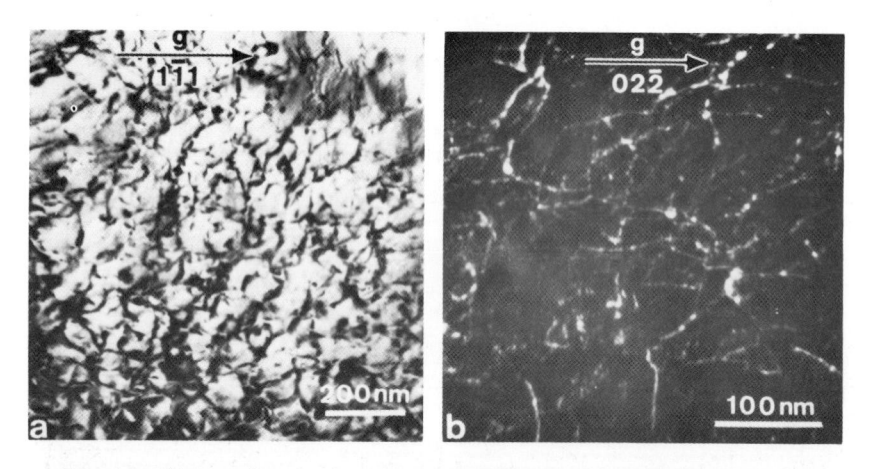

FIG. 1. Micrographs showing displacement damage as the main feature evident in aluminum as-implanted with 4.4 at. % Mo. (a) Bright-field. (b) Weak-beam dark-field.

In-situ annealing was performed for 10 min at temperatures increasing by 50°C intervals from 200°C in a Philips EM400T/FEG. Precipitation first occurred at 550°C. Specimens were then conventionally vacuum annealed at 550°C for 100 min and prepared for TEM examination by backthinning. Examples of the microstructure are shown in figure 2. The precipitates would probably form a lamellar structure in bulk material, but here the thickness (normal to the specimen surface) is only approximately 50 nm (the range of the implanted ions).

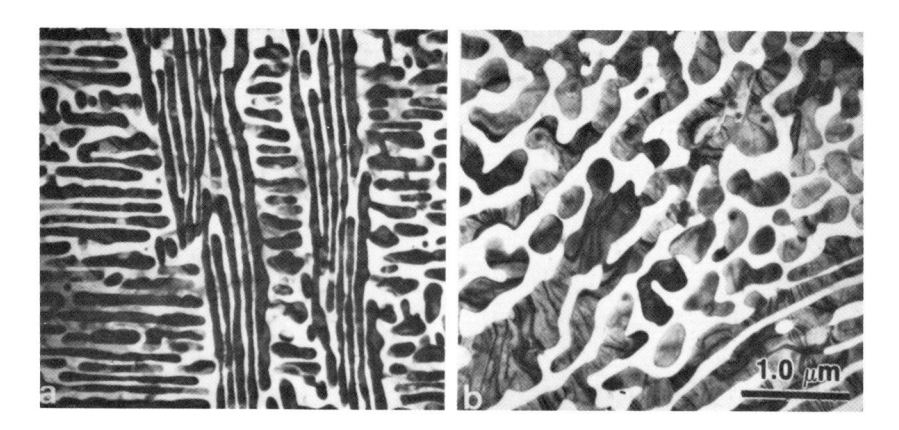

FIG. 2. Micrographs showing the pseudo-lamellar precipitates formed in aluminum implanted with 4.4 at. % Mo and annealed at 550°C for 100 min.

X-ray microanalysis of the precipitates in backthinned specimens indicated a composition of 6.7 ± 0.5 at. % Mo. Since the aluminum content is probably overemphasized by the presence of an Al_2O_3 surface film (identified by electron energy loss spectroscopy) and by contributions from the matrix as a result of beam spreading and secondary fluorescence, x-ray microanalysis was repeated on extraction replicas (Fig. 3). The indicated molybdenum content increased to 9.3 ± 1.0 at. % Mo. This increase was expected, but systematic errors in the standardless quantification procedure [4] may still exist. The composition is near $Al_{12}Mo$, the phase predicted from the equilibrium diagram.

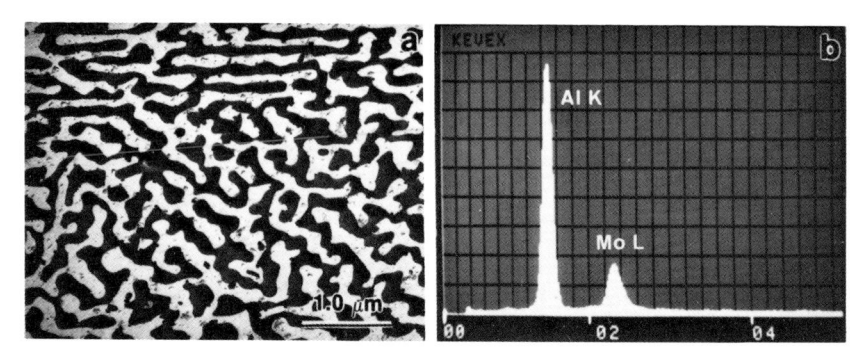

FIG. 3. (a) Micrograph of extraction replica showing the faithful replication procedure. (b) Energy-dispersive x-ray spectrum from an extracted precipitate.

A positive identification of the precipitates as $Al_{12}Mo$ was obtained from convergent beam electron diffraction (CBED). The lack of high-order Laue zone lines in the CBED patterns precluded the direct use of the tables of Buxton et al. [5] Nevertheless, space group determination was still possible. The projection symmetries along <111> and <001> were 6 and 2mm, respectively, (Fig. 4). From Table 3.7.1 of the International Tables for X-ray Crystallography [6] the point group is m3. The lattice was determined to be body-centered (cubic) with a lattice parameter of 0.75 nm. Since 0kl reflections with k and l both odd were observed on 0kl systematic rows the space group was identified as Im3. These data are in agreement with the structure of $Al_{12}Mo$ determined by Adam and Rich [7] using x-ray diffraction.

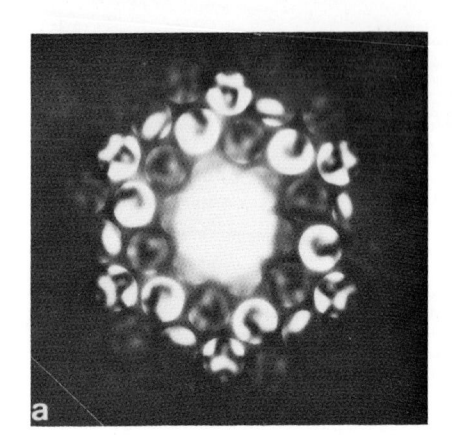

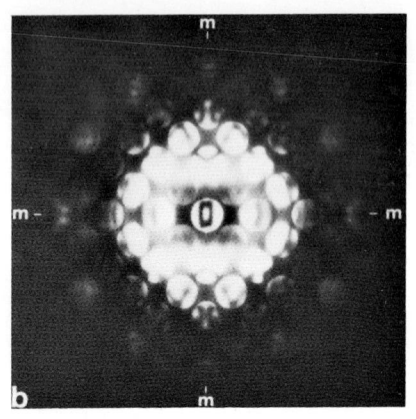

FIG. 4. CBED patterns from $Al_{12}Mo$ precipitates. (a) <111>, sixfold symmetry. (b) <001>, 2mm symmetry.

The orientation relationship (OR) between the matrix (m) and the precipitate (p) was determined by standard electron diffraction techniques and can be regarded as a relative rotation of approximately 40° around an approximately common <111>. The observed OR is consistent with that predicted from geometrical considerations. In the $Al_{12}Mo$ structure the molybdenum atoms are surrounded by 12 aluminum atoms (icosahedron); {111} contain triangular arrangements of aluminum atoms which are approximately congruent to triangular arrangements of atoms on {111} of fcc aluminum. Superposition of triangular arrangements of atoms occurs following a rotation of approximately 40° around <111>. With this OR there is also a close match between the d-spacings of the $\{100\}_m$ and $\{123\}_p$; this may determine the morphology of the pseudo-lamellar precipitates, but further work is required to confirm this.

Preliminary video recordings were made of precipitate growth during in-situ annealing experiments. The nucleation of misfit dislocations and the rearrangement of interfacial dislocations which occurred when precipitates impinged one another were clearly observed. Again further analyses are required to fully understand these results.

Aluminum implanted with 11 at. % Mo

The microstructure of the 11 at. % Mo material in the as-implanted state was similar to that of the 4.4 at. % Mo specimens. A metastable solid solution was retained and displacement damage in the form of dislocation loops and tangles was observed. Conventional vacuum annealing at 550°C resulted in the formation of a continuous film of $Al_{12}Mo$ in the ion-implanted surface region. As before, x-ray microanalysis and CBED were used to identify the precipitate.

In order to investigate further the formation of the continuous $Al_{12}Mo$ film, in-situ annealing experiments were performed. No visible changes in the microstructure were evident for annealing temperatures of up to 400°C, in contrast to the results of other workers. [1] Annealing at 425°C resulted in the formation of $Al_{12}Mo$. Video recordings were made of transformation fronts as they moved through the specimen at typically 0.2 µm/s. The concentration of dislocation loops was reduced by approximately one order of magnitude as the leading edge of the transformation front passed. The remaining loops shrank and disappeared within a few seconds, a process which is probably due to the thickening of the $Al_{12}Mo$. The dynamic behavior was more readily apparent in the video recordings than in a series of stills. The transformation fronts were observed to proceed most rapidly in a direction parallel to and some distance from the edge of the wedge-shaped thin region of the specimens. The lateral movement toward both thicker and thinner areas was more sluggish. It is probable that the transformation proceeds most rapidly where there is a high molybdenum content throughout the thickness of the foil, since the driving force is highest there and any constraints are a minimum. In thicker regions, the underlying, unimplanted material may impose restraints; in the thinnest regions, the molybdenum content is small. Fine dislocation tangles were visible in the aluminum matrix below the transformed layer. The microstructure of the resultant $Al_{12}Mo$ was complex; it exhibited a feathery appearance and although more-or-less the same crystallographic orientation exists over large areas, fine scale strain was present.

A transformation front was arrested during annealing by rapidly cooling the specimen from 425°C and the structure on either side of the interface was examined. The precipitate dark-field image of figure 5 reveals that a film of $Al_{12}Mo$ is forming on one side of the transformation front A-B. The $Al_{12}Mo$ film is not continuous near the transformation front, probably because this region transformed as the specimen was cooling. (In the video recordings the film appears to be continuous up to the transformation front). In regions which transformed at 425°C (i.e., further away from the transformation front) a continuous film was produced.

The Al-Mo phase diagrams give $Al_{12}Mo$ and Al_6Mo as the equilibrium phases expected for a composition of 11 at. % Mo. [2,3] However, Al_6Mo was not observed to form. The apparently preferential formation of the $Al_{12}Mo$ film is probably due to the peak concentration of molybdenum being reduced by diffusion such that the overall composition profile was near the optimum 7.7 at. % Mo required for $Al_{12}Mo$ formation. Diffusion in the fcc Al-rich matrix should be fairly rapid at temperatures >400°C (i.e., $T/T_m > 0.7$) and would be further enhanced by the presence of the high dislocation density in the implanted layer. Another factor may be that the formation of $Al_{12}Mo$ proceeds easily because of the orientation relationship which leads to a close match for several sets of lattice planes.

The fact that, for the processing conditions employed in this work, a continuous film of $Al_{12}Mo$ was produced in the ion-implanted surface layer could have significant effects on certain properties such as corrosion and selected mechanical properties, and these will be investigated in the near future.

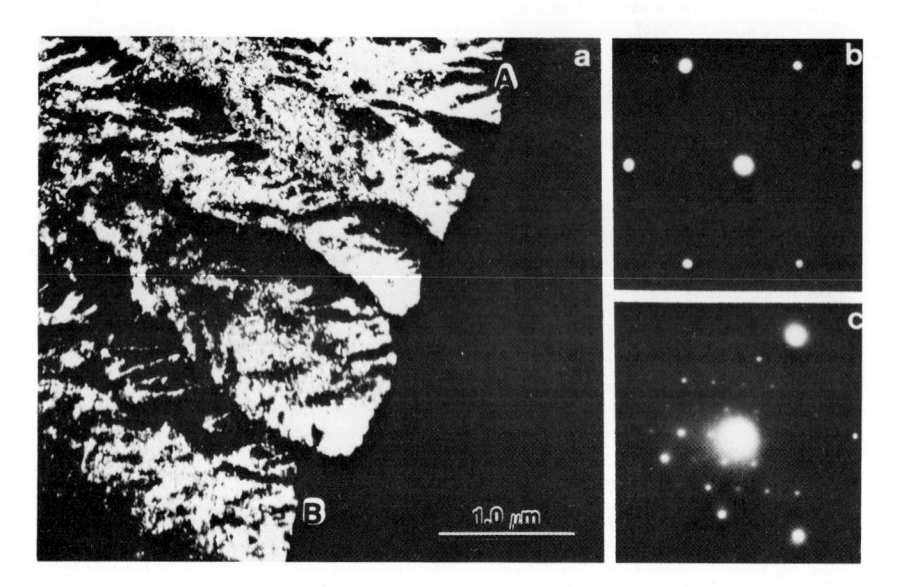

FIG. 5. (a) Precipitate dark-field image of the region near an arrested transformation front (A-B) in a specimen rapidly cooled from 425°C. (b) and (c) Selected area diffraction patterns from respectively the right and left sides of (a) showing $\{111\}_m \| \{111\}_p$.

SUMMARY

Analytical electron microscopy has shown that $Al_{12}Mo$ forms during post-implantation annealing of aluminum ion-implanted with molybdenum (peak concentrations of 4.4 and 11 at. % Mo). The morphologies of the precipitate phases and the transformation mechanisms in the two materials are quite different; a pseudo-lamellar structure is produced for the low molybdenum concentration, a continuous film is formed at the higher molybdenum concentration. Video recordings of in-situ annealing experiments have revealed the details of the transformation processes.

REFERENCES

1. H. Al-Saffar, V. Ashworth, A.K.D. Biorarnov, D.J. Chivers, W.J. Grant, and R.P.M. Proctor, **Corrosion Sci.** 20, 127 (1980).
2. M. Hansen, **Constitution of Binary Alloys** (McGraw Hill, New York, 1958).
3. W.G. Moffat, **The Handbook of Binary Phase Diagrams** (General Electric, Schenectady 1978), Vol. 1.
4. N.J. Zaluzec, **Introduction to Analytical Electron Microscopy**, (Plenum Press, New York 1979), eds. J.J. Hren, J.I. Goldstein, and D.C. Joy, pp. 121-167.
5. B.T. Buxton, J.A. Eades, J.W. Steeds, and G.M. Rackham, **Phil. Trans. Roy. Soc.** A281, 171 (1976).
6. **International Tables for X-Ray Crystallography**, (Kynoch Press, Birmingham 1969), Vol. 1, pp. 38-39.
7. J. Adam and J.B. Rich, **Acta Cryst.** 7, 813 (1954).

MODULATED STRUCTURES IN ION IMPLANTED Al-Fe SYSTEM

K.V. JATA, D. JANOFF, AND E.A. STARKE, JR.
Department of Materials Science, University of Virginia,
Charlottesville, VA 22901

ABSTRACT

The results of transmission electron microscopy studies of iron implantation into high purity aluminum foils are described. For both 50 and 100 + 50 keV incident ion energies, modulated structure has been detected in the as-implanted foils. Upon annealing at 793 K the modulated structure decomposes into the Al matrix and Al$_3$Fe precipitates for the 50 keV implantation. A similar annealing treatment for the 100 + 50 keV implantation indicates that the modulated structure is more stable, although some Fe$_3$Al precipitation occurs.

INTRODUCTION

Ion implantation offers the potential for developing alloys with a superior fatigue crack initiation resistance [1,2]. In structural alloy systems incoherent particles in the near surface region tend to disperse dislocations thereby homogenizing slip and improving fatigue crack initiation resistance. Aluminum alloys having solute additions that form nonshearable particles which are thermally stable are also desirable for high temperature services. Iron has a solubility limit of 0.05 wt. percent and a diffusivity of only 4.1×10^{-13} m^2/sec is a good candidate. This paper describes the microstructure resulting from ion implantation into aluminum. It is part of a program directed towards improving the fatigue resistance of aluminum and its alloys by surface modification using ion implantation.

The binary aluminum-iron alloys have been investigated by several workers over the past decade [3-5]. They have been processed either by continuous casting where the cooling rate is approximately 3 K per second or by rapid quenching from the melt with cooling rates of the order of 10^6 K per second. Alloys obtained by the latter method have superior mechanical properties primarily due to a refinement of the microstructure and lower segregation of embrittling elements to the grain boundaries. The rapidly quenched alloys often contain two characteristically different zones, designated Zone A and Zone B by Jones [3]. Zone A essentially consists of a supersaturated solid solution and Zone B contains both metastable and stable phases.

Ion implantation offers the possibility of obtaining alloys having: (1) a highly supersaturated solid solution with extended solid solubility over that predicted by the phase diagram; (2) a solid solution with stable and metastable phases; and (3) a disordered or amorphous structure. In many instances alloy systems produced by rapid quenching have been duplicated by ion implantation [6,7] and parallels have been drawn between the two techniques. The amorphous metal-metal or metal-metalloid systems produced by ion implantation have led researchers to again suggest that the thermal spike concept [8] where local heating of a solid takes place on the impingement of an ion with energy values of the order of 50-500 keV may apply. Although the temperatures in the small hot zone caused by the thermal spike could exceed the melting point of the solid, the time scale of the thermal spike is only a few **picoseconds resulting** in a ultra fast quench of the molten hot zone. Consequently, the diffusion required to form the metastable and stable phases cannot occur during the thermal spike. Precipitation is more likely to occur during and post implantation by radiation enhanced diffusion [6,9]. The radiation enhanced diffusion coefficient given by K/α, where K is the defect production rate and α is the number of dislocations per unit area, is many orders of magnitude higher than the thermal diffusion coefficient.

EXPERIMENTAL PROCEDURES

Discs of high purity aluminum of Marz grade quality with 3×10^{-3} m diameter and 2.54×10^{-4} m thickness were prepared. These discs were then thinned in a methanol-nitric acid electrolyte and subsequently implanted with iron on one face. Electrothinning was then completed by masking the implanted face with a lacquer. The thinned foils were subsequently examined in a transmission electron microscope at 100 kV. The ion-implantation parameters are shown below:

Energy (keV)	Dosage ions/m^2	Rp (nm)	ΔRp (nm)	∼ at. percent
50	5×10^{19}	12	5	8.5
100 + 50	5×10^{20}	23	8.5	36

The double dosage, 100 + 50 keV, implantation was performed with a time interval of one hour between the 100 and 50 keV implantations. The temperature rise of the discs during implantation is estimated to be less than 373 K.

RESULTS AND DISCUSSION

For both 50 and 100 + 50 keV dosages, extensively damaged substructure typical of ion implanted materials was observed. In Fig. 1(a) a bright field transmission electron micrograph for a 50 keV Al implanted foil is shown. The top part of the micrograph shows the implantation damage and the bottom part shows a typical modulated structure observed in these foils. The composition fluctuations in the modulation are found to be in <111> direction and 10 nm apart. The corresponding electron diffraction pattern of the modulated structure shows satellite spots near the Bragg reflections. These satellite spots are not observed due to the presence of dislocation defects produced by implantation. It was difficult to observe the modulated structure under two beam conditions since this structure was masked by the implantation damage which comes under view for maximum contrast conditions. Moire fringes were also observed in the areas of the modulated structure. As expected, these fringes shifted their relative position with respect to the modulated structure as the specimen was tilted in the stage around the axis of the transmitted electron beam path. Thus, it was clearly possible to distinguish between the modulated structure and the Moire fringes. The micrographs in Fig. 1 were taken using the conditions when the fringes and the damaged structure were completely out of view. A further confirmation of the presence of the modulated structure is obtained from the 100 + 50 keV foil which was given an annealing treatment at 793 K for 15 minutes. This treatment removed the extensive damage resulting from the double dosage ion implanta-

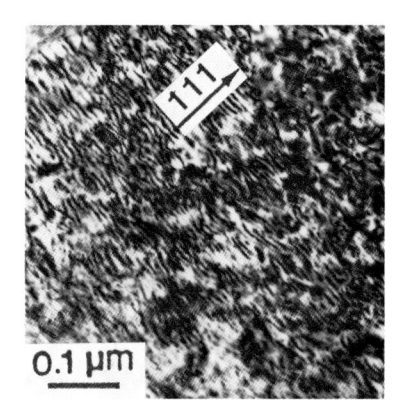

Fig. 1. TEM's of 50 keV as-implanted showing modulations in <111> direction and damage (a), SAD of modulations showing satellites (b).

tion. The associated micrographs, Fig. 2, clearly reveal rod-like precipitates and many areas of retained modulated structure. The diffraction patterns contain reflections, corresponding to Fe$_3$Al precipitates which are streaked due to the faults in the precipitate. The 50 keV foil was given an annealing treatment at 793 K for 30 minutes. The modulated structure has completely broken down and many precipitates are observed in the interior of the grain, Fig. 3. These precipitates are identified as Al$_3$Fe. Occasionally, a globular precipitate was observed along a grain boundary. Such globular precipitates have been observed before by rapid quenching an Al-7wt.%Fe and have been identified as metastable Al$_6$Fe precipitates [10]. However, in the present case Al$_6$Fe precipitates were not observed. By adopting a lower annealing temperature for a shorter time period the possibility exists for observing the intermediate Al$_6$Fe.

The majority of Al$_3$Fe precipitates observed in Fig. 3(a) are aligned in the <111> direction with respect to the matrix. The growth of these precipitates seems to bear an orientation relationship with the composition fluctuations found along <111> direction in Fig. 1(a). Fig. 3(a) might also imply that the modulations in regions where the subgrain boundaries are tilted at small angles prefer the same orientation.

In recent years, the thermal spike phenomenon has been invoked to explain the structure of quenched metastable alloys. Therefore, we have made an estimation of the temperature and lifetime of a thermal spike using the elastic-collision spike model proposed by Sigmund [8]. According to Sigmund the energy density Θ_0 or the temperature T_s within the spike depends on the mass of the ion (Fe$^+$) M_1, the target atom M_2 (Al), the density N of the target (Al) and on the primary ion energy E (50 keV). Assuming the ionized region to obey Boltzmann's Statistics Θ_0 could be written as,

$$\Theta_0 = \frac{G_2 N^2}{E^2} = \frac{3}{2}kT$$ and the lifetime τ of the spike as $\tau = M_2 \frac{E^3}{N^2}$ for an interatomic potential, $V \propto r^{-\frac{1}{2}}$ where r is the internuclear distance. For $V \propto r^{-1/3}$

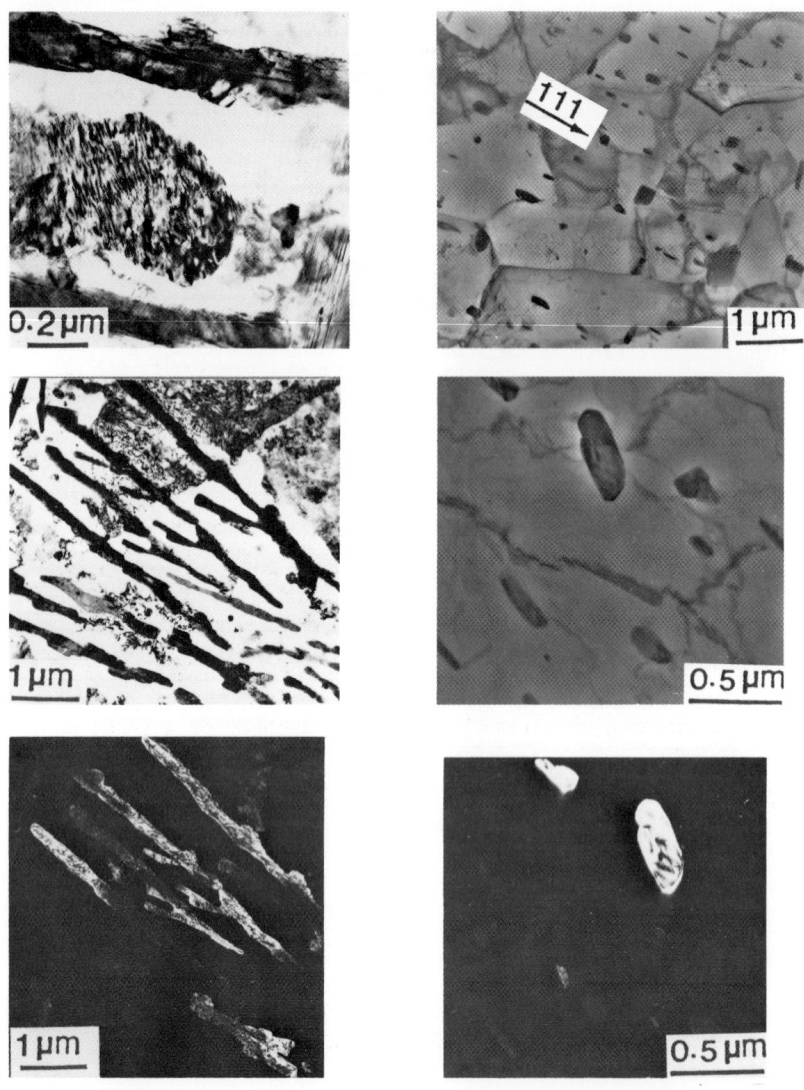

Fig. 2. TEM's of 100 + 50 keV + 15 min. at 793 K showing retained modulated structure and rod-like Fe₃Al precipitates (a), bright and dark fields of the Fe₃Al precipitates (b) and (c).

Fig. 3. TEM's of 50 keV + 30 min. at 793 K showing Al₃Fe precipitates aligned along <111> direction (a), bright and dark field of images of Al₃Fe (b) and (c).

similar expressions could be chosen from Sigmund's work. Although our calculations show temperatures of 1200 and 600 K, the corresponding lifetime of the spikes are only 10^{-11} and 10^{-10} seconds. The spacings of the modulations in Fig. 1(a) are of the order of 10 nm. At the temperature of 1200 K the thermal diffusion coefficient of Fe in Al is 3.8×10^{-15} m^2/sec. In order for Fe to migrate a distance of 10 nm in the hot zone 2.62×10^{-2} sec are needed. Even if one considers a radiation enhanced diffusion coefficient of two orders of magnitude more than the thermal diffusion coefficient, the time scale still falls short by a factor of 10^5. This suggests that the ultra fast quenching rate produced by a thermal spike may result in the jamming of Fe atoms in the aluminum lattice resulting in a disordered or an amorphous structure contrary to present observations. Our observations may be associated with normal spinodal decomposition resulting from a highly supersaturated structure produced by ion implantation.

We suggest that as ion implantation of Fe proceeds a critical composition is reached whereby the supersaturated Al-Fe solid solution becomes unstable to further compositional fluctuations. If our samples were heated to 373 K during implantation a time factor of 9.5 hours is needed for thermal diffusion of Fe to 10 nm (Fig. 1(a)). This time factor could be further reduced by a radiation enhanced diffusion coefficient. Poate and Cullis [6] have shown the effect of temperature on the radiation enhanced diffusion coefficient for various dislocation densities. Using their data for a defect production rate of 10^{-2} displacement/atom/second, we have chosen a four orders of magnitude increase in the diffusion coefficient. Then at 373 K, using a radiation enhanced diffusion coefficient of 2.7×10^{-17} m^2/sec, the time needed for iron to migrate would be only 3.7 seconds. We therefore propose here that the modulated structure observed in the Al/Fe foils for double dosage is due to the attainment of a critical composition due to implantation of Fe. The fineness of the spacing of the modulated structure is a function of the defect production rate which would dictate the radiation enhanced diffusion. The Al-Fe phase diagram shows a miscibility gap around 75 atomic percent of Fe. At this time, due to a lack of experimental evidence, we can only suggest that such high compositions (75 at. percent) of Fe could result from excessive sputtering of Al atoms during implantation.

CONCLUSIONS

1. The 50 keV and 100 + 50 keV implantations of Fe in Al result in modulated structures.

2. Some of the modulated structure produced by 100 + 50 keV implantation decomposes into Fe_3Al precipitates and matrix upon annealing at 793 K for 15 minutes.

3. The modulated structure produced by 50 keV implantation totally decomposes into Al_3Fe precipitates and matrix upon annealing at 793 K for 30 minutes.

ACKNOWLEDGEMENT

This research was sponsored by the Office of Naval Research under Contract N00014-78-C-0270, Dr. Philip A. Clarkin, Program Manager.

REFERENCES

1. K.V. Jata and E.A. Starke, Jr., J. Metals 35, 23 (1983).
2. N.E.W. Hartley in: Treatise on Materials Science and Technology, Ion Implantation 18, J.K. Hirvonen, ed. (Academic Press, New York

1980) pp.321-368.
3. H. Jones, Mater. Sci. Eng. 5, 1 (1969/70).
4. A. Tonejc, Met. Trans. 2A, 437 (1971).
5. H. Jacobs, A.G. Doggett and M.J. Stowell, J. Mater. Sci. 9, 1631 (1974).
6. J.M. Poate and A.G. Cullis, cf. 2, pp. 85-131.
7. S.P. Singhal, H. Herman and J.K. Hirvonen, Appl. Phys. Lett. 33, 25 (1978).
8. P. Sigmund, Appl. Phys. Lett. 25, 169 (1974).
9. S.M. Myers in Treatise on Materials Science and Technology, Ion Implantation 18, J.K. Hirvonen, ed. (Academic Press, New York 1980) pp. 51-82.
10. R. Yearim and D. Shectman, Met. Trans. 13A, 1891 (1982).

THE DEVELOPMENT OF THE AMORPHOUS PHASE IN NiTi DURING HEAVY ION OR ELECTRON BOMBARDMENT

J. L. BRIMHALL,* H. E. KISSINGER,* and A. R. PELTON**
*Pacific Northwest Laboratory, Richland, WA 99352; **Ames Laboratory, Ames, IA 50011

ABSTRACT

A supralinear dose dependence for the amorphous transformation was observed in NiTi during bombardment with 2.5 MeV Ni$^+$ ions. These results are consistent with a mechanism that requires cascade overlap to obtain a critical defect density for the amorphous transformation. Direct amorphization in the cascades was not resolvable. The temperature dependence of the minimum dose required for complete amorphous transformation had the same form as that observed for amorphization of silicon. Amorphization caused by electron bombardment required a higher dose than by ion bombardment. Different degrees of homogeneity of the damage state between ions and electrons can explain the dose dependence on particle type.

INTRODUCTION

This paper reports on a study of the mechanisms by which an intermetallic compound, NiTi, transforms to an amorphous phase during ion irradiation. Although there have been extensive studies on the mechanisms of amorphous transformation in silicon, [1] little attention has been devoted to other materials. In particular, the importance of displacement cascades and whether amorphous zones are formed directly in the cascades was investigated in this work.

The NiTi alloy was irradiated with 2.5 MeV Ni$^+$ ions at ambient temperature and the amorphous transformation followed by TEM observations. In addition, electron irradiation was carried out over a limited dose range and results compared with those from ion bombardment. The results are discussed in terms of current mechanisms for amorphization.

Experimental Procedure

The NiTi alloy was made by arc melting and had a nominal composition of 50% Ti - 50% Ni. Discs, 3 mm wide and 0.25 mm thick, were punched from rolled sheet and polished to a smooth finish. All discs were annealed at 900°C for 24 hrs prior to ion bombardment. The microstructure was predominantly NiTi phase with isolated particles of Ti$_2$Ni. The discs were bombarded with 2.5 MeV Ni$^+$ ions to doses ranging from 4 x 10^{13}/cm^2 to 5 x 10^{15}/cm^2, i.e., 0.020 to 3 displacements per atom (dpa) according to calculated damage rates [2]. The maximum damage occurred at a depth of ~0.6 μm. The ion current was normally 0.16 μamp/cm^2 but currents as high as 1.6 μamp/cm^2 were used. The same alloy was electron irradiated in the HVEM at Lawrence Berkeley Laboratory to doses >10^{22} e/cm^2 so as to produce damage states equivalent to ~1 dpa [3].

Standard microscopic techniques were used to record the microstructural changes. Selective electro thinning was used to observe the surface region and a region ~0.3 μm from the surface. The influence of the deposited Ni ion was negligible in these regions. The amorphous volume was determined by estimating a radius for the irregular amorphous regions and assuming a three dimensional shape. The foil thickness was also estimated but very thin foil regions were used to avoid overlaping of the amorphous regions in the two dimensional analysis. Because of these estimates, a factor of two is placed on the error spread. The degree of long range order was determined by measuring the ratio of the intensity of the [100] spot to the [200] spot as described by Howe and Rainville [4]. The long range order parameter S is a function of $(I_{[100]}/I_{[200]})^2$.

Results

The amorphous volume estimated from the micrographs is shown in Figure 1 as a function of the irradiation dose. Even though there are large error bands, the slope of the data is definitely greater than one, i.e., the dose dependence is supralinear. The slope decreases at the highest dose as the matrix saturates with amorphous material. The shape of the curve at lower doses could not be determined but presumably the overall shape is sigmoidal.

The minimum dose required to make an alloy fully amorphous increased as the ion current increased, Figure 2. This was a result of the increased beam heating at the higher ion currents. The temperature could only be measured on one specimen for each run but at the highest current, temperatures up to 180°C were recorded. At the lowest ion current, the temperatures did not exceed 40°C. The form of the curve in Figure 2 is also identical to that observed for the temperature dependence of ion induced amorphization in silicon [5].

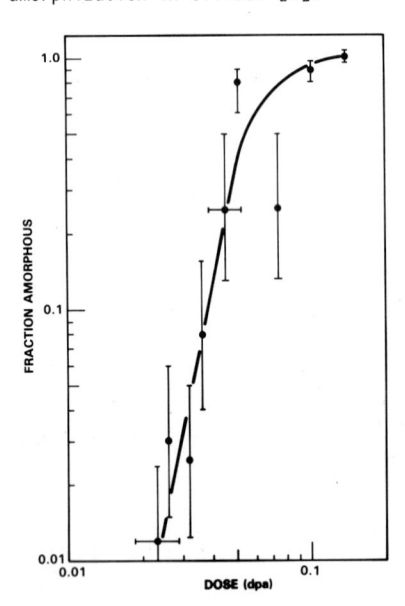

FIG. 1. Amorphous volume as a function of dose for NiTi irradiated with 2.5 MeV Ni$^+$ ions.

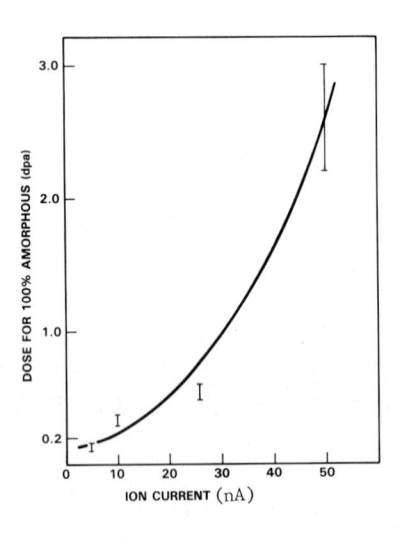

FIG. 2. The minimum dose required for 100% amorphous material as a function of ion current. Ion current is a measure of the beam heating.

Small, irregular regions with lighter or darker contrast from the matrix were characteristic of the damage at low doses, Figure 3a. These irregular zones expanded in volume with increase in dose but maintained their irregular shape, Figure 3b. At high doses, the amorphous nature of these zones was clearly evident as shown by a strong diffuse ring in the diffraction pattern.

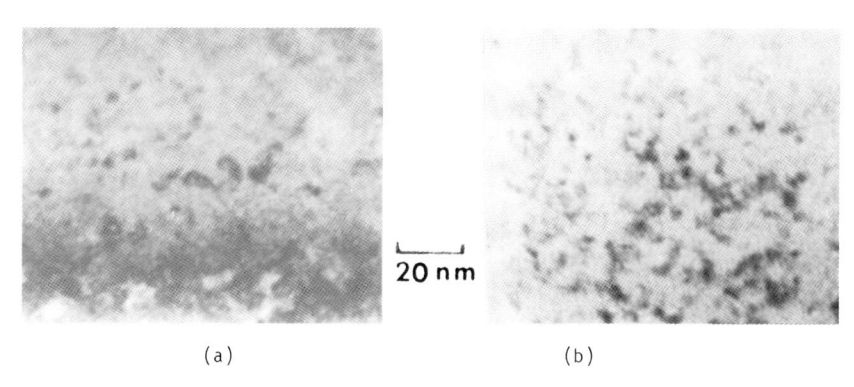

FIG. 3. Microstructure of NiTi bombarded to various doses (a) 0.025 dpa (b) 0.052 dpa. Dark regions are crystalline in (b).

The amorphous regions could also be identified by comparing bright field (BF) and dark field (DF) micrographs from a lattice reflection and DF micrographs from a portion of the diffuse ring. An example is shown in Figure 4 where the amorphous zones exhibit opposite contrast to the matrix in the BF and normal DF imaging conditions. In the DF using the diffuse

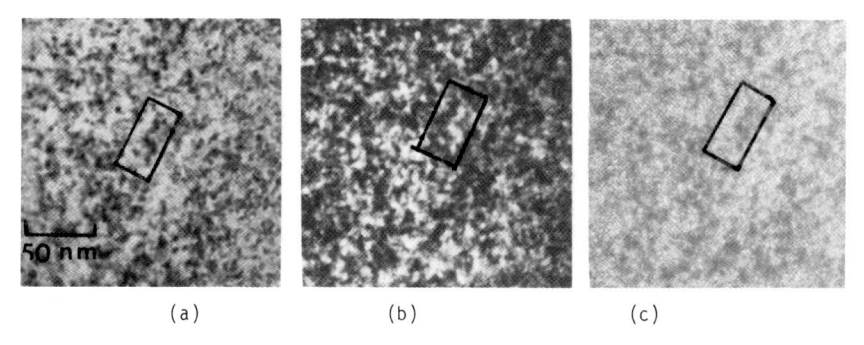

FIG. 4. Imaging of partially amorphous structure in NiTi (a) BF (b) DF, [200] reflection (c) DF using a portion of diffuse ring. The same area is delineated in each micrograph.

ring, the amorphous zones appear in somewhat stronger contrast since the matrix is not diffracting at all. The lowest dose at which these regions could reasonably be identified as amorphous was 0.025 dpa. It was not possible to determine if the smallest, isolated spots were actually amorphous regions. Strain contrast was observed in some cases for these small spots which indicated a defected but not amorphous region.

The matrix did not become completely disordered before becoming amorphous. The diffraction pattern from a region approximately 50% amorphous still revealed distinct superlattice (SL) spots. The relative intensity of the SL spots to lattice spots decreased with irradiation, but there was still considerable order in the matrix. Further, no additional disorder zones were observed when imaging in DF using a SL reflection.

The Ti_2Ni particles within the TiNi matrix also became amorphous during irradiation. The general appearance of the partly transformed Ti_2Ni was similar to that in NiTi although on a somewhat finer scale. The minimum dose required to make Ti_2Ni fully amorphous was approximately twice that for NiTi.

The electron irradiation in the HVEM also transformed NiTi to an amorphous state. For an equivalent damage rate, a dose between 1 and 3 dpa was required to make the alloy fully amorphous, Figure 5. This is considerably higher than for ion irradiation. The light region in the center of Figure 5b is the amorphous region. Although not noticeable in the micrograph, the onset of amophization occurs at ~1 dpa [3]. The matrix also became fully disordered before becoming amorphous during electron irradiation which is in contrast to the ion irradiation results.

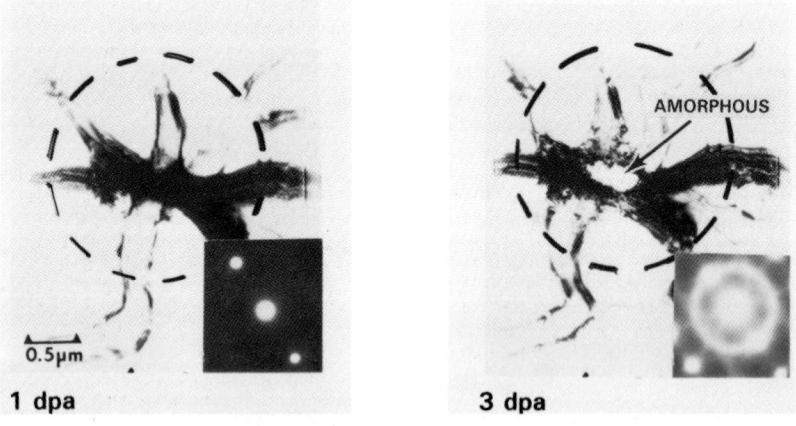

FIG. 5. Microstructure of NiTi irradiated with 1.5 MeV electrons in the HVEM. Dotted area delineates electron beam.

DISCUSSION

The supralinear form of the dose dependence for the amorphous transformation implies that the bombarding ions do not produce the amorphous phase directly within the displacement cascades. A linear dependence would be expected if each individual cascade created an amorphous zone. The decrease in the dose dependence at high doses occurs when a significant fraction of the crystalline phase becomes amorphous. Ions that bombard the previously transformed regions do not contribute any further to the amorphization process. Our observations are therefore consistent with the mechanism that assumes the cascades produce mainly regions of high defect density. Overlap of neighboring cascades is required to create the critical defect concentration necessary for the amorphous transformation [6-8]. Once a substantial fraction of subcritical defect regions is created, only a small increase in dose, hence defect concentration, can initiate the collapse of the lattice into an amorphous phase [1]. The density of cascades in NiTi was estimated using the primary knock-on (PKA) spectra for 4 MeV Ni$^+$ ions in nickel after Marwick [9]. The lowest dose of 4×10^{13} ions/cm^2 would produce $\sim 3 \times 10^{18}$ PKA/cm^3. This asssumes each PKA with E > 10 KeV produces at least one definable cascade. The average PKA spacing of 7 nm is therefore comparable to the size of cascades and considerable overlap in the cascades would be expected for these high densities.

The amorphization process is, however, expected to be a function of the ion mass based on results in silicon where heavy ions cause direct amorphization in the cascades and light ions create diffuse defect regions which must overlap before amorphization occurs [1,7]. Nickel is an intermediate mass ion in which some direct amorphization could occur for some of the higher energy cascades. The dose dependence may show some linear dependence at very low doses but it was not measurable in the technique used here. The data suggests that cascade overlap is the predominant mechanism to account for the rapid increase in amorphization with dose.

The temperature dependence of the dose required for full amorphization is also consistent with cascade overlap and the requirement of a critical defect concentration. As the temperature is increased, there is more annealing of defects within the cascade region. The net defect density per unit of ion dose is less; therefore a higher dose is required to attain the same critical defect density to form the amorphous structure. If the temperature is too high, an infinite ion dose will be required to reach the critical defect density as the curve in Figure 2 illustrates. A strong temperature dependence for the minimum dose has also been observed in electron irradiated NiTi [10].

The lower dose required for amorphization by ions compared to electrons is a result of the inhomogeneity of the damage in the case of ions. The fact that the remaining crystalline material between the amorphous regions is still ordered is a graphic illustration of the inhomogeneity of damage. A hypothetical representation of the defect density along an arbitrary direction in the bombarded material is shown in Figure 6 for both ions and electrons. C_D^{th} is the equilibrium defect concentration. Even though the average defect density, C_D, may be similar for ions and electrons, the localized defect density has attained the critical value, C_D^{crit}, at several points in the case of ions. These regions would become amorphous and continue to expand with increasing dose. The defect distribution is more uniform during electron irradiation and at no point has the defect concentration attained the critical value. A higher dose is required to raise the defect concentration above the critical level. This description implies that the amorphous transformation during electron irradiation should occur over a narrow dose range.

CONCLUSIONS

The amorphous transformation in NiTi caused by Ni$^+$ ion bombardment is consistent with a cascade overlap mechanism as opposed to amorphous formation directly in cascades. Electron irradiation in which there are no cascades also produces an amorphous phase in NiTi but a higher dose (dpa) is required than for ions.

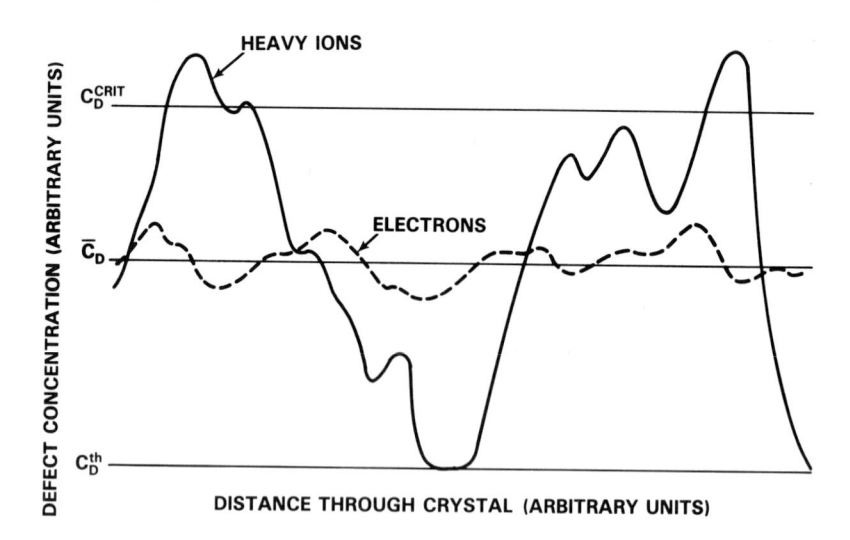

FIG. 6. Schematic illustration of the defect density along an arbitrary direction in a crystal. Solid line = ion bombardment, dashed line = electron bombardment. See text for definitions of C_D, C_D^{crit}, C_D^{th}.

ACKNOWLEDGMENT

This work was primarily supported by the Division of Materials Sciences, Office of Basic Energy Sciences of the Dept. of Energy under Contract DE-AC06-76RLO-1830. The assistance of Prof. R. Sinclair and support by the NSF - MRL program at Stanford University is gratefully acknowledged.

REFERENCES

1. D. A. Thompson, Rad. Eff., 56, 105(1981).
2. J. L. Brimhall, H. E. Kissinger and L. A. Charlot, Rad. Eff., 77, 237(1983).
3. A. R. Pelton, Seventh Intl. Conf. on High Voltage Electron Microscopy, R. M. Fisher, R. Gronsky and K. H. Westmacott eds., LBL-16031, 1983, (NTIS, Springfield, VA) p. 245.
4. L. M. Howe and M. H. Rainville, J. Nucl. Mat., 68, 215(1977).
5. J. R. Dennis and E. B. Hale, J. Appl. Phys., 49, 119(1978).
6. E. C. Baranova, V. M. Gusev, Yu. V. Martynenko, C. V. Starinin and I. B. Haibullin, Rad. Eff. 18, 21(1973).
7. D. A. Thompson, A. Golanski, K. H. Haugen, D. V. Stevanovic, G. Carter and C. E. Christodoulides, Rad. Eff., 52, 69(1980).
8. V. Bartuch and W. Karthe, Rad. Eff. Lett., 67, 187(1982).
9. A. D. Marwick, J. Nucl. Mat., 55, 259(1975).
10. H. Fujita, H. Mori and M. Fujita, ibid Ref. 3, p. 233.

HVEM AND INTERNAL OXIDATION STUDIES OF LITHIATED NICKEL[+]

K. SESHAN, P. BALDO* AND H. WIEDERSICH*
Department of Metallurgical Engineering, University of Arizona,
Tucson, Arizona 85721
*MST, Argonne National Laboratory, Argonne, Illinois 60439

ABSTRACT

Pure, polycrystalline nickel samples were implanted with lithium to doses up to 5×10^{17} lithium ions per square centimeter, at a temperature of 500°C, such that the implantation damage would anneal. These samples were then prepared for electron microscopy and examined at 1 MeV, in the Argonne National Laboratory HVEM facility. It was observed that compared to pure nickel, the lithium implanted nickel showed a different radiation damage behaviour. A plausible explanation for the difference in behaviour is presented in this paper.
The lithium implanted nickel, in the high dose samples, also showed an unusual form of precipitation. Electron microscopy revealed the precipitates to have truncated octahedral shapes with {111} planes for sides and {100} planes truncating the corners. They resemble voids and helium bubbles in nickel. The precipitates appear to be associated with dislocations. The lithium implanted nickel was internally oxidized in order to obtain evidence for the presence of lithium. Electron diffraction analysis of the internally oxidized lithiated nickel showed the presence of a topo-taxial compound being formed, with an ordered NaCl-structure. Possible interpretations of this diffraction pattern are discussed.

INTRODUCTION

The elements lithium and nickel are virtually impossible to alloy by conventional methods. It was, therefore, decided to implant the lithium into the nickel so that the properties of lithiated nickel could be studied. It has been reported that lithiation decreases the oxidation rate of nickel.[1-5] The object of this study was to investigate the radiation damage response of lithiated nickel using the Argonne National Laboratory HVEM facility and study the details of lithium precipitation and internal oxidation in nickel.

Experimental Procedure

High vacuum furnace annealed 3mm discs of five-nine pure nickel were implanted at the ANL facility, at 500°C, to doses in the range of 10^{15} to 5×10^{17} Li/cm^2. The samples were electropolished from the back to provide electron transparent specimens. Internal oxidation experiments were performed on the implanted nickel samples in a partially oxidizing atmosphere of CO/CO_2 (100:1). The pre-thinned transmission electron microscope samples used were observed before and after the oxidation process.

+Work supported by the U. S. Department of Energy

RESULTS

In nickel samples implanted to a dose of 7 x 10^{16} lithium/cm^2 at 500°C, electron microscopy revealed the presence of hexagonal-shaped defects arranged along dislocation lines. Tilting experiments, combined with weak-beam electron microscopy led to the conclusion that these defects have an octahedral shape with faces composed of {111} planes and corners truncated by {100} planes. It was not possible to obtain direct electron diffraction or energy-loss evidence for the presence of lithium inside these hexagonal features.

Electron Irradiation Experiments

Electron irradiation at 1 MeV was performed, on pure and lithium implanted nickel, using the ANL-HVEM fitted with the heating stage sample holder. In situ electron irradiation and specimen heating experiments were performed in the temperature range 300°K to 900°K. The sample was raised to a predetermined temperature, equilibrated, and then electron irradiated at that temperature for up to 40 minutes. Electron micrographs were obtained during the irradiation process.
An electron irradiation sequence of pure nickel is shown in Fig. 1. A similar sequence using lithium implanted nickel is shown in Fig. 2.
From the electron micrographs it is evident that the electron damage response of the lithium implanted nickel is completely different from that of pure nickel. The most striking difference is that in the temperature range 300°K - 673°K, the microstructure of the lithiated nickel is stable and does not change with electron damage to doses of 1 dpa. Pure nickel on the other hand damages quite readily in this temperature range.
The resistance to electron irradiation of the lithiated nickel could arise either from the precipitated lithium or the lithium in solution. It is not clear which of these two effects is operative. It is possible that the lithium in solution associates with the nickel interstitials produced during irradiation slowing them down, and enhancing their recombinations with vacancies. On the other hand, the resistance to irradiation could arise from an uniform, but unobservable distribution of sub-microscopic clusters of lithium. The interfaces of these clusters could act as unbiased sinks for electron-irradiation produced vacancies and interstitials.

Internal Oxidation Experiments

Internal oxidation experiments on the lithiated nickel were performed in an oxidation furnace where the CO/CO$_2$ atmosphere could be controlled. A CO/CO$_2$ ratio of 100/1 was used; samples were oxidized for 17 hours at 500°C. Under these conditions the lithium would, and the nickel would not, oxidize.
Routine bright field electron micrographs of the sample of the internally oxidized lithiated nickel did not reveal any noticeable difference from the un-oxidized specimen.
Electron diffraction patterns, obtained at the [001] zone were, however, remarkable different. An example of one such electron diffraction pattern is shown in Fig. 3. In order to distinguish more easily between matrix and precipitate spots, the diffraction pattern is re-drawn, to scale, in Fig. 3. The d-spacing obtained do not match Li$_2$O spacings. In order to explain this spot pattern it was hypothesized that an ordered Li$_x$Ni$_{1-x}$O compound is formed. This compound would be isostructural with Li$_x$Fe$_{1-x}$O which is reported to have an ordered NaCl structure.[6,7]
We attempted to synthesize the Li$_x$Ni$_{1-x}$O compound by co-precipitating and reducing a mixture of NiCO$_3$ and Li$_2$CO$_3$. XRD showed this compound as having a NaCl structure with d$_{100}$ and d$_{110}$ as 0.21 and 0.15 nm, respectively. If this were ordered then d$_{100}$ and d$_{110}$ would be 0.41 and 0.30 nm, respectively.

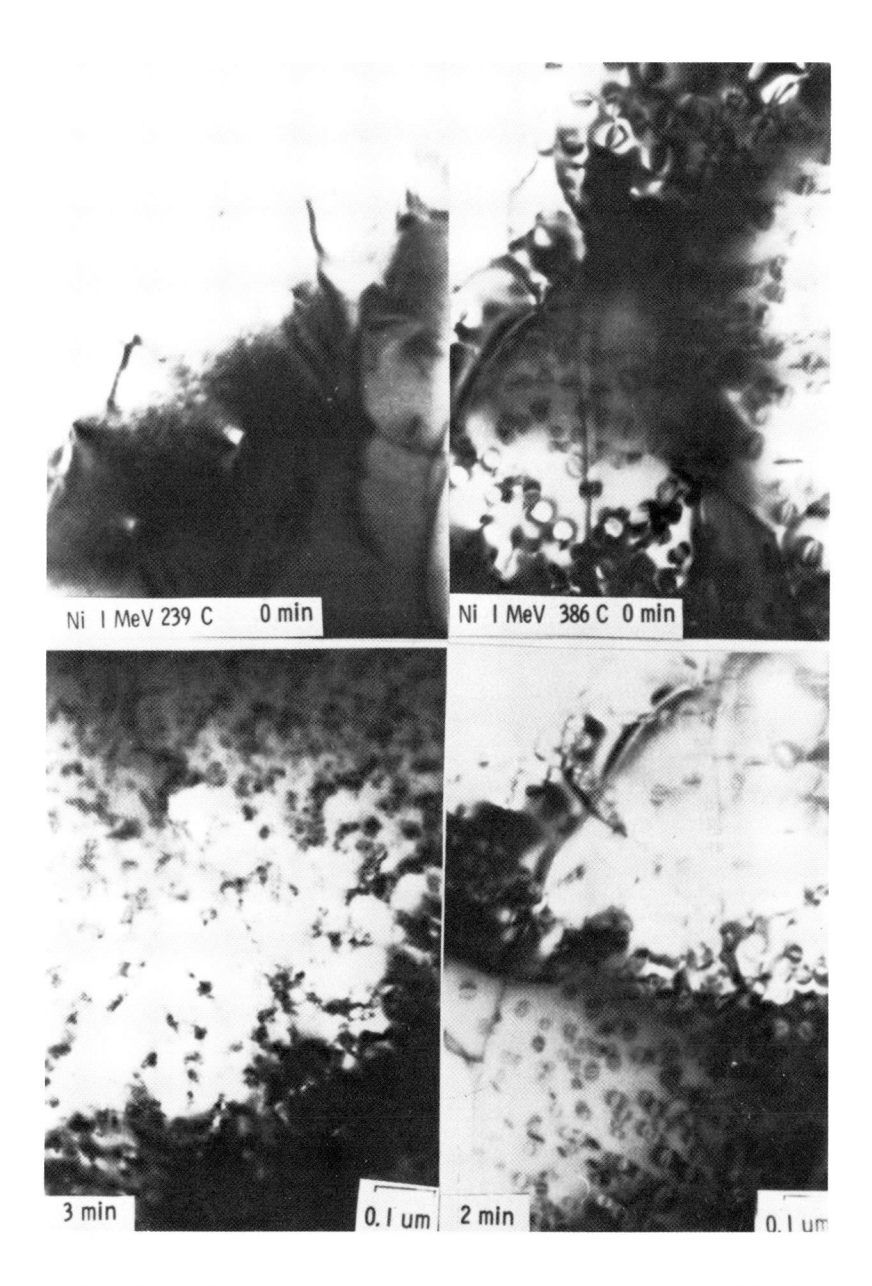

Fig. 1 A 1 MeV electron irradiation sequence of pure nickel at 239°C for
1-3 minutes, and at 386°C for 1-2 minutes. It is to be noted that
there is a rapid accumulation of radiation damage induced defects.
There also is observed a changeover in the damage mechanism between
239°C and 386°C.

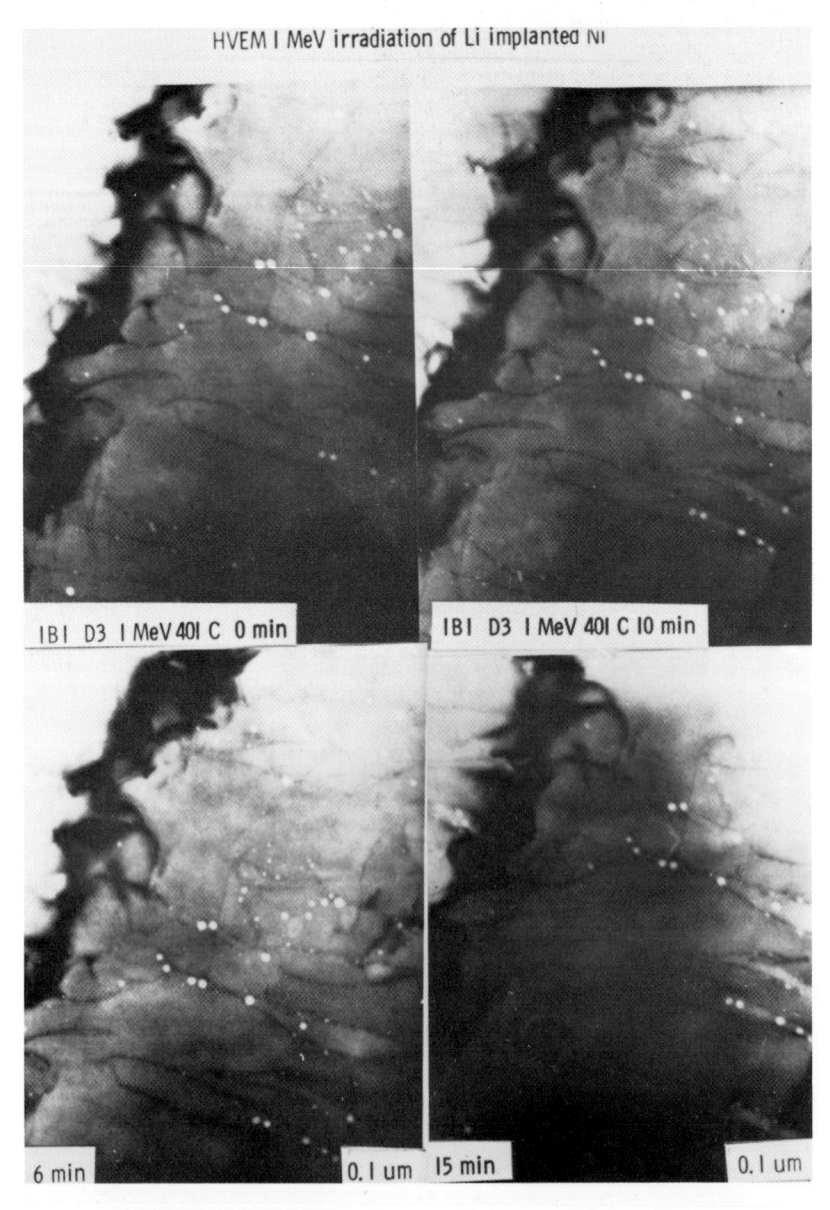

Fig. 2 A 1 MeV electron irradiation sequence of lithium implanted nickel. The white bubbles along the dislocations are lithium precipitates. The microstructure is stable and does not change during this sequence in which the sample was held at 400°C.

173

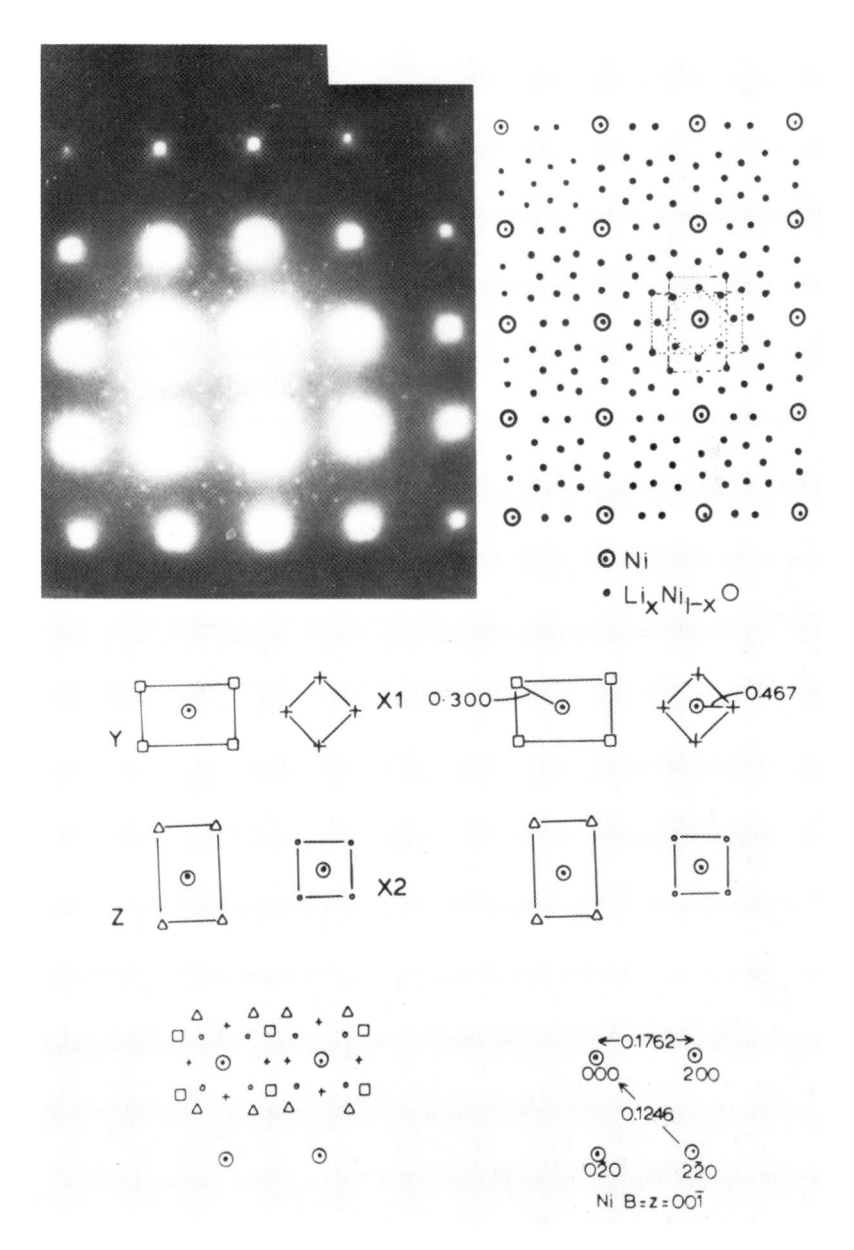

⊙ Ni
• $Li_xNi_{1-x}O$

Fig. 3 The electron diffraction pattern from internally oxidized, lithiated nickel is shown on the top left. The extra spots from the $Li_xNi_{1-x}O$ compound are identified and shown to arise from four varients x_1, x_2, y and z. When these varients are superimposed, and double-diffraction effects are taken into effect, all the extra spots can be accounted for.

These d- spacings are in good agreement with those obtained from the electron diffraction data on the internally oxidized lithium in nickel.

This experiment indicated that the compound being formed inside the lithiated nickel was ordering, whereas the compound fabricated by the co-precipitation and reduction was not ordered. The close agreements in the d-spacing shows that the compound formed in the lithiated nickel is of the $Li_xNi_{1-x}O$ type, with an ordered NaCl structure.

CONCLUSIONS

Implanting lithium in nickel to doses which exceed lithium solubility, causes a void-helium bubble-like precipitate of lithium. To observe these precipitates, the samples should be heated during implantation to anneal the implantation damage. These precipitates have a truncated octahedral shape. The litheated nickel is stable against 1 MeV electron-irradiation damage. Pure nickel, in contrast, has a constantly changing microstructure under similar radiation conditions. Internal oxidation of the lithium in the nickel cause the formation of a topotaxial compound with d_{100} = 0.46 nm and d_{110} = 0.30 nm.

A more detailed account of these experiments is in preparation and will be presented elsewhere.

ACKNOWLEDGEMENTS

It is a pleasure to acknowledge the ANL-HVEM facility staff for their unstinting cooperation. I, Krishna Seshan, am grateful to DOE-DBES for partial summer support, and to AUA for travel support. It is a pleasure to acknowledge the helpful conversations with Professor J. B. Wagner Jr. and Dr. K. H. Westmacott.

REFERENCES

1. H. Pfeifer and K. Hauffe, Z. Metalle. 43, 364 (1945).
2. E. J. W. Verwey, P. W. Haaijman, F. C. Romeijn, G. W. van Oosterhoot, Phillips, Res., Rep. 5, 173 (1950).
3. G. Dearnaley, Treatise on Materials Science and Technology 18, 257 (1980) and Nucl. Inst. Methods 182/183, 899 (1981).
4. P. D. Goode, Inst. Phys. Con. Ser. No. 28, 154 (1976).
5. Zhou Peide, F. H. Scott, P. P. M. Procter, W. A. Grant, Oxidation of Metals 16, (5/6), 409 (1981).
6. T. Matsui, J. B. Wagner Jr., J. J. Nuclear Materials 99, 213 (1981).
7. J. C. Anderson, M. Schnieber, J. Phys. Chem. Solids 25, 961 (1964).

AMORPHOUS SOFT MAGNETIC MATERIAL FORMATION BY ION MIXING

P. GERARD a), G. SURAN b), B. BLANCHARD c), J. DEVENYI d), M. DUPUY, d) and P. MARTIN a)
a) L.E.T.I-C.E.A. 85 X, 38041 Grenoble (F), b) Laboratoire de Magnétisme C.N.R.S. (92195) Meudon Principal CEDEX, c) DERDCA-CEA 85 X, 38041 Grenoble (F), d) LEPES-C.N.R.S. 166 X, 38041 Grenoble (F)

ABSTRACT

Ion mixing of FeNi-Si multilayers is performed at the same dose of 3.10^{15} Xe$^+$ cm^{-2} with either of the three energies 200, 300 or 400 keV. The irradiation is done at liquid nitrogen temperature (LNT) or at room temperature (RT). The mixed layers are characterized by reflection high energy electron diffraction (RHEED), transmission electron microscopy (TEM), Rutherford backscattering (RBS), secondary ion mass spectroscopy (SIMS) and ferromagnetic resonance (FMR). Films made at 400 keV are more reproducible and as homogeneous as the best amorphous films realized up to now, using sputtering techniques.

INTRODUCTION

Transition metal-metalloid amorphous soft magnetic alloys are at the origin of many investigations as much for a fundamental standpoint [1] as for numerous technological applications [2]. The theoretical approach is to look at the influence of the amorphous structure on the different magnetic parameters such as the magnetization ($4\pi M_s$), the Curie temperature (T_c), the anisotropy, etc., and their respective temperature dependence. Some of the amorphous soft magnetic materials have already found their use for certain applications, (e.g., transformers) because they have lower losses, lower coercivities, higher permeabilities, and are competitive in cost with some of their crystalline counterparts. There is in the literature much experimental data on the properties of these materials prepared by various techniques. Up to now, thin layers have only been made through coevaporation or sputtering and both have their advantages and their drawbacks. Here we propose the use of an ion mixing method, recently developed at Caltech [3], in order to make $(Fe_{20}Ni_{80})_{100-x}Si_x$ amorphous films, and to analyze by FMR the influence of certain irradiation parameters on the magnetic properties of these mixed layers. Indeed, magnetic measurements of amorphous transition metal-metalloid alloys are very sensitive to small variations of chemical composition and in depth homogeneity. Consequently such a system is particularly well adapted for the measurement of the ion mixing efficiency. The choice of the Permalloy composition ($Fe_{20}Ni_{80}$) as a starting material was made so as to have a final amorphous product, after the addition of the glass former dilutent Si, with no magnetostriction and little or no anisotropy. Under these circumstances, FMR spectra permit a straight-forward interpretation of the information giving access directly to $4\pi M_s$, the g factor, and T_c, and also permit the use of this method at low temperature on account of the narrow character of the resonance linewidth. The temperature dependence of the magnetic behaviour of certain of these layers has been studied elsewhere [4] and compared to results obtained on similar systems made by other techniques. Here, before describing the magnetic results, conventional characterizations by RHEED, TEM, RBS and SIMS of as-deposited and mixed layers under different mixing conditions are presented.

EXPERIMENTAL DETAILS

Alternate layers of $Fe_{20}Ni_{80}$ and Si were sequentially electron beam evaporated at room temperature with vacuum ranging from 1 to 3. 10^{-7} Torr on a 2 inch <111> oriented silicon wafer. The deposition rate was about 1 Å/sec. From 4 up to 6 alternate layers of FeNi-Si were deposited, resulting in a total thickness of the order of 800 Å. The thickness ratio between the FeNi and Si layers was chosen so as to have the desired composition after ion mixing. Once the layers were deposited onto the wafer, it was cut into small samples having a rectangular shape of 4.5 mm². The chosen compositions were obtained from the following multilayers: a) (30 Å Si, 100 Å FeNi).6; b) (60 Å Si, 100 Å FeNi).5; c) (100 Å Si, 100 Å FeNi).4 referred to hereafter as system (30-100), (60-100) or (100-100), respectively. Ion mixing was performed each time on 3 samples per composition. It was done either at LNT or at RT in a vacuum better than 1.10^{-6} Torr. The samples were Xe irradiated at the same dose of 3.10^{15} Xe^+ cm^{-2} with either of the three energies: 200 keV, 300 keV or 400 keV.

RESULTS AND DISCUSSION

Conventional Characterization

Fig. 1 a) and b) show RHEED diffraction patterns of system (60-100) before and after Xe irradiation at 400 keV.

It is important to note that RHEED only probes a 400 to 500 Å layer thickness. So, within the resolution of this technique, it can be ascertained that the dose of 3.10^{15} Xe is sufficient to amorphize the near surface layer of this multilayer system as seen from the diffuse half rings displayed in Fig. 1b. Similar results have been obtained with the two other compositions with either of the 3 energies and also with the two considered irradiation temperatures, i.e. LNT and RT. In Fig. 2, TEM cross section micrographs of system (100-100) are presented for; a) as deposited layers; b) and c) after a 200 keV and 300 keV Xe irradiation, respectively, both irradiations being done at LNT. From the bright field images b) and c) one can see that a 200 keV ion mixing only transforms a layer of about 500 Å to an amorphous alloy whereas a 300 keV irradiation is sufficient to make the whole layer amorphous. One can also notice in Fig. 2 a) and b), that there is a 30 Å thick film between the deposited layer and the substrate which is presumably silica. RBS spectra were taken with 1 MeV He ions at a scattering angle of 165° with the beam inclined at 60° from the sample normal (see scheme of Fig. 3) in order to increase the depth resolution. Fig. 3 shows spectra of the system (100-100) corresponding to the unmixed, 200 keV and 300 keV mixed system respectively. After a 200 keV irradiation the bottom of the multilayers is still unmixed, no stacking being left after a 300 keV Xe bombardment. RBS spectra of samples mixed at either RT or LNT were identical. The compositions of the 3 ion mixed amorphous alloys were determined from the RBS spectra heights relative to the backscattered energy corresponding to FeNi and Si given in Fig. 4.

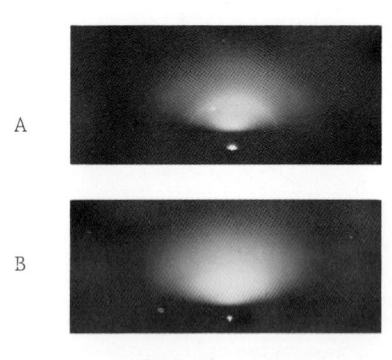

Fig. 1. RHEED diffraction patterns of system (60-100); a) before; b) after Xe irradiation at 400 KeV.

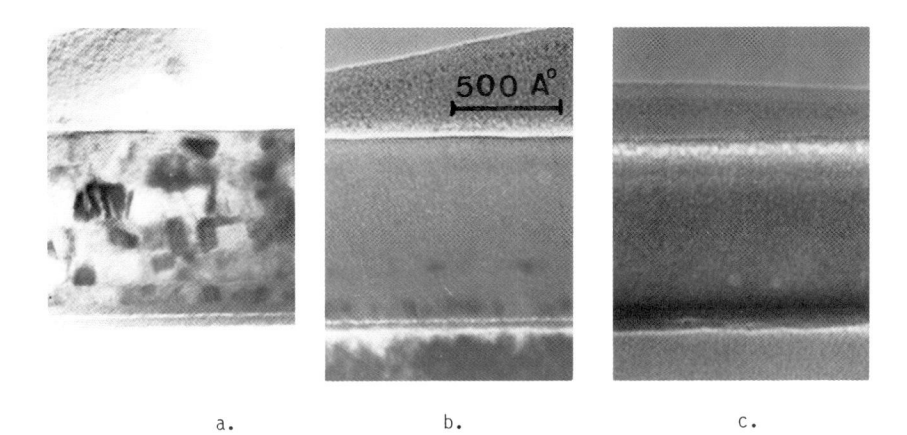

Fig. 2. Cross-section micrographs of system (100-100); a) as deposited layers; b) after a 200 keV and; c) after a 300 keV Xe irradiation.

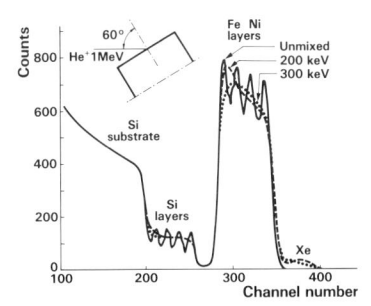

FIG. 3. RBS spectra of system 100-100 unmixed and ion mixed with Xe at 200 and 300 keV.

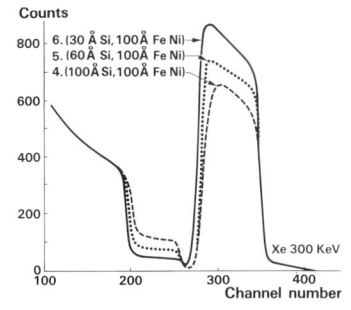

FIG. 4 RBS spectra of systems (30-100), (60-100) and (100-100) ion mixed with Xe at 300 keV.

These compositions as well as the projected range R_p and the projected standard deviation ΔR_p of Xe for 200, 300 and 400 keV are reported in Table I. R_p and ΔR_p were calculated by means of a program deduced from

TABLE I. Composition of the 3 systems, R_p, and ΔR_p for 200, 300 and 400 keV irradiation of each composition.

Composition	System (30-100) $Fe_{16}Ni_{64}Si_{20}$			System (60-100) $Fe_{14}Ni_{55}Si_{31}$			System (100-100) $Fe_{12}Ni_{49}Si_{39}$		
E_n (keV)	200	300	400	200	300	400	200	300	400
R_p (Å)	295	420	547	331	472	615	372	531	693
ΔR_p (Å)	80	113	149	87	127	167	98	142	188

L.S.S. theory. SIMS depth profile of Si distributions of system (100-100) before and after ion mixing are shown in Fig. 5 and 6, respectively. Fig. 5 displays the influence of Ne, Ar and Xe as primary ions, at an energy of 5.5 keV and at an incident angle of 65°, on the Si distribution. With Ne

FIG. 5. SIMS of Si in depth distribution obtained with Ne, Ar, and Se primary ions.

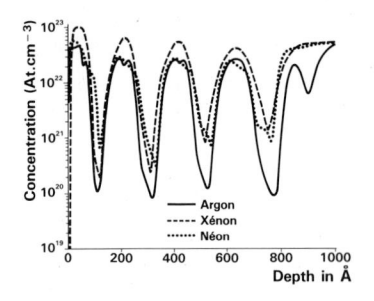

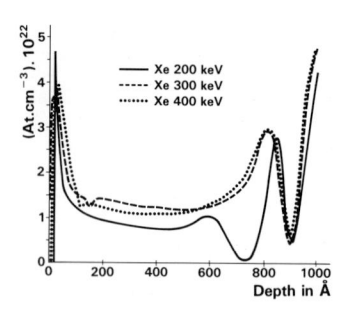

FIG. 6. SIMS of Si in depth distribution of system (100-100) after Xe irradiation at 200, 300 and 400 keV.

there is a knock-on effect of Si, which is less obvious in the case of Ar. This recoil effect disappears with Xe but one observes a degradation of the in depth resolving power which for Ar is about 20 Å for a signal variation of 50. It is also important to stress in the case of the curve obtained with Ar, the presence of the trough at 900 Å from the surface, which corresponds to a maximum of oxygen. This confirms that the 30 Å film on the substrate is silica (see Fig. 2). Fig. 6 shows the effect of ion mixing which is complete with energies of 300 and 400 keV but not in the case of 200 keV. The trough is still present at 900 Å after irradiaion. It appears that this silica layer acts as a barrier to ion mixing between deposited and substrate material.

Magnetic Characterizations by FMR

For reasons of convenience this technique is applied at room temperature. This restricts our studies to system (30-100) as it is, amongst the three ion mixed multilayers, the only composition to be ferromagnetic at this temperature. FMR is an ideal method for the determination of the magnetic parameters of a thin film, for the estimation of the magnetic loss (from the linewidth ΔH) and in particular for the analysis of the various magnetic (i.e., structural and chemical) in homogeneities within the layer thickness. In order to obtain the desired information it is necessry to take FMR spectra with the applied field perpendicular ($H_\perp$) and parallel ($H//$) to the film plane. Measurements were made by X-band (frequency f = 8.91 GHz) with a standard Varian spectrometer, the spectra having different typical features according to the film's homogeneity. So a homogeneous ferromagnetic film, in $H_\perp$ configuration is characterized by resonance spectra presenting one strongly excited mode (H_{1o}) followed by weakly excited modes (H_{1n}) which are stationary spin wave modes (SSW). These SSW modes obey a quadratic dispersion law [5]. If the film is heterogeneous, then there appears a certain discrepancy between the line position and the dispersion law and the linewidth of the uniform line (ΔH_0) is increased. In absence of anisotropy, $4\pi M_s$ and g are determined with the help of the following relations:

$$\frac{\omega}{\gamma} = H_{\perp o} - 4\pi M_s \quad (1) \quad \text{and} \quad \frac{\omega^2}{\gamma^2} = H_{//o}(H_{//o} + 4\pi M_s) \quad (2)$$

which correspond to the perpendicular and parallel configuration, respectively. ω is 2πf and γ is the gyromagnetic ratio (γ = g e/2mc). Fig. 7 shows spectra in $H_{//}$ and $H_\perp$ of the unmixed system (30-100).

There are five resonance lines in $H_\perp$ corresponding to five independent FeNi layers. Small variations of resonance field indicate that there is some small interdiffusion between Si and FeNi. Fig. 8 a and b are $H_{//}$ and $H_\perp$ spectra of 200 keV ion mixed layers at LNT. There are 5 resonance lines in $H_\perp$ and $H_{//}$ due to 5 uncoupled layers as drawn on the scheme of Fig. 8 b. The weakest $4\pi M_s$ belongs to the layer at the film-air interface, the largest $4\pi M_s$ corresponds to the deepest layer. The value of the latter is nearly that of Permalloy, proving that the FeNi layer next to the Si substrate is not mixed. ΔH varies from one mode to another. This shows that the ion mixing is inhomogeneous within each magnetic layer. The localization of the different layers was obtained through acid etching. Fig. 9 shows that after a LNT ion mixing at 300 keV there are typically two regions: a thick region at the film-air interface where the ion mixing is nearly complete (the layer being slightly inhomogeneous as established from SSW),

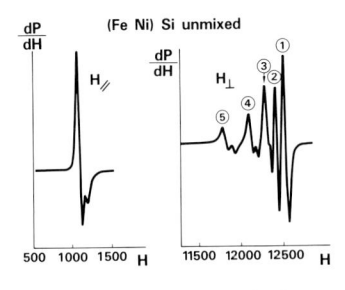

Fig. 7. FMR Spectra of the Unmixed System (30-100).

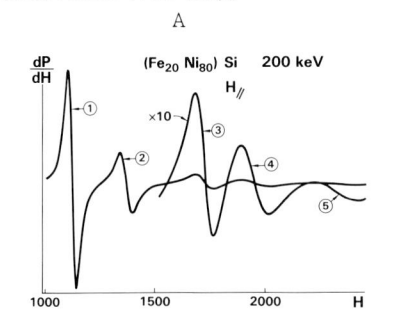

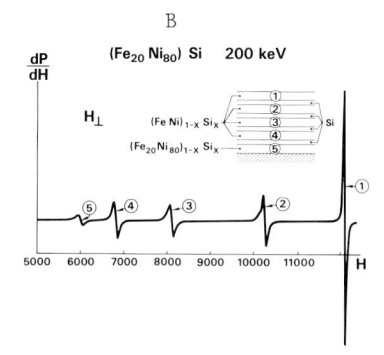

FIG. 8. FMR spectra of the system (30-100) ion mixed with Xe at 200 keV; a) in $H_{//}$ and; b) in $H_\perp$ configuration.

and a second depth region with a larger $4\pi M_s$. Fig. 10 shows $H_\perp$ and $H_{//}$ FMR spectra at LNT after ion mixing at 400 keV. These are typical spectra complying with all the criteria described earlier for an extremely homogeneous layer. The $4\pi M_s$ value of this layer is very close to the value obtained with bulk amorphous materials of the same composition. Table II displays the magnetic parameters of the system (30-100) mixed at 200, 300 and 400 keV. Several ion mixing experiments, performed under the same conditions at LNT and RT, on small rectangular samples coming from the same wafer have shown that: 1) good reproducible results are obtained whatever is the chosen ion mixing energy, and; 2) the temperature of the process has no influence on the final mixed product in the case of system (30-100).

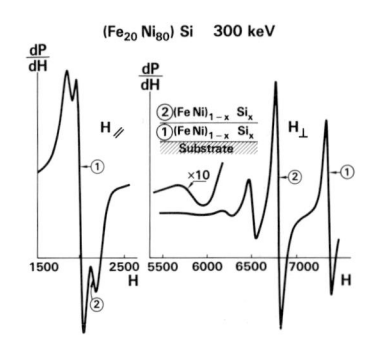

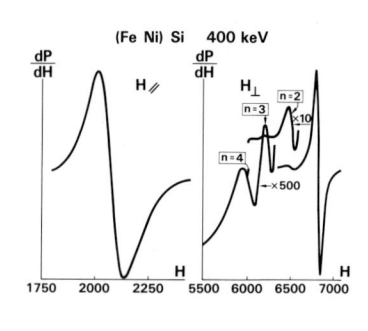

FIG. 9. FMR spectra of system (30-100) ion mixed at 300 keV.

FIG. 10 FMR spectra of system (30-100) ion mixed at 400 keV.

Table II. $4\pi M_s$ and g values of the system (30-100) after 200, 300 and 400 keV ion mixing at LNT.

layer	200 keV $4\pi M_s$	g	300 keV $4\pi M_s$	g	400 keV $4\pi M_s$	g
1	2620	2.066	3500-3800	2.08	3430	2.07
2	3520	2.13	3800	2.10		
3	4740	2.09				
4	6850	2.08				
5	8760	2.09				

CONCLUSION

Here for the first time an amorphous soft magnetic thin film of FeNiSi has been made by an ion mixing technique. The superiority of FMR over RBS and SIMS to characterize the materials has been demonstrated. Indeed, FMR has permitted the analysis of ion mixing effects in a much more sensitive fashion. The twofold interest of this study is to show that: 1) magnetic characterization of mixed magnetic films is an ideal tool for the understanding of ion mixing mechanism; 2) ion mixing can compare favourably with sputtering which has been up to now the best preparation technique for making amorphous magnetic layers. Indeed, ion mixed samples, at any energy, have revealed themselves to have reproducible characteristics.

ACKNOWLEDGMENT

The authors are indebted to H. Sibuet for making the vapor depositions and to A. Soubie for the irradiations.

REFERENCES

1. M. Manheimer, S.M. Bhagat, L.M. Kistler and K.V. Rao, J. Appl. Phys. 53, 2220 (1982)
2. H. Warlimont and R. Boll, J. Mag. and Mag. Mat. 26, 97 (1982)
3. B.X. Liu, W.L. Johnson, M.A. Nicolet and S.S. Liau, Nucl. Inst. and Meth. 209-210, 229 (1983)
4. G. Suran and P. Gerard, M.M.M. Int. Conf. Pittsburgh (Nov. 1983)
5. C. Kittel, Phys. Rev. 110, 1295 (1958)

HIGH-DENSITY CASCADE EFFECTS IN ION-IMPLANTED AG-AU ALLOY

F. R. VOZZO*
Physics Department, State University of New York at Albany, Albany, NY 12222
*present address: Naval Research Lab, Washington, DC 20375

ABSTRACT

High-purity, polycrystalline foils of 12 atomic percent gold Ag-Au alloy were implanted with polyatomic ions of arsenic and antimony at energies of 45 keV per atom. During fluence intervals before and after the steady state, sputtered material was collected on high-purity strips of aluminum foil. Subsequent backscattering analysis of the targets and collectors showed that significant redistribution and segregation occurred in the implanted layer, with relative depletion of silver consistently observed. The results suggested that preferentiality in sputtering is dependent on composition of an alloy and the ion beam used (even at low fluence), but there appears to be no major difference between the redistribution behavior of targets sputtered with atomic ions and equal-velocity molecular ions.

A model is presented which predicts the sputtering behavior and surface configuration of a binary alloy implanted to steady state with a third species.

INTRODUCTION

Certain conditions are known to promote preferential loss from the surface of one or more of the original species in an alloy subjected to sputtering. Sputtering and redistribution ultimately determine the final surface composition and the surface properties of implanted alloys.

Among the parameters on which redistribution effects depend are the amount of energy available for atomic displacements and the rate of production of mobile defects. High-density collision cascades, readily produced by molecular-ion bombardment [1], could be expected to produce changes in preferential sputtering and redistribution behavior accordingly. This paper presents the observed and modelled surface changes in targets of Ag-Au alloy, implanted by a third species to the steady state under conditions of both low and high density of energy in the atomic collision cascades.

EXPERIMENT

High-purity, $\langle 111 \rangle$-textured polycrystalline foils of silver, gold and 12 atomic percent gold Ag-Au alloy were implanted at normal incidence and at room temperature with monomer, dimer and trimer ions of arsenic and antimony at energies of 45 keV per atom. During fluence intervals before and after the steady state of implantation, sputtered material was collected on high-purity strips of aluminum foil mounted on a cylinder centered about the beam spot. Steady state was determined by monitoring the changes in intensity of light emitted by sputtered neutral atoms [2]. All implants were done in ultra-high vacuum (3×10^{-9} torr). The collection and analysis apparatus is shown schematically in figure 1.

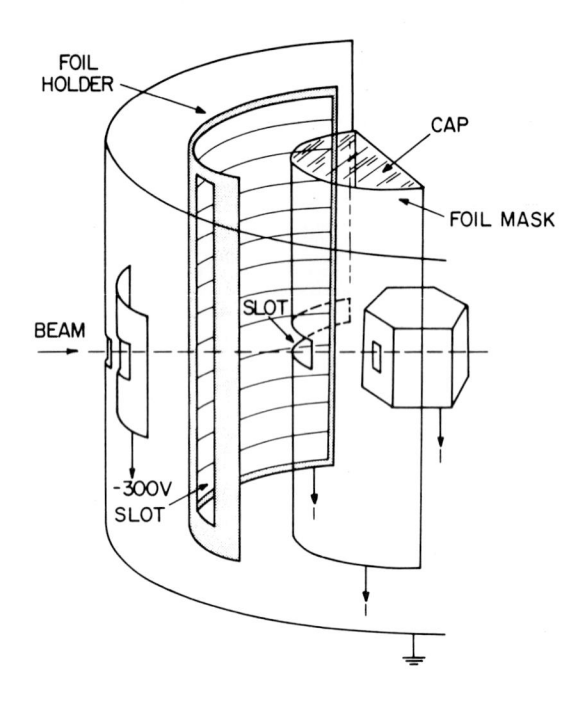

FIG. 1. Geometry of implantation
and sputtered product collection.

Sputtering yields and the relative angular distributions of sputtered atoms were determined from subsequent backscattering analysis of the collector foil strips. Steady-state surface fractions in elemental and alloy targets were determined by backscattering analysis of sputtered and unsputtered regions, using a subtraction technique where necessary.

RESULTS

A sampling of the partial yields of target atoms as a function of fluence for a few of the silver target cases is given in table I. The collection efficiency of the aluminum foil appeared to be inadequate for the determination of the beam atom partial yields. From these results and similar ones for gold and for other fluences, it was evident that high-density cascades (as produced by molecular ion impact) caused significantly less non-linearity of sputtering than had been previously measured [1] . In the limit of low incident mass (eg. As_1) there was good agreement with previous work for both silver and gold targets.

The alloy targets had initial surface composition ratios of 7.33 silver to gold, which were expected to change slightly due to preferential sputtering. At the steady state, stoichiometric sputtering was expected. Table II. shows the dependence on fluence for the composition ratio of the sputtered flux (ratio of partial sputtering yields). The ratios approached 7.33 as fluenced increased, except for Sb_3 and Sb where no change occurred.

TABLE I. Sputtering of elemental silver. Yields are given in number of silver atoms removed per incident atom.

ion	energy (keV)	fluence interval (fraction to s.s.)	Ag yield (± 5%)	Ref. [1] [a]
As	45	0 to 1/3	21	--
As	45	1/1 and up	19	18
As$_2$	90	1/1 and up	26	27
As$_3$	135	3/4 to 1/1	28	60
Sb	45	1/1 and up	19	32
Sb$_2$	90	1/1 and up	40	80
Sb$_3$	135	1/1 and up	47	270
Ar	90	0 to 1/4	6.3	--

[a] S.S. Johar and D.A. Thompson (1979)

TABLE II. Low and high-fluence sputtering of silver and gold from Ag-Au.

ion	energy (keV)	fluence interval	yields (± 5%) Ag	Au	yield ratio
As$_3$	135	1/1 and up	3.7	0.5	7.4
Sb	45	0 to 1/3	19.8	3.17	6.3
Sb	45	1/1 and up	9.76	1.56	6.3
Sb$_2$	90	0 to 1/3	28.6	4.76	6.0
Sb$_2$	90	1/1 and up	13.8	1.95	7.1
Sb$_3$	135	0 to 1/3	50.1	7.60	6.6
Sb$_3$	135	1/1 and up	19.8	3.07	6.4
Ar [b]	90	0 to 1/4	4.7	0.5	9.4
Ar	90	1/4 to 1/2	4.8	0.6	8.0
Ar	90	1/2 to 1/1	4.8	0.7	6.9

[b] all Ar data taken from Ref. [3], M.R. Weller (1980)

The magnitude and favor of preferential sputtering may be expressed as an enhancement factor f given by

$$f = \frac{s_{Ag}}{s_{Au}} / \frac{x^s_{Ag}}{x^s_{Au}} \quad (1)$$

where s_i is the partial sputtering yield of i atoms and x^s_i is the corresponding i-elemental "surface fraction." The x^s_i should be equal to the bulk atomic fractions in the low-fluence limit. An f value greater than one indicates silver sputters preferentially, and table II. shows that initially f was equal to 0.86 for antimony and 1.28 for argon bombardment. The yield ratios 6.3, 6.0, and 6.6 for monomer, dimer, and trimer (respectively) antimony implantation suggest no clear difference in overall preferential sputtering behavior from low and high-density cascades in Ag-Au.

The atomic "surface fractions" for targets implanted to the steady state were measured by backscattering spectroscopy, with the results shown in table III. The low-resolution backscattering surface fractions represent an average over a much greater depth than the ejection depth for sputtering, and thus may not be applicable to equation (1) above.

The relative angular distrubutions of collected silver and gold were used to infer the relative composition of the surface atomic layer, by the qualitative technique of Andersen, et al. [4]. The technique assumes forward-peaking of ejection from subsurface atomic layers. The results consistently showed the alloy surface layer to be slightly richer in gold than the layers below, even for the cases where the partial yield measurements indicated preferential loss of gold.

TABLE III. Alloy steady-state surface atomic fractions.

ion	x_o^s	$\pm .01$ x_{Ag}^s	x_{Au}^s	x_{Ag}^s/x_{Au}^s	s_{Ag}/s_{Au}	f
As$_3$	.17	.70	.13	5.4	7.4	1.4
Sb	.05	.83	.12	6.9	6.3	.91
Sb$_2$	.05	.83	.12	6.9	7.1	1.03
Sb$_3$	.03	.84	.13	6.5	6.4	.98

Discussion

The results generally indicate a strong dependence of preferential sputtering (even at low fluence) on the implanted ion, particularly when comparing argon (negligible retention) with antimony (non-negligible retention). Also, the three methods outlined above for determining the loss (relative) of one element lead to different conclusions; this can depend on whether the target or the flux is examined. Finally, only in one case (As$_3$) was stoichiometric sputtering ever observed. Some of the results may be partly explained by redistribution effects.

Three possible redistribution mechanisms are: (1) Gibbsian segregation = silver-rich surface due to reduction in free energy, (2) recoil implantation = inward flux of silver due to smaller mass, and (3) radiation-induced "de-mixing" [5] = outward flux of gold due to very slight solute size mismatch. That antimony sputters a gold-rich flux while producing a gold-rich or unchanged surface ratio violates mass conservation unless gold is transported from below; such an effect is illustrated in figure 2. The lack of any major difference in the observed effect with higher density of energy deposition, along with the fact that the targets were room-temperature metal, seem to rule out thermal processes such as enhanced diffusion or "spikes." Profiling of the targets with a sputter gun / Auger electron spectroscopy technique is planned.

FIG. 2. Redistribution in binary alloy "A-B" implanted to the steady state with ion "o". Region a corresponds to Gibbsian segregation, region b is the sputtered layer; recoil implantation transports from region c to d. ⟨Rp⟩ is the mean projected range for ion "o". Region f is the undisturbed bulk of the binary alloy.

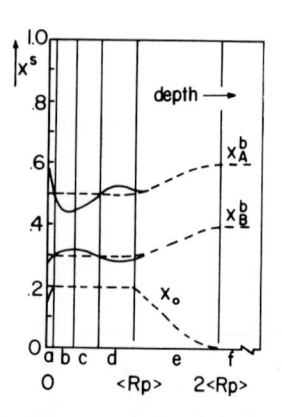

MODEL

The surface binding energy model proposed by Reynolds, et al. [6,7] has been extended to apply to ternary implantation alloys. Matrix binding energies, relative concentrations and displacement-energy partitioning (mass effects) have been incorporated into predictions of preferential sputtering, which allows determination of the steady-state composition if a fluence-independent, constant enhancement factor f (equation 1) is assumed. The assumption of a constant enhancement factor is reasonable if the dilution of the alloy surface by a third species does not result in preferential chemical effects (eg. formation of compounds with only one component). The assumption does not take redistribution effects into account.

The enhancement factor is derived from a sputtering yield expression of the form

$$S_{tot.} = \sum_i x_i (U_i/\overline{U}_i) S_{oi} = \sum_i s_i \quad (i=o,A,B) \quad (2)$$

where S_{oi} is the (elemental) sputtering yield from a pure surface of species i produced by a beam of o ions. The binding-energy ratio $(U_i/\overline{U}_i)$ effectively substitutes a matrix-dependent, quasi-chemical binding energy into each partial yield s_i of the form

$$\begin{pmatrix} \overline{U}_o \\ \overline{U}_A \\ \overline{U}_B \end{pmatrix} = Z_s \begin{pmatrix} U_{oo} & U_{oA} & U_{oB} \\ U_{oA} & U_{AA} & U_{AB} \\ U_{oB} & U_{AB} & U_{BB} \end{pmatrix} \begin{pmatrix} x_o \\ x_A \\ x_B \end{pmatrix} \quad (3)$$

where the U_{ij} are the pairwise matrix bond strengths and Z_s is the surface atom coordination number. From equations 1, 2 and 3 one obtains

$$f = \frac{U_A S_{oA} (U_{AB} x_A^b + U_{BB} x_B^b)}{U_B S_{oB} (U_{AA} x_A^b + U_{AB} x_B^b)} \quad (4)$$

where the x_i^b are the initial (bulk) atomic surface fractions. Applying steady-state boundary conditions,

$$\frac{x_B^s}{x_A^s} = f \frac{x_B^b}{x_A^b}, \quad s_o = 1 \quad (5)$$

the steady-state surface fractions are then derived:

$$x_A^s = \left[\frac{(U_o S_{oo} - Z_s U_{oo} + Z_s U_{oA}) + f(x_B^b/x_A^b)(U_o S_{oo} - Z_s U_{oo} + Z_s U_{oB})}{(U_o S_{oo} - Z_s U_{oo})} \right]^{-1}$$

$$x_B^s = f(x_B^b/x_A^b) x_A^s$$

$$x_o^s = 1 - (x_A^s + x_B^s) \,. \quad (6,7,8)$$

The S_{ij} may be determined from low-fluence measurements; this would incorporate relative energy loss characteristics into the model as well as non-linear sputtering yields produced by the molecular ions. For example, the sputtering of gold by trimer antimony ions produced a yield 2.4 times the yield for monomer antimony, compared with a value of 2.5 for silver (table I.); the fact that silver sputtered more non-linearly than gold would then be incorporated into the predictions.

Comparisons of modelled and experimental results are shown in table IV. The S_{oi} values came from measurements such as those shown in table I. The measured f values came from low-fluence measurements.

TABLE IV. Comparisons between the model and experimental results.

ion	model f	x_o	x_{Ag}	x_{Au}	experiment f	x_o	x_{Ag}	x_{Au}
As_3	0.78	.02	.89	.09	--	.17	.70	.13
Sb	0.56	.02	.91	.07	0.86	.09	.79	.12
Sb_2	0.52	.02	.92	.06	0.82	.05	.83	.12
Sb_3	0.74	.02	.89	.09	0.90	.03	.84	.13
Ar	1.03	--	--	--	1.28	--	--	--

The model was successful in predicting the favor of the preferential sputtering, but all of the values were lower than the experimental f values. The predicted ratios of silver to gold in the steady-state surfaces were all greater than experimental values, indicative of some process not accounted for which tended to bring extra gold to the surfaces (redistribution). The model is in need of further refinement, but the qualitative agreement is at least encouraging.

SUMMARY

Sputtering of Ag-Au alloy by polyatomic arsenic and antimony ions resulted in preferential loss of gold; the loss was apparently not limited to the near-surface region and hence some redistribution must have occurred. Dimer and trimer ion bombardment produced preferential sputtering similar to that of monomer ions; this result combined with very small observed non-linearity in sputtering yields suggest that high-density cascades do not produce significantly different sputtering in polycrystalline metals. Comparisons with argon sputtering suggest that preferential sputtering and redistribution depend on the ion used even at very low fluences; the dependence of recoil implantation on ion mass may contribute.

A model was developed to explain the dependence of preferential sputtering on the implanted species; good qualitative agreement was found.

REFERENCES

1. D.A. Thompson, Rad. Effects 56, 105 (1981).
2. G.W. Reynolds, A.R. Knudson and C.R. Gossett, Nucl. Instr. and Meth. 182/183, 179 (1981).
3. M.R. Weller, Yale University preprint 3074-703 (1982).
4. H.H. Andersen, J. Chevallier and V. Chernysh, Nucl. Instr. and Meth. 191, 241 (1981).
5. L.E. Rehn, Mat. Res. Soc. Symp. Proc. 7, 17 (1982).
6. G.W. Reynolds, Nucl. Instr. and Meth. 209/210, 57 (1983).
7. G.W. Reynolds, F.R. Vozzo, R.G. Allas, P.A. Treado and J.M. Lambert, in Proc. Tenth Int. Conf. on Atomic Collisions in solids (in press).
8. F.R. Vozzo and G.W. Reynolds, Nucl. Instr. and Meth. 209/210, 555 (1983).
9. G.W. Reynolds, F.R. Vozzo, et al., Mat. Res. Soc. Symp. Proc. 7, 51 (1982).

A NEW HAFNIUM-BERYLLIUM SYSTEM PRODUCED BY ION IMPLANTATION AND ANNEALING TECHNIQUES

J.C. SOARES*, A.A. MELO*, M.F. DA SILVA**, E.J. ALVES**, K. FREITAG***, AND R. VIANDEN***
*Centro de Física Nuclear da Universidade de Lisboa, Lisboa, Portugal;
**Laboratório Nacional de Engenharia e Tecnologia Industrial, Sacavém, Portugal; ***University of Bonn, F.R. Germany.

ABSTRACT

Low and high dose hafnium implanted beryllium samples have been prepared at room temperature by ion implantation of beryllium commercial foils and single crystals. These samples have been studied before and after annealing with the time differential perturbed angular correlation method (TDPAC) and with Rutherford backscattering and channeling techniques. A new metastable system has been discovered in TDPAC-measurements in a low dose hafnium implanted beryllium foil annealed at 500°C. Channeling measurements show that the hafnium atoms after annealing, are in the regular tetrahedral sites but dislocated from the previous position occupied after implantation. The formation of this system is connected with the redistribution of oxygen in a thin layer under the surface. This effect does not take place precisely at the same temperature in foils and in single crystals.

INTRODUCTION

Implanting atoms in a solid by bombardment with energetic ions (in the keV to MeV range) results normally in the creation of defects due to the energy loss by nuclear collisions. The high concentration of lattice defects in the region where the implanted ion comes to rest, the annealing behaviour of these defects and the diffusion coefficient of foreign atoms in a lattice are the most responsible parameters for the formation of a stable or metastable solid solution using the ion implantation technology. The use of nuclear techniques to study these solid solutions might be quite appropriate because they give complementary information about the same system. These techniques, like hyperfine interactions and RBS combined with the ion channeling, have shown quite interesting results in recent years. In particular it has been successfully applied to the study of ions implanted in beryllium metal [1,2]. It has been shown [1] that the implantation of beryllium single crystals with hafnium ions produces an interstitial solid solution where the hafnium ions occupy displaced tetrahedral sites in the beryllium lattice. The same results are obtained using beryllium commercial foils but in this case the samples must be annealed at 200 °C after the implantation [3].

In the present work the stability of the hafnium-beryllium system formed by ion implantation has been systematically investigated as a function of the annealing temperature in the range betwen 300 and 600 °C. The information of depth profiles of the implants and lattice location

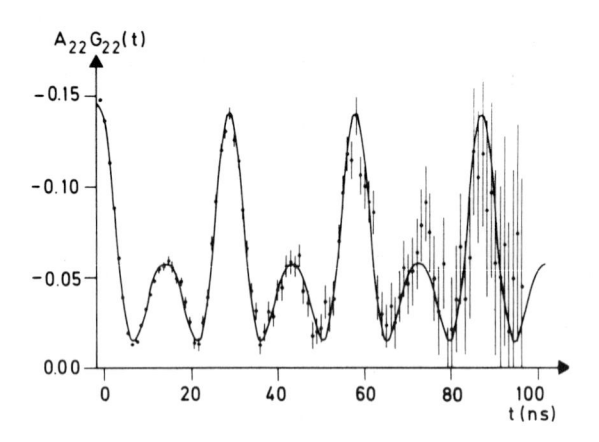

Fig.1 Quadrupole precession of ^{181}Ta in Be measured at room temperature using an Hf implanted foil.

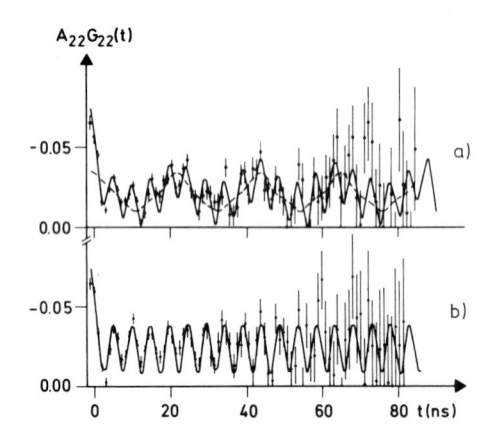

Fig.2 The same as in Fig.1 but measured:
  a) at 450 °C; b) at 500 °C. The new frequency remains after cooling the sample and is almost temperature independent.

A further experiment was carried out using the same isotope ^{181}Hf implanted in a beryllium single crystal which has been annealed in the same manner as the beryllium polycrystalline samples. After the observation of the same new frequency the measurements were repeated with the c axis of the beryllium single crystal oriented in the direction of one detector. The result of this measurement is shown in Fig.3. The EFG has the same direction as the c-axis of the beryllium crystal. This measurement shows also that all the hafnium atoms are in regular equivalent places in the beryllium lattice.

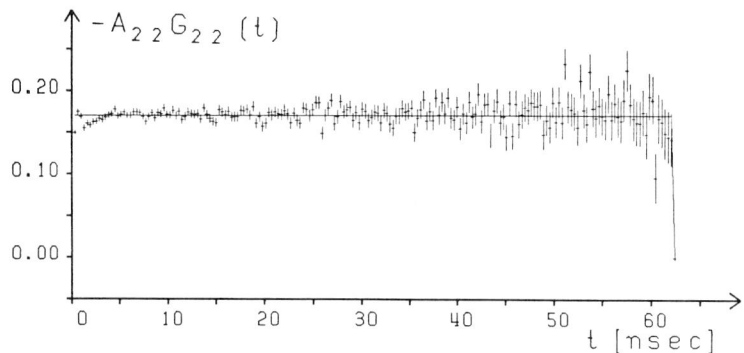

Fig.3 Angular correlation pattern of Ta in Be single crystal;c-axis parallel with the detectors.

Using the RBS-Channeling technique it has been shown [1] that this regular site of hafnium is the tetrahedral one but slightly dislocated. Our results deviate also from the ideal tetrahedral case.

The RBS analyses of the single crystals and of the foils have been very important to decide the mechanism producing the new charge distribution. RBS spectra of a Be single crystal annealed at 570°C taken in the implanted and in the not implanted region are shown in Fig. 4 and 5. These spectra show mainly that Hf does not diffuse into Be at this annealing temperature while oxygen does. In fact, in the implanted region it can be clearly observed that the oxygen peak is broad in a strong contrast with the oxygen peak of the not implanted region. This effect was more clearly observed using a beryllium foil implanted with a very high dose of hafnium. The depth profiles of hafnium and oxygen overlap very well.

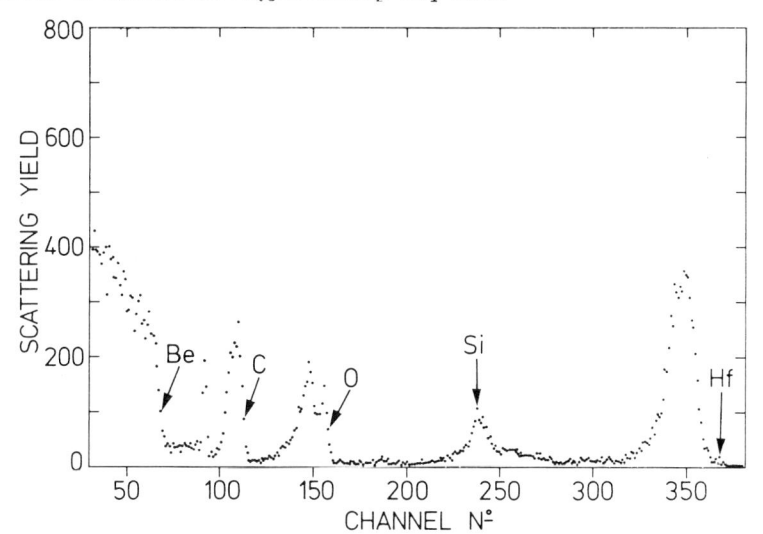

Fig.4 RBS spectrum of a Hf-implanted Be single crystal after annealing at 570 °C. The oxygen peak shows structure.

studies after the annealing of the samples are combined with the results obtained from the application of the quadrupole interaction (QI) technique. In fact, using the TDPAC technique it was possible for the first time to observe the formation of a new metastable system. Samples implanted with high (10^{15} atoms/cm^2) and very high (10^{16} atoms/cm^2) doses of hafnium, annealed at the same temperatures, showed very characteristic effects in the depth profiles of the oxygen peaks. These results can be explained with the formation of a Hf-O dumbbell in the implanted region. This effect has not been observed by other authors using other implanted impurities like Au [4] and Cu [5].

EXPERIMENTAL PROCEDURES AND RESULTS

The beryllium single crystals used in this work were cut approximately perpendicular to the [1010] axis and were carefully electropolished before the implantation.

The 125 μm thick beryllium commercial foils were used without further surface treatment. Using foils from different manufacturers, some differences in the impurity concentrations could be observed. However, the main impurities are accumulated in a thin film at the beryllium surface where the concentration in oxygen is always dominant. The main difference between the use of beryllium foils and of single crystals lies in the impurity content of the bulk material and in the oxide layer at the surface.

The samples were prepared by implantation of 80 keV Hf ions with the Bonn electromagnetic isotope separator. During the implantation the samples were kept at room temperature in vacuum better than 10^{-5} mbar. All the samples were implanted with the ion beam normal to the surface plane so that in the case of the single crystals channeling effect is not excluded. When radioactive samples were needed, the isotopes ^{175}Hf and ^{181}Hf were implanted.

After the implantation of the isotope ^{181}Hf, TDPAC measurements were carried out at 293 K to observe the spin rotation of the 482 keV state of ^{181}Ta [1,3]. This work was done by using the 133-482 keV - cascade of ^{181}Ta populated in the decay of the implanted isotope.

The observation of the characteristic spin rotation pattern shown in Fig. 1 indicates that all the hafnium atoms are in regular sites in beryllium. In fact the same QI frequency, ν_Q, measured for single crystals [1] is observed and the amplitude of the function does not show any damping. Structure in the foil is clear in this measurement since the amplitudes of the three harmonic components are not those expected for pure polycrystalline foils.

Spin rotation patterns of two annealing steps at 450 °C and 500 °C are shown in Fig.2. Clearly a new frequency appears and this new frequency is observed simultaneously with the previous one in the first annealing step. This very high, ν_Q = 1420(14) MHz, well resolved frequency means that the ^{181}Ta nuclei experience a much higher component V_{zz} of the electric field gradient (EFG). This effect must be caused by a new arrangement of the charge distribution surrounding the ^{181}Ta nuclei.

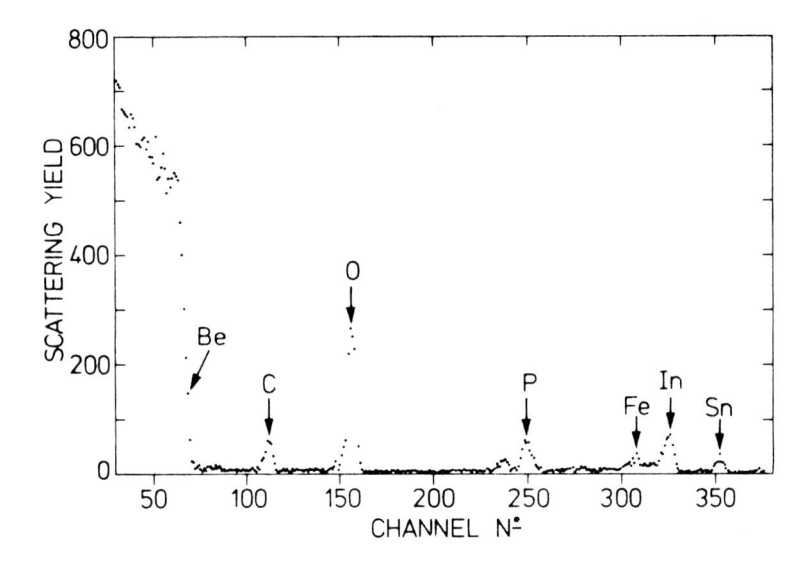

Fig.5 The same as Fig.4 but taken in the not implanted region.

CONCLUSIONS

Combining all the information obtained using these different techniques it is clear that hafnium implanted in beryllium traps oxygen during annealing at temperatures between 300 and 500 °C. This trapping however, forms a highly symmetric three component system where the hafnium and the oxygen keep the cylindrical geometry of the crystal lattice. Since hafnium remains settled in the tetrahedral site, the trapped oxygen atom must also occupy the nearer tetrahedral site in the c direction. Surprisingly, this system can be prepared in quite a pure state at an annealing temperature of 500°C using beryllium foils.

The most important information, however, one can get from the experiment shown in Fig. 3, is related with the geometry of the Hf-O-Be system. The most probable configuration is a Hf-O dumbbell connecting two adjacement tetrahedral sites.

The comparasion of our results with those published for Au [4] and Cu [5] implanted in beryllium shows that only in the case of the implantation of hafnium does the diffusion of oxygen in beryllium occur. This diffusion takes place through the bulk material where the hafnium atoms are located. Also the analysis of the samples implanted with very high dose does not show any diffusion of the Hf atoms like in the case of Au and Cu. Further details of these measurements will be published (6).

AKNOWLEGMENTS

The authors thank Prof. E. Bodensted for stimulating discussions.
This work was performed with support from JNICT under contract n° 426.82.83.
Financial support by the Alexander von Humboldt Foundation (J.C.S.) and the DAAD is acknowledged. The work at Lisbon was performed with equipment donated by the Federal Republic of Germany (GTZ).

REFERENCES.

1. E. N. Kaufmann, K. Krien, J. C. Soares and R. Freitag, Hyp. Int. $\underline{1}$, 485 (1976).
2. R. Vianden, E. N. Kaufmann and J.W.Rodgers Phys. Rev. B22(1980)63
3. K. Krien, J. C. Soares, K. Freitag, R. Tischler, G. N. Rao and H. G. Muller Phys. Rev. $\underline{B14}$, 4782 (1976).
4. S. M. Myers and R.A.Langley, J. Appl. Phys. $\underline{46}$, 1034 (1975).
5. S. M. Myers, S.T. Prevender, Phys. Rev. $\underline{B9}$, 3953 (1974).
6. J. C. Soares, A. A. Melo, M. F. Da Silva, D. Freitag, C. Herrmann, P. Herzog, H.J. Rudolph, K. Schoesser, R. Vianden, U. Wrede and D. O. Boerma, to be published.

PART III

SEMICONDUCTORS

ION INDUCED MORPHOLOGICAL INSTABILITIES IN GE*

B. R. APPLETON
Solid State Division, Oak Ridge National Laboratory, Oak Ridge, TN 37831

ABSTRACT

   The characteristics of an anomalous morphological instability initiated in amorphous Ge by heavy ion bombardment are reviewed and a model based on defect-production/defect-interactions is proposed.

INTRODUCTION

   A series of recent measurements have shown that heavy ion irradiation of Ge near room temperature induces a drastic morphological change in the ion bombarded surface [1-7]. For moderate doses ($<10^{15}$ ions/cm^2) the near surface of Ge single crystals are turned amorphous as with ion bombardment of Si. However, as bombardment continues ($>10^{15}$ ions/cm^2) a morphological instability is initiated in the amorphous phase of Ge which results in the formation of surface craters several hundred nanometers deep. This effect differs significantly from related defect-driven phenomena such as radiation-induced segregation, precipitation or phase instabilities in that those phenomena usually occur in binary or ternary alloys where preferential coupling of certain alloying elements to defect fluxes exist, or where the number of alternate phases available provide a variety of interaction channels for decay of the disordered system [8]. The Ge effect is an instability that occurs entirely in a single phase system, namely, amorphous Ge. Also the drastic nature of the morphological alteration has no direct parallel.
   In this paper we review experimental results which characterize this effect. Results published elsewhere are referenced but not reviewed in detail to allow space for speculation on the defect and materials interactions which could account for the observed morphological alterations.

CHARACTERISTICS OF THE EFFECT

   High purity Ge(100) and (111) single crystals from several different suppliers were implanted with He, Si, Ge, In, Sn, Sb, Tl, Pb, and Bi at doses ranging from 10^{13}-10^{17} ions/cm^2 and at Ge sample temperatures from ~10-800 K. Subsequent analysis included ion scattering, ion channeling, nuclear reaction analysis, optical and transmission electron microscopy, ESCA, SIMS, SEM, and surface profilometry measurements. Some implanted samples were exposed to air before any analysis, while others were analyzed in situ by ion beam analysis techniques to deduce the effects of air exposure [1-7].
   The anomalous behavior of Ge was discovered during studies comparing implantation and annealing effects in Ge and Si [1]. When Si is implanted with virtually any species to doses from 10^{14}-10^{17} ions/cm^2 the near surface of the single crystal is turned amorphous. The thickness of the amorphous layer is less for 300 K implantation compared to 70 K due to increased defect mobility and recombination at the higher temperature. Not surprisingly, there is no loss of dopant other than that from normal sputtering at either temperature.

----

*Research sponsored by the Division of Materials Sciences, U.S. Department of Energy under contract W-7405-eng-26 with Union Carbide Corporation.

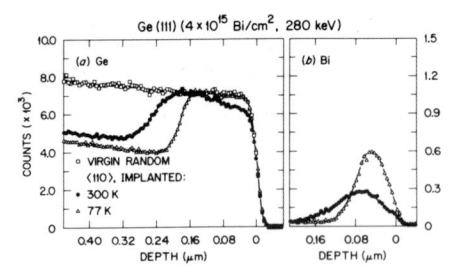

Fig. 1. Ion scattering/channeling analysis of Ge(111) implanted at 77 and 300 K.

In contrast, when Ge(111) is implanted with 280 keV, 4×10^{15} Bi/cm^2 at 300 and 70 K there are numerous differences from Si as Fig. 1 shows. After implantation the implanted Si (amorphous) surface has a slight milky appearance for both implantation temperatures. The Ge sample implanted at 77K has this milky appearance while the 300 K implant has a mat, black finish. Ion scattering/channeling analysis of the Ge (Fig. 1), after exposure to air, shows an amorphous surface layer that is apparently thicker for implantation at 300 K than 77 K, and the 300 K implant has a huge yield deficit in the near surface region. The implanted Bi distribution shows an altered depth profile and significant loss of dopant for the 300 K implant.

These differences are all related to the severe crater formation in the Ge surface. In the sections below some of the experimental characteristics of this process are reviewed in an effort to bound the possible origins of the effect.

Near-Surface Yield Deficit and Impurity Incorporation

One of the more striking differences between the Si and Ge results for implantations at 300 K is the huge near-surface yield deficit which is present in implanted Ge, such as in Fig. 1, but which is completely absent in Si and unimplanted Ge. A similar deficit in the scattered ion yield from a Ge crystal implanted with 40 keV, 3×10^{15} Pb/cm^2 was reported by Campisano et al. [9] and attributed to changes in the scattering and/or stopping cross sections due to the presence of the implanted Pb impurity. A series of measurements and calculations have shown that this is certainly not the explantation for the yield deficit observed in the present results [1,3-6]. The definitive experimental result is shown in Fig. 2 where "implanted impurities" were eliminated by implanting Ge$^+$ into Ge. The ion scattering spectra in Fig. 2 were taken under identical random conditions from a Ge single crystal half of which was implanted, the other half unimplanted

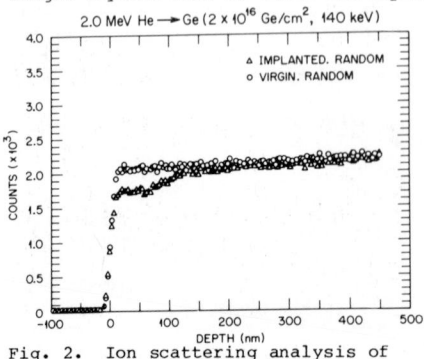

Fig. 2. Ion scattering analysis of virgin and Ge$^+$ implanted Ge.

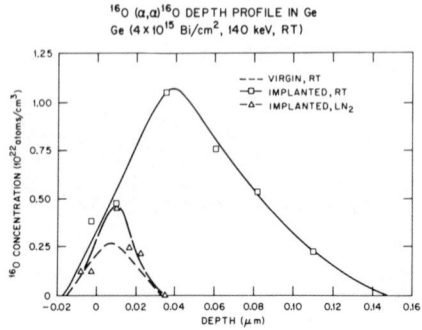

Fig. 3. Nuclear reaction analysis of the concentration of ^{16}O vs depth after Bi implantation in Ge.

(virgin). The implanted area was black in appearance and, after exposure to air, the ion scattering analysis shows a pronounced yield deficit. Since no impurities were implanted in this exposure this cannot be the origin of the yield deficit.

Nuclear reaction techniques were used to analyze a variety of virgin and implanted Ge single crystals for low mass impurities incorporated by some means other than implantation [1-3]. Results are reproduced in Fig. 3 where the $^{16}O(\alpha,\alpha)^{16}O$ nuclear resonance was used for a quantitative analysis of the oxygen concentration versus depth for virgin and Ge$^+$ implanted Ge. All Ge samples were exposed to air before analysis and scattered ion energies were converted to depth assuming standard stopping powers for He$^+$ in Ge. A similar reaction [$^{12}C(H,H)^{12}C$] was used to depth profile C in these same Ge samples and distributions comparable to those for O were detected [3]. The analysis in Fig. 3 shows about 2 monolayers of air-formed oxide on the surface of the virgin crystal and the equivalent of ~33 monolayers apparently incorporated over about 0.14 μm in depth for the room temperature implant. This corresponds to about 30 oxygen atoms per incident Ge$^+$ and, considering that a comparable amount of carbon is also present, it is difficult to conceive a mechanism where residual gas in 10^{-8} Torr vacuum could be incorporated to such depths with such efficiency. This skepticism was confirmed by performing room temperature implants with a variety of heavy ions including Ge$^+$ in vacuums ranging from 10^{-5} to 10^{-9} Torr, and analyzing the samples in situ using ion scattering/channeling. All samples apeared black after implantation, however, ion scattering showed little or no evidence for a near-surface scattering deficit. Upon prolonged exposure to air, subsequent analysis showed the reappearance of the yield deficit. This is detailed further in a later section.

Temperature Dependence and Microstructure

The microstructural relationships between crater formation and the experimental parameters were detailed using correlated ion scattering/channeling and cross sectional transmission electron microscopy. The dose dependence can be seen from the series of micrographs in Fig. 4 for implantation of 120 keV In$^+$ into (111)Ge at room temperature. Up to a

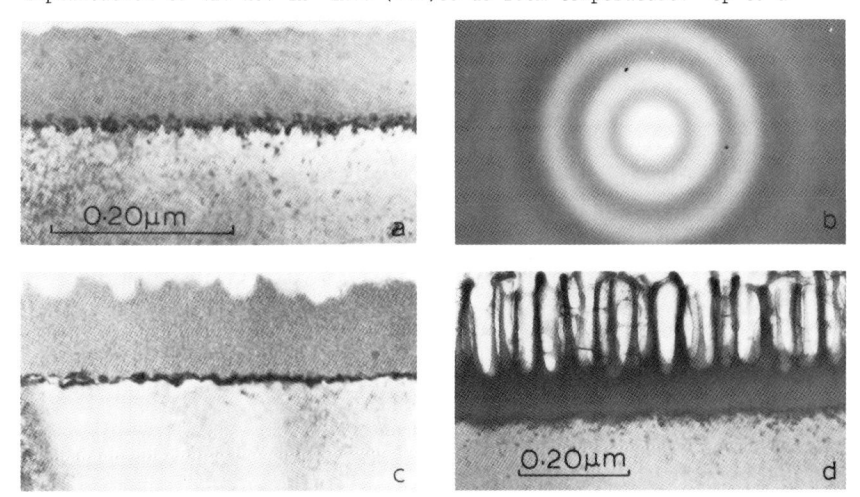

Fig. 4. Cross-sectional TEM micrographs of Ge(111) implanted at room temperature with 120 KeV In to a total dose of (a) 1 x 10^{15}, (c) 2 x 10^{15}, (d) 5 x 10^{15} In/cm^2; (b) microdiffraction pattern from region (a).

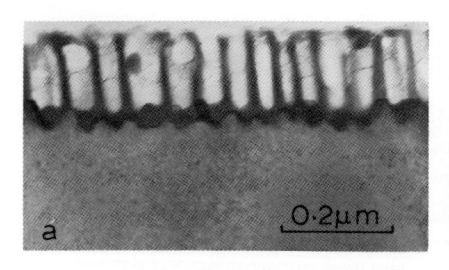

Fig. 5. Cross-section micrographs of Ge(111) implanted at (a) 300°C and (b) 350°C with Sn (100 keV, 5 x 10^{15} cm^{-2}).

fluence of 1 x 10^{15} In/cm^2 the implanted regions have a milky appearance; ion scattering/channeling analysis indicates an amorphous surface layer about 95 nm wide with no evidence of a yield deficit; and cross-sectional TEM (Fig. 4a) confirms that implantation to this dose has resulted in an amorphous surface layer from the single crystalline substrate separated by a narrow band (~15 nm) of dislocations. The surface and interface regions are planar with undulations < 5 nm, and microdiffraction from the surface layer shows characteristic amorphous rings corresponding to the first- and second-nearest neighbor distances in Ge. These results are analogous to those obtained for imlantation in Si. However, by 2 x 10^{15} In/cm^2 a darkening of the implanted Ge occurs and the corresponding micrograph (Fig. 4c) shows the presence of surface craters ~50 nm deep and the crystalline-amorphous interface has undulations similar to the surface craters. Following a total dose of 5 x 10^{15} In/cm^2 a drastic change in the surface morphology occurs as Fig. 4d shows. Regularly-spaced, columnar voids have developed which intersect the surface and extend ~170 nm into the amorphous layer. An amorphous region extends another 85 nm below this to the c-a interface. This unique structure has evolved within the amorphous phase of Ge and microdiffracion techniques showed only the amorphous phase present in the cratered region.

The strong suppression of this effect at low temperatures was demonstrated by ion scattering/channeling and TEM results [3] which revealed no evidence of crater formation for implantation of any species to fluences > 10^{17} ions/cm^2 into Ge crystals held at liquid nitrogen temperature. These low temperature implants produced only a simple amorphous-layer/dislocation-band structure. In contrast, crater formation is enhanced for elevated temperature implants. The micrographs in Fig. 5 show a cross-sectional view of high temperature, 300 and 350°C, implants of 125-keV, ^{120}Sn ions into (111) Ge to a fluence of 6 x 10^{15} cm^{-2}. The micrograph of the 300°C implant shows a 140-nm cratered region followed by a 25-nm amorphous layer. Structure like the columnar fingers which bound the void regions in the RT implants (see Fig. 4d) can easily be identified, but the region in between is no longer completely void of material. Microdiffraction analysis indicated that material in both regions is amorphous. The contrast in the micrograph between the regions is apparentely due to density variations in the amorphous phase. The less dense region, between the column structure, may contain a high density of vacancy clusters. The lattice of the 350°C implanted sample has remained crystalline, as seen in Fig. 5. The near-surface region (90 nm) is fairly free from damage with a wide band of dislocation loops (40 nm) behind. There is no evidence of crater or void formation.

The suppression of crater formation under ion irradiation at low temperatures and its enhancement at temperatures below 350°C suggests that "defect mobility" in the amorphous phase of Ge plays an important part in initiating the craters.

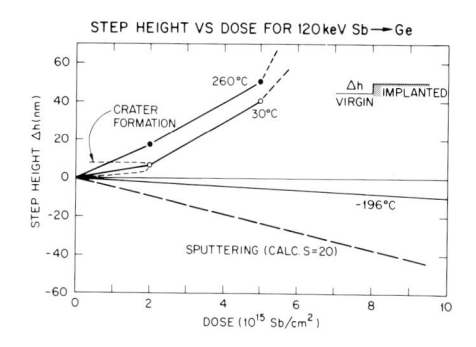

Fig. 6. Step height vs dose for Ge ion implanted at various temperatures.

Surface Profilometry

Additional chracteristics of the crater formation can be deduced from measurements of surface swelling and the associated TEM analyses. When Ge is bombarded with heavy ions a variety of synergistic effects occur which change the height of the surface relative to the original (unimplanted) surface. The implanted surface expands upon going from the crystalline to amorphous phase; it expands due to incorporation of implanted ions (a negligible effect); it contracts due to normal sputtering; and, as we shall see, its height increases due to crater formation. These cumulative effects were assessed by measuring the step height changes Δh relative to the unimplanted surface, and results are shown plotted in Fig. 6 for 120 keV Sb^+ implanted in Ge at various doses and temperatures. The line labeled sputtering was calculated assuming a sputtering yield of 20 Ge atoms per Sb^+ and it represents the maximum decrease expected from sputtering alone. Since only an amorphous surface layer is formed for liquid nitrogen implants, the data represented by the line labeled -196°C show the difference between surface swelling, due to the a-c phase change, and recession, due to sputtering. The remaining data show the changes in Δh observed when craters are formed. These results show the Δh increases rapidly with dose and temperature (for T $\lesssim$ 325°C), and from correlated TEM analysis it was found that the critical dose for crater formation varied roughly as the dashed line labeled crater formation in Fig. 6. Initially, for low doses an amorphous surface layer is formed over approximately the range of the incident ions. With increasingly higher doses, craters form. Since some ions bombarding a crater-laden surface experience no collisions until they reach the bottoms of the craters, a new amorphous layer is formed beneath the craters. Consequently, as the surface becomes more porous the amorphous/cratered region extends further into the surface. This is why altered surface layers in the micrographs of Fig. 4 can extend several times the ion range.

Loss of Implanted Dopant

Another consequence of crater formation, already eluded to in discussing Fig. 1, is the loss of the implanted dopant [1,3,4,6]. In Fig. 1 the retained Bi in the liquid nitrogen temperature implant (no craters) is the same as the measured implantation dose corrected for sputtering. However, in the equivalent room temperature implant (craters) 20-30% of the Bi is lost. Evaluation of dopant retention in room temperature implantations of ion species ranging from Si to Pb showed that, depending on conditions, as much as 80% of the dopant could be lost [1,3-6]. In all cases the loss of dopant in surfaces where craters formed was considerably greater, 2 to 15 times, that calculated for sputtering from a normal Ge surface.

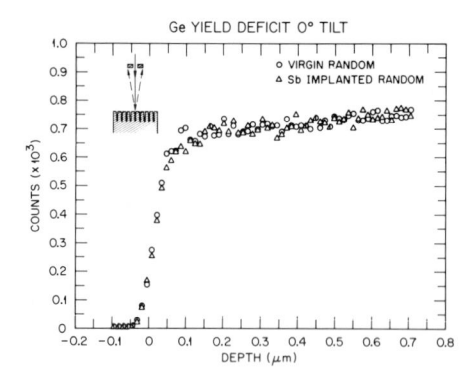

Fig. 7. Ion scattering analysis of implanted Ge surface with craters (see text).

An anomalous loss of dopant from Ge implanted at room temperature and 300°C and then thermally annealed was reported earlier by Johansson et al. [10]. They did not observe the yield deficit, crater formation, etc., however, this must be recognized in their interpretation of the measurements. They found that appreciable dopant loss started at annealing temperatures of approximately 350, 390, and 560°C for Hg, Tl, and Bi, respectively implanted at room temperature; and at about 590°C for Bi and Tl implanted at 300°C. Their implanted doses ranged from 2-6 x 10^{14} ions/cm^2. They concluded that the loss of dopant was due to preferential diffusion toward the surface followed by out diffusion, thermal etching or both.

From our measurements it is clear that craters form with increasing efficiency for implants from room temperature to 300°C so some craters were probably present in all the samples of Johansson et al. prior to annealing. Because of the increased surface area thermal diffusion should transport large amounts of the implant to a free surface. It should also be noted from our measurements, such as those in Fig. 1, that a significant loss of dopant occurred after ion implantation at room temperature and no annealing. However, our doses were usually > 10 times those of Johansson et al. (i.e., more craters and increased surface area) and it is possible that the nature of the cratered surface gave rise to poor heat conduction and thus surface temperatures above room temperature. A possible explanation of the dopant loss mechanism is offered in the next section.

Origins of the Yield Deficit and Dopant Loss

Correlation of the measurements discussed in the previous sections shows that the near-surface yield deficit, and the presence of low mass impurities are observed only when surface craters are present, and only when the implanted sample has been exposed to air prior to ion beam analysis. For example, the ^{16}O concentration retained in the Ge sample of Fig. 3 which was implanted at liquid nitrogen temperature is only slightly greater than for the virgin sample because no craters were formed in the low temperature implant, and the room temperature implant showed the large oxygen concentration only after prolonged exposure to air. The suggested model to explain the yield deficit and the apparent incorporation of low mass impurities is that once a heavily cratered surface is exposed to the air, atmospheric gases are adsorbed onto the crater surfaces. The overall increase in adsorbed impurities is due to the greatly increased surface area. The appearance that the low mass impurities are distributed in depth and the observed yield deficit are both due to the interplay of the ion scattering and stopping processes for ions penetrating through the honeycomb-shaped craters at an angle.

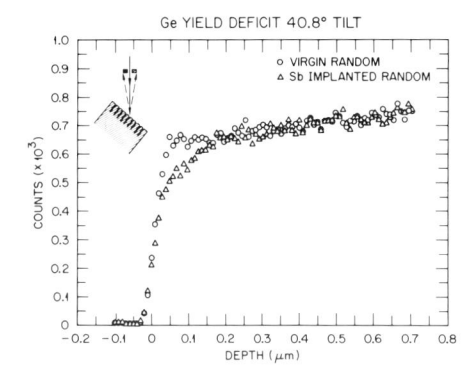

Fig. 8. Ion scattering analysis of implanted Ge surface with craters (see text).

This effect can be illustrated by comparing the two ion scattering measurements in Fig. 7 and Fig. 8 which were designed to exploit the geometry of the surface cratering effect. Both spectra were recorded for ions backscattered 180 ± 0.75° from a heavily cratered surface (90 keV, 6 x 10^{15} Sb/cm^2) which had been exposed to air. The geometry used to obtain the spectra in Fig. 7 was with the well-collimated (0.01°) ion beam incident normal to the surface craters as shown by the illustration on the figure. In this arrangement the incident ions backscatter parallel to the axes of the craters from a surface which appears in projection as a simple oxygen covered surface. Consequently, random spectra from the virgin and Sb implanted (cratered) surface appear the same with no evidence of a yield deficit. (This scattering geometry is unlike all others reported in this paper. The usual arrangement has the detector located at scattering angles from 140°-165° and the ion beam is usually incident in a non-normal direction.) The same conditions were used to obtain the spectra in Fig. 8 except the sample was tilted 40.8°. In this geometry the incident ion beam penetrates many craters and alternately scatters from an oxygen covered crater surface, pure Ge in the crater wall, etc. This gives the appearance in the ion scattering and nuclear reaction analysis spectra of low mass impurities distributed in depth in the Ge, and accounts for the near-surface yield deficit in the Sb implanted spectrum of Fig. 8 as well as the extended oxygen distribution in Fig. 3.

Dopant loss from Ge samples implanted at room temperature and 300°C has been observed for In, Sn, Sb, Tl, Pb, and Bi [1-6] and for Hg, Tl, and Bi [10]. All these specie induce crater formation and all have fairly low melting temperatures between -38 and 330°C, except Sb whose melting temperature is 630°C. It is conceivable that dopant loss occurs when the implanted impurities are transported to the surfaces of the craters during irradiation, either by thermal diffusion or some ion induced segregation process, and that the thin crater walls rise in temperature enough to have significant dopant loss in vacuum. This mechanism could account for both our results [1-6] and those of Johansson et al. [10]. An alternate or additional loss mechanism which could account for our results but not necessarily those of Johansson et al. is increased sputtering by the implant beam of dopants segregated to the crater surfaces. This hypothesis is currently being tested.

Pulsed Laser and Thermal Annealing Characteristics

The effects of surface craters and incorporated impurities retard solid phase epitaxial regrowth by thermal annealing as well as liquid phase recrystallization by pulsed laser melting [5]. These annealing characteristics have been discussed in detail elsewhere [1,3-7].

MODEL CONSIDERATIONS AND DISCUSSION OF RESULTS

The central remaining unknowns are the mechanisms responsible for inducing the morphological instabilities in amorphous Ge which result in a spatially periodic array of craters. Given the existance of craters, reasonable causes for the near-surface impurity incorporation and yield deficit, the thermal and laser annealing behavior, and even the loss of dopant can be identified; but these observations are all secondary to the crater formation mechanisms.

There is strong circumstantial evidence that the driving force for the instability is purely a defect-production/defect-interaction mechanism. The first supporting observation is that the instability is initiated in a single phase system. Thus, there are no obvious thermodynamic (chemical, metallurgical, etc.) driving forces which serve to nucleate the instability. In alloy systems the preferential nucleation of a phase of one of the alloying constituents can interact with the defect flux to form a different phase, or in a two phase system the defect fluxes can change the free energy of the system in favor of one phase [8]. As an example, Maksimov et al. [11] postulate that the stoichiometric inbalances induced in III-V compounds by ion implantation result from decomposition of the compound due to its low formation enthalapy, the loss of the more volatile constituent, and surface segregation of the other. However, in our case the instability appears to be initiated in the amorphous phase of Ge and the final cratered surface is, as best we can determine from microdiffraction techniques, still in the amorphous phase. This suggests some sort of periodic density variation arising from the interactions of the defect production and annihilation rates alone.

The second piece of evidence in favor of defects is the strong mass dependence of the effect which can be interpreted as a defect-production-rate effect. Systematic studies of crater formation at room temperature showed that craters were not formed by light implants such as He^+ and Si^+ but were formed with increasing efficiency by ions heavier than Ge^+. This suggests that some critical rate of defect production, which is greater in the damage cascades of heavy ions compared to light ions, may be necessary to nucleate the instability. We also observed, as did Wilson [2], that the craters formed normal to the surface independent of the angle of incidence of the bombarding ions.

Thirdly, the strong temperature dependence of the effect is reminiscent of defect phenomena. Crater formation is insignificant below 77 K but increases rapidly with temperature until near the a-c phase transition temperature (~325°C) in Ge. Above this transition temperature dynamic annealing of the damage from each cascade occurs, the amorphous phase never forms and neither do the surface craters. This is additional evidence that the amorphous phase is essential for nucleation of the instability and suggests that increased defect mobility also leads to accelerated nucleation.

A possible sequence of events leading to the observed ion induced morphological instability could be the following. As we have demonstrated, it is first necessary to produce the amorphous phase of Ge by ion bombardment. Additional irradiation creates "defects" in the amorphous phase which become increasingly mobile with temperatures up to the a-c transition temperature. Drawing an analogy with defects in crystalline solids, we can assume that the mobile defects are free to recombine and annihilate in the amorphous phase. For low mass ions (low defect production rates) these defect elimination mechanisms are in some sort of balance with the production rates and the amorphous phase is homogeneously preserved. For sufficiently heavy ions a critical defect production rate is exceeded and spatially periodic clustering of defects occurs in the amorphous phase. This evolves into a periodic density variation which leads to crater formation.

The speculations for this model description draw heavily on concepts proposed by Martain [12] in 1975 to explain phase instabilities produced in crystalline metals during ion irradiation. Martain argues that the energy introduced into materials systems in the form of ion induced defects can be dissipated in highly irreversible processes and can draw the irradiated solid far from its equilibrium state. He shows [12] that above a critical defect production rate g* the uniform point defect production rate in crystals may become unstable with respect to spatial point defect fluctuations, and this can lead to a spatially periodic point defect clustering. Although these general concepts may be useful in analyzing our results, it should be emphasized that the drastic morphological alteration we observed in Ge is significantly different than a spatially periodic point defect clustering. It is not clear that many of our observations can be accommodated in the same formalism.

One complication in modeling the interactions leading to a periodic density variation arises because we are dealing with defects in the amorphous phase. Defects in amorphous materials are little studied and poorly understood compared to crystalline solids. It would appear from the results in Fig. 5a that the assumption of a spatially periodic density variation as a precursor to crater formation is a good one. The contrast variation in the micrograph in Fig. 5a indicates, we believe, an altered surface layer with dense amorphous crater walls bounding less dense amorphous Ge in the regions which will become open craters. Detailed microdiffraction analysis of individual filled craters has shown what could be clusters of voids in these less dense regions. This suggests that during ion irradiation mobile vacancies agglomerate, form voids, and these voids coalesce to form the denuded areas in Fig. 5a. Annealing stages in crystalline Ge attributed to vacancy and multivacancy mobility have been identified for temperatures above liquid nitrogen temperature, and the defects associated with this stage were stable to 90-120 K [13]. If similar defects and mobilities are postulated for amorphous Ge, a qualitative description of the strong temperature dependence of the crater formation is possible.

The defect/materials interactions least understood at this stage of the investigation are those responsible for stabilizing the voids into a periodic array with the particular dimensions of the observed craters, and those which explain how such an array can erupt into a drastic morphological state such as craters. Given the relatively large spacings between the denuded craters it seems more probable that elastic interactions similar to those responsible for stabilized voids in a void lattice [14], rather than special electronic interactions are responsible for the stability. This would require a description of elastic interactions in an amorphous medium. A modified description analogous to spinodal decomposition or lamellar eutectics may be possible also [8]. Impurities or a second metastable phase could serve to nucleate and stabilize the defects produced during irradiation. Since we used ultrapure Ge and since the craters formed using a wide range of low-dose bombarding species including Ge$^+$, it is unlikely that impurities play a dominant role. Two amorphous states of Si with equivalent densities but different dangling-bond states have been identified by Hubler et al. [15] and Donovan et al. [16], using differential scanning calorimetry to measure the enthalpy of crystallization in Si and Ge made amorphous by LN$_2$ implants, observed that amorphous Ge relaxed continuously to an amorphous state of lower free energy, in contrast to Si, before crystallization started. These observations leave open the possibility that a second amorphous phase of Ge with lower free energy (and possibly less dense) becomes available under ion irradiation and that nucleation of this phase could serve as a sink for mobile vacancies. Several of these possibilities are presently being tested.

ACKNOWLEDGMENTS

I would like to acknowledge the collaboration of my colleagues O. W. Holland and J. Narayan throughout the course of these studies.

REFERENCES

1. B. R. Appleton, O. W. Holland, J. Narayan, O. E. Schow III, J. S. Williams, K. T. Short, and E. M. Lawson, Appl. Phys. Lett. 41, 711 (1982).
2. I. H. Wilson, J. Appl. Phys. 53, 1698 (1982).
3. O. W. Holland, B. R. Appleton, and J. Narayan, J. Appl. Phys. 54, 2295 (1983).
4. E. M. Lawson, K. T. Short, J. S. Williams, B. R. Appleton, O. W. Holland, and O. E. Schow III, Nucl. Inst. Methods 209/210, 303 (1983).
5. O. W. Holland, J. Narayan, C. W. White, and B. R. Appleton, p. 297 in Laser-Solid Interactions and Transient Thermal Processing of Materials, ed. by W. L. Brown, J. Narayan, and R. A. Lemons, North Holland, New York, 1983.
6. E. M. Lawson, J. S. Williams, D. J. Chivers, K. T. Short, and B. R. Appleton, Australian Atomic Energy Commission Report AAEC/E573.
7. J. S. Williams, D. J. Chivers, R. G. Elliman, S. T. Johnson, E. M. Lawson, I. V. Mitchell, K. G. Orrman-Rossiter, A. P. Pogany, and K. T. Short, these proceedings.
8. Phase Transformation During Irradiation, ed. by F. V. Nolfi, Applied Science Publishers, New York, 1983 and references therein.
9. S. U. Campisano, M. G. Grimaldi, P. Baeri, G. Foti, and E. Rimini, Appl. Phys. Lett. 22, 201 (1980).
10. N.G.E. Johansson, D. Sigurd, and K. Bjorkqvist, Radia. Eff. 6, 257 (1970).
11. S. K. Maksimov, V. L. Egorov, V. V. Kryuk, D. I. Piskunov, E. A. Pitirimova, and V. F. Veselov, Phys. Stat. Sol. (a) 73, K283 (1982).
12. G. Martain, p. 1084 in Fundamental Aspects of Radiation Damage in Metals, CONF-751006-P2, U.S. Department of Commerce, ed. by M. T. Robinson and F. W. Young, 1975. Also, Phil. Mag. 32, 615 (1975).
13. J. Bourgoin and F. Mollot, Phys. Stat. Solidi 43, 343 (1981). T. A. Callcott and J. W. MacKay, Phys. Rev. 161, 698 (1967).
14. A. M. Stoneham, J. Phys. F 1, 778 (1971) for example.

THE PRODUCTION OF POROUS STRUCTURES ON Si, Ge AND GaAs BY HIGH DOSE ION IMPLANTATION.

J.S. WILLIAMS*, D.J. CHIVERS*[†], R.G. ELLIMAN*[φ], S.T. JOHNSON*, E.M. LAWSON**
I.V. MITCHELL*[θ], K.G. ORRMAN-ROSSITER, A.P. POGANY*, AND K.T. SHORT*
*Microelectronics Technology Centre, RMIT, Melbourne 3000 Australia;
**AAEC Lucas Heights, Australia; † Present address: AERE Harwell, England
φJoint appointment with CSIRO Division of Chemical Physics, Clayton 3168
Australia: θ Permanent address: CRNL Chalk River, Canada.

ABSTRACT

This paper presents new data on the previously observed porous structures which can be developed in high dose, ion implanted Ge. In addition, we provide strong evidence to suggest that such porous structures can be formed in high dose, ion implanted Si and GaAs substrates under particular implant conditions. Comparison of the various systems using RBS analysis indicates that heavy ion doses as low as $10^{14} cm^{-2}$ can give rise to such structural modifications in GaAs, whereas doses of $10^{15} cm^{-2}$ are needed to observe an effect with Ge and doses usually exceeding $10^{16} cm^{-2}$ are required for Si.

INTRODUCTION

Recent studies [1 - 4] have shown that high dose implantation of heavy ions into Ge can lead to near-surface structural modifications which result in the formation of highly porous surface layers under certain implantation conditions. Such porous layers have been identified by both TEM [1,4] (where the porous layer is observed to be essentially amorphous) and SEM [2]. Furthermore, the surfaces of these specimens appear black to the naked eye, even while the samples are still under vacuum in the implantation chamber. Upon exposure to air, the porous structure is found to absorb high concentrations (>20 atom % in some cases) of light impurities such as oxygen and carbon. These porous, contaminated surface layers give rise to a Ge yield deficit in Rutherford backscattering spectra [1,3] and this yield deficit extends over a depth corresponding to the thickness of the implanted layer. Studies to date [1,3,4] have indicated that the degree of porosity is sensitive to the implant temperature, the implant mass and dose. These porous, contaminated layers cannot be recrystallised to good quality single crystal during subsequent furnace heating. In a recent report [5] we have indicated that the development of porous structures during high dose implantation of Si may also occur under certain bombardment conditions. No such effects have been reported in GaAs studies.

In this paper, we report on further experiments in Ge, designed to provide more detail on the temperature dependence of implantation-induced structural modifications and to investigate the subsequent annealing behaviour. We provide strong evidence to suggest that porous, amorphous structures can also occur in both Si and GaAs substrates during high dose implantation, under appropriate conditions.

EXPERIMENTAL

In the present study, p-type (100) Ge slices were implanted with 80 keV Sb ions to doses of $5 \times 10^{15} cm^{-2}$ at a current density of $< 1\mu A\ cm^{-2}$. Implantations were carried out at temperatures between $-196°C$ and room temperature. Selected implantations were annealed to $400°C$ in-situ within the implantation chamber without breaking vacuum, while other implants were annealed at temperatures up to $400°C$ in flowing nitrogen, following removal from the implantation chamber.

Si wafers of (100) orientation were implanted with $^{29}Si^+$ at 40 keV. Implants were undertaken at current densities of $< 1\mu A\ cm^{-2}$ to total doses in the range $5 \times 10^{15} cm^{-2}$ to $1 \times 10^{18} cm^{-2}$ with the substrates loosely clamped to a plate held at $-180°C$. Implantation of other ions (notably Sb^+ and As^+) was carried out at three target temperatures ($-196°C$, room temperature and $250°C$) to doses of $0.5-2 \times 10^{16} cm^{-2}$ at current densities of $< 1\mu A\ cm^{-2}$.

Finally, (100) GaAs was implanted with Ar^+, Te^+ or Sb^+ ions at 50 - 100 keV to doses up to $10^{16} cm^{-2}$. Most implants were undertaken at room temperatrue but selected samples were also implanted at temperatures between $-196°C$ and $250°C$. The dose rates were kept below $1\mu A\ cm^{-2}$ in all cases.

The implanted samples were analysed using ion beam techniques, most usually Rutherford backscattering and channeling with 2MeV He^{++} ions. TEM, SEM and optical microscopy were employed to examine selected implanted samples. This paper focusses on the ion beam analysis results.

GERMANIUM

Fig. 1 shows RBS and channeling spectra from (100) Ge implanted with 5×10^{15} Sb cm^{-2} at temperatures between $-196°C$ and room temperature. A near-surface yield deficit is clearly apparent for the room temperature implant (in comparison with a virgin Ge spectrum) and, from previous studies [1, 3], we interpret this behaviour in terms of a porous surface layer containing high concentrations of light-mass impurities. The magnitude of the yield deficits for $-196°C$ implants and implants just below $0°C$ are much smaller, which indicates the extreme sensitivity to temperature of the processes which give rise to the deficit. The yield deficit observed in the lower temperature implants does not substantially exceed the ~4% deficit which is calculated to be the result of incorporation of the retained Sb into the near surface of the Ge. Furthermore, the yield deficit for the room temperature implant extends over the damage depth, as shown by the aligned (dashed) spectrum in Fig.1. Other differences were apparent, namely that the room temperature implant appeared black to the naked eye, whereas the lower temperature implants had a light-grey 'amorphous' appearance.

In Fig.2, we illustrate the annealing behaviour of high dose, implanted Ge. The annealing temperature of $400°C$ is sufficient to completely recrystallise the amorphous Ge layer [3]. It is clear that the porous surface layer inhibits epitaxial growth, even when the annealing has been carried out in-situ before, presumably, uptake of light impurities from air exposure. However, substrates which were exposed to air prior to annealing resulted in even worse regrowth. In contrast, further results indicated that annealing of implants having the smallest yield deficits (e.g. liquid nitrogen implants or lower implant doses) exhibited the best epitaxial regrowth. These effects are attributed to lower levels of porosity.

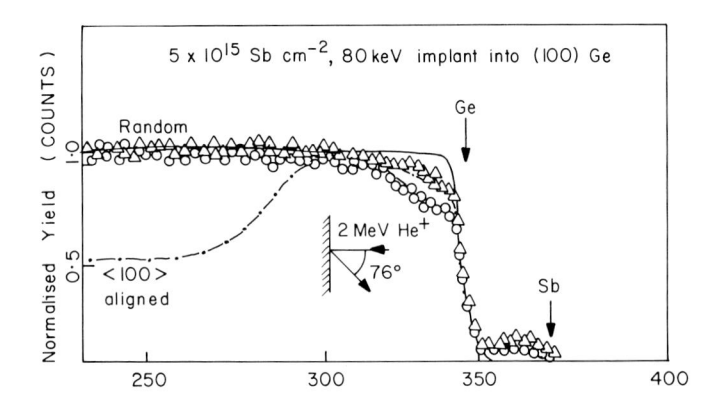

Fig. 1 Comparison of RBS spectra for virgin (100) Ge (—) and Ge implanted with 5 x 10^{15} Sb cm^{-2} at 80 keV at -196°C (Δ), at 0°C (-·-) and at 25°C (o), showing the temperature dependence of the yield deficit. The <100> channeling spectrum for 0°C implant conditions is also shown.

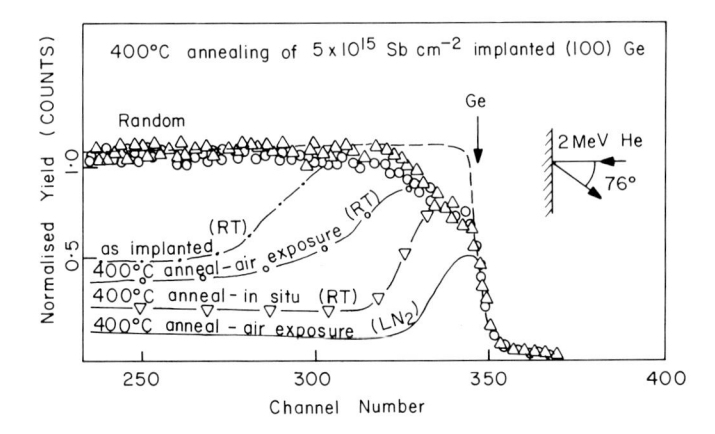

Fig. 2. Comparison of random and <100> channeling spectra for (100) Ge implanted with 5 x 10^{15} Sb cm^{-2} at 80 keV. Data are for virgin (--), as-implanted (-·-) 400°C in-situ vacuum annealed (Δ) and 400°C anneal in N$_2$, i.e. following exposure to air. Implant temperature is indicated in parentheses.

SILICON

The most dramatic evidence for porous structures in ion implanted silicon was obtained from high dose $^{29}Si^+$ implants. These implants showed visible differences in colour and surface reflectivity across the wafer for doses exceeding $10^{17}cm^{-2}$. These lateral variations presumably arise, from temperature gradients across the wafer, attributable to poor thermal contact during implantation. A particularly striking result is illustrated in Fig.3 for a dose of $10^{17}Si$ cm^2 (spectrum labelled region 1). In this case, a large yield deficit can be observed in the near-surface Si spectrum. This yield deficit correlates well with an observed high concentration of oxygen throughout the near-surface layers of the Si. Carbon and nitrogen RBS signals were not seen and hence we infer that their concentrations must be at least an order of magnitude lower than oxygen. Close to the surface, the average film stoichiometry (from Fig.3) is near $SiO_{1.2}$. The channeling spectrum (dashed curve) indicates a damaged layer extending some 1400Å below the surface. TEM observations taken from the same region of the sample (as that giving spectrum 1 in Fig. 3) indicated the presence of considerable oxide within a near-surface layer. This layer overlays an amorphous region which, in turn, overlays a region rich in crystalline defects. A 'cellular'-type microstructure (visible in TEM) within the near-surface layers is suggestive of a porous surface morphology.

The behaviour described above for Si, is consistent with that observed in Ge. We therefore suggest that the development of a porous structure has occurred in Si during implantation with the uptake of oxygen into this structure on exposure to air. We speculate that the high oxygen content is primarily contained in the form of surface oxide on internal surfaces within the porous structure. Other features of the near-surface microstructure have been revealed by RBS, TEM, SEM and optical microscopy examination and these results will be reported at a later date. It is, however, worth commenting here on the extreme sensitivity of the development of the porous structure to the implant temperature. This feature is illustrated in Fig.3, where the spectrum labelled 2 was taken on a neighbouring region (on the same $10^{17}cm^{-2}$ implanted sample) which did not exhibit a dark (coloured) appearance. For this region, no yield deficit is observed in the RBS spectrum, although the same concentrations of impurities, implanted along with the high dose Si^+ ions, are apparent. This result suggests that i) the oxygen corresponding to the yield deficit in the previous spectrum was not implanted: ii) the unintentionally implanted impurities do not play a role in the development of the porous structure and, most significantly, iii) small changes in the implantation temperature across the wafer dramatically alter the development of surface microstructure. Our observations indicate that implant temperatures somewhat above room temperature particularly favour the development of a porous surface morphology in Si. We have observed similar behaviour on higher dose Si^+ implants up to $10^{18}cm^{-2}$. In addition, Sb^+ and As^+ implants to doses $>10^{16}cm^{-2}$ appear to exhibit an anomalous yield deficit, although a surface layer of high porosity is not observed by SEM or TEM in these cases. Further details of microstructural changes to silicon under high dose implantation conditions and possible formation mechanisms will be reported elsewhere.

GALLIUM ARSENIDE

Fig.4 shows RBS spectra from Sb implanted into (100) GaAs at various temperatures. For the implant at room temperature, a clear yield deficit is observed in the near-surface yield from GaAs and, in light of similar observations in Ge and Si, we attribute this behaviour to a porous structure which has absorbed light-mass impurities on exposure to air. For comparison, an implant of the same Sb dose ($3 \times 10^{15}cm^{-2}$) at a temperature of 250°C does not give rise to a yield deficit, the spectrum in this case overlays the

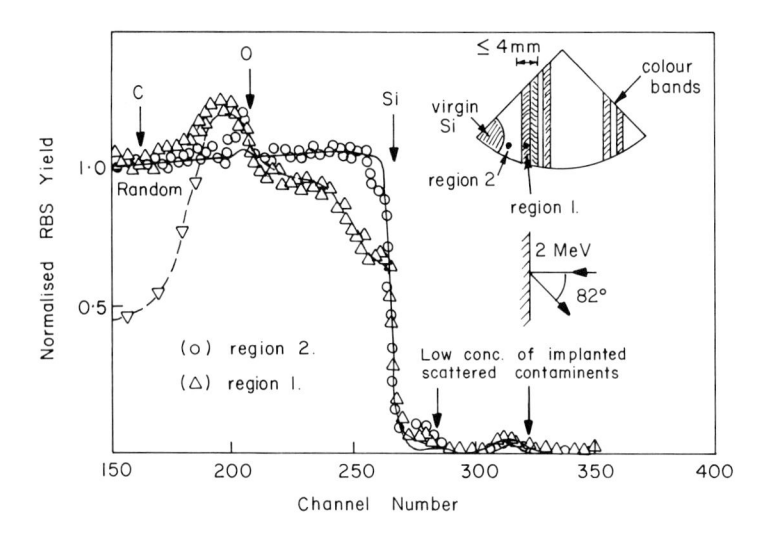

Fig. 3 RBS random spectra from two adjacent regions on a 1 x 10^{17}Si$^+$ cm^{-2} implant (40keV) into (100) Si (see text from implant details). The insert shows the sample region from which the spectra were taken The dashed curve illustrates an aligned spectrum on spot 1.

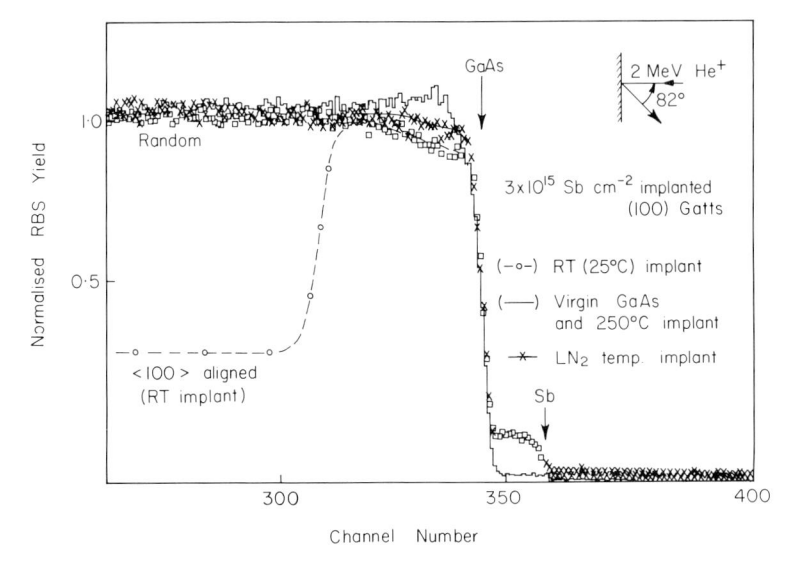

Fig. 4 Random RBS spectra from 3 x 10^{15}Sb cm^{-2} implanted into (100) GaAs at various implant temperatures. The dashed curve shows the aligned spectrum of the RT implant.

random spectrum taken from an unimplanted region on the GaAs samples. The Sb profiles of both the room temperature and 250°C implants are identical so that this Sb concentration alone cannot account for the yield deficit. In fact, incorporation of this amount of Sb into the GaAs near-surface is expected to lower the GaAs yield by less than 5% (see ref. 5 for details of such yield calculations). Indeed, within our counting statistics, no observable deficit is obtained from this Sb implant at 250°C.

Compared in Fig. 4 are the yield deficits obtained following liquid nitrogen, room temperature and 250°C implants of 3×10^{15} Sb cm^{-2}. Clearly, the maximum deficit is observed for the room temperature case, an observable but much reduced deficit is evident following liquid nitrogen temperature implantation, but no deficit occurs at 250°C. It is interesting to note that the extent of the yield deficit corresponded well with the damage width (of presumably amorphous GaAs) as determined from the channeling spectra of the two lowest temperature cases. In contrast, the 250°C implant did not form an amorphous layer but appeared to result in a large number of residual defect clusters within crystalline GaAs, as far as can be ascertained from channeling spectra.

Implants into GaAs of other species (Ar and Te) as well as further Sb implants at lower doses ($\sim 10^{14}$ cm^{-2}) also gave substantial yield deficits for room temperature implants. It is intriguing that such effects can appear in GaAs at heavy ion doses as low as 10^{14} cm^{-2} in contrast to Ge and Si substrates where much higher implantation doses are required. We are currently investigating the characteristics of the observed yield deficit in GaAs.

CONCLUSION

We have reported data in this paper which strongly suggests that near-surface porous structures can be developed in Ge, Si and GaAs under appropriate high-dose implantation conditions. Our results indicate that the development of such microstructures is most sensitive to implantation temperature and appears to be favoured in situations where the implant temperature is just low enough to ensure amorphisation. This suggests that mobile defects within the amorphised surface region may be responsible for the observed phenomenon.

ACKNOWLEDGEMENTS

The Australian Research Grants Committee and the Commonwealth Special Research Centres Scheme are acknowledged for financial support. One of us (RGE) would like to acknowledge support under the CSIRO postdoctoral fellowship scheme.

REFERENCES

1. B.R.Appleton, O.W.Holland, J.Narayan, O.E.Schow III, J.S.Williams, K.T.Short and E.M. Lawson, Appl. Phys.Lett 41, 711 (1982).
2. I.H. Wilson J. Appl. Phys. 53, 1698 (1982).
3. E.M. Lawson, K.T.Short, J.S.Williams, B.R.Appleton, O.W.Holland, and O.E.Schow III, Nucl. Instr. 209/210, 303 (1983).
4. O.W.Holland, J.Narayan and B.R.Appleton J.Appl. Phys. 54, 2295 (1983).
5. E.M.Lawson, J.S.Williams, D.J. Chivers, K.T.Short, and B.R.Appleton AAEC Report E573 (1983).

THERMODYNAMICS AND KINETICS OF CRYSTALLIZATION OF AMORPHOUS Si AND Ge
PRODUCED BY ION IMPLANTATION

E.P. DONOVAN,*[+] F. SPAEPEN,* D. TURNBULL,* J.M. POATE,** AND D.C. JACOBSON**
*Division of Applied Sciences, Harvard University, 29 Oxford Street,
Cambridge, MA 02138; **Bell Laboratories, 600 Mountain Avenue, Murray Hill,
NJ 07974

ABSTRACT

Amorphous Si and Ge layers, produced by noble gas (Ar or Xe) implantation of single crystal substrates, have been crystallized in a differential scanning calorimeter (DSC). This technique allows determination of the growth velocity (which is proportional to the rate of heat evolution, $\Delta \dot{H}_{ac}$), and the total enthalpy of crystallization ΔH_{ac}. Amorphous Ge was found to relax continuously to an amorphous state of lower free energy, with a total enthalpy of relaxation of 6.0 kJ.mole^{-1} before crystallization started. The regrowth velocity on (100) substrates, measured to be $4.2 \times 10^{17} \exp(-2.17 eV/kT)$ Å/sec, is compared to other determinations. The value of ΔH_{ac} was found to be 11.66 ± 0.7 kJ.mole, in good agreement with ΔH_{ac} for amorphous Ge produced by other methods. For Si, ΔH_{ac} was determined to be 11.95 ± 0.7 kJ.mole without any evidence of heat release due to relaxation. The kinetics of crystallization measured by DSC are compared with those determined by other techniques. The effects of the implant profile on the regrowth velocity could also be observed directly in the DSC signal. The more accurate value of ΔH_{ac} allowed a more precise determination of the melting temperature of amorphous Si: $T_{a\ell} = 1420$K.

INTRODUCTION

Ion implantation offers some unique advantages in the study of the thermodynamics and kinetics of crystallization of amorphous elemental semiconductors:

(i) Since the impurity content can be controlled very precisely, the amorphous material and the crystalline-amorphous interface can be kept very clean. This results in an increased interface mobility over that in deposited samples, and rapid crystallization therefore occurs at lower temperatures [1,2].

(ii) Since the amorphous-crystalline interface is planar, and remains so during regrowth, the kinetics of the regrowth process can be monitored *directly* by the rate of heat evolution, $\dot{H}$. This makes it possible to identify impurity effects and to distinguish between the thermal effects of regrowth and structural relaxation.

Evidence is accumulating [3,4] that a-Si melts at a temperature, $T_{a\ell}$, that is considerably lower than the melting temperature, $T_{c\ell}$, of crystalline Si. Given the higher chemical potential of the amorphous phase, this is to be expected. A quantitative check of $T_{a\ell}$, however, requires more accurate thermodynamic data. Our experiments [2] have led to a precise determination of the heat of crystallization; additional measurements of the specific heat, C_p, of the amorphous phase would be desirable for a more accurate estimate of $T_{a\ell}$.

EXPERIMENTAL METHODS

Single crystal wafers of Si and Ge were implanted at 77°K with noble gases, at dose rates not exceeding 0.15 μA.cm^{-2}. The minimum dose to form a fully amorphous buried layer was determined from RBS to be, for (110) Si and 1.7 MeV Ar$^+$ ions, 2×10^{14} cm^{-2}, and for (110) Ge and 1.6 MeV Ar$^+$ ions, 2×10^{13} cm^{-2}. The factor of ten difference in these results is attributed to more efficient damaging in Ge due to a higher cascade density.

For calorimetry, 2.0 μm amorphous layers were produced on (100) Ge substrates with Ar$^+$ ions of four energies from 0.5 to 3 MeV; the fluences were chosen to give a flattened impurity profile of about twice the minimum dose for 1.6 MeV. Amorphous layers 1.6 μm thick were produced on (100) Si by a flat profile of Xe of four energies ranging from 0.5 to 3 MeV and fluences from 0.58×10^{14} to 2.5×10^{14} cm^{-2}, respectively.

Effects of substrate temperature, dose, and dose rate dependence on the formation of amorphous Si layers are to be discussed in a forthcoming paper [5].

The calorimetry was performed with a Perkin-Elmer differential scanning calorimeter, model DSC 2. At a temperature scan rate of 40°K/min, the differential power, $\dot{H}$, needed to keep the implanted wafer and an unimplanted reference wafer at the same temperature was measured.

RESULTS FOR Ge

In scans of as-implanted samples, a constant heat release was observed followed by a peak corresponding to rapid crystallization between 660 and 740K. The total integrated heat for both effects was about 20 kJ/mole. Annealing at a temperature T_A prior to the scan eliminated the heat release below T_A; RBS showed that the interface moved only 0.26 μm up to 660K. The low temperature release was therefore attributed primarily to structural relaxation, a phenomenon well known in other amorphous materials, such as metallic and silicate glasses.

Since the amorphous-crystalline interface in epitaxial regrowth remains planar, the regrowth velocity, v, is proportional to the rate of heat release, $\dot{H}$: $v = \dot{H} \bar{V}/\Delta H_{ac} A$ ($\bar{V}$: molar volume of the amorphous phase, taken equal to that of the crystal [2,6], ΔH_{ac}: molar heat of crystallization, A: area of the wafer). An Arrhenius plot of $\dot{H}$ (Fig. 1) clearly shows that the high temperature part of the data corresponds to crystallization with an activation energy of 209 kJ.mole^{-1}; its extrapolation agrees well with the lower temperature crystallization velocity measurements of Csepregi et al. [7].

By fitting an exponential, corresponding to the activation energy for crystallization, to the high temperature calorimeter peak, the effects of relaxation can be eliminated; integration of this peak gives a heat of crystallization $\Delta H_{ac} = 11.66 \pm 0.7$ kJ.mole^{-1}. This agrees well with values observed for relaxed amorphous Ge produced by vapor deposition [8]: $\Delta H_{ac} = 11.5$ kJ.mole^{-1}, or sputter deposition [9]: $\Delta H_{ac} = 10.9$ kJ.mole^{-1}. The agreement indicates that the properties of relaxed amorphous semiconductors produced by ion implantation are representative of the intrinsic properties of the material.

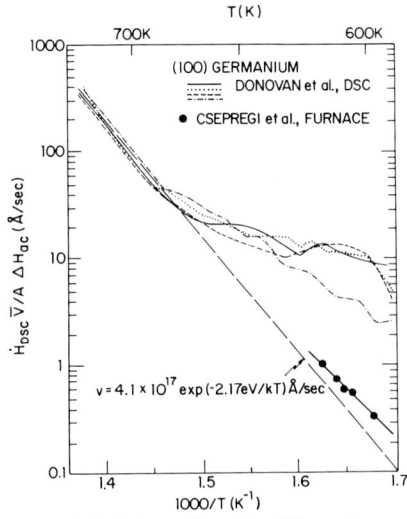

FIG. 1. Arrhenius plot of the calorimeter signal, $\dot{H}_{DSC}$ (normalized), for regrowth into amorphous Ge produced by ion implantation of a (100) substrate. Furnace regrowth data are from Csepregi et al. [7].

With this value of ΔH_{ac}, the regrowth velocity of (100) Ge can then be obtained from the previous equation: $v = 4.1 \times 10^{17} \exp(-2.17 \text{ eV}/kT)$ Å/sec. By taking into account the small amount of interface motion during structural relaxation below 660K, and the value of ΔH_{ac}, the heat of relaxation is found to be 6.0 kJ·mole^{-1}. Although the total distance moved by the interface below 640K is only 0.26 μm out of 2 μm, it is much greater than what would be expected from an extrapolation of the high temperature growth velocity. This could be a result of a large number of extra dangling bonds at the interface [10].

RESULTS FOR Si

In contrast to the results for amorphous Ge, no thermal effect of structural relaxation was observed. The calorimeter trace therefore corresponds entirely to crystallization and its Arrhenius plot is a straight line (see Figs. 2 and 3). The activation energy of the regrowth velocity in Fig. 3 is 216 kJ·mole^{-1}, and the extrapolation of the calorimeter data corresponds well to the regrowth measurements of Csepregi et al. [11,12] at lower temperatures. The values of the activation energies for all samples lay between 216 and 267 kJ·mole, which is within the range of other investigations [11-16]. Since the Si data were obtained near the upper temperature limit of the calorimeter, for some runs it was difficult to obtain a precise baseline, which is normally traced by rerunning the sample after crystallization. In such cases, the baseline was adjusted to give a straight Arrhenius plot. The average value of the heat of crystallization for all the high-quality runs was 11.95 ± 0.7 kJ·mole^{-1}. Based on this value, the growth kinetics of Fig. 3 can be written as:
$v = 1.77 \times 10^{14} \exp(2.24 \text{ eV}/kT)$ Å/sec. In a number of samples, with uneven concentration profiles, the effect of the noble gas impurities on the regrowth velocity could be observed directly from the calorimeter trace: minima in the trace were shown, with RBS, to correspond to maxima in the impurity profile, and vice versa [5].

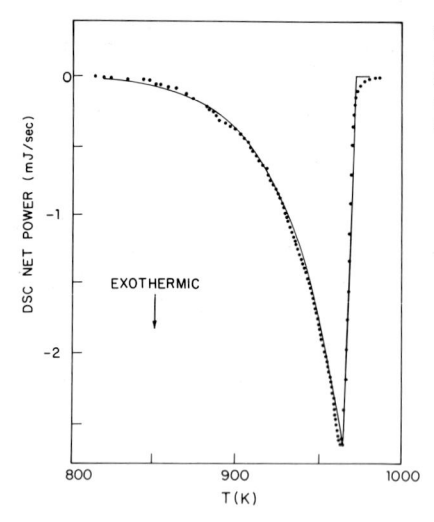

FIG. 2. Calorimeter signal for regrowth into amorphous Si produced by ion implantation of a (100) substrate. The smooth line is from the fit to the data on Figure 3.

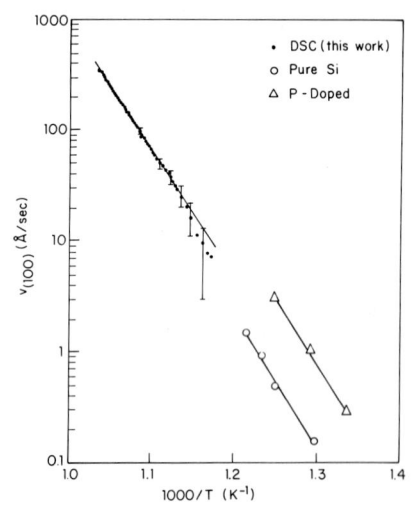

FIG. 3. Arrhenius plot of the data of Figure 2. The furnace regrowth data are from Csepregi *et al.* [11,12].

DISCUSSION

A comparison of the results for Si and Ge raises two surprising points: (i) the absence for a-Si of the thermal manifestation of structural relaxation observed for a-Ge, and (ii) the similarity in value between the two heats of crystallization: 11.66 kJ.mole^{-1} for Ge, and 11.95 kJ.mole^{-1} for Si.

The first point remains so far unexplained, and no specific information is available on systematic differences in relaxation behavior between a-Si and a-Ge.

The second point, however, has recently been somewhat clarified. The structure of both amorphous Si and Ge can be modeled by the same tetrahedrally coordinated continuous random network. Most of the difference in energy between such a network and the diamond cubic crystal can be attributed to the strain energy associated with the bond angle distortion. Since the force constant for this distortion is, depending on the method of estimation, 15-20% greater for Si than for Ge, the heat of crystallization was initially expected to scale accordingly. However, a recent more detailed investigation of this problem [17], using computer relaxation in a Keating potential [18], using the *specific* force constants for Si and Ge [19] gives a strain energy for Si that is only 4.5% greater than for Ge, which is much closer to the 2.5% difference observed experimentally. The reason for this improved agreement is that, although the structures of a-Si and a-Ge are *topologically* isomorphous, the bonds in Si are *relatively* more stretchable, so that a greater fraction of the bond angle distortion energy can be eliminated by stretching the bonds. Since the Keating potential has its limitations, the comparison of Si and Ge random networks is now also being pursued using other interaction potentials.

The precise determination of ΔH_{ac} for a-Si allowed a more accurate estimate of its melting temperature, $T_{a\ell}$ [2]. Figure 4 is a free energy diagram for amorphous and liquid Si relative to crystalline Si. The free energy of the amorphous phase was calculated using a model-based estimate of the residual entropy [20], and a specific heat difference, $\Delta C_{p,ac}$, either equal to zero (line a(1)), or scaled with the measured value of $\Delta C_{p,ac}$ for a-Ge [8] (line a(2)). It is clear that direct measurements of $\Delta C_{p,ac}$ would improve the accuracy of the free energy estimate. However, the present prediction of $T_a = 1420$ K is certainly an improvement over earlier predictions and is in reasonable agreement with experimental evidence from the various ultra-rapid heating experiments which indicate $T_{c\ell}-T_{a\ell}$ to be at least several hundred degrees [4].

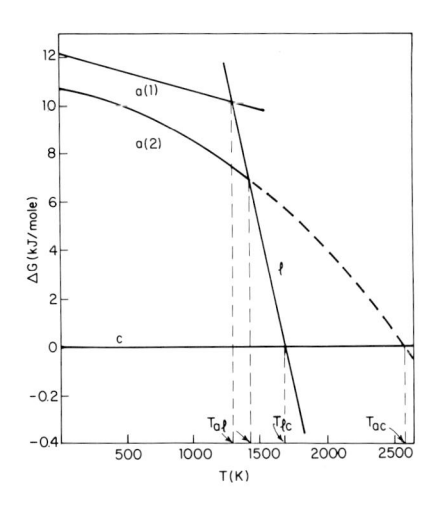

FIG. 4. Free energy of liquid (ℓ) and amorphous (a) Si, relative to crystalline (c) Si; lines a(1) and a(2) correspond to different estimates of the heat capacity of amorphous Si (see text).

CONCLUSIONS

1. Amorphous Ge produced by ion implantation exhibits a large heat release due to structural relaxation, which has so far not been observed in Si.

2. The heat of crystallization of well-relaxed a-Ge produced by ion implantations is in excellent agreement with that measured for relaxed samples

produced by deposition. This indicates that ion-implanted samples have representative properties.

3. The heat of crystallization of a-Si in only 2.5% greater than that of a-Ge.

4. The crystallization kinetics of a-Ge and a-Si obtained by calorimetry agree well with those observed by other methods.

5. The melting points of a-Si and a-Ge are estimated to be about 250K below those of the crystalline phases.

ACKNOWLEDGMENTS

Work performed at Harvard was supported by the National Science Foundation, Materials Research Laboratory, under Contract No. DMR 80-20247.

REFERENCES

1. J.C.C. Fan and H. Andersen, J. Appl. Phys. 52, 4003 (1981).
2. E.P. Donovan, F. Spaepen, D. Turnbull, J.M. Poate, and D.C. Jacobson, Appl. Phys. Lett. 42, 698 (1983).
3. P. Baeri, G. Foti, J.M. Poate, and A.G. Cullis, Phys. Rev. Lett. 45, 2036 (1980).
4. J.M. Poate, Mat. Res. Soc. Symp. Proc. 13, 263 (1983).
5. E.P. Donovan, F. Spaepen, D. Turnbull, J.M. Poate, and D.C. Jacobson, to be published.
6. D.E. Polk, J. Non-Cryst. Solids 5, 365 (1971).
7. L. Csepregi, R.P. Kullen, J.W. Mayer, and T.W. Sigmon, Sol. St. Comm. 21, 1019 (1977).
8. H.S. Chen and D. Turnbull, J. Appl. Phys. 40, 4214 (1969).
9. R.J. Temkin and W. Paul, Proc. 5th Int. Conf. on Amorphous and Liquid Semiconductors, J. Stuke and W. Brenig, eds. (Taylor and Francis, London 1974) p. 1193.
10. F. Spaepen and D. Turnbull, in: Laser Annealing of Semiconductors, J.M. Poate and J.W. Mayer, eds. (Academic, New York 1982) p. 15.
11. L. Csepregi, E.F. Kennedy, T.J. Gallagher, J.W. Mayer, and T.W. Sigmon, J. Appl. Phys. 48, 4234 (1977).
12. L. Csepregi, E.F. Kennedy, J.W. Mayer, and T.W. Sigmon, J. Appl. Phys. 49, 3906 (1978).
13. A. Lietoila, R.B. Gold, J.F. Gibbons, T.W. Sigmon, P.D. Scovell, and J.M. Young, J. Appl. Phys. 52, 30 (1981).
14. A. Lietoila, A. Wakita, T.W. Sigmon, and J.F. Gibbons, J. Appl. Phys. 53, 4399 (1982).
15. G.L. Olson, S.A. Kokorowski, J.A. Roth, and L.D. Hess, Mat. Res. Soc. Symp. Proc. 13, 141 (1983).
16. S.A. Kokorowski, G.L. Olson, J.A. Roth, and L.D. Hess, Phys. Rev. Lett. 48, 498 (1982).
17. E. Nygren and F. Spaepen, unpublished results.
18. P.N. Keating, Phys. Rev. 145, 637 (1966).
19. F. Spaepen, in: Amorphous Materials: Modeling of Structure and Properties, V. Vitek, ed. (TMS-AIME, New York 1983) p. 265.
20. F. Spaepen, Phil. Mag. 30, 417 (1974).

+ Present address: Naval Research Laboratory, Washington, D.C., 20375

PHYSICAL PROPERTIES OF TWO METASTABLE STATES OF AMORPHOUS SILICON

G.K. HUBLER*, C.N. WADDELL**, W.G. SPITZER**, J.E. FREDRICKSON***, AND T.A. KENNEDY*
*Naval Research Laboratory, Washington, D.C.; **Physics and Materials Science Departments, University of Southern California, Los Angeles, California; ***Physics-Astronomy Department, California State University, Long Beach, California

ABSTRACT

Characterization of the two metastable states of amorphous Si produced by ion implantation is extended to include electron paramagnetic resonance, fundamental absorption edge, and density measurements in addition to infrared reflection. It is found that the properties of the two a-Si states are not dependent upon the mass of the incident ion (^{12}C, ^{29}Si, ^{31}P, ^{120}Sn) or upon the anneal temperature for $400°\leq T_A \leq 600°C$. The dangling-bond density drops about a factor of 2, the absorption coefficient drops by more than a factor of 5, but the density does not change when the a-Si makes a transition between the two states.

INTRODUCTION

In previous work [1] a number of Si samples were implanted at either 200K or room temperature with Si or P ions which had incident energies between 200keV and 2.7MeV and fluences between 1.0×10^{16} and 10.0×10^{16} ions/cm^2. In all cases the amorphous (a-Si) produced by the implantation process had an infrared refractive index, $n_I(\nu)$ (see curve in Fig. 1) which was independent of the fluence and the implanted ion, and was 12% larger than the crystalline value at the frequency ν corresponding to a wave number of 4000cm^{-1}.

In subsequent studies [2,3] it was found that annealing the implanted samples for about 2 hours at 500°C caused the index of refraction to decrease to a new value, $n_{II}(\nu)$, also shown in Fig. 1. The ratio $n_{II}/n_I = 0.96$ at $\nu = 4000$cm^{-1} indicates that annealing causes the refractive index to drop about 1/3 of its original implantation induced increase from the crystalline value $n_c(\nu)$. The values of $n_{II}(\nu)$ were also independent of the implantation parameters and did not change with further annealing. These two optical states were identified as (I), the defect-saturated (as-implanted) and (II), the thermally-stabilized (annealed) states of a-Si [4]. For this work, we define these states as a-Si-I and a-Si-II, respectively.

In a recent study [5] the infrared refractive indices, the strength of the electron paramagnetic resonance (EPR) dangling-bond signal, and the changes in density of the a-Si were measured as a function of annealing time at $T_A = 500°C$ for ^{29}Si-implanted samples of Si. These measurements indicated that the dangling bond density behaved in a manner similar to that of the refractive index, i.e., a large drop in the early stages (first two hours) of the 500°C anneal, and only slight further changes resulting from prolonged anneal times. However, there was no corresponding change in density.

The results presented here extend those of Ref. 5 in two ways. Firstly, because there was an indication in the literature that the refractive index of a-Si-I depends upon the mass of the implanted ion [6], it is important to determine if these properties of a-Si-I and II are

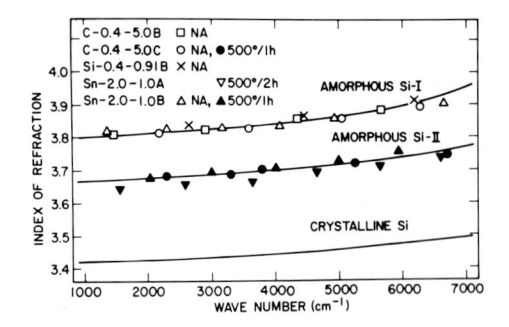

Fig. 1. Indices of refraction of crystalline Si, a-Si-I (as-implanted), and a-Si-II (thermally stabilized). The curves are from Ref. [1,2] and the data points are measurements of C, Si, and Sn implanted samples in the as-implanted state and after annealing at 500°C.

dependent upon the ion mass. Group IV (^{12}C, ^{29}Si, ^{120}Sn) ions were chosen to avoid complications of electrical doping effects. Secondly, since the transition between the two states had only been studied for $500 \leq T_A \leq 550°C$, we studied the annealing process over a wider temperature range ($300 \leq T_A \leq 600°C$) to determine whether the properties of the thermally stabilized state (a-Si-II) are dependent upon the anneal temperature employed. If the several properties of a-Si-II are independent of all implantation parameters and of the annealing temperature, then the results would suggest that a definable and unique thermodynamic state has been produced.

Experimental Procedures

The techniques used are essentially the same as those used previously [1-3,5,7] and therefore they will be reviewed here only briefly. The implanted samples are identified in Figs. 1, 2, and 3 by the designation of: implanted ion, incident ion energy in MeV, fluence in units of 10^{16} ions/cm^2, and if there is more than one sample, by A,B, etc. Thus, C-0.4-5-A is a C ion-implanted Si sample, where the C-ion energy is 400 keV, the fluence is 5×10^{16}/cm^2, and "A" indicates that this the first such implanted sample. The starting material was single crystal Si having an initial resistivity > 10 Ωcm. Multiple energy implantations between 25 and 400 keV were used in some cases to achieve more uniform ion and damage density profiles within the implanted layer. The sample temperature was maintained near 200K during implantation and, to reduce channeling, the incident ion beam direction was approximately 8° off of the high symmetry directions.

Room temperature, near-normal incidence, reflection measurements were made over the wave number range $400 < \nu < 7000$ cm^{-1}. The reflected intensity varies strongly with ν because of interference between the reflections at the air-amorphous and amorphous-crystalline interfaces. Computer analyses of the reflection spectra yields precise values for (i) the refractive-indices of the a-Si and the recrystallized region, (ii) the depth of the amorphous layer, and (iii) the width of the transition between the a-Si and c-Si regions [7].

During implantation a mask covered a small portion of the surface of each sample. The swelling of the implanted region produces a sharp step in surface height at the edge of the mask. Because the thickness of the implanted region is very small compared to the lateral dimensions of the implanted region, only the thickness of the sample is significantly affected by the radiation damage [8]. Using this model the change in density, Δρ, is given by Δρ = ρ$_c$ (step height)/(a-Si thickness) where ρ$_c$ is the crystalline Si density. The thickness of the a-Si region was determined from the infrared reflection measurements, and a Sloan Dektak profilometer was used to determine the step height to an accuracy of ~25Å per measurement. The latter measurements contributed most of the inaccuracy in the determination of $\Delta\rho$. For our implantation conditions we estimate that 5 to 11Å of Si was removed by sputtering for ^{29}Si implants, 18Å for C implants, and 38Å for Sn implants. Since step heights for the as-implanted samples were usually between 100 to 160Å, the correction to the density would be negligible for Si, 10% for C and about 20% for Sn. Since the step height measurements themselves are accurate only to about 20%, and the sputtering coefficient estimates are accurate only to 50%, we have not applied the correction.

The EPR measurements were made at room temperature with a Varian E3 spectrometer at 9GHz. The signal from a-Si, which has an isotropic g = 2.0055, has been attributed to dangling bonds [9]. To obtain relative spin density values, the EPR signal strength is divided by the a-Si volume which is obtained by using the sample surface area and the layer thickness determined from the infrared reflection measurements. Infrared band edge absorption measurements were made at 77K by directly comparing the infrared transmission of the implanted sample with that of an identical, non-implanted sample.

Results

The curves in Fig. 1 show the refractive indices $n_I(\nu)$ of a-Si-I (as-implanted), $n_{II}(\nu)$ of a-Si-II (thermally stabilized, 500°C), and $n_c(\nu)$ for crystalline Si, all as previously reported for Si and P implants. The data points are the refractive indices obtained for several of the samples from this study implanted with ^{12}C, ^{29}Si, or ^{120}Sn ions. Both the as-implanted and the 500°C anneal refractive indices for all samples are in excellent agreement with our previously published results [1-3,5]. Thus, there is no discernible dependence upon the mass of the incident ion (^{12}C→^{120}Sn), or the use of a single ion energy or multiple energies during amorphization.

For ^{12}C, ^{29}Si, and ^{120}Sn implanted samples, the measured absorption for non-annealed samples in the region of the absorption edge was close to $\alpha_I \sim 5 \times 10^{-13} \nu^4$ in agreement with the previous results for P-implanted Si [1], and there was no detectable dependence upon which ion is used for the implantation. The implanted samples were annealed for 2 hours at 500°C to produce the change from $n_I(\nu)$ to $n_{II}(\nu)$ and the absorption edge was remeasured. It was not possible to detect a difference between the absorption of the annealed implanted samples and the crystalline reference samples. Therefore, we were unable to obtain meaningful values for α_{II}, the absorption coefficient for the thermally stabilized state. We estimate that $\alpha_{II}(\nu)$ is less than $\alpha_I(\nu)/5$ or $\approx \alpha_c(\nu)$ for all cases.

EPR measurements for some of the ^{12}C, ^{29}Si, and ^{120}Sn implanted samples are presented in Table I. The values for ^{29}Si are the same as those given previously [5], and it was indicated there that the as-implanted value corresponds to a dangling bond density of $\sim 2 \times 10^{19}$cm^{-3}. When the samples are annealed (500°C/2h) the spin density drops about a factor of 2. There are variations in the values measured for both as-implanted and annealed samples, but no trend due to ion mass is discernible. Therefore, the decrease in the spin density is also attributed to the transition between the two states of the a-Si. The spin densities of a-Si-I and a-Si-II compare to within a factor of 3 for pure films evaporated onto substrates held at room temperature [9].

To characterize the changes in the indices of refraction $n(\nu)$ we define a single parameter $f_D = n(\nu)/n_I(\nu)$. The value of this parameter was determined at ν = 4000 cm^{-1}, the average frequency of the infrared measurements. The results of annealing ^{12}C, ^{29}Si, and ^{120}Sn implanted samples of at 400°C are shown in Fig. 2. The f_D measurements in Fig. 2a clearly show the change in the index of refraction that occurs when the a-Si makes a transition between the two states. There is no

TABLE I: Relative Dangling-Bond Densities[a]

Anneal/Ion	^{12}C	^{29}Si	^{120}Sn	Average
as-implanted	299	353[b] 325 320	252	310 ± 38
500°C/2h	150	136	148	
500°C/8h	163	~100		140 ± 22
			Ratio	0.45 ± 0.10

[a] (The isotropic g = 2.0055 EPR signal)/(a-Si volume).
[b] This value corresponds to a spin density of ~2×10^{19} cm.$^{-3}$

discernible difference due to the implanted ion specie or the duration of the annealing time. The average value of $f_D=0.965$ is in good agreement with the previously determined values [3].

The density measurements shown in Fig. 2b indicate that there is no change in the density of the a-Si layer when the transition in states occurs. Note that for an anneal time of two hours, large changes are observed for f_D and the EPR spin density. There is also no observed density effect due to the implanted ion specie. There is a correlation between the magnitude of $\Delta\rho/\rho_c$ and the calculated magnitude of the sputtering corrections, but the large errors mentioned earlier preclude an accurate sputtering correction. The density of a-Si-I and a-Si-II is 2.0±0.5% less than crystalline Si. Very similar results for f_D and $\Delta\rho/\rho_c$ to those in Fig. 2 were obtained for 500°C anneals except that the transition occurred much more rapidly.

Figure 3 shows the results obtained when ^{29}Si implanted samples are isothermally annealed at temperatures between 300°C and 600°C. Again the f_D values indicate similar values of index of refraction $n_{II}(\nu)$ for the annealed state for $T_A=400°$ to 600°C. However, at $T_A=300°C$ the transition rate is much slower, and the index of refraction remained at $f_D=0.974$ after 118h which is significantly greater than the value $f_D=0.965\pm0.005$ obtained at the higher temperatures. These results suggest that the annealing mechanism responsible for the transition between states may be different at the lowest anneal temperature.

Discussion

The experimental results presented here and in previous publications support the view that the physical properties being measured are intrinsic to two distinct states of a-Si. The refractive index of as-implanted amorphous silicon, $n_I(\nu)$, is independent of ion mass ($^{12}C - ^{120}Sn$), implantation dose and energy (3×10^{15} to $10^{17}Si$, P/cm^2; 200 to 2700 keV P) [1-3] and implantation temperature (100K to 300K for Si implants). The refractive index of thermally stabilized a-Si, $n_{II}(\nu)$, is independent of the above parameters and is unique for annealing temperatures $400 \leq T_A \leq 600°C$. For 300°C annealing the final $n_{II}(\nu)$ state is not quite achieved. If we assume the change in refractive index, $n_c(\nu) \rightarrow n_I(\nu)$, is caused by the introduction of ~2×10^{19} cm^{-3} dangling bonds and to the decrease in dipole moment/volume caused by the observed 2% decrease in mass density, one calculates that the dangling bond polarizability must increase by a factor of 2500 over that of a normal bond. This result is physically unreasonable and indicates that the dangling bonds are not responsible for the change in indices of refraction. Averaged over the entire lattice one finds that there is an

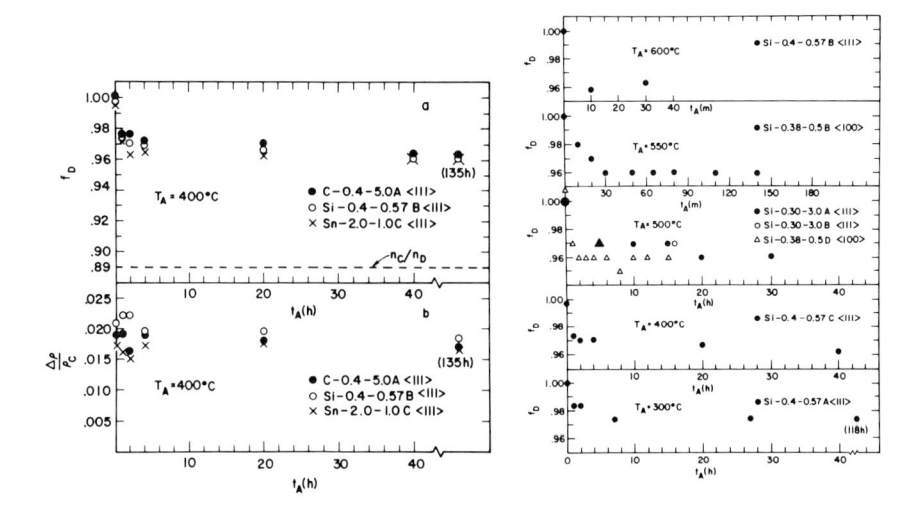

Fig. 2. The relative change in index of refraction, f_D, and mass density, $\Delta\rho/\rho_C$ for C, Si, and Sn implanted samples annealed at 400°C as a function of anneal time.

Fig. 3. Comparison of the changes in index of refraction, f_D, for anneal temperatures T_A = 300°, 400°, 500°, 550° and 600°C.

increase of ~28% in the average bond polarizability for the change from c-Si to a-Si-I, and a decrease of ~8% for the change from a-Si-I to a-Si-II. The observed decrease in the near band edge absorption after the a-Si-I⟶a-Si-II transition occurs appears to be consistent with the proposal of a network reorganization in which there is a reduction in disorder [5]. The absorption in this region is often associated with the fluctuations in the atomic configurations [10] which cause shifts in electronic states and give rise to band tails. The reduction in disorder by network reorganization thus appears to be the cause for the reduction in absorption and the ~8% decrease in average bond polarizability. The decrease in dangling bond density seems to be directly related to the annealing of defects within the amorphous structure.

If the transition to a-Si-II is a phase transition, one might expect a release of heat during the phase change. Using differential scanning calorimetry, Donovan et al. [11] has measured the heat of crystallization of a-Si and a-Ge prepared by ion implantation. For a-Ge, they observe a substantial heat of stabilization prior to crystallization, (1/2 of the heat released during crystallization) but no heat of stabilization is observed for a-Si prior to crystallization. Our preliminary refractive index data for as-implanted and thermally stabilized a-Ge [12] shows a similar relaxation to that occurring in Si. It is therefore puzzling that there is no heat of stabilization measured for Si and indicates that the mechanism producing the relaxation is different for the two materials.

Given that two distinct states of a-Si can be produced, it is appropriate to speculate on the effects that the two states may have on other material properties. For example, in pulse laser annealing, the coupling of the radiation prior to melting would change due to changes in the optical constants, especially if the light can penetrate to a-c boundry and back to the front surface to interfere with the incident wave. Also, any enthalpy decrease (probably small) between a-Si-I and a-Si-II will increase the melting temperature of a-Si-II; and/or, the apparent melting temperature of a-Si-II might be larger than a-Si-I because advantageous sites for heterogeneous and homogeneous nucleation of the liquid are reduced which allows the a-Si-II to be more readily superheated [13] in a fast pulse anneal. In addition, Csepregi et al. have observed that the crystalline quality of epitaxially recrystallized amorphous layers on (111) Si can be improved by performing a 550°C, 1h anneal prior to performing the normal 950°C, 1h anneal [14]. Our work shows that this pre-anneal should produce the a-Si-II state before recrystallization occurs. While we have no evidence that the production of a-Si-II is related to the results of Cspregi et al., it is possible that the nucleation of microtwins and polycrystallites [14] is suppressed in the relaxed "structure" of a-Si-II such that epitaxial regrowth and simple defect annealing can occur at 550 and 950°C without competition from polycrystallite and microtwin formation. This deserves further study.

In any case, it is clear that refractive index changes induced by the thermal annealing of a-Si provide a physical basis for systematic investigations of the effects of two-step thermal processing on the properties of recrystallized Si and other semiconductors as well [12].

This work was partially supported by the joint Services Electronics Program under contract No. F44620-76C-0061 monitored by the Air Force Office of Scientific Research. We would like to thank J.M. Poate and E.P. Donovan for useful discussions.

REFERENCES

1. G.K. Hubler, C.N. Waddell, W.G. Spitzer, J.E. Fredrickson, S. Prussin and R.G. Wilson, J. Appl. Phys. 50, 3294 (1979).
2. J.E. Fredrickson, C.N. Waddell, W.G. Spitzer and G.K. Hubler, Appl. Phys. Lett. 40, 172 (1982).
3. C.N. Waddell, W.G. Spitzer, G.K. Hubler and J.E. Fredrickson, J. Appl. Phys. 53, 5851 (1982).
4. M. Janai, D.D. Allred, D.C. Booth and B.O. Seraphin, Solar Energy Mat. 1, 11 (1979).
5. W.G. Spitzer, G.K. Hubler and T.A. Kennedy, Nucl. Instr. Meth. 209/210. 309 (1983).
6. W. Wesch and G. Gotz, Radiat. Eff. 49, 137 (1980).
7. G.K. Hubler, P.R. Malmberg, C.N. Waddell, W.G. Spitzer and J.E. Fredrickson, Radiat. Eff. 60, 35 (1982).
8. K.N. Tu, P. Chaudhari, K. Lai, B.L. Crowder and S.I. Tan, J. Appl. Phys. 43, 4262 (1972).
9. P.A. Thomas, M.H. Brodsky, D. Kaplan and D. Lepine, Phys. Rev. B 18, 3059 (1978.)
10. G.A.N. Connell, "Optical Properties of Amorphous Semiconductors", in Amorphous Semiconductors, ed. M.H. Brodsky, (Springer Verlag, Berlin, 1979).
11. E.P. Donovan, F. Spaepen, and D. Turnbull, these proceedings.
12. K-W. Wang, W.G. Spitzer, and G.K. Hubler, unpublished.
13. D. Turnbull, Metastable Materials Formation by Ion Implantation, eds. S.T. Picraux and W.J. Choyke (Elsevier, 1982) 103.
14. L. Csepregi, W.K. Chu, H. Muller, and J.W. Mayer, Rad. Eff. 28, 227 (1976).

IMPLANTATION INDUCED FILAMENTARY STRUCTURES

C. JAUSSAUD, B. MAILLOT[*], M. BRUEL
LETI - COMMISSARIAT A L'ENERGIE ATOMIQUE
85 X, 38041 GRENOBLE CEDEX, FRANCE
* Present address : IBM FRANCE
  224, boulevard Kennedy, 91102 CORBEIL, FRANCE

ABSTRACT

Heavy rare gases implantations have been performed in layered structures : Sb/In/Si bulk and Sb/Ga/Si bulk. Layer thicknesses lie in the range of 150 Å - 600 Å, ion doses in the range of 10^{14} cm^{-2} - 10^{16} cm^{-2}.

Ion implantation induces in such structures a change in the visual aspect from a metallic appearance to a dark black aspect. This modification corresponds to the creation of voids with, as a consequence, the expansion of the structure. This effect is so important that even for low and medium doses, the structure becomes filamentary. Expansions as high as 2,8 μm have been measured. SEM photographs illustrate this spectacular effect.

Relations between expansion and implantation parameters are given. A comparison between this effect and those observed in bulk InSb, GaSb and Ge is presented.

INTRODUCTION

In order to study recoil implantation of Antimony, multilayers composed of a layer of Indium covered by a layer of Antimony were deposited onto silicon and irradiated with rare gases ions (1). The Indium was used as a marker. For doses above 5×10^{14} ions x cm^{-2}, implantation resulted in a blackening of the surface. SEM observations of the surface showed a filamentary structure, with elevations of the surface up to 2,8 μm (2). Because of this unexpected effect implantations were done in other bilayer structures : Ga/Sb, Ga/In, In/Se, Sn/Sb.

EXPERIMENTAL PROCEDURE

Layers were deposited by vacuum evaporation, Joule heating was chosen for Antimony, electron bombardment for Indium evaporation.

Both layers were deposited in the same vacuum cycle at a pressure around or below 10^{-6} torr.

Layer thicknesses were monitored by a crystal oscillator.

Implantations were performed on our facilities with current densities low enough to avoid any temperature effects. Scanning Electron microscopy was used to see the changes in structure and to measure the amount of swelling.

RESULTS

Structure of the deposited layer

The deposited layer has a rough aspect (see photo n° 1). This rough aspect is due to the first (In or Ga) layer which does not form a continuous film, but which is composed of discrete In or Ga islands. The size of those islands is a few thousand Angstroms.

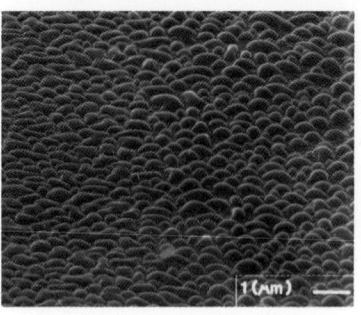

Photo n° 1 : 600 Å Sb over 300 Å In, non implanted.

Effects of ion implantation

The effects of ion implantation in such structures are described with the example of Xenon implanted into a 300 Å thick Indium layer deposited on Silicon and covered with a 600 Å thick Antimony layer (see fig. 1). The Xenon ions were implanted at 200 Kev, which corresponds to Rp = 500 Å.

200Kev ions

FIG. 1

For doses less than 10^{14} ions x cm^{-2} no change in the structure can be observed.

At 3.10^{14} ions x cm^{-2} we observe a measurable increase in the layer thickness (about 500 Å) which seems due to void formations (see photo n° 2).

At 10^{15} the increase in layer thickness reaches 1,7 µm and the structure has become filamentary (see photo n° 3) with some filaments extending about 1000 Å over the surface of the layer.

At 10^{16} ions x cm^{-2} the deposited layers are almost completely sputtered away (see photo n° 4).

Such evolutions in the structure of the deposited layers have been observed for Indium and Antimony layers ranging from 150 to 1000 Angstroms, and for implantations with Argon, Krypton, Xenon and Antimony ions. The increase in layer thickness varies linearly with ion dose up to a maximum dose at which sputtering begins to erode the layer. The maximum dose depends on the mass of the implanted ion (see fig. 2). Xenon and Antimony, which have similar masses give similar expansion.

An expansion of 2.85 µm was observed after 200 Kev Kr implantation into a multilayer (Sb/In/Sb) deposited onto silicon (see photo n° 5).

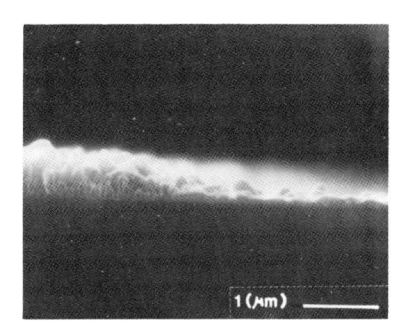

Photo n° 2 : 600 Å Sb on 300 Å In, implanted with Xe (200 Kev, 3×10^{14} ions x cm^{-2})
Formation of voids in the implanted zone (left)

Photo n°3 : 600 Å Sb on 300 Å In, implanted with Xe (200 Kev, 10^{15} ions x cm^{-2})
Formation of filaments in the implanted zone (left)

Photo n° 4 : 600 Å Sb on 300 Å In, implanted with Xe (200 Kev 10^{16} ions x cm^{-2})
Sputtering of the deposited layers.

Photo n° 5 : Sb 600 Å/In 300 Å /Sb 600 Å implanted with Kr (200 Kev 5×10^{15} ions x cm^{-2})
The expansion reaches 2.8 µm.

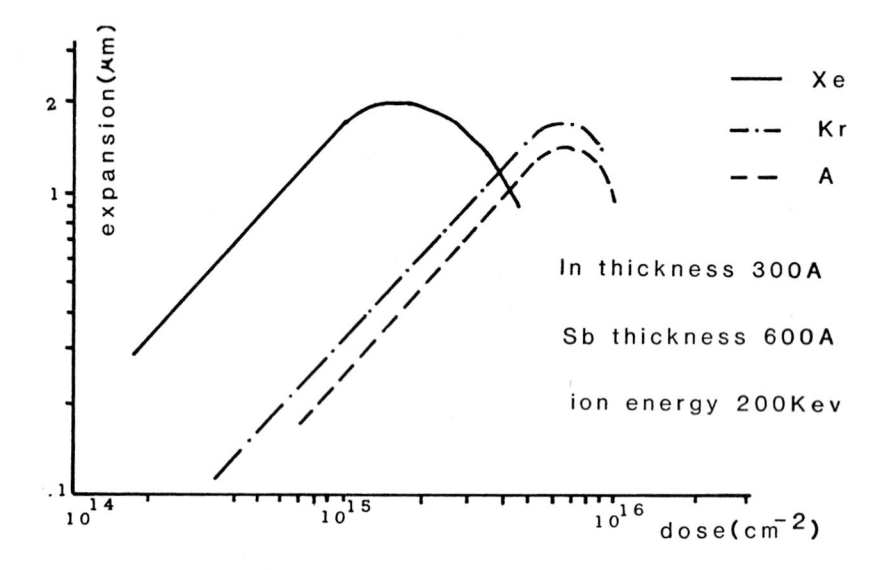

FIG. 2. Expansion as a function of ion dose

Two implantations were done at liquid nitrogen temperature with conditions (layer thicknesses, ion dose) giving a maximum increase in the layer thickness. No change in the layer structure could be observed after these low temperature implantations.

Implantations of rare gas ions have been done in other bilayers. Ga/Sb, Ga/In, In/Se, Sn/Sb and in Germanium layers deposited on silicon. Effects similiar to those described above for In/Sb have been observed for implantations in Ga/Sb layers. The maximum expansions are nearly the same for In/Sb and Ga/Sb layers, but higher doses are required in the case of Ga/Sb (see fig. 3). S.E.M. photos show that the structure obtained after implantation into Ga/Sb layers are quite similiar to those observed in In/Sb. For implantations in other layers (Ga/In, In/Se, Sn/Sb and Ge) no change in the structures could be observed by S.E.M.

thicknesses

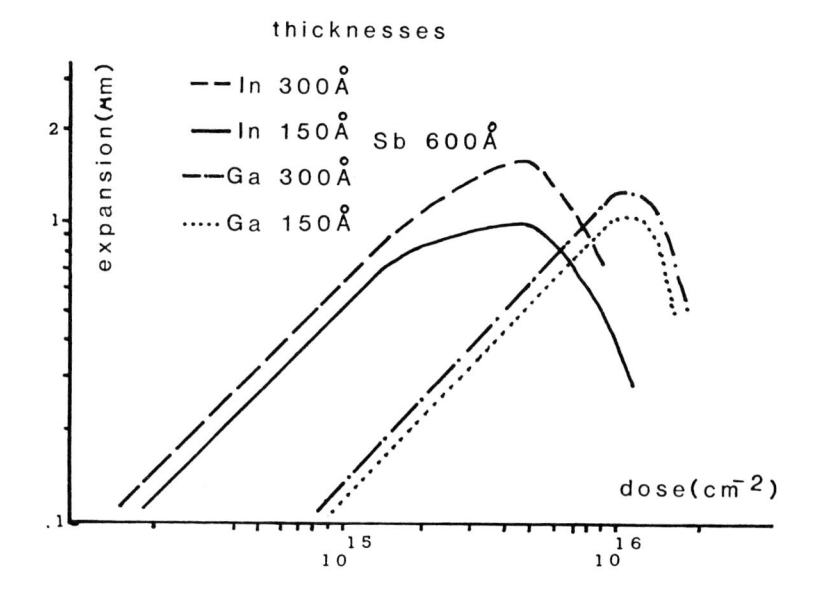

- - In 300 Å
— In 150 Å   Sb 600 Å
- - Ga 300 Å
..... Ga 150 Å

FIG. 3. Expansion as a function of Ion dose for Ga/Sb and In/Sb bilayers.

DISCUSSION

Elevation of the surface and void formation after implantation have been observed by several authors (3-8) in bulk Ga/Sb, In/Sb and Ge. When an In/Sb or Ga/Sb bilayer is implanted, ion mixing can lead to the formation of an InSb or GaSb phase. Further implantation into this phase can result in void formation and layer expansion as observed in bulk materials. In facts, for lower doses, we observe the formation of voids. The formation of a filamentary structure, observed only in the case of implantation into deposited layers is related to the geometry of the layer : for thicknesses less than 1000 Å, the In or Ga do not deposit as a continuous film, but as discrete islands. The size of the islands depends on the layer thickness. In figure 4 we have plotted the layer expansion as a function of mean island diameter. For small layer thicknesses, the expansion is limited by the amount of InSb phase. There is a maximum for an island diameter of about 3000 Å, and for larger diameters the layer expansion goes to zero. Another evidence of the essential role played by the geometry of the layer on layer expansion and the formation of a filamentary structure is given by the following experiment : Sb was deposited on bulk Si. The Sb layers always formed a continuous film. In was deposited on the Sb layers and those bilayers were implanted under conditions which give the maximum layer expansion in In/Sb layer. No expansion or formation of filamentary structure was observed in those bilayers. Thus, filamentary structure formation is observed only for Ga/Sb and In/Sb bilayer deposited under conditions which give a discontinuous intermediate Ga or In layer.

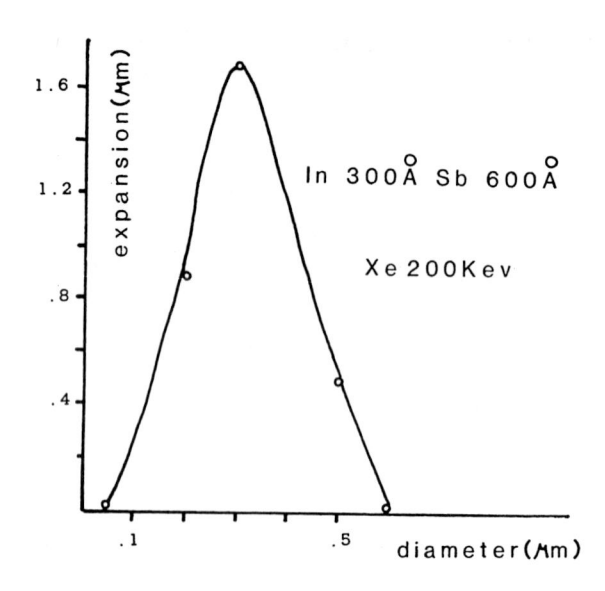

FIG. 4. Expansion as a function of mean Island diameter. This mean Island diameter increases with the In layer thickness.

REFERENCES

1. Implantation ionique au travers de couches d'antimoine déposées sur silicium. Etude des phénomènes de surface et d'interface.
   Brigitte MAILLOT - Thèse de 3ème cycle - 11.06.82. Univ. de Grenoble.
2. M. BRUEL and B. MAILLOT
   Proceedings ot the International Ion Engineering Congress ISIAT'83 and IPAT'83 12-14 sept. 83 KYOTO JAPAN P1773
3. U. GONSER and B. OKKERSE,
   Phys. Rev. 105 757 (1957)
4. D. KLEITMAN and H.J. YEARIAN
   Phys. Rev. 108 901 (1957)
5. Yu. A. DANILOV and V.S. TULOVCHIKOV
   Sov. Phys. Semicond. 14, 117 (1980)
6. I.H. WILSON
   Low Energy Ion Beams Bath 1980
   Conference series Number 54. The Institute of Physics Bristol and London. p.262
7. G.L. DESTEFANIS and J.P. GAILLIARD
   Appl. Phys. Lett. 36, 40 (1980)
8. O.W. HOLLAND, B.R. APPLETON and J. NARAYAN
   J. Appl. Phys. 54, 2295 (1983).

EPITAXIAL CRYSTALLISATION OF DOPED AMORPHOUS SILICON

R.G. ELLIMAN**, S.T. JOHNSON*, K.T. SHORT* AND J.S. WILLIAMS*
*Microelectronics Technology Centre, RMIT, Melbourne 3000 Australia;
**Joint Appointment with CSIRO, Division of Chemical Physics, Clayton 3168 Australia.

ABSTRACT

This paper outlines a model to account for the influence of doping and electronic processes on the solid phase epitaxial regrowth rate of ion implanted (100) silicon. In addition we present data which illustrates good quality epitaxial crystallisation of silicon at 400°C induced by He^+ ion irradiation. We tentatively suggest that electronic energy-loss processes may be responsible for this behaviour.

INTRODUCTION

Amorphous silicon layers produced in single crystal substrates by ion implantation are metastable and recrystallise epitaxially from the crystal-amorphous interface when annealed at temperatures of ~500°C [1]. For uncontaminated layers; i.e. those amorphised by implantation with silicon ions, the kinetics of regrowth have been extensively investigated [1-3]. Indeed, the substrate orientation dependence [1] is reasonably well understood in terms of structural models of the amorphous-crystalline interface, such as that proposed by Spaepen and Turnbull [4,5]. The regrowth of layers containing impurities is, however, less well understood [3,6,7]. In particular, it has been observed that the presence of electrically active impurities can enhance the epitaxial growth rate by more than an order of magnitude [3,6-8]. Furthermore, if equal concentrations of both n-type and p-type dopants are introduced into silicon by implantation, a compensating effect is observed in which the epitaxial growth rate is found to return to that of undoped silicon [3,8,9]. These results clearly indicate the importance of electronic processes in determining the recrystallisation kinetics. In this paper we discuss a phenomenological model to explain the role of electronic doping in enhancing the rate of epitaxial crystallisation. In addition, we present new data on helium ion-beam-induced epitaxy of amorphous silicon at temperatures as low as 400°C. These intriguing results suggest further experiments which can be devised to probe details of solid phase recrystallisation processes in semiconductors.

REGROWTH MODEL

The Spaepen and Turnbull structural model of the crystal-amorphous interface during recrystallisation is based on minimum free energy and bonding arguments [5]. They suggest that during regrowth the crystal-amorphous interface will facet to maximise the exposure of minimum free energy (111) planes and conclude that the interface consists of (111) terraces, bounded by <110> ledges. Crystal growth is envisaged to proceed along these ledges, initiated at appropriate nucleation sites. The number of ledges, and hence nucleation sites, is determined by the angle the interface plane subtends to the (111) plane. From geometric considerations, the ledge concentration, and hence the regrowth rate, is expected to be approximately proportional to sin (θ), where θ is the angle between the regrowth plane and the (111) plane. The predictions of this model agree well with the measured orientation dependence of solid phase epitaxial growth kinetics[5]. In order to account for the observed dependence of regrowth on doping, the present model extends the Spaepen and Turnbull model by assuming that the concentration of nucleation sites on ledges is influenced by doping in a manner analogous to that proposed for enhanced dislocation motion [10]. Furthermore, the nucleation sites are envisaged to be kink or jog sites on ledges, as illustrated in Fig. 1. These kink sites uniquely provide two 'crystalline' bonds to the adjacent atom in the amorphous phase (open circles in Fig. 1). Such sites are assumed to introduce bound electron states in the band gap of silicon, such that kinks may exist in either a charged or neutral state. The charged-to-neutral fraction is simply determined by Fermi-Dirac statistics, and hence by the doping level. Since the concentration of neutral kinks is a constant at a given temperature, as dictated by Boltzmann statistics, the total kink concentration, charged and neutral, becomes a function of the doping level. Furthermore, because the rate of crystallisation is directly proportional to the number of growth sites, from established growth rate theory [5], the rate of epitaxial recrystallisation is a similar function of doping level. To establish the form of this dependence, we follow an approach analogous to that given by Hirsch [10] for the enhancement of dislocation motion by doping.

Details of our theoretical approach are given elsewhere [11]. The treatment leads to the following expression for the ratio of the growth velocity in n-doped silicon to that in intrinsic material:

$$\frac{V_n}{V_i} = \frac{(N_D/N_c)\exp(E_G - E_K^A)/kT}{1 + g\exp(E_{Fi} - E_K^A)/kT} \quad \ldots\ldots\ldots (1)$$

where N_D and N_C are the active donor concentration and density of states in the conduction band, respectively; E_G is the band gap energy; E_K^A is the energy of kink acceptor level(s) in the band gap; g is the degeneracy factor; and E_{Fi} is the intrinsic Fermi energy. This can be further reduced to the form

$$\ln(V_n/V_i) = \ln N_D + \text{const} \quad \ldots\ldots\ldots (2)$$

Fig. 2 compares some experimentally determined growth rates as a function of impurity concentration with the predictions of equation 2. The results are particularly encouraging in view of the fact that no account has been taken of the influence of band bending at the interface or of the reduction of effective doping concentration by compensation and trapping effects at the interface. Indeed, it is interesting to conjecture about the nature of the band structure at a moving crystalline-amorphous interface in doped silicon. Fig.3 illustrates a possible interfacial band structure. We suggest that the position of the Fermi level at the interface (**right** portion of Fig. 3) is determined by both the implanted dopant concentration profile (**left** portion of Fig.3), which gives the Fermi level on the crystalline side of the interface, and the band bending at the interface necessitated by pinning of the Fermi level close to mid-band in the high resistivity amorphous silicon. Possible kink-related defect levels, both donor and acceptor, are also indicated.

Although the existence and magnitude of kink-related defect levels, as indeed the degree of interfacial band bending, may be extremely difficult to establish experimentally, there is some hope of establishing the credibility of the model by design of suitable experiments. To this end, the simple form of the expression in equation 2, coupled with possibilities for interpreting differences between regrowth rates for n-type and p-type dopants in terms of simple band pictures such as those in Fig3, suggest that further detailed regrowth measurements as a function of dopant concentration may be valuable. In addition, to establish further details of the electronic and bond-breaking processes at the interface, it may be fruitful to

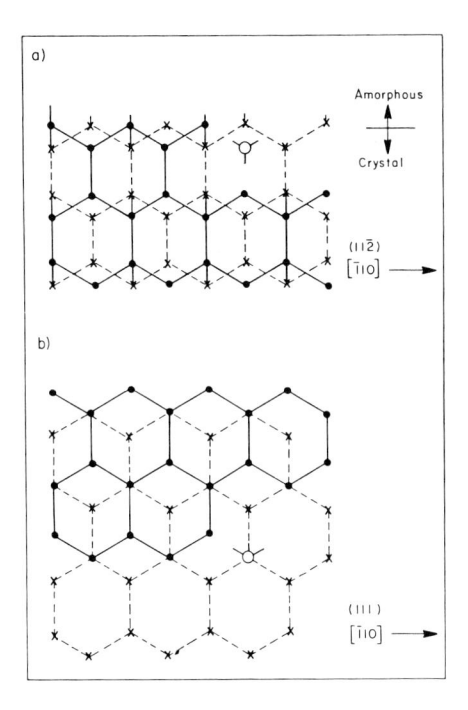

Fig. 1. Detailed schematic of the kink site at which crystal growth initiates. The figure depicts the crystalline side of the interface, where full circles show the top plane of atoms and the crosses the second plane. The open circle indicates the next atom to crystallise. A cross sectional view, along the <112> direction, is shown in part (a) and a plan view, along the <111> direction, is shown in (b).

Fig. 2. Comparison of theoretically predicted and experimentally observed regrowth rate versus impurity concentration behaviour. The broken line of unity slope has been fitted to the data using the theoretically predicted slope of 2. The arrow indicates the magnitude of the conduction band density of states.

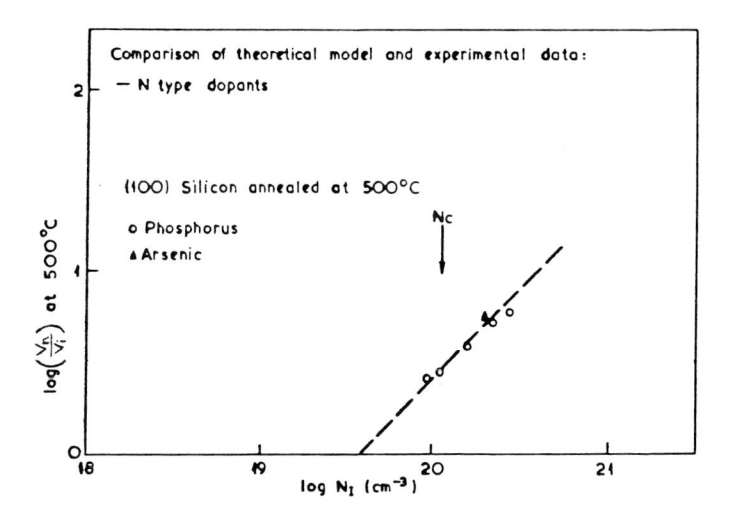

investigate possible regrowth enhancement under (non-thermal) experimental conditions which may aid bond breaking, such as the application of an external stress to the sample during regrowth or bombardment of the interfacial region with ionizing radiation in the form of light ions, electrons or, possibly, light. In this regard, the following section outlines some initial experiments on epitaxial regrowth in silicon under helium ion irradiation, the results of which suggest exciting possiblities.

HELIUM ION BEAM ANNEALING

In this initial series of experiments, amorphous layers were first produced in (100) silicon by cooling the sample to liquid nitrogen temperatures and irradiating with 80keV Sb ions to fluences of 10^{14} to 10^{15} Sb cm^{-2}. The substrate temperature following these implants was then raised to between 200°C and 400°C and selected regions of each implant irradiated with 80 keV He$^+$ ions to doses in the range 5×10^{14} to 3×10^{16} cm^{-2}.

The high resolution Rutherford backscattering and channeling spectra in Fig.4 clearly indicate He$^+$ bombardment-induced regrowth at 400°C for a silicon layer previously amorphised with 1×10^{15} Sb cm^{-2}. Indeed, the amorphous layer is observed to regrow epitaxially from the crystal-amorphous interface and the degree of regrowth increases with increasing He$^+$ dose. Following irradiation with 3×10^{16} He$^+$ cm^{-2}, the highest dose employed in the present experiments, the layer has regrown by ~320Å. The quality of the regrown crystal is good and the Sb atoms appear to occupy substitutional lattice sites, as measured by channeling. In contrast, He$^+$ irradiations at 200°C and 300°C did not produce measurable regrowth, consistent with a previous He$^+$ bombardment study [12]. Our regrowth observations at 400°C constitute the first report of He$^+$-induced epitaxial recrystallisation of silicon.

In Fig.5 the Rutherford backscattering and channeling spectra at 165° scattering reveal the deep damage generated by the nuclear energy-loss component of the irradiating helium, at the end of the helium ion range (~6000 - 7000Å). This deep damage in the silicon could also result from some agglomeration of the implanted helium into gas bubbles and attendant lattice strain. However, the surface region, where He$^+$ ions lose most of their energy by electronic loss, appears to be relatively free of He$^+$-induced damage. Interestingly, the extent of epitaxial recovery also appears to depend on the concentration of Sb since i) the regrowth width in Fig.4 does not scale with the He$^+$ irradiation dose as the growth front moves through regions of increasing Sb concentration, and ii) a sample amorphised with a lower Sb dose (10^{14}cm^{-2}) showed regrowth of < 100Å at 400°C for a He$^+$ fluence of 3×10^{16}cm^{-2}. These doping effects are analogous to those observed during thermal annealing and are sonsistent with the regrowth model proposed earlier.

He$^+$ induced annealing results raise some interesting questions as to the operative regrowth-enhancing mechanism. Although the production of mobile defects (notably vacancies) under He$^+$ irradiation cannot be overlooked as a possible cause of crystal growth, in view of the much higher electronic energy-loss component of the He$^+$ ions as they pass the interfacial region and the crystallisation of ~6 Si atoms per indicent He ion we tentatively suggest that electronic bond breaking processes may be responsible for the observed behaviour. Beam induced annealing of amorphous layers in silicon has been observed with heavier bombarding ions than helium [12-14], but we believe that this annealing may result from an entirely different effect, related to nuclear energy loss and spike effects [14]. We are currently involed in a series of experiments to test our suggestion that the He$^+$ results are attributable to electronic processes.

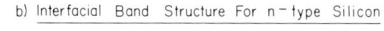

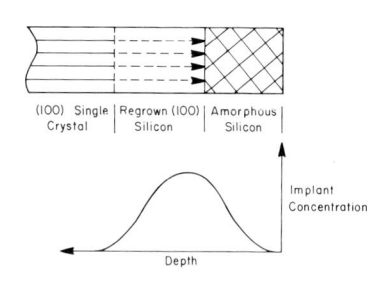

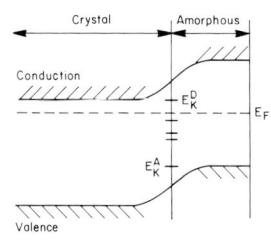

Fig. 3

    a) Schematic of partially regrown n-type doped silicon, showing the implantation profile.

    b) The proposed amorphous-crystalline band structure showing band bending and the existence of kink levels, donor E^D_K and acceptor E^A_K, within the band gap.

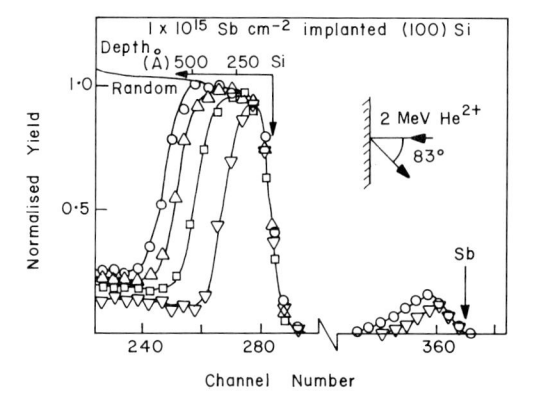

Fig.4. <100> channeling spectra showing He^+-induced epitaxial regrowth as a function of 80 keV He^+ dose, for a 1×10^{15} Sb cm^{-2} implanted (100) Si sample. Profiles correspond to $1 \times 10^{15} He^+ cm^{-2}$ (o), $3 \times 10^{15} He^+ cm^{-2}$ (△), $1 \times 10^{16} He^+ cm^{-2}$ (□) and $3 \times 10^{16} He^+ cm^{-2}$ (▽), all at 400°C substrate temperature.

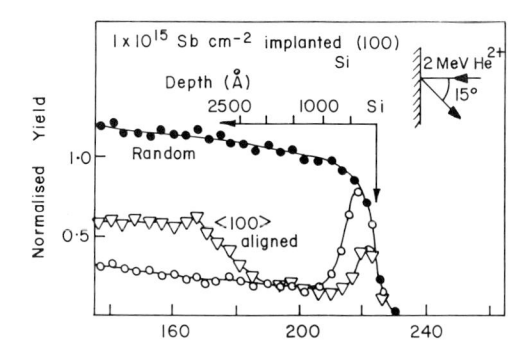

Fig.5 165° RBS and channeling spectra, showing deep crystalline damage generated by 80 keV He^+ irradiation of Sb-implanted (100) Si. Profiles correspond to random alignment (●) and < 100> channeling in as-implanted Si (o) and following irradiation with $3 \times 10^{16} He^+ cm^{-2}$ at 400°C (▽), respectively.

CONCLUSIONS

We have demonstrated that the influence of electronic dopants on the epitaxial growth rate of silicon can be understood in terms of simple electro-physical effects. In addition, it has been demonstrated that energetic helium ion bombardment can also induce epitaxial crystal growth at temperatures of 400°C in Sb-doped silicon. We tentatively suggest that this behaviour is a result of electronic energy loss processes at the crystalline-amorphous interface.

ACKNOWLEDGEMENTS

We acknowledge the financial support of the Australian Research Grants Committee and the Commonwealth Special Research Centres Scheme. One of us (RGE) would also like to acknowledge the financial support of the CSIRO postdoctoral fellowship scheme.

REFERENCES

1. L. Csepregi, E.F., Kennedy, J.W. Mayer and T.W. Sigmon, J. Appl. Phys. 49, 3906 (1978).
2. B. Drosd and J. Washburn, J. Appl. Phys. 51, 4106 (1980).
3. A. Lietoila, A. Wakita, T.W. Sigmon and J.F. Gibbons, J. Appl. Phys. 53, 4399 (1982).
4. F. Spaepen, Acta. Metall. 26, 1167 (1978).
5. F. Spaepen and D. Turnbull, in Laser and Electron Beam Processing of Semiconductor Structures, edited by J.M. Poate and J.W. Mayer (Academic, New York, 1981), p. 15.
6. L. Csepregi, E.F.Kennedy, T.J. Gallagher, J.W.Mayer and T.W.Sigmon. J. Appl. Phys., 48, 4234, (1977).
7. E.F.Kennedy, L. Csepregi, J.W.Mayer and T.W.Sigmon. J. Appl. Phys. 48, 4241 (1977).
8. I. Suni, G. Goltz, M.G.Grimaldi, M.A. Nicolet and S.S. Lau. Appl. Phys. Lett. 40, 269 (1982).
9. J.S.Williams and K.T.Short, Nucl. Instr. Meth. 209/210, 767 (1983).
10. P.B. Hirsch, J. Phys. (Paris), Colloq. 40, C6 (1979).
11. J.S.Williams and R.G.Elliman. Phys. Rev. Lett., 51, 1069 (1983).
12. B. Svensson, J. Linnros and G. Holmen. Nucl. Instr. Meth., 210, 755 (1983).
13. J. Nakata and K. Kajiyama. Appl. Phys. Lett. 40, 686 (1982).
14. K.T.Short, D.J. Chivers, J. Liu, R.G. Elliman, A.P. Pogany, H.K. Wagenfeld and J.S.Williams, these proceedings.

HIGH CURRENT DENSITY IMPLANTATION AND ION BEAM ANNEALING IN Si*

O. W. HOLLAND AND J. NARAYAN
Solid State Division, Oak Ridge National Laboratory, Oak Ridge, TN 37831

ABSTRACT

Annealing of amorphous layers in Si by high flux, self-ion irradiation will be discussed. The mechanism for the lattice recovery is presented and related to the structure of the residual damage. It will be shown that highly supersaturated, alloyed regions, free from extended defects, result from the annealing process.

The characteristics of high flux implantation in Si will be presented including a study of 'dynamic annealing'; a process whereby a high dose rate beam self-anneals its own displacement damage; and annealing of preexisting amorphous layers on crystal substrates by high flux irradiation. The mechanism for lattice recovery during ion bombardment and its relation to the morphology of the residual damage will be discussed. Various aspects of ion beam annealing (IBA) of Si were investigated including metastable phase formation and solid solubility under irradiation. One motivation for this study was to determine if new metastable materials or more highly supersaturated alloys could be produced by IBA; materials which cannot be formed by more conventional techniques such as furnace processing [1,2] or various transient thermal techniques including RTA [3] (rapid thermal annealing) and cw laser annealing [4,5]. The crystals were analyzed after implantation by ion channeling techniques using a 2.5 MV Van de Graaff accelerator. The channeling technique not only gives the depth distribution of the damage in the crystal but also the lattice location and distribution of the implanted dopant. Selected samples were also viewed in cross-section by transmission electron microscopy to determine the detail structure of the residual damage.

Shown in Fig. 1 are channeling spectra from single crystal Si(100) samples which were irradiated with a 100 μamp/cm^2, ^{28}Si (self-ion) beam at two different fluences. By comparison with the random spectrum in the figure, it can clearly be seen that at a fluence of 0.375 x 10^{16} Si/cm^2, a buried amorphous layer has formed as a result of the displacement damage produced by the incident ions. The rapid change in the scattered yield at the location of the crystal-amorphous interface on both sides of this buried layer indicates that the transition between the phases is sharp and the interface is planar. At an increased fluence of 0.75 x 10^{16} Si/cm^2, there is no evidence of an amorphous layer at any depth as seen in the aligned spectrum from this sample. It is clear that at this fluence a significant amount of the lattice damage has been dynamically annealed. The minimum yield (which is the ratio of the aligned to random yield) is 2.8% just behind the surface peak which is comparable to that of a virgin (nonimplanted) crystal. There are however, two distinct damage regions in this sample which can be identified in the aligned spectrum by an increase in the scattered ion yield. The damage peak centered at 290 nm is located near the original interface of the buried amorphous layer and crystal substrate. This damage consists of interstitial loops which form as a result of coalescence of point defects generated near the end of range of the ions [6]. There is also a damage region located at 160 nm which will

---

*Research sponsored by the Division of Materials Sciences, U. S. Department of Energy under contract W-7405-eng-26 with Union Carbide Corporation.

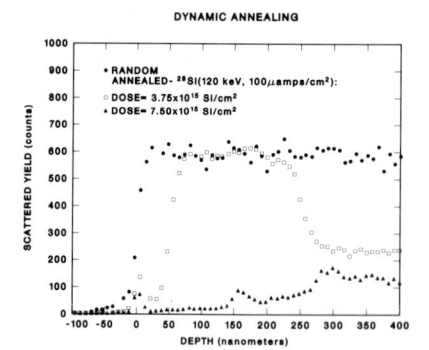

Fig. 1. Backscattering spectra from a Si(100) single crystal irradiated with a high flux ^{28}Si ion beam.

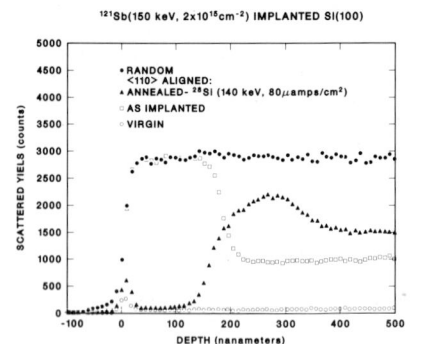

Fig. 2. Backscattering data from an ^{121}Sb (140 keV, 2x10^{15} cm^{-2}) implanted Si(100) single crystal which has been ion beam annealed.

be discussed shortly. Aligned spectra (not shown) from samples implanted with annealing doses intermediate to those discussed above showed that the thickness of the buried amorphous layer is reduced by the movement of both c-a interfaces. Since the amorphous region is bound on both sides by single crystal, epitaxial growth is possible from both the front and back side. Since no reduction in the scattered yield below the amorphous level was seen between the interfaces as they moved, it is concluded that dynamic annealing is dominated by interface-controlled epitaxial growth and not by a volume nucleation and growth process within the amorphous layer. As the moving interfaces come together, any residual defects (atoms which are in excess of what can be accommodated in the lattice) will tend to coalesce and be trapped. It is these defects which give rise to the damage peak at 160 nm in the aligned spectrum from the 'annealed' sample in Fig. 1. Also any incoherency in the interfaces as they join due to grown-in dislocations will contribute to the observed peak. No significant differences were observed in the growth rates of the two interfaces which is consistent with the location of the damage peak near the center of the original buried amorphous layer.

The relative importance of beam heating versus defect production on the epitaxial process during dynamic annealing was investigated. In an attempt to isolate beam heating effects, the rate of epitaxial growth of very deep amorphous layers, which were irradiated with a 90 keV, high flux ^{28}Si ion beam, was determined. Since the range of the ions was much less than the depth of the amorphous layer (formed by a low flux implantation of 175 keV, ^{30}Si ions at liquid nitrogen), it was felt that only beam heating could initiate epitaxial growth. For comparison, single crystals were irradiated identically and the recrystallization rates of the buried amorphous layers (which form early during 'dynamic annealing') determined. Various annealing doses were implanted and the growth monitored at each dose by channeling measurements. The deep amorphous layer grew at a faster rate than the buried layer leading to the conclusion that epitaxial growth is retarded in regions traversed by ion beams. The deep amorphous layer grew at a rate of 13.0 nm/s while the buried layer grew at 7.6 nm/s. These velocities were determined by the growth which occurred during an annealing dose range of 0.5-1.0 x 10^{16} Si/cm^2. Solid-phase-epitaxial regrowth velocities have been measured at high temperatures by means of cw laser heating and dynamic reflectivity techniques [4]. An Arrhenius dependence, $V = v_o \exp(-E_a/kT)$, was observed over a wide range of temperature; where $v_o = 3.07 \times 10^8$ cm/s and $E_a = 2.68$ eV. A regrowth velocity of 13.0 nm/s corresponds to a lattice temperature of 666°C which is consistent with thermocouple measurements during implantation.

Dynamic annealing requires implantation of a critical dose to raise the temperature of the lattice sufficiently to initiate the amorphous to crystalline transition ~525°C. High flux ion beams of group III,V dopants have been used to directly dope Si during implantation [7-10]. This method of implantation is limited since, for doses below the critical dose, no significant 'dynamic annealing' occurs, and for a higher dose, radiation induced precipitation of the nonequilibrium fraction of dopant would proceed. Additionally, the dislocation band at the end of range of the ions would intercept the dopant profile, contributing to the degradation of the electrical properties and also providing sites for precipitation. Because of these problems, it was felt that other ion beam techniques would have to be used to produce metastable alloys. Therefore, to overcome these problems, the desired dopant profile was implanted at a low dose rate so that a continuous amorphous layer formed at the surface extending over the damage range of the ions. Annealing of this layer was attempted by implantation of a high flux, self-ion beam with a range which substantially exceeded the depth of the amorphous layer. Channeling results are shown in Fig. 2 from a Si(100) sample implanted with ^{121}Sb (140 keV, 2 x 10^{15} cm^{-2}) ions and then ion beam annealed by a 150 keV, 80 μamp/cm^2, 2 x 10^{16} Si/cm^2 irradiation. Comparison of the aligned spectrum from the sample before and after annealing shows that the amorphous layer formed by the low flux Sb implantation was completely annealed. Little differences can be seen between the aligned spectrum from the annealed sample and that from a virgin (nonimplanted) crystal over the range of the original amorphous layer. There is evidence, however, of a very wide defect band center at 260 nm, near the end of range of the annealing beam. Cross-sectional transmission electron micrograph from this sample is shown in Fig. 3. The micrograph shows that a 'nearly defect-free' region extends 140 nm from the surface followed by a dense 150 nm band of dislocation tangles and loops. Also a single dislocation can be seen, extending between this band and the surface, which was grown-in during the epitaxy. The total and substitutional concentrations in Fig. 4 show that a supersaturated alloy had formed as a result of ion beam annealing. Greater than 80% of the implanted Sb was substitutional after annealing with a peak concentration of 2.2 x 10^{20} cm^{-3}, which is greater than the retrograde maximum by a factor of 3. For higher annealing doses, the substitutional fraction decreased dramatically due to irradiation induced precipitation of the supersaturated concentration of Sb.

During furnace annealing at 550°C of an ^{115}In implanted Si(100), accummulation of indium at the moving crystal-amorphous interface above a critical interfacial concentration has been reported [11]. Also considerable redistribution of In, attributed to fast diffusion of liquid, indium precipitates which form ahead of the moving interface, was observed. This precipitation and subsequent redistribution limits the concentration of indium which can be trapped onto lattice site during epitaxy. Ion beam annealing of In implanted layers was attempted to see if dopant redistribution could be limited and solid solubility increased. In Fig. 5, the total and substitutional concentration of In after IBA are shown. The maximum substitutional concentration 5 x 10^{19} cm^{-3}, while 60 times greater than the retrograde maximum, was essentially the same as reported for furnace annealing. A greater total concentration was trapped in the bulk by IBA, presumably in the form of small precipitates, than for furnace annealed samples, although the fraction on lattice site is less.

Therefore, it has been shown that high flux ion irradiation is capable of annealing amorphous layers on Si. Nearly defect free, supersaturated regions are produced with maximum substitutional concentrations which are comparable to those achieved by other nonequilibrium, annealing techniques. However, a rather extensive band of dislocation tangles and loops results from the coalescence of point defects generated at the end of range of the

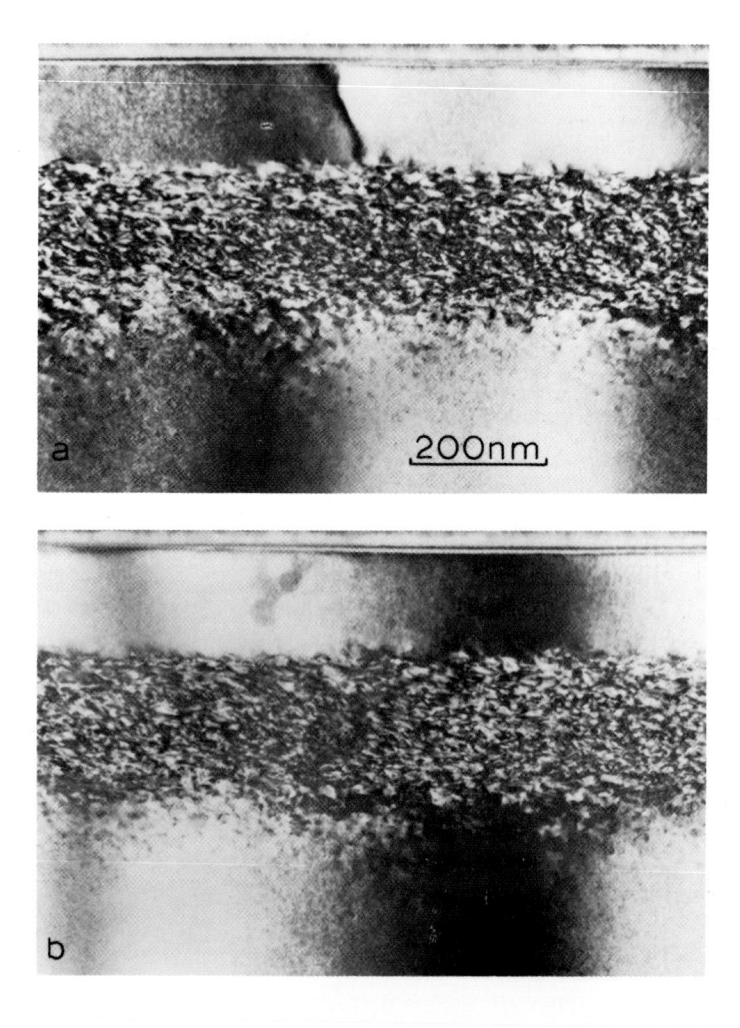

Fig. 3. Cross-section transmission electron micrograph of ^{121}Sb implanted sample which has been ion beam annealed. Part (a) shows a region which contains a single dislocation which extends to the surface and (b) a region which is free of any extended defects in the near surface.

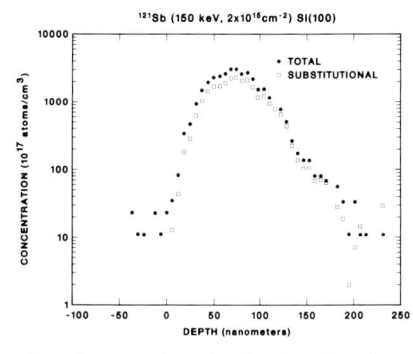

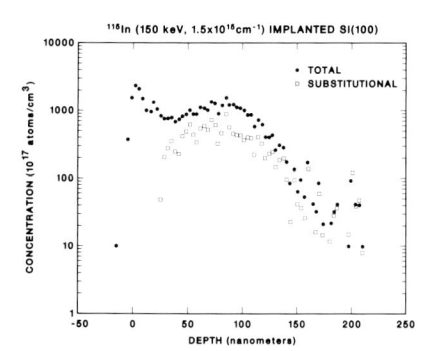

Fig. 4. Total and substitutional ^{121}Sb concentration in Si after ion beam annealing.

Fig. 5. Total and substitutional ^{115}In concentration in Si after ion beam annealing.

annealing ion beam. The basic annealing mechanism is epitaxial regrowth of the amorphous layer initiated by beam heating. Growth rates were found to be retarded in these regions.

REFERENCES

1. L. Csepregi, E. F. Kennedy, J. W. Mayer, and T. W. Sigmon, J. Appl. Phys. 49, 3906 (1978).
2. L. Csepregi, R. P. Kullen, J. W. Mayer, and T. W. Sigmon, Solid State Commun. 21, 1019 (1977).
3. T. O. Sedgwick, J. Electrochem. Soc. 130, 484 (1983).
4. G. L. Olson, S. A. Kokorowski, J. A. Roth, and L. D. Hess, p. 125 in Laser and Electron-Beam Solid Interactions and Materials Processing, ed. by J. F. Gibbons, L. D. Hess, and T. W. Sigmon, North Holland, New York, 1981.
5. S. A. Kokorowski, G. L. Olson, and L. D. Hess, J. Appl. Phys. 53, 921 (1982).
6. J. Narayan and O. W. Holland, Phys. Stat. Sol. (a) 73, 225 (1982).
7. M. Tamura, K. Yagi, N. Sakudo, K. Tokiguti, and T. Tokuyama, International Conference on Ion Beam Modification of Materials, Budgapest, 4-8 September 1978.
8. G. F. Cembali, P. G. Merli, F. Zignani, Appl. Phys. Lett. 38, 808 (1981).
9. G. Cembali, M. Finetti, P. G. Merli, and F. Zignani, Appl. Phys. Lett. 40, 62 (1982).
10. E. Gabilli, R. Lotti, P. G. Merli, R. Nipoti, and P. Ostoja, Appl. Phys. Lett. 41, 967 (1982).
11. J. Narayan and O. W. Holland in Metastable Materials Formation by Ion Implantation, ed. by S. T. Picraux and W. J. Choyke, North Holland, New York, 1982.

ION-BEAM PROCESSING OF ION-IMPLANTED Si

H.B. DIETRICH, R.J. CORAZZI, AND W.F. TSENG
Naval Research Laboratory, Washington, DC 20375

Abstract

Substrates can undergo major temperature excursions during ion implantation if they are not well heat sunk. At power densities on the order of 50 watts per cm^{-2} radiatively cooled Si will melt in a matter of seconds. Such power densities can be maintained over a few sq. cms with many of the beams produced by even the moderate current machines currently used for doping Si and the III-V's. We have made use of this fact to study pulsed ion-beam annealing of implanted Si. Two types of studies have been carried out. In the first, 5-20 sec proton irradiations were done at power densities of 3-35 watts cm^{-2} to produce sample temperatures of 500 to 1100°C. $2x10^{16}cm^{-2}$ 280 keV B, BF_2, As and P implants were annealed in this manner. Sheet resistances, ρ_s, versus power density curves were obtained for each ion and compared to ρ_s vs T data obtained for furnace annealed companion samples. In the second study the $2x10^{16}cm^{-2}$ 280 keV implants were carried out at progressively higher current densities so that the dopant beam itself raised the sample temperature to 500-1000°C. For each ion (other than B) it was possible to obtain power densities which resulted in self-annealing implants whose sheet resistances were as low as those obtained with the optimal furnace anneal. Details of the experiments, electrical and physical properties of the pulsed ion-beam annealed layers and device applications will be presented in this paper.

Introduction

Some years ago, Freeman[1] showed that if a Si wafer is restricted to radiative cooling its temperature will increase to something on the order of 1000°C, within seconds after it begins to absorb power at a level of 10's of watts/cm^2. This fact has been made use of to do transient annealing with furnaces, lamps and charged particle beams. It also explains why, if not well heat sunk, substrates undergo major temperature excursions during ion implantation. In fact, if some minimal care is taken to minimize conduction losses the dopant beam can be used to bring samples to temperatures typical of post implantation annealing, during the implantation process.

Merli and Zignani[2] first drew attention to the possibility of these self-annealing implants (SAI) and later Merli and co-workers published a series of papers showing the feasibility of this process.[3,4,5] In this paper, we report on the results of self-annealing As, P and BF_2 implants and we also give data on B, BF_2 and As implants annealed with 5-20 sec proton beam irradiations.

Experimental

For this work, 0.8 cm^2 Si samples (8.3x9.5 mm) were mounted free standing in a 1.3mm slot in a stainless bar and a 0.5 cm^2 area (7x7 mm) was irradiated with protons or the self-annealing implant. The active dopants were implanted into lightly doped substrates of the opposite sign. All proton irradiations and self-annealing implants were done at 280 keV. Sheet resistances were measured with a four-point probe. Damage and profile information were obtained from 2 MeV alpha particle RBS measurements. Additional profile information was obtained from SIMS profiling.

Results

280 keV 3×10^{15} cm^{-2} B implants were done at room temperature and and the Si substrate subsequently annealed with 20 second 280 keV proton beams of from 10 to 100 microamps. For the geometry used in these experiments, 7 watts/cm^2 corresponds to 20 microamps. The sheet resistance measured as a function of power density in the annealing proton beam is plotted in Fig. 1. Data were obtained for both back and implanted side proton irradiations. For the B implants, no significant difference was observed in the measured sheet resistances, or dopant profiles obtained from SIMS measurements, for these two anneal configurations. In Fig. 2, the B proton anneal data is replotted as a function of substrate temperature and compared to the values obtained subsequent to a 30 min furnace anneal. The temperature attained during the proton irradiation was measured with an optical pyrometer where possible and calculated from an extrapolation of the measured temperatures beyond the range of the pyrometer. The values are judged to be minimum values and could be in error by as much as 100°C. As can be seen in the plots, the transient anneal and furnace anneal curves are qualitatively similar and the minimum sheet resistances obtained are equivalent. At 35 watts/cm^2 the sheet resistance measured subsequent to a 5 second anneal was the same as that observed after a 20 second one.

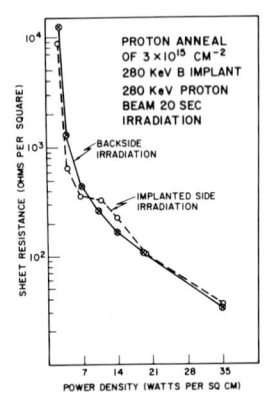

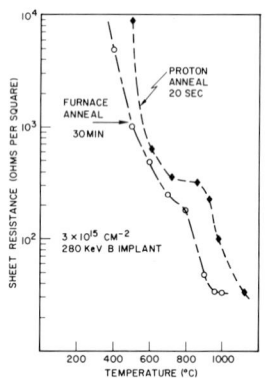

Fig. 1. Sheet resistance versus input power density for proton annealed B implanted layers.

Fig. 2. Sheet resistance versus temperature for proton annealed B implant data from Fig. 1.

In Fig. 3, the sheet resistances measured for proton annealed and self-annealing BF$_2$ implants are compared. At power densities of 35 watts/cm^2 and less, the self-annealing implants result in sheet resistances which are significantly higher than those obtained with the proton anneals. However, at power densities greater than this the two anneal methods result in similar sheet resistances and they compare favorably with the values obtained with furnace annealing. Figure 4 shows the sheet resistances measured subsequent to the proton anneal of 1×10^{16} cm^{-2} 190 keV As$_2$ implants. The curve is indicative of a furnace anneal of a heavy As implant where the amorphous layer crystallization results in high carrier concentrations even after a comparitively low temperature anneal. In Fig. 5 are data for self-annealing As and P implants. Sheet resistances as low as 15 ohms/square can be attained at power densities which correspond to a substrate temperature on the order of 600°C and the minimum values obtained again compare favorably with those obtained with furnace annealing.

A measure of the residual lattice damage can be obtained from dechanneling measurements. Spectra for proton annealed and self-annealed As implants are shown in Fig. 6. The energies of the two As implants are different, the proton annealed samples were 190 keV As$_2$ implants and the self-annealing implants were 280 keV, but the doses are the same and the power densities

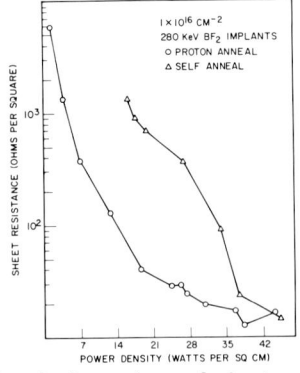

Fig. 3. Comparison of sheet resistances measured on proton-annealed and self-annealed BF$_2$ implanted layers.

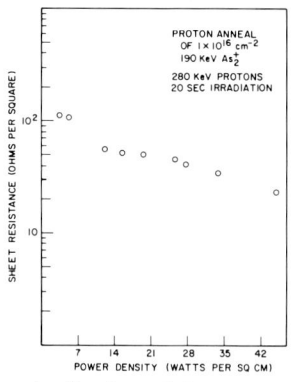

Fig. 4. Sheet resistance versus input power density for proton annealed As implanted layers.

similar. As can be seen in the figure, the lattice recovery is slightly better for the proton annealed sample and it is the same as is observed for a 900°C furnace anneal. The self-annealing implant took 15 seconds as compared to the 20 second proton anneal.

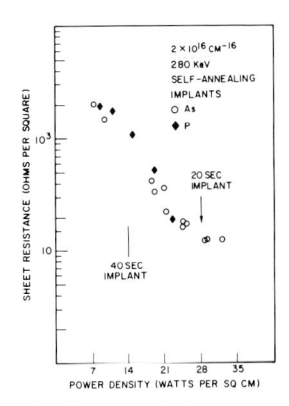

Fig. 5. Sheet resistance versus input density for self-annealing P and As implants.

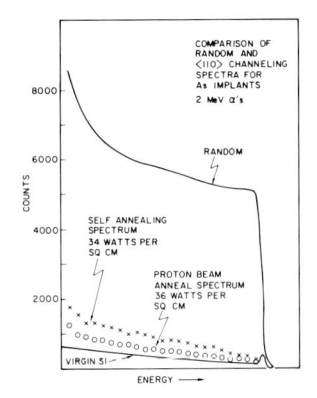

Fig. 6. Channeled and random RBS spectra for proton-annealed and self-annealed As implants.

Figures 7 and 8 show the measured As distribution for the proton annealed and self-annealed implants associated with the dechanneling spectra shown in Fig. 6. The As in the proton annealed layer shows no diffusion when compared to the as implanted profile and it is 91% substitutional. The As distribution in the self-annealed layer is strikingly different. First the As has diffused some 0.75 microns into the bulk and it is only 75% substitutional. Secondly, both the random and channeled distribution show a depletion of As from the region around R_p. The 900°C furnace annealed layers show little diffusion and no depletion in the random spectrum but there is a depletion at

R_p in the distribution measured with the channeled beam. The lower power self-annealing implants show less As redistribution, a smaller degree of depletion at the maximum range and lower substitutional fractions.

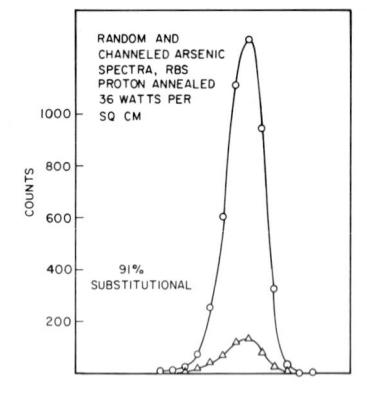

Fig. 7. Arsenic profiles for proton annealed layer taken from random and channeled RBS spectra.

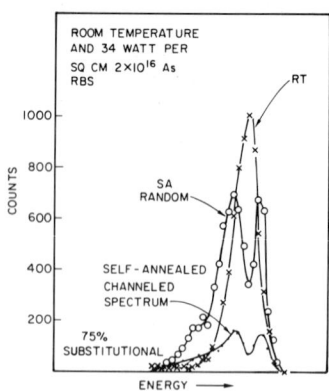

Fig. 8. Arsenic profiles for self-annealed layer taken from random and channeled RBS spectra.

Figure 9 shows the dechanneling spectra obtained for proton and self-annealing BF$_2$ implants. Comparison with Fig. 6 shows that the residual damage level in these layers is much higher than that found in the As implanted material and interestingly enough in the BF$_2$ case the self-annealed implants show a significantly lower degree of damage than the proton annealed sample. SIMS profiles show that the B and F redistribution are also different from the As case, in that there is redistribution in the proton annealed case which is measurably greater than that observed for the self-annealing implant.

In Fig. 10, a SIMS profile of an as implanted and proton anneal B layer is shown. There is a slight redistribution of the B during the 20 second anneal but it is similar to that observed during a 30 minute furnace anneal.

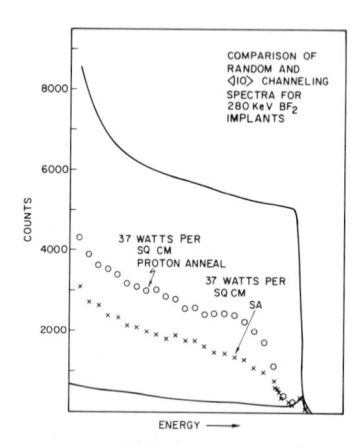

Fig. 9. Channeled and RBS spectra for proton annealed and self-annealed BF$_2$ implants.

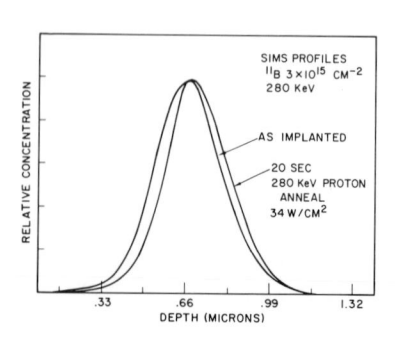

Fig. 10. SIMS profiles for as implanted and proton annealed B implanted layers.

Applications

The application of the type of self-annealing implants described here is under one major constraint. The dose associated with the required temperature rise is on the order of 10^{16} cm^{-2} at implant energies on the order of 300 kV, higher at lower energies and lower at higher energies. This restricts one to those applications which require high dose implants. The Italians[5] identified an excellent application in Si pn junction solar cells. Another possibility is the fabrication of self-aligned NMOS in which the thermal processing is supplied by a final self-annealing source-drain n$^+$ implant. We have fabricated transistors in this manner. The results of this work will be reported elsewhere.

Summary

In this work, we have shown that 5-20 second 280 keV proton irradiations can be used to anneal implanted layers in Si. The lattice recovery resulting from these anneals at power densities of 35 watts/cm^2 is typical of that seen with furnace anneals as are the sheet resistances and substitutional fractions. We have also shown that self-annealing As and P implants can be carried out at power densities of 20 watts/cm^2 or higher. In the case, although the sheet resistances obtained are approximately equal to those obtained with furnace annealing the lattice recovery is less complete and the diffusion is found to be greater for at least the As case. Self-aligned Si NMOS transistors can be fabricated with a final self-annealing As SD implant providing the required anneal process. The details of this work will be reported elsewhere.

References

1. Ion Implantation, edited by G. Dearnaley, J.H. Freeman, R.S. Nelson, J. Stephen, (North Holland Publishing Co., 1973), p. 421.
2. P.G. Merli and F. Zignani, Rad. Effects Lett. 57, 59 (1980).
3. G.F. Cembali, P.G. Merlie and F. Zignani, Appl. Phys. Lett., 38, 808 (1981).
4. G. Cembali, M. Finetti, P.G. Merli and F. Zignani, Appl. Phys. Lett., 40, 62 (1982).
5. E. Gabilli, R. Lotti, P.G. Merli, R. Nipoti and P. Ostoja, Appl. Phys. Lett., 41, 967 (1982).

ION-BEAM-INDUCED DAMAGING AND DYNAMIC ANNEALING PROCESSES IN SILICON

K.T. SHORT*, D.J. CHIVERS*[†], R.G. ELLIMAN*[φ], J. LIU**, A.P. POGANY*
H. WAGENFELD*, AND J.S. WILLIAMS*
*Microelectronics Technology Centre, RMIT Melbourne 3000 Australia
†Present address: AERE Harwell, England; φ Joint appointment with CSIRO
Division of Chemical Physics, Clayton 3168 Australia;
** Chinese Academy of Sciences, Beijing, China.

ABSTRACT

We have employed high resolution ion channeling and TEM methods to investigate the damage production and dynamic annealing processes which take place in (100) silicon bombarded at elevated temperatures. Two important observations have arisen from our results i) We have observed an amorphisation process for Sb-implanted silicon at 250°C which is more akin to amorphisation processes in metals, whereby the impurity (Sb) appears to influence the stability of amorphous zones associated with individual ion tracks. ii) We have demonstrated that previously amorphised layers in silicon can be recrystallised through a solid phase epitaxial process by subsequent bombardment with He$^+$, Ar$^+$ and Sb$^+$ ions at substrate temperatures of 300-400°C, which are significantly below normal thermal regrowth temperatures of >500°C.

INTRODUCTION

At room temperature and below, it is well known that the implantation of heavy ions into silicon generates stable displacement damage and leads to amorphisation [1]. For elevated temperature implants there is a clear competition between damage production and dynamic annealing during ion bombardment. The amount and nature of the residual damage following elevated temperature implantation is particularly sensitive to several implantation parameters including ion species, dose, dose rate [2], substrate orientation [3] and implantation temperature [4,5]. If the implantation temperature is sufficiently high (e.g.>300°C for Sb implantation) amorphous layers do not form and the residual damage consists of crystalline defect clusters and loops [3]. Furthermore, it has been recently demonstrated that it is possible to recrystallise previously amorphised silicon by ion bombardment of Ne$^+$[6] and As$^+$ [7] ions at temperatures of 300-400°C which are significantly below the normal thermal recrystallisation temperatures of >500°C [8].

In this paper we report on two aspects of damage production and dynamic annealing processes in (100) silicon during elevated temperature ion bombardment. Firstly, amorphisation processes are examined for Sb implanted silicon and an intriguing dose dependence of damage production is observed at 250°C. Secondly, we have observed epitaxial recrystallisation of previously amorphised silicon layers under He$^+$, Ar$^+$ and Sb$^+$ bombardment at substrate temperatures between 300° and 400°C.

EXPERIMENTAL

We have previously carried out Sb implantation into (100) silicon at implantation temperatures between LN$_2$ and 450°C. The temperature, substrate orientation and dose dependencies of the implantation-induced damage were studied by ion channeling and TEM and the results have been reported in detail elsewhere [3]. The first part of the present study concentrated on investigating the interesting damage structures which were observed at 250°C. In particular, 80keV Sb$^+$ ions were implanted into (100) silicon at 250°C for doses in the range $1 \times 10^{14} \text{cm}^{-2}$ to $5 \times 10^{15} \text{cm}^{-2}$. The amount of damage and the damage distribution were monitored using high resolution Rutherford backscattering and channeling of 2MeV He^{++} ions. The nature of the damage (i.e. crystalline defects or amorphous regions) was identified by TEM.

In the second part of this study, (100) silicon wafers were pre-amorphised with 80keV Sb ions to doses of $1 \times 10^{14} \text{cm}^{-2}$ and $1 \times 10^{15} \text{cm}^{-2}$ at LN$_2$ temperatures. These wafers were further bombarded with He$^+$, Ar$^+$ and Sb$^+$ ions to various doses at 80keV, with the wafers held at temperatures between 200-400°C. The dose rates of these elevated temperature bombardments were kept low ($\lesssim 1\mu\text{Acm}^{-2}$) to avoid any significant local heating of the wafer with the ion beams. High resolution channeling techniques were used to monitor beam-induced epitaxial recrystallisation of the previously-amorphised layers.

RESULTS AND DISCUSSION
Amorphisation at Elevated Temperatures
Fig. 1 shows high resolution channeling spectra indicating the build up of damage with dose for 80keV Sb implantation of (100) Si at 250°C. It is evident that doses of $< 2 \times 10^{14} \text{cm}^{-2}$ result in very low levels of disorder at 250°C despite the fact that a dose of only $4 \times 10^{13} \text{cm}^{-2}$ is sufficient to generate an amorphous layer of ~500Å thickness for implants carried out at room temperature and below [1,2]. However, from Fig.1, a dose of 5×10^{14} Sb cm^{-2} produces at 250°C a sharp increase in damage up to the random level as measured by channeling. This damage further increases in thickness with increasing dose above $5 \times 10^{14} \text{cm}^{-2}$ consistent with the generation of an amorphous layer continuous to the surface

In Fig.2, the disorder peak area, N_d, obtained from Fig.1 (normalised to the $5 \times 10^{14} \text{cm}^{-2}$ value), is plotted as a function of implant dose. The dramatic increase in disorder at doses between $2 \times 10^{14} \text{cm}^{-2}$ and $5 \times 10^{14} \text{cm}^{-2}$ is clearly shown. Furthermore, TEM has indicated that no amorphous disorder is associated with doses of $< 2 \times 10^{14} \text{cm}^{-2}$. The disorder consists, in these cases, of crystalline defect clusters and loops which have a low enough density so as not to be readily apparent in channeling spectra. In contrast, TEM showed clearly resolved amorphous layers for doses $>5 \times 10^{14} \text{cm}^{-2}$.

We interpret the dose dependence of implantation damage at 250°C with respect to the schematic representation of the damage structure in Fig.3. At doses $< 2 \times 10^{14} \text{cm}^{-2}$ our results suggest that individual Sb ions do not leave any residual amorphous zones in the silicon lattice. Overlap of many collision cascades, to a level where each silicon lattice atom has been displaced at least 6 times (at the peak of the damage for a $2 \times 10^{14} \text{cm}^{-2}$ implant), also results in the absence of any detectable amorphousness.

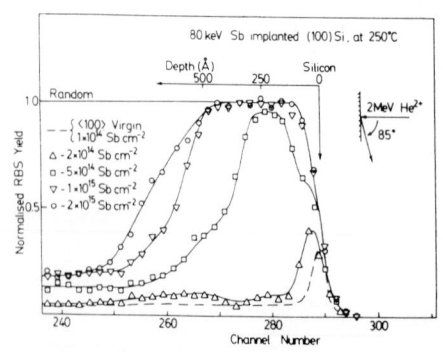

Fig.1. High resolution < 100> channeling spectra of 80keV Sb implanted (100) Si showing damage build up as a function of dose for implantation at 250°C.

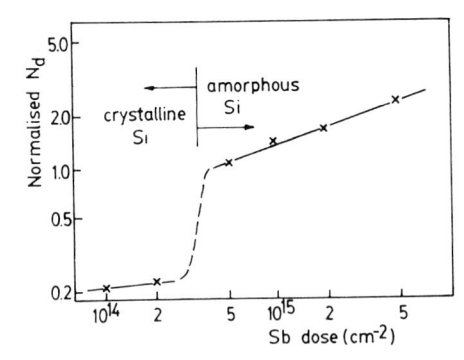

Fig.2. Plot of disorder (N_d) as measured by <100> channeling as a function of implant dose, for Sb implantation of (100) Si at 250°C.

We therefore suggest that the formation of an amorphous layer, rather abruptly at a dose of ~5 x 10^{14}cm^{-2}, is not consistent with previous models of amorphous layer formation in semiconductors under heavy ion bombardment [9,11]. These models contend that the displacement damage or amorphous (spike-generated) zones associated with individual implanted ions increase in density with increasing dose to ultimately produce an amorphous layer. As the implantation temperature is increased the residual displacement or amorphous damage associated with individual ion tracks is reduced in magnitude as a result of competition between damage production and dynamic annealing processes. Nonetheless, damage build up is envisaged to result from the cumulative effect of individual ion damage[11]. Indeed, the damage production process that we observe for Sb-implanted silicon at temperatures much less than 250°C [3] is consistent with such a mechanism. However, at 250°C we believe that an alternate mechanism exists for amorphisation in silicon. In this regime the implantation damage associated with individual ions is initially completely annealed leaving only crystalline defect clusters. We speculate that at a critical dose of between 2 x 10^{14}cm^{-2} and 5 x 10^{14}cm^{-2} stable amorphous zones begin to form about individual ion tracks. We can envisage two possible mechanisms for the sudden stability of these amorphous zones; i) the build up of lattice strain associated with crystalline defect formation during the initial stages of bombardment, and ii) the build up of a high, peak Sb concentration (~ 5 x 10^{19}cm^{-3} by a dose of 2 x 10^{14}cm^{-2}) which may nucleate subsequent amorphous zone formation. The lattice amorphisation process (impurity stabilization) is expected to be the more likely one in view of the models which have been suggested to explain the amorphisation of metals under ion bombardment [12]. Indeed, although our damage results at 250°C for Sb-implanted silicon are not consistent with the expected behaviour of semiconductors under ion bombardment, the amorphisation process we propose is entirely consistent with models to account for amorphisation of metals. As a consequence, we would expect the amorphisation behaviour in silicon at 250°C to be extremely sensitive to implant species of

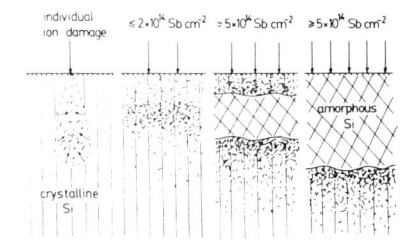

Fig.3. Schematic representation of implant damage build up as a function of dose for 250°C Sb implantation of Si.

the same mass but with different chemical properties. We are currently investigating this aspect.

Bombardment-Induced Annealing

Fig.4 shows high resolution ion channeling spectra which indicate the typical bombardment effects induced by 80 keV Ar$^+$ irradiation at 300°C, of a (100) silicon wafer previously amorphised using 80keV Sb$(1 \times 10^{14} cm^{-2})$. The as-implanted spectrum (open circles) indicates the amorphous layer thickness corresponding to a 1×10^{14}Sb cm^{-2} implant heated to 300°C: this spectrum is identical to that from an as-implanted sample not heated to 300°C, indicating that no significant annealing takes place at 300°C without simultaneous bombardment with Ar$^+$. The remaining spectra in Fig.4 show the effect of Ar bombardment to doses of 5×10^{14}cm^{-2} and 5×10^{15}cm^{-2} under irradiation conditions which keep the wafer at 300 ± 20°C. Clear epitaxial growth of the previously amorphised layer is observed, the extent of regrowth increasing with Ar dose. However, the Ar also generates deep damage close to the range of 80 keV Ar$^+$ in silicon (~ 1100Å). This damage consists of an agglomeration of crystalline defects (e.g. clusters, loops) as a result of the dynamic annealing of the displacement damage produced by the Ar ions at the end of their range.

The bimodal damage distribution observed in Fig. 4 for the highest dose Ar implant contains a residual but regrown surface amorphous layer and a deeper damaged layer produced by the Ar irradiation. This behaviour is typical of the damage structure observed for both Sb$^+$ and He$^+$ irradiation at temperatures of 300°C and 400°C, respectively. Although, the details of these studies are not shown here. The depth distribution and magnitude of the two components of the bimodal distribution vary dramatically with implant species and irradiation temperature. For example, for elevated temperature 80 keV Sb$^+$ irradiation of a preamorphised sample the near-surface amorphous layer and deeper crystalline layer overlap for low Sb fluences (~ 10^{14}cm^{-2}) and only a crystalline damaged layer remains following irradiation to fluences >2 $\times 10^{15}$cm^{-2}. In contrast, He$^+$ irradiation produces a small amount of very deep damage (>5000Å) well separated from the near-surface amorphous layer, which regrew epitaxially as a function of increasing He$^+$ dose. The He$^+$-induced epitaxial regrowth results are presented in detail elsewhere in these proceedings [13].

We describe the dynamic annealing processes presented above with reference to the schematic representations given in Fig. 5. We envisage that there may be two regimes applicable to the irradiation-induced recrystallisation processes. For heavy ions in which the displacement cascades overlap the pre-amorphized layer, we suggest a recrystallisation mechanism illustrated on the left hand side of Fig.5. In this case the individual ions produce

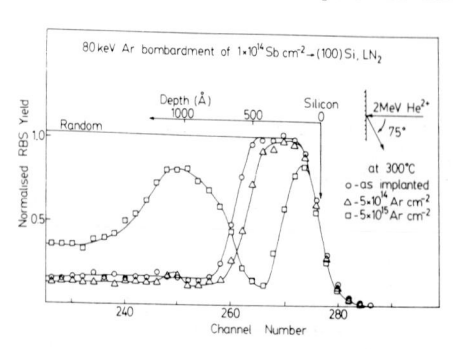

Fig. 4. High resolution <100> channeling spectra following 300°C Ar$^+$ bombardment of (100) Si, pre-amorphised with 1×10^{14}cm^{-2} Sb implants at LN$_2$

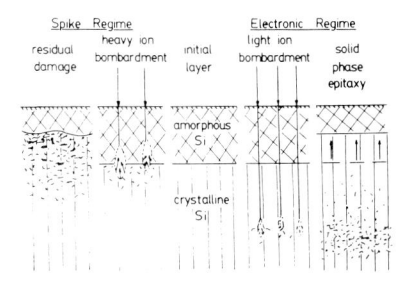

Fig. 5. Schematic representation of possible dynamic annealing processes occuring during >300°C ion bombardment of pre-amorphised Si.

displacement or thermal spikes [10] which completely recrystallise via dynamic annealing (consistent with Sb$^+$-implantation at 300°C [3]) and consequently recrystallise the pre-amorphised layer as the Sb$^+$ fluence increases. For lighter ions electronic energy loss processes are dominant to depths well below the pre-amorphised layer. As discussed in more detail elsewhere [13], two possible processes can be envisaged for beam-induced epitaxy. Firstly, mobile point defects (e.g. vacancies and interstitials) which are generated deeper in the crystalline silicon may migrate to the interface and stimulate epitaxial growth. Secondly, electronic energy loss may directly result in bondbreaking and consequential atom rearrangement at the amorphous-crystalline interface. We are currently undertaking further irradiation experiments which are aimed at distinguishing between these possible beam-induced epitaxial growth processes.

CONCLUSION

We have observed two intriguing aspects of damage production and dynamic annealing during ion bombardment of silicon at elevated temperatures. Firstly we have observed an amorphisation process in silicon for Sb-implantation at 250°C which is consistent with models for impurity stabilised amorphisation of metals. Secondly, we have demonstrated that pre-amorphised layers can be epitaxially recrystallised at temperatures of 300-400°C under He$^+$, Ar$^+$ and Sb$^+$ bombardment. Both of these observations clearly warrant further investigation.

ACKNOWLEDGEMENTS

The ARGC and The Commonwealth Special Research Centres Scheme are acknowledged for financial support. RGE also wishes to acknowledge the CSIRO Postdoctoral Award Scheme for financial support.

REFERENCES

1. J.W. Mayer, L. Eriksson and J.S. Davies: In "Ion Implantation in Semiconductors". Academic Press N.Y. (1970).
2. J.M. Poate and J.S.Williams: In "Ion Implantation and Beam Processing" J.S. Williams and J.M. Poate eds. Academic Press Sydney (1983).
3. I.F.Bubb, D.J. Chivers, J.R. Liu, A.P. Pogany, K.T. Short, H.K. Wagenfeld and J.S.Williams: Nucl. Inst. Meth., in press.
4. L. Csepregi, E.F. Kennedy, S.S. Lau, J.W. Mayer and T.W. Sigmon; Appl. Phys. Lett. 29, 645 (1976).
5. D.G. Beanland; Rad. Effects 33, 219 (1977).
6. B. Svensson, J. Linnros and G. Holmen; Nucl. Inst. Meth., 210, 755 (1983).
7. J. Nakata and K. Kajiyama; Appl. Phys. Lett. 40, 686 (1982).

8. L. Csepregi, E.F. Kennedy, T.J. Gallagher, J.W. Mayer and T.W.Sigmon. J. Appl. Phys. 48, 4234, (1977).
9. F.F. Morehead and B.L. Crowder: In "Ion Implantation", F.H. Eisen and L.T. Chadderton eds. Gordon and Breach London (1971).
10. D.A. Thompson, Rad. Effects 56, 105 (1981).

11. L.A. Christel and J.F. Gibbons: J. Appl. Phys. 52, 5050 (1981).
12. W.A. Grant: J. Vac. Sci. 17, 1663 (1978).
13. R.G. Elliman, S.T. Johnson, K.T.Short and J.S.Williams, these proceedings.

MEGAVOLT BORON AND ARSENIC IMPLANTATION INTO SILICON

P.F. Byrne[1], N.W. Cheung, S. Tam[2], C. Hu,
Y.C. Shih, J. Washburn, and M. Strathman[3]
University of California, Berkeley, California 94720
1. current address LSI Logic Corp., Santa Clara, CA
2. current address Intel Corp., Santa Clara, CA
3. current address Charles Evans and Assoc., San Mateo, CA

Abstract

Formation of buried p-type and n-type layers in (100) silicon has been accomplished by implanting with 4 MeV boron and 11 MeV arsenic ions respectively. The projected range (R_p) of 4 MeV boron is 5.2 microns with a straggle (ΔR_p) of .2 microns. The 11 MeV arsenic implant has a R_p of 4.37 microns with a ΔR_p of .37 microns. The 4 MeV boron implant was carried out to a dose of $1 \times 10^{15}/cm^2$ while the 11 MeV arsenic implant dose was $1.9 \times 10^{15}/cm^2$. For both dopants the target holder could be cooled with either liquid nitrogen (LN) or flowing room temperature water (RT). Buried amorphous regions are seen by cross sectional transmission electron microscopy (XTEM) for both boron and arsenic when LN cooling is used. Arsenic shows a buried amorphous region for the RT case as well. The extent of the buried amorphous regions are compared with the energy deposited into nuclear stopping as determined by computer simulation. Threshold levels are determined for the creation of these buried amorphous regions. The boron samples were annealed for 30 minutes at 900°C in a nitrogen ambient, and XTEM shows no residual damage for both cooling conditions. The arsenic samples underwent a two step annealing procedure; 545°C for 16 hours followed by a 945°C step for 15 minutes. Regions containing dislocation networks are observed by XTEM for both cooling conditions.

The 4 MeV implanted boron buried layer was applied to an NMOS process. Transistors fabricated above the p-type buried layer show channel mobility, threshold voltage, and sub-threshold leakage which are indistinguishable from transistors fabricated without the buried layer. Vertical npn bipolar transistors have been fabricated using the 11 MeV arsenic buried layer to form the collector. Electrical characteristics from both these devices indicate that megavolt ion implantation can be applied to silicon for active device geometries.

Introduction

Ion implantation in the conventional energy range of 100 to 400 keV has replaced silicon doping through thermal processes for many integrated circuit applications. The ability to directly monitor and control the ion beam dose and energy during implantation gives this technique precision that cannot be achieved in a diffusion furnace. It is this precision that makes ion implantation more attractive than a batch furnace process where many wafers at a time can be doped. Some applications such as threshold control are possible by ion implantation, but are not practical through diffusion.

Because of this precision in energy and dose control, ion implantation in the megavolt regime has potential to replace chemical-vapor-deposition (CVD) epitaxial growth for the formation of buried dopant layers. CVD epitaxy, like diffusion, is a batch process which is critically dependent on times, temperatures, and gas concentrations for uniformity and reproducibility. A further advantage of an ion implanted buried layer over a CVD epitaxial layer is

the freedom to do the implant after high temperature steps that would redistribute the dopant. This freedom is not available for an epitaxial buried layer where the epitaxial growth step must precede any further high temperature modification of the epitaxial silicon.

Ion implantation in the megavolt and near megavolt regime has made possible some novel and perhaps important device structures. These structures include a vertical JFET [1], bipolar transistors [2,3], a retrograde CMOS transistor [4], and a buried interconnect [5].

To determine the approximate profiles of boron and arsenic in the megavolt regime design curves were produced for R_p and ΔR_p from the theory of Lindhard, Scharff, and Schiott (LSS) [6]. These are presented in Figures 1 and 2. The electronic stopping power used by Brice [7] has been applied to the boron case since the LSS electronic stopping power is not valid for boron at energies where the incoming ion is stripped of its electrons. Also plotted on these Figures are the data obtained by Maby [8], Kostka and Kalbitzer [9], and Takahashi, Nakata, and Kajiyama [10]. Agreement is found between the data and experiment except for ΔR_p from reference [9]. This digression from the theory indicates need for more work in this area. An upper limit on the energy for applications in the integrated circuit area may be the coulomb barrier energy for these ions on silicon. This energy has been estimated to be 10.5 MeV for boron and 63.2 MeV for arsenic [11]. Above these energies residual radiation becomes an issue.

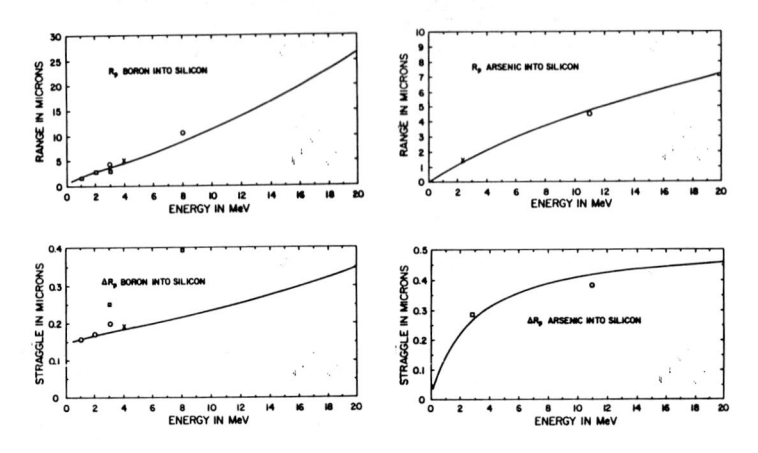

Figures 1 and 2. The calculated projected range (R_p) and parallel straggle (ΔR_p) for boron and arsenic into silicon as functions of energy. The data points are taken from references 5,8,9,10, and this work.

Residual Damage

When megavolt boron or arsenic is used to form a buried layer for a solid state device, the active device area is contained in a region that has been damaged by the passage of the ions. The annealing of damage and understanding the nature of the residual defects is therefore essential for device applications of megavolt implantation. Figures 3 and 4 show XTEM micrographs for the 4 MeV boron implant while Figures 5 and 6 show micrographs for the 11 MeV arsenic implant where both cooling conditions have been applied. Buried amorphous regions are seen for the LN case for both dopants while the RT case only shows a buried layer for the arsenic implant. It is important to note that even for arsenic at this high dose an amorphous region continuous to the surface is not formed for LN cooling. This implies that annealing of the damage will proceed from two fronts when a buried amorphous region is formed. The extent of these amorphous regions is compared with computer simulations for the amount of energy deposited into nuclear stopping. These simulations are based on the programs developed by Brice [7]. Plots of the energy deposited into nuclear stopping and the threshold energies for the formation of amorphous regions are shown in Figures 7 and 8. These threshold numbers correspond to energy densities of 2.9×10^{20} keV/cm^3 for boron at LN temperatures, and 4.56×10^{21} keV/cm^3 and 2.1×10^{21} keV/cm^3 for the RT and LN cases of arsenic.

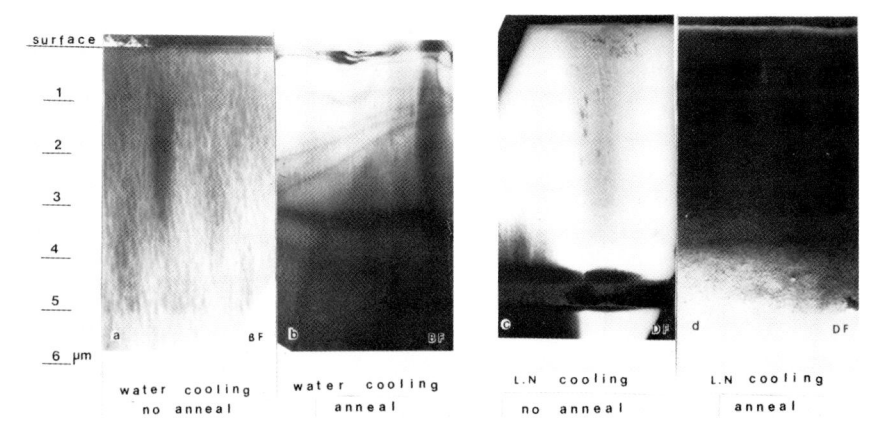

Figures 3 and 4. XTEM micrographs for boron implanted into silicon at an energy of 4 MeV and a dose of $1\times10^{15}/cm^2$. Micrographs a and b are dark field exposures while c and d are bright field exposures. A buried amorphous region is shown in micrograph c, and no other extended defects are observed in the other micrographs.

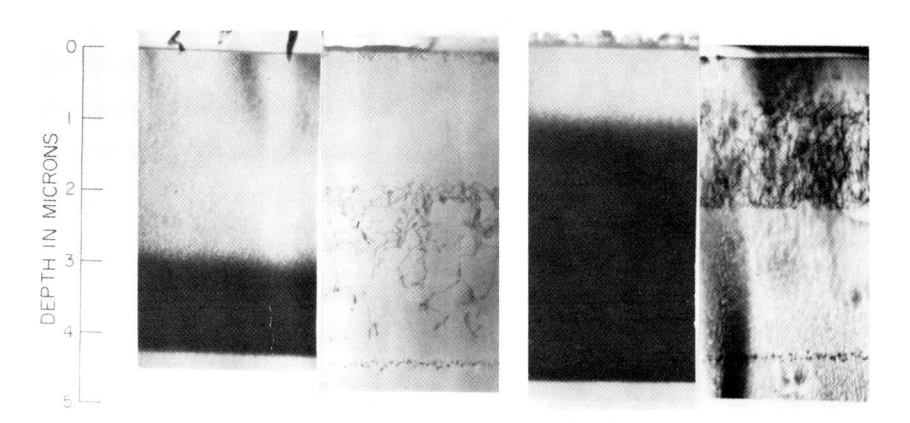

Figures 5 and 6. XTEM micrographs for arsenic implanted into silicon at an energy of 11 MeV and a dose of $1.9\times10^{15}/cm^2$. These micrographs are all bright field exposures. The annealing sequence is the same as Figure 3. Dislocation networks are observed for both annealed samples.

After implantation the silicon was furnace annealed in a nitrogen ambient. The boron was annealed at 900°C for 30 minutes. This anneal has been shown to produce minimal residual damage for megavolt boron implantation into silicon [9]. XTEM micrographs of the silicon substrates after this anneal are shown in Figures 3 and 4. There are no visible extended defects. The arsenic implanted samples were annealed by a two step process. The first step was at 545°C for 16 hours. It was then followed by a 945°C for 15 minutes. The XTEM micrographs for the annealed samples are shown in Figure 5 and 6. The dislocation lines show that annealing proceeded from both an upper and a lower interface. These dislocations would be detrimental for devices with junctions located at the depths where the dislocations occur. It is seen that for the arsenic RT case that the upper two microns of material is dislocation free after annealing. It is this upper two microns that is used to form the base region of a bipolar transistor that was fabricated using 11 MeV arsenic to form the base region. The results show that further work is needed to optimize the arsenic annealing.

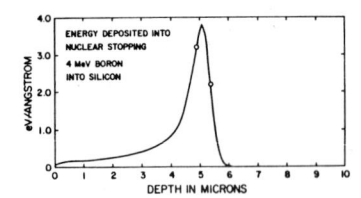

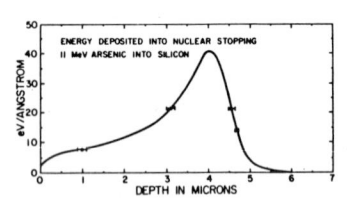

Figures 7 and 8. Energy deposited into nuclear stopping as a function of depth for 4 MeV boron and 11 MeV arsenic into silicon. The locations of the amorphous to crystalline interfaces are shown for the LN cooled boron implant, and the LN and RT cooled arsenic implant.

Masking for Megavolt Implantation

To define the lateral extent of an implanted region, a masking material needs to be deposited or thermally grown, and then patterned. It should be noted that masking is not necessary for some applications of megavolt ion implantation. For example, the NMOS structure described in the next section require a blanket implant over the whole wafer, and therefore does not need a masking step. Other applications such as the bipolar structure need lateral definition of the implanted region.

For conventional ion implantation; photoresist, silicon dioxide, or silicon nitride can be used to mask the implant. For implantation in the megavolt regime, impractical thicknesses of photoresist, silicon dioxide, or silicon nitride would be needed. Instead, for the 11 MeV arsenic implants tungsten was sputtered deposited and defined in a fluorine based plasma. Tungsten, being refractory, did not buckle or experience adhesion problems for the 11 MeV arsenic exposure which was performed at 1.2 μamps particle current. A gold-on-chromium mask under similar conditions buckled. Figures 9 and 10 contain plots of R_p and ΔR_p for boron and arsenic into tungsten. These curves were used to determine the required thickness for a tungsten mask.

A constraint on the masking of megavolt ion implantation comes from the transverse straggle of the beam. Figure 11 shows this straggle ΔR_t for boron and arsenic into silicon based on an LSS computer simulation. We see that in the megavolt regime ΔR_t is significant when compared with device dimensions. As can be seen in Figure 11, this is not the case for ions with energies of a few hundred keV.

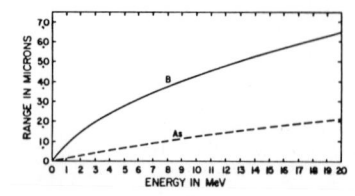

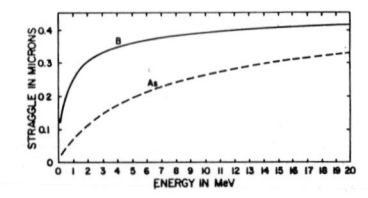

Figures 9 and 10. The calculated projected range (R_p) and parallel straggle (ΔR_p) for boron and arsenic into tungsten as functions of energy.

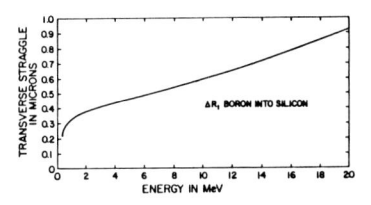

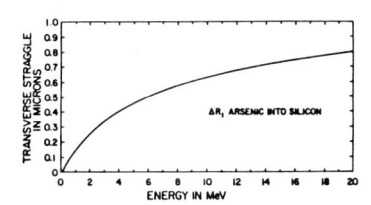

Figure 11. The calculated transverse straggle for boron and arsenic into silicon as functions of energy.

Devices

Using the 4 MeV boron implant, a modification to NMOS processing with a buried p-type layer was achieved. This buried layer should reduce the alpha particle susceptibility of the device [12]. The process began with a blank (100) silicon wafer which was implanted with 4 MeV boron in 1 cm^2 spots to a dose of 1×10^{15} / cm^2. The other areas of the wafer did not receive any 4 MeV boron. The wafer was then annealed at 900°C for 30 minutes, and a standard local oxide NMOS transistor structure was fabricated on the wafer. After fabrication the implanted and unimplanted transistors were compared. No differences were seen in the I-V characteristics, mobility, or threshold voltages for the two types of transistors. The I-V characteristic for the implanted transistor is shown in Figure 12. This transistor had a masked channel length of 10 microns and a width of 50 microns. Subthreshold leakage was also the same between both sets of transistors. This indicates that the residual damage in the surface region has been reduced to the point where it does not effect device performance. A plot of the subthreshold leakage for the implanted structure is shown in Figure 13.

The 11 MeV arsenic implant was used to form the collector for a vertical npn bipolar transistor. A schematic cross section for this device is shown in Figure 14. For the base doping, the background p-type concentration of 1×10^{17} / cm^3 was used. This transistor structure does not have the epitaxial growth step or the base doping step used in conventional bipolar transistors. The collector is implanted after the isolation wells are diffused, and therefore the collector does

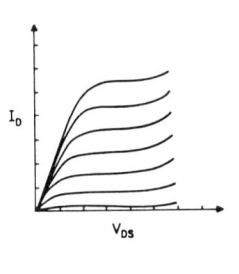

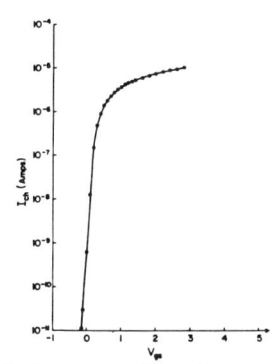

Figure 12. The drain current as a function of drain-source voltage for an NMOS transistor. The vertical scale is 2 mamps per division while the horizontal scale is 2 volts per division. The gate voltage was stepped in 1 volt increments.

Figure 13. The channel current is plotted as a function of gate-source voltage for the NMOS transistor. No significant sub-threshold leakage is observed.

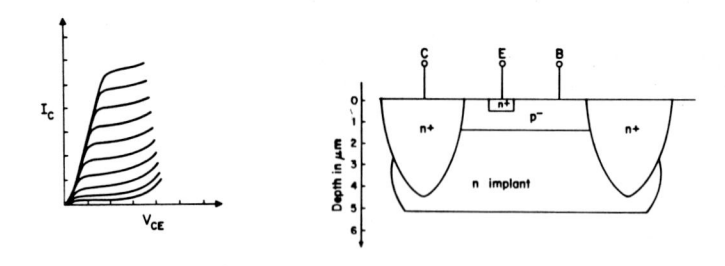

Figure 14. The collector current as a function of collector-emitter voltage for the npn bipolar transistor. The vertical scale is 2 mamps per division while the horizontal scale is .5 volts per division. The base current was stepped in 50 µamp increments.

Figure 15. Schematic cross section of the npn bipolar transistor.

not experience the the redistribution inherent with the epitaxial buried layer process. Figure 15 shows the I-V characteristic of this device. These devices showed a gain of forty. The low punch through voltage is due to the lack of an n$^-$ region in the collector that is available in the standard process. An optimized bipolar transistor would include an implant for this n$^-$ region as well as forming the isolation wells by multi-energy implantation.

References

1. D.P. Lecrosnier, and G.P. Pelous, IEEE Trans. Electron Devices, ED-21, (1974) 113.
2. J.F. Ziegler, B.L. Crowder, and W.J. Kleinfelder, IBM J. Res. Develop., Nov., (1971) 452.
3. M. Doken, T. Unagami, K. Sakuma, and K. Kajiyama, IEDM, (1981) 586.
4. R.D. Rung, C.J. Dell'Oca, and L.G. Walker, IEEE Trans. Electron Devices, ED-28, (1981) 1115.
5. P.F. Byrne, N.W. Cheung, and D.K. Sadana Thin Solid Films **95**, pg. 363 (1982).
6. J. Linhard, M. Scharff, and H.E. Schiott, Kgl. Danske Videnskab. Selskab., Mat-fys. Medd. **33**, (1963).
7. D.K. Brice, SAND75-0622, Sandia Laboratories, 1977.
8. E. W. Maby, Ph. D. thesis, Massachusetts Institute of Technology, (1979).
9. A. Kostka and S. Kalbitzer, Rad. Effects **19**, 77 (1973).
10. M. Takahashi, J. Nakata, and K. Kajiyama, Proc. of the 5th Symposium on Ion Source and Ion Assisted Technology, Tokyo (1981).
11. P.F. Byrne, J. Appl. Phys. **54**, pg. 1146 (1983).
12. A. Silburt, and R. Foss, Electronics, Nov. pg. 155 (1982).

LATTICE SITE LOCATION OF GROUP VII IMPURITIES IN SILICON

D.O. BOERMA, P.J.M. SMULDERS, AND T.S. WIERENGA
L.A.N. and Materials Science Centre, University of Groningen,
Westersingel 34, 9718 CM Groningen, The Netherlands

ABSTRACT

Chlorine- and iodine-implanted silicon have been investigated using the channeling/RBS technique after laser and oven annealing. Both annealing methods result in good epitaxial recrystallization of the implanted layer. A large near-substitutional I fraction exceeding 90% was found after oven annealing at 875°C. The near-substitutional fractions found after laser annealing for Cl and I, and after thermal annealing at 900°C for Cl are in the order of 50% only. The precise lattice site for these Cl and I fractions are determined by fitting the angular scans with simulated results. The results are interpreted in terms of vacancy association.

INTRODUCTION

Implantation of impurities is a widely used process in the semiconductor technology [1]. Implantation produces well-defined doped regions with steep profiles and has proved to be of great value for manufacturing a variety of semiconductor devices. Relatively little is known about implanted group VII impurities in silicon. The location of iodine in silicon was previously investigated by ^{129}I Mössbauer spectrocospy on ^{129m}Te-implanted and laser-annealed crystals. It was found that the I-atoms introduced into silicon via the decay of ^{129m}Te are either part of a cubically symmetric or a (Jahn-Teller) distorted configuration, dependent on the charge state [2,3]. Tellurium has been found to occupy mainly substitutional sites in silicon after laser annealing [4]. In previous channeling studies [4,5] on iodine-implanted silicon it was found that after laser annealing only 60% of the I-atoms are at near-substitutional sites, and a considerable redistribution and loss of impurities was observed. However, after thermal annealing, about 90% of the I-atoms were found to occupy near-substitutional sites. To our knowledge nothing is known about the site of other group VII impurities in silicon. It has been shown [6] that after thermal annealing to 1000°C chlorine-implanted silicon exhibits a donor character.

In this work we report results of an analysis of channeling data on iodine and chlorine in Si. A preliminary account of the iodine experiment was published earlier [5].

EXPERIMENTAL DETAILS

Single crystals of <100> orientation of pure silicon were implanted with ^{127}I ions or with ^{37}Cl ions. To circumvent ion channeling the implantations were done with the beam at an angle of 7° with the normal string. After implantation part of the samples were oven-annealed in a quartz tube, either in vacuum or in a flow of argon. Other samples were laser-annealed in air using a Q-switched ruby laser with a pulse width of about 30 ns and a power density of 1.5 Jcm^{-2}. A surface layer of ∿ 5 nm was removed from the iodine-implanted samples by anodic oxidation and subsequent etching in a dilute HF-solution. Further details on the sample preparation are given in table I.

Table I. Survey of preparation of samples used in this work.

Crystal	Implantation dose, energy depth ± width (nm)	Anneal	Experiment for which the sample was used.
#1	$5 \cdot 10^{14}$ I 110 keV 49 ± 15	875°C, 30 min vacuum	Displacement cross section <100> Angular scan <100>
#2	"	"	Angular scans <100>, <111>
#3	$2 \cdot 10^{15}$ Cl 115 keV 119 ± 42	875°C, 30 min argon flow	check of recrystallization of silicon
		L.A. 1.5 J/cm²	Angular scans <100>,<110>,<111> Displacement cross section <110>
#4	$1 \cdot 10^{15}$ Cl 54 keV 55 ± 25	900°C, 60 min argon flow	Angular scans <100>,<110>

For the channeling work a two-axes goniometer provided with an x-y translational system was used. The goniometer was mounted in a vacuum system (10^{-5} Pa) within LN_2 cooled shields to prevent carbon build-up. The samples were cooled to temperatures between 100 K and 120 K. The cooling was done to decrease the influence of the thermal vibrations of the atoms on the analysis and to reduce the association of beam-induced damage with the impurities. Beams of He⁺ ions of 2 MeV (iodine measurements) or 3 MeV (chlorine measurements) from the Groningen 5 MV Van de Graaff accelerator were used. The beam was collimated to 0.7 × 1.4 mm² with a beam divergence of 0.04°. Three cooled (-30°C) silicon detectors were used at backward angles. The beam current was stabilized to obtain a constant count rate and pile-up rejectors were applied. The resulting background under the I-peak on the spectra was negligible. The background was not small relative to the Cl-peak.

The depth profiles of the impurities and the crystallinity of the samples were measured before and after annealing. After annealing the increase in the minimum yield of silicon and iodine in the <100> direction and of silicon and chlorine in the <110> direction was measured as a function of the beam charge. The increase for silicon was about equal in both cases. For iodine the increase was large (4% per 10^{15} He⁺ atoms/cm²), whereas we found no increase for chlorine.

Angular scans were taken through the three major strings of silicon keeping the angle φ with respect to a plane through the string at a constant value. A new spot on the crystal was chosen after one or two points of the scan were measured (for iodine) or after a complete scan was done (for chlorine). With the latter method a fluctuation of the impurity concentration in different spots is not important. The frequent choice of a new spot was necessary to avoid the influence of crystal damage due to the probing beam, and particularly the displacement of the impurity due to the damage. For iodine about 20 spots were used for one scan. For chlorine 12 spots were used per scan.

DATA ANALYSIS

The yields χ_m for silicon were determined by summing the channel contents over a range corresponding to the depth of the impurity. The angular scans for the impurity were determined from the impurity peak content corrected for a (linear) background. The results were normalized to equal collected charge during measurement, taking into account the corrections for dead time and loss of pulses due to pile-up rejection. The results for the three detectors were added. For iodine values of χ_m for positive and negative values of the angle θ between the string and the scanning point were averaged to obtain "half scans". For chlorine in most cases only half of the scan was used in the analysis. It turned out that the shape of the second half of the scan was affected by beam-induced damage.

The theoretical shapes of the scans were calculated with a Monte Carlo program. For 11 points (θ,ϕ) of a scan trajectories of He^+ ions starting at random positions on the surface of a perfect crystal are traced along a string over a range corresponding to the depth of the impurity. The beam divergence and surface contamination are taken into account by introducing a two-dimensional Gaussian distribution of angles with a standard deviation of 0.06° around θ,ϕ. The trajectories are straight between traversed planes. At planes the direction is updated by taking into account the scattering angle due to a binary collision with the nearest atom, and also the deflection due to the field of the surrounding strings, treated in the continuum-string approximation. A potential due to Biersack [7] is used. The influence of the thermal vibration of the atoms, described by the Debye model with θ_D = 543 K is taken into account, as well as the energy loss and angular straggling due to interaction with electrons. The program generates the close-encounter yield of scattering with lattice atoms as a function of depth, and the particle flux as a function of lateral position, integrated over the depth with a weight given by the impurity depth profile.

In a second program this weighted flux is used to calculate the impurity yield χ_m for a given set of (equivalent) sites taking into account the thermal vibration amplitude of the impurity. The impurity yield χ_m calculated as a function of θ for many different sites chosen along the <100>, <110> and <111> directions serves as input for a third program, in which the measured impurity yield is fitted with a constant background plus one or more of the yields calculated for assumed positions. With this program the best fit as a function of impurity site(s) is searched. The silicon yield is fitted with the calculated close-encounter yield plus a constant background to take into account scattering by a disordered surface layer. In all fits the height of the measured scan is normalized to the theoretical value.

In the analysis it is assumed that shifts of impurities from substitutional sites are either in the <111>, <110>, or <100> directions. For displacements due to simple defects this may be realistic. To describe the thermal vibrations of iodine in silicon the Debye model was used. The Debye temperature θ_D was derived from a Mössbauer experiment in which the recoilless fraction of ^{125}I decaying to ^{125}Te was measured as a function of temperature in a single Si crystal along the three main crystal directions. Two different values for θ_D were found: θ_D = 160 K for vibrations in the <111> directions and θ_D = 213 K for directions perpendicular to <111>. This asymmetric vibration was taken into account by summing the yield for two positions displaced along <111> with respect to each other by 0.15 Å, the yields at the two positions being calculated for θ_D = 213 K. For chlorine in silicon an arbitrary value of θ_D = 300 K was assumed.

RESULTS

The depth profile of iodine and chlorine was found not to change after thermal annealing at 875°C. After this anneal complete recrystallization of the iodine-implanted silicon was observed. Only partial recrystallization of chlorine-implanted silicon was found. After thermal annealing for 60 min. at 900°C, and also after laser annealing of the chlorine-implanted wafers, com-

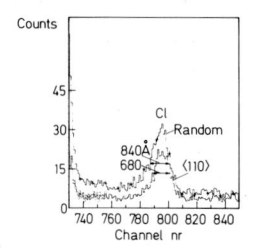

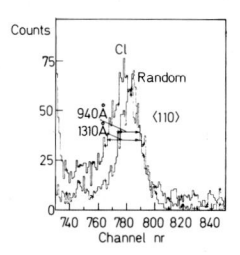

FIG. 1. Impurity part of RBS-spectra of 3 MeV α-particles incident in random and <110> aligned directions on a chlorine-implanted Si-crystal. Left: implantation energy 54 keV, thermal annealing at 900°C for 60 min. Right: implantation energy 115 keV, laser annealed.

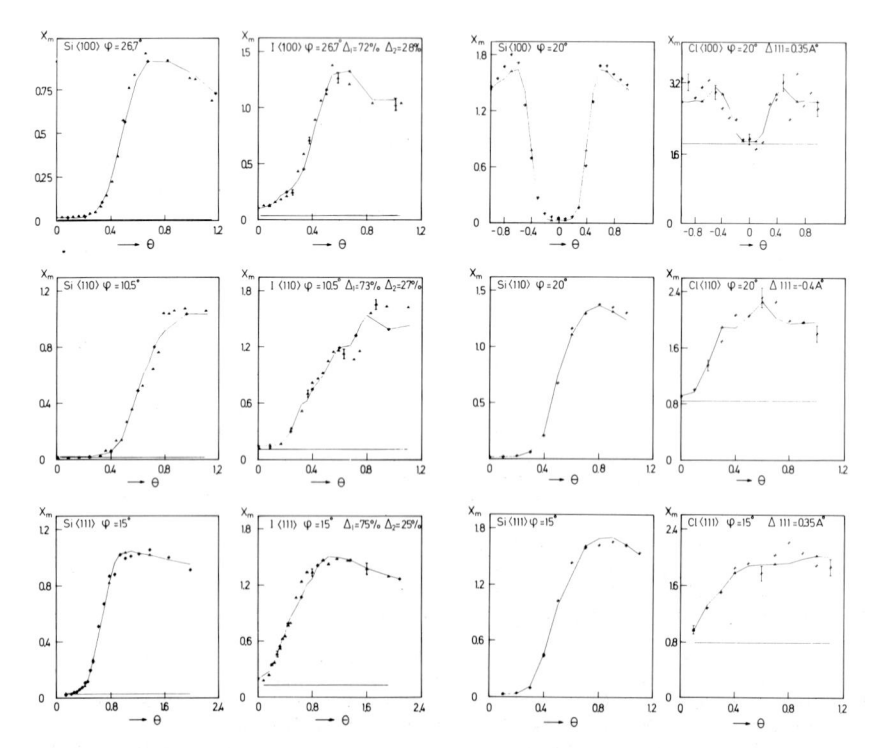

FIG. 2. Angular dependence of the normalized backscatter yield of 2 MeV He$^+$ ions for Si (left) and for I in Si (right). Experimental data are indicated by triangles, calculated points are connected by straight solid lines.

FIG. 3. Angular dependence of the normalized backscatter yield of 3 MeV He$^+$ ions for Si (left) and for Cl in Si (right). See figure caption 2 for further details.

plete recrystallization was observed, together with a redistribution of chlorine to greater depths, as is shown in fig. 1. After laser annealing a loss of chlorine of 30% was found. In previous work [4] a similar loss and redistribution of iodine was observed after laser annealing of iodine-implanted wafers. The loss of the impurity may be due to segregation at the recrystallization front.

In fig. 2 the angular scans for silicon and iodine are compared with simulated results. The agreement for silicon is rather good, but there are deviations. We contribute these deviations mainly to experimental errors in the determination of the scan positions (θ,ϕ) with respect to the string position. An error of $0.02°$ in the determination of the string position may lead to a deviation of $2.5°$ in ϕ at $\theta = 0.5°$. Such a deviation in ϕ can give rise to a different yield, as the ϕ-dependence of the yield is appreciable.

The iodine yields were fitted simultaneously with a random fraction R plus two fractions Δ_1 and Δ_2 at different sites. These fractions were chosen from all possible shifts S along the major axes. In the fit the fractions were adjusted to each scan separately. The result is

$\Delta_1 = 73\% \pm 4\%$, $S = 0.13 {}^{+ 0.08}_{- 0.03}$ Å along <111> towards the nearest neighbour

$\Delta_2 = 27\% \pm 4\%$, $S = 0.40 \pm 0.06$ Å along <100> or <110>.

The random fraction R is defined as a percentage to be added to the 100% of the fractions Δ_1 and Δ_2. For crystal 1 (see table I) R was found to be 5%; for crystal 2 we found 11% and 13%. In earlier work we also found a large variation of R after thermal annealing of iodine-implanted wafers. A second best solution, with Δ_1 shifted along <111> away from the nearest neighbour position, cannot be excluded entirely. No acceptable fit with other combinations of shifted fractions, or with one fraction only, was found. The present result is at variance with a preliminary result [5]. The quality of the present analysis is superior, due to the use of the search program and also due to a redetermination of the scan angle ϕ.

In figure 3 we present the results for chlorine in silicon after laser annealing. To obtain the experimental points for the <110> and <111> scans only the low energy part of the chlorine peak was taken into account. As can be seen in figure 1, the shadowed fraction of chlorine atoms at the larger depths is bigger. In comparison with scans obtained from the full chlorine peak, the only difference is a lower random fraction R. The chlorine scans were fitted with one displaced fraction and a random fraction. The inclusion of a second smaller fraction might improve the fit and change the result. The best fit was obtained for $S = 0.35 \pm 0.07$ Å along <111> towards the nearest neighbour. A shift of equal magnitude in the opposite direction along <111> could not be excluded. Assuming a unique impurity site, shifts in other directions are excluded. Depending on the depth of the chlorine more than 40% of the chlorine atoms were found to occupy random positions.

A similar result was found for the chlorine in silicon after thermal annealing (not illustrated). Here the scan could be fitted with a large random fraction plus one shifted fraction with $S = 0.35 {}^{+ 0.15}_{- 0.05}$ Å. The direction of the shift could not be established. Here the random fraction averaged over the full chlorine peak was twice as high as the shifted fraction. The random fraction was decreasing with the depth of the chlorine (see fig. 1).

DISCUSSION

There is a great similarity in Cl and I-implanted silicon. In both cases no substitutional fraction is found after thermal or laser annealing. It was found recently that the Mössbauer spectrum obtained from the decay of ^{125}I implanted in silicon contains after annealing to 875°C only a quadrupole split structure which possibly consists of two quadrupole split components. This indeed means that no purely substitutional I exists. After laser annealing the random fraction of both Cl and I is large and redistribution and loss of the impurity is observed. After laser or thermal annealing of Cl in silicon and after thermal annealing of I in silicon the non-random fraction is shifted mainly in a <111> direction towards the nearest neighbour. This might be due to the association of the impurity with one vacancy. This interpretation is also in agreement with the asymmetric vibration amplitude found in the Mössbauer work. The second fraction found for I might be due to association with two or more vacancies. The assumption that vacancy association takes place may explain the fact that after laser annealing only a moderate near-substitutional fraction is found. It would be hard to conceive that the short time the sample is at high temperatures during pulsed laser annealing is sufficient to associate all impurity atoms with one or more vacancies.

The differences in Cl and I-implanted silicon might be understood also. The trapping radius for iodine to form a one-vacancy plus impurity complex is probably larger than for chlorine. This larger trapping radius would offer an explanation for the larger near-substitutional I fraction in comparison with the Cl fraction after oven annealing. Diffusion of Cl seems to occur simultaneously with the formation of a near-substitutional fraction. This diffusion may be due to the mobility of Cl-vacancy complexes. A difference in atomic radii between I and Cl could be the cause for the differences in trapping radii and mobility.

The configuration in the lattice of impurity atoms forming a "random fraction" is not understood. If clustering were involved, one has to assume that such clusters break up again at the high temperatures used in the present anneal study.

This work is part of the research program of the Stichting voor Fundamenteel Onderzoek der Materie (Foundation for Fundamental Research on Matter) and was made possible by financial support from the Nederlandse Organisatie voor Zuiver-Wetenschappelijk Onderzoek (Netherlands Organization for the Advancement of Pure Research).

REFERENCES

1. P.J. Dean, A.G. Cullis and A.M. White, Handbook on Semiconductors, ed. S.P. Keller, Vol. 3 (North Holland Publ. Co., Amsterdam, 1980) p. 113.
2. G.J. Kemerink, J.C. de Wit, H. de Waard, D.O. Boerma and L. Niesen, Phys. Lett. 82A, 255 (1981).
3. G.J. Kemerink, H. de Waard, L. Niesen and D.O. Boerma, Hyperfine Interactions 14, 53 (1983).
4. G.J. Kemerink, D.O. Boerma, H. de Waard and L. Niesen, Rad. Eff. 70, 183 (1983) and Rad. Eff. 69, 83 (1983).
5. D.O. Boerma, L. Niesen, P.J.M. Smulders and F.H. du Marchie van Voorthuysen, Nucl. Instr. Meth. 209/210, 375 (1983).
6. G. Greeuw and J.F. Verwey, Solid State Electronics 26, 241 (1983).
7. J.P. Biersack and J.F. Ziegler, Nucl. Instr. Meth. 194, 93 (1982).

A REVIEW OF SILICON-ON-INSULATOR FORMATION BY OXYGEN ION IMPLANTATION

RUSSELL F. PINIZZOTTO
Materials Science Laboratory, Texas Instruments Incorporated,
P.O.Box 225936, M.S.147, Dallas, Texas, 75265, USA

ABSTRACT

Silicon-on-Insulator structures will be an important technological advance used in future VLSI, VHSIC and three-dimensional integrated circuits. The most mature SOI technology other than silicon-on-sapphire is SIMOX, or Separation by IMplanted OXygen. High energy oxygen ions are implanted into single crystal silicon until a stoichiometric buried silicon dioxide layer is formed. After implantation, the material is annealed at high temperature to remove implantation induced defects. The structure is completed by the growth of a thin epitaxial silicon layer. Devices and complex circuits have been successfully fabricated by several research groups. This paper reviews the development of this buried oxide SOI technology from 1973 to 1983. The five major sections discuss the advantages of SOI, the basics of buried oxide formation, the literature published between 1973 and 1983, key issues that must be solved before large scale implementation takes place and, finally, predictions of future developments.

INTRODUCTION

This paper is a review of silicon-on-insulator (SOI) formation by oxygen ion implantation. It covers research published in scientific periodicals from 1973 to mid-1983. It is divided into five sections. The first section is a brief overview of the advantages of SOI technology. It explains why SOI may be used in the production of ICs in the near future. The second section presents the basic principles of buried oxide formation. The third section reviews the literature from the classic nitrogen implantation paper of Dexter, Watelski and Picraux [1] to the most recent electric field shielding layer work of Nakashima, Akiya and Kato [2]. The fourth section discusses the issues that will determine whether or not buried oxide SOI is used in production. The last section is the weakest and most controversial. I will make some predictions about future developments and will go on record with specific dates for several major milestones. I hope that these predictions will appear reasonable to those with the enviable position of making postdictions on the same topic.

THE ADVANTAGES OF SOI

The advantages of SOI naturally fall into 4 categories. The first is the improved resistance to radiation induced failures. A typical SOI structure is shown in Figure 1. The devices are physically and electrically separated from one another and from the substrate. Any charge formed in the substrate by ionizing radiation cannot directly affect the active devices because it must pass through the dielectric isolation layer. Only that charge produced in the thin surface silicon film is detrimental. This is a small volume compared to the bulk volume, so SOI circuits should be radiation hard. In addition, soft errors will be minimized and the effects of secondary impact ionization will be reduced.

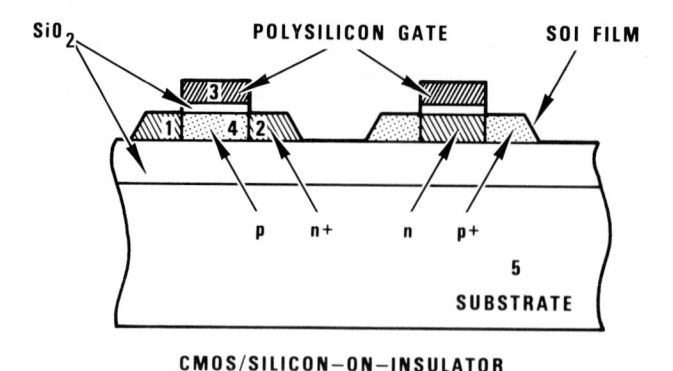

Figure 1. A typical SOI structure. The five terminals are: 1, source; 2, drain; 3, gate; 4, body; and 5, substrate.

The physical separation also leads to major improvements in device performance and circuit design. CMOS latch-up is avoided because there are no electrical connections between the p- and n- channel devices. Device to device high voltage isolation is increased which leads to greater packing densities. The devices have reduced parasitic capacitances and hence should have lower power dissipations and operate at higher speeds. The substrate can act as an easily accessible ground plane. This is an important "fifth" terminal (the five terminals are source, drain, gate, body and substrate) that does not exist in other traditional device designs. It may prove valuable in future applications.The thin silicon dioxide insulating layer used in buried oxide SOI can dissipate more heat than heterogeneous SOI technologies such as silicon-on-sapphire or silicon-on-spinel (SOS). This may be important for VLSI circuits. SOI materials on Si substrates also have greater electron and hole mobilities and longer carrier lifetimes than SOS and do not suffer from large interface defect densities. Finally, homogeneous substrate SOI may be cheaper to manufacture than SOS, the major SOI technology in production today. These materials are similar to conventional substrates and will not require any specialized fabrication techniques.

THE BASICS OF BURIED OXIDE FORMATION

Figure 2 illustrates the basic principles of the formation of an SOI structure by ion implantation. Figure 2a shows that ion implantation leads to a Gaussian distribution of the implanted impurity at a fixed depth below the surface of the sample. Undoped matrix material remains on both sides of the implant. For 150 keV oxygen atoms, the mean range is 371 nm with a Gaussian distribution width of 98 nm [3]. The peak of the damage distribution profile is slightly closer to the sample surface than the peak of the impurity concentration distribution (Figure 2b). During implantation, it is mandatory that the damage at the front surface remains less than the critical value needed for the formation of amorphous silicon. This is usually achieved through the sample heating caused by the ion beam itself. The elevated temperature results in *in situ* annealing which maintains the single crystal nature of the target. With increasing ion dose, the peak oxygen concentration reaches the stoichiometric limit for SiO_2. Increasing the dose further does not lead to the formation of super-stoichiometric oxides. Instead the width of the buried SiO$_2$ layer increases. Eventually, the width of the oxide layer becomes larger than the width of the implant and abrupt silicon/oxide interfaces are formed (Figure 2c). After implantation, the sample is annealed at high temperature to remove implantation induced defects and to adjust the stoichiometry of the buried layer. Most workers use 1150°C, since this is an easily accessible temperature using conventional furnaces. There are no fundamental reasons why higher temperatures cannot be used. Table I lists the experimental details of the entire buried oxide SOI formation process.

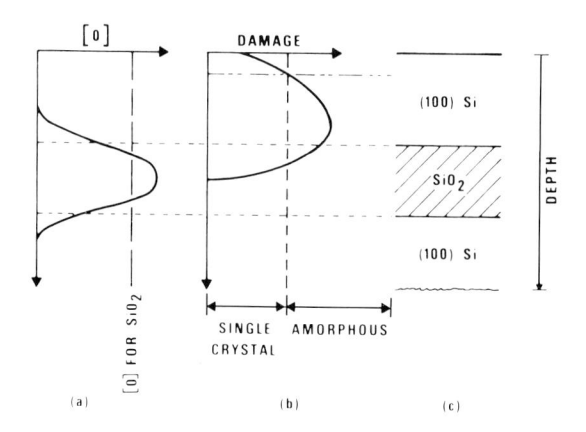

Figure 2. The basics of buried oxide SOI formation: a, the impurity distribution profile after implantation; b, the damage distribution profile; c, a schematic of the final structure.

Table I. Experimental Details for the Formation of Buried Oxide SOI.

Material:
    (100) Si
    p-type
    6-8 ohm-cm

Ion Implantation:
    300 keV O$_2^+$
    6-13 x 10^{17} O$_2^+$/cm^2
    150 μA beam current

Anneal:
    1150°C
    2-4 hours
    Ar or N$_2$

A REVIEW OF SOI FORMATION BY OXYGEN ION IMPLANTATION

The first paper that detailed the formation of a buried insulator by ion implantation was published in Applied Physics Letters in 1973 by Dexter, Watelski and Picraux of Texas Instruments and Sandia Laboratories [1]. They implanted nitrogen and used a 1200°C post-implantation anneal to form a buried nitride layer. An epitaxial silicon layer was grown on top of the resulting structure, proving that the surface does remain crystalline and that it would be possible to build devices in this material. While this is not really a paper on buried oxide, it contains all the basic steps still used today to form buried oxide SOI and is really the predecessor of all later work in ion implantation for buried layer formation as applied to ICs.

The first major paper to show that a buried oxide of high quality could be formed was published in 1977 by Badawi and Anand of the University of Kent [4]. They reported that buried oxides were formed when high implantation energies were used. This was confirmed by ellipsometry, infra-red spectroscopy and electron diffraction after etching. They also found that a high temperature anneal was necessary for the implanted oxide layer to have the same physical properties as thermal oxide. The as-implanted oxide was significantly inferior. They suggested that buried oxides could be used for dielectric isolation of ICs, but did not demonstrate it themselves.

The first paper which showed the electrical characteristics of devices built on buried oxide was by Izumi et al. of NTT in 1978 [5]. They implanted 1.2×10^{18} O atoms/cm^2 at 150 keV. The samples were then annealed for 2 hours at 1150°C in a N$_2$ ambient. These authors published the first sputter Auger electron spectroscopy (AES) data that directly proved that a buried layer was formed. The results agreed with LSS theory [6]. An epitaxial layer was grown and 19-stage CMOS ring oscillators were fabricated. The oscillators operated at twice the speed of the same circuits fabricated in bulk silicon. The increase in speed was attributed to the reduction of parasitic capacitances. In addition, Izumi et al. coined the acronym SIMOX for Separation by IMplanted OXygen. As is often the

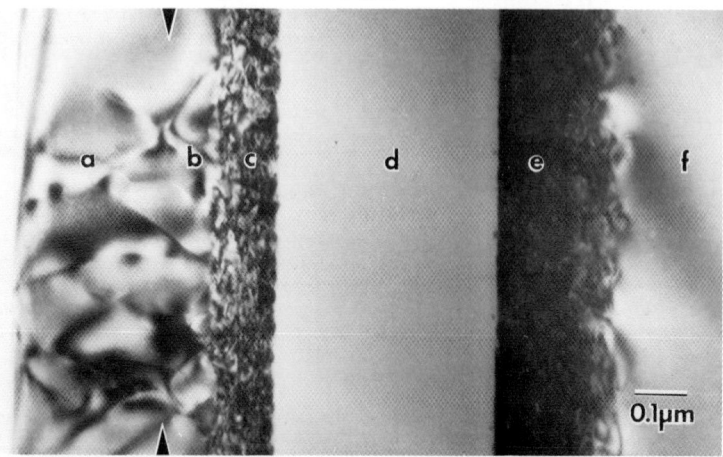

Figure 3. A cross-sectional transmission electron micrograph of buried oxide SOI. The original sample surface is marked with arrows. Note the abrupt silicon/oxide interfaces. a, the epitaxial layer; b, the top single crystal Si layer; c, a damaged polysilicon layer; d, the buried oxide; e, a damaged substrate layer; and f, the substrate.

case, each research group tends to use their own acronyms and no one title for this technology has been generally accepted. The others include B-IMPLOX [7] and "buried oxide" [8]. The latter term has been adopted for this paper.

In 1980, Hayashi et al. of NTT published several papers that presented detailed materials analyses of buried oxide SOI [9,10]. In their first paper, they observed abrupt interfaces between the buried oxide and both the top surface layer and the substrate. This was measured using sputter AES. They also found that the microstructure changed as a function of the oxygen dose. For very large doses, the top layer could be defect free. Figure 3 is a cross-sectional transmission electron micrograph of a sample implanted with 2.6 x 10^{18} O/cm^2 at 150 keV. A 0.3 μm thick epitaxial layer was grown after the sample was annealed for 2 hours at 1150°C in an Ar ambient. The original surface is marked with arrows. The interfaces are indeed abrupt, as described by Hayashi et al. The defects will be described in detail below.

In 1981, Pinizzotto et al. of Texas Instruments described a mechanism for the formation of the abrupt interfaces [11]. As oxygen is implanted into silicon, an oxygen concentration is eventually reached where stoichiometric SiO$_2$ is formed. If more oxygen is implanted into this SiO$_2$ layer, the excess oxygen acts like an internal oxidation source and diffuses to the outer edges of the buried layer. It reacts there with the remaining silicon to form more SiO$_2$. As the implantation proceeds, the width of the stoichiometric oxide layer exceeds the width of the Gaussian implant distribution and the interfaces become abrupt. SiO$_2$ can never be a super-stoichiometric oxide, e.g. SiO$_3$ or SiO$_4$, so the buried layer width increases by consuming parts of the top layer and the substrate. Pinizzotto et al. were also the first workers to use cross-sectional TEM to study the microstructures of buried oxide as a function of dose. They found that the primary defects were dislocations lying on the (111) slip planes (see Figure 3). This is different than SOS where the primary defects are microtwins and stacking faults. The dislocation density in the top silicon layer can vary from 10^6 to 10^{10} cm^{-2} depending on the particular implant and anneal conditions used. This paper also reported the use of dark field TEM to image the top silicon layer that remained in the crystalline state after implantation. For a dose of 2.4 x 10^{18} O/cm^2 implanted at 150 keV, the remaining single crystal material consists of a discontinuous layer of silicon islands that are about 40 nm thick. This would seem to be a very tenuous amount of material to use as seeds for epitaxial growth, but remarkably, epitaxial layers can be grown directly after implantation with no post-implantation annealing. After post-implantation annealing, the top silicon layer is a continuous single crystal about 100 nm thick. A polysilicon layer, also about 100 nm thick, remains between the top silicon layer and the buried oxide.

In 1980, Ohwada et al. of NTT reported that MOSFETs with submicron channel lengths could be built on buried oxide SOI [12]. They fabricated 21-stage ring oscillators. The SOI devices had a smaller threshold voltage shift as a function of channel length than devices fabricated in bulk silicon. This improvement was attributed to the elimination of the "punch through" effect by the SOI structure. The electron mobility was 750 cm^2/V-sec, the same as for bulk devices. These CMOS SOI circuits had low device leakage and small interface charge and the authors concluded that this technology could be important for VLSI.

Hayashi et al. published another materials characterization paper in 1981 that showed the existance of a second surface peak in the AES spectra [13]. This was confirmed by Kilner et al. using secondary ion mass spectroscopy in 1983 [14]. This second "oxide" layer between the surface silicon layer and the buried oxide could only be formed in a very narrow dose range. In addition, the substrate temperature was a very important parameter. Unfortunately, the substrate temperature and the dose rate are related and the two effects have not been independently studied even to this time.

Nakashima et al. used the existence of this second layer to great advantage as an "electric field shield" [2]. This layer resembles semi-insulating polycrystalline silicon (SIPOS) and hence can act as a means of decoupling the electric field produced in the bulk and the operation of devices built in the top silicon layer. Nakashima et al. showed that the threshold voltages of both p- and n- type devices were not affected by substrate biases up to 50 V. This could have an important role in improving the radiation hardness of buried oxide ICs. It implies that radiation induced charge build up at the buried oxide/substrate interface will not affect devices built in the top Si layer.

In 1981, Irita et al., again of NTT, published the first paper showing that three dimensional SOI structures with single crystal silicon layers interspersed between dielectric layers were possible [15]. They extended the ideas of implantation plus epitaxial growth to a repeated process. They reasoned that since a buried layer with a single crystal layer could be fabricated, it would be possible to grow a thick epitaxial layer that could itself be implanted with oxygen leaving a second buried oxide layer within the epitaxial layer. In fact, they repeated the implantation plus epitaxy processes three times to form a triple SOI structure. Each SiO_2 layer was stoichiometric and each silicon layer was a single crystal with exactly the same orientation as the substrate. Unfortunately, no electrical characterization was performed.

Das et al. of Middlesex Polytechnic published two papers in 1981 [7,16]. They were the first to publish an XTEM micrograph of buried oxide SOI and also the first to study epitaxial material grown on implanted, but unannealed substrates. This group uses 200 keV implants, unlike most other workers who use 150 keV. They feel that the increased thickness of the top silicon layer more than makes up for the longer implantation time. Thick top layers may make the epitaxial growth step unnecessary. As the implantation energy is increased, the implant dose necessary for stoichiometric oxide formation increases because the width of the Gaussian distribution increases with implant energy. Therefore, the implanted oxygen is spread out over a larger depth range and more oxygen is needed to form the buried layer. Das et al. use a H_2 ambient for post-implantation annealing. They report that very long anneals in H_2 can completely destroy the buried oxide layer. They propose that the oxygen reacts with the hydrogen to form water, which diffuses out of the material. Similar oxygen/hydrogen interactions have been found in solar cell grain boundary passivation experiments [17]. This could be an important effect if it is possible to reduce the residual oxygen concentration in the top silicon layer and sharpen the oxide/silicon interfaces even further.

In 1982, Maeyama and Kajiyama of NTT demonstrated that buried nitride formation, unlike buried oxide, is not limited at the stoichiometric level [18]. When more nitrogen than the amount needed for the formation of Si_3N_4 is implanted into the silicon, the excess remains at the implanted position, presumably forming Si_3N_4 plus N_2 after a 2 hr 1150°C anneal. It is not possible to form abrupt interfaces during nitrogen implantation. The excess nitrogen may be a threat to long term stability of the material since the gas could diffuse throughout the substrate. For these reasons, buried nitride SOI will probably not be used in production.

Fabrication of a large SIMOX circuit, a 1k CMOS SRAM, was reported by Izumi et al. in 1982 [19]. The devices had channel lengths of 1.5 µm. The IC had a chip-select access time of 12 nsec with a 5V supply voltage. The NTT group also fabricated a 400 gate phase-locked-loop logic chip that operated at 420 MHz. These devices are the most complex built with any SOI technology except SOS. They demonstrate that the buried oxide material is certainly production worthy and is the most advanced of all the SOI techniques.

Malhi et al. of Texas Instruments showed in 1982 that devices can be built directly in the top silicon layer [20]. The epitaxial layer is not a necessity. The electron and hole mobilites for n- and p- channel devices were 600 and 300 cm^2/V-sec, respectively.

Hamdi et al. of North Texas State University and Texas Instruments used Rutherford backscattering spectroscopy (RBS) to study the effects of various anneal ambients on the quality of the top silicon layer [21]. They found that Ar ambients yield better material than N$_2$ ambients. Impurities in the Ar gas cause a small amount of oxidation, which protects the top silicon layer from further degradation. In addition, it was shown that the 2 hour 1150°C anneals used by most groups are not sufficient to totally anneal the material. The minima in the RBS measurements do not occur until the material is annealed for 4 hours. Long anneals degrade the sample surface, however, so this group uses 3 hrs as a compromise.

Hamdi et al. also used RBS channeling to compare the buried oxide plus epitaxial layer SOI to SOS (see Figure 4) [22]. The buried oxide SOI has a much smaller minimum channeling yield than SOS indicating that it has a lower defect density. The minimum value for the buried oxide is 2.6% compared to 2.2% for unprocessed Si and 6% for SOS. The buried oxide values are only slightly greater than the unprocessed Si values for the first 300 nm, whereupon there is a large increase in the yield due to the damage layer above the oxide layer (see Figure 3). The SOS spectrum initially has a steeper slope which then increases to a much larger value than for the buried oxide. The primary defects in SOS are microtwins and stacking faults which are planar defects. In buried oxide, the defects are dislocations which are linear defects. The dislocations do not cause as much ion dechanneling.

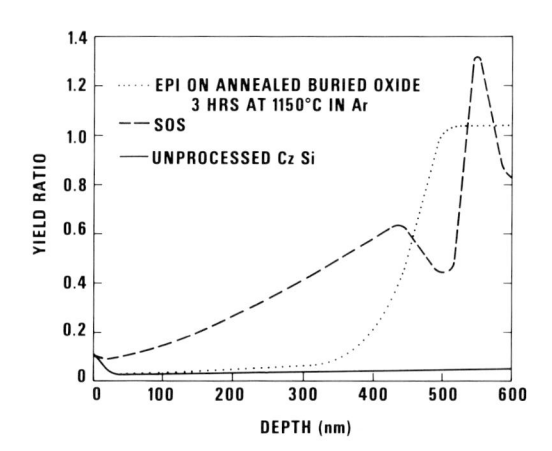

Figure 4. RBS <110> channeling spectra of unprocessed Si, SOS and buried oxide SOI plus an epitaxial layer.

In 1983, Kim et al. of General Electric published a paper that highlights one of the major SOI problems [23]. SOS could be used for widespread IC production only after it became possible to measure the quality of the incoming material. This was accomplished using ultraviolet reflectance (UVR) measurements correlated with x-ray diffraction data. Kim et al. demonstrated that the UVR technique is applicable to buried oxide.

KEY ISSUES

Scientific Issues

There are a number of basic effects that occur during buried oxide formation that have been documented, but not explained. The most basic question is the mechanism of buried oxide formation itself. There are early indications that the last oxygen implanted remains in the Gaussian distribution that is expected from LSS predictions [24]. This implies that the oxygen implanted first "diffuses" to the interfaces and reacts there to form SiO_2. There are no data on the diffusion kinetics or on the reaction rates. It is only known that a high temperature anneal is needed after implantation to complete the process. Elucidation of the formation mechanism will also play an important role in understanding other phenomena. The electric field shielding layer is only formed with very specific implantation dose rates and with substrate temperatures that are in a relatively narrow range. It is not known if the EFS layer is a dose rate effect or a sample temperature effect. At the present time, it is not possible to separate the two because the dose rate and the temperature are not independently variable. The substrate temperature is raised to 400 to 600°C using the ion beam and not external heaters. Future implanters are being designed to independently vary the two parameters. Understanding the oxide formation mechanism may also help explain why oxygen and nitrogen behave differently. It may be possible to separate the chemical effects from the physical effects. The mechanism of buried oxide film removal by hydrogen annealing needs far more study and confirmation. H_2 annealing could have important production benefits if the oxygen concentration in the top silicon layer can be controlled. Finally, there is no knowledge of what UV reflectance techniques measure. UVR requires further theoretical development to understand the empirical results. This technique is in its infancy and needs maturing before it gains widespread acceptance.

Technological Issues

The ideal experimental conditions for buried oxide fabrication can only be found empirically. In the past, long implantation times limited the amount of available material so that all of the parametric effects could not be studied. The implant temperatures and anneal temperatures need to be optimized. The regimes studied to date are only those easily accessible using standard equipment. The implant temperature is fixed by the ion current used in each individual implanter. The anneal temperatures are those used in conventional furnaces. Much is known about anneal ambient effects, but more experiments are needed for optimization. H_2 anneals must be studied in detail. Devices must be fabricated to determine whether or not abrupt interfaces are necessary for long term stability and device performance. The EFS layer must be evaluated in detail and its benefits balanced against process complexity. Finally, a technique of non-destructive evaluation of substrates must be developed. This is absolutely crucial. It was only when UVR was accepted by the SOS industry that it became possible for SOS substrate manufacturers and users to agree on the material quality needed for good device performance and high fabrication yields.

Production Issues

Cost is the major hurdle that buried oxide SOI must clear before it is applied to VLSI. The extra cost is caused by the extended implantation time needed to produce the material. At present, implantation of a full 3-inch diameter silicon wafer takes about 16 hours using a standard implanter and a slightly modified oxygen ion source. New equipment is being designed to implant a standard 100mm wafer in about 30 minutes. However, this design is based on batch processing. Twenty to 25 wafers are implanted simultaneously while traveling on a carousel. The actual batch implant time will be 14 hours. This estimate is based on the use of a high current pre-deposition implanter with a custom designed oxygen source. Eventually, it may be possible to build an implantation machine that produces a 150 keV, 200 mA oxygen beam. Such a machine has a total beam power of 90 kW! It will be an interesting design problem to insure that the machine cannot destroy itself under fault conditions. Other production issues include yield and reliability. Since buried oxide defect densities are of the same order as those of SOS, similar results can be expected. This must be tested in detail before it is known if the technology is production worthy. Radiation hardness testing is also needed. Finally, customers who are willing to pay the additional costs that will always be a part of SOI must be identified and enough sales predicted to warrant the huge investment in time, capital and talent.

PREDICTIONS

The following predictions are based on the data available in the literature reviewed above. The views stated here are my own and are not necessarily the same as those of my co-workers, Texas Instruments or contract personnel.

A large LSI circuit will be demonstrated in 1984. This step is a necessity to the continued development of the buried oxide process. The part will be a large memory (>4k SRAM equivalent) or a high voltage logic chip. Buried oxide will be the first SOI technology other than traditional dielectric isolation and SOS that is used in production. The other possibilities, for example, beam recrystallization or porous oxidized silicon, have too many severe problems. Based on the successes predicted above, it would be reasonable to expect a special product to be available within 4 years. This will occur because of the advent of large current implantation machines of 200 mA beams of oxygen at 150 keV. The implantation time will be on the order of 4 minutes for a 100 mm diameter wafer. Such a machine should be available for experimental laboratory applications within 3 years.

ACKNOWLEDGEMENTS

I thank all of my colleagues at Texas Instruments and North Texas State University without whose efforts and support this work would not have been possible. This research was supported in part by the United States Air Force Rome Air Development Center and the Defense Nuclear Agency through contract F19628-81-C-0136, P.J. Vail, contract monitor.

REFERENCES

1. R.J. Dexter, S.B. Watelski and S.T. Picraux, Appl. Phys. Lett., 23, 455 (1973).

2. S. Nakashima, M. Akiya and K. Kato, Elec. Lett., 19, 568 (1983).

3. J.F. Gibbons, W.F. Johnson and S.W. Mylroie, Projected Range Statistics, Dowden, Hutchinson and Ross, Stroudsburg, PA, 1975.

4. M.H. Badawi and K.V. Anand, J. Phys. D, 10, 1931 (1977).

5. K. Izumi, M. Doken and H. Ariyoshi, Elec. Lett., 14, 593 (1978).

6. J. Lindhard, M. Scharff and H.E. Schiott, Matt. Fys. Medd. Dan. Vid. Selsk., 33, 1 (1963).

7. K. Das, J.B. Butcher and K.V. Anand in Chemical Vapor Deposition 1981, edited by J.M. Blocher, Jr., G.E. Vuillard and G. Wahl, The Electrochem. Soc., Pennington, NJ, 81-7, 427 (1981).

8. H.W. Lam, R.F. Pinizzotto, H.T. Yuan and D.W. Bellavance, Elec. Lett., 17, 356 (1981).

9. T. Hayashi, H. Okamoto and Y. Homma, Jpn. J. Appl. Phys., 19, 1005 (1980).

10. T. Hayashi, S. Maeyama and S. Yoshi, Jpn. J. Appl. Phys., 19, 1111 (1980).

11. R.F. Pinizzotto, B.L. Vaandrager and H.W. Lam, Proc. Mater. Res. Soc., 7, 401 (1982).

12. K. Ohwada, Y. Omura and E. Sano, 1980 International Electron Devices Meeting Technical Digest, IEEE, New York, 1980, pg. 756.

13. T. Hayashi, H. Okamoto and Y. Homma, Inst. Phys. Conf. Ser., 59, 533 (1981).

14. J.A. Kilner, S. Littlewood, P.L.F. Hemment, E. Maydell-Ondrusz and K.G. Stephens, presented at the Sixth International Conference on Ion Beam Analysis, Tempe, AZ, May 1983, paper Th 7. To be published in Nucl. Instr. Meth. Phys. Res., (1984).

15. Y. Irita, Y. Kunii, M. Takahashi and K. Kajiyama, Jpn. J. Appl. Phys., 20, L909 (1981).

16. K. Das, J.B. Butcher, M.C. Wilson, G.R. Booker, D.W. Wellby, P.L.F. Hemment and K.V. Anand, Inst. Phys. Conf. Ser., 60, 307 (1981).

17. L.L. Kazmerski, J.R. Dick and R.J. Matson, American Vacuum Society Thirtieth National Symposium, Paper ESThA01, Boston, MA, November 1983.

18. S. Maeyama and K. Kajiyama, Jpn. J. Appl. Phys., 21, 744 (1982).

19. K. Izumi, Y. Omura, M. Ishikawa and E. Sano, in Digest of Technical Papers, 1982 Symposium on VLSI Technology, Kanagawa, Japan. Also see K. Izumi, Y. Omura and T. Sakai, J. Elec. Mat., 12, 845 (1983).

20. S.D.S. Malhi, H.W. Lam, R.F. Pinizzotto, A.H. Hamdi and F.D. McDaniel, 1982 International Electron Devices Meeting Technical Digest, IEEE, New York, 1982, pg. 107.

21. A.H. Hamdi, F.D. McDaniel, R.F. Pinizzotto, S. Matteson, H.W. Lam and S.D.S. Malhi, Appl. Phys. Lett., 41, 1143 (1982).

22. A.H. Hamdi, F.D. McDaniel, R.F. Pinizzotto, S. Matteson, H.W. Lam and S.D.S. Malhi, IEEE Trans. on Nucl. Sci., NS-30, 1722 (1983).

23. M.J. Kim, D.M. Brown and M. Garfinkel, J. Appl. Phys., 54, 1991 (1983).

24. M.A. Nicolet, C.D. Lien, R.F. Pinizzotto and S. Matteson, unpublished results.

FORMATION OF OXIDE LAYERS BY HIGH DOSE IMPLANTATION INTO SILICON

S.S. GILL and I. H. WILSON*
Royal Signals and Radar Establishment, St Andrews Road, Great Malvern, Worcestershire WR14 3PS, U.K.; *Department of Electronic and Electrical Engineering, University of Surrey, Guildford, Surrey GU2 5XH, U.K.

ABSTRACT

Single crystal silicon was implanted with 80, 120, 160 and 240 keV oxygen ions. Rutherford backscattering (RBS) analysis was used to obtain the implanted oxygen profile and the oxygen to silicon ratio in the implanted layer for doses in the range 10^{16} to 1.5×10^{18} O_2^+ cm^{-2} for room temperature implants. The depth and the thickness of the buried oxide layer has been measured as a function of implantation energy and oxygen dose. Chemical formation of stoichiometric SiO_2 was confirmed by infra-red (IR) spectroscopy. Both RBS and IR indicate that once a surface oxide layer is formed for very high dose levels, the layer thickness decreases with increasing implanted dose beyond a critical dose level.

INTRODUCTION

For higher speed, high density VLSI circuits it is desirable to have vertical isolation and low capacitance coupling to the substrate material. At present three choices are available (i) silicon-on-sapphire (SOS) (ii) silicon-on-insulator (SOI)[1] and (iii) silicon-on-implanted-oxide SIMOX[2]. The limitations of SOS are well known for its high cost, and low mobility of the lattice mismatched top silicon layer. The technique of SOI has been described by Geis et al[1] where they used the zone melting recrystallization technique. Using a moving graphite strip heater they obtained several square cm of single crystal silicon on deposited insulator. Silicon on insulator has also been reported where ∼ 150 μm long single crystal regions across a whole wafer were obtained using a dual-electron beam system[3]. Recently Hayafuji et al[4] have reported up to 100 μm wide and 2 cm long quasi-single-crystal Si films on SiO_2 obtained using a high-energy density strip electron beam. In this paper we shall report results on synthesizing silicon oxide layers in silicon (both buried and at the surface) by high dose oxygen ion implantation.

METHOD

Oxygen ions have been implanted into 4 Ωcm p-type <100> silicon at an incident angle of 7° from normal to minimize possibility of channeling during implantation. The dose range was 10^{16} to 1.5×10^{18} cm^{-2}, in most cases molecular oxygen was implanted, with energies per molecule of 80, 120, 160 and 240 keV at nominally room temperature (beam current density 10 μA cm^{-2}). The implanted samples were analysed using Rutherford backscattering (RBS) both in the aligned and non-aligned direction using 1.5 MeV He^+ ions. A dual beam Perkin-Elmer spectrophotometer was used to record the IR transmission spectra of the implanted samples.

RESULTS

In figure 1 is shown the peak oxygen to silicon ratio in the implanted layer as a function of oxygen dose for the four implant energies investigated in the present work. The oxygen to silicon ratio was estimated using the two relative yields of the implanted and non-implanted layers[5]. As can be seen from this figure the dose required to form SiO_2 increases with increasing implant energy; 4.5×10^{17}, 5.5×10^{17}, 6.3×10^{17} and 6.8×10^{17} O_2^+ cm^{-2} for the 80, 120, 160 and 240 keV implants respectively.

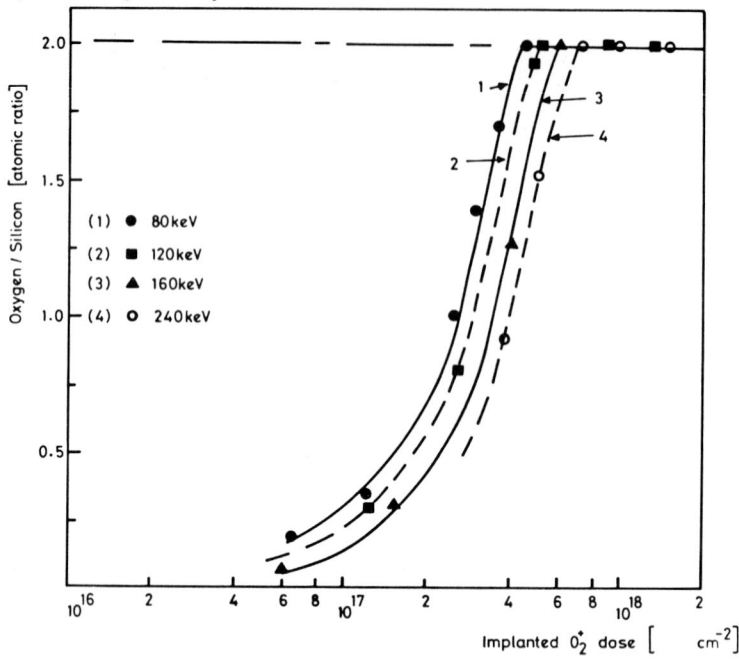

FIG. 1. Oxygen to silicon ratio as function of implanted oxygen dose for the four energies.

At lower doses the oxygen profile in gaussion[6] at higher doses this saturates once SiO_2 is formed to create a buried layer. This is clearly shown in figure 2, which also shows that the depth and the thickness of the buried layer increases with implant energy. As the implanted dose is increased further, the width of the implanted profile increases and at very high dose levels an oxide extending to the surface can be formed[6].

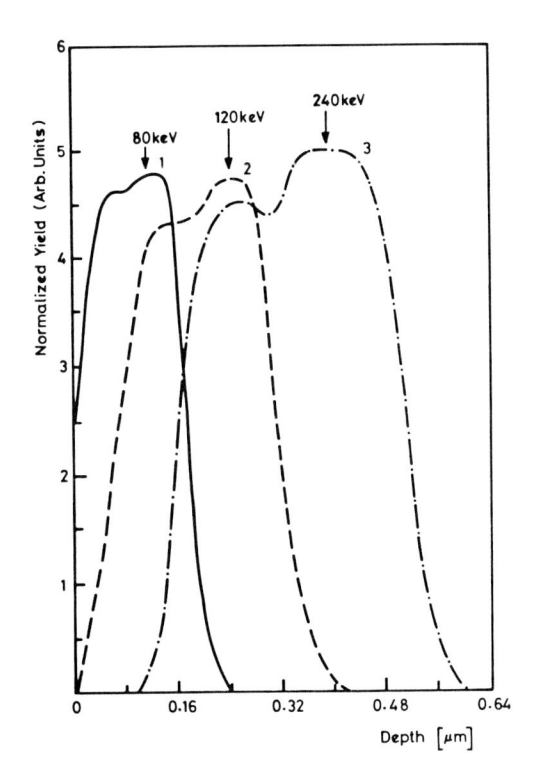

FIG. 2. Oxygen profiles obtained from RBS, showing both layer thickness and depth. (1) 80 keV, 5 x 10^{17} O_2^+ cm^{-2}; (2) 120 keV, 5 x 10^{17} O_2 cm^{-2}; (3) 240 keV, 7 x 10^{17} O_2^+ cm^{-2}.

Figure 3 demonstrates the variation of buried oxide layer depth and thickness with implanted oxygen energy. Our results agree well with other workers and link the low energy results of Badawi and Anand[7] with the higher energy results of more recent workers. There is also a good fit to the theoretical model of Maydell-Ondrusz and Wilson[12]. In the published work[7]-[11] many different implant conditions have been used (oxygen dose and implant temperature) however the general agreement is still very good.

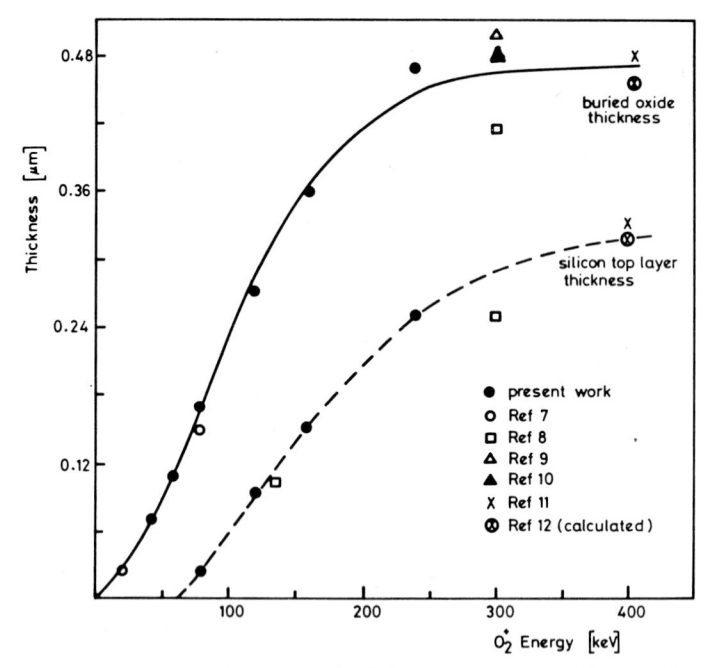

FIG. 3. Buried oxide thickness and silicon top layer thickness (depth) as function of implant energy. Results from published work are also shown for comparison.

In figure 4 we have plotted both the depth of buried oxide layer and thickness as a function of implanted dose for 160 and 240 keV implants. This figure enables the determination of an optimum dose of buried oxide thickness and depth. By way of comparison the results of Hemment et al[11] for 400 keV implants are also shown in figure 4. Considering the different experimental conditions employed for their work, the results are in reasonable agreement with the present work and show similar trends. This figure also demonstrates that in the case of the 160 keV implant the oxide layer thickness decreases with increase in implanted dose at very high doses. This has been more clearly shown by Gill[6] for 120, 160 and 240 keV implants.

The formation of stoichiometric SiO_2 can be confirmed by IR spectroscopy[13] however these results have to be interpreted with caution[6]. The oxide layers formed by ion implantation can give erroneous results due to high strain levels introduced during formation, causing the maximum IR absorption to occur at a lower wave number. However after heat treatment at 850°C for 30 minutes in flowing nitrogen, the IR measurement indicates a true result and this technique can then be used to identify chemical formation of SiO_2. In figure 5 are shown the IR spectra of 80 keV molecular implants as function of implanted oxygen dose. Maximum absorption for thermally grown SiO_2 occurs at a wave number of 1080 cm^{-1}, therefore as shown in figure 5 for an implanted dose of 5×10^{17} O_2^+ cm^{-2} a silicon dioxide layer is formed. For this low energy implant a surface layer is formed at this

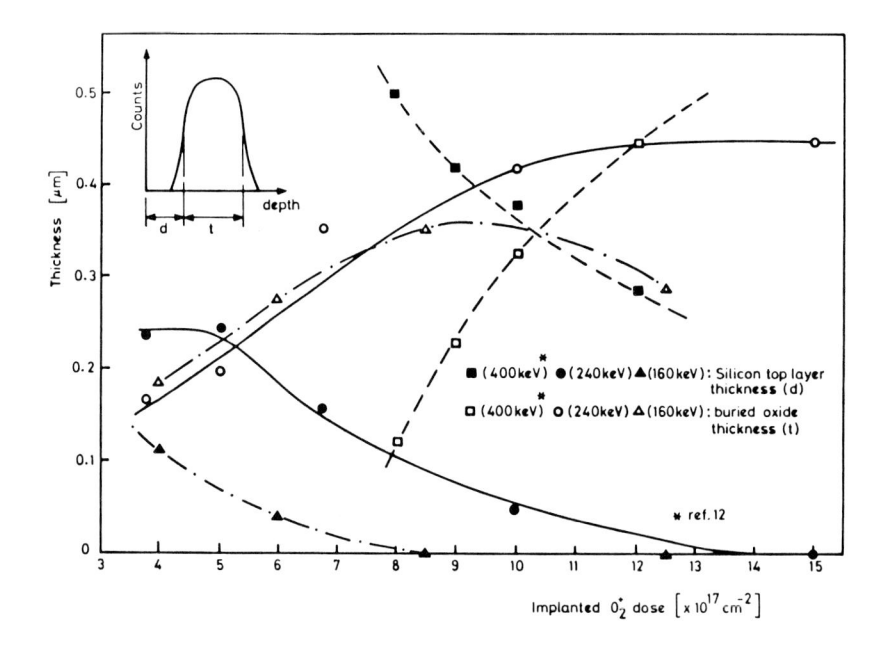

FIG. 4. Buried oxide thickness and silicon top layer thickness (depth) as function of implanted dose, for 160, 240 and 400 keV (reference 12) implants.

dose and hence any further increase in implanted oxygen dose causes the layer thickness to decrease due to sputtering[5], hence the reduction in intensity of the IR absorption for the highest dose.

DISCUSSION AND CONCLUSIONS

The saturation of the oxygen to silicon ratio once stoichiometric SiO_2 is formed is now a well documented phenomenon, our results provide data for energies intermediate between the early low energy work and that of current workers producing SIMOX layers. There appears to be a consistent trend in the increase in Si and SiO_2 layer thickness with oxygen implant energy. The data on the variation of these thicknesses with dose will be of use to those desirious of reducing the cost of the SIMOX process, by reducing the ion energy and thereby minimising the required ion dose. Silicon epitaxy would have to be used to grow a layer suitable for devices. The minimum oxygen implant energy for this purpose would appear to be in the region of 150 keV per molecule.

The IR and RBS results of this work and earlier low energy work[5] indicate that once a surface oxide is formed the layer thickness decreases with increasing oxygen dose. We attribute this to sputtering combined with an increase in the stopping power once SiO_2 is formed.

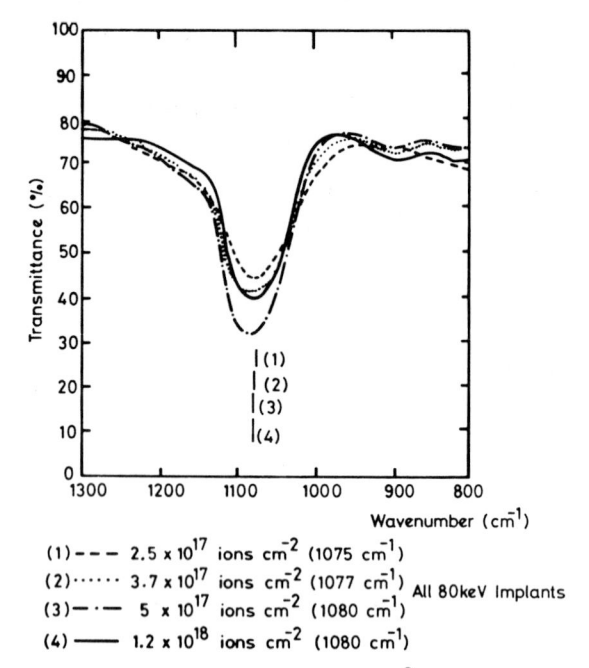

FIG. 5. IR spectra for 80 keV implants, after 850°C/30 min anneal in flowing nitrogen.

(1) - - - 2.5×10^{17} ions cm^{-2} (1075 cm^{-1})
(2) ······ 3.7×10^{17} ions cm^{-2} (1077 cm^{-1})   All 80keV Implants
(3) — · — 5×10^{17} ions cm^{-2} (1080 cm^{-1})
(4) ——— 1.2×10^{18} ions cm^{-2} (1080 cm^{-1})

REFERENCES

1. M. W. Geis, H. I. Smith, B-Y Tsaur, J. C. C. Fan, D. J. Silversmith, R. W. Mountain and R. L. Chapman, Mat. Res. Soc. Symp. Proc. 13, p 477-489 (1983).
2. I. H. Wilson, Int. Conf. on Rad. Effects in Insulators, (1983), Albuquerque, New Mexico. To be published in Nucl. Instrum. and Meth. (1984).
3. H. Ahmed and R. A. McMahon, Mat. Res. Soc. Symp. Proc. 13, p 653-664 (1983).
4. Y. Hayafuji, T. Yanada, S. Usui, S. Kawado, A. Shibata, N. Watanabe, M. Kikuchi and K. E. Williams, App. Phys. Lett. 43 (5) p 473-475 (1983).
5. S. S. Gill and I. H. Wilson, Thin Solid Films 55 p 435-448 (1978).
6. S. S. Gill, Ph.D. Thesis, University of Surrey, Guildford, UK (1980).
7. M. H. Badawi and K. V. Anand, J. Phys. D. Appl. Phys. 10, p 1931-1941 (1977).
8. S. Maeyama and K. Kajiyama, Jpn. J. Appl. Phys. 21 (5) p 744-751 (1982).
9. M. Akiya, K. Ohwada and S. Nakashima, Electronics. Lett. 17 (18) p 640-641 (1981).
10. Y. Omura, E. Sano and K. Ohwada, IEEE Trans. Elec. Dev ED-30 (1) p 67-73 (1983).
11. P. L. F. Hemment, E. Maydell-Ondrusz, K. G. Stephens, J. A. Kilner and J. B. Butcher, Low energy ion beams Conference, Loughborough 1983.
12. E. A. Maydell-Ondrusz and I. H. Wilson, submitted for publication in Thin Solid Films.
13. J. Dylewski and M. C. Joshi, Thin Solid Films, 35 p 327-336 (1976).

Ⓒ HMSO London 1983

CHARACTERISATION OF DEVICE GRADE SOI STRUCTURES FORMED BY IMPLANTATION OF HIGH DOSES OF OXYGEN

P.L.F. HEMMENT[*], E.A. MAYDELL-ONDRUSZ[*†], K.G. STEPHENS[*], R.P. ARROWSMITH[+], A.C. GLACCUM[+], J.A. KILNER[o] and J.B. BUTCHER[Δ]

[*]Dept. of Electronic & Electrical Engineering, University of Surrey, U.K.; [+]British Telecom Research Laboratory, Martlesham Heath; [o]Imperial College, London; [Δ]Middlesex Polytechnic, Bounds Green, London.

†Permanent address: Institute of Nuclear Physics, Krakow, Poland

ABSTRACT

SOI structures have been formed in (100) silicon by implanting 400 keV molecular oxygen to a dose of 1.8×10^{18} O atoms cm^{-2}. These samples were annealed at $1150°C$ for 2 hours with a SILOX cap. Oxygen depth profiles have been determined by SIMS and wafers implanted at about $500°C$ have been characterized by studying the regrowth kinetics, As drive in and oxidation rate in the top silicon overlay.

INTRODUCTION

Silicon on insulator (SOI) structures formed by the implantation of high doses of oxygen ions have been studied recently by a number of groups [1],[2],[3]. Very high doses of oxygen (> 10^{18} O cm^{-2}) are required to synthesise the buried oxide layer and, consequently, the supply of wafers has been very limited but, nevertheless, the material has been shown to be a suitable substrate for LSI circuits. As the material is compatible with current silicon processing technology it is expected to be suitable also for VLSI circuits. In this paper we report further results from experiments to characterize the material, the work being part of an "on going" study of this SOI technology [4].

EXPERIMENTAL DETAILS

Synthesised SOI structures have been formed by implanting 400 keV molecular oxygen ions to a dose of 1.8×10^{18} atomic O cm^{-2} into device grade (100) silicon wafers of diameter 50 mm and 75mm using the Heavy Ion Accelerator at the University of Surrey [3]. Wafers have been implanted within the temperature range $325°C$ to $600°C$. After implantation the wafers were furnace annealed at $1150°C$ for 2 hours in flowing N_2, using a SILOX cap.

Batches of samples (5mm x 5mm) have been cleaved from wafers implanted at about $500°C$ and further processed (i) by implantation with 1×10^{16} As$^+$ cm^{-2} at 40 keV and annealed and (ii) oxidized in dry oxygen at $950°C$. Samples have been surface analysed by the SIMS [5] and Rutherford backscattering (RBS) methods. Test batches of wafers have also been fully processed to 5 μm NMOS and 3 μm CMOS rules [6].

RESULTS AND DISCUSSION

Figure 1 shows SIMS depth profiles from samples implanted with 1.8×10^{18} O cm^{-2} at substrate temperatures of $325°C$, $400°C$, $500°C$ and $600°C$. Figure 1(a) shows the profiles from the as implanted samples. The profiles are smooth, asymmetric and show saturation of the oxygen volume concentration at the value corresponding to stoichiometric SiO_2. Independent analyses

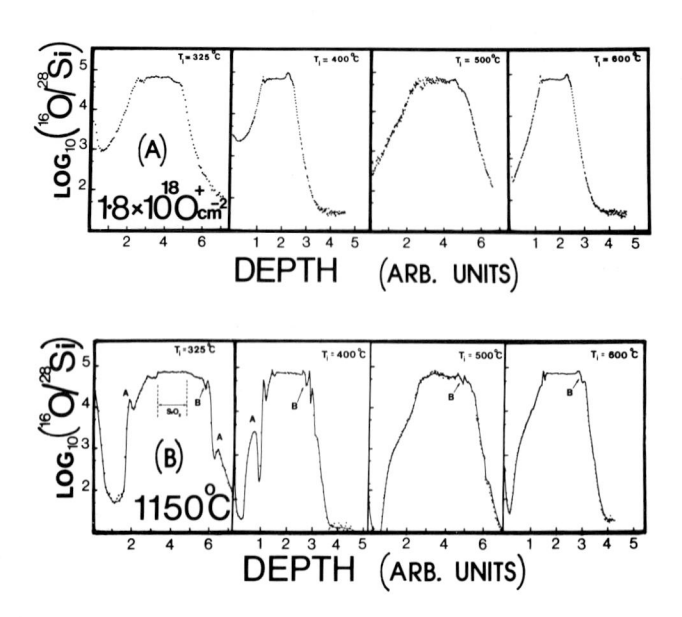

FIG. 1. Oxygen depth profiles from silicon samples implanted with 1.8 x 10^{18} O atoms cm^{-2} at 200 keV (a) as implanted and (b) after an anneal at 1150°C for 2 hrs. The implantation temperatures were 325°C, 400°C, 500°C and 600°C. The depth scale for each pair of samples, (a) and (b), are similar.

using Auger [7] and TEM [8] methods confirm that the central region of the buried oxide is stoichiometric, amorphous and continuous. The volume concentration at the surface is about 10^{20} O cm^{-2}.

Figure 1(b) shows the oxygen distributions after annealing the samples at 1150°C for 2 hours. Although stoichiometric SiO$_2$ is still present in the saturated region of the profiles, it is evident that significant redistribution of the implanted oxygen has occurred within the wings of the profiles. In all cases the near surface region is depleted of oxygen. The samples implanted at 500°C and 600°C still have smooth depth distributions but now regions may be identified by different concentration gradients. These correlate with structure sensitive measurements [3]. The samples implanted at the lower temperatures show pronounced structure in both wings of the distributions which is believed to be due to the growth of SiO$_2$ precipitates at specific depths beneath the surface [10]. All samples show characteristic structure ("B") at the back SiO$_2$/Si interface which correlates with the "banding" observed in TEM micrographs of samples implanted at the higher temperatures [8]. The "banding" is believed to be due to a coarse two phase (Si and SiO$_2$) microstructure [10].

The sample implanted at 500°C has the widest (~ 1000Å) oxygen denuded layer within which the minimum oxygen concentration is estimated to be about 10^{18} O cm^{-3}. This feature and the good single crystal quality of the top layer [3], which does not contain any large oxide precipitates, all point to the suitability of this material as a SOI substrate. Further experiments have been carried out, therefore, to characterize the substrates formed by implantation at about 500°C.

Regrowth

Arsenic ions (1x10^{16} cm^{-2}, 40 keV) have been implanted, at a power density of less than 0.01 W cm^{-2}, into samples cleaved from SOI substrates. During the implant the sample temperature did not exceed about 20°C. Under these conditions an amorphous layer of thickness 750Å is formed and the regrowth of this arsenic doped material, within the temperature range 440°C to 500°C, has been followed by RBS and channelling analysis [11]. The measured regrowth rates in these samples and also control samples of bulk silicon are plotted against reciprocal temperature in Figure 2. An activation energy of 2.7 eV ± 0.2 eV has been determined and this value compares favourably with values of 2.4 eV for pure silicon and 2.7-2.8 eV for highly doped material. Within experimental error the regrowth rates are similar in both bulk silicon and the SOI substrates, with a value of 3Å min^{-1} at 475°C. The composition and microstructure of these SOI substrates are clearly different from that of bulk silicon and yet from these measurements, no difference is detected in the kinetics of regrowth of amorphous layers formed by typical source/drain implants.

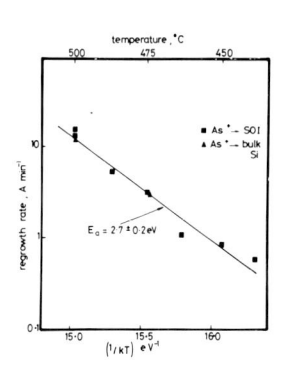

FIG. 2. Regrowth rate of the amorphised top silicon layer plotted against reciprocal temperature. [11]

Arsenic drive-in

Both SOI and bulk silicon samples have been implanted with 1x10^{16} As$^+$ cm^{-2} at 40 keV and the arsenic doped layer has been regrown by pulse annealing at 1100°C, under standard conditions. These samples were then further heat treated in a furnace at 950°C in flowing nitrogen, for times within the range 1 hr to 16 hr. Following these "drive-in" anneals the arsenic depth distributions were determined by RBS analysis. Figure 3 shows the arsenic signals in the non channelled and channelled energy spectra of 1.5 MeV He$^+$ backscattered from the thermally processed samples. For clarity, smooth curves have been drawn through the data points. Little change occurs during the regrowth, but after a "drive-in" anneal of 1 hr the arsenic is almost uniformly distributed through the top silicon layer. With further increases in time the arsenic gradually piles up in the vicinity of the Si/SiO$_2$ interface, at a depth of about 3000Å beneath the top surface. The channelled spectra are plotted as dotted lines and show that, with increasing time, arsenic atoms move off substitutional lattice sites.

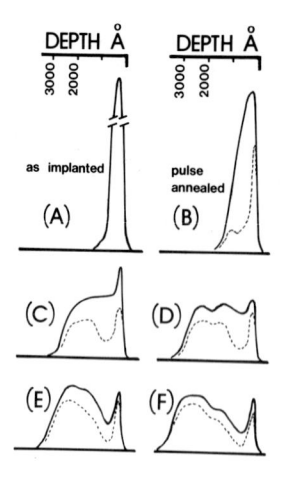

FIG. 3. Arsenic signals in the non channelled and channelled RBS spectra of 1.5 MeV He$^+$ from SOI material implanted with 1×10^{16} As$^+$ cm^{-2} at 40 keV and pulse annealed at 1100°C. (a) Reference, as implanted; (b) pulse annealed 1100°C; (c) 1 hr drive in at 950°C; (d) 2 hr at 950°C; (e) 8 hr at 950°C and (f) 16 hr at 950°C.

Figure 4 shows the dependence of the sheet resistance (R_s) of the arsenic implanted samples. Unexpectedly, R_s has a minimum value after about 6 hours at 950°C. The control samples show a constant value of 34 Ω/☐, which agrees with published data [12]. The ratio of the integrated arsenic yields in the channelled and non channelled spectra is plotted also. As the proportion of non substitutional arsenic atoms tends asymptotically to a maximum value with increasing time, it is concluded that the minimum in the value of R_s is due to annealing of scattering centres giving a higher free carrier mobility.

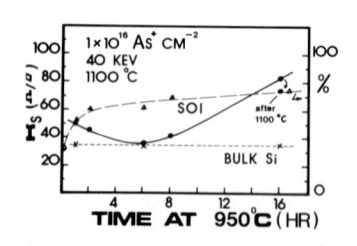

FIG. 4. Sheet resistance (R_s) of annealed layers implanted with 1×10^{16} As$^+$ cm^{-2} at 40 keV plotted against the drive-in time at 950°C. The dashed curve is a measure of the non substitutional arsenic.

A subsequent pulse anneal of the sample heat treated for 16 hours showed a small recovery of conductivity of only 10%.

Oxidation

Synthesised SOI (2.0×10^{18} O cm^{-2}) and bulk silicon samples have been oxidized at 950°C in dry oxygen at atmospheric pressure, for times within the range 3 hours to 45 hours. Both SIMS and RBS methods have been used to determine the thicknesses and compositions of the thermal oxides and no differences are evident, within the experimental uncertainty of not more than ± 2.5%. Figure 5 shows SIMS profiles of thermal oxides on the SOI material and bulk silicon. The oxide thickness of 0.12 μm was determined on the bulk sample by ellipsometry. Both samples have an abrupt thermal SiO$_2$/Si interface whilst in the SOI material the oxygen distribution merges smoothly with the distribution of implanted oxygen in the upper wing of the buried oxide. As discussed above, the small spikes in the wings of the buried oxide are believed to be due to SiO$_2$ precipitates which grow during the 1150°C anneal. This sample which was implanted in the temperature range 450°C to 480°C therefore provides further evidence for the growth of oxide precipitates in samples implanted at temperatures below about 500°C.

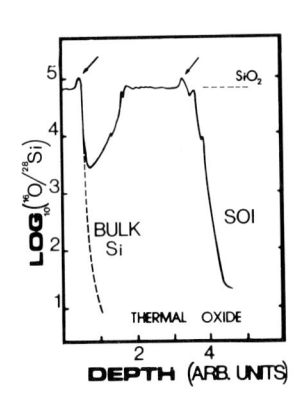

The oxide thickness shows the usual linear dependence with the square root of time.

FIG. 5. Oxygen depth profiles of thermal oxides grown on (a) processed SOI substrate and (b) bulk silicon.

Device performance

Batches of SOI substrates have been processed to 5 μm NMOS and 3 μm CMOS rules using bulk silicon processing schedules which have been modified only by the inclusion of higher energy (deeper) implants.

Measurements show vertical leakage currents to be as low at 10^{-10} A cm^{-2} at 5 volts and breakdown voltages to be much greater than 50 volts. Back channel leakage, between the source and drain contacts, is low and values of 25 pA have been measured on 10 μm x 50 μm n-channel devices. Gain factors within 6% of the values for devices on bulk material were measured and this indicates that the carrier mobility in the top layer of silicon, which is denuded of oxygen, is almost the same as in bulk silicon. However, measured sheet resistances were a factor of two higher than in bulk silicon, in both the n- and p-type implanted layers. This must be due to the dopants being deactivated by trapping on defects, as is evident in Figure 3.

The I_{DS} versus V_{DS} (common source) characteristics showed structure, in the form of a "kink", which is absent when the MOS devices are formed on bulk silicon but occurs to a smaller extent, for similar devices formed on silicon on sapphire (SOS) substrates [6]. The effect is believed to be due to impact ionization in the drain region and the more pronounced effect in these substrates is assumed to be due to better quality silicon in the top layer. The yield of chips containing 21 stage inverter chains was up to 95% with the major cause of failures being defects produced by the photo-lithography.

CONCLUSIONS

SOI structures have been formed by implantation of high doses of oxygen into silicon.

The regrowth and oxidation kinetics of the top silicon layer of thickness 1000Å-1500Å are found to be similar to bulk silicon. The presence of the buried oxide blocks the diffusion of implanted arsenic during 950°C "drive-in" anneals and the dopant gradually piles up at the Si/SiO_2 interface. The number of arsenic atoms on lattice sites decreases with increasing time.

The sheet resistance of n- and p-type implanted layers is higher than in bulk silicon, however, the good performance and high yields of 3 μm and 5 μm MOS devices further confirms the suitability of these SOI structures for silicon integrated circuits.

ACKNOWLEDGEMENTS

The authors are indebted to D.J. Godfrey, GEC, Hirst Research Centre, London and P.D. Scovell, STL, Harlow for carrying out some of the processing and for many useful discussions and to J.E. Mynard for assistance with the implantations. Acknowledgement is made to the Director of Research, British Telecom for permission to publish. This work is funded in part by the UK SERC.

REFERENCES

1. K. Izumi, M. Doken & H. Ariyoski, Electron Lett. 14 593 (1978).
2. H.W. Lam, R.F. Pinizzotto, H.T. Yuan & D.W. Bellavance, Electron Lett. 17, 356 (1981).
3. P.L.F. Hemment, E. Maydell-Ondrusz, K.G. Stephens, J.B. Butcher, D. Ioannou & J. Alderman, Nucl. Inst. & Methods 209/210, 157 (1983).
4. P.L.F. Hemment, E. Maydell-Ondrusz, K.G. Stephens, R.P. Arrowsmith, A.E. Glaccum, J.A. Kilner, M.C. Wilson & G.R. Booker, ISIAT83, Kyoto, September 1983.
5. J.A. Kilner, S. Littlewood, P.L.F. Hemment, E. Maydell-Ondrusz & K.G. Stephens, Sixth Inter. Conf. on Ion Beam Analysis, Arizona, May '83.
6. A.E. Glaccum, R.P. Arrowsmith, P.L.F. Hemment & J.D. Speight, ESSDERC83, Canterbury, September 1983.
7. C.G. Tupper & G.J. Davies. Fall Meeting of Electrochem Soc., Washington, October 1983.
8. M.R. Taylor, C.G. Tupper, R.P. Arrowsmith, R.M. Dobson, A.E. Glaccum, M.C. Wilson, G.R. Booker and P.L.F. Hemment, Microscopy of Semiconducting Materials, Oxford, March 1983.
9. E.A. Maydell-Ondrusz - to be published.
10. P.L.F. Hemment, E.A. Maydell-Ondrusz, J.E. Castle, R. Paynter, M.C. Wilson, R.G. Booker, J.A. Kilner & R.P. Arrowsmith, to be published.
11. P.L.F. Hemment, E. Maydell-Ondrusz, K.G. Stephens & P.D. Scovell; Electron Lett. 19, 483 (1983).
12. P.D. Scovell & E.J. Spurgin, J. Appl. Phys., 54, 5, 2413 (1983).

CHARACTERIZATION OF N-TYPE LAYERS FORMED IN Si BY ION IMPLANTATION OF HYDROGEN

S. R. WILSON, W. M. PAULSON AND W. F. KROLIKOWSKI[a]
Semiconductor Products Sector, Motorola, Inc., Phoenix, AZ 85008

and

D. FATHY[b]
Center for Solid State Science, Arizona State University, Tempe, AZ 85281

and

J. D. GRESSETT, A. H. HAMDI[c] AND F. D. MCDANIEL
Dept. of Physics, North Texas State University, Denton, TX 76203

ABSTRACT

Silicon wafers have been implanted with H^+ (90 keV) to doses of 5.0E15/ cm^2 and 2.0E16/cm^2. The wafers were annealed in nitrogen at temperatures between 450 and 700°C for times between 10 and 60 min. The electrically active carrier profiles were measured by capacitance voltage and spreading resistance techniques. The residual damage was measured by TEM and RBS. The electrical measurements were essentially the same in both FZ and CZ silicon implying that oxygen is not playing a role in the donor formation which was observed. The donor concentration peaks near the projected range of the hydrogen after annealing at temperatures between 450-500°C. As reported previously 1000 H^+ ions generate 1 donor in the implant peak. In addition, the donor concentration between the surface and R_p has increased more than a factor of 10 above the background concentration after a 450°C 10 min anneal. Anneals of 550°C for 30 min or more annihilates essentially all of the donors. The RBS results show small amounts of damage for the 5.0E15/cm^2 implant dose but considerable crystal damage with a dose of 2.0E16/cm^2, even after a 500°C, 30 min anneal. Cross-sectional TEM analysis of 500°C annealed samples showed a large number of small loops at depths corresonding to the depth of the peak electrical carrier concentration. The donors are directly correlated to the implant damage and resultant defects. SIMS data shows little diffusion for anneals of 500°C or less but after 550°C, 30 min the peak H concentration decreases by approximately a factor of 10.

INTRODUCTION

The limited implant range of As and P with conventional implanters (<200 keV) prevents the generation of n-type layers well below the surface of Si.

---

[a]Present address: Rockwell, Malibu Beach, CA
[b]Present address: Solid State Division, Oak Ridge National Lab., Oak Ridge, TN 37830
[c]Present address: Applied Physics Dept., Cal. Tech., Pasadena, CA

However, it has been reported that implanted hydrogen forms a shallow donor after low temperature anneals. Furthermore, hydrogen has a range of approximately 1 μm per 100 keV of implant energy. These two characteristics of proton implants may lead to applications for semiconductor device processing. G. H. Schwuttke et al.[1] bombarded Si with 1 MeV protons to a dose of 3E16/cm². They observed increases in resistance after the implant, but after 400°C anneals shallow donors were formed near the peak of the proton implant. The damage introduced in Si-diodes by protons was examined by Sigmon and Gibbons [2]. The number of positively changed defect centers increased with proton dose at 2.8 MeV. The damage decreased rapidly at anneal temperatures above 460°C. Deep traps in Si were reported by Onmura, Zohta and Kanazawa [3] following proton implants. These traps annealed out at 300°C leaving shallow donors, that disappeared for anneal temperatures above 700°C. The damage distributions were presented by Chu et al.[4] for proton implants from 50 to 250 keV to doses of 1E16 or 1E17/cm². The loss of donors at these doses results from the defects caused by displacement damage. They also reported dopant rejection because of radiation enhanced diffusion with a 900°C anneal. Wood et al.[5] implanted Si with H to dose of 1.0E16 to 8E17/cm² and observed the damage with TEM and RBS. Imai and Nakajima [6] utilized the shallow donors formed by implanted hydrogen to selectively anodize silicon and form silicon-on-insulator structures. The purpose of the present experiments is to implant hydrogen into silicon and correlate the resulting electrical measurements with the physical damage distribution from RBS and the defect structures observed with the TEM and the hydrogen profiles measured by SIMS.

SAMPLE PREPARATION

Silicon wafers, (100) orientation, n-and p-type, 1-10 Ω/cm, were used as substrates. We used both Czochralski (CZ) and Float Zone (FZ) wafers to see if oxygen donors which are generated between 400 and 500°C played a significant role in the results. The wafers were implanted with H⁺ (90 keV) to doses of 5.0E15/cm² and 2.0E16/cm². Samples were annealed in N_2 at temperatures of 450, 500, 550 and 700°C for times of 10, 30 and 60 min. The samples were analyzed by Capacitance-Voltage (CV) techniques, using a mercury probe, Spreading Resistance Probe (SRP), Transmission Electron Microscopy (TEM) and Rutherford Backscattering-Channeling (RBS-C) analysis. The hydrogen profiles were measured by Secondary Ion Mass Spectroscopy (SIMS).

RESULTS

Electrical Data

The electrically active carrier profiles were measured by CV and SRP techniques. Figure 1 is a plot of the donor concentration determined by CV measurements for a sample implanted to a dose of 5.0E15/cm² and annealed at 450°C for 30 min. This sample was a 1 Ω-cm, n-type, FZ wafer. The plot shows a peak in the donor concentration of ∼8.0E16/cm² at a depth of ∼.75 μm. Wafers which were implanted but received no anneal produced similar CV plots, but the uniformity across a wafer was not as good as after a 450°C anneal. This peak in the donor concentration is ∼3 orders of magnitude lower than the peak hydrogen concentration measured by SIMS. The donor concentration peaks at a depth of ∼.75 μm which is slightly closer to the surface than the peak in the hydrogen profile (∼.80 μm) measured by SIMS. As discussed in the following sections this depth agress with the maximum damage location. The peak donor concentration is nearly two orders of magnitude greater than the background donor concentration. The CV plot also indicates a donor concentration of $\sim 1.5 \times 10^{16}$/cm³ extends toward the surface from the peak. The zero bias depletion prevents a determination of the donor concentration at the surface by this technique. Similar results were obtained on a wafer

FIG. 1. Carrier concentration vs. depth measured by CV for H+ (90 keV, 5.0E15/cm²) implanted Si. The sample was annealed at 450°C for 30 min.

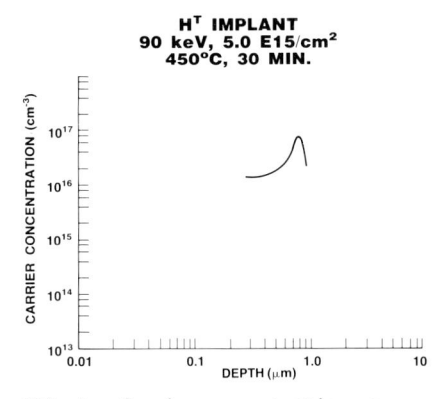

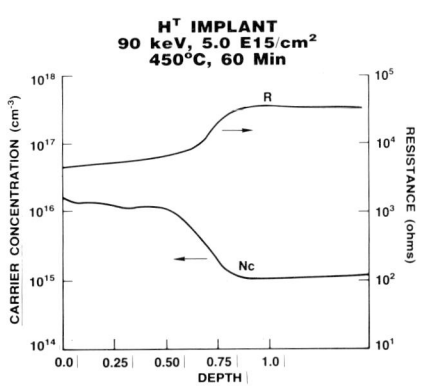

FIG. 2. Carrier concentration and probe to probe resistance measured by SRP for H+ (90 keV 5.0E15/cm²) implanted Si. The sample was annealed at 450°C for 60 min.

annealed at 450°C for 1 hr. We did not observe any significant differences between FZ and CZ wafers. When the implant was into p-type material compensation occurred and the implanted layer became n-type.

The CV results are supported by the SRP data shown in Fig. 2 for a sample annealed at 450°C for 60 min. The SRP shows an approximately constant carrier concentration of ~1E16/cm³ from the surface to a depth of ~0.65 μm. The carrier concentration decreases to the background level at depths greater than ~0.75 μm. The corresponding probe-to-probe resistance shows a lower resistance near the surface and an increase in resistance in the starting wafer level at depths greater than the implant. The peak in the concentration which was present in all of the CV plots was not observed in any of the SRP measurements. This is probably due to a difference in resolution in the two techniques. The SIMS data (not shown) indicates the hydrogen distribution is tightly peaked around 0.8 μm with very little hydrogen between 0.0 and 0.6 μm.

Figures 3a and 3b are CV plots for wafers annealed at 500 and 550°C for 10 min, respectively. The results for the 500°C anneal are essentially the same as the 450°C anneal, 30 min (Fig. 1). However, Fig. 3b shows four different CV plots taken from four different areas on a wafer which was annealed at 550°C for 10 min. These results imply that the donor concentration produced by the hydrogen implant is decreasing and the uniformity across a wafer is decreasing. A 30 min anneal at 550°C annihilates the donors generated by the hydrogen implant and the carrier concentration has decreased to the level present in the starting material prior to the H+ implant (Fig. 3c). Anneals at 700°C produce profiles identical to the starting material and will annihilate any donors which were generated by a 450 to 550°C anneal. The SRP plots actually show an increase of a factor of 2-3 in probe to probe resistance in the first 0.1 μm after a 550 or 700°C anneal. This resistance decreases to background at the depth of the implant.

Some wafers were implanted to a dose of 2.0E16/cm² to examine differences in doses. Figure 4 is a CV plot from a wafer implanted to the higher dose and annealed at 500°C for 30 min. The carrier concentration profile is nearly the same as the profile from the wafer that was implanted to a dose of 5.0E15/cm². The peak in the carrier concentration is slightly wider for the 2.0E16/

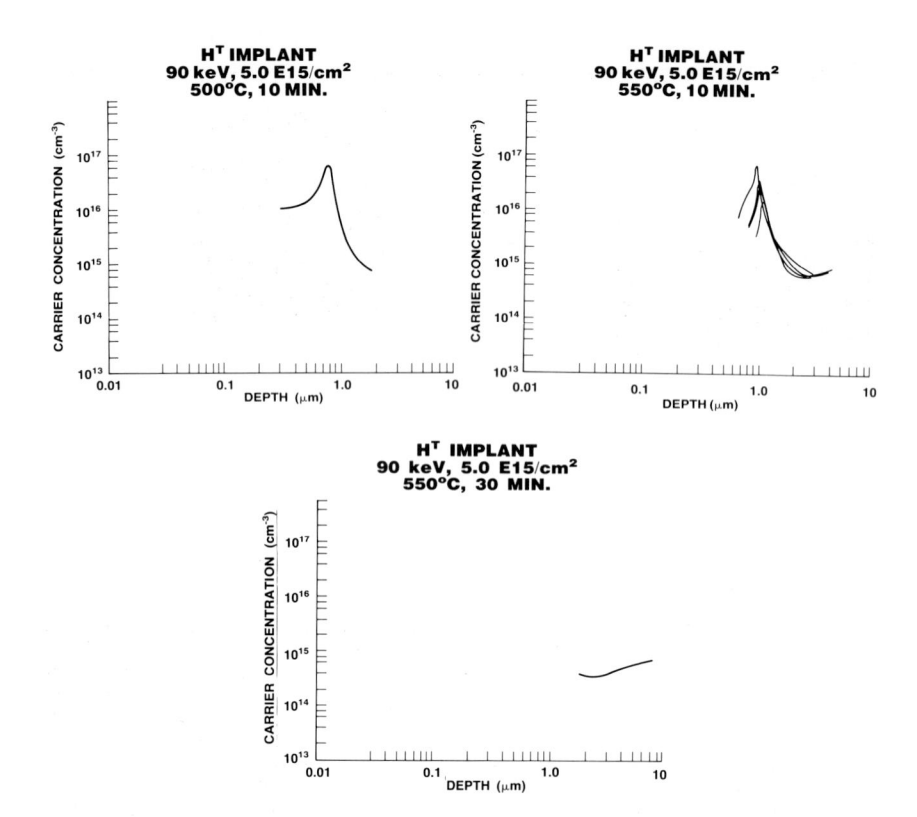

FIG. 3. Carrier concentration vs. depth measured by CV for H+ (90 keV, 5.0E15/cm²) implanted Si. The samples were annealed at (a) 500°C, 30 min and (b) 550°C, 10 min, (c) 550°C, 30 min.

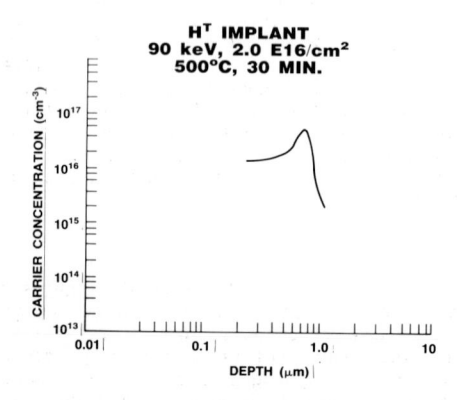

FIG. 4. Carrier concentration vs. depth measured by CV for H+ (90 keV, 2.0E16/cm²) implanted Si. The sample was annealed at 500°C for 30 min.

H⁺ (90keV, 5.0 E15/cm²) IMPLANTED Si 500°C 30 MIN.   H⁺ (90keV, 5.0 E15/cm²) IMPLANTED Si 550°C 30 MIN.

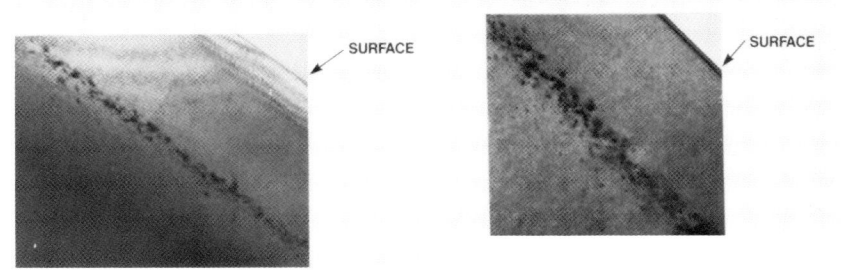

FIG. 5. Cross sectional TEM of samples implanted to a dose of 5.0E15/cm² and annealed at (a) 500°C and (b) 550°C.

cm² plot relative to the 5.0E15/cm² profile. This implies the donor concentration is not sensitive to changes in the implant dose in this range.

Structural Data

Cross-sectional TEM micrographs for wafers implanted to a dose of 5.0E15/cm² are presented in Figs. 5a and 5b. The anneal temperatures were 500 and 550°C respectively. In both cases there is a band of damage between .60 and .70 µm. The depth of this damage layer agrees fairly well with the peak observed in the carrier concentration plots (Fig. 1, 3a) determined by CV. The micrograph for the 500°C annealed sample shows that the defects are fairly small (<100 A°) and are predominantly loops. After the 550°C anneal the defects have increased in size and cluster density. The defects are now a combination of line defects and loops. The 700°C anneal (not shown) shows a further increase in the size of the defects.

Rutherford backscattering analysis was performed using a 2.0 MeV He⁺ beam aligned parallel to a <110> axis. The 5.0E15/cm² as implanted sample shows very little damage detectable by RBS-C. There is a slight increase in χ_{min} from 4.4% to 7.9% between 0.5 and 0.8 µm. This is in good agreement with the TEM, which shows is only a lightly damaged layer between 0.6 and 0.7 µm. However, at the higher dose, 2.0E16/cm², the surface layer is heavily damaged (χ_{min} = 40%) as shown in Figure 6. The χ_{min} increases rapidly between 0.6 and 0.7 µm to 85% at a depth of 0.8 µm. The 500°C 30 min anneal produces no change in the scattering yield near the surface, but causes χ_{min} to approach 100% at 0.8 µm. This implies the heavily damage layer is becoming more defective during the anneal. This is probably due to the high concentration of hydrogen.

CONCLUSION

We have shown that hydrogen implanted into Si produces donors after a 450-500°C anneal. Donors are generated from the surface to the depth of the implant. The peak donor concentration occurs at a depth which correlates to the maximum damage. The donor profiles (especially between the surface and R_p) do not correlate with the hydrogen profiles. An anneal at 550°C or 700°C annihilates the donors and actually causes an increase in the resistance of the surface layer. The 5.0 E15/cm² sample has a layer of small dislocation loops at the depth of the peak donor concentration after the 500°C anneal.

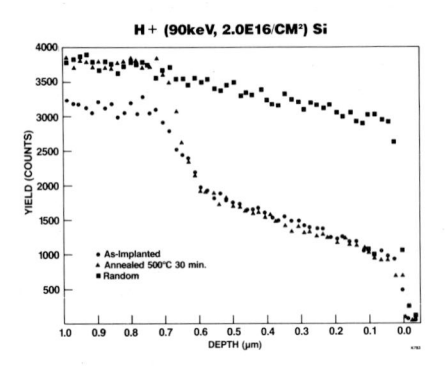

FIG. 6. RBS-C spectra from samples implanted to a dose of 2.0E16/cm^2 before and after annealing at 500°C for 30 min.

The higher temperature anneals cause the loops to grow in size. This indicates the electrically active donors may be related to point defects [7]. The point defects become larger dislocations with the higher temperature anneal, thus decreasing the number of donors. A higher implant dose (2.0E16/cm^2) produced little change in the donor concentration but substantially increased the residual damage. This result indicates that the type of damage present after the anneal is playing a significant role in the number of donors that are measured. This further supports the idea that only the point defect are active and the large defects are not. The fact no differences were observed between FZ and CZ material implies oxygen donors are not playing a major role.

ACKNOWLEDGMENTS

We would like to thank Bob Lorigan for performing the implants, O. W. Holland for assistance with the computer calculations of implantation damage and M. Scott for typing the manuscript. Partial support for NTSU provided by the Robert A. Welch Foundation and the NTSU Organized Research Committee.

REFERENCES

1. G. H. Schwuttke et al, in Ion Implantation, ed. F. H. Eisen and L. T. Chadderton, Gordon and Breach, London, 1971, p 139.
2. T. W. Sigmon and J. F. Gibbons, ibid, p 131.
3. Y. Ohmura, Y. Zohta and M. Kanazawa, Phys. Stat. Sol. (a) 15, 93 (1973).
4. W. K. Chu et al, Physical Review B16, 3851 (1977).
5. Susan Wood, J. Greggi, J. A. Spitznagel, N. J. Doyle, R. B. Irwin, J. R. Townsend, and W. J. Choyke, 1983 Oxford Conference on Microscopy of Semiconductors in press.
6. K. Imai and S. Nakajima, Proceeding of the IEDM, 1981, p 376.
7. L. M. Brown and D. Fathy, Phil. Mag. B, 43, 715 (1981).

POINT DEFECT SUPERSATURATION AND ENHANCED DIFFUSION IN SPE REGROWN SILICON.[*]

S. J. PENNYCOOK, J. NARAYAN, AND O. W. HOLLAND
Solid State Division, Oak Ridge National Laboratory, Oak Ridge, TN 37831

ABSTRACT

Transient, greatly enhanced diffusion has been observed on annealing solid-phase-epitaxial (SPE) grown Si-Sb alloys. This is shown to be due to a high concentration of interstitials being trapped during SPE regrowth. The migration enthalpy, for diffusion of Sb by an interstitialcy mechanism was measured as 1.8 ± 0.2 eV. The interstitials eventually condensed into loops, marking the end of the transient. In a SPE grown Si-Bi alloy a similar transient enhanced diffusion was observed, with an activation energy of 2.0 ± 0.2 eV, but no loops formed.

INTRODUCTION

Supersaturated Si alloys have recently been produced using ion implantation followed by new thermal processing techniques, such as furnace or pulsed laser annealing.[1,2] These techniques remove the implantation damage but leave the dopant trapped on substitutional sites to achieve electrically active dopant concentrations greatly exceeding the solubility limit. Such alloys have been characterized by ion channeling for determining the lattice location and concentration of dopant, and by transmission electron microscopy (TEM) to demonstrate the absence of extended defects. However, possible non-equilibrium concentrations of point defects could not be detected by these techniques. Here we demonstrate that alloys produced by solid-phase-epitaxial (SPE) growth contain a high concentration of trapped interstitials which gives rise to transient, greatly enhanced dopant diffusion during subsequent annealing.

Si-Sb Alloys

Samples of (100)Si were implanted with ^{121}Sb$^+$ (200 keV, integrated dose 4.4×10^{15} cm^{-2}) and recrystallized, either by SPE growth (furnace annealing under flowing dry He gas at 550°C for 20 minutes), which will be referred to as sample 1, or by LPE growth (pulsed ruby laser annealing, 1.4 J cm^{-2} and 15 ns pulse duration). Ion channeling analyses of these alloys are shown in Fig. 1a and b. We have also used a graded energy implant, SPE regrown at 550°C for 30 minutes, to give an approximately uniform concentration of 1.6 at. % down to a depth of about 100 nm as shown in Fig. 1c, which will be referred to as SPE sample 2. All alloys are >99% substitutional and exceed the retrograde maximum solubility of 0.16 at. %. Thin TEM samples were prepared by a chemical thinning procedure and studied in a Philips EM400 electron microscope. Annealing the thin samples at temperatures of 650°–750°C (SPE samples) or 850°–950°C (LPE samples) induced the precipitation of the dopant in excess of the solubility limit. The thin regions of the TEM samples did not contain the layer of residual implantation damage underlying the SPE regrown layer and thus this damage could not affect the precipitation process occurring within the regrown layer. The electrical junction had also been removed from these thin regions.

---

[*]Research sponsored by the Division of Materials Sciences, U. S. Department of Energy under contract W-7405-eng-26 with Union Carbide Corporation.

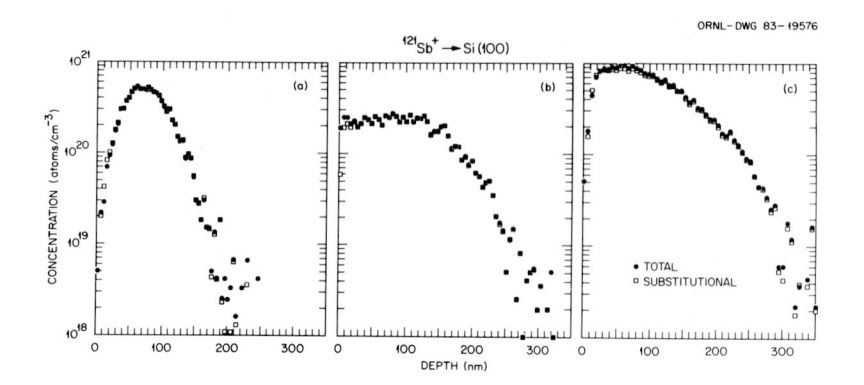

FIG. 1. Ion channeling analysis of (100)Si implanted with ^{121}Sb$^+$ (200 keV, 4.4 x 10^{15} cm^{-2}), (a) SPE regrown at 550°C/20 min. (b) pulsed laser annealed at 1.4 J cm^{-2} (c) graded energy implant SPE regrown at 550°C/30 min.

Figure 2 shows the results of 20 minute anneals of SPE sample 1 at various temperatures. At the lowest temperature the nucleation rate is very low but a few precipitates are visible (Fig. 2a). The nucleation rate and the mean precipitate size increase with higher temperature anneals (Fig. 2b, and c). After the 750°/20 min. anneal, loops are visible (Fig. 2d), which grow with further annealing and additional loops nucleate (Fig. 1e). The loops were found to be faulted, lying on {111} planes, having Burgers vectors of a/3<111> and to be interstitial in nature.[3] Figure 3 shows the results of annealing the LPE sample, where higher temperatures were required to induce the precipitation and no loops were observed.

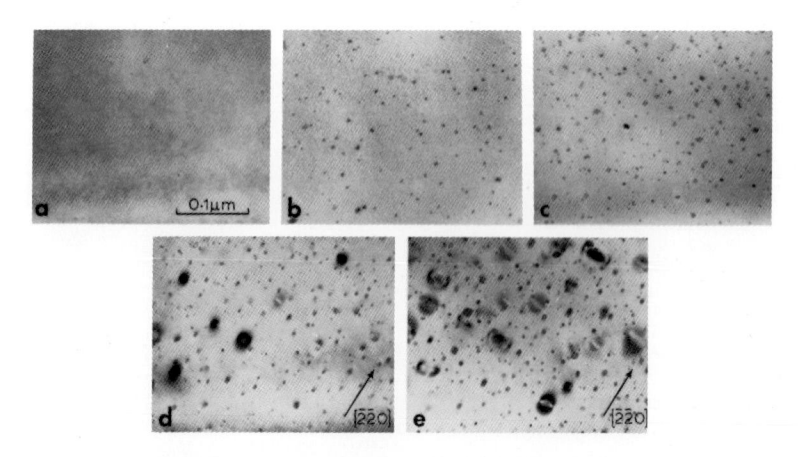

FIG. 2. Results of 20 min. anneals of SPE sample 1 at a) 650°C; b) 700°C; c) 750°C observed in a non-diffracting orientation; d) area of (c) observed in diffraction contrast showing onset of loop formation; e) as (d) but with additional 780°C anneal.

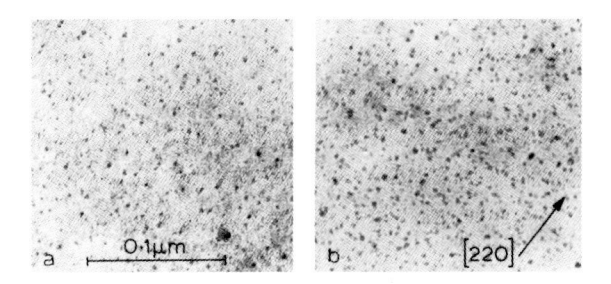

FIG. 3. Results of 20 min. anneals on the LPE sample at (a) 850°C, (b) 900°C.

The diffusion coefficients were obtained as follows.[4] For diffusion limited growth, and small precipitated fraction, we deduce,

$$D = \frac{C_p}{C_o} \frac{\overline{R_p^2}}{2\bar{t}} \quad (1)$$

where D = diffusion coefficient of Sb
 C_p = Sb concentration in the precipitate
 C_o = initial Sb concentration in the matrix
 $\overline{R_p^2}$ = mean square precipitate radius
 $\bar{t}$ = mean time of groth = t/2
 t = annealing time.

An isothermal annealing study of SPE sample 2 is shown in Fig. 4, which justifies the assumptions of the growth law. The nucleation rate was approximately constant for at least the first 20 minutes of annealing, and the mean-square precipitate radius increased linearly. Before the onset of loop formation the precipitated fraction was small (calculated from the precipitate size distribution observed in the micrographs). The diffusion coefficients can, therefore, be obtained from eqn. (1) and were enhanced over the tracer values[5] by about five orders of magnitude as shown on the Arrhenius plot in Fig. 5.

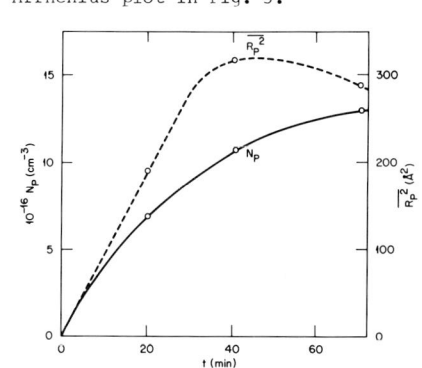

FIG. 4. Isothermal annealing study at 680°C to justify precipiate growth law. N_p = precipitate number density.

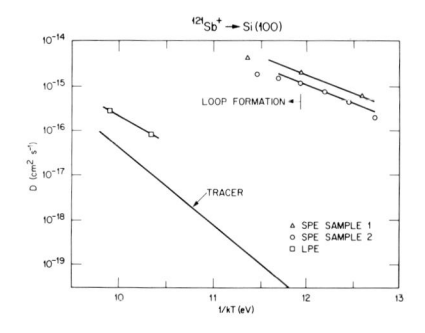

FIG. 5. Arrhenius plot of Sb diffusion coefficients deduced from 20 min. anneals.

Also shown are similar results from SPE sample 2, obtained at closer temperature intervals. Both samples show an activation energy for diffusion of 1.8 ± 0.2 eV during the initial transient. During the higher temperature anneals loops were formed by the condensation of the trapped interstitials. The concentration of interstitials and, therefore, the diffusion coefficients, reduced during these anneals so that the points deduced from eqn. (1) fall below the extrapolation of the initial transient. With the LPE sample, diffusion coefficients were obtained much closer to the tracer values. The small enhancement may indicate some point defect trapping in this case, even though no loops were observed in this alloy.

Si-Bi Alloys

The same procedure has been used to obtain diffusion coefficients in graded energy Bi implanted Si, SPE regrown at 575°C/30 min. An ion channeling analysis is shown in Fig. 6 indicating an approximately constant Bi concentration of 0.3 at. % to a depth of 150 nm with a 94% substitutional fraction. Higher temperatures were needed to precipitate the dopant than with SPE regrown Sb alloys, as shown in Fig. 7. No loops were observed even after annealing at 1000°C for 30 min. However, the diffusion coefficients were enhanced by four orders of magnitude over the tracer values as can be seen from the Arhennius plot in Fig. 8. The initial transient shows an activation energy of 2.0 ± 0.2 eV, similar to the Sb case, although less dopant is precipitated during it.

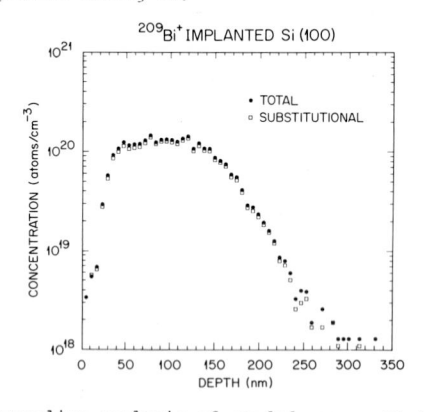

FIG. 6. Ion channeling analysis of graded energy Bi implanted Si.

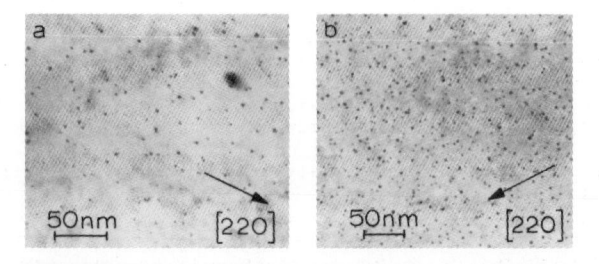

FIG 7. Results of 20 min. anneals on SPE regrown Bi implanted Si, at a) 800°C, b) 920°C.

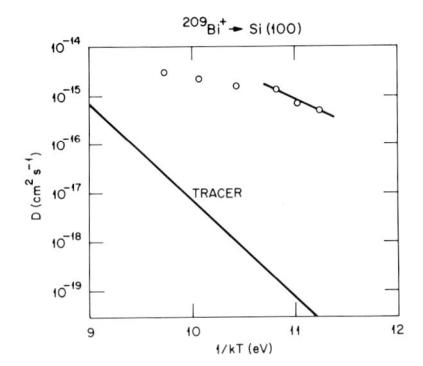

FIG. 8. Arrhenius plot of Bi diffusion coefficients.

DISCUSSION

Consider first the case of Si-Sb alloys. Transient enhanced diffusion was observed in SPE regrown alloys during which 30% of the dopant was precipitated out. Since only 1% of the dopant was nonsubstitutional following SPE regrowth, this cannot be due simply to fast diffusion of the nonsubstitutional fraction. Neither can it be due solely to the high dopant concentration since the diffusion coefficients were enhanced by five orders of magnitude over the tracer values. In the LPE alloy the dopant concentration was a third that of the SPE alloy, but the diffusion coefficients were enhanced by less than an order of magnitude. Interstitial dislocation loops formed only in the SPE samples, and marked the end of the transient enhanced diffusion. Therefore, the enhanced diffusion is due to interstitials trapped during SPE regrowth. These observations provide a clear demonstration of an interstitialcy mechanism of diffusion.[6] Furthermore, the activation energy of the transient enhanced diffusion must, therefore, correspond to the migration enthalpy of Sb by the interstitialcy mechanism.

The original concentration of trapped interstitials was estimated to be 3×10^{18} cm^{-3} from an isothermal annealing study at 680°C, in which the size and number density of loops eventually became constant. From the experimental diffusion coefficients, therefore, the Sb diffusivity by this mechanism can be estimated as[3] $D_I = 0.04 \exp(-1.8/kT)$ cm^2 s^{-1}. The concentration of trapped interstitials is less than 0.4% of the concentration of implanted Sb. The trapping site may be an Sb atom itself, but if so, on heating the complex must split up since about 30% of the implanted Sb is precipitated by these trapped interstitials.

In the case of Si-Bi alloys, the situation is not so clear. Although enhanced diffusion was clearly observed, the end of the transient was not marked by loop formation. Therefore, we cannot conclusively say that the enhanced diffusion is due to trapped interstitials, although it seems extremely likely since the activation energy during the transient is similar to that obtained in the Si-Sb alloys. The absence of loops also meant that there was no direct measure of the concentration of trapped interstitials. It seems likely that a lower concentration was trapped in the Si-Bi alloy since less dopant was precipitated during the transient. The nucleation of interstitial loops would, therefore, be considerably more difficult. In addition, the higher temperatures needed to cause the precipitation of Bi would greatly enhance competing processes for the removal of excess interstitials, namely, their migration to the free surfaces or recombination with thermally-generated vacancies. The activation energy for vacancy-interstitial recombination is only 1.4 eV and many results showing enhanced and retarded diffusion can be explained on the basis of a local equilibrium model.[7] Presumably, the supply of vacancies would be limited by thermal activation to concentrations much lower than the concentration of trapped interstitials. Therefore, it is unlikely that they can be efficiently removed by this mechanism and we presume that migration to the free surfaces is occurring. Loop nucleation, therefore, only occurs when the concentration of interstitials is very high, as in the Si-Sb alloys.

We now discuss the reason for the trapping of intersitials in SPE grown Si alloys. The effect of the lattice strain around the substitutional dopant can be ruled out since this is a compressive strain for Sb and Bi and would lead to vacancy trapping. Another possibility is the volume increase associated with the amorphous to crystalline transformation, which may lead to interstitial injection as found during Si oxidation. This seems unlikely, however, since the volume increase is much smaller, and the excess interstitial concentration is very much larger than in the case of oxidation. The most likely mechanism is electronic in nature. If a small fraction of the Sb or Bi atoms are fully saturated in the amorphous state, having five neighbors, they may remain so as the crystallization interface passes, creating a trapped interstitial silicon atom. This process could not occur in regrowth from the melt, since liquid silicon is metallic, consistent with the measurement of diffusion coefficients much closer to the tracer values in LPE grown alloys.

CONCLUSION

Supersaturated Si alloys produced using SPE regrowth contain a supersaturation of trapped interstitials. These give rise to transient greatly enhanced diffusion during subsequent heat treatment. Concentrations of trapped interstitials exceeding 10^{18} cm^{-3} will condense into loops following the initial transient.

REFERENCES

1. J. Narayan and O. W. Holland, Appl. Phys. Letts. 41, 239 (1982).
2. C. W. White, S. R. Wilson, B. R. Appleton and F. W. Young, Jr., J. Appl. Phys. 51, 378 (1980).
3. S. J. Pennycook, J. Narayan and O. W. Holland, J. Appl. Phys. 55, 837 (1984).
4. P. Haasen, p. 203 in Physical Metallurgy, Cambridge University Press, London (1978).
5. H. Reiss and C. S. Fuller, p. 222 in Semiconductors, ed. by N. B. Hannay, Reinhold, NY (1959).
6. J. R. Manning, p. 80 in Diffusion Kinetics for Atoms in Crystals, Van Nostrand, Princeton (1968).
7. T. Y. Tan, U. Goesele and F. F. Morehead, Appl. Phys. A31, 97 (1983).

CHARACTERIZATION OF ION IMPLANTED SILICON BY SPECTROSCOPIC ELLIPSOMETRY AND CROSS SECTION TRANSMISSION ELECTRON MICROSCOPY

P. J. McMARR, K. VEDAM AND J. NARAYAN*
Materials Research Laboratory, The Pennsylvania State University, University Park, PA 16802 and *Solid State Division, Oak Ridge National Laboratory, Oak Ridge, TN 37380

ABSTRACT

This paper deals with the application of spectroscopic ellipsometry (SE) and cross-section transmission electron microscopy (XTEM), to the characterization of damaged surface layers in ion implanted Si single crystal. Si samples of 2-6$\Omega\cdot$cm resistivity and <100> orientation were implanted with $^{28}Si^+$ ions in the dose range of $1.0 \times 10^{16} - 1.5 \times 10^{16}$ ions/cm^2 using ion energies of 100 and 200 keV. Ion current densities were varied from 6 to 200 µA/cm^2. Depth profiles of the implanted samples were evaluated from the spectroscopic ellipsometry data. These calculated profiles were compared with the TEM micrographs of the cross sections of the samples. Excellent agreement is obtained between the two characterization techniques. The characteristics of the depth profiles of the samples, as established by the two techniques, is shown to be the result of annealing occuring during implantation .

INTRODUCTION

It is well known that the surface layers of ion implanted specimens are highly damaged during the ion implantation process. The depth profile and the degree of damage at various depths from the surface are very strongly dependent on a number of parameters such as the type and the energy of the impinging ions, total dose as well as the dose rate, the orientation and the temperature of the substrate during implanation, etc. In fact, the degree of damage in the implanted region can vary over the entire range -- from total amorphization, to dislocation tangles, to a region totally free of damage. The depth profile of such a damaged region is usually obtained by various techniques such as RBS, ion channelling, XTEM, etc. The present paper shows that spectroscopic ellipsometry is another powerful technique that can yield this information in a quantitative and nondestructive fashion and can also be used as an in-situ technique in most situations.

The application of conventional null ellipsometry, for the characterization of damaged surface layers in ion implanted semiconductors is well documented [1]. Good comparisons between the results of these null ellipsometry measurements and RBS, ion channelling, as well as other techniques, have been made [2,3].

The advantages of a spectroscopic ellipsometry over that of a single wavelength null system are discussed in references [4-8]. In general, the spectroscopic capability adds the ability to characterize the depth profile of the material, utilizing the fact that the optical penetration (and hence the region where the reflection takes place) is strongly wavelength dependent. In particular in the case of Si, the high energy regions of the spectra principally contain information about the surface regions, while the low energy regions ($\lambda<2.5$eV) contain information about progressively deeper layers of the material.

EXPERIMENTAL

A spectroscopic ellipsometer of the rotating analyzer configuration was constructed and used in the present investigation. The configuration adopted is similar to that of Aspnes et al. [9-11]. The current spectral range of our instrument covers the energy range 1.5-4.5eV. The precision attainable with our system is comparable to the current state of the art [12].

The <100> oriented Si wafers of 2-6Ω·cm resistivity were implanted at room temperature with Si+ ions at normal (or almost normal) incidence. Table I presents the various relevant parameters of the ion implanted specimens used in the present study.

TABLE I. Si+ Ions Implanted in Si Single Crystal

Sample number	Orientation	Ion energy	Total dose $x-/cm^2$	Dose rate
561	(100)	100 kev	1.5×10^{16}	6 ± 0.6 μamp/cm^2
562	(100)	200 kev	1.5×10^{16}	6 ± 0.6 μamp/cm^2
626	(100)	200 kev	1.0×10^{16}	200 μamp/cm^2

Cross-section transmission electron micrographs (XTEM) were obained on samples prepared under the same conditions as those used in the ellipsometry studies. The specimens were thinned in cross-section using the ion milling technique described elsewhere [13].

RESULTS AND DISCUSSION

Figures 1-2 present the spectroscopic ellipsometric data obtained on the ion implanted samples. The data obtained on Sample No. 562 and hence their interpretation as well, resemble exactly those of sample No. 561 and hence will not be discussed any further. The ordinates in these figures show the calculated values of the ellipsometric parameters Δ and ψ from the measured cos Δ and tan ψ.

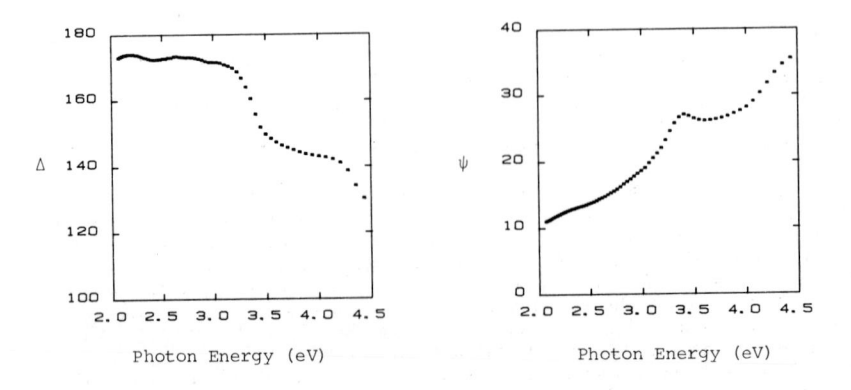

FIG. 1. Experimentally determined values of the ellipsometric parameters Δ and ψ. Sample No. 626.

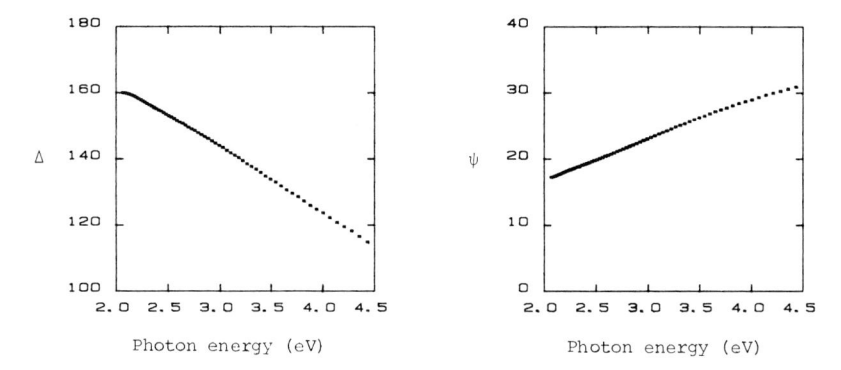

FIG. 2. Experimentally determined values of the ellipsometric paremters Δ and ψ. Sample No. 561.

In general, a number of parameters of the ion implanted Si can influence the measured values of cos Δ and tan ψ. These effects include possible overlayers, such as oxide or organic contaminant films, surface roughness caused by ion bombardment effects, as well as density and compositional changes in the bulk of the implanted sample [14-16]. Consequently the effective optical parameters such as the refractive index or the dielectric function of the implanted samples, as evaluated from the measured values of cos Δ and tan ψ, will be different from those of the intrinsic bulk silicon. But these modified dielectric responses can be used to obtain the depth profile information of the ion-implanted specimens.

In order to extract such information the following procedures was adopted. The ellipsometric parameters Δ and ψ were first calculated for various realistic n-phase multi-layer models (to represent the experimental sample) using exact ellipsometric equations at each wavelength studied. Then the calculated and experimental values of Δ and ψ were compared systematically, and objectively, by evaluating the mean-square deviation σ,

$$\sigma = \frac{1}{(n-p-1)} [\sum_{j=1}^{n} (\Delta_{expt} - \Delta_{calc})^2 + (\psi_{expt} - \psi_{calc})^2]^{1/2}$$

by standard least squares Linear Regression Analysis (LRA) techniques. Here n is the number of independent readings corresponding to different wavelengths at which the measurements were carried out and p is the number of unknown parameters used in the mode. Bruggemann's [17] effective medium approximation theory was used to evaluate the dielectric function of the medium if it contains another phase such as voids or a-Si or c-Si. For this computation the reported best values of the dielectric functions of c-Si and a-Si were obtained from Aspnes [18].

The results of such an analysis are presented in Figures 3-6. While Figures 5 and 6 show the models that yielded the best fit with the experimental data for the samples, Figures 3 and 4 show the agreement between the calculated and the experimental values of the observed dielectric function. The latter values are evaluated from the Δ, ψ data given in Figures 1-2. We find the plot of the dielectric function versus photon energy (rather) than tan ψ or cos Δ versus energy) to be more sensitive to bring out the disagreement between the calculated and experimental values.

Figures 7 and 8 present the cross-sectional TEM micrograph obtained on the samples with identical preparatory history. In fact, both the SE and XTEM were carried out on the same specimens, with one half used for XTEM while the other half was used for SE studies. On comparing the XTEM micrographs (Figures 7 and 8) with the corresponding models shown in Figure 5

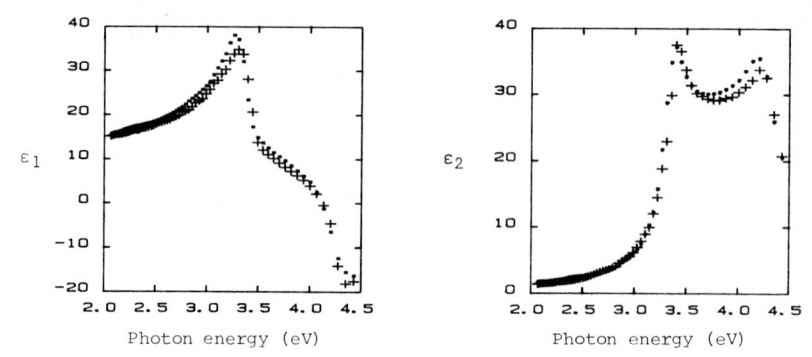

FIG. 3. Plot of real (ε_1) and imaginary (ε) parts of dielectric response for sample No. 626. (...) experimental, (+++) model.

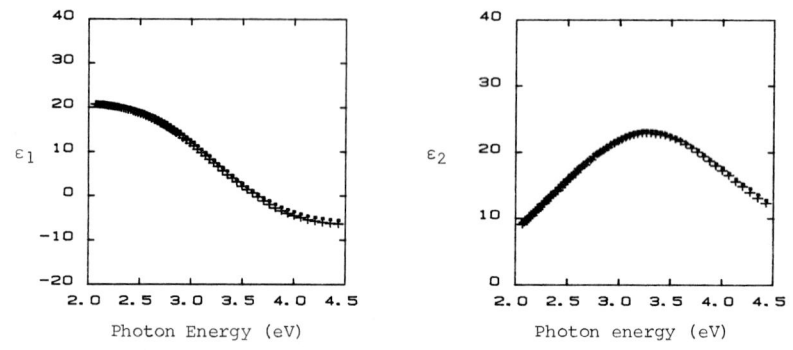

FIG. 4. Plot of real ( ) and imaginary ( ) parts of dielectric response for sample No. 561. (...) experimental, (+++) model.

and 6 respectively, it is seen that the agreement is very good. The numerical values of the various parameters of the models used for the interpretation of the SE data are entered in Figures 5 and 6 along with their 95% confidence limits. Thus it is clear the technique of spectroscopic ellipsometry can provide <u>quantitative</u> depth profile information <u>non-destructively</u>.

From Table I it is seen that the main difference between the sample Nos. 626 and 561 is the rate of ion implantation. In the former the ion current was 200µamp whereas for the latter it was 6µamp. Since the rate of implantation for sample No. 626 is quite high and since the relative orientation of the specimen with respect to the direction of the impinging ions was almost ideal with the result the ions are channelled in, the slight damages introduced in the surface layers are self annealed out even during the implantation stage. On the other hand when the rate of implantation is very slow as in the case of Samples 561 and 562, the temperature rise at the surface region during implantation is not sufficiently high to anneal out the damages. Linear regression analysis (LRA) of the SE data on Sample Nos. 561, revealed that the <u>outermost surface layer</u> of this sample exhibits the characteristics of <u>microroughness</u> [19] or equivalently that the outermost 50Å of the surface layer has 36±2% void region, as shown in Figure 6. The XTEM micrograph on this sample confirms this conclusion, as can easily be verified with

SiO$_2$ 23.0 ± 1.1Å

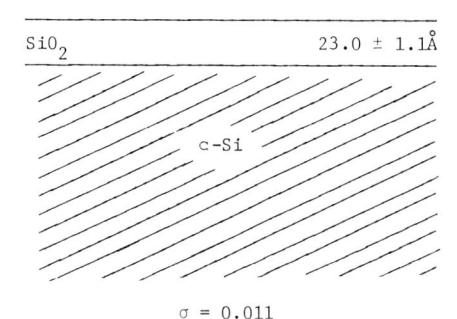

a-Si

σ = 0.011

FIG. 5. Schematic diagram of model for sample No. 626. Best fit parameters and their 95% confidence limits are also given.

FIG. 7. Cross section TEM micrograph of sample No. 626. OPD-optical penetration depth.

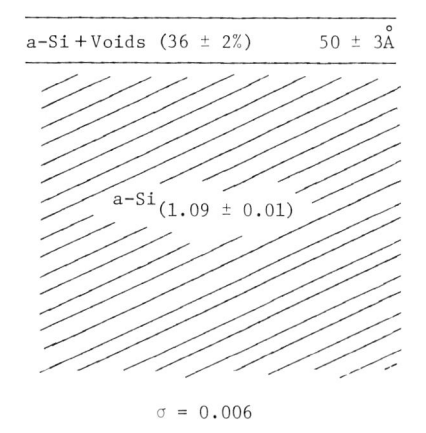

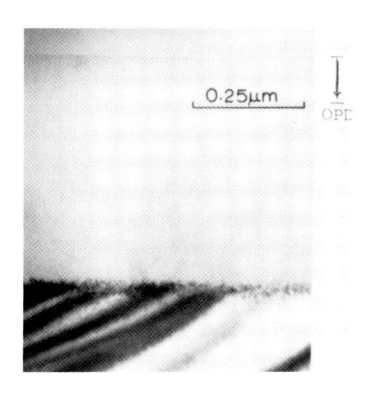

a-Si+Voids (36 ± 2%)   50 ± 3Å

a-Si (1.09 ± 0.01)

σ = 0.006

FIG. 6. Schematic diagram of model for sample No. 561. Best fit parameters and their 95% confidence limits are also given.

FIG. 8. Cross section TEM micrograph of sample No. 561. OPD-optical penetration depth.

the help of a straight edge placed on the outermost layer shown in Fig. 8.
From Fig. 6 it is seen that the a-Si region is densified by 9 ± 1% compared to the density of the best a-Si film (prepared by LPCVD) reported in the literature. This is not surprising since almost all the films prepared by CVD, evaporation or sputtering techniques contain columnar structures and voids [20] with the result the density of such films will be less than void-free a-Si. Similar increase in the density of ion bombarded (and amorphized) layer in germanium has been reported by Aspnes and Studna [21].
XTEM micrographs in Figures 7 and 8 show that the ion implantation damage extends far beyond the optical penetration depth of the present SE studies. However, if the SE measurements can be extended to the near IR range, this limitation can be overcome. Efforts are currently underway for increasing the spectral range of our present spectroscopic ellipsometer.

Lastly, it may be mentioned that similar SE studies on silicon crystals implanted with 80 KeV Si^+ ions reveal, that it can detect and characterize nondestructively, the buried amorphous layer beneath about 500A of crystalline silicon. These results will be presented elsewhere.

ACKNOWLEDGEMENT

The authors' sincere thanks are due to Dr. D.E. Aspnes of Bell Labs for giving us the exact values of the dielectric function of c-Si and a-Si, and to Dr. O.W. Holland of ORNL for valuable discussions.

REFERENCES

[1] J.R. Adams and N.M. Bashura, Surface Sci. 49 (1975) 441.
[2] D.K. Sadana, M. Stratham, J. Washburn and G.R. Booker, J. Appl. Phys. 51, 5718 (1980).
[3] T. Lohner, G. Mezey, E. Köta, F. Pâszti, A. Manuaba and J. Gyulai, Nuclear Instruments and Methods 199 (1982) 405.
[4] R.M.A. Azzam and N.M. Bashara, Ellipsometry and Polarized Light (North Holland, Amsterdam, 1977).
[5] J.B. Theeten in Proceedings of the Fourth Interantional Conference on Ellipsometry, edited by R.H. Muller, R.M.A. Azzam and D.E. Aspnes (North Holland, Amsterdam, 1980), Surf. Sci. 96, (1980), 275.
[6] D.E. Aspnes, G.P. Schwartz, G.J. Gualtieri, A.A. Studna and B. Schwartz, J. Electrochem. Soc., Vol. 128, No. 3, 590 (1981).
[7] M. Erman and P.M. Frimk, Appl. Phys. Lett. 43, 285 (1983).
[8] J.B. Theeten and D.E. Aspnes, Thin Solid Films, 60 (1979) 182-192.
[9] D.E. Aspnes and A.A. Studna, Appl. Opt. 14, No. 1 (1975) 220.
[10] D.E. Aspnes, J. Opt. Soc. Am. 64, 639 (1974).
[11] D.E. Aspnes, in Optical Properties of Solids/New Developments, B.O. Seraphin, Ed. (North Holland, Amsterdam) 800-816.
[12] R.H. Muller, Surface Sci. 56 (1976) 19-36.
[13] J. Narayan and O.W. Holland, Phys. Stat. Solidi 73, 242 (1982).
[14] K. Vedam, R. Rai, F. Lukes and R. Srinivasan, J. Opt. Soc. Am. 58, 526 (1968).
[15] D.E. Aspnes, J. Vac. Sci. and Technol., Vol. 17, No. 5 (1980) 1057-60.
[16] M. Ermen and J.B. Theeten, Surface and Interface Analysis, Vol. 4, No. 3, 98 (1982).
[17] S.A.G. Bruggemann, Ann. Phys. (Leipzig) 24, 636 (1935).
[18] B.G. Bagley, D.E. Aspnes, A.C. Adams and C.J. Mogab, Appl. Phys. Lett. 38, 56 (1981).
[19] D.E. Aspnes, J.B. Theeten and F. Hottier, Phys. Rev. 20 B, 3292 (1979).
[20] R.C. Ross and R. Messier, J. Appl. Phys. 52(8), 1981, pp. 5329-5339.
[21] D.E. Aspnes and A.A. Studna, Surface Sci. 96, 294 (1980).

SPATIAL CORRELATION INTERPRETATION OF EFFECTS OF As+ IMPLANTATION ON THE RAMAN SPECTRA OF GaAs

D.E. ASPNES,* K.K. TIONG**, P.M. AMIRTHARAJ**, AND F.H. POLLAK**
*AT&T Bell Laboratories, Murray Hill, NJ 07974; **Physics Department, Brooklyn College of CUNY, Brooklyn, NY 11210

ABSTRACT

The red shift and asymmetric broadening of the LO phonon mode of ion-implanted GaAs are both described quantitatively by a spatial correlation model based on a damage-induced relaxation of the momentum selection rule previously used by Richter, Wang, and Ley to describe similar effects in microcrystalline Si. The success of the model for a qualitatively different disorder microstructure suggests it may be possible to evaluate average sizes of crystallographically perfect regions in semiconductors from the phonon lineshapes of their Raman spectra.

Recently, Richter, Wang, and Ley [1] showed that the red shift and broadening of the LO phonon line in the first-order Raman spectrum of microcrystalline silicon could be described in terms of the lineshape and phonon dispersion data for the perfect crystal if it is assumed that the $\vec{q}$-selection rule is relaxed in the microcrystalline material. In this model the phonon wave function in a spherical crystallite was assumed to have the form of that of a phonon in an ideal crystal except for being attenuated by a spatial factor $\exp(-2r^2/L^2)$ representing the finite extent of the crystallite. With this assumption the phonon wavefunctions are no longer eigenstates of the crystal momentum $\vec{q}$ but become superpositions of such eigenstates with a weighting factor $\exp(-L^2(\vec{q}-\vec{q}_0)^2/8)$, where $\vec{q}_0$ is the wavevector of the initial state. Under this condition the observed first-order lineshapes become superpositions of lineshapes representing the individual eigenstates of the unperturbed crystal, and red shifts and broadenings arise because the energies of the unperturbed eigenstates are themselves functions of $\vec{q}$.

It is clear that the idea of the finite extent of a phonon wavefunction can be extended to microstructural geometries that are not microcrystalline, such as point or line defects, or sublattice disorder in the case of ternary or quaternary semiconductor alloys [2]. However, it is not clear that the spherical correlation model [1] should be applicable in these cases. To provide some insight into this question, we have measured the first-order Raman spectra of a series of single-crystal GaAs crystals of <100> surface orientation implanted 7° off axis with 270 keV As+ as described previously [3]. The damage distributions caused by ion implantation can be calculated from LSS theory [4]; the relevant parameters describing the damage resulting from each ion are the mean damage depth $<x_D>$ = 880Å, the mean straggle $<\Delta x_D>$=350Å, and the transverse straggle $<y>$ = 220Å. Thus, the implantation damage forms a relatively narrow, approximately conical region around the track of each As+ ion with an apex angle of about 30°. At small fluences these regions locally destroy the long-range order without introducing grain boundaries or other perturbations to disrupt the connectivity of the crystalline host matrix as a whole. At large fluences the long-range order is completely destroyed and the material is amorphous. Consequently, the microstructure of ion-implanted single-crystal material is qualitatively different from that of microcrystalline material and the relative fraction of damaged material can be controlled by varying the fluence.

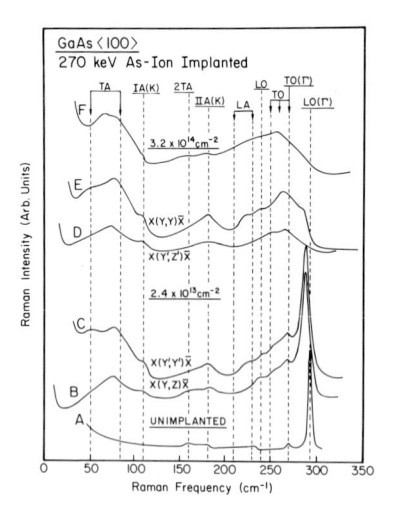

FIG. 1. Raman spectra for <100> GaAs: (A) before implantation; (B) (C) (D) (E), to 2.4 x 10^{13} cm^{-2}; (F), to 3.2 x 10^{14} cm^{-2}. The spectra are not corrected for instrument resolution.

Figure 1 shows unpolarized first-order Raman spectra for an unimplanted sample and one implanted with a fluence 3.2 x 10^{14} cm^{-2}. Several polarized spectra for a sample implanted to 2.4 x 10^{13} cm^{-2} are also shown. All spectra were measured at λ5145 where the penetration depth of light is approximately 1100Å. Similar spectra were obtained for λ4880 where the penetration depth is approximately 800Å, indicating that the Raman response is originating primarily from the near-surface region even though damage and light penetration depths are comparable. The positions of various zone center and zone edge phonons are shown at the top. The dominant feature of the unimplanted sample spectrum is the structure due to the $\vec{q} = 0$ LO phonon at 292 cm^{-1}. After correction for instrument resolution, the linewidth of this mode is found to be 3 cm^{-1}. At 3.2 x 10^{14} cm^{-2} the spectrum is that [5] of amorphous material, in agreement with dielectric function measurements on the same sample [3].

Results for several polarizations x(y,z)x̄ are shown for a fluence of 2.4 x 10^{13} cm^{-2}, where x, y, z, y', and z' denote [100], [010], [001], [011], and [01̄1] directions, respectively. The LO phonon is forbidden for x(y,y)x̄ and x(y',z')x̄ but is allowed for x(y,z)x̄ and x(y',y')x̄. As can be seen, these spectra contain features of both unimplanted and amorphous samples. However, the dominant LO phonon structure in the allowed configurations has clearly red-shifted and asymmetrically broadened relative to its counterpart for the unimplanted material. In contrast, the small $\vec{q} = 0$ TO phonon feature at 269 cm^{-1} is almost unchanged.

To interpret these observations, we use the model of Richter, Wang, and Ley [1]. Let $W(\vec{q})$ be the weighting factor describing the average $\vec{q} = 0$ phonon mode of the implanted material in terms of the momentum eigenstates of the perfect crystal, and let $\omega(\vec{q})$ represent the energy variation of the unperturbed states with $\vec{q}$. Then supposing that the phonon matrix elements for $\vec{q} \neq 0$ are approximately equal to those for $\vec{q} = 0$ it follows that the Raman intensity $I(\omega)$ can be written [6]

$$I(\omega) \propto \int_0^{2\pi/a_0} d\vec{q} W^2(\vec{q}) [(\omega-\omega[\vec{q}])^2 + (\Gamma_0/2)^2]^{-1}, \qquad (1)$$

where Γ_0 is the broadening parameter describing the linewidth of the unimplanted sample and a_0 is the lattice constant. To be specific, we take a spherically symmetric, analytic representation for $\omega(\vec{q})$:

$$\omega(\vec{q}) = A + B \cos(a_0 q/2), \quad (2)$$

where $A = 269.5$ cm^{-1} and $B = 22.5$ cm^{-1}. This represents quite well the actual dispersion of the LO mode [7].

Results of calculations in a three-dimensional spherically symmetric model using the weighting function $W(\vec{q}) = \exp(-L^2 q^2/8)$ of Ref. 1 are shown in Figs. 2 and 3. In these calculations L is a parameter having the meaning of

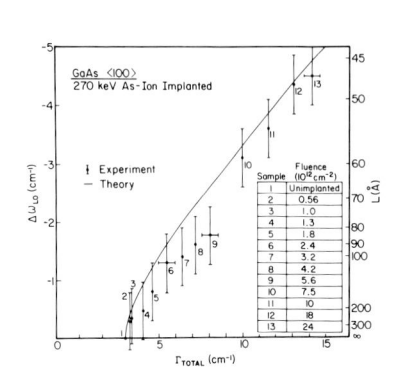

FIG. 2. Experimentally determined energy shift $\Delta\omega_{LO}$ and broadening Γ as a function of fluence. The curve shows the theoretical relationship obtained from Eqs. (1) and (2).

FIG. 3. As Fig. 2, but for $\Delta\omega_{LO}$ vs. linewidth asymmetry Γ_a/Γ_b.

a correlation length, and the parameters describing the calculated lineshapes $I(\omega)$ are $\Delta\omega$, the apparent shift of the phonon frequency, Γ, the full width at half maximum of the phonon line, and Γ_a and Γ_b, the division of Γ into components below and above, respectively, the energy at which $I(\omega)$ reaches its maximum value. Figures 2 and 3 show the two independent constraints that the model imposes on the data: the variation of $\Delta\omega$ with Γ, and the variation of $\Delta\omega$ with Γ_a/Γ_b. These comparisons depend only parametrically on L and thus represent self-consistency checks for the model.

As seen in these figures, both correlations are reproduced well by the model. Thus, both asymmetry and shift are consistent with the idea that the extent of the phonon is limited in three dimensions. Although entering only as a parameter in the above model, the values of L obtained are consistent with length scales evaluated from the fluence. For example, the mean distance between impact sites for a fluence of 2.4×10^{13} cm^{-2} is 20Å, which compares to the value 45Å deduced for L from the data of Figs. 2 and 3. The behavior of the TO mode can also be represented by the model. Here, there is only a small dependence of $\omega(\vec{q})$ on $\vec{q}$, leading to the conclusion that there should be very little change of this spectral feature with fluence, as observed.

The idea that a model initially developed to describe approximately spherical microcrystals should also describe the microstructural configuration of damage due to ion implantation is at first sight surprising, and we make some remarks on the more general problem of spatial extent and correlation lengths. The concept of a spherical correlation region defined by a Gaussian cutoff length L related to particle size is appealing for microcrystalline configurations because one intuitively expects zero vibrational amplitude outside the particle boundary (even though the amplitude function has physical meaning only when evaluated at lattice sites). However, it is clear that the surface vibrational amplitudes need not be zero at the surface but simply different from those in the interior (although the narrow energy width of phonon resonances ensures that the vibrational amplitudes of surface atoms, with bonding significantly different from that of bulk atoms, will be small compared to those of bulk atoms at the bulk phonon frequencies). Thus, the relationship between L and particle size cannot be exact even for microcrystals but will depend on the actual boundary conditions at the surface. This, in fact, is tacitly assumed in the Gaussian form since the cutoff is not sharp.

The present microstructure, consisting of conical defect trails in an otherwise perfect crystal, is superficially quite different and in principle should be described by a different assumption. Ruppin and Englman [8] have shown in the case of diatomic crystals that the amplitude $\vec{f}(\vec{r})$ of the relative motion coordinate for longitudinal waves (the only type that satisfies the LO phonon selection rules and that we need consider here) satisfies the condition $\nabla \times \vec{f}(\vec{r}) = 0$ in the long-wavelength limit. Therefore, we can write $\vec{f}(\vec{r}) = \nabla g(\vec{r})$, where $g(\vec{r})$ is any scalar function (including the Gaussian weighting function used in Ref. [1] and in the above calculation). Since other weighting functions also satisfy this rather general condition, we have investigated the possibility of two-dimensional averaging [$d^3q \rightarrow d^2q$ in Eq. (1)] which may be expected to more nearly represent the damage configuration of our samples where the tracks have not yet begun to overlap. Indeed, we found good agreement between our data and the predictions of a two-dimensional model and also values of L more nearly similar to the mean distance between impacts for fluences less than 10^{13} cm^{-2}. However, the overall agreement was poorer.

Spatial correlation effects due to point and line defects can be expected to show another form. In fact, more recent work [2] has shown that the spatial correlation model also describes compositional disorder on the separate cation and/or anion sublattices for which the microstructure may be expected to approximate that of point defects. It is becoming clear that the spatial correlation model is more general than the initial application implies and must therefore result from a rather fundamental property of the effect of disruptions in an otherwise perfect crystal lattice on phonon propagation. The continuum approximation [8] is probably too crude; a theoretical treatment to describe these effects must take into account the discrete nature of the lattice. Further theoretical work is needed in this area.

REFERENCES

1. H. Richter, Z. P. Wang, and L. Ley, Sol. St. Commun. 39, 625 (1981).
2. P. Parayanthal and F. H. Pollak, Phys. Rev. Lett. (submitted).
3. D. E. Aspnes, S. M. Kelso, C. G. Olson, and D. W. Lynch, Phys. Rev. Lett. 48, 1863 (1982).
4. K. B. Winterbon, Ion Implantation Range and Energy Distributions, Vol. 2 (Plenum, New York, 1975).
5. J. E. Smith, Jr., M. H. Brodsky, B. L. Crowder, and M. I. Nathan, in Light Scattering in Solids, ed. by M. Balkanski (Flammarion, Paris, 1971), p. 330.
6. K. K. Tiong, P. M. Amirtharaj, F. H. Pollak, and D. E. Aspnes, Appl.

Phys. Lett. (in press).
7. J. T. Waugh and G. Dolling, Phys. Rev. $\underline{132}$, 2410 (1963).
8. R. Ruppin and R. Englman, Rep. Prog. Phys. $\underline{33}$, 149 (1970).

HIGH RESOLUTION TRANSMISSION ELECTRON MICROSCOPY STUDY OF Se$^+$ IMPLANTED AND ANNEALED GaAs

D.K. SADANA*, T. SANDS AND J. WASHBURN
Materials and Molecular Research Divisions, Lawrence Berkeley Laboratory, University of California, Berkeley, CA 94720;*Now at Microelectronics Center of North Carolina, Research Triangle Park, NC 27709.

ABSTRACT

High resolution transmission electron microscopy (HRTEM) has been applied to the study of amorphization and recrystallization mechanisms in Se$^+$ implanted (100) $_o$GaAs. An Se$^+$ dose of 1×10^{14} cm^{-2} at 450 keV (projected range 1550A) produced an amorphous band in the depth range 250 to 2150A below the surface. Annealing at 400°C resulted in the epitaxial regrowth of the upper and lower transition region (0-250A and 2150 - 2500A, respectively). Regrowth of the amorphous layer was found to proceed by the nucleation and propagation of the dense network of stacking faults bundles. These bundles disappeared on higher ($\geq$600°C) temperature annealing. Amorphization and recrystallization mechanisms in Se$^+$ implanted GaAs are discussed in light of these HRTEM results.

INTRODUCTION

Conventional (bright-field and weak-beam) transmission electron microscopy (TEM) and Rutherford backscattering (RBS)/ channeling have been utilized extensively for structural examination of ion implanted and subsequently annealed GaAs[1-7]. It is known from these measurements that amorphous layers produced by ion implantation recrystallize at a much lower temperature (150-200°C)[1-7] than analogous amorphous layers of Si (450°C). Another important difference in the recrystallization behavior of these two materials is that regrowth of amorphous layers in (100) Si can result in essentially defect-free material, whereas regrowth of amorphous GaAs layers in (100) GaAs usually results in microtwins, stacking faults, misoriented crystallites and other irregular structures[1,4,6,7]. These differences are also reflected in the electrical behavior of recrystallized layers. For example, the optimum electrical activation (90-100%) of dopants in Si is achieved when the implanted region is completely amorphized before annealing. However, in GaAs such conditions result in the lowest activation of dopants ($\sim$10%)[4,8]. Better electrical results in GaAs have been reported for elevated temperature (>100°C) implants for which amorphization of the implanted region does not occur[9]. It is therefore apparent that more detailed characterization of amorphized and regrown GaAs is necessary. In this communication, the high resolution TEM (HRTEM) and conventional TEM results from Se implanted and furnace annealed GaAs are presented along with a phenomonological discussion of GaAs amorphization and recrystallization mechanisms.

EXPERIMENTAL

Selenium ions accelerated to 450 keV were implanted at nominal room temperature into semi-insulating Cr-doped (100) GaAs samples to a dose of 10^{14} cm^{-2}. The implanted samples were capped with Al and then annealed in a furnace in the temperature range 400-800°C. For TEM analysis, specimens of both plan and cross-sectional geometries were used. The former were prepared by chemical (methanol + 3% Cl$_2$) thinning from the unimplanted side. The latter were prepared by Ar^{+2} ion milling at LN$_2$

temperature (4.5kV, 20 µA specimen current, 14° tilt). Room temperature ion milling caused in situ annealing of the amorphous layer. A JEOL 200 CX HRTEM (Cs ≃1.2 mm, top entry stage) was used to obtain HRTEM images in <011> zone axis orientation from cross-section specimens.

RESULTS

Figure 1 is a set of cross-sectional TEM micrographs from the unannealed (Fig. 1a) and 400°C annealed (Fig. 1b) samples. The unannealed sample contained three distinguishable damage regions: (I) an amorphous layer between 250 and 2150A below the surface, (II) Two heavily damaged amorphous/crystalline transition regions in the depth ranges 0-250A and 2150-2500A and, (III) a lightly damaged crystalline region in the depth range 2500-3000A.

The micrograph of the annealed sample shows that region I of the unannealed sample was converted into a dense crystalline band of microtwins and fault bundles (on {111}) which grew from both sides of the amorphous region and met at a depth of 1100A. The inset diffraction pattern (Fig. 1b) also shows spots from the [$\bar{2}$110] zone of wurtzite, suggesting that some of the faults were bundled in such a sequence as to produce thin plates of wurtzite (e.g., an extrinsic fault adjacent to an intrinsic fault is equivalent to 4 layers of wurtzite. The damage in regions II, III and beyond consisted of small dislocation loops which decreased in density with depth.

HRTEM of regions I, II and III (see Figures 2, 3 and Reference 10) showed the following:
i) Region I was found to be completely amorphous (except near the boundary with regions II).
ii) The heavily damaged material in regions II of the unannealed sample was found to consist of an intimate mixture of small (50-150A) in diameter) amorphous and crystalline pockets (Figure 2). Many microtwin nuclei and small (~50A) stacking faults were present in the regions II. These nuclei were found at amorphous-crystalline interfaces which were micro-faceted on {111}.[11]
iii) Rod-shaped defects approximately 30A long and lying along <110> directions were found in the regions II and III in the unannealed sample. Figure 2 shows several of these defects near the boundary between regions II and III. The density of these defects in region III was estimated to be $5 \times 10^{11} cm^{-2}$ (±50%), which is numerically equal to 0.5% of the total Se$^+$ dose. We propose that these defects are damage cascades due to either channeling of Se$^+$ ions at the end of their tracks or channeling of recoiled Ga or As.
iv) The annealed sample showed that region II had recrystallized epitaxially but was not defect-free. Both perfect dislocations and isolated stacking faults bounded by Shockley partials were found in this region.[10]
v) The annealed sample also confirmed that region I had completely recrystallized in the form of fault bundles and microtwins up to 10 atomic layers (≃33A) thick. These faults and twins nucleated at both top (~250A) and bottom (~2150A) interfaces and grew until impingement at a depth of 1100A.[10]
vi) After annealing, no evidence of channeled cascades was found. Instead, small (35A in diameter) extrinsic Frank dislocation loops ($\vec{b}$ <11>) were found in the depth regions 3000 to 3500A (see Figure 3).

313

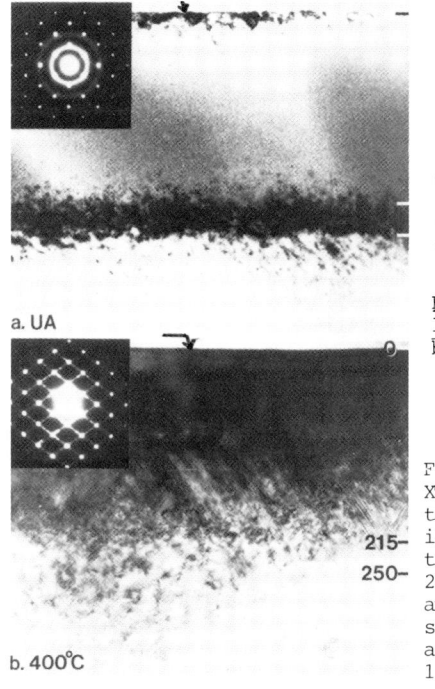

Figure 3: An isolated extrinsic Frank loop at a depth level of ~3000Å with $\vec{b} \parallel \langle 111 \rangle$ (defocus - 1100Å).

Figure 1: (a) Low magnification (1$\bar{1}$0) XTEM of unannealed sample with crystalline pockets at surface and heavily damaged amorphous-crystalline transition region between 2150 and 2500Å (depth indicated in nanometers) and (b) 400°C annealed sample with stacking fault bundles initiating at 250Å and 2150Å and meeting at 1100Å. Arrows point to the surface.

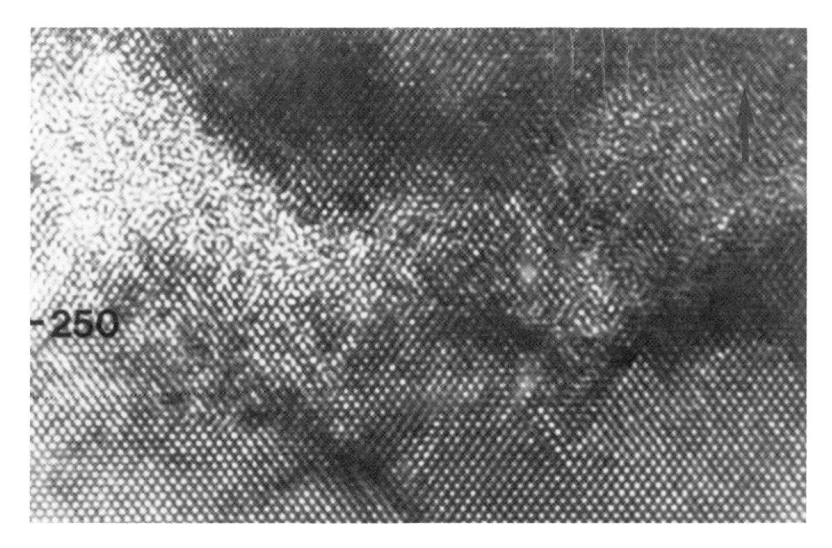

Figure 2: High resolution image of lower edge of transition region (at intersection of regions II and III of text). Note the amorphous pockets (which are partially buried in the TEM specimen). Several "channeled cascades" are visible below the interface. The level at 2500Å below the surface is indicated at left (defocus -900Å). Arrow points toward the surface.

DISCUSSION

Based upon the HRTEM results discussed above, a schematic diagram of damage distributions in the unannealed and annealed samples is shown in Figure 4. We propose the following model for the amorphization and recrystallization of GaAs. During the early stages of implantation, while the GaAs is still crystalline, either a small fraction of Se$^+$ ions (probably 0.5%) or a fraction of the recoiled Ga and As are channeled into <110> directions resulting in the cylindrical damage zones observed in region III (Fig. 2).

The distinct two-phase nature of the transition region II[11] suggests that the amorphous state is formed by relaxation of crystalline material when the defect density reaches some critical value and that there is a competition between formation of amorphous zones and their epitaxial regrowth by dynamic annealing process. A similar model for the crystalline to amorphous transformation has been proposed for silicon[12].

Dynamic annealing during the nominal room temperature implant results in the formation of microfacets on the crystalline pockets in the transition region II. In addition, dynamic regrowth of the crystalline pockets during implantation leads to the formation of stacking fault nuclei[11].

Recrystallization during the 400°C annealing treatment occurs in two stages. The first stage involves the complete epitaxial regrowth of the transition region. The fault nuclei present in the unannealed sample grow until they reach the growth front of a neighboring crystallites. Formation of large area stacking faults or twins is impossible in this region due to the large number of growth fronts in various orientations. The rough nature of the boundary between regions I and II and the fact that <11> is the slowest growing direction eventually results in a microfaceted, and simply connected growth front at a depth of approximately 2150A (a similar process is occuring near the upper surface). Stage two in the regrowth process begins at this microfaceted growth front (facets on {111} have dimensions on the order of 50 to 100A as judged from the size of the stacking fault bundles.[10]

The {111} facets provide favorable sites for accommodation of dopant concentration above the solid solubility limit and non-stoichiometry which was produced by the implantation and accumulated during the last part of stage one. Schockley partial dislocation cores form on the defected facets, thereby forming the nuclei for stacking fault bundles. Once a bundle of stacking faults has been nucleated, the propogation of the bundle is likely to be rapid because excess Ga or As and dopant atoms can be efficiently rejected to the amorphous material. Stage II is complete when stacking fault bundles from both interfaces meet in the middle of region I. Excess point defects and non-stoichiometry in region III and below is accommodated by the formation of the small extrinisic Frank loops (Figure 3).

Microtwin growth due to stacking errors is also encountered in the regrowth of amorphized (111) Si. However, such morphologies are not generally found in recrystallized (100) silicon. It is likely that many observed differences between the behavior of (100) GaAs and (100) Si are a result of the binary nature of GaAs. For example, the different displacement properties of Ga and As result in deviations in local stoichiometry which cannot be fully eliminated by low temperature annealing (400°C). These deviations in stoichiometry, along with impurities present

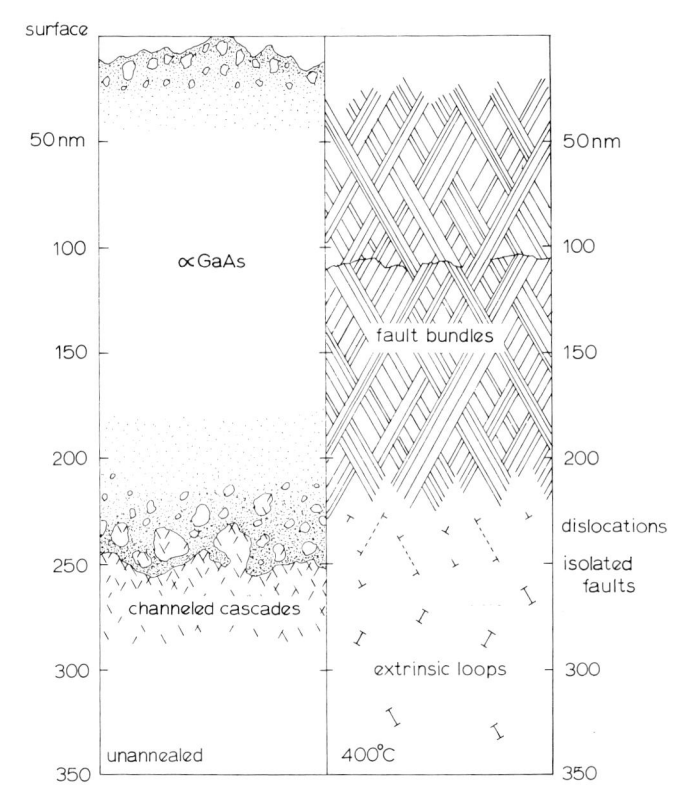

Figure 4: A schematic diagram of damage distributions in the unannealed and 400°C annealed samples based on the HRTEM results.

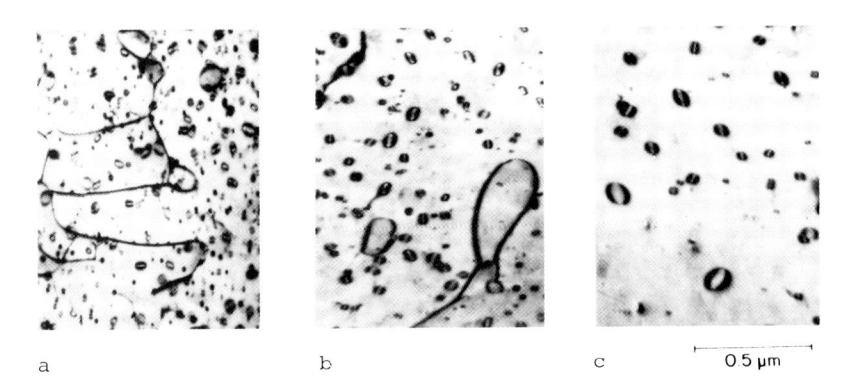

Figure 5: Low magnification view TEM micrographs showing annealing sequence of the damage in Fig. 1a. (a) 600°C, (b) 700°C and (c) 800°C. The microtwins, stacking faults and their bundles annealed out. Only dislocation loops are present.

above the solid solubility, are probably accommodated by extrinsic Frank loops, Shockley partial cores, perfect dislocation cores and defects at regrowth front intersections. At higher temperatures however, due to increased mobility of point defects, constituent atoms and impurities, local non-stoichiometry can be minimized. Under these circumstances, the microtwins and bundles of stacking faults may anneal out. Figure 5 shows an annealing sequence of the damage in the plan view in the temperature range 600-800°C. The microtwins and stacking fault bundles were no longer present; only dislocation loops remained. The effect of solid solubility of impurities on the nucleation and annealing out of microtwins/stacking fault bundles in (100) GaAs will be discussed later.[11]

CONCLUSIONS

In summary, HRTEM examination of medium dose (10^{14} cm^{-2}) Se implanted GaAs revealed the following:

1. A buried amorphous layer that was bounded by two relatively broad (~300Å) transition regions.
2. The amorphous-crystalline interface within the transition regions were sharp within an atomic distance.
3. During low temperature annealing (400°C), recrystallization recurred by the nucleation of bundles of stacking faults and microtwins at both interfaces that met in the middle of the original amorphous layer. The wurtzite phase of GaAs was also present in this region.
4. The lower transition region converted into isolated extrinsic stacking faults and Frank loops.
5. The bundles of stacking faults and microtwins annealed out at higher temperatures (≥600°C).

ACKNOWLEDGMENTS

The authors wish to acknowledge Dr. Ronald Gronsky of the National Center for Electron Microscopy, Lawrence Berkeley Laboratory, for valuable discussions and continued encouragement. We would also like to thank Dr. Brian Sealy of the University of Surrey, England for the Se^+ implantation. This work was supported by the Director, Office of Energy Research, Office of Basic Energy Sciences, Materials Science Division of the U.S. Department of Energy, under Contract No. DE AC03-76SF00098.

REFERENCES

1. G.H. Narayanan and A. Kachare, Phys. Status Solidi 26 657 (1974).
2. K. Gamo, T. Ineda, J.W. Mayer, F.H. Eisen and G.G. Rhodes, Rad. Eff. 33 85 (1977).
3. D.K. Sadana and G.R. Booker, Rad. Eff. 42 35 (1979).
4. S.S. Kular, B.J. Sealy, K.G. Stephens, D. Sadana and G.R. Booker, Solid State Electron 23 831 (1980).
5. J.S. Williams and M.W. Austin, App. Phys. Lett. 36 994 (1980).
6. M.G. Grimaldi, B.M. Paine, M-A. Nicolet and D.K. Sadana, J. App. Phy. 53 1803 (1982).
7. R.S. Bhattacharya, A.K. Rai, P.P. Pronko, J. Narayan, S.C. Ling and S.R. Wilson, J. Phys. Chem. Solids 44 61 (1983).
8. S.S. Kular, B.J. Sealy and K.G. Stephens, Elec. Lett. 14 22 (1977).
9. D.K. Sadana, G.R. Booker, B.J. Sealy, K.C. Stephens and M.M. Badawi, Rad. Eff. 49 183 (1980).
10. D.K. Sadana, T. Sands and J. Washburn (paper submitted to App. Phys. Lett.).
11. T. Sands, D.K. Sadana, J. Washburn and R. Gronsky, (unpublished).
12. J. Washburn, C.S. Murty, D. Sadana, P. Byrne, R. Gronsky, N. Cheung and R. Kilaas, Nucl. Inst. & Meth. 209/210 345 (1983).

ION IMPLANTATION IN GALLIUM INDIUM ARSENIDE

M. ANJUM, M. A. SHAHID, S. S. GILL*, B. J. SEALY AND J. H. MARSH**
Department of Electronic and Electrical Engineering, University of
Surrey, Guildford, Surrey GU2 5XH, U.K.; *Royal Signals and Radar
Establishment, St Andrews Road, Great Malvern, Worcestershire WR14 3PS,
U.K.; **Department of Electronic and Electrical Engineering, University
of Sheffield, Mappin Street, Sheffield, Yorkshire, U.K.

ABSTRACT

We have studied the formation of heavily doped n-type layers in LPE GaInAs using ion implantation. 400 keV selenium ions have been implanted in dose ranges of 5×10^{13} to 1×10^{15} cm^{-2} at room temperature. For the high dose implants we have reproducibly achieved activities of 20-40% and sheet Hall mobilities of 700-1000 cm^{-2} V^{-1} s^{-1} and peak carrier concentrations of about 10^{19} cm^{-3}. TEM and RBS results indicate that for long time anneals residual damage persists in the implanted layers, however, anneals at 800°C for 30 seconds perfectly recrystallize the implanted layers.

INTRODUCTION

Investigation into the use of ion implantation to dope III-V semiconducting materials other than GaAs and InP is becoming increasingly important. There have been, however, only a few papers published which describe preliminary results and most of these are concerned with acceptor ion implants[1]-[4]. Although the formation of p$^+$ layers by ion implantation is very important, we decided to concentrate initially on the implantation of selenium ions, since this has not yet been treated elsewhere. This paper therefore describes our preliminary work using 400 keV selenium and highlights a number of problems which still require adequate explanation.

EXPERIMENTAL METHOD

The $Ga_{0.47}In_{0.53}As$ layers were grown on <100> InP in an automated LPE system. The growth was isothermal at 659°C with ≃ 8°C supersaturation, in a Pd purified H_2 atmosphere. The layers were Mn doped with p ≃ 10^{17} cm^{-3} and 2-4 μm thick. The InP substrates were Zn doped with p ≃ 2×10^{18} cm^{-3}. The melts were made up from weighed quantities of In, InAs and GaAs, with Mn being obtained from a master melt containing 0.01% Mn in In. The substrate was protected with an InP cover plate prior to growth and no etch melt was used.

400 keV selenium ions were implanted into substrates at room temperature to doses of 5×10^{13} to 1×10^{15} cm^{-2}. Following implantation samples were annealed either in a furnace with an overpressure of arsine or with an rf furnace or a graphite strip heater. AlN was used as an encapsulant for the latter two methods and was depositied either by reactive evaporation[5] or by sputtering. Annealing temperatures were 700, 750 and 800°C, the annealing time being reduced to 30 s at the highest temperature.

The Hall coefficient and sheet resistivity were measured on clover-leaf shaped samples using the van der Pauw method. Contacts to the layers were made by alloying tin dots at about 250°C. Carrier concentration and mobility profiles were obtained by differential Hall effect measurements using a chemical etch of composition 1 H_2O_2:1 H_2SO_4:125 H_2O.

Samples for transmission electron microscopy (TEM) were prepared by cleaving into 2 mm x 2 mm squares. The implanted surface was protected with lacomit, and the thinning was done by 10% bromine in methanol jet polishing from the back on a rotating pad. After dissolving lacomit in acetone the samples were examined in bright and dark field modes in a JEOL 200 CX scanning transmission electron microscope operated at 160 or 200 kV. Rutherford backscattering (RBS) was also used for analysis both in the aligned and non-aligned direction using 1.5 MeV He^+ ions.

RESULTS AND DISCUSSION

Our initial results for a dose of 1×10^{15} Se^+ cm^{-2} annealed at 700°C for 10-15 minutes gave activities of about 20% with good mobilities (table I). The electrical profiles showed that peak carrier concentrations exceeding 10^{19} cm^{-3} were achieved (figure 1a). Similar results were obtained also for a dose of 5×10^{14} Se^+ cm^{-2} (table 1,

TABLE I Sheet electrical properties of Se^+ implanted GaInAs for various annealing schedules

Dose	Annealing Conditions Temp(°C)	Time(sec)	ρ_s ($\Omega/\square$)	μ_s ($cm^2 V^{-1} s^{-1}$)	N_s (cm^{-2})	CAP*
1×10^{15}	700	1500	35.1	507	3.5×10^{14}	AlN(e)
1×10^{15}	700	1500	45.6	884	1.5×10^{14}	AlN(e)
1×10^{15}	700	1500	34.2	855	2.1×10^{14}	AlN(e)
1×10^{15}	700	600	21.3	845	3.4×10^{14}	AlN(e)
1×10^{15}	700	600	25.3	790	3.1×10^{14}	AlN(e)
1×10^{15}	700	600	22.6	720	3.8×10^{14}	AlN(e)
1×10^{15}	700	900	32.0	1152	1.6×10^{14}	AsH_3
1×10^{15}	700	900	31.1	1057	1.9×10^{14}	AsH_3
1×10^{15}	700	900	40.1	787	1.9×10^{14}	AsH_3
1×10^{15}	700	900	32.7	1061	1.8×10^{14}	AsH_3
1×10^{15}	700	900	38.6	836	1.9×10^{14}	AsH_3
1×10^{15}	700	600	33.7	1650	1.1×10^{14}	AsH_3
5×10^{14}	700	900	33.5	560	3.3×10^{14}	AsH_3
5×10^{14}	700	900	30.8	976	2.0×10^{14}	AsH_3
5×10^{14}	700	700	31.6	931	2.1×10^{14}	AsH_3
5×10^{14}	700	700	34.3	885	2.0×10^{14}	AsH_3
5×10^{14}	750	600	33.7	1650	1.1×10^{13}	AlN(s)
1×10^{14}	800	30	187.2	1345	2.6×10^{14}	AlN(s)
1×10^{14}	800	30	36.2	1587	1.0×10^{14}	AlN(s)
1×10^{14}	800	30	36.1	1478	1.1×10^{14}	AlN(s)
1×10^{14}	800	30	35.8	1483	1.1×10^{14}	AlN(s)
5×10^{13}	800	30	25.3	2217	1.11×10^{14}	AlN(s)

* e ≡ evaporated AlN, s ≡ sputtered AlN.

figure 1b). However, the sheet carrier concentration for doses of 1 x 10^{14} and 5 x 10^{13} cm^{-2} does not agree with the integrated area of the electrical profiles (figure 1c and 1d). The 5 x 10^{13} cm^{-2} dose apparently has an activity of about 200% from the sheet carrier concentration measurement and yet the area enclosed by the profile (figure 1d) suggests that the net activity is only about 30-40%. This profile indicates also that there is a large and unexpected surface electron concentration. In a similar way the profile for the 1 x 10^{14} cm^{-2} dose suggests that the net activity is about 30% whereas the sheet

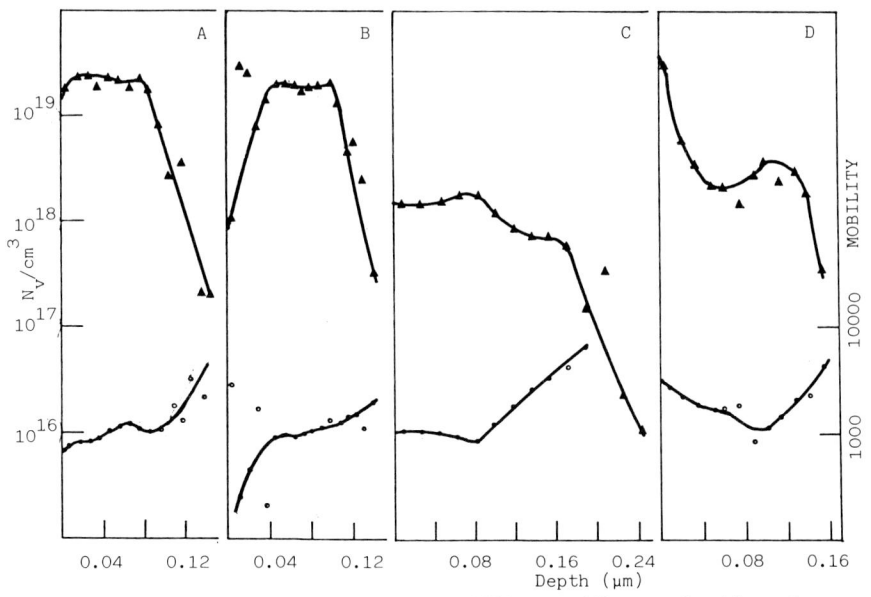

FIG. 1. Electron concentration and mobility profiles as function of selenium dose; (a) 1 x 10^{15} cm^{-2}, 700°C/900s (AsH$_3$); (b) 5 x 10^{14} cm^{-2}, 700°C/900s (AsH$_3$); (c) 1 x 10^{14} cm^{-2}, 800°C/30s (AlN-sputtered); (d) 5 x 10^{13} cm^{-2}, 800°C/30s (AlN-sputtered).

carrier concentration indicates that an activity above 100% was obtained. For one sample, however we did measure a sheet carrier concentration corresponding to an activity of 26% (table 1) which is in agreement with that obtained from the profile of figure 1c. We believe this discrepancy in the electrical properties is associated with a surface conducting layer around the edges of the samples. The effect can be eliminated by careful placement of electrical contacts and by appropriate masking using black wax in our case.

Figure 2a shows a TEM bright field micrograph of untreated GaInAs, the inset shows the diffraction pattern. The salient features are (a) non-uniform contrast consisting of black and white spots and (b) straight fringes of black-white contrast running along the <100> directions. The diffraction pattern, however shows, that the material is single crystal. After implanting with Se$^+$ ions at an energy of 400 keV, amorphous layers have been observed that result in diffraction rings centred at the strongly diffracted spots in the diffraction pattern, when the electron beam is incident along the [100] direction.

The micrograph of figure 2b shows a sample implanted at room temperature with a dose of 1×10^{15} Se$^+$ cm^{-2} and subsequently annealed at a temperature of 700°C for 900 seconds. It shows that the surface layers contain heavily strained regions and very complicated dislocation complexes. The diffraction pattern in the inset confirms this conclusion. Another sample implanted with 5×10^{13} Se$^+$ cm^{-2} and annealed at 800°C for 600 seconds is shown in figure 2c. This micrograph also shows features similar to those described in figure 2b. The bright field micrograph of figure 2d has been produced from a

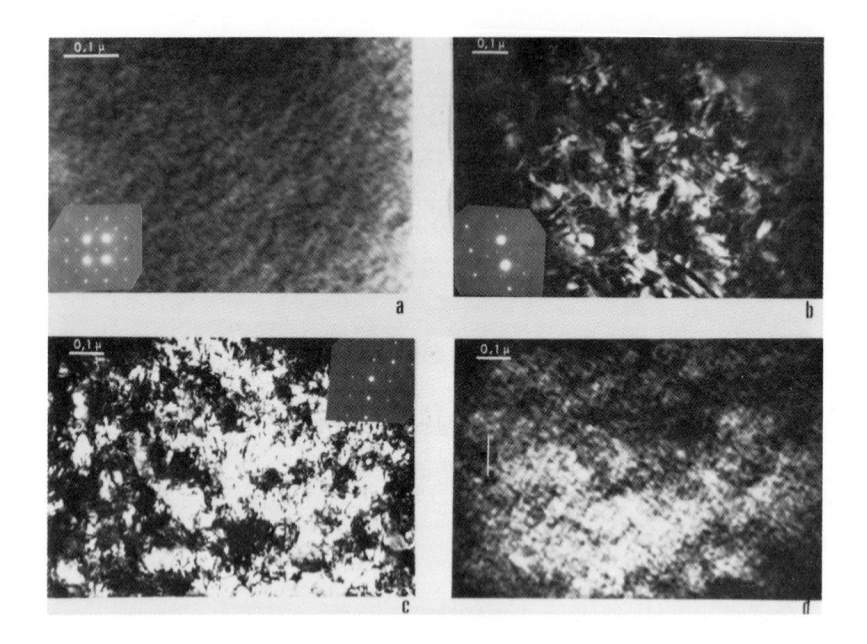

FIG. 2. Bright field TEM micrographs; (a) as grown GaInAs; (b) 1×10^{15} cm^{-2}, 700°C/900s (AsH$_3$); (c) 5×10^{13} cm^{-2} 800°C/600s (AlN-sputtered); (d) 1×10^{14} cm^{-2} 800°C/30s (AlN-sputtered).

sample implanted with 1×10^{14} Se$^+$ cm^{-2} and annealed at 800°C for 30 seconds. This shows perfect annealing, without any gross damage as seen above for the samples annealed for longer times. The only features observed are similar to those seen with the as grown and untreated material of figure 2a. Similar results have been achieved for samples implanted with 1×10^{15} Se$^+$ cm^{-2} and annealed at 800°C for annealing times of 60, 90 and 120 seconds.

RBS studies confirm the TEM results in that samples implanted with doses of 5×10^{14} cm^{-2} and above are amorphous and that residual damage is found following anneals at 700 and 750°C. For example a sample implanted with 1×10^{15} Se$^+$ cm^{-2} and annealed at 750°C for 600 seconds had a X_{min} of 22% compared with a value of about 12% for good crystalline layers. RBS has been used also to check if decomposition occurs during annealing at 700°C for 900 seconds using sputtered AlN as the cap (figure 3). This figure indicates that no significant changes have taken place.

In summary despite the large amount of residual damage, it is possible to obtain carrier concentrations of 2-3 x 10^{19} cm^{-3} with mobilities of 700-1000 cm^2 V^{-1} s^{-1} for high dose selenium implants. However, all the profiles (figure 1) are shallower than expected, the calculated range being 0.14 μm which corresponds approximately to the tail of the electrical profile. We suggest that the radiation damage is in some way affecting the activation of the selenium ions, the measured profiles being assciated with the region where most damage was introduced.

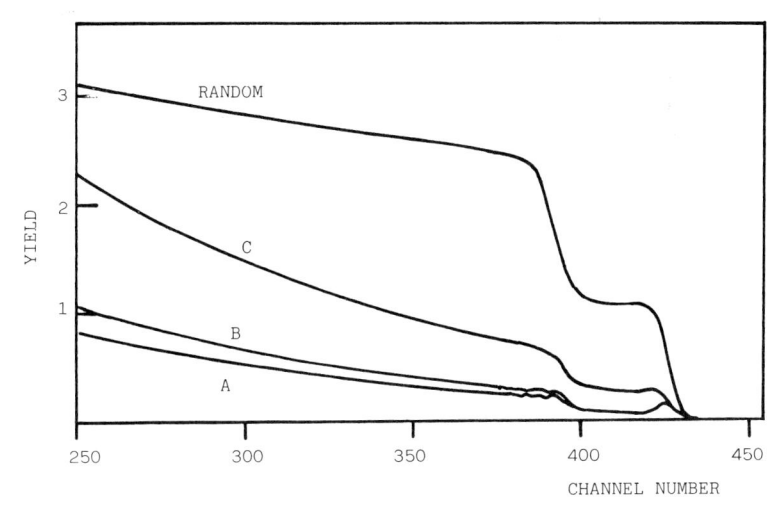

FIG. 3. RBS spectra of: (a) as grown GaInAs; (b) unimplanted and heat treated, 700°C/900s (AlN-sputtered); (c) 1 x 10^{15} Se$^+$ cm^{-2} implanted, annealed at 750°C/600s (AsH$_3$). a, b and c are aligned spectra. A typical non-aligned spectrum is also shown.

One of the aims of this work was to find an annealing schedule to remove completely the radiation damage. It seems that conditions used in the published work[1]-[4], that is 650-700°C for about 15 minutes, does not produce good crystalline material. We suggest that temperatures of the order of 800°C are required with times as short as 30 seconds. We have made a preliminary study of 400 keV magnesium ions implanted to a dose of 1 x 10^{15} cm^{-2} and these also have no residual damage following a pulse anneal of 800°C for 30 seconds. Our results show also that it is possible to use both sputtered and reactively evaporated AlN or an arsine overpressure to anneal selenium implanted GaInAs to temperatures up to 800°C. We conclude that transient annealing is a useful way of removing damage in ion implanted GaInAs layers.

ACKNOWLEDGEMENTS

The authors thank SERC for the financial support and the staff of the accelerator laboratory and microstructural studies unit (University of Surrey) for technical assistance. They also acknowledge Mr P Wells (University of Sheffield) for help with growth of material.

REFERENCES

1. N.J. Slater, A.N.M.M. Choudhury, K. Tabatabaie-Alavi, W. Rowe, C.G. Fonstad, K. Alavi and A. Y. Cho, Inst. Phys. Conf. Ser. No. 65 (1983) 627.
2. K. Tabatabaie-Alavi, A.N.M.M. Choudhury, N.J. Slater and C.G. Fonstad, Appl. Phys. Lett. 40 (1982) 517.
3. K. Tabatabaie-Alavi, A.N.M.M. Choudhury, K. Alavi, J. Vlcek, N.J. Slater, C.G. Fonstad and A.Y. Cho, IEEE Elec. Dev. Lett EDL-3 (1982) 379.
4. L. Vescan, J. Solders, H. Krautle, W. Kutt and H. Beneking, Elect. Lett. 18 (1982) 533.
5. R. Bensalem, N.J. Barrett and B.J. Sealy, Elect. Lett 19 (1983) 112.

© HMSO London 1983

ION IMPLANTATION DAMAGE IN CdS CRYSTALS USING RBS/CHANNELING AND TEM

N.R. PARIKH, D.A. THOMPSON, R.BURKOVA[*] AND V.S. RAGHUNATHAN[**]
Department of Engineering Physics and Institute for Materials Research
McMaster University, Hamilton, Ontario, Canada, L8S 4M1
*Short term visitor from Inst. for Semiconductor Technology, Botevgrad, 2140 Bulgaria; **On leave from the Reactor Research Centre, Kalpakkam, India

ABSTRACT

Implantation damage in single crystal of CdS produced by 60 keV Bi$^+$ and 45 keV Ne$^+$ at 50 K and at 300 K has been studied. Measurements of Cd disorder and dechanneling behaviour have been made by means of RBS/channeling for He$^+$ ions ranging in incident energy from 1.0 to 2.8 MeV either along $\langle 0001 \rangle$ or $\langle 11\bar{2}0 \rangle$ axial channeling directions. The amount of disorder measured were two orders of magnitude lower than the calculated Cd disorder. Damage when analysed along the $\langle 0001 \rangle$ axis is larger than when analysed along the $\langle 11\bar{2}0 \rangle$ axis. χ_{min} values for the implanted crystals decreases as the E_0 increases, when analysed along $\langle 0001 \rangle$ direction. TEM observations of Bi implanted samples show that the dislocation loops of $\bar{b}= 1/3 \langle 11\bar{2}0 \rangle$ are produced. Attempts have been made to correlate the RBS/channeling results with the defect structures observed in microscopy.

INTRODUCTION

Cadmium sulphide, being a wide and direct band gap (2.42 ev) semiconductor, is potentially attractive for use in opto-electronic devices. It is normally either n-type or intrinsic and attempts to produce p-type semiconductor by conventional doping methods have not been successful, mainly due to a self-compensation effect which occurs at high temperatures. Attempts have been made to use ion implantation techniques to produce p-type conversion [1,2]. However, most of the work has concentrated on implanting different ions to improve the opto-electronic characteristics of the junction formed. Some studies of heavy ion implantation damage in CdS crystals using Rutherford Backscattering (RBS)/channeling [3] and TEM [4] have been carried out. However, the results of these two techniques have not been correlated.
In this work, we have studied implantation damage due to 60 keV Bi$^+$ and 45 keV Ne$^+$ at 50 K and at 300 K analysed using the RBS/channeling technique with 1.0 to 2.8 MeV He$^+$ ions, E_0, incident either along the $\langle 0001 \rangle$ or the $\langle 11\bar{2}0 \rangle$. TEM observations have been made of unimplanted CdS and CdS crystals implanted with 60 keV Bi$^+$. The results of these two techniques have been correlated to identify defects produced by ion implantation.

EXPERIMENTAL

CdS crystal (wurtzite structure - see Fig. 1) of high resistivity were obtained commercially, with surfaces parallel to either {0001} or {11$\bar{2}$0}. The crystals were polished chemically with dilute HCl (1HCl:1H$_2$O) on acid resistant cloth for ~5 min. to remove damaged layers caused either by mechanical polishing or prior implantation. The value of χmin (the ratio of the aligned to the nonaligned yield of RBS Spectra) was ~5% at 300 K. A detailed description of the experimental set-up is described in ref. [5]. The base pressure in the target chamber was $< 10^{-6}$ Torr. The sample was mounted on a two axis goniometer which has an angular resolution of $0.05°$ in the tilt axis.

Crystals were implanted at $\sim 20°$ off the axial channeling direction to minimize any channeling of the implanted ions. The crystals were either bombarded at 300 K or at 50 K. In all implants, a shield surrounding the target was mounted at 30 K so that the effective pressure near the target was reduced to $< 10^{-8}$ Torr which minimized surface contamination of the target.

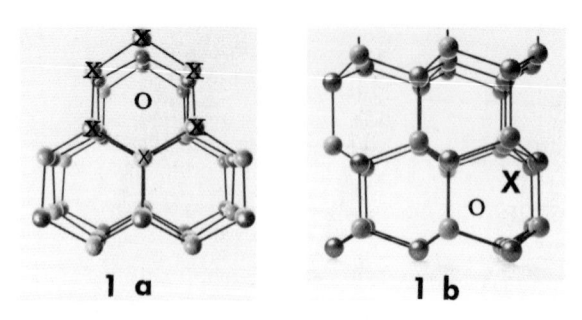

Fig.1. Model of Wurtzite Structure, a) along c-axis ($<0001>$) b) along a-axis ($<11\bar{2}0>$).

In-situ analysis was performed on the damaged sample by means of RBS/channeling with MeV He^+ ions. The ion fluence was determined to an accuracy of $\pm 5\%$ from the integrated electric charge received by the target. Lattice disorder was calculated only for Cd atoms, as analyzing RBS damage peaks of S involved large inaccuracies due to high background counts. The Cd disorder i.e. the number of displaced atoms, N_D, was calculated using the linear dechanneling method. The experimental stopping powers were used to establish depth scale on the backscattering spectra [6]. This enabled values of $\Delta \chi_{min}$ (which is the change in the normalized minimum yield) to be determined at the same depth for different E_o. TEM observation of implanted and unimplanted crystals were carried out in a Phillips EM 300 transmission electron microscope.

RESULTS AND DISCUSSIONS

(1) RBS/Channeling

Figs. 2 and 3 show the variation in N_D as a function of ion dose, Φ, for 60 keV Bi^+ and 45 keV Ne^+ implants at 300 K and at 50 K, analyzed along $<0001>$, c-axis, and $<11\bar{2}0>$, a-axis. Fig. 4 shows a plot of $\Delta \chi_{min}$ vs. E_o for 60 keV Bi^+ implants of various doses at 50 K analyzed along the c and a axes. The following observations can be made from the results:
(a) For both ion species; the observed Cd disorder initially increases with increasing ion dose. In the case of 60 keV Bi^+, the observed disorder tends to saturate at dose of $\sim 1-2 \times 10^{15}$ $Bi^+.cm^{-2}$. At a dose of 6×10^{15} $Bi^+.cm^{-2}$, the damage peak only reaches to $\sim 40\%$ of the nonaligned backscattering level which indicates that the damaged region is not amorphized. No such saturation of disorder is observed for 45 keV Ne^+ implants up to the maximum dose investigated of 1×10^{16} $ions.cm^{-2}$.
(b) For both ion species studied, the damage observed at 50 K along the c-axis was greater than the damage observed along the a-axis.

TABLE I: Comparison of Calculated and Measured Damage Rates

Ion	E_i/keV	Temp/K	$dN_D/d\phi$ experimental c-axis	$dN_D/d\phi$ experimental a-axis	(calculated)*
Bi$^+$	60	50	19	7	2400
		300	50		
Ne$^+$	45	50	5.5	0.7	1300
		300	9		

* Calculated using Ed for cadmium as 7.3 eV [8]

TABLE II: Effect of Dislocation loops on RBS/Channeling

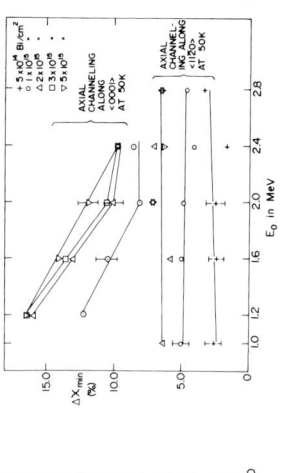

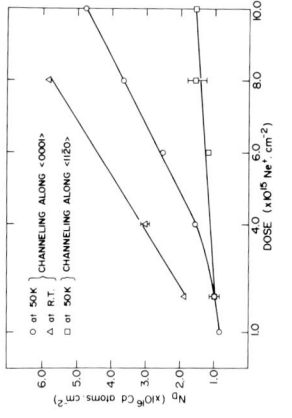

Plane of Dislocation Loop	$\bar{b}$	RBS/Channeling effects c-axis N_D	Dech	a-axis N_D	Dech
Basal	1/2[0001]	S	W	W	S
	1/3[11$\bar{2}$0]	S	S	W	M
Prism	1/2[0001]	W	W	M	M
	1/3[11$\bar{2}$0]	S	S	W	M

S-Strong, W-Weak, M-Moderate

Fig. 2. N_d vs Dose for 60 keV Bi$^+$

Fig. 3. N_d vs Dose for 45 keV Ne$^+$

Fig. 4. $\Delta\chi_{min}$(%) vs E_o for various Doses 60 keV Bi$^+$ at 50 K

(c) For both the ion species studied, the damage observed at 300 K along the c-axis appears to be greater than that observed at 50 K, which is contrary to the normally expected behaviour due to the thermal annealing of defects.
(d) Table I gives the measured and calculated [7] values of $dN_D/d\phi$ extrapolated to low dose. It is seen that the calculated values are approximately two orders of magnitude larger.
(e) There is no observable systematic dependence of N_D on E_o over the range of 1.0 to 2.80 MeV.
(f) Fig. 4 shows that $\Delta\chi_{min}$ at 50 K is independent of E_o within the experimental uncertainty when the analyzing beam is parallel to the a-axis, but when the beam is parallel to the c-axis, the $\Delta\chi_{min}$ decreases with increasing E_o.

(2) Transmission Electron Microscopy

Fig. 5 shows the electron-micrograph (EM) from a CdS crystal after an implantation dose of 1×10^{14} ions.cm^{-2} of 60 keV Bi$^+$ at 300 K. At this ion dose, the predominant defects observed have a characteristic black-white dot contrast (average size ~5 nm). This type of contrast is seen from either very fine precipitates or voids or possibly from dislocation loops with very small diameter.

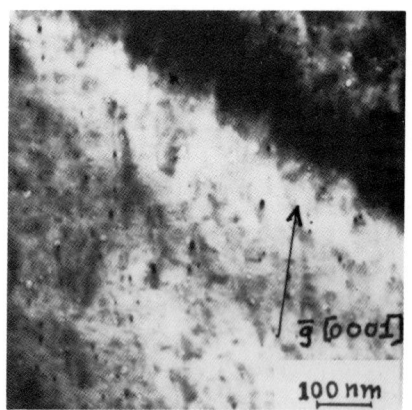

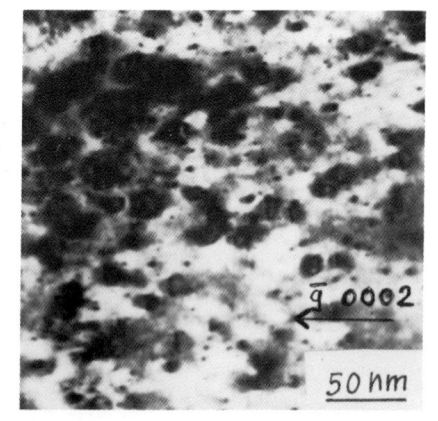

Fig. 5. Black-white dot contrast

Fig. 6. Dislocation Loop Constant

The same sample was further implanted to a total dose of 1×10^{15} Bi+.cm^{-2}; the EM is shown in Fig. 6. TEM contrast experiments confirmed the existence of dislocation loops whose average diameter is ~20 nm. It was noticed that when $\bar{g} = (11\bar{2}0)$, a large number of loops went out of contrast. Figs. 7 (a and b) show the EM from the same region with $\bar{g}= (0002)$ and $\bar{g}= (11\bar{2}0)$ after a dose of 2×10^{15} Bi$^+$.cm^{-2}. Diffraction patterns from the region in both the contrast conditions are inset. It was concluded that the Burgers vector of the loops was predominantly of the $1/3\langle11\bar{2}0\rangle$.

Two types of self interstitial sites can be considered for the wurtzite structure, namely octahedral sites (O) and tetrahedral sites (X). There are six sites of each type in each unit cell, the projections of these sites are shown in Fig. 1. If the X sites are preferentially occupied, then a basal plane projection indicates that the interstitial site is shadowed by rows of atoms when the analyzing beam is along c-axis. Hence there should not be any increase in backscattering yield due to occupancy of such sites. On the other hand, if O sites are preferentially occupied, they are in the center of the channel when the analyzing beam is along c-axis and would give rise to strong backscattering. For the case of beam parallel to a-axis, a similar argument leads to the fact that both O and X sites will give rise to moderate backscattering probability.

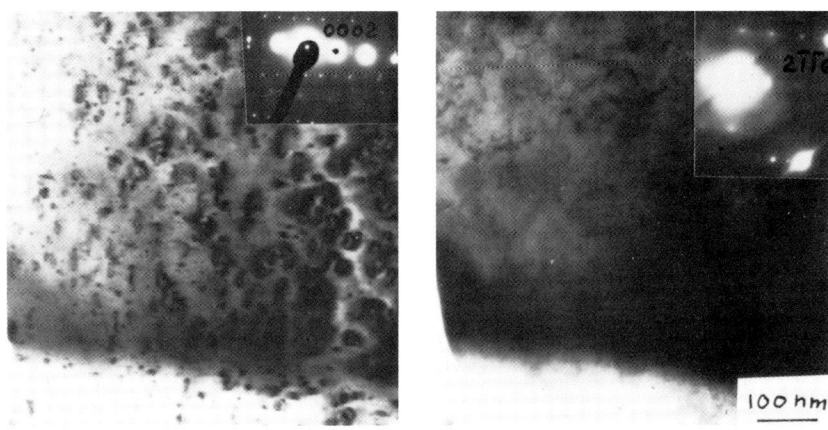

Fig. 7. EM of CdS implanted to 2×10^{15} $Bi^+.cm^{-2}$ under two different contrast conditions. a) $\bar{g} = [0002]$ and b) $\bar{g} = [11\bar{2}0]$

From the results shown in Figs. 2 and 3, it is reasonable to conclude that if the defects giving rise to the surface peak (and hence N_D) are interstitials, than the O site occupancy probability would be high in the CdS structure.

When the interstitial concentration increases with the dose, the dislocation loops grow in size. Table II lists the possible dislocations in the CdS structure which have been observed [9,10] with their appropriate Burgers vectors. When a dislocation loop of interstitial type lies either in the basal or in the prism plane with $\bar{b} = 1/3 \langle 11\bar{2}0 \rangle$ (which can dissociate in two partials), atom in the loop will change the sequence of stacking from ABABA to ABCBAB. The atoms of position C are in the center of the channel for c-axis channeling but only slightly extended into the channel for a-axis channeling. Therefore, there should be a high backscattering yield when analyzing along c-axis but a weak backscattering will result when analyzing along a-axis.

In general, the atomic displacement around the dislocation is parallel to the Burgers vector. Accordingly, the displacement vectors would be perpendicular to c-axis, it is either $\langle 10\bar{1}0 \rangle$ or $\langle 11\bar{2}0 \rangle$ type. These displacements are small and decrease with distance from the dislocation line. However, considering that the displacement field would extend to significant distances away from the dislocation line and also along the dislocation line, there should be high dechanneling along c-axis for the two Burgers vectors considered. There is a small but finite probability of backscattering due to this displacement. Similar arguments are used to evaluate the effects of dislocation loops with Burgers vectors on the backscattering and dechanneling behaviour along the two channeling directions. The results are shown in Table II.

RBS/channeling results considered along with Table II suggest that the dislocation loops are of the type with $\bar{b} = 1/3 \langle 11\bar{2}0 \rangle$ and lying either in the basal plane or in the prism plane. The consequence of the defects produced and dechanneling behaviour is illustrated as an energy dependence of $\Delta \chi_{min}$. The results are similar to the dependence of the point defect scattering centers when the damage is analyzed along c-axis.

The larger damage observed at 300 K compared to that observed at 50 K by RBS/channeling along c-axis is an unexpected effect. Further experiments and analysis are necessary to explain this result.

CONCLUSION

i) Dislocation loops with Burgers vector $\bar{b} = 1/3 \langle 11\bar{2}0 \rangle$ are predominantly produced by heavy ion implantation. These grew in size with implantation dose suggesting interstitial migration at 300 K.

ii) In the heavy ion implanted CdS crystals, the Cd disorder, N_D, when analyzed along c-axis ($\langle 0001 \rangle$) is higher than when analyzed along a-axis ($\langle 11\bar{2}0 \rangle$). This could result from the preferential occupation of octahedral interstitial sites and/or from the interstitial dislocation loops.

iii) Higher dechanneling levels observed along c-axis than along a-axis is explained on the basis of projections into the c and a-axes of the displacement field associated in the dislocation loops.

iv) A qualitative correlation between the nature of defects indicated by RBS/channeling data and the structural defects observed by TEM has been realized.

ACKNOWLEDGEMENTS

The authors would like to thank Dr. J.A. Davies for stimulating discussions and encouragement during the course of this work. One of the authors (N.R.P.) would like to thank the Canadian International Development Agency and Secretaria de Tecnologia Industrial, Ministério da Indústria e Comercio, Brazil for their financial support.

REFERENCES

1. F. Chernow, G. Eldridge, G. Ruse and L. Wahlin, Appl. Phys. Lett. 12, 10, 339 (1968).
2. Y. Shiraki, T. Shimada, and K.F. Komatsubara, J. Appl. Phys. 43, 2, 710 (1972).
3. D. Baxter, Ph.D. Thesis, Dept. of Electrical Enginnering, Univ. of Salford, England (1977).
4. P.K. Govind and F.J. Fraikor, J. Appl. Phys. 42, 6, 2476 (1970).
5. R.S. Walker and D.A. Thompson, Nucl. Instr. and Meth. 135, 489 (1976).
6. W.E. Miller, J.A. Hutchby, J. Appl. Phys. 46, 10, 4479 (1979).
7. P.Sigmund, Rad. Effect. 1, 15 (1969).
8. B.A. Kulp, Phys. Rev. 125, 6, 1965 (1962).
9. T. Yishiie, H. Iwanga, N. Shibata, K. Suzuki, M. Ichihara, and S. Takeuchi, Phil. Mag. A, 47, 3, 315 (1983).
10. D.J.H. Cockayne, A. Hons, and J.C.H. Spence, Phil. Mag. A, 42, 6, 773 (1980).

COMPARISON OF HEAT-PULSE AND FURNACE ISOTHERMAL ANNEALS OF Be IMPLANTED InP

B. MOLNAR,* G. KELNER,* G.L. RAMSEYER,* G.H. MORRISON,** S. C. SHATAS,***
*Naval Research Laboratory, Washington, D.C. 20375; **Cornell University, Ithaca, N.Y. 14853; and ***AG Associates, Palo Alto, CA 94303

ABSTRACT

Annealing in the 600-900°C temperature range, using either a halogen lamp for periods of seconds or a furnace for periods of minutes, has been applied to activate Be implanted InP samples. The Be was ion-implanted at room temperature into InP substrates. The substrates were uncapped and in close contact with another smooth surface during annealing.

It was found that in the 10^{17}-10^{19}/cm^3 range, the acceptor concentration increased with the temperature of anneal. It was also found that redistribution effects decreased with decreasing anneal time. For short anneals, the optimum condition for an 80-90% activation in the 10^{17}-10^{18}/cm^3 range was estimated to be 10 seconds at 950°C. It was also found that after rapid anneal, the carrier concentration profile closely approximated the as-implanted Be profile. In the case of 10^{18}-10^{19}/cm^3 implants, which were rapidly annealed, there was a low concentration component (10^{16}-10^{17}/cm^3) to the redistribution; this ancillary component to the main active distribution was detected by SIMS and was electrically inactive. For long term annealing this electrically inactive Be component was partly converted to substitutional Be and became electrically active. Rapid thermal annealing eliminated this conversion.

INTRODUCTION

The development of controllable acceptor doping of InP by ion implantation is important for several electronic device applications. Because it creates less damage than other heavier p-type ions, low mass Be is particularly attractive for p-layer formation. Be has already been used for p-n junction formation in InP diodes [1,2,3]. Such devices require precise control of the impurity concentration. For furnace annealed cases, appreciable spreading of Be has been observed [4,5]. During the past few years several short time anneal techniques [6] have been explored for annealing ion implantation damage while producing minimal dopant diffusion. Such techniques use a variety of energy sources and can thermally process materials in the 10^{-8}-10^2 s range. This is short compared to a standard 15-20 min thermal anneal. Annealing of ion implanted InP with pulsed laser beams [7], pulsed electron beams [8], and different lamps [9,10,11] has been successfully applied over a broad range of time. It has been recently been demonstrated [11] that a few second lamp anneal of 10^{18}/cm^3 Be, results in dopant activation and mobilities comparable with the best furnace anneal. The lamp anneal technique is simple, relatively inexpensive, and minimizes the effect of thermal gradients.

We have established [5] that for Be implantation at a level of 10^{18}/cm^3, or higher, a significant tail develops during the thermal anneal. The comparison of atomic (SIMS) and electronic (C-V) carrier distribution profiles revealed that most of the Be in the tail is in electrically inactive sites. The tail is developed by a fast moving Be component. In order to understand the p-layer formation by Be implantation, we have studied the kinetics of Be redistribution in the range from a few seconds to 20 min by using two isothermal anneal procedures. In order to avoid doping anomalies, often found with the use of dielectric encapsulation, the samples were annealed without an encapsulant by the close contact method [12].

EXPERIMENTAL

The InP substrates were cut with a (100) orientation from several Fe doped and undoped LEC grown crystals. The polished surfaces were etched before implantation. All the implants were done at room temperature and the wafers were positioned off axis to reduce channeling. Shallow (4-5x10^{3}Å) or extended (1x10^{4}Å) Be layers in the range of 10^{17}-10^{19}/cm^3 were examined.

Thermal activation was performed by a capless annealing method between 600-900°C in a conventional furnace system [12] for 2-20 min range or in a HEATPULSE processing system [13] for the 1-100 s range. A detailed description of the tungsten halogen system was presented elsewhere [14]. The 6x6 mm^2 samples flat implanted surface faced a flat Si surface in close proximity during annealing. The thin enclosed space, on the order of a micron, limits the InP dissociation. The Si provided a mechanical support, and received equal energy on its two sides during the heat pulse. The temperature of the Si was measured by a thermocouple located adjacent to the sample. A typical anneal cycle of 10 s duration at peak temperature (850°C) is presented in Fig. 1.

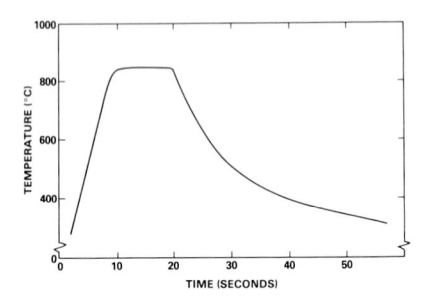

Fig. 1. The time-temperature cycle for an InP sample annealed for 10 s at 850°C.

Because of the presence of the Si support, the sample temperature was uniform throughout its bulk during the heat cycle. The rise and fall times of the temperature transient were, however, influenced by the supporting Si thickness and its thermal properties.

The majority carrier profiles corresponding to the substitutional Be were determined from C-V measurements using Al Schottky diodes. In order to profile the implant, the Al was removed, the InP etched, the Al pattern was reapplied and the C-V measurement was repeated. The carrier concentration and Hall mobility were determined from van der Pauw type Hall measurements.

The atomic profiles of Be were determined by using a CAMECA IMS-3f ion microanalyzer (SIMS). Depth measurements, made with a TALYSTEP stylus device, were used to convert the sputtering time to depth. The Be secondary ion count data were calibrated by integrating the area under the profile and setting it equal to the implanted dose.

RESULTS

Representative low level 6.6x10^{12}/cm^2, 140 keV; 3.9x10^{12}/cm^2, 55 keV; and 7.1x10^{11}/cm^2, 20 keV Be implantation results for a conventional furnace anneal at 750°C for 22 min are presented in Figure 2.

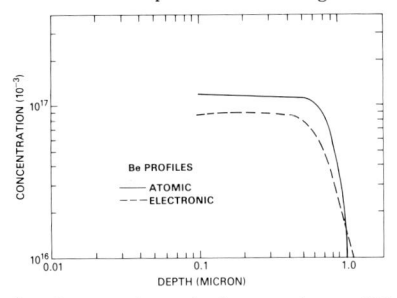

Fig. 2. Be atomic and electronic profiles after a 22 min furnace anneal at 750°C.

The Be atomic (SIMS) (solid line) and the Be electronic (dashed line) profiles are compared in Figure 2. The annealed profile measured by SIMS is a slightly diffused LSS profile. The carrier profile agrees reasonably well with the Be atomic profile. The activation is about 80% and the effective Hall mobility is approximately 120 cm^2/Vs. The activation of the Be is somewhat dependent upon the quality of the semi-insulating InP ingot.

The temperature dependence of the activation of the same Be implants is demonstrated in Figure 3 for short time, 10 s lamp anneals.

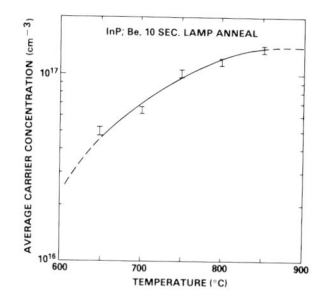

Fig. 3. Carrier concentration as a function of temperature for 10 s lamp annealing.

Fig. 4. Carrier concentrations after 650°C and 800°C, 10 s lamp annealing.

The average hole concentration as shown in Fig. 3, represents the substitutional Be concentration, increased with annealing temperature, and should reach its maximum at approximately 900-950°C.

The carrier concentration as a function of depth was derived from repeated C-V measurements and is represented in Fig. 4. The as-implanted Be LSS profile is compared with the carrier profiles at two temperatures; 650°C and 800°C. The Be activation increased with the annealing temperature, and with higher activation the carrier profile was in close agreement with the implanted profile. The experiments with 10^{17}/cm^3 Be implants have shown that rapid isothermal annealing can produce an activation comparable or better than the furnace anneal and the profiles indicated no diffusion.

These results are in contrast to the results found for Be concentrations higher than $3-5 \times 10^{17}/cm^3$. In an earlier study [5] of the Be redistribution using conventional furnace anneals, we have established that at these higher concentrations the Be atomic (SIMS) distribution is different from the acceptor profiles. These results are recapitulated in Figures 5 and 6. Shown in Figure 5 are the Be atomic profiles of an $8.8 \times 10^{13}/cm^2$, 50 keV Be implant (dashed line), and the redistribution profile after a 10 min furnace anneal (solid line) at 700°C.

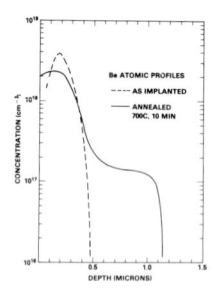

Fig. 5. The as implanted and 700°C 10 min furnace annealed atomic profiles of Be.

Fig. 6. Atomic and carrier profiles after 2 and 15 min furnace anneals at 750°C.

The as-implanted profile has been found to agree with the Gaussian approximation of an LSS [15] calculation to within 10%. It is important to notice that there is no tailing in this profile. Therefore, any channeled Be is below the SIMS sensitivity. The annealed profile is characterized by the slightly diffused version of the implanted Gaussian profile followed, after a short concave region, by a rapidly diffusing low concentration tail section which has a steep leading edge. The leading edge of the plateau was at around 1.15 micron. The leading edge of the profile would be at approximately 0.65 micron if only the slight diffusion were present. The development of the tail section indicated an enhanced diffusion for a few percent of the implanted Be concentration.

The kinetics of the carrier development after furnace anneal are presented in Figure 6. In Figure 6 the majority carrier profiles of a $2.2 \times 10^{14}/cm^2$, 50 keV Be implant are compared with the Be atomic (SIMS) profile of the 750°C anneal. The Be SIMS profile (solid line) represent the atomic Be distribution after a 2 min anneal. It has a fully developed tail section, and its leading edge is at about 1.57 micron. The corresponding carrier profile (dashed line) is a slightly diffused version of the unannealed profile with decreased activation. After a 15 min anneal the Be activation clearly increases as can be seen from the carrier profile. However, in this case the carrier profile also has a tail which is less developed than the tail in annealed atomic profile associated with the 2 min anneal. The differences between the Be atomic and carrier profiles are significantly larger than the errors associated with the depth determination. The tail section of the Be atomic profile is substantially deeper than the majority carrier profiles; this indicates the presence of neutral Be. The tail formation occurs by the rapid diffusion of electrically inactive Be during the initial stage of annealing. As the annealing progresses the inactive Be relaxes and becomes electrically active, substitutional Be. In Figure 6 it can be seen that after a 2 min of annealing there already exists some conversion of the neutral Be tail. This slight conversion can be eliminated by using short, 10 s, anneals.

In Figure 7, the results of 6.6×10^{13}/cm^2, 140 keV; 7.9×10^{13}/cm^2, 55 keV, and 7.1×10^{12}/cm^2, 20 keV Be implants into an undoped InP substrate are presented. These data were obtained for 600°C and 800°C 10 s anneals. The Be atomic (SIMS) profile (solid line) after a 10 s anneal at 800°C, clearly demonstrats the presence of the fast diffused tail. The Be atomic profile is deeper than the carrier profiles. Some electrically active Be is already present after the 600°C anneal and the activation is almost complete after the 800°C anneal. The electrical carrier profile after the 800°C anneal is a sharp profile profile coinciding with the as-implanted Be profile and shows no diffusion tail. This is to be contrasted with the Be atomic profile, for the same anneal conditions which exhibits the low concentration, rapidly diffusing tail.

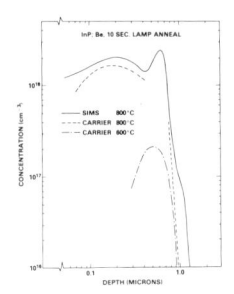

Fig. 7. Atomic and electronic profiles after 10 s lamp anneals at 600°C and 800°C.

SUMMARY

We have examined Be implants below 10^{19}/cm^3 in order to avoid the complications arising from either the formation of an amorphous layer or from exceeding the Be solid solubility. Conventional furnace (2-30 min) and halogen lamp (1-100 s) anneals have been used for isothermal annealing. For both types of anneal we have avoided the use of dielectric encapsulation. While capless annealing [12] eliminated possible adverse effects such as thermal stress and impurity incorporation from the capping film, it also set a limit for the maximum anneal temperature. Consequently, we had to limit furnace annealing to below 780°C and also to restrict the anneal time to less than 1 s for lamp anneals above 850°C. Within these limitations we were able to maintain a surface quality sufficient for device processing.

From an examination of the Be atomic and electronic profiles we have determined that, if the implanted Be concentration is below $\approx 3-5 \times 10^{17}$/cm^3, the Be outdiffusion can be described with a single diffusion component as discussed by Seidel and MacRae [16]. Applying their method of approximation, we estimated the Be diffusion coefficient to be D $\approx 10^{-13}$ cm^2/s at 700°C. During conventional furnace anneal times (10-30 min) the implanted Be redistributed slightly, but redistribution was not detected after 10 s anneals. With short time anneals, as the annealing temperature approached 900-950°C, Be on the substitutional In lattice sites (acceptors) approaches the total Be concentration present in the InP. Finally with Be implantation levels below $\approx 3-5 \times 10^{17}$/cm^3, no low concentration, fast diffusing Be component was detected.

From the investigation of the 10^{18}-10^{19}/cm^3 implanted Be redistribution during annealing we have observed a second, rapidly redistributing component first reported by Oberstar et al. [4]. This second component leads to a low concentration (10^{16}-10^{17}/cm^3) tail which has a steep leading edge. We have established that this rapidly diffusing second component consists of neutral electrically inactive Be. In addition, we have shown that this tail formation occurs within a 10 s anneal period. From these results we estimated that the second component redistribution is at least 100 times faster than the normal Be component. The comparison of the Be atomic and the associated carrier profiles after short time (seconds) anneals revealed that only the normal component Be is activated. As the anneal time is increased, as with conventional furnace annealing, the neutral Be is partly converted to substitutional Be. Consequently, after a 10-30 min of furnace anneal the carrier profile also has a tail. However, this latter tail is less extensive than the tail in the Be atomic profiles.

In conclusion, we have show that the complete description of the redistribution of implanted Be requires characterization of both the atomic and electronic profiles. Using a high power lamp for post implant short time anneals, the carrier profile showed no spreading and showed a maximum activation of 80-90% at about 900°C. For high dose implants a rapidly developing tail exists. This tail is primarily composed of electrically neutral Be. As the anneal time is increased the neutral Be is partly converted to substitutional Be. Rapid thermal annealing eliminates this conversion.

This work was supported by the Office of Naval Research (GOR and GHM).

REFERENCES:

1. W.J. Devlin, K.T. Ip, D.L. Leta, L.F. Eastman, G.H. Morrison, and J. Comas, Inst. Phys. Conf. Ser. 45, 510 (1979).
2. J.P. Donnelly and C.A. Armiento, Appl. Phys. Lett. 34, 96 (1979).
3. J.B. Boos, H.B. Dietrich, T.H. Weng, K.J. Sleger, S.C. Binari, and R.L. Henry, IEEE Electron Device Lett. EDL-3, 256 (1982).
4. J.D. Oberstar, B.G. Streetman, J.E. Baker, and P. Williams, J. Electrochem. Soc. 129, 1312 (1982).
5. B. Molnar, G. Kelner, G.O. Ramseyer, and G.H. Morrison, Electrochem. Soc. Meeting, Extended Abstrat #361, May 1983, to be published.
6. T.O. Sedgewich, J. Electrochem. Soc. 130, 484 (1983).
7. B. Tell, J.E. Bjorkholm, and E.D. Bebe, Appl. Phys. Lett. 43, 655 (1983).
8. D.E. Davis, E.F. Kennedy, J.J. Comer, and J.P. Lorenzo, Appl. Phys. Lett. 36, 922 (1980).
9. S.S. Gill, B.J. Sealy, P.J. Topham, N.J. Barrett, and K.G. Stephens, Electron. Lett. 17, 623 (1981).
10. D.L. Lile, D.A. Collins, and C.R. Zeisse, IEEE Electron Device Lett. EDL-4, 231 (1983).
11. A.N.M. Masum Choudbury, K. Tabatabaie-Alivi, and C.G. Fonstad, Appl. Phys. Lett. 43, 381 (1983).
12. B. Molnar, Appl. Phys. Lett. 36, 927 (1980).
13. HEATPULSE system of AG Associates.
14. A. Gat, IEEE Electron Device Lett. EDL-2, 15 (1981).
15. I. Manning and G.P. Mueller, Computer Phys. Commun. 7, 85 (1974).
16. T.E. Seidel and A.U. MacRae, Trans. Met. Soc. A.I.M.E. 245, 491 (1969).

A CHANNELING STUDY ON Mg IMPLANTED InSb SINGLE CRYSTALS

H.W. ALBERTS
Department of Physics, University of Pretoria, Pretoria, South Africa

ABSTRACT

Proton and α-particle channeling were used to study the radiation damage caused by the implantation of 160 keV Mg ions in InSb. The implantations took place at various substrate temperatures ranging from room temperature to temperatures just below the melting point and doses ranging from 5.10^{13} to 1.10^{16} Mg$^+$ cm^{-2}. The isochronal annealing of the room temperature implanted crystals started at 200°C and damage could not be completely removed even at temperatures just below the melting point. For crystals implanted at elevated substrate temperatures no annealing effects during implantation occured up to 400°C. Above 400°C a sharp reduction of damage indicates that the rate of formation of more complex defect configurations during the implantation process becomes smaller than the annihilation rate of the vacancy-interstitial pairs. A non-linear dependence exists between the degree of radiation damage in the InSb lattice and the implanted dose.

INTRODUCTION

Laser annealing techniques can be applied to obtain improved electrical properties of InSb devices [1-3]. However, information about radiation defects in ion-irradiated InSb is important because radiation defects exert a strong influence on the properties of the implanted layers. Therefore, the radiation damage produced in InSb single crystals has been studied for a wide variety of bombarding particles at different energies. The radiation damage due to electron beam irradiation was the most extensively studied [4, 5 and other references therein] and seems to be well understood.
Studies were also performed with light particles where the radiation damage is stable at room temperature. It has been reported that p-type InSb can be converted by irradiation with protons [6, 7] and neutrons [8] to n-type. The radiation damage produced in InSb at liquid nitrogen temperature with 9 MeV [9] and 12 MeV [10] deuterons seems to anneal below 150°C.
Since He$^+$-particles are normally used as an analyzing beam in RBS and channeling techniques [11-15] it is of utmost importance to know what structural changes may be introduced by these particles in order to avoid misinterpretation of the results. Bogatyrev and Kachurin [7] implanted 300 keV He$^+$ ions at room temperature and found that n-type InSb was formed after annealing for doses $\geq 10^{14}$ cm^{-2}. Bogatyrev et. al [16] implanted 300 keV He$^+$ ions at elevated substrate temperatures and found that when the irradiation temperature did not exceed 200°C, subsequent isochronous annealing (up to 400°C) produced n-type layers and no evidence of amorphization was found. However, when the substrate temperature was higher than 250°C no n-type layers were formed. Also Chernysheva et. al [21] found no evidence of amorphization after 300 keV implants at room temperature. Friedland et. al [15] implanted 50 keV α-particles at 20°, 200° and 250°C as well as 2 MeV particles at room temperature. For both energies appreciable damage is produced in the room temperature implanted crystals with fluences of about 10^{17} cm^{-2}. However, no damage was observable in the crystals implanted at 250°C.
With the implantation of a wide variety of heavy ions (ranging from Be to Xe) [3, 7, 11-14, 16-23] severe damage is produced in InSb. For many types

of implanted ion species subsequent annealing at temperatures just below the melting point does not remove the radiation damage or move an appreciable number of implanted ions onto electrically active lattice sites. Most studies have been devoted to the measurement of the Hall effect. The RBS and channeling techniques were used in only a few cases [11-15] to study the defects induced by the implantation process.

Amorphization of the surface layer is observed after implantations of Ar [21] and S [17, 20] ions at room temperature. However, if the implantations were done at higher substrate temperatures (> 200°C), the amorphous layers did not form. Pronounced damage peaks are observed in the aligned backscattering spectra after the implantation of 300 keV Cd and Xe ions for doses up to 5×10^{13} cm^{-2} [12]. The damage peak reaches the random level only at the high dose rate (40 nA/cm^2).

With the implantation of Bi ions for doses up to 10^{15} cm^{-2} [11] complete amorphization was not found. For higher doses a powderlike surface layer formed which did not anneal at temperatures up to 450°C.

Although no dose rate is given Gamo et. al [13] also did not reach the random level after implantation of 70 keV Cd ions at room temperature at a dose of 2.10^{15} cm^{-2}.

Because various implantation conditions such as the dose, dose rate and substrate temperature can have a marked effect on the radiation damage and annealing conditions of ion implanted InSb, it is considered important to study these factors in Mg implanted InSb.

EXPERIMENTAL

The InSb crystals were mechanically polished and then chemically etched with CP4A. Mg ions at an energy of 160 keV were implanted into <111> n-type InSb with a Varian-Extrion ion implanter with doses ranging from 5×10^{13} to 1×10^{16} ions cm^{-2}. During implantation the crystals were kept at temperatures ranging from room temperature to about 500°C. In order to avoid axial channeling the samples were tilted about 10° relative to the normal direction.

The samples were analyzed at energies ranging from 0.5 to 2.5 MeV by means of the channeling technique using a 3-axis goniometer. Protons and α-particles from a 2.5 MV Van de Graaff accelerator were backscattered and detected at 160°.

RESULTS AND DISCUSSION

As pointed out by Friedland et. al [15] InSb single crystals can be severely damaged with high doses of He ions. Therefore, in order to minimize the influence of the He ions on the structure of InSb during the analysis, the orientation of the samples and the angular scans were performed with proton beams. Care was taken to ensure that the analyzing beam current did not exceed 1.5 nA.

Typical channeling spectra for 1.5 MeV α-particles backscattered from a <111> oriented InSb single crystal are given in figure 1 as a function of the implanted dose. Although not shown in figure 1, the normalized yield of the channeled He ions for a virgin crystal was approximately 5%, indicating that the original InSb crystal was almost free from defects.

At the lower doses (up to 1.10^{14} Mg$^+$ cm^{-2}) no well defined damage peak is observable and the damage produced during the implantation process shows only as an enhanced dechanneling in the aligned spectra, indicating that the atoms were displaced from their crystal lattice sites to distances exceeding the Thomas-Fermi screening radius. At doses exceeding 1.10^{14} cm^{-2} a well defined damage peak appears extending deeper into the crystal at higher doses. As is the case with various other implanted ions [12, 21] the damage peak almost reached the random level after a dose of 10^{16} Mg$^+$ cm^{-2}, starting right at the surface. This indicates that contrary to Bi$^+$ bombardment with doses up to

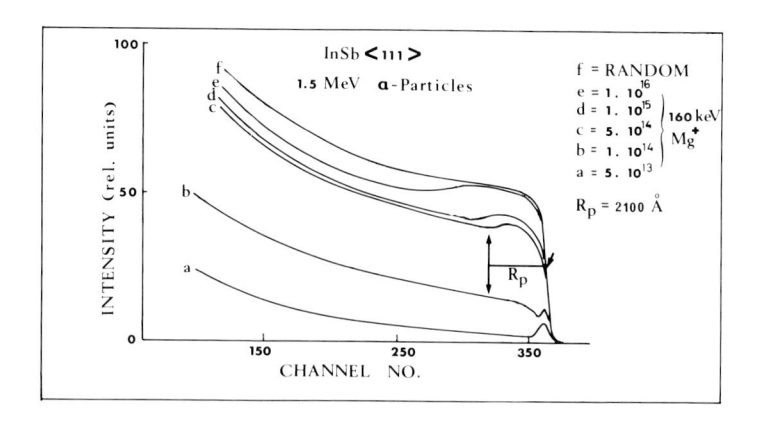

FIG. 1. Backscattering yield of 1.5 MeV α-particles from <111> aligned InSb as function of the implanted dose.

10^{15} cm^{-2} [11], amorphization took place after implantation at room temperature. Gamo et. al [13] implanted different ions into various compound semiconductors to the same dose of $1-3 \times 10^{15}$ cm^{-2}. Under equivalent implantation conditions, an amorphous layer was formed in some of the compound semiconductors but not in InSb. However, their conclusion that a certain critical implantation temperature is necessary in order to avoid the formation of an amorphous layer may not be valid for higher doses.

As shown in figure 2 the energy dependence of the dechanneling on the analyzing beam energy was studied for α-particle beams with energies between 1.0 and 2.5 MeV for a crystal implanted with 1×10^{15} Mg$^+$ cm^{-2} at 160 keV. This normally reveals some information on the major type of defect structure present [24]. Because of the damage peak that is usually present in semiconductors caution must be exercised in the interpretation of such results. However, the analysis was performed just behind the damage peak and the results are presented in figure 3. Dechanneling decreases with energy according to $E^{-\frac{1}{2}}$ indicating that interstitial atoms are mainly responsible for the dechanneling, which is supported by the tendency of InSb to form polycrystals because of the existence of a great number of In and Sb in a non-stoichiometric state [21, 23].

The isochronous annealing behaviour of room temperature implantations at a dose rate of 0.1 μA/cm^2 is shown in figure 4 for crystals implanted with a dose of 1×10^{15} Mg ions cm^{-2}. The annealing takes place in two phases, the first starting at 200°C and the second at about 300°C. The first annealing phase at 200°C was also reported for Bi implanted InSb [11]. According to a few papers dealing with the ion doping of InSb, p-type layers are formed if the implantation takes place below 200°C and inversion of the type of conduction occurs after heating above 200°C [7, 16, 20]. The p-type conduction is attributed to disordered regions formed during the implantation process. After annealing at temperatures ≥ 200°C relatively simple defect complexes are formed. Chernysheva et. al [21] also found that the amorphous layer formed by 5×10^{15} Ar$^+$ cm^{-2} becomes polycrystalline after annealing at 200°C and 300°C.

Complete recrystallization of damaged crystals does not occur even at temperatures just below the melting point [19, 22]. This was the case for all the crystals implanted at beam intensities of 0.05, 0.1, 0.5 and 1.0 μA cm^{-2} at a dose of 1×10^{15} Mg$^+$ cm^{-2}. However, Skakun et. al [14] reported complete annealing between 300°C and 400°C for crystals implanted at an energy of 70 keV.

Usually implantation into InSb is carried out at room temperature which requires subsequent annealing of defects [1, 7, 11, 12, 14, 17-19, 23]. In a few

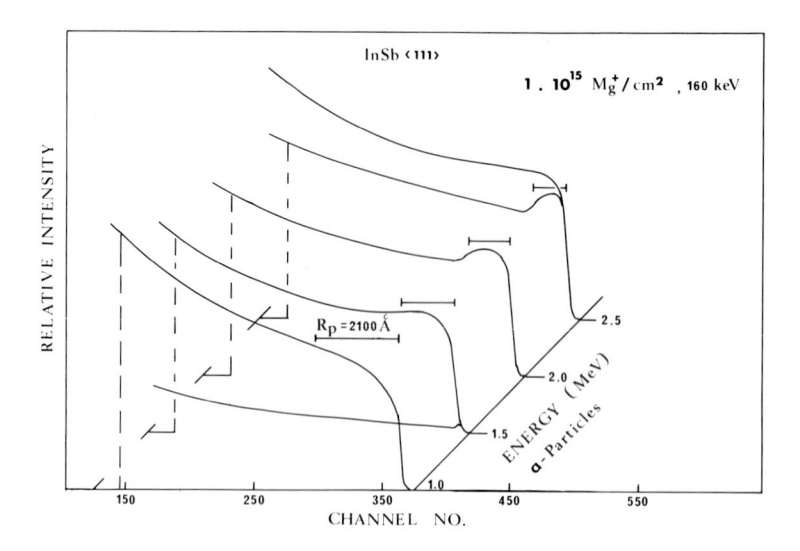

FIG. 2. Dependence of dechanneling on analyzing beam energy for $1 \cdot 10^{15}$ Mg$^+$ cm^{-2} implanted InSb.

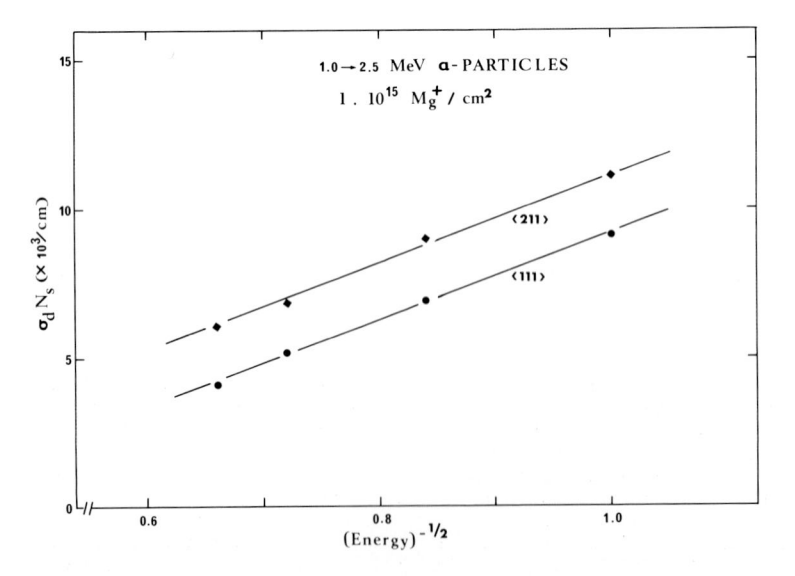

FIG. 3. Dependence of defect cross section on energy of the α-particle analyzing beam for room temperature implanted InSb. ($\phi = 10^{15}$ Mg$^+$ cm^{-2})

cases the radiation damage was studied after implantation at elevated temperatures [16, 20-22]. It was found that at irradiation temperatures above ~ 200°C only polycrystalline layers formed [21] but temperatures up to 400°C were insufficient for complete annealing of defects. The annealing behaviour for 1×10^{15} Mg$^+$ cm^{-2} is shown in figure 5 as a function of the implantation

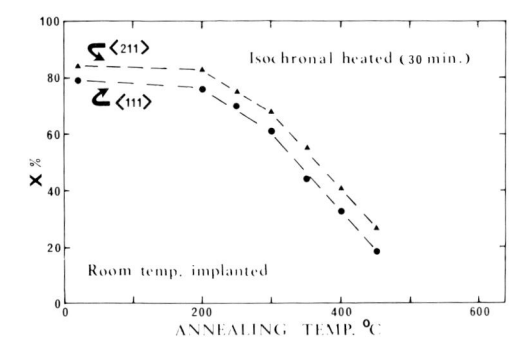

FIG. 4. Radiation damage as function of temperature for room temperature implanted InSb ($\phi = 10^{15}$ Mg$^+$ cm^{-2}) after isochronal annealing.

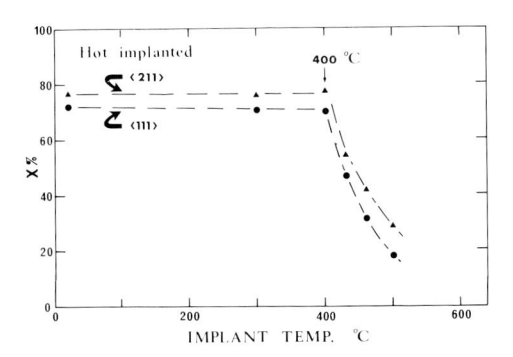

FIG. 5. Radiation damage as a function of implantation temperature for $\phi = 10^{15}$ Mg$^+$ cm^{-2} at 160 keV.

temperature. No annealing was reached up to 400°C followed by a very sharp annealing stage. This indicates that the polycrystalline phase only starts at 400°C and that there might be an increase in the mobility of vacancies preventing the creation of complex and stable defect structures. At 500°C the crystals were not restored to their initial condition before implantation.

The radiation damage was measured as a function of the implanted dose at room temperature and this is presented in figure 6. The nonlinear dependence of the total number of defects on the dose was also reported by Skakun et. al [14] and was ascribed to a competition between the formation of complexes and the annealing of defects during irradiation. Friedland et. al [15] considered the rate at which point defects are created and their annihilation due to migration or recombination of vacancy-interstitial pairs. With increasing dose the number of stable defect structures increases due to an increase in the number of interstitial crystal nuclei in the paths of the mobile point defects. At doses between 1×10^{14} to 5×10^{14} cm^{-2} an accumulated high level of point defects does not appear to be possible resulting in an equilibrium between the rate at which point defects are created and the mobility of vacancies forming more complex and stable defects. At higher doses the individual disordered zones merge to form an amorphous layer.

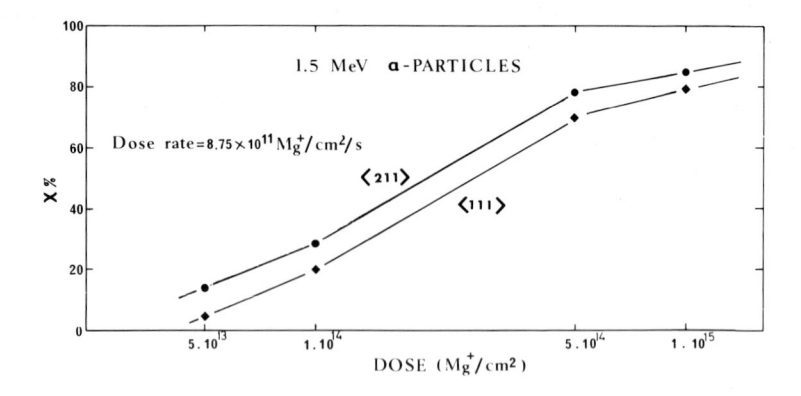

FIG. 6. Radiation damage as function of the dose for room temperature implanted InSb.

REFERENCES

1. V.A. Bogatyrev and G.A. Kachurin, Sov. Phys. Semicond. 11 56 (1977).
2. D.H. Lee, G.L. Olson and L.D. Hess, Mat. Res. Soc. Symp. Proc. 1, 281 (1981).
3. A.B. Korshunov, E.K. Dubrovskaya, V.I. Sokolov and O.V. Tikhonova, Sov. Phys. Semicond. 16 964 (1982).
4. F.H. Eisen, Phys. Rev. 148 828 (1966).
5. S. Myhra, Rad. Effects 59 1 (1981).
6. A.G. Foyt, W.T. Lindley and J.P. Donnelly, Appl. Phys. Letters 16 335 (1970).
7. V.A. Bogatyrev and G.A. Kachurin, Sov. Phys. Semicond. 11 798 (1977).
8. J.W. Cleland and J.H. Crawford Jr., Phys. Rev. 95 1177 (1954).
9. D. Kleitman and H.J. Yearian, Phys. Rev. 108 901 (1957).
10. U. Gonser and B. Okkerse, Phys. Rev. 105 757 (1957).
11. G. Langguth, E. Lang and O. Meyer, Ion Implantation in Semiconductors (Springer-Verlag Berlin 1971) 228.
12. G. Bahir and R. Kalish, Rad. Effects 33 199 (1977).
13. K. Gamo, M. Takai, H. Yagita, N. Takada, K. Masuda, S. Namba and A. Mizobuchi, J. Vac. Sci. Technol. 15 1086 (1978).
14. N.A. Skakun, I.G. Stoyanova, N.P. Dikil, P.A. Svetashev, A.S. Trokhin and A.L. Chapkevich, Sov. Phys. Semicond. 15 1112 (1981).
15. E. Friedland, B.J.E. van Tonder and J.F. Prins, Nucl. Instr. and Meth. (In press) (1983).
16. V.A. Bogatyrev, G.A. Kachurin and L.S. Smirnov, Sov. Phys. Semicond. 12 57 (1978).
17. P.J. McNally, Rad. Effects 6 149 (1970).
18. C.E. Hurwitz and J.P. Donnelly, Solid State Electr. 13 753 (1975).
19. K.-H. Wiedeburg, H. Betz and H. Kranz, Phys. Stat. Sol. 31 K69 (1975).
20. M.I. Guseva, A.N. Mansurova, V.G. Tikhonov and S.N. Khorvat, Sov. Phys. Semicond. 10 872 (1976).
21. N.Y. Chernysheva, G.A. Kachurin and V.A. Bogatyriov, Phys. Stat. Sol. 47 K5 (1978).
22. Yu. A. Danilov and V.C. Tulovchikov, Fiz. An. Tekh. Polup. USSR 14 (1) 197 (1980).
23. G.L. Destefanis and J.P. Gailliard, Appl. Phys. Lett. 35 (1) 40 (1980).
24. Y. Quéré, Rad. Effects 28 253 (1976).

STUDY[†] OF NEAR SURFACE STRUCTURE AND COMPOSITION FOR HIGH DOSE IMPLANTATION OF Cr[+] INTO Si

F. NAMAVAR,* J. I. BUDNICK,* H. C. HAYDEN,* F. A. OTTER,** AND V. PATARINI**
*The University of Connecticut, Storrs, CT 06268;
**United Technologies Research Center, East Hartford, CT 06108

ABSTRACT

The dependence of the implanted layer composition on total dose, dose rate and target chamber environment for Cr[+] implanted Si have been studied by means of Rutherford Back Scattering (RBS) and Auger Electron Spectroscopy (AES). Implantation of Cr[+] for doses up to 2×10^{18} ions/cm^2 and a fixed dose rate and energy were carried out in an ultra high vacuum (UHV) system as well as in a diffusion pumped vacuum (DPV) system. For the former, the maximum Cr concentration was about 42%. On the other hand, implantation of Cr in a DPV system resulted in a much higher peak concentration (86%) and retention.

Both the RBS and AES results positively demonstrate the existence of extensive surface carbon for a Si-rich surface and a chromium oxide layer for the Cr-rich surface. This result suggests that the interaction of oxygen or carbon occurs preferentially and depends on the surface composition.

No surface compositional variation could be observed by the RBS experiments for Cr implanted in a UHV system for different dose rates. In contrast, for implantation in a DPV system, higher concentrations can be achieved for lower dose rates.

INTRODUCTION

In this work we are interested in: (i) investigating silicide formation by means of the high dose implantation technique for the Cr-Si system, (ii) obtaining information concerning the saturation concentration of implanted Cr into Si in an ultra high vacuum as well as in a DPV system, and (iii) determining the nature of the final products for different vacuum conditions as well as understanding the effect of oxygen and carbon impurities on compound formation or precipitation.

The maximum achievable concentration in ion implanted systems is governed by the sputtering of the near surface region of the target system [1]. Implantation in a vacuum system with a pressure of the order of 10^{-6} Torr may, in some cases, result in the production of an oxide or carbide layer where generally such a layer can reduce the sputtering rate. Armenian et al. [2] reported the production of a very high concentration of Cu for Cu implantation in Al as a result of the creation of an oxide layer on their target.

Introduction of elements of residual adsorbed gases by recoil implantation [3] or any other means [4] has two facets: (1) a detrimental effect, i.e., the target becomes contaminated with unwanted impurities; and

(2) a beneficial effect, i.,e., the production of a protective layer from sputtering. Indeed, because of this protective layer, we have been able to achieve a very high concentration (86%) of Cr for the Cr-Si system.

EXPERIMENTAL PROCEDURES

Polished single crystals of Si with (111) and (100) orientations were implanted with energetic Cr^+ ions ranging from 40 to 160 keV by means of the high current implanter of The University of Connecticut. During implantation three parameters were varied:
   (i)   dose, ranging from 1×10^{16} to 2×10^{18} ions/cm^2
   (ii)   dose rate, ranging from 5×10^{12} to 3×10^{14} ions/cm^2 sec
   (iii) vacuum conditions, DPV system with and without liquid nitrogen trap $(2-4) \times 10^{-6}$ Torr and an UHV system $(2-4) \times 10^{-8}$ Torr
During implantation and RBS, samples were tilted to 10° from the normal incidence of ion beam and were fastened to the heat sink with conducting materials (silver paint). The temperature of some high dose rate implantations was monitored and no rise in target temperature was observed.

EXPERIMENTAL RESULTS

Fig. 1 demonstrates the RBS spectra of implanted Cr into Si(100) with 100 keV energy and a dose rate of 6.2×10^{13} ions/cm^2 sec and doses ranging from 2.5×10^{16} to 10^{18} ions/cm^2 in a UHV system operating at a pressure of about $(2-4) \times 10^{-8}$ Torr. Cr concentration at the maximum for each individual implanted sample was calculated [5] and is shown versus dose of implantation in Fig. 2. It is clear that a steady state concentration (42%) was achieved for a dose of about 5×10^{17} ions/cm^2. Implantations with higher doses do not result in higher Cr concentrations.

Fig. 3 demonstrates the projected range (R_p) and straggle (ΔR_p) determined from the RBS data of Fig. 1 along with estimates of stopping cross sections for He^+ in Cr-Si.

Fig. 4 shows RBS spectra of Cr implanted with 100 keV energy and a dose rate of 6.2×10^{13} ions/cm^2 sec into a Si(100) target, in a DPV system operating at a pressure of $(2-4) \times 10^{-6}$ Torr. Cr concentration as well as Cr retention for implantation in a chamber with a pressure in the order of 10^{-6} Torr are plotted in Fig. 2.

Fig. 5 shows AES data at the surface and at a depth of 50Å for those samples whose RBS was shown in Fig. 4. RBS and AES results are in very good agreement.

Curves a and b of Fig. 6 demonstrate the Cr concentration curves versus implantation for samples prepared in a DPV system operating at pressures of about 10^{-6} Torr and a dose rate of 6.2×10^{14} and 1×10^{13} ions/cm^2 sec respectively. Curve c in Fig. 6 is plotted for purposes of comparison and demonstrates the Cr concentration for implantation in a UHV system. From these results it is obvious that when implantation is carried out at a pressure of about 10^{-6} Torr a higher concentration of implanted atoms can be achieved for a low dose rate implantation.

DISCUSSION AND CONCLUSION

An examination of the first four data points of Fig. 3 suggests that depth decreases linearly with dose. For these lower doses, Cr lies beneath

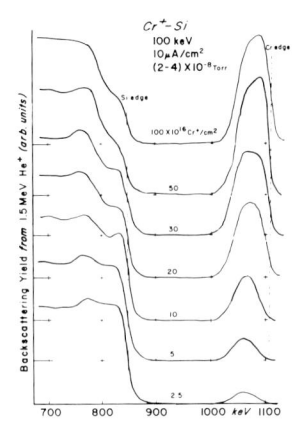

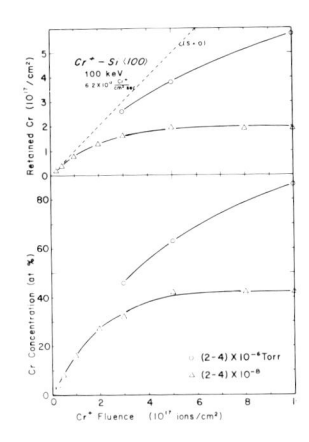

FIG. 1: RBS spectra of 1.5 MeV He$^+$ for implanted Cr$^+$ into Si with 100 keV energy.

FIG. 2: Concentration and retention curves for implanted Cr$^+$ into Si
△ : Implanted in UHV system
○ : Implanted in DPV system

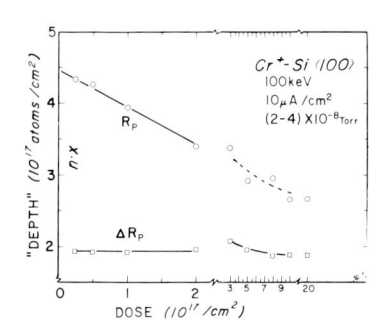

FIG. 3: Projected range (R_p) and straggling (ΔR_p) determined from the data of Fig. 1 (UHV).

the surface (Fig. 1), and the concentration is less than 27% (Fig. 2). The Cr-Si phase diagram [6] does not suggest the existence of any compounds with Cr concentration of ≤27%. It therefore seems possible that the zero order approximation model [7] can be applied to the first four data points. In zero order approximation one assumes that: (i) range distribution of implanted ions is not influenced by atoms previously introduced, (ii) surface sputtering is defined by a constant factor, (iii) the rest position

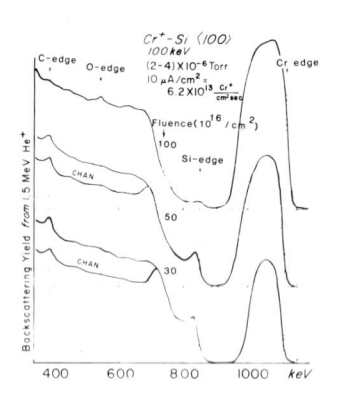

FIG. 4: RBS spectra of Cr implanted with 100 keV energy and a dose rate of 6.2 x 10^{13} ions/cm^2 into Si in a DPV system (2-4) x 10^{-6} Torr.

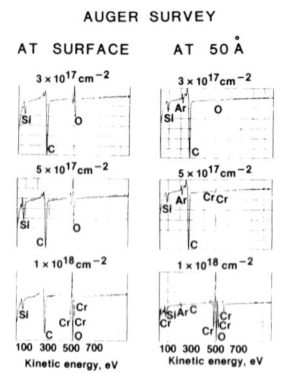

FIG. 5: AES data for the surface and for a depth of 50 Å for those samples whose RBS data are shown in Fig. 4.

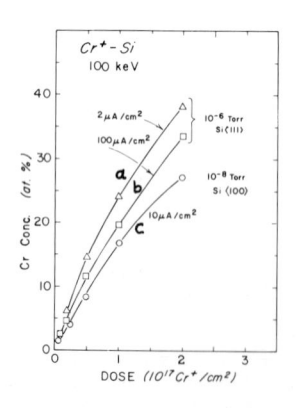

FIG. 6: Cr concentration curve versus implantation dose.

of an implanted atom does not change by subsequent implantation, and (iv) enhanced diffusion does not occur. During implantation, as a result of surface sputtering, the depth profile of the implanted atoms moves toward the instantaneous surface where the amount of the shift is equal to half the thickness of sputtered layers for symmetrical range distribution [7].

Assuming the validity of this model, twice the absolute value of the slope of the fitted line gives the sputtering factor. We can then conclude that the sputtering factor for a normal incident Cr^+ beam with 100 keV energy and a dose rate of 6.2×10^{13} ions/cm^2 sec is about $S = 1.06$. This value can be considered as the lowest limit for the sputtering factor, if range shortening [8] due to previously implanted atoms influences the range distribution of implanted ions.

In Fig. 1 the profile of Cr for implantation with a dose of 3×10^{17} ions/cm^2 is relatively flat. Fig. 3 indicates that, although no change in R_p can be observed, the value of straggling increases. Considering Fig. 2, one can realize that the Cr concentration is comparable with $CrSi_2$.

Fig. 2 shows that by increasing the system pressure by two orders of magnitude, concentration and retention of implanted atoms increases by a factor of 2 and 3 respectively. The RBS results shown in Fig. 4 demonstrate that an increase in the concentration and retention of Cr atoms is always accompanied by the presence of a carbon or an oxygen peak. These peaks indicate that surface of these samples possesses large amounts of carbon or oxygen depending on the surface composition which in turn is controlled by the total dose. Fig. 4 demonstrates in addition that the near surface region of samples implanted with doses of 3 and 5×10^{17} ions consists predominantly of Si atoms. Moreover, about 1.1×10^{17} and 1.4×10^{17} carbons/cm^2 are present in the surface region as a result of implantation of 3 and 5×10^{17} ions/cm^2 respectively. On the other hand, for implantation with a total dose of 1×10^{18} ions/cm^2, the carbon peak has disappeared and a small oxygen peak emerges which corresponds to about 3.3×10^{16} oxygens/cm^2.

Two mechanisms can be responsible for the introduction of oxygen or carbon into the surface sample: (i) recoil implantation of elements of adsorbed residual gas and/or (ii) enhancement of the chemical reaction of surface atoms with the element, of adsorbed residual gas as a result of ion implantation. It is well known that during implantation the surface impurity is recoil-implanted into the target (3). However, as can be seen from the result (Fig. 4), if recoil implantation is responsible for the introduction of oxygen and carbon in our samples, one should expect both oxygen and carbon to be present in large quantities at the same time.

On the other hand, Fig. 4 suggests that when surface targets are predominantly constituted of Si atoms, a carbon build-up occurs. However, when the dose of implantation was increased to 1×10^{18} ions/cm^2, mostly Cr atoms reached the surface and the carbon peaks disappeared while an oxygen peak emerged. AES results of the above sample at a depth of 50Å clearly support the conclusion derived from RBS data.

At this time we believe that the mechanisms involved in this process are of a chemical nature because the accumulation of oxygen or carbon occur preferentially and depend upon surface composition. Judging from our results, it seems that during implantation there is a strong affinity between Si and C as well as between Cr and oxygen which determines the process of carbon or oxygen gettering.

Understanding the transformation of implanted samples from a carbon rich surface (with a dose of 5×10^{17} ions/cm^2) to an oxygen-rich surface (with a dose of 1×10^{18} ions/cm^2) merits serious consideration. By implanting Cr with a dose of 5×10^{17} ions/cm^2 a surface layer rich in C and Si is observed (see Figs. 4 and 5). At this stage of implantation very few Cr atoms are present on the C-Si/Si interface. Indeed, the latter can be verified by the experimental results shown in Fig. 4. Further implantation causes Cr atoms to move toward the interface (as well as deeper) and/or the C-Si/Si interface also moves to the Cr-rich region because the surface Si is consumed by the Si-C reaction as well as by sputtering.

At certain doses which obviously should be less than 1×10^{18} ions/cm^2, a large number of Cr atoms will reach the interface and come in contact with the carbon present in the C and Si region. One would expect that further implantation should initiate a reaction between Cr and carbon. Very recently it was reported that chromium carbide was formed by implantation of Cr$^+$ into the CrFe system [9]. In addition, so-called vacuum carbonization [4] is cited to occur for pure Fe, Ni and Ta by a variety of ions such as Ti$^+$, Ta$^+$ and Cr$^+$. Our RBS and AES indicate that carbon occurs more readily at a surface rich in silicon than for one rich in Cr.

However, we believe that, if in both high and low dose rate implantation a uniform C-Si layer is created, then there should be no difference in Cr concentration or in retention for a relatively high dose implantation because sputtering is a surface phenomenon. On the other hand, if islands of C-Si are formed, we believe that a reduction of sputtering or an increase in retention of implanted atoms will depend on the surface concentration of these islands. More islands are probably created during low dose rate implantation because of a longer implantation time. The low dose rate implantation time was about 50 times that of the high dose rate implantation (Fig. 6).

Because of the lack of required resolution RBS and AES techniques cannot be applied in order to verify the lateral uniformity of implanted surfaces. Therefore, we have studied samples implanted with doses of 3 and 5×10^{17} ions/cm^2 by scanning electron microscopy. SEM micrographs indicate the presence of light spots which could be associated with exposed Si surfaces.

†Supported by the Office of Naval Research.

REFERENCES

1. Z.L. Liau and J.W. Mayer, J. Vac. Sci. Technol. **15**(5) 1-29 (1978).
2. E. Arminen, A. Fontell, and V.K. Lindroos, Phys. Stat. Sol. (a) **4** 663 (1971).
3. R. Kelly and J.B. Sanders, Surface Science **57** 143 (1976).
4. I.L. Singer, J. Vac. Sci. Technol. **A1**(2) 419 (1983).
5. W.K. Chu, J.W. Mayer and M.A. Nicolet, Backscattering Spectroscopy (Academic Press, New York 1978).
6. M. Hansen, Constitution of Binary Alloys (McGraw-Hill, New York 1958).
7. F. Schulz and K. Wittmaack, Radiation Effects **29** 31 (1976).
8. P. Blank, K. Wittmaack and F. Schulz, Nuclear Instruments and Methods **132** 387 (1976).
9. W.K. Chan, C.R. Clayton, R.G. Allas, C.R. Gossett and J.K. Hirvonen, Nuclear Instruments and Methods **209/210** 857 (1983).

ACKNOWLEDGMENTS

We are very grateful to Professor Quentin Kessel for his help with the Van de Graaff accelerator, to Professor D. I. Potter for allowing us to use his ultra high vacuum implantation chamber and to M. Ahmed for his assistance during implantation. In addition, we would like to thank R. Roser, P. Clapis, John Gianopoulous and C. H. Koch for their technical assistance.

THE INFLUENCE OF IMPLANTATION CONDITIONS AND TARGET ORIENTATION IN HIGH DOSE IMPLANTATION OF Al[+] INTO Si[#]

F. NAMAVAR,* J. I. BUDNICK,* A. FASIHUDDIN,* H. C. HAYDEN,* D. A. PEASE,* F. A. OTTER,** AND V. PATARINI**   *The University of Connecticut, Storrs, CT 06268; **United Technologies Research Center, East Hartford, CT 06108

ABSTRACT

We report the preliminary results of a study to determine the dependence of the near surface composition and structure on total dose, dose rate, vacuum condition and substrate orientation for Al implantation into Si (111) and Si (100) with doses up to 2×10^{18} ions/cm^2. Our studies include the results of Rutherford Back Scattering (RBS), Auger Electron Spectroscopy (AES) and x-ray diffraction measurements on samples implanted with a 100 keV energy in a diffusion pumped vacuum (DPV) system (10^{-6} Torr) with and without a LN$_2$ trap and in an ultra high vacuum (UHV) system $(2-4) \times 10^{-8}$ Torr.

Results of high dose rate (50 μA/cm^2) implantation into Si (111) in an untrapped DPV system indicate that Al segregates with a preferred (111) orientation. For a dose of 1×10^{18} ions/cm^2 the surface is Al-rich to a depth of 2500Å while for lower doses the surface is silicon-rich. A carbon build-up occurred for samples prepared by low dose rate (5 μA/cm^2) implantation. However, no Al segregation could be observed for doses of less than 10^{18} ions/cm^2. A similar behavior has been observed for Si (100) except that Al segregation occurs with a polycrystalline structure. Moreover, the segregated Al is present at depths greater than the projected range.

When implantation was carried out in a DPV system with a LN$_2$ trap, no carbon peaks could be observed by RBS regardless of the dose rate. For these conditions, as well as for the implantation of Al in an UHV system, we find Al segregation with a polycrystalline structure independent of the dose rates and target orientations we used. Al is observed at a depth greater by a factor of two than the expected value from the R_p calculations. The Al depth penetration increases with the dose of implantation.

INTRODUCTION

We have recently shown [1] that, under certain conditions, the implantation of energetic Xe through the interface of deposited Al thin films on Si substrates can produce an Al-Si alloy layer which is uniform in texture. No segregation was observed for a long period of room temperature annealing. Our subsequent experiments demonstrated [2] that Al-Si alloy layers are stable under heat treatments of up to 450°C where at this temperature a thin film of Al is formed at the sample surface. We have been very interested in understanding the reasons for this stability and in determining the dependence of stability on the presence of undersized and oversized impurities such as oxygen and Xe.

Since the production of an oxygen-free Al-Si interface is normally very difficult, we proceeded to prepare Al-Si mixtures by direct high dose implantation of Al into silicon. Our interest in this work was enhanced because of recent stimulating work on metal-semiconductor eutectic systems [3-6]. For some of the metal-semiconductor eutectic systems such as Au-Si and Au-Ge, metastable phases have been produced by splat cooling, laser and ion beam experiments [3, 7].

Therefore, it seemed very interesting to study by high dose implantation techniques the possible existence of metastable phases in the Al-Si system and to determine the dependency of formation of such a phase upon contaminating elements such as oxygen. Here, we report preliminary results of the first part of this study, namely, the influence of implantation conditions and target orientation for high dose implantation of Al into Si and the results of studies by RBS, x-ray diffractometry, and AES. We are also in the process of applying nuclear resonance profiling (NRP) using the ^{27}Al (P,γ) ^{28}Si reaction in order to obtain additional information on the depth of profile for Al.

EXPERIMENTAL PROCEDURES

Polished single crystals of Si with (111) and (100) orientations were implanted with 100 keV Al$^+$ ions by means of the high current implanter of The University of Connecticut. During implantation three parameters were varied:
 (i) dose, ranging from 1×10^{17} to 2×10^{18} ions/cm^2
 (ii) dose rate, ranging from 3×10^{13} to 3×10^{14} ions/cm^2 sec
 (iii) vacuum conditions, diffusion pumped vacuum system with and without liquid nitrogen trap $(2-4) \times 10^{-6}$ Torr and an ultra high vacuum (UHV) system $(2-4) \times 10^{-8}$ Torr
During implantation and RBS, samples were tilted to 10° from the normal incidence of the ion beam and were fastened to the heat sink with conducting materials (silver paint). The temperature of some high dose rate implantations was monitored and no rise in target temperature was observed.

EXPERIMENTAL RESULTS

Fig. 1 demonstrates the RBS results for Al implanted Si (111) samples for an untrapped diffusion pumped system (UDPV), with 100 keV energy and doses of 3 and 10×10^{17} ions/cm^2 and a dose rate of 2.4×10^{14} ions/cm^2 sec. As is shown for the sample implanted with a dose of 3×10^{17} ions/cm^2, there is a substantial reduction in the backscattered signal for depths of up to 2,000Å. However, Si constitutes about 70% of the surface sample.

More reduction of scattering signal was observed for doses such as 5×10^{17} ions/cm^2 as compared with ones with a dose of 3×10^{17} ions/cm^2. It appears that the backscattered signal from the (Si) surface is reduced as the implantation dose is increased. Eventually, at a dose of 1×10^{18} ions/cm^2 we observe a predominance of Al to a depth of about 2,500Å.

Sputtered depth profiles by Auger electron spectroscopy of samples whose RBS is shown in Fig. 1 have been carried out and are demonstrated in Fig. 2. From these data, the formation of a high concentration Al surface region, which was suggested by RBS at a dose of 1×10^{18} ions/cm^2 is verified. Although we cannot accurately estimate the effect of preferential sputtering on the AES data, the RBS and AES results are in agreement. The above samples were studied by means of an x-ray diffractometer. By this means, a pronounced (111) Al reflection peak has

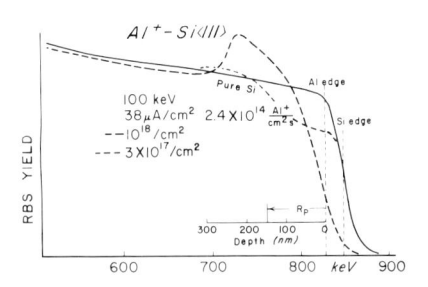

Fig. 1: RBS spectra of 1.5 MeV He$^+$ for Al implanted Si (111) samples (UDPV).

been observed for all Al implanted samples into Si (111) substrates with doses ranging from 3 to 10 x 10^{17} ions/cm^2 and at a dose rate of 2.4 x 10^{14} ions/cm^2 sec. However, no x-ray peaks due to reflections from other planes could be observed in our diffractometer, in contrast to the Al implanted Si (100) samples which do not show such preferential orientation.

The RBS spectra for a dose rate of 3.3 x 10^{13} ions/cm^2 sec into Si (111) (UDPV) at a dose of 3 x 10^{17} ions/cm^2 is shown in Fig. 3. A thin carbon film on the surface is observed. No elemental Al peaks could be observed by x-ray diffractometry for the low dose rate implanted samples.

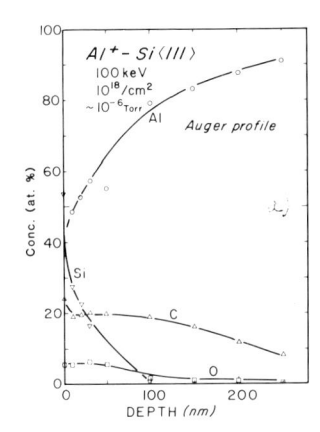

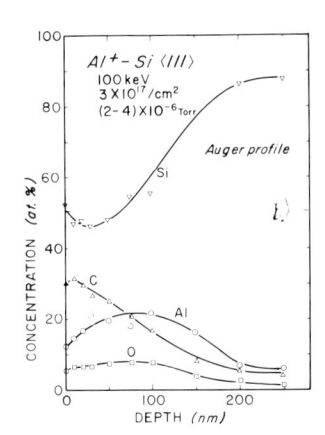

Fig. 2: Depth profiles by AES of samples whose RBS spectra are shown in Fig. 1, (a) implanted with a dose of 1 x 10^{18} ions/cm^2
(b) implanted with a dose of 3 x 10^{17} ions/cm^2

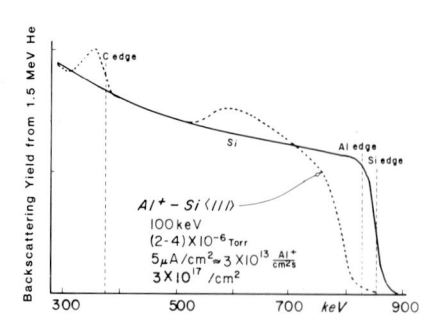

Fig. 3: RBS spectrum of Al implanted into Si (111) (UDPV).

Fig. 4 demonstrates RBS spectra of implanted Al in Si (100) in a trapped diffusion pumped system with a dose rate of 3×10^{14} ions/cm^2 sec and different doses. These data suggest that as the dose of implantation increases, the Al concentration profile shifts to a greater depth. For doses of 2×10^{18} ions/cm^2, Al penetrates to a depth which is almost twice that of the projected range. Two x-ray lines which correspond to (111) and (200) Al reflections were observed at intensities expected for polycrystalline aluminum.

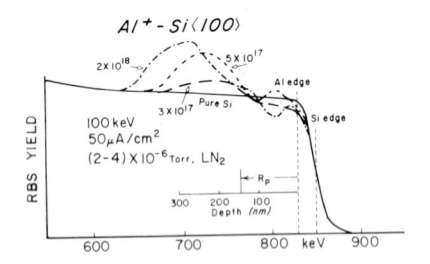

Fig. 4: RBS spectra of implanted Al into Si (100) (DPV).

Fig. 5 demonstrates RBS results of Al implanted with 100 keV energy, a dose of 1×10^{18} ions/cm^2 and a dose rate of 6.2×10^{13} and 3.1×10^{14} ions/cm^2 sec into Si (100) and Si (100) targets in an UHV system operating at $(2-4) \times 10^{-8}$ Torr. As is shown, Al penetrates much deeper than the expected value for the projected range. A higher concentration of Al could be achieved for implantation into Si (111) than into Si (100). Al reflections with an intensity ratio corresponding to polycrystalline aluminum were observed for all four samples whose RBS results are demonstrated in Fig. 5.

DISCUSSION AND CONCLUSION

Implantation of Al in a DPV system into Si (111) with 100 keV energy

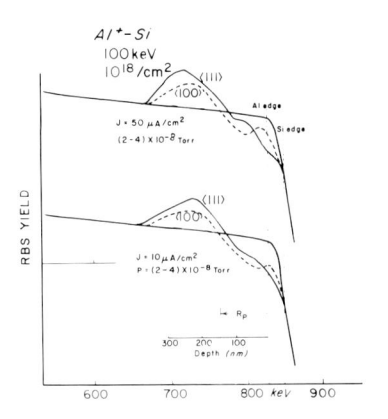

Fig. 5: RBS spectra of Al implanted into Si (111) and Si (100) (UHV).

and a dose rate of 2.4×10^{14} ions/cm^2 and doses ranging from 3 to 10 x 10^{17} ions/cm^2 resulted in Al segregation where Al is preferentially oriented in a (111) plane parallel to the (111) plane of the Si substrate. On the basis of RBS Fig. 1, AES Fig. 2 and x-ray results, we may conclude that for implantation of Al with a dose of 3×10^{17} ions/cm^2 Al islands are formed where they occupy about 30% of the surface sample. However, as the supply of Al increases by increasing the dose of implantation to 1×10^{18} ions/cm^2, the surface density of these islands increases and eventually an Al-rich layer with a thickness of approximately 2500Å is created. It is expected that the near surface region of Si targets becomes highly disordered (amorphized) because of high dose implantation. However, it is intriguing that Al precipitates possess a preferred (111) orientation.

Very recently Kaufmann et al. [8] reported that as a result of implantation at room temperature of silver into a single crystal of beryllium, silver precipitates at room temperature such that the (111) axis in silver is parallel to the (0001) axis of the beryllium target. Furthermore, Tsaur et al. [9] reported the formation of (111) preferred orientation in ion beam mixed Au-Ni, Ag-Cu and Au-Co systems.

On the other hand, our x-ray diffraction measurements, of implanted Al into Si (111) in an ultra clean vacuum system, suggests that, although Al segregates, it possesses a polycrystalline structure. Since implantation conditions for samples prepared in UHV and UDPV systems were the same except for the pressure, we believe that the formation of highly oriented Al films is probably related to impurities, possibly oxygen and carbon, which were introduced during implantation in the UDPV system. At high doses, for implantation in a clean system, Al is present at a greater depth than that expected from the projected range calculation (Fig. 5). For poor vacuum conditions, though, the segregated Al remains at the near surface region.

Results shown in Figs. 4 and 5 suggest that for the implantation of Al in Si in an ultra clean or moderately clean vacuum system, implanted Al redistributes during implantation or during post-implantation room temperature annealing. Implanted atoms have a tendency to penetrate deeper as the dose of implantation increases (see Fig. 4). There is a similarity between the results of this work and those reported by Rusbridge [4] for implanted Al into AlGe at 160°C.

At the present time, interpretation on the redistribution of Al in our work in comparison to the interpretation of AlGe results [10] is hindered because of the lack of detailed information on depth profiles for implanted Al in Si.

No Al segregation was observed for Al implanted Si (UDPV) with a low dose rate and for doses of up to 5×10^{17} ions/cm^2. In contrast, implantation under similar conditions, even in an oil trapped DPV system, resulted in Al segregation. From these results, it seems that during high dose rate implantation in the UDPV system, fewer contaminated elements were available because of the sputtering effects and the short times required for implantation. Secondly, precipitated nucleation for implanted Al occurred as soon as the Al concentration at the damage peak surpassed the solubility limit and before any contaminant has a chance to diffuse to the damage peak. On the other hand, for low dose rate implantation in UDPV systems, contaminant atoms such as carbon may reach damage peaks faster than implanted Al reached its solubility limit. Probably the presence of very chemically active small impurities such as carbon and oxygen could act as a sink for defects which as a result delay or prohibit Al precipitation.

#Supported by the Office of Naval Research.

REFERENCES

1. F. Namavar, J.I. Budnick and F.A. Otter, Thin Solid Films 104 (1983) p. 31.
2. F. Namavar, J.I. Budnick and F.A. Otter, unpublished data, 1983.
3. S.S. Lau, B.Y. Tsaur, M. von Allmen, J.W. Mayer, B. Stritzker, C.W. White and B. Appleton, Nuclear Instruments and Methods 182/183 (1981) pp. 97-105.
4. K.L. Rusbridge, Nuclear Instruments and Methods 182/183 (1981) p. 521.
5. K.L. Bertram, F.J. Minter, J.A. Hudson and K.C. Russell, Journal of Nuclear Materials 75 (1978) p. 42.
6. M.R. Mruzik and K.C. Russell, Journal of Nuclear Materials 78 (1978) p. 343.
7. H. Jones and C. Suryanarayana, Journal of Materials Science 8 (1973) p. 705.
8. E.N. Kaufmann and L. Buene, Nuclear Instruments and Methods 182/183 (1981) p. 327.
9. B.Y. Tsaur, S.S. Lau, L.S. Hung and J.W. Mayer, Nuclear Instruments and Methods 182/183 (1981) p. 67.
10. A.D. Marwick in Surface Modification and Alloying, edited by J.M. Poate, G. Foti and D.C. Jacobson (Plenum Press, New York and London, 1983), p. 211.

ACKNOWLEDGMENTS

We are very grateful to Professor Quentin Kessel for his help with the Van de Graaff accelerator, to Professor D. I. Potter for allowing us to use his ultra high vacuum implantation chamber and to M. Ahmed for his assistance during implantation. In addition, we would like to thank R. Roser, P. Clapis, John Gianopoulous and C. H. Koch for their technical assistance.

EFFECTS OF DOUBLE-IMPLANT ON THE EPITAXIAL GROWTH OF AMORPHOUS SILICON

L.J. CHEN, AND C.W. NIEH
Department of Materials Science and Engineering, National
Tsing Hua University, Hsinchu, Taiwan, ROC.

ABSTRACT

Investigation has been made on the effects of double implant of BF_2^+-As^+, BF_2^+-P^+, B^+-As^+ and B^+-P^+ on the thermal annealing behaviors of silicon. For specimens with As^+ or P^+ as the major dopants, the annealing behaviors were similar to those singly implanted with As^+ or P^+. For samples with BF_2^+ or B^+ as the major impurities, drastic changes in annealing characteristics were found in comparison with BF_2^+ or B^+ single implant specimens. High density twins, including primary twins on all four {111} planes and twelve types of secondary twins were observed in BF_2^+-As^+ and BF_2^+-P^+ double implant samples.

I. INTRODUCTION

The annealing behaviors of ion-implanted silicon have been extensively studied for its wide applications in solid state devices. The collision of energetic ions can cause considerable damage to the crystalline substrate. Continuous amorphous layer may be formed for implantation above a critical dose. To reduce the disorder, samples must be annealed either thermally or by high intensity energy beam. During annealing, the amorphous layer regrows epitaxially on substrate silicon. It has been well known that the impurity has strong influence on the regrowth behaviors of amorphous silicon [1,2]. Arsenic, boron and phosphorus, having an impurity concentration near 0.5 at%, can increase the growth rate by factors of 6-25. On the other hand, carbon, oxygen and nitrogen at the level of about 0.5 at% strongly retard the growth rate. For amorphous layer double implanted with equal amount of As and B or P and B, to concentrations of about 0.5 at%, the growth rate was decreased 2-3 order of magnitude [3]. The dominant defects and their density were also dependent on the type and amount of impurities. For (111) oriented silicon wafers, implanted either by 5×10^{14} As^+, 1×10^{16} Au, 5×10^{15} Ar, 3×10^{15} Pb or 8×10^{15} Si/cm^2, post annealing defects were dominated by twins [4,5]. While implanted by 1×10^{16} B^+ or P^+/cm^2, dislocation lines were the major defects [6]. For (100) substrates implanted with either 8×10^{15} Ne, 7×10^{14} Kr, 2×10^{15} Ar, 5×10^{15} O or 5×10^{15} N/cm^2, the dominate defects after thermal annealing were twins [7,8], whereas implanted by B^+, P^+ or As^+ to about 1×10^{16}/cm^2, the dominant defects were dislocation lines and loops. The general trend for twin formation was above a critical dose (D_C), twin will be formed on both (111) and (100) specimens. D_C for (100) substrate was higher than that for (111) substrate and was different for each implant species [8].
Atomic size difference between impurity and host atoms leads to the generation of misfit dislocations. The atomic radius of Si is 1.17 Å, if it was doped with B or P, having an atomic radius of 0.88 Å and 1.10 Å, respectively, misfit dislocation will be formed. If the silicon substrate was doped with both P and Ge (atomic radius 1.22 Å) or B and Sn (atomic radius 1.40 Å), the double impurity effect was to reduce the dislocation density [9].
In recent years, BF_2^+ ion has appeared to be a good candidate for producing shallow implanted layer in device fabrication. A number of

studies have been conducted to characterize the electrical properties and microstructures of BF_2^+-implanted silicon [10,11]. In this paper, we report the results of the investigation of the effects of double implant of BF_2^+-As^+, BF_2^+-P^+, B^+-As^+ and B^+-P^+ on the epitaxial regrowth of amorphous silicon.

II. EXPERIMENTAL RPOCEDURES

Silicon wafers, phosphorus or boron doped, 10 Ω-cm, (100) or (111) oriented were first partly implanted with one type of dopant to a dose of $5 \times 10^{15}/cm^2$, using thermal oxide as selective implantation mask. The oxide mask was subsequently removed and the second implantation of another dopant to a dose of $1 \times 10^{15}/cm^2$ followed. Fig. 1 shows an optical micrograph of a typical patterned specimen. Implantation energies were 25 KeV, 100 KeV, 65 KeV and 100 KeV for B^+, As^+, P^+ and BF_2^+, respectively, so that the projected ranges for each dopant are approximately equal ($\sim$600 Å). Substrate normal was kept 7º off the incident ion beam direction to minimize channeling effect. After implantation, the specimens were annealed at 1000ºC for 30 mins. All the annealings were performed in N_2 flowing diffusion furnace. Post-implantation annealed specimens were then examined by transmission electron microscopy.

III. RESULTS AND DISCUSSION

For the regions double implanted with 5×10^{15} BF_2^+ (or B^+)/cm^2 and 1×10^{15} As^+(or P^+)/cm^2, the annealing behaviors were drastically different from those of single implanted specimens. Tangled dislocations were found in samples implanted by 5×10^{15} BF_2^+/cm^2 followed by 1000ºC annealing as shown in Fig. 2. Few twin was observed. Double implants promoted twin formation significantly and high density twins were found in both (111) and

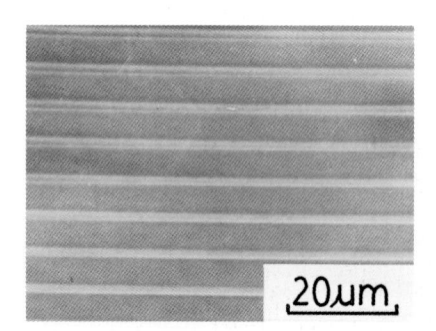

Fig. 1 Optical micrograph of a typical patterned sample.

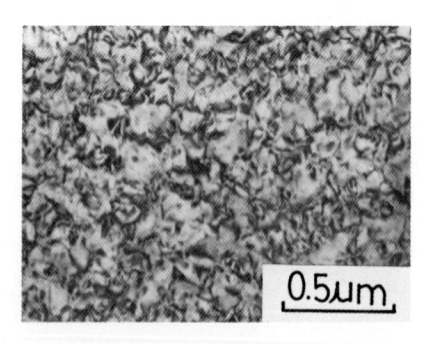

Fig. 2 Bright field micrograph (B.F.), (111), 5×10^{15} BF_2^+/cm^2.

(100) samples. Fig. 3 shows a dark field image of twins formed on inclined ($\bar{1}11$) plane for a (111) sample annealed at 1000ºC, viewed near [110] direction. The D region was implanted with 5×10^{15} BF_2^+ and 1×10^{15} As^+/cm^2, S region was irradiated with 1×10^{15} As^+/cm^2. The twin density was much higher than that of S region. For other (111) specimens as well as (001) samples with the major dopant being BF_2^+ or B^+, the characteristics of twin formation were similar.

For the regions double implanted with 5×10^{15} As^+ (or P^+)/cm^2 and

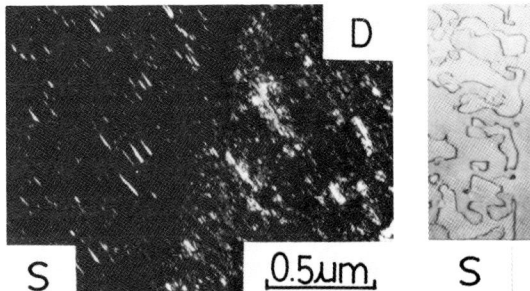

Fig. 3 Dark field micrograph (D.F.), primary twins on (1$\bar{1}$1) plane, (111), D: 5x10^{15} BF$_2$$^+$ + 1x10^{15} As$^+$/cm^2, S: 1x10^{15} As$^+$/cm^2.

Fig. 4 B.F., (001), D: 5x10^{15} As$^+$ + 1x10^{15} BF$_2$$^+$/cm^2, S: 1x10^{15} BF$_2$$^+$/cm^2.

1x10^{15} BF$_2$$^+$ (or B$^+$)/cm^2, the annealing behaviors were similar to those of the specimens singly implanted with As$^+$ (or P$^+$). Fig. 4 shows an example of a (001) sample. The double implant region was implanted with 5x10^{15} As$^+$ and 1x10^{15} BF$_2$$^+$/cm^2 and the single implant region was irradiated with 1x10^{15} BF$_2$$^+$/cm^2. The D region was nearly defect free, whereas irregular dislocations were observed in S region. The microstructure in D region was similar to that in 5x10^{15} As$^+$/cm^2 single implant specimen. For (001) samples with 5x10^{15} P$^+$ and 1x10^{15} BF$_2$$^+$/cm^2 in D region and 1x10^{15} BF$_2$$^+$/cm^2 in S region, see Fig. 5, similar results were obtained.

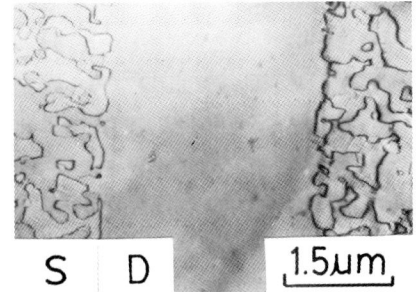

For (111) specimen, D-region with 5x10^{15} As$^+$ (or P$^+$) and 1x10^{15} BF$_2$$^+$/cm^2, dislocations and twins were observed. The results were similar to those found in 5x10^{15} As$^+$ (or P$^+$) single implant samples. An example is shown in Fig. 6.

High density twins were observed for samples implanted with As$^+$-BF$_2$$^+$ and P$^+$-BF$_2$$^+$ samples. Figs. 7(a) to (c) are typical diffraction patterns of

Fig. 5 B.F., (001), D: 5x10^{15} P$^+$ + 1x10^{15} BF$_2$$^+$/cm^2, S: 1x10^{15} BF$_2$$^+$/cm^2.

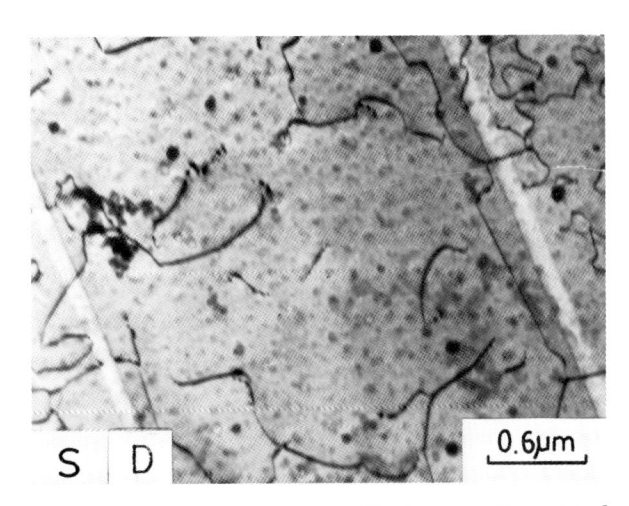

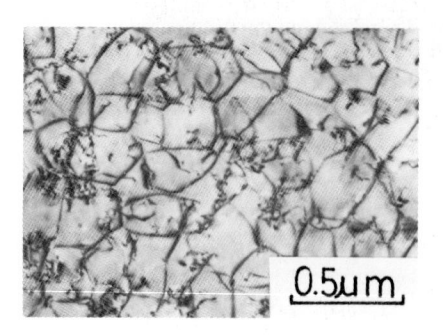

Fig. 6 B.F., (111), 5x10^{15} As$^+$ + 1x10^{15} BF$_2^+$/cm^2.

silicon single crystal containing twins taken near [111], [110] and [001] directions, respectively. Analysis of the diffraction patterns indicate that primary twins on all {111} planes and twelve possible types of primary twins will all present in both (111) and (001) wafers.

Previous studies have atributed the generation of structural defects in post-implantation annealed samples to the stresses induced by amorphization of crystalline substrate and atomic size difference between dopant and host atoms. Strain compensation was invoked to explain the reduction of defect density in P$^+$-Ge$^+$ or B$^+$-Sn$^+$ double doped samples compared to singly doped samples [9]. In the present investigation, all the samples were initially rendered amorphous and the As atom was known to have almost identical atomic size as Si (1.18 Å vs. 1.17 Å). The drastic changes of microstructures in post-implantation annealed BF$_2^+$-As$^+$ compared to that in BF$_2^+$-implanted samples are therefore not explained satisfactorily by these two effects. Other factors, such as changes in bonding at amorphous/crystalline interface and chemical effect may exert strong influence on the epitaxial growth of amorphous silicon.

V. SUMMARY AND CONCLUSIONS

Transmission electron microscopy has been applied to study the effects of double implant of BF$_2^+$-As$^+$, BF$_2^+$-P$^+$, B$^+$-As$^+$, B$^+$-P$^+$ on the thermal annealing behaviors of silicon. For specimens with As$^+$ or P$^+$ as major dopants, the annealing behaviors were similar to those singly implanted with As$^+$ or P$^+$. For samples with BF$_2^+$ or B$^+$ as the major impurities, drastic changes in annealing characteristics were found in comparison with BF$_2^+$ or B$^+$ single implant specimens. Changes in bonding at amorphous/crystalline interface or chemical effect may be involved in the epitaxial growth of amorphous silicon. High density twins, including primary twins on all four {111} planes and twelve types of secondary twins, were observed in BF$_2^+$-As$^+$ and BF$_2^+$-P$^+$ double implant specimens.

ACKNOWLEDGEMENTS

The research was supported in part by ROC National Science Council through Grant No. NSC72-0414-P-007-03.

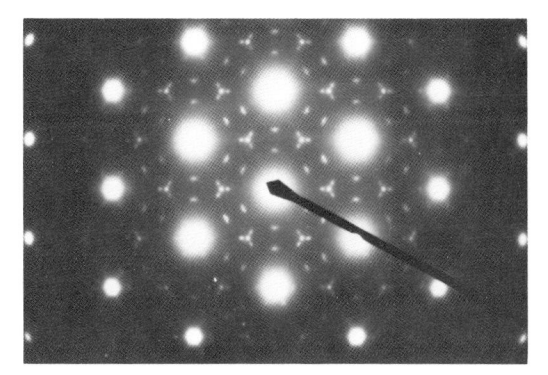

Fig. 7(a)

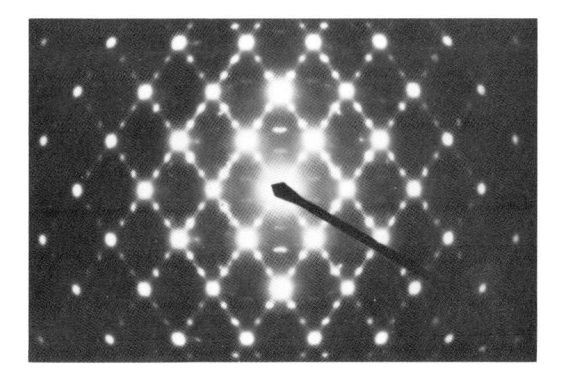

Fig. 7(b)

Fig. 7 Typical diffraction patterns of silicon single crystal containing primary and secondary twins, taken along (a) [111], (b) [110] (c) [100] directions.

Fig. 7(c)

REFERENCES

1. L. Csepregi, E.F. Kennedy, T.J. Gallagher, J.W. Mayer, and T.W. Sigmon, J. Appl. Phys. 48, 4234 (1977).
2. E.F. Kennedy, L. Csepregi, J.W. Mayer, and T.W. Sigmon, J. Appl. Phys. 48, 4241 (1977).
3. I. Suni, G. Göltz, M.G. Grimaldi, M-A Nicolet, and S.S. Lau, Appl. Phys. 40, 269 (1982).
4. I. Ohdomari and N. Onoda, Phil. Mag. 35, 1373 (1977).
5. M.D. Rechtin, P.P. Pronko, G. Foti, L. Csepregi, E.F. Kennedy, and J.W. Mayer, Phil. Mag. A, 37, 605 (1978).
6. M. Tamura, Phil. Mag. 35, 663 (1977).
7. M. Wittmer, J. Roth, P. Revesz, and J.W. Mayer, J. Appl. Phys. 49, 5207 (1978).
8. E.F. Komarov, V.S. Solov'yev, V.S. Tishkov, and S. Yu Shiryayav, Rad. Eff. 69, 179 (1983).
9. G. Nakamura: in Ion Implantation in Semiconductor, S. Namba, ed. 557 (1975).
10. L.J. Chen, and I.W. Wu, J. of Appl. Phys. 52, 3310 (1981).
11. L.J. Chen, Y.J. Wu, and I.W. Wu, J. of Appl. Phys. 52, 3520 (1981).

A COMPARATIVE STUDY OF NEAR-SURFACE EFFECTS DUE TO VERY HIGH FLUENCE H+ IMPLANTATION IN SINGLE CRYSTAL FZ, CZ, AND WEB SI*

W.J. CHOYKE,* R.B. IRWIN,** J.N. MCGRUER,** J.R. TOWNSEND,** N.J. DOYLE,*** B.O. HALL,*** J.A. SPITZNAGEL,*** AND S. WOOD***
*University of Pittsburgh and Westinghouse R&D Center, Pittsburgh, PA 15235; **University of Pittsburgh, Pittsburgh, PA 15260; ***Westinghouse R&D Center, Pittsburgh, PA 15235

ABSTRACT

Crystals of float-zone, Czochralski, and Web Si having widely differing oxygen concentrations were implanted at 300°K with large fluences of hydrogen. Experiments using RBS/channeling, profilometry and microscopy are reported. For room temperature implants the step-height h, the distribution of displaced atoms $N_D(x)$ and the total number of displaced atoms/cm^2 (a_D) are all essentially independent of the oxygen content of the Si.

INTRODUCTION

The role of oxygen in silicon has been of interest for a long time but despite this it remains a very active area of investigation. J.R. Patel [1] has given a concise review of the current problems on oxygen in silicon covering the field up to 1981. While experiments using large fluence implants of hydrogen in various forms of Si are constantly being reported, we have seen no systematic study in the literature of the effect of the oxygen concentration in the implanted Si.

Recently [2] we reported on the mechanical response of float-zone <111> and <100> Si at room temperature when implanted with H+ from 2×10^{16} H+/cm^2 to 1.6×10^{18} H+/cm^2 in a range from 20 keV to 1 MeV. A short discussion of older results may be found there.

In the present paper we report a comparison of RBS/channeling, profilometry and optical microscopy on hydrogen implanted Si crystals grown in a number of ways and having different oxygen concentrations.

EXPERIMENT

Three types of samples were used in these experiments:
(i) Monsanto N-type (FZ) float zone grown Si with a conductivity of 30-60 Ω-cm. The oxygen concentration in these crystals is much less than 1 appm [3].
(ii) Westinghouse P-type dendritic web Si [4] with a conductivity of 2-8 Ω-cm. The estimated oxygen concentration is 30 appm.
(iii) Monsanto N-type (CZ) Czochralski-grown Si with a stated interstitial oxygen concentration of 37 appm.

All samples were cleaned, prior to implantation, by a 16 step cleaning procedure used for integrated circuit device fabrication. The dimensions of samples for channeling and step-height measurements were 1x1 cm.

---

*Supported in part by NSF Grant DMR-81-02968.

The thickness of the FZ and CZ samples cut from large wafers was between 200 and 300 μm whereas the dendritic web was 150 μm thick and originally 3 cm wide. Each implanted crystal was covered with a tantalum mask over less than one-half its area so as to provide a step between the implanted and non-implanted surface and a virgin-sample reference for the channeling measurements. Implants were made with a 200 keV implanter or a 2 MeV Van de Graaff, both machines being equipped with post acceleration magnetic mass separation. Samples were implanted at pressures of ∼10^{-7} Torr and hydro-carbon build-up was reduced to a minimum with cryogenic shielding. To conduct the channeling experiments we used a 1.5 MeV He beam from the 2 MeV Van de Graaff. Beam currents were normally 15 nA and the spot size was 0.8 mm. The beam divergence was less than 0.03° and backscattering measurements were made at 168°.

In Figure 1, we show a typical channeling spectrum for a dendritic web Si sample with a <111> face implanted at room temperature, with 75 kV and a fluence of 8.0 x 10^{17} H$^+$/cm^2. The large backscattering peak observed for the implanted section is roughly confined between channels 110 and 175. Note that the backscattering yield from the surface of the implanted region varies very little from the non-implanted region. This is consistent with our cross-section TEM results, previously reported [2] for FZ Si <111> at room temperature, where electron diffraction patterns from the surface zone indicated good quality crystalline material.

Step-height measurements were made with a Dektak II. Normally, the step was measured at several places along the implanted and non-implanted interface and averaged. We did not attempt to obtain step-heights less than 10 nm in height since reproducibility becomes very poor at smaller steps.

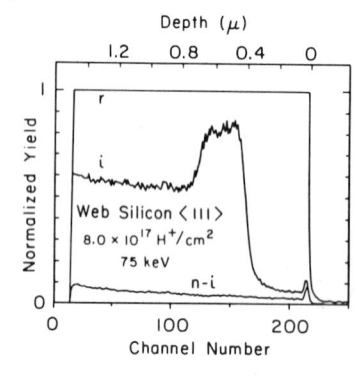

Fig. 1. Typical channeling spectrum for a dendritic web silicon <111> sample (∼30 appm Oxygen) implanted with 8 x 10^{17} H$^+$/cm^2 at 75 keV

RESULTS AND DISCUSSION

As the fluence is increased in an implantation of a Si sample the channels of the Si crystal become increasingly "littered" with displaced Si atoms and implanted H atoms. This disorder is expected to be at a maximum in a region centered either about the projected range R_p or the maximum, x_m, of the calculated damage distribution $S_D(x)$. We calculated $S_D(x)$ (the energy loss/μm to atomic displacements) using a modified EDEP-1 code [5] and hydrogen stopping powers given by Anderson and Ziegler [6]. As a consequence of the kinematics of the backscattering

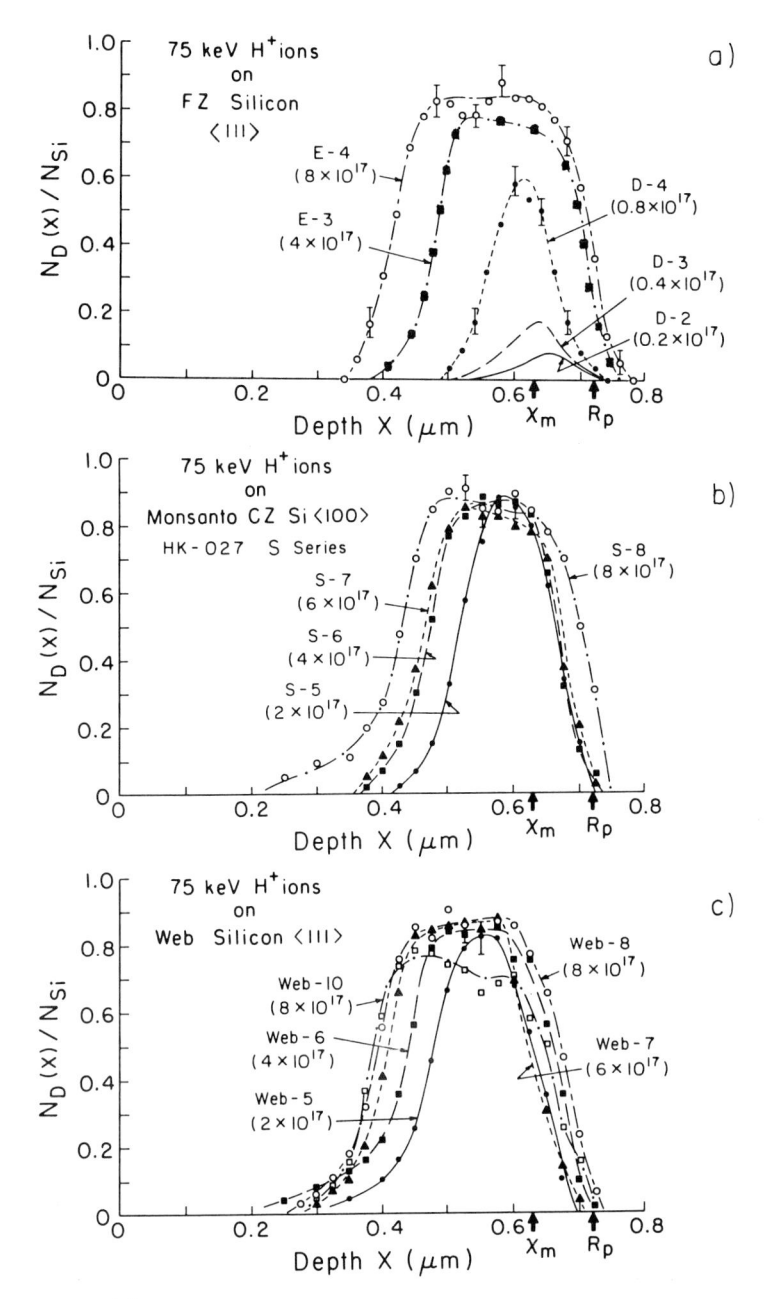

Fig. 2a. $N_D(x)/N_{Si}$ for FZ Silicon <111> (<1 appm Oxygen)
2b. $N_D(x)/N_{Si}$ for CZ Silicon <100> (37 appm Oxygen)
2c. $N_D(x)/N_{Si}$ for web Silicon <111> (~30 appm Oxygen)

process, the hydrogen atoms make no direct contribution to the backscattered He ions used for the channeling measurements. Hence, the observed backscattering is determined entirely by the distribution of displaced Si atoms. We denote by $\chi_R(x)$ that fraction of the He channeling beam which has been dechanneled when the sample is positioned in an aligned, low index direction. The backscattered yield from such an aligned crystal divided by the backscattered yield from the same crystal in a randomly oriented position is denoted by $\chi(x)$. Following Rimini [7] we can determine N_D, the number of "equivalent displaced atoms" per unit volume from

$$\frac{N_D(x)}{N_{Si}} = \frac{\chi(x)-\chi_R(x)}{1-\chi_R(x)}, \qquad (1)$$

$N_D(x)$ represents the equivalent density of Si atoms (as seen by the channeled He ions) required to produce the observed direct backscattering. N_{Si} is the density of Si atoms.

While $\chi(x)$ is directly measured, the determination of $\chi_R(x)$ normally involves a numerical procedure [7]. In the present work, however, the direct backscattering peak is so prominent that use of a graphically interpolated $\chi_R(x)$ or a single scattering model is adequate.

Figures 2a, b, c show the resulting $N_D(x)$ distributions which we obtained in this manner using various fluences of 75 keV H^+ ions on FZ, CZ, and web Si.

The integral of $N_D(x)$ will give the number of equivalent displaced atoms projected per unit area of the sample surface; this projected number is called a_D. Figure 3a shows how a_D is calculated from the usual channeled backscattering spectrum. Figures 3b, c, d show the variation of a_D with Ψ_H (the hydrogen fluence) for the 75 keV hydrogen implants of FZ, CZ, and web material.

Figures 4a, b, c give the results of the step-height measurements for the 75 keV implants as a function of hydrogen fluence.

Our data indicate that the step-height (h), the distribution of displaced atoms ($N_D(x)$) and the total number of displaced atoms/cm^2 (a_D) are all essentially independent of the oxygen content of the Si crystal. This lack of dependence may reflect the limited mobility of the oxygen or the defects at room temperature. At higher implant temperatures, one might expect the role of O on h, N_D and a_D to be increased.

In comparing h versus Ψ with a_D versus Ψ at room temperature, one is struck by the saturation in a_D which begins around 4×10^{17} H/cm^2. This behavior is not reflected in the step height data which shows a linear dependence on Ψ up to much larger fluences. In fact, no departure from linearity is seen for $h(\Psi)$ below the fluence where surface failure begins (2×10^{18} H$^+$/cm^2). This extended linearity implies that each H ion produces the same volume change in the implant region whether at low or high fluence. This volume change may be either the result of displacement damage, or inserting a H atom or a combination of both. This volume change per implanted ion does not change when the displacement production seems to saturate; this probably indicates that the H atom insertion is the dominant cause of the volume change.

As discussed previously [2], our evidence suggests that about 1/2 of the implanted H is trapped in the implant region, the other half pre-

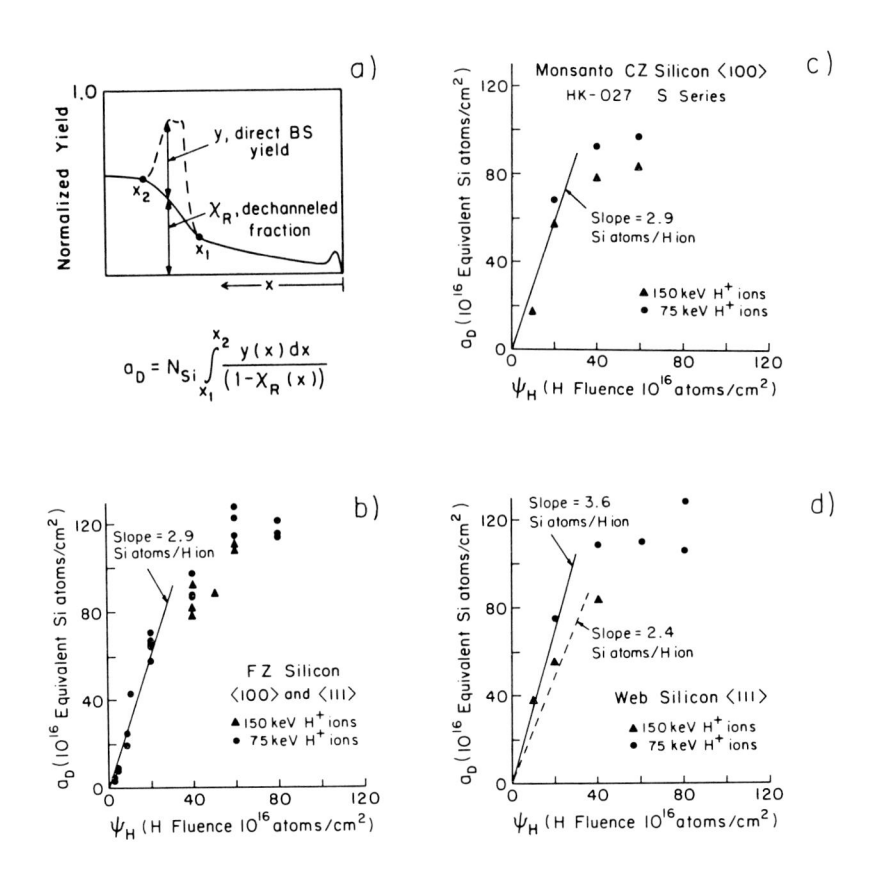

Fig. 3a. Calculation of a_D from $N_D(x)/N_{Si}$
3b. a_D versus ψ_H for FZ Silicon (<1 appm Oxygen)
3c. a_D versus ψ_H for CZ Silicon (37 appm Oxygen)
3d. a_D versus ψ_H for web Silicon (~30 appm Oxygen)

sumably diffusing into the base material. The observed volume change, based on the $h(\psi)$ data, is 24 Å³ per H ion implanted. If only 1/2 of the implanted ions are effective in producing $h(\psi)$, then the volume per trapped hydrogen atom will have to be substantially larger.

The different oxygen levels in the FZ and web Si may be linked to the observed enhancement or blister formation at high hydrogen fluences in the web & CZ materials. Surface profilometry measurements have shown a definite shift in preferred surface failure mode for the three materials. For comparable implant conditions, the FZ material shows a marked preponderance of craters while blisters are the preferred failure mode at 300 K for the web & CZ Si. Annealing experiments are in progress to determine the effects of oxygen denuded zone formation and nucleation and growth of oxide precipitates on mechanical response.

Fig. 4a. Step height versus Ψ_H for FZ Silicon (<1 appm Oxygen)

4b. Step height versus Ψ_H for CZ Silicon (37 appm Oxygen)

4c. Step height versus Ψ_H for web Silicon (~30 appm Oxygen)

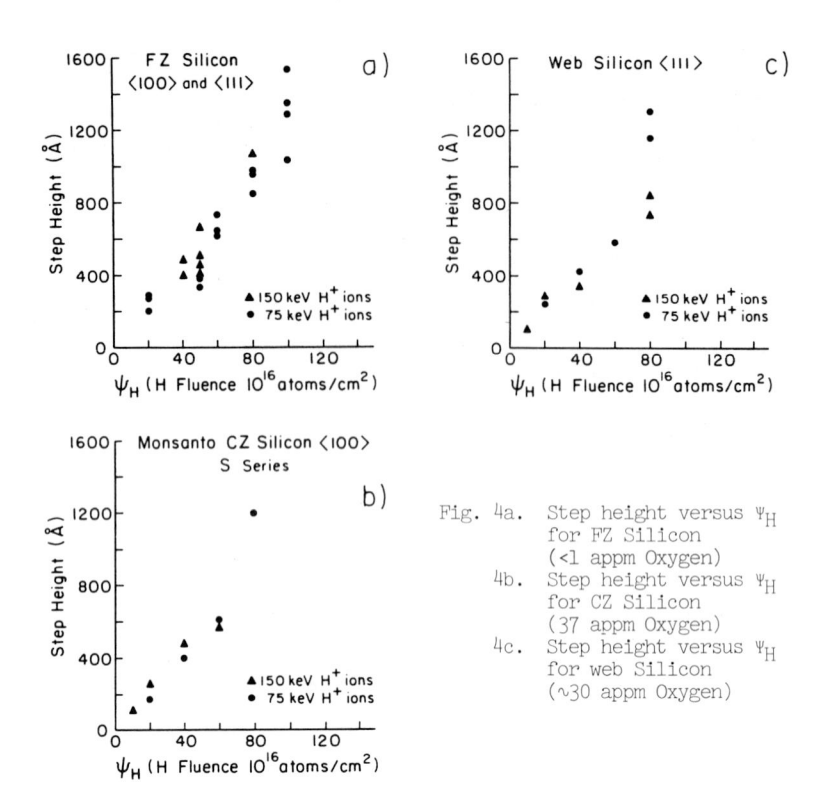

ACKNOWLEDGEMENTS

It is a pleasure to thank J. Moody of Monsanto for the CZ Silicon samples and R.G. Seidensticker of the Westinghouse R&D Center for the dendritic web Si samples.

REFERENCES

1. J.R. Patel, in "Semi-conductor Silicon 1981" edited by H.R. Huff, R.J. Kriegler, and Yoshiyuki Takeishi, Proceedings Volume 81-5 (The Electrochemical Society, Inc.) 189 (1981).
2. W.J. Choyke, R.B. Irwin, J.N. McGruer, J.R. Townsend, N.J. Doyle, B.O. Hall, J.A. Spitznagel and S. Wood, Nucl. Inst. and Meth. 209/210, 407 (1983).
3. K. Graff, E. Grallath, S. Ades, G. Goldbach, and G. Tölg, Solid State Electronics 16, 887 (1973).
4. R.G. Seidensticker, in "Crystals Vol. 8", J. Grabmaier, Editor (Springer-Verlag, New York) 145 (1982).
5. I. Manning and G.P. Mueller, Comput. Phys. Commun. 7, 85 (1974).
6. H.H. Anderson and J.F. Ziegler, "Hydrogen Stopping Powers and Ranges in all Elements", Vol. 3 (Pergamon Press, New York) (1977).
7. E. Rimini, in "Materials Characterization Using Ion Beams", Editors, J.P. Thomas, A. Cachard (Plenum Press, London-New York) (1978).

CHARACTERIZATION OF ION IMPLANTATION DAMAGE IN CAPLESS ANNEALED GaAs

H. KANBER,* M. FENG,* and J. M. WHELAN**
*Torrance Research Center, Hughes Aircraft Company, Torrance, CA 90509;
**Materials Science Dept., University of Southern California, Los Angeles, CA 90089

ABSTRACT

Arsenic and argon implantation damage is characterized by Rutherford backscattering in GaAs undoped VPE buffer layers grown on Cr-O doped semi-insulating substrates and capless annealed in a H_2-As_4 atmosphere provided by AsH_3. The damage detected in the RBS channeled spectra varies as a function of the ion mass, the implant depth and the annealing temperature of the stress-free controlled atmosphere technique. This damage is discussed in terms of the stoichiometric disturbances introduced by the implantation process. The as-implanted and annealed damage characteristics of the Ar and As implants are correlated to the electrical activation characteristics of Si and Se implants in GaAs, respectively.

INTRODUCTION

Ion implantation causes collision damage in compound semiconductor crystal lattices which can become sufficiently extensive to reach the amorphous level depending upon the fluence and energy of the incident ions. For most applications in GaAs discrete devices and integrated circuits, ion implantation followed by an annealing step is necessary to permit the reordering of the near surface damage and to electrically activate the implanted channel layer. Ion beam backscattering and ion channeling can be used to study induced damage in ion implanted samples. The study of disorder by the ion channeling technique [1] is based on the principle that channeled ions detect displacements from regular lattice sites. Thus, ion channeling is sensitive to interstitial host and impurity atoms or changes in the host periodicity, e.g., stacking faults, but is insensitive to isolated vacancies. One of the capabilities of ion beam backscattering is that the depth scale can be established by the analysis of the energy of the backscattered ions. The channeling spectrum from the implanted sample can be converted to relative damage with respect to the random yield versus depth distribution of the damage.

This paper compares the damage detected by Rutherford backscattering spectrometry (RBS) of channeled He^+ ions resulting from ion implantation and annealing of semi-insulating (SI) GaAs. The upper temperature used, 850°C, is in the range required for substantial electrical activation of donor implants. The samples were implanted with Ar^+ and As^+ ions with projected ranges (R_p) of 850 and 1350 Å to illustrate the effects of damage caused by different ion masses and energies. The dose of 5×10^{14} cm^{-2} was used to avoid full amorphization at any depth. Results are interpreted in terms of the stoichiometric disturbances resulting from ion implantation of compound semiconductors as proposed by Christel and Gibbons [2] and used by Magee et al [3]. According to their model, implantation creates a region at a depth exceeding R_p, where the recoiled Ga and As atoms produce excess Ga or As distributions in the form of interstitials. Thus, the near surface region has a net concentration of Ga and As vacancies and the net host atom excess concentration maximum is between the projected range (R_p) and $R_p + \Delta R_p$.

EXPERIMENTAL

All implants were made into undoped epitaxial layers with thicknesses of 3.8 μm grown on semi-insulating Cr-O doped substrates with dislocation densities under 5000 cm^{-2}. Impurity concentrations in the buffer layers were ≤5x10^{15} cm^{-3}. The (100) surfaces were tilted 7° relative to the Ar$^+$ and As$^+$ beams to minimize channeling. Substrates were held at room temperature during implantation in an end station which did not have cooling provisions. The ^{40}Ar$^+$ ion energies were 130, 200 and 400 keV corresponding to R_p's of 850, 1330 and 2730 Å, respectively, whereas the ^{75}As$^+$ beam energies were 240, 380 and 750 keV with corresponding R_p's of 850, 1350 and 2720 Å. In all cases the dose was 5x10^{14} cm^{-2} so as to avoid amorphization at any depth and to provide a range of damage detectable by RBS. The annealing of the implanted wafers was done using the controlled atmosphere technique (CAT) [4,5] without encapsulants. Surface integrity of the GaAs wafers is maintained by using a high purity H$_2$ carrier stream containing 0.021 atmospheres of As$_4$ which is introduced as AsH$_3$ in the annealing apparatus. All anneals were for 30 minutes at temperatures of 500, 700 and 850°C. The upper temperature was selected as corresponding to the optimum for electrical activation [6] of shallow ^{28}Si implants and less than the temperature at which impurity migration from the substrate becomes significant [7,8]. The 850°C anneal is not high enough for the full electrical activation of ^{80}Se implants with a mass similar to ^{75}As.

The implanted and annealed samples were analyzed [8] by backscattering spectrometry of 2.275 MeV ^{4}He ions channeled in the <100> direction and were detected at scattering angles of 169° and 99°, the latter for improved depth resolution. The backscattered counts from the random yield were normalized to 100% and the following data are presented as percent of the random yield which is a direct measure of the amount of damage in these crystals. The energy spectrum was converted to a depth scale through the use of standard RBS calculations involving tabulated physical constants.

RESULTS

Fig. 1 shows the percent amorphization versus depth for the 240 keV As$^+$ implant with R_p of 850 Å before and after capless anneals at 500, 700 and 850°C. The shallow near surface peaks are artifacts of RBS caused by surface scattering. It is interesting to note that a dose of 5x10^{14} cm^{-2} was not enough to fully amorphize the implanted layer, implying that mixed ordered (crystalline) and disordered regions exist within this layer. Significant points are the large amount of damage with a broad peak at a depth of 550 Å in the unannealed sample with no shift at the depth of the peak position after the 500°C anneal but a reduction in the amount of remaining damage. The 700 and 850°C anneals caused the peak to shift to 750 Å and closer to R_p with further reductions in residual damage. It is noteworthy that damage remains even after the 850°C anneal. These results are in marked contrast to those for the 380 keV As$^+$ implant shown in Fig. 2 where R_p is 1350 Å. The extent of damage was less with a poorly defined peak at ≃1000 Å and a long tail in the as-implanted sample. All anneals at 500, 700 and 850°C effectively removed the RBS detectable damage. No measurable RBS damage was observed for the 750 keV As$^+$ implant.

Fig. 3 illustrates the consequences of reducing the implanted ion mass (130 keV Ar$^+$), but maintaining the same R_p, 850 Å, of the As implant in Fig. 1. The peak damage region was closer to the surface and lesser in extent. Upon the 500°C anneal the damage peak depth shifted from 450 to 550 Å. Both the 700 and 850°C anneals effectively removed the RBS detectable damage. Increasing the Ar$^+$ ion energy to 200 keV (R_p = 1330 Å) introduced less detectable damage which essentially is removed by all three anneals as shown in Fig. 4. No RBS damage was observed for the 400 keV Ar$^+$ implant.

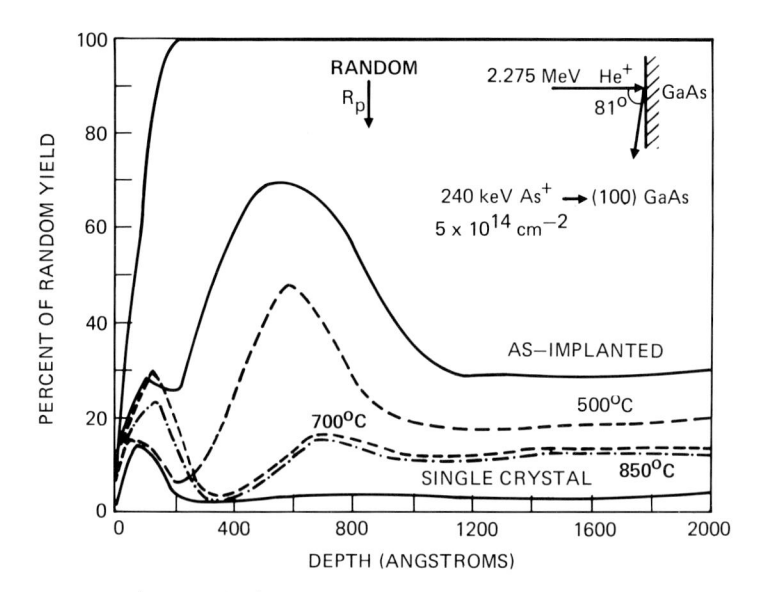

FIG. 1 High resolution backscattering spectra for 240 keV As+ implanted (100) GaAs with a dose of 5×10^{14} cm^{-2} illustrating the annealing out of damage at 500, 700 and 850°C using capless H$_2$/AsH$_3$ overpressure for 30 min. The single crystal and the random spectra are also shown.

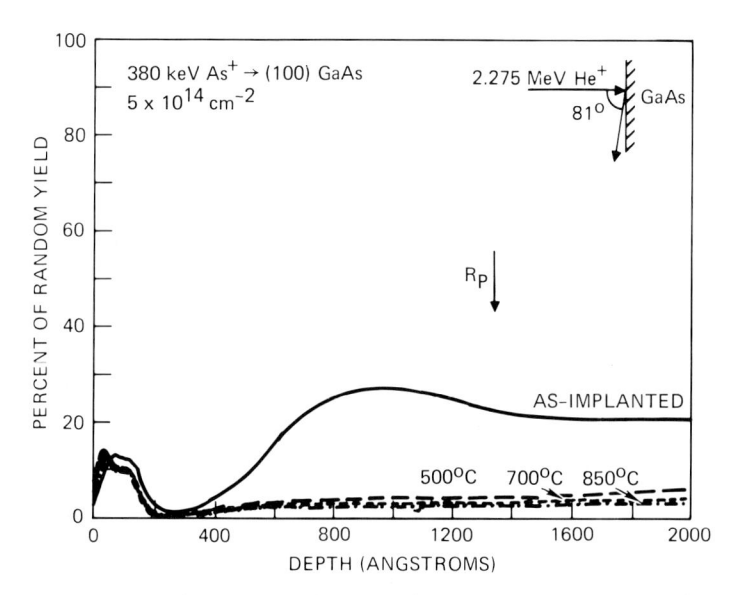

FIG. 2 Channeling spectra illustrating the damage and annealing behavior (30 min. flowing H$_2$/AsH$_3$ gas) for 380 keV As+-implanted (100) GaAs with a dose of 5×10^{14} cm^{-2} and capless annealed at 500, 700 and 850°C.

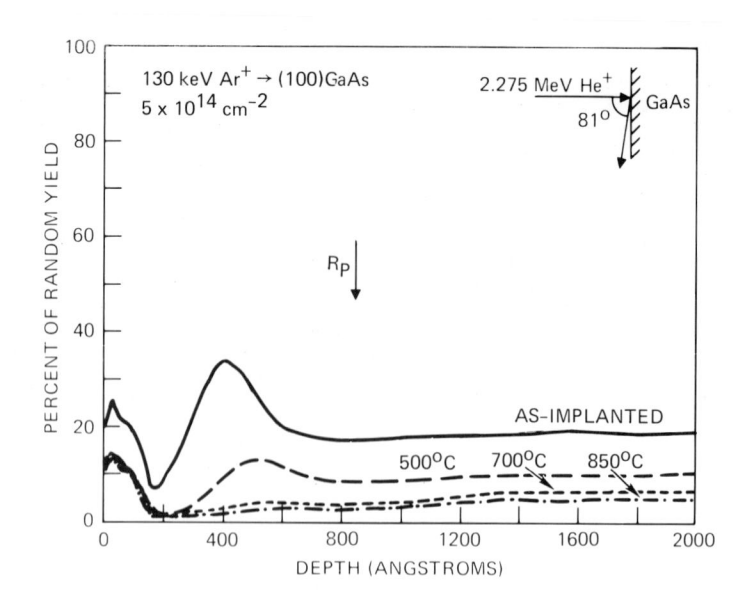

FIG. 3 Channeling spectra for 130 keV Ar+ implanted (100) GaAs with a dose of 5x10^{14} cm^{-2} and capless annealed for 30 min. at 500, 700 and 850°C.

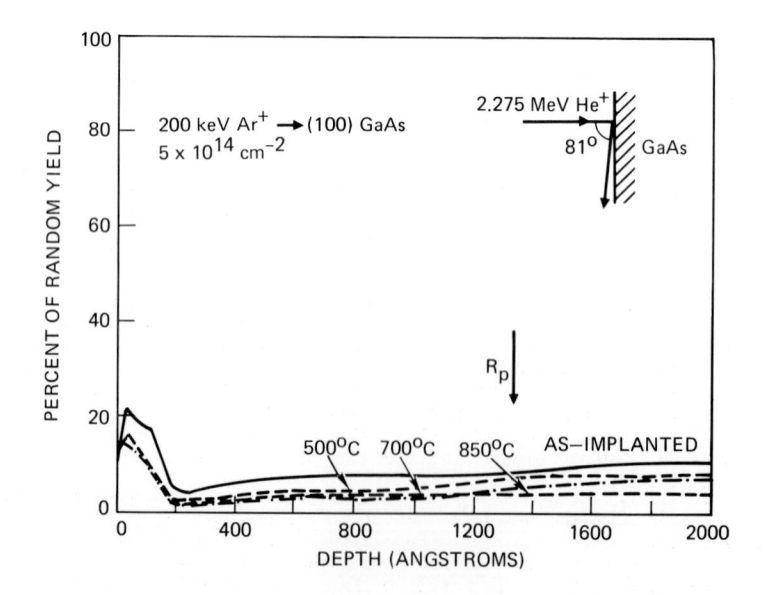

FIG. 4 High resolution backscattering spectra for 200 keV Ar+ implanted (100) GaAs with a dose of 5x10^{14} cm^{-2} illustrating the annealing behavior at 500, 700 and 850°C for 30 min. in H$_2$/AsH$_3$ overpressure.

DISCUSSION

Christel and Gibbons [2] have calculated relative concentrations of net excess host vacancy concentrations and net excess host atom concentrations as functions of depth resulting from ion implantation into compound semiconductors. They showed that the net excess vacancy concentration is highest at the surface and monotonically decreases as the depth, R_p, is approached; whereas the net excess atom density has its maximum value between R_p and $R_p + \Delta R_p$. For given doses and values of R_p, the concentrations of both the net vacancies and net excess atoms increase nonlinearly with the mass of the implant for identical doses and values of R_p. With Magee [3], Christel and Gibbons also calculated that for a given implant mass and dose, the net vacancy concentrations and net excess atom concentrations at depths measured in fractions of R_p increase as the implant energy (and R_p) increases. This model agrees with the earlier observations by Sadana et al [10] that for moderately high dose hot implants two bands of dislocations formed upon annealing. A very shallow band at the surface consisted of poorly defined dislocation loops and a well defined band of extrinsic dislocation loops was located between R_p and $R_p + \Delta R_p$. We also found this model to be consistent with the accumulation of rapidly diffusing Mn, which then occupies the vacancies that peak upon annealing near the surface without precipitation and the accumulation by precipitation of Cr at the extrinsic loop region [8,11].

The above damage model also accommodates the present RBS detectable damage if one makes several plausible assumptions. These are: (1) interstitial atoms are the primary defects detected by RBS at the relatively low implant dose used herein; and (2) these interstitials anneal most readily at a given depth where the net vacancy concentration is high; (3) our room temperature implants are not artificially cooled to lower the wafer temperature and some annealing occurs during the implant itself; and (4) less readily detectable damage consisting of poorly defined dislocation loops is more difficult to remove than interstitials and less difficult to remove than well-developed dislocation loops. Returning to Fig. 1, the primary damage peak at 550 Å for the as-implanted sample decreases after annealing at 500°C without shifting and at higher anneal temperatures shifts deeper without complete recovery because the initial net vacancy concentration is relatively low compared to the more energetic As^+ implant shown in Fig. 2. Much more complete RBS detectable damage recovery was achieved by all anneals shown in Fig. 2 because of the higher initial net vacancy concentrations generated by the ion implantation process. Less than complete recovery can be attributed to the annealing characteristics of complex extended defects, i.e., dislocation networks and/or stacking faults. Coalescence of the point defects into more complex, extended ones (i.e., interstitial loops) will make it more difficult for these to anneal out as readily and could be the reason the anneals are incomplete for the lower energy implants. Lesser RBS detectable damage associated with the lighter Ar ion is common to all theories as well as the above views. However, the same trend is seen with both the shallow As implant (Fig. 1) and the Ar implant in Fig. 3. Since the damage is less for Ar, the peak shifts even for the 500°C anneal and the channeled RBS spectrum is reduced to the virgin crystal spectrum level by the 700 and 850°C anneals.

The above results have a bearing on lighter device type implant doses of 1 to 10×10^{12} cm^{-2} for selecting a preferred donor implant in GaAs, i.e., Si or Se. We conclude that the small residual damage associated with the shallow Ar^+ anneal compared to the As^+ one is one reason why 850°C anneals are satisfactory for Si implants and inadequate for Se implants in GaAs.

ACKNOWLEDGEMENT

The authors would like to thank R. C. Rush and W. B. Henderson for their technical help, and Michael Strathman of Charles Evans and Associates for the RBS data. We also wish to thank Dr. T. A. Midford for his comments and support.

REFERENCES

1. D.S. Gemmell, Rev. Mod. Phys. 46, 129 (1974).
2. L.A. Christel and J.F. Gibbons, J. Appl. Phys. 52, 5050 (1981).
3. T.J. Magee, H. Kawayoshi, R.D. Ormond, L.A. Christel, J.F. Gibbons, C.G. Hopkins, C.A. Evans Jr. and D.S. Day, Appl. Phys. Lett. 39, 906 (1981).
4. R.M. Malbon, D.H. Lee and J.M. Whelan, J. Electrochem. Soc. 123, 1413 (1976).
5. H.B. Kim, J.M. Whelan, V.K. Eu and W.B. Henderson in: Proceedings of the Seventh Biennial Cornell Electrical Engineering conference (Cornell University, Ithaca, NY, 1979) p. 121.
6. H. Kanber, M. Feng, V.K. Eu, R.C. Rush and W.B. Henderson, J. Electron. Mater. 11, 1083 (1982).
7. M. Feng, V.K. Eu, H. Kanber and W.B. Henderson, J. Electron. Mater. 10, 973 (1981).
8. H. Kanber, M. Feng and J.M. Whelan, Appl. Phys. Lett. 40, 960 (1982).
9. Charles Evans and Associates, San Mateo, California.
10. D.K. Sadana, G.R. Booker, B.J. Sealy, K.G. Stephens and M.H. Badawi, Radiation Effects 49, 183 (1980).
11. V.K. Eu, M. Feng, W.B. Henderson, H.B. Kim and J.M. Whelan, Appl. Phys. Lett. 37, 473 (1980).

STOICHIOMETRIC DISTURBANCE IN InP MEASURED DURING ION IMPLANTATION PROCESS

D. HABERLAND, P. HARDE, H. NELKOWSKI
Technical University Berlin, D-1000 Berlin 12, West-Germany

and

W. SCHLAAK
Heinrich-Hertz-Institut, D-1000 Berlin 12, West-Germany

ABSTRACT

To measure the sputtered ions during implantation a specially designed UHV-target chamber with a SIMS apparatus was set up. Quantitative analysis are possible with an Auger spectrometer. Disturbances in the stoichiometry in InP are measured during implantation of Sn. The enrichment of the doped surface of InP with the lighter component phoshorus will be discussed in consideration of preferential sputtering and recoil effects during implantation. Measured depth profiles of Sn in InP will be compared with calculated distributions on condition that sputtering takes place. The sputtering yield of InP bombarded by 120 keV Sn^+ is 17 ± 5.

INTRODUCTION

With the growing interest of the electronic industry in III-V compound semiconductors the use of ion implantation to dope such materials has become quite important in the last few years. Indium phosphide and its ternary and quaternary compounds are used to manufacture electro-optical and electronic devices. As in silicon technology the advantages of ion implantation include the good control over carrier concentration, uniformity, and reproducibility. However, apart from radiation damage new problems arise in compound semiconductors. Preferential sputtering and recoil effects will lead to stoichiometric disturbances in the implanted layer (1). These effects could influence the electrical activation of dopants, or in case of amphoteric dopants, changes the probability of occupancy of the suitable lattice sites. Particularly for InP we expect these disturbances because of the large difference in masses of the components. In this paper we will report on the decomposition of the doped surface of InP during implantation and present calculations which include the sputtering of the surface during implantation and explain quite successful the depth profile of secondary ion mass spectroscopy measurements.

EXPERIMENTAL

The reported measurements were carried out in a specially designed UHV surface analysis chamber connected to an implantation plant (2). This system makes it possible to observe sputter effects during implantation with almost any element at energies up to 170 keV. Figure 1 shows a schematic drawing of the surface analysis chamber. It consists of a SIMS system with an ion source, a quadrupole mass analyzer and an AES system (CMA with an integral electron gun) for quantitative surface analysis. A differential pumping stack with built-in deflecting plates connects the implantation system to the UHV target chamber, so a pressure is maintained at 10^{-6}-10^{-7} Pa during implantation. The quadrupole is mounted 90° to the implantation beam and to the beam of the SIMS ion gun. Thus it was achieved that measurements

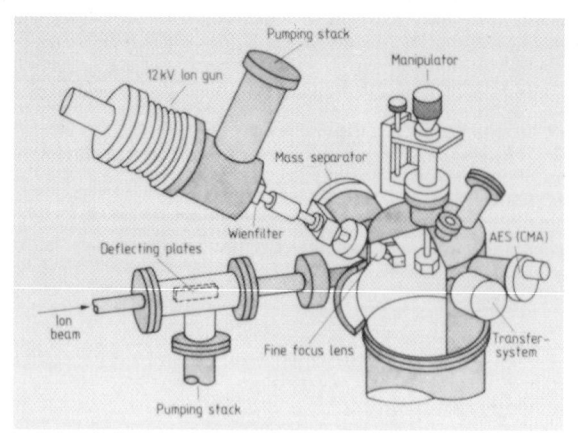

FIG. 1:
Schematic drawing of the UHV-chamber for surface analysis

of the flux of the sputtered ions during implantation are possible as well as the recording of depth profiles with the low energy beam of the SIMS ion gun. The manipulator for 8 samples allows rotations and any linear movement. To shorten the time for exchanging samples into the UHV chamber a vacuum lock transfer system is provided.

RESULTS AND DISCUSSION

The InP (100) samples were polished and have a size of 400 μm x 9 mm x 9 mm. They were cleaned and chemically etched with a mixture of $H_2SO_4/H_2O_2/H_2O$. The prepared crystals were bombarded with 80 to 120 keV Sn^+ ions at room temperature with doses up to 1×10^{16} ions/cm².

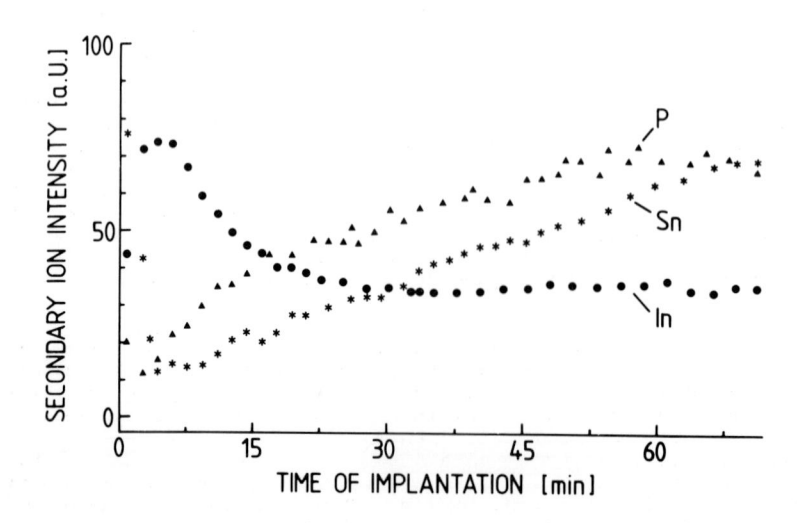

FIG.2: Secondary ion intensities for In^+ (●●●●), P^+ (▲▲▲), and Sn^+ (∗∗∗∗) during the implantation of 1×10^{16} Sn^+/cm².

Figure 2 shows exemplary the secondary ion intensities for In$^+$, P$^+$ and Sn$^+$ during the implantation of 1×10^{16} Sn$^+$/cm² ions. During the first three minutes of the implantation there are instabilities of the secondary ion yield of all detected elements which will not be considered in the following discussion. Then a continuous rise of the Sn$^+$ signal can be noticed. These Sn$^+$ secondary ions are not backscattered but are predominantly originated from the increasing doped InP surface. Simultaneously we observe a distinct decrease of the In$^+$ signal and an increase of the P$^+$ intensity. These characteristics indicate an enrichment of the lighter species phosphorus at the surface during implantation.

	In [at%]	P [at%]	Sn [at%]
before implantation	50	50	-
after implantation of 1x10¹⁶ Sn⁺/cm²	36	61	3

TABLE 1: Composition of InP before and after implantation of 1×10^{16} Sn$^+$/cm².

AES measurements of samples just before and right after the implantation allow a quantitative analysis of this process. In table 1 you see the concentration of In, P and Sn at the surface before and after implantation. Due to the cleaning of the InP samples with organic solvents the surface will always be contaminated with carbon, sulfur and chlorine. Therefore it is necessary to calculate the concentration with Chang's method (3), who considered the screening effect of other elements at the surface.

Because of the sputtering during the bombardment with Sn$^+$ the surface will be free of contamination so there is no need for this correction procedure. Both, the measurements of secondary ion intensities during implantation and the AES analysis before and after implantation show consistently an impoverishment of the heavier component indium in the surface of the substrat. This does not agree with the theoretical prediction of an enrichment of the heavier component as reported by Christel et al. (4). But their calculation only considers recoil implantation without any sputtering. However, calculations by Sigmund (5) show in additional consideration of sputtering and with a high sputter yield that you will get an inversion of the composition of the components at the surface i.e. an enrichment of the lighter component. Following his model and with our experimental data the inversion of concentration of indium to phosphorus should take place at about 40 min of the time of implantation. Because of the monotonous course of secondary ion intensity (Fig. 2) which is not in agreement with Sigmund's model we assume that preferential sputtering of indium causes the enrichment of the lighter component.

That we can measure a considerable intensity of secondary ions during implantation with Sn$^+$ ions and that we have an ion etching effect, you may see on the photograph of the scanning electron microscope (Fig. 3a). The measurement of the step between implanted and virgin area was realized mechanically with an α-step (Tencor Instruments) (Fig. 3b). Considering the thickness of the sputtered layer and the dose of 1×10^{16} Sn$^+$/cm² the sputter yield is 17 ± 5. This is in good agreement with the value calculated by the model of Sigmund (6). The depth d = 40 nm is of the same order as the projected range of Sn$^+$ ions (120 keV) R_p = 45 nm and the standard deviation ΔR_p = 14 nm. This results in a considerable change of the implantation profile.

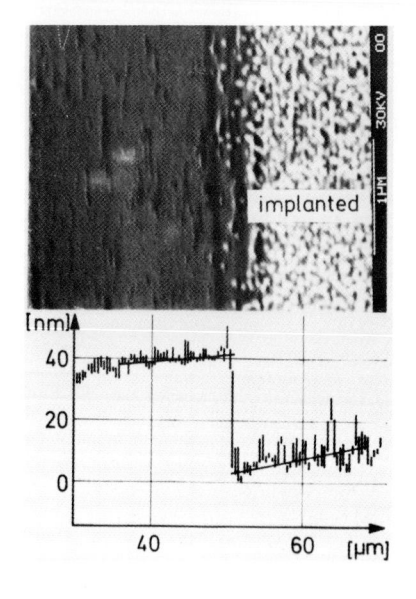

FIG. 3:
a) Scanning electron micrograph of the sputtered surface by bombardment of 10^{16} Sn$^+$/cm^2
b) Mechanically measured depth profile (depth d ≈ 40 nm).

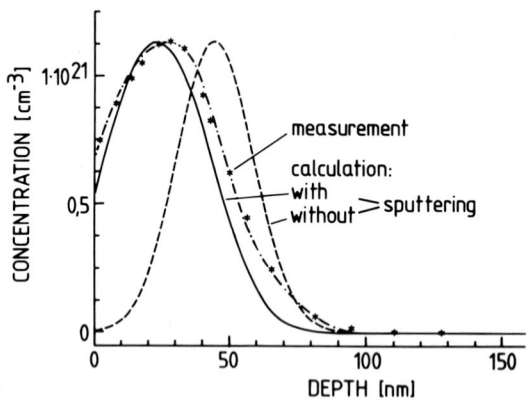

FIG 4:
Depth profiles of 120 keV Sn$^+$ into InP. Measured *--*, calculated with ——, and without sputtering ----.

Fig. 4 shows the experimental depth profile for Sn measured with SIMS and the calculated depth profiles without and with considering the sputter effect. The dashed line corresponds to a Gaussian distribution. The projected range and the standard deviation are calculated with the PRAL algorithm proposed by Biersack (7). Considering the sputter effect we assume that the total implantation N_D will be subdivided in n steps, so that in each step 1/n of the total dose of 1×10^{16} Sn$^+$/cm^2 will be implanted. Further on we will assume that with each step a layer of the thickness d/n will be removed (Fig. 5). The range distribution C(z) of the total implantation process (solid curve) will result in a summation of the n single steps.

$$C(z) = \sum_{i=0}^{n-1} \frac{N_D}{\sqrt{2\pi} \cdot n \cdot \Delta R_p} \exp\left(\frac{-(z-R_p+i\frac{d}{n})^2}{2\Delta R_p^2}\right)$$

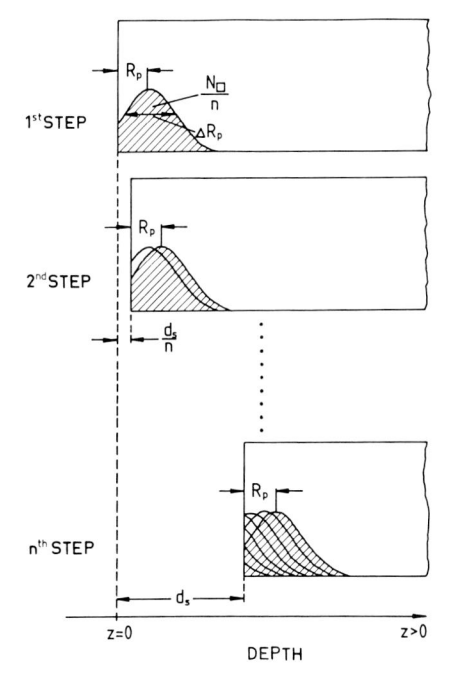

FIG. 5:
Theoretical model considering the sputtering effect during implantation. (R_p = projected range, ΔR_p = standard deviation, $N_\square$ = total dose, d_s = sputtered depth.

The depth profile measurements show that the sputtering during implantation with high doses of heavy ions has a non-neglectable influence on the range distribution of the dopant. We can see an increase in surface concentration and also a shift of the maximum of the profile to the surface.

CONCLUSION

The analysis of the processes during implantation of Sn into InP shows a stoichiometric disturbance of the substrate. The enrichment of phosphorus, which was measured with the secondary ion intensity during implantation and with AES, is interpreted as a preferential sputtering of the heavier component indium. Considering the ion etching during implantation the range distribution is calculated which results in good agreement with the measured depth profiles.

We are grateful to the Heinrich-Hertz-Institute for supplying the InP crystals which made the measurements possible.

REFERENCES

1. Tong He Zheng, R.G. Elliman, G. Carter: Nucl. Instr. Methods 209/210, 761-766 (1983)
2. D. Haberland, P. Harde, H. Nelkowski, W. Schlaak: (to be published).
3. C.C. Chang: Surface Science 48, 9-21 (1975).
4. L.A. Christel, J.F. Gibbons: J. Appl. Phys. 52 (8), 5050-5055 (1981).
5. P. Sigmund, J. Appl. Phys. 50 (11), 7261-7263 (1979).
6. P. Sigmund, Topics in Appl. Phys. 47, pp. 1-71 (Springer-Verlag 1981).
7. J.P. Biersack, Z. Phys. A 305, 95-101 (1982).

DAMAGE DISTRIBUTION STUDIES IN PROTON-IMPLANTED GaAs

H. A. JENKINSON, M. O'TOONI, J. M. ZAVADA, T. J. HAAR
U. S. Army Armaments Research and Development Center, Dover, NJ 07801

and

D. C. LARSON
Drexel University, Philadelphia, PA 19104

ABSTRACT

Samples of n^+-GaAs implanted with 300 keV protons have been examined using high resolution electron microscopy, capacitance-voltage profilometry, and infrared reflectance. In contrast to previously reported results, electron microscopic examination of the as-implanted samples revealed the presence of dislocation loops and/or precipitates both near the wafer surface and at the bottom of the implanted layer. These results are corroborated by electrical and optical measurements.

INTRODUCTION

Ion implantation is a technique which has been used to alter the refractive index of transparent dielectrics and semiconductors in order to form optical waveguides. The mechanisms by which the implanted ions achieve this effect are generally classified as either changes to the density or polarizability of the initial material, chemical doping, or carrier compensation [1]. Although all of these processes can occur simultaneously, usually one will predominate depending on the selection of target material and ion species. In highly doped semiconductors, the refractive index is depressed from that of the lattice by the free carrier plasma contribution to the dielectric function. The damage created in such a material during ion implantation can be used to trap the free carriers, allowing the refractive index in the damaged layer to rise towards the value it would attain in intrinsic material.

In an effort to understand the mechanisms involved in modifying the refractive index of n-GaAs through proton implantation, irradiated samples have been examined using high resolution electron microscopy, capacitance-voltage profilometry, and infrared reflectance. Although the point defects believed to be responsible for carrier compensation are too small to be resolved by electron microscopy, examination of the samples using this technique was performed to determine the degree of crystallinity of the implanted layer and to look for defect agglomerations, such as loops and precipitates, which might indicate that excess damage was induced. The actual carrier compensation achieved was determined electrically by profiling the carrier concentration within the implanted layer using standard capacitance-voltage profiling techniques. Differential infrared reflectance measurements were performed to assess the effects of the implantation on the optical properties of the samples.

SAMPLE PREPARATION

The GaAs wafers used in these experiments were obtained from Laser Diode Laboratories. They were doped with Si to produce n-type material with a carrier concentration of $3-4\times 10^{18}/cm^3$ and a relatively high mobility of 1700 cm^2/V-sec. The wafers were oriented to expose (100) surfaces, which were optically polished by the manufacturer. The implantations, done at the Naval Research Laboratory, were performed at room temperature with 300 keV protons. During implantation, the beam current was held at 10^{-6} amps/cm^2 to minimize substrate heating and the sample was oriented at an 8° angle with resepect to the beam to minimize the effects of ion channeling. A dose of $10^{15}/cm^2$ was achieved. From the LSS projected range calculations [2], it was expected that the proton distribution would be nearly Gaussian with a projected range of 2.4 μm and a standard deviation of 0.26 μm. It was expected that the damage distribution would be similar.

ELECTRON MICROSCOPIC EXAMINATION

Prior to electron microscopic examination, the samples were mechanically polished from the back to reduce the thickness to about 25 μm while leaving the implanted region intact. The samples were then further thinned by the ion-milling technique to produce electron-transparent specimens. To accurately determine the observed regions within the samples, intermittant ion millings from the top and bottom surfaces were performed. The locations of the sections finally examined were calculated from the time and rate of the ion-milling erosion.

High resolution electron micrographs from two separate depths from the surface are shown in Figures 1 and 2. The image in Figure 1 was obtained

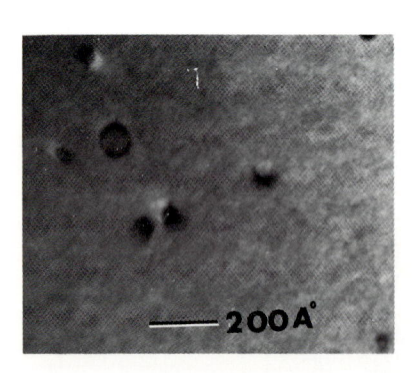

FIG. 1. Electron micrograph of damage at 1.5 μm depth.

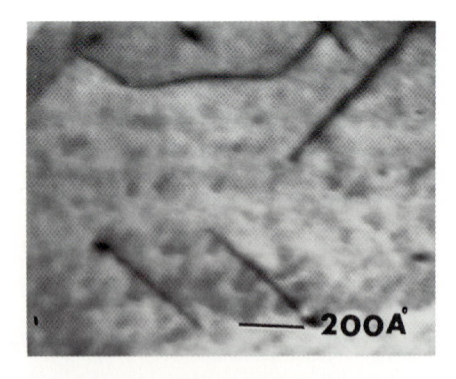

FIG. 2. Electron micrograph of damage at 3 μm depth.

at a depth of approximately 1.5 μm below the original surface. It is characterized by a high density distribution, 5×10^{15} def/cm^2, of randomly distributed dislocation loops and spherical precipitates. The image of Figure 2 was obtained at a depth of approximately 3 μm from the surface. This region is characterized by precipitates and dislocations, single or tangled, with a density of about 4×10^{13} def/cm^2. The corresponding (100) and (110) electron diffraction patterns are given in Figures 3 and 4.

FIG. 3. (100) Electron diffraction pattern at 1.5 μm depth

FIG. 4. (110) Electron diffraction pattern at 3 μm depth.

CAPACITANCE-VOLTAGE PROFILING

Schottky barriers formed on semiconductors behave in many respects like p^+-n junctions in which the metal electrode acts as the p^+ region. It can be shown that such a junction has a capacitance and depletion width which depend on the carrier concentration of the semiconductor [3]. By applying a reverse bias to the junction, the depletion width is increased and the capacitance is decreased. Measuring the rate at which the capacitance decreases as the bias is increased in magnitude allows the carrier concentration to be parametrically profiled according to the equations:

$$N(V) = (2/K\varepsilon_0 eA)[\, d(1/C^2)/dV \,]^{-1}$$

$$t = K\varepsilon_0 A/C(V)$$

In these equations, N is the carrier concentration, K is the low frequency dielectric constant, ε_0 is the vacuum permittivity, e is the electronic charge, A is the area of the Schottky barrier, V is the bias, C is the capacitance, and t is the depletion width.

Schottky barrier diodes were formed on the implanted surfaces of these samples by electron beam evaporation of Au through a stainless steel mask to produce dots 0.8 mm in diameter and about 2000 Å in thickness. Ohmic contact was made to the back surface of the samples by depositing an Ag layer followed by a thin coating of Au to prevent sulfiding.

The capacitance-voltage measurements were made using phase-sensitive detection techniques to measure the reactive current through the sample when a 10 mv, 20 kHz modulation was superimposed on the dc bias. As these samples were heavily doped, the conductance component of the admittance became much larger than the susceptance as the sample was reverse biased, limiting the depth which could be accurately probed to about 0.5 μm. In order to reach an area more affected by the implantation, a layer of about this thickness was removed from the surface of the sample using an argon ion sputtering system. The sample was thermally bonded to a water cooled target to keep heating to a minimum. New Schottky barriers were then applied and the C-V measurements repeated. The results of these measurements are shown in Figure 5, which presents the composite carrier concentration profile

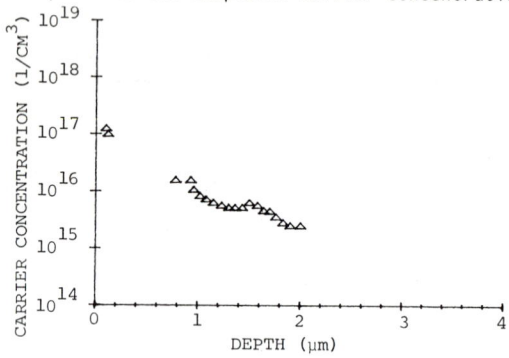

FIG. 5. Carrier concentration profile of as-implanted sample (1x10^{15} H$^+$/cm^2 @ 300 keV, RT)

for this sample. At the surface of the sample the implantation produced enough damage to trap over 90% of the free carriers while further into the material, over 99% were trapped. The carrier concentration was still decreasing when the limit of the second series of measurements was reached. Presumably this curve would dip to a minimum near the proton projected range and then rise sharply to the initial doping level of the material.

INFRARED REFLECTANCE

The optical properties of the implanted sample were characterized directly from differential reflectance measurements made over the range of 800 cm^{-1} to 4000 cm^{-1}. As shown in Figure 6, this procedure yielded interference fringes typical of a thin film/substrate structure. The interesting

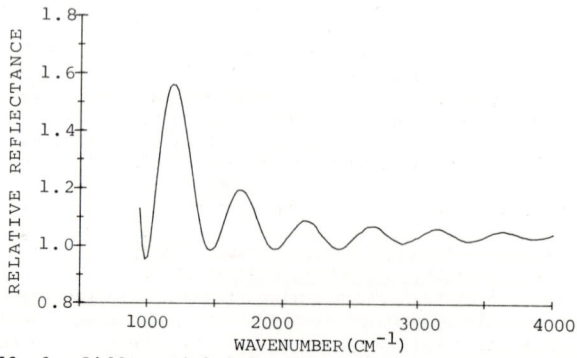

FIG. 6. Differential infrared reflectance spectrum of as-implanted sample (1x10^{15} H$^+$/cm^2 @ 300 keV, RT)

features of this curve are the periodicity of the fringes and their increase in amplitude as the measurement progressed to smaller wavenumbers. The basic features of this curve can be explained with the aid of an approximation to the reflectance formulae for such a slab dielectric structure. In this model [4] losses are neglected and the implanted layer is considered to be a film of refractive index n_f sitting on a substrate whose index has been depressed a small amount Δn by its higher concentration of free carriers. The reflectance reference sample is considered identical to the substrate. Expanding the relevant reflectance equations in a McLaurin series to first order in Δn yields

$$R_D(\sigma) = \frac{R_{sample}}{R_{reference}} = 1 + \frac{4\sigma_p^2(1-N_f/N_s)}{n_f(n_f-1)}\left(\frac{\sin(2\pi t n_f \sigma)}{\sigma}\right)^2$$

In this equation R_{sample} and $R_{reference}$ are the power reflectances of the sample and reference respectively, R_D is their ratio, N_f and N_s are the carrier concentrations in the film and substrate respectively, and σ is the wavenumber. The plasma term, σ_p^2, is given by

$$\sigma_p^2 = \frac{N_s e^2}{(2\pi c)^2 m^* \varepsilon_o}$$

where m* is the effective mass of an electron in GaAs and c is the speed of light. From this analysis, it can be shown that the thickness of the implanted layer can be determined by the separation of the fringe minima and the carrier concentration in the film from the fringe amplitudes.

These measurements were performed on a Perkin-Elmer Model 180 double beam infrared spectrophotometer fitted with dual specular reflectance attachments. Characterization of this sample using the analysis described results in a thickness of 3.35 µm and a carrier concentration of 2.8×10^{17} carriers/cm^3.

DISCUSSION

The samples used in this study were implanted with a monoenergetic beam of protons. According to the energy loss models used in the LSS projected range calculations, these ions should interact with the GaAs lattice primarily when they have lost most of their energy through electronic interactions. Thus the damage is expected to occur with a distribution which approximates the distribution of the protons in the material. For protons implanted in GaAs at 300 keV, this distribution is expected to be nearly Gaussian with a mean projected range of 2.4 µm and a standard deviation of 0.26 µm. Previous studies of the damage induced in similarly irradiated GaAs have indeed found that the defect density does approximate the proton distribution, but only after the samples were annealed so that the defects were large enough to be resolved by the electron microscope [5].

The results of this study indicate that significant damage is induced in the material from the surface to well below the projected range of the protons. The electron micrographs show that large defect clusters were produced during the implantation both at a depth at which the primary energy loss mechanism is electronic (1.5 µm) and at a depth near the peak of the proton concentration (3 µm). No such evidence of defects was found in a piece of unimplanted material obtained from the same wafer. The corresponding electron diffraction patterns show that the material is still highly crystalline throughout this whole region. The density of the dislocations found near the damage peak is two orders of magnitude smaller than that of the loops and precipitates found closer to the surface. One could tentatively hypothesize that the energy required for the formation of these defects is higher than that for prismatic loops, a theory which agrees with the energy loss mechanisms of the implantation process.

The existence of defects nearer the surface and beyond the projected range is also indicated by the carrier concentration profile and the reflectance spectrum of the sample. The carrier concentration over the region profiled is very low and uniform in comparison to the implanted proton distribution. This indicates damage sufficient to trap carriers has occured even at the surface. The reflectance spectrum, however, provides the best indication of the optical uniformity of the implanted region. The uniformity of the periods and the gradual growth in amplitude are adequately explained by the model of a single layer dielectric structure sitting on an infinite substrate where the refractive index difference is due to the free carrier plasma effect. Although not discussed here, extensive modeling studies with multilayer structures have not significantly improved the interpretation of this spectrum. The spectrum indicates that optical changes, and hence carrier compensation and the damage producing it, extends about four standard deviations beyond the projected proton range.

In conclusion, the results reported here indicate that the mechanisms of the carrier compensation process may be more complicated than generally believed. The defects responsible for compensation need to be identified as well as a mechanism for their formation during the implantation process. The effect that the larger defects found here have on the refractive index and electrical properties of the implanted layer is also a subject for further investigation.

REFERENCES

1. G.L. Destefanis, B.W. Farmery, J.P. Gailliard, E.L. Ligeon, A. Perez, P.D. Townsend, and S. Valette, J. Appl. Phys. $\underline{50}$, 7898 (1979).
2. J.F. Gibbons, W.S. Johnson, and S.W. Mylroie, Projected Range Statistics, 2nd Edition (Dowden, Hutchinson, and Ross, Stroudsburg, PA 1975).
3. G. Dearnaley, J.H. Freeman, R.S. Nelson, and J. Stephen, Ion Implantation (North-Holland, Amsterdam, 1973).
4. H.A. Jenkinson and D.C. Larson, Proc. of NASA Conf. on Optical Information Processing for Aerospace Applications, NASA Conference Publication 2207, 231-240 (1981).
5. H.C. Snyman and J.H. Neethling, Rad. Eff. $\underline{60}$, 147-154 (1982).

PART IV

CERAMICS, POLYMERS AND GRAPHITE

ION BEAM MODIFICATION OF CERAMICS*

C. J. McHARGUE, C. W. WHITE, B. R. APPLETON, G. C. FARLOW, J. M. WILLIAMS
Oak Ridge National Laboratory, Oak Ridge, Tennessee 37831

ABSTRACT

Alterations to the structure and properties of ceramics are complex due to the range of bonding types encountered and the necessity for maintaining local charge balance. Ion damage can occur as a result of ionizing effects as well as displacement collisions. Ion species, implantation temperature, implantation energy, and the specific bonding characteristics of the host are important parameters in determining the structure and properties of implanted ceramics. Some of these effects will be illustrated for Al_2O_3 implanted with chromium or zirconium and silicon carbide implanted with chromium.

INTRODUCTION

Although ion implantation doping has had its greatest success in semiconductor technology, it has been utilized in recent years to alter the physical and chemical properties of metals [1,2], and the optical [3,4] and electrical [5,6] properties of insulators. Relatively little work has been reported on changes in the mechanical and chemical properties of ceramics as a result of ion implantation.

Implantation for metallurgical purposes requires implanted concentrations of a few to several atomic percent (fluences of 10^{16}–10^{17} ions·cm^{-2}). At such high fluences, effects such as sputtering and composition-dependent phase stability become important considerations. Similar concentrations are required to alter the surface mechanical properties of ceramics.

Implantation and radiation damage in ceramics is much more complex and less studied than in semiconductors or metals. In the displacement cascade, one must deal with at least two sublattices that may have different displacement energies. The types of defects that can be produced are strongly influenced by the requirements of local electrical charge neutrality, the local stoichiometry, and the nature of the chemical bonding of the particular lattice. Similarly, the local structure is strongly sensitive to the chemical and electrical nature of impurities or dopants. In addition, ionizing effects may introduce lattice defects, whereas, in metals such effects are generally unimportant.

This paper reports the changes in structure and mechanical properties which occur during implantation in two ceramic materials, α-Al_2O_3, and α-SiC. These materials have similar hexagonal crystal structures but markedly different bonding, varying from an intermediate ionic nature to highly directional covalent.

---

*Research sponsored by the Division of Materials Sciences, U.S. Department of Energy, under contract W-7405-eng-26 with Union Carbide Corporation.

EXPERIMENTAL PROCEDURE

Single crystals of Al_2O_3 (obtained from Crystal Systems, Inc., Salem, MA) of high purity (<100 ppm total) and low dislocation density (10^3–10^4 cm^{-2}) were cut to within ±2° of <0001> and <1$\bar{2}$10>, and polished and annealed at 1400°C in air for 120 h to produce damage-free samples. One half of each sample was retained as virgin reference material and the remaining one half was implanted with various cations at room temperature, liquid nitrogen temperature, or an elevated temperature with the ion beam incident at 7° or 3° from the crystal normal at fluences ranging from 10^{15} to 10^{17} cm^{-2} and energies from 100 to 300 keV.

Single crystal [0001] samples of α-SiC were obtained from the Carborundum Company as platelets produced in an Acheson furnace. These were implanted with nitrogen or chromium to fluences of 10^{14} to 10^{17} cm^{-2} at energies from 62 to 280 keV at room temperature.

The specimens were examined using Rutherford backscattering-ion channeling techniques (RBS-C) with 2.0 MeV $^4He^+$ to determine the depth profile of the implanted species, the depth distribution of damage in the host lattice, and the lattice location of the impurity. Transmission electron microscopy specimens (TEM), prepared by ion milling, with the plane of observation both parallel and perpendicular to the implantation beam were employed to determine the structural characteristics of the implanted zone. Surface profilometry gave data on the volume changes introduced by implantation. Raman spectroscopy gave important information regarding the structure of implanted SiC.

The hardness of the samples was determined by the Knoop microhardness technique. With a 15-g load (0.147 N), this procedure made indentations which were approximately 250 to 300 nm deep, a distance greater than the range of the implantations. The hardness values obtained represent a composite response of the implanted layer and the underlying unmodified lattice. The hardness values are reported only as relative values, expressed as the ratio of the hardness of the implanted area to that of an unimplanted region on the same crystal, indicating magnitudes of hardness changes rather than absolute hardness values.

An evaluation of the response of the surface to simulated mechanical abrasion was made by means of a scratch test [7]. In this test, a stylus was slowly translated across the surface under normal loads of 0.098 to 0.49 N while the tangential force on the stylus was continuously measured. A qualitative evaluation of the deformation characteristics was obtained from scanning electron microscopy (SEM) of the grooves.

RESULTS AND DISCUSSION

Aluminum Oxide

Figure 1 shows typical backscattering spectra of 2 MeV $^4He^+$ ions from Al_2O_3 crystal implanted with ^{52}Cr (300 keV) ions to doses of 10^{16} and $10^{17}/cm^2$. Due to the large mass difference between chromium, aluminum, and oxygen, the kinematics of the RBS process produce characteristic steps in the yield of particles scattered from the matrix elements (aluminum and oxygen) as well as an isolated peak for scattering from implanted chromium.

Comparing the <0001> aligned yields from the implanted and virgin regions at energies below 1.1 MeV shows that substantial disorder has been introduced into both the aluminum and oxygen sublattices as a result of implantation. It should be noted, however, that the near-surface region was

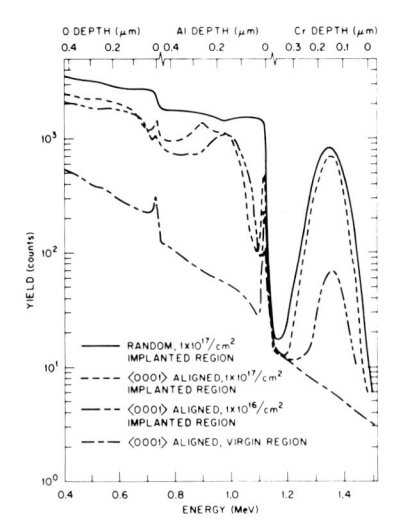

FIG. 1. Backscattering spectra of 2 MeV He$^+$ from ^{52}Cr (300 keV, 1×10^{16} and 1×10^{17}/cm^2) implanted α-Al$_2$O$_3$.

not turned amorphous by the implantation of chromium. For room temperature implantation, we have not observed a completely disordered surface region up to fluences of 1×10^{17}/cm^2 of chromium, nickel, and iron [8–12] or to fluences of 4×10^{16}/cm^2 of titanium, indium, gallium, copper, manganese, or niobium. However, as discussed below, the results for zirconium implantation were strikingly different.

The relative damage to the aluminum and oxygen sublattices reached saturation for implantation doses from 5×10^{15} to 1×10^{17} Cr·cm^{-2} with the oxygen sublattice exhibiting the greater amount of damage. Utilizing the ion-channeling geometry to detect preferential lattice locations of the implanted species, it was found that chromium atoms showed only a slight bias toward substitutionality in the samples, ranging from about 29% at 4×10^{16} to about 45% at 1×10^{17} Cr·cm^{-2}.

The Al$_2$O$_3$ specimen implanted to 2×10^{16} Cr·cm^{-2} was examined in the ORNL Hitachi 1000 microscope. The diffraction pattern showed the implanted region to be crystalline despite the large amount of damage. The TEM images contained a high density of "black spots," suggestive of point defect clusters. Attempts to determine the character of these clusters were unsuccessful due to large residual stresses in the thinned foils [12].

In order to explore the effects of implantation temperature upon the structure and properties of Al$_2$O$_3$, specimens were implanted with 4×10^{16} Cr·cm^{-2} at 77, 300, and about 640 K. The backscattering spectra for these conditions are shown in Fig. 2. The liquid nitrogen implant gave a scattering curve typical of an amorphous layer extending from the surface to a depth about 120 nm, well beyond the peak concentration of chromium. The elevated temperature implant sustained about the same amount of damage to the aluminum and oxygen sublattices at the depth of the peak concentration as at room temperature. However, there was considerable recovery in the immediate surface layers as indicated by the aluminum surface peak in Fig. 2. The degree of substitutionality of the chromium for the 640 K implantation was only slightly greater than the room temperature specimen (34 versus 29% for this fluence).

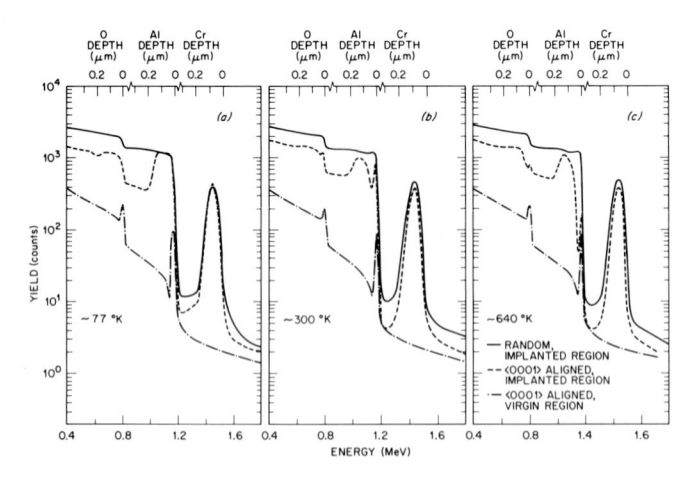

FIG. 2. Backscattering spectra from α-Al$_2$O$_3$ implanted with ^{52}Cr (280 keV, 4 × 10^{16}/cm^2) at 77, 300, and 640 K).

As also recently observed by Burnett and Page [15], room temperature implantation of zirconium caused more damage per unit fluence than did other species studied. We found an amorphous subsurface layer to be present after a fluence of 4 × 10^{16} Zr·cm^{-2}. Figure 3 shows the spectra for room temperature implantation of 2 × 10^{16} Zr·cm^{-2}. The channeling data gave no evidence of any substitutionality, indicating an interstitial solid solution to have formed. At the higher fluence (4 × 10^{16} ions·cm^{-2}), the spectra of Fig. 4 indicate a random layer in the <0001> aligned specimen that extended from about 40 to 80 nm from the surface. Transmission electron microscopy confirmed that a subsurface amorphous layer existed.

As might be expected from the large amounts of damage to the aluminum and oxygen sublattices and the formation of nonequilibrium interstitial solid solutions, the surface mechanical properties were markedly affected by the implantation. Figure 5 shows the relative hardness (implanted versus unimplanted) as a function of fluence for the chromium implantations. Because of the distribution of the implanted ions and defects and the depth of the 0.147 N (15 g) indentations the hardness values are for some ill-defined composite region. They show the direction of hardness changes but probably underestimate the true values. Thus, the room temperature implanted samples show hardness increases of at least 50% for a fluence of 1 × 10^{17} Cr·cm^{-2} (corresponding to a cation concentration of about 10% Cr). Conventionally prepared solid solutions of Cr$_2$O$_3$-Al$_2$O$_3$ exhibit hardness increases of approximately 10% for this concentration [16,17].

The hardness of the amorphous Al$_2$O$_3$ (4 × 10^{16} Cr·cm^{-2}, 77 K) was less than the crystalline specimens. The measured ratio shown in Fig. 5 is 0.85. It is believed that the true value is considerably less.

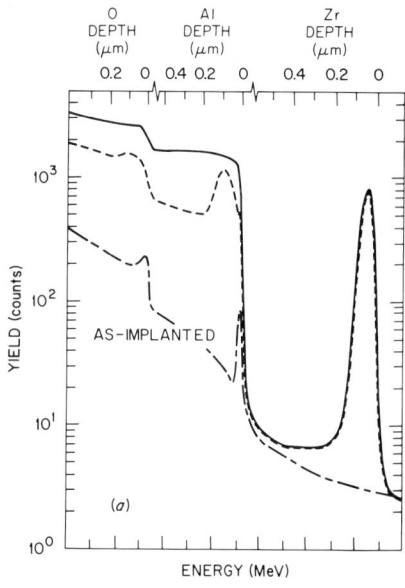

FIG. 3. Backscattering spectra from α-Al$_2$O$_3$ implanted with ^{90}Zr (150 keV) to a fluence of 2 × 10^{16} ions/cm^2 at room temperature.

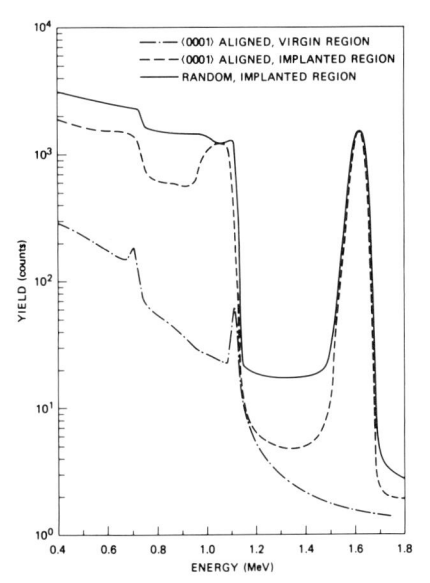

FIG. 4. Backscattering spectra from α-Al$_2$O$_3$ implanted with ^{90}Zr (150 keV) to a fluence of 4 × 10^{16} ions/cm^2 at room temperature.

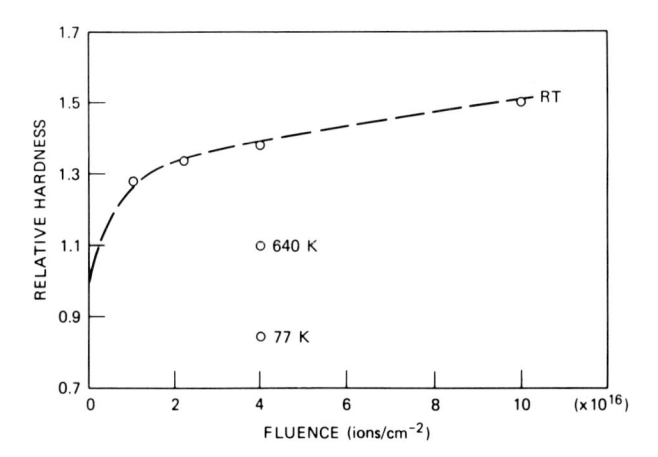

FIG. 5. Relative hardness (implanted to unimplanted) of chromium-implanted α-Al$_2$O$_3$.

The specimen implanted hot (640 K) had a hardness less than the room temperature specimen. This reflects the recovery of the near surface region as indicated by the backscattering spectra of Fig. 2.

Because of the change from a crystalline to an amorphous structure at the surface, the deformation characteristics underwent dramatic changes at this interface. Figure 6 shows SEM of a typical scratch in each region for the specimen implanted at 77 K (4×10^{16} Cr·cm^{-2}). In the unimplanted region, the groove edge contains fractures, as does the base of the groove. In the implanted region fracturing in and adjacent to the groove disappears. The value of the tangential force upon passing from the unimplanted to the implanted part of the single crystal decreased by 20 to 30% for the very light normal loads. A significant amount of pile-up along the sides and in front of the diamond indenter were noted for the amorphous samples.

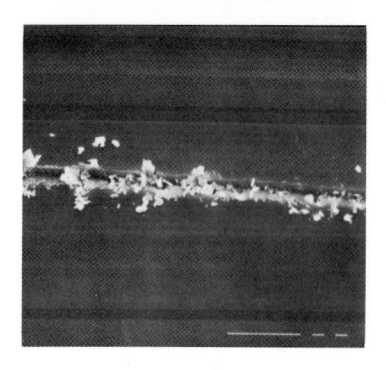

(a)　　　　　　　　　　　　　　(b)

FIG. 6. Scanning electron micrographs of scratches made with a diamond stylus under a load of 0.049 N (5 g) in the (a) unimplanted region and (b) implanted region of α-Al$_2$O$_3$ implanted with ^{52}Cr (4×10^{16} ions/cm^2, 280 keV, 77 K).

We have noted that the nature of the scratch and the degree of cracking remain altered even for loads sufficient to completely penetrate the amorphous (implanted) layer. In these cases, the indenter is pulled through a virgin crystalline-damaged crystalline-amorphous composite. Subsurface cracks which initiate in the crystalline region are markedly influenced by the amorphous overlayer. The SEM examinations suggest that the cracks stop at the crystalline-amorphous interface. Page and co-workers have made similar observations for implanted silicon and silicon carbide [18] and recently in implanted Al$_2$O$_3$ [15].

Silicon Carbide

Ion-channeling analyses of single crystals of SiC implanted with nitrogen or chromium to various fluences showed that the channeled ion-scattering yields reached the random yield when implanted doses corresponded to about 0.2 dpa [12,13]. Figure 7 contrasts [0001] channeling spectrum for a crystal implanted with 2.9×10^{14} Cr·cm^{-2} with the [0001] channeling and rotating random reference spectra taken from an unimplanted (virgin) portion of the crystal. For this fluence, the damage induced by the chromium ions has "randomized" the crystal in a region 0.02 to 0.2 μm from the surface. This is the region where the damage energy was a maximum and brackets the range

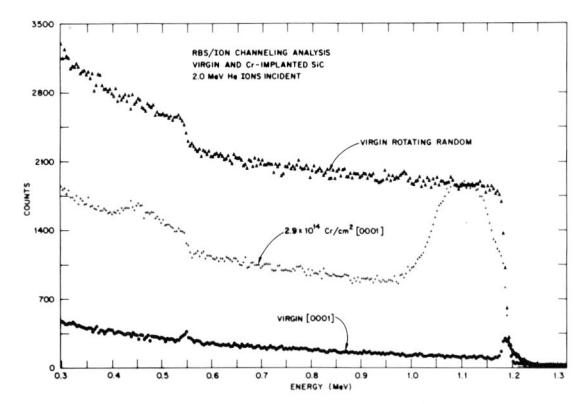

FIG. 7. Backscattering spectra of 2 MeV He$^+$ from ^{52}Cr (280 keV, 2.9 × 10^{14}/cm^2) implanted α-SiC.

where it exceeded the critical value of 0.2 dpa. At higher fluences, the width of the random region spreads in both directions. The dose dependence of randomization for each ion is given in Ref. 13.

It is generally assumed that an overlapping of the aligned spectrum with the random spectrum indicates an amorphous structure. In order to determine if this was the case for the present study, TEM and Raman spectroscopy were used to examine the chromium-implanted silicon carbide specimens. The TEM micrographs (Fig. 8) showed halos in the diffraction patterns characteristic of amorphous material to a depth of 0.25 μm and crystalline patterns at greater depths. The range of 280 keV chromium ions in SiC is about 0.25 μm. The Raman spectra for the virgin region contained peaks at 768.13, 784.51, 796.15, and 959.95 cm^{-1} which are characteristic of crystalline silicon carbide [19]. After implantation to 2 × 10^{15} Cr·cm^{-2}, these crystalline modes were completely absent [20], confirming the ion channeling and TEM observations on the amorphous nature of the implanted region.

A large volume increase was associated with the crystalline to amorphous transformation. A step-height of 40 nm was measured for the single crystal specimen implanted to 2 × 10^{16} Cr·cm^{-2}. The depth of the amorphous layer as measured by TEM and calculated is about 200 nm. If the material expanded only in a direction perpendicular to the free surface, this measurement corresponds to a volume increase of 20 to 25%.

The measurement of the hardness of SiC, as a function of irradiation dose, revealed that the relative hardness decreases 0.7 at the highest implantation level. This is consistent with the fact that the hardness measurement in the implanted layers is a composite measure of the layer and the underlying bulk; as the layer thickness increases, the influence of the implanted layer hardness is proportionately greater.

The response to the scratch wear test was similar to the amorphous Al$_2$O$_3$. Photographs of SEM scratches were similar to Fig. 6 and the tangential force decreased upon passing from crystalline to amorphous region by 20 to 30%.

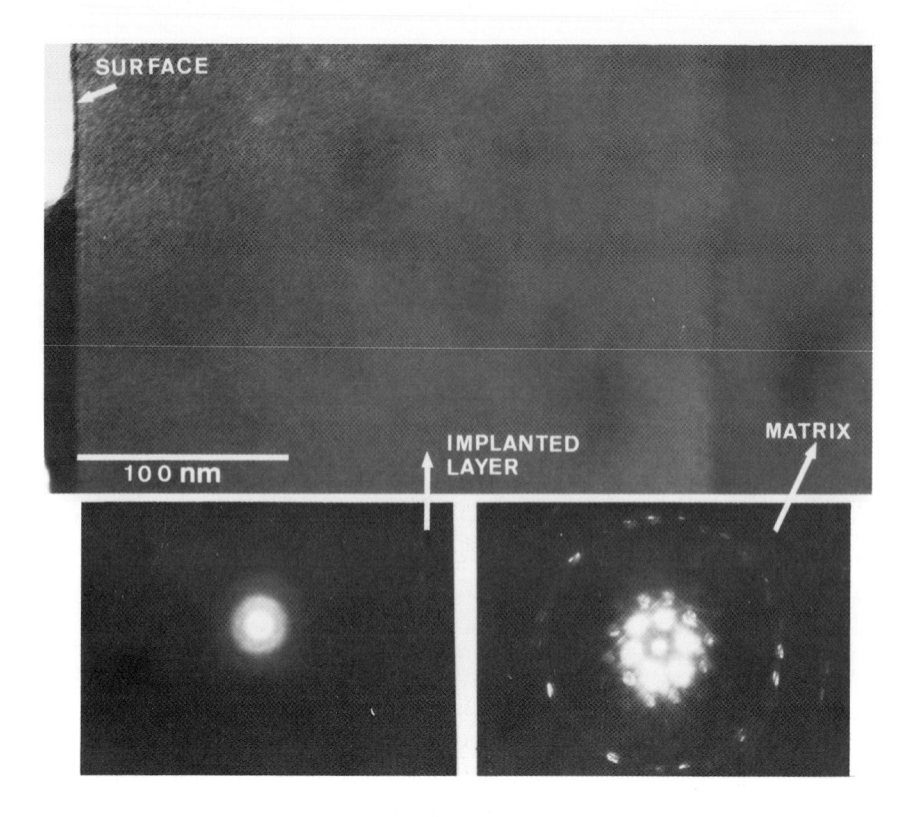

FIG. 8. Transmission electron micrograph and electron diffraction patterns for α-SiC implanted with ^{52}Cr.

SUMMARY

Both highly damaged but crystalline lattices and amorphous layers have been produced by ion implantation of Al_2O_3. The fluence required to turn a region amorphous is species-dependent in some unknown manner. Low temperature (77 K) implantation promotes the formation of the amorphous phase during implantation. An amorphous SiC was produced by all our implantation conditions.

The surface hardness of all implanted crystalline samples was increased and the amount of the increase was fluence-, temperature-, and species-dependent. All amorphous specimens showed a decrease in hardness relative to their crystalline state.

ACKNOWLEDGEMENT

The author gratefully acknowledges the aid of a number of colleagues including B. C. Leslie and S. B. Waters for sample preparation, and C. S. Yust and J. S. Pullium for the scratch-wear tests.

REFERENCES

1. Treatise on Materials Science and Technology, Vol. 18, Ion Implantation, J. K. Hirvonen, ed. (Academic Press, New York, NY 1980).
2. Metastable Materials by Ion Implantation, S. T. Picraux and W. J. Choyke, eds. (North Holland, New York, NY 1982).
3. P. D. Townsend and S. Valette in: Treatise on Materials Science and Technology, Vol. 18, Ion Implantation, J. K. Hirvonen, ed. (Academic Press, New York, NY 1980).
4. G. W. Arnold, G. B. Krefft, and C. B. Norris, Appl. Phys. Lett. 25, 540—42 (1974).
5. W. J. Choyke, L. Patrick, and P. J. Dean, Phys. Rev. B 10, 2554—60 (1974).
6. D. A. Thompson, M. C. Chan, and A. B. Campbell, Can. J. Phys. 54, 626—32 (1976).
7. C. J. McHargue, H. Naramoto, B. R. Appleton, C. W. White, and J. M. Williams in: Metastable Materials by Ion Implantation, S. T. Picraux and W. J. Choyke, eds. (North Holland, New York, NY 1982), pp. 147—53.
8. A. B. van Groenou, N. Maan, and J.D.B. Veldkamp, Philips Research Report 30, 320 (1975).
9. H. Naramoto, C. W. White, J. M. Williams, C. J. McHargue, O. W. Holland, M. M. Abraham, and B. R. Appleton, J. Appl. Phys. 54, 683—98 (1983).
10. H. Naramoto, C. J. McHargue, C. W. White, J. M. Williams, O. W. Holland, M. M. Abraham, and B. R. Appleton, Nucl. Instr. and Methods 202/210, 1159—66 (1983).
11. C. J. McHargue, H. Naramoto, C. W. White, J. M. Williams, B. R. Appleton, P. S. Sklad, and P. Angelini in: Emergent Process Methods for High Technology Ceramics, R. F. Davis, H. Palmour III, and R. L. Porter, eds. (Plenum Press, New York, NY in press).
12. C. J. McHargue, M. B. Lewis, B. R. Appleton, H. Naramoto, C. W. White, and J. M. Williams in: The Science of Hard Materials, R. K. Viswanadham, D. J. Rowcliffe, and J. Gurland, eds. (Plenum Press, New York, NY 1983) pp. 451—66.
13. C. J. McHargue and J. M. Williams in: Metastable Materials by Ion Implantation, S. T. Picraux and W. J. Choyke, eds. (North Holland, New York, NY 1982), pp. 303—9.
14. J. M. Williams, C. J. McHargue, and B. R. Appleton, Nucl. Instr. and Methods 209/210, 317—23 (1983).
15. P. J. Burnett and T. F. Page, J. Mater. Sci. (to be published).
16. R. C. Bradt, J. Am. Ceram. Soc. 50, 54—55 (1967).
17. B. B. Ghate, W. C. Smith, C. H. Kim, D.P.H. Hasselman, and G. E. Kane, Ceram. Bull. 54, 210 215 (1975).
18. S. G. Roberts and T. F. Page in: Ion Implantation into Metals, V. Ashworth, W. A. Grant, and R.P.M. Procter, eds. (Pergamon Press, New York, NY 1982), pp. 135—46.
19. R. B. Wright, R. Varma, and D. M. Gruen, J. Nucl. Mater. 63, 415—21 (1976).
20. G. Begun and C. J. McHargue, unpublished work at Oak Ridge National Laboratory.

BEHAVIOR OF IMPLANTED α-Al$_2$O$_3$ IN AN OXIDIZING ANNEALING ENVIRONMENT*

G. C. FARLOW,[+] C. W. WHITE, C. J. McHARGUE,[++] AND B. R. APPLETON
Solid State Division, Oak Ridge National Laboratory, Oak Ridge, TN 37831

ABSTRACT

The transition metal ions Fe, Mn, Ni, Ti, and Cr as well as the group III-A ion Ga were implanted in a random direction close to the c-axis of α-Al$_2$O$_3$ single crystals. These were subsequently annealed in flowing oxygen at temperatures between 600 and 1500°C. The impurity and damage distributions were determined by He backscattering and channeling techniques. The normally trivalent ions, Fe, Ga, and Cr show strong tendency to become incorporated into substitutional sites, whereas the normally divalent ions, Cu and Mn show no such tendency, even at high temperatures. Most of the impurities tend to move toward the surface with only Fe, Ga, and Cr showing tendency towards bulk diffusion. In all cases the Al sublattice recovers at lower temperatures than the oxygen sublattice.

INTRODUCTION

This paper describes an extension of work done on the implantation of metal ions into α-Al$_2$O$_3$ [1,2]. In addition to Cr [1] and Ti [2] we will consider the first row transition metals Mn, Fe, Cu, Ni, and also Ga. The other published work on implantation of metals into Al$_2$O$_3$ concerns Pt [3] and Pb [4].

These studies were undertaken in an effort to determine whether implantation techniques could harden the surface of Al$_2$O$_3$, or otherwise improve wear properties as has been possible with metals and alloys [5]. In the process we also hope to relate surface changes to fundamental properties of the sample such as lattice damage, microstructure, and the lattice interactions of the implanted species.

The details of implantation and annealing are described elsewhere in these proceedings [6]. Briefly, each species was implanted to a dose typically $4 \times 10^{16}/cm^2$ and to a depth of 500-1000 Å. Sections of each sample were annealed in flowing oxygen at temperatures between 600 and 1600°C, analyzed for lattice damage using He backscattering and channeling techniques, and measured for hardness with a Knoop microhardness device. No particular evidence of unwanted impurity deposition during implantation or annealing can be seen in the backscattering spectrum.

LATTICE ANNEALING EFFECTS

Lattice damage recovery during annealing is similar in all cases. An example is shown in Fig. 1 for the case of ^{56}Fe (150 keV, $4 \times 10^{16}/cm^2$).

First, annealing of the lattice damage begins during implantation. Since we know that at low temperatures the lattice is amorphized over the entire implantation range [6] it is clear that the implantation beam induces annealing in the as implanted specimen. Presumably this is produced by

---

*Research sponsored by the Division of Materials Sciences, U. S. Department of Energy under contract W-7405-eng-26 with Union Carbide Corporation.
[+]ORAU Postdoctorate
[++]Metals and Ceramics Division.

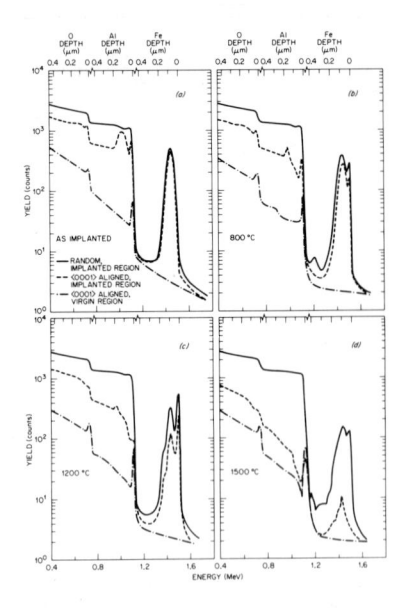

Fig. 1. ^{56}Fe (160 keV, 4 × 10^{16}/cm^2) in α-Al$_2$O$_3$ Annealed in Air.

kinetic interactions of the beam with existing lattice damage and/or conversion of the beam energy to thermal energy as the ions come to rest. The damage distribution of samples implanted at high temperatures is different only in the very near surface region [6] suggesting that the impurities may be stabilizing the damage in the region where the impurity distribution is relatively high.

Second, in all cases the Al sublattice recovers at lower temperatures than the oxygen sublattice. This is shown most clearly in Fig. 2 which consists of plots of χ_{min} (the ratio of the aligned yield to the random yield) vs. temperature for the Fe and Cu implanted specimens. These plots show two very different patterns of annealing, yet in both cases the damage recovery in the Al sublattice is greater than in the oxygen sublattice at temperatures below 1200°C. One will note too that the oxygen sublattice is still very nearly amorphous in the as implanted state and thus relatively unaffected by beam induced annealing. Thus, even though the annealing paths are very different, we infer that the Al interstitial becomes mobile at much lower temperatures than the oxygen. These figures also show that there are two stages in the annealing of the oxygen sublattice and probably two in the Al sublattice.

Third, in most cases (the Cu implanted sample being a notable exception) the Al density in the vicinity of the impurity distribution peak is depressed. This deficit tends to anneal out and is generally gone after a 1500°C anneal. In the Ni implanted specimen however the deficit becomes more pronounced after a 1200°C anneal. We have not associated any physical process with this feature but it might represent a swelling of the lattice.

IMPURITY ANNEALING EFFECTS

Certain effects are common to all implanted species that we have studied. As implanted the impurities have a roughly gaussian distribution showing little or no substitutionality at all for 4 × 10^{16}/cm^2 doses. Specimens implanted with Cr and Ti at lower doses do show some substitutional fraction in the as implanted state [1,2]. If the impurity exhibits any mobility during annealing, there is a definite preference to move

Fig. 2a.

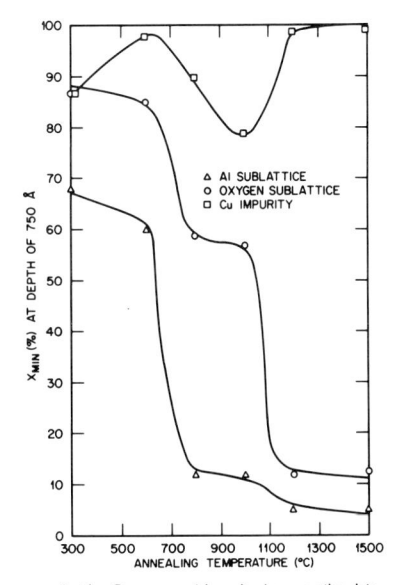

Lattice Recovery and Impurity Incorporation into the Lattice for implants of ^{63}Cu (130 keV, 4×10^{16}/cm^2) in α-Al$_2$O$_3$ Annealed in Air

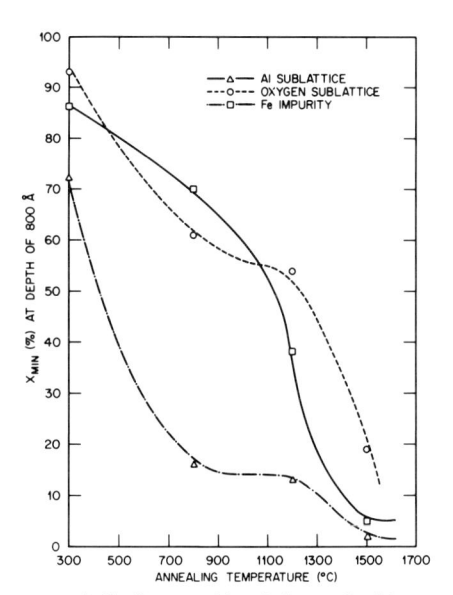

Fig. 2b.

Lattice Recovery and Impurity Incorporation into the Lattice for implants of ^{56}Fe (160 keV, 4×10^{16}/cm^2) in α-Al$_2$O$_3$ Annealed in Air

toward the geometrical surface rather than towards the bulk. This suggests that the presence of oxygen in the atmosphere during annealing induces a chemical gradient affecting the diffusion. In several cases there is a significant loss of the impurity after anneals at the highest temperatures. We suspect either evaporation of the oxide or leaching of the metal by the flowing oxygen gas.

Lattice incorporation of the impurity may be divided into three groups. The first consists of Cr [6], Fe, and Ga which are characterized by a strong tendency to go substitutional during annealing at temperatures above 1200°C and are better than 95% substitutional after annealing at 1500°C. Additionally, Fe (Fig. 1) and Ga (Fig. 3) have diffused into the bulk after 1500°C anneals, and Cr [1] is known to begin diffusing at about 1600°C. The expected oxidation state of these elements is 3$^+$ and each has the same crystal structure as Al$_2$O$_3$. It is known that the Cr is in the 3$^+$ state after a 1600°C anneal [1]. Ga and Cr do not seem to be so strongly bound to the surface as the Fe. Neither do the former show the degree of surface precipitation as does the Fe.

The second group consists of Cu (Fig. 4) and Mn (Fig. 5) which show no evidence of substitutionality nor diffusion into the bulk. It is questionable whether the Mn even reaches the surface in spite of the surface-side hump observed after a 1200°C anneal. The oxides of these elements are expected to be divalent, MnO having cubic symmetry and CuO having tetragonal symmetry. These samples also show what appear to be incoherent precipitates at 1200°C and above.

The third group consists of Ni (Fig. 6) and Ti (Fig. 7) which show partial substitutionality at 1200°C and above, and which exhibit some bulk diffusion after 1500°C anneals. If being in the 3$^+$ state is important to lattice incorporation, as the data of group I above indicates, the fraction that is substitutional may well reflect the portion of Ti in the 3$^+$ state on the way to the preferred 4$^+$ state. Burnett and Page [7] report that the fine precipitates (as distinguished from the large platelet shaped

TABLE I. Relative hardness of Fe and Cu implanted specimens vs. annealing temperatures.

Annealed	Cu	Fe
As Implanted	1.18	1.24
800°C	1.27	1.27
1200°C	1.12	1.17
1500°C	1.00	1.20

precipitates) which form around 1300°C are not Al_2TiO_5. These might contain trivalent Ti. Nickel is normally divalent and NiO has cubic symmetry, thus one would expect the Ni implanted specimens to behave like the ones implanted with Cu and Mn. We do not know why it does not, but, on the basis of the behavior of the other implanted species, we suspect the chemistry of the ion is responsible. Optical absorption and EPR studies are under way to try to identify the exact chemical states of the implanted ions.

HARDNESS

Knoop microhardness measurements have been made on the Cu and Fe implanted samples. Both show an increase in the relative hardness of the implanted region as compared to the unimplanted region of between 20 and 30%. The Cu hardness quickly diminishes after annealing at 1200°C and

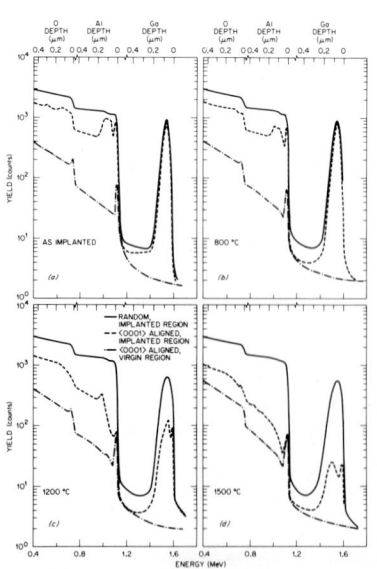

Fig. 3. ^{69}Ga (130 keV, $4 \times 10^{16}/cm^2$) in α-Al_2O_3 Annealed in Air.

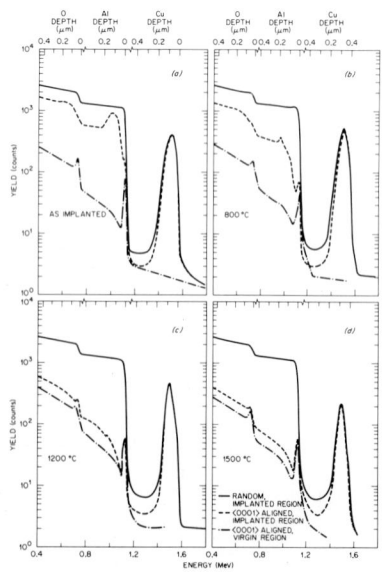

Fig. 4. ^{63}Cu (130 keV, $4 \times 10^{16}/cm^2$) in α-Al_2O_3.

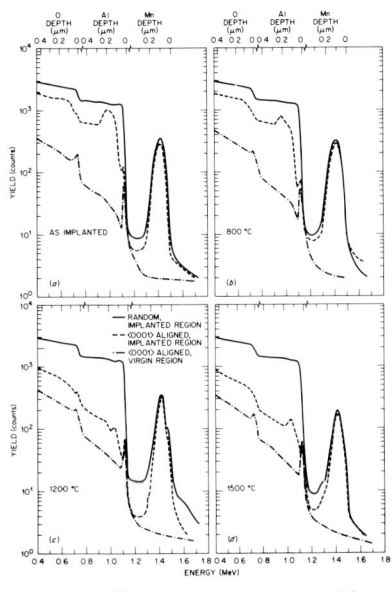

Fig. 5. ^{55}Mn (150 keV, 4 × 10^{16}/cm^2) in α-Al$_2$O$_3$ Annealed in Air.

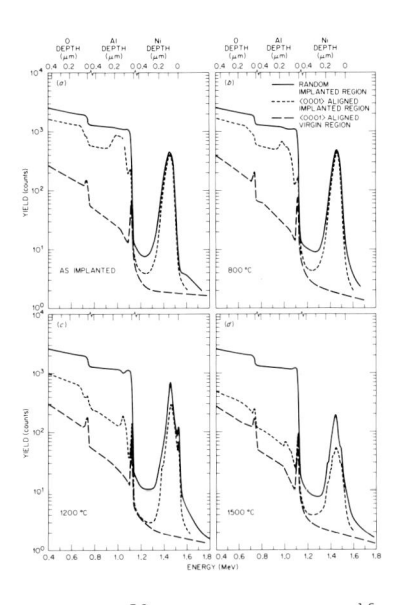

Fig. 6. ^{58}Ni (160 keV, 4 × 10^{16}/cm^2) in α-Al$_2$O$_3$.

above. The lattice damage has also diminished to nearly the unimplanted state. The damage in the Fe implanted sample however shows considerably more residual damage at these temperatures and the hardness also remains high. The microstructure of the Cu implanted samples shows a very coarse and nonuniform precipitate at these temperatures. The Fe implanted samples by contrast exhibit a very fine and uniform precipitate. We suspect that the finer precipitate structure is stabilizing the damage and its contribution to the hardening as well as contributing directly to the hardening.

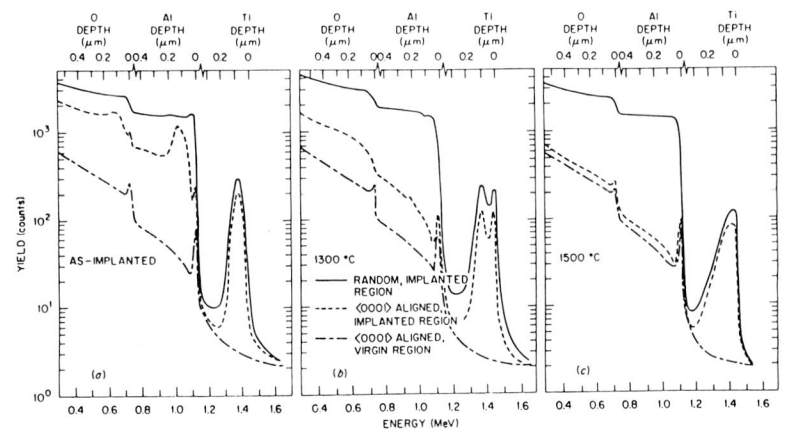

Fig. 7. ^{48}Ti (150 keV, 3×10^{16}/cm^2) in α-Al$_2$O$_3$

CONCLUSIONS

Our principal conclusions are as follows:

1. Lattice damage is partially annealed during implantation.
2. The Al ion become mobile at lower temperatures than the oxygen ion.
3. Many implanted impurities appear to be bound to the surface after thermal annealing in oxygen.
4. Lattice incorporation of implanted species is correlated with valence states.
5. Hardness as a function of annealing temperature correlates reasonably well with residual lattice damage.

REFERENCES

1. H. Naramoto, C.W. White, J.M. Williams, C.J. McHargue, O.W. Holland, M.M. Abraham, and B.R. Appleton, J. Appl. Phys. 54, 683 (1983).
2. H. Naramoto, C.J. McHargue, C.W. White, J.M. Williams, O.W. Holland, M.M. Abraham, and B.R. Appleton, Nucl. Instr. and Methods 201/202, 1159 (1983).
3. A. Canera, A.V. Drigo, and P. Mazzoldi, Radia. Eff. 49, 29 (1980).
4. A. Turos, H.J. Matzke, and P. Rabette, Phys. Stat. Sol. A 64, 565 (1981).
5. Treatise on Materials Science and Technology, Vol. 18, Ion Implantation, ed. by J.K. Hironven, Academic Press, New York, (1980). Also see articles in the Proceedings of the Second International Conference on Ion Beam Modification of Materials, Albany, New York, July 1980. Proceedings published in Nucl. Instrum. and Methods 182/183 (1981).
6. C.J. McHargue, C.W. White, B.R. Appleton, G.C. Farlow, and J.M. Williams, these proceedings.
7. P.J. Burnett, T.F. Page, J. Mat. Sci., to be published.

CHANGING THE INDENTATION BEHAVIOUR OF MgO BY ION IMPLANTATION

P.J. BURNETT AND T.F. PAGE
University of Cambridge, Department of Metallurgy and Materials Science, Pembroke Street, Cambridge, CB2 3QZ, U.K.

ABSTRACT

Cleaved {100} MgO single crystal surfaces have been implanted with 300keV Ti^+ and Cr^+ ions to doses in the range 10^{16}-10^{18} ions cm^{-2}. Localised plasticity and fracture behaviour has been investigated using microhardness indentation tests at loads of between 10gf and 500gf. Significant surface hardening has been observed at lower doses with amorphisation related softening occurring at higher doses. Post-implantation heat treatments produce precipitation reactions with associated hardness changes.

INTRODUCTION

By changing the chemistry, structure and properties of a highly-localised thin surface layer ion-implantation is capable of modifying materials behaviour in areas such as corrosion, fatigue and wear [e.g. 1]. This paper arises from a study of the means by which implantation might affect the tribological properties (e.g. hardness, friction, wear) of hard ceramics. We have already reported [2-4] that implantation-induced stresses can modify near-surface crack genesis and growth, thus altering the indentation fracture behaviour of brittle solids. Also, implantation can lead to significant solid solution and/or radiation hardening effects, while at higher doses (typically $\sim 4 \times 10^{17}$ ions cm^{-2}) amorphisation of the surface can result in surface softening. Further we have proposed a simple quantitative model which allows the position and extent of this amorphous layer to be predicted [3]. Thus, it has been shown that amorphisation begins at the peak of the damage profile (e.g. for MgO this occurs at $\sim 8 \times 10^{16} Ti^+ cm^{-2}$ @ 300keV [5]), initially forming a sub-surface amorphous layer. With increasing dose the layer extends to the surface (e.g. for MgO this occurs at $\sim 5 \times 10^{17} Ti^+ cm^{-2}$ @ 300keV [5]). Precipitation hardening has also been reported during the post-implantation heat-treatment of sapphire [4,6].

This paper presents the results of an exploration of the extent to which similar phenomena might be induced by implanting MgO with aliovalent species.

EXPERIMENTAL

High purity (99.99%) {100} habit MgO single crystals, supplied by W & C Spicer, were cleaved into (100) surface slices $\sim 10 \times 5 \times 2$mm and implanted in the as-cleaved state. Implantation with 300keV Ti^+ and Cr^+ ions to doses in the range 10^{16}-10^{18} ions cm^{-2} were carried out in the Cockcroft-Walton facility at AERE Harwell. These ion species were chosen on the criteria of ionic radius and, more importantly, valency as those likely to cause solid-solution hardening in MgO [7]. The doses were determined by post-implantation Rutherford backscattering (RBS).

Microhardness tests (Leitz "Miniload") were used to determine materials response to surface contacts. All tests were performed under ambient conditions at room temperature. Knoop microhardness tests at loads of 10gf and 25gf were performed on all specimens, the long axis of the indenter being aligned along a <100> direction in the surface. Six indentations per specimen were made and the hardness calculated from the mean long diagonal value. These loads produce indentations which lie substantially (∿0.3-0.6µm penetration) within the implanted layer. Vickers indentations (diagonals parallel to <100>) were made at 100gf, 200gf and 500gf (corresponding to depths of ∿1-2µm) were used to study the modifications to deformation and fracture morphologies. Five indentations per load per specimen were made and both the indentation diagonal lengths and the radial crack trace lengths measured and subsequently used to evaluate the apparent K_{IC} values as in [2].

Two specimens (MT4 and MC4, 6.8 x 10^{16} ions cm^{-2} Ti$^+$ and Cr$^+$ respectively) were isothermally annealed at 800°C in air using a resistance element furnace. A section of MT4 not previously annealed was given a 1hr anneal at 1150°C in order to produce over-aged precipitates ameanable to easy identification by transmission electron microscopy (TEM).

RESULTS AND DISCUSSION

Hardness Behaviour

The effect of Ti$^+$ and Cr$^+$ implantations on the low-load Knoop microhardness of MgO is shown in Fig. 1. The trend is for hardness to increase with dose to a peak at ∿8 x 10^{16} ions cm^{-2}, thereafter the hardness falls to a value less than that of the unimplanted control. For both Ti$^+$ and Cr$^+$ the 10gf Knoop hardness (H_{10}) is initially less than the 25gf hardness (H_{25}). As the surfaces become hardened by the implantation process, then H_{10} becomes greater than H_{25}. For Ti$^+$ implantation this order then reverses again at high doses. For unimplanted MgO the observation that $H_{10}<H_{25}$ is contrary to the published load-variant microhardness behaviour of MgO [e.g. 8]. This reverse effect is almost certainly due to a water-absorption-affected layer [9]. The hardening observed at low doses in the as-implanted state may be attributed to solid-solution and/or radiation hardening. Cr and Ti may be expected to harden MgO as a result of being aliovalent ions. This hardening is generally attributed to the formation of vacancy complexes around the solute ion, these being required to maintain charge neutrality within the crystal [7]. Additional hardening may occur due to misfit between the solute and solvent anions, but this effect should be much smaller than that due to charge effects [7]. Since no significant difference is seen in the hardening capability of Ti and Cr implanted into MgO, and since these ions are of different sizes, it may be assumed that they both harden as a result of charge effects rather than strain effects. Perez et al. [10] implanted Fe$^+$ into MgO and found Fe^{2+} (in preference to Fe^{3+}) present in the dose range used here. Thus, Ti is probably present as Ti^{3+} but is expected to change to Ti^{4+} during annealing in air. A similar Fe^{2+}→Fe^{3+} transformation has also been reported [10] for the annealing of Fe$^+$ implanted MgO.

Correlation of the hardness data with dose shows that, while hardening is initially observed, subsequent softening is apparent around the dose at which amorphisation first occurs. However, some hardening is retained whilst a sub-surface amorphous layer is present, but, at higher doses, when the surface itself becomes amorphous, an absolute softening may occur. Similar behaviour for the case of implantation into Al_2O_3 has been discussed elsewhere [4].

The initial crossing over of the H_{10} and H_{25} values as a result of changing hardness is due principally to implantation hardening effects. However, it is expected that, as the amorphization induced softening of the surface progresses, the lower load hardness, H_{10}, would again become less than H_{25}, as the indentation is contained more completely within the softened layer. This behavior is seen for Ti$^+$ into MgO (Fig. 1) but not for

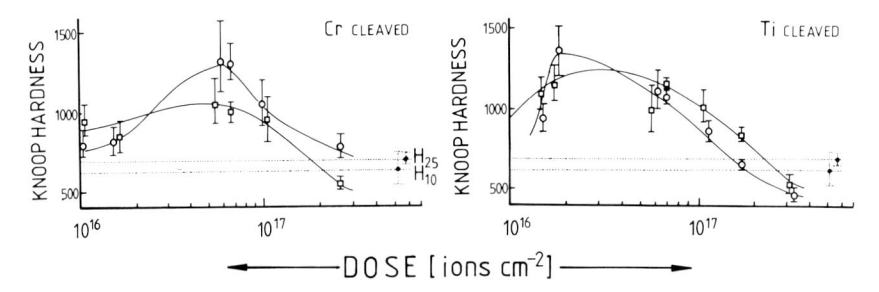

Fig. 1. The experimental variations of 10gf (o) and 25gf (■) Knoop hardness with dose for Ti$^+$ and Cr$^+$ implantations into MgO @ 300keV. The broken horizontal lines represent the 10gf and 25gf hardness values (H_{10}, H_{25}) of the unimplanted control.

Cr$^+$. This may be due to changes in the chemical nature of the surface resulting in a decreased affinity for water, i.e. implantation may render the material either insensitive to, or differently sensitive to, chemomechanical effects.

The Effect of Annealing on Hardness

For the two specimens annealed at 800°C in air, the expected stable phases in the composition range (∿0-10 at% solute) of the implanted surface are MgO+Mg$_2$TiO$_4$ (spinel) and MgO+MgCr$_2$O$_4$ (spinel) (as shown on the equilibrium phase diagram [11]). Precipitation should thus occur during annealing. A series of isothermal anneals were performed sequentially, hardness tests being undertaken between anneal stages, and the results shown in Fig.2. Both specimens showed broadly similar behaviour, the hardness decreasing initially then rising rapidly at ∿3hr for MT4 and ∿2hr for MC4. In both cases, this sharp change in hardness occurred over the range in which radiation induced colour centres were removed (i.e. the removal of most of the point defects). The hardness rose to a peak before falling off with increasing time. Observations made using reflected light microscopy and transmission electron microscopy showing that substantial precipitation had occurred in unison with the re-hardening observed. Attempts at identifying precipitates formed in MT4 at 800°C and 1150°C by electron diffraction (TEM) indicates that there is probably more than one phase present. Possibilities include MgTiO$_3$ (orthorhombic a=9.58, b=9.69 c=3.62Å) and Mg$_2$TiO$_4$ (spinel a=8.45Å). Fig.3a

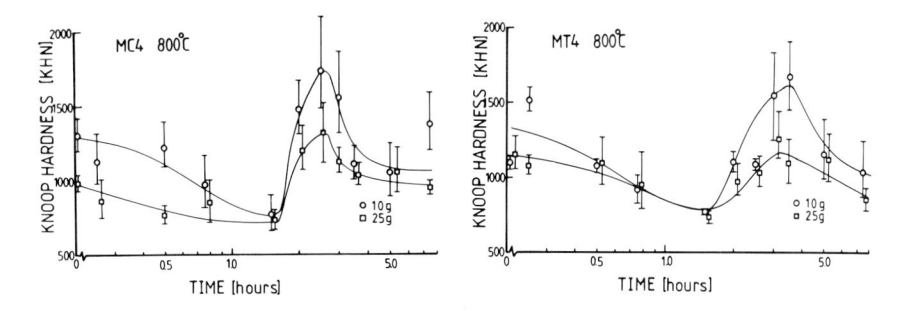

Fig. 2. The observed variation of 10gf (o) and 25gf (■) Knoop hardnesses with anneal time for MT4 and MC4 at 800°C.

shows the larger (∼1µm) precipitates found in MT4 after annealing at 1150°C
for 1 hr. EDX analysis performed in a Philips 400T TEM shows that these are
of equal Mg and Ti content, thus indicating that they are probably MgTiO$_3$. At
higher magnification a finer dispersion of precipitates can be seen (Fig.3b).
These are probably responsible for the re-hardening observed. The precipitate
phases observed are metastable, their rapid formation (2-3 hrs. @ 800°C) being
due to the initial high diffusion rates resulting from the thermodynamic
excess of point defects introduced by the implantation process. The initial
softening is probably due to the annealing out of radiation damage and/or the
depletion of solute prior to the formation of the finer precipitate phase
later in the anneal. These results are analagous to the annealing behaviour
found in sapphire [4,6].

Indentation Fracture and Deformation Behaviour

The most striking effects of implantation into MgO are the changes in
fracture morphology and deformation geometry. It is believed that the <110>
radial crack traces seen on the {100} surface of unimplanted MgO after indentation (e.g. Fig. 4) result from the action of the indentation stress
field upon microcracks formed by reactions between dislocations on different
slip planes [12]. It can be seen from Fig. 4 that these cracks only extend
to the limit of the square slip trace array. Fig. 5 shows that ion-implantation can markedly alter the observed fracture geometry and dislocation etch-
pit patterns. Since the surface has become hardened it is reasonable to expect the extent of the dislocation motion to decrease, as is observed. The
most marked difference between the unimplanted and implanted etch-pit rosettes
is the shortening of the <110> etch-pit rays and the disappearance of the
smaller <100> rays (see Figs. 4,5). Associated with this effect is the shortening of the <110> bands of diffuse scattering observed when using polarised
light microscopy. The square slip trace arrays are also noticably less extensive after implantation, disappearing altogether at higher doses (e.g.Fig.
5a,d). However, despite the absence of slip steps, the troughing (as described for pure MgO in [13]) around the indentation is still apparent (Fig. 5d).
This suggests firstly that the sub-implanted layer deformation mode is unchanged (as witnessed by the surface topography) and, secondly, that the
strengthened surface layer is perhaps deforming by some mechanism other than
dislocation motion (as suggested by the disappearance of the etch-pit.rosett-
es). Indeed, the implanted layer may simply be deforming elastically. Fig.5
also shows a modified fracture geometry, but again the cracks do not extend
beyond the confines of the square slip trace array. After implantation the
crack traces no longer exclusively follow the <110> trajectories. In the
higher dose specimens (Fig. 5a) the <110> form completely disappears and the

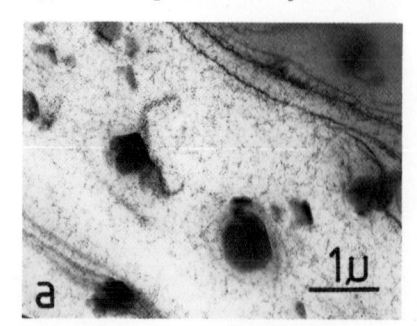

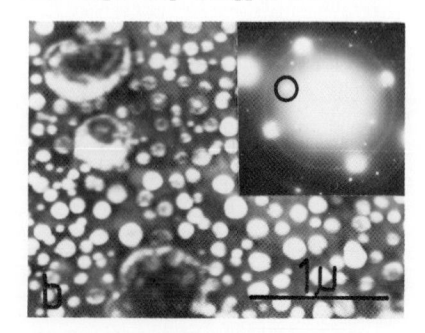

Fig. 3(a). TEM bright field image of MT4 annealed for 1hr @ 1150°C showing
large angular precipitates with a finer distribution between them;(b) TEM CDF
micrograph (imaging beam marked) of the smaller rounded precipitates believed
responsible for the rehardening that occurs during annealing.

cracks, rather than originating from the halfway along the sides of the indentations, now grow from the corners of the indentation. This implies that the crack nucleation mechanism has altered from being slip induced (e.g. microcracking etc.) to being initiated at the points of highest stress concentration during indentation (i.e. the corners of the indentations). When the K_{IC} values are calculated as in [2] a large (∼80%) increase in toughness is observed. Further modification to the fracture morphology may arise from implantation induced stresses [e.g. 2].

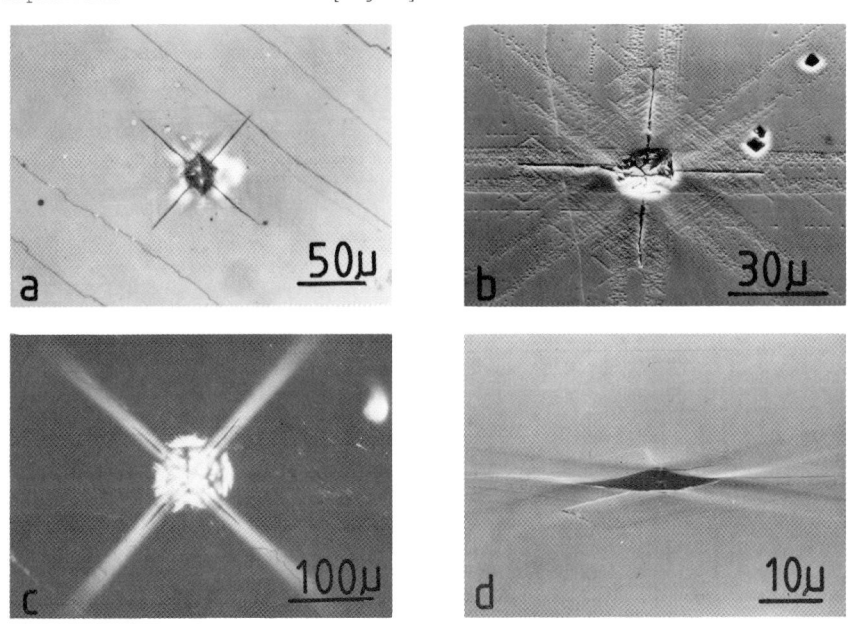

Fig. 4. Vickers indentations on unimplanted MgO, (a) 500gf; reflected light; (b) 200gf, after etch-pitting SEM secondary electron image; (c) 500gf, reflected polarised light; (d) 200gf, 70° tilt, SEM secondary electron image.

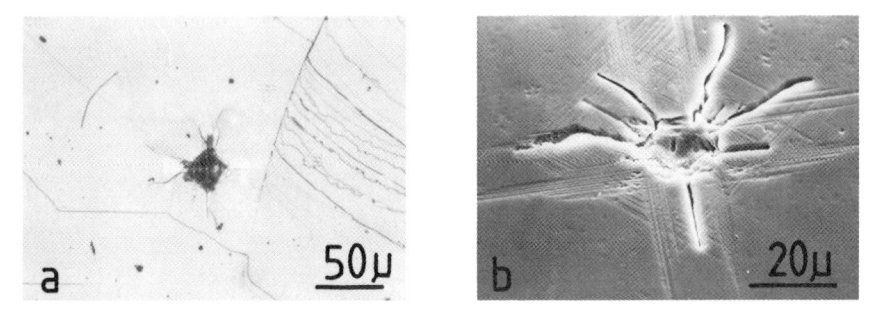

Fig. 5. Vickers indentations in implanted MgO, (a) 500gf, reflected light, dose 1.7×10^{16}Ti$^+$ cm^{-2}; (b) 200gf, after etch-pitting SEM secondary electron image, dose 10^{16}Cr$^+$cm^{-2}; (c) 500gf, reflected polarised light, dose 10^{16} Cr$^+$ cm^{-2}; (d) 200gf, 70°tilt, SEM secondary electron image, dose 6.8×10^{16}Cr$^+$ cm^{-2}.

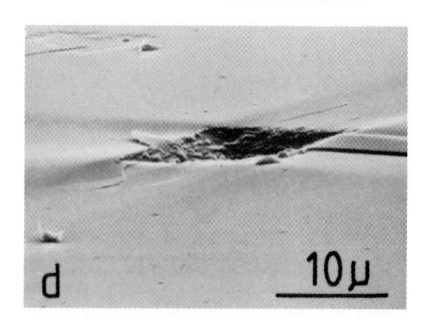

Fig. 5. contd.

CONCLUSIONS

Ion-implantation can effect the mechanical properties of {100} MgO single crystals in a number of ways: (i) implantation to doses below those required to amorphise MgO have been found to harden the surface, (ii) implantation to doses that result in amorphisation cause a softening of the surface to occur, this being less for Cr implants than Ti implants, (iii) post-implantation annealing results in precipitation hardening by the formation of a variety of as yet unidentified second phases (iv) a change in indentation fracture morphology has been observed after implantation and this has been associated with the decrease in the size of the dislocation rays along <110> believed to be linked with the hardening in (i).

ACKNOWLEDGEMENTS

The authors wish to thank Prof. R.W.K. Honeycombe, F.R.S., for provision of laboratory facilites, SERC and AERE Harwell for support and Mrs. E. Palmer for kindly typing this manuscript.

REFERENCES

1. G. Dearnaley, Mats. in Eng. Applications, 1, 28 (1978).
2. P.J. Burnett and T.F. Page, submitted to J. Mater. Sci., (1984).
3. P.J. Burnett and T.F. Page, J. Mater. Sci., in press.
4. P.J. Burnett and T.F. Page, Plastic Deformation of Ceramic Materials, eds. R.C. Bradt and R.E. Tressler (Plenum Press, New York, in press).
5. P.J. Burnett, unpublished work.
6. H. Naramoto, C.W. White, J.M. Williams, C.J.McHargue, O.W. Holland and M.M. Abraham, J. Appl. Phys., 54 683 (1983).
7. T.E. Mitchell and A.H. Heuer, Mats. Sci. Eng., 28, 81 (1977).
8. P.M. Sargent and T.F. Page, Proc. Brit. Ceram. Soc., 26, 209 (1978).
9. A.R.C.Westwood, J.S.Ahearn and J.J.Mills, Colloids and Surfaces,2,1 (1981).
10. A. Perez, M. Treilleux, P. Thevenard, G. Abouchacra, G. Marest, L. Fritsch and J. Serughetti, Proc. Mat. Res. Soc., 7, 159 (1982).
11. C.M. Levin,C.R. Robbins and H.F.McMurdie, Phase Diagrams for Ceramists, (Am. Ceram. Soc., Ohio, 1964).
12. A.N. Stroh, Proc. Roy. Soc. A, 232, 548 (1955).
13. R.W. Armstrong and C. Cm. Wu, J. Am Ceram. Soc., 61, 102 (1978).

MICROSTRUCTURAL DEVELOPMENT OF TiB$_2$ ION IMPLANTED WITH 1 MEV NICKEL[*]

P. S. SKLAD, P. ANGELINI, M. B. LEWIS, AND C. J. McHARGUE
Metals and Ceramics Division
Oak Ridge National Laboratory, Post Office Box X,
Oak Ridge, TN 37831

ABSTRACT

An Analytical Electron Microscopy (AEM) investigation of polycrystalline TiB$_2$ implanted with 1 MeV Ni$^+$ to 1×10^{21} ions/m^2 has shown that the implanted region remained crystalline and showed no evidence of precipitation. A region containing tangled dislocations extended from the implanted surface to ~500 nm. Between ~500 and 750 nm, the microstructure was more complicated and could be indicative of a high density of 5 to 10 nm diam defects. The maximum nickel concentration determined by energy dispersive spectroscopy (EDS) occurred at ~450 nm, slightly deeper than the calculated depth of 390 nm. Observations after in situ annealing revealed cavities and nickel-rich precipitates. Radiation damage models are invoked to explain the microstructures observed.

INTRODUCTION

It is well known that many of the properties of ceramics are controlled by the condition of the specimen surface. For example, fracture toughness can be improved with the application of compressive surface stresses, oxidation resistance may depend critically on the presence of a protective surface layer, and wear resistance may be enhanced by improving the response of the surfaces to contact stresses. Surface modification techniques which have been successfully applied to a number of metals and semiconductors in recent years, offer the potential for overcoming some of the surface related limitations on the use of ceramics. Specific applications include structural materials in advanced energy conversion systems where high operating temperatures and highly reactive environments may be encountered.

The research emphasis at ORNL is on Al$_2$O$_3$ implanted with Cr, Zr, Ti, Ni, and Fe, SiC implanted with Cr and Zr, and TiB$_2$ implanted with Ni and Zr. The present study is concerned with the microstructure produced by implanting TiB$_2$ with 1 MeV Ni$^+$ ions. The experimental program includes not only characterization of the as-implanted microstructure but also studies of the stability of the microstructure during exposure to elevated temperatures with the purpose of determining the mechanisms by which these microstructures are altered. In this respect, comparison of the experimental observations with predictions of radiation damage models of microstructural development has been useful. Such information is important to the development of surface modified ceramics for use in elevated temperature applications and will form a basis on which ion implantation techniques may be used to successfully tailor surface characteristics of this class of materials.

---

[*]Research sponsored by the Division of Materials Sciences, U.S. Department of Energy under contract W-7405-eng-26 with the Union Carbide Corporation.

MATERIALS AND PROCEDURES

Specimens of TiB_2 were prepared from powders obtained from Stark Company, West Germany, by vacuum hot pressing at ~2000°C and a pressure of 25 MPa for 4 hours. The resulting compacts of 98.4% theoretical density had an average grain size of 23 µm, although grain growth produced some grains as large as 100 µm. Specimens of appropriate size were cut from these compacts and polished mechanically to a surface finish of <1 µm.

The implantation of 1 MeV Ni^+ ions was carried out at ambient temperatures using the ORNL 5MV Van de Graaff facility. The fluence was 1×10^{21} ions/m^2. The implanted specimens were cut into pieces approximately 1 mm × 2 mm × 5 mm and pairs of these pieces were glued together with implanted surfaces facing each other. Slices ~250 µm thick were then cut from this composite, mechanically polished to ~75 µm, mounted on copper washers, and argon ion milled. This technique allowed examination of the implanted layer in cross section while at the same time maintaining (in most cases) the original implanted surface.

The TEM specimens were examined in a Philips EM 400T/FEG equipped with 6585 STEM and EDAX 9100 x-ray EDS system. In addition, a Gatan double tilt liquid nitrogen cooled specimen holder with a beryllium cup was used during EDS analysis. Areas were selected for analysis where specimen thickness did not require absorption corrections. The STEM probe size was ~2 nm. Subsequent quantitative analyses were performed using integrated peak intensities and standardless analysis routines using programs developed by Zaluzec [1]. In situ heating was carried out in the EM 400 using a Philips PW6592 heating holder operating at 850°C. Conventional bright-field, weak-beam dark-field, centered-beam dark-field, and selected area diffraction techniques were used to analyze the microstructures.

RESULTS AND DISCUSSION

The effect of implanting TiB_2 with energetic nickel ions is illustrated in Fig. 1. The micrographs are a bright-field and weak-beam dark-field pair which show the change in the nature of the damage as a function of distance from the implanted surface. It can be seen that the implanted layer remained crystalline and that there is a reasonably uniform, moderate density of tangled dislocations in the region extending from the surface to a depth of ~500 nm. The areas between the dislocations appear to be relatively free of defects. In the region between 500 and 750 nm from the surface the microstructure is noticeably different and may contain a high density of 5 to 10 nm diam defects. Because of the contrast from the small defects, it is impossible to tell whether any tangled dislocations of the type seen nearer the surface are present. Such contrast could also be caused by a significant increase in thickness. At a depth of ~750 nm observable damage ends abruptly.

Measurements of the nickel content of the implanted layer were made using x-ray EDS. In this case the measurements were obtained in STEM mode. The Ni Kα, Ti Kα, and respective background regions were recorded at 40 discrete intervals as the probe was moving continuously from a depth of 1000 nm within the specimen towards the surface. Due to the nature of the scan, the spatial resolution was approximately 30 nm. After quantification, with appropriate background and hole count subtraction, the results are plotted as the ratio of nickel to titanium, Ni/Ti, versus depth in Fig. 2. As can be seen there is a maximum in the nickel content at a depth of approximately 450 nm. Although nickel is present in fairly high levels there is no evidence of precipitation either in the bright-field images or in the

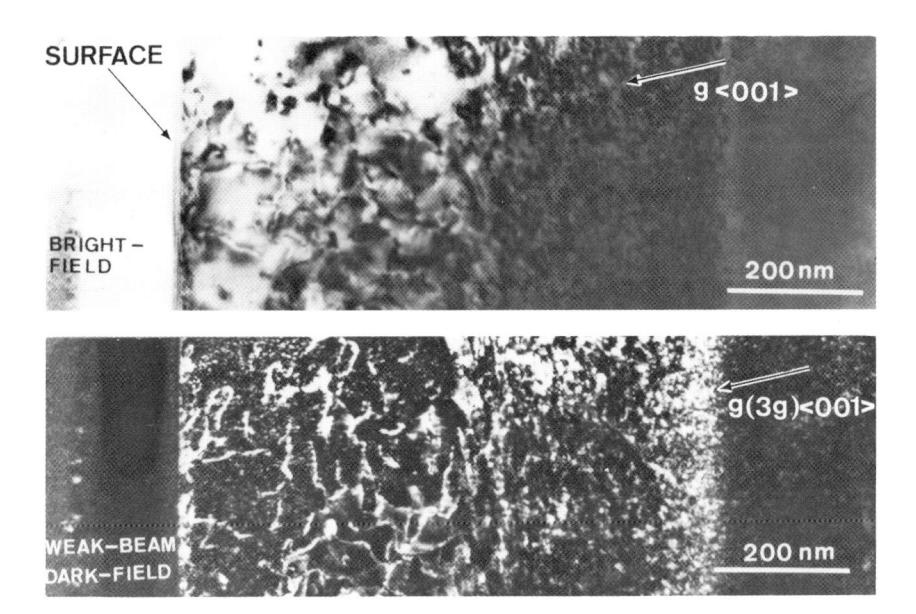

FIG. 1. Bright-field (top) and weak-beam dark-field (bottom) pair showing the implanted layer in cross section. Note the dislocation network near the surface and the fine damage near the end of the implanted layer. No precipitates were observed.

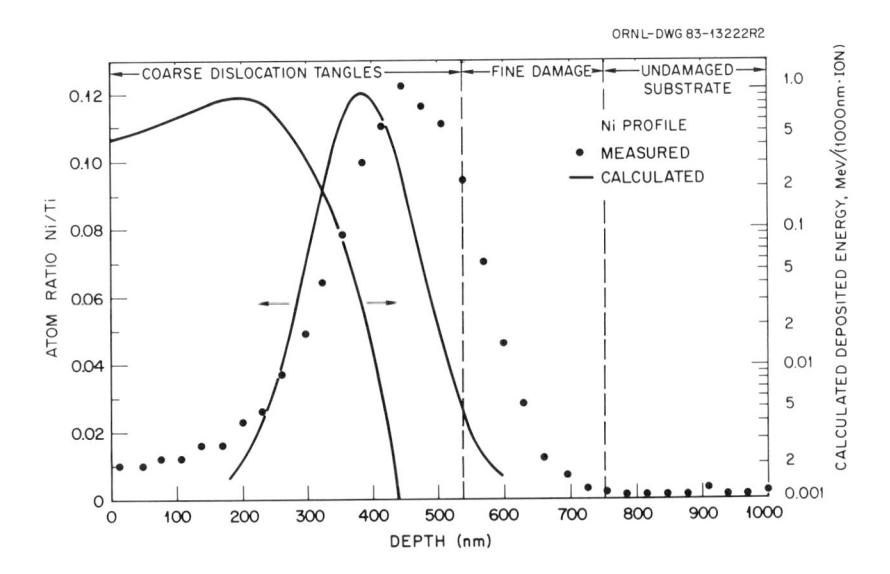

FIG. 2. The Ni/Ti atom ratio depth profile in as-implanted TiB_2 as determined by x-ray EDS. The results agree well with the calculated profile.

diffraction pattern from the modified layer. Further, the nickel concentration in the region deeper than 650 nm is low despite the obvious presence of a modified microstructure.

Since the atomic interactions which occur during surface modification by ion beam techniques are the same as those which occur during neutron or ion irradiation of materials, radiation damage models provide insight into the results of this study. One such computer model, the E-DEP-1 code of Manning and Mueller [2], was used to calculate the deposited energy and ion profiles for 1 MeV Ni$^+$ in TiB$_2$. The results of these calculations are also plotted in Fig. 2. There is reasonable agreement with the measured profiles, i.e., the peak in the deposited energy profile corresponds to a highly damaged region as indicated by the dislocation tangles and the peak in the deposited nickel ion profile occurs at a depth of ~390 nm, a difference of only 15% from the observed peak. Although this agreement is good, other calculations using the same code and taking into account oscillations in electronic stopping power of both the bombarding and the target species result in even better agreement. In this respect the Marlowe Binary Collision Program [3], which is based on detailed atomic structure and therefore accounts for oscillations in stopping power, also results in good agreement.

The observation of an obviously modified microstructure at depths where the nickel ion concentration and the deposited energy are low suggests that another phenomena may also be involved. Calculations indicate that there is a reasonable probability for nickel ions to produce displaced boron knock-on atoms. The range of these boron knock-on atoms is calculated to be approximately 800 nm, which is in reasonable agreement with the observed end of the modified layer, 750 nm. However, other effects such as diffusional spreading [4], as observed in other radiation damage experiments, could also be involved. At present it is not possible to distinguish between these two possible explanations. Future experiments may resolve this question.

In situ annealing for 30 minutes at 850°C produced a number of significant changes in the microstructure in the implanted layer. As the temperature was increased to ~500°C, the dislocations in the near-surface region were observed to move and in some cases to escape from the specimen. At higher temperatures precipitate formation occurred. Some of these

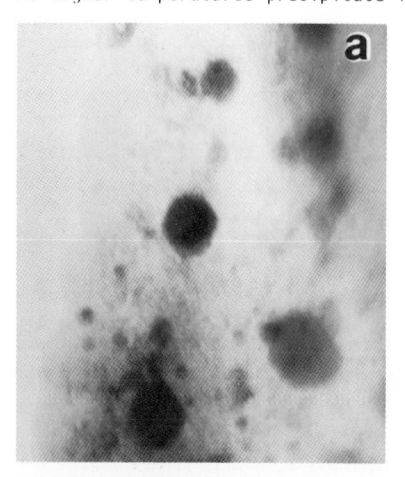

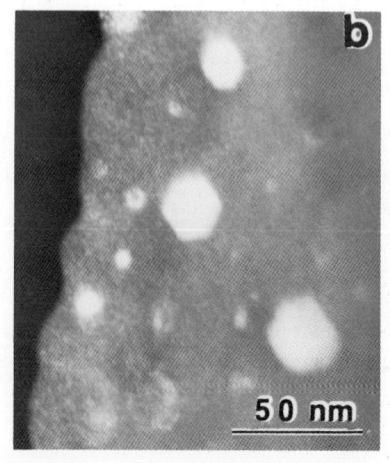

FIG. 3. Bright-field (a) and centered-beam dark-field using precipitate reflection (b) showing precipitates which formed during in situ annealing.

precipitates are shown in Fig. 3 which is a bright-field, centered-beam dark-field pair. The particles range in size from about 3 nm to about 40 nm. It is also clear from the micrographs that some of the larger precipitates are faceted. The EDS measurements made in the areas between the precipitates reveal only negligible amounts of nickel while measurements made on the precipitates themselves reveal that they are rich in nickel, Fig. 4. Since the precipitates are contained within the TiB_2 matrix it is difficult to determine the nickel content accurately. However, estimates based on the results of quantification indicate that the minimum nickel content of the precipitates is approximately 50 at. %. Identification of the precipitates based on crystal structure determination is not yet complete.

Observation of the implanted layer after annealing also revealed the presence of cavities distributed throughout the material, Fig. 5. Qualitatively the cavities appear to be distributed bimodally with the number density of cavities in the size range 1.5 to 2.5 nm approximately equal to $1-2 \times 10^{23}$ cavities/m^3, and a few cavities as large as 12.5 nm diam. In general the larger cavities are faceted. Stereo-microscopy

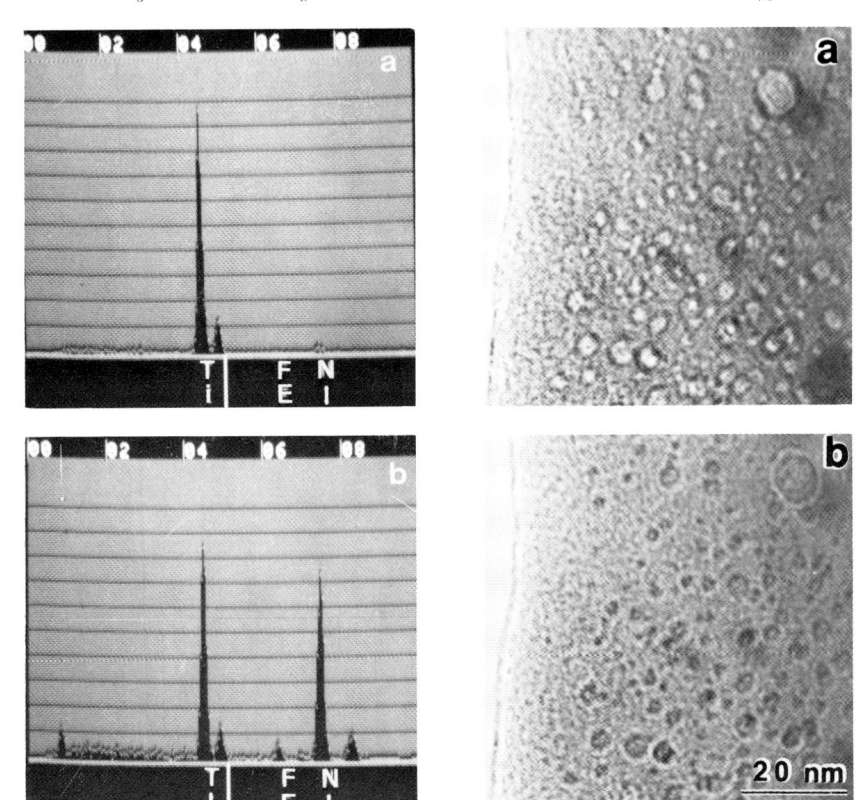

FIG. 4. X-ray spectra from (a) matrix, and (b) precipitate in implanted layer after in situ annealing.

FIG. 5. Cavities produced in the implanted layer during in situ annealing. (a) Underfocused. (b) Overfocused.

measurements confirm that the cavities are distributed throughout the thickness of the specimen and are not surface related. These cavities are similar in appearance to those seen in other materials (metals as well as ceramics) irradiated with energetic particles at higher homologous temperatures.

There are also some examples of cavity formation during elevated temperature annealing of materials which had been previously irradiated at low temperatures. For example Evans et al. [5] observed cavities which formed during annealing of molybdenum which had been neutron irradiated at 60°C. In this work low level gaseous impurities were shown to play an important part in stabilizing the cavities. Mazey and Francis [6] have reported on the evolution of small cavities in the microstructure of 316 stainless steel following helium implantation at ambient temperatures and subsequent annealing. Since oxygen is undoubtedly a contaminant in the TiB_2 used in this study, a similar mechanism may be operative here. Future work will seek to quantify this effect by measuring changes in the cavity size and distribution as a function of depth, annealing temperature, etc. Experiments to measure the effect of the observed microstructural changes on surface properties are also planned.

SUMMARY

Analytical electron microscopy has been used to characterize the near surface microstructure of TiB_2 implanted with 1 MeV Ni^+ ions. It was found that while the surface region was modified to a depth of ~750 nm, it remained crystalline and showed no evidence of precipitation. The microstructure observed was characterized by one region extending from the surface to ~500 nm containing dislocation tangles and a second region between ~500 and 750 nm possibly containing a high density of small defects. The maximum nickel concentration as measured by EDS occurred at 450 nm. Calculation using various radiation damage models agreed well with experimental observations. Observations made after in situ annealing revealed cavities and nickel-rich precipitates. Further experiments are planned to investigate the nature of these microstructural features and to determine their effect on surface properties.

REFERENCES

1. N.J. Zaluzec, "Introduction to Analytical Electron Microscopy," J.J. Hren, J.I. Goldstein, and D.C. Joy, eds. (Plenum Press, New York 1979) p. 121.
2. I. Manning and G.P. Mueller, Comp. Phys. Comm. 7, 85 (1974).
3. M.T. Robinson and I.M. Torrens, Phys. Rev. B9, 5008 (1974).
4. L.K. Mansur and M.H. Yoo, J. of Nucl. Mater. 85 and 86 523—532 (1979).
5. J.H. Evans, S. Mahajan, and B.L. Eyre, Phil. Mag. 26, 813 (1972).
6. D.J. Mazey and S. Francis, Proc. of the Consultant Symposium, "The Physics of Irradiation-Produced Voids," Harwell (1974).

ION IMPLANTATION OF POLYMERS

M.S. DRESSELHAUS,[#*+] B. WASSERMAN,[#+] AND G.E. WNEK[#θ]
[*]Department of Electrical Engineering and Computer Science;
[#]Center for Materials Science and Engineering; [+]Department of Physics;
[θ]Department of Materials Science and Engineering;
Massachusetts Institute of Technology, Cambridge, MA 02139, USA

ABSTRACT

Ion implantation provides a mechanism for radically modifying the electronic and transport properties of a variety of polymers that are normally insulating. By using masks and tailoring the implanted species and ion energies, conducting paths in an insulating medium can be fabricated between specific reference points, an application of obvious relevance to the microelectronics industry. Specific results are reported for modification of the structure, electrical conductivity, thermoelectric-power, optical transmission and electron spin resonance for several polymers under a variety of implantation conditions. The temperature and frequency dependence of the conductivity suggest a one-dimensional variable range hopping mechanism for conduction along the polymer chains. Comparison is made between implantation in the 200 keV and 2 MeV energy ranges.

INTRODUCTION

Most polymeric materials can be regarded as good insulators and many have electrical resistivities of $> 10^{12}$ Ωcm. The electronic structure of these insulating polymers consists of saturated bonds and tangled and interrupted chains which result in few if any conduction mechanisms. The low cost and easy processibility of some polymers make them attractive as electronic materials. The general instability of chemically doped conducting polymers presents problems with regard to their practical utilization. In contrast, the alternate doping scheme by ion implantation produces stable conducting materials. However, the conducting mechanism for ion implanted polymers is more difficult to determine because of the physical damage associated with the implantation process. Extensive studies of ion implantation in graphite have been reported [1] and it is found that the extensively damaged graphite material bears many similarities to highly damaged ion-implanted polymeric materials with regard to their electrical properties and lattice modes [2].

Recent developments in microelectronics technology make extensive use of energetic beams of radiation (electron, x-ray, and ion) for patterning. Of significance is the higher sensitivity of the polymer (by ~2 orders of magnitude) to ion beams relative to electron beams, so that ion beam lithography is feasible at fluences of less than 10^{13} cm^{-2} [3]. For positive resists (e.g., PMMA), ion implantation results in scission of the molecular chains, while for negative resists (e.g., polystyrene), implantation results in crosslinking of polymer chains. In order to understand this technology in a more fundamental way, the investigation of radiation-induced defects in polymers has become an area of interest. The use of ion beams to define conducting paths (for example as interconnects) through an insulating polymer film has been demonstrated and is of potential interest in very large scale integrated circuit technology. Furthermore, microelectronic device structures have been applied to sensitively measure transport properties of ion-implanted polymer films, using an interdigitated electrode configuration upon which a polymer film was spun and subsequently implanted using charge flow transitor technology [4,5].

A band-like conduction process has been discussed for the observed electronic properties of polyacetylene. Ion-implantation and chemical doping studies on polymers have been carried out on very disordered materials, so that a hopping process involving transitions between localized states forms the basis for any model of the electronic properties. Polymers typically have electronic band gaps $\geq$ 2 eV so that the localized states formed by the doping or implantation of polymers typically lie in this band gap. This paper focuses on the conduction mechanism in ion implanted polymers and on the nature of implantation-induced defects, including cross linking and scission of polymer chains, double bond, localized carrier and free radical formation and gas evolution.

ION IMPLANTATION AND SAMPLE CHARACTERIZATION

Ion implantation has been successfully applied to a number of normally insulating polymers to make them conducting, including the polymers poly(p-phenylene-sulfide) (PPS), poly(2,6-dimethylphenylene-oxide) (PPO), polyacrylonitrile (PAN) [4-6], poly(methyl methacrylate) (PMMA), poly(vinyl-chloride) (PVC) and other photoresists [2], and related non-polymeric organic thin films [7,8], 3,4,9,10-perylene-tetracarboxylic dianhydride (PTCDA), 1,4,5,8-napthalene-tetracarboxylic dianhydride (NTCDA), and Ni-phthalocyanine (NiPc). It has also been demonstrated that ion implantation can enhance the conductivity of polymers that are already conducting [9-11] such as polysulfurnitride $(SN)_x$ and polyacetylene $(CH)_x$. From this listing (see Fig. 1), it is seen that both simple chain (planar zig-zag) polymers and aromatic polymers with helical conformations can be rendered conducting by ion implantation.

Ion implantation studies on polymers are normally carried out on thin polymer films. These films are often prepared by dissolving the desired polymer in a solvent, which is then spun on to a substrate and heated to ~100°C to remove the solvent. Some solvents that have been used are: diphenyl ether or n-methyl-2-pyrrolidone for PPS [4,5]; chloroform for PPO [6]; and N,N dimethylformamide for PAN [6].

A wide range of ions have been implanted, including: ^{19}F, ^{36}Cl, ^{80}Br, ^{84}Kr, ^{127}I into PPS; ^{14}N and ^{80}Br into PPO; ^{14}N, ^{75}As, ^{80}Br, and ^{84}Kr into PAN; ^4He and ^{40}Ar into PTCDA, NTCDA and NiPc; ^4He and ^{40}Ar into PMMA and

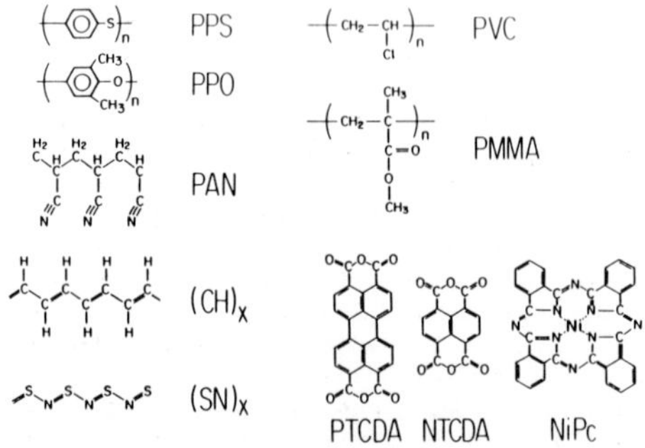

Fig. 1. Monomer units of polymers and related materials for which ion implantation has resulted in enhanced conduction.

PVC; ^{80}Br and ^{106}Pd into (SN)$_x$; and ^{19}F, ^{36}Cl, ^{80}Br, ^{127}I into (CH)$_x$. Some workers [9-11] have used relatively low energy implants $25 < E < 90$ keV to confine the implants to near-surface regions of the polymer substrate, while other workers [4-6] have used intermediate ion energies $35 < E < 230$ keV and fluences ϕ in the range $3 \times 10^{14} < \phi < 5 \times 10^{16}$ cm^{-2}, and yet other workers [2,7,8] have used sufficiently high energy implants (E = 2 MeV) to completely penetrate the polymer films. Approximately uniform ion irradiation can be achieved in this case, since R_p for a 2 MeV ^{4}He$^+$ ion exceeds 3μ, which is significantly larger than the film thickness of ~0.5 μ that was used [12]. The work in the various groups has not overlapped significantly either with respect to polymer substrate or ion energy range. In comparing the observed properties, consideration must be given to the difference in implantation conditions and choice of polymer films used by different workers.

The theory of implantation of ions in materials was originally developed for random or disordered solids [13]. In this form the theory is applicable to disordered polymeric materials with regard to the depth distribution of both the implanted ions and the structural damage. From this theory, a good estimate for the mean ion penetration depth R_p and the half width of the Gaussian distribution ΔR_p for ion implantation into bulk polymers can be obtained by assuming the density of the polymer to be that of amorphous carbon. Typical values of $R_p \sim 800$ Å and $\Delta R_p \sim 200$ Å are given here for ^{75}As implantation at 100 keV into PPS.

For the low and intermediate energy implants, a Gaussian distribution of implants is expected [13], and this has been confirmed experimentally [4-6]. For example Auger electron spectroscopy has been used to measure the spatial profile of the implanted ions in a polymer. The results verify the Gaussian distribution, and the magnitudes of R_p and ΔR_p are consistent with LSS theory; specific results have been reported in terms of the arsenic Auger line at 376 eV for ^{75}As implanted into PAN at 200 keV to a fluence of 1.0×10^{16}/cm^2 [6]. In addition, Auger spectra for elements of the host polymer material have been measured and the results indicate a depletion of volatile components such as sulfur in PPS in the near-surface region (d < R_p) and a corresponding relative increase in carbon concentration in PPS implanted with ^{84}Kr at 100 keV and a fluence of 1×10^{16}/cm^2. In contrast to the Gaussian distributions discussed above for the low and intermediate energy ion implants, the ions completely penetrate the thin film samples (1000 < d < 3000 Å) implanted at high energies (E = 2 MeV) and high fluence [2,7,8].

Structural Damage and Chemical Modifications

Ion implantation damage to polymers tends to be more destructive and irreversible in comparison with other types of targets. Some polymers (particularly those with simple unbranched chains) tend to crosslink and conjugate, while others (especially those with complex side chains) tend to degrade. For a given implant, the type of implantation-induced structural damage and chemical modification depends sensitively on the ion energy and mass. The extent of the structural damage depends sensitively on the fluence. Specifically, larger ionization density along the track of a high mass ion results in much larger structural damage in the case of high mass ions compared to lower mass ions with comparable energy and fluence. Evidence for irreversible structural changes for intermediate and high energy implants is provided by a variety of techniques.

Of particular significance has been the monitoring of the gas evolution during the implantation process, using an in-situ residual gas analyzer (quadrupole mass spectrometer) within the ion implantation chamber. For the case of ^{80}Br implanted into PAN at 200 keV, large amounts of hydrogen gas are emitted during the implantation process [6] with a very large initial increase in gas evolution followed by a decrease

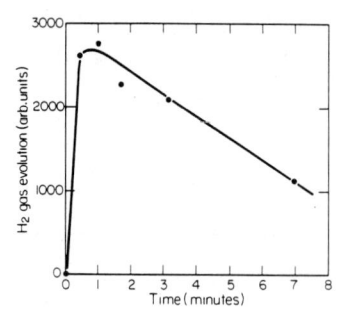

Fig. 2. Gas evolution of a polymer resulting from ion implantation. Here hydrogen gas evolution (arbitrary scale) is plotted as a function of time of implantation (proportional to ion fluence) obtained with an in-situ residual gas analyzer and a PAN film implanted with ^{80}Br at 200 keV.

at later times as shown in Fig. 2; here the time scale is proportional to the ion fluence. From the massive gas evolution we infer that the energy dissipation of the incident ions by the electronic energy loss mechanism occurs through the formation of large densities of free radicals (as confirmed by ESR measurements on the same samples [14]) and the molecular recombination of side chain and backbone constituents in the vicinity of free radical sites. We further infer that the implantation-induced structural changes result in conjugation of the polymer chains, the extent of the conjugation increasing with increasing ion energy. From Fig. 2, we see that as the fluence is initially increased, the gas evolution also increases, suggesting the release of loosely bound hydrogens from the sidechains. As diffusion proceeds, the gas evolution increases until a significant fraction of the hydrogens are removed, after which the probability of further gas evolution is reduced.

Copious gas evolution was also observed for high energy (2 MeV) implants into polymers. Massive quantities of CO_2, CO, formic acid, acetic acid and other species were detected by a quadrupole mass spectrometer [12] upon implantation of PMMA by 2 MeV ^{40}Ar ions. By monitoring the time dependence of the irradiation-induced emission using a 50 msec pulsed ion beam, and identifying the emission delays with molecular diffusion through the polymer films, a determination of the diffusion coefficients of various gases in polymers was made [12]. Radiation-enhanced oxidation commonly occurs in polymers such as $(CH)_x$, PPO and PVC [15].

For certain organic films (e.g., PTCDA), high energy, high fluence ion implantation is found to reduce the films mainly to carbon, with a large depletion of the original nitrogen content. These stoichiometric determinations were sensitively measured using Rutherford backscattering spectrometry (RBS) [8]. The RBS results are consistent with the copious gas evolution of volatile constituents in ion implanted PAN [6]. These stoichiometric modifications are also consistent with the shrinkage of the PTCDA film thickness from 3000 Å to 1500 Å upon ^{40}Ar implantation at 2 MeV and a fluence of 10^{16} cm^{-2} [8].

Even for low energy implants (e.g., 25 keV ^{106}Pd and ^{80}Br into $(CH)_x$), the broadening of the 1s carbon XPS (ESCA) lines indicates extensive structural damage. A comparison of the intensities of the 1s nitrogen ESCA peak with the 2p sulfur peak further indicates implantation-induced depletion of nitrogen in $(SN)_x$ in the near-surface region for low energy (25 keV) ^{80}Br ions at a fluence of $\phi = 5 \times 10^{15}$ cm^{-2} [9]. Evidence has also been presented for the chemical binding of the implanted bromine to a single site of the $(CH)_x$ backbone. Even for high fluences of implantation in the low energy regime, no evidence for implantation-induced ion diffusion was reported, based on RBS, ESCA, and infrared and ESR spectroscopies. From these measurements it is inferred that low energy implantation of ^{80}Br into $(CH)_x$ results in the formation of free radicals and some neutral hydrogen which is subsequently released as hydrogen gas, in addition to bonding some of the halogens to the polymer backbone.

Qualitative evidence for implantation-induced changes in the structure and properties of polymer targets is provided by color changes of the films upon implantation. For polymer films that are colorless prior to implantation, irradiation with the ion beam often imparts a brownish tinge upon implantation at low fluences. The color change deepens with increasing fluence and the effect is more pronounced for high energy implants. For example, PTCDA films exhibit a darkening for $\phi \geq 10^{12} cm^{-2}$; this darkening increases until a fluence of $5 \times 10^{15} cm^{-2}$ where the irradiated stripe was reported to become a dense shiny black. These color changes, the gas evolution and the stoichiometric changes all suggest the formation of carbonaceous material as a result of implantation of polymers. For high energy and high fluence implants (E = 2 MeV, $\phi \geq 10^{16} cm^{-2}$), evidence from the transport measurements described below suggest that this carbonaceous material bears a close resemblance to coal or amorphous carbon.

Electrical Conductivity

Dramatic increases in electrical conductivity σ (by 14 orders of magnitude) have been reported to result from ion implantation into a variety of normally insulating polymers and organic films under various conditions of implantation [4-8]. The general dependence of the conductivity on ion fluence ϕ seems to be qualitatively similar for several different ion species implanted into several host polymers under different implantation conditions. Examples of the observed dependence of log σ on log ϕ are shown in Fig. 3(a) for ^{84}Kr, ^{75}As and ^{80}Br into PPS [5] and in Fig. 3(b) for ^{40}Ar into various organic films [7,8]. Similar results have also been obtained for various implants into other polymers [6]. Measurements in all cases are made in the ohmic region; in Fig. 3(a), microelectronics devices (based on charge flow transistor technology) are used in the conductivity measurements, while the measurements in Fig. 3(b) were made with conventional 4-probe techniques, except for the high resistivity regime where 2-probe techniques had to be used.

For normally insulating polymers and organic films that can be made conductive by implantation, the lowest significant conductivity values shown in Fig. 3 ($\sim 10^{-9}$ to 10^{-10} $(\Omega cm)^{-1}$) correspond to a fluence of $\sim 10^{14} cm^{-2}$, where extensive structural modifications have already occurred, as indicated by optical transmission measurements [7,16]. While as-deposited organic films show a rich spectra of absorption bands, ion doses as low as $\phi \sim 10^{12} cm^{-2}$ are effective in attenuating and broadening absorption bands; at a fluence of $\sim 10^{14} cm^{-2}$ and above, a featureless absorption spectrum is obtained, suggesting the formation of amorphous carbon[7]. In the fluence range $10^{14} < \phi < 5 \times 10^{15} cm^{-2}$, a rapid increase in conductivity is observed, followed by a saturation behavior above $\sim 5 \times 10^{15} cm^{-2}$. It is significant that the behavior shown in Fig. 3 is common to a wide variety of implanted polymers over a similar fluence range. From these measurements we conclude that the dominant effect of ion implantation is the generation of defects such as scission and cross linking of polymers chains, the generation of broken bonds associated with the liberation of volatile species, and the formation of free radicals, double bonds and charge carriers. In the range $10^{14} < \phi < 5 \times 10^{15}$ cm^{-2}, chemical modifications to the polymer chains seem to be less important than the formation of structural defects, since little dependence on ion species is found. The results however suggest that the absolute magnitude of the conductivity achieved at the highest available fluences increases with increasing ion energy.

The correspondence of the maximum conductivity for the high energy implants with that of amorphous carbon suggests that ion implantation plays a role similar to high temperature heat treatment in the pyrolysis

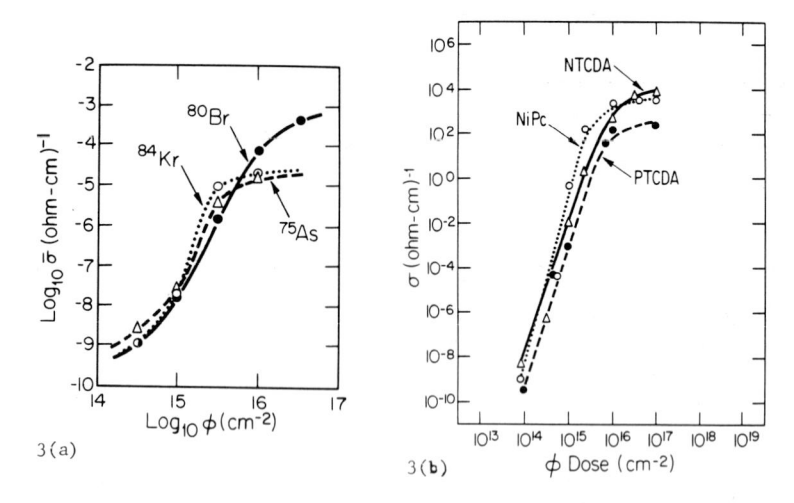

Fig. 3. Log-log plots of the dependence of the room temperature conductivity on the fluence of implantation. (a) Conductivity plot of ^{84}Kr (at 100 keV), ^{75}As (at 100 keV) and ^{80}Br (at 200 keV) implanted into PPS. (b) Resistivity plot of ^{40}Ar (at 2 MeV) implanted into PTCDA (closed circles), NTCDA (triangles) and NiPc (squares).

of precursor materials to form carbons. In this context, an analogy can be made between higher ion energies and higher heat treatment temperatures, so that as the incident ion energy is increased, the implantation process becomes more effective in selectively removing non-carbon atoms from the original polymer material. It is of interest to note that the maximum conductivity that has been achieved by ion implantation at high energies (2 MeV ^{40}Ar ions with $\phi \sim 10^{16}$ cm^{-2}) is at least an order of magnitude greater that for amorphous carbon [7], which implies that high energy ions promote conjugation and formation of local sp^2 carbon bonds. The higher graphitic content of the ion-implanted materials relative to amorphous carbon prepared by evaporation is also supported by the Raman spectra obtained from similar ion-implanted samples as were used for the conductivity measurements [2,7].

The rapid increase in conductivity with increasing ion fluence can also be related to other characterization measurements such as transmission electron microscopy (TEM) and electron spin resonance (ESR), discussed below. With regard to TEM measurements, sharp electron diffraction rings from as-deposited PTCDA films (indicating polycrystalline grain sizes ranging between 100 Å and 500 Å) were transformed by 2 MeV ^{40}Ar ions at a fluence of 1.0 x 10^{15} cm^{-2} into a broad diffraction ring, characteristic of amorphous carbon [7,8].

The saturation effect above 5 x 10^{15} cm^{-2} is very pronounced for chemically inert implants (e.g., ^{40}Ar and ^{84}Kr) and also for ^{75}As, while reactive halogens may continue to exhibit modest increases in conductivity at higher fluences. The different behaviors in the saturation region between the reactive halogen implants and the chemically inert implants suggest the formation of halogen bonds to the polymer side chains and backbones. A similar bonding arrangement for implanted halogens has been proposed [9] for halogens implanted (CH)$_x$ and (SN)$_x$.

Valuable information on the conduction mechanism is provided by studies of the temperature and frequency dependence of the ion-implanted

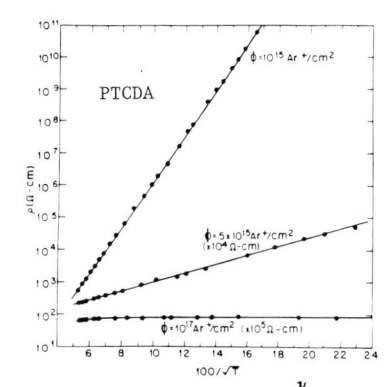

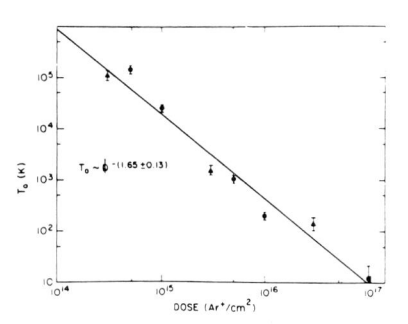

Fig. 4. Resistivity vs. $T^{\frac{1}{2}}$ for 3000 Å PTCDA films at various doses of 2 MeV ^{40}Ar ions (from Ref. 7).

Fig. 5. Dependence of characteristic T_o vs. 2 MeV ^{40}Ar ion dose for PTCDA (circles) and NiPc (triangles) (from Ref. 7).

polymers. Measurements of the temperature dependence of the conductivity of the ion-implanted polymers PAN and PPO [6], PMMA and other photoresists [2] and of related organic films [7,8] all yield a functional form of $\sigma = \sigma_0 \exp(-T_0/T)^{\frac{1}{2}}$ for a wide range of fluences and energies of implantation (see Fig. 4). It is of interest to observe that the same functional form is applicable to a large number of different implants and polymers, and over a wide range of fluences [2,6-8], though the values of σ_0 and T_0 are strongly fluence dependent (see Fig. 5).

Two models have been used to interpret conductivity data in ion implanted polymers. A conducting grain model has been proposed by Forrest et al. [7] to explain the conductivity behavior for high energy implants. According to this model the polymer is considered to consist of conducting particles or grains (idealized as spheres) separated by narrow insulating regions [17] and the current is controlled by the quantum mechanical tunneling between the grains. The fluence dependence is related by percolation theory to the number of paths through the polymer. In the context of the charge grain model, the material is assumed to be uniform, consisting of conducting grains or islands which are identified with crystallites or with amorphous carbon particles of diameter d separated by non-conducting barriers of width w. According to this model, T_0 becomes $T_0 = 4 \chi$ w E/k_B where $\chi = (2m^*\Psi/\hbar^2)^{\frac{1}{2}}$, and m^* is the effective mass, Ψ is the energy barrier height over which the charge must hop and E is the electrostatic energy necessary to remove a charge from an uncharged island and place it on an adjacent previously uncharged island, thereby producing two adjacent oppositely charged islands separated by a dielectric layer with permittivity η, so that assuming cylindrical symmetry, E is written as $E = 2q^2w/[\eta d^2(1/2 + (w/d))]$. From analysis of their data, Forrest et al. [7] determine w/d as a function of ϕ, and find for low fluences ($\phi \sim 10^{14}$ cm^{-2}), that (w/d) ~ 0.14, which they relate to conducting grains in the range 200 < d < 500 Å separated by w in the range 15 < w < 20 Å. With increasing ϕ, the w/d ratio is found to decrease as $\phi^{-0.83}$, so that for high fluences, w becomes less than the nearest neighbor interatomic distance, indicating the formation of bridges between conducting grains.

A one-dimensional hopping model has been proposed to explain transport phenomena in polymers implanted with ions of intermediate energies

50 < E < 250 keV [6]. The H_2 gas evolution suggests that the saturated bonds along the polymer chains are broken by collision with the energetic ions, and double bonds, free radicals and free carriers are generated. However, a hopping mechanism is necessary to complete the conduction path through the bulk material, as implied by the transport measurements of Wasserman et al. [6] and summarized below. This model relates to conduction in the semiconducting chalcogenide glasses, which are built from helical chains of Se atoms, that become highly disordered by doping the Se with other elements, thereby modifying and breaking the chains. In the ion implanted polymers, the one-dimensional character of the material is identified with the disordered polymer chains, with charge transfer occurring via hopping along, across and between the chain fragments.

Since ion implanted polymers have a morphology similar to that of chalcogenide glasses, it is suggestive to apply models which have been developed for these materials to the ion implanted polymers. On the basis of the one-dimensional thermally activated hopping model, the conductivity is expected to follow the relation $\ln(\sigma/\sigma_o) = -(T_o/T)^{\frac{1}{2}}$, in agreement with experimental results on ion implanted polymers. The model defines T_o by $T_o = 4\alpha/k_B N(E_F)$, where α denotes the exponential decay of localized states, k_B is the Boltzmann constant and $N(E_F)$ the density of states at the Fermi energy.

According to this one-dimensional variable range hopping model, a significant frequency dependence is expected. The frequency dependence of the real part of the conductivity is normally written in the form $\sigma_\omega = \sigma_o + A\omega^s$. For the charged grain model s is expected to be between $0.8 \leq s \leq 1$, while the experimental results [6] indicate a frequency dependent s rising above s = 2 for $\omega \geq 50$ kHz in PAN implanted with 200 keV ^{80}Br ions to a fluence of 2 x 10^{16} cm^{-2}. For the one-dimensional variable range hopping model a frequency dependence of $(\omega \ln\omega)^2$ is predicted [18]. Thus the frequency dependent conductivity appears to favor the one-dimensional variable range hopping model, though much quantitative work remains to be done.

DISCUSSION

Other transport and related measurements, such as thermoelectric power, Hall effect, optical transmission and electron spin resonance (ESR) also relate to the conduction mechanism. A summary of results obtained with these techniques is given here and related to the conduction model.

The temperature dependent thermopower measurements on ion implanted PAN and PPO exhibit the form $S = S_o + \beta T$ which is consistent with the one-dimensional variable range hopping model [6]. For example for PAN implanted with ^{80}Br to a fluence of 2 x 10^{16} cm^{-2}, n-type behavior is found with $S_o = -2.4 \mu V/K$ and $\beta = 0.018$ $\mu V/K^2$ for temperatures 200 < T < 300 K. The small magnitude of the thermopower implies a large carrier density of very low mobility carriers [6]. A lower limit on the carrier mobility μ is found from attempted Hall effect measurements on the same sample as for the thermoelectric power, yielding $\mu \leq 10^{-3}$ cm^2/V sec [16]. Very low mobilities are also characteristic of chalcogenide glasses.

Important information relevant to the conduction process in ion implanted polymers has also been obtained from analysis of the visible spectra of PPO and PPS in the wavelength range 310 < λ < 850 nm [16]. The results show a slowly increasing absorption with a nearly exponential dependence near the absorption edge. The short wavelength portion of the spectra can be interpreted in terms of a plasma resonance absorption which is fitted to obtain an estimate for the carrier density. An approximation to Mie's general formula for light scattering by small particles has been given by Kawabata and Kubo [19] where the transmission coefficient γ is related to the plasma resonance wavelength λ_R by the Lorentzian form

$\gamma = \gamma_0 \varepsilon_2 / [\beta^2 (\lambda-\lambda_R)^2 + \varepsilon_2^2]$ where $\gamma_0 = 18\pi\alpha(\varepsilon_m{}^3)^{1/2}/\lambda_R$ in which α is the volume density of the conductive part of the sample, ε_m is the dielectric constant of the non-conducting medium in which the conducting particles are embedded, ε_1 and ε_2 are the real and imaginary parts of the measured dielectric constant and β is the dispersion $\beta = \partial\varepsilon_1/\partial\lambda$ near the resonant wavelength λ_R. A good fit to this Lorentzian function for γ has been obtained for PPO implanted with 1×10^{16} cm^{-2} ^{80}Br ions at 200 keV for the wavelength range 310 < λ < 730 nm. A small deviation at higher wavelengths is attributed to the exponential shape expected near the absorption edge of disordered materials (Urbach rule [20]). The analysis carried out to relate the carrier density to λ_R employs the Bruggeman effective medium theory [21] by which λ_R and the bulk plasma frequency ω_p are related by $\lambda_R = 2\pi c/\omega_p^*$ where $\omega_p^* = (\{\omega_p^2/f\}-\tau^{-2})^{1/2}$ and $f = 1 + n^2(2+\alpha)/(1-\alpha)$ and $\varepsilon_2(\lambda_R) = [f\lambda_R/\pi\tau c]$ where n is the refractive index, τ is the scattering time and α is the volume density of the conducting medium. The analysis of the transmission data yields an estimate of the carrier density (~3.5 x 10^{22}/cm^3) and an estimate for the length of the dipole moment (~5 Å) responsible for free carrier optical absorption. Qualitatively, the broadness of the Lorentzian lineshape suggests a very short dipole moment which is readily interpreted by the one-dimensional variable range hopping model in terms of long thin charged particles (chains) with short lateral separations (interchain distances). In contrast, it is difficult to explain the broad Lorentzian lineshape in terms of a charge grain model.

Information on the density of free radicals and free carriers is provided by ESR measurements carried out on similar polymer samples [14]. In contrast to the absence of any ESR signal for the unimplanted polymers, the implanted polymers show a single, narrow Lorentzian ESR line, indicative of homogeneous broadening and non-metallic behavior, consistent with the conductivity model described above. As the fluence is increased, the ESR linewidth is observed to decrease, indicating motional narrowing and an increased delocalization of the carriers along the polymer chains. The decrease in the ESR linewidth $\Delta\omega$ with increasing ϕ indicates the formation of C=C double bonds [22] to a density n_d where $\Delta\omega \sim (n_d)^2$. The ϕ dependence of the ESR signal intensity shows a rapid increase up to a fluence of ~5 x 10^{15} cm^{-2} above which saturation is observed. It is significant that a similar behavior for the ESR signal intensity is found for both PAN and PPO and that the saturation behavior for the ESR signal and conductivity occur at the same fluence, suggesting an overlap in carrier wavefunctions between adjacent chains in this limit.

To separate the contribution of the localized spins (free radicals) and conduction electrons, the temperature dependence of the ESR signal was investigated in the range 90 < T < 300 K [14]. Since the density of localized spins n_s follows a Curie law, while the density of free carriers n_f make a temperature-independent Pauli contribution to the ESR signal intensity, by fitting the temperature dependence of the ESR intensity n_s and n_f can be separately determined. Investigation of the temperature dependent ESR signal intensity for a range of fluences for a variety of implanted PPO and PAN samples shows that for low fluences (~6 x 10^{14} cm^{-2}) the density of free carriers is about an order of magnitude smaller than the density of free radicals, while at high fluences (~1 x 10^{16} cm^{-2}), the densities of free carriers and free radicals are about equal [14].

The total density of spins in PPO implanted with 200 keV ^{80}Br ions is found to be 6 x 10^{17} spins/g (for ϕ = 6 x 10^{14} cm^{-2}) and 2.7 x 10^{19} spins/g (for ϕ = 1 x 10^{16} cm^{-2}). Taking into account the mass density of the polymer film (~3) and the fraction of the ~2 μm film that contains unpaired spins (~1/7) leads to a rough estimate of ~5 x 10^{20} spins/cm^3 for PPO implanted to 10^{16} cm^{-2} with 200 keV ^{80}Br ions. A comparison of this spin density with the carrier density deduced from the optical transmission data suggests that only a fraction of these carriers have unpaired

spins [14]. Although the one-dimensional variable range hopping model
account for many of the observed phenomena, many questions about the natture of the charge transport in ion implanted polymers remain unanswered.

ACKNOWLEDGMENTS

We gratefully acknowledge Drs. K. Sugihara, T. Venkatesan and G. Dresselhaus for many discussions, and the NSF-MRL grant #DMR-81-19295 for support.

REFERENCES

1. B.S. Elman, M.S. Dresselhaus, G. Braunstein, G. Dresselhaus, T. Venkatesan, B. Wilkens, and J.M. Gibson, Proceedings of this Symposium.
2. T. Venkatesan, S.R. Forrest, M.L. Kaplan, C.A. Murray, P.H. Schmidt, and B.J. Wilkens, J. Appl. Phys. 54, 3150 (1983).
3. T.M. Hall, A. Wagner and L.F. Thompson, J. Appl. Phys. 53, 3997 (1982).
4. J.S. Abel, H. Mazurek, D. R. Day, E.W. Maby, S.D. Senturia, G. Dresselhaus, and M.S. Dresselhaus, in Metastable Materials Formation by Ion Implantation, Proceedings of the MRS Symposium on Ion Implantation, edited by S.T. Picraux and W.J. Choyke, North Holland, New York 1982, Vol. 7, p. 173.
5. H. Mazurek, D.R. Day, E.W. Maby, J.S. Abel, S.D. Senturia, G. Dresselhaus, and M.S. Dresselhaus, J. Polymer Science, Polymer Physics Edition 21, 539 (1983).
6. B. Wasserman, M.S. Dresselhaus and G. Wnek, Proceedings of this Symposium.
7. S.R. Forrest, M.L. Kaplan, P.H. Schmidt, T. Venkatesan, and A.J. Lovinger, Appl. Phys. Lett. 41, 708 (1982).
8. M.L. Kaplan, S.R. Forrest, P.H. Schmidt, T. Venkatesan and A.J. Lovinger (unpublished).
9. W.N. Allen, P. Brant, C.A. Carosella, J.J. DeCorpo, C.T. Ewing, F.E. Saalfeld, and D.C. Weber, Synthetic Metals 1, 151 (1980).
10. D.C. Weber, P. Brant, C.A. Carosella, L.G. Banks, J.C.S. Chem. Commun. 198, 522 (1981).
11. D.C. Weber, P. Brant, and C.A. Carosella, in Metastable Materials Formation by Ion Implantation, Proceedings of the MRS Symposium on Ion Implantation, edited by S.T. Picraux and W.J. Choyke, North Holland, New York 1982, Vol. 7, p. 173.
12. T. Venkatesan, W.L. Brown, C.A. Murray, K.J. Marcantonio, and B.J. Wilkins (unpublished); T. Venkatesan, D. Edelson and W.L. Brown (unpublished).
13. J. Lindhard, M. Scharff, H.E. Schiott, Mat. Fys. Medd. Dan. Vid. Selsk. 33, 14 (1963).
14. G.E. Wnek, I.-H. Loh and B. Wasserman, (to be published).
15. M.C. Wintersgill, 2nd Int. Conf. on Radiation Effects in Insulators, Albuquerque, NM (1983) p. 92.
16. B. Wasserman (unpublished).
17. P. Sheng, B. Abeles and Y. Arie, Phys. Rev. Lett. 31, 44 (1973); B. Abeles, P. Sheng, M.D. Coutts, and Y. Arie, Adv. in Phys. 24, 407 (1975).
18. A.A. Gogolin, Physics Reports 86, 1 (1982).
19. A. Kawabata and R. Kubo, J. Phys. Soc. Jpn. 21, 284 (1966).
20. N.F. Mott and E.A. Davis, Electronic Processes in Non-Crystalline Materials, (Clarendon Press, Oxford, 1979) p. 287-293.
21. Y.K. Szeto, BS Thesis, Univ. of Toronto, 1980 (unpublished); R. Landauer, AIP Conference Proceedings, No. 40, edited by J.C. Garland and D.B. Tanner, American Institute of Physics (1978), p. 2.
22. S. Onishi, Y. Ikeda, S. Sugimoto and I. Nitta, J. Polymer Science 47, 503 (1960).

IMPLANTATION-INDUCED CONDUCTIVITY OF POLYMERS

B. WASSERMAN,[*,+] G. BRAUNSTEIN,[*,+] M.S. DRESSELHAUS,[*,+,#] AND G.E. WNEK[*,θ]
[*]Center for Materials Science and Engineering; [+]Department of Physics;
[#]Department of Electrical Engineering and Computer Science;
[θ]Department of Materials Science and Engineering;
Massachusetts Institute of Technology, Cambridge, MA 02139, USA

ABSTRACT

Ion implantation causes an increase by ~14 orders of magnitude in the electrical conductivity of normally insulating polymers such as polyacrylonitrile (PAN), poly(2,6 dimethyl phenylene-oxide) (PPO), and poly (p-phenylene sulfide) (PPS), after ion implantation with Br at fluences of 3×10^{15} ions/cm^2. The temperature dependence of the dc conductivity was measured in the range $23 < T < 293$ K and results show an exponential law $\sigma \sim \exp(-T_o/T)^\alpha)$ for PAN, PPO and PPS with $\alpha = \frac{1}{2}$, suggesting a one-dimensional hopping mechanism. The temperature dependence of the thermoelectric power (TEP) identifies the sign of the dominant carrier type. The TEP exhibits linear metallic behavior, with small magnitudes (~3μV/K), and shows Br implantation to yield p-type material in PPS and n-type material in PAN with extremely low values of mobility ($< 10^{-3}$ cm^2/V sec) and correspondingly very high values of the carrier concentration estimated to be 5×10^{22} cm^{-3}. Results are also reported for the frequency dependence of the AC conductivity and of similarly implanted PPS and PAN samples.

INTRODUCTION

In this work we present the results of studies on transport properties of three ion implanted polymers; polyacrylonitrile (PAN), poly(2,6 dimethyl phenylene-oxide) (PPO), and poly (p-phenylene sulfide) (PPS). The observed tremendous (~14 orders of magnitude) increase of conductivity through implantation leads to the possibility of creating easily processed, low cost materials with selective conductivity, interconnections for 2D and 3D integrated circuits, and of constructing p-n junctions.

EXPERIMENTAL

To achieve easily measurable conductivity for typical ion penetration depths ($\lesssim 3000$ Å), thin polymer films were used for the host matrix. These thin polymer film (2000Å to 1μm) samples were prepared by a spinning technique, using a thin microscope slide as a substrate. The spinning rate varied from 0 to 2500 R.P.M. Chloroform was used as a solvent for PPO and N,N-dimethylformamide was used to dissolve PAN. The best films were obtained using solution temperatures T_s between $100 < T_s < 130°C$. The films were cured in a vacuum oven for a few hours at $100°C$ to remove the remaining solvent. Electrical contacts were made using EPO-TEKH20E conductive epoxy, cured at $120°C$, and good ohmic contacts to the polymer were thus achieved. The implantation was performed on an Accelerators Incorporated 300 MP model ion implanter for doses ranging between 10^{14} and

$10^{17}/cm^2$, and with an ion beam energy of 200 keV. The samples were implanted with ^{79}Br, ^{75}As, and ^{14}N ions at room temperature.

The depth distribution of the ions was determined by Auger spectroscopy as a function of Ar ion sputtering time and typical results are presented in Fig. 1 for a PAN sample implanted with ^{75}As at 200 keV and a fluence of $1.0 \times 10^{16}/cm^2$ based on analysis of the 376 eV arsenic Auger transition. A crude estimate of the sputtering rate for the polymer was made by comparison with the sputtering rate of a chemically grown Ta_2O_5 layer of 1000 Å thickness, which was sputtered under the same conditions. The results of Fig. 1 give a Gaussian profile, as expected from LSS theory [1], and the measured ion penetration depth ($R_p \sim 2500$ Å) and halfwidth ($\Delta R_p \sim 360$ Å) are in good agreement with theoretical values, assuming the average mass of the polymer atoms corresponds to that of carbon [1].

During the implantation process large amounts of hydrogen gas were emitted from the polymer and the gas evolution was monitored by an in-situ residual gas analyzer mounted in the implantation chamber. Although other gas species could also have been emitted, the geometry of our residual gas analyzer permitted only the detection of hydrogen. The massive gas evolution is indicative of modifications to the polymer bonding and hence relates to the conduction mechanism.

The temperature dependence of the DC conductivity and thermoelectric power (TEP) were measured by means of a 2-probe technique (see inset to Fig. 2), due to the very high sample resistance at low temperatures. A schematic diagram of the experimental arrangement for each of the measurements is shown as an inset to the experimental results presented in the text. The measured voltages were continuously fed into a PDP 11/23 computer at very fine temperature increments, so that the measurement accuracy (between 5 and 10%) was defined by the load resistor. The temperature dependence of the conductivity for implanted PAN is shown in Fig. 3. The frequency dependence of the AC conductivity was measured using a Gen Rad 1689 Precision RLC Digibridge in the frequency range 300 Hz $< \omega <$ 100 kHz and the equivalent circuit for the experimental arrangement is shown in the inset to Fig. 4. The real part of the conductivity is given by

$$Re[\sigma_\omega] = G \omega (C - C_o) (1 + D^2)/D \qquad (1)$$

where G is the geometrical factor, ω is the frequency, C is the measured capacitance and C_o is the capacitance of the sample holder and the static contribution of the sample, and D is the measured loss factor.

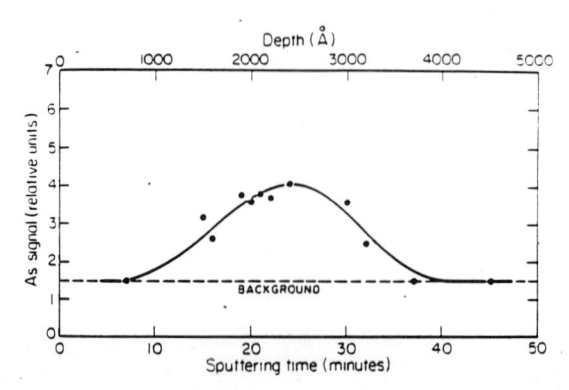

Fig. 1. Auger electron spectroscopy profile of an ^{75}As implanted PAN sample (E = 200 keV, $\phi = 1 \times 10^{16}/cm^2$). The curve is a Gaussian fit to the experimental points.

RESULTS

Conductivity measurements were made as a function of fluence, temperature and frequency to gain insight into the conduction mechanism. To establish the fluence range of interest, measurements of the room temperature conductivity were made as a function of fluence ϕ and the results are presented in Fig. 2. The results for PAN implanted with ^{80}Br at 200 keV are plotted in terms of $\log_{10}\sigma$ vs $\log_{10}\phi$ and show a rapid rise in σ until a fluence of $\phi_c \sim 5 \times 10^{15}/\text{cm}^2$ is reached, above which saturation occurs. Measurements repeated on the same polymer (PAN) with different ions (^{14}N, ^{75}As, ^{80}Br, ^{84}Kr) also exhibited the same general behavior and a similar value for the threshold fluence ϕ_c. These results are also similar to those previously obtained on other implanted polymers by Mazurek et al. [2] on PPS, and by Venkatesan et al. [3] on poly(methylmethacrylate) (PMMA), poly(vinyl chloride) (PVC), polyimide and others. It should however be noted that not all polymers that show increased conductivity upon doping, also show increased conductivity upon implantation, as for example our findings for poly(N-vinylcarbazole) PVK.

From Fig. 2, we see that the interesting range of fluence for conductivity measurements is $10^{14} < \phi < 10^{16}/\text{cm}^2$. To provide information on the conduction mechanism, the temperature dependence of the DC conductivity was measured over a wide (4.2 < T < 370 K) temperature range for ion-implanted PAN and PPS. A similar T dependence of $\ln \sigma/\sigma_0 = -(T_0/T)^{\frac{1}{2}}$ was obtained for the various polymers and implants used in this work, and typical experimental results are shown as points in Fig. 3 for ^{80}Br ions implanted into PAN at 200 keV and $\phi = 2 \times 10^{16}$ cm^{-2} for 80 < T < 280 K; for this sample $T_0 = 6 \times 10^3$ K. The inset shows results in the low temperature regime 4.2 < T < 50 K for another sample nominally implanted under the same conditions, and again the functional form $\ln \sigma/\sigma_0 = -(T_0/T)^{\frac{1}{2}}$ was obtained above 10 K, but with $T_0 = 2 \times 10^3$ K. The temperature independent conductivity below 10 K can be related to the loss of sensitivity of the apparatus rather than to an intrinsic effect. It is generally found that T_0 decreases with increasing fluence ϕ. However, at a constant fluence (e.g., $\phi = 2 \times 10^{16}$ cm$^{-2} > \phi_c$), similar values of T_0 (i.e., $6 \times 10^3 < T_0 < 8 \times 10^3$ K) were generally obtained for PAN and PPO implanted at 200 keV with several different ions but similar masses (^{75}As, ^{80}Br,

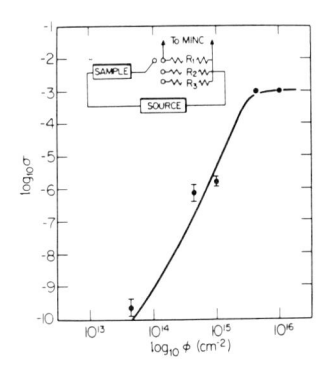

Fig. 2. Log-log plot of fluence dependence of the DC conductivity for ^{80}Br implanted PAN (E = 200 keV). The inset shows the experimental configuration.

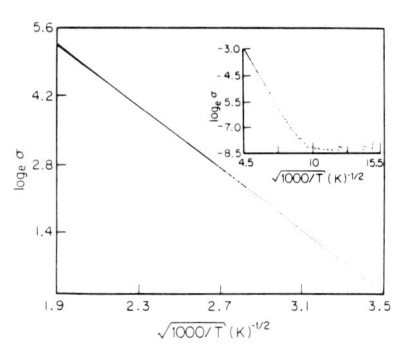

Fig. 3. Temperature dependence of the conductivity of PAN implanted with ^{80}Br at $\phi = 2 \times 10^{16}/\text{cm}^3$ and E = 200 keV (see text). The inset shows the low temperature behavior of a "similar" ion-implanted PAN sample.

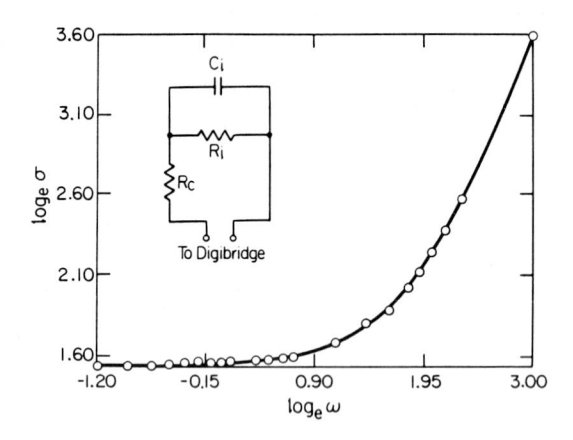

Fig. 4. Frequency dependence of the real part of the conductivity for ^{80}Br ion implanted PAN (E = 200 keV, ϕ = 2 x 10^{16}/cm^2). The inset shows the equivalent circuit.

^{84}Kr), though significant scatter in the T_o values were obtained for the same polymer implanted under nominally similar implantation conditions.

The frequency dependence of the electrical conductivity from a PAN sample implanted with ^{80}Br at a fluence of 2 x 10^{16}/cm^2 and 200 keV is displayed in Fig. 4. The results follow the traditional law for hopping conductivity in amorphous materials $\sigma = \sigma_o + A \omega^s$ with a frequency dependent exponent ranging from 0 < s < 1.5 in the frequency range $\omega \lesssim$ 20 kHz. It is significant that the frequency dependent term increases faster than linear and in fact exceeds 2 above 50 kHz (not shown in the figure).

Thermoelectric power (TEP) measurements were carried out to determine the carrier type dominating the conductivity. The results exhibit striking differences between PAN and PPO samples implanted under identical conditions. Specifically, PAN exhibits n-type behavior while PPO and also PPS exhibit p-type behavior. The magnitude of the thermopower is very small (S ~ 3 µV/K), consistent with a large concentration of low mobility carriers ($\mu \lesssim 10^{-3}$ cm^2/Vsec). Superimposed on a background temperature-independent term S_o, the TEP shows a slight linear dependence βT. For example for PAN implanted with ^{80}Br to a fluence of 2 x 10^{16}/cm^2, we find S = S_o + βT where S_o = -2.4 µV/K and β = 0.018 µV/K^2 in the temperature range 200 < T < 300 K. The corresponding Fermi level as deduced from TEP measurements is 0.9 eV relative to the mobility edge.

DISCUSSION

We model the implantation-induced conductivity as follows. When the initially insulating polymer is implanted, scission of the polymer chains occurs along the ion path, leaving free radicals and ions behind. Furthermore, we interpret the extensive hydrogen gas evolution in terms of the breaking of C-H bonds and the concomitant formation of a large concentration of charged centers along the polymer backbone and side chains. To maintain charge neutrality a large concentration of charged carriers must be simultaneously generated. Evidence for the large carrier concentration is provided by several independent experiments, such as the magnitude of the thermopower, reported in the present work, and Hall effect and optical absorption measurements to be reported elsewhere [4]. Ion implantation thus results in the formation of a large concentration of charged centers and low mobility carriers which give rise to electrical conductivity.

Of the various measurements reported in this work, the temperature dependence of the conductivity provides the most direct insight into the conduction mechanism. Following the work of Bloch et al. [5], we consider our system as a quasi-one dimensional conductor, where charge transport occurs via a variable range hopping of carriers along isolated chains as well as between backbone chains through side chain paths. The dominance of one-dimensional thermally activated hopping is supported by the observed $\exp(-T_o/T)^{\frac{1}{2}}$ temperature dependence of the conductivity.

We propose that the inhomogeneous nature of the host polymer and consequently also of the implanted polymer can be treated as a function of fluence according to the model of Cohen and Jortner [6] (CJ), who consider an inhomogeneous conductor in a pseudo-semiconducting regime. The temperature dependence of both the conductivity and the TEP for the ion implanted polymers in this work is consistent with the CJ model. We identify the polymer residual with the quasi-metallic component of the inhomogeneous composite material in the pseudo-semiconducting regime. The partial metal-like delocalization of electrons along the polymer chains is also consistent with our ESR measurements on the same implanted samples as were used in the present conductivity measurements [4]. We find that for all the polymer hosts and implanted ion species that were investigated in this work, the conductivity and ESR linewidths [4] saturate above a fluence of $\phi_s \sim 5 \times 10^{15}$ ions/cm^2. We explain this observation in terms of the formation of connected networks of conducting chains above ϕ_s. Because of the inhomogeneous nature of the samples, we would not expect a minimum metallic conductivity at low temperatures.

According to this model we infer that above ϕ_s we have strongly overlapping states where the intermolecular distance is smaller than the characteristic wave function decay length at the Fermi level. Following Ambegaokar et al. [7], the temperature-dependent conductivity is given by

$$\sigma = \sigma_o \exp(-T_o/T)^{\frac{1}{2}} \qquad (2)$$

where $T_o = 4 \alpha/k_B N(E_F)$ and $N(E_F)$ is the density of states at the Fermi level, α is the coefficient of exponential decay of the localized states and k_B is the Boltzmann constant. It follows that the energy spread at the Fermi level is given by $2 \Delta E_{max} = k_B (T\, T_o)^{\frac{1}{2}} = 2 k_B T_{max}$ where T_{max} is the corresponding temperature spread. Substituting for typical values of $T_o \sim 10^4$ K and $T = 300$ K, we find $T_{max} \sim 850$ K. It is significant that for $\phi > \phi_c$ the experimental values of T_o fall within a narrow range for all polymers and implants that were measured (i.e., $2 \times 10^3 < T_o < 8 \times 10^3$ K). It is also significant that characteristic T_o values for conducting polymers such as quinolinium-TCNQ ($T_o \sim 6 \times 10^3$ K) [5] are also similar to the values obtained in the present work.

The effect of ion implantation into PPO and PAN is to introduce delocalization of π orbitals resulting from bond cleavage and rearrangements. This model is supported by the narrowing of the ESR signal and the formation of a brown coloring of the implanted films. Interchain hopping via bipolarons is suggested as a conduction mechanism in chemically oxidized (doped) PPO and PPS [8].

We attribute the n-type behavior observed in the TEP measurements for the implanted PAN to the formation of dangling carbon bonds. These dangling bonds result from the broken hydrogen bonds, implied by the voluminous hydrogen evolution during the implantation process. We identify the dangling bonds with the formation of a donor band in the mobility gap.

In terms of the model of Chaikin et al. [9], the observation of p-type thermopower for PPO and PPS indicates that the defect band is more than half full. We note that according to the model of Chaikin for a single one-dimensional band, the thermopower has a weak linear temperature dependence which is identified with scattering processes.

The one-dimensional thermally activated hopping model [5-7] is also supported by the frequency dependent conductivity results. The observation of a ω^s dependence with $s > 1$ favors the one-dimensional hopping model following the work of Gogolin [10] who developed some qualitative suggestions of Mott, and showed that for a one-dimensional disordered lattice the real part of the conductivity exhibits a $(\omega \ln\omega)^2$ dependence for the delocalized electron states within the mobility gap. In contrast, the charge grain model for the conductivity, which also follows a $\ln(\sigma/\sigma_o) = -(T_o/T)^{\frac{1}{2}}$ dependence, implies a frequency dependence of ω^s with $s \leq 1$, which is inconsistent with our observations (see Fig. 4). Three dimensional interchain hopping would also contribute a term in ω^2 to the real part of the conductivity arising from the existence of a lower limit on the scattering time [11], implied by $\tau^{-1} < E_{max}/\hbar$, where τ^{-1} in this limit is replaced by ω_{phonon}. Substitution of this limit into Pollak's theory [11], and using the nearest neighbor distance for carbon atoms (1.4 Å), a carrier concentration of $\approx 10^{22}/cm^3$ (consistent with ESR and optical measurements [4]), and assuming a density of states of 10^{21} states/cm^{3}eV, we obtain a value of $\sigma \sim 0.1$ $(\Omega cm)^{-1}$ at 20 kHz, which is consistent with the order of magnitude measured experimentally.

CONCLUSIONS

The temperature dependence of the DC conductivity, together with the frequency dependence of the AC conductivity and TEP measurements, are consistent with a one-dimensional variable range hopping model for the carriers from one polarizable site to another along the chains, with some three dimensional contributions coming from hopping between the chains. The formation of a defect band in the mobility gap is suggested to explain the n- and p-type conductivities which result from ion implantation in PAN and PPO, respectively.

We gratefully acknowledge Drs. K. Sugihara and G. Dresselhaus for many enlightening discussions, Dr. A.S. Hay of the General Electric Company for the PPO, and the NSF-MRL grant #DMR-81-19295 for support.

REFERENCES

1. J. Lindhard, M. Scharff, H.E. Schiott, Mat. Fys. Medd. Dan. Vid. Selsk. 33, 14 (1963).
2. H. Mazurek, D.R. Day, E.W. Maby, J.S. Abel, S.D. Senturia, G. Dresselhaus and M.S. Dresselhaus, J. Polymer Science, Polymer Physics Edition 21, 539 (1983).
3. T. Venkatesan, S.R. Forrest, M.L. Kaplan, C.A. Murray, P.K. Schmidt, and B.J. Wilkens, J. Appl. Phys. 54, 3150 (1983).
4. M.S. Dresselhaus, B. Wasserman and G.E. Wnek, Proceedings of this Symposium.
5. A.N. Bloch, R.B. Weisman and C.M. Varma, Phys. Rev. Letters 28, 273 (1972).
6. M.H. Cohen and J. Jortner, Phys. Rev. Lett. 30, 699 (1973).
7. V. Ambegaokar, B.I. Halperin and S.J. Langer, Phys. Rev. B4, 2612 (1971).
8. R.R. Chance, D.S. Boudreaux, J.L. Bredas, and R.J. Silbey, Handbook on Conducting Polymers, edited by T. Skotheim (Marcel Dekker, New York) (to be published).
9. P.M. Chaikin, R.L. Greene, S. Etemad and E. Engler, Phys. Rev. B13, 1627 (1976).
10. A.A. Gogolin, Physics Reports 86, 1 (1982).
11. M. Pollak, Phil. Mag. 23, 519 (1970).

CHEMICAL AND PHYSICAL INTERACTIONS IN COVALENT POLYMERS IMPLANTED
WITH TRANSITION METALS

PEHR E. PEHRSSON,* D. C. WEBER,** N. KOONS,** J.E. CAMPANA,** S.L. ROSE**
*Catholic Univ. of America, Wash., D.C.
**Naval Research Laboratory, Wash., D.C.

ABSTRACT

Multiple analytical techniques including FABMS, XPS/Auger, magnetic susceptibility, and conductivity are used to probe the effects of transition-metal implantation on the structural and physical properties of covalent polymers and HOPG. Post-implantation chemistry and solid-state interactions between the implanted atoms and the polymer are investigated.

Introduction and Review

Ion implantation is a technique not usually applied to organics, particularly polymers. However, its unique ability to introduce large nonequilibrium quantities of atomically dispersed species into a solid matrix has recently demonstrated potential for unique and useful chemistries; viz, implantation of halogens into certain polymers to greatly increase their conductivity[1].

Beam damage to the original polymeric matrices in the energy and fluence regimes used in our experiments is substantial and irreversible. High kinetic energy ions striking the surface of the polymer target are slowed by electronic stopping mechanisms, resulting in polymer chain scission, crosslinking, and bond disruption[2]. Different polymers exhibit varying resistance to such damage, primarily as a result of variations in chain substituents and the degree of unsaturation. As the implanted atoms slow, nuclear stopping displaces large numbers of atoms in the polymer chains[3], destroying their long-term periodicity, and ultimately obliterating the molecular structure of the original monomeric units. In addition, preferential sputtering and the diffusion of volatile reaction products alter the sample stoichiometry.

Once in the matrix, the implanted species are subject to a variety of possible interactions. They may react chemically with the surrounding matrix analogous to the formation of carbides by implantation of carbon into certain bulk metals[4]. Alternatively, they may be physically or electrostatically trapped and isolated by the matrix, as seen in the vapor deposition of metal vapors on defect-saturated surfaces of Ar^+-sputtered graphite[5]. The metal atoms stick at the point of initial contact with the surface, and so are slow to nucleate and develop the bulk metal electronic structure. Or, the implanted atoms may diffuse either into the sample bulk or to the surface, with the accompanying prospect of nucleation and precipitation, similar to metal atoms vapor deposited on a graphite surface, which behave as a two-dimensional gas and diffuse on the surface until encountering a defect or nucleation site[5-6].

In this report, we describe the implantation of transition metal ions into a variety of covalent polymer matrices, followed by analysis and characterization of the resulting products.

Experimental

The polymers used were free-standing films of Mylar(polyethylene terephthalate), Nylon, polyethylene, PVF$_2$ (polyvinylidene fluoride), and polyacetylene. Polystyrene films were prepared by spin-casting onto an Al substrate from a toluene-based solution of the solid polymer. HOPG was cleaved in a He-filled glovebox and was transferred along with the other samples to a modified Varian/Extrion Ion Implanter. All transfers were accomplished under vacuum or N$_2$. The samples were implanted with a 25 KeV beam of Cr+, Fe+, Cu+, or Ag+, at a current density of 125 namps/cm² or less, to minimize heating damage. Fluences ranged from 4×10^{15} to 1×10^{17}/cm². Samples were stored in a glovebox at 2 ppm of O$_2$ and H$_2$O.

Analytical techniques included Fast-Atom Bombardment Mass Spectrometry (FABMS), X-ray Photoelectron (XPS) and Auger Spectroscopies, and fixed-and-variable-temperature conductivity and magnetic susceptibility measurements.

FABMS was performed on a ZAB-2F Reverse Geometry Mass Spectrometer, using a 4 KeV Ar+ sputtering beam. XPS spectra were taken on a McPherson Model 3600, with dedicated PDP12 computer, and on a PHI XPS/Auger equipped with a Model 255G Cylindrical Mirror Analyzer(CMA). Auger data was also taken on the latter instrument. Magnetic data was taken in a vibrating sample magnetometer with the capability of sample cooling to 2K. Room temperature sample conductivities were measured using a linear 4-point probe configuration, while the variable temperature studies were done with a van der Pauw 4-point probe configuration.

Results and Discussion

The samples acquired a silvery metallic color on implantation with metals, while Ar beams of equal fluence resulted in a brown color which varied in intensity with the fluence. Some polymers, particularly Mylar, exhibited superior resistance to gross manifestations of damage, while others, notably polyethylene, blistered and curled when implanted.

XPS spectra were taken both before and after implantation. The beam energy resulted in large implantation depths, necessitating the use of sample surface sputtering to obtain a metal XPS signal.

Implantation caused drastic changes in the C(1s) spectrum. Structure corresponding to carbon in different chemical and stuctural environments was lost, to be replaced with a single broad peak, slightly skewed to the high binding energy side. Binding energies converged at a value a fraction of an eV lower than that for hydrocarbons adsorbed on a metal standard in the spectrometer. Examination of the changes in the spectrum of PVF2 illustrate the extent of the beam damage. The virgin film consists of a polyethylene chain with the hydrogen substituents on every other carbon replaced by fluorine atoms. The C(1s) spectrum reflects these two markedly different bonding environments as two distinct peaks[7]. Implantation fluences of 1×10^{16} destroy the doublet, as well as distinguishing structure in the C(1s) signal of the other polymers (Fig.1). Subtle differences in the high binding energy shoulders suggest that some structure still remains in the carbon matrix, even at high fluences. Shakeup satellites characteristic of unsaturated systems[8] are also lost, indicating the destruction of the highly stable aromatic structures of the polymer. The valence band spectra of the samples also reflect the destruction of the original polymer, as once characteristic structure is replaced by the spectrum of amorphous carbon[9].

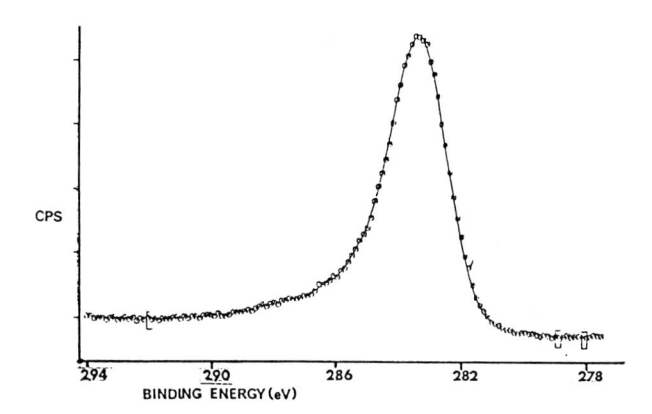

Fig 1. C(1s) spectrum of PVF_2, implanted with 2×10^{16} Ag^+/cm^2.

Auger spectra of graphite, amorphous carbon and metal carbides are quite distinctive[10]. Our spectra were clearly not carbidic, but rather resembled that of sputtered graphite, i.e. the random trigonal and tetrahedral bonds of amorphous carbon. Subtle changes occured in the fine structure of these spectra, particularly in the spectra where the metal XPS signal was strong. More definitive interpretation of the changes is underway, as are the metal spectra, using sophisticated lineshape-analysis techniques.

In the FABMS technique, the sample surface is sputtered with a beam of atoms generated by neutralizing a beam of accelerated ions. The ejected fragments are mass analyzed spectrometrically, providing a depth profile of the implanted species, as well as information on matrix stoichiometric and structural changes resulting from the implantation. Although the probing beam results in damage to the sample, experimental conditions, e.g. low current densities, can be chosen to minimize these effects. It is then possible to dislodge large fragments from the sample, up to and including monomeric units of the original polymer chain. Most of our spectra were obtained with higher beam densities, so that even the mass spectra from unimplanted films were composed primarily of small-chain fragments and recombination species. These were often still sufficient for at least partial identification of the original polymer.

The spectra of implanted PVF2 (Fig. 2) reveal the depletion of fluorocarbon fragments in the metallized region. The presence of higher mass fragments in the spectrum from the outer layers suggests that damage is incomplete in this region where electronic stopping mechanisms predominate. As the beam sputters down to the region where the metal concentration is low, but cascade damage from the implantation is still high, the abundance of the lower fluorocarbon fragments again increases, demonstrating the utility of FABMS for distinguishing even heavily damaged materials.

Fig. 2a

Fig. 2b

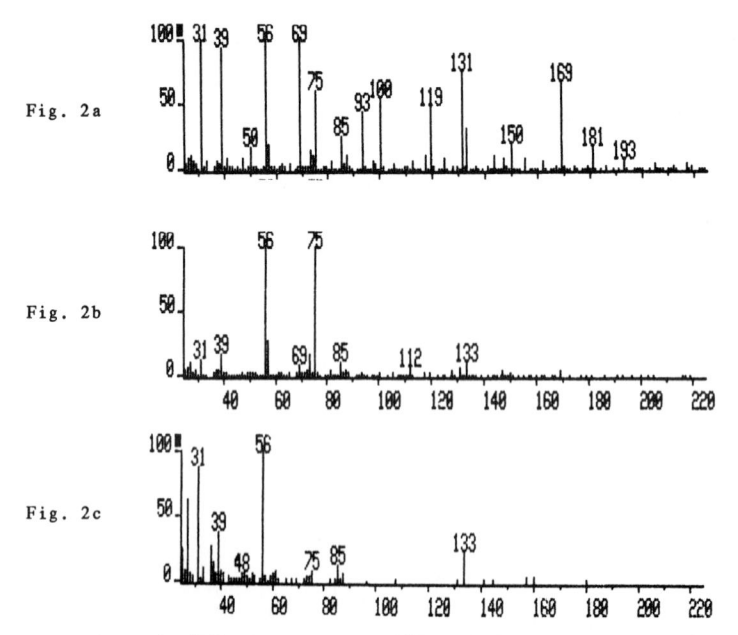

Fig. 2c

Fig. 2(a,b,c) FABMS:Intensity vs. M/Z of PVF_2 implanted with $3x10^{16}$ Fe+/cm² a) Scan 6-outermost sample layers b) Scan 20-region of peak metal intensity c) Scan 90-region of deepest implant penetration(low metal concentration) Comparative intensities of three scans are 4:20:1, respectively.

Two significant trends appeared in the positive ion FABMS spectra. In unimplanted, Ar⁻-implanted, and low fluence(10^{16}/cm²) metal-implanted samples, the profile of the total number of fragment ions ejected(TIC) vs. scans was comparatively low and flat, while samples implanted with higher metal fluences showed a large increase in TIC when the zone containing the metal was being sputtered. Metal species contributed a large part of the increased signal, but ejection of carbon species, particularly smaller clusters of C_x^+ and $C_xH_y^+$ also increased significantly. The ratio of carbon monomer to that of the dimer varied with the metal concentration. The outermost layers of material, where the metal concentration was low, as well as the underlying unimplanted polymer, had a consistently higher monomer to dimer ratio than did the metal-rich zone. For example, HOPG implanted with $1x10^{16}$/cm² Fe⁺ had a stronger monomer then dimer peak, while the reverse was true for a sample implanted to $6x10^{16}$/cm². The change in this ratio was weaker, although still evident in PVF2 samples implanted with high fluences of Fe⁺ compared to low fluence implants and the virgin film.

Fluence-dependent changes in the negative ion spectra were also noted. The pure HOPG spectrum consisted mostly of peaks corresponding to negative carbon clusters from 1 to 9 atoms, dominated by the dimer. The spectrum of PVF2 implanted with $1x10^{17}$/cm² Fe⁺ changes with sputtering depth. In the surface and unimplanted regions the higher order clusters are significantly less abundant than in the metal-rich region, where they increase, and the spectrum resembles that of HOPG. (Fig. 3)

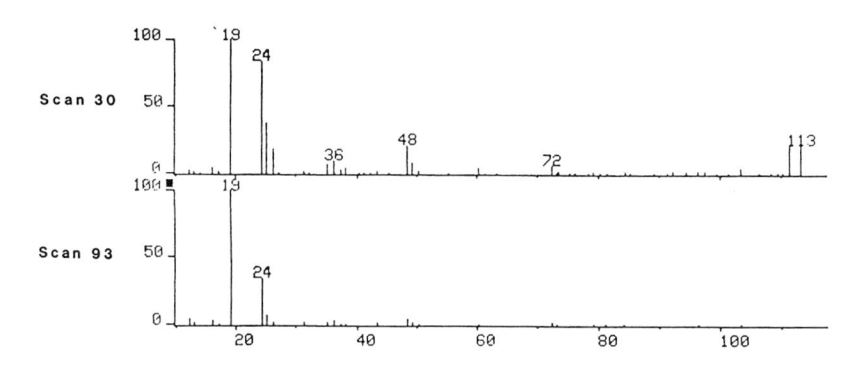

Fig.3(a,b) Negative ion FABMS spectrum of PVF$_2$ implanted with 1×10^{17}/cm² Fe+. a) Scan 30-Region of highest metal concentration b) Scan 93-underlying unimplanted polymer.

The positive ion spectra of the metallic species are consistent with the sputtering of isolated atoms and small clusters, as in pure metal. The strongest peak is that of the unipositive metal ion. Peaks corresponding to the dimer and trimer are seen only at fluences of 2×10^{16}/cm² or greater, and then in different proportion to those in bulk metal. For example, with a Ag standard the trimer is more intense than the dimer, whereas in our samples, if the trimer exists at all it is very weak. Mixed clusters of metal and carbon atoms equivalent to those seen in analysis of chromium carbides[11] were seen in the high fluence implants of Fe and Cr. The Fe-based species were quite scarce, while the Cr-based species were somewhat more abundant, although less so than in pure lower carbides. The spectra generally revealed little, if any, evidence of metal oxidation.

It is possible that these trends are artifacts of the plasma density of sputtered species above the sample surface, or of the stability of the fragment during its flight to the mass analyzer. Therefore they are not self-sufficient evidence of variations in cluster size or changes in chemical bonding within the polymer bulk.

Significant changes occured in the results of the other analytical methods in the same fluence regimes as in the FABMS data.

Magnetic susceptibility data was taken on samples of HOPG, polyethylene, and PVF2 implanted with fluences of 1,3,6,and 10×10^{16} Fe$^+$/cm². The samples were sealed in evacuated quartz tubes, and analyzed over the temperature range from 2 to 120 °K. All demonstrated similar characteristics, undergoing a transition from paramagnetic to ferromagnetic behavior between the two lowest fluences. The possibility exists that the Curie Point for at least some of the samples occurs below room temperature. Samples with fluences of 3×10^{16}/cm² had a weak magnetization temperature dependence, suggesting that they were intermediate between ferromagnetic and superparamagnetic behavior.

Room temperature measurements for polymers implanted with 1×10^{16} /cm² metal ions, or with Ar^+, revealed small or nonexistent conductivity increases. Higher fluence metal implants showed substantial fluence-dependent conductivity increases with little correlation to the implanted species or the polymer. Samples implanted with 1×10^{17} Fe⁺/cm² exhibited increases of 8 orders of magnitude after correction for the width of the implanted region. Sample conductivity decreased as the temperature fell from 298 to 78 °K, suggesting a nonmetallic conduction pathway[12]. Attempts are underway to clarify the conduction mechanism.

Conclusions

In this work, we have attempted to determine whether implantation of transition metals into covalent polymers results in (1) products from unique chemical reactions, (2) metastable solid solutions of metal in carbon, or (3) a collection of chemically discrete phases. The results indicate that polymers implanted to fluences of 1×10^{16}/cm² and greater lose much of their original molecular structure, albeit at somewhat different rates. At low fluences, metal atoms and small clusters are trapped in the carbon network, although their exact chemical disposition is unclear. Careful examination of the XPS metal spectra should provide information in this regard. As the metal concentration increases, larger clusters form, with concomitant changes in electronic structure evidenced by the onset of ferromagnetic behavior in the Fe⁺-implanted samples. Metallic clusters and bulk metal surfaces sometimes function as reaction sites for hydrogenation, graphitization and other transformations of simple carbon species[13]. In our samples, the presence of larger metallic clusters appears to promote double-bonding and conjugation in the damaged carbon matrix. This conclusion is consistent with the behavior of the conductivity of the samples with respect to the fluence, and with the increase in conductivity accompanying a rise in temperature. The results, although encouraging, do not completely reveal the nature of the changes occuring in the samples. Progress has been made, but much work clearly remains to be done to definitively characterize these unique materials.

REFERENCES

1. W.N. Allen, P. Brant, et. al., Synthetic Metals 1, 151-159, 1979/80
2. A.J. Swallow, Radiation Chemistry of Organic Compounds, (Pergamon Press, N.Y., 1960)
3. K.B. Winterbon, Radiation Effects 13, 215-226, 1972
4. I. Singer, Journal of Vacuum Science and Technology A1(2) 419-422 (1983)
5. R.C. Baetzold, Journal of Applied Physics 47 (9), (1976)
6. William F. Egelhoff, Gary G. Tibbetts, Physical Review B 19(10) 5028-5035, (1978)
7. D.T.Clark, H.R. Thomas, Journal of Polymer Science, Polymer Chem. Ed. 16, 791-820 (1978)
8. D.T. Clark, Advances in Polymer Science 24, 126-188 (1978)
9. F.R. McFeely, S.P. Kowalczyk, et.al., Physical Review B 9(12), 5268-5278 (1974)
10. T.W.Haas, J.T. Grant, G.J. Dooley, J. Applied Physics 43(4) 1853-1860 (1972)
11. R. Colton, T. Barlak, unpublished results
12. C. Kittel, Introduction to Solid State Physics, 5th ed. (J. Wiley and Sons, N.Y., 1976) Ch. 7-8
13. H.J.Krebs, H.P. Bonzel, Surface Science 99 570-580 (1980)

STRUCTURE/MAJORITY CARRIER RELATIONSHIPS IN ION-IMPLANTED POLYMER FILMS

G. E. WNEK, B. WASSERMAN† AND I.-H. LOH
Department of Materials Science and Engineering
†Department of Physics
Massachusetts Institute of Technology, Cambridge, MA 02139, U.S.A.

ABSTRACT

Ion implantation of selected polymer films leads to the formation of semiconductive ($\sigma \sim 10^{-4}$ S/cm) derivatives, which are presumably partially carbonaceous products derived from gross structural rearrangements. In support of this contention we find that Br-implanted polyacrylonitrile, PAN, exhibits a narrow ESR signal with a g-value of 2.0033, consistent with the presence of free radicals delocalized within a π-electron system. Thermopower measurements reveal that the sign of the majority carrier is dependent upon the molecular structure of the parent polymer. For example, implantation of PAN with Br$^+$ affords n-type derivatives while under similar conditions poly(p-phenylene sulfide), PPS, and poly(2,6-dimethylphenylene-oxide), PPO, yield p-type semiconductors. It is suggested that the majority carriers are the ions (carbenium ions or carbanions) which are best stabilized by the parent polymer structure, remnants of which presumably exist after implantation. Resonance and/or inductive effects are invoked to explain the data. For example, the rather electron-rich backbone of PPO is expected to stabilize holes more effectively than electrons, and the observed p-type behavior is consistent with this prediction.

INTRODUCTION

Ion implantation of several low molecular weight organic solids and high polymers results in rather large increases in electrical conductivity [1-4]. Of particular interest is the use of this technique for the fabrication of environmentally stable semiconductor devices from thin organic films. Thus, the ability to control the sign of the majority carriers is of great importance. In this article we present preliminary data concerning the majority carriers from thermopower measurements for ion-implanted poly(acrylonitrile), PAN, poly(p-phenylenesulfide), PPS, and poly(2,6-dimethylphenylene oxide), PPO, and suggest a simplistic (but potentially useful) rationalization of the data.

EXPERIMENTAL

Thin films of PAN, PPS, and PPO (solvents: DMF, diphenyl ether and chloroform, respectively) were spun (0-2500 R.P.M.) onto ~2 cm. glass squares cut from microscope slides. Electrical contacts were made with conductive epoxy (EPO- TEKH20E; cured at 120°C). Ion implantation was performed on an Accelerators Incorporated 300 MP implanter using doses between 10^{14} and 10^{17}/cm^2, and an ion beam energy of 200 keV. The samples were implanted with ^{79}Br and ^{75}As. The sample temperature during implantation was monitored by a thermocouple attached to the glass substrate. It was found not to exceed ca. 120°C and therefore the effects of implantation cannot be attributed to pyrolysis.

EPR spectra were recorded on a Varian E-9 spectrometer. Spin concentrations were determined [5] using a CuSO$_4$ standard.

Thermopower measurements were performed on the implanted samples by measuring the voltage gradient resulting from an applied temperature gradient ($\Delta T \sim 5°C$). Contacts to the samples were made with conductive epoxy. The thermopower coefficients, S, were determined from $S = \Delta V/\Delta T$.

RESULTS

It is found that ion-implanted PPO and PAN display intense EPR signals. The intensity increases and the linewidth decreases with increasing fluence. The spin concentrations reach an asymptotic value of ca. 10^{20} spins/g at the highest fluences (10^{16}/cm) where rather narrow linewidths (ca. 2-3 Oe) are observed. The g-value of Br^+-implanted PAN (g = 2.0033; BDPA [g = 2.0026] used as reference) at 3×10^{15} ions/cm^2 is higher than the free electron value (g = 2.0023).

Table 1 gives the sign of the majority carrier from thermopower measurements for PPO, PPS, and PAN implanted with ^{79}Br and/or ^{75}As. Implanted PPO and PPS are consistently found to be p-type, while implanted PAN exhibits n-type behavior. Thermopower coefficients were ca. 3μV/K in each case.

Table 1

Polymer	Ion	Fluence	Sign of Carriers
PAN	As, Br	2×10^{16}/cm^2	−
PPO	As, Br	2×10^{16}/cm^2	+
PPS	Br	2×10^{16}/cm^2	+

DISCUSSION

A chemical description of the complex consequences of ion implantation may be as follows. The energetic ions cause both homolytic bond cleavage affording free radicals and heterolytic cleavage yielding carbenium ions (positively charged carbons) and carbanions (negatively charged carbons), as well as expulsion of small molecules such as H₂ [4]. Bond cleavage is accompanied by significant rearrangement to π-electron systems (most likely fused rings) as suggested by the EPR data. The g-values of ca. 2.0033 for implanted PAN suggest spin-orbit coupling to heteroatoms (presumably N in this case) as is found in many samples of coal [6]. (In fact, the term 'coal' may be a reasonable descriptor for the material formed upon implantation.)

The fused, π-electron ring systems are amenable to thermal generation of carriers (electron-hole pairs). Which of the two is the more mobile will be the majority carrier. We rationalize the data of Table 1 by suggesting that there is a similarity between conduction in 'chemically doped' polymers [7] and ion-implanted polymers. In the former, carriers are primarily transported by intermolecular (interchain) redox processes. Any electron donor functionalities attached to π-systems can best stabilize carbenium ions (holes). It is well know that sulfur and oxygen atoms can donate electron density to π-systems via resonance. In this connection, it is noteworthy that poly(p-phenylene sulfide) is much more amenable to p-type doping (oxidation) than n-type doping (reduction) [7]. Thus, residual oxygen or sulfur in ion-implanted PPO and PPS could enhance hole mobility. In the case of implanted PAN, the π-system is probably similar in some respects to that of pyrolyzed PAN, which contains N atoms in conjugated imine structures [9]. Such nitrogens cannot donate electron density and, since nitrogen is more electronegative than carbon, these atoms can actually withdraw electron density via induction. This should lead to greater stabilization of electrons (carbanion centers) during intra- and intermolecular redox processes and allow for more facile transport compared with holes. Interestingly, pyrolyzed PAN is typically n-type [8]. That the majority carriers are dependent on polymer structure is further supported by the apparent independence of the nature of the ion employed (see Table 1).

Future studies include implantation of several other electron-rich and electron-deficient polymers to test the ideas discussed above. These studies, in conjunction with more detailed structural characterization of implanted polymers, will hopefully lead to the rational design of semiconductor devices derived from synthetic organic polymers.

ACKNOWLEDGEMENTS

We acknowledge Prof. M. S. Dresselhaus and Dr. G. Dresselhaus for stimulating discussions, Dr. A. S. Hay (General Electric) for the gift of PPO, and support from NSF-MRL grant #DMR-81-19295 and the MIT Materials Processing Center.

REFERENCES

1. S. R. Forrest, M. L. Kaplan, P. H. Schmidt, T. Venkatesan and A. J. Lovinger, Appl. Phys. Lett., 41, 708 (1982).
2. D. C. Weber. P. Brant, C. A. Carosella and L. G. Banks, J. Chem. Soc. Chem. Comm., 522 (1981).
3. H. Mazurek, D. R. Day, E. W. Maby, J. S. Abel, S. D. Senturia, G. Dresselhaus, J. Polym. Sci. Polym. Phys. Ed., 21, 539 (1983).
4. M. S. Dresselhaus, B. Wasserman and G. E. Wnek, Proceedings of this Conference.
5. J. E. Wertz and J. R. Bolton, "Electron Spin Resonance. Elementary Theory and Practical Applications", McGraw-Hill, New York, p. 502 (1972).
6. H. L. Retcofsky, in "Coal Science. Vol. 1", M. L. Gorbaty, J. W. Larsen and I. Wender, eds., Academic Press, New York, p. 56 (1982).
7. R. H. Baughman, J. L. Bredas, R. R. Chance, R. L. Elsenbaumer and L. W. Shackletter, Chem. Revs., 82 (2), 209 (1982).
8. H. Meier, "Organic Semiconductors", Verlag Chemie, Weinheim, Ch.7 (1974).

SYNTHESIS OF HARD SI-C COMPOSITE FILMS BY ION BEAM IRRADIATION OF POLYMER FILMS

T. VENKATESAN, T. WOLF, D. ALLARA, B. J. WILKENS and G. N. TAYLOR
Bell Laboratories, Murray Hill, New Jersey 07974

G. FOTI
University of Catania, Catania, Italy

ABSTRACT

We propose a novel technique to convert polymer films into useful inorganic films by ion beam irradiation. Along the track of an ion the polymer is dissociated into smaller fragments. Volatile fragments diffuse through the film and escape. Any element which is not removed in the form of volatile species is subsequently enriched with respect to the other elements. We demonstrate this effect in a polymer poly(dimethylsilylene-co-methylphenylsilylene), which initially has a C:Si ratio of 4.5:1. Upon irradiation with 2 MeV Ar$^+$ ions at a dose of 10^{15} ions/cm^2 the C:Si ratio changes to 3.4:1 as verified by Rutherford backscattering spectrometry. We believe that the effect of the ion beam irradiation is to produce more Si-C bonds at the expense of the C-H and Si-Si bonds, with $\leq 10\%$ of the original hydrogen being present in the film at high doses. The loss of the H atoms is further confirmed by a nuclear reaction technique. The IR spectra of the film as a function of the irradiation dose shows a progressive loss of fine molecular features with significant increase of the refractive index. The IR spectrum at the high doses appears to be due to a mixture of various Si and C bonds. However, the irradiated films are very hard and scratch resistant (knoop value $\gtrsim 1300$) suggesting an increase in the number of silicon carbide bonds.

Introduction

Ion beam interaction with polymers has been studied with the intention of using ions to expose resists for high resolution lithography.[1] Recently, ion beam irradiation has also been shown to reduce the resistivity of organic[2] and polymeric[3] films. The dynamics of the degradation of polymers by ion beam irradiation also has been studied[4] and in this study significant emission of molecular species has been observed from the irradiated polymer films. In general, most polymers tend to carbonize with increasing radiation dose and the reason for this is simply that not all the carbon atoms form volatile species during or after the irradiation process. This idea can be generalized to elements other than carbon. One could in principle design a polymer with elements A, B, C, etc. (Fig. 1) such that when the film is irradiated, of the various molecules formed, A and B are not removed significantly as volatile molecules, while the other elements are completely volatilized. One can then form an inorganic composite with a large number of A-B bonds. Coupled with the ability to pattern using ion beams, this may enable the patterned synthesis of an inorganic compound from a polymer film.

To demonstrate this idea we chose the polymer poly(dimethylsilylene-co-methylphenylsilylene), which is rich in silicon. The polymer is prepared (Fig. 2) by the method of West and co-workers.[5] A toluene-sodium mixture was heated to the reflux temperature resulting in a sodium metal dispersion. After cooling the dispersion to 70°C, a mixture of 52.6g of freshly distilled methylphenyldichlorosilane and 35.5g of freshly distilled dichlorodimethylsilane was added gradually. The mixture was heated for 6 hours at 110°C. After cooling, methanol and subsequently water were added to the mixture to ease the separation of the organic layer. Toluene was subsequently removed and the product was dissolved in tetrahydrofuran (THF). The high molecular weight fraction was precipitated from solution by the slow addition of the THF-polymer solution to methanol. After vacuum drying poly(dimethylsilylene-co-methylphenylsilylene) was obtained. The molecular weight of this fraction was 125,000 with a polydispersity of ~ 10. Nuclear magnetic resonance measurements indicated the polymer to be a 50:50 mixture of the two monomers. The polymer was dissolved in chlorobenzene and spun onto silicon wafers. The polymer films were baked in an oven at 120°C for 30 minutes to evaporate the solvent. Typical film thicknesses of $\sim 6000\text{\AA}$ were used in the experiments to be described.

The films were then irradiated by a 2 MeV Ar$^+$ beam from a Van de Graaff generator. Irradiated regions typically 5×5 mm were produced on these films. To study the composition of the film after

$A_a B_b C_c D_d$

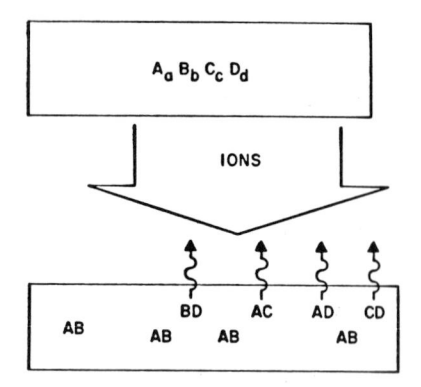

IONS

AB BD AB AC AB AD AB CD

Fig. 1 Schematic representation of formation of an inorganic film of molecules AB by ion beam, irradiation of a film containing atoms A, B, C and D.

$A_m B_n$

irradiation we prepared the polymer films on Be substrates. After irradiating several regions with varying doses, the films were analyzed by Rutherford scattering spectrometry (RBS) using a 2 MeV He$^+$ ion beam. The backscattering spectra are shown in Fig. 3. The spectrum of the pristine polymer shows strong Si and C peaks besides the Be continuum from the substrate. The oxygen peak is from a layer of beryllium oxide on the substrate, which can be consistently subtracted from the spectra. The measured areas under the peaks for C and Si gives C:Si = 4.5:1, as expected from the molecular structure. There is no backscattering peak of H as the atoms are too light to backscatter the He ions. however, the H atoms do contribute to the energy loss of the He ions and the height of the C and Si peaks would be affected by the amount of H atoms present in the film. The height of the silicon peak is proportional to

$$F_{Si} = \frac{f_{Si}\epsilon_{Si}}{f_{Si}\epsilon_{Si} + f_C\epsilon_C + f_H\epsilon_H} \quad (1)$$

where f_I and ϵ_I are the atomic fraction and the stopping cross section of element I respectively. Putting in known values[6] of ϵ in the unirradiated material (with C:Si:H=9:2:14) we get $f_{si} = 0.19$. In the material dosed with 10^{16} Ar$^+$/cm^2, the Si peak height increases (with negligible decrease in area) by a factor of 40.4% implying an $f_{si} = 0.27$. Putting in the measured atomic fractions for C and Si (from the areas under the peaks) into equation 1 we get

C:Si:H = 7.2:1.38 which implies

90%, 10% and ~5% loss of H, C and Si respectively. By using a residual gas analyzer we have directly measured the evolution of molecules from the polymer film. The emission consist of mostly H_2 and to a smaller extent CH_4 consistent with the measured loss of elements by RBS. Thus there is a significant enrichment of Si in the film with the final C:Si ratio = 3.4:1. Interestingly, an increasing oxygen peak is observed for doses >10^{13} ions/cm^2, but this begins to decrease at the very high doses. Oxygen incorporation is common in most organic films irradiated by energetic ions with the oxygen from the atmosphere reacting with the free radicals produced in the film by the irradiation. At high doses the free radicals in the film recombine and much less oxygen is incorporated.

Fig. 2   Synthesis of poly(dimethylsilylene-co-methylphenylsilylene).

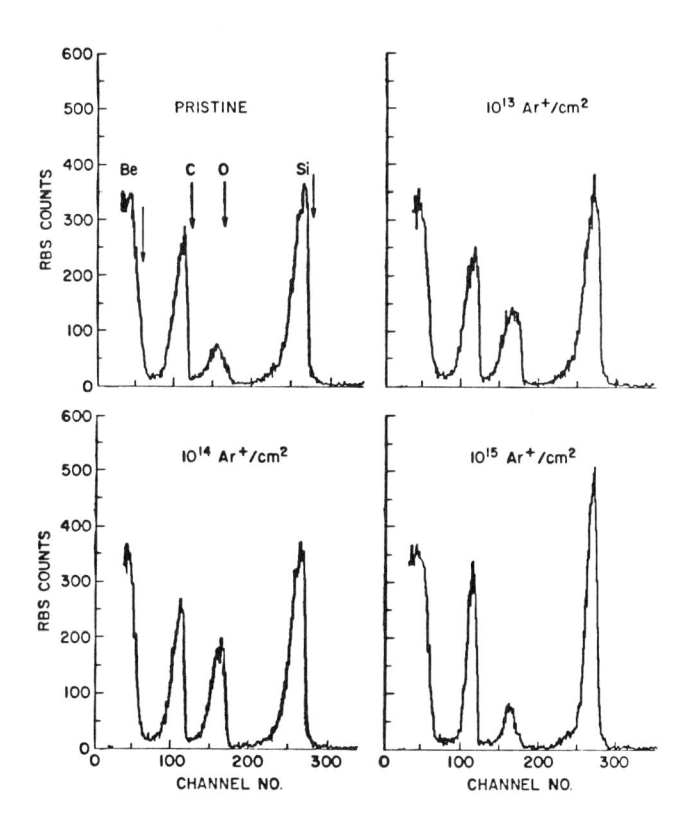

Fig. 3   Rutherford backscattering spectra as a function of Ar^+ ion dose for the Si-C polymer film, on Be substrate.

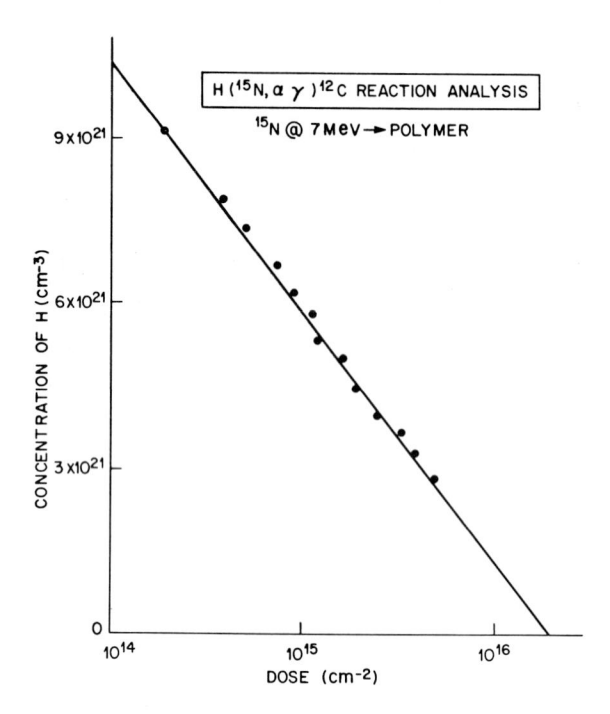

Fig. 4 ^{15}N nuclear reaction measurement of the H concentration as a function of the irradiation, dose.

The loss of H atoms is further confirmed by a nuclear reaction analysis technique[7] using the reaction H(^{15}N, αγ)C^{12}, which has a resonance at 6.385 MeV with a width of 0.9 keV. The reaction produces a 4.43 MeV γ-ray with the γ-ray yield proportional to the hydrogen concentration. Using an incident 7 MeV ^{15}N$^+$ ion the resonance is expected at a depth of 2000Å in the film over a region of 500Å width. We assumed a density of 1 gm/cc for the polymer and an electronic energy loss of 1.0 MeV/μm. The measured hydrogen concentration in the film as a function of ion dose (Fig. 4) exhibits a steady exponential decrease and by extrapolating the curve the hydrogen content of the film seems to be negligible by an ion dose of 2×10^{16} cm^{-2}. The energy loss of a 2 MeV Ar$^+$ ion is comparable if not higher than that of the 7 MeV N$^+$ ion and the loss of hydrogen with increasing Ar$^+$ irradiation dose is further brought out by these results.

The regions irradiated beyond a dose of 5×10^{14} ions/cm^2 are very hard. The film could not be scratched by very sharp stainless steel blades, quartz or iron carbide or garnet with a hardness[8] of >1300 knoop values indicating presumably a significant formation of silicon carbide bonds. The films were electrically insulating with resistivity >10^6 Ω cm. In comparison, other polymer films carbonized by similar irradiation[3] are highly conductive (ρ ~ 10^{-3} Ω cm) and nowhere near as hard and are easily scratched by stainless steel blades or quartz. This proves that the hardness of the irradiated organo silicon polymer is not due to structural rearrangement of the C atoms (e.g., diamond-like carbon).

The chemical structure of the film was studied using IR spectra as a function of irradiation dose. The spectra of the heavily dosed sample was compared with that of an amorphous film of SiC and carbon. The IR samples were typically 3000Å thick and spun onto doubly polished, undoped silicon wafers. The IR data are shown in Fig. 5. Sharp lines are observed in the pristine film at ~3000

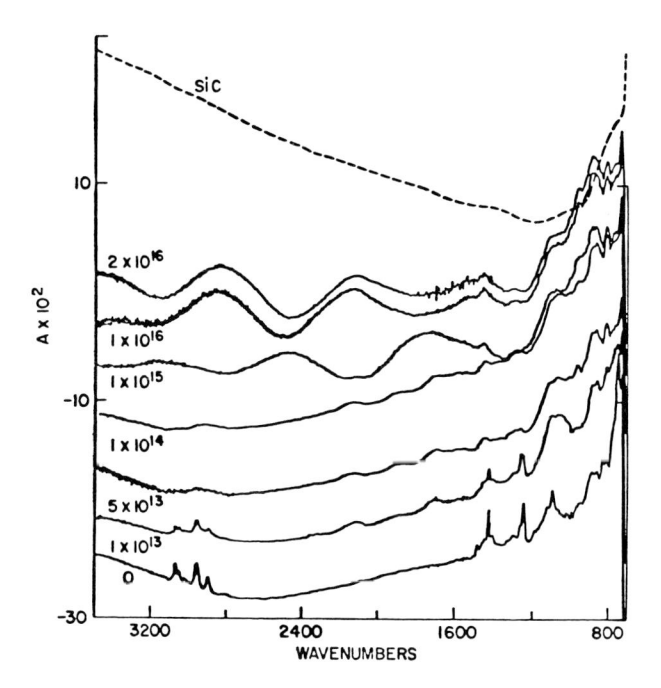

Fig. 5   The infrared spectra as a function of irradiation dose. The dotted line is the spectrum of an, evaporated silicon carbide film.

wavenumbers (corresponding to C-H stretching modes), and ~1400 wavenumbers (corresponding to phenyl group modes). The broad structure at 800 wavenumbers is associated with Si-C stretches. As the films are irradiated the sharp features are completely eliminated by doses of ~10^{14} ions/cm². However, at higher doses oscillation occurs due to interference in the film which indicates an increase in the refractive index. In fact the refractive index increases by a factor of 2 in going from 10^{13} to 2×10^{16} ions/cm². SiC has a strong absorption feature at 800 cm^{-1} and an e-beam evaporated film of SiC did show a broad band at 800 cm^{-1} (Fig. 4). The highly ion irradiated polymer does show prominent features at 800 cm^{-1} in addition to absorption features at higher frequencies. This is not surprising, as the irradiated film, owing to the excess carbon, is presumably a mixture of SiC and Si regions in a carbon host. That is the reason why the film has lower hardness than that of SiC[7] (knoop value of ~2480).

In conclusion, polymers are a rich source of starting materials for the patterned synthesis of interesting inorganic films utilizing energetic ion beam irradiation. The adhesion enhancement[9,10] of the film to the substrate, when the ions are made to traverse the film-substrate interface, is a further benefit of ion beam irradiation. Polymers could be synthesized with other inorganic elements besides Si. Many other novel compounds could be formed using ion beam irradiation.

We would like to thank R. Hamm for e-beam evaporation of the SiC film and Prof. J. Choyke for the bulk SiC material.

REFERENCES

[1] W. L. Brown, T. Venkatesan and A. Wagner, Rev. of Nucl. Instrum. and Meth., *191*, 157 (1982) and ref. therein.

[2] S. R. Forrest, M. L. Kaplan, P. H. Schmidt, T. Venkatesan and A. Lovinger, Appl. Phys. Lett. *41*, 708, (1982).

[3] T. Venkatesan, S. R. Forrest, M. L. Kaplan, C. A. Murray, P. H. Schmidt and B. J. Wilkens, in J. Appl. Phys., *54*, 3150 (1983).

[4] T. Venkatesan, D. Edelson and W. L. Brown, Appl. Phys. Lett. *43*, 364 (1983).

[5] R. West, L. D. David, P. I. Djurovich, K. L. Stearley, K. S. V. Srinivasan, and H. Yu, Jour. of Am. Chem. Soc., *103*, 7352 (1981).

[6] W. K. Chu, J. W. Mayer, M. Nicolet, p. 362, Back Scattering Spectrometry, Academic Press, NY (1978).

[7] W. A. Landford, H. P. Trautvetter, J. F. Ziegler and J. Keller, Appl. Phys. Lett. *28*, 567 (1976).

[8] Handbook of Chemistry and Physics, edited by R. C. West and M. J. Astle, CRC Press, Inc., Boca Raton, Florida, (1982-83).

[9] J. C. Griffith, Y. Qiu and T. A. Tombrello, Nucl. Instr. Meth., *198*, 607 (1982).

[10] B. T. Werner, T. Vreeland, Jr., M. H. Mendenhall, Y. Qiu, and T. A. Tombrello, Thin Solid Films, *104* (1983) In press.

MAGNETIC PROPERTIES OF IRON IMPLANTED POLYMERS AND GRAPHITE

N. C. KOON, D. WEBER, P. PEHRSSON, AND A. I. SCHINDLER
Naval Research Laboratory, 4555 Overlook Avenue,
Washington, DC 20375

ABSTRACT

We have measured the magnetic properties of highly oriented pyrolytic graphite (HOPG), polyethylene (PE), and polyvinylidine fluoride (PVF_2), implanted with fluences of 25 keV iron atoms ranging from 10^{16} to 10^{17} atoms/cm². The lowest fluence specimens were paramagnetic down to 2 K, with evidence for clusters of only a few spins, while the highest fluence specimens were clearly ferromagnetic, with magnetization curves resembling those of a set of randomly oriented soft magnetic planes. The critical fluence for formation of a ferromagnetic state appears to be between 1 and 3 x 10^{16} atoms/cm² at 25 keV. These results can be qualitatively understood based on the critical density for percolation of near neighbor exchange interactions.

INTRODUCTION

In recent years the modification of crystallographically layered materials by intercalation of foreign species, principally by chemical means, has been the subject of intense interest. Many species, however, cannot be intercalated by chemical means. Ion implantation offers an alternative method of inserting foreign species into host materials. For a variety of reasons it cannot really be considered equivalent to chemical intercalation. For ions implanted into polymers and graphite, one of the common results is an amorphous surface layer (1-3) which typically occurs for fluences greater than 1-5 x 10^{15} ions/cm². The purpose of the present work was to explore the magnetic properties of iron implanted into graphite and the polymers PE and PVF_2. Because of the typically strong nature of the exchange coupling between iron spins, even in amorphous materials, it was expected that the magnetic propeties would provide useful information about the structure and separation of the implanted atoms.

EXPERIMENT

The samples were prepared by implantation at a fixed energy of 25 keV normal to the graphite face or free standing polymer films. The penetration depth of iron atoms in graphite at that energy is expected to be 147 Å. The 1/e width of the distribution is expected to be 47 Å, assuming no sputtering at the surface. The penetration into PE and PVF_2 is approximately 500 Å. Samples were prepared with fluence levels of 1, 3, 6, and 10 x 10^{16} atoms/cm². Peak concentrations of the iron atoms in the HOPG, for example, were expected to be 6.5, 17.3, 29.5, and 41.1 atomic percent, respectively.

The ion implanted samples were removed in an inert atmosphere and sealed in quartz tubes under vacuum. Magnetic measurements were made using a vibrating sample magnetometer and a superconducting solenoid. For experimental reasons the data were limited to temperatures less than 150 K.

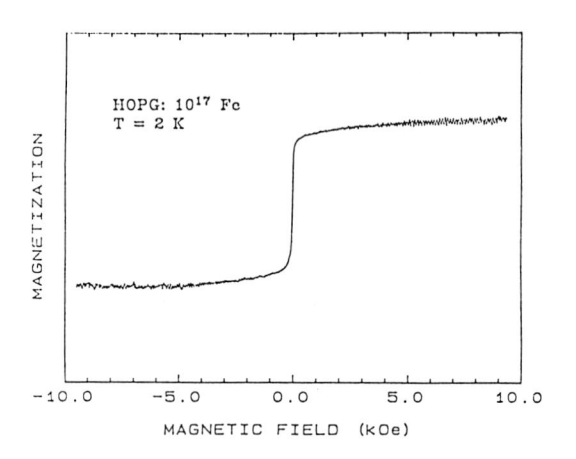

FIGURE 1. Magnetization vs. applied magnetic field at 2 K for HOPG implanted with 10^{17} atoms/cm² of iron. The diamagnetic signal from the graphite has been subtracted.

RESULTS

The fluence levels chosen for investigation resulted in samples whose magnetic properties spanned the range from clearly ferromagnetic at the highest fluence level (10^{17} atoms/cm²) to paramagnetic at the lowest (10^{16} atoms/cm²). Magnetization curves taken at 2 K for the highest fluence HOPG sample (Figure 1), after the linear diamagnetic signal from graphite has been subtracted, show the striking feature of a rapid change in magnetization at low fields. This is consistent with a randomly oriented arrangement of thin magnetic films with low coercive forces. Because of the low demagnetizing factor in the plane, there would be an initial sharp rise in magnetization to 78.5% of saturation as the moment of each plane rotated in the plane to maximize the component parallel to the applied field. There would then be a slower increase with applied field as the moments rotate out of the plane. Similar results were observed for a fluence of 6 x 10^{16} atoms/cm², except that the coercive force was somewhat greater. The coercive force at 2 K reached a maximum of 550 Oe at a fluence of 3 x 10^{16} atoms/cm², which is probably close to the critical fluence for formation of a ferromagnetic state.

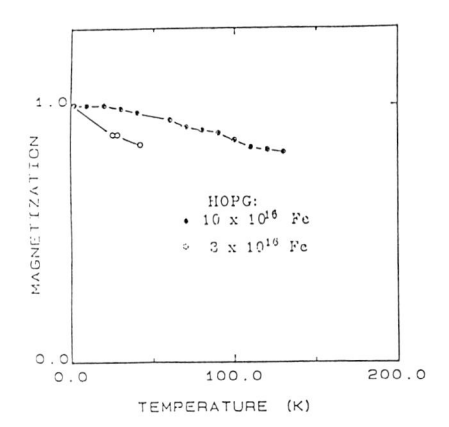

FIGURE 2. Relative spontaneous magnetization vs. temperature for HOPG implanted with 3×10^{16} (open circles) and 10^{17} (solid dots) atoms/cm² of iron.

The temperature dependence of the spontaneous moments for HOPG (Figure 2) for the 3 and 10×10^{16} atoms/cm² samples, also suggest that 3×10^{16} is approaching the critical fluence for the formation of a ferromagnetic state. The two highest fluence samples had a very weakly temperature dependent spontaneous moment at low temperatures, while the specimen with 3×10^{16} atoms/cm² has a magnetization which drops off much more rapidly with increasing temperature, suggesting a much larger percentage of weakly coupled spins.

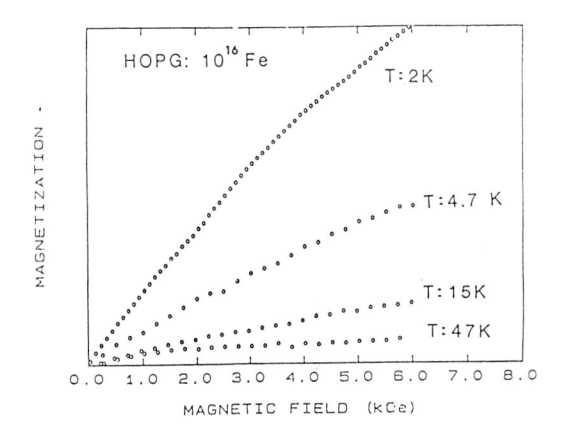

FIGURE 3. Magnetization vs. applied magnetic field at various temperatures for HOPG implanted with 10^{16} atoms/cm² of iron. The diamagnetic signal from the graphite has been subtracted.

The magnetization of the lowest fluence specimen of implanted HOPG (10^{16} atoms/cm^2) (Figure 3) is strikingly different from the other three in that there is no evidence for a spontaneous moment even at the lowest temperature measured. The susceptibility varies more slowly than 1/T, but still increases rapidly with decreasing temperature. Higher field data suggests that the average moment size per spin is larger than two Bohr magnetons, indicating there is exchange coupling between spins, but the average number of spins in each exchange coupled cluster is rather small.

DISCUSSION

The results of this study can be explained by assuming only neighboring iron spins in the heavily radiation damaged (and probably amorphous) layer are strongly exchange coupled. Variation in magnetic properties with concentration then depends mainly on how this strong exchange propagates through the lattice. In this model true ferromagnetism only occurs at iron concentrations greater than the percolation concentration, where the exchange interactions can propagate to infinity. The percolation problem has been treated extensively for a variety of physical processes (4, 5). Sher and Zallen (6) defined a critical density for percolation close to 15 volume percent for a wide variety of lattice types. In the present case the high fluence specimens for each system are all well above the iron percolation density and are clearly ferromagnetic, while the lowest fluence specimen is well below the iron percolation limit and is clearly paramagnetic. The magnetic properties observed in the present cases therefore seem to correlate well with the structural characteristics expected for iron implanted into graphite, PE, and PVF$_2$.

REFERENCES

1. B. L. Crowder, J. E. Smith, Jr., M. H. Brodsky, and M. I. Nathan, Proceedings of the Second International Conference on Ion Implantation in Semiconductors (Garmisch-Partenkirchen, Federal Republic of Germany, 1970), p. 255; J. E. Smith, Jr., M. H. Brodsky, B. L. Crowder, and M. I. Nathan, J. Non-Cryst Solids, 8-10, 179 (1972).

2. R. B. Wright, R. Varma, and D. M. Gruen, J. Nucl. Mater., 63, 415 (1976).

3. B. S. Elman, M. S. Dresselhaus, G. Dresselhaus, E. W. Maby, and H. Mazurek, Phys. Rev. 24, 1027 (1981).

4. H. L. Frisch and J. M. Hammersley, J. Soc. Ind. Appl. Math, 11, 894 (1963).

5. R. J. Elliott, B. R. Heap, D. J. Morgan, and G. S. Rushbrooke, Phys. Rev. Letters, 5, 366 (1960).

6. H. Shur and R. Zallen, J. Chem. Phys., 53, 3759 (1970).

COMPARISON OF CONDUCTIVITY PRODUCED IN POLYMERS AND CARBON FILMS BY PYROLYSIS AND HIGH ENERGY ION IRRADIATION

T. Venkatesan, R. C. Dynes, B. Wilkens, A. E. White,
J. M. Gibson, and R. Hamm

Bell Laboratories
Murray Hill, New Jersey 07974

INTRODUCTION

The electrical properties of pyrolyzed polymers have been studied recently.[1,2] It has been shown that organic, polymeric[3] and non-polymeric[4] films can be made conductive ($\rho \sim 10^{-3} \Omega cm$) by ion beam irradiation. Common to all of the films was the presence of carbon as a constituent element and both pyrolysis and ion beam irradiation[3] was shown to increase the relative carbon content of the films. The ion beam irradiated organic films[3,4] exhibited a temperature dependence of their resistivity of the form $\rho(T) = \rho_\infty e^{-(T_o/T)^n}$, where ρ is the ion-induced resistivity, ρ_∞ and T_o are constants and T is the temperature. At very high doses of irradiation ($10^{17} cm^{-2} Ar^+$ @ 2MeV), the film resistivity was temperature independent. Very similar transport properties were observed in the pyrolyzed polymers[1] as well, though the lowest resistivities achieved were higher than the resistivity values observed in the ion irradiated[3] polymer films. In both the pyrolysis and ion-irradiation experiments the temperature dependence has been explained by a model due to Sheng and Abeles,[5] which involves charge transport by hopping between conducting islands embedded in an insulating matrix. Such striking similarities between two distinctly different modes of energy deposition in the films, prompted us to compare the effects of pyrolysis and ion irradiation in different carbon containing films. We compared both a polymer (HPR-204*) and a film of electron beam evaporated carbon film. While in the former case one would observe chemical degradation as well as structural modification, by studying pure carbon films the physical nature of the processes could be clarified. We report metallic carrier densities in both films and evidence for significant structural rearrangement. We conclude that pyrolysis and ion beam irradiation have similar effects on both polymer and carbon films.

EXPERIMENTAL

Polymer films were spun onto various substrates (quartz, silicon and sapphire) to a thickness of ~8000Å and were baked at 120°C for 30 minutes to evaporate the solvent. Carbon films were prepared by electron beam evaporation of 99.9999% purity carbon source material onto various substrates at room temperature. We found the electrical properties of the pristine and modified films to be independent of the substrates used. Approximately 3000Å thick films on quartz substrates were used for the four-probe resistivity measurements whereas Hall measurements were made on thinner films (~300Å) on quartz, sapphire or high resistivity silicon substrates. The pristine carbon films had resistivities of $\sim 5 \times 10^{-2} \Omega cm$. (Hauser[6] has demonstrated that by sputter depositing C onto low temperature substrates films of resistivity as high as $10^2 \Omega cm$ can be prepared, though our experiments show that the initial state of the pristine film is not important). To study the effect of pyrolysis, the samples were introduced into a tubular furnace in flowing argon gas for about one hour which was found to be adequate for equilibration. In the ion irradiation experiments, 2 MeV Ar^+ ions from a Van de Graaff generator impinged on the samples through a contact mask to provide the necessary electrical lead configuration (Fig. 1) for the Hall measurement. The lead configuration was critical for the measurement of the films with large carrier densities, because the normal magnetoresistance contribution due to misalignment must be minimized. In all cases the range of the ion was larger than the film thickness. For the Hall measurements, the samples were immersed in liquid helium (4.2K). A magnetic field was applied normal to the plane of the substrate with a superconducting solenoid capable of 50 kG. In order to compare the structure of the pristine and irradiated films, thin films (~200Å) were deposited on salt substrates, ion beam irradiated and then floated off to be examined in a 200 keV transmission electron microscope (TEM).

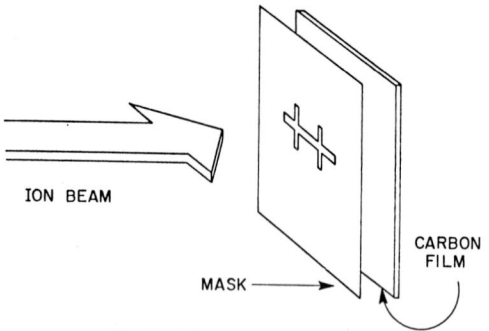

Fig. 1 The irradiation setup.

RESULTS

The resistivity of the films as a function of the ion dose, for various substrate temperatures during irradiation are shown in Fig. 2. Compared to polymers carbon films undergo modification at lower ion doses and the saturation values are also much lower ($\rho \sim 5 \times 10^{-4} \Omega$cm). It is noticeable that these values are comparable to those found in disordered metals or metallic glasses,[7] where the carrier mean free paths are a few lattice constants, and because of the metallic carrier densities (10^{22}–10^{23}cm^{-3}) the resistivity values lie between 100 and 500 $\mu\Omega$cm. Since we expected the carbon films to have short carrier mean-free-paths, based on the resistivity values we suspected large carrier densities. Increasing substrate temperature during irradiation helps in the reduction of the film resistivity. The effect of pyrolysis on the carbon and polymer films is shown in Fig. 3. The results are qualitatively similar to the ion irradiated films. The decrease in the resistivity of evaporated carbon films has been reported earlier by Morgan.[8] In addition, we find that the resistivity exhibits a thermally activated characteristic, with an

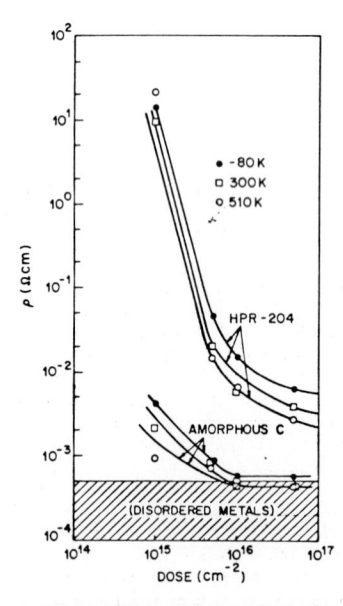

Fig. 2 The resistivity vs. dose with samples held at different temperatures.

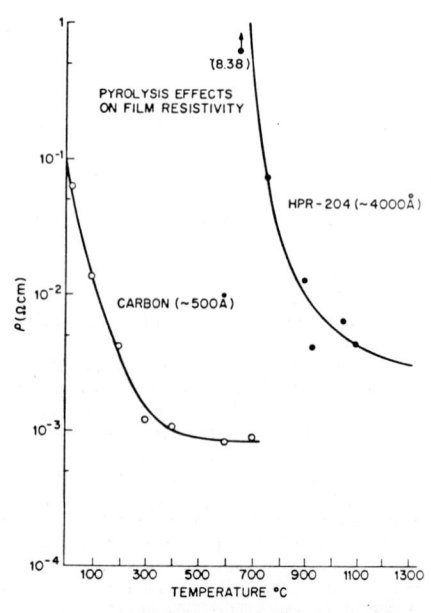

Fig. 3 Resistivity vs. annealing temperature.

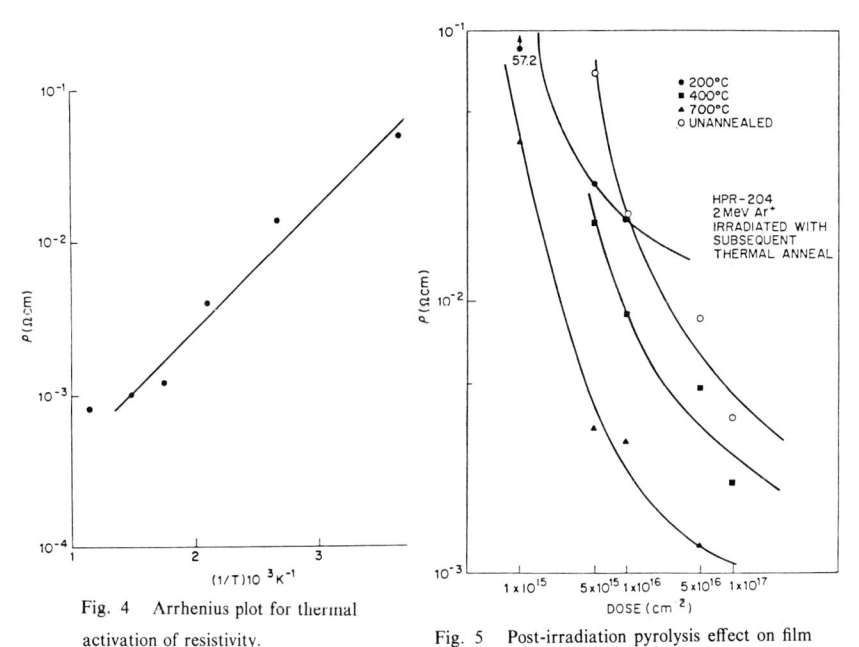

Fig. 4 Arrhenius plot for thermal activation of resistivity.

Fig. 5 Post-irradiation pyrolysis effect on film conductivity. The curve drawn through the data is only an approximation.

activation energy of $\sim$ 30 meV (Fig. 4). The resistivities of the thermally annealed films subsequent to ion beam irradiation are shown in Fig. 5 and one sees further significant decreases in the film resistivity after the thermal anneal.

Carrier densities derived from Hall measurements are summarized in Table 1. In thicker films we could only estimate a lower limit owing to the very small Hall voltages at these carrier densities. However, in the thinner films the signals were well above the noise and we estimate the stated carrier densities to be accurate to within a factor of two. Therefore, the observed carrier densities (10^{22}–10^{23} cm^{-3}) are larger by several orders of magnitude than in graphite. As one would expect ion-irradiated polymer films, which are eventually carbonized also exhibit large carrier densities. Even the thermally annealed carbon films show high carrier densities but not as high as those in the ion-irradiated films.

A film irradiated with 10^{17} Ar$^+$/cm^2 was compared with an unirradiated region of the same specimen in a TEM. The electron diffraction patterns of different regions are shown in Fig. 6. The electron beam is normal to the film in Fig. 6a (unirradiated) and Fig. 6b (irradiated) and a slight sharpening of the characteristic diffuse rings of amorphous carbon subsequent to irradiation is seen. This implies some increase in the atomic ordering within the film subsequent to the irradiation. When the electron beam is incident on a region where the film has folded over, the beam samples the film in a direction parallel to the film surface and a series of diffuse, diffraction spots are obtained (Fig. 6c). These arise from irregular stacking of graphitic-like planes, parallel to the film surface, with mean spacings of $\sim 3.5 \pm 0.5$ Å. These findings are consistent with the results of Sander et al,[9] who proved that ion-irradiated and thermally annealed C films are structurally equivalent. These results are further corroborated by the ultra-high resolution images of the same regions shown in Fig. 7. The high resolution image with the beam normal to the film (Fig. 7a) shows the "random" image typical of amorphous films. However, in the region where the film is folded over, one gets a "cross-sectional" view (Fig. 7b). Here a graphitic-like stacking parallel to the film surface is evident with a correlation length of ~ 20 Å. This is reminiscent of 'turbostatic' carbon structures and is consistent with earlier Raman measurements in ion-irradiated polymer[3] films.

TABLE I

Film	Thickness (Å)	Resistivity (Ω–cm) (T=300°K)	Electron Density (cm^{-3}) T=4°K
Evaporated Carbon Film ($10^{17}Ar^+/cm^2$)	3000	5×10^{-4}	$> 10^{22}$
Evaporated Carbon Film ($10^{17}Ar/cm^2$)	300	5×10^{-4}	1.8×10^{23}
Evaporated Carbon Film ($5 \times 10^{15}Ar^+/cm^2$)	300	1×10^{-3}	1.1×10^{23}
Evaporated Carbon Film (pristine)	300	5×10^{-2}	Carrier Freeze Out
HOPG (pristine)	30,000	1.5×10^{-4}	4.8×10^{18}
Polymer (HPR-204) Film ($10^{17}Ar^+/cm^2$)	4,000	3×10^{-3}	$> 10^{22}$
Evaporated Carbon Film (Thermal Annealed) T = 300°C	400Å	2×10^{-3}	1.4×10^{22}

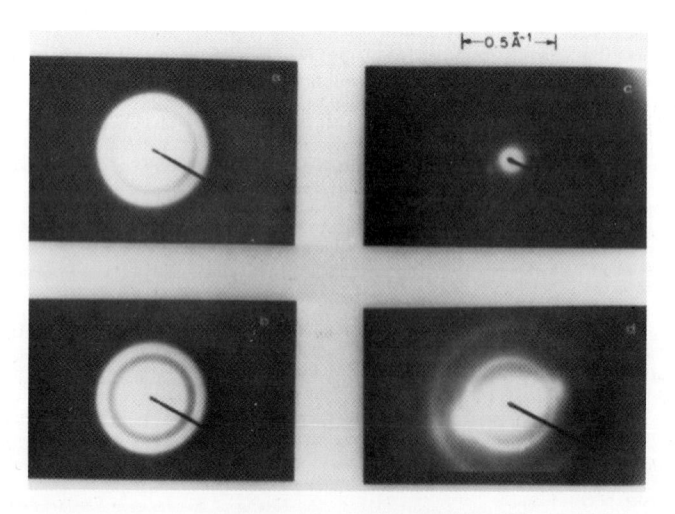

Fig. 6 The electron diffraction patterns with beam
    (a) perpendicular to pristine film
    (b) perpendicular to irradiated film
    (c) almost parallel to irradiated film (short camera exposure)
    (d) almost parallel to irradiated film (long camera exposure)

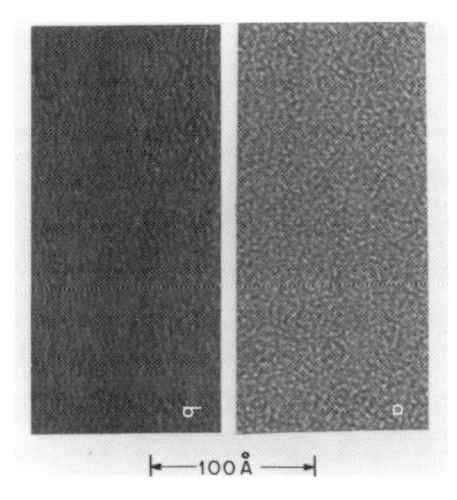

Fig. 7 Ultra high resolution TEM images with beam
(a) perpendicular to irradiated film
(b) almost parallel to irradiated film.

DISCUSSION

The polymer films required higher doses for modification of the resistivity probably because the film is carbonized first in addition to a subsequent rearrangement of the carbon atoms. It is possible that the polymeric films degrade to a more defective, loosely packed carbon film and so the saturation resistivities are higher than those in pure carbon film. However, organic precursers[4] that carbonize readily exhibit lower saturation resistivities ($\rho \sim 5\times 10^{-4}\,\Omega\,cm$) upon irradiation. The effect of increasing temperature is to reduce the resistivity of the film, with or without ion-irradiation. This suggests either annealing of traps, or increase in the crystalline order of the film. The chemical degradation of the polymer would also be favoured by increasing the sample temperature. The Arrhenius characteristic of the resistivity produced by thermal annealing of C films suggests ~ 30 meV barrier energy for the traps that localize the electrons of the carbon atom.

Even though pyrolysis and high energy ion irradiation yield similar end results, there is one significant difference between the two processes. The ion beam modification occurs even at low beam currents where there is negligible ohmic heating of the film. In the case of the high energy ion, huge amounts of electronic energy (~ 30 eV/atom) are deposited in the track of the ion (of diameter of ~ 10Å). Further, ion beam interaction with material gives a cumulative effect of individual ion tracks which do not overlap in time (as the lifetime[10] of the collision induced processes is $\sim 10^{-11}$sec, much faster than the arrival rate of ions in typical interaction regions). As a result, the maximum correlation length of the crystalline order induced by the ion is likely to be comparable with the transverse extent of the energy deposited in the ion track. However, this need not be true for pyrolysis where there is no limit to the transverse extent of the crystalline order induced in the annealing process.

In the case of graphite, the band structure allows negligible overlap of the valence and conduction bands resulting in very few carriers at the Fermi level. The TEM observation of some graphitic ordering in the irradiated film and the Hall measurement of carrier densities four to five orders of magnitude larger than in graphite are not easily reconcilable. It is possible that for such small, distorted "crystallites" the electronic structure is significantly different from that of graphite. (One could expect numerous dangling bonds at the "crystallite" boundaries but this would not be adequate to explain carrier densities $>10^{23}\,cm^{-3}$.) Because of the uncertainty principle the short scattering times ($<10^{-15}$ sec) of the carriers would lead to a large spread of the carrier energy (~ 1 eV) at the Fermi level. While in crystalline graphite the overlap of the conduction and valence band is small (~15–30 meV), resulting in low carrier densities (~10^{18}cm^{-3}), for tiny crystallites of graphite (~10Å) the band overlap would be significantly higher because of the short carrier scattering times and a resultant increase in the energy spread of the carrier at the Fermi level, resulting in large carrier densities. Hence, the smaller the graphite crystallites, the larger the carrier density is likely to be.

However, pristine carbon films do not exhibit metallic carrier densities at low temperatures (4K). In fact the carriers freeze out at low temperatures. Thermal annealing at relatively low temperatures (<400°C) delocalized the electrons. With increasing temperature one expects increased graphitic ordering and an eventual decrease in the carrier density. Ion irradiation can successfully delocalize the electrons with a limit to the maximum graphitic order (~20Å) in the films. Hence the highest carrier densities are expected in the ion irradiated rather than the pyrolyzed film and the experimental data supports this fact.

One consequence of the limit to maximum crystalline order induced by ions is that after heavy doses of ion irradiation graphite,[11] diamond,[12] carbon[12] and polymers[13] appear to show structurally the same end product, as verified by Raman scattering. In this context we would like to point out that Hauser[13] had studied the ion beam (6×10^{16}C$^+$/cm^2 @ 20-70 keV) induced conductivity in diamond and measured a resistivity of 3×10^{-3} Ωcm, which he attributed to overlapping localized states.

In conclusion, from our work on widely disordered carbon containing films it is found that extended states are produced by ion beam irradiation. Further, in the experiments described we have found the role of the ion in modifying the transport properties is in the deposition of energy in the film rather than as a dopant, and is qualitatively similar to thermal annealing. It is notable that high energy ion irradiation or low temperature pyrolysis leads to a carbon phase which exhibits metallic behaviour similar to metallic glasses.

REFERENCES

1. E. K. Sichel and T. Emma, Solid State Comm. (1982) *41*, 747.
2. Carbon Black - Polymer Composites, edited by E. K. Sichel, Published by Marcel Decker Inc., NY (1982).
3. T. Venkatesan, S. R. Forrest, M. L. Kaplan, C. A. Murray, P. H. Schmidt, and B. J. Wilkens, J. Appl. Phys., June 1983.
4. S. R. Forrest, M. L. Kaplan, P. H. Schmidt, T. Venkatesan and A. Lovinger, Appl. Phys. Lett. (1982) *41*, 708.
5. B. Abeles and P. Sheng, M. D. Coutts and Y. Arie, Advances in Phys., *2A*, 407 (1975); P. Sheng, A. Abeles and Y. Arie, Phys. Rev. Letts. *31*, 44 (1973).
6. J. J. Hauser, Solid State Comm. (1975) *17*, 1577.
7. J. H. Mooij, Phys. Stat. Sol. (1973) *17*, 521.
8. M. Morgan, Thin Solid Films, (1971) *7*, 313.
9. U. Sander, H. H. Bukow and H. Von Buttlar, J. Phys. Colloq. (1979) *1*, 301.
10. L. M. Howe, M. H. Rainville, H. K. Harigan and D. A. Thompson, Nucl. Instrum. and Meth. (1980) *170*, 419.
11. B. S. Elman, M. S. Dresselhaus, G. Dresselhaus, B. W. Maloy and H. Mazurele, Phys. Rev. *B24*, 1027 (1981).
12. C. A. Murray, private communications.
13. J. J. Hauser, J. R. Patel and J. W. Rodgers, Appl. Phys. Lett. *30*, 129 (1977).

DEPTH PROFILING OF HYDROGEN IN ION-IMPLANTED POLYMERS

J. DAVID CARLSON
Lord Corporation Research Center, P.O. Box 1107, Cary, NC 27511 USA

PETER P. PRONKO and DAVID C. INGRAM
Universal Energy Systems, 4401 Dayton-Xenia Road, Dayton, OH 45432 USA

ABSTRACT

Depth profiling of hydrogen in polymeric materials poses special problems. Backscattering methods are ruled out because of kinematics. Nuclear reaction methods are undesirable because small reaction cross sections necessitate large fluences of high mass projectiles and result in unacceptable levels of radiation damage. We have used a helium-induced proton-recoil technique with 3 MeV ^{4}He particles to measure the hydrogen distribution in pristine and ion-implanted polyvinylidene fluoride (PVDF) films. The incident ^{4}He particles stopped in the 25 micron PVDF films while the recoiling protons were detected after passing through the polymer film. Large changes in the hydrogen content of PVDF films implanted with modest fluences of 6 MeV carbon, oxygen and nickel ions were observed.

INTRODUCTION

Ion implantation has become a well established and powerful technique for modifying metals and semiconductors [1-4]. Recently, interest has developed in the application of ion-implantation and ion-irradiation to the modification of polymers [5-9]. Implantation and irradiation effects in polymers present a number of contrasts to those in semiconductors and metals. Not only can the presence of an implanted species affect the chemistry of the implanted polymer, but the transferred energy alone may drastically alter the entire character of the polymer through a variety of processes. These include crosslinking, chain scission, free radical formation, and chemical decomposition into small organic molecules.

The evolution of molecular species during the course of ion irradiation of polymer films [10] and the eventual formation of residual carbonaceous films [9,11] indicate that large changes in the elemental composition of polymers are induced by implantation or irradiation with MeV heavy ions. We have found that, with care, Rutherford backscattering (RBS) can be used to quantify changes in the carbon and heavier element composition of polymers. Depth profiling of hydrogen, however, poses special problems and conventional analytical tools are found to be undesirable.

Backscattering methods are ruled out because of kinematics. Nuclear reaction analysis (NRA), on the other hand, which works quite well for profiling hydrogen in metals is ruled out because the relatively small reaction cross sections necessitate large fluences of heavy ions and result in unacceptable levels of damage in radiation sensitive polymers. Heavy ion fluences sufficient for NRA of hydrogen in organic polymers are typically in the range of 10^{14} to 10^{16} ions/cm^2. These are significantly larger than the 5×10^{13}/cm^2 fluences of similar mass ions we have found sufficient to cause changes of nearly a factor of two in polymer elemental composition (see below). NRA defeats its own purpose by causing changes larger than those we wish to quantify.

We have used an elastic recoil detection analysis (ERDA) [12,13] technique to measure the depth profile distribution of hydrogen in unimplanted and ion-implanted polyvinylidene fluoride (PVDF) films. In this method the energy spectra of helium-induced proton recoils which leave the polymer sample in a forward direction are measured. We have used a transmission geometry in which protons recoiling at 5 degrees to the incident beam direction were detected after being knocked through the nominally 25 micron thick PVDF films. The incident 3 MeV ^{4}He ions stopped within the PVDF films. The 5 degree detector angle was chosen to maximize recoil energy while maintaining a margin of protection for the detector in case of target failure.

Depth information comes primarily from the energy loss suffered by the ^{4}He ions as they penetrate the polymer film and is carried through into the detected proton energy spectrum. This method can be used as a compliment to conventional RBS for the complete characterization of the elemental composition of the polymer material. For thick polymer targets a glancing incidence geometry can be used. Advantages of the helium-induced proton-recoil analysis technique include: (a) low mass projectiles to minimize radiation damage, (b) large cross sections which minimize necessary ^{4}He fluences, and (c) the ability to obtain the entire hydrogen depth profile at one time rather than from a series of individual measurements at different depths.

BASIC PRINCIPLES

The basic arrangement for the helium-induced proton-recoil analysis method with a transmission geometry is illustrated in Fig. 1. A beam of ^{4}He ions of initial energy $E_\alpha(0)$ enters the polymer film and penetrates to a depth d where it suffers an elastic collision with a proton from the host material. The proton recoils at an angle θ, penetrates the remaining distance s through the polymer film, and is detected by a particle energy spectrometer, e.g., silicon surface barrier detector. The energy of the detected recoil proton depends upon: (a) $E_\alpha(0)$, the incident ^{4}He energy, (b) dE_α/dx, the energy lost by the ^{4}He before producing the recoil proton, (c) K_θ, the kinematic factor, and (d) dE_p/dx, the energy lost by the recoil proton as it leaves the sample.

Fig. 1. Basic arrangement for transmission mode helium-induced proton-recoil analysis.

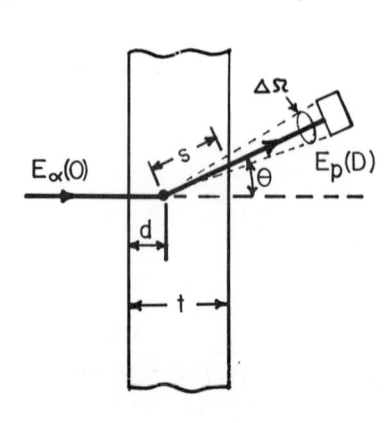

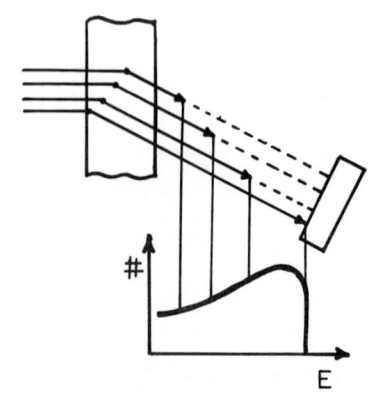

Fig. 2. Relationship between scattering depth and recoil proton energy.

The detected recoil proton energy, $E_p(D)$, is a function of d as given by

$$E_p(D) = K_\theta \left[E_\alpha(0) - \int_0^d \frac{dE_\alpha(x)}{dx} dx \right] - \int_0^s \frac{dE_p(x)}{dx} dx \qquad (1)$$

where $s = (t-d)/\cos\theta$. The relationship between detected proton energy and elastic scattering depth is illustrated schematically in Fig. 2. In order to convert the detected proton energy spectrum to a depth profile distribution one needs to obtain a depth scale by solving (1) for d in terms of $E_p(D)$. This can be done numerically. For polymers of known elemental composition, e.g., unimplanted materials or implanted materials having only slightly altered compositions, stopping power values for carbon, hydrogen and other elements are combined using Bragg's Rule. Heavily implanted and modified polymers not having compositions which are known a priori may require an iterative approach and complimentary RBS measurements for heavy element content.

EXPERIMENTAL RESULTS

Helium-induced proton-recoil energy spectra obtained for unimplanted and implanted PVDF films are shown in Fig. 3. The 3 MeV ^{4}He^{++} analysis beam was produced by a General Ionex Tandetron accelerator located at Universal Energy Systems. The PVDF films were implanted with 6 MeV C, O and Ni ions to a fluence of $5 \times 10^{13}/cm^2$ using the same accelerator. Recoil protons were detected using a 6 mm diameter silicon surface barrier detector located 10 cm from the target at an angle of 5° to the incident beam direction. The spectra shown here were each obtained in approximately 600 sec. with a 1nA beam.

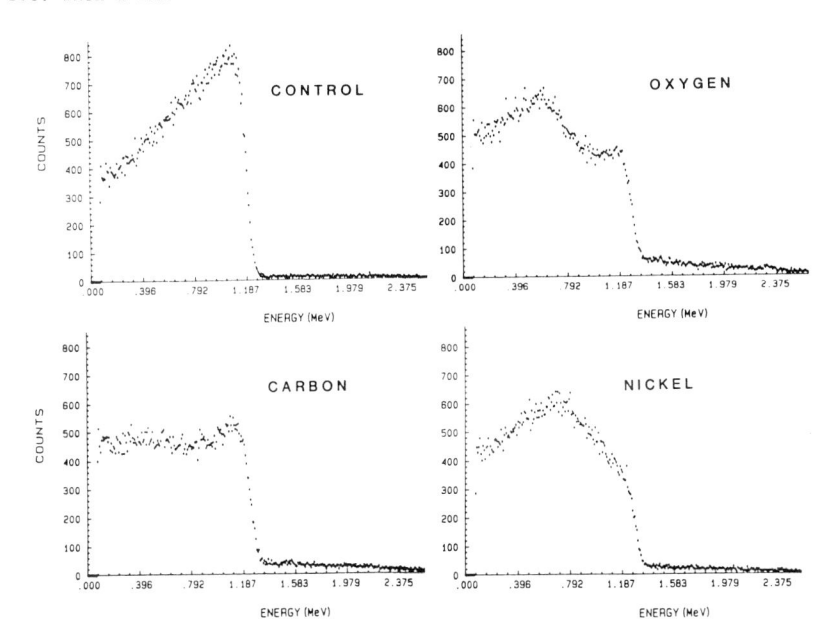

Fig. 3 Helium-induced proton recoil energy spectra for unimplanted and implanted (6 MeV C, O, and Ni at $5 \times 10^{13}/cm^2$) polyvinylidene fluoride.

HYDROGEN DEPTH PROFILES

The recoil proton energy spectra were converted to hydrogen depth profile distributions using tabulated stopping powers for He and H [14]. The following steps were followed in order to obtain a depth scale and perform this conversion for the PVDF spectra. First, the residual ^{4}He energy as a function of distance into the PVDF films was calculated using stopping powers for C, H and F, the unmodified composition of PVDF ($C_2H_2F_2$) and the following expression:

$$E_p(d) = E_\alpha(0) - \int_0^d \frac{dE_\alpha(x)}{dx} \quad (2)$$

Next, the initial proton-recoil energy for protons originating a distance d inside the PVDF film was calculated using the kinematic factor for elastic recoil at 5°.

$$E_p(0)d = K\theta E_\alpha(d) \quad (3)$$

The energy of the proton-recoil after passing through the remainder of the PVDF film was then calculated using proton stopping powers for C, H and F with the following

$$E_p(D)_d = E_p(0)_d - \int_0^s \frac{dE_p(x)}{dx} dx \quad (4)$$

where $s = (t-d)/\cos\theta$. This calculation is rather sensitive to the polymer film thickness, t. We have found that nominal or even measured film thickness values are not entirely satisfactory due to normal, small variations in polymer film thickness. However, the actual polymer thickness at the spot where the analysis beam strikes can be determined from the proton-recoil data itself since the calculated energy for recoils originating at the film front surface, $E_p(D)_o$, must equal the maximum recoil energy in the measured spectrum. Thus, the calculation of $E_p(D)_d$ is forced to meet this condition with t as a free parameter.

The depth scale, d versus $E_p(D)$, which is obtained from this process is illustrated in Fig. 4. Note that the depth scale is not linear. This non-linearity is primarily responsible for the fact that the homogeneous, pristine PVDF control sample gives a non-flat recoil energy spectrum.

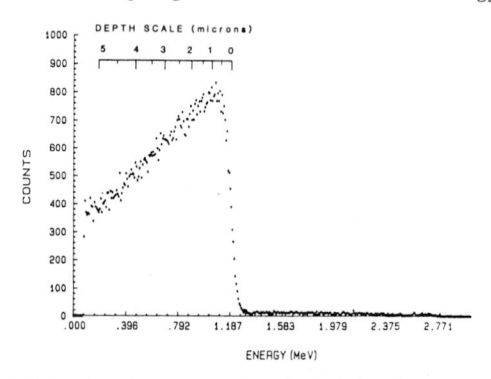

Fig. 4. Recoil proton energy spectrum for unimplanted, homogeneous PVDF film showing the corresponding non-linear depth scale.

Finally, the recoil-proton energy spectra are converted to depth profile distributions by redistributing the particle counts into equal width depth bins as shown in Fig. 5. Here we see a flat hydrogen distribution for the unimplanted, homogeneous control PVDF and the loss of hydrogen in the first 4 microns for the carbon implanted PVDF. For the distributions shown here, no correction was made for any variation in the ^{4}He-p differential cross section with energy. This is reasonable as a first approximation since this cross section is highly non-Rutherford and is not strongly dependent upon energy in the range near 3 MeV [13].

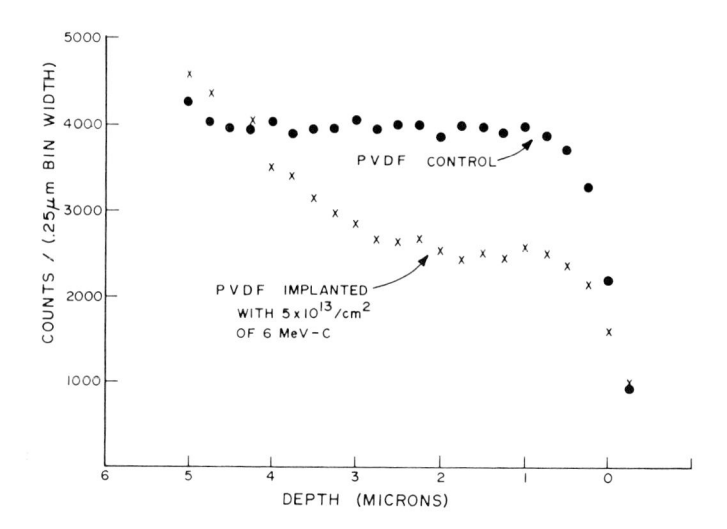

Fig. 5. Hydrogen depth profile distributions for unimplanted and implanted PVDF films.

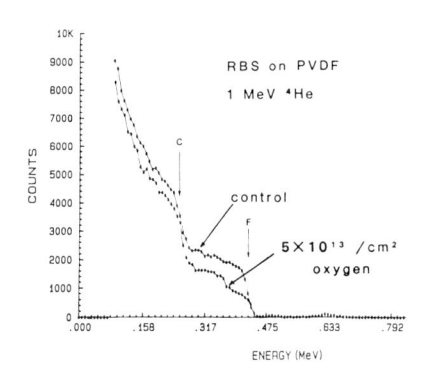

Fig. 6. RBS spectra for PVDF films corresponding to films used in the p-recoil analysis.

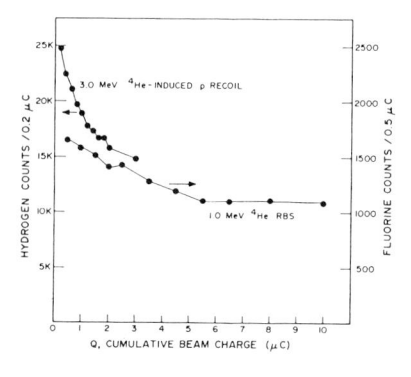

Fig. 7. Effect of analysing beam on PVDF film during RBS and p-recoil analysis.

The recoil-proton energy spectra in Fig. 3 and the depth profile distributions in Fig. 5 show that implantation with only very modest fluences (5×10^{13}/cm^2) of C, O and Ni can significantly decrease the hydrogen content to depths on the order of the implanted ion range. Corresponding changes in the fluorine content of these same implanted PVDF films have also been observed using conventional RBS as shown in Fig. 6.

EFFECT OF ANALYSING BEAM

The effect that the 4He^{++} analysing beam can have on the composition of a polymer sample is illustrated in Fig. 7. Shown here are the fluorine and hydrogen count rates as a function of cumulative beam charge for 1 MeV RBS and 3 MeV p-recoil analysis of PVDF. In both cases significant decreases in count rate are observed as the analysis proceeds indicating loss of F and H due to the He beam. For both of the cases illustrated, the beam spot had an area of 0.02 cm^2 so that each microcoulomb represents a fluence of approximately 3×10^{14}/cm^2. Typical p-recoil analyses require 0.25-0.5 μC and use a 0.1 cm^2 beam spot.

CONCLUSION

Helium-induced proton recoil analysis can give quantitative information on the distribution of hydrogen in polymeric materials which compliments RBS information on heavy element composition. This technique has been used to measure the depth profile distribution of hydrogen in MeV heavy ion implanted PVDF films and shows that large decreases in hydrogen content occur at relatively modest fluences.

REFERENCES

1. J.K. Hirvonen, ed., Treatise on Materials Science and Technology, Vol. 18 (Academic Press, New York, 1980).
2. G. Dearnaley, J.H. Freeman, R.S. Nelson and J. Stephan, eds., Ion Implantation, (North-Holland, Amsterdam, 1973).
3. P.D. Townsend, J.C. Kelly and N.E.W. Hartley, Ion Implantation, Sputtering and Their Applications, (Academic Press, London, 1976).
4. Proceedings of the Second International Conference on Ion Beam Modification of Materials, Nucl. Instrum. Methods 182/183 (1981).
5. D.C. Weber, P. Brant, C. Carosella and L.G. Banks, J. Chem. Soc: Chem. Comm., 522 (1981)
6. M.C. Wintersgill, 2nd International Conference on Radiation Effects in Insulators, Albuquerque, 1983, to be published.
7. T. Venkatesan, W.L. Brown, C.A. Murray, K.J. Marcantonio and B.J. Wilkens, Polymer Eng. Sci., to be published.
8. H. Mazurek, D.R. Kay, E.W. Maby, J.S. Abel, S.D. Sentura, M.S. Dresselhaus and G. Dresselhaus, J. Poly. Sci: Poly. Phys. 21, 537 (1983).
9. T. Hioki, S. Noda, M. Sugiura, M. Kakeno, K. Yamade and J. Kawamoto, Appl. Phys. Lett. 43, 30 (1983).
10. T. Venkatesan, D. Edelson and W. L. Brown, Appl. Phys. Lett. 43, 364 (1983).
11. T. Venkatesen, S.R. Forrest, M.L. Kaplan, C.A. Murray, P.H. Schmidt and B.L. Wilkens, J. Appl. Phys. 54, 3150 (1983).
12. B. Terreault, J.G. Martel, R.G. St.-Jacques and J. L'Ecuyer, J. Vac. Sci. Technol. 14, 492 (1977).
13. L.S. Wielunski, R.E. Benenson and W.A. Lanford, to be published.
14. L.C. Northcliffe and R.F. Schilling, Nuclear Data Tables A7, 233 (1970).

TWO-DIMENSIONAL ORDERING OF ION DAMAGED GRAPHITE

B.S. ELMAN,[*,θ] M.S. DRESSELHAUS,[*,θ,#] G. BRAUNSTEIN,[θ] G. DRESSELHAUS,[+]
T. VENKATESAN,[$] B. WILKENS,[$] AND J.M. GIBSON[$]
[*]Center for Materials Science and Engineering; [θ]Department of Physics;
[#]Department of Electrical Engineering and Computer Science; [+]Francis Bitter National Magnet Laboratory; Massachusetts Institute of Technology, Cambridge, MA 02139; [$]Bell Laboratories, Murray Hill, NJ 07974, USA

ABSTRACT

Post implantation annealing of ion-damaged, highly oriented pyrolytic graphite (HOPG) has been studied by Raman spectroscopy, the ion channeling technique and Transmission Electron Microscopy. Complementary information obtained by these methods provides confirmation for the completion of the first step of graphitization of ion-damaged graphite at annealing temperatures of ~2300°C. This is manifested by the formation of carbon planes with two dimensional ordering but no correlation in the third (c-axis) dimension.

INTRODUCTION

Radiation damage introduced into graphite during the process of ion implantation can be removed by subsequent annealing and recrystallization (graphitization) of the damaged region. Bulk graphitization of different types of carbons has been widely studied [1]. The graphitization is generally considered to be a two step process. The first step involves in-plane ordering which starts above 1500°C. This is followed by the second step of graphitization above 2000°C, where, in addition, three dimensional ordering takes place [1]. Generally, x-ray diffraction experiments are performed to measure the interplanar distance from the linewidth of the (002) reflection, from which the degree of graphitization is determined. Diamagnetic susceptibility measurements of the magnitude of χ and the anisotropy $\chi_3 - \chi_1$ are also carried out to monitor the graphitization process as a function of heat treatment temperature [1].
Raman spectroscopy was successfully applied to follow annealing experiments on ion implanted HOPG [2,3] and the heat treatment of graphite fibers prepared by the pyrolysis of benzene [4]. The technique is sensitive to disorder in the crystalline substrate, since for disordered materials Raman scattering is not limited to contributions from Raman-active zone-center modes, and contributions from other phonons in the Brillouin zone become possible. The first-order Raman spectra are highly sensitive to the first step of graphitization, while the onset of the second step of graphitization can be determined from the second-order Raman spectra [2,3,5]. For most carbon blacks, the heat treatment temperature for the onset of the second step of graphitization is found to be between 2300°C and 2500°C; for cokes it is ~2000°C [1].
Rutherford backscattering spectrometry (RBS) in the channeling geometry is widely applied to obtain information on lattice disorder and recrystallization of ion-damaged semiconductors [6-8] and to HOPG [9]. In the present work we demonstrate how the complementary information from Raman scattering and RBS measurements can be supplemented with TEM data to provide strong evidence for the establishment of 2D ordering of graphite planes with no correlation in the third (c-axis) direction after subsequent annealing of ion-damaged graphite at temperature T_a ~ 2300°C.

EXPERIMENTAL

Samples of HOPG were implanted with a variety of ions at different energies. The implantation was performed with an Accelerators Inc. 300R instrument at ambient temperatures. The energies of implantation for different ions were chosen to obtain a broad disordered region near the surface such that the disordered region would not be less than the skin depth of light at the 4880 Å laser line for the Raman measurements and would be wide enough for observations by the RBS channeling technique. High fluences of implantation have been chosen in order to achieve a homogeneous extension of the highly disordered region up to the surface of the samples that were examined [3].

Specimens were prepared for transmission electron microscopy (TEM) by affixing the implanted surface to a glass slide and removing the bulk material by repeated cleaving with adhesive tape. The glue was then dissolved and thin flakes were picked up on fine mesh copper grids. Observations were made on a JEOL 200CX TEM operating at 200 keV.

The Raman scattering experiments were performed at ~300 K using a Brewster angle back-scattering geometry. Incident radiation at 4880 Å and ~100 mW was provided by a CW argon-ion laser. The scattered radiation was collected at 90° to the sample surface, analyzed by a double grating monochromator and detected by a cooled photomultiplier tube [2].

To observe channeling along the c-axis, the c-face of the HOPG sample was accurately oriented perpendicular to the ion beam using a two angle goniometer. The goniometer rotation angles could be controlled with a reproducibility of ±0.02° and ±0.04° for the θ and ∅ directions, respectively [9]. For the channeling studies, a beam of 2.0 MeV ^{4}He$^+$ ions from a Van de Graaff generator was used typically with a current 20 nA through a 1mm diameter aperture. Backscattered particles were detected by a surface-barrier detector with energy resolution of < 20 keV (FWHM) at the scattering angle of ~175°. Energy analysis was performed using a computer based data acquisition system.

The ion-implanted samples were isochronally annealed subsequent to implantation in an argon atmosphere in the 1100 < T < 2700°C temperature range to study the graphitization of the ion damaged region. Different samples were generally used for each annealing temperature.

RESULTS

Raman Spectroscopy

The Raman spectra of highly disordered graphite consist of a very broad asymmetric line in the first order spectrum and a very broad structure in the second order spectrum. In previous work, it was shown that several lines contribute to the structure in the first order spectra [3].

First we will consider the annealing behavior of HOPG implanted with ^{75}As at 230 keV and a fluence of 1 x 10^{15} cm^{-2}. The Raman spectrum for the implanted (unannealed) sample is shown in Fig. 1a. When post implantation annealing at a sufficiently high temperature is performed, we are able to resolve the first order Raman structure, so that the process of graphitization can be monitored. For example, the change in the first order Raman spectrum of this sample resulting from annealing at 1900°C for 20 min is shown in Fig. 1b. Three lines are resolved in Fig. 1b. The strongest line is peaked at ~1580 cm^{-1} and corresponds to the Raman-allowed zone-center high frequency mode of HOPG. The other two are disorder-induced lines appearing due to the high density of phonons at ~1360 cm^{-1} and ~1620 cm^{-1} [3,4]. Raman spectra corresponding to Fig. 1b were taken for a number of annealing temperatures and the relative intensities, positions and linewidths of the three lines were determined. A very convenient estimate of the in-plane crystalline size L_a of ion damaged HOPG is provided by R the ratio of the intensities of the disorder-induced ~1360 cm^{-1} line to the Raman-allowed ~1580 cm^{-1} line ($R = I_{1360}/I_{1580}$) which was calibrated by x-ray analysis from carbons with different in-plane crystallite dimensions L_a [10]. Our experimental results for the parameter R vs. annealing temperature for the sample of Fig. 1 is presented in Fig. 2. On the right-hand side, the vertical scale relates R to the in-plane crystallite size L_a, for reference. It is seen that as the annealing temperature T_a increases above ~1300°C, the in-plane crystallite size increases very rapidly until about 2300°C where L_a reaches a value of ~1000 Å. At this temperature R becomes very small, the intensity of the disorder-induced ~1360 cm^{-1} line approaches zero, and the Raman spectrum approaches that of crystalline graphite.

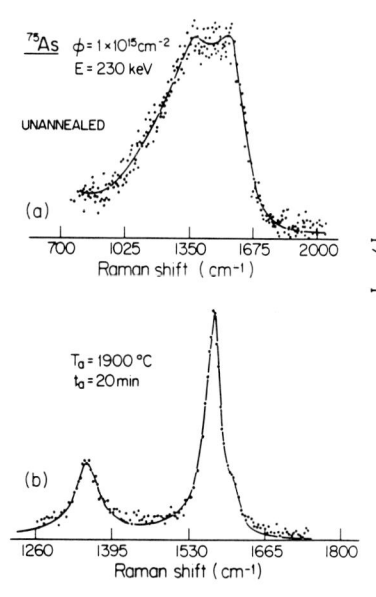

Fig. 1. Raman spectrum of HOPG ion implanted with 230 keV ^{75}As at a fluence of 1 x 10^{15} cm^{-2} before (a), and after (b) subsequent annealing at 1900°C for 20 min.

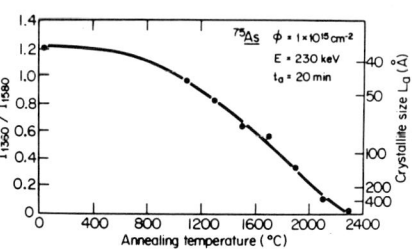

Fig. 2. Ratio of intensities of the disorder-induced ~1360 cm^{-1} line to the Raman-allowed ~1580 cm^{-1} line vs. annealing temperature for samples of HOPG implanted with ^{75}As ions under the same conditions as in Fig. 1 and annealed for 20 min. at various annealing temperatures.

Essentially the same information as we get by measuring the intensity ratio R can be deduced from the linewidth behavior [11], by looking at the linewidths of the disorder-induced ~1360 cm^{-1} and Raman-allowed ~1580 cm^{-1} lines vs. annealing temperature. The linewidths of both the ~1360 cm^{-1} and ~1580 cm^{-1} lines are greatly decreased upon annealing at T_a = 1600°C and above. In particular, near ~2300°C where the ~1360 cm^{-1} line disappears (Fig. 2), the linewidth of the ~1580 cm^{-1} line reaches the HOPG value.

In summary, the Raman spectroscopy results indicate that by annealing the highly disordered regions of HOPG to temperatures of ~2300°C and above, all the parameters of the first order spectrum are essentially indistinguishable from those of pure HOPG.

RBS Channeling

Information on ordering in the c-axis direction can be conveniently obtained by the RBS channeling technique, which has been widely applied to studying disordered layers formed on a substrate by ion implantation [6-8,12]. The channeling effect appears as a reduction of the backscattering yield when the beam is aligned along the channeling direction. In some cases, the interpretation of the aligned RBS spectra is very difficult and one has to make assumptions about the microscopic nature of the disordered region. These assumptions for the case of the graphitization of ion damaged HOPG are confirmed by electron microscopy measurements and the results are presented here.

The possibility of channeling along the c-axis of pristine HOPG has previously been demonstrated [9]. In the present work we apply this technique to study ion implanted graphite and focus on the observation of implantation-induced disorder and subsequent restoration of 2D order in the graphite planes by post implantation annealing. Detailed studies of the regrowth kinetics and the process of three-dimensional graphitization will be reported elsewhere [12].

In the RBS spectra, the backscattered yield is plotted vs. energy of the backscattered ions, and this energy can be related to the sample depth where the scattering event occurred. When the RBS probing beam is in an aligned direction, and the backscattering yield within the energy range ΔE reaches a value characteristic of a random direction, the corresponding region of the sample is considered to be highly disordered [13].

Random and aligned spectra from a set of HOPG samples ion implanted and subsequently annealed are shown in Fig. 3. It is seen that for a depth range of ~2600 Å, the backscattering yields for the aligned (unannealed) spectra reach the random backscattering value. This indicates that the highly disordered region produced by implantation extends all the way to the surface; the surface is indicated in the figure by zero depth.

As a function of increasing annealing temperature, the spectra in Fig. 3 clearly shows that the interface between the disordered region and the undamaged HOPG moves towards the sample surface but the backscattering yield (from disordered regions, ΔE) still remains at the "random" value until an annealing temperature of ~2300°C is reached. Above this annealing temperature, the spectra are qualitatively different, as discussed below in detail in Ref. 12.

We have observed the same behavior in the RBS spectra (see Fig. 3) as a function of annealing temperature from a number of other samples implanted with several different ion species under a variety of implantation conditions, including the set of samples that was studied in this work by Raman spectroscopy (^{75}As at E = 230 keV and ϕ = 1 x 10^{15} cm^{-2}). In all cases, the annealing temperature above which the qualitative differences in spectra are observed is $T_a \cong 2300°C$. The observation of the first step of graphitization in the same temperature range, independent of the implant or implantation conditions supports our identification of $T_a \cong 2300°C$ with the completion of the first step of the graphitization of ion damaged graphite. Even if the aligned spectra from the unannealed and 2300°C-annealed samples look qualitatively similar (taking into account the difference in the depth of the disordered layers), we know from the Raman data that there are significant differences in the microstructure of these disordered layers.

<u>Transmission Electron Microscopy (TEM)</u>

To provide more complete information on the first step of the graphitization process, we report transmission electron microscopy (TEM) results for ^{12}C and ^{75}As ion implantation into graphite. Figure 4 shows selected-area diffraction patterns from regions within ≲1000 Å of the surface of ion implanted HOPG before (Fig. 4(a)), and after annealing at 2300°C (Fig. 4(b)). The diffraction pattern before annealing is seen to have diffuse rings close to the position of the ⟨100⟩ and ⟨110⟩ spacings

in graphite. This is characteristic of amorphous carbon formed by
evaporation [14]. After annealing, the diffraction pattern of Fig. 4(b)
clearly indicates that the ⟨100⟩ and ⟨110⟩ spacings have become very
well-defined; however the observation of the ring patterns indicate a
random orientation of the crystallites about the c-axis (sample normal).
This identifies the presence of well-ordered graphite planes with no
positional correlation between the planes.

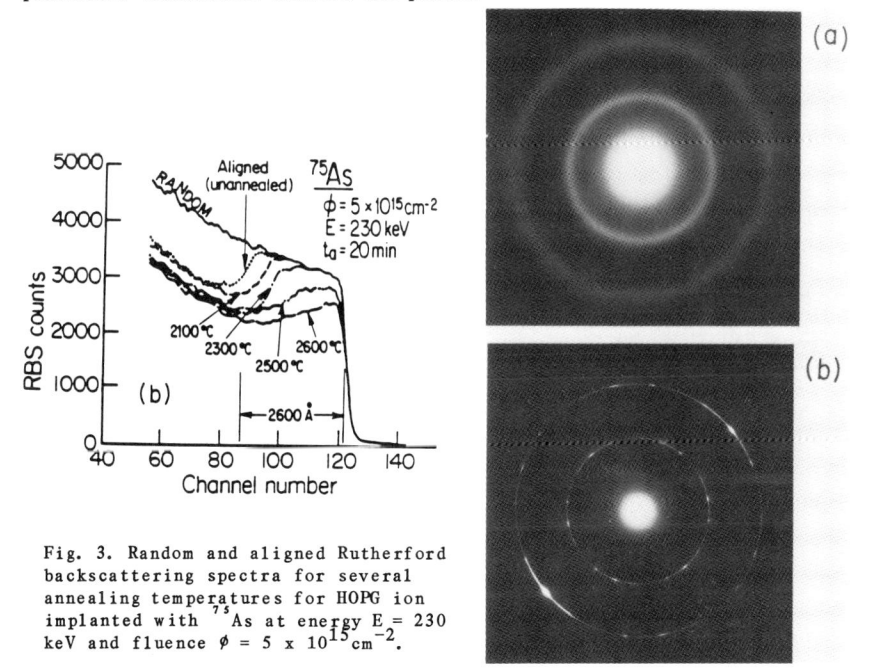

Fig. 3. Random and aligned Rutherford
backscattering spectra for several
annealing temperatures for HOPG ion
implanted with ^{75}As at energy $E = 230$
keV and fluence $\phi = 5 \times 10^{15} cm^{-2}$.

Fig. 4. Selected-area transmission electron diffraction patterns
taken at 200 keV from: a) an unannealed sample of HOPG ion implanted
with ^{12}C at an energy of $E = 100$ keV and a fluence of $\phi = 5 \times 10^{15}$ cm^{-2},
and b) a sample of HOPG ion implanted with ^{75}As at an energy of $E = 230$
keV, a fluence of $\phi = 5 \times 10^{15}$ cm^{-2} and annealed at 2300°C for 20 min.
Note the coincidence of the diffuse rings before annealing in (a) with
the sharp ⟨100⟩ and ⟨110⟩ graphite rings for the c-axis parallel to the
foil normal, but the a-axes are randomly distributed after annealing in
(b).

DISCUSSION

Our Raman spectroscopy data show that the first-order spectra obtained from samples of HOPG ion implanted and subsequently annealed at
2300°C are indicative of the formation of graphitic crystallites
extending up to the implanted surface. On the other hand, the RBS data
from the same samples are consistent with the existence of a highly
disordered region near the surface. There is no contradiction between
these data. The link between the Raman and RBS data is provided by the
TEM measurements, which confirm that the near-surface region considered
by all three techniques consists of 2D-planes of graphite with no
correlation in the third (c-axis) direction. The in-plane crystallite
size in this case is > 1000 Å. The absence of three-dimensional ordering

between the graphitic planes blocks the channels parallel to the c-axis, thereby explaining the high backscattering yield observed in the "aligned RBS spectra" from the samples annealed at 2300°C in Fig. 3. As is also seen in Fig. 3, the backscattering yield from samples annealed at temperatures T > 2300°C is lower than that for the "random" spectrum, providing clear evidence for the ordering of graphite in the third (c-axis) direction as indicated by the unblocking of c-axis channels.

SUMMARY

In summary, at sufficiently high fluences of implantation, a highly disordered region is formed near the surface of HOPG which is observed by Raman scattering, RBS spectrometry and electron diffraction studies. The complementary information obtained by these techniques clearly indicates that the first step of graphitization of the highly disordered region produced by ion implantation is complete at the annealing temperature $T_a \sim 2300°C$ where the 2D-ordering of graphite planes takes place with no correlation in the third (c-axis) dimension. Our results for the establishment of 2D order by high temperature annealing of ion damaged HOPG are consistent with previously reported data on the graphitization of a variety of disordered carbons.

The MIT authors gratefully acknowledge support from ONR grant #N00014-77-C-0053.

REFERENCES

1. A. Marchand and A. Pacault, Nouveau Traite de Chemie Minerale, vol. VIII, No. 1, 457 (1968); D. Fishbach, Chemistry and Physics of Carbon, vol. 7, P.L. Walker, Jr., ed. (Marcel Dekker, New York 1971), p. 1.
2. B.S. Elman, M.S. Dresselhaus, G. Dresselhaus, E.W. Maby and H. Mazurek, Phys. Rev. B24, 1027 (1981).
3. B.S. Elman, M. Shayegan, M.S. Dresselhaus, H. Mazurek and G. Dresselhaus, Phys. Rev. B25, 4142 (1982).
4. T.C. Chieu, M.S. Dresselhaus and M. Endo, Phys. Rev. B26, 5867 (1982).
5. P. Lespade, R. Al-Jishi, and M.S. Dresselhaus, Carbon 20, 427 (1982); A. Marchand, P. Lespade (private communication); A. Marchand, P. Lespade and M. Covzi, Extended Abstracts of the 15th Biennial Conference on Carbon, University of Pennsylvania, p. 282 (1981).
6. J.W. Mayer, L. Eriksson and J.A. Davies, Ion Implantation in Semiconductors, (Academic, New York 1970).
7. W.-K. Chu, J.W. Mayer, M.-A. Nicolet, Backscattering Spectrometry, (Academic Press, New York 1978), and references therein.
8. L.C. Feldman, J.W. Mayer, S.T. Picraux, Materials Analysis by Ion Channeling, (Academic Press, New York 1982).
9. B.S. Elman, M.S. Dresselhaus, G. Dresselhaus, T. Venkatesan and B. Wilkens, Extended Abstracts of the 16th Biennial Conference of Carbon, San Diego, CA (1983), p. 459; B.S. Elman, G. Braunstein, M.S. Dresselhaus, G. Dresselhaus, T. Venkatesan and B. Wilkens, (to be published).
10. F. Tuinstra and J.L. Koenig, J. Chem. Phys. 533, 1126 (1970).
11. B.S. Elman, M. Hom, E.W. Maby and M.S. Dresselhaus, Intercalated Graphite, M.S. Dresselhaus, G. Dresselhaus, J.E. Fischer, M.J. Moran, eds. (North-Holland Elsevier, New York 1983), p. 341.
12. T. Venkatesan, B.S. Elman, G. Braunstein, M.S. Dresselhaus and G. Dresselhaus (to be published).
13. Ref. 7, p. 247.
14. J. Kakinoki, K. Katada, T. Tanawa and T. Ino, Acta. Cryst. 13, 171 (1960).

CHANNELING STUDIES OF THERMAL REGROWTH IN ION DAMAGED GRAPHITE

T. VENKATESAN[+]
AT&T Bell Laboratories, Murray Hill, New Jersey 07974

B. S. ELMAN, G. BRAUNSTEIN, M. S. DRESSELHAUS and G. DRESSELHAUS
Center for Materials Science and Engineering, MIT, Cambridge, MA 02139

ABSTRACT

The crystallization of disordered surface layers on highly oriented pyrolytic graphite (HOPG) have been studied by Rutherford backscattering spectrometry (RBS) and channeling techniques. Disordered layers (~1000–3000Å thick) are produced on the surface of HOPG by the implantation of various ions. The disordered layers are regrown by thermal annealing of the samples in an inert environment. Isochronal anneals reveal two distinct regrowth processes: one, a rapid process of low activation energy ($E_a \sim 0.15$ eV) which is observed primarily in regions where the disorder is sufficient to prevent the channeling of the ions but insufficient to totally destroy the graphitic structure. This low activation energy may indicate annealing of the damage by migration of interstitials where the interstitials are the knock-on carbon atoms produced by the primary ions. A regrowth process with higher activation energy ($E_a \sim 2.0$ eV) occurs primarily in regions where the disorder is close to the saturation-disorder produced by ion implantation. Both the regrowth processes are epitaxial in nature and the epitaxial nature of the process may explain the much lower activation energy for 3D AB stacking as measured in ion implanted graphite when compared with results on the bulk graphitization of pyrocarbons.

INTRODUCTION

Crystal growth usually occurs at a solid-liquid or solid-vapor interface at which the perfection of the crystal is controlled by the crystallinity of the interface. The growth of graphite crystals from the vapor has been achieved in the formation of graphite whiskers. The normal synthetic graphites are obtained at a crystal-amorphous solid interface at which small crystallites grow as a result of solid state diffusion. This graphite crystal growth is difficult to monitor and characterize. The present work reports on a series of experiments related to the regrowth of graphite under more controlled conditions making use of ion implantation to introduce disorder into the substrate in a controlled manner. Specifically, the implantation process produces a crystal-disordered solid interface which we monitor by Rutherford backscattering spectrometry (RBS) (with depth resolution of ~400Å) during a series of isothermal and isochronal anneals during which the crystalline regrowth occurs.

The graphite recrystallization studies reported in this paper are to be compared and contrasted with the graphitization studies which have been reported with various precursor carbons.[1] Carbons are classified into graphitizable and ungraphitizable precursors. The graphitizable carbons generally start with either large or small "crystallite" planes which are to some extent aligned. In the graphitization process these planes grow in lateral extent and decrease the mosaic spread of their c-axis orientations until they obtain the AB registry of three-dimensional graphite. The ungraphitizable carbons generally have precursor morphologies which could best be described as tangled ribbons of graphite-like planes. Upon heat treatment these planes grow but because of the constraints imposed by the original tangle they often cannot fully form 3-dimensional graphite.

The conventional wisdom about the characteristic temperatures associated with graphitization are as follows. The carbon atoms in the disordered (amorphous) region migrate to the edges of the planar crystallites and form the trigonal bonds with the planar edge atom starting at about 1500°C. The energy required for this process (~0.4 eV) is much less than that associated with a σ-bond and the rate limiting step may possibly be associated with carbon atom migration. In the temperature between 1500° and 2300°C the separation between the planes in graphitizable carbons is observed to decrease from

---
[+] Central Services Organization of the Regional Bell Operating Companies.

~3.45Å down to about 3.35Å. This decrease in interplanar separation is associated with plane flattening which must occur when the crystalline platelets increase in area. Finally at ~2500°C when the crystallite sizes as inferred by Raman and x-ray measurements are greater than 300Å the three dimensional AB order sets in and larger 3-dimensional crystals are formed in the temperature range of 2500—3000°C. The three dimensional registration provides the closest packing of the layer planes with the separation of 3.35Å.

The behavior described above is typical for pyrolytic carbons. The recrystallization associated with ion implantation is unique in that the regrowth occurs from the crystalline substrate which is undamaged by implantation. The rate limiting step associated with this growth is inferred in this work from the activation energies which we determine by a series of isothermal and isochronal annealing experiments on highly oriented pyrolytic graphite (HOPG) implanted with carbon and impurity ions under a variety of carefully selected implantation conditions.

ION IMPLANTATION AND CHANNELING

Ion implantation provides a convenient means to damage a surface layer of a solid in a clean and reproducible manner. Sharp amorphous to crystalline interfaces have been demonstrated in Si by ion implantation[2] and the regrowth kinetics of silicon have been studied[3] by subsequent thermal annealing of such disordered layers. In ion implantation, ions (As^+, C^+, Si^+, etc.) with energies of ~30—300 keV impinge on a solid and disorder is induced in the solids by means of direct nuclear displacement collisions or by the generation of dense electronic ionizations. The thickness and extent of damage are controlled by the mass, energy and the fluence of the ions. The literature on ion implantation in solids is an extensive one[2] and we will not delve into the details of the implantation process at this juncture.

Rutherford backscattering[4] (RBS) and channeling[5] spectrometry of high energy ions is a routine technique for studying surface layers of solids. The technique provides depth dependent information on the chemical composition, stoichiometry and crystallinity of the surface region (~ few 1000Å) of the solid with depth resolutions of typically a few 100Å. When MeV α-particles impinge on a sample, these probe ions backscatter with energies characteristic of the element and its depth in the target. In addition, if the solid is crystalline and the crystal axis lines up with the ion direction, a significant reduction in the backscattering of the ion occurs due to a process called "channeling" by which the ions are coherently guided down the crystal channels by the atomic potentials. However, if disordered layers intercept the beam, then the probe ions backscatter with energies characteristic of the depth. The backscattering yield is proportional to the scattering centers and depends upon the atomic number of the scattering nucleus. Thus the backscattering signal would contain information about the extent of damage and the location of the damaged layer.

Though the channeling technique has been applied primarily to the studies of crystalline materials, recently there have been some study of textured silicides[6,7] and polycrystalline graphite.[8] In polycrystalline materials as long as the crystallites are large enough (≥ 1000Å) and the mosaic spread of the crystal axes with respect to the surface is small compared to the critical angle to observe the channeling of ions, RBS is a viable technique to study disordered layers. An extensive study of the channeling phenomenon in HOPG will be reported elsewhere.[9]

EXPERIMENTAL

The stress annealed HOPG samples obtained from Union Carbide typically exhibit a mosaicity of $0.4 \pm 0.1°$ FWHM as measured by x-ray diffraction. The mosaic spread of these crystal as measured by the channeling technique[9] was $0.48 \pm 0.05°$. The samples used in the present work (10×5×0.1 cm) were cut from a large HOPG block and were ion implanted in vacuum with various species at different energies and fluences in an Accelerators 300R ion implanter. Subsequent to the implantation, identical samples were subjected to ioschronal anneals of ~20 minutes at temperatures ranging from 300—3000°C in a tubular carbon furnace in a flowing argon ambient. Series of isothermal anneals 'at a few temperatures were also carried out to study the time dependence of the regrowth process. The samples were then analyzed by the backscattering study of 2 MeV He ions obtained from a Van de Graaff generator. The samples were mounted in a goniometer providing adequate tilt controls to enable the alignment of the c-axis of the crystal with the beam direction. Typical signals required He ion fluences of ~4×10^{15} ions/cm^2, but at such high energies (2 MeV) the damage produced by a light ion like He was not significant.

RESULTS AND ANALYSIS

Typical spectra of channeling from ion implanted graphite surfaces are shown in Fig. 1. The curve corresponding to "random" represents the spectrum obtained with the c-axis of the crystal not aligned with the beam direction. In this mode there is no distinction between the implanted or pristine graphite

469

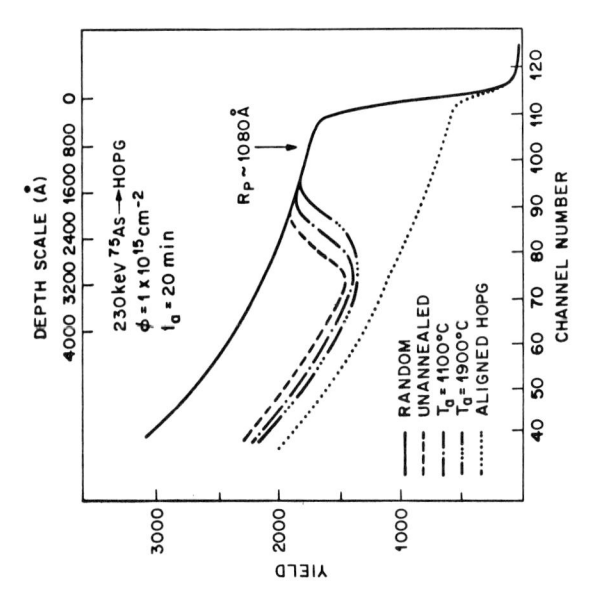

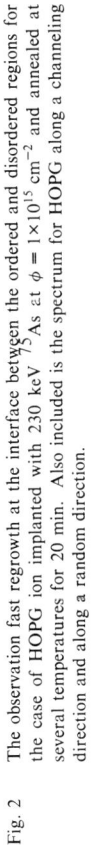

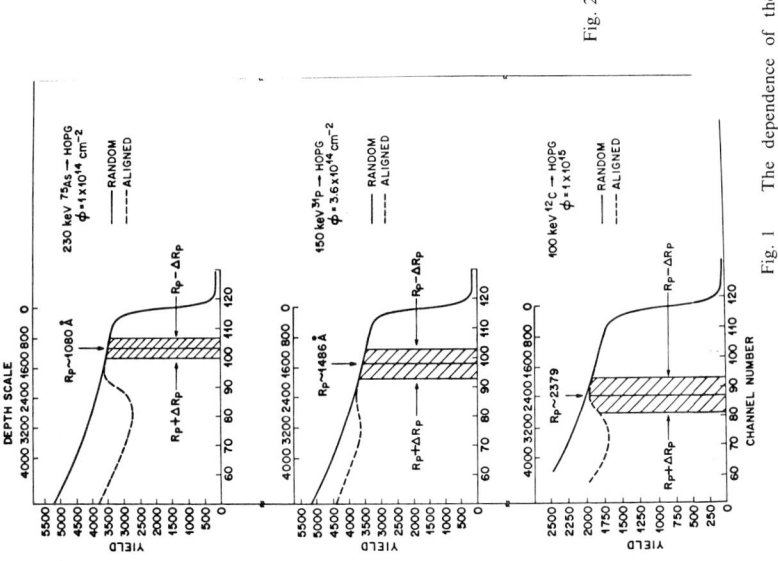

Fig. 2 The observation fast regrowth at the interface betyween the ordered and disordered regions for the case of HOPG ion implanted with 230 keV ^{75}As at $\phi = 1\times10^{15}$ cm^{-2} and annealed at several temperatures for 20 min. Also included is the spectrum for HOPG along a channeling direction and along a random direction.

Fig. 1 The dependence of the disorder on the mass of the implanted ion. The conditions of implantation and the distribution parameters calculated from LSS theory ($R_p \pm \Delta R_p$) are indicated. A comparison of spectra taken along a random and channeling directions is given.

as far as the C signal is concerned. The rest of the curves show the signal obtained in the channeling mode and the effect of the implantation induced disorder. For 100 keV C atoms the penetration depth (R_p) and the straggle (ΔR_p) in graphite calculated from theory[10] is ~2379.0 and ~462.0Å respectively. By varying the ion energy and fluence the extent and thickness of the disordered layer could be varied.

The dependence of the disorder on the mass of the projectile is also shown in Fig. 1. We compare the RBS spectra for three ions (As, P and C) at energies of 230 keV for ^{75}As, 150 keV for ^{31}P and 100 keV for ^{12}C. The fluences are chosen to be 1×10^{14} cm^{-2}, 3.6×10^{14} cm^{-2} and 1×10^{15} cm^{-2} respectively. This is done in accordance with our Raman spectroscopy results[11] which show that for a lighter ion mass, one must use a higher fluence of implantation to produce a highly disordered region extending all the way to the surface. In Fig. 1 the regions of $R_p \pm \Delta R_p$ are also indicated for each ion at a particular energy for reference. Clearly it is seen, that the width of the damaged region is larger compared to $(R_p + \Delta R_p)$, the more massive the ion. Considering again the cross sectional view of ion implanted HOPG (Fig. 3 of Ref. 11), we conclude from Fig. 1 that the region beyond PP′ (Fig. 3 of Ref. 11) is not simply the pristine crystalline material, but also contains a disordered region the extent of which increases with increasing ion mass. The presence of this disordered region can be explained by the effect of "knock-on" collisions, whereby the more massive the implanted ion, the higher the energy where the maximum ion-nuclear stopping occurs and hence the larger the range of the displaced carbon atoms. In all three cases demonstrated in Fig. 1, the secondary "knock-on" atoms are the carbon atoms which cause further damage to the lattice before coming to rest at random positions. In terms of the annealing kinetics, the damaged regions produced by the secondary carbon atoms would only require the migration of interstitials and hence could be activated with very low activation energies.[1] At the same time, the regions closer to the damage peak of the primary ion beam would require much larger activation energies, since these regions are severely damaged.

In Fig. 2 we show the regrowth of the back of the disordered region by annealing. Specifically, we show an HOPG sample implanted with 1×10^{15} cm^{-2} ^{75}As ions at 230 keV for several annealing temperatures. The aligned spectrum from HOPG is shown as a reference. It is seen that the width of the disordered region decreases as the temperature of annealing increases. This is a clear indication of an epitaxial regrowth of the damaged graphite layers at the order-disorder interface. The regrowth is observed to start at temperatures as low as 300°C, which is understandable from the origin of the near-interface disorder as discussed above. The area under the disordered peak of the RBS spectrum (ΔA) is proportional to the number of blocked channels and provides a measure of the lattice damage in the sample.

We now present the results for the movement of the interfaces in Fig. 2 by plotting this movement (Fig. 3) in terms of Δz and ΔA with respect to $(1/T)$, where Δz is the thickness of the regrown disordered region. The data for samples implanted under different conditions are seen to yield a reasonable Arrhenius fit. It is interesting to note that the areas (ΔA) under the disordered peaks in Fig. 2 are also directly proportional to the number of carbon atoms displaced from their lattice sites by a distance larger than the Thomas-Fermi screening distance.[5] The activation energies obtained for all samples are very close to an average value of ~0.15 eV. As discussed earlier, these regrowth processes are most likely due to the motion of interstitial atoms blocking the channels, and the low activation energy associated with this process justifies this interpretation.[1,12] The same value for the activation energy obtained from Fig. 3 for samples implanted under different conditions, supports the interpretation that the regrowing of the implantation-induced disordered regions has a similar origin, and the annealing process is also similar.

Now we will again consider isochronal annealing (t = 20 min) at temperatures ($T_a > 2300$°C) high enough so that the 3D ordering of the graphite planes can take place. Figure 4 shows the behavior of the damaged layer of HOPG implanted with ^{12}C at $\phi = 5\times10^{15}$ cm^{-2} and E = 100 keV over a wide range of annealing temperatures ($T_a < 3000$°C). The "random", aligned channeling spectra for HOPG and the spectrum for the as-implanted (unannealed) sample are given for comparison. For annealing temperatures in the range $T_a < 2300$°C an interfacial regrowth occurs is shown in Fig. 4. In contrast, at higher annealing temperatures, the backscattering yield decreases concomitant with the advancement of the interface. This implies one of two possibilities: either the formation of well ordered two dimensional regions with partial three dimensional registry or the formation of three dimensionally ordered islands in a disordered host. In both cases, however, there is a decreasing registry of the layers as we move from the undamaged bulk region towards the surface. This implies that independent of the above scenarios, the three dimensional registry occurs in an epitaxial manner.

Typical estimated velocities of the establishment of 3D ordering are ~1 Å/s at a temperature of 2600°C. In Fig. 5 we show Arrhenius plots of the regrowth of interface (Δz) as well as the fraction of

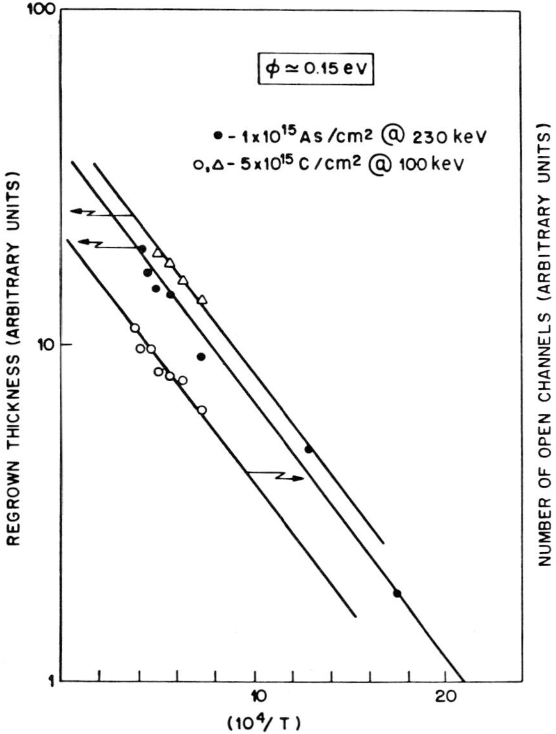

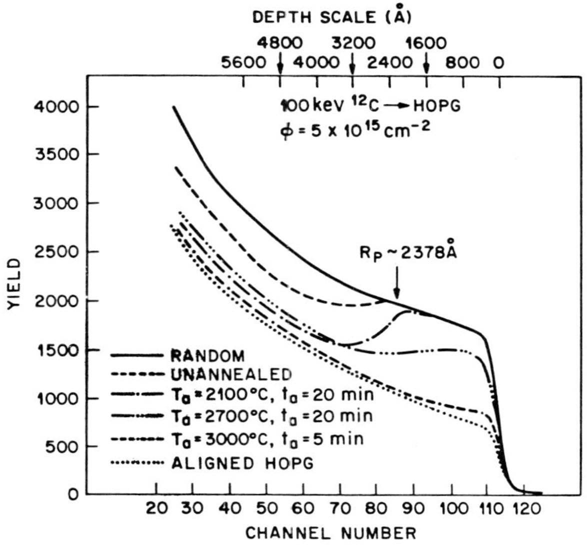

471

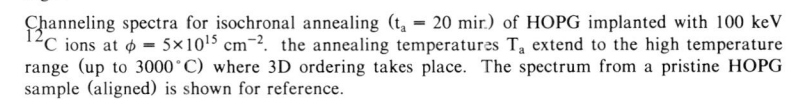

Fig. 4

Channeling spectra for isochronal annealing ($t_a = 20$ min) of HOPG implanted with 100 keV ^{12}C ions at $\phi = 5 \times 10^{15}$ cm^{-2}; the annealing temperatures T_a extend to the high temperature range (up to 3000°C) where 3D ordering takes place. The spectrum from a pristine HOPG sample (aligned) is shown for reference.

Fig. 3 Arrhenius plots of the fast regrowth of the disordered layers of HOPG initially implanted under various conditions (as indicated). The areas under the curves in the channeling spectra (Fig. 4) are also plotted and show the same behavior. In all cases, the activation energy is $E_a \simeq 0.15$ eV.

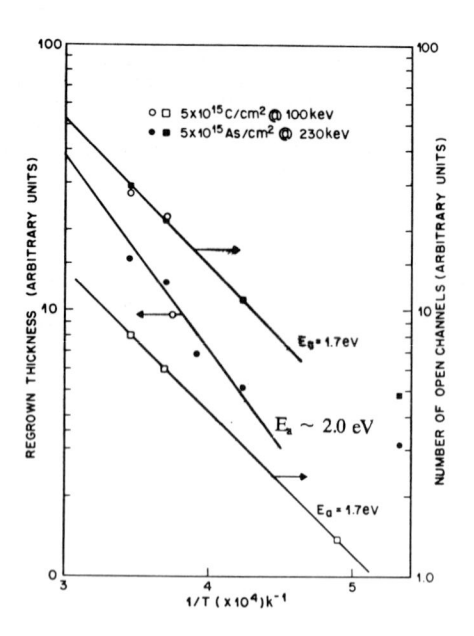

Fig. 5   Arrhenius plots the interface regrowth and the number of open channels in the slow regrowth regions in the HOPG implanted under various conditions. In all cases the source activation energy is obtained ($E_a \simeq 2$ eV) over the limited available temperature range.

open channels for two different conditions of implantation (indicated in the figure). The activation energies obtained from the analysis of both of these parameters appears to be the same with the value of $E_a \sim 2$ eV. This value of E_a is considerably less than the typical energies of 6-11 eV measured for the bulk graphitization.[1,12] The epitaxial nature of the regrowth process for the case of the ion implanted samples may be responsible for such a low value for the activation energy. The crystalline substrate provides a large area on which the adjacent disordered layers can be aligned with a nucleation of the three dimensional registry occurring at the undamaged substrate interface. Another possible difference between these experiments and the earlier ones on the graphitization of carbons is the extent of amorphicity of the disordered layer. It is known[11,13] that even after very heavy doses of ion bombardment, the graphite surfaces exhibit partly oriented crystallites of dimensions of $\sim 15-20$ Å. Such crystallites at the interface may further be responsible for the lower activation energy.

The model developed in this paper to explain the epitaxial regrowth has relevance to the graphitization process for carbons. That is, a graphitizable carbon could be initially represented by a "quasi-amorphous" collection of carbon atoms with 1,2 and 3 fold coordination. The initial stages of graphitization consist of slow growth of graphite platelets by acceration. After the platelet lateral dimension L_a grows to a size of ~ 100Å, then three dimensional order sets in and the interplanar separation c drops to its final value of 3.35Å. The (002) x-ray lines originally broad and shifted, become narrow and displaced[14] indicating an increasing L_c and L_a and an appropriate graphitic interplanar separation of 3.5Å. In contrast to the usual graphitization described above the epitaxial regrowth occurs at the interface between the crystalline substrate and the disordered graphite. An interface is clearly defined in the RBS channeling data and implies a discontinuous change in the parameters c, L_a and L_c which are traditionally used to characterize the degree of graphitization. Our Raman experiments[14] indicate that the microcrystallite parameters L_a and L_c grow during the heat treatment to a maximum size of ~ 300Å. The disordered region is probably most clearly defined by the interplanar parameter, $\Delta c/c_o \sim \dfrac{(c-c_o)}{c_o} \simeq 0.03$ which changes to zero in the bulk crystalline material. The distance over which this change occurs defines the width of the interface. Thus the usual graphitization process which shows

L_a and L_c increasing monotonically with heat treatment temperature behaves identically with the disordered layer on the graphite. However, the deviation of the interplanar distance from the ideal, i.e., $\Delta c/c_0$, occurs discontinuously at the interface for the epitaxial regrowth. Our results suggest that graphitization could occur at a lower temperature if a pyrolytic carbon were deposited on a graphite substrate. It has been observed that a few layers of carbon deposited on diamond can epitaxially grow to form 4-fold coordinated diamond bonds. The measurements reported here suggest that a similar process occurs on graphite to an even greater degree.

SUMMARY

In conclusion, ion implantation and channeling techniques are ideally suited for the study of regrowth phenomena on the surfaces of graphite. Such studies can yield quantitative measure of the thermodynamics of the regrowth processes in graphite. These studies not only enhance our understanding of the physics and chemistry of carbon but also associated phenomena like intercalation and radiation induced damages.

We would like to acknowledge J. M. Poate for stimulating ideas and encouragement.

REFERENCES

[1] B. T. Kelly, "*Physics of Carbon*", Applied Science Publishers, London and New Jersey, 1981.

[2] J. W. Mayer, L. Erikson and J. A. Davies, Ion Implantation in Semiconductors, Phys. Rev. 180.

[3] L. Csepregi, J. W. Mayer and T. W. Sigman, Appl. Phys. Lett. 29, 92 (1976).

[4] W. K. Chu, J. W. Mayer, and Marc-A. Nicolet, Backscattering Spectrometry, Academic Press New York, 1978.

[5] L. C. Feldman, J. W. Mayer and S. T. Picraux, Materials Analysis by Channeling, Academic Press, New York, 1982.

[6] H. Ishiwara, and S. Furukawa, J. Appl. Phys. 47, 1686 (1976); H. Ishiwara, K. Hikosaka, M. Nagatom and S. Furukawa, Surface Sci. 86, 711 (1979).

[7] D. Sigurd, R. W. Bower, W. F. Van Der Weg and J. W. Mayer, Thin Solid Films, 19, 328 (1973).

[8] T. Iwata, K. Komaki, H. Tomimitsu, K. Kawatsura, K. Orawa and K. Doi, Rad. Effects. 24, 63 (1975).

[9] B. S. Elman, G. Braunstein, M. S. Dresselhaus, G. Dresselhaus, T. Venkatesan and B. Wilkens, to be published.

[10] J. Lindhard, Kgl. Danske Videnskal. Selskab., Mat-Fys. Medd. 34, No. 14 (1965).

[11] B. S. Elman, M. Shayegan, M. S. Dresselhaus, H. Mazuzek and G. Dresselhaus, Phys. Rev. *B25*, 4142 (1982).

[12] D. Fishbach, "*Chemistry and Physics of Carbon*", vol. 7, ed. by P. L. Walker, Jr., (Marcel Dekker, New York, 1971) p. 1; A. Pacault, "*Chemistry and Physics of Carbon*", vol. 7, ed. by P. L. Walker, Jr., (Marcel Dekker, New York, 1971), p. 107.

[13] T. Venkatesan, R. C. Dynes, B. Wilkens, J. M. Gibson, A. E. White and R. Hamm, Proc. of Second International Conf. On Radiation Effects on Insulators, Albuquerque, NM, May 1983. To be published.

[14] B. S. Elman, M. S. Dresselhaus, G. Dresselhaus, T. Venkatesan and J. M. Gibson. To be published.

HIGH TEMPERATURE IMPLANTATION IN GRAPHITE

G. BRAUNSTEIN,[*] B.S. ELMAN,[*] M.S. DRESSELHAUS,[+*] G. DRESSELHAUS[#]
[*]Department of Physics; [+]Department of Electrical Engineering and Computer Science; [#]Francis Bitter National Magnet Laboratory;
Massachusetts Institute of Technology, Cambridge, MA 02139, USA

T. VENKATESAN
Bell Laboratories, Murray Hill, NJ 07974, USA

ABSTRACT
　　In previous studies it was found that when highly oriented pyrolytic graphite (HOPG) is implanted at room temperature, the damage caused by the implantation could be completely annealed by heating the sample to temperatures higher than ~ 2500°C. However at these high temperatures, the implanted species was found to diffuse out of the sample, as evidenced by the disappearance of the impurity peak in the Rutherford backscattering (RBS) spectrum. If, on the other hand, the HOPG crystal was held at a high temperature ($\geq$ 600°C) during the implantation, partial annealing could be observed. The present work further shows that it is possible to anneal the radiation damage and simultaneously to retain the implants in the graphite lattice by means of high temperature implantation ($T_i \geq$ 450°C) followed by annealing at 2300°C.

INTRODUCTION

　　Graphite is an interesting prototype material for ion implantation studies because of the high anisotropy in its structure and physical properties [1]. For example, the layered nature of the graphite lattice gives rise to a regrowth behavior completely different from that observed in widely studied semiconductors [2,3]. In addition, interesting applications of ion implanted graphite can be envisaged in terms of modification to the electrical, thermal and magnetic properties of the host material [4]. There are two main ways to change the physical properties of graphite: a) by substitutional doping or b) by intercalation. In both cases the number of species which can be introduced into the graphite lattice by thermal or chemical means is very limited, while in principle any ion can be introduced into the graphite host by means of ion implantation.
　　Initial studies on radiation damage in ion implanted highly oriented pyrolytic graphite (HOPG) [2,3] have shown that the recrystallization (graphitization) of the disordered lattice is a two step process, as also occurs in the graphitization of pyrocarbons [5,6]. The first step involves two-dimensional or in-plane ordering, which occurs for annealing temperatures in the approximate range 1500°C $\leq T_a \leq$ 2300°C [2]. The second step of graphitization occurs at higher temperatures where three dimensional or c-axis ordering takes place [2]. Unfortunately when the implantation is done at room temperature, the amount of implanted species retained by the graphite lattice decreases with increasing annealing temperature, until, at a temperature of approximately 2300°C (slightly dependent on implantation dose), all the impurities diffuse out of the substrate [7]. If on the other hand, the HOPG crystal is held at about 600°C during the implantation, significant preservation of crystallinity can be achieved. This behavior had already been observed in another variety of carbon (i.e., diamond [8]) as well as in silicon [9].

The object of the present work is to investigate if it is possible by means of high temperature implantation to achieve both recrystallization of the lattice and retention of the implants by the graphite host. For this purpose we have carried out: a) Raman scattering and RBS-channeling measurements in which the damage accompanying ^{75}As implantation into HOPG has been studied as a function of implantation temperature T_i as well as after subsequent annealing at 2300°C. b) RBS-channeling studies of the evolution of the impurity distribution under the same implantation and annealing conditions.

EXPERIMENTAL

The graphite host material in the present work is HOPG which is polycrystalline with crystallites exhibiting a mosaic spread in their c-axis orientations of less than 0.5°. ^{75}As ions were implanted into the HOPG samples at 230 keV, to a dose of 1×10^{15} cm^{-2} at different implantation temperatures. The experimental set up used for the high temperature implantations was previously described [10]. The Raman scattering experiments were performed at room temperature using a Brewster angle back-scattering geometry. Incident radiation at ~4880 Å and ~100 mW was provided by a CW argon-ion laser. The scattered radiation was collected at 90° to the sample surface, analyzed by a double grating monochromator and detected by a cooled photomultiplier tube. RBS-channeling analysis was carried out using a probing beam of 2 MeV He$^+$ ions. Backscattered particles were detected at ~175° by an annular surface-barrier detector with energy resolution $\leq$ 20 keV (FWHM). Post-implantation annealing was performed for 20 minutes at 2300°C in a carbon furnace containing an argon atmosphere.

RESULTS

Figure 1 shows RBS-channeling spectra of the HOPG samples implanted at different temperatures T_i with 230 keV ^{75}As ions to a fluence of 1×10^{15} cm^{-2}. It is seen that as T_i is progressively increased, the

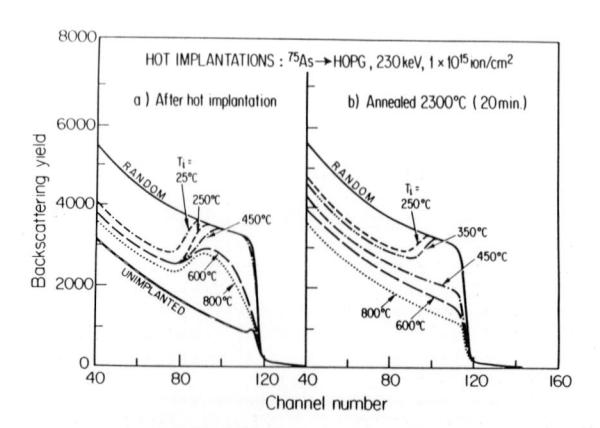

Fig. 1. RBS-channeling spectra of hot implanted ^{75}As into HOPG, showing the dependence of the residual damage on the implantation temperature T_i: a) after high temperature implantation, b) after subsequent annealing for 20 minutes at 2300°C.

radiation damage gradually diminishes. After implantation at $T_i \leq 450°C$, a highly disordered region is produced which extends all the way from the surface to a depth between 2500Å < d < 3000Å (depending on T_i), as indicated in Fig. 1 by the fact that the aligned spectra reach the random level in that region. However after implantation at $T_i = 600°C$ and $T_i = 800°C$, the observed damage peak is much lower than the random level, indicating significant preservation of crystallinity after the hot implantation. To demonstrate the healing effect of high temperature implantation, the RBS-channeling spectrum for the room temperature implantation is included.

In the case of ion damaged HOPG, analysis of the backscattering results alone can give rise to misleading conclusions due to the particular annealing characteristics of graphite, where it was found that 2-dimensional ordering is well established before 3-dimensional ordering proceeds [2]. It was previously shown [10] that Raman spectroscopy is a useful technique for monitoring the residual amount of lattice damage accompanying ion implantation as well as after subsequent annealing. In particular the Raman spectrum is especially sensitive to the in-plane ordering (e.g., 2D ordering) of HOPG [2]. We have therefore also used the Raman scattering technique to characterize the hot implanted samples.

Figure 2 shows 1st order Raman spectra of the same HOPG samples whose RBS-channeling spectra appear in Fig. 1. The 1st order Raman spectrum for room temperature implantation ($T_i \sim 25°C$) consists of a very broad asymmetric line; however, as the implantation temperature is gradually increased, we are able to resolve a first order Raman structure consisting of mainly 3 lines: I) the ~1580 cm^{-1} Raman-allowed line corresponding to the zone-center high frequency mode of crystalline HOPG, and II) and III) the ~1360 cm^{-1} and 1620 cm^{-1} disorder-induced lines appearing because of the high density of phonon states at those frequencies.

Tuinstra and Koenig [11] have shown that the relative intensity of the disorder-induced ~ 1360 cm^{-1} line relative to the Raman-active ~ 1580 cm^{-1} line $R = I_{1360}/I_{1580}$ varies inversely with the crystallite size L_a of the graphite sample, a result which we use to monitor the 2D reordering of the crystalline structure [2]. Previous studies have shown that the linewidths (FWHM) of the ~ 1360 cm^{-1} and 1580 cm^{-1} lines also contain information about the degree of disorder induced by ion implantation [7].

A summary of the results for the first-order Raman spectra of the

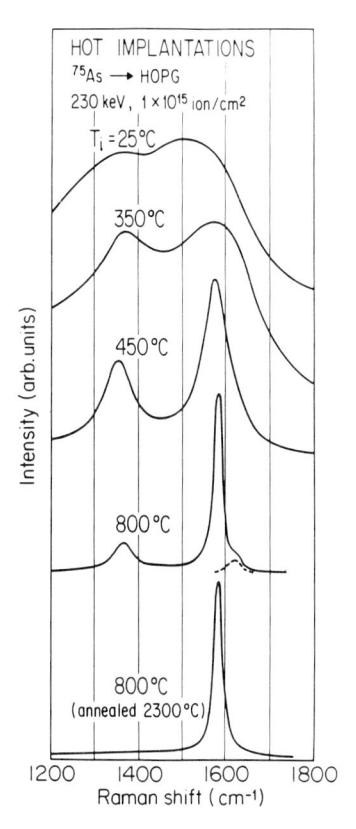

Fig. 2. First order Raman spectra of HOPG samples implanted at various temperatures T_i with 230 keV ^{75}As ions to a dose of 1×10^{15} ion/cm^2.

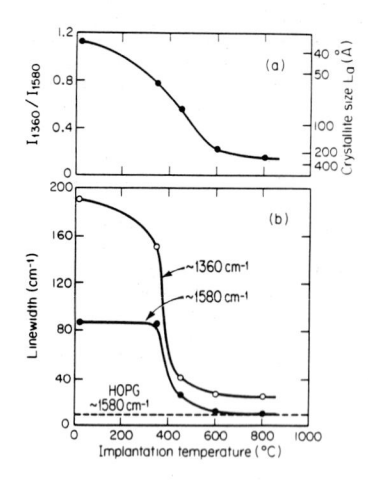

Fig. 3. Results of the analysis of the first order Raman spectra of ^{75}As hot implanted HOPG:
a) Ratio of the intensities of the disorder-induced ~ 1580 cm^{-1} line to the Raman-allowed ~ 1580 cm^{-1} line as a function of implantation temperature, b) plot of linewidth (FWHM) of the disorder-induced ~ 1360 cm^{-1} line (o), and Raman-allowed ~ 1580 cm^{-1} line (•) as function of implantation temperature.

hot implanted HOPG is given in Figs. 3a and b where the intensity ratio I_{1360}/I_{1580} as well as the linewidths of the two lines are, respectively, plotted as a function of the temperature of implantation T_i. It is seen that I_{1360}/I_{1580} decreases with increasing T_i, while the crystallite size L_a gradually increases. On the other hand, the linewidths of both lines show a rather discontinuous behavior near T_i ~ 400°C. With respect to the retention of the implanted As ions by the graphite lattice, Fig. 4 shows that indeed most of the implants remain in the sample after the high temperature implantations. The results described above show that although the amount of implantation-induced damage can be greatly reduced by implanting at elevated temperature, even for T_i ~ 600°C, the crystalline structure is not completely restored.

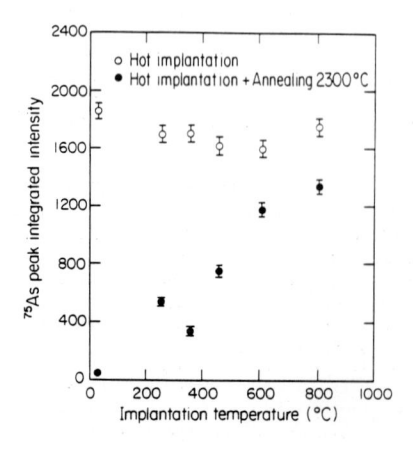

Fig. 4. Integrated intensity under the As peak in the RBS-channeling random spectra of HOPG samples implanted at various temperatures T_i, showing the relative amount of impurity retained by the lattice:
o) after hot implantation,
•) after subsequent annealing at 2300°C.

To see whether it is possible to completely heal the radiation damage by subsequent annealing, the samples were annealed for 20 minutes at 2300°C in an argon atmosphere. This particular temperature was chosen because previous studies [2,7] of room temperature implanted HOPG have shown that 2D ordering was established at 2300°C, although in that case the As implants were found to leave the substrate at this elevated annealing temperature.

RBS-channeling and Raman spectroscopy results for the subsequently annealed samples are shown in Figs. 1b, 3b, and 4. The RBS-channeling spectra of the subsequently annealed samples shown in Fig. 1b reveal that the lattice damage of the samples implanted at 450°C, 600°C and 800°C is almost completely healed. In contrast, the spectra of the samples implanted at 350°C, 250°C and room temperature show some epitaxial regrowth, but the backscattering yields still reach the random level, indicating a heavily disordered region near the surface (from the RBS-channeling point of view). The variation of the backscattering minimum yield χ_{min} among the samples implanted at 450°C, 600°C and 800°C may be due to differences in the mosaic spread resulting from ion implantation and subsequent heat treatment or simply due to differences in the degree of perfection of the starting HOPG material. The first order Raman spectra of all the samples show only a very narrow ~ 1580 cm^{-1} line, which is taken as evidence that the first step in the graphitization process, the 2D ordering, has indeed been completed.

Figure 4 reveals the important result that most of the As implants remain inside the sample after post-hot-implantation annealing. The amount of As retained by the lattice appears to increase approximately linearly with increasing implantation temperature. Furthermore, a comparison between the As peaks obtained in channeling and random geometries for T_i = 450°C, 600°C and 800°C indicates that about ~15% of the implants occupy sites along the c-axis rows, possibly at substitutional or commensurate interstitial positions.

DISCUSSION

The present study of high temperature implantation into HOPG suggests that analogous to the two stage graphitization process (2D and 3D reordering) found in regrowth studies of room temperature ion implanted HOPG [2,3,7], it is possible to define a two stage disordering process. The first stage corresponds to a partial or severe disordering of the 2D or in-plane structure but the layered structure of the lattice (along the c-axis) is still conserved. In the present case the first stage of disordering corresponds to implantation temperatures above 400°C. The second stage corresponds to the destruction of the layered structure, and occurs for T_i < 400°C. It is interesting to note that we cannot distinguish one stage of disorder from the other simply by monitoring when the aligned RBS-channeling yield reaches the random level. Instead these two stages can be distinguished by the rapid broadening of the Raman linewidths in the vicinity of T_i = 400°C. The annealing of the 2D or in-plane disorder by heating the sample to 2300°C [2], leads to the complete recovery (2D and 3D ordering) of the crystalline structure for those samples implanted at temperatures higher than 400°C, that is, samples with a degree of damage corresponding to the first stage in the disorder process.

The behavior of the implanted ions observed in high temperature implantation and subsequent annealing is markedly different from that observed in room temperature implantation followed by annealing. In the latter case, the impurity peak in the RBS-channeling spectrum starts to decrease at annealing temperatures T_a > 1600°C, and almost no impurity peak is left by T_a = 2300°C, the temperature at which only 2D reordering occurs. During this annealing process the position of the impurity peaks do not change while the FWHM linewidths and the backscattering yields decrease with increasing annealing temperature, implying a lateral diffusion of the arsenic atoms, rather than diffusion along the c-axis [7]. On the other hand, the As implants are retained by the lattice after post-hot-implantation annealing at 2300°C. The fraction of As atoms which remain in the sample is seen in Fig. 4 to increase almost linearly from a vanishing amount for the room temperature implantation to about 70% of the implants embedded in a perfectly recovered crystalline structure for the T_i = 800°C implantation. For T_i > 400°C, the depth profile of the implant does not change significantly after annealing at 2300°C and indeed the depth profile parameters are also similar (except for the impurity concentration) to those measured for the unannealed room temperature implantations. These results suggest that the magnitude of T_i is the critical factor which determines whether or not the implant is retained after high temperature annealing. Further studies are under way to understand in more detail the behavior of the impurities in high temperature implantation and subsequent annealing processes.

CONCLUSION

We have shown in the present work that it is possible to introduce As into a recovered graphite lattice by means of high temperature implantation (T_i > 450°C) and subsequent annealing at 2300°C. This result opens the possibility for studies of the modification of the electronic properties of graphite by means of ion implantation as well as the possibility for new materials synthesis of elements that cannot be introduced into the graphite host by thermal or chemical means.

ACKNOWLEDGMENT

The work at MIT was supported by ONR grant #N00014-77-C-0053. We wish to thank Dr. J.M. Poate for helpful discussions, Mr. M. Rothman for assistance with the ion implantation and Dr. A.W. Moore of Union Carbide Corporation for the HOPG material. Two of us (G. Braunstein and B.S. Elman) also acknowledge Bell Laboratories for their hospitality when carrying out the RBS measurements.

REFERENCES

1. B.T. Kelly, Physics of Graphite (Applied Science Publishers Ltd., Essex, 1981).
2. B.S. Elman, M.S. Dresselhaus, G. Braunstein, G. Dresselhaus, T. Venkatesan, B. Wilkens and M. Gibson, Proceedings of this Symposium.
3. T. Venkatesan, B.S. Elman, G. Braunstein, M.S. Dresselhaus and G. Dresselhaus, Proceedings of this Symposium.
4. M.S. Dresselhaus and G. Dresselhaus, Advances in Physics 30, 139 (1981).
5. A. Marchand and A. Pacault, "Nouveau Traite de Chemie Minerale" vol. VII, No. 1, 457 (1968).
6. D. Fishbach in: Chemistry and Physics of Carbon vol. 7, P.L. Walker Jr. ed. (Marcel Dekker, New York 1971) p.1.
7. B.S. Elman, Ph.D. Thesis, M.I.T., 1983 (unpublished).
8. G. Braunstein and R. Kalish, Nucl. Inst. and Meth. 209/210, 387 (1983).
9. J.R. Dennis and E.B. Hale, J. Appl. Phys. 49, 1119 (1978).
10. B.S. Elman, M. Hom, E. Maby and M.S. Dresselhaus, Mat. Res. Soc. Symp. Proc. 20, 341-346 (1983).

STOICHIOMETRIC DETERMINATION OF GRAPHITE INTERCALATION COMPOUNDS USING
RUTHERFORD BACKSCATTERING SPECTROMETRY

L. SALAMANCA-RIBA[*], B.S. ELMAN[*], M.S. DRESSELHAUS[*+]
[*]Department of Physics; [+]Department of Electrical Engineering and Computer
Science; Center for Materials Science and Engineering;
Massachusetts Institute of Technology, Cambridge, MA 02139, USA

T. VENKATESAN
Bell Laboratories, Murray Hill, NJ 07974, USA

ABSTRACT

Rutherford backscattering spectrometry (RBS) is used
to characterize the stoichiometry of graphite intercalation compounds (GIC). Specific application is made to
several stages of different donor and acceptor compounds
and to commensurate and incommensurate intercalants. A
deviation from the theoretical stoichiometry is measured
for most of the compounds using this non-destructive
method. Within experimental error, the RBS results agree
with those obtained from analysis of the (00ℓ) x-ray
diffractograms and weight uptake measurements on the same
samples.

INTRODUCTION

Rutherford backscattering spectrometry (RBS) provides a very useful
non-destructive technique for studying the stoichiometry of multi-elemen-
tal and layered compounds [1] and yields information averaged over an area
corresponding to the ~1 mm diameter of the ^{4}He$^+$ ion beam. Analysis of the
energy distribution of the backscattered ions provides information on the
stoichiometric dependence on depth from the near-surface region (~1 µm) of
the sample; this information is generally not available using other non-
destructive techniques.
Graphite intercalation compounds (GIC) are layered materials with a
large variety of structural orderings [2]. In general, the measured
stoichiometry differs significantly from the theoretical in-plane density
deduced from the intercalate lattice constants. At present, it is not known
the extent to which these deviations from the theoretical stoichiometry are
intrinsic to a given intercalant and stage, or are dependent on the details
of the preparation procedure. Thus, a non-destructive method for
stoichiometric determination with the capabilities of measurement as a
function of depth and lateral dimensions is expected to have a major impact
on the field of intercalated graphite. The applicability of the RBS
technique to GICs has already been successfully demonstrated on the SbCl$_5$-
GIC system [3]. In the present work we show that the method can be
generally applied to a wide variety of GICs including intercalants that are
not air-stable and are not commensurate with the graphite. Specifically,
application of the technique is made to acceptor and donor intercalants
forming commensurate (KHg and SbCl$_5$) and incommensurate (FeCl$_3$ and CuCl$_2$)
compounds. In this work we compare the RBS results with stoichiometric
determinations obtained on the same samples either from analysis of the
(00ℓ) x-ray diffraction peak intensities, or from weight uptake measurements.

EXPERIMENTAL DETAILS

The preparation methods for the $SbCl_5$-GIC samples [3] and the KHg-GIC samples [4,5] were reported earlier. The acceptor compounds with intercalants $FeCl_3$ and $CuCl_2$ were prepared by the two-zone method, by placing the intercalant (~1.5 mg) at one end of the tube and the graphite at the other end under a partial pressure (~300 Torr) of Cl_2 gas. The different stages were obtained [6] by keeping the graphite temperature constant and varying the intercalant temperature, T_I.

All the samples used in this experiment were characterized for stage index and stage fidelity using (00ℓ) x-ray diffraction and techniques previously described [3]. The stage infidelity was less than 5% for all the samples used in this experiment. No special handling technique was employed for the air stable compounds ($FeCl_3$, $SbCl_5$ and $CuCl_2$-GIC) that were used in these experiments. On the other hand, the reactive KHg-GIC samples were encapsulated under a partial pressure of argon gas in capillary pyrex tubes (0.1 mm wall thickness). The (00ℓ) x-ray diffractograms for the KHg system were obtained using a 1 mm diameter collimator and the K_α radiation from a Mo source.

The RBS measurements were carried out by using a primary beam of 2 MeV $^4He^+$ ions from a Van de Graaff generator with a typical current of 20 nA through a 1 mm diameter aperture, and using a set-up previously described [3]. To carry out RBS measurements on the air sensitive KHg-GIC samples, a glove bag was placed around the sample holder of the RBS set-up with a constant flow of N_2 gas. The glass ampoules containing the samples were opened inside the glove bag and the samples were mounted on the sample holder with vacuum grease.

RESULTS AND DISCUSSION

The main technique used for the stoichiometric determination of the composition of the GIC samples employed Rutherford backscattering spectrometry (RBS). In previous work [3] we applied this technique to the $SbCl_5$-GIC system and the method of data analysis is described therein. The essence of the RBS technique is in the analysis of the energy spectrum of the $^4He^+$ particles backscattered from the atoms of the substrate. Simply considering a process of hard sphere collisions, it is clearly understood that the heavier the atom of the substrate from which the incoming particle is backscattered, the less energy will be transferred to this atom during the collision process. Thus, the higher energies of backscattered helium ions correspond to heavier masses present in the substrate. Moreover, for each particular atomic mass in the substrate M_i, one expects to see a step in the energy spectrum at a characteristic energy E_i, corresponding to the scattering from these atoms of mass M_i located at the surface of the substrate. At energies lower than E_i we expect to have a continuous spectrum corresponding to the in-depth distribution of species i. Typical RBS spectra for some of the intercalants studied in this work are shown in Figs. 1 and 2 in terms of the backscattering yield (counts) vs. energy of the backscattered ions. The inset in Fig. 1(a) shows the set-up for the RBS experiment.

Writing a general expression for the stoichiometry of these compounds as $C_{\xi n}MN_x$ where MN_x denotes the intercalate stoichiometry, and n the stage index, we can express our results obtained from analysis of the RBS spectra in terms of ξ and x. The analysis of the RBS spectra is done by relating the heights of the steps H_i in the RBS spectrum to the relative concentrations C_i/C_j of the elements present in the sample by performing an analysis of the raw data based on an iteration of the RBS yield equations [3]. The relation between the relative concentrations and the heights in the RBS spectrum is given by

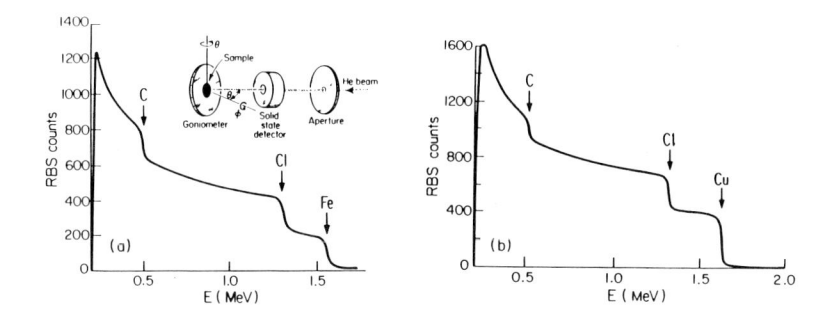

Fig. 1. Typical RBS spectra of (a) a stage 2 FeCl$_3$-GIC and (b) a stage 2 CuCl$_2$-GIC. The inset to (a) shows the experimental geometry.

$$C_i/C_j = [H_i/H_j] \; [\sigma_j/\sigma_i] \; [\Psi_i/\Psi_j] \qquad (1)$$

where σ_i is the differential scattering cross-section of atomic species i and Ψ_i is its stopping cross-section factor which can be calculated from tabulated values [1]. The application of this iteration method to the analysis of RBS spectra of GIC has been discussed in reference [3].

The analysis of the RBS spectra was carried out for different commensurate and incommensurate GICs. The results from this RBS analysis are

Table I. Summary of the measured stoichiometries for the graphite intercalation compounds reported in this paper. The parameters are for the compound $C_{n\xi}MN_x$. Experimental weight uptake $W_u(exp)$ is also reported.

Intercalant MN	Stage n	ξ RBS	ξ x-ray	ξ ref.	x RBS	x x-ray	x ref.	$W_u(exp)$ %
KHg	1	3.0	4.6	4[a]	0.7	0.75±0.05	1[a]	–
	2	3.5	–	–	0.7	–	–	–
	3	3.3	–	–	0.6	–	–	–
SbCl$_5$	2-4,6	14.1[b]	–	14[d]	4.4[b]	–	5[d]	–
		13.5[c]	12.9	12[e]	4.6[c]	4.9±0.3	5[e]	–
FeCl$_3$	2	7.3	5.9	8.5[f] 9.0[g] 6.2[f]	2.4	2.6±0.5	3[f] 3[g] 3[f]	47.6
CuCl$_2$	2	4.6	6.5	6.0[g] 4.9[h]	2.0	2.2±0.4	2[g] 2[h]	57.74

a) From Ref. 4. b) From uncleaved samples. c) From cleaved samples.
d) From Ref. 7 using x-ray analysis. e) From Ref. 8 using chemical analysis. f) D.G. Onn, M.G. Alexander, J.J. Ritsko and S. Flandrois, J. Appl. Phys. 53, 2751 (1982). g) M.E. Vol'pin, Yu.N. Novikov, N.D. Lapkina, V.I. Kasatochkin, Yu.T. Struchkov, M.F. Kazakov, R.A. Stukan, V.A. Povitskii, Yu.S. Karimov and A.V. Zvarikina, J. Amer. Chem. Soc., 97, 3366 (1975). h) From Ref. 6.

summarized in Table I in terms of ξ and x. Table I also contains the values for ξ and x obtained from analysis of the integrated intensities under the (00ℓ) x-ray diffraction peaks, and the values are listed under the columns labeled x-ray. The table also includes under the columns labelled ref. the values ξ and x reported in the designated references. A discussion of the results for each system studied in this work is given below. No dependence on host material (HOPG vs. Kish single crystal) was found in the stoichiometry for $FeCl_3$, $CuCl_2$, KHg, and $SbCl_5$-GIC.

The KHg-GIC system has been found to form three coexisting phases commensurate with the graphite lattice: a (2 x 2)R0°, a (2 x √3)R(0°,30°) and a (√3 x √3)R30° superlattice [4,5,9]. For the (2 x 2)R0° and (2 x √3)R(0°,30°) superlattices, the expected stoichiometry is C_{4n}KHg where n denotes the stage index, and for a (√3 x √3)R30° superlattice it is C_{3n}KHg. Our RBS results on the KHg-GIC system show a ratio of Hg to K atoms x < 1 in all the samples we have studied (see Table I, and Figs. 2(a) and 2(b)). This mercury deficiency is possibly associated with deintercalation of the mercury occurring during the process of mounting the samples for the RBS experiment. This is consistent with the temperature dependent in-situ x-ray experiment [9] performed on a stage 1 KHg-GIC sample for 300 < T < 500 K, where it was shown that the mercury leaves the graphite host at a temperature lower than that where potassium leaves. The results for ξ and x reported here for the KHg-GIC system were obtained from analysis of one RBS spectrum for every stage and therefore a sample dependence could produce a scatter from these reported values. The RBS results for the KHg-GIC system for stage 1 (see Table I) are consistent with the results obtained from an analysis of the integrated intensities of the (00ℓ) x-ray diffractograms taken from the same samples.

The lateral distribution of the intercalant of a stage 3 KHg compound showed a very interesting depth dependence of the stoichiometry. The regions at the edges of the sample showed an equilibrium value of x = 0.62 for the ratio of Hg to K atoms with a uniform depth distribution for x (see Table I). In contrast, the central region of the sample shows a decrease of Hg and K with depth (see Figs. 2(a) and 2(b)). The analysis of the spectrum in Fig. 2(b) gave ξ = 4.76 and x = 0.79. This is in agreement with the proposed mechanism for intercalation whereby intercalation starts from the a-face edges and from the sample surface planes. To our knowledge, there has been no independent measurement of x for stage 3 KHg-GIC. It is interesting to note that for the preparation of stage 3 KHg-GIC the intercalant has a Hg to K ratio of 2.5 prior to intercalation.

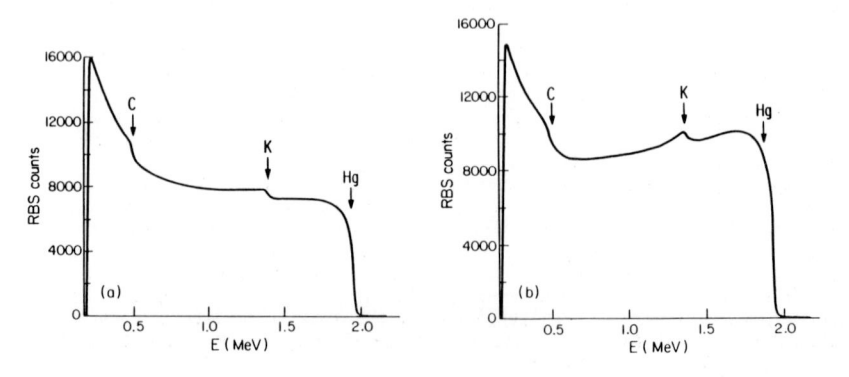

Fig. 2. RBS spectra of a stage 3 KHg-GIC from (a) the edge of the sample and (b) from the center of the ~1.5 x 1.5 mm^2 sample.

Our study of the RBS spectra for $SbCl_5$-GIC (stages 2,3,4 and 6) has been previously reported [3]. Table I contains the results for ξ and x obtained for this system and we summarize the results below. The $SbCl_5$ system has been found to form a ($\sqrt{7}$ x $\sqrt{7}$)R19.1° commensurate superlattice as the principal phase [7,8,10]. We have found a small but significant deviation from x = 5 for cleaved and uncleaved samples, by taking two spectra, one before and one after cleaving the samples with a razor blade [3]. We have also found that x is different within the bulk and at the surface of 'as-grown' samples. Within experimental error, the value of ξ for the $SbCl_5$-GIC system obtained from analysis of the RBS spectra agrees with the reported values and with the expected value ξ = 14 for a ($\sqrt{7}$ x $\sqrt{7}$)R19.1° commensurate superlattice [7,8,10]. It also agrees with the result obtained from analysis of the (00ℓ) x-ray diffractograms. The deviation from x = 5 is in agreement with the Mossbauer results on $SbCl_5$-GIC which have shown that there is a disproportionation of sites ($SbCl_5$, $SbCl_3$, $SbCl_6^-$) in this system [11]. The results obtained from analysis of the (00ℓ) x-ray diffractograms agree with the results obtained from analysis of the RBS spectra for the cleaved samples because the penetration depth for x-rays (> 10 µm) is larger than the penetration depth for the $^4He^+$ ions (~1 µm). No change in the stoichiometry (within experimental error) was found in the lateral direction for cleaved and uncleaved samples for the $SbCl_5$ system, consistent with a high in-plane diffusivity.

The results obtained from analysis of the RBS spectra from several cleaved stage 2 $FeCl_3$-GIC samples are summarized in Table I. A typical RBS spectrum for this system is shown in Fig. 1(a). Based on the measured lattice constants, the theoretical values for ξ and x, for the $FeCl_3$-GIC system are ξ ~ 5.3 and x = 3. Our results show a statistically significant deviation from x = 3 in this system. Electron diffraction patterns and bright and dark field studies from these samples using the TEM have shown that on a microscopic scale these compounds are not completely homogeneous. Some regions of the samples are homogeneous and show continuous intercalate regions. The electron diffraction patterns taken from these regions show rings with superimposed spots whose wave numbers agree with the reported values for interplanar distances for pristine $FeCl_3$ (d_s = 5.25 Å). There are other regions where the homogeneous intercalate layer is interrupted by large islands (~2000 Å diameter) that scatter electrons strongly and give rise to very intense spots in the diffraction pattern that show hexagonal symmetry. The measured wave numbers for these bright spots are in agreement with reported interplanar spacings for $FeCl_2$ (3.10 Å). The TEM results thus suggest the coexistence of $FeCl_2$ with $FeCl_3$ in the intercalate layer, consistent with Mössbauer results previously published [12]. Our results do not agree with the stoichiometry $C_n^+Cl^-FeCl_2 3FeCl_3$ suggested by Dzurus and Hennig [13]. From the stoichiometry obtained from analysis of the RBS spectra for this system we have calculated the percent weight uptake (W_u(RBS)) for the samples used in this experiment. Our calculated values for W_u(RBS) agree with the experimental weight uptake values W_u(exp) to within 7% for all the samples we have studied (see Table I).

RBS spectra from the $CuCl_2$ system were obtained from three stage 2 samples. The results from the analysis of the RBS spectra are summarized in Table I. Figure 1(b) shows a typical RBS spectrum for this system. Contrary to the case of $FeCl_3$-GIC and $SbCl_5$-GIC, we found a ratio of Cl to Cu atoms of x ~2, suggesting that there is no disproportionation of sites in this system. Our results for ξ and x, obtained from analysis of the RBS spectra, agree with reported stoichiometric values (see Table I). The calculated W_u(RBS) for this system are in agreement with the experimental values of W_u(exp) to within 5%.

ACKNOWLEDGMENTS

The MIT authors gratefully acknowledge support from ONR grant #N00014-77-C-0053 (B.S.E., M.S.D.) and from AFOSR contract #F49620-83-C-0011 (L.S.-R.). We also wish to thank Dr. G. Dresselhaus for valuable discussions.

REFERENCES

1. W-K. Chu, J.W. Mayer, M.A. Nicolet, Back Scattering Spectrometry, Academic Press, New York (1978).
2. M.S. Dresselhaus and G. Dresselhaus, Adv. Phys. 30, 139 (1981).
3. B.S. Elman, L. Salamanca-Riba, M.S. Dresselhaus and T. Venkatesan, J. Appl. Phys. (in press).
4. M. El Makrini, P. Lagrange, D. Guerard, and A. Herold, Carbon 18, 211 (1980).
5. G. Timp, B.S. Elman, R. Al-Jishi and G. Dresselhaus, Solid State Commun. 44, 987 (1982).
6. A. Herold, in Physics and Chemistry of Materials with Layered Structures, edited by F. Levy (Reidel, Dordrecht, Holland, 1979), 6, 323.
7. P.C. Eklund, G. Giergiel and P. Boolchand, in Physics of Intercalation Compounds, Vol. 38 of Springer Series in Solid-State Sciences, edited by L. Pietronero and E. Tosatti, (Springer, Berlin, 1981), p. 168.
8. J. Melin and A. Herold, Carbon 13, 357 (1975).
9. A. Erbil, G. Timp, A.R. Kortan, R.J. Birgeneau, and M.S. Dresselhaus, Synthetic Metals 7, 273 (1983).
10. R. Clarke, M. Elzinga, J.N. Gray, H. Homma, D.T. Morelli and C. Uher, Phys. Rev. B26, 5250 (1982).
11. P. Boolchand, W.J. Bresser, D. McDaniel, and K. Sisson, Solid State Commun. 40, 1049 (1981).
12. S.E. Millman, Solid State Commun. 44, 23 (1982).
13. M.L. Dzurus and G.R. Hennig, J. Am. Chem. Soc. 79, 1051 (1957).

TRANSPORT PROPERTIES AND ELECTRON MICROSCOPY STUDIES OF ION IMPLANTED GRAPHITE

T.C. CHIEU[*], B.S. ELMAN[+], L. SALAMANCA-RIBA[+], M. ENDO[&] AND G. DRESSELHAUS[#]
[*]Department of Electrical Engineering and Computer Science; [+]Department of Physics; [#]Francis Bitter National Magnet Laboratory;
Massachusetts Institute of Technology, Cambridge, MA 02139, USA;
[&]Faculty of Engineering, Shinshu University, Nagano, Japan.

ABSTRACT

Graphite fibers with high structural perfection and small diameters (~1μm) provide a sensitive medium to study the effect of ion implantation on the transport properties of graphite and to observe the defect structure associated with the implantation process. Graphite fibers prepared from the thermal decomposition of benzene and subsequent heat treatment to high temperatures (~2900°C) have been shown to achieve the high structural perfection necessary to carry out such experiments. Implantation-induced changes in the fiber resistivity are reported and are found to be larger at low temperature, as expected on the basis of residual resistance arguments. Using the lattice fringe imaging technique of high resolution electron microscopy, the implantation-induced defect structure can be observed directly. The results show local expansion of the interlayer graphite planes, with an interlayer separation ranging up to 3.9 Å in the implanted region, compared with ~3.4 Å for the well-ordered layers beyond the ion penetration depth.

INTRODUCTION

Graphite fibers with the highest structural perfection yet achieved have been successfully fabricated by thermal decomposition of a mixture of benzene and hydrogen gas [1-3] at a temperature of approximately 1100°C. The fiber axis is oriented along a basal plane direction and the c-axis is radial [4]. When heat treated to temperatures as high as 2900°C, the fibers exhibit graphite crystallites with dimensions larger than 1000 Å and an unusually high degree of structural order. These fibers also have the highest electrical conductivity, bulk modulus and tensile strength yet found in fibrous graphite materials [4-8]. Furthermore, for fibers with diameters less than 0.8 μm, the c-axis lattice image and defect structures can be seen directly using high resolution electron microscopy without any thinning and sectioning or special sample preparation [9]. Thus, benzene-derived fibers (BDF), heat treated to 2900°C and above, provide an excellent matrix for the study of structural order associated with the implantation process.

The focus of this work is the direct observation by the lattice fringe imaging technique of the defect structures created by the implantation of various ions into these well-ordered graphite fibers. The results show a variation in the defect density with depth and an enlargement in the c-axis graphite spacing from d = 3.36 Å to d as large as 3.9 ± 0.1 Å. A temperature dependent resistivity curve for a boron implanted benzene-derived fiber (T_{HT} = 3500°C) is compared to that of the original BDF, showing a degradation in conductivity upon implantation because the decrease in mobility due to ion beam-induced structural defects dominate over the increase in carrier density for room temperature implantation.

EXPERIMENTAL

The pristine graphite fibers used in this study were derived from a benzene precursor material [1-3], and subsequently heat treated to temperatures of 2900°C and 3500°C. Because of their small diameters (~1 μm), the fibers require no special sectioning or microtoming for electron microscopy observations. To investigate the defect structure as a function of fluence, a set of individual fibers was implanted with ^{31}P, ^{75}As, ^{122}Sb$_2$ and ^{209}Bi ions at an energy of 30 keV and at fluences of 5×10^{12} cm^{-2}, 1×10^{14} cm^{-2} and 1×10^{15} cm^{-2}. The implanted fibers were subsequently annealed in a flowing argon atmosphere at temperatures up to 2500°C in order to study the graphitization process of the ion damaged layer during annealing.

The lattice images were obtained using a JEOL 200 CX transmission electron microscope with high resolution pole pieces (C_s = 2.8 mm) and a LaB$_6$ filament. The observed point to point resolution with this stage was 2.9 Å. Because the LaB$_6$ filament is a bright electron source, exposure times used for recording the image could be a short as 5 seconds at a magnification of 530,000 X. The images were recorded on Kodak SO-163 electron microscope film. A typical acceleration voltage was 200 keV with a first condensor aperture of 20 μm.

Suitable areas of the fiber specimen for lattice imaging are the edge regions of the fiber, within several hundred Å from the fiber surface. In this region, we selected an area of the sample where the sample thickness was less than ~500 Å. The graphite layer planes in these regions are oriented such that the (00ℓ) diffraction pattern is always observed under axial illumination. The lattice images were obtained with the circular objective aperture positioned to encompass, at minimum, the (000) and (002) reflections observed in the back focal plane of the objective lens. The truncation of the Fourier series corresponding to the selected area of the diffraction pattern enhances the contrast of the spatial frequencies of interest in the observed real-space image. The interlayer spacing of lattice fringes was determined by taking an optical interferogram [10] of the electron micrograph, and using the pristine fiber interlayer spacing of 3.36 Å as a reference.

The electrical resistivity of single fibers was measured by a conventional four terminal method [7,8]. For this experiment the fibers were implanted with ^{11}B ions at 100 keV and a fluence of 2.8×10^{15} cm^{-2}.

We emphasize that low energy and heavy mass ions were used to prepare samples for electron microscopy studies in order to have the damaged region as close to the surface as possible. To enhance the changes in the electrical transport properties, we used light mass ions and intermediate accelerating voltages to damage homogeneously as thick a near-surface layer as possible. In all cases the implantation was carried out at room temperature and no corrections for ion beam heating were made.

RESULTS AND DISCUSSION

Figure 1(a) shows a typical (002) lattice fringe image obtained from a pristine BDF sample heat treated at 2900°C. Straight defect-free (002) graphite fringes have been found to extend for over 1000 Å along both the a-axis and c-axis directions [9]. The interplanar spacing for the pristine fibers is 3.36 Å as measured from the optical diffractogram taken from the negative of Fig. 1(a).

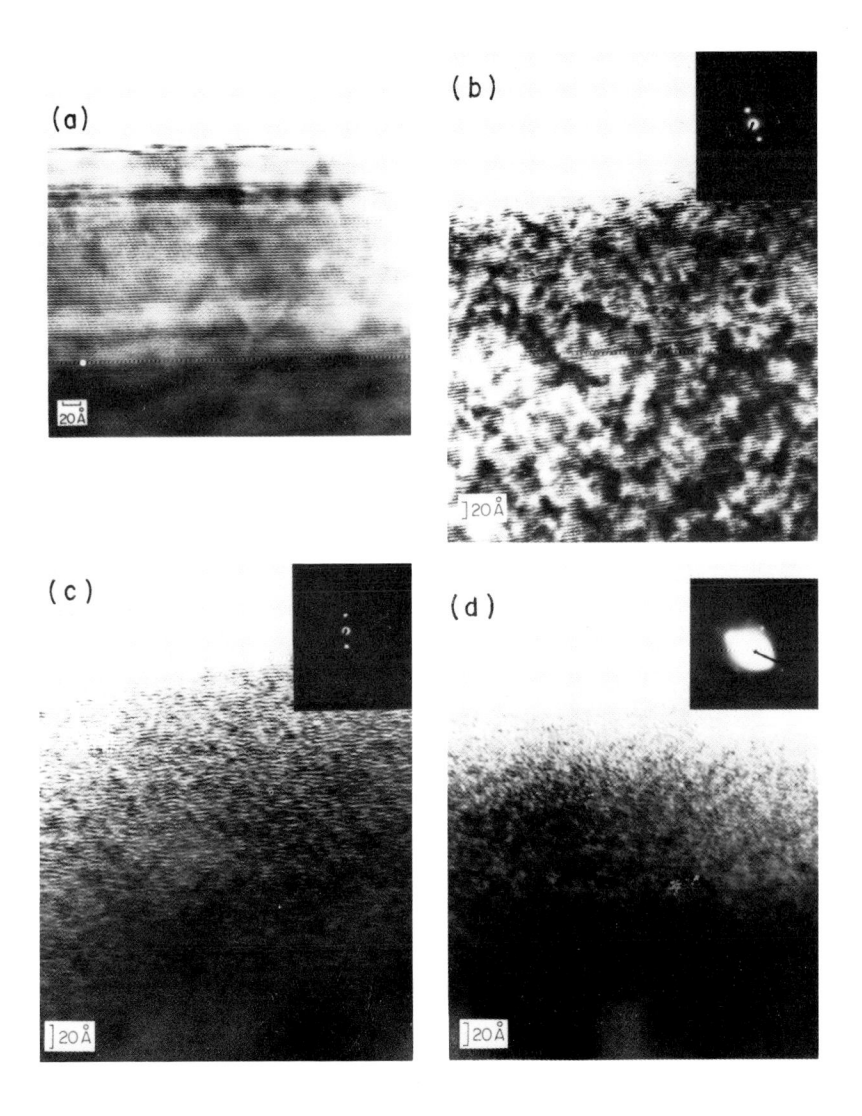

Fig. 1. Lattice fringe images obtained using the (002) beam for (a) pristine BDF sample heat treated at 2900°C. The inset shows the optical diffractogram taken from the negative. Fibers inplanted with ^{122}Sb ions are shown at a fluence of (b) 5×10^{12} cm^{-2}, (c) 1×10^{14} cm^{-2}, and (d) 1×10^{15} cm^{-2}.

For the ion-implanted BDF samples, we studied the effect of the ion fluence and mass on the graphite structure. The effect of fluence is shown in Figs. 1(b-d), where (002) lattice fringe images of ion implanted BDF samples are shown with three different fluences of ^{122}Sb ions: 5×10^{12} cm^{-2}, 1×10^{14} cm^{-2}, and 1×10^{15} cm^{-2}. It is clearly seen from

490

these figures that the distances (along both a-axis and c-axis directions) over which the fringes are straight get shorter as the fluence is increased. In fact, at the highest fluence (1×10^{15} cm^{-2}) shown in Fig. 1(d), the area close to the surface looks amorphous in texture. This conclusion can also be reached from viewing the respective electron diffraction patterns shown in the insets of Figs. 1(b) - 1(d), where the (00ℓ) diffraction spots become broadened and the basal plane order is reduced with increasing fluence. The crystallite sizes (L_a and L_c) were obtained directly from the lattice images by measuring the distances along the a-axis and c-axis over which the fringes are parallel.

The results of these measurements are difficult to summarize quantitatively, because of the uncertainty of the axis of the TEM column with respect to the direction of implantation for a particular fiber sample. However, qualitative conclusions can be reached readily. It is noticeable on the lattice fringe patterns that the in-depth crystallite size L_c is smallest near the surface and increases slightly as a function of depth. Values for L_c for zero depth and at a depth of ~200 Å from the fiber surface being ~20 Å and ~35 Å for the lowest fluence of implantation and ~ 6 Å and ~15 Å for highest fluence, respectively. However the in-plane crystallite size L_a behaves in a different way, but still consistent with

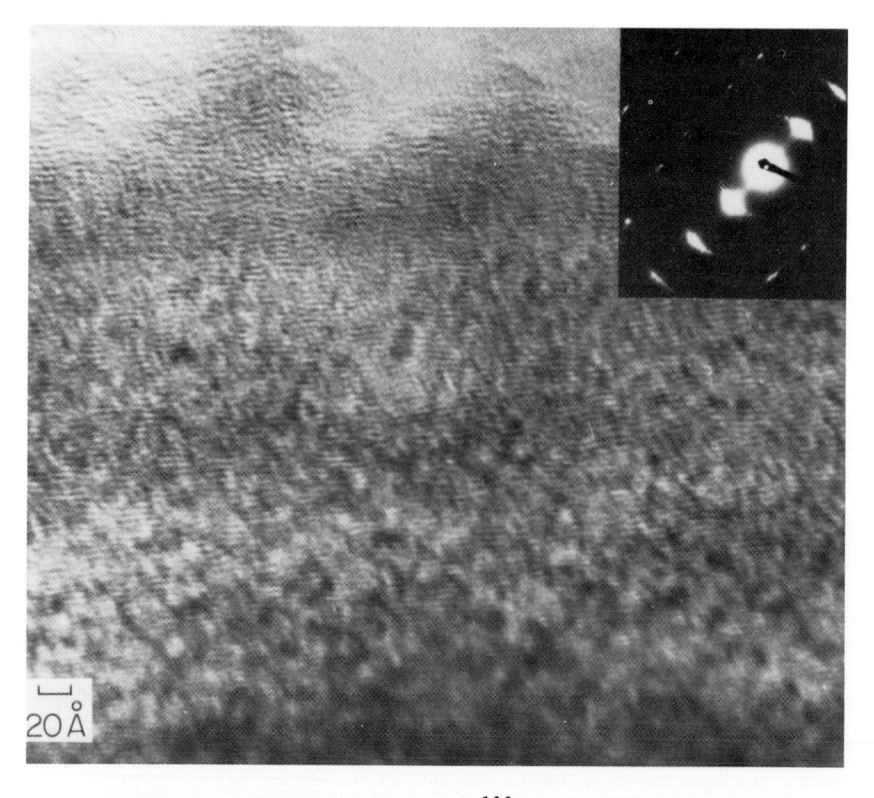

Fig. 2. (002) lattice fringe images of a ^{122}Sb ion-implanted benzene-derived fiber ($\phi = 1 \times 10^{13}$ cm^{-2}) annealed at 2500°C for 45 min.

the observed depth distribution of damage. This is particularly noticeable for the case of lowest fluence where the following depth distribution for L_a is found. In the depth range, $0 < d < 75$ Å, we find $L_a \sim 20$ Å; for $75 < d < 150$ Å, we find the greatest damage with $L_a \sim 15$ Å; and for the depth range, $150 < d < 200$ Å, the value of L_a is ~ 35 Å. It is of interest to note that the penetration depth of antimony atoms at an accelerating voltage of 30 KeV is $R_p \sim 150$ Å with a spread of $\Delta R_p \sim 35$ Å. Thus the depth distribution of L_a at the fluence of 1×10^{15} cm^{-2} directly shows that the maximum damage to the substrate is produced before the implanted ions are stopped by the target. The extent of the damaged region beyond $R_p + \Delta R_p$ is largely due to secondary effects ("knock-on" carbon events).

The interplanar distances of the lattice fringes were obtained by taking optical diffractograms of the negatives of Figs. 1(b) - 1(d) and are shown as insets of these figures. Our analysis indicates an expansion of the graphite layers from 3.36 Å for pristine fiber to ~ 3.6 Å after implantation at 1×10^{15} cm^{-2} (Fig. 1(d)).

Preliminary observations have been made on the effect of post-implantation annealing at temperatures of $\sim 2500°C$ of the ion implanted fibers mentioned above. The results show an increase in both L_a and L_c as measured from the lattice images taken of these samples; these results are in agreement with their respective electron diffraction patterns which show sharper (hkℓ) spots for the annealed as compared to the unannealed fibers. Figure 2 shows (002) lattice images of a ^{122}Sb ion-implanted fiber ($\phi = 1 \times 10^{15}$ cm^{-2}) annealed at 2500°C for 45 min. The comparison of these lattice fringe images (Fig. 2) with those for the unannealed sample (Fig. 1d) shows that the crystallite size grows after annealing at 2500°C, but L_a and L_c still remain small compared to values for the pristine fiber. A small decrease in the interplanar distances is found upon annealing, but large separation distances (~ 3.5 Å) are still found.

We have also studied the effect of ionic mass on the generation of lattice damage for a fixed fluence of implantation ($\phi = 1 \times 10^{15}$ cm^{-2}). The optical diffractograms taken from the lattice fringe images of BDF samples

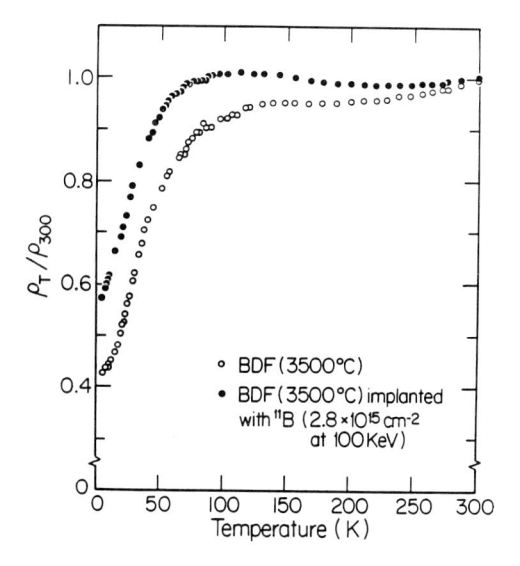

Fig. 3. Resistivity vs. temperature for heat treated and implanted fibers.

implanted with ^{31}P, ^{75}As, ^{122}Sb and ^{209}Bi ions indicate that the (002) graphite interplanar spacings get larger as the mass of the implanted ion increases, with the largest interplanar separation distances in a given sample varying from ~3.5 Å for ^{31}P implantation to ~3.9 Å for implantation with ^{209}Bi ions. It is important to note that the layer stacking in the interior part of the fiber was observed to remain unaffected by ion implantation, the lattice fringe pattern for the interior of the fibers showing the same stacking texture as the unimplanted fibers.

The effect of ion implantation on the electrical transport properties of these fibers has also been examined. Explicit results for the temperature dependence of the electrical resistivity are given in Fig. 3 for an unimplanted benzene-derived fiber ($T_{HT} = 3500°C$) and a similar fiber heavily implanted with 2.8×10^{15} ^{11}B ions/cm^2 at 100 keV. It is seen that upon implantation of the fiber, the room temperature resistivity increases by about 5% and the residual resistivity ratio defined as $(R = \rho(300K)/\rho(4K))$ decreases from R = 2.33 (before implantation) to R = 1.78 (after implantation). Also the temperature dependence of the resistivity changes its functional form and gradually approaches that characteristic of fibers with a lower heat treatment temperature. For example, the implanted fiber in Fig. 3 exhibits a behavior similar to an unimplanted fiber with $T_{HT} = 2900°C$, showing a flat maximum in resistivity at about 120 K and a shallow minimum at about 220 K. The results are consistent with the introduction of structural defects by ion implantation, as evidenced by the direct observation of the defect structures in Fig. 1.

In conclusion, the high structural order and small fiber diameters that can be achieved with the benzene-derived fibers facilitate structural studies of the implantation process using high resolution lattice fringe imaging. For this process, lattice fringe imaging provides direct information on the induced structural defects; such information can be directly related to the electrical transport properties of these fibrous materials.

ACKNOWLEDGMENTS

We take pleasure in acknowledging support by NSF grant #DMR-81-19295 for T.C.C and M.E., and ONR grant #N00014-77-C-0053 for B.S.E. and G.D. and AFOSR contract # F49620-83-C-0011 for L.S.-R.

REFERENCES

1. T. Koyama, Carbon 10, 757 (1972).
2. T. Koyama, M. Endo and Y. Onuma, Japan J. Appl. Phys. 11, 445 (1972).
3. M. Endo, K. Komaki and T. Koyama, International Symposium on Carbon, (Toyohashi, 1982), p. 515.
4. M. Endo, Y. Hishiyama and T. Koyama, J. Phys. D15, 353 (1982).
5. T. Koyama and M. Endo, Japan J. Appl. Phys. 13, 1175 (1974).
6. T. Koyama, M. Endo and Y. Onuma, Japan J. Appl. Phys. 13, 1933 (1974).
7. T.C. Chieu, M.S. Dresselhaus, and M. Endo, Phys. Rev. B26, 5867 (1982).
8. T.C. Chieu, G. Timp, M.S. Dresselhaus, M. Endo, and A.W. Moore, Phys. Rev. B27, 3686 (1983).
9. M. Endo, T.C. Chieu, G. Timp, M.S. Dresselhaus, and B.S. Elman, Phys. Rev. B28, (1983).
10. G. Timp, M.S. Dresselhaus, L. Salamanca-Riba, A. Erbil, L.W. Hobbs, G. Dresselhaus, P.C. Eklund and Y. Iye, Phys. Rev. B26, 2323 (1982).

MAGNETOREFLECTION IN ION-IMPLANTED GRAPHITE*

L.E. MCNEIL,$ B.S. ELMAN,$ M.S DRESSELHAUS,#$ G. DRESSELHAUS,+
#Department of Electrical Engineering and Computer Science; $Department of Physics; +Francis Bitter National Magnet Laboratory, Massachusetts Institute of Technology, Cambridge, MA 02139

T. VENKATESAN
Bell Laboratories, Murray Hill, NJ 07974

ABSTRACT

The use of a hot stage (T ~ 600°C) for ion implantation into graphite permits the introduction of foreign species into the host material while eliminating most of the lattice damage associated with ion implantation at room temperature. This permits the use of the magnetoreflection technique for examination of changes in the electronic band structure induced by implantation. Samples of graphite implanted with ^{31}P and ^{11}B at various energies and fluences are examined, and the in-plane and c-axis disorder are characterized using Raman spectroscopy and Rutherford Backscattering Spectrometer (RBS) techniques. Implantation-induced changes in the electronic band structure are interpreted in terms of the Slonczewski-Weiss-McClure band model. Small changes are found relative to the band parameters that describe pristine graphite.

INTRODUCTION

Ion implantation makes possible the physical introduction of foreign species which, for various reasons, cannot be chemically introduced into the host material. This mechanism induces a number of changes in the host material, including displacement of atoms of the host material (radiation damage), changes in the carrier concentration of the host, and modification of the electronic structure of the host material [1].
The relationship between the implantation-induced damage and changes in the electronic structure of the semimetal graphite is the subject of this paper. The electronic structure is affected both by lattice damage due to the incoming ions and the secondary carbon ions, and by the presence of the foreign ions once the ions come to rest. Implantation-induced damage was characterized by two complementary techniques: Raman spectroscopy, which is sensitive mainly to the degree of in-plane disorder within the optical skin depth (δ ~ 800 Å); and Rutherford Backscattering Spectrometry (RBS), which is sensitive to the c-axis disorder to a depth of a few microns. Changes in the electronic structure were measured using the magnetoreflection technique, in which the relative changes in the infrared reflectivity from the sample surface are measured as a function of applied magnetic field. This technique is especially suited to the examination of ion-implanted materials since the infrared skin depth (δ) is comparable to the penetration depth of the implanted ions (R_p), so that the region of maximum sensitivity corresponds to the region of physical interest.
Necessary conditions for the observation of magnetooptical effects include $\omega_c\tau > 1$ and $\hbar\omega_c > kT$. The condition $\omega_c\tau > 1$ (where ω_c is the angular cyclotron frequency $\omega_c = eH/m_c^*c$, m_c^* is the cyclotron effective mass and τ is the relaxation time), implies that an electron or hole completes at least one cyclotron orbit before being scattered. The magnetoreflection experiment is sensitive to the degree of implantation-induced damage within

a depth δ from the surface, since δ is comparable to R_p in magnitude. The condition $\hbar\omega_c > kT$ implies that the characteristic magnetic energy $\hbar\omega_c$ is large compared with thermal energies, so that distinct Landau levels can be resolved when in competition with thermal broadening effects.

We analyze the experimental magnetoreflection spectra from ion-implanted graphite in terms of the Slonczewski-Weiss-McClure (SWMcC) model which is based on the crystal symmetry of the perfect crystal and gives an analytical expression for the electronic states near the Fermi energy E_F [2,3]. The model expresses the electron dispersion relations in terms of seven parameters (Δ, γ_i, $i = 0,1,...,5$) which are related to π band overlap integrals and are in practice determined experimentally.

EXPERIMENTAL DETAILS

Magnetoreflection data have been taken on c-faces of pristine and ion-implanted highly oriented pyrolytic graphite (HOPG) samples [4]. Sample preparation and characterization techniques are described below.

To achieve the resonant condition $\omega_c\tau > 1$, the HOPG samples were implanted at elevated temperatures (implantation temperature $T_i \sim 600°C$). Elevated temperatures of implantation were used to reduce lattice damage in the region where the implanted ions pass and stop, and to reduce the mechanical damage to the HOPG surface. Ions of ^{31}P and ^{11}B were implanted at a variety of energies E and fluences ϕ in the range $30 < E < 200$ keV and $8.5 \times 10^{13} < \phi < 1.0 \times 10^{15}$ ions/cm^2 respectively. The penetration depth (R_p) and the corresponding Gaussian halfwidth of the implanted ions (ΔR_p) in HOPG for this range of ion energies are in the range $670 Å < R_p < 1980$ Å and $220 Å < \Delta R_p < 470$ Å when the implantation is done at room temperature [5].

The implanted samples were characterized using Raman spectroscopy to determine the in-plane microcrystallite size within the optical skin depth. It has been shown [6] that the crystallite size L_a can be deduced from the ratio of the intensities of the disorder-induced line at 1360cm^{-1} (I_{1360}) to the Raman-allowed line at 1580cm^{-1} (I_{1580}) in the first-order spectra. RBS data were taken using $^4He^+$ ions at 2 MeV and an experimental apparatus described elsewhere [7]. The magnetoreflection experiments were made in an experimental arrangement also described elsewhere [8,11]. The selected photon energies were in the range 99 meV $< \hbar\omega <$ 300 meV and the data were taken using (+) and (−) circularly polarized light to accommodate the different selection rules for the two polarizations. Magnetic fields of up to 15 Tesla were used to record the spectra.

RESULTS AND ANALYSIS

After the samples had been implanted, Raman scattering experiments were performed to determine the in-plane crystallite size. It was found that for hot-stage implantation ($T_i \sim 600°C$) the ratio I_{1360}/I_{1580} was typically ~ 0.08, giving an in-plane crystallite size $L_a \sim 400$ Å. Samples with I_{1360}/I_{1580} as large as 0.125, or $L_a \sim 300$ Å, were shown to give magnetoreflection spectra with well resolved resonant structure.

Analysis of the disorder introduced during the process of ion implantation was performed by the RBS channeling technique. Figure 1 shows the backscattering yield as function of depth for HOPG implanted at various energies with ^{31}P to a fluence of 1×10^{15} ions/cm^2, with a "random" direction spectrum included for comparison. The backscattering yield is sensitive to the number of phosphorus or secondary carbon atoms blocking the c-axis channels. The spectrum does not distinguish one type of displaced atom from another. The loss of crystalline order would cause the backscattered intensity to rise to that of the "random" direction value, and this is the case when the implantation is done at room temperature [9]. For hot-stage implantation the disorder is considerably less, as can be seen in Fig. 1. The disorder distribution is broader and reaches its maximum value deeper into the sample for higher implantation energies. The excess scattering region is not coincident with the region where the implanted ions reside, but extends deeper into the sample where the secondary carbon atoms come to rest. Figure 2 shows the distribution of disorder as function of depth for ^{31}P ions implanted into HOPG at three different energies of implantation.

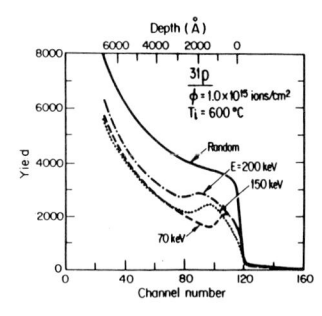

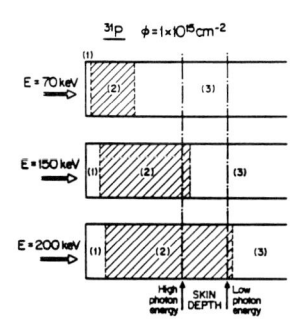

Fig. 1 RBS backscattered yield as a function of depth for HOPG implanted at $T_i \sim 600°C$ with ^{31}P to a fluence of 1.0×10^{15} ions/cm^2 at various energies.

Fig. 2 Schematic view of the spatial distribution of the damage produced in the substrate for different energies of implantation (see text).

The samples so characterized were examined using the magnetoreflection technique. The analysis of the oscillations in the infrared reflectivity as a function of magnetic field were carried out using pristine HOPG as a reference material to derive changes in the electronic structure near the Fermi level of graphite. These changes are presented in terms of changes in the band parameters near the HK axis of the Brillouin Zone, using the complementary information provided by the Landau level transitions at the H-point and at the K-point. The band parameters used are those of

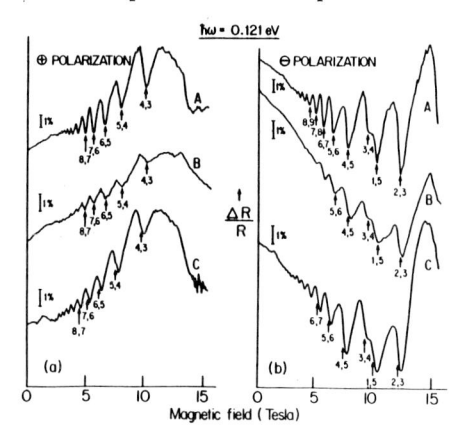

Fig. 3 Magnetoreflection traces for (a) (+) polarized and (b) (−) polarized light taken on pristine HOPG and HOPG implanted with ^{31}P ions (E = 70 keV) and ^{11}B ions (E = 30 keV). A) pristine HOPG; B) ^{31}P, $\phi = 8.5 \times 10^{13}$ ions/cm^2; C) ^{11}B, $\phi = 1.0 \times 10^{14}$ ions/cm^2. For all the traces, the photon energy $\hbar\omega$ = 121 meV. The K-point transitions are labeled with quantum numbers indicating the initial and final levels.

the well-established Slonczewski-Weiss-McClure (SWMcC) model [2,3].

Figure 3 shows typical magnetoreflection spectra at $\hbar\omega$ = 0.121 eV for unimplanted HOPG (A), and for graphite implanted with ^{31}P at E = 70 keV to a fluence of 8.5 x 10^{13} ions/cm^2 (B), and ^{11}B at E = 30 keV to a fluence of 1.0 x 10^{14} ions/cm^2 (C). It is clear from the figure that implantation induces a broadening and weakening of the resonances, and that these effects are more pronounced in the sample implanted with ^{31}P than in that implanted with ^{11}B. The degradation of the spectra implies that the relaxation time τ is lower in the implanted samples than in HOPG due to the scattering centers introduced by implantation, and that the ^{11}B ions introduce less lattice damage than the ^{31}P ions because of their smaller mass.

The analysis of the H-point spectra follows the work of Toy et al. [10] and uses a relationship between the transition energies at the H-point and the resonant magnetic field which allows the determination of γ_0, the nearest neighbor intraplanar overlap integral. The values of γ_0 were determined by a fit to the H-point spectra for each of the implanted samples [11]. The results obtained for relative changes in γ_0 for the samples implanted with ^{31}P at 70 keV, γ_0^{imp}, are shown in the solid circles in Fig. 4 with respect to γ_0^{HOPG} = 3.11 eV for the unimplanted sample. The decrease in γ_0 with increasing fluence (shown on a log scale) gives direct information on the expansion of the graphite in-plane honeycomb lattice between $R_p - \Delta R_p$ and $R_p + \Delta R_p$ due to the presence of the implanted ions.

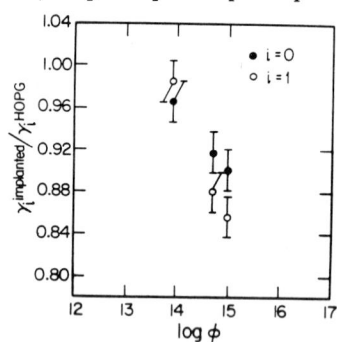

Fig. 4 Values for the γ_0 and γ_1 parameters with respect to those for unimplanted HOPG vs fluence ϕ of ^{31}P ions implanted at energy E = 70 keV and T_i ~ 600°C. Note the log scale for ϕ. [Ref. 11]

Unfortunately, the data are not extensive enough to distinguish reliably between different possible functional dependences of ($\gamma_0^{imp}/\gamma_0^{HOPG}$) on fluence. A similar analysis was carried out for a sample implanted with ^{11}B at 30 keV to a fluence of 1 x 10^{14} ions/cm^2, but no measurable shift in γ_0 was detected, consistent with the relatively small size of the ^{11}B ion.

Information on the interplanar interactions is provided by the K-point transitions which are most sensitive to the parameter combination γ_0^2/γ_1, where γ_1 is the nearest-neighbor interplanar overlap integral. The analysis of the K-point transitions was carried out using a simple two-band model and is described in detail elsewhere [11]. The band parameter combination γ_0^2/γ_1 is determined from the slope of resonant photon energy vs. magnetic field. By combining this with the values for γ_0 obtained from the H-point transitions we obtain the values for γ_1 which are shown as open circles in Fig. 4. The decrease in γ_1 with increasing fluence for the samples implanted with ^{31}P implies an expansion along the c-axis of the graphite lattice due to the presence of the foreign ions. The K-point transitions for the above-mentioned ^{11}B-implanted sample again showed no detectable differences from those of pristine graphite.

To confirm the results obtained from the simple two-band model, the values derived for γ_0 and γ_1 were used in a fit to the experimental points using the full SWMcC model, and when satisfactory fits were obtained [11]. It was thus concluded that the results obtained from the simple model are consistent with the more sophisticated analysis of the SWMcC model.

DISCUSSION

It has been demonstrated [12] that under the conditions of hot stage implantation, the concentration and spatial distribution of the ions implanted into graphite are nearly the same after hot stage implantation as after room temperature implantation. This implies that the effect of implantation is manifested not only through implantation-induced defects but also through long range strain effects due to the implanted ions resident in the samples.

It should also be noted that the magnitude and sign of the shifts observed in γ_0 and γ_1 are in agreement with the expected consequences of a long-range strain induced by implantation. The strong in-plane coupling in graphite implies that only a very small in-plane expansion, and therefore a very small decrease in γ_0, is expected. Most of the excess volume introduced by the implanted ions should be accommodated by expansion along the c-axis, leading to a decrease in the interplanar coupling and therefore in γ_1. However, even in pristine graphite the interplanar coupling is very weak, and so a large expansion along the c-axis leads to only a small (~ 15%) change in γ_1.

To prevent fracture of the bulk sample, an in-plane lattice expansion must also occur in the regions containing no implanted ions unless dislocations are present. Since the magnetoreflection spectra exhibit relatively small line broadening compared with the unimplanted samples, we infer that the variation of $\Delta a_0/a_0$ along the c-axis is long-range.

In order for resonant oscillations to be observed in the magnetoreflection signal in a polycrystalline material such as HOPG, it is necessary for the in-plane crystallite size to be at least as large as the classical cyclotron radius $r_c = v_F/\omega_c$, where v_F is the Fermi velocity. For pristine graphite $r_c \sim 350$Å at $H = 10$ Tesla, and this requirement is well satisfied since typically the in-plane crystallite size is $L_a \sim 1000$ Å [4]. For the implanted samples considered here m_c^* is only ~ 6% larger than the value for unimplanted HOPG [11], so r_c is only slightly larger in the implanted materials. The Raman spectra of these samples give values of L_a in the range of $300 \lesssim L_a \lesssim 500$ Å, and therefore the in-plane crystallite size is adequate for cyclotron oscillations to occur, at least in the near-surface region sampled by Raman scattering.

The implanted samples can be modeled as consisting of several regions at various depths, as shown in Fig. 2. Ions incident upon the surface pass through region (1) without initially causing much lattice damage as the energy loss is primarily from electron plasma oscillations. As the ions slow and stop (in region (2)), the lattice damage increases markedly. In region (3) the lattice damage is minimal, but long range strain effects should be present due to the foreign ions in region (3) and the disorder produced by the secondary carbon ions. The damage seen in the RBS spectrum in region (3) anneals out with a very low activation energy [13], which implies that the scattering is due to secondary carbon atoms which have come to rest between the graphite planes and which readily diffuse along the planes. These defects would have a limited effect upon the electron relaxation time within the planes, and so $\omega_c\tau$ would remain high in region (3). In region (1) the excess scattering is small and presumably the defects induced by implantation consist primarily of vacancies and interstitials within the planes, which would degrade $\omega_c\tau$ in this region. Thus, although the optical density is lower in region (3) than in region (1), because $\omega_c\tau$ is larger in region (3), the primary contribution to the magnetoreflection signal should come from this region.

The volume sampled by the magnetoreflection signal extends only to a depth δ from the surface, where δ is the frequency-dependent skin depth.

As can be seen in Fig. 3, for low energy implants the sampled region includes a substantial portion of the undamaged region (3) for all photon energies. As the energy of implantation is increased, region (3) is sampled only at low photon energies; and for the highest energy implants region (3) is not sampled at all.

The observed behavior of the magnetoreflection spectra is consistent with this model. For the sample implanted at 70 keV, clear oscillations are visible at all photon energies. For the sample implanted at 150 keV the strength of the oscillations diminishes with increasing photon energy and vanishes at $\hbar\omega \sim 140$ meV. For the 200 meV sample, the oscillations were reduced to the noise level of the measurements over the entire available photon energy range.

The failure of the sample implanted with ^{11}B to a fluence of $\phi = 1.0 \times 10^{14}$ ions/cm^2 to show shifts in γ_0 and γ_1 may be attributed to several factors. The boron ion is smaller than that of either phosphorus or carbon and would therefore introduce less strain into the graphite lattice at a given fluence. In addition, boron is known to enter substitutionally when chemically doped into graphite [14], and the presence of some substitutional boron would reduce the strain still further.

It would be of interest to extend these studies to higher fluences where larger changes in the band parameters would be expected. At very high values of ϕ it may be possible to observe the onset of lattice relaxation phenomena and a consequent saturation in the values of the band parameters. Further studies of ^{11}B and other implants would be useful to determine the dependence of the damage and strain effects on ion size.

REFERENCES

1. J.W. Mayer, L. Eriksson, and J.A. Davis, Ion Implantation in Semiconductors, (Academic Press, New York, 1970).
2. J.C. Slonczewski and P.R. Weiss, Phys. Rev. 109, 272 (1958).
3. J.W. McClure, Phys. Rev. 108, 612 (1957); Phys. Rev. 119, 606 (1960).
4. A.W. Moore, Chemistry and Physics of Carbon, edited by P.L. Walker and P.A. Thrower, (Dekker, New York, 1973), vol. 11, p. 69.
5. Calculated from the LSS theory [J. Lindhard, M. Scharff, and H.E. Schiott, Dan. Vidensk. Selsk., Mat. Fys. Medd. 33, 14 (1963)] and electronic stopping-power tables [L.C. Northcliffe and R.F. Schilling, Nucl. Sect. A7, 233 (1970)].
6. F. Tuinstra and J.L. Koenig, J. Chem. Phys. 53, 1126 (1970).
7. W.-K. Chu, J.W. Mayer, and M.-A. Nicolet, Backscattering Spectrometry, (Academic Press, New York, 1978); B.S. Elman, G. Braunstein, M.S. Dresselhaus, G. Dresselhaus, T. Venkatesan, and B. Wilkens, (to be published).
8. M.H. Weiler, Semiconductors and Semimetals, edited by R.K. Willardson and A.C. Beer, (Academic Press, New York, 1981); M.H. Weiler, Ph.D. Thesis, Massachusetts Institute of Technology, 1977 (unpublished).
9. G. Braunstein, B.S. Elman, G. Dresselhaus, and T. Venkatesan, (to be published).
10. W.W. Toy, M.S. Dresselhaus, and G. Dresselhaus, Phys. Rev. B15, 4077 (1977).
11. B.S. Elman, L. McNeil, C. Nicolini, T.C. Chieu, M.S. Dresselhaus, and G. Dresselhaus, Phys. Rev. B28, (in press).
12. B.S. Elman, M. Hom, E.W. Maby, and M.S. Dresselhaus, Intercalated Graphite, edited by M.S. Dresselhaus, G. Dresselhaus, J.E. Fischer, M.J. Moran, (North-Holland Elsevier, New York, 1983), p. 341.
13. T. Venkatesan, B.S. Elman, G. Braunstein, M.S. Dresselhaus, G. Dresselhaus, (to be published).
14. J.A. Turnbull, M.S. Stagg, and W.T. Eeles, Carbon 3, 387 (1966); C.E. Lowell, J. Am. Ceram. Soc. 50, 142 (1967).

*The work at MIT was supported by ONR Grant #N00014-77-C-0053.

PART V

NOVEL PROCESSING TECHNIQUES

DEVELOPMENT OF NEW MATERIALS BY IONIZED-CLUSTER BEAM TECHNIQUE

T. TAKAGI
Ion Beam Engineering Experimental Laboratory, Kyoto University, Sakyo Kyoto 606, Japan

ABSTRACT

In the Ionized-Cluster Beam (ICB) technique, deposition by macroaggregates consisting of 500 to 2000 atoms loosely coupled together involves low ratios of charge to mass. Consequently, high mass density beams at low equivalent energy per atom in an optimum range for film formation can be transported without problems due to space charge repulsion forces and deposition onto insulating substrates is easily possible due to low accumulation of ion charge. The presence of ionic charge has great influence upon film formation mechanisms in spite of low content of ions in the total flux. Also, because of the kinetics of cluster breakup upon impact, enhancement of migration of adatoms upon a substrate surface can be achieved by increasing the acceleration voltage. It is possible to control the mechanical, crystallographic, optical and magnetic properties of films over three dimensions by variation of acceleration voltage or ion content in the total flux. Films of many materials have been formed at low temperatures with well-controlled characteristics. Among examples of interest are included metal and semiconductor material films for functional devices and VLSI applications, intermetallic compound films for magnetic or thermoelectric uses and organic material films. Results suggest that ICB offers exceptional potential for applications involving formation of new materials.

INTRODUCTION

Clusters are created by condensation of supersaturated vapor atoms produced by adiabatic expansion through a small nozzle into a high vacuum region. Once formed, the cluster beam can be partially ionized to positive single charge state by electron impact. Energy can then be added to ionized clusters by use of acceleration potentials [1,2].

Unique capabilities of ICB deposition are attributed to the properties of the cluster state. Enhancement of the adatom migration effect is one of the most significant properties of cluster beam deposition. A number of other properties of ICB, such as extremely low charge to mass ratios which prevent space charge problems and high equivalent current, low energy ion beam transport in a range from above thermal energy up to a few hundred eV, are also important characteristics of this technology for film formation purposes.

The importance of low energy ion beams for film formation can be easily understood when we recognize that binding energies of the atoms in solids are of the order of 10eV. Atomic kinetic energies corresponding to thermal energies given by kT/e are only 0.01-0.1eV in the temperature range below 1000°C and are consequently much less than the binding energies. A strong effect can be expected however as the result of bombardment using accelerated ion beams, even at energies of only a few eV approximately corresponding to binding energies. In the cluster beam, the cluster is an aggregate of about 1000 atoms. Therefore, even when the clusters themselves are accelerated to kilovolts, energies of individual constituent atoms of the clusters are in a range from a few to a few tens of eV and are therefore comparable to the binding energies of the atoms in solids.

This paper will review studies done on the ICB technique and will discuss applications to development of new materials. Recently this technique has broadened from laboratory investigations toward applications studies. Several types of ICB deposition apparatus have been developed for experimental work and for applications use. These are now commercially available in Japan as well as in the United States. This review will also discuss the newest data relative to industrial applications of ICB.

FORMATION OF CLUSTERS AND PROPERTIES OF THE CLUSTER

Presented within this paper is an overview of the properties of the cluster beam so as to distinguish the unique features of ICB technology which can make possible development of new materials.

It is reported that clusters of different sizes have different structures[3]. This is easily understood by taking surface effects into account. In clusters with sizes less than 20 to 50 atoms, all constituent atoms are situated on the cluster surface. In clusters of sizes between 50 and 500 atoms most, but not all, of the atoms are situated on the surface. Therefore, in these two cases, structure and properties are strongly affected by the surface atoms[4]. On the other hand, in clusters with more than a few thousand atoms, surface contributions become lower and the properties of these clusters become closer to those of the bulk[5].

It is commonly believed that metals should have little tendency to condense because they are characterized by high surface tension coefficients[6]. However, recent calculations of nucleation rates for metals show that they have values comparable to those of alkali metals and water[7]. Since the classical nucleation rate is expressed as a function of σ/kT, even when a metal has a high value of surface tension, σ, the ratio will not increase significantly. This is due to the fact that a metal in a crucible must be heated to high temperature in order to get sufficiently high vapor pressure for expansion through the crucible nozzle.

A cluster beam can be created by condensation of supersaturated vapor atoms produced by adiabatic expansion through a small nozzle into a high vacuum region[8]. The source of atoms which form the clusters can be simply a solid heated to sufficient temperature to produce the pressure difference for expansion through the nozzle (pure-expansion method) or can be a mixture of a vaporized solid with a pressurized carrier gas, such as argon, ejected together through a nozzle (gas mixture method)[9]. To utilize the ICB technique, clusters can be formed by the pure expansion method. The resulting type of cluster has a size of 500 to 2000 atoms per cluster and is in a form of liquid phase even though it is close to vapor phase because interatomic distances between atoms are much longer than in the liquid state.

Requirements for and construction of an ICB source have been described elsewhere[2]. The measurement of cluster size was made by collecting clusters on an electron microscope mesh which was cooled to liquid nitrogen temperature [10]. As shown in Fig 1, the distribution of cluster diameters was found to be 20 to 50Å. Energy analysis of the cluster beam was also made to determine the cluster size [11]. By combining the results with ejection velocity measurements, the cluster size has been calculated to be 500 to 2000 atoms per cluster.

We analyzed the structure of the clusters using an electron diffraction technique [12]. The vaporized solid material cluster source was mounted in the diffraction chamber of an electron microscope. Diffraction patterns of an antimony cluster beam showed a typical amorphous signature. Microdensitometer traces of the diffraction patterns on the electron microscope films were Fourier transformed and a radial distribution function was calculated. The results were compared with those from crystal and amorphous films. A comparison of the interatomic distances in a cluster with those of bulk crystal and amorphous films is shown in Fig. 2. In the amorphous film, the interatomic distance between layers is longer than that in the crystalline films.

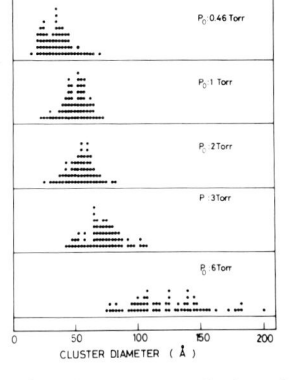

Fig. 1. Histrograms of the cluster diameter distributions. P_o is the pressure in the crucibles.

However in our clusters, a layer structure which is seen in crystal and amorphous films is not observed and further spread in the interatomic distance was apparent.

It became obvious that the structural properties of clusters influence film formation kinetics. This is in contrast to the situation with other film formation methods, such as vacuum evaporation and sputter deposition, in which atomic and molecular particles are used. We have explained that the clusters have different structure which causes them to exhibit weak binding energy [13], low melting temperature [14], and other unique physical properties. These properties of the cluster influence film formation processes by enhancement of adatom migration and allow low temperature epitaxy, formation of preferentially oriented films and effective compound film formation.

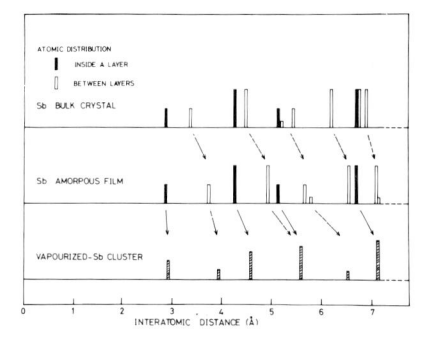

Fig. 2. Comparison of interatomic distances of the Sb clusters with those in amorphous thin film and bulk crystal.

DEPOSITION EQUIPMENT

In 1972, we suggested the basic construction of an ICB ion source for film formation [1]. Recently, we have designed different types of cluster ion sources for different purposes. In these various sources, crucible heating, cluster ionization and acceleration components are designed so as to be controlled independently. The crucible can be heated to temperatures higher than 2000°C. The ratio of ionized clusters to total clusters can be adjusted by variation of the ionization electron current. For example, the degree of ionization is 5 to 7 percent at an ionization current I_e of 100mA, 7 to 15 percent at I_e of 150mA and 30 to 35 percent at I_e of 300mA [15]. Theoretical calculations of the ionization cross-section have been done by using hydrogen clusters[16]. For small clusters of less than 50 atoms, the cross-section increases with the number of atoms per cluster. For larger clusters only the surface atoms can be ionized and the cross-section becomes proportional to the 3/2 power of the cluster size.

Several types of ICB apparatus have been developed. Typical construction of a source with single crucible and nozzle is shown in Fig. 3. One type of ICB system now commercially available has four crucibles which can be moved sequentially into the ion source by remote control.

Deposition conditions such as crucible temperature, deposition rate, acceleration voltage, electron current for ionization, substrate temperature, and film thickness can be controlled automatically by system computer [17]. A dual cluster beam system having two groups of crucibles has also been developed. A single crucible system with multiple nozzles which forms a curtain beam with high uniformity is shown in Fig. 4 [18].

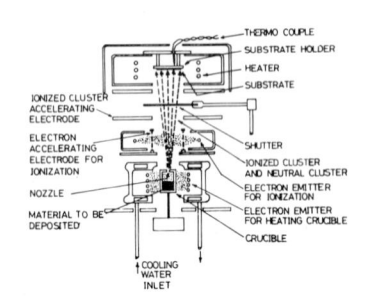

Fig. 3. Ionized cluster beam source with single crucible and single nozzle.

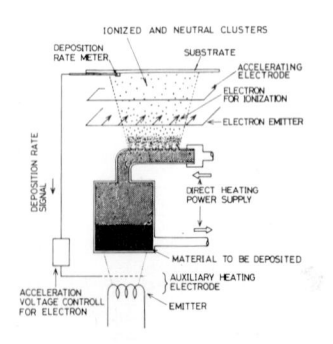

FILM FORMATION KINETICS

When considering film growth mechanisms by ICB, factors arising from the use of ionized clusters, such as special effects due to the cluster state, the existence of electric charge and the kinetic energy, etc., must be taken into account. Growth mechanisms are generally enhanced by the kinetic energy of the accelerated ionized clusters. Electric charge of the cluster also influences the critical parameters in the condensation process, such as nucleation, coalescence, etc., and causes enhancement of chemical reactions [19]. It is important to note that, even when only a few percent of the incoming particles are ionized, these effects are strong.

Fig. 4. Ionized cluster beam source with single crucible and multiple nozzle (curtain beam).

Unique film formation kinetics due to controlled atomic processes by ICB make it possible to accurately control film characteristics to nanometer scale resolution. In actual depositions, a number of aspects of ICB processes which are listed below are of importance. (a) In-situ surface cleaning by ICB bombardment just prior to deposition and during deposition. (b) Low temperature crystal growth. (a) and (b) are largely the

the results of enhanced surface mobilities of adatoms. (c) Controlled sticking coefficients. Although the sticking coefficient is determined by the combination of depositing atoms and substrate materials, surface conditions such as cleanliness, presence of defects, introduction of defects, etc., also have influence. These factors can be controlled by ion bombardment. (d) Enhanced adatom migration. (e) Control of film formation kinetics. These include the rate of formation of nucleation sites, the density of nucleation sites, the energy associated with deposition which is related to the adatom migration effect. (f) Capability to form large area films at adequate deposition rates on an industrial scale.

EXAMPLES OF FILM APPLICATIONS

Many experiments have been undertaken to prepare various kinds of films by ICB and by reactive ICB (RICB). The films deposited by these methods and their features are summarized in Table 1. Because of the ability to properly control the deposition kinetics, it has been possible to form films with special qualities and planned characteristics. Several important examples will be discussed.

TABLE I. Film properties and their features

Films/Substrate	Accel. Volt. (kV)	Subs. Temp. (°C)	Advantages (Applications)
		METALS	
Au,Cu/glass, Kapton	1-10	RT	Strong adhesion, high packing density, good electrical conduction in a very thin film (high resolution flexible circuits, optical coating)
Pb/glass	5	RT	Controllable crystal structure, strong adhesion, smooth surface, improved stability on thermal cycling (superconductive device)
Ag/Si(n-type) Ag/Si(p-type)	5	RT 400*	Ohmic contact without alloying Ohmic contact at low temperature (semiconductor metallization)
Ag-Sb/GaP	2	400	Ohmic contact, strong adhesion (semiconductor metallization)
Al/SiO$_2$,Si	0-5	RT-200	Controllable crystal structure strong adhesion, ohmic contact at low temperature, electro-migration resistant (semiconductor metallization)
c-axis preferentially oriented MnBi/glass	0**	300	Spatially uniform magnetic domain, high density optical memory (magneto-optical memory)
GdFe/glass			Thermally stable and uniform, amorphous (magneto-optical memory)
c-axis preferentially oriented PbTe/glass	0-3	200	High Seebeck coefficient, low thermal conductivity (high efficiency thermo-electrical converter)
Preferentially oriented ZnSb/glass	1	140	Thermally stable amorphous, high Seebeck coefficient (high efficiency thermoelectric converter
ʃ-FeSi$_2$/glass	0-5	150	Thermally stable, amorphous, p- or n-type controllable, high Seebeck coefficient (high efficient thermo-electric converter)

TABLE I. Film properties and their features (continued)

Films/Substrate	Accel. Volt. (kV)	Subs. Temp. (°C)	Advantages (Applications)
SEMICONDUCTORS			
Si/(111)Si Si/(100)Si Si/(1$\bar{1}$02)sapphire	6	620	Low temperature epitaxy in a high vacuum (10^{-7}-10^{-6} Torr) shallow and sharp p-n junction (semiconductor devices)
Amorphous Si/glass	2	200	Thermally stable (solar cell, thin film transistor)
Ge/(100)Si Ge/(111)Si	0.5-1	300-500	Low temperature epitaxy in a high vacuum (10^{-7}-10^{-6} Torr) shallow and sharp p-n junction (semiconductor devices)
GaAs/Cr-doped GaAs	6	550	Epitaxy in a high vacuum (10^{-7}-10^{-6} Torr) (semiconductor devices)
GaP/GaP GaP/Si	4	550 450	Low temperature epitaxy in a high vacuum (low cost LED)
ZnS/NaCl ZnS:Mn/glass	1 1	200 200	Single crystal formation (optical coating) (ac-dc electroluminescent cells of low impedance)
CdTe/glass	2	250	Improved monocrystalline domains (infrared detector)
InSb/sapphire	3	250	Controllable crystal structure (magnetic sensor)
OXIDES, NITRIDES, AND CARBIDES			
Li-doped ZnO/ (1$\bar{1}$02) sapphire	0.5-1	230	Single crystal formation (optical wave guide)
C-axis preferentially oriented ZnO/glass	0	150	Controllable crystal structure Controllable optical transmission (optical wave guide, SAW)
BeO/(0001) sapphire	0	400	Single crystal formation, transparent film (semiconductor device, heat sink, electrical insulator)
c-axis preferentially oriented BeO/glass	0	400	High electrical resistivity and high thermal conductivity coating, low temperature growing (semiconductor device, heat sink electrically insulating, and thermal conduction)
PbO/glass	3	RT	Low temperature growth, smooth surface, accurately controllable thickness, good adhesion (superconductor)
SnO$_2$/glass	0	400	Transparent low resistivity film, strong adhesion (NESA glass)
SiO$_2$/Si	0-2	200	Extremely low temperature growth (Si devices)
FeO$_x$/Si,glass	3	250	Controllable optical bandgap (photovoltaic cell)
GaN/ZnO/glass	0	450	Low temperature growth of GaN film on amorphous substrate (low cost LED, photocathodic electrode)
SiC/Si,glass	0-8	600	Controllable crystal state (energy converter, surface protective coating)
ORGANIC MATERIALS			
Anthracene/glass	0-2	-10	Controllable crystal structure (detectors)

TABLE I. Film properties and their features (concluded)

Films/Substrate	Accel. Volt. (kV)	Subs. Temp. (°C)	Advantages (Applications)
Cu-Phtalocyanine/ glass	0-2	200	Controllable crystal structure (photovoltaic devices)
Polyethylene/glass	0-1	0-110	Controllable crystal structure (photovoltaic devices)

* Annealing temperature
** Ejection velocity only

Materials for Semiconductor Devices

To extend the limits of miniaturization of semiconductor devices, ICB techniques can offer low-temperature process capabilities and make it possible to control dimensions of devices to atomic scale accuracy. This will become important for VLSI and three dimensional semiconductor devices.

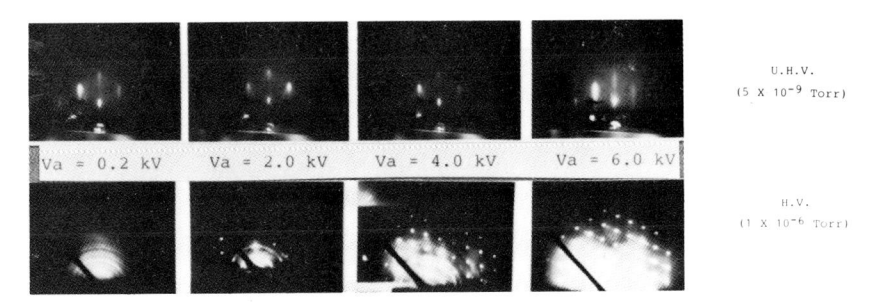

Fig. 5. Electron diffraction patterns of epitaxial silicon films by ICB different conditions.

Silicon epitaxial films can be formed by ICB [20-22]. Electron diffraction was used to evaluate the silicon films. The results show that an amorphous or polycrystalline structure was formed in a range of 0-4kV and that the crystallinity was improved at higher acceleration voltages. On the other hand, epitaxial growth of Si on an atomically clean and well-ordered silicon surface in ultra-high vacuum chamber was obtained at only 200V acceleration with substrate temperature of 500°C [23]. The film formed in this method was examined by in-situ electron diffraction at 5keV. The results are shown in the upper part of Fig. 5. In the case of standard high vacuum, at 10^{-7} to 10^{-6} Torr, acceleration voltage required for epitaxial deposition was 6 to 8kV because it was necessary to sputter the native oxides from the surface and to remove impinging residual gas atoms during the deposition. On the other hand, in the deposition on atomically clean surfaces under ultra-high vacuum, a single crystalline pattern was ob- served at 200V acceleration because it was not necessary to sputter the oxides and residual gas atoms during the deposition. At increasing acceleration voltages, improvement of the crystalline quality could be observed. This might be due to enhanced adatom migration effect. The Hall mobilities of the films grown at substrate temperatures of 730°C and acceleration voltage of 5kV in a HV chamber was 800cm^2/V.sec and increased with increasing acceleration voltage. The p-n diodes formed by deposition of an n-Si film onto a p-Si substrate had good and abrupt junctions. By changing deposition conditions, thermally stable hydrogenated amorphous silicon films could be formed. This is due to the fact that the film could be formed in a relatively low gas pressure of the range of 10^{-5} to 10^{-4} Torr and bonding of hydrogen atoms to silicon atoms could be controlled by changing the acceleration voltage [24].

Low temperature semiconductor film formation will require complementary low temperature oxide [25,26], nitride [27] and carbide [28] film formation abilities to complete the overall low temperature approach. Some of the results show possibilities for these materials.

Aluminum deposition is also very important in VLSI metallization. Low temperature ohmic contact formation, interconnections with improved electromigration lifetime, smooth surfaces and good step coverage and corrosion resistance were achieved by ICB deposition [29]. Mean-time-to-failure (MTF) of the interconnections is related to the crystallite size, the size distribution of the grains and the crystalline structure. X-ray diffraction measurements show that an increase of acceleration voltage results in an increase of (111) axis preferential orientation. Grain size becomes larger in the depositions at higher acceleration voltages. Narrow distribution of grain sizes was also observed. These factors contribute to increased lifetime.

Semiconductor films on insulators are becoming important for three dimensional device fabrication. We have attempted to grow GaN films on a c-axis oriented ZnO layer deposited onto a glass substrate [25]. The ZnO layer was also deposited by ICB. Figure 6 shows an x-ray diffraction pattern and the fractured edge of the GaN film on the ZnO layer grown at I_e=300mA, V_a=0kV and T_s=450°C. These diffraction and SEM data show that the GaN film has successfully grown on the ZnO film. This result shows that ZnO film by ICB offers an adequate seed nucleation surface for growing GaN films. The optical absorption characteristics of GaN films grown on ZnO/glass substrate are comparable to those of GaN films grown on sapphire substrates by using a CVD method. This seed-epitaxy method may be one approach to development of fabrication technologies for three-dimensional VLSI devices.

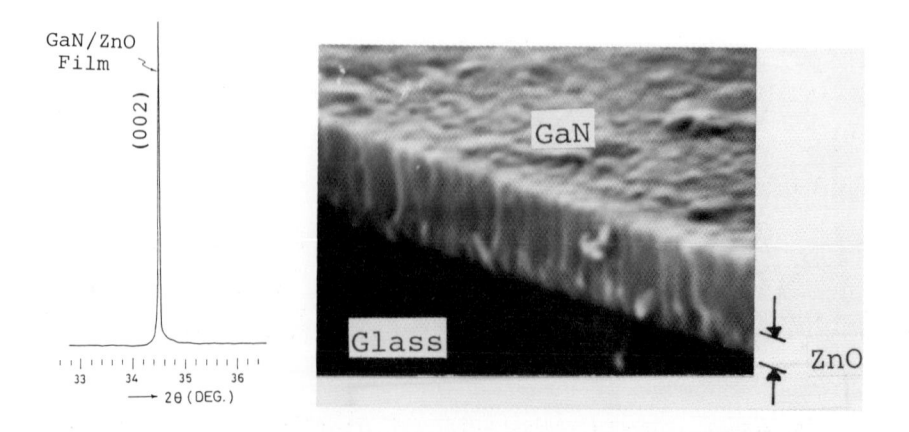

Fig. 6. X-ray diffraction patterns: (a) and scanning electron microstructure of a GaN/ZnO/Glass prepared at Va=0kV, Ie=300mA and Ts=450°C.

Intermetallic Compound Materials

In the investigation of films for thermoelectric energy conversion devices, films formed by ICB exhibited exceptionally high Seebeck coefficients. In case of ZnSb films, the maximum observed Seebeck coefficient of 600 μV/deg at 500°K is twice as high as the highest value previously reported [30]. The films were formed by co-deposition using an ionized and accelerated Zn cluster beam and a neutral Sb cluster beam. The film showed (102) preferentially oriented structure. In the case of amorphous iron disilicide, Seebeck coefficients of 10mV/deg for a p-type film and -20mV/deg for an n-type film have been obtained [31]. The figure of merit for these films is on the order of $Z \cong 10^{-2}$ deg C^{-1}. This value is approximately 100 times higher than that of polycrystalline materials. To form the p-type film, deposition was made using an ionized silicon monoxide cluster beam and a neutral iron cluster beam in an oxygen atmosphere of 10^{-5} to 10^{-4} Torr. Manganese was evaporated from a heated basket as doping material. For n-type film, an ionized Fe-cluster beam and neutral Si clusters were used in an oxygen atmosphere of 2×10^{-4} Torr. Figure 7 shows temperature dependencies of the Seebeck coefficient as a function of manganese concentration. Maximum value of the Seebeck coefficient was obtained for the films doped with 0.5 atomic percent Mn. The figure also shows higher Seebeck coefficients than previously observed even without manganese doping.

As one possible application for photovoltaic energy conversion devices, deposition of CdTe and PbTe films has been studied. Particularly interesting results have been observed when one of the constituent materials is ionized. In case of CdTe when ionized Cd clusters and neutral Te clusters were used, the film showed cubic structure. However when ionized Te clusters and neutral Cd clusters were used the film became of hexagonal structure. A superlattice structure of cubic-PbTe and cubic-CdTe films with 100Å period on a (111) silicon substrate surface could be formed by ICB. Optical absorption which corresponds to the n=1 quantum state was observed [32].

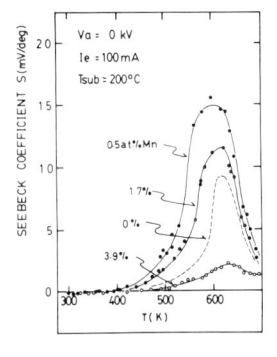

Fig. 7. Temperature dependence of the Seebeck coefficient S for manganese doped Fe (SiO)$_2$.

Organic Materials

Although the plasma discharge technique, in which electrons, ions, radicals and energetic neutrals of a few eV move randomly, is one of the conventional methods for organic film formation, the particle energy cannot be controlled by changing discharge conditions and the composition of the products is too complicated to allow identification. This is due to difficulties of sustaining the discharge in the high vacuum region. By using ICB, deposition of organic materials with controlled film quality has been attempted.

Anthracene, copper-phthalocyanine and polyethylene films were deposited by ICB. Experiments on the formation of the organic clusters have been reported elsewhere [33]. Figure 8 is an example indicating the controllability of the crystal system of copper-phthalocyanine by ICB.

Films deposited at lower acceleration voltage showed (001) preferentially oriented α-CuPc structure. But β-CuPc structure, which cannot be formed by conventional vacuum deposition at low substrate temperatures, appeared at acceleration voltages higher than 200V. The ICB technique was found to have a capability for depositing polymer compounds as well as molecular crystal organic materials with controlled crystallinity [34]. The general characteristics of these films indicate that the crystallinity, grain size, film adhesion strength and photoluminescent intensity could be improved by controlling acceleration voltage and ion beam density.

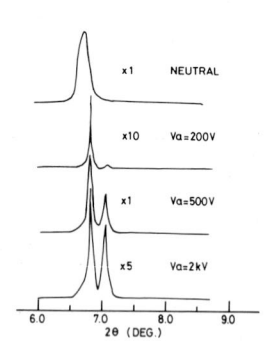

Fig.8. X-ray diffraction patterns of Cu-Pc films deposited at different acceleration voltages at 200°C substrate temperature.

CONCLUSIONS

The ICB technique allows growth of elemental or compound material hydride, oxide, carbide and nitride films. It has been shown that films could be deposited at lower substrate temperatures than are required in other deposition methods. Vacuum pressures in the chamber during the deposition of films was in the range from 10^{-7} to 10^{-6} Torr, a range which can be obtained by conventional vacuum equipment. In the case of reactive ICB deposition, the gas pressure in the vacuum chamber is not required to be higher than 10^{-5} to 10^{-4} Torr. The remarkable features of ICB relative to conventional techniques are due to unique film formation mechanisms supported by distinctive properties of the clusters and by controlled film formation kinetics.

REFERENCES

1. T. Takagi, I. Yamada, M. Kunori, and S. Kobiyama, proc. 2nd Int. Conf. Ion Sources, 1972, Vienna (Ostereichiche Studiengesellshaft fur Atomonergie, Vienna, 1982), p.790.
2. T. Takagi, I. Yamada and A. Sasaki, Thin Solid Films 45, 569 (1977).
3. J.-P Borel and J. Buttel (ed.), "Small Particles and Inorganic Clusters", Suf. Sci. 106 (1981).
4. J.K. Lee, J.A. Barker, and F.F. Abraham, J. Chem. Phys. 58, 3166 (1973).
5. A. Yokozeki and G.D. Stein, J. Appl. Phys., 49, 2224 (1978).
6. P.G. Hill, H. Witting, and E.P. Demetri, Trans. Am. Soc. Mech. Eng. 85, 303 (1963).
7. I. Yamada, in T. Takagi (eD.), Proc. Int. Ion Eng. Congr. -ISAT'83 & IPAT83-, 1983, Kyoto (IEEJ, Tokyo, 1983), p. 1117.
8. P.P. Wagener, J.A. Clumpuer and B.J.C. Wu, Phys. Fluids 15, 1869 (1972).
9. G.D. Stein, Phys.s Teach. 503 (1979).
10. H. Usui, H. Takaoka, I. Yamada, and T. Takagi, Proc. 4th Symp. Ion sources Ion-Assisted Technology, 1981, Tokyo (IEEJ, Tokyo, 1981), p.175.
11. I. Yamada and T. Takagi, Thin Solid Films 80, 105 (1981).
12. I. Yamada, G.D. Stein, H. Usui, and T. Takagi, Proc. 6th Symp. Ion Sources Ion-Assisted Technology, 1982, Tokyo (IEEJ, 1982), p.47.

REFERENCES (Concluded)

13. A.B. Anderson. J. Chem Phys. 64, 4046 (1976).
14. J.Buttet and J.-P.Borel, Phys. Rev. A 13, 2287 (1976).
15. T. Takagi, I. Yamada and A. Sasaki, Inst. Phys. Conf. Ser.38, 142 (1978).
16. F. Bottiglioui, J. Coutant and M. Fois, Phys. Rev. A 6, 1830 (1972).
17. Technical Data, Eaton Corporation, 133 Brimbal Avenue, Beverly, MA U.S.A.
18. T. Takagi, I. Yamada and H. Takaoka, Surface Science 106, 544 (1981).22.
19. I. Yamada, H. Takaoka, H. Inokawa, H. Usui, S.C. Cheng and T. Takagi, Thin Solid Films 92, (1982).
20. T. Takagi, I. Yamada and A. Sasaki, J. Vac. Sci. Techno. 12, 1128 (1972).
21. T. Takagi, I. Yamada and A. Sasaki, Thin Solid Films 39, 207 (1976).
22. I. Yamada, F.W. Saris, T. Takagi, K.Matsubara, H. Takaoka and S. Ishiyama, Jap. J. Appl. Phys. 19, L181 (1980).
23. T. Takagi, Preprint of Ion Assisted Surface Treatments, Techniques and Processed, 1982, Coventry (The Metal Society, London, 1982), p.1.1.
24. I. Yamada, I. Nagai, M. Horie and T. Takagi, J. Appl. Phys. 54, 1583 (1983).
25. T. Takagi, I. Yamada and K. Matsubara, Thin Solid Films, 58, 9 (1970).
26. T. Takagi, K. Matsubara and H. Takaoka, J. Appl. Phys. 51, 5419 (1980).
27. K. Matsubara, T. Horibe, H. Takaoka and T. Takagi, Proc. 4th Symp. Ion Sources Ion Applic. Technology, 1980, Tokyo (IEEJ, Tokyo, 1980), p.137.
28. K. Mameno, K. Matsubara and T. Takagi, Proc. 6th Symp. Ion Sources Ion-Assisted Technology 1982, Tokyo (IEEJ, Tokyo, 1982), p.341.
29. H. Inokawa,K. Fukushima, I. Yamada and T. Takagi, Proc. 6th Symp. Ion Sources Ion-Assisted Technology, 1982, Tokyo (IEEJ, Tokyo, 335 1982), p.355.
30. T. Koyanagi, K. Matsubara, H. Takaoka and T. Takagi, Proc. 6th Symp. Ion Sources Ion-Assisted Technology, 1982, Tokyo (IEEJ, Tokyo, 1982), p.409.
31. T. Takagi, K. Matsubara, M. Oura and T. Koyanagi, Proc. 6th Symp. Ion Sources Ion-Assisted Technology, 1982, Tokyo (IEEJ, Tokyo, 1982), p.371.
32. H. Takaoka, K. Matsubara and T. Takagi, in T. Takagi (eD.), Proc. Int. Ion Eng. Congr. -ISAT'83 & IPAT'83-, 1983, Kyoto (IEEJ, Tokyo, 1982), p. 1241.
33. H. Usui, N. Naemura, I. Yamada and T. Takagi, Proc. 6th Symp. Ion Sources and Ion-Assisted Technology, 1982, Tokyo (IEEJ, Tokyo, 1982), p.331.
34. H. Usui, I. Yamada and T. Takagi, in T. Takagi (eD.), Proc. Int. IonEng. Congr. -ISAT'83 & IPAT'83-, 1983, Kyoto (IEEJ, Tokyo, 1983), p.1427.

DYNAMIC RECOIL MIXING FOR THE PRODUCTION OF SILICON NITRIDE FILMS

H. KHEYRANDISH, J.S. COLLIGON AND A.E. HILL
Thin Film and Surface Research Centre, Department of Electronic and
Electrical Engineering, University of Salford, Salford M5 4WT, UK

ABSTRACT

In Dynamic Recoil Mixing (DRM) a film of constant thickness is sputtered on to a substrate by using a broad low energy (1keV) ion beam and is subsequently bombarded by a high energy (10 keV) ion beam. During the bombardment process a dynamic balance is maintained between the back-sputtering and the deposition of the film, thus providing an 'unlimited' source of the dopant material. In this way very high surface dopant concentrations may be achieved which otherwise cannot be reached by more conventional ion beam mixing techniques. Furthermore, by choosing reactive ion species, films of novel chemical compositions may be produced. This technique has been employed to produce silicon nitride films on polished commercial mild steel samples. Measurements of knoop hardness of these films indicate an increase of over 200% whilst the wear and friction properties have shown considerable improvement. Conventionally sputtered silicon films of similar thickness, recoil mixed silicon using argon, or implantation of 10 keV argon and nitrogen to similar doses, indicate only a modest change in tribological characteristic compared to the reactive mixed films.

INTRODUCTION

In recent years ion implantation has played an important role in altering the hardness, friction and wear of steels and has been the subject of extensive study {1}. An alternative to ion implantation is the deposition of thin films of borides, carbides and nitrides of certain transition metals which are hard and resistant to both wear and corrosion in bulk form. Such films may be produced by reactive sputtering {2}, plasma and ion assisted film formation {3}, chemical vapour deposition {4} as well as reactive ion beam mixing. In comparison with ion implantation the tribological and corrosion characteristics of such overlayers have been the subject of limited study {5}.

It has been established that atomically mixed layers have considerably greater adhesion and endurance {6,7} and the inclusion of silicon in pure iron has been demonstrated to have a significant effect on the oxidation resistance of pure iron {8}. The purpose of the present work has been to investigate the tribological properties of reactive ion beam mixed films of silicon on commercial mild steel.

EXPERIMENTAL

The Apparatus

The dynamic recoil mixing (DRM) apparatus has been described in detail elsewhere {9} and hence only a brief description of it is presented here. A schematic diagram of this apparatus is shown in Figure 1. It consists of a

Kaufman-type multiray ion source and produces currents of ~ 1 mA/cm^2 of 1 keV ions and this is used as a sputtering source to deposit thin films. There is also a higher energy (10 keV) recoil implanting ion source positioned perpendicular to the sputtering source, capable of producing currents of ~ 20 μA/cm^2 uniformly over 1 cm^2. This source is used to bombard the deposited films. Both or either of these sources may be used with inert gases (for example, argon or krypton) or reactive gases (such as nitrogen or oxygen for **reactive ion beam mixing**). After a film of desired optimum thickness has been deposited it is bombarded with the implanting source during which time a dynamic balance is maintained between the backsputtering and deposition of the film (i.e. controlled simultaneous bombardment and deposition), thus providing in effect, an unlimited source of the dopant (film) material. The state of this balance is detected on a quartz crystal microbalance which ensures that the backsputtered and deposited masses of the film are equal at any time. Alternatively the microbalance may be used to monitor the rate of growth of a film under simultaneous ion bombardment and deposition. The same quartz crystal is also used to measure the initial film thickness deposited prior to mixing.

DRM offers distinct advantages over more conventional ion beam mixing techniques for certain applications where extremely high surface concentrations or well-bonded films are required. One inherent advantage of DRM over conventional mixing is that even after bombardment to high fluences the mixed layer is not eroded by sputtering.

Sample Preparation and Treatment

Samples of commercial mild steel (30 x 18 x 3 mm) and flat to 0.1 μm were used in the present experiments. Si films of 150 Å thickness were sputter deposited on to these specimens using a 1 keV nitrogen beam at typical deposition rates of 25 Å/min. These films were then dynamically recoil mixed at room temperature with 10 keV ~ 10 μm/cm^2 nitrogen ions using fluences in the range 5 x 10^{16} - 10^{17} ions/cm^2. Samples were also prepared with Si films deposited using argon sputtering and recoil mixed with 10 keV argon to a fluence of 5x10^{16} ions/cm^2. For comparison, further mild steel samples were implanted with 10 keV argon or nitrogen to similar fluences.

A summary of the various sample treatments is given in Table 1. However, it should be noted that since the ion beams used in this system are not mass analysed it is not possible to comment on the ratio of ion currents $I_{N_2^+}$ to I_{N^+}. This may influence the recoil mixing since molecular ions produce overlapping collision cascades which effect radiation damage processes. Also the two atomic N particles created from dissociation of N_2^+ at the surface have half the energy of the N^+ ion so that the mixing is occuring in the presence of two energetic groups of N atoms {10}.

Measurements of Wear, Friction and Hardness

The knoop microhardness of the specimens was measured using a Leitz microhardness tester. Loads of 10 and 25 g were used in order to minimize any contribution from the substrate.

Friction and wear tests were conducted by using a pin-on-disc machine with pins of ball bearing steel (BS 535A99, SAE 52100) of 6.35 mm dia. A load of 1.47 N was applied to the pin and the specimen rotated at a sliding speed of 19.3 mm/sec under dry and unlubricated conditions. A force transducer was used to measure the frictional force and, in the absence of a vertical displacement transducer, the specimen and the pin wear were **measured (after completing 11.55 m of travel) by using a Sloan Instruments** Dektak and a microscope. The worn and unworn regions of the specimens were monitored by the Dektak which indicated some plastic deformation as a result of wear. To quantify these results, the net amount of material lost has been estimated by integrating the profiles. Similarly the wear volume of the pin has been calculated by measuring the diameter of the worn region using a microscope. From these measurements the wear coefficient K, for both the specimen and the pin have been estimated using

$$K = \frac{\text{Wear volume}}{\text{Force} \times \text{Distance travelled}}.$$

RESULTS AND DISCUSSION

The values of knoop microhardness H_K, kinetic coefficients of friction μ and the wear coefficients K for both the specimen and pin are given in Table 1 which also contains a summary of the sample treatment parameters.

TABLE 1. A summary of the various sample treatments and the results.

Deposition species and initial film thickness (Å)	Bombardment species and fluence (ions/cm^2)	Colour of specimen	H_K Kg/mm^2	μ	$K_{specimen}$ x 10^{-13} m^2/N	K_{pin} x 10^{-13} m^2/N
(N_2^+, N^+), 150	—	Transparent yellow	552	0.50	1.06	0.95
" "	$(N_2^+, N^+) 5 \times 10^{15}$	"	569	0.48	1.27	0.98
" "	" 10^{16}	"	653	0.42	0.94	0.86
" "	" 3×10^{16}	"	791	0.35	0.89	0.73
" "	" 5×10^{16}	"	815	0.32	0.59	0.75
" "	" 7×10^{16}	"	975	0.34	0.47	0.73
" "	" 10^{17}	"	890	0.32	0.41	0.74
——	5×10^{16}	Metallic	491	0.51	0.98	0.87
——	$(Ar^+), 5 \times 10^{16}$	"	522	0.53	1.19	0.91
(Ar^+), 150	——	Blue	430	0.52	1.40	1.17
(Ar^+), 150	$(Ar^+), 5 \times 10^{16}$	"	510	0.45	1.38	1.14
Control (untreated) specimen		Metallic	430	0.56	1.37	1.10

It is noted that in all cases sputtering of silicon with nitrogen resulted in deposition of transparent yellow insulating films resembling silicon nitride whilst in corresponding cases using argon a formation of deep blue films results. This is in agreement with Weissmantel {11} who has reported deposition of Si$_x$N$_y$ films by reactive ion beam sputtering under similar conditions. The formation of stoichiometric Si$_3$N$_4$ have only been reported for low deposition rates ($\simeq$100 Å/min); our own deposition rates being 25 Å/min. RBS and nuclear reaction studies of our films have revealed considerable nitrogen concentrations as well as an increasing degree of Si-Fe mixing with increasing bombardment fluence. The present discussions are, however, confined to the tribological characteristics of these films whilst the atomic mixing, depth profiling and stoichiometry of similar films on pure Fe are the subject of a present investigation which will be reported elsewhere {12}.

The values of hardness for all samples are shown in Figure 2. It is noted that at a bombardment fluence of 5×10^{16} ions/cm^2 the hardness increases by $\simeq$200% whilst the as-sputtered films of similar thickness or recoil mixed Si films using argon and direct implants of 10 keV argon and nitrogen to similar fluences show only a 20-30% increase in hardness. These values are in agreement with previously published results {13,14}.

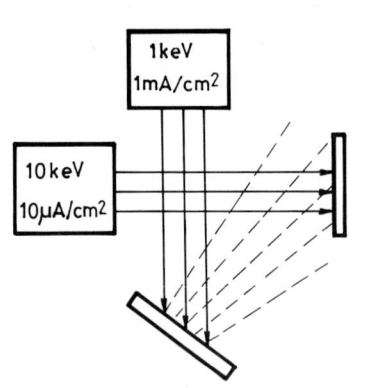

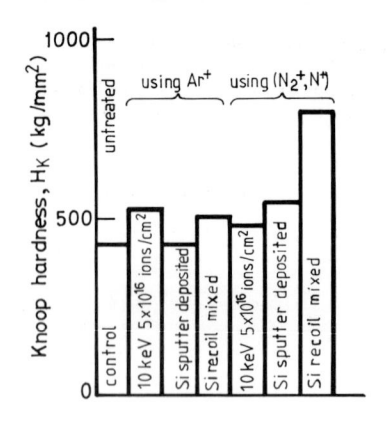

FIG. 1. A schematic of the apparatus.

FIG. 2. Knoop hardness of the specimens.

The relative change in hardness with nitrogen fluence is shown in Figure 3.

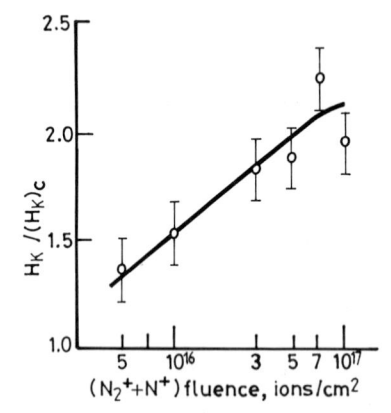

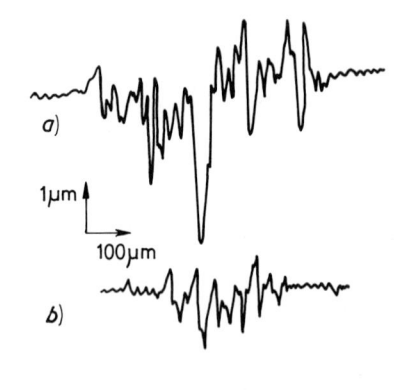

FIG. 3. The relative change in hardness with nitrogen fluence.

FIG. 4. The Dektak profiles for (a) the control (untreated) sample and (b) the recoil mixed sample using 7×10^{16} nitrogen ions/cm^2.

The rate change of hardness with nitrogen fluence is estimated to be 30 kg/mm^2 per 10^{14} ions in the range $5 \times 10^{15} - 10^{17}$ ions/cm^2 representing an increase in excess of 200% at maximum dose. However in these measurements the penetration depths of the indentor ($\sim$0.5 μm) is approximately an order of magnitude higher than the total modified depths and hence the reported values of hardness relate to those for the film together with a contribution from the substrate. It is therefore reasonable to assume that the actual hardness of the film is considerably larger than the presently measured values.

The wear profiles for the control (untreated) and the recoil mixed 7×10^{16} ions/cm^2 specimen are shown in Figure 4. These suggest varying degrees of plastic deformation of the substrate, considerable reduction in wear and, in the latter case, and average wear depth of $\sim$0.7 μm which is over an order of magnitude larger than the treated depths. The relative wear coefficients of the nitrogen bombarded specimens K_c/K (where K_c is the wear coefficient of the control sample) together with corresponding values for the pins are plotted as a function of the nitrogen fluence and are shown in Figure 5. From this a steady improvement in the

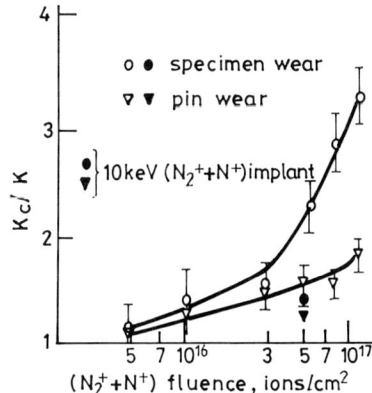

FIG. 5. The relative change in the wear coefficient with nitrogen fluence.

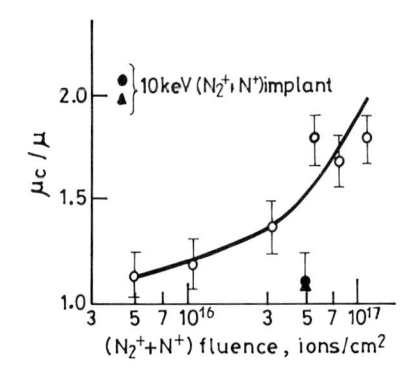

FIG. 6. The relative change in the coefficient of friction with nitrogen fluence.

relative wear coefficient of the specimens is observed up to a fluence of 3×10^{16} ions/cm^2 (K_c/K = 1.5) after which there appears an enhanced improvement up to a fluence of 10^{17} ions/cm^2 where K_c/K is 3.3. The presence of nitrogen in the substrate alone will not explain such wear improvements as a 10 keV, 5×10^{16} ions/cm^2 nitrogen implant resulted in a wear coefficient of 0.98×10^{-13}m^2/N (K_c/K = 1.4) whilst the recoil mixed specimen bombarded to a similar fluence indicated a wear coefficient of 0.59×10^{-13}m^2/N (K_c/K = 2.3). The pin wear coefficient is also observed to improve with nitrogen fluence though the improvement rate is not as great as that for the specimens, the highest value of K_c/K being 1.8 for a dose of 10^{17} ions/cm^2. It is also important to note the absolute values of wear coefficients as listed in Table 1. For the control (untreated) specimen (H_K = 430 kg/mm^2), the harder pin (H_K = 900 kg/mm^2) displays the lower wear coefficient of 1.10×10^{-13}m^2/N compared to that for the specimen of 1.37×10^{-13}m^2/N, whilst for the 10^{17} ion/cm^2 bombarded specimen the now "hardened" specimen (H_K = 890 kg/mm^2) has the lower wear coefficient of 0.41×10^{-13}m^2/N compared to that for the pin of 0.74×10^{-13}m^2/N.

The relative coefficients of friction which are shown in Figure 6 show a similar trend to the wear properties. These show a moderate improvement up to a fluence of $\sim 3 \times 10^{16}$ ions/cm^2 (μ_c/μ = 1.35) after which an enhanced improvement is observed up to a fluence of 10^{17} ions/cm^2 where μ_c/μ = 2.6. Again, direct implantation of nitrogen cannot account for such changes in μ, since a 10 keV, 5×10^{16} nitrogen implant appears to have only a slight effect on the coefficient of friction (μ_c/μ = 1.1) compared to the recoil mixed specimen where μ_c/μ = 1.75.

It is also interesting to see that the sputter-deposited 150 Å films of silicon using argon and nitrogen, recoil mixed layers of silicon using argon or 10 keV direct implants of argon have relatively little or no effect on the tribological characteristics of these samples.

CONCLUSIONS

Using the dynamic recoil mixing technique and nitrogen ions for the sputtering and recoiling beams, mixed layers of silicon and nitrogen on commercial mild steel have been formed. The hardness of the layers is increased by 200%, wear properties are improved by a factor of 3, and the coefficient of friction is significantly reduced.

Results presented here are of a preliminary nature but show conclusively that the dynamical recoil mixing process using nitrogen ions is essential to produce the changes in mechanical properties. Use of argon for recoil mixing or even direct implantation of nitrogen into silicon does not appear to produce similar improvements.

Work is proceeding to compare other overlayers, the effect of ion-assisted film formation, and the chemical state of the intermixed layer and the film.

ACKNOWLEDGEMENT

The author wishes to thank Dr. R.D. Arnell for many valuable discussions, during the course of this study.

REFERENCES

1. G. Dearnaley and N.E.W. Hartley, Proc 4th Conf on Scientific and Industrial Applications of Small Accelerators (IEEE, New York), 20 (1976).
2. G.K. Wehner, Thin Film Dielectrics, Electrochemical Society, New York, 117, (1969).
3. J.M.E. Harper, J.J. Cuomo and H.R. Kaufman, Ann Rev Mat Sci 13, 413-439 (1983).
4. P.J.M. van der Straten and G. Verspui, Philips Tech Rev 40(7), 204-210 (1982).
5. N.E.W. Hartley and J.K. Hirvonen, Nucl Instrum and Methods 209/210, 933-940, (1983).
6. L.E. Collins, J.G. Perkins and P.T. Stroud, Thin Solid Films 4, 41-45 (1969).
7. M.L. Hitchman, J.S. Colligon, A.E. Hill, C. Southway, H. Kheyrandish and P. Argyokastritis, European Conf on Sensors and their Applications, UMIST, Manchester, UK, 20-22 September 1983.
8. A. Galerie and G. Dearnaley, Nucl Instrum and Methods 209/210, 823-829 (1983).
9. G. Fischer, A.E. Hill and J.S. Colligon, Vacuum 28, 277 (1978).
10. Cui Fu-Zhai, L.I. Heng-De and Zhang Xiao-Zhong, Nucl Instrum and Methods 209/210, 881-889 (1983).
11. Chr Weissmantel, Thin Solid Films 32, 11-18 (1976).
12. H. Kheyrandish and J.S. Colligon. To be published.
13. S. E. Donnelly, Vacuum 28, 163 (1978).
14. H.T. Li, P.S. Liu, S.C. Chang, H.C. Lu, H.H. Wang and K. Tao, Nucl Instrum and Methods 182/183, 915-917 (1981).

STRUCTURE OF Al-N FILMS DEPOSITED BY A QUANTITATIVE DUAL
ION BEAM PROCESS

H.T.G. HENTZELL*, J.M.E. HARPER and J.J. CUOMO
IBM Thomas J. Watson Research Ctr.
P.O. Box 218, Yorktown Heights, NY 10598

ABSTRACT

We describe the structure of Al-N films deposited with N/Al ratios varying from 0 to 1. A dual ion beam system supplies the Al flux by inert ion sputtering, and the N flux by low energy (100-500 eV) N_2^+ ion bombardment of the growing film. For N/Al < 1, a cermet structure forms with large Al grains mixed with AlN. Above the composition N/Al = 1, excess N is rejected from the AlN. The AlN films show a pronounced change in preferred orientation from c-axis perpendicular to the film surface, to c-axis parallel to the film surface with increasing N_2^+ energy.

INTRODUCTION

Thin film nitrides have exceptional electrical properties and high temperature and mechanical stability. AlN is interesting because of its wide band gap (6.3 eV), high decomposition temperature (2490°C) and chemical stability. AlN is also one of the promising thin film piezoelectric materials owing to its high ultrasonic velocity and fairly high piezoelectric coupling factor.

In this work, thin films of AlN were deposited at room temperature in a dual ion beam system (Fig. 1) where the Al atom flux (deposition rate = 2.5Å/s) is supplied by inert ion beam sputtering (1500 eV Ar^+, P_{AR} = 2 x 10^{-4} Torr) and the reactive flux (P_{N_2} = 1x10^{-4}Torr) is supplied by a low energy (100-500 eV) ion beam directed at the growing film. Insitu monitors of atom and ion flux enable the fundamental deposition parameters of ion and atom arrival rates, ion energy and direction to be measured and controlled [1,2]. The accumulated flux, i.e. the composition, is analyzed by Rutherford Backscattering (RBS). In the present paper, which is an extended abstract of Ref. 3, the deposition parameters are related to the microstructure of AlN films across the substrate holder. We also report on the influence of the N_2^+ ion beam energy on the microstructure of AlN films.

The crystallographic structure and microstructure of

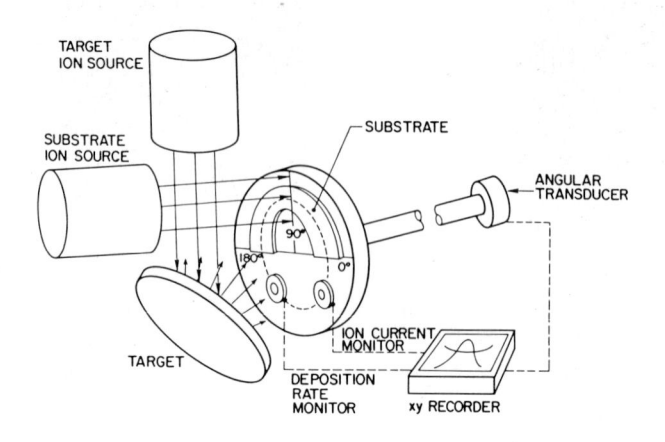

FIG. 1 Diagram of dual ion beam deposition system.

the films were analyzed as deposited, on Si_3N_4 windows and after floating off NaCl, in a TEM. The size and shapes of the grains were measured from bright field images for the Al-AlN films and from high magnification dark field images for the AlN films. Inaccuracies in measurements of grain size are small for Al-AlN films but may be as high as 25% in the AlN films.

DEPENDENCE OF STRUCTURE ON COMPOSITION

In this section we describe the changes in crystallographic structure and microstructure as the arrival flux of 500 eV N_2^+ ions onto the film surface increases. Fig. 2a shows a TEM micrograph of a pure Al film, ion beam sputter deposited with Ar. The microstructure consists of small equiaxed grains. The diffraction pattern indicates that the film has grown without any texture. In films deposited in the presence of 1×10^{-4} Torr N_2 but no N_2^+ ion beam, no indication of AlN formation is observed from the electron diffractograms. Careful measurements of the lattice parameter show, however, that the Al lattice is expanded ($a_0 = 4.08$Å) as compared to pure Al ($a_0 = 4.05$Å), consistent with the presence of about 9% N dissolved in Al.

The influence of the 500 eV N_2^+ ion beam on the structure of the film is shown in Fig. 2b. When nitrogen ions impinge on the growing film, all nitrogen reacts with Al to form AlN [2]. The microstructure consists of large uniformly distributed Al grains surrounded by very fine-grained AlN. The lattice parameter of Al has relaxed to

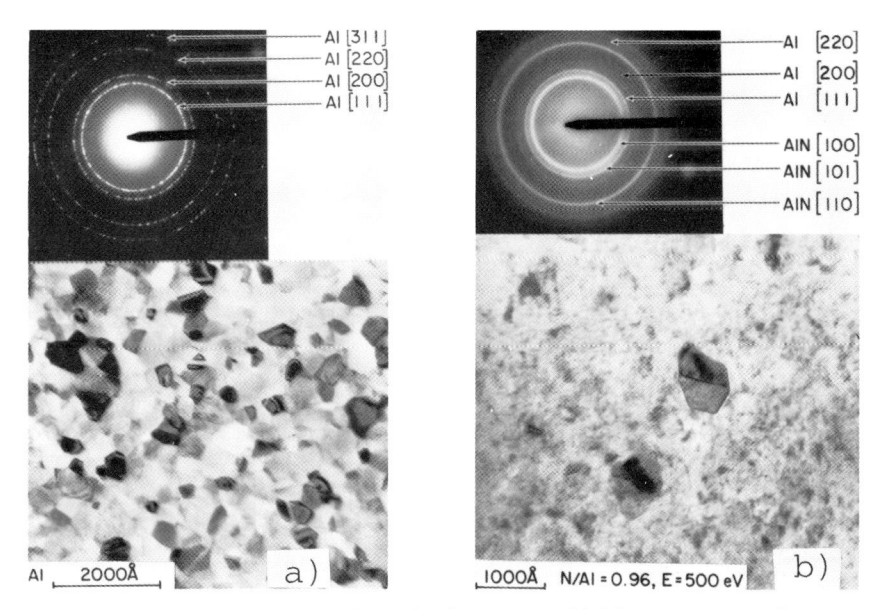

FIG. 2 TEM micrograph and electron diffractogram from
a) pure Al
b) cermet Al-AlN film of composition N/Al = 0.96.

the bulk value as compared to Al + N films prepared with a
N_2^+ ion beam indicating that all nitrogen has reacted
with aluminum to form AlN instead of dissolving into the
aluminum lattice. The a-axis and c-axis of the hexagonal
AlN structure in the films (a = 3.11Å, c = 4.98Å) agree
very well with table values for AlN (a = 3.111Å and c =
4.979Å[4]). In the cermet Al-AlN films where 0.54 < N/Al <
0.82 the films have a dull appearance due to surface
roughness, because the Al grains have grown faster than
the surrounding AlN even though Al is preferentially
sputtered [2,3]. The electron diffractogram (Fig. 2b)
shows that the AlN (002) reflection is missing, indicating that the AlN has grown with the c-axis perpendicular
to the film surface.

The range of grain sizes observed in pure Al, cermet
Al-AlN and single phase AlN films is shown in Fig. 3. The
average and maximum aluminum grain size increases when
AlN forms. Further increase in the nitrogen to aluminum
ratio causes a decrease in the Al grain size but even in
films with N/Al = 0.96 (Fig. 2b) the Al grain size is similar to that in pure Al. The microstructure of AlN in the
cermet Al-AlN films is too distorted for any measurements to be made. The grain size in single phase AlN is

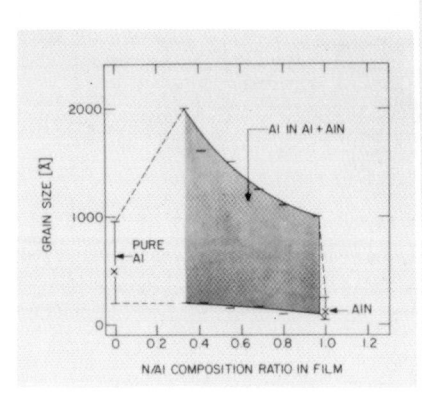

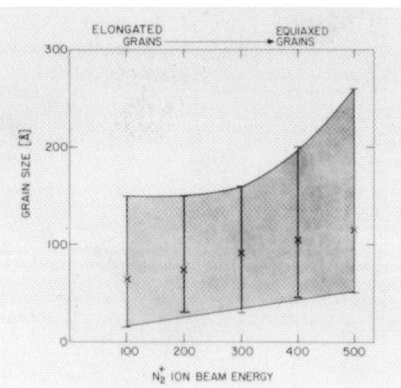

FIG. 3 Maximum and minimum grain sizes of pure Al films, Al in cermet Al-AlN films, and single phase AlN films.

FIG. 4 Maximum and minimum grain size of AlN thin films as a function of N_2^+ ion beam energy (Arrival rate ratio N/Al = 1.5).

considerably smaller than that of aluminum in the cermet films.

Two important factors that will influence the growth of the Al grains are enhanced diffusion of the species due to the low energy bombardment [5] and a driving force for phase separation because of the difference in free energy between Al dissolved in AlN and in the form of pure α-Al [3].

DEPENDENCE OF ALN STRUCTURE ON N_2^+ ION ENERGY

Single phase AlN films are produced when the arrival rate of nitrogen (as N_2^+) is equal to or exceeds the arrival rate of aluminum [2]. These films are electrically insulating and transparent. We have studied the effect of the N_2^+ ion energy (100-500 eV) on the microstructure and crystallographic orientation, using an arrival rate ratio N/Al = 1.5 for all energies. Fig. 4 shows the grain size measured in the plane of the film as a function of the ion beam energy. The grain size increases as the energy increases and the microstructure changes from predominantly of a fiber type at 100 eV to equiaxed at 500 eV. Fig. 5 shows the microstructure and crystallographic structure of an AlN film deposited with an ion energy of 400 eV. The microstructure in films prepared with high ion ener-

FIG. 5

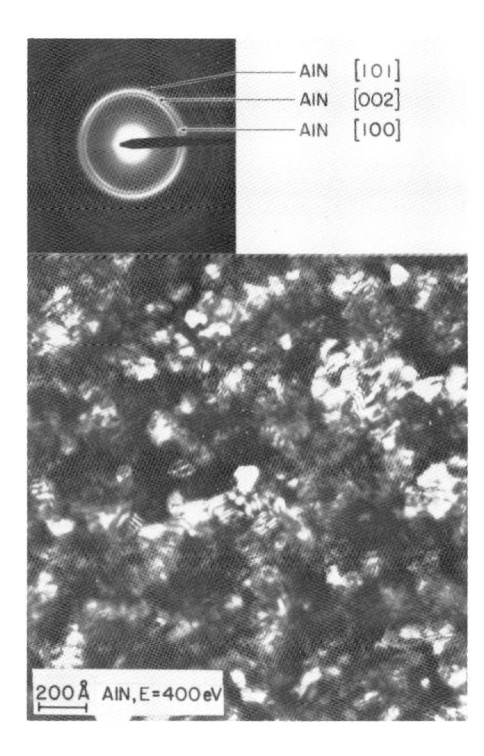

TEM micrograph and electron diffractogram for single phase AlN ($E_{N_2}+ = 400 eV$).

gies (400-500 eV) consists of equiaxed and overlapping grains as compared to the microstructure in films deposited at low ion energies (100-200eV) which are more of a fiber type.

The measured relative peak intensities from electron diffractograms of AlN (100) and (002) as a function of ion beam energy are shown in Fig. 6. Also shown is the relative peak intensity in a sample with random orientation [4]. In films prepared with ion beam energies of 100 and 200 eV the relative intensity of AlN (002) is very weak indicating that the film has grown with the c-axis perpendicular to the film surface. The same was observed for AlN in Al + AlN cermet films prepared with 500 eV N_2^+. Therefore, higher ion energy but low flux produces the same preferred growth orientation as high flux and low energy. Increasing the N_2^+ ion energy in forming AlN causes a change in texture to random for E_{N_2} of 300 eV and with the c-axis in the plane of the film for E_{N_2} of 400eV. We may therefore conclude that an increase in the ion energy from 100 to 500 eV causes a change from small fiber grains with the c-axis along the fiber direction to equiaxed grains with the c-axis more in the plane.

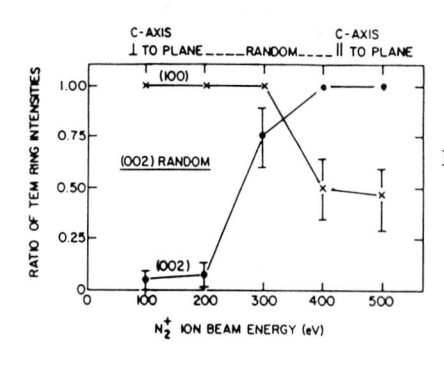

FIG. 6 Ratio of TEM ring intensities for AlN (100) and (002) diffraction rings as a function of N_2^+ ion beam energy.

ACKNOWLEDGMENTS

We thank P. Saunders for the RBS measurements. One of us (H.H.) thanks K.N. Tu of IBM Yorktown Heights for making it possible to participate in this work.

*Permanent address: Department of Physics and Measurement Technology
Linköping University of Technology, S-58183
Linköping, Sweden

REFERENCES

1. J.M.E. Harper, J.J. Cuomo and H.T.G. Hentzell, Appl. Phys. Lett., 43, 547 (1983).

2. J.M.E. Harper, H.T.G. Hentzell and J.J. Cuomo, to be submitted to J. Appl. Phys.

3. H.T.G. Hentzell, J.M.E. Harper and J.J. Cuomo, to be submitted to J. Appl. Phys.

4. "Inorganic Index to the Powder Diffraction File" Joint Committee on Powder Diffraction Standards, Pennsylvania (1972).

5. A.H. Eltoukhy and J.E. Greene, J. Appl. Phys. 51, 8 (1980).

SURFACE MODIFICATION BY ION BEAM ENHANCED DEPOSITION

R. A. KANT AND B. D. SARTWELL
U.S. Naval Research Laboratory, Washington, D.C., 20375

ABSTRACT

Copper films given multiple sequences of Ta implantation and Cu depositions were analyzed using electron microscopy, backscattering, and Auger spectroscopy. Ta retention is 92% following direct implantation, and 100% retention was achieved for the same Ta dose if sputtered Cu is replaced during implantation. Lateral migration of Ta and microroughness were observed for all cases studied. Evidence for TaC formation is presented.

INTRODUCTION

The use of concurrent ion irradiation and vapor deposition to achieve enhanced thin film properties and to improve coating adherence has, for the most part, been studied using ion beams with energies less than 5 keV [1-3]. The reasons for using low-energy ions as opposed to beams of moderate-energy ions that are obtained from ion implantation systems include (1) energy deposition will be very near the surface, (2) low energy ion sources are readily available and reasonably inexpensive and (3) sputtering is reduced at low energies. If, however, the negative aspects of increased sputtering could be overcome, then there would be several advantages to using higher energy ion beams available from an implanter. Principle among these are that the ion beam can consist of any element and is not limited to gaseous species as is frequently the case for low energy ion guns. Because of the increased energy available, the dose required to produce the same energy deposition is reduced. In addition, the combined ion implantation and deposition technique offers a new approach for the study of sputtering phenomena. Ordinarily the measurement of a sputtering rate is complicated by an increase in concentration of the implanted element at the surface. If, however, the target material is continuously being replenished by vapor deposition during the sputtering process, the surface fraction of the implanted element can be kept arbitrarily low. Another interesting potential application of the combined technique is to use vapor deposition to precisely offset sputtering. If this can be done, then the maximum concentration of an implanted specie should no longer be fixed by the "sputtering limit" and the fraction of the implanted dose that is retained could be increased significantly. In the process, a completely new range of material compositions would become accessable via ion implantation.

To establish the feasibility of combining vapor deposition techniques and moderate energy ion implantation, a long range systematic study has been initiated. The preliminary results of that study are reported here. An initial goal is to determine the extent to which conventional models for implanted systems are applicable to ion beam enhanced deposition (IBED) systems. For this initial phase of the study, copper was implanted with tantalum while copper was being deposited to replace the sputtered atoms. This system was selected because the sputtering rate for Ta on Cu is high, and because Ta is insoluble in and forms no compounds with Cu.

EXPERIMENTAL APPARATUS AND TECHNIQUES

For these treatments, the sample is positioned in the center of a chamber 30 cm above an electron beam evaporator and can be rotated about a horizontal axis in order to expose it to either the evaporant from below or the ion beam from the side. The samples are affixed to a cylinder of circular or square cross section or in some cases are themselves cylinders. In order to maintain a high degree of control over the relative arrival rates of the evaporant atoms and the beam ions, the treatment process is broken into two completely isolated steps. First a small fraction of the total intended dose is implanted. Then the sample is rotated into the evaporant stream where a correspondingly thin coating of the desired element is deposited. This process is then repeated until the desired film thickness or ion dose is achieved. If small variations in the relative arrival rates can be ignored, or if an external means of controlling the two fluxes is provided, then the sample is simply rotated continuously through the two beams. The chamber is equipped with two variable apertures which will eventually be used to adjust the relative arrival rates of ions and evaporant atoms, but for this work, these apertures were used to mask different portions of the sample from each beam. The evaporator is diffusion pumped and is isolated from the sample by a set of differential pumping apertures. A cryogenic pump is mounted directly opposite the sample which is itself surrounded by a cold wall held at 77 K.

In order to evaluate the extent to which the effects of the sputtering of copper by tantalum can be offset by concurrent evaporation of copper, three types of samples were prepared. For all three cases investigated, the substrate was a 12 mm diameter aluminum disk on which 63 nm of Cu had been deposited. The first treatment (case I) consisted of normal incidence implantation of 50 keV Ta to a fluence that was believed to be sufficient to sputter 48 nm of Cu or four times the range of the Ta ions. Assuming a sputtering rate of 18 atoms/ion for 50 keV Ta on Cu, this required a fluence of 2.26×10^{16} ions/cm^2 to remove the necessary amount of Cu. For the second treatment (case II), the same Ta fluence was implanted, but this time in 12 equal increments. Between each increment, 4 nm of Cu were deposited, which was approximately the thickness required to replace the Cu sputtered during each implant increment, assuming a sputtering yield of 18 atoms/ion. For case III, the sample received a total of 48 increments, each identical to those used in case II. Thus the total fluence of Ta for this sample was 9.04×10^{16} ions/cm^2 and the total equivalent thickness of deposited copper during the implantation was 188 nm (47 increments).

These samples were evaluated using Rutherford backscattering (RBS), scanning electron microscopy (SEM), and Auger electron spectroscopy (AES) combined with low-energy argon ion bombardment. For the RBS measurements, 2 MeV alpha particles were normally incident on the samples, with the backscattered ions being detected at an angle of 135 degrees. Scanning electron microscopy was performed in an ISI model SX30 equipped with energy dispersive X-ray analysis and a Coates and Welter Cwicscan 106A. Auger electron spectra were obtained using a single-pass cylindrical mirror analyzer, with an incident electron beam energy of 3 keV and current of 0.05 mA. Depth profiles of the films were obtained by sequential Auger analysis and sputtering with 3 keV argon ions incident at an angle of 60 degrees with respect to the sample normal.

RESULTS AND DISCUSSION

Analysis of the RBS spectrum of the direct Ta-implanted Cu film (case I) (Fig. 1A) indicated that the average thickness of copper removed was 43 nm (nearly the equivalent of four times the calculated range of the implanted Ta) and that the average sputtering yield was 16.4. It is, therefore, remarkable that the RBS spectrum also indicated that approximately 90% of the

527

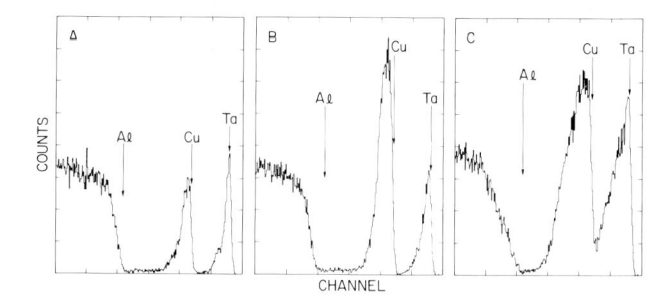

FIG. 1. Rutherford backscattering spectra of Cu films on Al following A) 2.26 x 10^{16} Ta/cm^2; B) same Ta dose delivered 12 increments separated by 11 Cu depositions of 4 nm each; and C) 9 x 10^{16} Ta/cm^2 in 48 increments separated by 47 Cu depositions.

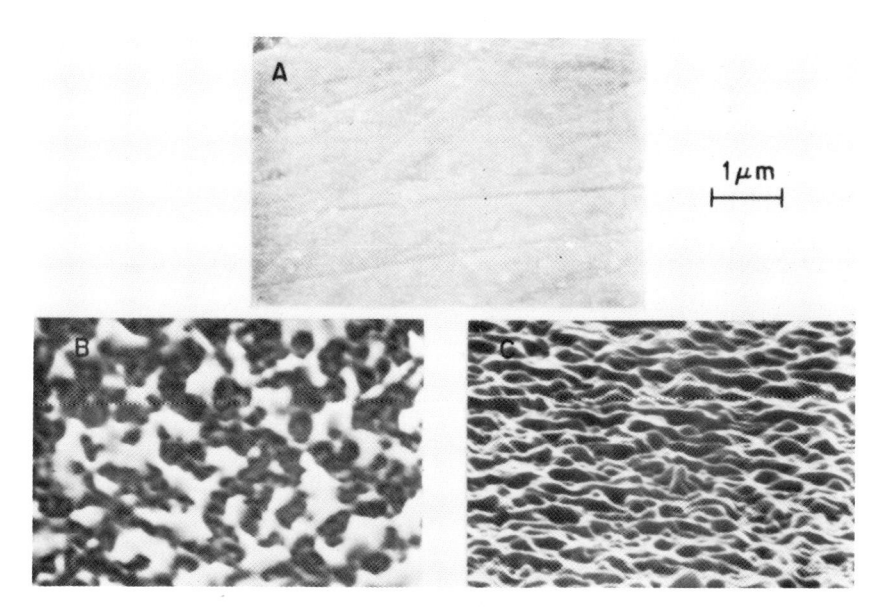

FIG. 2. Scanning electron micrographs: A) unimplanted Cu film; B) following 48 Ta implantations with Cu deposited; and C) same as B but viewed at 79 degrees to surface normal.

implanted Ta had been retained. The RBS spectrum for the second case, 4 nm of Cu deposited between each of the 12 incremental doses of Ta (Fig. 1B), showed that essentially 100% of the implanted Ta had been retained. For this case, nearly exact replacement of the sputtered Cu had been achieved and the apparent sputtering rate was 18.3. It should be noted that these determinations of the amount of Ta and Cu present are uncertain to 10 to 15%; thus the difference between the amounts of retained Ta for the two cases should not be viewed as significant. The amount of Cu on this second type of sample was essentially identical, within experimental uncertainty, to the amount present before implantation. For the third case (48 increments of Ta implantation plus 47 evaporations of 4 nm of Cu), the RBS revealed that the system no longer was retaining all of the implanted Ta and that the Cu depositions were increasing the film thickness more rapidly than the Cu was being sputtered away. The sputtering yield for this case was 18.3, 16% of Cu deposited during implantation remains, and retained Ta is 84% of that implanted.

Scanning electron microscopy of these samples (Fig. 2) revealed that considerable roughening of the surface had taken place in all three cases. Fig. 2C shows the surface for case III viewed at an angle of 79 degrees with respect to the surface normal. Energy dispersive X-ray analysis was used to examine the light and dark areas shown in Fig. 2B. The ratios of the X-ray yields in the light areas to that in the dark areas for Ta, Cu and Al are 1.6:1, 3.5:1 and .5:1 respectively.

The relative peak heights of the derivative Auger spectra plotted as a function of sputtering time for the three samples are shown in Fig. 3. An Auger depth profile of the unimplanted 63 nm thick copper film (not shown) indicated that it required approximately 20 minutes of sputtering to remove the film and observe the Al KLL peak. Small amounts of oxygen and carbon (less than 5 at.%) were incorporated into the film. For each of the profiles shown, it is assumed that approximately 2-3 minutes of sputtering was required to remove the surface contamination produced when the samples were exposed to the atmosphere. Thus, the 2-3 minute Auger peak heights should be considered to represent the "surface" as it existed before being removed from the IBED vacuum chamber.

When viewed collectively, the above results suggest the following interpretation. The RBS data indicate that there is Al very close to the surface. The gradual slope of the near surface Al signal in the RBS spectra (particularly for case III) suggests that the thickness of the film is not constant, a result that is certainly supported by the SEM pictures. These results, together with the X-ray data, indicate that the ridges that have developed are rich in Ta and have resisted sputtering while the valleys between the ridges have been preferentially sputtered, thereby exposing the Al substrate. The nonuniformity of the Ta indicates that significant lateral migration has occurred. In addition, the AES data indicate that sigificant Cu concentration occurs at the surface. This supports the view that the Ta has segregated on the surface leaving exposed regions of Cu. Analysis of the C KLL Auger line shape indicates that the C was present as a carbide. Applying sensitivity factors to the C and Ta Auger peak heights indicated that there was approximately a 1:1 correspondence between the concentrations of those two elements throughout the implanted region. This supports the contention that a carburization process has resulted in the formation of TaC. The amount of sputtering that has taken place during the Ta implantation is well illustrated in Fig. 3A. Ta is still observed after all the Cu had been removed by sputtering or during AES analysis, indicating that some Ta was implanted into the Al substrate. Also, the fact that the ratio of the Al Auger peak at the "surface" (after 3 minutes of sputtering) to that obtained at saturation is 0.3 indicates that 30% of the aluminum substrate was exposed during the implantation. This uncovering of Al was reduced substantially for case II sample but reappears for case III.

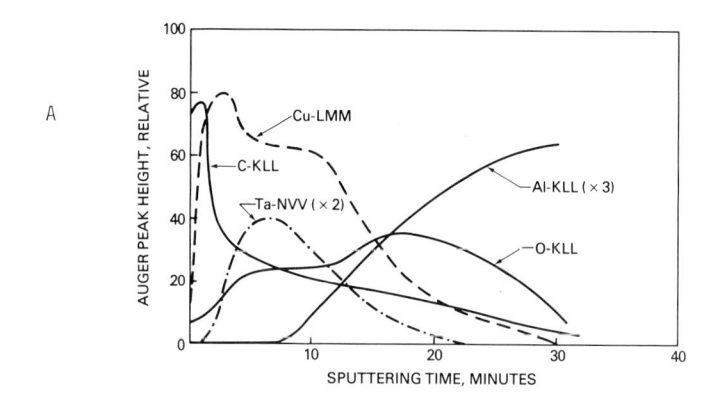

FIG. 3. Auger electron peak heights for Ta, Cu, C, and O versus Ar sputtering time. A) Ta implantation only; B) and C) sequential deposition and implantation of 2.26×10^{16} and 9×10^{16} Ta/cm^2 respectively.

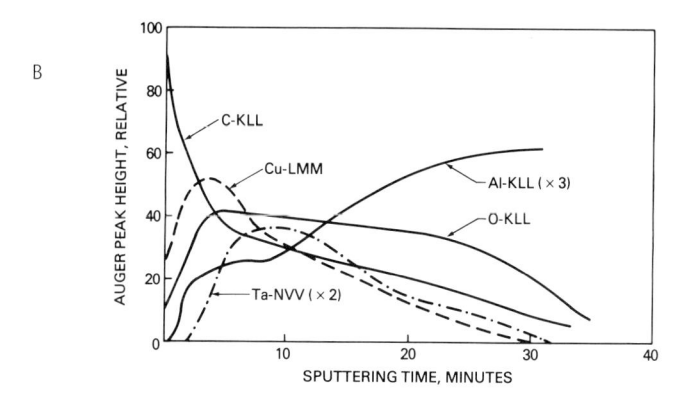

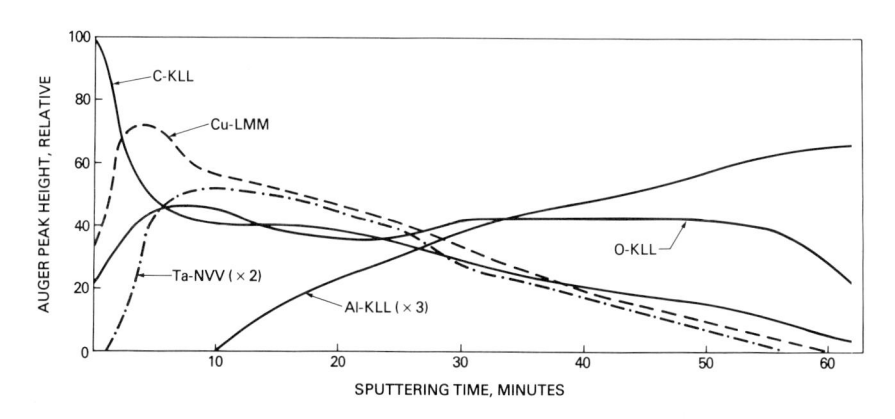

CONCLUSIONS

The interpretations presented above lead to the following conclusions. During the early stages of the Ta implantation, two mechanisms may be contributing to development of the roughened topography which subsequently dominates the evolution of the system. First, there is the usual development of hills and valleys due to sputtering. This development is enhanced by the segregation of the exposed Ta which forms TaC layers over the hills which further reduces the sputtering rate in these regions. It is apparent that the Cu deposited between Ta implantations has had an influence on the total retained Ta, but this added Cu has not significantly altered the development of surface roughness. This result has significant implications for the use of the combined technique for generating thick layers. Since the development of micro-roughness has not been substantially mediated by the deposition of additional Cu during implantation, it may be necessary to avoid applying the technique to systems with high sputtering rates. At the very least, such roughness severely complicates the interpretation of various surface analytical techniques such as AES. For example, without topographical information, one might conclude from the Auger depth profiles that intermixing between the Cu and Al had occurred. The results of these experiments also demonstrate the potential value of the IBED technique as a tool for the investigation of sputtering processes. Note, for example, that small changes in the sputtering rates become magnified if deposition is employed during implantation. This is illustrated by comparing the average thicknesses of the Cu film following 12 increments of implanted Ta with that following 48 such implants. A change in the sputtering rate of less than ten percent has lead to a readily detectable accumulation of several hundred angstroms of copper over the course of the 48 Ta implants.

ACKNOWLEDGEMENTS

We would like to thank Dr. A. R. Knudson for generating the RBS spectra and R. A. Walker for his skillful operation of the ion implanter.

REFERENCES

1. C. Weissmantel, Thin Solid Films 58, 101-105 (1979).

2. L. Pranevicious, Thin Solid Films 63, 77-85 (1979).

3. J. J. Cuomo, J. M. E. Harper, C. R. Guarnieri, D. S. Yee, L. J. Attanasio, J. Angllello, and C. T. Wu, J. Vac. Sci. Technol., 20(3), 349-354 (1982).

MASKLESS PATTERNING OF Cr FILMS USING FOCUSED ION BEAMS

K. GAMO, K. MORIIZUMI, T. MATSUI AND S. NAMBA
Faculty of Engineering Science, Osaka University,
Toyonaka, Osaka, Japan

ABSTRACT

Characteristics of maskless patterning of Cr films using focused Sb^+ ion implantation have been investigated. Dose and depth dependence of the etching rate of Sb-implanted layers during plasma etching using CCl_4 were measured. Sb profiles were also measured by Rutherford backscattering techniques. It was found that a sharp threshold dose exists to form an etch-resistant layer by Sb implantation. It was also found that a latent image of an Sb implanted pattern at a dose $\geq 3.8 \times 10^{15}/cm^2$ was developed by the plasma etching, and that Cr patterns with a thickness of a few hundred nanometers were formed by the present maskless patterning technique.

INTRODUCTION

Focused ion beam techniques have attracted much interest in applications for various maskless processes such as scanning ion beam lithography, maskless etching and maskless ion implantation. For maskless etching or patterning, use of chemical effects induced by high energy ion beams is crucial because simple physical sputter etching provides very low etching rates. For example, a dose of 10^{18}-$10^{19}/cm^2$ is required to etch a few hundred nanometers.

Ion beam induces various chemical effects through kinetic energy or doping[1]. These effects are applied as radiation enhanced etching and ion beam modification techniques. For radiation enhanced etching, a dose of 10^{14}-$10^{15}/cm^2$ is enough, which is a significant reduction compared to a physical sputter etching. This technique is useful for various materials and has been applied to various materials such as Si, SiO_2, [2,3], garnet[4] or $LiNbO_3$[5].

For patterning of metal films, however, use of radiation enhanced etching is difficult because metals are generally highly radiation-resistant and amorphous layers are rarely formed. In this case, use of doping effects is necessary. For example, it was found that Cr films implanted with Sb or As at a dose of $\leq 10^{16}/cm^2$ become resistant to a plasma etching using CCl_4 gas[6]. For Al films, it was observed that ion implanted regions become less soluble in acids and patterning is possible[7]. Recently it was reported that Zn doped Al films, at a concentration of a few percent, by laser photochemical doping become soluble in acetic acid[8].

These results can be applied for maskless patterning using focused ion beams. Implanted latent patterns are developed by a proper etching process after maskless implantation.

In the present paper, focused ion beam techniques were combined with ion beam modification techniques and characteristics of maskless patterning of Cr films were investigated.

EXPERIMENTAL PROCEDURES

Fig. 1 shows a maskless pattern fabrication process of Cr films using focused ion beams. To modify the chemical properties of the film surface,

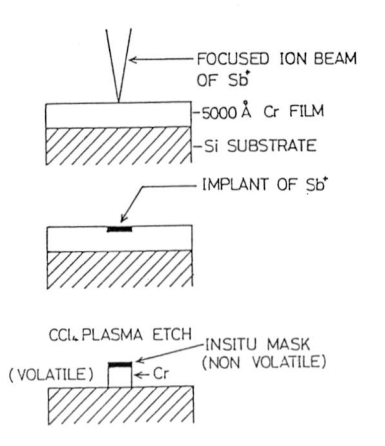

Fig. 1 Maskless pattern fabrication process of Cr films by Sb implantation and plasma etching.

50keV, about 10μA, focused Sb+ ions were implanted in Cr films at a dose ranging from 1×10^{14} to $1\times10^{16}/cm^2$ and at room temperature. The focused Sb beam was obtained using Sb-Pb-Au liquid metal alloy ion sources and a mass-separated focused ion beam system. Their characteristics are reported elsewhere[9,10]. To obtain a necessary dose, the focused Sb beam was singly or multiply scanned. The Sb dose and the profile were measured by Rutherford backscattering techniques using 2MeV He ions.

About 500nm thick Cr films were deposited on Si wafers by electron beam evaporation at a vacuum pressure of about 2×10^{-6} Torr.

The implanted latent image was developed by plasma etching using CCl_4 gas. The plasma etching was done using a barrel type plasma etching system at an rf power of 80w and at a pressure of 0.6 Torr.

RESULTS AND DISCUSSION

Fig.2 shows the etching rate for the plasma etching of Sb+ implanted Cr films as a function of dose (a) and depth (b.) In these Figures, the etching rate is normalized to the rate observed for unimplanted Cr films, and in Fig. 2a, the normalized etching rate corresponds an average rate over a removed depth indicated by a horizontal bar. The present result indicates that a sharp threshold dose exists to form an etch-resistant layer. For example, the etching rate obtained for an implantation at a dose of $3.0\times10^{15}/cm^2$ is reduced by a factor of ≥ 4 from the unimplanted sample which has an etching rate of 4.3 nm/min. Increasing the dose by 30% gives almost a perfect reduction of the etching rate and the etching is suppressed even after etching for 35 min. The results shown in Fig. 2b indicate that the etching rate depends on the depth or the concentration of Sb. After etching a depth of 100 nm, the etching rate becomes almost equal to the rate for unimplanted samples.

The observed dose to form an etch-resistant layer is smaller by a factor of about 3 than the reported value[6]. The reason is not clear but may be due to the difference in the etching process or the film properties.

The concentration profile of implanted Sb measured before and after the plasma etching for a depth of 35 nm is shown in Fig. 3. In this Figure, the observed Rutherford backscattering energy spectra were converted to a

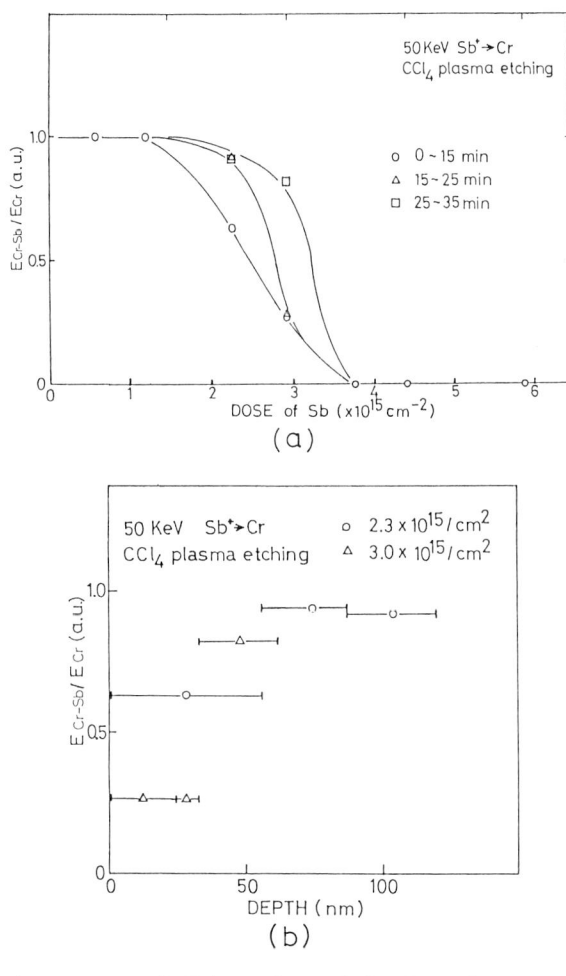

Fig. 2 Etching rate of Sb+ implanted Cr for plasma etching using CCl₄ gas. (a) dose dependence and (b) depth dependence.

depth profile by using a stopping power tabulated by Ziegler and Chu[11]. The observed projected range and range straggling were about 11 and 8.8 nm, respectively. The projected range is in good agreement with a theoretical prediction[12]. However, the projected range straggling is about 2 times larger[12].

By comparing the two spectra shown in Fig. 3, it was confirmed that the total area of the Sb peak or the total number of Sb atoms are unchanged before and after the etching. For another sample, which was implanted with a dose of $1.2 \times 10^{15}/cm^2$, all of the implanted Sb still remains after the plasma etching for a depth of 48 nm. These results suggest that Sb atoms are not removed by the plasma etching. From the backscattering spectrum for a sample after the etcing as shown in Fig. 3, it is clear that the Sb contained in the etched-off layer piles up at the Cr film surface.

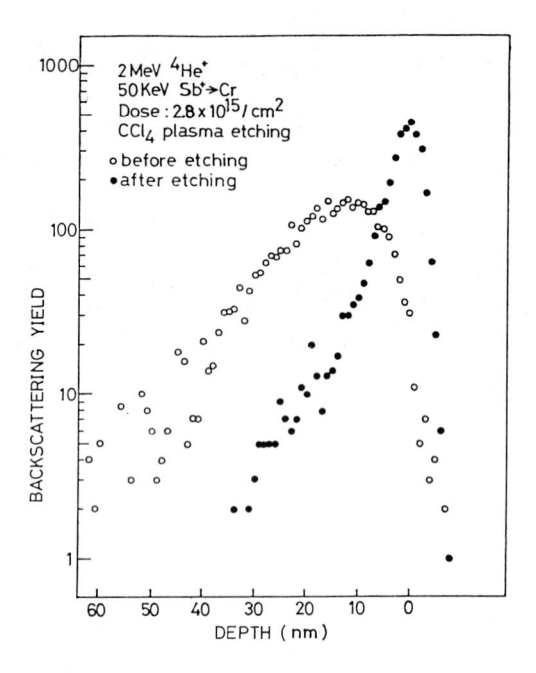

Fig. 3 Concentration profile of 50keV Sb+ implanted in Cr films observed before and after the plasma etching for a depth of 35nm.

From the concentration profile observed before the etching as shown in Fig. 3, the concentration at a depth of 33 nm is estimated to 20 % of the maximum concentration and corresponds to the maximum concentration for samples implanted at a dose of $0.56 \times 10^{15}/cm^2$. Therefore, if the etching rate is proportional to the surface concentration, samples implanted at a dose of $2.8 \times 10^{15}/cm^2$ and etched for a depth of 33 nm should show the same etching rate observed for the initial surface layer for a sample implanted at a dose of $0.56 \times 10^{15}/cm^2$. However the observed etching rate for the former and the latter samples was 0.27 and 1.0, respectively. Similar differences were also observed for a sample implanted at a dose of $2.3 \times 10^{15}/cm^2$. Therefore, these results suggest that Sb piled up at the surface partially contributes to a reduction of the etching rate. The fraction of the reaction-inhibiting Sb decreases with increasing depth and the piled-up Sb is ineffective after etching to a depth larger than 100nm.

SEM photographs of line patterns formed by the maskless patterning is shown in Fig. 4a and b. Fig. 4b indicates lines B and C viewed at larger magnification to show the surface smoothness after the plasma etching. Samples were implanted at various line doses by a single (for line E) or multiple (for lines A-E) scanning of the focused Sb beam. Therefore, the dose across the line is not uniform but may be similar to a gaussian profile. On average, a line dose of $6 \times 10^{11}/cm$ roughly corresponds to a dose of $3 \times 10^{15}/cm^2$. It was observed that the film thickness and the surface smoothness at the implanted region is unchanged after the plasma etching for Sb implantation at a dose more than $6 \times 10^{11}/cm$. The line E is composed of two lines. These are due to unresolved $^{121}Sb^+$ and $^{123}Sb^+$ beams.

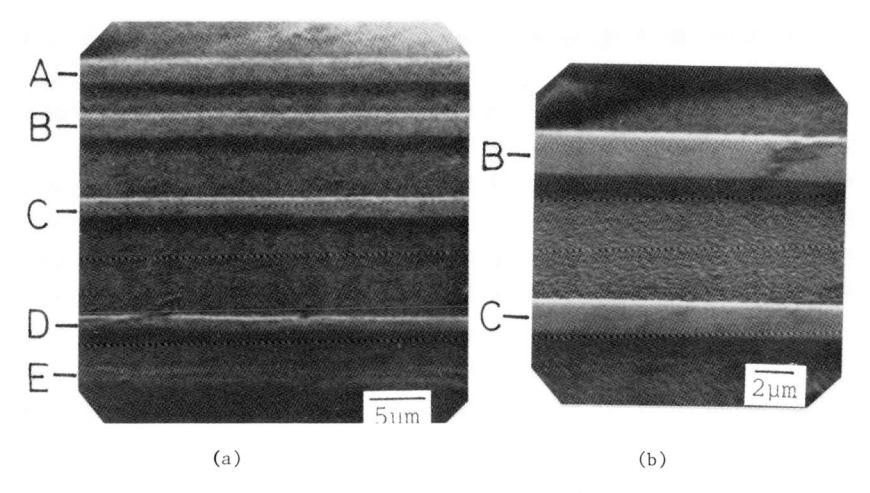

Fig. 4. SEM photograph of Cr line patterns formed by the maskless patterning process.

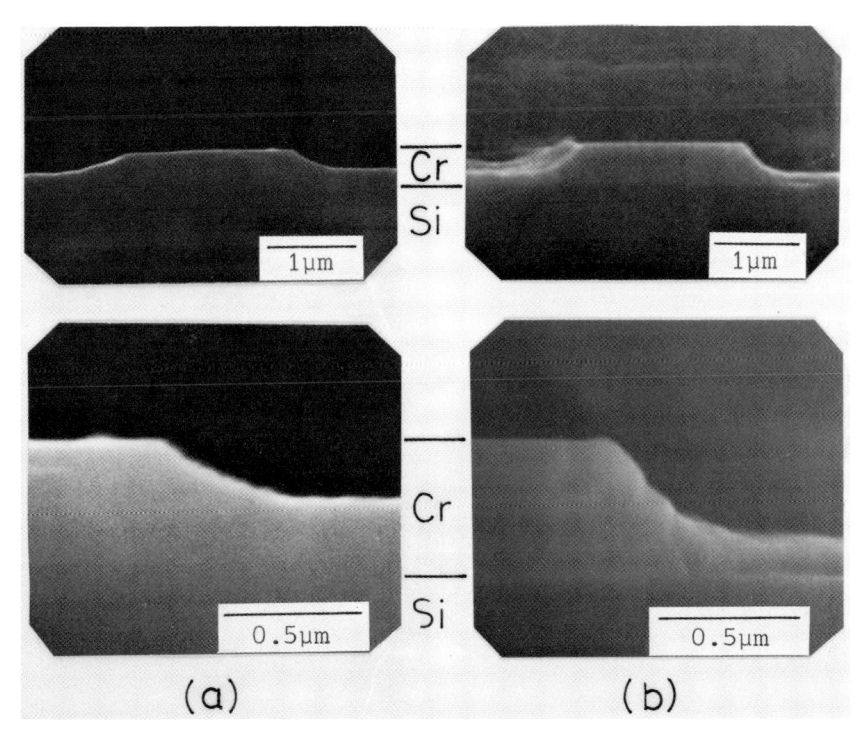

Fig. 5. Cross-sectional view of Cr line pattern. Plasma etching was done for 90 min (a) and 140 min (b).

It was observed that the line width became wider for lines formed at higher dose. This is because the etch-resistant region becomes wider for higher dose implants. From the result shown in Fig. 2, it can be said that the dose at the line edge corresponds to about $3.8 \times 10^{15}/cm^2$. This value is in reasonable agreement with a value estimated from a line dose.

Fig. 5 shows a cross-sectional view of the Cr line B shown in Fig. 4. The figures at the bottom show edge profiles viewed at a larger magnification. The plasma etching was done for 90 min in Fig. 4a and 140 min in Fig. 4b. It was observed that the line width and the film thickness was unchanged and the slope of the side wall became steeper after a longer time of the etching.

The observed profile can be explained by the dose dependence of the etching rate shown in Fig. 2. The line width and the film thickness should be unchanged because a sharp threshold dose exists to form an etch-resistant layer. The plasma etching is normally isotropic and induce a large undercut. However the observed profiles shows no undercut. This may be partially due to the lateral distribution profile of Sb and partially due to Sb atoms piled up at the etched surface.

Summary

Maskless patterning characteristics of Cr films using Sb implantation and a plasma etching were investigated. It was found that a few hundred nanometer thick patterns in Cr are formed by the present maskless process. The present technique makes use of an ion beam modification of material surface chemical properties. Similar modification effects have been found for other materials such as Al and, therefore, the present technique may be applicable for other materials.

REFERENCES

1. G. K. Wolf, Treatise on Materials Science and Technology Vol. 18 ed. J. K. Hirvonen (Academic Press, New York, 1980) pp. 373-414
2. J. F. Gibbons, F. O. Hechfl and T. Tsurushima, Appl. Phys. Lett. 15, 117 (1969)
3. K. Moriwaki, K. Masuda, H. Aritome and S. Namba, Jpn. J. Appl. Phys. 19, 491 (1980)
4. T. Tsurushima, H. Tanoue and H. Funabashi, Inst. Electron. Commu. Eng. 527 (1976)
5. M. Kawabe, M. Kubota, K. Masuda and S. Namba, J. Vac. Sci. Technol. 15, 1096 (1978).
6. T. Yamazaki, Y. Suzuki and N. Nakata: J. Vac. Sci. Technol. 17, 1348 (1980)
7. J. I. Pankove and C. P. Wu, Proc. IEDM 1980 (Dec. 8-10, Wahington D. C.)
8. D. J. Ehrlich, R. M. Osgood, Jr and T. E. Deutsch, Appl. Phys. Lett. 38, 399 (1981)
9. K. Gamo, T. Ukegawa, Y. Inomoto, Y. Ochiai and S. Namba, J. Vac. Sci. Technol. 19, 1182 (1981)
10. K. Gamo, Y. Inomoto, Y. Ochiai and S. Namba, Jpn. J. Appl. Phys. 21 Suppl. 21-1, pp.415 (1981)
11. W. K. Chu, J. W. Mayer and M-A Nicolet, Backscattering spectroscopy (Academic Press, New York, 1978)
12. K. B. Winterbon, Ion Implantation Range and Energy Deposition Distributions Vol. 2 (IFI/Plenum, New York, 1957)

CARRIER LIFETIME REDUCTION BY ION IMPLANTATION INTO SILICON

A. MOGRO-CAMPERO AND R.P. LOVE
General Electric Research and Development Center, Schenectady, NY 12301

ABSTRACT

Ion implantation is emerging as a versatile tool for the control of carrier lifetime in silicon. We present results on the use of argon and proton implantations to reduce lifetime. Particle energies used imply submicron penetration depths into the silicon. We find that within the first micrometer from the surface, the generation lifetime can be reduced by several orders of magnitude, and varies inversely with dose. A change in doping concentration was found only at the highest dose used (10^{14} cm^{-2}). Effects of the implantations on surface generation velocity and oxide charges were also studied.

INTRODUCTION

Ion implantation is used extensively in silicon device processing. The most common application at present is probably the introduction of dopants into silicon, but other uses of ion beams are also practiced or are being developed [1]. Ion implantation can be used indirectly to increase carrier lifetime by gettering impurities. In this case, impurities which decrease the lifetime are removed from the region of interest and gathered at a damaged portion of the silicon wafer. This damaged region can be created by ion implantation [1]. This paper is not directly concerned with this effect of implantation.

Our study involves the use of ion implantation to decrease carrier lifetime in silicon; we present here some results on the effects of proton and argon ion beams. At low carrier injection levels the dominant mechanism for carrier generation and recombination is phonon-assisted events via deep level defects [2]. These defects can be crystalline imperfections, impurities (e.g., gold atoms), and/or a combination of both such as the oxygen-vacancy complex. In the case of ion implantation all of these can be important, since crystalline imperfections result from ion implant damage, and the implanted species itself may act as a deep level impurity.

The techniques in common use to reduce carrier lifetime in silicon (the diffusion of gold or platinum, and electron irradiation) result in a reduction of lifetime throughout the wafer. On the other hand, due to the well defined range of ions, ion implantation makes possible the spatial localization of regions of reduced lifetime. Masking can be used effectively to provide planar selectivity. Depth control can be achieved by varying the energy of the incident ion or by burying the implanted region by epitaxial overgrowth. The effect of the ion implantation on other material properties is also of concern. The thermal stability of the ion implantation effects is important depending on thermal requirements of subsequent processing steps or device operating temperatures.

EXPERIMENTAL PROCEDURES

Silicon wafers of orientation (111) and doping density of 10^{15} phosphorus atoms cm^{-3} were used. The backs of the wafers were implanted with 10^{15} cm^{-2} phosphorus ions of 100 keV to provide an Ohmic contact to the backside aluminum. Argon implantation at 200 keV (203 nm range in silicon [3]) was performed prior to 60 nm oxide growth at 1000°C. A set of aluminum dots was then evaporated on top to form MOS capacitors. Backside aluminum was evaporated after oxide removal. In the case of samples implanted with protons, the implantation was performed as the last step, i.e., the 80 keV protons traversed through 0.3 μm of aluminum and 0.1 μm of SiO$_2$, resulting in a penetration of 0.5 μm into the silicon [3]. In all cases a portion of the sample was left unimplanted to serve as a control area.

Generation lifetime was measured at room temperature by the Zerbst technique [4,5]. A MOS capacitor is pulsed into deep depletion, and the formation of the inversion layer is observed via the capacitance transient. By this method it is possible to deduce the generation lifetime (τ_g) in the depleted zone (~1μm for the samples in our experiment), and the parameter s due to minority carriers whose current is independent of the time-varying depth of the depletion zone. The parameter s is the sum of the surface generation velocity and a term corresponding to diffusion of carriers into the depletion zone from the quasi-neutral bulk [6]. A typical variation by a factor of 2-3 was found in the values of τ_g and s across a given sample. Capacitance was measured at 1 MHz.

RESULTS

Argon Implantations

Figure 1 shows the generation lifetime as a function of argon dose. These results are consistent with those reported by others using similar

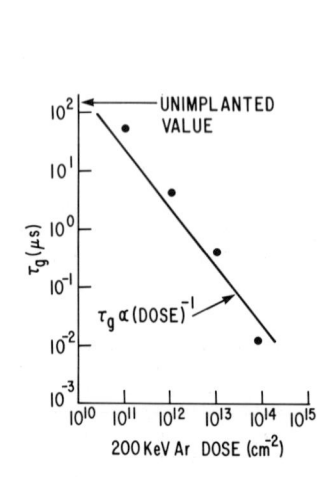

FIG. 1. Generation lifetime vs. argon dose.

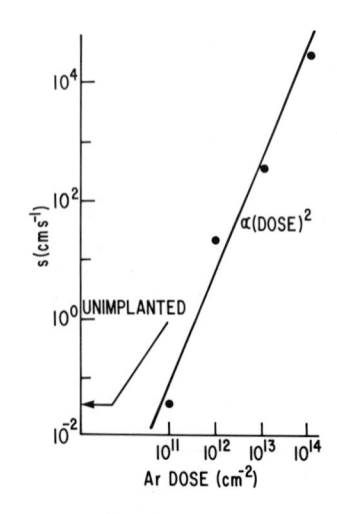

FIG. 2. Parameter s vs. argon dose.

experimental conditions [7]. The inverse dose relationship seems reasonable, since carrier lifetime is inversely proportional to the density of deep levels [2], which are being introduced by the implantation. Figure 2 shows the parameter s from the Zerbst analysis as a function of argon dose. By studying the temperature dependence of the parameters s, we have concluded in a separate study that in the dose regime where the lifetime is determined primarily by the implantation (argon or protons), the dominant term in the parameter s is the near-surface generation velocity (i.e., the diffusion component is negligible at room temperature). The surface generation velocity is proportional to the surface density of deep levels [2], and these would increase as a result of surface damage related to the implantation. The observed dependence of s (proportional to the square of the dose, see Fig. 2), could be due to a combination of two irradiation-related defects which have become paired, possibly as a result of the post-irradiation heat treatment at $1000°C$ to grow the oxide. Capacitance-voltage measurements were used to determine the total charge density associated with this oxide. The measured charge density ($10^{11} cm^{-2}$) was found to be independent of argon dose.

C-V measurements were also used to measure doping density within the depleted zone by the maximum-minimum method [8]. Doping density ($10^{15} cm^{-3}$) was found to be unaltered within the implanted regions, except for the highest dose used ($10^{14} cm^{-2}$), where it was found that doping density increased by a factor of 5. Within a standard deviation from the peak of the implanted argon, its average density for a dose of $10^{13} cm^{-2}$ is $4 \times 10^{17} cm^{-3}$, far in excess of the shallow dopant concentration of $10^{15} cm^{-3}$. Therefore, the effective electrically active fraction of the implanted argon atoms seems to be low. Donor generation by implantation of argon ions at 25 keV has recently been reported [9]. A change in doping density was found at $10^{15} cm^{-2}$, but not at $10^{14} cm^{-2}$. The difference between these results and ours is probably due to the fact that the doping density of their unimplanted wafers ($4 \times 10^{15} cm^{-3}$) was higher than in our case.

Annealing treatments were carried out in nitrogen prior to metallization. Detailed results are presented elsewhere [10]. The lifetime was observed to recover to within a factor of about 100, and the surface generation velocity to within a factor of 1000, after an hour at $1200°C$. Complete recovery was not observed even after 5 hours at $1200°C$. These samples were cooled rapidly (hundreds of degrees per second) in order to postpone recovery. A change in cooling rate to a slow two degrees per minute showed that complete recovery of lifetime and surface generation velocity was observed after a few hours at $1200°C$.

In order to detect implanted argon by secondary ion mass spectroscopy (SIMS), doses much higher than the ones used here are necessary ($10^{15}-10^{16} cm^{-2}$). After such implantations into silicon samples, the SIMS analysis showed unexplained erratic behavior in different runs: the shapes of depth profiles were sometimes inconsistent with tabulated values.

Proton Implantations

Generation lifetime and surface generation velocity as a function of proton dose are shown in Figures 3 and 4. Functional dependence on dose is seen to be close to linear. At $10^{14} cm^{-2}$, analysis by the Zerbst technique was no longer possible, but the duration of the capacitance transient continued to follow an inverse dose dependence as shown by the dotted line extension into the high dose regime. The superlinear dependencies (reverse leakage current of diodes approximately proportional to dose squared) for helium and carbon implantations within the range of doses in our experiment [11] were not observed.

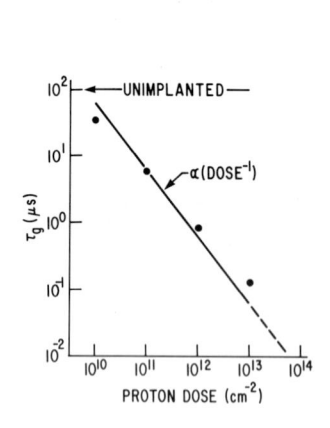

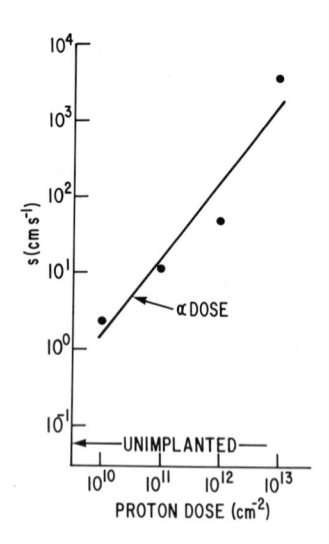

FIG. 3. Generation lifetime vs. proton dose.

FIG. 4. Parameter s vs. proton dose.

The generation lifetime as a function of proton dose has been reported for similar experimental conditions [12]. We find no evidence for the precipitous drop in lifetime (more than three orders of magnitude in the dose range 10^{12}-10^{13} cm^{-2}). In that experiment, the lifetime was also deduced from a capacitance transient analysis, but the surface contribution was not taken into account. It has been shown that for irradiated samples, this procedure can lead to large errors [10].

Capacitance-voltage curves of the samples implanted with protons show stretch-out characteristic of an increase in interface traps [13]. In addition, hysterisis appears in these curves of the type which also indicates the existence of carrier trapping in the oxide near the silicon [14]. From the flatband voltage shift of the C-V curves, one can compute the total charge associated with the oxide [13]. This charge increases linearly with the logarithm of the proton dose (from 1.5 x 10^{11} cm^{-2} to 4.5 x 10^{11} cm^{-2} for doses between 10^{10} cm^{-2} and 10^{14} cm^{-2}). The resulting shift in threshold voltage for inversion is 2 volts for a proton dose of 10^{14} cm^{-2}. Proton dose and range were checked by SIMS analysis on samples implanted with 5 x 10^{15} cm^{-2}.

Using the same method as with the argon samples, an average doping density within the depletion zone was calculated. There was no change in doping density with proton dose, except at the highest dose of 10^{14} cm^{-2}, where an increase in doping by a factor of 3.5 was detected. Doping concentration changes after proton implantation into n-type silicon have been reported; carrier concentration decreases were found after implantation, but increases appeared after annealing at around 300°C [15]. This indicates a competition between the dominance of deep and shallow levels, which should depend on the energy level spectrum within the silicon bandgap. Since various point defect complexes and/or defect-impurity associations are probably responsible for the levels, the details in a particular situation are bound to be different. It is important to note in this respect that we have used Czochralski samples, whereas the other experimenters [15] have worked with float-zoned material. It is well known that there is a large difference in impurity concentrations in these two types of crystals, notably with respect to oxygen concentration, which in Czochralski samples is many orders of magnitude higher (in the 10^{18} cm^{-3} range).

DISCUSSION

In this section we group some comments which compare or combine results obtained by both the argon and proton implantation experiments. In the last paragraph we alluded to various defect combinations which could contribute to levels within the silicon band gap; some of these may form directly as a result of implantation, but the mobility of vacancies even at room temperature [16] gives added opportunities. Proton implantation can give rise to damage clusters, but these are much more abundant for heavy ion implantation such as argon, and the cluster portion of the damage seems to be better related to lifetime degradation [17].

The relative implantation damage in the silicon samples for the conditions of this experiment can be calculated in the usual way [18] to obtain the number of displaced target atoms expressed in the displacements-per-atom (dpa) unit used in radiation damage studies. For the conditions of this experiment, and a dose of 10^{15} cm^{-2} argon, one obtains 1.7 dpa near the silicon surface, rising to a maximum of 2.8 dpa near the end of range. For the same dose of protons, the damage is 1.7×10^{-3} dpa near the silicon surface, increasing to 4×10^{-3} dpa at 0.35 μm into the sample.

A variety of energy levels have been reported after proton [19], and argon [19,20] implantations. These lie throughout the bandgap. Since the samples implanted with argon have been annealed at temperatures of up to 1200°C in this experiment, the energy levels reported after annealing treatments for various noble element implantations [19] are of particular interest. The deep level at 0.5 eV below the conduction band for argon may be a true impurity level rather than one associated with the implantations, since no similar deep level was found in the cases of neon nor xenon. In addition, in a study of the relative effectiveness of argon and silicon implantations into silicon, after a 1000°C heat treatment, it was found that silicon implantation had no effect on the lifetime at doses < 10^{14} cm^{-2} [7], whereas their results as well as ours show lifetime reductions at considerably lower argon doses. Stabilization of defect structures after argon implantation and annealing may also contribute to lifetime degradation; detectable bubbles of 3 nm diameter were observed at a dose of 10^{14} cm^{-2} under similar experimental conditions [7].

CONCLUSIONS

Ion implantation is emerging as a versatile tool for lifetime control in silicon. Carrier lifetime reduction can be varied by several orders of magnitude. Well-defined range and damage profiles allow localization of low lifetime regions within silicon wafers. Resistance to high temperature treatment (as in the case of samples implanted with argon) allow the introduction of regions of low lifetime early in the device processing sequence; the quality of oxide grown after implantation is not compromised as judged from C-V measurements at 1 MHz. In addition, epitaxial overgrowth may be used to bury regions of low lifetime deeper into the silicon. For lifetime tailoring after device completion, protons offer the possibility of least damage near the surface. Some of the effects observed in our experiments on MOS capacitors (charge trapping near the oxide, and threshold voltage shifts) may be removable by low temperature annealing. For both proton and argon implantations, an increase in doping concentration was observed at 10^{14} cm^{-2}, the highest dose used.

ACKNOWLEDGMENTS

We are indebted to P. Cerniglia, G. Gidley, B. Hatch, and M. Lazzeri for carrying out various wafer processing steps; argon and phosphorus implantations were performed by R. Riehl and W. Whitney, and proton implantations at the State University of New York at Albany with Prof. H. Bakhru. We thank P. Chow for making us aware of reference 12, and W. Katz for the SIMS runs.

REFERENCES

1. H. Ryssel, Adv. Electron. and Electron Phys. 58, 191 (1982).
2. S.K. Ghandhi, The Theory and Practice of Microelectronics (John Wiley and Sons, New York 1968).
3. B. Smith, Ion Implantation Range Data for Silicon and Germanium Device Technologies (Research Studies Press, Inc., Forest Grove, Oregon 1977).
4. M. Zerbst, Z. Angew. Phys. 22, 30 (1966).
5. D.K. Schroder and J. Guldberg, Solid-St. Electron. 14, 1285 (1971).
6. D.K. Schroder, IEEE Trans. Electron Devices ED-19, 1018 (1972).
7. H.F. Kappert, G. Sixt, and G.H. Schwuttke, Phys. Stat. Sol. (a) 52, 463 (1979).
8. B.E. Deal, A.S. Grove, E.H. Snow, and C.T. Sah, J. Electrochem. Soc. 112, 308 (1965).
9. G. Greeuw and J.F. Verwey, Solid-St. Electron. 26, 241 (1983).
10. A. Mogro-Campero and R.P. Love in: Defects in Silicon, W.M. Bullis and L.C. Kimerling, eds. (The Electrochemical Society, Inc., Proc. Vol. 83-9, Pennington, NJ 1983) p. 595.
11. K.A. Pickar and J.V. Dalton, Rad. Eff. 6, 89 (1970).
12. Y. Wada and M. Ashikawa, Japan J. Apl. Phys. 14, 1405 (1975).
13. S.M. Sze, Physics of Semiconductor Devices (John Wiley and Sons, Inc., New York, 1981).
14. P.V. Gray, Proc. IEEE 57, 1543 (1969).
15. Y. Ohmura, Y. Zohta, and M. Kanazawa, Phys. Stat. Sol. (a) 15, 93 (1973).
16. J.W. Corbett, Electron Radiation Damage in Semiconductor and Metals, Solid State Physics, Supplement 7 (Academic Press, New York 1966).
17. V.A.J. van Lint and R.E. Leadon in: Lattice Defects in Semiconductors 1974, p. 227, Inst. Phys. Conf. Ser. No. 23 (1975).
18. W.G. Johnston, A. Mogro-Campero, J.L. Walter, and H. Bakhru, Mater. Sci. and Eng. 55, 121 (1982).
19. J.-W. Chen and A.G. Milnes, Ann. Rev. Mater. Sci. 10, 157 (1980).
20. J. Garrido, E. Calleja, and J. Piqueras, Solid-St. Electron. 24, 1121 (1981).

MODIFYING POLYCRYSTALLINE FILMS THROUGH ION CHANNELLING

R. B. IVERSON AND R. REIF
Department of Electrical Engineering and Computer Science,
Massachusetts Institute of Technology,
Cambridge, MA 02139

ABSTRACT

A novel low-temperature process to enhance the grain size of a polycrystalline film on an amorphous substrate has been previously reported. In this process, ion implantation is used to selectively amorphize the film, and undamaged grains act as seed crystals in a subsequent low-temperature anneal. In this work, a 120 nm polycrystalline silicon film was implanted from three angles with phosphorous at $150°K$. The total dose was $1.0 \times 10^{15}/cm^2$. Transmission electron micrographs after a partial anneal ($700°C$ for 30 minutes) indicate that some crystallites survived implantation due to ion channelling in the (111) plane. After a 60 minute anneal at $700°C$, 7 μm grains were observed.

INTRODUCTION

In crystals, the rate of amorphization due to ion implantation decreases as the ion beam becomes parallel to a plane or axis of symmetry of the crystal [1]. By utilizing this ion channelling effect, we hope to produce a large-grain device-quality film on an amorphous substrate. The fabrication process, reported earlier [2,3], has three steps. First, a thin polycrystalline film is deposited. Next, the film is implanted to amorphize all grains except where significant ion channelling occurs. Surviving grains act as seed crystals in the final step, a low-temperature anneal.

Combined with existing fabrication technologies for silicon devices, advantages of this process include:
  (a) the ability to build devices on mesas, eliminating parasitic capacitance associated with the reverse-biased PN junctions commonly used to isolate devices;
  (b) a much cheaper substrate than used in silicon-on-sapphire technology;
  (c) a technology which allows stacking of active layers of silicon on insulating or conducting films, making possible the fabrication of three-dimensional integrated circuits; and
  (d) the ability to fabricate efficient and inexpensive solar cells.

The most successful techniques reported in the literature (laser recrystallization [4], graphoepitaxy [5], strip-heat recrystallization [6]) require melting of the silicon film. In this technique, the recrystallization is achieved by solid-phase epitaxial growth, fundamentally a low-temperature process requiring no melting of the silicon film.

In this paper, the process is reviewed and some theoretical and experimental results are presented and evaluated.

PROCESS OVERVIEW

The fabrication procedure presented here consists of three steps: deposition, implantation, and anneal.

Deposition of Polycrystalline Film

Any method of deposition of the polycrystalline film can be used as long as it allows sufficient control of the size and orientation of crystals in the film. (Grain size and orientation will affect the density and quality of the seed crystals remaining after implantation.)

In this experiment, a 120 nm film of polycrystalline silicon was deposited on thermal SiO_2 using LPCVD. A transmission electron micrograph of the as-deposited film is shown in Fig. 1. Grains are approximately 10 nm in diameter. Samples of as-deposited film were prepared for transmission electron microscopy (TEM) by scoring the surface and placing it in HF until the underlying SiO_2 was removed and the film floated to the surface. The film was then transferred to a TEM grid for imaging.

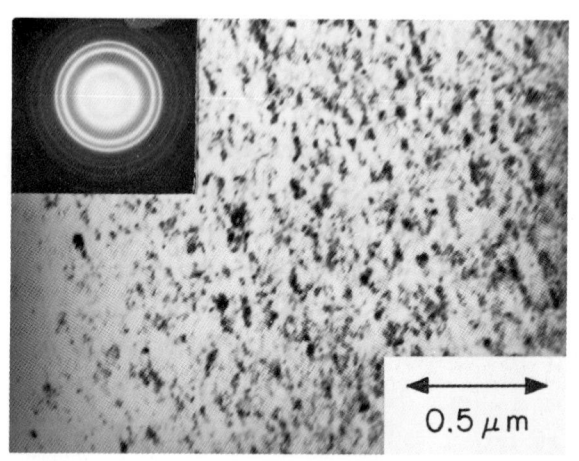

FIG. 1. TEM diffraction pattern and bright-field image for a 120 nm polycrystalline silicon film, as-deposited.

Selective Amorphization through Implantation

The crux of this process is the channelling of implanted ions through "aligned" grains, i.e. crystals with a plane or axis of symmetry parallel to the ion beam. Several implant angles must be used to preserve only grains in a single orientation. For each angle, the dose and energy must be selected so that any unaligned grains in the film are amorphized while aligned grains are not critically damaged. The range of doses for which this occurs is the "dose window". This window is wider at lower temperatures where atomic vibrations are reduced. At low temperatures, less self-annealing occurs so that the minimum dose required to amorphize unaligned grains is lower [7] while, at the same time, effective channel widths are increased [8]. For these reasons, lower implant temperatures are preferred.

In this experiment, three implant angles ($-73.2°$, $0°$, and $+73.2°$) were used. These are the angles between the $\langle\bar{1}21\rangle$, $\langle 110\rangle$, and $\langle 2\bar{1}\bar{1}\rangle$ directions, which are co-planar. These angles were chosen to preserve

(110) grains with equivalent three-dimensional orientations. The implant temperature was 150 K. For the 0° implant, 100 keV phosphorous ions were used at a dose of $5 \times 10^{14}/cm^2$. For the two 73.2° implants, 225 keV phosphorous ions were used, each at a dose of $2.5 \times 10^{14}/cm^2$. The beam current density was 0.2 to 0.4 $\mu A/cm^2$.

Low-Temperature Anneal

The final step is an anneal which should recrystallize the film. This step has been well characterized for amorphous silicon on a single-crystal silicon substrate [9]. Low temperature anneals, less than 650°C, have been used [10].

In this experiment, samples were annealed in a nitrogen ambient at 700°C for 30 minutes and for 60 minutes.

ANALYSIS

The implanted film was imaged in a transmission electron microscope at three stages: as-implanted, after a 30 minute anneal, and after a 60 minute anneal.

The As-Implanted Film

The diffraction pattern (TED) in Fig. 2 suggests that the implanted film is amorphous, but it is not conclusive. In other experiments, where lower doses were used, the as-implanted film appears to be amorphous; however, after a 30 minute anneal, the film could not be distinguished from as-deposited film. From this evidence, even though the film appears amorphous, enough crystalline order can remain to recover the original structure after a short anneal.

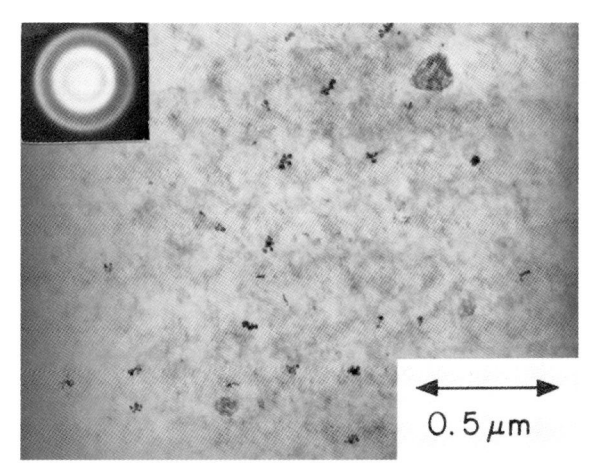

FIG. 2. TEM diffraction pattern and bright-field image for as-implanted silicon film.

The Film After a 30 Minute Anneal

The TED in Fig. 3 indicates a ⟨110⟩ orientation. Figures 4 and 5 show bright and dark-field images at a lower magnification. The dark-field image is from a (111) spot in the diffraction pattern, so bright crystals in this image have mutually-parallel (111) planes within 18° (aperture size) which are perpendicular to the surface of the film within 1°. The chance that three grains in a field of ten randomly oriented grains are oriented within this tolerance is less than 0.1%.

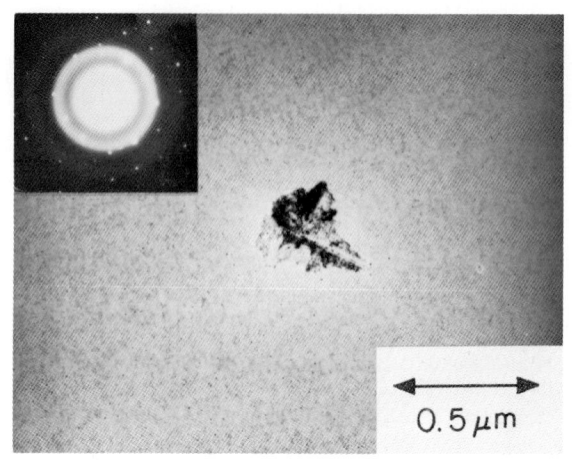

FIG. 3. TEM diffraction pattern and bright-field image for implanted silicon film after a 30 minute anneal at 700°C.

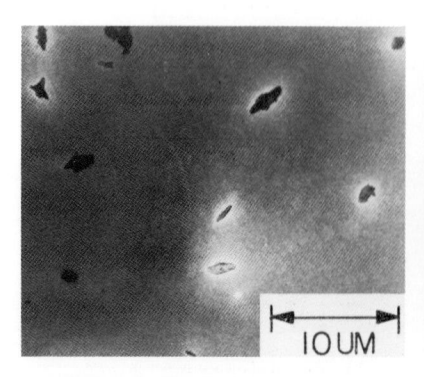

FIG. 4. TEM bright-field image for implanted film after a 30 minute anneal at 700°C.

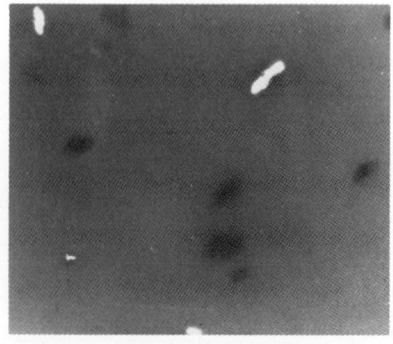

FIG. 5. TEM dark-field image for implanted film after a 30 minute anneal at 700°C. A (111) diffracted beam was used for this image.

Another TED of this sample was taken at a small tilt (a few degrees). In that pattern, three orientations are apparent: ⟨110⟩, ⟨431⟩, and ⟨541⟩. Evidenced from the diffraction pattern, the (1$\bar{1}$1) planes for these orientations were mutually parallel within 3° (including digitizing error) and perpendicular to the surface within 1°. The probability that three grains in a field of ten randomly oriented grains are oriented within this tolerance is around 3×10^{-5}. The correlation between these orientations is more than coincidence. It is strong evidence of planar channelling. In other words, some grains survived implantation because each of the implant directions was in the (111) plane.

The Film After a 60 Minute Anneal

From Fig. 6, channelling-induced orientations are not apparent after a 60 minute anneal. This could be explained by either twinning or spontaneous nucleation. Frequent twinning of the seed crystals can be expected, reducing the effect of aligned seeds on the final orientation. Crystallites which spontaneously nucleate will be randomly oriented, also reducing the effect of the aligned seeds.

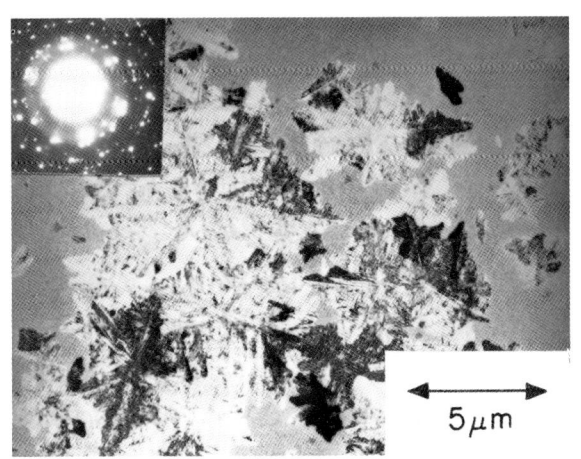

FIG. 6. TEM diffraction pattern and bright-field image for implanted film after a 60 minute anneal at 700°C.

The variety of grain sizes indicate that spontaneous nucleation is occurring, though the extent is not known. Data concerning spontaneous nucleation in silicon [11] indicate, after analysis, that the amount of spontaneous nucleation is not a strong function of anneal temperature. Though both nucleation rate and growth rate are strong functions of temperature, the data available for silicon indicate that changing the anneal temperature basically changes the time scale for each of these two processes by the same amount. From these considerations, spontaneous nucleation appears to impose an upper limit on the grain size that may be achieved using this procedure. Though grain growth in polycrystalline silicon can be greatly enhanced by implanting heavily with phosphorous [12], effects on the nucleation rate are not known. On basis of available data, spontaneous nucleation in 120 nm silicon films limits the grain-size to 8 or 10 microns.

SUMMARY

A novel low-temperature process incorporating ion channelling was used to modify the structure of a 120 nm polycrystalline film. The film originally had 10 nm grains. After processing, grains measured 7 μm. Observations of a partially-recrystallized film indicate planar channelling played a major role in selecting the seed orientations. Spontaneous nucleation and/or twinning of oriented seeds precluded observation of oriented seeds in a more fully recrystallized film.

It was hoped that axial channelling would be the dominant factor in determining the orientations of the seed crystals. Using larger implant doses may prevent planar channelling from occuring. If this does not solve the problem, a double-tilt stage for implantation may be required so that the implant directions are not co-planar.

Effects of twinning and spontaneous nucleation could be reduced by a series of implants and anneals.

This work was supported by the National Science Foundation, Grant No. 83-03450ECS.

REFERENCES

1. R.S. Nelson and D.J. Mazey, J. Mat. Sci. 2, 211-216 (1967).
2. R. Reif and J.E. Knott, Electron. Lett. 17, 586-588 (1981).
3. P. Kwizera and R. Reif, Appl. Phys. Lett. 41, 379-381 (1982).
4. A. Gat, L. Gerzberg, J.F. Gibbons, T.J. Magee, J. Peng, and J.D. Hong, Appl. Phys. Lett. 33, 775-778 (1978).
5. M.W. Geis, D.C. Flanders, and H.I. Smith, Appl. Phys. Lett. 35, 71-74 (1979).
6. M.W. Geis, D.A. Antoniadis, D.J. Silversmith, R.W. Mountain, and H.I. Smith, Appl. Phys. Lett. 37, 454-456 (1980).
7. F.H. Eisen and B. Welch, P. C. Atom. Coll., 111-127 (1969).
8. J.H. Barrett and D.P. Jackson, Nuc. Instr. Meth. 170, 115-118 (1980).
9. D. Drosd and J. Washburn, J. Appl. Phys. 53, 397-403 (1982).
10. S.S. Lau, J. Vac. Sci. Tech. 15, 1656-1661 (1978).
11. K. Zellama, P. Germain, S. Squelard, and J.C. Bourgoin, J. Appl. Phys. 50, 6995-7000 (1979).
12. Y. Wada and S. Nishimatsu, J. Elchem. Soc. 125, 1499-1504 (1978).

THIN POLYMERIC FILMS PRODUCED BY ION IMPLANTATION FROM FROZEN ORGANIC
MOLECULES

L. CALCAGNO, K. L. SHENG* AND G. FOTI
Instituto Dipartimentale de Fisica, Università di Catania,
Corso Italia 57, I95129, Catania, Italy

ABSTRACT

The synthesis of polymeric films under ion bombardment of
frozen benzene has been investigated for the dependence of the
thickness on the fluence and beam energy of keV proton and argon
ions. Chemical analyses have been performed to characterize the
films.

INTRODUCTION

The physical and chemical properties of polymers can be heavily changed
when they are irradiated with ion beams. Electrical resistivity [1] or
solubility [2] of organic films has been modified by several orders of
magnitude after implantation of keV ions in the fluence range
10^{14}-10^{16}/cm^2. Polymeric films can be produced by irradiation of
monomers deposited on cold substrates in high vacuum environments. Such an
approach is a potential method for ion beam lithography [3] because of the
low scattering of the incoming beam.

In this work we show the polymerizing process induced by energetic ion
beams, with frozen benzene acting as the monomer.

EXPERIMENTAL METHOD

A solid layer of benzene was formed by condensation from the gas phase
on a cold substrate (77K) in a vacuum chamber ($\sim 10^{-7}$torr). The
thickness of the deposited layer was directly measured with X-ray emission
induced by a 100 keV proton beam on the gold substrate [4]. This technique
is sensitive to the energy loss of the proton beam and the thickness of the
film will be given as carbon atoms/cm^2 because the energy loss for
hydrogen is negligible. After growing the film, the target was bombarded
at low temperature with proton and argon ions at different energies. The
beam flux was maintained below 10^{+2} nA/cm^2 and the fluence ranged
between 10^{14} and 10^{16} ions/cm^2. After bombardment the sample was
warmed-up and the volatile molecules which came out from the target were
measured with a gas chromatography system.

The thickness of the stable organic film at room temperature was
measured by using the X-ray emission technique and chemical composition was
analyzed with Auger Electron Spectroscopy.

*Permanent address: Institute of Nuclear Research, Academia Sinica,
Visiting scientist on the basis of INFN - Academia Sinica cultural
exchange plan.

RESULTS AND DISCUSSION

Fig.1 reports the thickness of synthesized residue as a function of fluence for proton irradiation. All the experimental points are obtained with the same starting thickness of the deposited film ($T_o=1.5 \times 10^{18}$ carbon atoms/cm^2), which is small compared with the range of 100 keV protons (8.6×10^{18} carbon atoms/cm^2) in carbon. The residual thickness T is proportional to fluence (ϕ) for low values, while for high fluence it reaches a saturation level which coincides with the initial thickness of deposited film T_o.

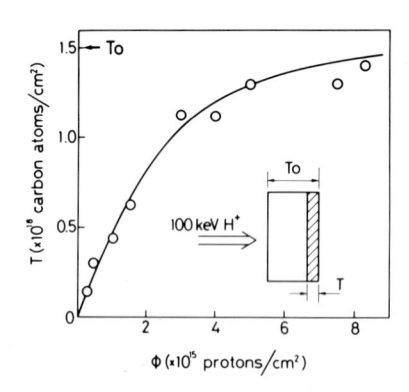

Fig.1. Number of bonded carbon atoms in the film obtained with different doses for 100 keV protons.

The experimental points are well fitted with a simple exponential curve (full line in the Fig.1):

$$T = T_o (1 - e^{-\sigma\phi})$$

where σ is the cross-section (cm^2) for the polymerization of benezene molecules during the interaction with proton beam. We found that the value of σ, which reproduces the experimental data very well is 4×10^{-16} cm^2 for 100 keV H$^+$; this value is very close to the geometrical cross-section of the irradiated molecules.

Some measurements have been performed with argon beams, but choosing an initial thickness of deposited film much larger compared with the range of the ions which is typically 1.1×10^{18} carbon atoms/cm^2 for 100 keV Ar$^+$. The experimental data are given in Fig. 2 where the residual thicknesses T are reported as function of the fluence for Ar$^+$ beams of different energies. The trend is very similar to the proton results but the maximum thickness changed considerably, giving 0.41×10^{18}, 0.55×10^{18}, and 1.2×10^{18} carbon atoms/cm^2 for 20, 50 and 100 keV, respectively. The value of the cross-section is estimated as 1.2×10^{-14} cm^2 for 100 keV Ar$^+$, which is larger than for a 100 keV proton beam. It is surprising that there is agreement between the maximum thickness of synthesized films and the range of argon ions evaluated for a carbon target (given as arrows in Fig.2). These results support the idea that the polymerization process occurs along the ion track, where the atomic bonds of benzene are rearranged in a new stable configuration. Similar results have been obtained for frozen CH$_4$ [5] and C$_2$H$_6$ but with temperatures below 10K.

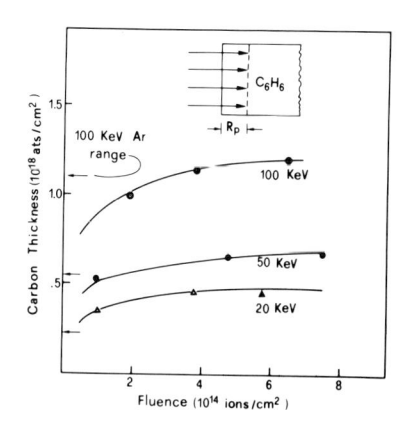

Fig. 2. Relationship between the number of carbon atoms and the fluence of Ar$^+$ ions of different energies.

The polymer films were analyzed with an Auger spectrometer and compared with bulk graphite as reported in Fig. 3. The Auger signals recorded between 200 and 300 eV are very similar in shape and the spectrum of the synthesized films exhibits a small energy shift ($\Delta E=1.54$ eV) compared with the graphite; this is typical for polymeric material. The main component of the film is carbon. However, a small amount of oxygen in the ratio of 6:1=C:O, is also detected. It is not clear if the oxygen is incorporated during the bombardment or the warm-up cycle with subsequent exposure to the air. We expect that this film is quite reactive because many dangling bonds are not saturated after implantation. During warming-up of the target, a large amount of volatile compounds have been released from the sample. The main component is, of course, benzene but analyses with gas chromatography show that more complex molecules are produced during the ion bombardment [6].

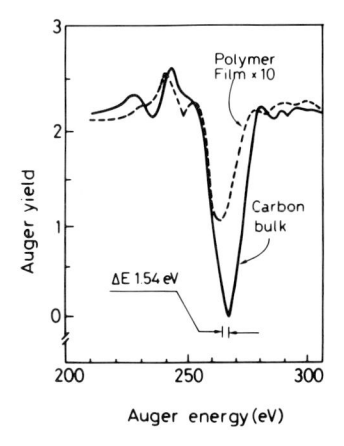

Fig. 3. Auger spectra for carbon bulk and polymer film, for 4.0 keV electron beam.

To increase the sensitivity, the analysis has been performed after several depositions and implantations to accumulate about 200 mg of irradiated benzene for a total fluence of 60 µC of 100 keV Ar$^+$. A typical spectrum of the released molecules is given in Fig.4 with a corresponding gas chromatography trace for unimplanted benzene film (dashed line). Most of the new molecules exhibit prominent peaks in the 8-12 min range of retention time and came from dimers, peaks 1, 2, and 3 in Fig.4, labelled as C_{12} compounds because twelve carbon atoms are contained. The C_{12} products have been reported in several studies, where solid benzene has been irradiated with energetic beams of electrons or alpha particles. The formation of C_{12} compounds is well known in radiation chemistry and involves the presence of H$^\cdot$ and $C_6H_5^\cdot$ radicals. The yield of molecular volatile components ranges between 10^2-10^3 molecules/ion.

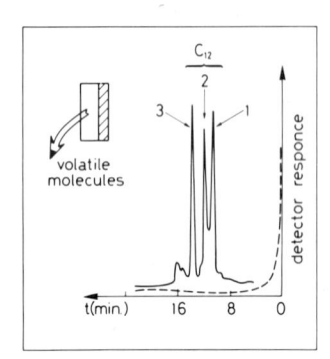

Fig.4. Gas chromatogram analysis of the benzene after exposure to the argon beam (full line) and without bombardment (dashed line).

CONCLUSION

Ion beams cause the formation of stable organic films starting from frozen benzene. The thickness can be controlled by changing the energy of the incoming ion in the range of 0.1 µm to 0.6 µm assuming a film density of 1g/cm^3. Chemical analysis show that the main component of the film is carbon with a small amount of oxygen.

The authors are greatly indebted to Mr. F. Arriva and Mrs. G. Giuffrida for their assistance in preparing this manuscript. Work supported in part by Progetto Finalizzato Chimica Fine e Secondaria.

REFERENCES

1) T. Venkatesan, R.S. Forrest, M. L. Kaplan and B. J. Wilkens, J. Appl. Phys. 53, 6 (1983).
2) T. M. Hall, A. Magner, and L. F. Thompson, J. Appl. Phys. 53, 3997 (1983).
3) W. Brown, T. Venkatesan, and A. Wagner, Nucl. Instr. and Meth. 191, 157 (1981).
4) L. Calcagno, G. Strazzulla and G. Foti, Rad. Effects Lett. 75/76, 157 (1983).
5) W. Brown, (1983) Private Communication.
6) G. Foti, L. Calcagno and O. Puglisi, Nucl. Instr. and Meth., 209/210, 87 (1983).

MAGNETOREFLECTION OF ION-IMPLANTED BISMUTH

E. M. KUNOFF,[*] B. S. ELMAN,[*] M. S. DRESSELHAUS[*+]
[*]Department of Physics; [+]Department of Electrical Engineering and Computer Science; Massachusetts Institute of Technology, Cambridge, MA 02139 USA

ABSTRACT

Bi single crystals have been implanted with isoelectronic ions (As, Sb and Bi) and the electronic structure of these implanted materials has been studied using the magnetoreflection technique. Since the ion penetration depth and optical skin depths are of roughly the same magnitude, this technique provides a sensitive test for implantation-induced changes in the electronic structure. Explicitly, the magnetoreflection spectra show changes in lineshape, resonant frequency and in some cases the introduction of Landau level transitions forbidden in unimplanted bismuth. In particular, implantation-induced changes in the resonance lineshapes indicate an increase in plasma frequency as either the fluence of the implants or the ion size is increased. Further analysis of the data shows that the Lax model, which accounts for the magnetoreflection spectra of unimplanted bismuth, is equally applicable to bismuth implanted with isoelectronic ions. Our results yield measurable changes in the L-point band gap and smaller relative changes in the band parameter combination E_g/m^*. The mechanism responsible for these changes in the electronic structure of bismuth is suggested.

INTRODUCTION

The magnetoreflection technique has been used successfully to study changes in the electronic structure of bismuth induced by pressure [1] and doping [2]. Since ion implantation strains the lattice by introducing a foreign species into the crystal, we expect this experiment to give similar information for the case of ion implanted bismuth. In this paper, we discuss the results obtained using the magnetoreflection technique to study the effect of ion implantation on the electronic properties of bismuth.

We have chosen isoelectronic implants, As, Sb and Bi, to isolate one of the possible modifications due to the presence of impurity atoms in the lattice; we consider the effect of strain and not charge transfer. Fluence was also varied to allow study of the effect of different concentrations of implanted atoms in the lattice. Results show that increasing both the size and fluence of the implanted atoms causes a small decrease in the band gap at the L-point, slightly increases the momentum matrix element coupling the valence and conduction bands, and significantly increases the number of carriers.

Some of the theory necessary to analyze our results is presented in section II. A description of the experimental details comprises section III and we present our experimental results and analysis in section IV. The implications of those results are discussed in section V, especially in relation to the model for the bismuth electronic band structure proposed by Abrikosov and Falkovskii [3] and Abrikosov [4] and modified by McClure [5] which is based on the deviation of the bismuth lattice from cubic symmetry.

THEORETICAL BACKGROUND

We analyze our data using the two band model [6,7,8] as modified by Vecchi [9,10]. We follow early convention [2,7,8,9] motivated by the large spin-orbit coupling in bismuth, and label energy levels by the quantum number $j = n + \frac{1}{2} - s$, where n is the Landau-level quantum number, $s = \pm \frac{1}{2}$ is the spin, and mirror conduction and valence bands are assumed. Measured from the center of the L-point gap, the energy of the $j \neq 0$ levels for both conduction and valence bands at the L point is then:

$$E^{c,v}_{j\neq 0, s} = \pm[(\varepsilon^2 + j\gamma H)^{\frac{1}{2}} - 2sG\beta^* H] \quad (1)$$

where $\varepsilon = E_g/2$ is half the gap energy E_g and H is the magnetic field. The coefficient γ can be expressed in terms of the effective Bohr magneton, $\beta^* = \beta_o m^*/m_o$ or the square of the momentum matrix element between the two bands [11]:

$$\gamma = E_g \beta^* = [e \hbar/2m^2 c] \langle c|p|v\rangle \langle v|p|c\rangle. \quad (2)$$

G is the effective g-factor and because the product $G_o \beta^*$ is small, the pairs of levels $E(n,-\frac{1}{2})$ and $E(n+1,+\frac{1}{2})$ are nearly degenerate. The $j \neq 0$ energy bands are nondegenerate and interact strongly with each other at high magnetic fields because of the small band gap. Using the notation of Vecchi et al. [9], the energy of the $j = 0$ bands at the L point is:

$$E^{c,v}_{j=0, s} = \pm[(\varepsilon - |G_o \beta^*|H)^2 + (Q\beta^* H)^2]^{\frac{1}{2}} \quad (3)$$

where G_0 is the effective g-factor for the j=0 levels, slightly different from G, and Q is a parameter introduced by Baraff [6,9,10] to account for coupling to other bands.

In pure bismuth, the $j \to j + 1$ transitions obey the selection rule [2,7,9] $\Delta j = \pm 1$ which includes both the spin selection rule $\Delta s = \pm 1$ and orbital selection rule $\Delta n = 0, \pm 2$ for nonparabolic bands. In the inset of Fig. 1, some possible transitions are labeled. The labels A, B, C and D, denoting the 0←1 transitions, are used to mark transitions in the experimental traces (see Fig. 2). A and C are interband transitions, the former conserving spin, B is the cyclotron resonance transition and D the spin flipping intraband transition.

EXPERIMENTAL DETAILS

Magnetoreflection measurements were taken on the binary faces of pristine and ion implanted single crystal bismuth samples [12]. These were oriented using Laue back reflection and then cut by a string saw to approximately 10x10x2 mm^3. The surfaces were mechanically polished using alumina grit of varying sizes from 9 μ down to 0.05 μ and then electro-polished to remove surface damage. Orientation was subsequently re-checked using an x-ray diffractometer. The samples were then implanted with As to fluences $2.2 \times 10^{15}/cm^2$, $5.5 \times 10^{15}/cm^2$ and $1.0 \times 10^{16}/cm^2$ and Bi to fluences $2.2 \times 10^{15}/cm^2$ and $5.5 \times 10^{15}/cm^2$. All implantation was done in vacuum at room temperature at incident beam energy of 150 keV. The ion beam was incident on the sample at an angle of 7° to avoid channeling. Ion penetration depths, R_p, calculated using the theory of Lindhard, Scharff and Schiott [13] are 484 Å and 268 Å for As and Bi implants, respectively. The corresponding spreads in R_p in the direction of implantation, ranging from ΔR_p, are respectively 289 Å and 154 Å. Since the optical skin depth, ~260 Å for our highest incident photon energies to ~470 Å at the lowest incident photon energies, is the same order of magnitude as R_p, the experiment reported in this paper probes the modification to the region extending from the sample surface up to the region of implantation, $R_p - \Delta R_p < x < R_p + \Delta R_p$ and, for the heavier implants, a small distance beyond.

The magnetoreflection measurements were made in the Faraday geometry using a cold finger dewar containing liquid helium. A globar provided infrared radiation. Using a single-pass monochomator, we selected photon energies ranging from 95 to 300 meV. Magnetic fields of up to 150 kG were produced by a two-inch-bore Bitter magnet. Further details relevant to the magnetoreflection experiment are available elsewhere [2,9,10].

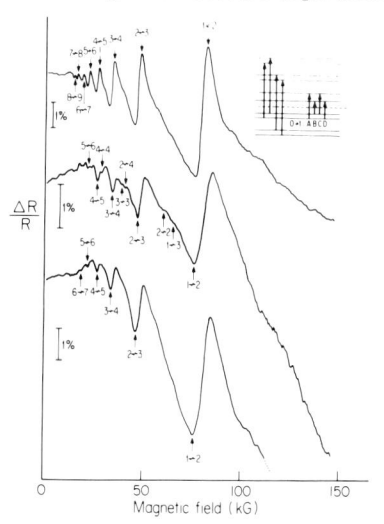

Fig. 1. Experimental magnetoreflection traces taken at photon energy 200 meV for pristine bismuth (top), bismuth implanted with As to fluence $5.5 \times 10^{15}/cm^2$ (middle), and bismuth implanted with Bi to fluence $5.5 \times 10^{15}/cm^2$ (bottom). Arrows mark resonant transitions. Inset depicts magnetic energy level scheme of pristine bismuth near the L point.

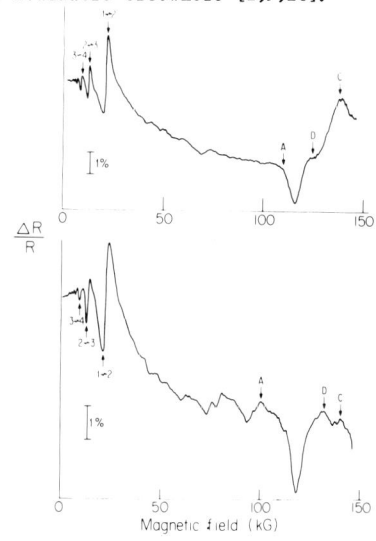

Fig. 2. Experimental magnetoreflection traces taken at photon energy 105 meV for (a) pristine bismuth and (b) bismuth implanted with As to fluence $1.0 \times 10^{16}/cm^2$. Low field resonances are main series $j \to j+1$ transitions; the higher field resonances, marked A, C and D are 0←1 low quantum limit transitions (see inset Fig. 1).

EXPERIMENTAL RESULTS

Figure 1 shows three typical traces taken at photon energy 200 meV in the high energy regime [2,9,10] for pristine bismuth, and bismuth implanted with As and Bi to fluence $5.5 \times 10^{15}/cm^2$ respectively. Much is learned from the qualitative differences between these traces, as well as those in Fig. 2, which compare results in the low energy regime for As implanted to a high fluence to pristine bismuth results. The most obvious change affected by implantation is the reversal in lineshape [14], obtained in all samples except the lowest fluence As implant of $2.2 \times 10^{15}/cm^2$ (not shown). This indicates that the plasma frequency has shifted to a value higher than the photon energy at which the trace was taken [14]. Based on observed lineshapes, the plasma frequency is shifted to greater than 300 meV in all implanted samples except the lowest fluence As sample, where no change in lineshape is observed, and the As implanted to fluence 5.5 x $10^{15}/cm^2$, where we observed a lineshape reversal only at incident photon energies below 225 meV. Since the plasma frequency is proportional to the square root of the number of carriers, we conclude that ion implantation increases the number of electrons at the L-point in bismuth. This is discussed further in the next section.

Next we note that in addition to the lineshape reversal, there are extra transitions appearing in the traces, especially prominent in the sample implanted with As to fluence 5.5 x $10^{15}/cm^2$. These are found to correspond to transitions not allowed by bismuth symmetry by fitting the resonant energy at different fields to the theoretical expression derived in Vecchi et al. [9] as described in the previous section. This additional structure was also observed by Misu et al. [2] in tin doped bismuth and was attributed to the lower symmetry of a tin atom at a substitutional site. In our case, because the implanted bismuth and arsenic atoms have the same valence as the atoms of the lattice, the occurrence of these resonances indicates the presence of atoms at interstitial sites in the bulk.

The final point to note regarding this figure is that the resonant field positions have been shifted to lower values as the fluence is increased for a particular ion. For all samples, we fit the energy vs. field curves in the high photon energy regime to

$$\hbar\omega = E^c(j,s) - E^v(j+1,s-1) \qquad (4)$$

where the energies are given by Eq. (2), letting the parameter γ vary. One such fit to the theory, also known as a fan chart, is given in Fig. 3. We find that γ increases with increasing fluence of a given implant. This dependence is shown in the graph of Fig. 4.

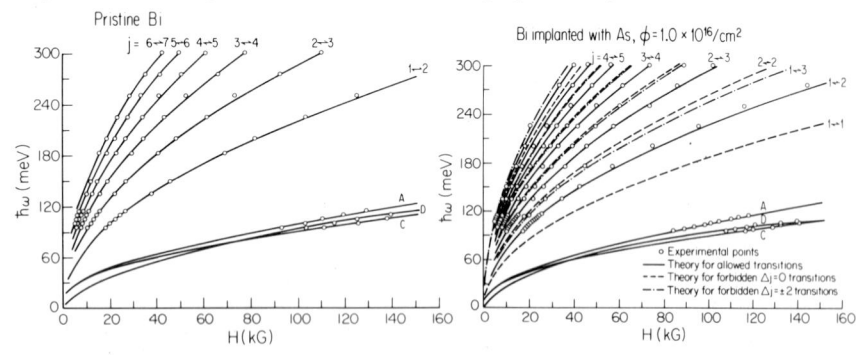

Fig. 3. Plot of resonant energy vs. magnetic field for (a) pristine bismuth and (b) bismuth implanted with As to fluence 1.0 x $10^{16}/cm^2$.

In Fig. 2 we present traces taken at 105 meV, the low photon energy regime, for (a) pristine bismuth and (b) bismuth implanted with As to a fluence of 1.0 x $10^{16}/cm^2$. Again, we note the lineshape reversal in the main series $j \to j+1$ transitions and point out the differences in the lineshapes of the 0↔1 transitions observable in the implanted samples. The labeling of the transitions is discussed in section II and illustrated in the insert of Fig. 1. The transition B is not observed because in our setup the lowest obtainable energy is not low enough. Comparison of the two traces of Fig. 2 reveals that the intensities of transitions A and D are greatly enhanced in the implanted samples while the C resonance intensity is somewhat diminished. The great intensity of the D resonance with C still significantly strong is indicative of interband coupling between the four 0↔1 transitions [15]. In addition, we find that all resonances are shifted to lower magnetic field as is the field at which the C and D resonances cross [6]. This is evident in the fan chart of Fig. 3b. To fit our data to the theory for the low quantum limit transitions, we allow three parameters, ε and G of Eq. (1) and Q of Eq. (3) to vary, leaving the value of the final parameter G_0 fixed at its pristine bismuth value. The fluence dependence of the energy gap E_g is shown in Fig. 5. Table I gives the fluence dependence of all 4 parameters.

TABLE I: Electronic Structure Parameters For Ion Implanted Bi

Implant	Fluence ($\times 10^{15}$)	γ (meV²/kG)	E_g (meV)	m^*/m_0 ($\times 10^{-3}$)	G ($\times 10^{-5}$)	Q ($\times 10^{-5}$)	$\hbar\omega_p$ (meV)
Pure Bi		83.0	13.8	192	756	252	30.9
As	2.2	83.0	12.0	167	708	391	< 95
	5.5	86.1	10.8	145	796	353	~ 200
	10.0	88.1	9.9	130	834	214	> 300
Sb	5.5	86.6	10.6	146	774	29	> 300
Bi	2.2	85.7	10.2	137	755	253	> 300
	5.5	86.2	9.7	131	755	240	> 300

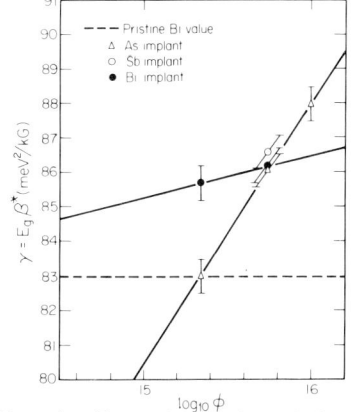

Fig. 4. Plot of γ vs log of fluence for bismuth implanted with As, Sb and Bi. The lines are drawn by eye.

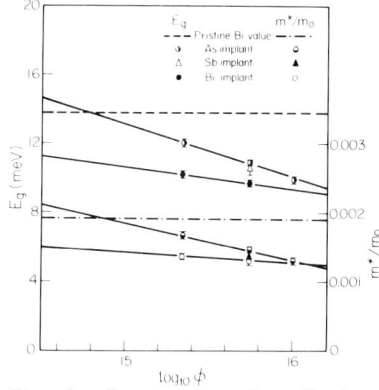

Fig. 5. Upper curve gives E_g dependence on log of fluence for bismuth implanted with As, Sb and Bi. Below is log fluence dependence of effective mass, $m^*/m_0 = \beta_0 E_g/\gamma$.

CONCLUSIONS

The variation of the parameters E_g, γ, G and Q with fluence of implanted ions can be related to strain on the bismuth lattice using the model of Abrikosov and Falkovskii [3,4] as modified by McClure [5] (AFMcC model). This model considers the bismuth rhombohedral lattice [16] as a perturbation of the simple cubic lattice, characterized by two parameters, α the angle between two primitive translations of the rhombohedral face centered lattice, and u, half the distance between adjacent atoms along the (111) direction in units of the rhombohedral diagonal. In the simple cubic lattice, these parameters are 60° and 0.25, respectively, as compared to 57.24° and 0.237 in bismuth. Increasing the perturbation to u increases E_g while making the lattice more rhombohedral causes the L and T point band overlap to increase.

The increase in plasma frequency with increased fluence, indicating additional charge carriers, can also be explained in terms of strain to the lattice. On the basis of pseudopotential calculations, Cohen et al. [17] conclude that increasing the perturbation to u away from 0.25 causes an increase of the energy overlap between the maximum at the T point and minimum at the L point, resulting in a larger Fermi surface. We note also that the variation of γ and E_g with increasing fluence is opposite to the results of Mendez et al. [1] for variation of these parameters with pressure. This, too, indicates that ion implantation causes an internal strain sufficiently small, and spread out over a large enough region of the crystal, to act as a coherent perturbation to the pristine bismuth energy bands within the optical skin depth.

It is important to note as well that the fluence dependence of all the band parameters seems to saturate at sufficiently high fluence, this being lower for the larger implanted atoms. This seems to indicate that the strain propagates through a larger region of the crystal for higher fluences, since the effect of several times as many implanted atoms is only slightly greater.

We have observed changes to the bismuth band structure caused by ion implantation that are consistent with an internal strain on the lattice. This strain causes the unit cell to become less rhombohedral but further separates the two interpenetrating face centered lattices, resulting in a larger number of charge carriers and a smaller energy gap.

ACKNOWLEDGMENTS

We gratefully acknowledge E.J. Alexander for help with sample preparation, M. Rothman for implantations and the staff of the Francis Bitter National Magnet Laboratory for experimental help. Helpful discussions with Profs. J.W. McClure and Y.-H. Kao are greatly appreciated. This work was supported by ONR contract #N00014-77-C-0053.

REFERENCES

1. E.E. Mendez, A. Misu and M.S. Dresselhaus, Phys. Rev. B24, 639 (1981).
2. A. Misu, T.C. Chieu, M.S. Dresselhaus and J. Heremans, Phys. Rev. B25, 6155 (1982).
3. A.A. Abrikosov and L.A. Falkovskii, Sov. Phys. JETP 16, 769 (1963).
4. A.A. Abrikosov, Sov. Phys. JETP 38, 1031 (1974).
5. J. McClure, J. Low Temp. Phys. 25, 527 (1976).
6. G.A. Baraff, Phys. Rev. 137, A842 (1965).
7. P.A. Wolff, J. Phys. Chem. Solids 25, 1057 (1964).
8. M.H. Cohen and E.I. Blount, Phil. Mag. 5, 115 (1960).
9. M.P. Vecchi, J.R. Pereira and M.S. Dresselhaus, Phys. Rev. B14, 298 (1976).
10. M.P. Vecchi, Ph.D. Thesis, MIT, unpublished (1975).
11. M.H. Cohen, Phys. Rev. 121, 387 (1961).
12. Crystal was graciously supplied by Prof. Yi-Han Kao.
13. J. Lindhard, M. Scharff and H.E. Schiott, Kgl. Danske Videnskab. Selskab, Mat. Fys. Medd. 33, 14 (1963).
14. C. Nicolini, T.C. Chieu, G. Dresselhaus and M.S. Dresselhaus, Solid State Commun. 43, 233 (1982).
15. P.R. Schroeder, Ph.D. Thesis, MIT, unpublished (1969).
16. R.W.G. Wyckoff, Crystal Structures, Vol. 1 (Interscience, NY, 1964).
17. M.H. Cohen, L.M. Falicov and S. Golin, IBM J. Res. Dev. 8, 215 (1964).

ION IRRADIATION SMOOTHING AND FILM BONDING FOR LASER MIRRORS

P. P. PRONKO,* A. W. MCCORMICK,* D. C. INGRAM,* A. K. RAI,*
J. A. WOOLLAM,** B. R. APPLETON+ AND D. B. POKER+
*Universal Energy Systems, 4401 Dayton-Xenia Road, Dayton, OH 45432,
**University of Nebraska, Lincoln, NB, +Solid State Division, Oak Ridge National Lab, Oak Ridge, TN 37830

ABSTRACT

Irradiation with high energy heavy ion beams has been investigated as a technique for improving the quality of highly reflecting metallic surfaces to be used as laser mirrors. Properties such as reflectivity, corrosion resistance, film bonding, and threshold to laser surface damage have been examined. Modifications of composition and microstructure of the material associated with the heavy ion irradiation have been measured with RBS, TEM, SEM, Auger, and ESCA. Reflectivity and extinction coefficient measurements were made using ellipsometry techniques. Observations indicate that keV heavy ion irradiations in the fluence range of 10^{15} to $10^{16} cm^{-2}$ produce significant surface smoothing. Additionally, MeV implants of heavy ions into films of Cu, Ag, Au and Al deposited on molybdenum substrates resulted in improvements to both tarnish resistance and structural bonding integrity.

INTRODUCTION

In the present study we have directed our attention to the application of energetic (keV and MeV) ion beam technology to the useful modification and improvement of metallurgical surfaces designed for application in laser mirror technology. In order to understand and appreciate the potential of such a processing technique it is necessary to first have a grasp of where the limitation of existing laser mirrors lie, and which features need to be improved.

Molybdenum mirrors, because of their attractive high temperature thermal properties, are widely used in applications involving high energy lasers. As a refractory metal, however, molybdenum has disadvantages in that it is difficult to machine as well as having a very modest surface reflectivity for optical radiation. Current technology is unsuccessful at precision machining molybdenum by point diamond turning thereby limiting the ultimate smoothness that can be achieved on complex surfaces. Additionally, the optical reflectivity of Mo for wavelengths below 1 μm is on the order of 60%, compared to much higher reflectivities for such metals as Cu, Ag, Au, or Al. In laser applications, therefore, one often finds high reflectivity coatings being applied to molybdenum mirrors. These coatings, however, are subject to mechanical degradation, tarnishing, corrosion, and laser damage in normal use under atmospheric environments resulting in serious loss of reflective efficiency.

Objectives

Ultimately, the most important aspect of a treated optical surface is the way in which it interacts with incident light. In the present work, we address the problems of improving surface finish, homogeneity, and film bonding. Laser mirrors must meet the standards required for adequate finish and homogeneity in addition to retaining very high reflectivities that approach intrinsic materials properties with regard to the onset of laser damage. Such damage usually occurs at local non-uniformities as a result of preferential absorption. All of the above properties are related to the ability of a manufacturing process to fabricate smooth surfaces with rugged non-tarnishing coatings.

Current knowledge relating to the effects associated with ion beam irradiation of solids suggests that this process could be a potentially useful adjunct to the existing surface finishing procedures for laser mirrors. Implantation of heavy ions into a surface has the ability to dramatically change the surface properties of that material. This is a consequence of the chemical composition change that can be introduced by the implanted ions as well as the radiation enhanced kinetic processes that occur on an atomic level.

When an ion beam impacts a surface it produces a high density of interstitial and vacancy pairs that are free to migrate and recombine. In general, this is referred to as a radiation damage process, however, it is not "damage" in the sense of being destructive. The large concentrations of atomic point defects can have, through their kinetic mobility, a number of important beneficial effects. These effects are primarily associated with the enormous enhancements to normal diffusion processes that can occur at surfaces and interfaces. In principle, it is possible for these diffusion processes to result in reduction of surface roughness and improved bonding at film interfaces. These effects in combination with surface chemical composition changes could result in improved corrosion and tarnish resistance as well. There are examples from the basic research literature demonstrating that these results can and do occur [1]. Additional discussion of these points will be presented in the sections that follow.

Experimental Procedure and Results

1. Mirror Preparation - Two sets of samples were used for this study. One set consisted of 1 cm x 1cm or 1 cm x 2 cm samples cut from 1 mm thick rolled sheet stock molybdenum. The second set was five 1.5 inch diameter commercial molybdenum laser mirrors, highly polished on both faces, obtained from Laser Power Optics in California.

The rectangular samples were mechanically polished using a sequence from 240 to 800 grit silicon carbide paper and then further polished with 6 μm diamond paste on a cloth lap. The purpose of these samples was to provide test specimens for the film deposition procedures, and to provide a data base for comparison with the highly polished commercial mirrors. Prior to film deposition, measurements were made on some of the test samples to estimate the thickness of any surface oxide. Resonant elastic collision cross-section enhancement indicated no residual oxide, i.e., very much less than 2 nm. Auger analysis indicated a very thin oxygen layer and ESCA showed that this oxide was MoO_2. Ellipsometry analysis again indicated an oxide layer very much less than 2 nm. The conclusion is that normal metallographic preparation and storage techniques result in only one or two monolayers of oxide formation over a period of months on the surface of these molybdenum samples.

Calculations of the film thickness required for the optical properties of the film to be dominant indicated that 130 nm would be adequate. The films were deposited by evaporation of 0.999999 pure material from a resistively heated boat. This work was done, at the Electronic Materials Department of the University of Nebraska, using a vacuum system pumped by a liquid nitrogen trapped oil diffusion pump. The pressure during evaporation was maintained to 10^{-6} torr. The evaporant charge was premelted with the target shielded. The thickness of the films was 130 $\pm$ 10 nm, and was determined by a quartz film thickness monitor during deposition and checked by weight change afterwards. During implantation of the films, to be described in detail later, blistering occurred on the gold, silver, and aluminum films but not on the copper. It was thought that these blisters were due to gas trapped in the film as it was deposited, and that the ion irradiation resulted in blistering through possible heating and radiation damage. Attempts were made to reduce this problem by heating the substrates prior to and during evaporation. This proved successful for the gold film when the substrate was heated to 200° C. The aluminium at 170° C and the silver at 200° C, however, had a milky appearance under these conditions. Oxygen analysis, by resonant elastic enhanced scattering, indicated a small quantity of oxygen in the aluminium film, but much less than in aluminium which had been left to oxidize in the atmosphere. Of the results reported here, only those for the gold and aluminium films on commercial mirrors are from substrates heated during deposition. The silver and copper films on the commercial mirrors were deposited at room temperature. The films on the rectangular samples were all deposited at room temperature.

The choice of ion species, energy and dose, with which to produce the film interfacial damage, is dictated by the type of mixing required. To produce ballistic cascade mixing requires 1-10dpa. Production of radiation enhanced diffusion mixing may only require 0.1 to 1 dpa. Thus, it was decided to perform experiments at three doses on the rectangular samples (See Table I). The lowest dose would correspond to less than 1 dpa. The medium dose would correspond to approximately 1 dpa, and the high dose to the 1-10 dpa range. The incident beam species chosen was nickel. This would yield dense cascades and would also be chemically compatible with the film even in low concentrations. An ion energy of 1 MeV was chosen following calculations with the Monte Carlo damage simulation code, TRIM. In all cases, these simulations indicated that 1 MeV energy would produce sufficient defects at the film interface. The version of TRIM used is a form modified to simulate ion implantation in multi-film targets and is based on the simple target form published by Biersack [2]. An example of the output from this modified version is shown in Fig. 1.

TABLE 1

Implantation Conditions used for 12 Molybdenum Test Samples

1MeV Nickel Ions

	Fluence 1 (Cm^{-2})	Fluence 2 (Cm^{-2})	Fluence 3 (Cm^{-2})
Au	4.3x10^{15}	1.28x10^{15}	4.3x10^{14}
Ag	8.6x10^{15}	4.3x10^{15}	8.6x10^{14}
Cu	8.6x10^{15}	2.5x10^{15}	8.6x10^{14}
Al	8.6x10^{15}	2.5x10^{15}	8.6x10^{14}

Implantations were performed on the UES Tandetron accelerator facility. The beam was approximately 5mm in diameter, and was swept electrostatically over the target. The beam current was 1 μA and the pressure in the target chamber was less that 2 x 10^{-5} torr. The doses used for the rectangular samples are given in Table 1. The mirrors were all implanted to a dose of 5 x 10^{15} ions cm^{-2}. All samples had half of their surface masked during implantation to provide an unimplanted reference for each.

Microstructural and Metallurgical Effects of Ion Irradiation

The effects of heavy ion irradiation on films deposited on substrates could result from interfacial effects associated with ion beam mixing or chemical effects associated directly with the implanted species. Radiation enhanced diffusion could also change the surface topography of these materials. All three of these physical property effects have been investigated in this study.

In cases where interfacial ion beam mixing occurs, there are two types of processes to be considered; ballistic mixing and radiation enhanced diffusion. Ballistic mixing results from the dynamic motion associated with direct collisions whereas radiation enhanced diffusion results in atomic transport through motion associated with gradients in defect fluxes. All the systems being studied in the present work are fcc-bcc combinations with either immiscible or limited thermal mixing characteristics. This leads us to suspect that gross mixing from a radiation enhanced diffusion process will not be the dominating mechanism but rather that ballistic mixing will be the more significant effect.

RBS analysis was performed with the same accelerator as used for the implants except that the beam was changed to 3 MeV He ions. No gross mixing of the type associated with radiation enhanced diffusion was observed. Some slight indication of ballistic mixing was seen for the case of silver films on Mo substrates, however, in most other cases the level of mixing was too small to be seen with normal RBS analysis. An interesting result was found for the gold film in that, for the highest fluence almost 200 A° of the film was lost (Fig. 2). Since sputtering did not occur for any of the other films at these energies and fluences, it is assumed that gold atoms migrated into the substrate after displacement and were trapped therein at concentrations below the detection limits of the backscattering analysis.

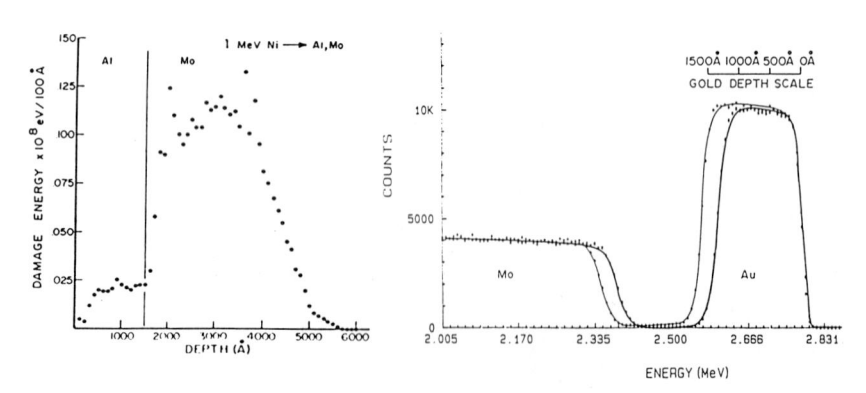

(Fig. 1) Trim Output; Damage Energy vs Depth

(Fig. 2) RBS For Au On Moly Substrate Before And After Irradiation With 4.3 x 10^{15} Ions/cm

Transmission electron microscopy (TEM) was performed with the UES Hitachi H-600 microscope. Samples were prepared by mechanical polishing and jet thinning from the backside using 20% H_2SO_4 and 80% methanol as a thinning solution. An optical sensing cutoff was used to terminate the jet process. In general, it was observed that all deposited films experenced some increase in average grain size as a result of the heavy ion irradiation. These increases were as large as 50% for the gold films. It is unlikely that this result was caused by thermal effects since adjacent non-irradiated material did not experience similar grain growth. This effect, therefore, is probably related to defect enhanced atomic mobility produced by the Ni$^+$ irradiation.

All observations were generally consistent with no significant alloy formation taking place in the bulk of the evaporated surface films. It is clear from some of the ion scattering results and the adhesion tests, however, that interfacial mixing of some kind occurred at the boundary between the films and the Mo substrate. Additional work will be required to resolve the details concerning the character of these interfacial effects.

Adhesion and Tarnish Resistance of Deposited Films

Adhesion experiments were performed on those films deposited on the rectangular test samples. It is anticipated from the earlier experiments of Tombrello and co-workers[3] that enhanced film adhesion will result from high energy heavy ion irradiation. The first test consisted of applying and then removing a 2 mm wide strip of adhesive "scotch" tape to the sample along a strip containing both the implanted and unimplanted regions (Fig. 3). This strip was confined to an edge of the rectangular specimen. A cardboard mask was used to define the area of the test. Good adhesion of the tape to the film was achieved by pressing the tape to the film with a cotton tipped probe.

For all the high and medium dose samples, the film adhered well to the implanted region. It was easily removed from the unimplanted region, revealing a sharp line at the boundary of the regions (Fig. 3). The low dose samples did not exhibit good adhesion on any parts of the film, both implanted and unimplanted.

Electron microscopy specimens were taken from these samples by removing 3 mm diameter disks with an electric core drilling tool. After this was accomplished, a second destructive test of the film was performed. This second test involved polishing the samples for a few minutes with 1 μm diamond paste on a cloth lap. It was seen that the implanted gold film resisted this abrasive polishing better than the other films. In all cases, no material was left on the unimplanted region.

The copper films exhibited striking tarnish resistance (Fig. 3). Similar but less vigorous improvements were visible on the other films as well.

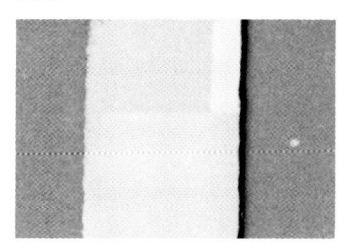

(Fig. 3)
Comparison of Irradiated (Lower) and Non-Irradiated (Upper) Copper Film on Polished Mo Substrates

Ion Irradiation Smoothing

As pointed out earlier, laser mirror surfaces are usually fabricated by mechanical polishing or diamond point turning. A cetain level of microstructural and microtopographic features will result from these treatments. Rough surfaces, being in a higher free energy state than flat surfaces will, under ion irradiation, couple with the defect fluxes to produce a smoother surface [4, 5]. In order to evaluate this effect Mo samples were irradiated with Mo$^+$ beams at 150 keV. This energy will maximize the defect production near the surface. Implants were performed at room temperature and at elevated temperatures. The surface microtopography was evaluated using SEM analysis and optical reflectivity.

The effect of ion irradiation on the surface roughness of these Mo samples was determined using the UES Hitachi H-600 electron microscope in the secondary electron surface scanning (SEM) mode. SEM pictures are reproduced in Figure 4 comparing the polished Mo surfaces before and after irradiation at room temperature. These results show conclusively that significant smoothing occured as a consequence of the ion irradiations. The photographs compare the polished Mo surface before (72377) and after (72378) 150 keV bombardment with 1×10^{15} Mo^+ cm^{-2} at room temperature. Smoothing has occurred to the degree that the polishing scratches are barely visible after implantation.

(Fig. 4) SEM micrographs showing the effects of ion irradiation smoothing (1×10^{15} Mo^+ ions/cm^2 at 150 keV)

Reflectivity, measurements using ellipsometry techniques [6] were made to confirm the quality of these smoothed surfaces before and after irradiation. The results of these tests, which were performed as a function of fluence and substrate temperature, confirmed the SEM observations through increases in optical reflectance.

REFERENCES

1. G. Dearnaley, Rad. Effects 63, 25 (1982).
2. S. Biersack, Nucl. Instrum. & Methods 174, 257 (1980).
3. B. T. Werner et al, Thin Solid Films 104, 163 (1983) and J. E. Griffith et al, Nucl. Instrum. & Methods 198, 607 (1982).
4. W. Marth, Z. Angew. Phys. 13, 224 (1961).
5. R. Sizmann, J. Nucl. Materials 69 & 70, 386 (1978) and R. Sizmann and V. Daeunent, Rad. Damage in Solids, 1, 351 (1962), IAEA, Vienna.
6. R. M. A. Azzam and N. M. Bashara, Ellipsometry and Polarized Light, North Holland Pub. Co., Amsterdam 1977.

Acknowledgments

This work was performed under NSF contract PHY-8260333 with Universal Energy Systems Inc. and also received partial DOE support under contract W-7405-eng-26 with Union Carbide Corporation.

ENHANCEMENT IN ADHESION OF Pt FILMS ON CERAMICS BY HELIUM ION AND ELECTRON IRRADIATION, AND A STUDY OF THEIR ELECTROCHEMICAL BEHAVIOUR.

D.K. SOOD*, P.D. BOND* AND S.P.S. BADWAL**.
*Microelectronics Technology Centre, Royal Melbourne Institute of Technology, Melbourne 3000 Australia.
**CSIRO Division of Materials Science, Advanced Materials Laboratory, P.O. Box 4331, Melbourne 3001 (Australia)

ABSTRACT

The adhesion of Pt films (10-330nm thick) sputter deposited on prepared substrates of yttria stabilized zirconia (YSZ) and alumina has been observed to be remarkably enhanced after irradiation with 2 MeV He^{++} ions and 5-30 keV electrons. The adhesion enhancement has been studied as a function of beam energy, dose and film thickness. Thermal stability of adhesion enhancement and of the microstructure of 'stitched' films has been investigated. The electrochemical behaviour of 'stitched' Pt electrodes on YSZ has been studied by complex impedance spectroscopy.

INTRODUCTION

Bombardment of thin films with low energy (up to a few hundred keV) heavy ions has been known [1,2] to lead to dramatic improvements in their adhesion to the substrate surface. The ion beam energies employed in such experiments have been in the nuclear stopping region and the adhesion enhancement can be attributed to atomic mixing effects at the interface. Similar enhancements in adhesion of metallic films on metal, semiconductor and insulator substrates were recently obtained [3-5] on bombardment with MeV/amu heavy ion beams in the electronic stopping region. It has been suggested [4] that the mechanism for this enhanced adhesion ('stitching') effect involves electronic energy loss at the film-substrate interface with no mass transport or atomic mixing effects. A direct proof of purely electronic origin of this effect has been provided by enhancement in adhesion of thin Au films on Si produced by bombardment with electrons of 5 - 30 keV energies which are well below the atomic displacement thresholds [6]. This paper presents a comparative study on adhesion enhancement of Pt films on ceramics by helium ion (in the electronic stopping region) and electron bombardment. Yttria stabilized zirconia and alumina have been chosen as substrate materials for this study. YSZ is an important electrolyte for several solid state electrochemical devices such as oxygen sensors, fuel cells and steam electrolysers. Alumina is widely used in electronic and refractory industries. It is highly desirable, though often a difficult task, to produce robust and adherent metallic contacts on these ceramics. Ion/electron beam induced adhesion may be a promising technique. The thermal stability and electro-chemical behaviour of 'stitched' Pt electrodes on YSZ have therefore been studied.

EXPERIMENTAL

Stabilized zirconia discs (9 mm diam, 1.5 mm thick, nominal density ~93%) of 7 mol % Y_2O_3 + 93 mol % ZrO_2 were prepared, polished to mirror finish and cleaned as described in ref.7. Alumina samples (19 mm square) were commercial substrates with coarse surface finish (~5μm) and were cleaned with organic solvents. Pt films (10-330nm thick) were deposited using a sputter coater operating at $\lesssim 8\times10^{-2}$ Torr pure Ar pressure. Repeated flushing with Ar was conducted prior to deposition to ensure clean interfaces. Film thickness was determined by RBS of 2MeV He^{++} beam at $170°$ and $108°$ scattering angles, using glancing exit angle geometry. Electron bombardment was done with a rastered beam from a scanning electron microscope (SEM), operated at selected voltages between 5 and 30 kV. Beam currents were maintained at <10 nA as measured in a Faraday cup mounted on a specimen stage, adjacent to samples. The irradiated area was typically 60 - 150μm square. He bombardments were done with a 2 MeV He^{++} beam obtained from a Tandetron accelerator. Circular apertures of 1 and 2mm diameter were used to select uniform regions from a defocussed beam of much larger size. Beam currents were 10 to 140 nA. Secondary electron suppression was achieved by a surrounding, biased Cu cage which was also cooled to liquid nitrogen temperature to eliminate any hydrocarbon cracking onto the target. He and electron bombardments were usually done on the same sample surface to ensure similar interface conditions. After irradiation, sample surfaces were rubbed with a dry cotton bud (Q-tip) until the unbombarded regions were free from the film. Pt residues replicating the bombarded area were left behind in regions which received particle fluences $\gtrsim$ a threshold dose for enhanced adhesion.

Complex impedance measurements[7] were made on two YSZ discs having identical Pt electrode films (325 nm thick) on both sides. One (cell A) had as-deposited electrode films and the other (cell B) had both electrode surfaces 'stitched' with 2 MeV He^{++} ions to a uniform dose of 2×10^{15} He^{++}/cm^2 (obtained by using a large number of 2 mm diameter beam spots juxtaposed to ensure 20% overlap between neighbouring spots). Measurements were made both during the heating and cooling (after holding at $600°C$ for 35-40 hours) cycle over the temperature range $400-600°C$, in a flowing atmosphere of pure oxygen. The frequency range used was 1 mHz-1 MHz. For thermal stability studies, several well separated 2 mm He beam spots at different doses were located on the same YSZ sample surface, which was then Q-tip treated and then heated at $600°C$ in air for 35h or in O_2 for 75h. A small region of unstitched film was retained along-side the stitched region as a control.

RESULTS AND DISCUSSION

A typical result on adhesion enhancement of a 170 nm Pt film on YSZ is shown in Fig.1. The sample was bombarded with electron and He beams over selected areas at the indicated doses. The Pt residues remaining on the surface after rubbing with a cotton bud appear as bright regions in Fig.1. The Pt film was readily removed from unirradiated areas and also from those areas which received very low doses. Partial residues of Pt are observed as the dose increases towards a threshold dose, D_{th} (square labelled 5.5 in Fig. 1) at, and above which, an essentially complete replica of the bombardment area is left as a residue. Thus, from a series of such bombardment spots, the threshold dose for adhesion enhancement can be readily measured within about 25%. At doses $>D_{th}$, the Pt residues are similar and persist even up to 6 times D_{th}, showing no evidence of any de-adhesion at the highest doses used. The Pt residues are highly resistant to prolonged rubbing with a Q-tip and are almost exact replicas of the original bombardment areas as shown in Fig.2 for an 80 nm Pt film on YSZ before and after Q-tip treatment. Small dose non-uniformity near the left corner of the electron beam spot can be noticed. A careful RBS depth profiling analysis of 100 nm Pt films on YSZ before and after Q-tip tests over several He bombardment spots (2 mm diameter) revealed that; a) the Pt film over unbombarded regions was completely removed; b) regions bombarded with doses $>D_{th}$ had residual Pt film thicknesses identical with the as-deposited value. It is thus clearly established that the Q-tip test is a true measure of

10 keV Electron Irradiation
x10^{16} e/cm^2

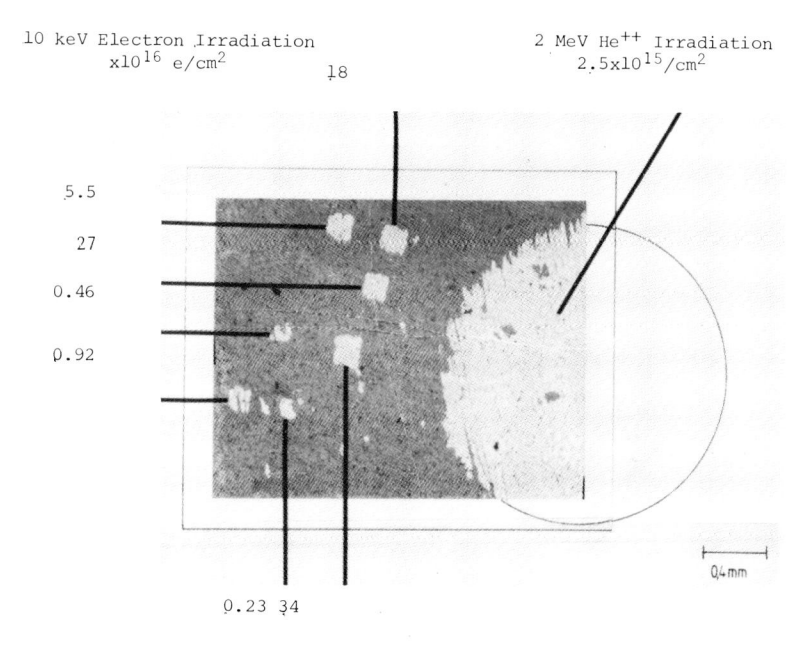

2 MeV He^{++} Irradiation
2.5x10^{15}/cm^2

Fig. 1 An optical micrograph showing He beam and electron beam 'stitching' of a 170 nm, thick film of Pt on YSZ.

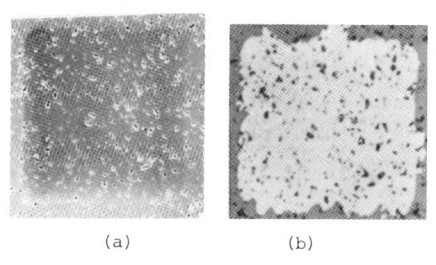

Fig. 2 A 5 keV electron beam spot on 80 nm Pt film on YSZ. a) SEM micrograph of as-bombarded spot, b) an optical micrograph of the same spot after Q-tip treatment. Spot width is 145μm.

adhesion (and not cohesion) of the films since continued rubbing either completely removes the surface film or leaves the original film thickness intact as a residue.

Pt films can be 'ion beam stitched' equally well on rough substrates as shown in Fig.3 for a 282 nm Pt film on an alumina substrate which had surface coarseness ~5μm. The as-deposited film was highly adherent and required very hard and persistent rubbing with a Q-tip before it peeled off (central frame in Fig. 3, where some broken film can be seen in side view). Residues over bombarded regions are identical in thickness (confirmed by RBS) with the as-deposited film.

All the measured threshold doses are listed in Table 1. The He ion induced D_{tn} values are dramatically (by an order of magnitude) lower than those reported by Tombrello[4] for Pd and Ag films on Al$_2$O$_3$, by 20 MeV Cl bombardment. These values are also similarly lower than D_{tn} ~1-2 E15/cm^2 [3] for 2MeV He^{++} bombardment of Pt films on Si. This observation is thus in direct disagreement with the general conclusion of the CALTECH work [3-5] that D_{tn}

increases as one goes from metal-metal to metal-semiconductor to metal-ceramic systems.

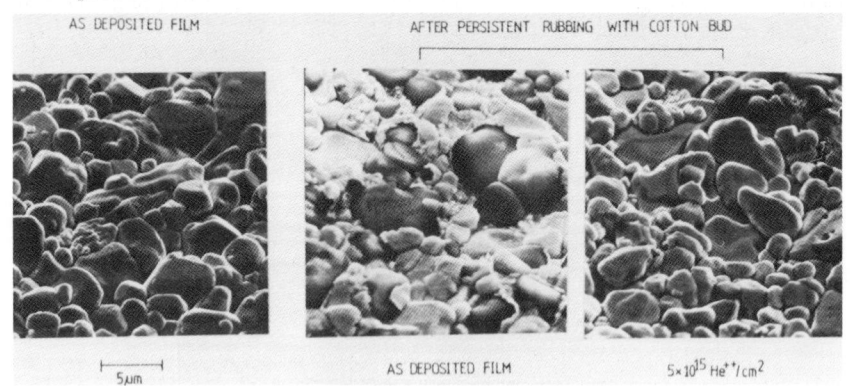

Fig. 3. SEM micrographs showing 282 nm Pt/Al$_2$O$_3$ (l. to r.) as-deposited, after rubbing (as-deposited), and after rubbing (5x10^{15} He^{++}/cm^2).

TABLE 1. Threshold doses for adhesion enhancement of Pt films by bombardment with He^{++} ions and electrons. Numbers preceded by "<" indicate the lowest dose successfully tested, but the threshold may be lower; ">" indicates highest dose studied without obtaining complete residues, threshold may be higher. Film thicknesses are determined by RBS.

Substrate	Pt film Thickness (nm)	Threshold Dose, D_{th} (cm^{-2})				
		He^{++}	Electrons			
		2MeV	5keV	10keV	20keV	30keV
YSZ	10	2E14	-	-	-	-
	55	-	1E17	1-4E17	-	>1.5 E17
	80	2E14	1.3E17	2.4-3E17	8.2E16	1-1.5E17
	100	< 5E14	-	-	-	-
	170	1-5E14	-	5.5E16	-	-
	186	< 1E15	-	-	5E16	5-8E16
	330	< 1E15	-	-	-	-
Al$_2$O$_3$	83	2E14	-	-	-	-
	282	1E14	-	-	-	-

Adhesion enhancement following electron bombardment occurs at D_{th} about 2 orders of magnitude greater than those for He^{++} ions. This difference could arise from changes in stopping powers, dE/dx. The particle range in μm (electronic stopping power in ev/Å) in Pt for 2MeV He [9], and 30,20 and 10 keV electrons [10], respectively, are 2.75 (68.2), 2.02 (0.94), 1.06 (1.21) and 0.37 (1.88). Tombrello has suggested [4] that D_{th} α (dE/dx)$^{-2}$. It is interesting to compare results for 2MeV He and 30keV electrons which have similar ranges in Pt. For an 80 nm Pt film on YSZ, the experimental ratio of (D_{th} for 30 keV electron)/(D_{th} for 2 MeV He ions) is 600 in great disagreement with the ratio of 5000 obtained from Tombrello model. The variation of D_{th} with film thickness and electron beam energy is moderate and much less clear because of small changes in stopping power for the range of energies used here. Furthermore, the adhesion of films is known to increase rapidly [11] with film thickness and this change in prebombardment adhesion could mask the bombardment - induced effects.

Fig. 4. SEM micrographs showing effect of heat treatment at 600°C for 35h in air on a 321 nm Pt film on YSZ (a) as deposited film, (b) 2E15 He^{++}/cm^2 and (c) 2E16 He^{++}/cm^2. Bar = 1.5μm.

The thermal stability of the adhesion enhancement (at doses between 1E15 - 2E16 He^{++}/cm^2) of 100 and 321 nm Pt films on YSZ is found to be very good. Even after heating at 600°C for 75h, the Pt residues of 100 nm films were unaffected by persistent and hard Q-tip tests. The microstructure of the film, however, undergoes considerable changes (Fig.4). The smooth surface of the as-deposited film transforms into large Pt crystallites (Fig.4a). The 'stitched' film shows quite different behaviour - Pt crystallites reside on a thin and perforated underlying Pt film (Fig. 4b,c). The size and number density of Pt crystallites reduces as the ion dose is increased. It appears as if a thin Pt film near the interface (where the 'stitching' effect would dominate) is highly resistant to thermal growth into large Pt crystallites. The obvious reduction in volume fraction of Pt crystallites with ion dose (Fig. 4a-c) could be a consequence of increasing extent of this 'resistant' film.

The electrochemical data were analysed[7] in the complex impedance plane assuming a Cole-Cole[12] distribution of relaxation times. The response of both cells at 600°C is shown in Fig.5 where each arc is due to the electrode reactions which involve transfer of oxygen across the electrode/electrolyte interface according to the overall reaction $O_2 + 4e = 2O^{2-}$. The difference of intercepts of this arc on the X-axis (real) represents the total electrode resistance, Ro, and the angle of depression[12] of this arc below the real impedance axis is a measure of the spread of the time constants. The left intercept of the electrode arc denotes the electrolyte resistance. The electrode arcs for noble metals are normally depressed below the real axis, the angle of depression varying between 15-45°[13,14]. The distortion of this arc on the high or low frequency side of the impedance spectra or the presence of two arcs are also not uncommon and this behaviour is believed to be associated with the electrode microstructure [15].

In the case of unstitched Pt, the electrode arc was invariably skewed on the low frequency side of the impedance spectrum and the angle of depression was around 23-24° during cooling cycle. For stitched Pt, the electrode arc was symmetrical and had a smaller (14-17°) angle of depression. The Arrhenius plots of the resistivity for both electrodes after the 600°C heat treatment are shown in Fig.6. Despite the differences observed in the shape of the electrode arc for stitched and unstitched Pt the Arrhenius plots had the same general characteristics. Both showed a change in slope around 500°C. At temperatures above 500°C the activation energy was 285± 10 kJ mol^{-1} for stitched and 278 ± 10 kJ mol^{-1} for unstitched Pt electrodes. These values are similar to those reported for the dissociative adsorption of molecular oxygen on the surface of Pt[16]. From these observations it can be inferred that the oxygen transport across the electrode/electrolyte

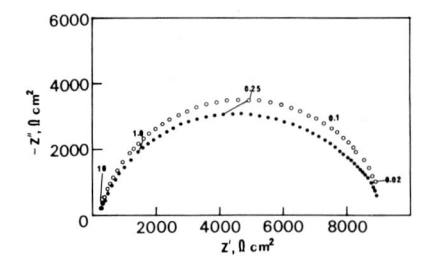

Fig. 5. Complex impedance spectra at 500C in pure oxygen for ● - cell A, o - cell B (with He stitched Pt electrodes). Numbers on the arcs are frequencies in Hz.

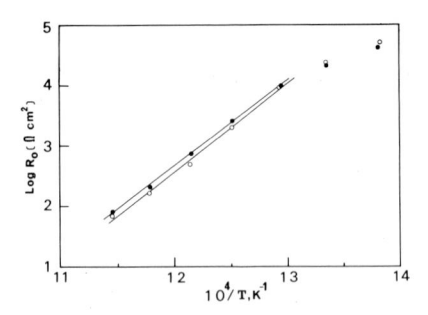

Fig. 6. Arrhenius plots for electrode resistance of ● - Cell A and o - Cell B

interface follows the same mechanism with and without stitching. The observed differences in the shape of the arc may therefore have been caused by different electrode microstructure near the electrode/electrolyte interface as evident from Fig.4. It must be emphasised here that any conclusions drawn at this stage are preliminary. Further work is in progress.

REFERENCES

1. L.E. Collins, J.G.Perkins and P.T.Stroud, Thin Solid Films 4, 41 (1969).
2. K.P. Padmanabhan and G. Sorensen, Thin Solid Films 81, 13 (1981).
3. J.E. Griffith, Y. Qiu and T.A. Tombrello, Nucl. Instrum. Meth. 198, 607 (1982).
4. T.A. Tombrello, Nucl. Instrum. Meth. (in press).
5. M.H. Mendenhall, Ph.D. Thesis (Cal.Inst. of Technology , 1983).
6. I.V.Mitchell, J.S.Williams, P. Smith and R.G.Elliman, Appl. Phys. Lett. (in press).
7. S.P.S. Badwal, J. Electroanal. Chem. (in press).
8. I.V.Mitchell et al (these proceedings).
9. J.F. Ziegler, Helium Stopping Powers and Ranges in All Elements (Pergamon Press, New York, 1977).
10. O.C.Wells, Scanning Electron Microscopy (McGraw Hill, New York, 1974).
11. K.L. Chopra, Thin Film Phenomena (McGraw Hill, New York, 1969)p. 320.
12. J.S.Cole and R.H. Cole, J. Chem.Phys. 9,341 (1941)
13. D. Braunshtein, D.S. Tannhauser and I. Riess, J. Electrochem. Soc. 128, 82 (1981).
14. M. Kleitz, H. Bernard, E. Fernandez and E. Schouler in Advances in Ceramics, Vol. 3, A.H. Heuer and L.W.Hobbs, eds. (Am Ceramics Soc. Inc., Columbas,Ohio, 1981), p.310.
15. S.P.S. Badwal, F.T. Ciacchi and A.S. Kashmirian, Sixth Australian Electrochemistry Conf. Feb. 1984. Geelong, Australia.
16. W.H. Weinberg and R.P. Merrill, Surface Sci. 39, 206 (1973).

H IMPLANTATION IMPROVES SUPERCONDUCTIVITY IN NON-TRANSITION METALS

F. OCHMANN and B. STRITZKER
Institut für Festkörperforschung, Kernforschungsanlage Jülich D-5100 Jülich
D-5170 Jülich, FRG

ABSTRACT

The superconducting transition temperature, T_c, of a series of non-transition metals was improved by implantation of H and D into these metals kept below 10 K. In all cases the increase of T_c was considerably larger for H and D implantation with respect of He implantation causing only radiation effects. It could be shown that the enhancement of T_c depends on the electron-phonon coupling constant of the host metal, suggesting that mainly phonon effects dominate the improvement of T_c.

INTRODUCTION

During the last decade many metal-hydrogen systems have been discovered showing interesting superconducting properties, i. e. hydrogenated $Pd_{55}Cu_{45}$ alloys with a $T_c \approx 17$ K [1]. Although the general superconducting properties of the widely studied Pd-H system are rather well understood, a detailed understanding of its inverse isotope effect or the influence of noble metals is still missing. This is even more true for the other superconducting metal-hydrogen systems.

In this paper we performed a systematic study of the influence of H on the superconducting properties of non-transition metals. We chose non-transition metals with broad s-p bands in order to avoid considerations about the influence of small changes within a complicated electronic band structure due to the H interstitials. Thus one should be able to extract a well defined H influence.

EXPERIMENTAL AND RESULTS

Since H and D are almost insoluble in all non-transition metals we used the method of low temperature implantation to introduce H and D. The target temperature < 10 K hinders the diffusion of H and D. Thus the hydrogen isotopes can be accumulated also in the non-transition metals. However, this method has the disadvantage that not only H and D but also lattice defects are introduced into the target material. It is known from a large variety of experiments on vapor quenched materials [2] that lattice disorder increases the superconducting transition temperature of non-transition metals due to a "weakening" of the phonon spectrum. In order to simulate the effect of radiation damage we implanted inert He atoms with the appropriate energy. For each non-transition metal we implanted H, D and He in three different experiments using similar starting materials.

The implantation of H_2^+, D_2^+ and He^+ was performed with energies between 7 and 100 KeV depending on the sample type. The implantations yield Gaussian type of concentration profiles in a depth of 50 to 300 nm and a width of 100 to 400 nm.

The elements under investigation were Be, Zn, Cd, In, Sn, Tl and Pb.
In the case of Zn, In and Pb we used both evaporated thin films of
100 to 200 nm and rolled foils of 10 to 20 µm thickness. We obtained
the same results for films and foils of the same material. Only foils
were used for the experiments on Be and Cd. In the case of Tl thin
films were evaporated in situ in the vacuum system of the cryostat
to prevent oxidation. In general the samples were about 25 mm long
and ~2 mm wide and deposited or mounted on sapphire or quartz substrates.
The substrates were attached to a substrate holder of an implantation
cryostat allowing temperatures between 0.1 and 300 K. The superconducting
transition was measured resistively using a standard four-point-probe
technique. This method easily allows the detection of increasing T_c
due to the implanted layer. The measured T_c corresponds to the bulk
value as long as the implanted layer is thicker than the superconducting
coherence length, a condition which is mostly fulfilled in our case.
However, a decreasing T_c can only be measured for thin evaporated films
which have been homogenously implanted. Thus, no surrounding higher T_c-
material has remained unimplanted.

The results of maximum T_c after implantation of H, D and He into the
non-transition metals under investigation are shown in Table 1 in comparison with the T_c values of the pure crystalline metal. In addition
the results for the heavily disordered metals as prepared by vapor
quenching onto cold substrates (<6 K) are included [3]. Except for
Pb the T_c values for the disordered non-transition metal are considerably
higher than for the crystalline metal. The T_c values of the disordered
metals are not very much different if the lattice disorder has been
produced by vapor quenching or by low temperature He implantation (with
the exception of Be). In all cases we achieved a substantial increase
of T_c after implantation of H or D which is considerably higher than
T_c of the disordered material (except Be). Thus we have to conclude
that H and D excert a special influence on the superconductivity of
non-transition metals besides plain radiation damage.

Table. 1

element	T_c / K				
	pure element		maximum after implantation of:		
	crystalline	vapor quenched	He$^+$	H$_2^+$	D$_2^+$
Be	0.026	8.60	1.20	1.64	1.64
Zn	0.85	1.51	1.37	2.28	2.09
Cd	0.52	0.91	0.57	1.73	1.73
In	3.41	4.65	4.24	5.72	5.72
Tl	2.39	2.90	2.06	3.13	3.73
Sn	3.72	4.50	4.80	6.21	5.33
Pb	7.20	7.03	7.10	7.60	7.81

The specific influence of hydrogen could be demonstrated most convincingly for thin evaporated Pb films. Fig. 1 shows the change of T_c as a function of implantation dose for both He and D implantation. For Pb it is known that radiation damage decreases T_c slightly [3] in agreement with the He results (dashed-dotted line) in Fig. 2. For low doses of implanted D T_c decreases similar to the He implantation. However, for doses exceeding $1.5 \cdot 10^{14}$ ions·cm^{-2} the T_c values after D implantation no longer decrease but increase substantially. After $5 \cdot 10^{14}$ D-atoms·cm^{-2} the starting T_c of Pb is restored. Further implantation up to doses of 10^{17} D-atoms cm^{-2} yields an increase $\Delta T_c = 0.61$ K over the value of pure Pb.

Fig. 1. Change of T_c of Pb for small concentrations of implanted D (solid curve) and He (dashed-dotted curve).

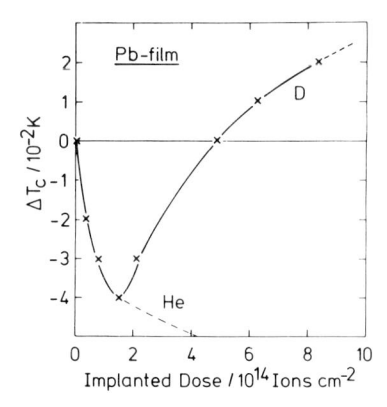

The results shown in Table 1 and Fig. 1 demonstrate clearly that the H (D) atoms excert a special positive influence on the superconducting properties of non-transition metals. However, the influence of the two isotopes is quite different for the different metals. No isotope effect is observed for Be, Cd and In. In contrast Zn and Sn show a positive but Tl and Pb a negative isotope effect. For the case of Pb we have strong indications that the apparent negative isotope effect is due to diffusion of H and D even at 4.2 K. The slower diffusion of the D would lead to higher concentrations and give rise to a higher T_c compared to the H case.

DISCUSSION

The superconducting properties of a material are determined by its phonon spectrum and its electronic bandstructure as well as by the electron-phonon interaction itself. In the present case of non-transition metals it is reasonable to assume that changes of T_c due to H or D are not caused by dynamic changes of the structure of the broad electronic bands. Therefore we applied considerations by McMillan [4] for non-transition metals. He assumes that T_c is only a function of a middle phonon energy $\langle \omega \rangle$. Thus one can optimise T_c by somehow varying $\langle \omega \rangle$. The resulting maximum T_c^{max} normalized to the starting value of T_c^o is plotted in Fig. 2 (dashed curve) as a function of the initial electron- phonon coupling constant λ_o. The result for more advanced theories by Rainer [5] and by Allen and Dynes [6], which yield a saturation behavior of T_c, is shown as solid curve in Fig. 2 and is not much different from McMillian's considerations.

For comparison we have included our results after hydrogen implantation, T_c^H, nomalized to T_c^o, in Fig. 2 (crosses for the different elements). As one can see the experimental points coincide very well with the simple theoretical considerations. The implanted H atoms seem to change the phonon spectrum in such a way, that the optimum T_c^{max} values are reached.

Fig. 2: T_c enhancement factor for H implantation as a function of the initial T_c^o value (crosses). Al data from ref. [7]. Included are theoretical curves for optimized T_c values (dashed curve after ref. [4], solid curve after ref. [5,6]).

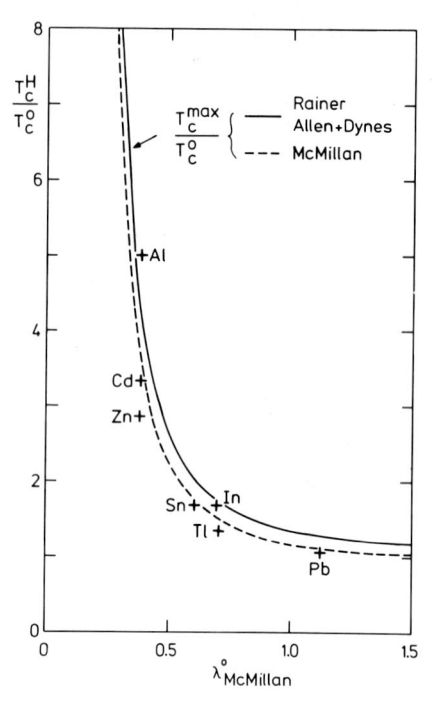

The comparison in Fig. 2 cannot give information on how this optimization of the phonon spectrum is done. There are mainly two possibilities:

1) The acoustic phonon modes of the host non-transition metal are altered. This assumption is supported by the known data of the T_c enhancement in amorphous superconductors which is in good qualitative agreement with the curves in Fig. 2. However, the quantitative T_c enhancement in amorphous non-transition metals is generally smaller than after H implantation (see Table 1). On the other hand He implantation yields even lower values than in the amorphous case. Thus one has to assume that implanted H atoms introduce a higher degree of lattice disorder than He implantation or vaper quenching; perhaps by stabilising more effectively a disordered phase.

2) The optic phonon modes due to H (D) interstitials take part in the electron-phonon coupling. This assumption is supported by tunneling experiments on Pb-H and In-H [8]. However, in the latter experiments

where only small amounts of H were built-in a slight decrease of T_c was observed in contrast to our results on highly concentrated H-metal systems.

In order to show which of the two alternatives or any combination of them is valid, tunneling experiments should be performed on these implanted metal-H systems. In addition, photoemission experiments would yield information about the validity of our basic assumption that the electronic band structure remains unchanged during the H (D) addition. Furthermore, detailed x-ray diffraction measurements are in progress to measure changes of lattice parameters in order to determine H and D concentrations. Thus we hope to obtain a better understanding of the puzzling isotope effect.

REFERENCES

1. For a recent review see: B. Stritzker, in "Electronic Structure and Properties of Hydrogen in Metals", P. Jena and C.B. Satterthwaite eds. (NATO Conf. Ser. VI, 6, Plenum Press, New York, 1983) p. 309 and M. Gupta, therein, p. 321.
2. For a review see: G. Bergmann, Phys. Rep. 276, 161 (1976).
3. B. W. Roberts, J. Phys. Chem. Ref. Data 5, 581 (1976) and references therein.
4. W.C. McMillan, Phys. Rev. 167, 331 (1968).
5. D. Rainer, Symp. on Superconductivity and Lattice Instabilities, Gatlingburg, USA (1973), unpublished.
6. P.B. Allen and R.C. Dynes, Phys. Rev. B 12, 905 (1975).
7. A. M. Lamoise, J. Chaumont, F. Mernier and H. Bernas, J. Phys. Lett. 36, 271 (1975).
8. B.W. Nedrud and D.M. Ginsberg, Physica 108 B, 1175 (1981).

ION IMPLANTATION INTO Nb/NbO/PbAuIn JOSEPHSON TUNNEL JUNCTIONS

G. J. CLARK and S. I. RAIDER
IBM Thomas J. Watson Research Center, Yorktown Heights, New York 10598

ABSTRACT

Boron ions were implanted into completed planar and edge Nb/Nb oxide/PbAuIn Josephson tunnel junctions to directly trim the Josephson pair currents, I_o. The implantation caused an increase in I_o and in the junction subgap conductance and a decrease in the junction energy gap. For a fixed junction fabrication procedure, the variations were observed to be monotonic with [11]B implant dose. Optimum trimming was found when the [11]B ions were implanted such that the peak in the depth distribution occurred at the tunnel barrier. The implantation caused an increase in I_o and in the junction subgap conductance and a decrease in the junction energy gap. The implanted junctions are stable at 80°C and under storage.

INTRODUCTION

The Josephson junction, a device based on the physical phenomena of superconducting and electron tunneling offers significant potential for application in computer technology [1]. If Josephson tunnel junctions are to be used in digital integrated circuit applications, then the Josephson current, I_o, must be controlled within tight margins. Despite a strong dependence of I_o on tunnel barrier processing parameters when fabricating Nb/Nb oxide/PbAuIn tunnel junctions, good I_o control has been obtained. Only small additional adjustments in I_o are generally necessary to meet design requirements. This paper addresses the possibility of providing this fine trim through the use of ion implantation.

The current density, J_1, in a Josephson tunnel junction is the Josephson current, I_o, normalized to the junction area, A

$$\text{ie.,} \quad J_1 = I_o/A \quad . \tag{1}$$

In a Josephson tunnel junction, J_1 is linearly dependent on the superconducting energy gap, E_g, and is exponentially dependent on $-d \times (\phi \times m^*)^{1/2}$, where d is the tunnel average barrier thickness, ϕ is the average barrier height, and m^* is the effective mass of the tunneling electron

$$\text{ie.,} \quad J_1 = kE_g \ell^{-d(\phi \times m^*)^{1/2}} \quad . \tag{2}$$

An I_o trim procedure is required to alter one or more of these parameters in a reproducible, controllable manner without significantly affecting other device properties. Attempts had previously been made to trim I_o by post-processing Josephson junctions by both thermal annealing [2] and by electron beam irradiation [3]. Junctions formed with Nb/Nb oxide/PbAuIn structures are more thermally stable than either Nb/Nb oxide/Nb or Pb alloy/oxide/Pb alloy junctions. Although annealing in N_2 at 210°C causes the junction current densities to change by about 20% after 1 hour and by 50% after 18 hours with little change in junction subgap conductance, the low melting point Pb-alloy counterelectrode makes this an unattractive I_o trim procedure. Electron beam irradiation of Nb/oxide/Nb and Pb-alloy/oxide/Pb-alloy junctions cause current densities to increase 20 to 30% but irradiation of Nb/oxide/Pb-alloy junctions do not change more than 2 to 3% at e-beam currents of 0.1 mA and voltages to 30 keV. Addition of an active element [4], such as In, to the counterelectrode metallurgy was used to alter I_o but this approach is not applicable to our structures which already contain In in the counterelectrode.

We have initiated a program to study whether ion implantation will provide an alternate scheme to directly trim I_o being aware of the sensitive dependence of the electrical properties of tunnel junctions on the junction chemistry and structure.

EXPERIMENTAL

The preparation of Josephson tunnel junctions was described previously [5]. Deposited Nb films were about 2000Å thick (M1) and the oxide barrier, formed by rf plasma oxidation of the Nb surface, was about 20Å thick (M2). A PbAu(4wt.%)In(12wt.%) counterelectrode (M3), 3500Å to 5000Å thick, was deposited over the oxide tunnel barrier. A 25 Å NbO_xC_y transition region separates the Nb base electrode from the Nb_2O_5 tunnel barrier. The planar junctions were defined by SiO windows with areas typically $6.6 \ 10^{-6} cm^2$. Edge junction areas ($7.1 \ 10^{-8} cm^2$) were defined by SiO windows and were formed on the side of an anodized Nb line. The edge junctions were inclined at about 45° angles from the oxidized Si substrate (2.54 cm diameter) surface. Implanted junctions were located within a 0.645 cm^2 area. Some edge junctions are positioned orthogonally to others. Samples with both edge and planar junctions were used.

Junctions were implanted with magnetically analyzed ^{11}B ions of energies from 50 keV to 2300 keV. Wafers were mounted in chambers that were pumped to 10^{-6} Torr. Ion implant doses and dose rates were monitored by direct current integration from the wafer. Implants were made at room temperature. The ^{11}B beam was incident at 7° to the wafer normal. Ion beam irradiations were spatially uniform to ± 2% over an area greater than 6.5 cm^2. The dose rates were typically 200 nA cm^{-2} and the integrated 11_B doses were 1.10^{13} to 1.10^{15} cm^{-2}.

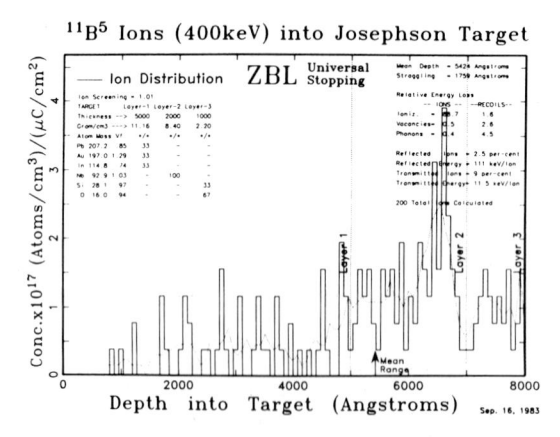

Fig. 1 Range distribution of 400 keV ^{11}B ions implanted into a Josephson junction having an PbAuIn counterelectrode (M3) of 3000Å thickness, an oxide layer (M2) of 50Å thickness and a Nb electrode (M1) of thickness 2000Å.

The range distribution of ^{11}B ions implanted into the Josephson junctions was determined using the Monte Carlo code of Biersack and Ziegler [6] for a multi-layered composite target. In this program the interactions between the incident ion and the target atoms are separated into binary collisions between the ions and the screened nuclei of the target atom and into inelastic collisions by the ion with the entire electron system of the Josephson junction. Figures 1 and 2 show the results from these calculations for 400 keV ^{11}B ions injected into a Josephson junction whose structure is as shown in Fig. 1. Figure 1 shows the range distribution of the implanted ions. The interfaces between the M3 and the M2 layers and the M2 and the SiO_2 substrate are shown. Figure 2 shows 2-D maps for the same calculation.

$^{11}B^5$ Ions (100keV) into Josephson Target

Particle Distribution   Vacancy Production

Total Energy Loss   Ionization Energy Loss

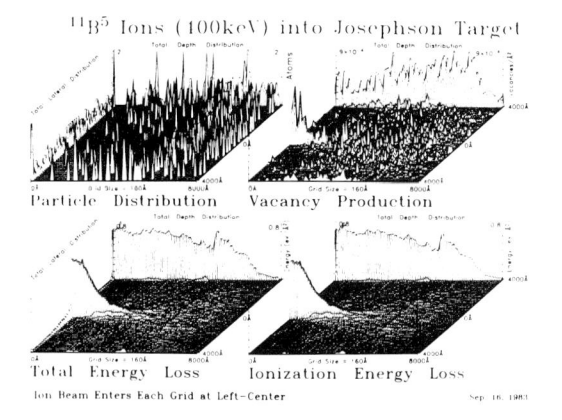

Ion Beam Enters Each Grid at Left-Center

Fig. 2   2-D plots showing the range distribution, total energy loss, vacancy production and ionization energy loss as a function of depth for 400 keV ^{11}B ions implanted into the Josephson junction described in the Fig. 1 caption.

The upper left plot of Fig. 2 shows the final distribution of the 400 keV ^{11}B ions injected into the center of the front edge of the plot. It is educational to note that the distribution is broad with the width of the distribution being nearly as broad as the depth distribution. Along the left and back of the plot are the summed longitudinal and lateral distributions.

The lower left plot of Fig. 2 shows the total energy loss of the injected ions. The energy loss is approximately constant for about half the ion track length.

The upper right plot shows the generation of vacancies in the target. The production of a vacancy is defined as a collision with a target atom which transfers more than a minimum amount of kinetic energy to the atom, i.e. the displacement energy which is here assumed to be 25 eV. Vacancies produced by recoiling atoms are also taken into account in this calculation using the Kinchin-Pease formalism.

The lower right plot in Fig. 2 shows the energy loss due to ionization processes in the Josephson junction. These processes may be important in the NbO_x region of the junction because in insulators, ionization effects can be more significant then vacancy production in modifying the target, both chemically and physically.

Changes in the junction properties due to ion implantation were determined from the junction current-voltage (I-V) characteristics. The I-V characteristics were measured before and after implantation. The current density, J_1, the gap voltage, V_g, and the subgap conductance were each measured. The design parameter, V_m, which is exponentially dependent on V_g/T where T is the absolute temperature, is inversely proportional to the subgap conductance. V_m is defined as $I_o \times R_j$, where R_j is the subgap resistance. R_j is empirically defined as a linear resistor which crosses the I-V curve at V = 1.7 mV.

RESULTS AND DISCUSSION

Only those ^{11}B ions implanted and stopped in the Nb/Nb oxide structure affected the junction properties. This is illustrated in Fig. 3 in which the change in current density, ΔJ_1, is plotted against the implant energy. When the ions are stopped in the PbInAu alloy or when they go through the junction and are stopped in the substrate, no change in the I-V characteristics of the junction were observed. The maximum change in the I-V characteristics for a given dose occurred when the ^{11}B range distribution was centered around the Nb oxide tunnel barrier. When the ^{11}B range distribution was shifted from the tunnel barrier, the changes in J_1 and V_m decreased but the relative change between J_1 and V_m were found to be independent of the location of the implant distributions.

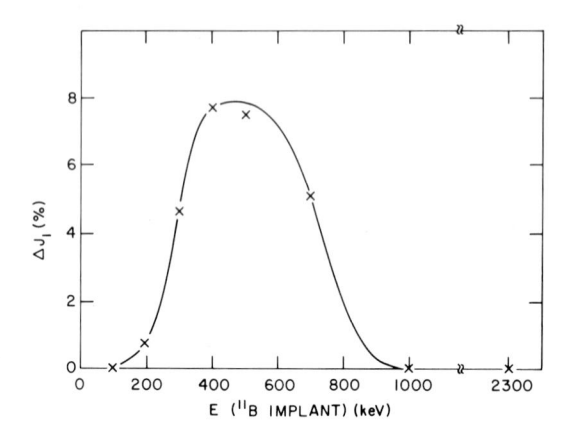

Fig. 3   Plot showing the change in the current density ΔJ_1 as a function of the ^{11}B implant energy.

In Fig. 4, the change in the current density, ΔJ_1, following 400 keV ^{11}B implants, is plotted as a function of the ^{11}B dose for both edge and planar junctions. Two planar and nine edge junctions were used in the measurements. The ΔJ_1 was observed to monotonically increase with ^{11}B dose and exhibited only a small dependence on whether the junction was edge or planar despite the differences in J_1, in subgap conductance, in V_g, in junction area or in junction inclination. After the total implant of 1.10^{14} ^{11}B/cm^{-2}, I_o increased by about 15%, a change that is in a range of practical interest for trimming. A small increase of about 2% in I_o spread for edge junctions is observed after the total ^{11}B implantation. Edge junctions facing orthogonal to each other did not differ in the I_o change due to implantation.

Following the initial electrical tests, the junctions were implanted and retested over a 6 week period. There was no observed change in any of the electrical properties indicating room temperature thermal stability of the modified junctions. After a 4 1/2 hour thermal anneal at 80°C of implanted junctions, no change in the electrical properties were observed.

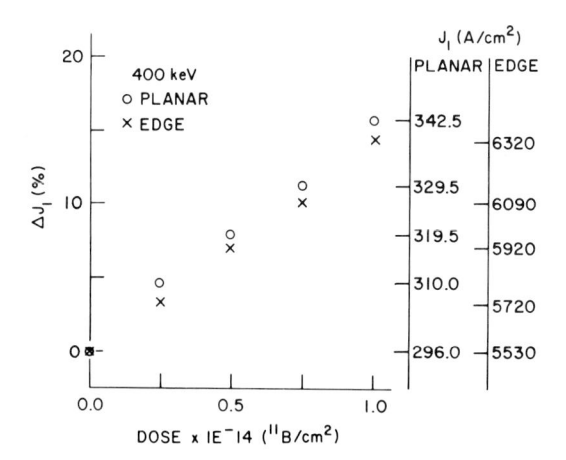

Fig. 4   Plot showing the change in the current density as a function of the ^{11}B dose for both edge and planar junctions. The absolute values for J_1 are also shown.

The value of I_o increases with the ^{11}B implant dose. This increase in tunneling is equivalent to a decrease in oxide film thickness or a lowering of the barrier height. The decrease in V_g cannot account for the change in I_o. We have established that the modification of the electrical properties by ^{11}B ion implantation are independent of the source of the Nb base electrode and of the PbAuIn counterelectrode. Doping of the tunnel barrier, which is about $1.10^{18} cm^{-3}$ for a ^{11}B implant of $1.10^{14} cm^{-2}$ centered at the tunnel barrier, is dependent on the depth distribution and the dose is not strongly dependent on the detailed plasma processing to form the tunnel barrier. A comparison of edge and planar junctions indicates that ion-induced changes are independent of the initial J_1 and the tunnel barrier thickness. This implies that doping of the oxide tunnel barrier is not the mechanism producing the electrical changes. Rather, the changes in junction electrical properties caused by ion implantation are probably due to modifications of the Nb/Nb oxide interfacial regions. This modification is possibly some radiation enhanced ion beam mixing effect. ^{11}B implantation thus offers a direct method for trimming I_o. The usefulness of this procedure as an I_o trim technique is limited only by the decrease in V_m and V_g.

SUMMARY

1. Ion implantation provides a direct I_o trim of completed Josephson tunnel junctions.
2. These changes are predictable, reproducible, have storage stability at room temperature and are stable at 80°C.
3. Changes are independent of whether we implant into an edge or planar junction.
4. The implanted ions must be distributed in the region of the oxide tunnel barrier.

REFERENCES

1. Special Issue on Josephson-Junction Devices, IEEE Transactions on Electron Devices ED-27 1855 (1980).
2. S. I. Raider (unpublished data).
3. Y. H. Lee and P. R. Brosious, Appl. Phys. Letters $\underline{40}$, 347 (1982).
4. S. S. Pie, T. A. Fulton, L. N. Dunkleberger and R. A. Keane, IEEE Trans. Magn., MAG-19, 820 (1983).
5. S. I. Raider and R. E. Drake, IEEE Trans. Magn., MAG-17, 299 (1981).
6. J. F. Ziegler, J. P. Biersack, U. Littmark, "The Stopping and Range of Ions in Matter," Volume 1, Pergamon Press (1984).

PART VI

APPLICATIONS: MECHANICAL

TRIBOMECHANICAL PROPERTIES OF ION IMPLANTED METALS

IRWIN L. SINGER, Naval Research Laboratory, Chemistry Division,
Code 6170, Washington, D.C. 20375

ABSTRACT

A review of tribomechanical studies supported by surface analysis finds ion implantation capable of increasing the sliding wear resistance of ion implanted metals in two ways. First, it can reduce friction by modifying the surface composition (e.g. Ti^+ into steel) or by promoting the growth of low friction oxide layers (e.g. N into Ti). Second, it can modify the subsurface composition and structure to resist fracture and debris formation. These modifications harden the surface, change its work-hardening behavior and/or increase residual stresses. Microindentation hardness measurements indicate that many but not all of the wear resistant surfaces are hardened by implantation; thus, surface hardness is a contributing but not necessarily a controlling factor in wear resistance. These mechanisms of wear reduction and the chemical and microstructural modifications responsible for them are discussed. Evidence for wear reduction through the migration of N during wear is critically reviewed. It is concluded that the principal benefit of ion implantation is to prevent or delay the formation of wear particles, thereby changing the wear mode during run-in and permitting metals to reach load-carrying capacities up to their elastic limits.

INTRODUCTION

Ion implantation has been highly touted by the material research community as a surface processing treatment for protecting metals against wear. So far, however, it has been regarded with healthy skepticism by many tribologists who are waiting for experimentally verifiable explanations of how a thin (100 nm) surface alloy can increase the wear resistance of metals. These explanations have been coming, slowly but steadily, from simple friction and wear tests supported by microscopic analyses of wear scars, and from surface analytical studies aimed at identifying the compositions and microstructures responsible for wear resistance. While these tests adequately describe the response of implanted layers to sliding contact, they are too crude to examine the tribomechanical properties of the thin (~100nm) implanted layer. More direct measures of these properties have been obtained from two very surface sensitive techniques, microindentation hardness and polishing wear studies. This paper reviews many of these studies and provides evidence for four mechanisms by which ion implantation improves the sliding wear resistance of metals.

The second section describes two ways that ion implantation can affect sliding wear processes: it can alter the deformation behavior of a metal surface under stresses transmitted during sliding, and it can reduce the stresses transmitted. The third section examines the tribomechanical effects of Ti- and N-implantation on the friction and sliding wear behavior of selected metals. Studies involving high speed sliding have been excluded because of the uncertainties associated with heating effects. Surface hardness and abrasion resistance studies are also presented in the context of identifying mechanisms of sliding wear resistance. Evidence for N migration during the wear process is also discussed. The fourth section then considers a variety of ion implantation treatments and the chemical and microstructural changes which reduce friction and delay or prevent wear during sliding contact.

Sliding and Wear Processes Affected by Ion Implantation

Detailed description of the wear processes occurring when two metals are placed in sliding contact can be found in references [1-6]. Briefly, wear is a complex interplay between adhesion, deformation and friction. Although relatively independent processes before sliding, adhesion and deformation become coupled during sliding by friction, which transmits high shear stresses to the uppermost surface layers and causes them to deform. Is this an adhesive wear mode or a deformation wear mode?

Two of the three processes, adhesion and friction, are very sensitive to the surface composition. For example, it is well known that oxide films can eliminate severe adhesion and reduce the coefficient of friction [1,3]. Steels sliding in a vacuum lose their oxides and experience severe and rapid adhesive wear, whereas the same steels sliding in a good lubricant have their oxides replenished and wear more slowly by deformation-controlled processes [3]. Therefore, it is easy to understand how ion implantation can affect adhesion and friction. By modifying the composition of the uppermost layers of metals, ion implantation can reduce the chemical affinity of surfaces in contact, promote oxide growth and/or strengthen the metal oxide/metal interface. We will see, later, how reducing friction also decreases the deformation mode of wear.

However, to understand how implantation can affect the deformation mode of wear directly, it is necessary to examine how a metal surface deforms. Initially, deformation is elastic, but goes plastic when the shear stress exceeds the yield stress of the metal or when it encounters stress concentrators such as inclusions or microcracks. Plastic deformation, the accumulation of plastic strain, may be accompanied by a variety of effects including work hardening, buildup of residual stresses [7], development of texture [5,6] and phase transformation [8]. When plastic deformation processes are exhausted, cold working ends and fracture begins [4-6]. In ductile metals, flow-induced cracks nucleate and grow in the subsurface; when they reach the surface, plate-like particles are formed. In hardened metals or coatings, fatigue cracks nucleate and grow; then fine debris particles are produced by brittle fracture at the edges of surface cracks. In addition, oxide particles can form by repetitive shear and regrowth of thin oxide films.

Ion implantation can affect the deformation mode of wear by modifying the composition and microstructure of the near-surface layer. The flow strength (hardness) of the surface can be increased by solid-solution strengthening or precipitate formation. The work-hardening behavior can be altered by stabilizing the microstructure, possibly by changing the stacking-fault energy [5]. In addition, a more homogeneous defect structure can be produced.

Implantation may also protect against wear by introducing residual compressive stresses into the surface. Compressive stresses can blunt crack propagation and resist the lifting off of plate-like particles by opposing sliding-induced tensile stresses. Residual stress produced by treatments such as shot peening have been used for years to increase resistance against rolling contact fatigue. Recently, Ho et al. [9] have demonstrated that residual stresses can also increase sliding wear resistance so long as they exceed the stresses induced during sliding.

Finally, implantation can reduce the deformation wear rate indirectly by reducing the coefficient of friction. Lowering the coefficient of friction reduces the intensity of the stresses transmitted to the surface. Moreover, it relocates the maximum shear stress component from the uppermost layer for $\mu > 0.4$ to a depth well below (10-20% of the contact diameter) the surface for $\mu < 0.2$ [7].

The load carrying capacity, as might be anticipated, also depends on friction. The load at which the elastic limit is reached can be calculated from P and τ, where P is the peak Hertzian pressure and τ is the yield stress in pure shear. For ideal elasto-plastic materials in the important ball-on-flat geometry, P is given by [7]

$$P = 3L/2\pi a^2 \qquad (1)$$

$$\text{and } a = (\frac{3}{4} RL)^{\frac{1}{3}} [\frac{(1-\nu_1)^2}{E_1} + \frac{(1-\nu_2)^2}{E_2}]^{\frac{1}{3}} \qquad (2)$$

where a is the radius of contact under normal load L, R is the radius of curvature of the ball, $E_{1,2}$ and $\nu_{1,2}$ are the elastic modulus and Poisson's ratio of the ball and flat, respectively. If we apply Tresca's yield criterion τ = Y/2, where Y is the yield strength, then the pressure-to-yield ratio at the load limit, L_{max}, may be approximated by

$$P/Y_{max} \simeq 2.3 (1 - \mu) \qquad (3)$$

for μ < 0.8 [7]. The dependence of L_{max} on μ follows by substituting eqs. 1 and 2 into 3, giving

$$L_{max} \propto (1 - \mu)^3 \qquad (4)$$

Eq. 4 shows that reducing μ can greatly increase the load carrying capacity. Eq. 3 will be used to predict the pressure-yield-ratio during sliding contact with friction; it will then be compared with observed P/Y values. Two caveats, however, should be mentioned. First, Eq. 3 underestimates P/Y_{max} when a lubricant is present; a lubricant tends to distribute the load over a larger area than would be calculated from Eq. 2. Second, when calculating observed P/Y values, it should be realized that the yield strength of a worn surface layer can be much higher (2-7 times) than the bulk yield strength [10]. Therefore, predictions made using Eq. (3) should be considered semiquantitative.

TRIBOMECHANICAL PROPERTIES OF ION IMPLANTED METALS

This section reviews the tribomechanical properties of a variety of ion-implanted metals and alloys. Compositions, microstructures and bulk hardness for many of these alloys are given in Table I. All of the investigations reviewed in this paper employed one or more surface analytical tools to ascertain composition and/or microstructures of the implanted surfaces. In several cases the compositions or microstructures responsible for wear resistance have been identified.

TABLE I. Composition, microstructure and bulk hardness of selected metals

Metal Designation	Composition (wt %)	Microstructure	Hardness (GPa)
Steels			
52100	Fe 1.5Cr-1C	Martensite	8.0
Fe1C	Fe-1C	Tempered Martensite	1.7-4.0
1018	Fe-0.2C	Ferrite + Fe Carbide	2.8
440C	Fe-18Cr-1C	Martensite + Cr Carbides	7.8
304	Fe-18Cr-8Ni	Austenite	2.9
Ti-6Al-4V	Ti-6Al-4V	α and β Ti	6.0
Stoody 3	Co-31Cr-13W-2.2C	Co(Cr)+Cr_7C_3+W_6C	6.6

Sliding wear studies permit several tribomechanical properties of ion implanted layers to be measured. These include friction coefficients and load-carrying capacities. In addition, the wear mode can be determined by analyzing the topography of the wear track and the debris generated. In the investigations reviewed, experiments were performed in a ball-against-flat geometry, with (initially) smooth surfaces, and at low speeds (<0.1 m/sec).

The hardness of implanted surface layers has been measured by Pethica [11] using a specially designed microindentation hardness (MIH) apparatus. Hardness numbers were obtained with a diamond pyramid indentor penetrating to depths as shallow as 50 nm.

Polishing wear measurements have been obtained by Bolster and Singer [12,13] with depth resolutions of 20 nm. Relative wear resistance (RWR) vs. depth profiles for implanted vs. non-implanted flats, shown in Fig. 1, illustrate the depth resolution. They also show that the steady state RWR values for two (304 and Stoody 3) of the four metals are higher than might be predicted from their bulk hardness values. RWR values, initially believed to measure surface hardness [12], are now thought to be related to the work-hardening rates of metals.

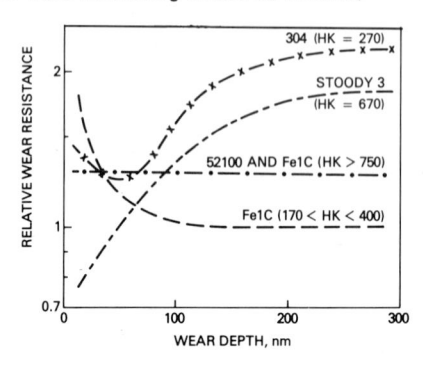

FIG. 1 RWR vs. depth profile for three steels and Stoody implanted with N. Bulk Knoop hardness, HK, is given in kgf/mm².

Implantation of Ti

Implantation of Ti into steels and Stoody 3 to high fluences has produced dramatic reductions in both friction and wear. In a hardened bearing steel, type 52100, it reduced the dry sliding friction coefficient from $\mu = 0.6$ (steel-vs-steel) to $\mu = 0.3$ and prevented the formation of oxide debris normally generated during dry sliding contact [14-16]. In soft (annealed) steels it again reduced friction and thereby delayed the onset of ductile fracture and plate-like debris formation [17,18]. Similar effects were achieved in a carbide strengthened Co-based alloy, Stoody 3 [19,20]. In that study equally low friction, wear resistant surfaces were obtained by covering nonimplanted Stoody 3 with a monolayer of a lubricious fatty acid, thereby confirming that low friction alone could produce the observed wear resistance [19].

A unique surface alloy, composed of Fe+Ti+C, has been found responsible for the low friction on steels [15,16]. When implanted to high fluences, Ti ions assist in the ''vacuum carburization'' of the surface, causing C atoms to be absorbed from gas molecules in the vacuum chamber [21]. In steels the Fe+Ti+C layer, and those produced by the intentional implantation of C, have been shown to be amorphous [15,22]. Analytical studies have found that low friction and wear resistance continues so long as the Fe+Ti+C layer is retained. This layer has remained intact even after being depressed to 1 μm below the non-worn surface [17,18,23]. In Stoody 3, a ''vacuum carburized'' layer was found in the Co matrix and the carbide phases [19].

The load-carrying capacity of dual implants of Ti and C into a variety of stainless steels (soft and hard) and Fe has been investigated by Pope, et al. [24,25] under the rather severe conditions of dry sliding contact at low speed (15mm/sec). This treatment reduced friction by approximately a factor of two and prevented wear at loads up to the maximum load-carrying capacity of the steels. For example, in a hardened 440C steel whose yield strength was Y = 1.8 GPa, the low friction (μ = 0.30) surface resisted wear up to a Hertzian pressure of P = 2.9 GPa [25]. The observed pressure to yield ratio P/Y = 1.6 is identical to that predicted from eq. 3.

MIH studies of these wear resistance Ti and Ti + C layers found no increase in surface hardness [26,27]. Polishing wear studies, however, showed dramatic increases (from x3 to x10) for Ti-implanted hard and soft steels [13,28] and Stoody 3 [20]. These increases have not been explained.

Implantation of N

N into steel is the implantation treatment most commonly used to improve its wear resistance. Four important conclusions about its tribological behavior can be drawn from studies on simple friction and wear apparatus. First, the friction coefficient does not decrease significantly in any of the steels, even those showing increased wear resistance [14,17,18,25,29,30]. In fact, one low speed sliding study of steel against N-implanted steel [17] showed increased wear resistance despite a slight increase in the friction coefficient. Therefore, the improved wear resistance cannot be attributed to a reduction in friction, as is the case for Ti-implanted steel. Second, large increases in wear resistance are obtained in ductile steels, but essentially no changes have been found in the four hardened (martensitic) steels examined. Third, the wear resistance has been attributed to changes in the wear mode and not simply to a "slowing down" of the wear rate [18,25,29]. Fourth, wear resistance has persisted up to loads that reach the load-carrying capacity of the steels. I have calculated pressure-to-yield ratios from wear data for 304 steel from two different investigations [29,30] and find values from 1.2 to 2.5 times larger than predicted from eq. 3 (see Table II). Therefore since friction isn't reduced, implantation must affect the way steel surfaces respond to deformations induced by the stresses transmitted during sliding.

TABLE II. Load carrying capacity of selected ion-implanted metals.

Implanted Species	Ref.	Metal Substrate	Slider	Y (GPa)	Frict. Coef.	P/Y Observed	P/Y Predicted
Ti + C	22	440C	440C	1.8	0.3	1.6	1.6
N	29	304	304	0.6*	0.7	1.8	0.7
N	30	304	Al_2O_3	0.6*	0.15	2.5	2.0
N	41	Ti6A14V	Al_2O_3	0.83	0.17	2.9	1.9
N	41	Ti	Al_2O_3	0.28	0.17	3.1	1.9

*Estimated as Y=H/3 where H is the hardness [10].

The SEM photos of wear tracks in Fig. 2 show clearly that N-implantation altered the flow behavior of 304 and 1018 surfaces of steels during sliding. In the early "run-in" stage of wear, tracks in non-implanted steels show more ductile fracture than those in N-implanted surfaces [17, 18]. Ductile fracture in these steels produces shear marks transverse to sliding direction, plates uplifted from the tracks, and plate-like debris smeared along the track. This mode of wear was delayed in the N-implanted tracks, despite having the same friction history as the non-implanted track [18]. Profilometry has shown that surfaces of tracks formed during sliding at high load can be depressed from 0.1 to 1 μm below the nonworn surface. Auger analysis of these tracks found no change in composition of worn and non-worn surfaces, indicating that only the underlying metal was deformed during sliding [17,18,23].

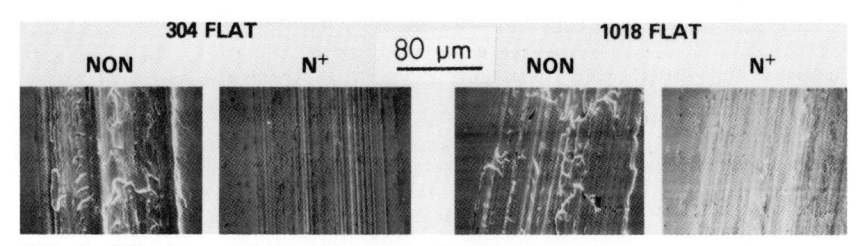

FIG. 2 SEM micrograph of wear tracks formed during dry sliding. (Left) after 3 passes on 304 steel; (right) after 7 passes on 1018 steel.

Microindentation hardness and polishing wear studies showed no changes in the surface hardness or abrasion resistance (see Fig. 1) of N-implanted hardened steels which, to repeat, showed no increased sliding wear resistance. In contrast, N-implantation was able to increase the MIH value of Fe by 50-100% [29] and the RWR value of a ductile steel (tempered Fe1C) by about that amount (see Fig. 1). Presumably N strengthens ferritic Fe by forming nitrides, solution strengthening and pinning dislocations [31]. Another strengthening mechanism for steels recently proposed by Fischer, et al. [17] is that microprecipitates of carbides form from C atoms that segregate to steel surfaces during high fluence implantation [17,18]. Since surface segregation of solute C has been observed for Ar and Fe implants as well as N implants, this mechanism should be independent of implant species. Microindentation hardness results for the austenitic 304 steel have been much less definitive [27]. N-implantation clearly hardened electropolished 304. But in polished 304, already hardened by abrasion, N-implantation did not substantially increase the hardness and may have softened it. Polishing wear studies showed that the RWR values of 304 steel were lowered by N-implantation [12,13].

Austenitic 304 is known to have a high work-hardening rate when it undergoes a phase transformation during abrasion [32]. N-implantation is believed to interfere with this phase transformation, thereby lowering its capacity to work-harden. Vardiman, et al. [33] have identified the microstructures associated with polishing then N-implanting 304 steel. Polishing strain-hardened γ(fcc)-austenite phase transformed it to an α'(bcc)-martensite phase. N-implantation reverted the martensite back to a N-dilated austenite [34]; in addition, the austenite was stabilized against transforming to martensite during further polishing.

Stabilization of the austenite by N-implantation can increase the wear resistance of 304 steel during sliding. In its unstable condition, the near surface layer work-hardens and becomes a hard, brittle layer attached to a softer, ductile substrate [8]. With a N-implanted layer, the near surface remains austenitic, work-hardens less and thereby resists the deleterious sliding condition of a hard layer on a soft substrate. TEM studies have confirmed that martensite forms in wear tracks of 304 steel [8,35], but so far the microstructure of a wear track on N-implanted 304 steel has not been identified.

Finally, it appears that steels containing nitride formers (e.g., Cr,V,Al) are benefited more by N-implantation than low alloy steels. This may be due the chemical pinning of implanted N to these atoms [36] and its effect on maintaining implantation induced residual stresses. It has been shown that implanted N redistributes in metals without nitride formers but maintains a Gaussian profile in the presence of these alloys [13,23] up to fluences causing the surface to blister [37].

<u>Implantation of N into Ti Alloys.</u> Ti and Ti alloys have notoriously poor resistance to delamination in sliding wear. Several studies have shown that N implantation into Ti can impart wear resistance and lower friction [38,39]. (C implants have provided similar benefits but required post-implantation heat treatments [40]). The delay in severe wear of Ti-6Al-4V as

a function of N fluence has been demonstrated convincingly by Hutchings and Oliver [39]. Their results were obtained with a stationary ruby ball (d = 5mm) against a Ti-6Al-4V disk (speed = 3mm/sec) sliding in alcohol. Severe wear occurred immediately in a non-implanted disk, but was delayed from 1 to 10^4 cycles as implant fluences increased from 0.5 to 4×10^{17} N/cm². Before wear, the tracks remained smooth and the coefficient of friction stayed low, $\mu = 0.17$. Surface analysis showed that a thin $Ti(O_{1-x}N_x)$ layer formed during sliding and that low friction persisted so long as this layer remained intact. After removal of the layer, the tracks resembled non-implanted wear tracks, the wear rate increased a hundred-fold and the friction coefficient jumped to $\mu = 0.4$. This is a clear case of N-implantation delaying wear by changing the wear mode.

Surface hardness studies were performed on the same specimens. At fluences from 0.5 to 4×10^{17} N/cm² MIH values increased from 60% to 180%, leveling off at fluences greater than 1×10^{17} N/cm². Since no TiN precipitates were observed at 1×10^{17} N/cm², solid solution hardening was indicated. Moreover, since most of the hardening had occurred at lower fluences, it could not account for the nearly logarithmic increase in wear resistance with higher fluence. Hutchings and Oliver postulated that N promotes the growth of a wear resistant, low friction oxide layer. Alternatively, N-implantation may be changing the work-hardening behavior of Ti. Although no measurements have been performed on N-implanted Ti, RWR measurements on C-implanted Ti show increases up to a factor of 6.

Oliver, et al. [41] have reported the load-carrying capacities of both pure Ti and Ti-6Al-4V implanted with N. Using their data, I have calculated pressure-to-yield ratios ~ 50% greater than the ratios predicted by eq. 3 (see Table II). The calculated ratios, however, should be regarded as upper limit values since it is likely that the alcohol contributed to the load-carrying capacity and the actual yield strength of the surface was probably much higher than the bulk value [10].

<u>N-implantation into Electroplated Cr</u> has also changed the wear behavior of this hard metallic coating. Oliver et al. [26,29] showed that wear of the non-implanted Cr began when sliding induced transverse cracks. Next, fine debris generated at the edges of these cracks got trapped between the track and the slider and abraded the Cr surface. This mode of wear was prevented by N-implantation, and without changing the friction coefficient, resulting in a reduced wear rate (by a factor of 20-50 in the particular test) of the Cr plate. Surface hardness measurements showed N-implantation increased MIH values up to 30% [26].

<u>N-implantation into Stoody 3.</u> Low speed, dry sliding studies on N-implanted Stoody 3 showed the same friction and similar plastic deformation as on non-implanted Stoody 3 [20]. However, Co-debris particles were found in non-implanted wear tracks but only slider debris (steel, Stoody, Ni alloys) could be found in N-implanted tracks. Polishing wear measurements showed N-implantation reduced the RWR of Stoody 3. N-implantation appears to interfere with the work-hardening of Stoody 3, the mechanism responsible for giving it high abrasion resistance and suspected of reducing the sliding wear resistance.

N Migration and Wear

So far, no mention has been made of the mobile interstitial model which depicts N migrating below the wearing surface [42]. As intriguing and plausible as the idea may be, to date there is no irrefutable evidence for it in the literature. The often referred to results of Lo Russo, et. al. [43] should not be cited as evidence for N migration because no microprobe analysis of the wear scar was provided. A broad-beam nuclear reaction technique was used which detected 20% of the original N content in a region worn under reciprocating sliding. A likely source of N is debris that was worn from the implanted surface then entrapped in the scar. Cui Fu-Zhai, et. al. [44] provided a more detailed wear scar analysis, by secondary ion mass spectroscopy and profilometry, to support their finding that approximately 10% of the implanted N was detected in a track nearly 2µm below the original surface. However, their analyzing beam was also wider than 2 of the 3 tracks analyzed. In addition, all tests were performed at loads high enough to depress the track by plastic deformation. To date, the more detailed microprobe analyses of wear tracks on implanted surfaces have shown that implanted ions do not migrate below the worn surface [17,18,22, 23,24,25,29,35,39].

COMPOSITIONS AND MICROSTRUCTURES THAT INCREASE SLIDING WEAR RESISTANCE

Surface Compositions that Reduce Friction

Ti-implantation to high fluences ''carburizes'' the surface, thereby producing a low friction surface on metals; too low a fluence creates a highly adhering, high friction surface [16]. The fluences and energies needed to produce this surface have been described by Singer and Jeffries [45]. This implantation treatment is particularly beneficial for the harder, more wear resistant alloys such as 52100 [14] and 440C [25] steels, for which N-implantation does not appear to be very effective. For more ductile steels, dual implants of Ti + N or Ti + C are recommended. Singer and Jeffries [18] have shown that the two act synergistically: Ti reduces friction and N increases resistance to deformation-induced wear. Other strong carbide forming implants such as Ta [23], Hf, Nb, and probably Zr and V, should provide similar friction and wear benefits as Ti, with the possibility of increased corrosion resistance as well [46].

Ion implantation can also promote the formation of lubricious (i.e. low friction) oxides on metals. As indicated earlier, Hutchings and Oliver [39] have suggested that implanted N atoms in Ti may have stimulated the growth of a wear resistant Ti oxynitride layer. Moreover, thick oxynitride layers which can form during N-implantation might decrease friction during run-in. The wear resistant, low friction films that Shephard and Suh [38] observed on N-implanted annealed Fe and Ti were probably oxynitrides, not metal nitrides.

Microstructures that Increase Fracture Resistance

In order to resist wear, the implanted layer must be able to withstand the stresses induced during sliding without fracturing. One way ion implantation has been able to accomplish this is by hardening the surface. The implant species can harden the surface directly by forming second phases, solution strengthening and pinning dislocations [31]. It can also contribute, indirectly, by dissolving phases already present, causing solutes in the lattice to redistribute and by creating defects. However, as indicated by the hardness and wear studies of Ti-6Al-4V, hardening of the surface alone cannot account for the wear improvements [39]. Moreover, Shephard and Suh [38] have concluded from theoretical studies that a hardened implanted layer cannot increase the load carrying capacity during sliding unless it also reduces friction, which is not the case for most N-implanted metals.

A second way in which implantation can protect against wear is to produce microstructures that resist excessive work hardening. Results from polishing wear studies demonstrate that ion implantation can modify the work hardening behavior of metals [12,13]. The N stabilized austenitic phase of type-304 steel is one example of a microstructure less prone to sliding wear failure [33]. Other examples considered in the literature are the stabilization of the fcc phase of Co in Stoody 3 (against the fcc to hcp transition) by N-implantation [20], the amorphization of steel surfaces by

Ti-implantation [14,22], and the hexagonal lattice expansion (i.e. c/a ratio increase) of Ti to modify its operating slip system by N-implantation [38].
There has recently been a proliferation of studies on microstructures formed in ion-implanted metals. The number of phases identified in steels, alone, by conversion electron Mossbauer spectroscopy [47,48] is overwhelming. Deciding which microstructures are responsible for any tribomechanical effect is a challenging task.

A third, more general way in which implantation may protect against wear is by introducing residual stresses in the surface. Residual compressive stresses introduced by high fluence implantations (of virtually any atom) have been shown to be extremely high, possibly in excess of the bulk yield strength of the metal itself [49]. In ductile metals, these stresses may counter those produced during sliding and may prevent cracks from growing and plate-like particles from lifting off the track. The photomicrographs of Fig. 2 show clearly that N-implantation into the two softer steels has delayed incipient plate formation. In harder, more brittle metals such as electroplated Cr [29] or nitrided steel [50], a net compressive stress at the surface may prevent cracks from opening and the surface from fracturing.

CONCLUSIONS AND RECOMMENDATIONS

A survey of the tribomechanical properties of ion implanted metals has suggested four mechanisms by which an ion implantation treatment can reduce sliding wear in metals. It can 1) produce a low friction surface by altering the surface chemistry; 2) modify the microstructure to harden the surface; 3) stabilize microstructures against deleterious work-hardening effects; and 4) introduce residual stresses to combat fracture. A summary of the effects on selected ion-metal substrates is given in Table III. By preventing or delaying the formation of wear particles, implantation alters the wear mode during the early ''run-in'' stage of wear. The change in wear mode has permitted several of the ion-implanted metals to withstand Hertzian pressures that reach the elastic-deformation limit of the metal (see Table II). Most implantation treatments have not yet been optimized, so it is likely that future treatments (e.g. dual implants, high temperature implants, ion beam enhanced deposition), will provide longer lasting, higher load carrying capacities for a wide variety of metals.

Table III. Tribomechanical properties affected by ion implantation
(O) no change; (I) increase; (D) decrease; (N.D.) no data

Implanted Species	Substrate	Friction Coef.	Wear Scar Analysis	Surface Hardness	Polishing Resistance
Ti	52100	D	No oxide debris	O	I
N	Ti-6Al-4V	D	Submicron oxide debris; delayed plates	I	N.D.
N	ferritic steels	O	Delayed plates	I	I
N	austenitic steel	O	Delayed plates	I/D	D
N	martensitic steel	O	No change	O	O

This review has been limited to investigations of polished surfaces, implanted at or near room temperatures then subjected to low speed sliding. In practice, engineering surfaces have rougher finishes due to machining, grinding or lapping; ion beams can easily heat a substrate above 500°C at typical beam power densities of 2W/cm² or greater; and components may be run at high speeds or in high temperature environments. Future investigations must address these issues and tackle the difficult but important task of characterizing the surface.

To illustrate the second point, there are several ion beam/surface interactions, besides implantation alloying, that can affect the tribological behavior of surfaces. Poor vacuum conditions during implantation can contaminate surfaces with carbon, which may be lubricious [26]. Excessive target heating during implantation in a poor vacuum can lead to the growth of extremely thick oxide layers. In alloys, heating can also promote decarburization and surface segregation of impurities (e.g. Si, S, B). Sputtering, especially common with heavy atoms implanted to high fluences, can remove native oxide layers, may alter the near surface composition [15] and can produce unexpected surface textures [19]. Each of the above implantation effects may change the tribological behavior of implanted metals, but not by the expected mechanism. Surface analytical studies should be carried out in conjunction with wear studies so that unanticipated surface films are not overlooked.

The author acknowledges his collaborators at NRL for their ongoing contributions, Jim Murday, Ron Vardiman and Fred Smidt for critical comments on the manuscript, and ONR for supporting the research.

REFERENCES

1. F.P. Bowden and D. Tabor, Friction and Lubrication of Solids (Oxford Press, Oxford). Part I (1950); Part II (1964).
2. H. Czichos, Tribology (Elsevier, Amsterdam, 1978).
3. D.H. Buckley, Surface Effects in Adhesion, Friction, Wear and Lubrication (Elsevier, Amsterdam, 1981).
4. N.P. Suh, The Delamination Theory of Wear, Wear 44 (1977).
5. Fundamentals of Friction and Wear of Materials, D.A. Rigney, ed. (ASM, Metals Park, OH, 1981).
6. R. Glardon and I. Finnie, ASME Trans 103, 333-340 (1981).
7. D.A. Hills and D.W. Ashelby, Wear, 75, 221-240 (1982).
8. K.L. Hsu, T.M. Ahn and D.A. Rigney, Wear 60, 13 (1980).
9. J.W. Ho, C. Noyan, J.B. Cohen, V.D. Khanna and Z. Eliezer, Wear 84, 183-202 (1983).
10. R.C.D. Richardson, Wear 10, 353-382 (1967).
11. J.B. Pethica: in Ion Implantation into Metals, V. Ashworth, ed. (Pergamon Press, Oxford, 1982) pp 147-156.
12. R.N. Bolster and I.L. Singer Appl. Phys. Letts. 17, 327 (1980).
13. R.N. Bolster and I.L. Singer ASLE Trans. 24, 526 (1981).
14. C.A. Carosella, I.L. Singer, R.C. Bowers and C.R. Gossett, in: Ion Implantation Metallurgy, C.M. Preece and J.K. Hirvonen, eds. (AIME, Warrendale, PA 1980) p. 103.
15. I.L. Singer, C.A. Carosella and J.R. Reed, Nucl. Instrum. Methods 182/183, 923 (1981).
16. I.L. Singer and R.A. Jeffries, J. Vac. Sci. Technol. A1, 317 (1983).
17. T.E. Fischer, M.J. Luton, J.M. Williams, C.W. White and B.R. Appleton, ASLE Trans. 26, 466 (1983).
18. I.L. Singer and R.A. Jeffries, ''Friction, Wear and Deformation...'', these proceedings.
19. S.A. Dillich and I.L. Singer, Thin Solid Films 108, 219-227 (1983).
20. S.A. Dillich, R.N. Bolster and I.L. Singer, these proceedings.
21. I.L. Singer and T.M. Barlak, Appl. Phys. Lett. 43, 457-459 (1983).
22. D.M. Follstaedt, F.G. Yost and L.E. Pope, these proceedings.
23. I.L. Singer, Appl. Surface Sci. 18, (1984).
24. L.E. Pope, F.G. Yost, D.M. Follstaedt, J.A. Knapp and S.T. Picraux, in: Wear of Materials - 1983, K.C. Ludema, ed. (ASME, New York, NY 1983), 280.
25. L.E. Pope, F.G. Yost, D.M. Follstaedt, S.T. Picraux and J.A. Knapp, these proceedings.

26. J.B. Pethica, R. Hutchings and W.C. Oliver, Nucl. Instrum. Methods., 209/210, 995-1000 (1983).
27. W.C. Oliver, R. Hutchings, J.B. Pethica, I.L. Singer and G.K. Hubler, these proceedings.
28. I.L. Singer, R.N. Bolster and C.A. Carosella, Thin Solid Films, 73, 283 (1980).
29. W.C. Oliver, R. Hutchings and J.B. Pethica, Metall. Trans, A (to be published).
30. F.G. Yost, S.T. Picraux, D.M Follstaedt, L.E. Pope, and J.A. Knapp, Thin Solid Films, 107 287-295 (1983).
31. H. Herman, Nucl. Instrum. Methods, 182/183, 887-898 (1981).
32. C. Allen, A. Ball and B.E. Protheroe Wear, 74, 287-305 (1981).
33. R.G. Vardiman, R.N. Bolster and I.L. Singer, MRS Symp. 7, 269 (1982).
34. R.G. Vardiman and I.L. Singer, Material Letts. 2, 150-154 (1983).
35. D.M. Follstaedt, F.G. Yost, L.E. Pope, S.T. Picraux and J.A. Knapp, Appl. Phys. Lett. 43, 358-360 (1983).
36. I.L. Singer and J.S. Murday, J. Vac. Sci. Technol. 17, 327-329 (1980).
37. W.M. Bone, R.J. Colton, I.L. Singer and C.R. Gossett, J. Vac. Sci. Technol. A2 (1984).
38. S.R. Shephard and N.P. Suh, J. Lub. Technol., 104, 29-38 (1982).
39. R. Hutchings and W.C. Oliver, Wear 92, 143-153 (1983).
40. R.G. Vardiman, these proceedings.
41. W.C. Oliver, R. Hutchings, J.B. Pethica, E.L. Paradis and A.J. Shuskas, these proceedings.
42. N.E.W. Hartley, in: Ion Implantation, J.K. Hirvonen, ed. (Academic Press, NY, 1980) p. 321.
43. S. Lo Russo, P. Mazzoldi, I. Scotoni, C. Tosello and S. Tosto, Appl. Phys. Letts. 34, 627 (1979).
44. F.Z. Cui, H-D. Li, X-Z. Zhong, Nucl. Instrum. Methods 209/210, 881-887 (1983).
45. I.L. Singer and R.A. Jeffries, "Processing Steels by Ti-implantation", these proceedings.
46. G.K. Hubler, P. Trzaskoma, E. McCafferty and I.L. Singer, in: Ion Implantation into Metals, V. Ashworth, ed. (Pergamon Press, Oxford, 1982), 24-34.
47. G. Marest, C. Skoutarides, Th. Barnavon, J. Tousset, S. Fayeulle and M. Robelet, Nucl. Instrum. Methods 209/210, 259-265 (1983).
48. E. Ramons, G. Principi, L. Giordano, S. Lo Russo, and C. Tosello, Thin Solid Films, 102, 97-106 (1983).
49. N.E.W. Hartley, J. Vac. Sci. Technol. 12, 485 (1975).
50. G.M. Ecer, S. Wood, D. Boes and J. Schreurs, Wear 89, 201-214 (1983).

EFFECTS OF NITROGEN AND HELIUM ION IMPLANTATION ON UNIAXIAL TENSILE
PROPERTIES OF 316 SS FOILS

J. A. SPITZNAGEL,* B. O. HALL,* N. J. DOYLE,* RAMAN JAYRAM,**
R. W. WALLACE,** J. R. TOWNSEND,** AND M. MILLER***
*Westinghouse R&D Center, Pittsburgh, PA 15235; ** University of
Pittsburgh, Pittsburgh, PA 15260; *** Oak Ridge National Laboratory,
Oak Ridge, TN 37830

ABSTRACT

Implantation of nitrogen into steels is known to affect surface sensitive mechanical properties. Tensile properties of thin foils implanted with either nitrogen or helium at 300 K have been measured. Fluences greater than 1×10^{16} ions/cm^2 raise the yield stress and fracture stress and reduce the plastic strain to failure. Both nitrogen and helium give comparable stress-strain responses for equal average concentrations of implanted ions. The mechanical response is discussed in terms of plastic flow of laminated structures and hardening mechanisms. Initial results of atom probe field ion microscopy examinations of nitrogen implanted Fe-15 wt.% Cr-12 wt.% Ni alloy are described.

INTRODUCTION

Implantation of nitrogen into steels often results in altered mechanical and electrochemical properties. The effects have been attributed to the formation of various nitride phases, implantation induced residual stresses and decoration of mobile dislocations by the implanted solute atoms [1]. Auger spectroscopy and T.E.M. studies [2,3], for example, have indicated that CrN may precipitate in 304 and 316 type austenitic stainless steels during implantation at ∼300 K when implanted concentrations exceed ∼2-8 atomic percent. Cantilever beam deflection experiments, however, suggest that large residual stresses approaching the yield stress may be introduced at much lower fluences [4,5].

Theoretical treatments of surface hardening or softening presuppose a knowledge of the effective volume change per implanted ion, which determines the magnitude of the residual stresses and of the nature and distribution of barriers to dislocation motion, e.g., precipitate particles, small dislocation loops, etc.[6]. Such information has been very difficult to obtain experimentally because the microstructural changes occur on a very fine scale for room temperature implants. In this paper we examine the feasibility of using simple uniaxial tension tests and atom probe field ion microscopy to deduce the origin(s) of mechanical property changes arising from implantation of a chemically active species (nitrogen) or an inert gas ion (helium) in polycrystalline 316 SS.

EXPERIMENTAL

Thin foils of polycrystalline 316 SS, five microns thick, with grain sizes larger than the foil thickness and initial nitrogen concentrations ∼1600 appm, were prepared by cold rolling bulk strip with intermediate anneals. This same material had been used in a previous extensive study of the stress-strain behavior of unimplanted micron thickness foils [7]. As in the earlier study, tensile specimens were stamped from the foil in hard

Supported in part by NSF grant DMR-81-02968.

rubber dies with the specimen axis parallel to the rolling direction. Specimen dimensions (gage section 9.4 mm long by 2.5 mm wide) and heat treatment after stamping (1173K for 0.2h in vacuum) were chosen because they provided highly reproducible, bulk representative flow properties in uniaxial tension [7].

The specimens could thus be considered as thin sections of bulk material and handbook values of Young's Modulus and yield stress were obtained for unimplanted foils using the thin foil test apparatus shown in Figure 1.

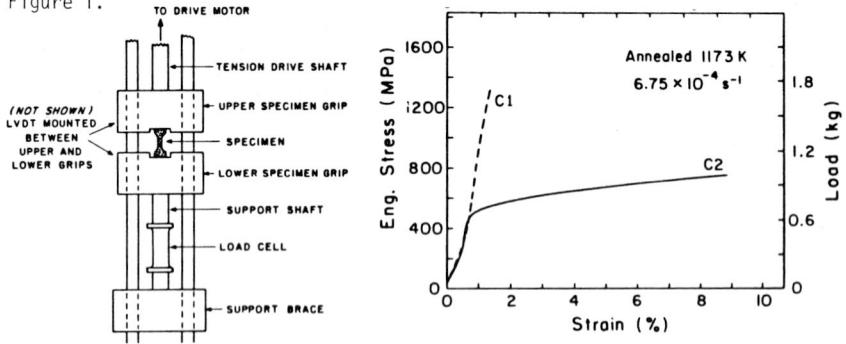

Figure 1. Schematic of Thin Foil Tensile Test Apparatus.

Figure 2. Effect of Annealing on Stress-Strain Curves of Unimplanted Foils.

Foils were implanted on both sides with He^+ or N^+ ions having energies of 0.75 MeV and 1.9 MeV, respectively. The calculated [8] projected range, R_p, for both ions was 1.3 μm or approximately one quarter of the foil thickness. Calculated rms straggle (ΔR_p) was ~0.07 μm for helium and 0.12 μm for nitrogen. Ion fluxes of 6.25 x 10^{12} $He^+/cm^2 s$ or 2.3 x 10^{12} $N^+/cm^2 s$ maintained a constant beam loading of 0.75 W/cm² on the specimen during implantation. The ion beam was rastered to cover the gage section and a portion of the grip sections on each specimen. Foils were attached to an aluminum holder with silver paste. Care was taken in mounting so that no paste was exposed to the beam. Samples were demounted by soaking in acetone and subsequently cleaned with an ultrasonic cleaner to completely remove any traces of silver paste. The temperature of the sample holder was maintained at ~300K by active cooling during implantation. Table 1 summarizes the implantation conditions. A minimum of four foils were implanted at each condition and subsequently pulled to failure at a strain rate of 6.75 x 10^{-4} per second. The fluences of He and N were chosen to provide a comparison based on equal average concentrations of implanted solute, equal maximum damage levels (dpa) or equal integrated damage. Range statistics and damage energy values were calculated with the EDEP-1 Code [8].

TABLE 1. Implantation Conditions

Ion	Energy (MeV)	Fluence, Ψ, (ions/cm²)	Condition
N^+	1.9	1.0 x 10^{16}	4300 appm
He^+	0.7	7.1 x 10^{15}	<He>=<N>~ 4300 appm
He^+	0.75	1.0 x 10^{17}	$dpa_{He}=dpa_N$ (at peak damage depth)
He+	0.75	1.4 x 10^{17}	$\int_0^{R_p + \Delta R_p} S_D(x)_{He^+} dx = \int_0^{R_p + \Delta R_p} S_D(x)_{N^+} dx$

Average Concentration $<> = \dfrac{\psi \times 10^6}{2.35\,(\Delta R_p\,N_0)}$; N_0 = atomic density of stainless steel

$\mathrm{dpa} = \dfrac{0.8\,S_D(x)\,\psi}{2E_d\,N_0}$; $S_D(x)$ = energy deposited into atomic displacements
E_d = threshold displacement energy $\sim 40\,\mathrm{eV}$

Initial attempts at producing needles suitable for atom probe field ion microscopy (APFIM) from the implanted foils have been unsuccessful. Consequently it was decided to implant preformed tips of a simple ternary alloy of similar major element content to the 316 SS. Ten needles of the Fe-15 W/o Cr-12 W/o Ni alloy prepared by electropolishing thin wires were implanted with 70 keV N^+ at constant ion flux to concentrations of ~ 5 and 15 atomic percent in a copper holder. Temperature of the holder was maintained at ~ 300K although the actual temperature of each tip (radius of curvature 50-100 nm) was undoubtedly higher. It was assumed that this would not pose a major complication since the phases Cr_2N and CrN should be stable up to ~ 1173K [9]. As will be shown, this assumption was probably incorrect. The specimen holder was placed inside a cryoshield during implantation to avoid hydrocarbon contamination. Neon imaging of the tips and time-of-flight mass spectrometry were conducted at 78K in an atom probe at the U.S. Steel Research Laboratory. A detailed description of APFIM techniques can be found elsewhere [10].

RESULTS AND DISCUSSION

Typical results of tensile tests on the implanted foils are given in figures 3 and 4. Average values and maximum/minimum values of yield stress, fracture stress, plastic strain and work hardening rate for all conditions are given in table 2. Yield stress was determined by a simple intercept method using straight line approximations for the elastic and plastic regions. Stresses and strains have been obtained by dividing the load and elongation by the initial cross-sectional area and gage length, respectively.

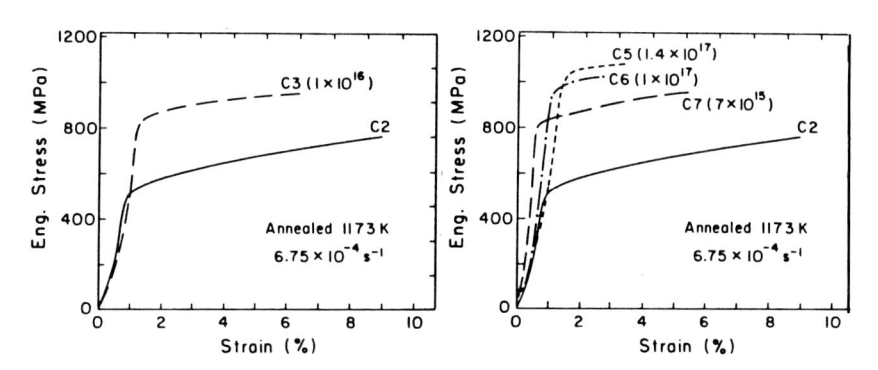

Figure 3. Effect of nitrogen implantation on stress-strain behavior of 5 μm thick foils.

Figure 4. Effect of the helium fluence on stress-strain behavior of 5 μm thick foils.

TABLE 2. Tensile properties of unimplanted and nitrogen or helium implanted 5 μm thick foils of 316 SS

Curve	Species	ψ (ions/cm²)	$\sigma_y \pm \Delta\sigma_y$ (MPa)	$\sigma_f \pm \Delta\sigma_f$ (MPa)	$\varepsilon_p \pm \Delta\varepsilon_p$	$\eta \pm \Delta\eta$ (MPa)
C1	Cold Worked	---	---	1341 ± 91	---	---
C2	Annealed	---	560 ± 50	760 ± 50	8.05 ± 0.89	24.84 ± 3.40
C3	N⁺	1.0 × 10¹⁶	880 ± 52	960 ± 18	4.25 ± 0.57	17.95 ± 3.15
C7	He⁺	7.1 × 10¹⁵	820 ± 33	950 ± 36	4.83 ± 0.19	26.91 ± 4.64
C6	He⁺	1.0 × 10¹⁷	970 ± 40	1020 ± 30	1.83 ± 0.19	27.32 ± 5.78
C5	He⁺	1.4 × 10¹⁷	1050 ± 53	1030 ± 43	2.00 ± 0.44	15.00 ± 10.2

ψ = fluence
σ_y = yield stress
σ_f = fracture stress
ε_p = plastic strain to fracture
η = work hardening rate ($= \frac{d\sigma}{d\varepsilon_p} \sim \frac{\sigma_f - \sigma_y}{\varepsilon_p}$)

The principal effect of implanting either N⁺ or He⁺ is to raise the yield stress and the fracture stress ($\sim$ tensile strength) and to reduce the amount of plastic strain to fracture. A general discussion of the tensile properties of thin foils is beyond the scope of this paper and may be found elsewhere [12]. The main conclusion from figures 3 and 4 and tables 1 and 2 is that the stress-strain response for foils implanted with either N or He is almost identical for equal average concentrations of implanted ions. When the implanted concentrations are normalized to give equal displacements per atom at the peak damage depth or equal amounts of energy deposited into displacement processes over the range of the ions, helium produces a much greater increase in strengthening and loss of ductility. The tensile properties change smoothly with increasing helium ion fluence. SEM observations of fracture surfaces have shown that specimens implanted with $\sim$4330 appm N or He neck down smoothly through the thickness to produce almost a chisel point failure. At higher helium concentrations cleavage failure is observed in the outer implanted layers and demarcations between damaged, damaged and doped and unimplanted regions are clearly seen. The results suggest that these 5 μm foils, implanted to approximately one quarter of their thickness from either side, can be considered perfectly bonded laminate structures, and the stress-strain curves can be analyzed using the general theories for the uniform deformation of multilayer bodies [12]. The analysis is complex and will be presented in a later paper. We will simply summarize a few general conclusions. (1) Even with this simple loading geometry, a uniform deformation coincides with a nonuniform stress distribution across the multilayer body under longitudinal stress. (2) The residual stresses resulting from implantation must be taken into account. (3) The measured instantaneous-yield strength lies between the maximum and minimum values of the instantaneous yield strength of the components. Thus either the first zone containing both displacement damage and/or the atoms has a yield stress higher than that measured for the composite. (4) Preliminary estimates of the hardening due to dislocation loops in the first zone and dislocation loops plus nitride precipitates or small bubbles ($\leq$ 1 nm diameter) in the second zone [6] account for a factor of $\sim$1.5 increase in yield stress. This is true even if relative zone widths and residual stress effects are taken into account. It appears necessary to consider hardening of the unimplanted core due to pinning of glide dislocations extending into the second zone to account for measured effects of implantation on the stress-strain curves.

Examination of the nitrogen-implanted Fe-15Cr-12 Ni needles in the atom probe has revealed a darkly imaging blunted surface that resists field evaporation. The underlying metal contains 0.5-0.8 atomic percent nitrogen uniformly distributed over the tip cross-section. This matrix concentration does not change when the nitrogen ion fluence is increased by a factor of three. No small precipitate particles have been imaged in the matrix, but some clustering was observed. These results suggest that the implanted nitrogen has considerable mobility. The invariant matrix concentration indicates an effective nitrogen solubility limit for the Fe-Cr-Ni alloy under the implant conditions. Although no direct evidence for Cr_2N or CrN has been obtained yet, the constant matrix concentration suggests that a nitride phase must be present in thermodynamic equilibrium with the matrix. Because of beam heating the nitride particles in the needles could be large but present in small number density ($<10^{15}$ cm^{-3}), and would probably escape detection with the atom probe. Future experiments will concentrate on TEM observations of the tips to monitor sputtering effects and formation of coarse nitride particles. Lower incident ion fluxes will be used to maintain lower tip temperatures during implantation.

ACKNOWLEDGEMENTS

The authors wish to thank R. B. Irwin for assistance with the implantation; J. Sheehan and J. N. McGruer for assistance with the uniaxial test apparatus and J. Rabel and M. Bruno for assistance with software development.

REFERENCES

1. H. Herman, Nucl. Instrum. & Methods 182/183, 887 (1981).
2. I. L. Singer and J. S. Murday, J. Vac. Sci. Technol. 17 (1), 59 (1980).
3. M. Baron, A. L. Chang, J. Schreurs and R. Kossowsky, Nucl. Instrum. & Methods 182/183, 531 (1981).
4. E. P. Eernisse and S. T. Picraux, J. Appl. Phys. 48, 9 (1977).
5. N.E.W. Hartley, J. Vac. Sci. Technol. 12 (1), 485 (1975).
6. B. O. Hall, J. Nucl. Mater. 116, 123 (1983).
7. D. L. Harrod and D. A. Kaminski, Mech. Res. Comm. 5 (6), 319 (1978).
8. I. Manning, G. P. Mueller, Computing Physics Communications 7, 85 (1974).
9. Diagrammy Sostoianiia Metallicheskikh Sistem; Nauka, Moscow p. 135 (1971).
10. S. S. Brenner and M. K. Miller, J. of Metals 35, (3), 54 (1983).
11. A. Lawley and S. Schuster, Trans. AIME 230, 27 (1964).
12. G. E. Arkulus, Compound Plastic Deformation of Layers of Different Metals, Daniel Davey & Co., Inc., New York, NY (1965), p. 44.

HARDNESS AS A MEASURE OF WEAR RESISTANCE

W.C. OLIVER[1], R. HUTCHINGS[2], J.B. PETHICA[3], I.L. SINGER[4], AND G.K. HUBLER[5]
[1] United Technologies Research Center, E. Hartford, CT 06108;
[2] Brown Boveri Research Center, CH-5405 Baden, Switzerland;
[3] Cavendish Lab., Cambridge, CB3 OHE, U.K.;
[4,5] Naval Research Laboratory, U.S.A.

ABSTRACT

One measure of the surface mechanical properties of materials can be obtained through microhardness data. The success of microhardness in predicting the improvements in wear resistance of ion implanted metals has been mixed. In this paper the cases of N implantation into 304 S.S. and Ti implantation into 52100 bearing steel will be examined. Microhardness data indicates little or no hardness changes whereas large wear rate changes are observed. From these two examples it is clear that the wear mechanism, the chemical nature of the surface, the ductility, and the toughness can be more important than the hardness changes.

INTRODUCTION

Two tribological systems of technological interest that have been studied extensively are N implanted 304 S.S. and Ti implanted 52100 bearing steel. Much of the early work on these systems involved extensive wear testing and little evidence was collected concerning the mechanism by which the improvements might occur. General explanations involving increased hardness of the implanted surface as giving rise to better wear resistance have been offered.

The hardness of a material does affect its wear rate; however, other parameters can be more important. Hardness gives some indication of strength. Other mechanical properties of a material that help determine its wear rate are ductility, toughness and the temperature dependence of these quantities. Hornbogen [1] and Atkins [2] consider the effects of some of these parameters on wear rates. The mechanical structure of the surface is also important. For example, a very hard surface on a very soft substrate does not result in high wear resistance if high contact stresses occur. Finally, chemical effects between each component of the wear couple and the atmosphere and between the two components themselves can also be extremely important.

Many of these properties are very difficult to measure. Ease of measurement has led to hardness being used extensively when considering wear experiments and modeling of wear processes. In the case of ion implanted surfaces, even hardness is very difficult to measure.

A hardness tester that is capable of sampling the very thin implanted layers has been constructed and used to measure the hardnesses of implanted surfaces [3-5]. The results of these tests, wear data, and microanalysis results for a range of implanted metals have been presented [6].

By comparing hardness and wear data, it has been shown that indentation hardness is qualitatively correct in predicting pin on disc wear property improvements. The magnitude of the improvements is much less predictable from the hardness data. In fact, in the cases where dramatic improvements in wear properties are observed, the improvements are too large to be explained through the effect of hardness on a given wear mechanism. For the cases of Ti-6Al-4V and hard electroplated chromium, N implantation can cause a change in the dominant wear mechanism [6]. The important point is, if the hardness of a surface is increased through implantation, some increase in wear resistance can be expected; however, other properties of the surface may also effect the wear rate.

With this discussion as background, we shall now proceed to two specific cases for which the evidence appears to be conflicting. These are nitrogen implanted 304 S.S. and titanium implanted 52100 bearing steel. The data in question concerns the indentation hardness, the surface abrasion resistance, and pin on disc type wear data that has been collected for these two systems.

The surface abrasion, relative wear resistance (R.W.R.) test was introduced by Bolster and Singer [7] as a relatively simple way to get some indication of the hardness of thin surface layers. This is a reasonable approach if all the parameters, other than hardness, that effect abrasive wear are held constant. The apparent inconsistencies arise when actual indentation hardness tests, taken using the machine mentioned above, do not match the R.W.R. results.

The hardness data presented here will highlight these apparent inconsistencies. The discussion of the results will help explain them.

EXPERIMENTAL

The 52100 specimens were implanted at the Naval Research Laboratory (N.R.L.). The conditions for the Ti implantation of the 52100 steel specimens have been previously reported [8]. The carbon was implanted at 40 keV to a dose of 2×10^{21} ions/m^2. The Ta was implanted to a dose of 1.8×10^{21} ions/m^2.

One 304 stainless steel specimen was polished to a 9 μm diamond abrasive finish. Part of the surface was electropolished and part was ion milled with 1 keV Ar ions. Part of the resulting surface was then implanted with 1×10^{21} N$^+$/m^2 at 40 keV. Thus the sample had surfaces treated as shown in Figure 2 available for hardness testing.

The other 304 S.S. specimen was polished to a 1 μm diamond finish and one section of the surface was implanted with N_2^+ at 90 keV to a dose of 2.5×10^{21} ions/m².

RESULTS AND DISCUSSION

First let us consider 52100 bearing steel. This material is used in the through hardened and slightly tempered condition. It has been shown that nitrogen implantation is not effective at hardening the surface of this material or improving its wear resistance [6,9]. For this reason extensive work has been carried out at the Naval Research Laboratory on Ti, Ta and Ti+C implantations of this material [8-13]. Positive improvements in wear life have been achieved. In addition the surface sensitive abrasive wear test used at N.R.L. indicated a factor of 6 increase in the relative wear resistance (R.W.R.) of the Ti implanted surfaces [8]. A dose of 5×10^{21} ions/m² titanium is known to produce an amorphous film on 52100 bearing steel [11]. It has been established that the titanium implantations are most effective when a significant amount of carbon is gettered from the atmosphere. Auger results have shown that the carbon is bound to the titanium in the same way as in TiC; however, there is no evidence that TiC exists as a second phase.

52100 STEEL

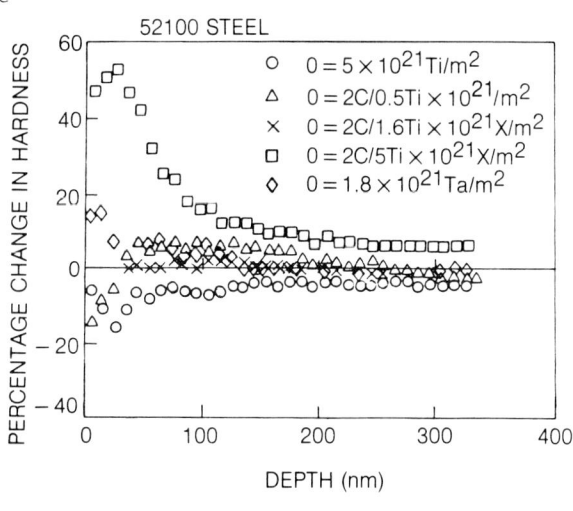

○ $\varnothing = 5 \times 10^{21}$ Ti/m²
△ $\varnothing = 2C/0.5Ti \times 10^{21}$/m²
✕ $\varnothing = 2C/1.6Ti \times 10^{21}$ X/m²
□ $\varnothing = 2C/5Ti \times 10^{21}$ X/m²
◇ $\varnothing = 1.8 \times 10^{21}$ Ta/m²

Figure 1

To confirm that the surface of 52100 bearing steel is hardened by Ti and Ta implantation, five samples were tested for indentation hardness versus depth. This data is presented in Figure 1. The first surprising result shown in Figure 1 is that Ti or Ta implanted by themselves did not increase the hardness of 52100 steel. In fact, the only effective hardening process was implanting with 2×10^{21} ions/m² C and 5×10^{21} ions/m² of Ti. This dose level of titanium corresponds well with the critical level for positive wear results which have been reported; however the wear results were determined without C implantation. It is clear that a minimum amount of both carbon and titanium are required for positive hardness results. In the presence of 2×10^{21} ions/m² of C the threshold for hardness increases is between 1.6 and 5×10^{21} ions/m² of Ti. The wear improvements demonstrated in both the abrasive wear test (R.W.R.) and the pin on plate tests were from samples that according to Figure 1 should

not have significantly hardened surfaces; thus, hardness changes cannot
be responsible. This is a clear example of the dangers of interpreting
abrasive wear resistance directly in terms of hardness. The dose level
associated with wear property improvements correlates with the dose required to obtain an amorphous layer. It may be some property of that layer
that is important. Some possibilities are ductility or toughness. The
hardness data indicates that further wear property improvements might be
obtained if the surface was hardened by dual implantation of C and Ti.

The apparent inconsistency in the data for this material, (that is
improvements in R.W.R., no hardness increase, and pin on plate wear
improvements), are rationalized when two facts are considered. First,
R.W.R. is actually a wear test. Second, properties other than hardness
can be important in determining wear rates.

The second material we will discuss is 304 S.S. Significant increases
in pin on disc wear resistance have been reported through N implantation
of this material [6,14]. Figure 2 shows the indentation hardness versus
depth curves for a 304 S.S. surface modified in several ways. The base
line (0% change) is the hardness of electropolished, ion milled, unimplanted
surface. The error in the measurements at less than 40 nm depth is
large, particularly for this specimen which was only polished with 9 μm
diamond. The data in this region should be considered with caution. It
is clear that implanting electropolished 304 S.S. increases the near surface hardness. This has been measured to cause an increase of at least
50% at 40 nm on two separate specimens. The change in hardness on implantation of the mechanically polished 304 S.S. shown in Figure 2 is
more difficult to interpret due to a lack of a mechanically polished base
line. The change in hardness due to N implantation of a 1 μm diamond
polished 304 S.S. sample is shown in Figure 3. It is clear that N implantation increased the surface hardness of this specimen by about 40-60%.
The change for the mechanically polished material shown in Figure 2 is
certainly no greater than 25% and perhaps significantly less. However,
if one interprets the previously reported R.W.R. results for mechanically
polished 304 S.S. [7] in terms of hardness one would expect a 50% decrease.
This does not agree with the 40-60% increase shown in Figure 3 or the
data presented in Figure 2. It is clear from Figure 2 that the hardness
at 400 nm is greatly increased (∿ 50%) by mechanical polishing. It seems
that the easiest way to harden the surface of 304 S.S. is through cold work.

The decrease in relative wear resistance near the surface of the implanted material has been explained by a reversion of a work-hardened
surface martensite to softer austenite [7]. It is clear that strain induced martensite does revert to austenite due to nitrogen implantation [14];
however, it is not clear under what polishing conditions the surface will
contain a large fraction of martensite. The samples used to measure the
R.W.R. and indentation hardness results were polished using vibratory
and wheel techniques, respectively. This in itself could explain the
discrepancy between these results. It is also possible the implanted
austenite is as hard or harder than the martensite from which it is formed.
Other properties than hardness may cause the decrease in R.W.R. and increase in pin on disc wear properties.

BASE LINE: ELECTRO POLISH AND ION MILLED
○ BASE LINE + 1 × 10^{21}N$^+$/m^2 AT 40 KeV
△ 9μm MECH POLISH + 1 × 10^{21}N$^+$/m^2 AT 40 KeV

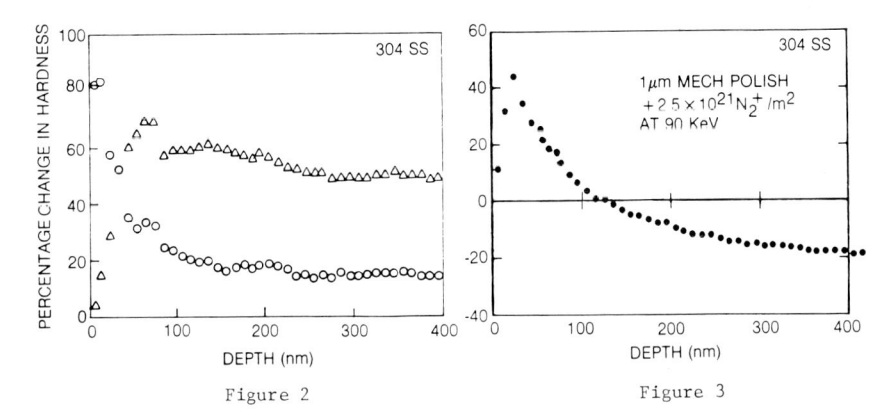

Figure 2

Figure 3 (304 SS, 1μm MECH POLISH + 2.5 × 10^{21}N$_2^+$/m^2 AT 90 KeV)

The two examples discussed above illustrate the errors obtained if on interprets surface abrasive wear results in terms of hardness changes. The R.W.R. test is an important tool for learning about the abrasive wear properties of implanted layers. Any consistent model of the behavior of such layers should be able to explain R.W.R. and pin disc results; however, what effect hardness has on the results can easily be masked by other effects.

CONCLUSIONS

1. Hardness is not the important property through which the wear properties of 304 S.S. or 52100 are improved by N and Ti implantation, respectively.

2. The surface hardness of 52100 steel can be increased by dual implantation of C and Ti.

3. The surface hardness of electropolished 304 S.S. can be increased by at least 50% through nitrogen implantation; however, the surface of mechanically polished 304 S.S. is only increased at most by 10-20%.

4. Indirect estimates of hardness through abrasive wear tests can yield misleading results.

REFERENCES

1. E. Hornbogen, Wear 33, 251 (1975).
2. A.G. Atkins, Wear 61, 183 (1980).
3. J.B. Pethica and W.C. Oliver, "Metastable Materials Formation By Ion Implantation", S.T. Picraux and W.J. Choyke, eds., Elsevier, New York (1982) p. 373.

4. J.B. Pethica: "Ion Implantation into Metals", V. Ashworth, W.A. Grant, and R.P.M. Proctor, eds., Pergamon Press, New York (1982) p. 147.
5. J.B. Pethica, R. Hutchings and W.C. Oliver, Phil. Mag., in press.
6. R. Hutchings, W.C. Oliver, and J.B. Pethica, to be published in Proc. of NATO/ASl on "Surface Engineering", Les Arcs, France, 1983.
7. R.N. Bolster and I.L. Singer, ASLE 24, 526 (1981).
8. I.L. Singer, R.N. Bolster and C.A. Carosella, Thin Solid Films 73, 283 (1980).
9. I.L. Singer and R.A. Jeffries, J. Vac. Sci. Technol. A1, 317 (1983).
10. C.A. Carosella, I.L. Singer, R.C. Bowers, and C.R. Gossett, "Ion Implantation Metallurgy", Eds., C. Preece and J.K. Hirvonen, AIME, (1980) p. 103.
11. I.L. Singer, J. Vac. Sci. Technol. A1, 419 (1983).
12. I.L. Singer, C.A. Carosella, and J.R. Reed, Nucl. Inst. and Meth. 182/183, 923 (1981).
13. I.L. Singer and R.A. Jeffries, Appl. Phys. Lett. 43, 925 (1983).
14. R.G. Vardiman and I.L. Singer, Mater. Lett. 2, 150 (1983).

MODELING OF HIGH FLUENCE TITANIUM ION IMPLANTATION AND VACUUM CARBURIZATION IN STEEL

D. Farkas[*], I. L. Singer[**], and M. Rangaswamy[*]
*Department of Materials Engineering, VPI & SU, Blacksburg, VA 24061
**Chemistry Division, Code 6170, Naval Research Laboratory, Washington, DC 20375.

ABSTRACT

Concentration vs. depth profiles have been calculated for Ti and C in 52100 Ti-implanted steel. A computer formalism was developed to account for diffusion and mixing processes, as well as sputtering and lattice dilation. A Gaussian distribution of Ti was assumed to be incorporated at each time interval. The effects of sputtering and lattice dilation were then included by means of an appropriate coordinate transformation. C was assumed to be gettered from the vacuum system in a one-to-one ratio with the surface Ti concentration up to a saturation point. Both Ti and C were allowed to diffuse. A series of experimental (Auger) concentration vs. depth profiles of Ti implanted steel were analyzed using the above-mentioned assumptions. A best fit procedure for these curves yielded information on the values of the sputtering yield, range and straggling, as well as the mixing processes that occur during the implantation. The observed values are in excellent agreement with the values predicted by existing theories.

INTRODUCTION

Ions implanted to high fluences (> 10^{17}/cm^2) in metals are capable of creating unique alloys with remarkable mechanical and chemical properties. An example of such an alloy is the wear-resistant amorphous layer formed when Ti is implanted into 52100 steel (1). This layer forms by adsorbing carbon from residual gases in the vacuum chamber (2), assisted by Ti atoms which reach the surface by sputter erosion during implantation.
The present paper describes a computational method for modeling high fluence implantation and presents calculated Ti and C depth profiles which mimic those observed for the above mentioned Ti implantation into 52100 steel. The model accounts for ion collection, sputtering and lattice dilation in a manner similar to the earlier treatments by Schultz and Wittmaak (3) and Krautle (4).
It also considers the diffusion-like transport processes which affect the shape of the evolving profiles and incorporates the vacuum carburization process elucidated by Singer (2). The formalism is based on a numerical solution of the coupled diffusion equations for implanted Ti and adsorbed C. Effective diffusivities for the two species and sputtering yield by Ti ions were obtained by comparison of the calculated and experimental profiles. The experimental data modeled in the present work were obtained by Singer (2)

THEORETICAL CONSIDERATIONS

The profile of Ti implanted to high fluences is affected by four processes:
o Ion collection, with a Gaussian distribution

- Sputter erosion of the surface
- Lattice dilation as a result of ion collection
- Diffusion-like broadening resulting from the collision cascades or radiation-enhanced diffusion.

The adsorbed C profiles are affected by two processes:
- Surface buildup of C as a function of time
- Diffusion-like penetration

The term "diffusion-like" is used here to denote transport processes that obey Fick's second law (5) which represents thermal and radiation-enhanced diffusion, as well as cascade mixing (6). The diffusion equations were solved for D_{Ti} and D_C values which best fit all the experimental curves. In these calculations D_{Ti} and D_C were approximated as constants, as suggested by the calculations of Eltoukhy et al. (7). Non-zero values for D_{Ti} indicate the relative importance of diffusion-like mixing under the conditions of implantation studied. We emphasize that both D_C and D_{Ti} should be considered as effective diffusivities, since they may result from processes other than thermally activated diffusion. The boundary condition is the surface carbon concentration as a function of time, taken from the model for vacuum carburization presented by Singer (2). The model predicts a surface concentration of C that is proportional to the amount of Ti exposed at the surface. As a first approximation it can be assumed that every exposed Ti atom adsorbs a C atom. This assumption is justified by the experimental data for surface contents up to concentrations around 16 at% where there appears to be a saturation of the adsorbed carbon. Therefore, the boundary condition used is that, at the surface, the C and Ti concentrations are equal.

The diffusion equations were solved by a finite difference technique, the Cranck-Nicholson method (5,8). At each time step of the calculation a Gaussian distribution of collected Ti was added to the existing profile, which was initially zero. Also, the C concentration at the surface, $[C]^s$, was made equal to the Ti concentration for every time increment. In addition, at each time step coordinate transformations were performed that accounted for the sputter erosion of the surface and the lattice dilation.

RESULTS

The formalism described requires the input of the following parameters:
- Range and range straggling of Ti ions in steel. These values were obtained from calculations done using LSS theory as described in the Manning and Mueller procedure (9).
- Flux of the incoming ions, f, which can be obtained from the experimental conditions as fluence per unit time. The Ti flux is converted to added thickness per unit time, using the atomic density of the 52100 steel. (f = 9.26 x 10^{13} at/sec cm^2, n_o = 8.21 x 10^{22} at/cm^3)
- Effective diffusivity of C, which was varied in order to obtain a good fit with the experimental C profiles.
- Effective diffusivity of Ti, which was varied in order to obtain the best fit to the experimental Ti profiles.
- Sputtering yield (S), which can be obtained directly from the experimental data by considering the total area under the profiles. As an additional confirmation the sputtering yield was allowed to vary in a best fit procedure to the experimental profiles. This was done simultaneously with the variation of the Ti diffusivity.

The effect of lattice dilation was found to be very important. Calculations were done without lattice dilation, and it was observed that the only way to get agreement with experimental values was to use a range and range straggling that are very much different from the ones predicted by the

Manning and Mueller formalism. For 190 kev a good fit without the lattice dilation effect required values of R_p = 900 A° and ΔR_p = 500 A°, which are almost double of the LSS values (R_p = 590 A° and ΔR_p = 230 A°). Since this discrepancy is unlikely it was concluded that the lattice dilation effect is absolutely necessary to describe high fluence ion implantation profiles.

The next series of calculations included the lattice dilation effect but did not consider a diffusion-like process for Ti. It was observed that the calculated curves were consistently narrower than the experimental ones. This suggested that a diffusion-like process is indeed necessary to account for the observed results. When both diffusion and lattice dilation were included in the calculations, only one combination of S and D_{Ti} values resulted in Ti profiles that fit all the seven measured profiles. These values were D_{Ti} = 6x10^{-15} cm^2/sec. and S = 2.0.

The sensitivity of the profiles to the value of S can be seen in Fig. 1 for six of the seven experimental curves studied. (The profile for 190 kev at lowest fluence of 5x10^{16} was found to be insensitive to the variation of the parameters of interest and therefore is not shown). Fig. 1 shows calculated and experimental Ti vs. depth profiles for 3 fluences at each of two energies. As expected, the effect of increasing sputtering yield is to move the profiles closer to the surface. The value of S that best describes the experimental curves seems to decrease from a value somewhat greater than 2.0 to a value less than 2.0 as the fluence increases. This is consistent with the value of 1.9 derived directly from experimental data. The results of Fig. 1 were calculated for a Ti effective diffusivity of 6x10^{-15} cm^2/sec.

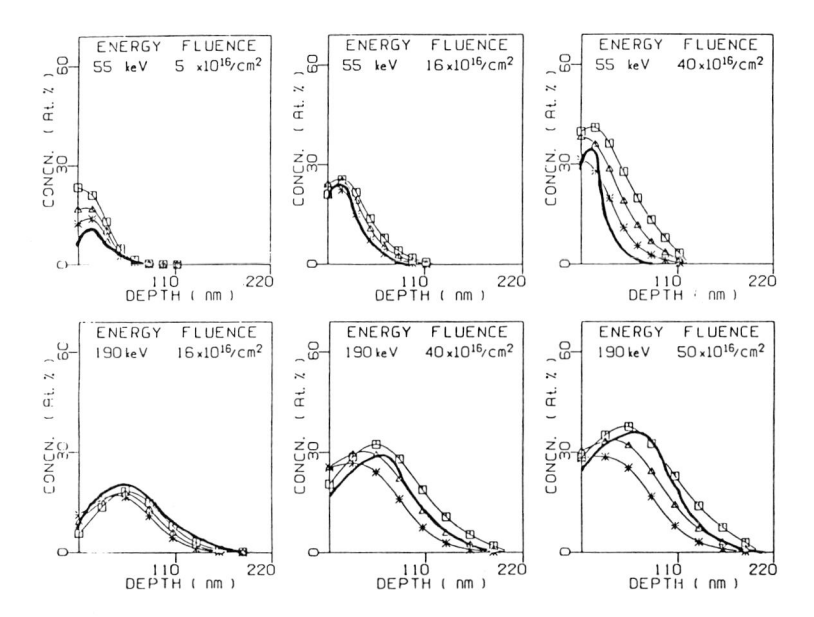

Fig 1 Experimental (heavy line) and calculated Ti profiles for several values of S [□ - 1.0, ∆ - 2.0, * - 3.0], with D_{Ti} = 0.6x10^{-14}(cm^2/s).

The sensitivity of the calculated profiles to the value of D_{Ti} is shown in Fig. 2. This figure shows calculations done for a sputtering yield of 2.0. These results include the ones obtained in the absence of diffusion broadening, showing that the experimental data are consistently broader. A value of D_{Ti} different from zero results in better agreement with experiment. An order of magnitude estimate of the transport process present can be obtained from these figures which is approximately, $D_{Ti} = 6 \times 10^{-15}$ cm^2/sec.

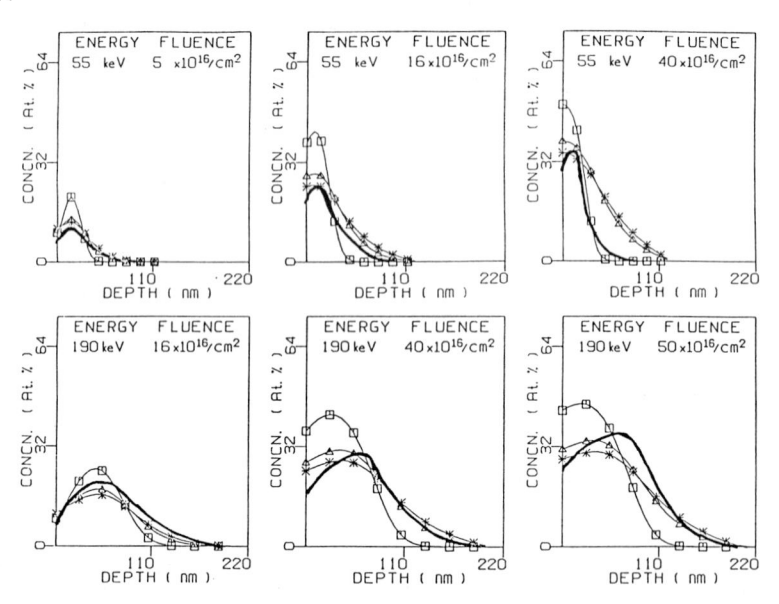

Fig 2 Experimental (heavy line) and calculated Ti profiles for several values of D_{Ti} [$(\times 10^{-14}$ cm^2/s) □ - 0.0, △ - 0.6, * - 1.0], with S=2.0.

Fig. 3 shows the calculated carbon concentration profiles compared to experiment. In this figure calculations are presented for different values of D_C, with $[C]^s$ restricted to ≤ 16 at%. As an order of magnitude estimate it may be concluded that the value of the effective diffusivity of carbon is around, $D_C = 6 \times 10^{-15}$ cm^2/sec.

DISCUSSION

The calculated composition vs. depth profiles are in good agreement with the experimentally determined profiles for high fluences of Ti implanted into steel. The computational method described takes into account the effect of sputtering and lattice dilation, as done by previous investigators. In addition, it is able to describe diffusion-like processes. Other phenomena may also be present in high fluence ion implantation, namely preferential sputtering and radiation induced segregation. The agreement between experimental and calculated profiles that was obtained in the present work suggests that these effects are not very significant in the case studied.

Several conclusions about the implantation alloying process can be drawn from the results of the present work. First, the lattice dilation and

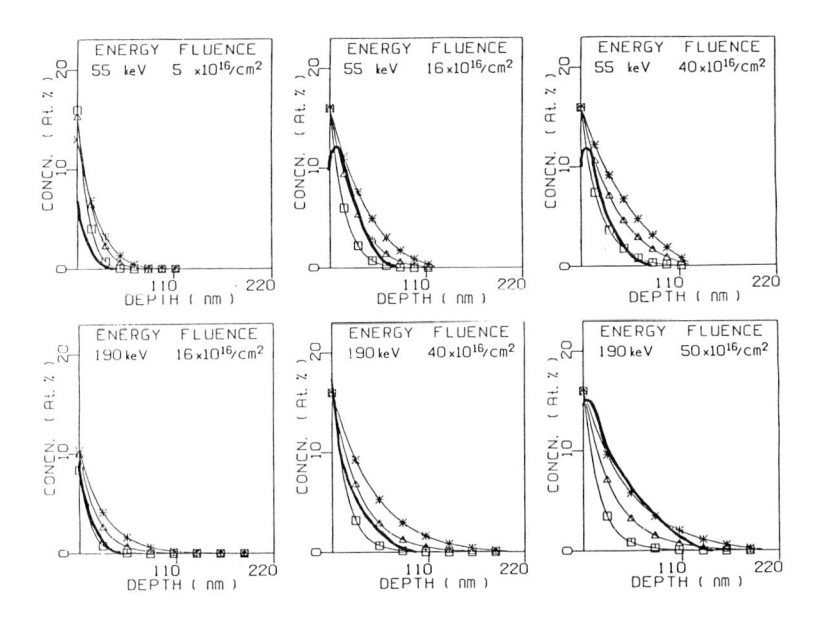

Fig 3 Experimental (heavy line) and calculated C profiles for several values of D_C [($\times 10^{-14}$ cm^2/s) □ - 0.3, Δ - 0.6, * - 1.0], with S=2.0.

the sputter erosion of the surface must be included in order to obtain reasonable agreement with experiment based on the values for range and straggling given by LSS theory.

Second, the sputtering yield of steel by Ti atoms is approximately 2.0 at fluences below those needed to form a carburized surface, but less than two at higher fluences. A sputtering yield of 2.5 ± .5 has been measured for Ar in Fe and stainless steel targets (10). It is possible that the sputtering yield of Ti is lowered when it is bonded to C. It is even possible that the anomalous dips in the self sputtering yield of reactive metals (V, Ti, Zr, Nb, Hf, Ta) presented by Almen and Bruce (11) over twenty years ago are attributable to carburization effects.

Third, the broadened profiles obtained experimentally can be understood in terms of an effective diffusivity for Ti of $D_{Ti} = 6 \times 10^{-15}$ cm^2/sec. Since this diffusivity occurs thermally at 580°C (12) it is clear that the diffusivity observed in the present problem is implantation-induced. For typical implantation conditions the effective diffusivity due to collision cascades is known to be of this order of magnitude (13). Furthermore, if following Myers, the effective diffusion coefficient is calculated on the basis of the Kinchin-Pease relation (6) the value obtained is of the same order of magnitude as the one obtained by using the parameters of the present problem.

Finally, a diffusion-like process accounts nicely for the inward migration of C into steel during Ti implantation. The assumption of a saturation value for the C adsorbed on the surface, which resulted in better agreement with experiment, is reasonable for a sputtered surface.

The present model cannot definitely identify the mechanism of the observed effective diffusion of C, 6×10^{-15} cm^2/sec. Three processes can contribute to this effective diffusivity value. These are thermal (non-enhanced) diffusion, radiation enhanced thermal diffusion and cascade mixing. The thermal diffusion of C in Fe at the sample temperature (40°C) is 10^{-16} cm^2/sec. (14), and therefore contributes only 2% of the migration process. A recent Montecarlo type calculation for the collision cascade effect during the carburization of Cr implanted Cr could not account for the majority of the C adsorbed as observed experimentally (15). It appears that in the present case both collision cascades and radiation-enhanced diffusion may contribute to the inward migration of C, although the latter may be more significant. The chemical affinity of C for Ti might also enhance C diffusion.

CONCLUSIONS

The formalism developed permits quantitative evaluation of several of the dynamic processes present in high fluence reactive ion implantation. For the case of Ti implanted into 52100 steel the following can be concluded: the sputtering yield decreases from a value greater than 2.0 to a value smaller than 2.0 as fluence increases, probably due to the incorporation of carbon with increasing fluence.

It is necessary to account for some mixing process for Ti in order to explain the experimental results. Alternatively a range and straggling that are very different from those predicted by LSS theory have to be assumed. The order of magnitude of this mixing process is characterized by a value of $D_{Ti} = 6 \times 10^{-15}$ cm^2/sec., which is consistent with a cascade mixing mechanism.

The mixing process responsible for C penetration can be described by Fick's law and is characterized by an effective diffusion coefficient of $D_C = 6 \times 10^{-15}$ cm^2/sec. It appears that both radiation enhanced diffusion and cascade mixing contribute to the inward migration of C. The amount of C adsorbed at the surface can be adequately described by a one to one relation with the amount of Ti at the surface up to a certain saturation limit around 16 at.% carbon.

REFERENCES

1. I. L. Singer, C. A. Carosella, and J. R. Reed, Nucl. Instr. Methods, 182/183, 923 (1981).
2. I. L. Singer, J. Vac. Sci. Technology. A, 1, 419 (1983).
3. F. Schulz and K. Wittmaak, Radiat. Eff., 29, 31 (1976).
4. H. Krautle, Nucl. Instrum. Methods, 134, 167 (1976).
5. Cranck, "The Mathematics of Diffusion" (Clarendon press, Oxford, 1975)
6. S. M. Myers, Nucl. Instr. Methods, 168, 265 (1980).
7. A. H. Eltoukhy and J. E. Greene, J. Appl. Phys., 51, 4444 (1980).
8. A. R. Mitchell, "Computational Methods in Partial Differential Equations",(J. Wiley and Sons, New York 1976) Chapt. 2.
9. I. Manning and G. P. Mueller, Comp. Phys. Comm., 7, 84 (1974).
10. H. H. Andersen and H. L. Bay, in "Sputtering by Particle Bombardment I", edited by R. Behrisch (Springer Verlag 1981), p. 173.
11. O. Almen and G. Bruce, Nucl. Instr. and Meth., 11, 257 (1961).
12. Swisher, Trans. AIME, 242, 2438 (1968).
13. G. Dearnaley, J. H. Freeman, R. S. Nelson, and J. Stephen, "Ion Implantation", (North Holland, Oxford, 1973), p. 228.
14. Smithells, "Metals Reference Book", 3rd edition (Butterworths, Washington, 1962), p. 594.
15. R. H. Bassel, K. S. Grabowski, M. Rosen, M. L. Roush, and F. Davarya, to be published.

RETENTION OF IONS IMPLANTED AT NON-NORMAL INCIDENCE

K.S. GRABOWSKI*, N.E.W. HARTLEY**, C. R. GOSSETT*, and I. MANNING*
*Naval Research Laboratory, Washington, D.C. 20375, **Georgetown
University, Washington, D.C. 20007 (Present address: AERE Harwell, UK)

ABSTRACT

Future applications of ion implantation require a knowledge of how the retention of implanted ions varies with angle of ion incidence. In this work the retention of 150-keV Ar,Ti,Cr,and Ta ions in AISI-M50 and 52100 bearing steels was measured for incidence angles up to 60° off normal. Fluences between 3×10^{15} and $3 \times 10^{17}/cm^2$ were used, typically on 3/8" diameter cylindrical samples. Retention was measured for Ar, Ti, and Cr by ion induced x-ray emission and for Ta by backscattering of He ions. Range and sputtering parameters needed for model calculations were experimentally determined from Ta-implanted thin Fe film samples. Generally, at the low-fluence limit a near-cos θ dependence obtained while at the high-fluence limit a $(\cos \theta)^{8/3}$ dependence applied where θ is the angle between the sample normal and the beam direction.

INTRODUCTION

It is clear that in practical applications it will be necessary to ion implant some metal workpieces at non-normal angles of incidence. Not surprisingly, there is concern that large angles of incidence may limit the benefits obtainable by line-of-site ion implantation. Singer and Jeffries [1] have provided recent evidence that the angle of incidence can be an important factor in determining the friction and wear properties of Ti-implanted steel, substantiating some of these concerns.

Since the benefits of ion implantation are typically related to the concentration of the ions at the surface, and the durability of this improvement is likely related to the depth of the distribution, it is appropriate to consider the integrated quantity of implanted ions remaining in the target (the retained dose). This work addresses how the retained dose of Ar, Ti, Cr, and Ta ions implanted into steel depends on the incidence angle of the ion beam.

EXPERIMENTAL PROCEDURE

The angle of beam incidence was controlled by implanting a scanned ion beam into symmetrically curved surfaces and into flat samples inclined to the beam. In the majority of this work, Ar, Ti, and Ta ions were implanted into stationary 3/8" diameter cylinders of AISI-M50 and 52100 martensitic bearing steels. M50 has a composition of Fe-4Cr-4Mo-1V-0.8C (in weight percent) and 52100 has a composition of Fe-1.5Cr-1.0C. Both steels were in a hardened and polished condition. A Ta slit 3/16" wide limited the 150-keV ion beams to a specific band across each cylinder. The slit was translated along the cylinder's axis to obtain separate bands for fluences between 3×10^{15} and $3 \times 10^{17}/cm^2$ (all fluences specified are for normal incidence). Typically, the chamber pressure was 3×10^{-6} torr and the beam current density was between 1 and 15 $\mu A/cm^2$ during these implantations.

Other steel samples were implanted under similar conditions. Flat disks of 52100 steel were inclined at various angles up to 40° to the beam and collectively implanted with 145-keV Ta ions to a fluence of 2 x 10^{17}/cm^2. An M50 steel inner-ring raceway of a ball bearing (2-cm bore diameter, 7.94-mm ball diameter) was implanted with 150-keV Cr ions to a fluence of 2 x 10^{17}/cm^2. This was the only concave surface examined. Additionally, 150-keV Ti ions were implanted to a fluence of 1 x 10^{18}/cm^2 into a 1/4" diameter half cylinder of 304 stainless steel (Fe-18 Cr-8Ni).

Fe thin-film samples were also Ta-ion implanted. An Al cylinder (3/8" diameter) was coated with 1600Å of evaporated Fe and implanted with 1 x 10^{17}Ta/cm^2 at 145 keV in a 3/16" wide band. This sample enabled measurement of both Ta retention and sputtering yields for Ta and Fe as a function of incident angle. Sputtering yields and retention were also examined as a function of fluence at normal incidence using 1900Å-thick Fe films on a saphire substrate. Fluences of 1 - 18x10^{16}/cm^2 were examined.

The retention of implanted Ar, Cr, and Ti ions was determined by proton or alpha-induced x-ray emission (PIXE), and of implanted Ta ions by 2-MeV Rutherford backscattering (RBS) of He$^+$. During the analysis, the 3/8"-diameter cylinders were rotated about their axis to select a specific incidence angle of the various implanted ions. The M50 bearing race was rotated during analysis so that the angle of Cr incidence was examined in the ball raceway at one radial section of the inner ring. The incidence angle on the 1/4"-diameter 304 s.s. half cylinder was selected by translating the cylinder diameter across the beam during analysis. This latter technique was somewhat less accurate and was therefore abandoned in later work. It was not possible to analyze implanted Ti in M50 due to overlap with the V signal from the bulk. For RBS analysis a scattering angle of 135° was used, and for the PIXE work an angle of 90° between the detector and beam direction was used, except for the Cr in M50 bearing analysis where an angle of 65° was used. The PIXE determinations of retained dose were all normalized to the retention at 0° incidence, whereas the absolute retention (in ions/cm^2) was obtained from the RBS measurements. As mentioned RBS measurements also provided the Fe and Ta sputter yields from the Fe thin-film samples.

RESULTS AND DISCUSSION

Theoretical

A rather simple dependence of the retained implantation dose on angle of incidence can be derived at both low-dose and high-dose (or steady-state) limits, using the model of Schulz and Wittmaack [2] for the evolution of the concentration-versus-depth profile from high-dose ion implantation. This model assumes that the implanted ion occupies no volume (i.e. no swelling occurs), the sputter coefficient and projected range remain constant, and no ion mixing or diffusion occurs. Although these assumptions appear quite restrictive, they do not prevent the model from identifying to first order the functional dependence of ion retention on incidence angle.

Fig. 1 shows the evolution of an implantation profile with increasing fluence based on the Schulz and Wittmaack model. R_p and σ represent the projected range and straggling of the incident ion, respectively, S the constant sputter coefficient, F_{ion} the atomic fraction of ions in the target, and ω the thickness of material sputtered off (S Φ/N) relative to R_p after an implantation fluence Φ(ions/cm^2). N is the constant target atomic density (at./cm^3). This figure represents implantations for which $R_p/\sigma = 6\sqrt{2}$.

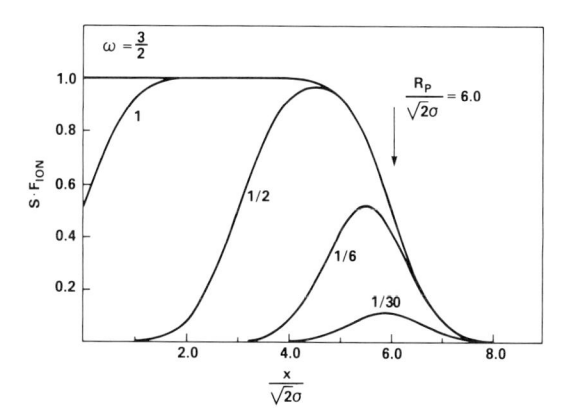

Fig. 1 Development of implanted ion distribution with increasing fluence for case of $R_p/\sigma = 6\sqrt{2}$. ω is the amount of material sputtered away relative to the projected range, R_p.

In the low-fluence limit for this value of R_p/σ, all implanted ions are retained beneath the surface, as is evident from Fig. 1. For an incident fluence, Φ_0, specified for normal incidence, the actual fluence is $\Phi_0 \cos\theta$ where θ is the angle between the beam direction and the surface normal. Therefore, the retention, $R(\text{ions/cm}^2)$, at low fluences should be

$$R_0 = \Phi_0 \cos\theta \tag{1}$$

In the high-fluence limit, the retention of ions is related to the area under the curve shown in Fig. 1 with $\omega=3/2$. Approximating this area by a step function of height $S F_{ion}$ with the step at $X=R_p$ leads to $R=NR_p/S$. At non-normal incidence, R_p shortens to $X_p=R_p \cos\theta$ and S increases. Sigmund [3] has estimated that S depends on θ according to $S=S_0/\cos^f\theta$ with f approximately equal to 5/3 for target masses less than about 3 times the ion mass, a condition met in all of the present work. Combining these angular dependences leads to a high-fluence limit of

$$R_\infty = \frac{NR_p}{S_0} (\cos\theta)^{8/3} \tag{2}$$

At intermediate fluences, R should vary with θ so as to fall between the two limits set by Eqs. 1 and 2.

A more realistic description of the evolution of an implantation profile is shown in Fig. 2, where $R_p/\sigma = \sqrt{2}$. This value of R_p/σ is more in line with the values of 1.7 - 2.8 expected for normal incidence and the ions and targets studied here. Even in Fig. 2, however, close examination reveals that in the high fluence limit ($\omega=3$), the area under the curve can be closely approximated by a step function of height $S F_{ion}$ and width $X=R_p$. However, conditions may worsen at non-normal incidence since X_p/σ is smaller. Similarly, the retention at low fluences may deviate from the $\cos\theta$ dependence due to an increase of ion loss from backscattering at large incidence angles.

Fig. 3 shows the extent to which Eqs. 1 and 2 deviate from the low and high-fluence limits expected from the complete Schulz and Wittmaack model of retention during implantation of 145-keV Ta into Fe. The model calculation used measured values for R_p, σ, and S_0 of 106Å, 122Å, and 6.0, respectively, and assumed $X_p = R_p \cos \theta$. The dashed lines in Fig. 3 represent values at low fluence from Eq. 1, normalized at $\theta=0°$, and at high fluence directly from Eq. 2. In the low-fluence limit, even though ω is only 0.1 at 0°, retention falls off more quickly than $\cos \theta$ with increasing angle due to an increase in ion backscattering. In the high-fluence limit, however, the simple approximations of Eq. 2 seem to hold quite well. This was found to be true for all cases of the present work.

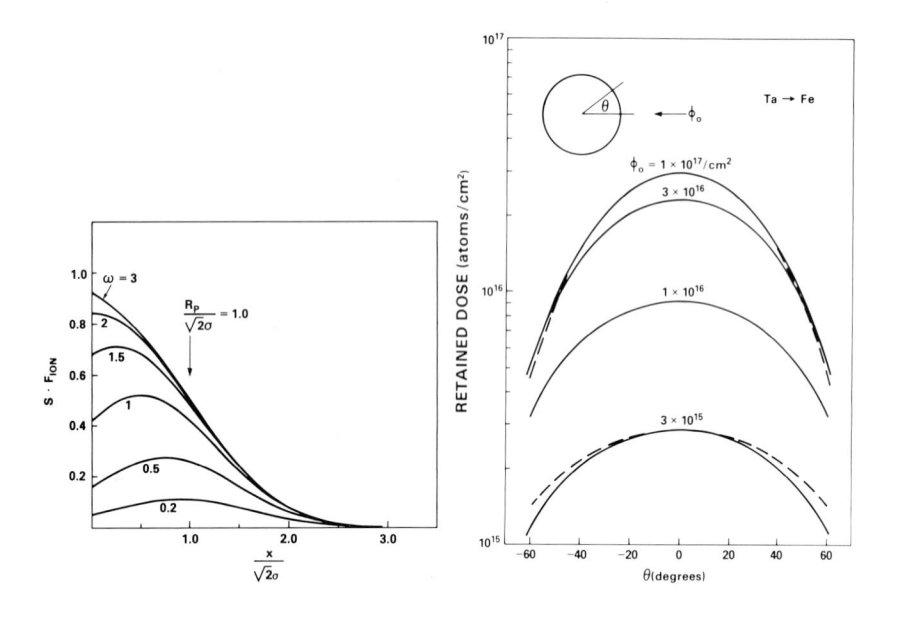

Fig. 2. Development of implanted ion distribution with increasing fluence for $R_p/\sigma = \sqrt{2}$.

Fig. 3. Theoretical retention of implanted Ta ions as a function of incidence angle and normal fluence. Dashed lines are simplified versions of theory for low and high-dose limits. Measured values for R_p, σ and S_0 were 206Å, 122Å, and 6.0, respectively.

Experimental

Because theory predicted a more general result for retention in the high-fluence limit, comparison with experiment was first conducted in that regime. Fig. 4 shows the retention of various ions normalized to unity at $\theta=0°$. All fluences were selected to produce a steady state distribution in the target. For Ar- and Cr-implanted samples good agreement between theory and experiment was obtained. The slight discrepancy near 60° for the 304 stainless steel target implanted by Ti is believed caused by difficulties in the experimental measurement. These were associated with the small cylinder diameter and the translation approach to selecting θ. Conversely, the slightly narrower shape to the distribution observed in Ti-implanted 52100 and somewhat more so in Ta-implanted M50 is believed to be real. A similar narrower shape was also observed from Ta-implanted 52100 flat samples, the Ta-implanted 52100 cylinder, and the Ta-implanted Fe film on Al cylinder.

To assess where the model was failing in Ti and Ta-implanted steel, the absolute retention of Ta in M50 and 52100 cylinders was determined as a function of incident angle and fluence. Only results for the 52100 cylinder are shown in Fig. 5, although analogous results were obtained from the M50 cylinder. For fluences up to about $1\times10^{16}/cm^2$, theory agreed within experimental accuracy with the data. However, at higher fluences where sputtering becomes important, much more retention was measured at all angles than was predicted by the Schulz and Wittmaack theory, even though the input values for R_p, σ and S_0 had been measured in Fe thin-film samples.

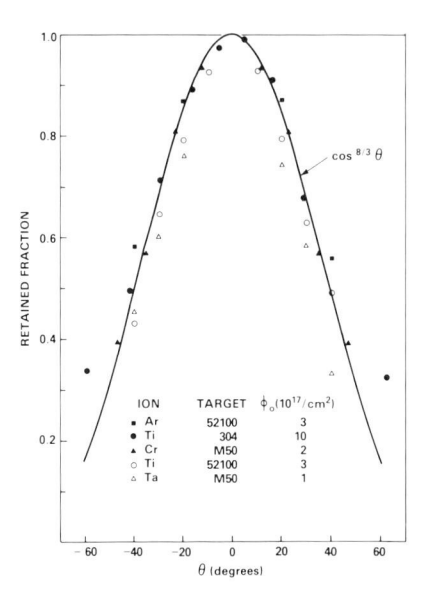

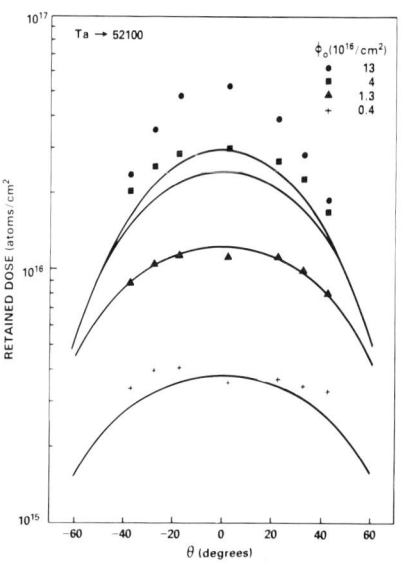

Fig. 4. Normalized retained dose in the high-dose limit from simplified theory and from measurements on steel cylinders. Data points at normal incidence were not plotted.

Fig. 5. Retention of implanted Ta in 52100 cylinder as function of fluence and incidence angle for theory and experiment. Theory used measured R_p, σ and S_0 as input and same fluences as experiment.

A more elaborate calculation of retention versus fluence at normal incidence was performed for Ta implantation, based on the model of Krautle [4]. This model includes the effects of target swelling, range shortening, and preferential sputtering, but not migration. This model, like the Schulz and Wittmaack model, predicts a plateau level for the atomic fraction of beam atoms beneath the surface. This level is equal to 1/S where S is the total sputtering coefficient of both ion and target atoms. Target swelling and range shortening alone reduced the saturation retained dose of Ta by about 5%. Including preferential sputtering (which alters S) could drive theoretical retention towards the experimental results, but the change in S required was unrealistic when compared to the values measured from Fe thin-film samples.

A plausible explanation of the discrepancy shown in Fig. 5, then, is that preferential removal of Fe combined with ion mixing resulted in the increased retention of Ta, distributed over about the range of the incident ion. This is in fact the model discussed by Liau and Mayer [5]. From this model it can be shown that the steady state distribution of implanted ions can again be represented by a step function of width about R_p and height (in atom fraction of beam atoms) equal to $1/S_{ion}$, where S_{ion} is the sputter coefficient for ions incident on themselves as a pure target. Therefore, this new model predicts approximately the same dependence on angle as described by Eq. 2, with the amplitude scaled by approximately the ratio of S to S_{ion}. In the present work this implies that S_{ion} for Ta should be about 3. Ongoing research at NRL will attempt to determine whether this assessment is correct. Considering this latter model to be correct, perhaps the slight discrepancy between experiment and Eq. 2 in Fig. 4 is caused by variations in the preferential removal of Fe or in ion mixing, with changing angles of incidence. This might be related to the carburization process which is known to occur during the implantation of Fe by reactive metal ions like Ti^+ and Ta^+.

CONCLUSIONS

The experimental data suggest that the preferential removal of Fe combined with ion mixing during implantation with Ta is necessary to explain the fluence dependence of Ta retention observed in steels. It is not clear from the present results whether these processes also occur during the implantation of Ar, Ti, or Cr ions into these same steels.

Nevertheless, in the high-fluence limit the retention of all these ions follows quite closely a $(\cos \theta)^{8/3}$ dependence, where θ is the angle between the beam direction and the surface normal. A slightly stronger decay with increasing angle of incidence does appear to occur for Ta and sometimes Ti-ion implantations. At lower fluences the retained dose falls off more gradually with increasing θ but may never quite approach the $\cos \theta$ dependence expected at low fluences from simple arguments, since an increase in ion backscattering occurs at large angles of incidence.

ACKNOWLEDGEMENTS

The authors are indebted to R.A. Jeffries for early x-ray measurements, C.M. Davisson for her model calculations, and to F.D. Correll for his help with some RBS measurements.

REFERENCES

1. I.L. Singer and R.A. Jeffries, these proceedings.
2. F. Schulz and K. Wittmaack, Rad. Eff. 29, 31(1976).
3. P. Sigmund, Phys. Rev. 184, 383(1969).
4. H. Krautle, Nucl. Instrum. Methods 134, 167(1976).
5. Z.L. Liau and J.W. Mayer in: Treatise on Materials Science and Technology, Vol. 18, Ion Implantation, J.K. Hirvonen, ed. (Academic, New York 1980) ch. 2, pp. 17-50.

INDUSTRIAL APPLICATIONS OF ION IMPLANTATION

J.K. HIRVONEN
Zymet, Inc., 33 Cherry Hill Drive
Danvers, MA 01923

ABSTRACT

   The use of ion implantation for non-semiconductor applications has evolved steadily over the last decade. To date, industrial trials of this technology have been mainly directed at the wear reduction of steel and cobalt-cemented tungsten carbide tools by high dose nitrogen implantation. However, several other surface sensitive properties of metals such as fatigue, aqueous corrosion, and oxidation, have benefitted from either i)direct ion implantation of various ion species, ii)the use of ion beams to "intermix" a deposited thin film on steel or titanium alloy substrates, or iii)the deposition of material in conjunction with simultaneous ion bombardment.

   This paper will concentrate on applications that have experienced the most industrial trials, mainly high dose nitrogen implantation for reducing wear, but will present the features of the other ion beam based techniques that will make them appear particularly promising for future commercial utilization.

INTRODUCTION

   It has been approximately ten years since the first publications by Hartley and colleagues at Harwell of the effect of high dose nitrogen implantation on wear reduction [1]. Since that time, numerous laboratories have gotten involved in R&D applications involving both mechanical and chemical surface-sensitive properties, many of which are presented in this proceedings. In spite of the large number of properties (e.g., fatigue, aqueous corrosion, high temperature oxidation) that have shown significant improvements due to ion implantation, to date, the vast majority of industrial trials has remained in the area of wear.

WEAR REDUCTION OF INDUSTRIAL TOOLING

   Wear improvements have been seen in both steel as well as Co-cemented WC tools under a variety of applications. It is important to note that nitrogen implantation has been found ineffective for those applications involving high operating temperatures such as found in chip forming applications. This is because of the instability of the nitrogen-defect structures that are thought to be responsible for improving the wear resistance. There is direct evidence from several groups [2,3,4] that there is an anomalous retention of nitrogen in the wear scar of implanted wear samples and a persistence of wear resistance after the removal of material corresponding to 10-50 times the original range of nitrogen. There is however still considerable debate as to what mechanism or mechanisms are responsible for this persistence. Dearnaley et al. [2] have suggested that the nitrogen decorates dislocation structures and migrates under the locally high temperatures produced at interacting asperities during wear. Baumvol and associates [5] have seen evidence for nitride and carbo-nitride formation following implantation and subsequent dissolution at elevated temperatures using conversion electron Mossbauer spectroscopy. These results suggest that such precipitates may pin dislocations, yet still be able to dissolve so that nitrogen migrates inward during wear. Another possibility for this anomolous retention is that nitrogen may be continuously transferred from one wear surface to another as the wear front advances.

Despite these uncertainties as to cause there are enough industrial tests to demonstrate the efficacy of the process. To date Harwell has worked with over a hundred companies in the U.K. involving industrial trials of nitrogen implanted tools and now several U.S. companies are initiating trials. As several other papers in this proceedings show, other ion species such as high dose Ti implantation can also have a marked effect on wear reduction. However because few of these have yet been tested industrially they won't be discussed here.

Implantation is an attractive technique for treating industrial tooling where one or more of the following criteria are met: i)there is a need to avoid dimensional changes of the tooling, ii)the tooling can't accomodate the elevated temperatures associated with other surface treatments such as PVD or CVD because of possible distortion, or iii)conventional antiwear coatings have delamination problems during use.

Table I is a compilation of several applications of nitrogen implantation for wear resistance. The doses used range from $2X$ to $6X10^{17}$ ions/cm^2 at energies of 50keV-100keV. In addition to the criteria given above most of the successful usages involve relatively expensive tooling or high costs of lost throughput associated with downtime of a production line. These points are illustrated by the first examples listed in Table I.

An example of a precision tool costing several hundred dollars is shown in Fig. 1. It is a D2 steel scoring die used for the production of flip tops for aluminum beverage cans. One large company currently evaluating implantation uses up to a hundred of these in a week before rejecting them due to wear of the fine pattern on the face of the tool. This particular tool normally produces between three to five million tops before wearing out whereas initial tests of N-implanted tools have produced seven to seventeen million tops before wearing out. Another tool shown in Fig. 2 is also made of D2 tool steel and is used to form the bottom of aluminum beverage cans. It is very highly polished and normally tends to pick up aluminum from the can surface thereby weakening the can as well as requiring time consuming repolishing several times during its lifetime. An initial trial of a nitrogen-implanted tool has shown markedly reduced metal pick-up during operation with further tests underway.

TABLE I. IMPROVEMENTS IN WEAR LIFETIME OBTAINED BY NITROGEN IMPLANTATION

APPLICATION	MATERIAL	LIFETIME INCREASES, ADDITIONAL BENEFITS [REF]
Scoring Die for Aluminum Beverage Can Lids	D2 Tool Steel	3X
Forming Die for Aluminum Beverage Can Bottoms	D2 Steel	Lowered wear, markedly lowered material pickup
Wire Guides	Hard Cr Plate	3X without significant wear
Finishing Rolls for Cu Rod	H-13 Steel	Negligible wear after 3X normal lifetime; Improved surface finish of product[6]
Paper Slitters	1.6% Cr 1% C Steel	2X[7]
Punches for acetate sheet	Cr-plated steel	Improved product[7]
Taps for phenolic resin	M2 high speed steel	Up to 5X[7]
Tool inserts	4% Ni 1% Cr Steel	Reduced tool corrosion by 3X[7]
Forming Tools	12% Cr 2% C Steel	Greatly reduced adhesive wear[7]
Fuel injectors and metering pump	Tool steel	>100X in engine tests[8]
Cam followers	Steel	Improved lifetime[8]
Plastic Cutting	Diamond tools	2X - 4X lifetime[9]
Hip joint prostheses	Ti/6Al/4V	100X in laboratory tests[10]

Application	Material	Result
Printed Circuit Board Drills	WC - Co	3X, Lowered smearing with 50% lower operating temperature, reduced breakage [6]
Dental Drills	WC - Co	2X - 3X, significantly lower cutting force required
Precision punches for electronic parts	WC - 15% Co	Greater than 2X
Punch and die sets for sheet steel laminations	WC - Co	4X-6X, improvement remains after resharpening [6]
Swaging dies, press tools for wheels	WC - Co	2X [11]
Dies for copper rod	WC - 6% Co	5X throughput, improved surface finish [7]
Dies for steel wire	WC - 6% Co	3X [7]
Deep drawing dies	WC - 6% Co	Improved lifetime, markedly reduced material pickup [7]
Sheet steel chopper blades	WC - Co	Greater than 3X, reduced chipping [6]
Injection molding nozzle, molds screws, gate pads for glass and mineral filled plastics	Tools steels and hard Cr plated steels	4X - 6X [11]
N into thermally nitrided steel molds	Tool steels	Combination better than either process alone [7]
Plastic Calibrator Die	Nitrided H-13 tool steel	2X [12]
Profile Hot Die for plastic extrusion	P-20 Tool steel	4X [12]

Fig. 1 Scoring die used for producing lids of flip top beverage cans

Fig. 2 D2 Steel nose punch used for forming bottoms of aluminum beverage cans.

Precision Co-WC punches and dies have previously been reported to wear longer following N-implantation by Westinghouse researchers[6]. In industrial trials with a major electronics manufacturing company, we have also found a 2X-4X improvement in the lifetime of N-implanted precision WC-Co punches such as used in the production of the so called lead frames that hold and provide electrical connections to integrated circuit chips as shown in Fig. 3. The cost of an individual punch as shown in Fig. 3 is approximately $100-200 and the cost of an entire progressive punch and die set for this application is typically $50K-100K. Downtime is costly in terms of both resharpening time (typically 1 shift) and lost production.

Fig. 3 Precision WC-Co punch and lead frame used in integrated circuits.

In all of these examples, a 2X improvement in tool lifetime would give a return-on-investment time of less than a year for a dedicated implanter which is less than the two to three year return on investment time for capital equipment that most industries require. In addition to reduced wear, other side benefits may be equally important to introducing implantation technology into these industries. The use of nitogen implantation for increasing the lifetime of Co-cemented WC printed circuit board (PCB) drills is another application previously reported by Westinghouse researchers [6]. In some recent tests we have done in conjunction with Precision Carbide, a major manufacturer of PCB drills, the surface temperature of N-implanted drills was measured to be approximately 175°F as compared to 350°F for untreated drills. This is another observation of reduced friction of N-implanted Co-WC tools reported initially by Dearnaley [2]. It is expected that lowered drill temperatures would significantly decrease or eliminate the smearing of the resin in the PCB on the walls of the drilled holes, which presently requires a separate "desmearing" or cleaning process utilizing hot sulphuric acid to remove the smeared resin. The possiblility of eliminating this toxic and corrosive process is expected to be another strong driving force for adoption of the implantation process in this industry. We have also observed extended life and reduced friction of N-implanted WC dental burs. Clinical evaluation of this application by several groups is currently underway. Here again, reduced temperatures (i.e., reduced trauma) may be as important or more important than reduced wear for acceptance.

It is obvious from a perusal of Table I that the diversity of applications is extreme and acceptance will be based not only by longevity increases but also by economics and industrial attitudes. For example, circular cross section wire drawing dies cost only a small fraction of shaped dies thereby significantly influencing the economics of treating them. A two inch diameter round die typically costs 3 to 5 dollars which is comparable to implantation costs using dedicated equipment and therefore wouldn't necessarily justify treatment costs whereas shaped dies would. At this time, it is too early to accurately gauge whether implantation technology for metals will spread predominatly through service centers (e.g., heat treating facilities) or via inhouse equipment. The first alternative will obviously be most appealing to low volume users and many of the industries who have traditionally had tooling both made and surface treated out-of-house. It is interesting to note that the initial purchasers of metals implanters in the U.S. are firms intending to offer service work. Other industries have great misgivings about having their expensive tooling out of their direct supervision and also have concern about fast turn around times for treatment of tools used in high volume production runs. The combination of these concerns, coupled with higher volume and concern with proprietary usage and development, warrants buying dedicated equipment.

The majority of the applications listed in Table I involve production tooling but the advantage of allowing processing on an as-finished part without altering an existing production line is also pertinent to consumer items. A particularly promising application of this sort involves surgical implants such as hip

joint prostheses, for which titanium alloys are being
increasingly used. The metal femoral component bears against an
acetabular cup of high molocular weight polyethylene. The
rubbing of the metal ball against the polyethylene pocket is
thought to remove protective oxides from the metal surface so
that it is attacked by body fluids thereby releasing
undesireable (metal) ions into the body. The wear rate is such
that a typical surgical implant will last less than 10 years and
surgical replacement poses significant problems. Dearnaley [10]
has reported a dramatic reduction in the wear rate of Ti/6Al/4V
alloy used in surgical implants by both direct nitrogen
implantation, which gives a 100X reduction in wear, or by ion
beam mixing of Sn, which gives a 1000X reduction. These results
were found in laboratory tests simulating the geometry and fluid
conditions found in the body. Williams et al. [13] have
reported similar findings using nitrogen implantation.
Considering there are approximately 100,000 operations of this
kind each year in the U.S. alone, even a small wear reduction
would have an immense benefit to recipients.

ION BEAM MIXING AND ION BEAM ENHANCED DEPOSITION

Due to the additional cost and complexity of implantation
equipment incorporating mass analysis, it is likely that future
commercial applications requiring nongaseous species will be
done using ion beams in conjunction with conventional thin film
deposition, such as in ion beam mixing, as an alternative to
direct high dose implantation.

The question of how effective ion beam mixing is versus
direct implantation must presently be addressed on a case to
case basis, but some studies have shown that higher alloy
concentrations can be achieved than in sputter-limited high-dose
implantation. An extension of this idea involves simultaneous
ion bombardment and thin film deposition. This technique has
been studied by a number of researchers [14,15,16,17] for the
purpose, at least partially, of understanding the role ions play
in conventional PVD processes. This combination allows thicker
alloyed regions to be produced than by either direct
implantation or ion beam mixing. It has been recognized for
some time that the presence of energetic ions has a synergistic
effect on thin film growth, and this topic has been recently
reviewed by Takagi[18]. There are several aspects of film
growth that have been beneficially influenced by ion bombardment
during deposition including i) film nucleation and growth[15],
ii) adhesion[15], iii) internal stress[19], iv) morphology, v)
density[20], and vi) composition.

One of the most striking changes in structure produced by
ion beams has been the deposition of carbon films with
"diamond-like" properties observed either after ion bombardment
of carbon films or the collection of energetic ionized
hydrocarbons. It was first seen in 1971 by Aisenberg and
Chabot[21] and later by many other groups. The carbonaceous
films produced by these methods (termed i-C denoting the role of
ions) appear transparent with a high refractive index,
are quasi-amorphous, are very hard, and have high electrical
resistivity. Weissmantel et. al.[16] have reviewed this area of
producing hard coatings by ion beam techniques, including the
production of cubic boron nitride, another extremely hard

material second only to diamond in hardness and of obvious interest for cutting surfaces. Weissmantel et. al.[22,23] evaporated pure boron in a residual atmosphere of nitrogen, and subsequently ionized and accelerated the resultant species at 0.5 - 3.0 keV onto various substrates. IR absorption spectra confirmed that B-N bonding states predominate while transmission electron microscopy and selected area electron diffraction patterns indicate a quasi-amorphous structure; however high energy deposits contained small crystallites of about 100 Angstrom diameter which had a lattice constant corresponding to cubic BN. Shanfield and Wolfson[24] present hardness, x-ray, and Auger analysis data supporting the production of cubic BN by using an ion beam extracted from a borazine ($B_3N_3H_6$) plasma. Satou and Fujimoto have also recently published results suggesting the formation of cubic boron nitride by the deposition of elemental boron and the simultaneous implantation of 30 keV nitrogen ions[25].

These examples serve to show the extremely beneficial effects ion beams can have when used in combination with conventional deposition processes. Clearly, we are only at the beginning of developing wear, corrosion and oxidation resistant coatings. This development work will be accelerated now that commercial equipment is available for such processing.

REFERENCES

1. N.E.W. Hartley, W.E. Swindlehurst, G. Dearnaley, and J.F. Turner, J. Mater. Sci. **8**, 900-904 (1973).

2. G. Dearnaley, Rad. Effects **63**, 1 (1982).

3. S. Fayeulle, D. Treheux, P. Guiraldenq, T. Barnavon, J. Tousset, and M. Robelet, Scripta Met. **17**, 459-461 (1983).

4. F.Z. Cui, H. D. Li, and X. Z. Zhang, Nucl. Instr. and Meth. **209/210**, 881-887 (1983).

5. C.A. Dose Santos, M. Behar, J.P. De Souze, and I.J.R. Baumvol, Nucl. Instr. and Meth., **209/210**, 907-912 (1983).

6. R.E. Fromson and R. Kossowsky in: Metastable Materials Formation by Ion Implantation, S.T. Picraux and W.J. Choyke, eds. (Elsevier, New York 1982) pp. 355-362.

7. N.E.W. Hartley, in: Ion Implantation Case Studies-Manufacturing Applications, Harwell Report AERE-R 9065 (1978).

8. G. Dearnaley, private communication.

9. N.E.W. Hartley in: Metastable Materials Formation by Ion Implantation, S.T. Picraux and W.J. Choyke, eds. (Elsevier, New York 1982) pp. 355-362.

10. G. Dearnaley in: Proc. of NATO Advanced Study Institute on Surface Engineering, Les Arcs France, July, 1982. R. Kossowsky and S. Singhal, eds.

11. G. Dearnaley in: Ion Implantation Metallurgy, C.M. Preece and J.K. Hirvonen, eds. AIME, Warrendale, PA (1980).

12. A.H. Deutchman and R.J. Partyka in: Proc. of NATO Advanced Study Institute on Surface Engineering, Les Arcs France July, 1982. R. Kossowsky and S. Singhal, eds.

13. J.J. Williams, G.M. Beardsley, R.A. Buchaman, and R.K. Bacon, This Proceeding.

14. J.M.E. Harper, and R.J. Gambino, J. Vac. Sci. Technol. 16 1901-1905 (1979).

15. L. Pranevicius, Thin Solid Films 63, 77-85 (1979).

16. C. Weissmantel, G. Reisse, H.J. Erler, F. Henny, K. Bewilogua, U. Ebersback and C. Schurer, Thin Solid Films 63, 315-325 (1979).

17. J. Colligon, and A.E. Hill. 6th National Conference on Interaction of Atomic Particles with Solids (Minsk USSR, Sept. 1981).

18. T. Takagi, Thin Solid Films 92, 1-17 (1981).

19. D.W. Hoffman, and M.R. Gaerttner, J. Vac. Sci. Technol. 17 425-428 (1980).

20. R.F. Bunshah, J. Vac. Sci. Technol., 11, 633 (1974).

21. S. Aisenberg and R. Chabot. J. Appl. Phys. 42, 2953-2961 (1971).

22. C. Weissmantel, K. Bewilogua, D. Dietrich, H.J. Erler, H.J. Hinneberg, S. Klose, W. Nowick and G. Reisse, Thin Solid Films 72, 19-31 (1980).

23. C.J. Weissmantel, Vac. Sci. Technol. 18, 179 (1981).

24. S. Shanfield, S. and R. Wolfson, J. Vac. Sci. Technol. (1983).

25. M. Satou and F. Fujimoto, Jap. Jour. of Appl. Physics. 22(3) L171-L172, (1983).

THE CHARACTERIZATION OF NITROGEN IMPLANTED WC/Co

DANIEL W. OBLAS
GTE Laboratories Incorporated, Waltham Massachusetts, 02254, USA

ABSTRACT

Although the mechanical properties of nitrogen implanted cemented carbides show significant improvements in performance over unimplanted specimen, the chemical or physical mechanisms giving rise to these improvements are not well understood. Furthermore, the mechanical properties of the implanted samples under thermal stress may not be stable. This study presents analytical results obtained for a series of nitrogen implanted cemented carbides that were subsequently characterized by Auger Electron Spectrometry (AES) and Thermal Desorption Mass Spectrometry (TDMS). Results show that, in the fluence range of 0.5 to 2.5×10^{17} N_2/cm^2, peak nitrogen concentration is linear and reaches a maximum concentration of 25 a/o for the highest fluence and 180 keV energy. During vacuum heating, the implanted nitrogen is evolved rapidly between 600-700°C, which corresponds to an activation energy of approximately 7 kcal/mole (.30 ev). A small quantity of nitrogen is still present after heating to 1000°C, at an estimated average concentration of 3 a/o with little diffusional broadening as a result of post-implant heating.

INTRODUCTION

The improved performance of cemented carbide tools such as punches, drills and wire drawing dies after nitrogen implantation has been widely reported (1-3). Unimplanted cemented carbides have excellent mechanical properties that are sustained to temperatures as high as 800°C. However, the improvements induced by ion implantation may not be retained at high operating temperatures. In general, the physical or chemical mechanisms giving rise to these improvements are not well understood (4,5) and the temperature range over which these changes are stable has not been clearly established. In a recent study (6), N^+, Ar^+ and other species were implanted into WC/Co at 50 keV. Improvements in microhardness were observed for all species, and nitrogen showed the best results. Post heating the implanted specimen up to 600°C produced small improvements in hardness and heating to higher temperatures caused the hardness to decrease rapidly. The delineation of the effects of radiation damage and chemistry on microhardness is still not clear.

The purpose of this study is to present some initial data on nitrogen implanted WC/Co using Thermal Desorption Mass Spectrometry (TDMS) and Auger Electron Spectrometry (AES) to characterize the implanted specimen. The nitrogen concentration, distribution, and thermal stability were determined as a function of fluence in the range of 0.5 to 2.5×10^{17} N_2/cm^2.

EXPERIMENTAL

The samples for this study were WC/Co tool inserts (0.5"x0.5"x0.25") 94%/6% by weight, respectively. The WC grain size was between .5 and 2.5 microns with an average size of 1.25 microns. The specimen was polished to a 3 micron finish with diamond paste and cleaned. Subsequently, it was heated in hydrogen for one half hour at 800°C prior to implantation to remove surface contaminants, principally hydrocarbons.

The implantations were done at 180 keV with an average, scanned, current density of 10 microamps/cm² of N_2, corresponding to a surface energy density of 1.8 watts/cm². The pressure in the target chamber was approximately 2×10^{-6} torr during implantation, with H_2O and CO the main species. The temperature of the tool insert during implantation, measured on the edge of the sample, was approximately 65°C.

TDMS analyses began at room temperature and extended to 1000°C in 100°C increments. For example, after placing the sample in the quartz chamber, the oven temperature was set to 100°C. It took about 3 minutes to come to temperature and the sample was kept at temperature for another 1 minute. The evolved gases were collected in a calibrated volume, and subsequently introduced into a high resolution mass spectrometer for identification. This gas was pumped away while gases from the next higher temperature were being collected. Carbon monoxide is a significant portion of the desorbed gas and, therefore, a mass spectrometer with sufficient mass resolution was used to separate the CO and N_2 mass peaks. A fluence of 2.5×10^{17} N_2/cm² implanted into the cemented carbide specimen is equivalent to 11.2 microliters of N_2 gas, which, if it is all desorbed, corresponds to a pressure in the collection volume of about 3 microns.

AES, in conjunction with argon ion sputtering (1 keV at 39 nanoamps), was used to determine the nitrogen concentration and distribution after ion implantation (7). Profiles obtained with standard AES operating parameters showed a distorted nitrogen distribution. The surface energy density of the analyzing electron beam for standard AES operating parameters was 25 watts/cm². Under less severe operating conditions, 1.0 watt/cm² (3 keV at 2 microamps, spot size = .011"), the nitrogen concentration profiles were characteristic of implanted ions. The elemental sensitivity of AES decreased by about a factor of 10, and the minimum detectable nitrogen concentration is approximately 1.0 a/o.

Talysurf measurements, coupled with known sputtering times, were used to estimate the sputtering rates for the 3 keV argon ions and resulted in sputtering rates of approximately 30 Å/min. The projected ranges discussed here are, therefore, only approximations.

RESULTS

Nitrogen profiles by AES, for a series of implanted WC/Co specimens, are shown in Figure 1 for five fluences from 0.5 to 2.5×10^{17} N_2/cm². Nitrogen concentration is in atomic % and is plotted against the sputtering time. The data points for the highest fluence sample show a saturation effect, however a smooth curve has been drawn through the points. Table 1 tabulates the peak nitrogen concentrations, the average nitrogen concentrations, projected range, R(p) and ΔR(p), the standard deviation. The average nitrogen concentration was determined by integrating the nitrogen distribution curve and dividing by the width of the distribution. The R(p), calculated from the E-DEP-1 program (8) for an average sample of WC/Co, is approximately 850Å with a ΔR(p) of 480Å, somewhat lower than the measured values. The peak nitrogen concentrations and the total area under the distribution curve (representing the total nitrogen content) are shown as a function of fluence in Figure 2.

Nitrogen was not detected by TDMS from an unimplanted WC/Co. However, NO is observed along with significant quantities of CO, H_2 and some CO_2. AES analysis of an unimplanted specimen showed significant surface concentrations of nitrogen and oxygen consistent with the detection of NO in the gas phase by TDMS. Figure 3 shows cumulative nitrogen evolution from two separate samples of WC/Co implanted with 2.5×10^{17}/cm² as a function of temperature. Most of the implanted nitrogen is rapidly

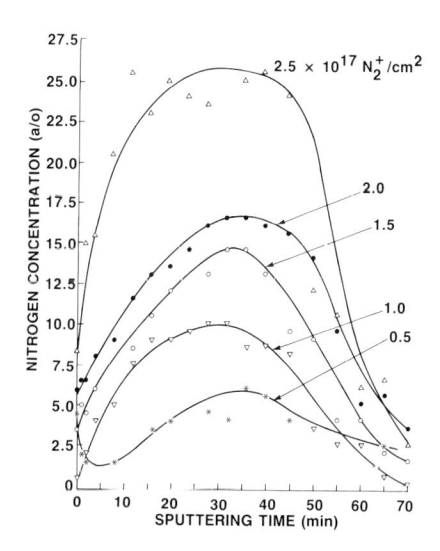

Fig. 1. Nitrogen distribution in implanted WC/Co as a function of fluence.

Table I. Peak Nitrogen and Average Nitrogen Concentrations, R(p) and $\Delta R(p)$ as a Function of Fluence.

FLUENCE	PEAK CONC.	AVG. CONC.	R_p^*	ΔR_p
5.0×10^{16} N_2^+/cm^2	5.0 a/o	3.2 a/o	1056 Å	570 Å
1.0×10^{17}	9.0	4.8	855	630
1.5×10^{17}	13.0	7.0	1020	525
2.0×10^{17}	16.5	9.6	1020	660
2.5×10^{17}	25.0	14.0	729	690

* Based On a Sputtering Rate of 30 Å/min
Specimen Surface Cobalt Rich

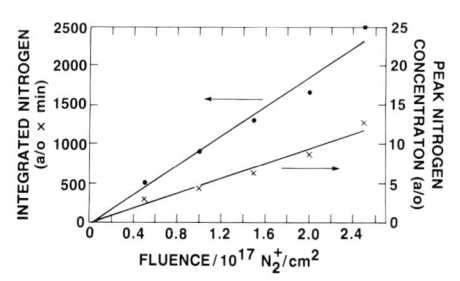

Fig. 2. Integrated and maximum nitrogen concentrations as a function of fluence, x - maximum concentration, a/o; ● - integrated concentration X2

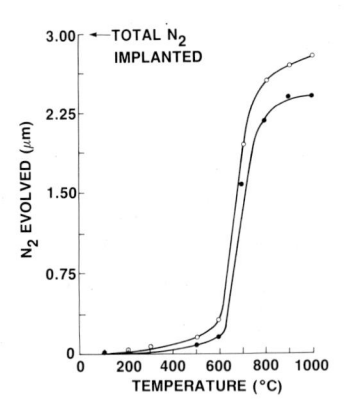

Fig. 3. Cumulative nitrogen evolution from implanted WC/Co, $2.5\times10^{17} N_2^+/cm^2$, as a function of outgassing temperature.

evolved between 600-700°C and the curves indicate the reproducibility of the experimental procedure. In addition, significant amounts of CO and H_2 and lesser quantities of NO and CO_2 are also measured as a function of outgassing temperatures. Most of the CO is evolved rapidly in the 700-800°C range. The arrow in Fig. 3, at 3 microns, corresponds to the quantity of implanted nitrogen that would be measured, if it were all desorbed. It is apparent that there are still small amounts of nitrogen remaining after the vacuum heating. In order to determine the concentration and the distribution of the remaining nitrogen, AES analyses were carried out on several outgassed samples. The nitrogen profile in WC/Co after vacuum heating to 1000°C is shown in Figure 4

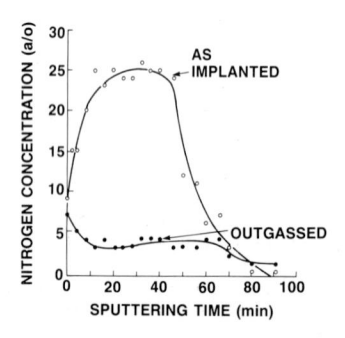

Fig. 4. Nitrogen distribution in nitrogen implanted WC/Co, before and after vacuum heating to 1000°C for a fluence of $2.5 \times 10^{17} N_2^+/cm^2$.

for a specimen implanted to $2.5\times10^{17}/cm^2$. Small amounts of nitrogen are still present at an average concentration of 3-4 a/o and there is no significant broadening of the profile due to nitrogen diffusion into the sample bulk. The nitrogen distribution from an implanted sample of the same fluence is included for comparison.

An Arrhenius plot of the data in Figure 3, gives an approximate activation energy for nitrogen desorption of 7 kcal/mole, or 0.30 ev. More precise values for activation energy can be obtained by using narrower temperature intervals for the TDMS measurements or a ramp heating technique with continuous nitrogen monitoring.

DISCUSSION

The cemented carbide is a complex, composite material, consisting of a hard, crystalline metal carbide, WC, and a relatively soft, ductile binder, Co, with a certain amount of alloying at the grain boundaries (9). The physical and the chemical effects of nitrogen implantation therefore, are extremely difficult to interpret. Furthermore, sample preparation will have a significant effect on the composition of the WC/Co surface. For example, relief polishing or smearing of the Co will leave the surface with a deficit or an enrichment of Co. These factors will have an effect on the consistency of the experimental results. However, collation of the data presented here attempted to minimize these effects.

Thermal desorption studies of nitrogen implanted iron have suggested that nitrogen is entrapped, partly as a gas and partly as a nitride, depending upon the chemical reaction between the implant and the substrate, and that the favorable mechanical changes observed are the result of the nitride formation (10). The results for implanted WC/Co however, are more difficult to interpret.

The large amounts of CO and H_2, and small amounts of NO, thermally desorbed from the implanted and unimplanted WC/Co samples, originate from oxide type films that are native to the sample surface. There is no indication of any interaction between the residual gases in the target chamber and the implanting nitrogen beam.

The low activation energy for thermal desorption of the implanted nitrogen is consistent with the view that the implanted nitrogen is trapped at defects within the damaged structure. The nitrogen is rapidly evolved in the 600-700°C temperature range, by moving with or from defects to the surface. The existence of stable chemical compounds or the presence of conventional diffusion phenomena, is unlikely, because of the higher activation energies associated with these phenomena. Nitrogen remaining in the sample after the vacuum heating however, is more strongly trapped in an undetermined way. In addition, for the fluence range of $1-2.5 \times 10^{17}/cm^2$ the quantity of nitrogen remaining after TDMS analysis is approximately the same, i.e., equivalent to an average nitrogen concentration of 3-4 a/o.

The total implanted nitrogen, determined by integrating the nitrogen (AES) profile curves (Fig. 2), was compared with values derived from the fluence and found to be less than the expected values by about 30-40%. If all of the implanted nitrogen is initially retained by the sample, it follows that the nitrogen concentrations determined by AES are too low. This could result from inappropriate Auger sensitivity factors for nitrogen, small thermal desorption effects during AES analysis or from Ar ion sputtering effects during AES profiling.

CONCLUSIONS

Although the low activation energy for nitrogen desorption indicates that most of the implanted nitrogen is trapped at near surface defects resulting from radiation damage, the formation and subsequent thermal decomposition of a loosely bound chemical system cannot be ruled out. The nitrogen that remains after post-implant heating is trapped more strongly within the same general region of the original nitrogen profile, shows little or no diffusion into the bulk of the sample and attains an average concentration of 3-4 a/o, throughout this region.

The nitrogen concentration is linear with fluence over the range of 0.5 to 2.5 x 10^{17} N_2/cm^2 for an implantation energy of 180 keV. The nitrogen profiles however, are broader than expected and, the maximum nitrogen concentrations are less. Furthermore, there is some indication that the nitrogen distribution, for the highest fluence, may be reaching a point of saturation. It should be noted that both of these results may be related to the analytical techniques used to characterize the samples rather than fundamentals. Characterization of the sample by a nondestructive method such as Nuclear Reaction Analysis should provide data to resolve these issues.

ACKNOWLEDGEMENTS

I should like to thank Dr. D. Dugger, GTE Labs, for his critical comments, and to the following members of the Materials Evaluation Group for their analytical support: M. Ames for TDMS and F. Kiluk for AES. A special thanks to J. Cambriello for bringing the ion implanter back to life and keeping it operational during the course of the work.

REFERENCES

(1) G. Dearnaley, Radiation Effects, 63, 1, 1982.

(2) S. LoRusso, P. Mazzoldi, I. Scotoni, C. Tosella and S. Tosto, Appl. Phys. Lett., 34, No10, 627, 1979.

(3) R. Kelly, J. Vac. Sci. & Tech., 21, No3, 778, 1982.

(4) J. Greggi and R. Kossowsky, International Conference on the Science of Hard Materials, Wyoming, August 1982.

(5) J. Greggi, Scripta Metallurgica, 17, 765, 1983.

(6) A. Kolitsch and E. Richter, Crystal Res. and Tech., 18, K5-K7, 1983.

(7) M. Baron, A. Chang, J. Schreurs and R. Kossowsky, Nuclear Instr. and Meth., 182/183, 531, 1981.

(8) I. Manning and G.P. Mueller, Computer Physics Communications, 7, 85, 1974.

(9) V. Jayaram and R. Sinclair, J. Amer. Ceramics Soc., In press.

(10) M. Guglielmi, A. Oliana and S.Tosto, Appl. of Surf. Sci., 10, 466, 1982.

FRICTION AND WEAR BEHAVIOR OF A COBALT-BASED ALLOY IMPLANTED WITH Ti OR N

S. A. DILLICH,* R. N. BOLSTER AND I. L. SINGER
Code 6170, Chemistry Division, Naval Research Laboratory,
Washington, D.C. 20375; *National Research Council Research
Associate

ABSTRACT

Dry sliding friction tests and relative abrasive wear measurements were used to investigate the effects of ion implantation on the tribology of a centrifugally cast Co-Cr-W-C alloy. Titanium implantation was found to significantly reduce the friction and wear of the alloy. Auger spectroscopy showed that vacuum carburization of the surface occurred during Ti implantation. Similar Ti and C profiles were seen in both carbide and (Co-rich) matrix phases. The abrasive wear resistance of the alloy decreased as a result of N-implantation. Nitrogen implantation did not reduce the friction during dry sliding, however, a change in the wear mode was observed.

INTRODUCTION

Cobalt-based superalloys show superior resistance to erosion, abrasion and galling wear owing to their microstructures, which consist of very hard Cr and W carbides dispersed in softer, Co-rich solid solutions [1-4]. Co-Cr-W solid solutions normally have hcp structures at low temperatures. However, in order to insure ductility during practical application, these alloys are designed to have fcc matrices.

Because of the technological and strategic importance of superalloy hardface materials, surface treatments to still further improve their mechanical and tribological properties should be investigated. The present study examines the effects of ion implantation on the dry sliding friction and abrasive wear resistance of Stoody 3, a centrifugally cast, cobalt-based alloy (50Co, 31Cr, 12.5W) (51-58 R_C). Nitrogen implantation has been reported [5-6] to result in large reductions in wear of many metals and cemented carbides. Similarly, high fluence implantation of Ti has been found to reduce the friction [7], and abrasive [8] and adhesive [9] wear in steels such as AISI 52100, a high carbon chromium steel. Our investigations therefore have concentrated on observing the effects of implantation of these two atomic species.

Dry sliding friction tests and relative abrasive wear measurements were made on nonimplanted, Ti-implanted, and N-implanted Stoody 3 disks. Wear tracks on the Stoody 3 surfaces were examined using Scanning Electron Microscopy (SEM) and optical microscopy with Differential Interference Contrast (DIC). The compositions of debris in the wear scars were identified by Energy Dispersive X-ray Analysis (EDX).

EXPERIMENTAL PROCEDURES

Friction measurements were made during low speed (0.1 mm/sec) dry sliding, using a ball on flat (disk) test geometry. Tests were conducted in ambient air (30 - 50% RH), at room temperatures, with a normal load of

9.8 N. The Stoody disks were tested against a variety of alloy balls which have been described elsewhere [10]. Tests were limited to twenty unidirectional passes of the ball (1.27 cm diam.) over the same path on the disk (1.27 cm diam.).

Prior to testing, balls and metallographically polished disks were ultrasonically cleaned in organic solvents [10]. Stoody 3 disks were implanted in a modified Model 200-20A2F Varion/Extrion ion implanter with a hot cathode arc discharge type ion source. The samples were mounted onto a water cooled holder and the target chamber was cryogenically pumped to pressures of about 3×10^{-6} Torr or better. The disks were implanted with ^{48}Ti ions at 190 keV beam energy to a fluence of 5×10^{17} Ti/cm², or with N_2 molecular ions at 100 keV to a fluence of 2×10^{17} N_2/cm² (i.e., 4×10^{17} N/cm² at 50 keV).

An abrasive wear technique [8,11] was used to determine the wear resistance of implanted disks relative to that of nonimplanted Stoody 3. Nine disks, 3 implanted and 6 reference, were simultaneously abraded for given intervals on a vibratory polisher charged with diamond paste (1-5 μm) in paraffin oil. The atmosphere was air dried to a frost point of 200 K. Wear rates were determined from weight losses measured with microgram precision. The relative wear resistance of a Ti or N-implanted disk was calculated by dividing the mean depth change of the two reference disks by the depth change of the implanted disk in each holder.

RESULTS

Friction Tests

Coefficients of kinetic friction (μ_K) measured on nonimplanted disks were initially quite low ($\mu_K \sim 0.25$, 1st pass), rising with successive passes to $\mu_K \sim 0.6$ as wear debris collected on the tracks. Optical microscope and SEM analysis revealed that considerable scratching and plastic deformation of the Stoody 3 alloy occurred even during the first pass [10]. Plowing of the Co-rich matrix phase was the most prominent feature of the wear tracks. However, carbide scratching and cracking, as well as separation from the matrix phase, were also observed. In all cases, the debris in the tracks, as determined by EDX analysis, had the composition of the softer member of the contact couple. Results of friction tests against two of the ball materials used, a hard bearing steel (AISI 52100) and a softer 302 SS, are listed in Table I for comparison.

Much lower (50% to 70%) coefficients of friction were observed on the Ti-implanted disks when tested against balls of hardness approximately equal to or greater than that of Stoody 3 [10]. As was the case for the nonimplanted disks, ball material was transferred to the tracks only when the bulk hardness was less than that of the disk [Table 1]. Some plastic deformation was apparent on tracks after 20 passes; however, for the harder ball materials, much less debris was found on the Ti-implanted disks than on the nonimplanted disks.

Nitrogen implantation did not reduce the coefficient of friction of the Stoody alloy. As was the case for the nonimplanted disks, high coefficients of friction ($\mu_K \sim 0.6$, 20th pass) were accompanied by considerable surface damage to the disk (Fig. 1c). However, unlike nonimplanted and Ti-implanted disks, the wear debris generally consisted of material transferred from the ball, regardless of the relative bulk hardnesses of the contacting alloys. This can be seen in Table I for AISI 52100 and 302 SS balls, and in Fig. 2 for the case of hard 440C SS balls.

TABLE I. Coefficients of kinetic friction of AISI 52100 and 302 SS balls sliding against Stoody 3 disks.

Ball	Surface	μ_K 1st pass	20th pass	Debris in track[a]
52100	Nonimplanted	0.32	0.56	Co-Cr-W
(Fe 1.5Cr 1C)	Ti-implanted	0.14	0.19	none
(60-65 R_C)	N-implanted	0.30	0.55	Fe
302 SS	Nonimplanted	0.24	0.58	Fe-Cr-Ni
(Fe 18Cr 8Ni)	Ti-implanted	0.17	0.51	Fe-Cr-Ni
(29-35 R_C)	N-implanted	0.22	0.57	Fe-Cr-Ni

[a] identified by EDX

Abrasive Wear Tests

Abrasive wear data on Ti-implanted and N-implanted Stoody 3, relative to that of nonimplanted Stoody 3, are shown in Fig. 3. The wear resistance of Ti-implanted disks was higher by a factor of about 4 at the surfaces, falling off gradually as indicated by the broken line in Fig. 3. Beyond 200 nm the wear resistance was the same as that of the nonimplanted disks (solid line, Fig. 3).

In contrast, N-implantation reduced the abrasion resistance of the Stoody 3 surfaces by about 50%. The wear resistance of the implanted disks increased steadily with wear depth, reaching that of the non-implanted disks at about 200 nm. The effects of both Ti-implantation and N-implantation on the wear resistance persisted to depths corresponding approximately to the depths of implantation i.e., ~150 nm [10].

NONIMPLANTED ($\mu_k = 0.56$)   Ti IMPLANTED ($\mu_k = 0.19$)   N IMPLANTED ($\mu_k = 0.55$)

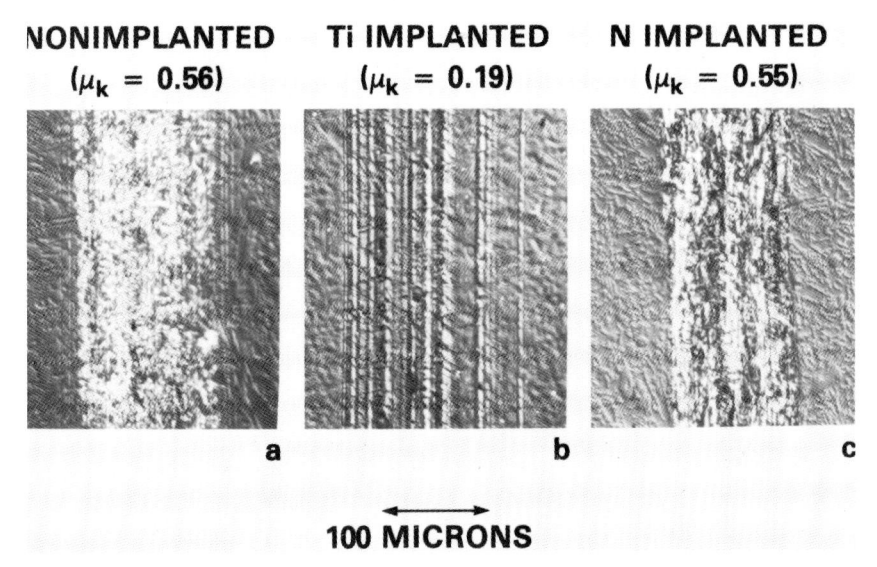

a   b   c

◄— 100 MICRONS —►

FIG. 1. Wear tracks of 52100 steel balls against Stoody 3; 20 passes.

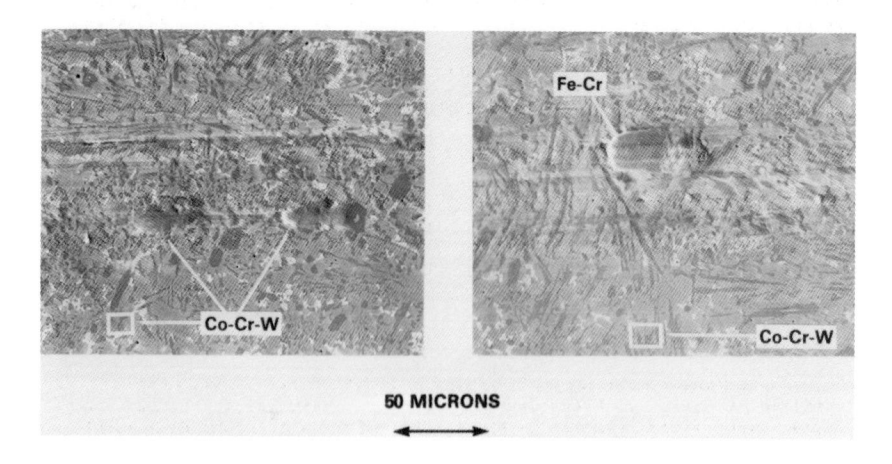

FIG. 2. SEM/EDX analysis of wear debris produced after 20 passes of a 440C ball (58-62R$_C$) on a Stoody 3 disk; (left) nonimplanted, (right) same disk after N-implantation. Note that the composition of wear debris differs for the two cases.

DISCUSSION AND CONCLUSIONS

High fluence Ti-implantation produced an abrasion resistant layer in Stoody 3 which also showed reduced friction and surface damage under dry sliding conditions. Scanning Auger microscope analysis revealed that implant assisted, vacuum carburization of both the carbide and matrix phases occurred during Ti-implantation [10]. Carburization of the surface, also observed in steels implanted with Ti under identical conditions [7-8], results when implanted Ti at the sample surface reacts with carbon from residual gases in the implanter vacuum chamber [12]. In 52100 bearing steel, reduced friction and wear following Ti-implantation is contingent upon carburization of the surface during implantation [9], and is associated with the presence of an amorphous surface layer [7]. A disordered phase is also believed to exist in Ti-implanted Stoody [13].

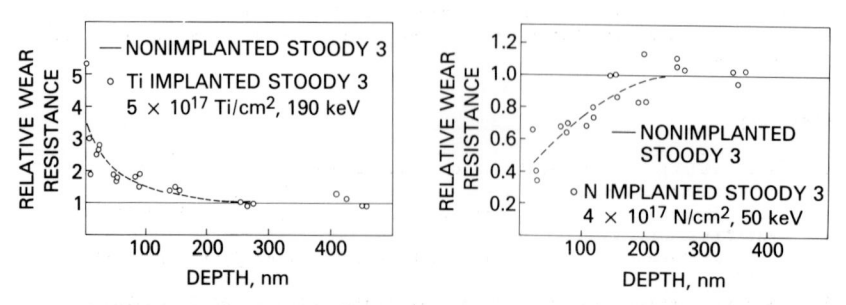

FIG. 3. Relative wear resistance of Ti-implanted Stoody 3 disks (left) and N-implanted disks (right), abraded with 1-5µm diamond, vs. depth.

however, its contribution to the establishment of low friction and wear is not fully understood. It is probable that, as is the case for 52100 steel, the improvements cannot be attributed to increased indentation hardness [14]. More likely, they are characteristic of the toughness and chemical nature of the carburized surface layer which prevents debris formation by inhibiting brittle fracture and/or adhesive welding at the contacts.

Nitrogen implantation into Stoody 3 reduced the relative wear resistance of the alloy surface to about half the bulk value. A possible mechanism for the observed behavior can be inferred from comparison with that of 304 SS, an alloy which also showed an approximately 50% reduction in relative wear resistance following N-implantation [11]. Surface martensite, produced by rolling deformation or abrasion of austenitic (γ,fcc) 304 SS, was found to revert back to the softer austenite after N-implantation [15].

The superior wear properties of cobalt based alloys are often credited, in part, to their low stacking fault energies which allow low temperature martensitic fcc $\rightarrow$ hcp transformations to occur with strain [1-4]. During wear these transformations increase the work hardening rate of an alloy, thereby providing the bulk fcc material with a hard wear resistant surface. Nitrogen generally has little or no solid solubility in Co. However, carbon in cobalt acts as a stabilizer for the fcc phase [16]. If implanted nitrogen played a similar role (i.e., stabilized the fcc phase and prevented the formation of hcp platelets at the contact surface during abrasion), the low wear resistance of the implanted disks could be attributed to the absence of the work hardened layer. As the implanted layer was removed, work hardening brought the wear resistance up to that of the nonimplanted disks.

The predominance of ball debris in the N-implanted tracks after dry sliding suggests an increase in the ductility of the alloy consistent with the above supposition of a stabilized fcc matrix phase. Evidently, shear at the contacts occurred primarily in the ball or at the junction interfaces, rather than in the Stoody 3 disk. Other possible influences on the tribology of N-implanted Stoody 3 which remain to be investigated include implantation induced changes in the microstructure and mechanical properties of the carbide phases, and W and Cr nitride formation.

ACKNOWLEDGEMENTS

We thank the Surface Modification and Materials Analysis Group at NRL for their cooperation with implantation. We would especially like to thank the Stoody Company, WRAP Division, for their generous donations of the materials used in this investigation. This project was supported in part by ONR.

REFERENCES

1. K.C. Antony, J. of Metals, Feb. 1973, 52-60.
2. K. J. Bhansali and A.E. Miller, Wear, 75 (1982) 241-252.
3. R.I. Blomberry and C.M. Perrot, J. Aust. Inst. Met. V19 #4 (1974) 254-258.
4. C.J. Heathcock, A. Ball and B.E. Protheroe, Wear 74 (1981-1982) 1126.
5. N.E.W. Hartley, in Treatise on Materials Science and Technology, V18 Ion Implantation, ed. J.K. Hirvonen (Academic Press, N.Y. 1980) 321-368.

6. G. Dearnaley, Radiation Effects (1982) V63, 1-15.
7. I.L. Singer, C.A. Carosella and J.R. Reed, Nucl. Inst. and Meth. 182/183 (1981) 923-932.
8. I.L. Singer, R.N. Bolster and C.A. Carosella, Thin Solid Films, 73 (1980) 283-289.
9. I.L. Singer and R.A. Jeffries, J. Vac. Sci. and Technol. A1 (1983) 317-321.
10. S.A. Dillich and I.L. Singer, Thin Solid Films, 108 No. 2, 1983, 219-227.
11. R.N. Bolster and I.L. Singer, Appl. Phys. Lett. 36 (1979) 208-209.
12. I.L. Singer, J. Vac. Sci. Technol. A1 (1983) 419-422.
13. S.A. Dillich, J.A. Sprague and I.L. Singer, to be published.
14. W.C. Oliver et. al., these proceedings.
15. R.G. Vardiman, R.N. Bolster and I.L. Singer, in Metastable Materials Formation by Ion Implantation, S.T. Picraux and W.J. Choyke, ed. (Elsevier Science Publ. Co. Inc., 1982) 269-273.
16. Cobalt Monagraph, Centre D'Information Du Cobalt, Brussels, (1960) 167.

THE MECHANICAL PROPERTIES OF Si$^+$ and Pb$^+$ IMPLANTED Al

P.B. MADAKSON
Department of Mechanical Engineering, King's College, Strand, London, WC2R 2LS, England.

ABSTRACT

Commercially pure Al was implanted with 300 keV Si$^+$ and 200 keV Pb$^+$ to doses of between 10^{11} and 10^{17} ions/cm^2. Changes in friction, wear, oxidation and hardness were investigated. Silicon increased the hardness and wear resistance of Al and significantly decreased friction and the oxidation of the implanted surface. These changes were observed to be almost proportional to the implanted dose. The implantation of Pb$^+$ resulted in a linear increase in hardness and a decrease in surface oxidation with dose. Friction decreased and wear resistance increased but the changes were not dose dependent. The implantation of Si$^+$ did not significantly alter the distribution of impurities, such as Fe and Cu within the Al matrix, but Pb$^+$ resulted in a diffusion of Fe to the implanted surface. Formation of precipitates was observed and the improvements in the surface properties studied are considered to result from precipitation hardening, which involves the impediment of dislocation movement by the precipitates during plastic deformation of the implanted Al.

INTRODUCTION

Silicon is always present as an impurity in Al and like lead it imparts fluidity in casting and welding. Lead is usually added to the cast aluminium to improve bearing properties. It produces only very slight changes in the mechanical properties of Al [1]. The presence of Si impurities in cast aluminium gives high mechanical properties through the formation of precipitates. Vandelli [2] observed that a fine dispersion of silicon needles, evenly distributed within the Al matrix, gives maximum improvement in wear resistance. Similar observations were made by Sarker and Clarke [3]. The work of Okabayashi et al [4] suggests however that both the distribution and the content of Si in Al are important for large improvements in mechanical properties.

Only lead had previously been implanted into aluminium. Thackery and Nelson [5] reported the formation of Pb precipitates after implantation of Al. The role of these precipitates on the mechanical properties of Al was not investigated. It is reported in this paper that implanted Pb$^+$ and Si$^+$ produce significant changes in the mechanical properties of Al. The hardness, friction and wear of the implanted Al were measured and compared with values from the unimplanted Al. The experimental method has been described elsewhere [6]. Changes in surface oxidation were studied by the Rutherford back-scattering (RBS) technique. Silicon and lead were implanted into Al at 300 keV and 200 keV respectively.

RESULTS

Figure 1 shows that Pb$^+$ and Si$^+$ significantly increase the hardness of aluminium. Silicon produces the maximum increase in hardness. The increase in hardness is almost proportional to the Si$^+$ dose. Lead implantation into Al only produces about 20% increase in hardness.

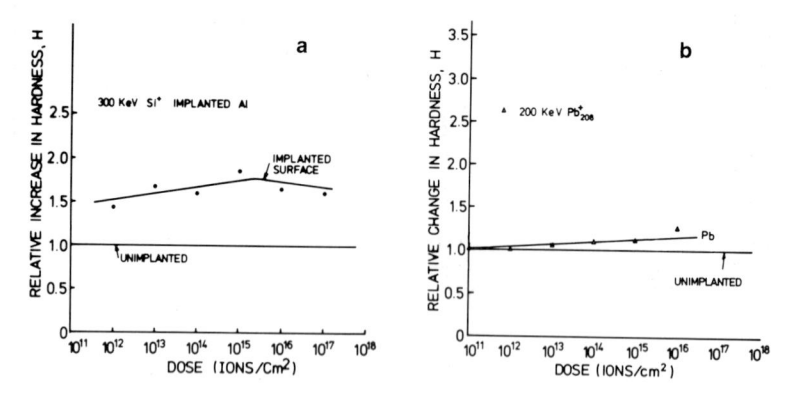

Fig. 1 Plot of hardness against implanted dose for Al implanted with (a) 300 keV Si^+ and (b) 200 keV Pb^+

Both lead and silicon reduced the friction of aluminium (fig. 2). In the case of silicon the reduction in friction is almost proportional to the Si^+ dose. The reduction in friction after 200 keV Pb^+ implantation of Al is not dose dependent. The maximum reduction in friction (50%) occurs at a dose of $10^{14} Pb^+/cm^2$ after which friction decreases with dose (fig. 2b).

Fig. 3 shows the variations in wear resistance and incubation time with dose for Al implanted with Si^+ (fig. 3a) and Pb^+ (fig. 3b). The incubation time is the period taken before observable wear commences on the implanted surface. It can be seen that for both ion species the incubation period is not dose dependent. In the case of Si^+ implantation of Al, the increase in wear improvement factor, W (taken as the wear depth on the unimplanted surface divided by that on the implanted surface) is proportional to the implanted dose. For Pb^+ implantation of Al the wear improvement factor is not dose dependent (fig. 3b). The maximum improvement in wear resistance occurs at doses of 10^{11} and 10^{13} Pb^+/cm^2.

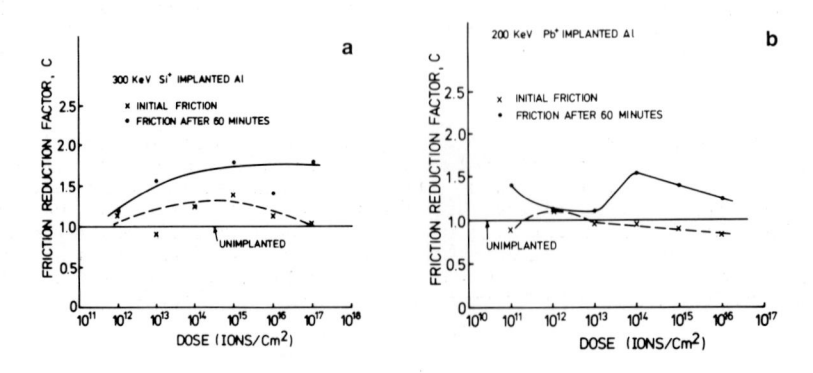

Fig. 2 Change in friction with dose for sliver steel pin sliding on Al implanted with (a) 300 keV Si^+ and (b) 200 keV Pb^+. The factor C is the ratio of the friction of the unimplanted to implanted Al

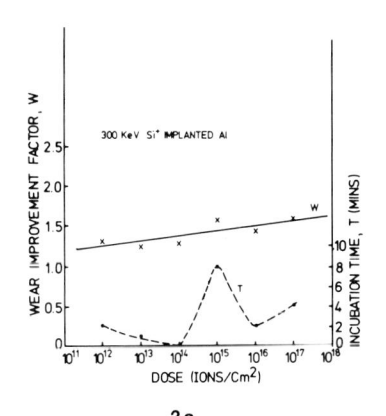

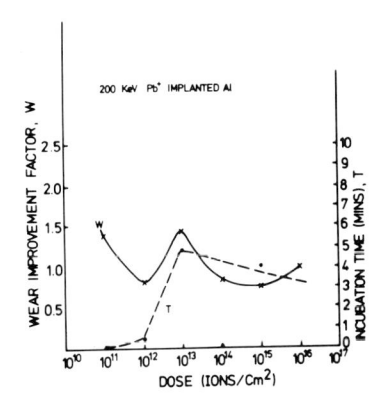

Fig. 3 Variations in wear resistance and incubation time with dose for Al implanted (a) 300 keV Si$^+$ and (b) 200 keV Pb$^+$

Dose (ions/cm^2)	unimplanted	10^{11}	10^{12}	10^{13}	10^{14}	10^{15}	10^{16}
Pb conc. ± 0.01%	–	–	–	–	0.009	0.030	0.310
Fe conc. ± 0.01%	0.016	0.035	0.060	0.035	0.050	0.070	0.070
O conc. ± 0.2%	1.4	1.3	0.7	–	–	–	–

Table A The surface concentration of impurities (Fe and O) and Pb after 200 keV Pb$^+$ implantation of Al

Table A gives the variation in the concentration of lead, iron and oxygen with Pb$^+$ dose. The method of measuring the surface concentration of these elements is given elsewhere [7]. Commercial aluminium contains iron and oxygen impurities [6]. Table A indicates that the concentration of these impurities at the implanted region varies with the implanted dose of Pb$^+$. The concentration of oxygen decreases with dose but the Fe concentration does not depend on the Pb$^+$ dose.

DISCUSSION

The observed changes in hardness are considered to be very significant in view of the fact that the method employed for measuring hardness is not sensitive to the implanted region. The hardness measurements were made on a Vickers microhardness tester which consists of an indenter that, on the application of a fixed load penetrates the implanted specimen to a depth of about 1µm. Such depths are beyond the ion penetration range. That there is an observable change in hardness after ion implantation suggests that the implantation effects extend beyond the ion range. This fact is well known and it has been reported that the depth of implanted ions can be increased by a thousand times or more [8] by radiation enhanced diffusion.

The linear dependence of hardness, H, on ion dosage, ϕ_r had previously been reported by Madakson [6,7] who showed that

$$H = \frac{3KE\alpha_r \phi_r R_p^2}{(1 - 2v)} \qquad (1)$$

or $H = K_1 \phi_r$

where K and K_1 are constants, E is Young's Modulus of elasticity, α_r is the coefficient of volumetric expansion (cm^3/ions), ϕ_r is the ion concentration (ions/cm^3), R_p is the ion projected range (cm) and v is the Poisson ratio [6]. Hardness was also reported to be proportional to the compressive stress σ generated at the implanted region [6], i.e. $H = K_2 \sigma$.

The implantation of both Si$^+$ and Pb$^+$ into Al resulted in a significant reduction in friction (fig. 2) and wear (fig. 3). In the case of Si$^+$ implantation these changes are almost proportional to the dose of silicon. This suggests that the role of implanted silicon in Al is similar to silicon introduced in the as-cast Al. It is the concentration of silicon that determines the magnitude of the changes in the mechanical properties of Al. A high concentration probably results in a fine dispersion of Si precipitates which, if evenly distributed in the Al matrix, would impede the movement of dislocation networks thereby hardening the implanted region.

There is no dose dependency in the friction and wear behaviour of Pb$^+$ implanted Al. This is probably because these properties do not only depend on the presence of lead atoms in the Al but the lattice damage created by Pb$^+$ during the implantation process. The damage created is manifested in sputtering of the implanted surface and the extensive migration of the existing impurities in Al. Table A shows that Fe migrates to the implanted region while the concentration of oxygen decreases. The reduction in oxygen concentration with Pb$^+$ dose is considered to be due to sputtering.

The concentration of Fe does not depend on the dose of lead ions (Table A). This is considered to be due to the creation and subsequent collapse of vacancies as the Pb$^+$ dose increases. The number of vacancies created during the implantation process would initially increase with Pb$^+$ dose. Since the Fe atoms migrate along these vacancies their concentration would also increase with the dose of lead ions. At a given Pb$^+$ dose most of these vacancies would collapse thereby terminating further migration of Fe atoms. There would be no further increase in the concentration of these atoms until more vacancies are created. The implantation of silicon into Al did not create similar changes in the distribution of the existing impurities in Al because Si$^+$ created negligible damage of the Al matrix.

CONCLUSIONS

Silicon and lead implantation of commercially pure Al produce up to 70% reduction in friction, 60% increase in wear resistance and 70% increase in hardness. In the case of Si$^+$ implantation these changes are due almost entirely to the presence of silicon in Al, their magnitude depends therefore on the Si$^+$ dose. Lead produces a significant damage of the Al lattice and as such alters the distribution of the existing impurities in Al. Changes in the mechanical properties investigated are determined by the surface concentration of these impurities. Migration of these impurities during ion implantation had previously been reported [6,7,9] and it seems to occur when heavy ions are used.

The changes in the surface oxidation of Si^+ and Pb^+ implanted Al are considered to be due to sputtering of the Al surface. The reduction in friction is considered to be associated with a decrease in the surface energy [10] while the increase in wear resistance is considered to result from the work hardening arising from the interaction of line and point defects with the implanted ions and precipitates.

ACKNOWLEDGEMENTS

The author gratefully acknowledges the award of grants from the Science and Engineering Research Council and the University of London Central Research Fund. He would like to thank Professor S.W.E. Earles for the provision of travel and subsistence funds and Dr. C. Jeynes, University of Surrey, for implantation of the specimens.

REFERENCES

1. L.E. Mondolfo, Aluminium Alloys Structure and Properties (Butterworths, London, 1976).
2. G. Vandelli, Alluminio and Nuova Metallurgia, 37 121 (1968).
3. A.D. Sarker and J. Clarke, Wear, 62 157 (1980).
4. K. Okabayashi, Y. Nakatani, H. Notani and M. Kawamoto, Keikinzoku (light metals, Tokyo), 14 57 (1964).
5. P.A. Thackery and R.S. Nelson, Phil. Mag. 19 169 (1969).
6. P.B. Madakson, Ph.D. Thesis (University of London, 1983).
7. P.B. Madakson, J. Appl. Phys., to be published.
8. S.M. Myers, Report, SAND-75-5386, 7 (1975).
9. P.B. Madakson, Proc. Intern. Conf. Ion Beam Analysis, Arizona 1983, in Nucl. Instr. Method, to be published.
10. P.B. Madakson, Wear, 87 191 (1983).

WEAR BEHAVIOR OF FLAT AND GRADED PROFILE BORON-IMPLANTED BERYLLIUM

K. KUMAR*, H. NEWBORN*, AND R. KANT**
*The Charles Stark Draper Laboratory, Inc., Cambridge, MA 02139
**Naval Research Laboratory, Washington, D.C. 20375

ABSTRACT

Pin-on-disk tests were performed for comparative friction and wear behavior on flat and graded profile boron implanted beryllium samples. Peak, intended boron concentrations of 10, 20, 30 and 40 atom percent were investigated. Auger Electron Spectroscopy was used to determine the boron concentration as a function of depth. Preliminary work was performed to study the effects of (1) a low temperature (450°C, 1-1/2 hours) heat treatment of the implanted specimens and (2) a change in the pin material. All of the boron implanted beryllium samples showed significant improvement versus unimplanted beryllium and an anodized beryllium surface. Graded samples showed comparable friction coefficients but inferior wear resistance with respect to the flat profile samples.

INTRODUCTION

The feasibility of implanting boron atoms into beryllium parts to significantly improve their surface hardness[1] and wear resistance[2] has been previously demonstrated. Variations in concentration of implanted boron as a function of depth are achieved through control of: (1) the boron ion flux, (2) the accelerating voltage, and (3) the duration for which the implantations are performed. Samples may be heat treated after implantation to form hard, beryllium borides and to change the boron concentration depth profile.

This paper reports upon the friction and wear behavior of implanted beryllium samples for intended peak boron concentrations of 10, 20, and 30 atom percent. In addition, the effect of pin material on friction and wear is discussed. In particular, a beryllium-titanium diboride composite[3] which is being examined as a potential gas bearing surface, was used as a pin. A comparison is, also, made between the wear resistance of the implanted beryllium with that of an anodized beryllium surface.

EXPERIMENTAL

Samples with intended peak boron concentrations of 10, 20, and 30 atom percent, denoted as samples 1, 2, and 3, respectively, were compared with previously discussed[2] 40 atom percent graded and flat profile samples, (identified here as 4 and 5), as well as with a lubricated anodized beryllium specimen. The energies used for this work were 25, 43, 67, 100, 140, and 192 keV and the fluences used for the 10 atom percent peak boron concentration sample corresponded to 7.1, 8.6, 8.8, 8.0, 6.3, and 3.8 x 10^{16} B/cm^2. The fluences used for the 20 and 30 atom percent samples, respectively, were two and three times these values.

The samples were analyzed by AES for concentration and depth profiles. Samples 1, 2, and 3 were additionally heat treated at 450°C for 1-1/2 hours and examined for wear behavior, designated 1HT, 2HT, and 3HT, respectively. Auger evaluation of the implanted materials was performed on a Physical Electronics instrument operated at 5 kV with a 1μm-diameter spot size.

Wear testing was performed using the pin-on-disk technique with a flex pivot wear-tester[4]. The tests were performed at 100 rpm for 10 minutes duration. Pins used were a 3.2-mm sapphire ball installed in a steel cylinder and a beryllium-titanium diboride composite cylinder with its end ground to the same radius of curvature. The samples were wear-tested with progressively higher loads applied to the pin to give a relative measure of performance. Optical micrographs were taken of the wear tracks generated at the increased loads.

RESULTS

AES Studies

As with the earlier 40 atom percent samples,[2] the peak boron concentrations (Figure 1) were consistently higher than intended. The profile of sample 1 shows a peak boron content of about 20 atom percent (versus an expected value of 10 percent). Sample 3 shows a peak value exceeding about 45 atom percent (versus an expected value of 30 percent boron). Sample 2 likewise had a peak value of about 35% boron. Of these three samples, only sample 3 had a smooth graded composition profile (similar to the 40 atom percent graded sample in Reference 2). Samples 1 and 2 showed a fluctuating composition profile as indicated in Figure 1.

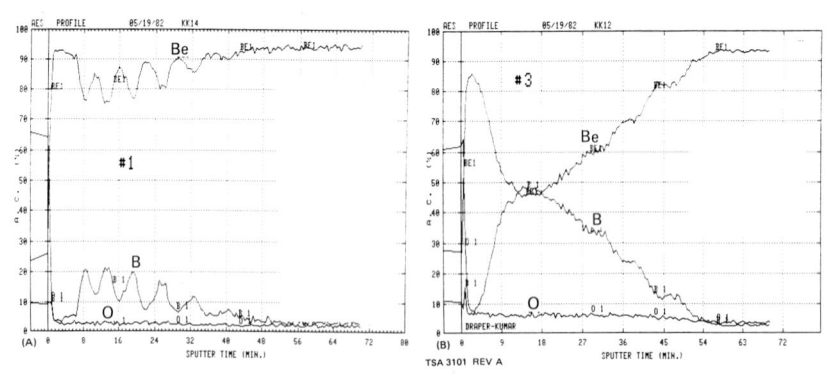

Figure 1. Plots of atom concentration versus sputter time (depth) for (A) Sample No. 1, and (B) Sample No. 3.

Friction and Wear Studies

Table I summarizes the friction coefficient data obtained with a 20g applied load. The peak intended boron concentrations of the different samples are shown in parentheses in the second column of this table.

The most repeatable values for μ were obtained at loads below which no evidence of a wear track could be detected with a Dektak surface profilometer. In most instances, applied loads of about 20g showed little or no evidence of a wear track. The data using the sapphire pin, indicated that even at the lowest (10 percent) boron levels, the initial friction coefficient was significantly lower than that measured for unimplanted beryllium,[6] and that this value was relatively constant for short duration (~0.1 sec). Figure 2 illustrates the observed effect of increasing the applied load on the coefficient of friction.

TABLE I. Measured values of the friction coefficients at different time for the several wear runs.

Speci-men No.	Actual (Intended) Peak %B	μ (20 g load) Sapphire .1 Min	μ (20 g load) Sapphire 1 Min	Composite 1 Min	Load Causing Severe Damage in Wear Region
1	20(10)	0.14(3)**	0.21(3)**	-	30g
2	35(20)	0.15(2)**	0.20(2)**	-	>40g
3	45(30)	0.14(3)**	0.26(3)**	-	30g
*4	55(40)	0.16(4)**	0.21(4)**	-	50g
*5	55(40)	0.15(6)**	0.15(6)**	0.46	60g with sapphire; up to 110g with composite
Be (anodized)	-		0.66	0.30	50g
1HT				0.53	30g
2HT				0.39	60g
3HT			0.22	0.63	60g

* The behavior of these samples was discussed in Reference 2. (Samples 4 and 5 had graded and flat profiles respectively.

**Number in parentheses indicates that the reported friction value was measured and averaged over that number of runs. Other data reported is based on a single run.

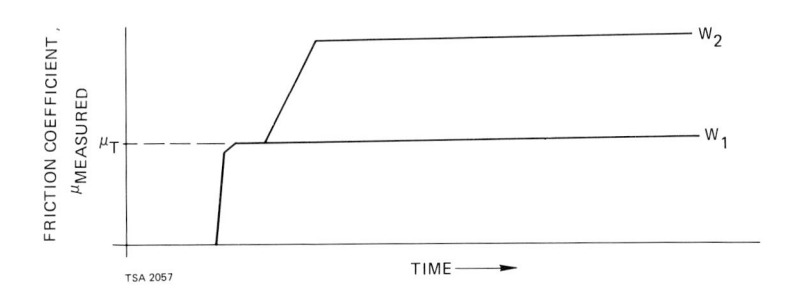

Figure 2. Schematic showing observed effect of increases in applied load on characteristics of friction trace of implanted beryllium.
$W_1 < W_2$.
$\mu_{measured} = \mu_T + \mu_W$, where μ_W is dependent on wear of surface, and
μ_T = true value of the friction coefficient.

This observed effect of the load on the friction traces was explained in the following way. The boron depth profiles indicated that little or no boron existed in the region near the surface of the implanted samples. The initial low and constant μ (for all the compositions that were investigated) were, therefore, possibly associated with increased wear resistance imparted to the surface by radiation damage from the penetrating boron ion flux. This reasonably boron-free beryllium surface layer was destroyed with increase in time of test and/or increased load, during which time the friction coefficient registered a reasonably rapid increase. Additional wear was very gradual through the boron rich layers. This latter effect appeared in the form of a very slowly increasing value of the friction coefficient with time. Somewhat surprisingly, this rapid stepped increase in μ was not noticed with the composite pin tests. The higher friction coefficient values that were measured were considered an indication that the microscopically rough surface of the composite pin possibly removed the boron-free beryllium surface quite rapidly to be sensed by the instrumentation.
Inconsistency in μ at the 0.1-minute interval with the composite pin required tabulation at the 1-minute duration; the sapphire data obtained at 1 minute is, therefore, also included for comparison.

Optical

Although the initial (0.1 second duration) μ values, for the sapphire pin, were roughly the same for all of the specimens, the wear resistance over full 10-minute intervals was quite different for various implanted conditions. Figure 3 depicts the wear tracks observed on

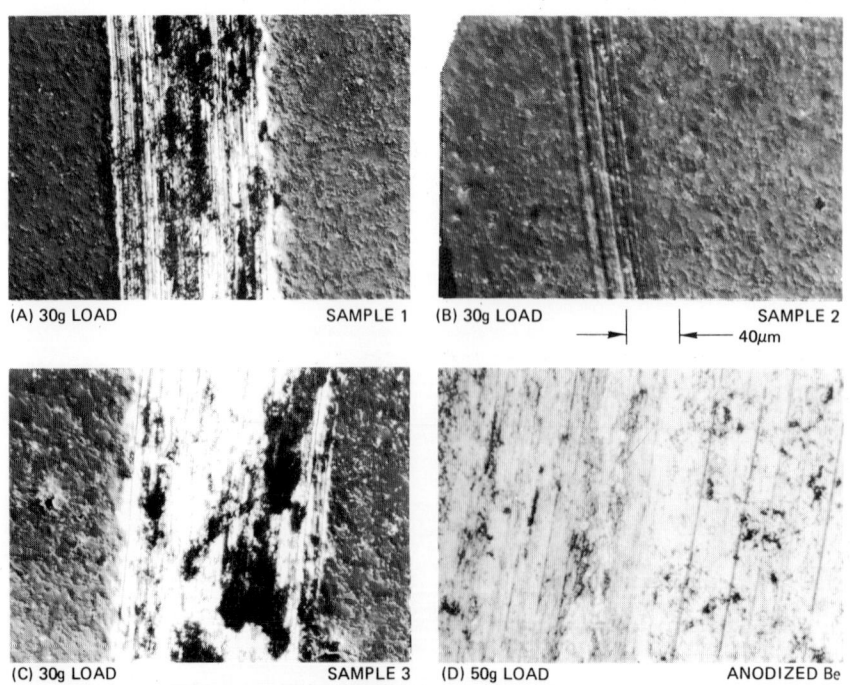

Figure 3. Wear tracks observed at indicated loads. Samples 1,2,3 contacted a Sapphire Pin. The anodized Beryllium sample was run against a Be-TiB$_2$ composite pin.

samples 1, 2, and 3 after 10 minutes of test using a sapphire pin at 30-gram load. Severe wear (i.e. significant debris, rugged wear track) has been initiated in samples 1 and 3. However, even at 40 grams, sample 2 showed no such signs. The reasons for this anomalous behavior of sample 2 are not clear. As indicated in Table I, after heat treatment, sample 1HT showed severe wear at the same 30g level with the composite pin; samples 2 and 3 HT survived to the 60-gram load level. This is an improvement after heat treatment for sample 3. It is possible that the heat treatment flattened the highly graded profile of sample 3. It could be concluded, similar to earlier work[2], that flat profile samples show superior wear resistance to graded samples. This study also demonstrates that the composition of the layer, a given distance below the actual wearing surface, determines the resistance of that surface to further wear. Further study on 40 intended atom percent flat profile samples has shown inconsistent results as to resistance to severe wear. One sample showed absence of severe wear even after a 110-gram loading with the composite pin.

Experiments performed with the composite pin on an anodized beryllium surface exhibited signs of severe wear after a 50-gram loading (Figure 3D). A low µ value was measured for this sample. However, the surface was lubricated prior to the measurement. This is believed to be responsible for the low value.

Microhardness Measurements

The hardness of a surface is generally considered an indication of its wear resistance. Therefore, the higher the hardness, the greater would be the expected wear resistance. Microhardness values in Knoop Hardness Numbers, KHN (expressed in kg/mm^2) made on several of the samples are shown in Table II. It is difficult to measure the hardness of such thin surface regions accurately. At high applied indentation load the substrate contribution is excessive. At very low loads the surface contribution dominates but questions arise with respect to "the actual" as compared to "the intended" load. We have, therefore, devised a procedure whereby microhardness measurements are made in a range of applied loads. The variation of microhardness as a function of applied load gives a more reliable assessment of the hardness of the surface.[5] Conventional thinking notwithstanding, the data in Table II show that the wear behavior of these samples cannot be explained on the basis of the measured microhardness data.

TABLE II. Average microhardness readings KHN (Kg/mm^2) on several implanted samples.

Applied Load (g) \ Sample	1	2	3	4	5
50				341	356
25	372	381	390		
10	703	643	686	762	542
5	1123	1017	1184	1211	1153
2	1736	1682	1622	2233	1965

CONCLUSIONS

1. Boron concentrations as low as 10 atom percent result in a substantial lowering of the friction coefficient. Higher concentrations appear to be needed for high resistance to wear. The behavior of sample 2 was anomalous and could not be explained.

2. The beryllium subsurface composition (of the implanted layers) plays a strong role in determining the wear resistance of the surface.

3. Microhardness and initial pin-on-disk friction coefficients do not correlate with resistance to longer-term severe wear.

4. Friction traces obtained using sapphire and beryllium titanium diboride composite pins were significantly different. It was suspected that this difference was related to rapid removal of the relatively boron free beryllium surface layer versus the composite pin material.

ACKNOWLEDGMENT

This work was supported by the Office of Naval Research.

REFERENCES

1. R.A. Kant, J.K. Hirvonen, A.R. Knudson, and J. Wollam, Thin Solid Films, 63, 27-30 (1979).
2. R.A. Kant, K. Kumar, A.R. Knudson, 1981 Annual MRS Meeting, Boston. Published in Proc of Symp E: Metastable Materials Formation by Ion Implantation; Eds: S.T. Picraux, W.T. Choyke, Elsevier Science Publ. Co. Inc., N.Y. 1982, p. 253.
3. K. Kumar, D. Das, Am. Ceram. Soc. Bull., 62, No. 2, 249-252, (1983).
4. K. Koehler, Technical Report C-5413, The Charles Stark Draper Laboratory, Inc., June 1981.
5. D.K. Das and K. Kumar, Thin Solid Films, 83, 53-60 (1981).
6. D. Das and K. Kumar, Thin Solid Films, 108, 181-188 (1983).

MICROSTRUCTURES OF STAINLESS STEELS EXHIBITING REDUCED FRICTION AND WEAR
AFTER IMPLANTATION WITH Ti AND C*

D. M. FOLLSTAEDT, F. G. YOST, AND L.E. POPE
Sandia National Laboratories, Albuquerque, NM 87185

ABSTRACT

Implantation of Ti and C into stainless steel discs of Types 304, 15-5 PH, Nitronic 60 and 440C has previously been reported to reduce wear depths by up to ~ 85% and friction by ~ 50% in unlubricated pin-on-disc tests. Our earlier studies relating microstructure to friction and wear results in Type 304 are first summarized; these indicate that the improvements in the surface mechanical properties are due to an amorphous surface layer, similar to the amorphous layer observed in pure Fe implanted with Ti and C. We have now examined the other three implanted steels and found similar amorphous layers. These results strongly suggest that the amorphous surface alloy is responsible for reduced friction and wear in all the steels.

INTRODUCTION

Ion implantation has been studied for about a decade as a method to reduce friction and wear on metal surfaces. Many studies have examined the effects of N implantation, but recently we and others have investigated steels implanted with Ti or Ti and C. This treatment has produced improvements in a wide variety of tests, including unlubricated pin- or ball-on-disc tests [1-4], lubricated tests [3,4], and tests of abrasive wear by diamond particles [5]. Here we examine implanted discs from our unlubricated tests, which gave reductions in maximum wear depth of up to ~ 85% and reductions in friction of ~ 50% [1]. Based on our observations in pure Fe implanted with Ti and C [6], we expected an amorphous surface alloy. Below we show that the amorphous phase forms in all the steels tested, and we propose that it is responsible for the reduced friction and wear in types 440C, Nitronic 60, and 15-5, as previously shown for 304 [7,8].

The details of our pin-on-disc tests are given elsewhere [1]. Mechanically polished discs were implanted with 2×10^{17} Ti/cm^2 (5×10^{16} Ti/cm^2 at 180, 150, 120 and 90 keV) and 2×10^{17} C/cm^2 (30 keV). Samples for transmission electron microscopy (TEM) were obtained from the wear test specimens by jet electropolishing from the unimplanted side to produce electron-transparent areas which included the surface alloy. Energy dispersive (x-ray) spectroscopy (EDS) was used with TEM analysis at 100 and 120 kV and with scanning (secondary) electron microscopy (SEM) examination of wear tracks at 20 keV.

SUMMARY OF TYPE 304 RESULTS

Our microstructure observations on 304 are discussed in detail elsewhere [7]. Unimplanted areas of the disc consisted of mostly fcc material of 20-30 μm grain size with a high density of defects. The TEM sample from implanted but unworn areas showed a free-standing amorphous surface layer which could be examined isolated from the crystalline substrate, just as in pure Fe [6]. EDS indicated the depth-averaged composition of the amorphous layer

---
*This work performed at Sandia National Laboratories supported by the U. S. Department of Energy under contract #DE-AC04-76DP00789.

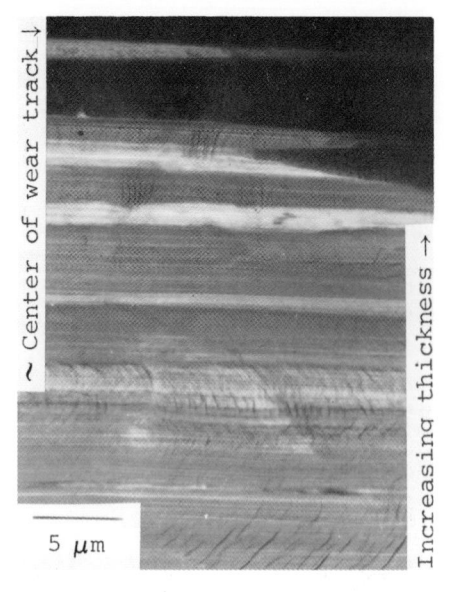

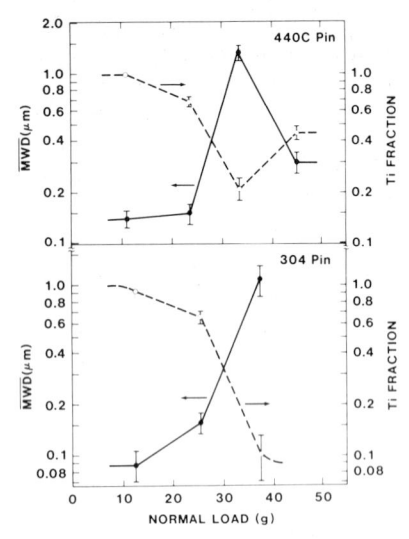

FIG. 1. TEM micrograph of wear track on 304 implanted with Ti and C, made with 440C pin and 23.6g load.

FIG. 2. $\overline{MWD}$ and fraction of Ti remaining in wear track as a function of pin load [8].

to be $(Fe_{0.495}Ti_{0.277}Cr_{0.146}Ni_{0.066}Si_{0.016})_{1-y}C_y$, where we have allowed for the presence of C which goes undetected in our spectrometer. The C concentration is believed to be approximately equal to that of Ti, y = 0.20-0.25. Electron diffraction patterns from the amorphous layer showed diffuse rings at atomic spacings d_1 = 2.18±0.05 Å and d_2 = 1.25±0.03 Å, which can be related to scattering parameters s by s = $4\pi\sin\theta/\lambda$ = $2\pi/d$.

Two tests have identified the amorphous layer as responsible for the observed reductions in friction and wear for Type 304. First, we have examined the wear tracks with TEM, SEM and EDS [8]. Figure 1 is a TEM micrograph from within an implanted wear track made with a light pin load (23.6 g) for which the mean maximum wear depth ($\overline{MWD}$) was reduced from ~ 1 μm before implantation to 0.15 μm, and the coefficient of friction was reduced from 0.8 to 0.35. The micrograph shows parallel wear grooves which are deepest at the edge of the thicker (darker) sample area, which is in the center of the track. Some groove areas show ridges roughly perpendicular to the wear grooves, which suggest that material was piled up ahead of the pin which then skipped over it. Electron diffraction shows that this wear surface is a nearly continuous amorphous layer, even within the deepest groove, which is ~ 0.15 μm below the original surface. These observations demonstrate that during the wear test the pin was sliding over a wear track with an amorphous surface, which can be inferred to be directly responsible for the reduction in friction and wear. This observation is very significant since devitrification could have resulted from heating or deformation during the test, and groove formation to depths greater than the implanted layer thickness might have disrupted continuity of the layer. Furthermore, the Hertzian stress calculated for this load with our pin (radius = 0.79 mm) is 900 MPa, which greatly exceeds the yield strength of 304 (290 MPa). The presence of the amorphous layer at depths greater than the original layer thickness demonstrates that the wear depth was at least partially generated by depressing the layer into the substrate.

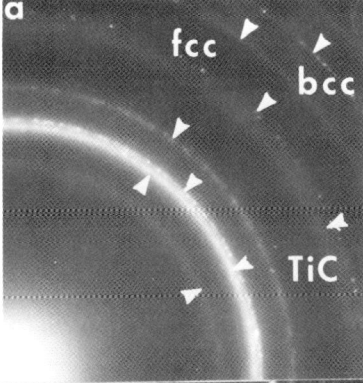

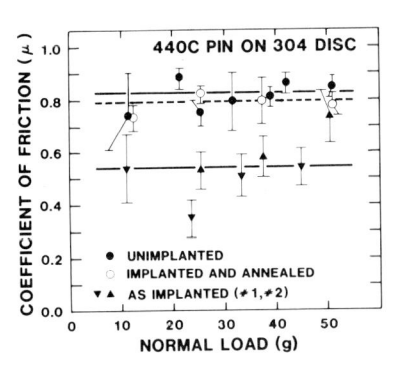

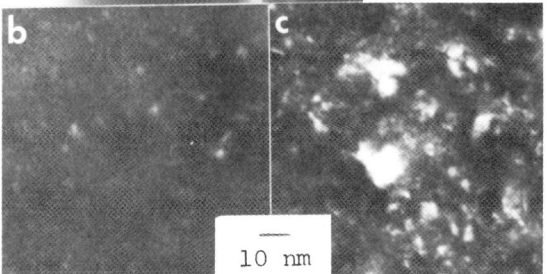

FIG. 4. Coefficient of friction versus pin load for unimplanted and implanted Type 304 [7].

FIG. 3 (left). a) Electron diffraction pattern from 304 implanted with Ti and C and annealed. b) TiC particles. c) TiC, bcc and fcc particles.

All wear tracks exhibiting reduced $\overline{\text{MWD}}$ (~ 0.15 µm) showed mild, abrasive wear scars in SEM and nearly continuous amorphous layers. Those implanted tracks exhibiting no reduction in $\overline{\text{MWD}}$ (~ 1 µm) showed coarse wear topographies; TEM revealed no amorphous wear surface in the tracks. As an additional means to monitor the amorphous layer in the wear process, EDS was used to measure the amount of implanted Ti remaining within ~ 20 µm square central areas of the ~ 100 µm wide wear tracks [8]. In tracks with reduced $\overline{\text{MWD}}$ most of the implanted Ti remained, while in tracks showing no wear reduction only a minor amount (10-20%) of the Ti remained, as shown in Fig. 2. The track in Fig. 1 contained 70%; correspondingly, 30% of the Ti was removed, which demonstrates that material removal also plays a part in generating the wear scar. We have found that Ti is transferred to the pin wear surface during our tests [1]. Thus the presence of the amorphous layer and large remaining Ti fractions correlate well with reduced wear depth [8].

The second test used a devitrified surface alloy [7]. In situ annealing of the amorphous layer in the microscope showed a microcrystalline surface layer after 15 min. at 650°C, as shown in Fig. 3. In Fig. 3a, the electron diffraction pattern indicates that three phases are present: bcc and fcc Fe-based phases, and TiC. We used the innermost TiC ring to dark-field image the TiC particles separately and found them to be 2-5 nm in diameter, as seen in Fig. 3b. Imaging with the fcc and bcc rings necessarily includes rings from all three phases; dark-field imaging with the innermost bcc and fcc rings yields the micrograph in Fig. 3c. Here particles up to 20 nm are seen, and the larger particles are taken to be due to the Fe-based phases.

As a test of the importance of the amorphous structure, we implanted a wear test specimen as before, annealed it 15 min. at 650°C to produce devitrification, and then did pin-on-disc testing. The results showed comparable reductions in wear depth to those obtained on as-implanted discs with amorphous surfaces, but as seen in Fig. 4 the friction coefficients increased to near those of unimplanted 304 material [7]. This result indicates that the

FIG. 5 Amorphous layer and $M_{23}C_6$ precipitates with surrounding fine-grained carbides in 440C.

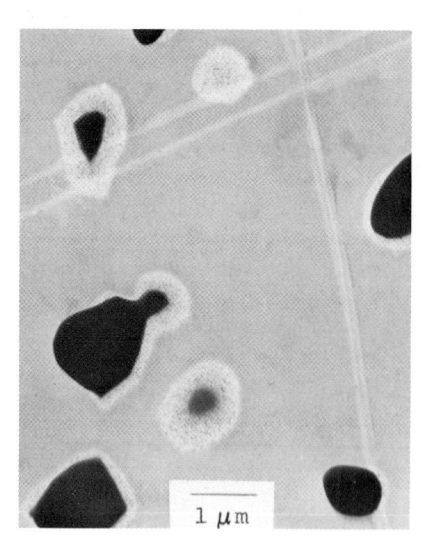

FIG. 6. Amorphous layer, crystalline substrate with bcc particles, and precipitate in Nitronic 60.

amorphous structure is required for reduced friction, which is consistent with our observation (above) of an amorphous wear surface. The reduced wear after devitrification is apparently due to the hard TiC particles.

We conclude that the amorphous surface layer is directly responsible for the reductions in both friction and wear on Type 304 implanted with Ti and C.

TEM RESULTS ON TYPES 440C, NITRONIC 60, AND 15-5 PH

Type 440C consisted of a body-centered matrix of $\sim$ 5 µm grains before implantation, with a high density of dislocations. Weak reflections from austenite were also observed. For the C concentration in 440C (1.0 wt.% $\approx$ 4.6 at.%), we presume that the body centered phase is a tetragonally distorted martensite. Within the martensite matrix, we also found $\sim$ 1 µm particles which indexed to the $M_{23}C_6$ carbide phase. These are expected from the high C level which exceeds the solubility of C in the austenite phase of 440C. Upon initial solidification to austenite, precipitation of these primary carbides occurs. Heat treating 440C to maximum hardness produces martensite and retains the primary carbides, which we then observe. EDS analysis indicates that the metal content in the carbides is $Cr_{0.57}Fe_{0.42}Mo_{0.01}$.

The implanted microstructure of 440C is shown in Fig. 5. Large areas of the amorphous layer (light area) were observed isolated from the substrate, and the primary carbide particles (dark areas) can be seen suspended upon this layer. Surface scratches apparently produced during mechanical polishing are also seen crossing the micrograph. The diffuse ring diffraction pattern from the amorphous layer gives d_1 = 2.18±0.05 Å and d_2 = 1.19±0.03 Å. EDS analysis of the amorphous layer in areas away from the carbides gives $(Fe_{0.633}Ti_{0.234}Cr_{0.110}Si_{0.020}Mo_{0.003})_{1-y}C_y$. The very light area surrounding the dark carbides shows a fine-grained crystalline structure; this was not seen in the unimplanted material and is apparently an implantation-induced change seen at the thin edge ($\sim$ 0.1 µm) of the carbides. The center of the carbide is usually thicker than the ion range and retains the $M_{23}C_6$ structure. The fine-grained areas show a metal content of $Cr_{0.48}Fe_{0.37}Ti_{0.14}Mo_{0.01}$; electron diffraction shows an fcc pattern with $a_o = 4.16\pm0.04$ Å. This result is consistent with most of the Cr within the implanted region of the original $M_{23}C_6$ particle being incorporated into a TiC-like lattice structure but with a smaller lattice constant than pure TiC ($a_o = 4.33$ Å) because Cr is a smaller atom. Others have found that chromium carbide dissolves into the TiC lattice at concentrations (~ 24 mol.% Cr_3C_2) which are needed to account for our lattice parameter [9].

We have also used TEM to examine a wear track produced on implanted 440C with a 304 pin and a 202 g load. The $\overline{MWD}$ was reduced by ~ 85% to 0.045 μm from 0.3 μm [1]. The amorphous layer was again found to be present across the wear track. This load is calculated to produce a Hertzian stress of 1850 MPa, which is higher than the stresses on the amorphous wear surface on implanted 304, and which is approximately equal to the yield strength of 440C.

Nitronic 60 had an fcc structure of ~ 50 μm grains before implantation, with a high density of stacking faults. It also contained precipitates (2-10 μm) of a cubic phase to which we tentatively assign $a_o = 6.6$ Å. After implantation, the TEM sample showed large areas of an amorphous surface layer (Fig. 6) with diffuse rings at $d_1 = 2.16\pm0.05$ Å and $d_2 = 1.21\pm0.03$ Å. The thicker (darker) area in Fig. 6 includes the crystalline substrate which contains ~ 0.5 μm bcc particles (small darker areas in the substrate). The transformation to bcc may have been induced by implanted ions passing through the amorphous layer, as discussed for fcc 304 [7]. Precipitates similar in size to those in unimplanted material are also observed after implantation, like the one shown suspended on the amorphous layer in Fig. 6. However, while some diffraction patterns from these precipitates indexed to $a_o = 6.6$ Å, other patterns indexed to an fcc structure with $a_o = 4.2$ Å.

Type 15-5 PH was found to consist of ~ 10 μm bcc grains before implantation with a high defect density. Weak reflections indexing to austenite were also observed. Implanted but unworn areas showed an amorphous surface layer with diffuse rings at $d_1 = 2.11\pm0.05$ Å and $d_2 = 1.20\pm0.03$ Å.

CONCLUSIONS

Listed in Table I are the steels which showed reduced friction and wear after Ti and C implantation and have now been determined to have amorphous surface layers. Also given are the composition, microstructure and hardness for each steel. Values for pure Fe [6], which showed lesser reductions in friction and wear [1], are given for reference. Included as well is 52100, which has been observed by workers at the Naval Research Laboratory to be driven amorphous when Ti is implanted to 28 at.% and C is simultaneously incorporated at the sample-vacuum interface to ≈ 15 at.% C [2,3]. These steels cover a wide range of composition, initial microstructure and hardness. The steels with high concentrations of alloy additions (a total of almost 40 at.% for Nitronic 60) still give amorphous structures very similar to amorphous Fe(Ti,C), as indicated by the similarity in their diffuse ring patterns summarized in the last two columns of Table I. These results suggest that other steels similarly implanted with Ti and C should also be amorphous, provided they do not contain alloy additions which cause the Ti and C to precipitate out of solution as observed at the $M_{23}C_6$ particles. Furthermore, every steel has shown some reduction in friction and wear in unlubricated testing, which suggests that implantation of Ti and C will probably give reductions for other steels as well. Amorphous layers on two very different steels (52100 and 304) also gave reduced wear in abrasive wear tests [5].

We have good evidence that the amorphous surface layer is directly responsible for reducing friction and wear on 304. We have also shown that the amorphous layer is present across a wear track exhibiting reduced friction

TABLE I. Summary of steels implanted with Ti or Ti and C

Type	Composition (wt.%)	Structure	Knoop Hardness	Diffuse Rings $S_1(\text{Å}^{-1})$	S_2/S_1
Pure Fe	99.99 Fe	bcc	95	3.06	1.73
304	19Cr-9Ni-2Mn-1.0Si-0.08C-Fe	fcc	180	2.88	1.75
Nitronic 60	17Cr-8.5Ni-8Mn-0.1C(Max)-4Si-0.08-0.18N-Fe	fcc + ppt	304	2.91	1.79
15-5 PH	15Cr-4.5Ni-3.5Cu-0.07C(Max)-Fe	bcc	382	2.98	1.76
440C	17Cr-0.75Mn(Max)-1.1C-Fe	bct + $M_{23}C_6$	789	2.88	1.83
52100 (NRL)[2,3]	1.5Cr-0.3-4Mn-0.24Si-1.0C-Fe	bct + carbides	750	3.01	1.6

and wear on 440C. These two steels are at opposite extremes in hardness; Nitronic 60, 15-5 PH and 52100 have intermediate values. Since we know that all the steels are initially amorphous after implantation, we can infer that the wear surface on tracks exhibiting reduced friction and wear in the steels not directly examined is also amorphous. Our findings strongly suggest that the amorphous surface alloy is responsible for reducing friction and wear in all the steels studied.

REFERENCES

1. L. E. Pope, F. G. Yost, D. M. Follstaedt, J. A. Knapp and S. T. Picraux, in Wear of Materials 1983, K. C. Ludema, ed., (Book No. H00254, ASME, New York, 1983), p. 280; also in Mat. Res. Soc. Symp. Proc. 7, 261-268 (1982).
2. I. L. Singer, C. A. Carosella and J. R. Reed, Nucl. Inst. and Meth. 182/183, 923 (1981).
3. C. A. Carosella, I. L. Singer, R. C. Bowers and C. R. Gossett, in Ion Implantation Metallurgy, eds. C. M. Preece and J. K. Hirvonen (Met. Soc. of AIME, Warrendale, PA, 1980), p. 103.
4. T. E. Fischer, M. J. Luton, J. M. Williams, C. W. White and B. R. Appleton, ASLE Trans. 26, 4, 466 (1983).
5. R.N. Bolster and I.L. Singer, ASLE Trans. 24, 4, 526 (1981); also I.L. Singer, R.N. Bolster and C.A. Carosella, Thin Sol. Films 73, 283 (1980).
6. D. M. Follstaedt, J. A. Knapp and S. T. Picraux, Appl. Phys. Lett. 37, 330 (1980); also in Ref. 3, p. 152.
7. D. M. Follstaedt, L. E. Pope, J. A. Knapp, S. T. Picraux and F. G. Yost, to appear in Thin Solid Films 107, 259 (1983).
8. D. M. Follstaedt, F. G. Yost, L. E. Pope, S. T. Picraux and J. A. Knapp, Appl. Phys. Lett. 43, 358 (1983).
9. W. B. Pearson, Handbook of Lattice Spacings and Structures of Metals, Vol. 1, (Pergamon Press, New York, 1958) p. 918.
10. F. G. Yost, S. T. Picraux, D. M. Follstaedt, L. E. Pope and J. A. Knapp, to appear in Thin Solid Films 107.
11. L. E. Pope, F. G. Yost, D. M. Follstaedt, S. T. Picraux and J. A. Knapp, proceedings this symposium.

FRICTION AND WEAR REDUCTION OF 440C STAINLESS STEEL BY ION IMPLANTATION*

L. E. POPE,* F. G. YOST, D. M. FOLLSTAEDT, S.T. PICRAUX and J. A. KNAPP
*Sandia National Laboratories, P. O. Box 5800, Albuquerque, New Mexico 87185 U.S.A.

ABSTRACT

Friction and wear tests on ion-implanted 440C stainless steel discs have been extended to high Hertzian stresses ($\leq$ 3150 MPa). Implantation of 2×10^{15} Ti/mm^2 (180-90 keV) and 2×10^{15} C/mm^2 (30 keV) into 440C reduces friction (~40%) and wear (> 80%) for Hertzian stresses as large as 2900 MPa, stresses which significantly exceed the yield strength of 440C (~1840 MPa). Implantation of 4×10^{15} N/mm^2 (50 keV) into 440C reduces friction slightly (~25%) for Hertzian stresses < 1840 MPa but provides little or no reduction in wear. The amount of Ti remaining in the wear tracks correlates with the reductions in friction and wear. The implantation of Ti and C produces an amorphous surface layer which is believed to reduce friction and wear, whereas N implantation is expected to produce hard nitride particles which probably do not modify the hardness of 440C (KHN = 789) significantly.

INTRODUCTION

Surface modification by ion implantation has been found in recent years to reduce friction and wear of metal surfaces [1-2]. For unlubricated wear couples, beneficial results have been obtained consistently for the implantation of Ti and C into steel surfaces [3-7]. The addition of Ti and C reduced friction coefficients by ~50% and decreased wear by up to ~85%. In contrast, beneficial results for unlubricated wear couples have not been found consistently for N implantation into steels. Reduced wear seems to be limited to tests with short sliding distances, low loads or abrasive wear by diamond particles for 304 and 15-5 PH stainless steels [8-10]. N implantation into 52100 steel did not always reduce wear [2,6]. Implantation of N into 304, 15-5 PH and 52100 steels did not reduce friction coefficients [4,9].

Our earlier studies have shown that the implantation of Ti and C into 440C stainless steel reduced friction and wear [3]; however, the lighter loads used previously produced relatively shallow wear depths and less easily discernable reductions in wear due to implantation. The effect of N implantation into 440C has not been reported. Presented here are the unlubricated pin-on-disc wear results on 440C after Ti and C implantation and after N implantation. Hertzian stresses up to 1.75 times the bulk yield strength are evaluated. Wear tracks are analyzed to determine wear mechanisms and the role of the implanted surface alloy in friction and wear reduction.

---

*This work was supported by the U. S. Department of Energy (DOE) under Contract Number DE-AC04-76-DP00789.

EXPERIMENTAL PROCEDURE

The 440C discs were prepared and heat treated to maximum hardness (KHN = 789) with a bulk yield strength of ~1840 MPa as discussed previously [3]. One group remained unimplanted for reference. A second group was implanted with 5×10^{14} Ti/mm^2 at each of four energies (180, 150, 120 and 90 keV, in that order) for a total fluence of 2×10^{15} Ti/mm^2, followed by 2×10^{15} C/mm^2 at 30 keV; these were expected to produce Ti and C concentrations of ~20 at. % each to a depth of ≤ 0.10 μm [11]. A third group was implanted with 4×10^{15} N/mm^2 at 50 keV, which has been shown to give a N concentration of ~40 at. % to a depth of ≤ 0.15 μm [9]. During implantation, the sample temperature was maintained below 313 K by water-cooling the sample holder.

The pin-on-disc configuration was used for friction and wear testing as described earlier [3]. Applied normal loads ranged from 200.3 to 1000 g (1840-3150 MPa Hertzian stress). Results from an earlier study for loads between 50 and 200 g have been included (labeled Ti and C #1 in the figures). Unimplanted 304 and 440C stainless steel pins were used for each load. Pins were polished to a 0.79 mm radius and were 8.28 ± 0.05 mm in length. Friction coefficients were measured continuously, but only the steady state values reached near the end of 1000 unidirectional cycles are plotted in the figures. Wear was determined after 1000 cycles by measuring the mean maximum wear depth (MWD), defined as the distance from the original surface to the deepest part of the profile averaged over four traverses of the wear track at approximately 90 degree increments. A Cameca MBX microprobe was used to determine relative chemical compositions across the wear tracks and to make chemical element maps inside and outside of the wear tracks. A scanning electron microscope (SEM) was used to obtain topographic micrographs of the wear tracks.

FRICTION AND WEAR RESULTS

Friction coefficients for 440C and 304 stainless steel pins are shown in Figure 1 for (Ti and C)-implanted and for N-implanted 440C discs. Friction results are similar for both pins. In comparison with the unimplanted discs, implanting Ti and C reduced the friction coefficient ~50% for loads ≤ 600 g (2650 MPa) for the 440C pin and ≤ 800 g (2920 MPa) for the 304 pin. Implantation of N into 440C reduced friction coefficients ~25% for loads 0 <

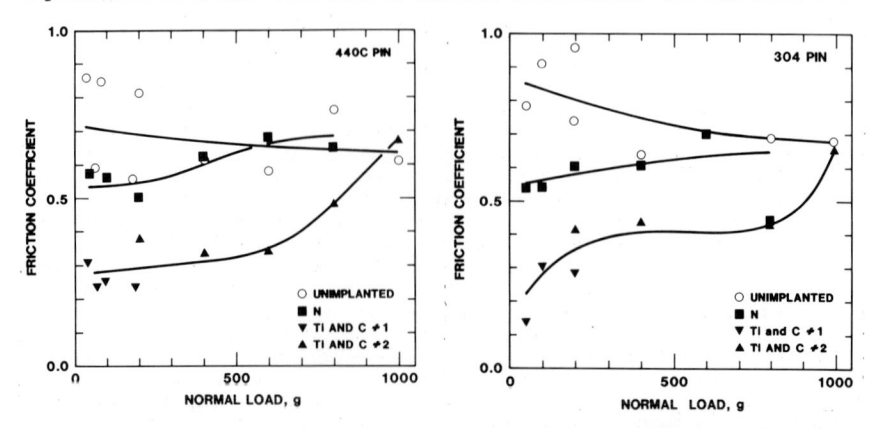

Figure 1. Friction coefficient versus normal load for implanted 440C; the average fluctuation in friction coefficient during one revolution for all tests was ± 0.06.

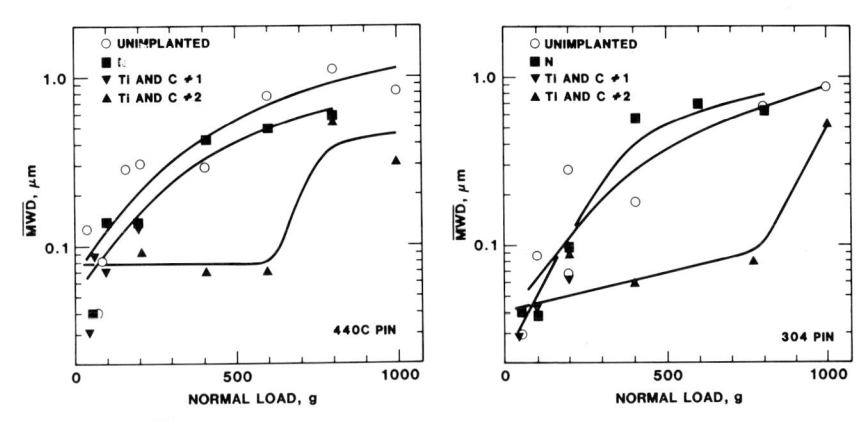

Figure 2. MWD versus normal load for implanted 440C stainless steel. The average coefficient of variation for the data is ± 0.3.

200 g (1840 MPa); the reductions disappeared for loads ≥ 400 g (2320 MPa) for the 440C pin and ≥ 600 g for the 304 pin.

Shown in Figure 2 are MWD results for 440C discs implanted with Ti and C and implanted with N. Wear results are similar for 440C and 304 pins. For loads less than 100 g (1460 MPa) implanted and unimplanted discs had similar wear. Between 100 and 600 g the MWD was relatively constant for (Ti and C)-implanted discs at ~0.08 μm, whereas the MWD increased continuously for un-implanted and N-implanted samples up to ~1 μm at 1000 g. Wear reductions were less at high loads (1000 g for the 304 pin and ≥ 800 g for the 440C pin). Results for unimplanted and N-implanted discs were similar; no clear reduction in wear was observed with the 304 pin, but a slight reduction (~30%) might have been present with the 440C pin.

CHEMICAL COMPOSITION AND TOPOGRAPHY RESULTS

At the completion of friction and wear tests, an electron microprobe was used to measure the Ti concentration profile across each of the wear tracks at two locations 180 degrees apart. A 15 kV beam was used to excite a 5 μm square area in steps 5 μm apart. All major elements were counted. The Ti concentration profiles were smooth with gradual changes across the wear tracks with one exception; the average of the two profiles was calculated after superimposing the midpoints of the profiles. Results are shown in Figure 3 for the highest load with significant wear reduction along with the next higher load which showed no wear reduction. This transition in wear (MWD reduction to no MWD reduction) correlates with the Ti remaining in the wear tracks. The Ti in the center of the wear tracks is 30-50% of that outside of the wear tracks for all loads where reduced MWD was measured, whereas little or no Ti remained when loss of MWD reduction occurred. For the 304 pin at 1000 g the wear track is extremely wide (see scale change) with large changes in the Ti concentration; in order to show the excursions to zero Ti for this case averaging over more than one track was not done.

The microprobe was used to prepare elemental area maps for Cr, Fe and C inside and outside the wear tracks. Primary carbides rich in Cr and C relative to bulk values were found with the same number, size and distribution inside the wear tracks as that outside the wear tracks. These carbides are of the $M_{23}C_6$ type and are nominally 0.2-1.5 μm in diameter [12]. Elemental maps of Ti and C showed that both elements were being worn away within wear tracks.

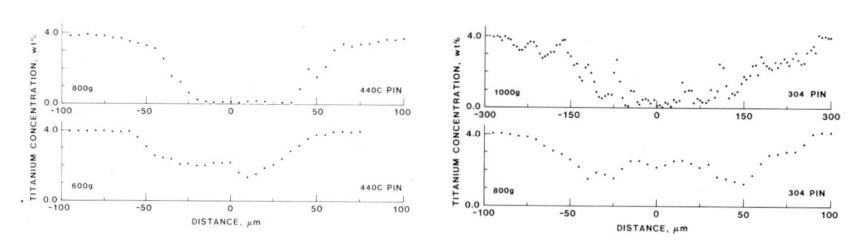

Figure 3. Ti concentration profiles across wear tracks with loads bracketing the wear transition.

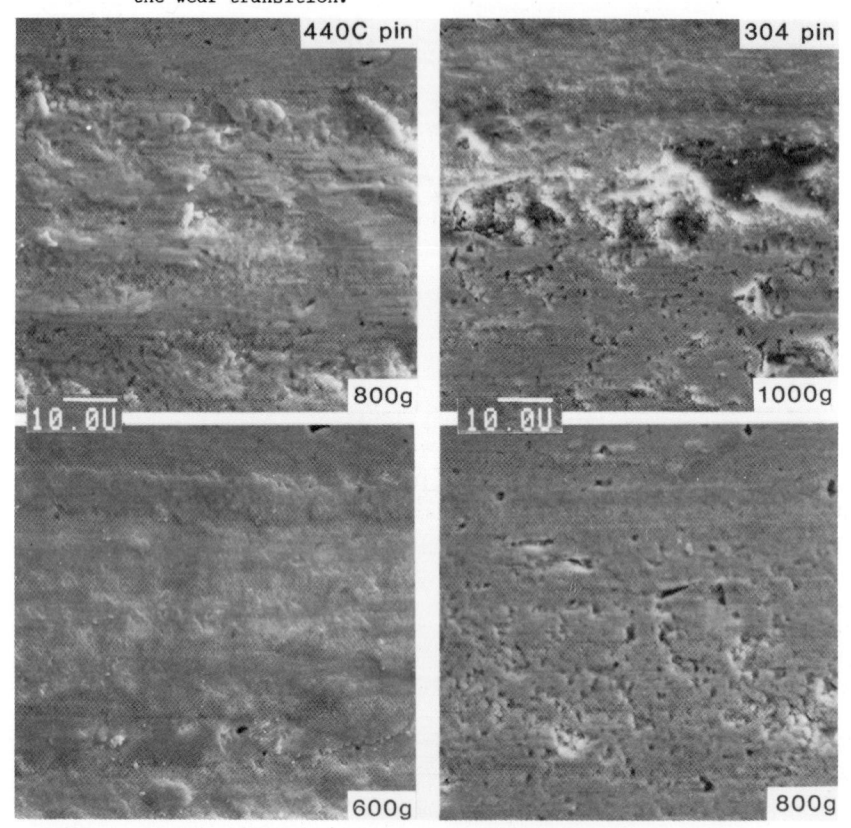

Figure 4. Scanning electron micrographs of wear tracks.

Scanning electron micrographs are shown in Figure 4 for the wear tracks corresponding to the Ti concentration profiles plotted in Figure 3. For the lighter loads the wear scars show mild abrasive wear which suggests that the surface layer is intact. At the heavier loads evidence of adhesive wear is seen which suggests that the implanted surface layer has been penetrated and disrupted. These results are in agreement with the presence and loss of Ti shown in Figure 3.

To compare the fraction of Ti remaining in the wear tracks with the MWD, the Ti and Fe signals obtained during the Ti concentration profile measurements were averaged over the center part of each track. Two areas 5 μm x

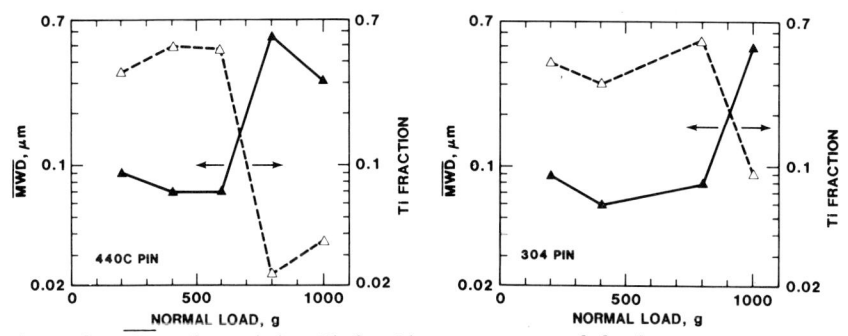

Figure 5. MWD and remaining Ti fraction versus normal load.

55 μm were used for each track. All of the data (over 200 points) taken in unworn areas were averaged to establish a reference for comparison. The average Ti and Fe concentrations (with standard deviations) in unworn areas were 4.0 ± 0.3 and 77.3 ± 4.8 wt. %, respectively. For each load the Ti/Fe ratio was calculated, and this ratio was divided by the Ti/Fe ratio in unworn areas to determine the Ti fraction remaining in the wear tracks. Shown in Figure 5 are MWD and Ti fraction for applied loads between 200 and 1000 g. The MWD curves have an abrupt increase in depth at 800 g for the 440C pin and at 1000 g for the 304 pin, and the Ti fraction shows a concomitant abrupt drop. Thus, the loss of Ti in the wear tracks correlates with an increase in wear at the highest loads.

SUMMARY AND DISCUSSION

The dual implantation of Ti and C into 440C stainless steel reduces friction coefficients and wear for Hertzian stresses significantly larger than the bulk yield strength. There is good correlation between the Ti fraction remaining in the wear tracks and reduction in friction and wear. Titanium concentration profiles are smooth and nonzero implying that the ion-implanted layer is continuous across the track.

Implantation of N into 440C reduced friction slightly for loads < 400 g (2320 MPa) with no reduction at loads > 600 g (2650 MPa). The reduced friction coefficients were > 0.6 and are not considered very advantageous. The MWDs were not reduced significantly by N implantation, as might be expected, since 440C is a wear resistant alloy with a microstructure of hard primary $M_{23}C_6$ carbides dispersed in a martensite matrix. The formation of nitride particles, as was found for N implantation into 15-5 PH [9], and the presence of dissolved N in 440C would not be expected to change the hardness of the alloy significantly. These observations are supported by surface hardness measurements which show that N implantation did not increase the hardness of 52100 steel [13]. Hence, it seems reasonable that N implantation should have little effect in 440C. These results of reduced friction but not wear are opposite to our findings on 304 and 15-5 PH, thus demonstrating that changes produced by N implantation depend upon the specific steel being treated.

An amorphous surface alloy results from the dual implantation of Ti and C with overlapping concentration profiles into pure Fe and 304, 440C, 15-5 PH and Nitronic 60 stainless steels [11-12,14-15]. The implantation of Ti alone into 52100 and 304 steels also produces an amorphous layer when C is incorporated into the alloy at the surface during Ti implantation [4-7,10]. The presence of the amorphous layer was shown recently to be responsible for reduced friction coefficients and reduced wear for 304 [15-16]. In particular, the amorphous layer was found to be continuous across tracks showing reduced friction and wear, and it did not devitrify during our unlubricated testing. We believe that the reductions in friction and wear observed for 440C after

Ti and C implantation are due to a similar amorphous layer, which we have observed with TEM across a wear track showing reduced friction and wear [16].

The wear transition from reduced MWD to no reduction in MWD occurs when the amorphous film is penetrated. After the amorphous film is penetrated, adhesive wear predominates in comparison with abrasive wear while the amorphous film is intact. The reduced wear due to the amorphous layer is probably attributable to increase hardness of the surface alloy, which is supported by two experimental observations on 52100 steel. An amorphous layer decreased abrasive wear by a factor of 5 suggesting higher hardness [17]. Implantation of Ti and C (but not Ti or C alone) at higher Ti concentrations than ours increased the surface hardness, though for Ti concentrations similar to ours a higher hardness was not detected (probably because the effect was not large) [13].

ACKNOWLEDGEMENTS

The contribution of P. F. Hlava in taking and interpreting chemical composition measurements in the microprobe is acknowledged. The contribution of M. M. Sturm in taking scanning electron micrographs and for performing energy dispersive spectroscopy measurements is recognized.

REFERENCES

1. H. Herman, Nucl. Instrum. Methods 182/183, 887 (1981).
2. J. K. Hirvonen, J. Vac. Sci. Technol. 15, 1662 (1978).
3. L. E. Pope, F. G. Yost, D. M. Follstaedt, J. A. Knapp and S. T. Picraux, in: Wear of Materials - 1983, K. C. Ludema, ed. (ASME, New York, NY 1983), p. 280.
4. C. A. Carosella, I. L. Singer, R. C. Bowers and C. R. Gossett in: Ion Implantation Metallurgy, C. M. Preece and J. K. Hirvonen, eds. (Metallurgical Soc. of AIME, Warrendale, PA 1980), p. 103.
5. I. L. Singer, C. A. Carosella and J. R. Reed, Nucl. Instrum. Methods 182/183, 923 (1981).
6. I. L. Singer, R. N. Bolster and C. A. Carosella, Thin Solid Films 73, 283 (1980).
7. I. L. Singer and R. A. Jeffries, Proc. 29th Nat'l Amer. Vac. Soc. Sym. Baltimore, MD (Nov. 1982).
8. A. K. Suri, R. Nimmgadda and R. F. Bunshah, Thin Solid Films 64, 191 (1979).
9. F. G. Yost, S. T. Picraux, D. M. Follstaedt, L. E. Pope and J. A. Knapp, to appear in Thin Solid Films, 107.
10. R. N. Bolster and I. L. Singer, Am. Soc. Lubr. Engr. Trans. 24, 526 (1981).
11. D. M. Follstaedt, J. A. Knapp and S. T. Picraux, Appl. Phys. Lett. 37, 380 (1980).
12. D. M. Follstaedt, F. G. Yost and L. E. Pope, these Proceedings.
13. J. B. Pethica, R. Hutchings and W. C. Oliver, Nucl. Instrum. Methods, 209/210, 995 (1981).
14. J. A. Knapp, D. M. Follstaedt and S. T. Picraux in: Ion Implantation Metallurgy, C. M. Preece and J. K. Hirvonen, eds. (Metallurgical Soc. of AIME, Warrendale, PA, 1980), p. 152.
15. D. M. Follstaedt, L. E. Pope, J. A. Knapp, S. T. Picraux and F. G. Yost, to appear in Thin Solid Films, 107, 259 (1983).
16. D. M. Follstaedt, F. G. Yost, L. E. Pope, S. T. Picraux and J. A. Knapp, Appl. Phys. Lett. 43(4), 358 (1983).
17. I. L. Singer, R. N. Bolster and C. A. Carosella, Thin Solid Films, 73, 283 (1980).

FRICTION, WEAR AND DEFORMATION OF SOFT STEELS IMPLANTED WITH TI AND N

I.L. SINGER,* AND R.A. JEFFRIES**
*Naval Research Laboratory, Code 6170, Chemistry Division, Washington, DC 20375, **Geo-Centers, Inc., 4710 Auth Place Suitland, MD 20746

ABSTRACT

Friction and wear measurements were performed with a hardened steel ball sliding against unhardened steels (AISI 1018, 304 and M2) implanted with either N^+ or Ti^+ or both $N^+ + Ti^+$ ions. Dry and lubricated sliding studies found that: Ti-implantation reduced friction and thereby delayed wear; N-implantation did not reduce friction but delayed wear by increasing the resistance of the surface to shear; and dual implants of $N^+ + Ti^+$ showed effects of both implants. Auger sputter depth profiles found the 1018 and M2 surfaces enriched in implanted N and bulk C. No evidence of N migration could be found in heavily deformed surfaces.

INTRODUCTION

Ti and N implantations have been shown to affect both the wear rate and the friction behavior of steels, but in different ways. Ti-implantation has been found effective at high fluences in reducing friction and wear in a variety of steels of different microstructures (e.g. ferritic, martensitic, austenitic, precipitation hardened) [1-6]. N-implantation has been found to reduce wear in all but through hardened steels, but not to reduce friction [1,6-8]. These studies suggest that during sliding contact Ti-implantation reduces adhesion whereas N increases the surface's resistance to deformation. Similar effects for Ti and N implants have also been found on a Co-based carbide alloy [10,11].

The present study was designed to investigate the effects of implanting both N and Ti on the friction and wear behavior of the three unhardened steels listed in Table I. First, N ions were implanted deep, then Ti ions shallow so that the effects of the two implant species might be separate. Ti ions were implanted to a fluence capable of producing a low friction surface on a hardened bearing steel [9]. Implantation parameters for N^+ and Ti^+ ions are listed in Table II. Friction and wear measurements were performed on nonimplanted, N-, Ti-, and dual (N + Ti) implanted steels with a hardened steel ball slider under dry and lubricated sliding conditions (Hertzian stress = 0.85 GPa). Wear tracks were analyzed by optical and scanning electron microscopes, and surface compositions were determined by Auger analysis.

TABLE I. Composition, microstructure and hardness of steel disks.

Steel Type (AISI)	Composition (at.%)	Microstructure	Knoop Hardness (20N load; in GPa)
304	Fe 20Cr-8Ni	Austenite	2.9
1018	Fe-1C	Ferrite and Cementite	2.6
M2	Fe-4.4Cr-3.0Mo-2.3V 1.9W-4.1C	Ferrite and Alloy Carbides	2.5

TABLE II. Implantation parameters for N and Ti ions.

Implanted Ion	Fluence ($10^{17}/cm^2$)	Energy (keV)	Range (nm)
N^+	4	180	190
Ti^+	2	50	18

EXPERIMENTAL

All steel samples were cut from as received bar stock, then were polished to a metallographic finish. The ion implanter and associated procedures have been described elsewhere [1,2].

The friction testing apparatus and procedures have been described previously [1,2]. Briefly, friction coefficients are measured as a ball is slid unidirectionally (speed=0.1 mm/sec) against an implanted disk from 1 to 20 passes along the same track. Surface compositions were obtained from Auger derivative spectra using the following sensitivity factors: $S_N = 0.31$; $S_C = 0.40$; S_{Fe} (650) = 0.17.

RESULTS

Dry Sliding

The dry sliding friction values obtained for 304 steel are shown in Fig. 1. Average μ_k values per pass for the N-implanted surface rose somewhat slower than nonimplanted surfaces. However, during most passes on these two surfaces, μ_k values rose (↑), fell (↓) or displayed stick-slip (↕). Ti-implantation significantly reduced the coefficient of friction and, initially, the tendency to stick-slip relative to the nonimplanted and N-implanted surfaces. The dual implant (N + Ti) produced the same average μ_k values as the Ti-implant and eliminated stick-slip.

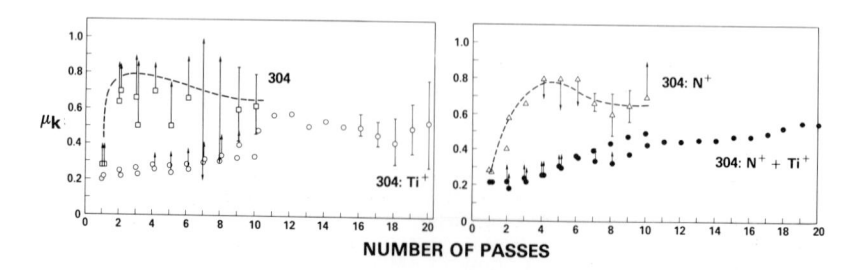

FIG. 1. Friction coefficients vs. pass number for nonimplanted (□), Ti (o), N(△), and N+Ti(●) implanted 304 steel.

Implantation dramatically altered the texture of wear tracks in dry sliding. The nonimplanted tracks were severely damaged after the first pass as large plates were lifted, sheared off, then smeared along the track. In the N-implanted tracks, shearing was delayed despite similar friction history. Textures of a nonimplanted and a N-implanted track, both after 3 passes, are seen in the SEM micrographs in Fig. 2 (left). The Ti-implanted tracks initially wore much less than either the non-implanted or N-implanted surfaces. But by the 20th pass, when stick-slip prevailed, large (25-50μm) plate-like patches were found distributed randomly over the track. The dual implanted tracks showed very little damage relative to the other samples, with no plate-like patches. Thus the dual implants of Ti and N lowered friction and prevented delamination in a way that neither could do alone.

In the ferritic 1018 steel, Ti-implantation again reduced friction and delayed severe wear, while N-implantation did not reduce friction but did eliminate stick-slip found in nonimplanted and Ti-implanted surfaces. N-implantation also strengthened the ferritic steel surface, as witnessed by the absence of smeared plates on the seventh pass track. (Compare tracks on right side of Fig. 2.)

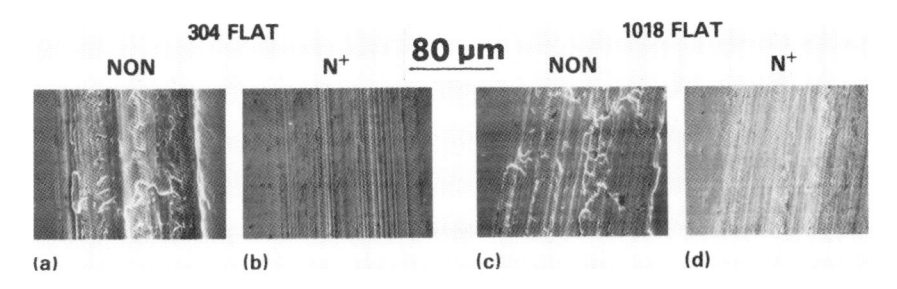

FIG. 2. SEM micrographs of wear tracks. (left)non- and N-implanted 304 steel after 3 passes; (right) non- and N-implanted 1018 steel after 7 passes.

The nonimplanted M2 steel showed a slow rise in μ_k from 0.3 to 0.6, and suffered gross deformation during sliding but only minimal wear. Ti-implantation lowered μ_k values and delayed the onset of microtexturing (described below). N-implantation caused μ_k values to rise sooner and caused the surface to wear more rapidly.

Lubricated Sliding

Under lubricated sliding conditions (Shell 2190 TEP), all three implanted 304 surfaces had friction coefficients of ~0.12. The non-implanted 304 surface displayed a higher coefficient of friction (μ_k ~ 0.16) initially, but decreased to a value of μ_k ~ 0.10 after 10 passes, and remained there for the next 10 consecutive passes. None of the lubricated surfaces had severely worn tracks, but 3 of the 4 showed the microtextures depicted by the interferograms in Fig. 3. After 15 passes, the dual implanted track was virtually undeformed, the N-implant showed light deformation and the Ti-implant slightly more deformation. By contrast, after 6 passes the nonimplanted track was highly deformed. Clearly, N was more effective than Ti, in the absence of high friction, in resisting surface deformation.

No obvious trends in the μ_k values or texturing of the 3 implants were observed for lubricated sliding against the two ferritic steels. In the 1018 tracks, the dual implant had a higher steady-state friction coefficient (μ_k = 0.14) than the other two steels (μ_k = 0.11). The dual implant, however, showed no evidence of microtexturing, whereas the N-implant and the Ti-implant showed about the same degree of microtexturing as the Ti-implanted 304 in Fig. 3. In M2 steel, the steady state friction coefficient was ~0.10 for all 4 samples, and none of the tracks showed microtexturing.

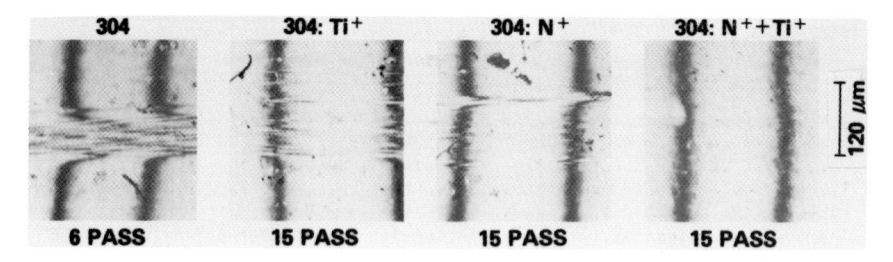

FIG. 3. Interferograms of wear tracks produced during lubricated sliding against 304 steel (fringe spacing = 273 nm).

The M2 steel, unlike the 304 or 1018, exhibited macroscopic plastic deformation during lubricated sliding even though its bulk hardness is almost identical to that of the other steels (see Table I). The inteferogram in Fig. 4 shows that the surface of a 10 pass lubricated track in N-implanted M2 was depressed ~ 140 nm. The Ti and dual implanted tracks had the largest deformation (~200 nm) and the non-implanted track, the least (~100 nm) deformation. Most deformation occurred during the first few passes.

Composition Analysis

Composition vs. depth profiles of implanted surfaces, worn and as-implanted, were taken by Auger sputter depth profiling using microprobes with 1 μm and 30 μm spot sizes. Identical profiles were obtained inside and outside the lubricated tracks of the three steels, confirming [6] that the majority of N atoms do not migrate as a result of low speed, high load sliding. The as-implanted profiles in the two steels having a ferrite matrix, M2 and 1018, showed evidence for outdiffusion of N; the profile for N-implanted M2 is shown in Fig. 4. Evidence for N outdiffusion has been reported previously for martensitic [12], and ferritic [6] steels and recently confirmed by nondestructive nuclear reaction profiling [13]. The profile of the N-implanted M2 surface in Fig. 4 also shows that C from the steel outdiffused along with the N. Similar N and C redistributions were observed in N-implanted 1018 steel but not in the 304 steel where the N bonds preferentially to the Cr in the lattice [4]. No surface carbon enrichment was found in nonimplanted disks. Ti-implanted surfaces showed the previously reported vacuum carburized Fe + Ti layer [1-3].

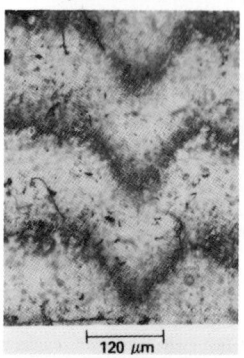

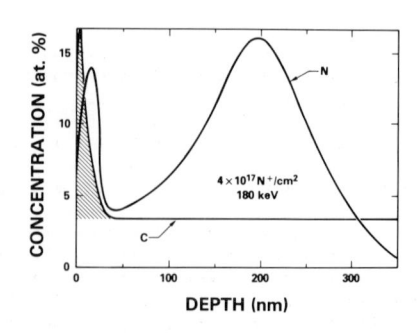

FIG. 4. (Left) Interferogram of a tenth pass lubricated track on N-implanted M2 steel (fringe space = 273 nm). (Right) Auger sputter depth profile of N and C in N-implanted M2 steel.

DISCUSSION

Ti-implantation reduced friction by nearly a factor of 2, initially, and delayed the onset of severe wear during dry sliding of a hardened steel against implanted unhardened steels. Friction reduction is attributed to lower adhesion of the vacuum carburized Fe + Ti layer, and wear resistance to the less severe subsurface deformation accompanying lower friction [15]. The microstructures of the three Ti-implanted surfaces are presumed to be amorphous, based on TEM results for other Ti-implanted Fe alloys [2,16]. The effect of the amorphous layer on deformation is yet unknown.

N-implantation did not alter significantly the friction coefficients of the steels, in agreement with previous studies [6-8], but did delay the onset of severe wear during dry sliding. The austenitic 304 steel was most noticeably improved by N-implantation. Surface damage during both dry and lubricated sliding was decreased significantly. Austenitic steels such as 304 steel have notoriously poor dry sliding wear behavior which, according to Hsu et. al. [17], is attributable to a brittle strain-induced α'-martensite layer formed during sliding. N-implantation has been shown to transform α'-martensite back to austenite [18] and to stabilize austenite against transforming under wear to martensite [19]. Thus, one way N-implantation could have changed the wear mode of 304 steel was by preventing the deleterious sliding condition of a hard "skin" on a soft substrate.

N-implantation also decreased the surface damage caused by sliding against the ferritic 1018 steels. This improvement could be due to surface hardening [8] and to the extremely high compressive stresses (>1GPa) residing in the implanted surface [20]. Residual compressive stresses would reduce the cyclic fatigue damage caused during sliding by decreasing the net tensile stress operating at the surface. The surface could also be strengthened by a microstructure incorporating defects and solutes (e.g. C,N) segregating to the surface. Fischer, et al. [6] found both C enrichment and wear resistance in a ferritic steel implanted with each of three different species: Fe, Ar, and N. They speculated that surface strengthening might have resulted from a combination of residual stresses and microprecipitates promoted by surface segregated C, neither of which is implant specific.

N-implantation was deleterious in an annealed M2 steel whose microstructure differed from 1018 steel mainly by the presence of large (5-10 μm) alloy carbides. Implantation may have degraded the wear resistance of the carbide phase.

Dual implants of N + Ti proved to be the most beneficial treatment for friction and wear reduction in 304 and 1018 steels. Dual implants of Ti + C have shown similar benefits on stainless steels [4,5,16]. We predict that dual implants of N + Ti or C + Ti will be more beneficial than either implant alone for many steels subjected to plastic deformation, but further work must be done to optimize the Ti fluence [21] and N or C energy for a given steel.

The composition profiles of as-implanted surfaces and worn tracks add a new dimension of complexity to the N-implantation and wear picture. First, they show clearly that most N atoms do not migrate under stresses large enough to induce plastic deformation, consistent with the results of Fischer et. al. [6]. Secondly, they emphasize the influence of alloy constituents (e.g. C in the ferritic steels and Cr in the 304 steels) on the composition of the implanted layer and, by implication, on the surface mechanical properties. Solute species can migrate, if not chemically bound, during high fluence implantation. High Cr steels, on the other hand, can ''pin'' implant solutes, to the point where the surface may crack and oxidize for lack of stress relief [13]. These composition effects should be taken into account when interpreting phases detected by Mossbauer analyses of N-implanted C steels. [22].

CONCLUSIONS

Ti-implantation reduced friction thereby delaying wear during sliding of a hard steel against implanted soft steels. N-implantation did not reduce friction but reduced wear by increasing the shear resistance of the 304 and 1018 steel surfaces. Dual implants of N + Ti acted additively and greatly benefited 304 and 1018 steel. The surfaces of the two ferritic steels were enriched in N and C by N-implantation. No evidence for N migration was found in a plastically deformed wear track.

REFERENCES

1. C.A. Carosella, I.L. Singer, R.C. Bowers and C.R. Gossett, Ion Implantation Metallurgy, Metallurgical Society of AIME, Warrendale, PA, 103 (1980).
2. I.L. Singer, C.A. Carosella, and J.R. Reed, Nucl. Instrum. Methods, 182/183, 923 (1981).
3. I.L. Singer, R.N. Bolster, C.A. Carosella, Thin Solid Films, 73, 283 (1980).
4. F.G. Yost, L.E. Pope, D.M. Follstaedt, J.A. Knapp, and S.T. Picraux, Metastable Materials Formation by Ion Implantation (Elsevier, Amsterdam), 261 (1982).
5. L.E. Pope, F.G. Yost, D.M. Follstaedt, J.A. Knapp, S.T. Picraux, Proc. Int. Conf. on Wear of Materials, Reston, VA, April 11-14, 1983, American Society of Mechanical Engineers, New York, 1983; also these proceedings.
6. T.E. Fischer, M.J. Luton, J.M. Williams, C.M. White and B.R. Appleton, ASLE Trans., 26, 31 (1983).
7. F.G. Yost, S.T. Picraux, D.M. Follstaedt, L.E. Pope, J.A. Knapp, Thin Solid Films, 107, 287 (1983).
8. R. Hutchings, W.C. Oliver, and J.B. Pethica, NATO/ASI Conference on Surface Engineering, Les Arcs, France, July, 1983 (edited by R. Kossowsky and S. Singhal).
9. I.L. Singer and R.A. Jeffries, Appl. Phys. Lett. 43, 925 (1983).
10. S.A. Dillich and I.L. Singer, Thin Solid Films, 108, 219 (1983).
11. S.A. Dillich, R.N. Bolster, and I.L. Singer, in these proceedings.
12. I.L. Singer and R.N. Bolster, Ion Implantation Metallurgy, Metallurigical Society of AIME, Warendale, PA, 116 (1980).
13. W.M. Bone, R.J. Colton, I.L. Singer and C.R. Gossett, presented at AVS Meeting, Boston, MA, November 1983.
14. I.L. Singer and J.S. Murday, J. Vac. Sci. Technol. 17, 327 (1980).
15. S. Jahanamir and N.P. Suh, Wear, 44, 17 (1977).
16. D.M. Follstaedt, F.G. Yost and L.E. Pope, in these proceedings.
17. K.L. Hsu, T.M. Ahn, and D.A. Rigney, Wear, 60, 13 (1980).
18. R.G. Vardiman, R.N. Bolster, and I.L. Singer, Metastable Materials Formation by Ion Implantation, (Elsevier, Amsterdam) 1982.
19. R.G. Vardiman, R.N. Bolster, I.L. Singer (to be published).
20. N.E.W. Hartley, J. Vac. Sci. Technol., 12, 485 (1975).
21. I.L. Singer and R.A. Jeffries, in these proceedings.
22. C.A. dos Santos, M. Behar, J.P. de Souza and I.J.R. Baumvol, Nucl. Instrum. Methods 209/210, 907 (1983).

PROCESSING STEELS FOR TRIBOLOGICAL APPLICATIONS BY TITANIUM IMPLANTATION

I.L. SINGER,* AND R.A. JEFFRIES**
*Naval Research Laboratory, Code 6170, Chemistry Division
Washington, DC 20375; **Geo-Centers, Inc., 4710 Auth Place
Suitland, MD 20746

ABSTRACT

Titanium implantation into steels has been shown to produce superior tribological surfaces. The fluence required to produce a wear resistant surface increases from 2 to 5×10^{17} Ti/cm² as the energy increases from 50 to 200 keV/ion. On curved surfaces (e.g., bearings, cutting tools, etc.) higher fluences are necessary due to effects of implantation at angles off normal incidence (i.e. the combined effects of higher sputtering rates, decreased range, and changes in the carburization process associated with duty cycles). Significant improvements in friction and wear have also been observed for surfaces which have been abraded by 600 and 120 grit SiC prior to implantation. Optimal benefits of Ti-implantation are associated with the formation of a modestly thick (>20nm) fully carburized layer.

INTRODUCTION

One of the most important tasks of surface processing technology will be to improve the tribological performance of load-bearing steels. Steel components such as balls, rollers, races and gears usually have sufficient bulk hardness to support the load but lack the needed "strength" at the surface to avoid adhesive, abrasive or fatigue failure. One of the most successful candidates for reducing friction and wear of bearing steels is Ti implantation [1-4]. Investigations of Ti-implantation in the most common of all bearing steels, AISI 52100 steel (Fe-1.5Cr-1C by wt.%), have found reduced friction under dry and poorly lubricated sliding [1,2], increased abrasion resistance [5] and increased scuffing resistance [2,3]. However, the benefits are obtained only when implantation is performed to high fluence, where an amorphous Fe-Ti-C layer forms in the surface [2] of this layer. Recent investigations [6-9] of what we call the implant-assisted vacuum carburization process have established a chemical basis for producing this superior tribological surface in 52100 steel by Ti-implantation.

The intent of this study was to establish the processing parameters that optimize the tribological benefits of Ti-implantation. The processing parameters investigated were fluence, energy and angle of ion implantation. Dry sliding friction and wear measurements were performed on polished and abraded 52100 steel disks. Auger sputter profiling was used to give a chemical basis for the observed tribological behavior.

EXPERIMENTAL

Friction measurements were carried out in air (~50% RH) at room temperatures at a sliding velocity of 0.1 mm/s. The sliders were 1.27 cm diameter AISI 52100 steel balls (R_c = 60) which were in contact with steel disks 0.95 cm in diameter and 0.32 cm thick. The applied force was 9.8N, giving a Hertzian stress of 0.85GPa. The first passes were 5 mm in length; subsequent passes over the same track were 3 mm each.

Disks were hardened (R_c = 60) AISI 52100 steels polished to a metallographic finish. Abraded samples were then prepared by regrinding the polished surface on 600 or 120 grit SiC paper. During implantation in NRL's modified Varian/Extrion implanter, disks were heat sunk onto a water-cooled holder and kept near room temperature (<40°C). Samples were implanted with a ^{48}Ti$^+$-beam rastered to give current densities of J ~ 10-20 μA/cm². The target chamber was cryogenically pumped to base pressures of 7 mPa (~5x10^{-7} Torr). Auger depth profile analyses have been described in detail elsewhere [2,6].

RESULTS

Fluence - Energy Requirements

Friction and wear studies were performed on polished 52100 steel disks implanted at normal incidence with Ti ions to fluences ranging from 5 to 50x10^{16}Ti/cm² at energies from 25 to 190 keV. Projected ranges of selected implant energies are given in Table 1 [10]. Disks implanted at or above a certain fluence, which depends on the implantation energy, had a superior tribological surface. Friction coefficients measured on the first pass of dry sliding contact were μ_k = 0.2, the same value obtained for a 52100 ball/TiC substrate couple. Friction coefficients rose to steady state values of μ_k = 0.3, 1/2 that of non-implanted couples (μ_k = 0.6), and the wear tracks showed virtually no adhesive wear.

TABLE 1. Projected Range (R_p) vs. energy of Ti$^+$ ions implanted into 52100 steel [10].

Energy (keV)	25	50	100	200
Rp (nm)	11	19	33	59

The minimum fluence required at a given energy for obtaining this superior surface is indicated as the carburization limit in Fig. 1. For example, at an energy of 50 keV, a fluence of 2x10^{17}Ti/cm² is sufficient, whereas at 190 keV, nearly 5x10^{17}Ti/cm² is required. The designation of the minimum fluences by the term ''carburization limit'' is based upon surface composition studies which have shown that fully carburized layers were produced by Ti implantation at or above these fluences for a particular energy [6,7].

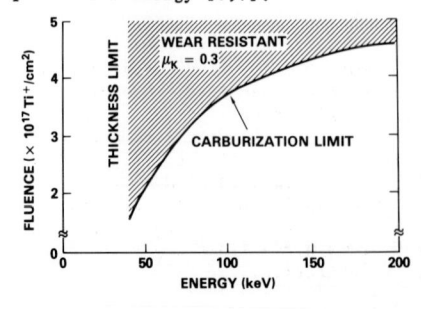

FIG. 1. Fluence of Ti ions vs. energy for producing wear resistant, low friction surfaces in 52100 steel.

At fluences below this limit, the benefits of Ti implantation on dry sliding wear decreased. Just below the limit, the coefficient of friction had a steady state value of $\mu_k = 0.4$ and the wear track showed mild adhesive damage and occasional debris flakes. At lower fluences, the friction coefficients reached values up to 50% higher than steel on steel and wear tracks showed considerable damage and debris accumulation. Surface composition studies showed that the tribological properties were degraded at lower fluences because a fully carburized layer did not develop and contact was made, instead, to an Fe-Ti alloy layer [6]. There also may be a ''thickness limit'' for Ti implanted steel. Disks implanted at 25 keV to fluences of 1.6 and 3.6×10^{17} Ti/cm² did not show the beneficial effects of Ti implantation even though a fully carburized layer was formed.

Implantation Off of Normal Incidence

Three hardened 52100 disks were implanted at essentially the same fluence and energy (4×10^{17} Ti/cm² at 100 keV). Two disks were mounted on a water-cooled target ladder such that one disk was normal to the beam, and the other at an angle of 45° relative to the beam. The third disk was mounted on a bearing race that was water-cooled and rotating during implantation such that a 90° sector ($\pm 45°$ off normal) was being implanted as the bearing rotated at 2 rpm (see insets of Fig. 3). As can be seen from Fig. 2, the sample implanted at normal incidence (0°) displayed low friction ($\mu_k = 0.3$). Its wear track was nearly featureless. The 45° sample exhibited stick slip behavior, with the average friction value increasing from $\mu_k \sim 0.3$ to $\mu_k \sim 0.45$ after 15 passes. Patches of debris were found in the track. The rotated sample had stick slip behavior, initially, and an average friction value of $\mu_k = 0.4$ from 7 to 15 passes. Some debris was found after 5 passes, but was subsequently removed by the slider. No measurable wear was observed.

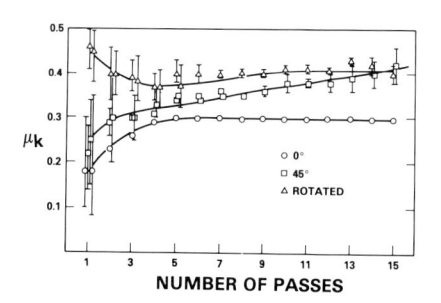

FIG. 2. Coefficient of friction vs. number of passes for samples implanted at 0°, 45°, and rotated. Bars on data points indicate stick-slip amplitude.

Compositions of the three disks, obtained by Auger sputter depth profiling, are shown in Fig. 3. The 0° implanted disk had a high surface Ti concentration, $[Ti]_s \sim 22\%$, and a fully carburized layer > 50 nm thick. Its composition is typical of disks alluded to in Fig. 1 which give superior tribological performance. The 45° implanted disk had a thinner (~25 nm), fully carburized layer but a much reduced $[Ti]_s \sim 13\%$. This layer was either too dilute in Ti+C or too thin to form a fully protective layer. A shortened range, a decreased fluence and enhanced sputtering contribute to altering the composition of off-normal implants, as will be discussed later. The rotated sample contained ample Ti, but less C than expected for so high a Ti concentration.

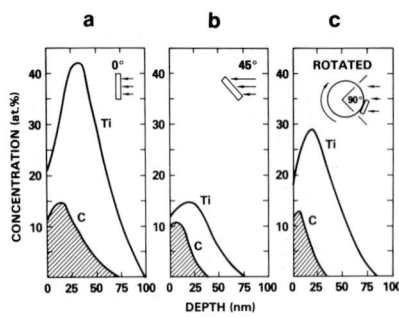

FIG. 3. Auger Composition vs. Depth profiles for samples implanted at 0°, 45°, and Rotated.

Surface Roughness

Polished disks were abraded unidirectionally on 600 grit paper then implanted to $2 \times 10^{17} Ti^+/cm^2$ at 50 keV or $5 \times 10^{17} Ti^+/cm^2$ at 190 keV. Fig. 4 presents the friction coefficients obtained with dry sliding, the sliding direction perpendicular to the abrasion direction. For the nonimplanted disk, μ_k rose to a steady state value ($\mu_k = 0.6$) by the second pass. By contrast, μ_k values for the implanted disks rose slowly from $\mu_k \sim 0.2$ up to $\mu_k \sim 0.4$ over 20 to 30 passes. Similar effects were observed for rougher surfaces (120 grit) implanted under identical conditions.

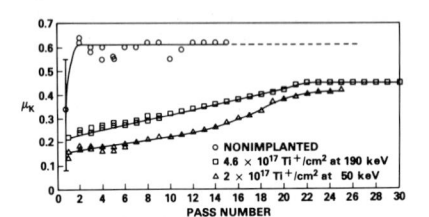

FIG. 4. Coefficient of friction vs. pass number for 600 grit SiC abraded disks.

This remarkable retention of low friction can be understood in terms of the wear processes occurring during the dry sliding of the 52100 ball against a roughened 52100 flat surface. The roughened surface of the nonimplanted disk, from the first pass on, behaved much like a file. Debris generated from the ball was deposited in the grooves of the abraded surface, then smeared along the track (see the DIC micrographs in Fig. 5). The friction coefficient rose to the steady-state value of a polished surface ($\mu_k = 0.6$) as the track became covered by the smoothed out debris. The tracks on the implanted surfaces showed virtually no wear and only slow accumulation of debris with each successive pass. Thus, the implanted roughened surface retained its initial low friction by reducing its wear against steel.

DISCUSSION

Titanium implantation at high fluences dramatically improves the dry sliding friction and wear behavior of 52100 steel. However, in order to achieve this superior tribological surface, one must tailor the implant fluence and energy to suit the geometry and surface condition of the specific component. If it has a flat surface, rough or smooth, the fluence may be chosen directly from Fig. 1. The choice of $2 \times 10^{17} Ti/cm^2$ at 50 keV instead of 5×10^{17} at 200 keV, for example, can be very cost effective. The current density of the 50 keV implant can be increased by

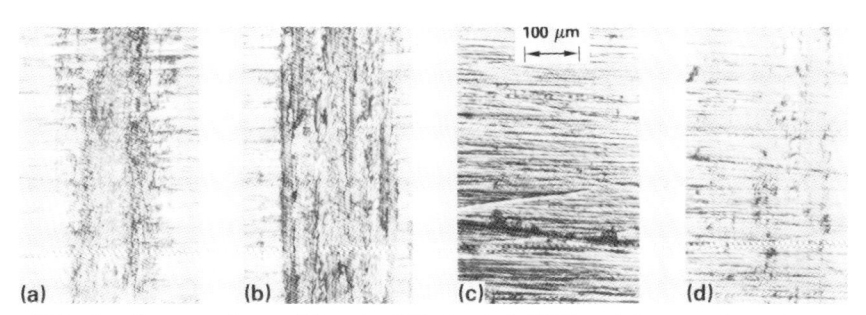

FIG. 5. Wear tracks on 600 grit SiC abraded 52100. (a) nonimplanted, 5 passes; (b) nonimplanted, 15 passes; (c) 4.6x10^{17}Ti$^+$/cm² at 190 keV, 10 passes; (d) 4.6x10^{17}Ti$^+$/cm² at 190 keV, 30 passes.

a factor of four, to achieve the same power density used in the 200 keV implant, and the processing time will be reduced by a factor of ten.

If irregular, curved or faceted surfaces are to be implanted, then energy and fluence must both be considered separately because of angular implantation effects. Implantation at an angle (see inset of Fig. 3) to the surface i) shortens the ion range to $R_p(\theta) = R_p \cos\theta$; ii) decreases the fluence to $f(\theta) = f \cos\theta$ (by increasing the intercepted surface area); and iii) increases the sputtering yield to $S(\theta) = S \cos^n \theta$ where $n = -5/3$ according to Sigmund's theory [11]. Implantations off-normal, therefore, at the least require higher energies to avoid encountering the thickness limit (Fig. 1).

A shorter range $R_p(\theta)$ compensates fairly well for decreased fluence $f(\theta)$, according to the fluence-range relations obtained from Table 1 and Fig. 1, and were it not for $S(\theta)$, one could probably obtain high Ti concentrations and fully carburized layers for wide ranges of angles at a fixed fluence. Unfortunately, the sputtering yield $S(\theta)$ fixes the maximum surface concentration of Ti at $[Ti]_s = 1/S(\theta)$. For large values of $S(\theta)$, $[Ti]_s$ and hence $[C]_s$ will be small, and the Ti+C layer may be too dilute to be effective. The surface implanted at $\theta = 45°$, while fully carburized, had only 1/2 as much $[Ti]_s$ as the 0° implant, consistent with the calculated ratio of sputtering coefficients at 0° and 45°. Therefore, angular implantations may be subjected to a third limiting factor, a surface Ti concentration limit, which is determined by sputtering.

Implantation of a rotating cylinder can be considered as angular implantation averaged over the sector ($-\theta_m$ to $+\theta_m$) (see inset of Fig. 3c). Average range and fluences are given by $\langle R_p \rangle = R_p^m \sin\theta_m/\theta_m$ and $\langle f \rangle = f \sin\theta_m/\theta_m$ and the average sputtering yield by

$$\langle S \rangle = \frac{1}{\theta_m} \int_0^{\theta_m} \cos^{-5/3}\theta \, d\theta.$$

A 90° sector ($\theta_m = 45°$) was chosen for the present study to minimize the time a rotating surface would be outside the beam. Preliminary studies have indicated that the carburized layer may be degraded by oxidation effects during the time period it spends outside the beam. The friction and wear results indicate that a 90° sector is a good compromise for a rotating cylinder or bearing race, but we recommend raising the fluence to 5x10^{17}Ti/cm² at 100 keV to assure adequate carburization.

Finally, it has been suggested that since the incorporation of C from the vacuum system is not easily controlled, the implantation of C (in addition to Ti) might produce the same tribological characteristics as obtained by the vacuum carburization process. Our investigations of dual implants of C and Ti into hardened 52100 steel have shown this is not true, and that a vacuum carburized layer is required for the low friction, wear resistant surface on hardened 52100 steel [9].

However, implanted C in dual Ti+C implants may benefit softer steels in the same way as N does e.g. by solid solution strengthening. The Sandia group [3] has reported significant reductions in friction and wear of soft stainless steels implanted with Ti+C. Our recent work on dual implants of Ti and N into soft steels also indicate beneficial effects [12]. Also, other strong carbide formers such as Ta or Hf can also produce low friction, low wear surfaces by forming carburized surface layers. Judicious choices of carbide formers and interstial implants may permit tailoring of surfaces to optimize several properties e.g. tribological and corrosion, at once.

In conclusion, we have enumerated the processing parameters required to optimize the beneficial effects of Ti implantation into 52100 steel. The steel requires a sufficiently thick, fully carburized layer to provide lowest wear at lowest friction values. A limited number of lifetime wear studies performed on Ti-implanted 52100 steel confirm the behavior described here [1,4], but more extensive testing of implanted components is required.

ACKNOWLEDGMENT

We thank Bruce Sartwell and Fred Smidt for their assistance and encouragement and NAVSEA for providing funds for the project.

REFERENCES

1. C.A. Carosella, I.L. Singer, R.C. Bowers, and C.R. Gossett, Ion Implantation Metallurgy, C.M. Preece and J.K. Hirvonen, eds., (Metallurgical Society of AIME, Warrendale, PA, 1980), p. 103.
2. I.L. Singer, C.A. Carosella and J.R. Reed, Nucl. Instrum. Methods, 182/183, 923 (1981).
3. F.G. Yost, L.E. Pope, D.M. Follstaedt, J.A. Knapp, and S.T. Picraux, Metastable Materials Formation by Ion Implantation, S.T. Picraux and W.J. Choyke, eds., (Elsevier, Amsterdam, 1983), p. 261.
4. N.E.W. Hartley and J.K. Hirvonen, Nucl. Instrum. Methods., 209/210 933 (1983).
5. I.L. Singer, R.N. Bolster, C.A. Carosella, Thin Solid Films, 73, 283 (1980).
6. I.L. Singer and R.A. Jeffries, J. Vac. Sci. Tech., A1 ,317 (1983).
7. I.L. Singer, J. Vac. Sci. Tech., A1, 419 (1983).
8. I.L. Singer and T.M. Barlak, Appl. Phys. Lett., 43, 457 (1983).
9. I.L. Singer and R.A. Jeffries, Appl. Phys. Lett., 43, 925 (1983).
10. I. Manning and G.P. Mueller, Comp. Phys. Comm., 7, 85 (1974).
11. P. Sigmund, Phys. Rev., 184, 383 (1969).
12. I.L. Singer and R.A. Jeffries, these proceedings.

THE REDUCTION OF WEAR AND WEAR VARIABILITY UNDER LUBRICATED SLIDING BY ION IMPLANTATION

J.J. AU* AND P. SIOSHANSI**
*Sundstrand Aviation Operations, 4747 Harrison Avenue, Rockford, IL 61101;
**Spire Corporation, Patriots Park, Bedford, MA 01730

ABSTRACT

Block-on-ring wear tests were performed on Ti-implanted AISI 52100 steel sliding against three commercial alloys: CDA 673 brass, AISI 4140 low alloy steel, and AISI 0-6 tool steel. A turbine oil, meeting MIL-L 23699 specifications, was used as the lubricant. The 52100 wear rings were ion implanted to a fluence of 2×10^{17} Ti/cm^2 at 55 KeV or 125 KeV. Comparison with non-treated 52100 wear rings indicated that ion implantation reduced the amount of wear and wear variability. It is hypothesized that reduced adhesion and improved lubricant film formation due to the presence of the implanted layer was responsible for the reduction in wear and wear variability.

INTRODUCTION

It has generally been recognized that ion implantation is most suitable for light load, lubricated sliding conditions owing to the shallowness of the implanted layer (.02 µm to .4 µm). It has also been shown that implantation of ions such as nitrogen or boron could significantly improve the lubricated sliding wear resistance of many steels [1-4], although titanium appeared to be the only ion species which was effective in reducing friction and wear of AISI 52100 steel [5]. Despite the large body of literature on tribology, very little has been reported on the technologically important problem of wear variability which manifested itself as infantile wear failures or as large scatter in the wear data. The lack of attention to studying wear variability in part could be attributed to the many variables involved in any wear process. For example, any attempt to alleviate wear variability by conventional coating techniques would encounter added variability from differences in coating quality and coating adhesion. Ion implantation appears to be a surface treatment technique which does not have the quality and adhesion problems inherent in other methods.

The current paper describes wear experiments carried out to study the effect of ion implantation, not only to reduce wear, but also to reduce wear variability. The LFW-1 block-on-ring wear apparatus was selected as the main wear testing tool owing to its capability to produce scatter in the wear data and thus simulate wear variability. The block-on-ring geometry also simulates a certain engineering applicaton at Sundstrand. For this reason, the wear blocks were conformed to a fixed contact area of 0.1 cm^2 (0.016 in^2). The wear rings were 3.5 cm in diameter and 0.87 cm wide, with a wear track of 0.5 cm. Three material combinations were selected as candidate materials for the intended engineering application: (1) AISI 52100 steel vs. CDA 673 (Mueller 602) brass, (2) AISI 52100 steel vs. AISI 4140 steel, and (3) AISI 52100 steel vs. AISI 0-6 graphitic tool steel. All three steels had a martensitic structure with the following hardness values: 58-62 HRC for AISI 52100 steel, 30-35 HRC for AISI 4140 and AISI 0-6 steels. The CDA 673 brass had a hardness of 75-80 HRB. Both Ti-implanted 52100 steel and non-treated 52100 steel were used as the rotating rings. The other three materials were used to make the conforming blocks.

Experimental Procedure

Titanium implantation was performed after ultrasonic cleaning in freon by two laboratories under two different conditions, as noted below:

Condition	Energy	Fluence	Ring Rotation
A	55 keV	2×10^{17} Ti/cm^2	Continuous
B	125 keV	2×10^{17} Ti/cm^2	Fixed, 120° apart

The variation in the accelerating energy is expected to change the penetration depth, while the nonuniformity of the implanted layer would be effected by the manner the ring was rotated relative to the ion beam. Higher accelerating voltage would give a deeper implanted layer, and continuous ring rotation would produce a more uniform implanted layer.

Both the implanted and the non-treated wear rings have an as-machined surface finish of 6-12 RMS. The wear blocks were conformed by rubbing on an oscillating ring covered with a 600 grit abrasive paper. The final polish used a 1200 grit alumina powder. The wear samples were ultrasonically cleaned in freon and then dried with hot air prior to installation into the LFW-1 wear machine. All wear tests were conducted at room temperature with the wear ring partially immersed in a bath of synthetic polyester lubricant which met MIL-L-23699 specifications. The applied normal load was 801 N (180 lbf) and the sliding speed was 55.9 cm/sec (22 in/sec). Continuous linear wear measurements were made with a LVDT pickup and recorder. The total linear wear result was verified with weight loss measurements before and after each test using a micro-balance having a sensitivity of 0.00001 gm and an accuracy of 0.00005 gm.

The wear samples were examined after the wear tests using scanning electron microscopy (SEM) and Auger spectroscopy in an attempt to elucidate the mechanisms which would explain the effect of ion implantation on wear and wear variability.

Experimental Results

Wear Test Results. The linear wear was converted to wear volumes and plotted as a function of sliding distance for both Ti-implanted and non-treated AISI 52100 steel rings sliding against the three materials CDA 673 brass, AISI 4140 steel and AISI 0-6 steel. These wear curves are shown in Figures 1-3. Wear test results of titanium implantation only are presented in Figure 4. The following legend is given to aid interpretation of the wear curves.

Legend	Description of AISI 52100 Ring
————	Non-treated
- - - - - -	Ti-implanted, condition A
— — —	Ti-implanted, condition B
—·—·—·—	Ti-implanted, condition A; ring used once in a previous test, new block.

Without Ti-implantation on the AISI 52100 steel ring, the wear and wear variability of CDA 673 brass and AISI 4140 steel blocks was large. On the other hand, wear of AISI 0-6 steel blocks was generally low with the exception of one test, indicating statistically large wear variability even with this low wear material. Condition A Ti-implantation on the AISI 52100 ring markedly reduced the wear and variability of wear of all three material combinations studied. Condition B Ti-implantation generally reduced wear but its effect on wear variability was less pronounced.

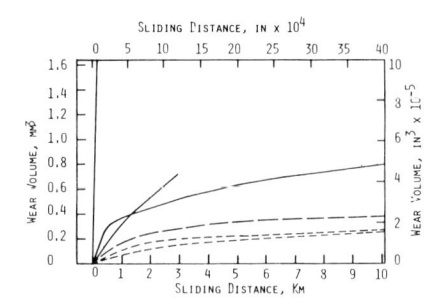

FIG. 1. Wear curves of non-treated and Ti-implanted AISI 52100 steel vs. CDA 673 brass.

FIG. 2. Wear curves of non-treated and Ti-implanted AISI 52100 steel vs. AISI 4140 steel.

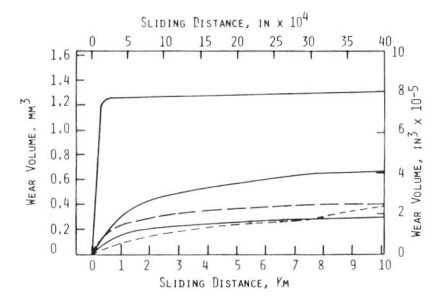

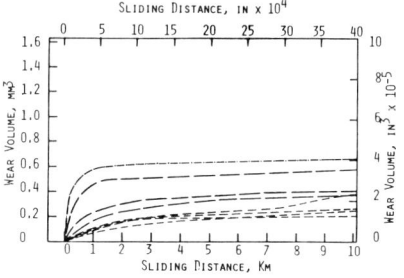

FIG. 3. Wear curves of non-treated and Ti-implanted AISI 52100 steel vs. AISI O-6 steel.

FIG. 4. Wear curves of Ti-implanted AISI 52100 steel vs. CDA 673 brass, AISI 4140 steel and AISI O-6 steel.

In addition, the wear curves of Figures 1-4 showed that condition A Ti-implantation in general is superior to condition B Ti-implantation. This may be indicative of a stronger dependence of wear behavior on the uniformity of the implanted layer than on the thickness of the implanted layer or it could reflect differences in surface chemistry. In fact, with condition A Ti-implantation, the wear behavior may be quite insensitive to the block materials, as noted in Figure 4. It is also worth noting that the beneficial effects of Ti-implantation persisted even with repeated use, as noted in Figure 2. The slightly higher wear of a once-used ring on a new block could be attributed to break-in of a new wear surface.

Scanning Electron Microscopy Results. Typical appearance of the ring and block before and after the wear test is shown in Figure 5. Without Ti-implantation, the original machining grooves were flattened and the surface appeared more smooth. Wear apparently occurred by deformation and fracture of asperities and subsequent polishing by entrapped wear particles, in accordance with the friction model advanced by Suh and Sin [6].

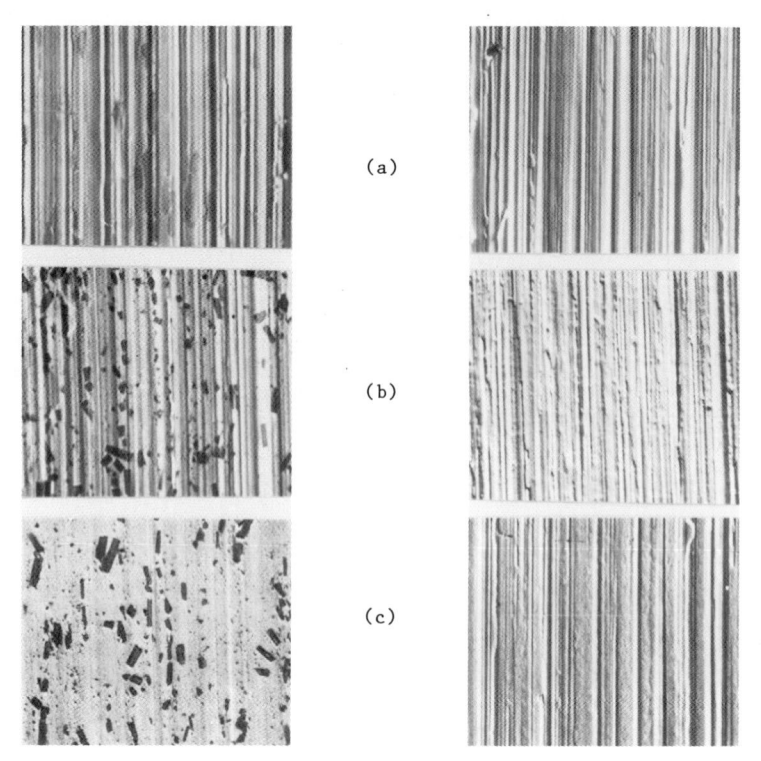

FIG. 5. Surface appearance of CDA brass block (left) and AISI 52100 steel ring (right). (a) Before test. (b) After test, condition B Ti-implanted ring. (c) After test, non-treated ring. Mag. 500X. Note: Similar results were observed in AISI 4140 and AISI 0-6 steels.

On the contrary, the condition B Ti-implanted AISI 52100 ring and its mating block did not show much change in its surface appearance after the wear test, compare (a) and (b) of Figure 5. It is postulated that the Ti-implanted layer promoted lubricant film formation and reduced adhesion such that wear occurred only by the slow oxidative mechanism. Follstaedt et.al. [7] and Singer et.al. [8] have shown that Ti-implantation of iron and steel produced an amorphous layer. It is possible that this amorphous layer changes the chemisorption or wetting characteristics of the surface. This lubrication model is consistent with the observation that wear behavior is almost independent of contact material under condition A Ti-implantation where positive identification of an amorphous layer has been observed.

Suh et. al. [9] have proposed a different model to account for the effect of ion implantation. In their model, reduction in friction and wear by ion implantation was due to decreases of the plowing component of friction in the presence of the thin implanted layer. Their model predicted that the decrease in friction in turn substantially reduced subsurface deformation and hence delamination wear. As shown in Figure 5, we have not observed any decrease in plowing by Ti-implantation. Nor have we found any evidence of delamination wear. We have also not found any correlation between the reduction in friction and reduction in wear [10].

The model of Suh et.al. thus cannot account for the current observations. It should be pointed out that geometry difference (block-on-ring vs. pin-on-disk) or differences in surface finish of the implanted member (as-machined vs. highly polished) may be responsible for the different observations reported by Suh et. al.

AUGER SPECTROSCOPY RESULTS

The auger electron spectroscopy (AES) analysis of condition A Ti-implantation is presented in reference [5] and shows a peak concentration of 28% Ti (at. %) at a depth of 750 Å. Only the AES results of condition B Ti-implantation is discussed herein.

Figure 6 presents the Auger spectrum on the wear track of the condition B Ti-implanted AISI 52100 ring which slid against a CDA 673 brass block. The absence of any copper or zinc (that is, transfer of CDA 673 brass) is consistent with our lubricant film model wherein wear occurred only by an oxidative mechanism. While a carbon peak has been observed, this carbon peak is attributable to hydrocarbon residue from the vacuum pump or the solvent used to clean the sample. This carbon peak was rapidly reduced to the value of the base metal carbon after removal of approximately 100 Å by sputtering. The major difference of condition A Ti-implantation and condition B Ti-implantation therefore lies in the C-content on the surface. A Ti+C implanted layer is present with condition A Ti-implantation [5], whereas only a Ti implanted layer exists with condition B Ti-implantation. The carbon is perhaps responsible for the difference in wear behavior noted in Figure 4.

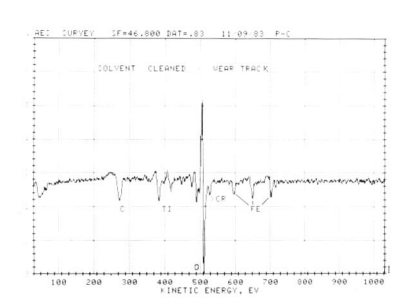

FIG. 6. Auger electron spectra of wear track of condition B Ti-implanted AISI 52100 ring.

In agreement with our proposed lubricant film model, Figure 7 shows that the implanted layer was only slightly reduced from about 1600 Å to about 1000 Å after wear testing. The maximum titanium concentration was not significantly affected. This retention of most of the implanted layer predicts that Ti-implantation is useful even after repeated use, as indeed observed in Figure 2.

CONCLUSIONS

Lubricated sliding wear of Ti-implanted AISI 52100 steel against CDA 673 brass, AISI 4140 steel and AISI O-6 steel has been studied. The conclusions are:

(1) Ti-implantation of AISI 52100 steel reduces wear and wear variability for each of the material combinations studied.

(2) A uniform Ti+C implanted layer is better than a thicker, non-uniform implanted layer of titanium only.

(3) With uniform Ti+C implantation, the wear behavior may be independent of mating material under the current test conditions.

(4) The beneficial effects of Ti-implantation persist with repeated use.

(5) Under lubricated sliding wear conditions, the surface micro-topography does not change significantly with Ti-implantation.

(6) It is hypothesized that the reduction of wear and wear variability may be attributed to reduced adhesion and improved lubrication due to the presence of an amorphous implanted layer (notably in condition A).

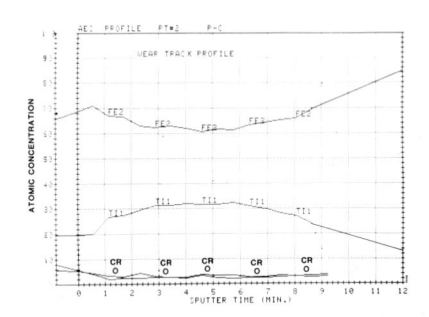

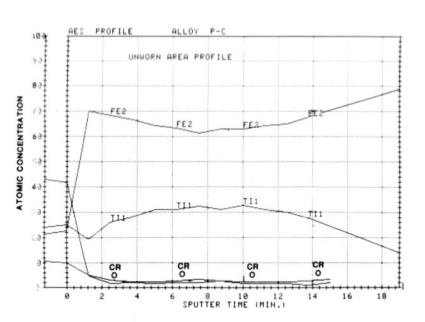

FIG. 7. Auger depth profiles of condition B Ti-implanted ring in wear track and in adjacent unworn area of ring (sputter rate approximately 100 Å/min.).

REFERENCES:

1. N. E. W. Hartley, G. Dearnaley, J. F. Turner, and J. Saunders in: Applications of Ion Beam to Metals, S. T. Picraux, and E. P. ErNisse, eds. (Plenum Press, NY 1974) pp. 123-138.
2. G. Dearnaley and N. E. W. Hartley, Thin Solid Films, 54, pp. 215-231 (1978).
3. J. K. Hirvonen, J. Vac. Sci. Technol., 15, pp. 1662-1668 (1978).
4. J. K. Hirvonen, C. A. Carosella, R. A. Kant, I. Singer, R. Vardiman, and B. B. Rath, Thin Solid Films, 63, pp. 5-10 (1979).
5. C. A. Carosella, I. L. Singer, R. C. Bowers, and C. R. Gossett in: Ion Implantation Metallurgy, C. M. Preece and J. K. Hirvonen, eds. (The Metallugical Society of A.I.M.E., Warrendale, Pennsylvania, 1980), p. 103.
6. N. P. Suh and H. C. Sin, Wear, 69, pp. 91-114 (1981).
7. D. M. Follstaedt, J. A. Knapp, and S. T. Picraux, Appl. Phys. Lett., 37, pp. 330-333 (1980).
8. I. L. Singer, C. A. Carosella, and J. R. Reed, Nucl. Inst. and Methd., 182/183, p. 923 (1981).
9. N. P. Suh and S. R. Shepard, Progress Report to the Office of Naval Research, Contract No. N00014-80-C-0255, (1981).
10. J. J. Au, Sundstrand Materials Project Report No. 19321, (1982).

EFFECTS OF ION IMPLANTATION ON THE ROLLING CONTACT FATIGUE OF 440C STAINLESS STEEL

F. M. KUSTAS,* M. S. MISRA,* AND P. SIOSHANSI**
*Martin Marietta Denver Aerospace, P.O. Box 179, Denver CO 80201; **Spire Corporation, Patriots Park, Bedford, MA 01730

ABSTRACT

Cylindrical 440C stainless steel specimens implanted with N and Ti were examined for their fatigue resistance and wear behavior by rolling contact fatigue (RCF) testing. The results obtained from RCF testing showed a 40% increase in the B-10 fatigue lifetime for N implanted and a 17% increase for Ti implanted 440C specimens compared to baseline, unimplanted 440C. Quantitative surface analysis by Auger Electron Spectroscopy (AES) was performed to determine the effects of process parameter optimization and ion beam masking on the elemental concentration profiles.

INTRODUCTION

Corrosion resistant stainless steels, such as 440C, are currently used for many aerospace bearing applications where good corrosion resistance and high load carrying capacity are primary requirements. Even though 440C steel exhibits excellent corrosion resistance it suffers from a low fatigue and wear resistance, especially in high load/high RPM applications. Premature failure of critical bearing components, before the useful design lifetime, has resulted in a reduced reliability of aerospace systems. Increasing the fatigue resistance of 440C stainless steel bearing components by 50-100% would have a significant impact on the operating efficiency of Space Transportation Systems.

Conventional surface treatments suffer from inadequate coating adhesion, and processing restrictions such as high temperature application and final grinding for dimensional accuracy. A novel surface processing technique which avoids all of the conventional coating deficiencies is ion implantation. This technique is useful for modifying surface dependent properties and its inherent advantages include, formation of highly alloyed, amorphous surface layers with no definable interface.

It has been established that ion implantation can provide wear and cyclic fatigue resistant surfaces by careful selection of processing parameters [1-3]. For example, Ti implantation of 52100 and 440C steels has shown a reduction in the kinetic coefficient of friction by up to 39% for 52100 [1] and 63% for 440C [2] steels. For the more severe condition of cyclic fatigue, an improvement in the fatigue endurance life of 1018 steel has been observed for N implantation followed by an annealing heat treatment [3].

Therefore ion implantation with Ti, for modification of the coefficient of friction, and N, for intermetallic compound formation, was selected as an approach to improve the rolling contact fatigue resistance of 440C steel.

EXPERIMENTAL

Test Specimens

Cylindrical 0.375 in. diameter by 3.25 in. long rods, fabricated from 440C stainless steel (18 wt% Cr, 1.02 wt% C, 0.56 wt% Mo), were used for the rolling contact fatigue (RCF) testing. After rough machining to a preliminary dimensional size, the cylindrical rods were heat treated to provide the required surface hardness of Rockwell C 58-63 and then given a final surface grind to a 6-8 micro-inch rms finish.

Ion Implantation

The implantation technique used for processing of the 440C RCF test specimens employed stationary specimens with an x-y rastered ion beam coverage in four overlapping angular increments of ~110° of the specimen circumference. Masking the ion beam was performed on selected specimens (see Figure 1) to prevent excessive sputtering of the ion beam processed area by ensuring near-normal beam incidence. Decreasing the ion beam width increases the degree of beam normality with the specimen surface. As shown in Figure 1, the specimens were firmly mounted to a Freon-cooled copper block which acted as a heat sink to prevent tempering of the substrate.

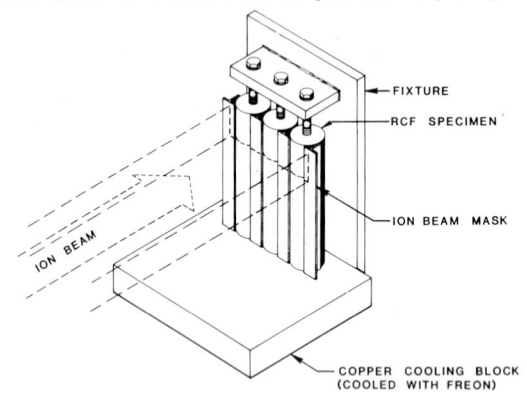

FIG. 1. Ion beam masking of stationary RCF specimens.

An approximate 0.5 inch length of each end of the RCF specimen was ion implanted for subsequent fatigue endurance testing. The process parameters listed in Table I, were selected from previous experience and a literature survey concerning the effects of alloy additions on material wear behavior.

TABLE I. Ion implantation process parameters.

Implanted ion	Accelerating voltage (keV)	Total integrated dose (ions/cm^2)
Ti	150	10^{17}
N	40	10^{17}
Ti[1]	40	2×10^{17}
N[1]	40	2×10^{17}

[1] Ion beam masking used during implantation.

Rolling Contact Fatigue (RCF) Testing

In order to evaluate the effects of ion implantation on the fatigue resistance of 440C stainless steel, a test technique was adopted that used continuous rotation of 2 M-50 steel discs against the test specimen to simulate rolling contact fatigue loading conditions. The RCF test technique is discussed in detail elsewhere [4]. The tests were performed under ambient conditions using a turbine engine oil lubricant, with a rotational velocity of 10,000 rpm at a maximum Hertz stress of 700 ksi.

Three fatigue endurance tests are conducted over each implanted section for a total of 6 tests per specimen, enabling the determination of a statistical fatigue life, termed the B-10 lifetime. This is defined as the number of cycles at which 10% of the specimens will fail by a fatigue spall mechanism.

RESULTS AND DISCUSSION

N-Implanted Specimens

Quantitative Surface Analysis - Ion concentration-depth profiles were determined by Auger Electron Spectroscopy (AES) and the results are shown in Table II for the unmasked and masked RCF specimens. The profiles were obtained from as-implanted, unworn areas of the RCF specimens.

TABLE II. Computer code prediction and AES quantitative analysis for N implanted 440C.

Type of analysis	Peak N a/o at given depth (μm)	Total depth (μm) of N penetration
a. Computer Model[1,2]	40 @ 0.02	0.05
b. Implanted[1] - AES	10 @ 0.02	0.09
c. Implanted[3] - AES	24 @ 0.07	0.15
d. Implanted[3,4] - AES	28 @ 0.05	0.15

Note:

a/o = atomic percent (%)

[1] N implantation parameters: 10^{17} ions/cm^2 @ 40 keV without masking.
[2] Polished flat steel substrate.
[3] N implantation parameters: 2×10^{17} ions/cm^2 @ 40 keV with ion beam masking (see Figure 1).
[4] AES analysis displaced ~45° from analysis c.

The concentration profile, for the unmasked N implanted specimen, showed a much lower peak concentration than predicted from the computer model. The discrepancy undoubtedly lies in the fact that the computer predictions were made for N implantation on a flat, polished steel substrate. It appears excessive sputtering of the ion beam processed areas occurred during implantation. The excessive sputtering is attributed to non-normal beam incidence on a portion of the cylindrical surface and surface-related effects due to the rough surface from grinding.

Using ion beam masking to optimize beam normality (see Figure 1), an enhanced peak N concentration of 24 atomic percent (a/o) was obtained. An additional AES concentration-depth profile was performed, rotated ~45° of the specimen circumference with respect to the original AES profile. A slightly higher peak N concentration (~17% enriched or 28 a/o) at a shallower depth (~0.05 μm) was observed for the alternate site. However, excellent correlation was found between total N penetration, namely 0.15 μm for both AES locations. This demonstrates that fairly good reproducibility can be obtained over a given cylindrical surface area.

Energy distribution information is useful in determining the phase or compound that individual elements can be expected to form. For N implanted 440C steel, it was not possible to discern the compound state of N due to interfering energy profiles. Previous studies have indicated the formation of iron nitride compounds after N implantation of a low alloy steel [5] and it appears nitride formation for N implanted 440C is also likely.

These higher concentration specimens have not been RCF wear tested as yet, however, it is anticipated that increasing the available nitrogen for iron nitride formation will provide a more wear resistant surface. In addition, increasing the total penetration depth will enable nitrogen translation to greater depths during the fatigue/wear process, an effect observed by other investigators [3].

Ti Implanted Specimens

Quantitative Surface Analysis - Implantation of cylindrical RCF specimens with Ti resulted in a near Gaussian distribution, with a peak atomic concentration of 4% for unmasked specimens, compared to the computer code prediction of 21% (see Table III). The penetration depths were comparable, while the lower concentration levels, determined by AES, were an effect of non-normal ion beam incidence as previously discussed. Attempts to increase the implanted Ti ion concentration and depth levels by employing the previously discussed ion beam masking technique (see Figure 1) were successful.

AES concentration-depth profiles for Ti implanted 440C RCF specimens are shown in Figure 2 with the peak Ti concentration being increased to 23 a/o at a depth of 0.03 μm as a result of masking. The observation of a carbon peak has been attributed to system contamination; however, its presence has been shown to be beneficial for carbide formation as well as an important element in amorphous layer production [6,7].

TABLE III. Computer code prediction and AES quantitative analysis for Ti implanted 440C.

Type of analysis	Peak Ti a/o at given depth (μm)	Total depth (μm) of Ti penetration
Computer Model[1,2]	21 @ 0.04	0.11
Implanted[1] - AES	4 @ 0.03	0.12
Implanted[3] - AES	23 @ 0.03	0.09

[1]Ti implantation parameters: 10^{17} ions/cm^2 @ 150 keV without masking.
[2]Polished flat steel substrate.
[3]Ti implantation parameters: 2×10^{17} ions/cm^2 @ 40 keV with ion beam masking (see Figure 1).

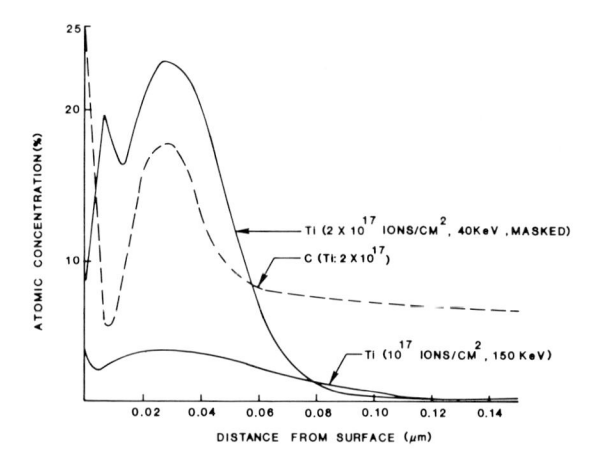

FIG. 2. AES concentration - depth profiles for Ti implanted 440C RCF specimens.

The theoretical explanation for carbide formation related Ti reaction with carbon containing gases, forming an excess surface concentration of TiC, which dissociates, allowing for C diffusion into the substrate [6]. The formation of TiC particles is desirable due to their excellent wear resistance. For all of the implantation parameters studied during the Ti implantation of 52100 steel, the peak carbon concentration was always smaller in magnitude and directly related to the Ti implantation parameters.

The results shown in Figure 2, for the high dose (2×10^{17} ions/cm^2) Ti implanted 440C RCF specimen support this observation, suggesting enhanced TiC formation by a similar mechanism.

In essence, the observation of this TiC layer is beneficial for wear resistance due to the formation of dispersed TiC particles, and has been adopted as an intentional processing approach with sequential implantation of Ti and C followed by an annealing treatment to stimulate TiC formation [7].

For 52100 steel, in addition to the formation of TiC particles, an amorphous Fe-Ti-C layer was found [1] and is also suspected for Ti implanted 440C.

The combination of an Fe-Ti-C amorphous zone along with dispersed TiC particles should provide an extremely wear resistant surface.

B-10 Fatigue Lifetime Test Results

The results of rolling contact fatigue (RCF) testing for baseline, N implanted and Ti implanted 440C specimens are shown in Table IV. There is a definite improvement in the B-10 lifetime for the N implanted specimens, approaching 40%, while the Ti implanted specimens showed a moderate improvement of ~17%. These initial results are encouraging, since the ion beam masking technique was not employed for these implantations, which was shown to be an effective method to maximize the surface ion concentration.

TABLE IV. B-10 fatigue lifetimes for baseline and ion implanted 440C RCF specimens.

Ion Implanted	Accelerating voltage (keV)	Total integrated dose (ions/cm^2)	B-10 lifetime (cycles)
Baseline	---	---	3.6×10^6
N	40	10^{17}	5.0×10^6
Ti	150	10^{17}	4.2×10^6

The noteworthy improvement in the fatigue lifetime for N and Ti implanted 440C stainless steel appears to be due to (1) enhanced formation of wear resistant iron nitride intermetallic compounds for N implanted specimens, and (2) production of an amorphous layer with a dispersion of TiC particles for Ti implanted specimens.

CONCLUSIONS

1. Successful enrichment in the N and Ti concentrations and enhanced ion penetration was observed for implanted 440C RCF specimens after increasing the total integrated dose and using an ion beam masking technique.

2. A significant improvement in the B-10 fatigue life by up to 40% for N implanted and a moderate gain of 17% for Ti implanted 440C stainless steel was observed by rolling contact fatigue endurance testing.

REFERENCES

1. I.L. Singer, C.A. Carosella and J.R. Reed, "Friction Behavior of 52100 Steel Modified by Ion Implanted Ti," in The Use of Ion Implantation for Materials Processing, NRL Memorandum Report 4527, June 24, 1981, pp.53-69.

2. F.G. Yost, et al., "Friction and Wear of Stainless Steel Implanted with Ti and C," in Proceedings of Boston Materials Research Society Meeting, Nov. (1981).

3. J.K. Hirvonen, "Ion Implantation in Tribology and Corrosion Science," J. Vac. Sci. Technol. 15(5), 1662-1668, (1978).

4. E.N. Bamberger and J.C. Clark "Development and Application of the Rolling Contact Fatigue Test Rig," Rolling Contact Fatigue Testing of Bearing Steels, ASTM STP 771, 85-106, (1982).

5. M. Carbucicchio, L. Bardani and S. Tosto, "Surface Mössbauer Analysis of 38NCD4 Steel Ion Implanted with Nitrogen," J. Appl. Phys. 52(7), 4589-4592, (1981).

6. I.L. Singer, R.A. Jeffries, "Surface Chemistry and Friction Behavior of Ti-Implanted 52100 Steel," J. Vac. Sci. Technol. A1(2), April-June 317-321, (1983).

7. G.C. Nelson, L.E. Pope and F.G. Yost, "Summary Abstract: Correlation of the Surface Compositon of C and Ti-Ion Implanted Layers in Iron and Stainless Steel with Reduced Friction and Wear," J. Vac. Sci. Technol. A1(2), 496, (1983).

ENHANCEMENT OF FERROUS ALLOY SURFACE MECHANICAL PROPERTIES BY NITROGEN IMPLANTATION

JOHN T.A. POLLOCK, MICHAEL J. KENNY*, PETER J.K. PATERSON,**
*CSIRO Division of Chemical Physics, Lucas Heights Research Laboratories, NSW, 2232, Australia. ** Applied Physics, R.M.I.T., Melbourne, VIC. 3000, Australia.

ABSTRACT

Clarification of the relationship between nitrogen profiles and wear behaviour has been sought by studying nitrogen implanted mild steel. Implant energy and dose were in the ranges 25 - 65 keV and $0.9 - 3 \times 10^{17}$ ions cm^{-2}, respectively. Wear characteristics were measured with a ball-on-disc system followed by interferometric analysis. Nitrogen distributions before and after wear were determined by Auger electron spectroscopy and Rutherford backscattering methods, and compared with wear track profiles. On balance, the data offers qualified support for nitrogen migration at low loads, although nitrogen was not detected for wear depths >2 times the implant depth. Observations of wear on the abrader ball-bearing surface and the role of oxygen in the wear process are reported.

INTRODUCTION

The effectiveness of nitrogen implantation as a wear-reducing agent is well established for many steels, particularly those which are not highly hardened by prior heat treatment (1-4). Improved properties are usually attributed to nitride formation and/or Cottrell atmosphere hardening. Evidence for these structural changes has been reasonably well established in recent years (5,6,7). However, the mechanism by which wear enhancement persists beyond the original implantation depth has yet to be clearly resolved. By inference, nitrogen retention by microstructural modification must occur during the wear of the implanted layer. Dissolution of nitrides due to localised heating at shearing asperities, together with thermal diffusion down the accompanying steep temperature gradient is the most accepted proposition (8).

In contrast with the former assertion of deep diffusion (8,9), recent experiments have shown that migration occurs at best over short distances (4,10,11).

The present work compares Auger electron spectroscopy (AES) nitrogen profiles to interferometric profiles of the same tracks and show in some detail, the relationship between wear condition and residual nitrogen.

EXPERIMENTAL PROCEDURE

Disc samples (12 mm dia.) of bright mild steel were polished metallographically to 1 μm diamond before implanting with unanalysed nitrogen at energies of 25 - 65 keV to nominal fluxes of $0.9 - 6 \times 10^{17}$ ions cm^{-2}, using beam current of $20 - 25 \times 10^{-6}$ A cm^{-2}. Implanted and unimplanted discs were subsequently annealed at 400 K for 6 h.

Wear measurements were made with a machine based on the pin-on-disc principle, but using a ball-bearing abrader. Sample speed and contact time between abrader (a 1.0 cm dia. carbide ball-bearing) and disc were normalised for various wear conditions so that tracks could be worn at various radii using one sample. Contact velocity was 0.8 m s^{-1} with paraffin oil continuously dripped onto the sample surface to provide light lubrication. Most wear measurements were made at loads in the range 20 - 100 g, however, loads up to 1000 g were employed.

Wear tracks were examined metallographically using interference microscopy; the wear depth profile was determined by interferometry using both white and monochromatic light. The samples were then depth profiled for N, O, C and Fe by AES focussed inside and outside wear tracks using a Varian 10 keV auger spectrometer equipped with a single pass cylindrical mirror analyser and an integral electron gun. The beam had a current of 5 x 10^{-6} A, a beam energy of 5 keV and a diameter of 10 x 10^{-6} m. Sputtering was achieved with 1 keV Ar ions from a Varian ion gun; the argon background pressure was 6.0 x 10^{-5} torr, and typically the sputtering current was 50 x 10^{-6} A cm^{-2}. Sputtering steps were in the range 5 - 20 nm, depending on the implant energy and wear condition of the area under examination.

Rutherford backscattering (RBS) analysis was performed on unworn implanted and unimplanted samples using 2 MeV He ions. The implant concentration/depth profile was determined from the difference between implanted and unimplanted spectra using the iterative procedures for high dose implants outlined by Chu et al. (12).

RESULTS AND ANALYSIS

Wear and Hardness

Wear volume ratios (implanted:unimplanted) were measured from interferogram profiles for mild steel implanted under a range of conditions and tested across a range of load and distance. The data displayed the trends reported in other studies (e.g. ref.1,2) using pin-on-disc systems. The maximum wear enhancement, recorded at the lightest load (20 g) was about 15:1 and contrasts with ≈30:1 or higher reported in other work. This reflects the tougher wear conditions produced by a ball tip compared with an approximately 1 mm dia. flat-topped pin.

Metallographic examination of wear tracks produced at loads of 50 g or less revealed a sliding wear mechanism with a very noticeable "burnishing" effect in the implanted tracks. Associated interferograms revealed significant deformation, in the form of material rollover at the track edge, occurring in both the implanted and unimplanted samples. In the case of the implanted sample, this may represent flow of the unimplanted material due to Hertzian stresses, which will be greatest just beyond the implanted layer.

With a ball-on-disc system, ball wear acts as a useful monitor of wear track depth relative to the implant zone. Low energy, high dose implants produce very hard near-surface zones which, at low wear distance and moderate load (<100 g) produce circular wear patterns on the ball. This pattern occurs for other implant conditions, but is not so marked. With increased wear time (distance), the ball breaks through the implanted layer and then wears less dramatically, with an increasing tendency to produce ellipsoidal scars. Of course, such scarring is produced during the wear of unimplanted samples reflecting the much reduced wear of the hard ball-bearing surface. Under these circumstances, ball markings are the result of interactions with hard inclusions and take the form of gouges.

Nitrogen Profiles

Representative sections of RBS spectra are shown in Figures 1a and 2a. The high dose implant (Figure 1a) has a clear dip at the top right of the spectrum indicating displacement of Fe atoms by N atoms. A corresponding displacement is not so discernable in the low dose spectrum (Figure 2a). The difference between the implanted and unimplanted spectra reveals the implant profiles more clearly. Atom concentration versus depth profiles were calculated from the backscattered data (Figures 1b and 2b). The nitrogen peak concentration of ≈34 at. %, measured on the high dose sample indicates the formation of Fe_2N (5).

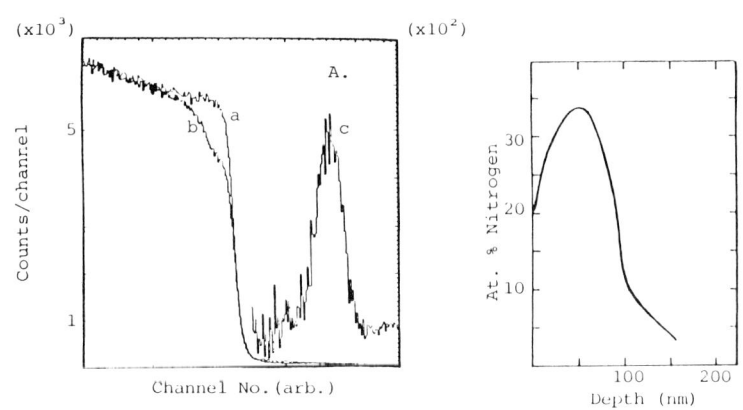

Figure 1. A. RBS spectra (a) unimplanted (b) implanted, 65 keV, 2.5×10^{17} ion cm^{-2} N^+/N_2^+ (c) difference between (i) and (ii). B. Nitrogen concentration/depth profile calculated from RBS data.

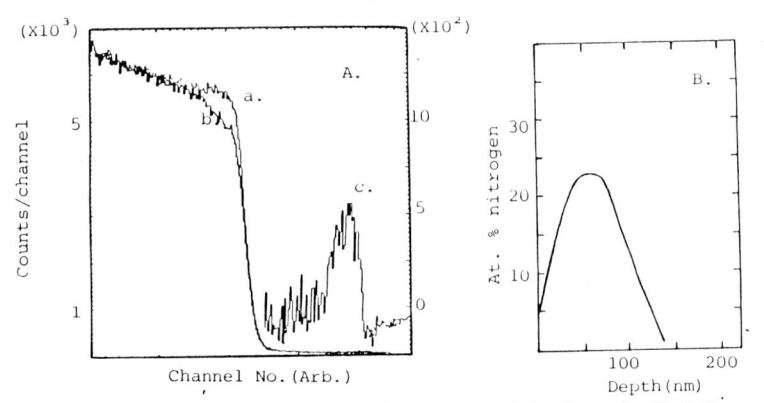

Figure 2. A. RBS Spectra (a) unimplanted (b) implanted, 65 keV, 0.9×10^{17} ions cm^{-2} (c) difference between (a) and (b). B. nitrogen depth/concentration profile calculated from RBS data.

AES ANALYSIS

Typical examples of between track and in-track AES concentration/depth profiles are shown in Figure 3. Comparison of the as-implanted AES profile (Figure 3a) with that produced by RBS analysis (Figure 1b) reveals a significant broadening of the AES profile. This could result from cascade mixing (13), heating effects during sputtering and the collection of the AES data (14) and contaminants, particularly oxygen, which can cause differential sputtering rates and the retention of nitrogen (13). For these reasons, the in-track AES profiles are discussed in terms of their relation to as-implanted standard profiles.

Oxygen is present in significant near-surface concentrations in both as implanted (Figure 3a) and in-track profiles (Figure 3b). However, in the as-implanted profile it falls to a small concentration within 20 nm, whereas it persists to >50 nm within the wear track. This behaviour is typical of sliding wear in these steels, and is likely to have an oxidation mechanism (15).

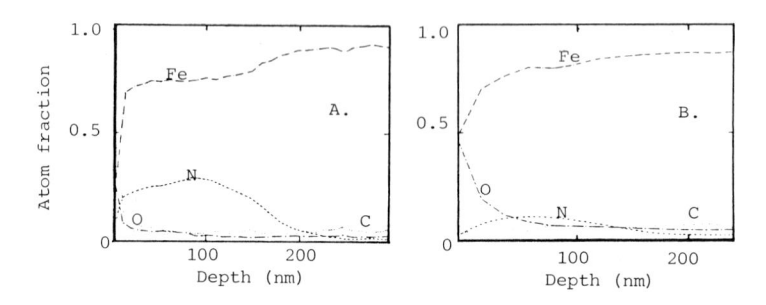

Figure 3. AES depth/concentration profiles for Fe,N,O and C.
A. As-implanted, 65 keV, 2.5×10^{17} ions cm^{-2}
B. In-track, 20 g load, 450 m distance.

Figure 4 summarises the AES nitrogen profile data for a sample examined in detail. A position for the wear track depth before sputtering has been used to allow comparison with the as-implanted profile. This depth was estimated by averaging the roughness peaks in the interferogram and taking 80% of the maximum depth. AES spectra were generated at the centre of the track with a beam size at least one order of magnitude smaller than the track width.

Details of the tracks examined and the results of AES examination for nitrogen are presented in Table I. Nitrogen was detected at the bottom of two wear tracks cut twice as deep as the original implant. Nitrogen was not detected within two tracks, each cut deeper than three times the implant depth. The remaining tracks, cut within the implant depth, contained nitrogen equal to or about 25% greater than that expected on the basis of material removed.

Implant Conditions Energy (keV) /Dose (cm^{-2})	Wear Conditions Load (g)/ Time (m)	Track Depth (nm)	% Nitrogen Observed (*)	% Nitrogen Expected (**)
65/2.5x10^{17}	20/45	100	47	46
"	20/90	150	24	18
"	20/180	400	10	0
"	100/10	200	60	40
"	100/45	650	0	0
"	200/10	200	23	15
"	300/10	500	0	0
65/10^{17}	20/45	150	40	40
45/2.5x10^{17}	20/45	140	23	22
25/2.5x10^{17}	20/45	200	16	0

TABLE I. Summary of AES nitrogen measurements.

(*) Nitrogen observed, as % of as-implanted profile.
(**) Nitrogen expected, as % of as-implanted profile.

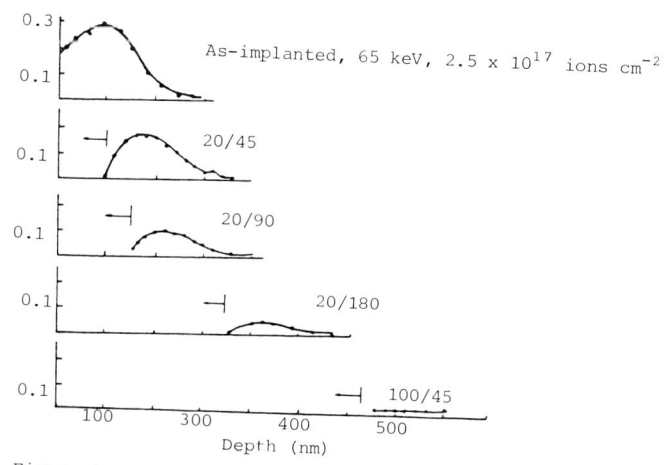

Figure 4. AES nitrogen concentration/depth profiles within wear tracks. Arrow marks track depth before profiling. Load (g)/Time (min) wear conditions indicated.

DISCUSSION

Although the carbide ball-bearing used in the present work is a more severe abrader than that used in earlier studies (1-4), wear characterisation is in general agreement. In this section, the nature of the wear process and the continuing presence of nitrogen at depths beyond that of the original implant are discussed.

The present work does not provide evidence of deep nitrogen migration during sliding wear. Comparison of the AES in-track and as-implanted profiles is conservative since RBS analysis indicates a broadening of the as-implanted profile during the AES characterisation. However, support for nitrogen migration is qualified by the possibility that surface roughness, transfer of nitrogen rich debris or push-in of the implanted layer (significant edge deformation was observed) could explain the variation in results presented in Table I. Bolster and Singer (11) reported short distance, ≈50 nm migration of nitrogen, during abrasive polishing wear of 304 SS discs, despite poorer wear characteristics of the implanted steel compared to the unimplanted standard. In the same study, a carbon steel, displaying improved wear resistance after nitrogen implantation, gave no indication of nitrogen movement. More recently, Fayeulle et al. (10), using a similar polishing technique, reported nitrogen movement over short distances (≈60 nm). Cui Fu-Zhai et al. (4), examined tracks produced by sliding wear and reported small quantities of the original implant (<13 vol.%) at depths three to four times greater than the initial implant depth. The present study is in general agreement with these reports and indicates the presence of nitrogen at the bottom of sliding wear tracks which were twice the depth of the initial implant depth. This contrasts with the original claim (8) supported by Lo Russo et al. (9) that nitrogen is present in considerable quantities (>30% of initial amount) within wear tracks cut more than 10 times deeper than the original implant depth. However, the wide variation in wear conditions and wear couples of these investigations and the current work, makes strong statements about nitrogen migration difficult.

The present work offers interesting observations on the oxidative nature of the wear process and suggests that the nitrogen implant stabilises the oxide film and is indirectly a wear-reducing agent. The oxygen levels measured with AES strongly confirm that in implanted mild steel sliding wear occurs via an oxidation stage (15). Indeed, the absence of nitrogen at the near-surface of wear tracks implies that nitrogen is not involved in the asperity/asperity wear contact process and oxidation wear of nitrogen implanted mild steel may be as follows. Initial contact between the abrader and disc is restricted to a few asperities. The stress concentration at these small areas, even for low applied loads, is sufficient for plastic deformation and eventual shearing or delamination of the weaker mild steel asperities. During this process, sufficient heat is generated for the formation of an oxide film and nitrogen diffuses away from the contact areas.

Drako and Gumanskji (16) have reported that Fe_2N is stable in the electron microscope at temperatures up to 900 K. Nitrogen migration during wear implies that such temperatures are reached and, more important, are acting at the oxide/steel interface following formation of the oxide film. It is possible, of course, that only free nitrogen is involved in the diffusion process and much lower temperatures will be sufficient. Quinn et al. (15) have recently reported a computer study of mild oxidational wear of steels and obtained oxide-forming temperatures in the range 500 to 900 K. It is possible that the nitride or nitrogen cluster base formed during ion implantation inhibits the growth of the oxide film to thicknesses at which delamination can occur.

ACKNOWLEDGEMENTS

The authors gratefully acknowledge the assistance of R.A. Clissold, M.D. Scott, M. Farrelly and J. Wells.

REFERENCES

(1) N.E.W. Hartley, Wear, 34 (1975) 427.
(2) T. Varjoranta, J. Hirvonen and A. Antilla, Thin Solid Films, 75 (1981) 241.
(3) H. Herman, Nucl. Instr. Meth. 182/183 (1981) 887.
(4) Cui Fu-Zhai, U. Heng-De and Zhang Xias-Zhong, Nucl. Inst. Meth., 209/210 (1983) 881.
(5) G. Longworth and N.E.W. Hartley, Thin Solid Films, 48 (1978) 95.
(6) R. Frattini, G. Principi, S. Lo Russo, B. Tiveron and C. Tosello, J. Mat. Sci. 17 (1982) 1683.
(7) C.A. Dos Santos, M. Behar, J.P. De Souza and I.J.R. Baumuol, Nucl. Instr. Meth., 209/210 (1983) 995.
(8) G. Dearnaley and N.E.W. Hartley, Thin Solid Films, 54 (1978) 215.
(9) S. Lo Russo, P.L. Mazzoldi, I. Scotoni, C. Tosello and S. Tosto, Appl. Phys. Lett., 34 (1979) 627.
(10) S. Fayeulle, D. Treheux, P. Guiraldenq, T. Barnavon, J. Touseet and M. Robelet, Scripta Met., 17 (1983) 459.
(11) R.N. Bolster and I.L. Singer, App. Phys. Lett., 36 (1980) 208.
(12) W.K. Chu, J.W. Mayer, M.A. Nicolet, "Backscattering Spectrometry", Academic Press, N.Y. 1978.
(13) P. Williams and J.E. Baker, Nucl. Instr. Meth., 182/183 (1981) 15.
(14) S. Hofman and A. Zalar, Thin Solid Film, 56 (1979) 337.
(15) T.F.J. Quinn, D.M. Rowson and J.L. Sullivan, Wear, 65 (1980) 1.
(16) V.M. Drako and G.A. Gumanskij, Radiation Effects, 61 (1982) 111.

WEAR IMPROVEMENT IN Ti-6Al-4V BY ION IMPLANTATION

R.G. VARDIMAN
Naval Research Laboratory
Washington, D.C. 20375

ABSTRACT

The friction and wear of Ti-6Al-4V are found to be sharply reduced by carbon implantation followed by heat treatment. Optimum wear resistance is developed at 400°C, at which the microstructure of the implanted layer shows a dense array of TiC precipitates up to 60 nm in size. The implanted layer in this case is worn through in a few thousand cycles of the ball on disc test, but by implanting at two energies to achieve a deeper carbide layer, negligible wear was found even after 20,000 cycles. No wear improvement was found for nitrogen implantation.

INTRODUCTION

Titanium alloys have always presented a particular problem with regard to wear, yet many applications involve fretting or wear situations. Most standard coating methods do not work well on titanium usually because of adhesion problems [1]. It is not always possible to lubricate the surfaces, and under unlubricated conditions rapid material transfer and severe wear can occur with even moderate loads [2].

Ion implantation creates a surface layer with no adhesion problems, since the affected layer is continuous with the underlying material. Wear improvement has been achieved in a variety of materials by implantation, usually with nitrogen [3]. Two recent studies [4,5] have reported substantial wear improvement in Ti-6Al-4V with nitrogen implantation.

The present work uses carbon implantation with post-implantation heat treatment to produce a low friction, wear resisting surface. Variations in the implanted layer microstructure are seen to correlate with wear behavior. A dual energy implant, increasing the depth of the affected layer, produces a larger than expected improvement in wear rate. A nitrogen implanted specimen tested for comparison did not show any wear improvement.

EXPERIMENTAL METHODS

The starting material was α-β processed (mill annealed) Ti-6Al-4V. This microstructure is characterized by equiaxed α grains and smaller, elongated β grains [6]. Wear discs were cut approximately 2.6 cm square and 0.45 cm thick, and the surface polished to 3 μm diamond. For transmission electron microscopy (TEM), 3 mm discs were mechanically polished to 100-150 μm thickness, then electropolished on one side before implantation. Final thinning was from the opposite side, with the implanted surface masked.

For implantation, wear specimens were clamped to an aluminum plate, while TEM specimens were attached with silver paint. Heat sinking was good in both cases, and maximum temperature during implantation is not expected to exceed 150°C. Two types of carbon implantation were used. First, 2×10^{17} at./cm^2 was implanted at 75 keV. All TEM specimens received this treatment, and one set of wear discs. One wear disc was first implanted to 3×10^{17} at./cm^2 at 175 keV, followed by the above treatment, in order to give a much deeper layer of carbon. Nitrogen was implanted to 2×10^{17} at./cm^2 at 100 keV, following the procedure of ref 4.

All material was heat treated in ultra-high vacuum. The cold vacuum was normally 1×10^{-9} torr, rising to roughly 2×10^{-8} torr at temperature. Temperature regulation was typically ± 2°C. Heat treatments were for one hour at 300, 400, or 500°C.

TEM examination was performed on a 200 kV instrument equipped with an electron energy loss spectrometer (EELS). Both EELS analysis and the observation of the highly strained appearance characteristic of an implanted layer confirmed that the material under examination was from the implantation layer.

The wear test used here was a simple ball on disc test. The ball was 12.7 mm diameter 440C stainless steel. The position of the ball was adjusted to give wear tracks with diameters between 14 mm and 23 mm. The rotation rate of the disc was adjusted to give a constant speed of 1.35 cm/sec of the ball relative to the disc. The friction force was recorded continuously, and calibrated before and after each test. The disc surface was covered with hexadecane, a poor lubricant which was used to give a more constant surface environment than laboratory air. The normal load was 100 g, with one run done at 300 g.

RESULTS

Microstructural Characterization

The microstructure of the 75 keV implants as seen in TEM shows the presence of TiC in all cases. The electron diffraction pattern shows in addition to the titanium spots diffuse rings corresponding to a FCC phase of the expected lattice parameter for TiC. The evenness of the ring intensities indicates the carbides are incoherent with the matrix. Dark field micrographs using portions of the (111) and (200) rings are shown in Fig. 1. The particles in the as implanted case have a maximum size of about 10 nm. After one hour at 300°C, there is only a small change in maximum size, but the density appears greater. One hour at 400°C produces a large change, with the maximum particle size increased to roughly 60 nm, while few particles of 10-20 nm size are seen. The overall density of carbides can be more clearly seen in the bright field micrograph of Fig. 2(a). After the 500°C heat treatment, few small particles are visible (Fig. 1(d)), while precipitates of the size 0.5-2.0 μm are found every two or three grains. A typical large carbide is shown in Fig. 2(b).

Wear Behavior

Friction and wear of the unimplanted material were found to be as expected [2]; the friction coefficient varying between approximately 0.4 and 0.5 (see Fig. 4(a)), heavy titanium transfer to the steel ball so that the wear couple was essentially titanium on titanium, and rapid removal of material from the disc wear track even at the relatively light load of 100 g.

Wear of the carbon implanted surface showed a low friction coefficient initially but after about 100 cycles changed to that of the unimplanted material, and after 1000 cycles no difference in total wear could be discerned. After the 300°C heat treatment, the low friction coefficient persisted for approximately 500 cycles, and for the 400°C specimens, for 2500 cycles. Prior to the transition in friction, the amount of wear was not discernable; that is, the wear track was indistinguishable from the surface roughness when measured on a surface profilometer. Wear of the

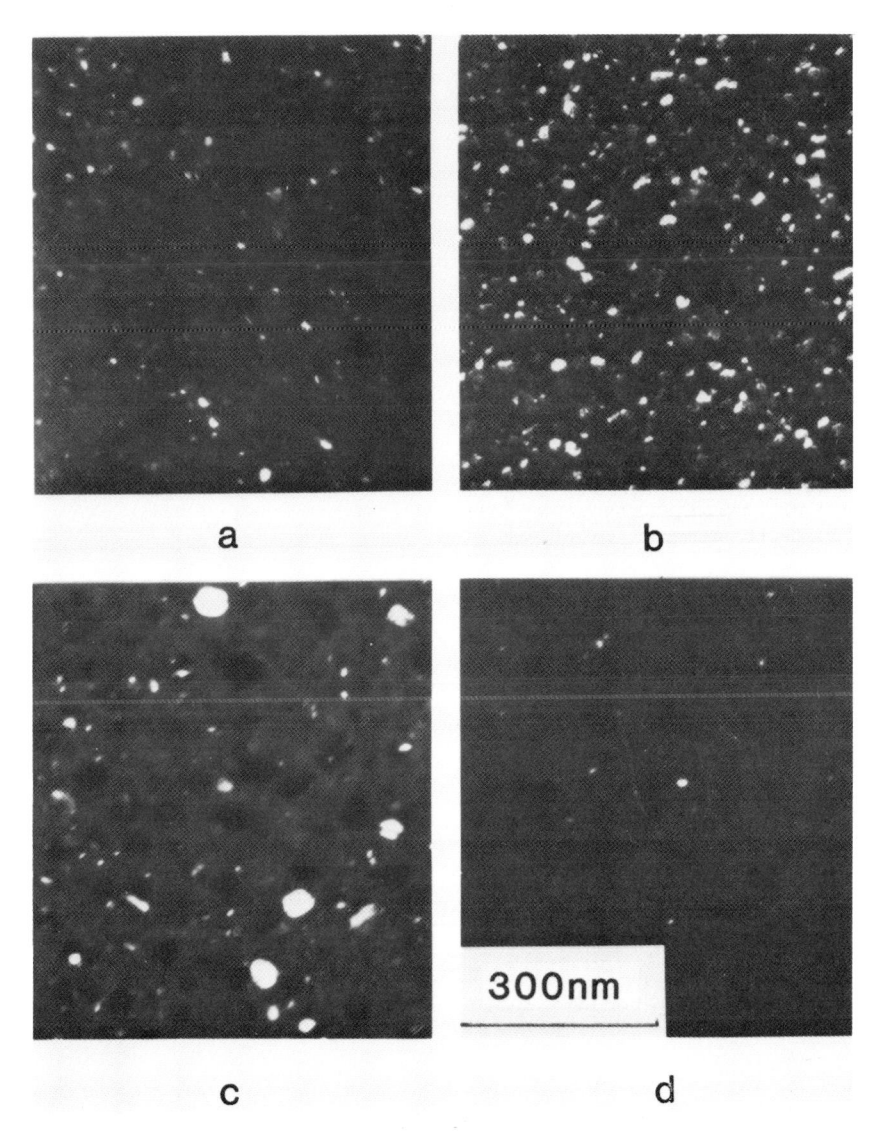

FIG. 1. Dark field TEM showing carbides in the implanted layer: (a) as implanted, (b) one hour at 300°C, (c) one hour at 400°C, and (d) one hour at 500°C.

500°C specimen was similar to the as implanted surface; that is, a transition at around 100 cycles. The number of cycles before the transition is plotted in Fig. 3 as a function of heat treatment temperature.

The friction coefficient in the low wear mode was sometimes constant around 0.17, a value similar to that found in other investigations of implantation improved wear in Ti-6Al-4V [5,6]. In some cases, however, the friction coefficient varied around the wear track from 0.17 to values as

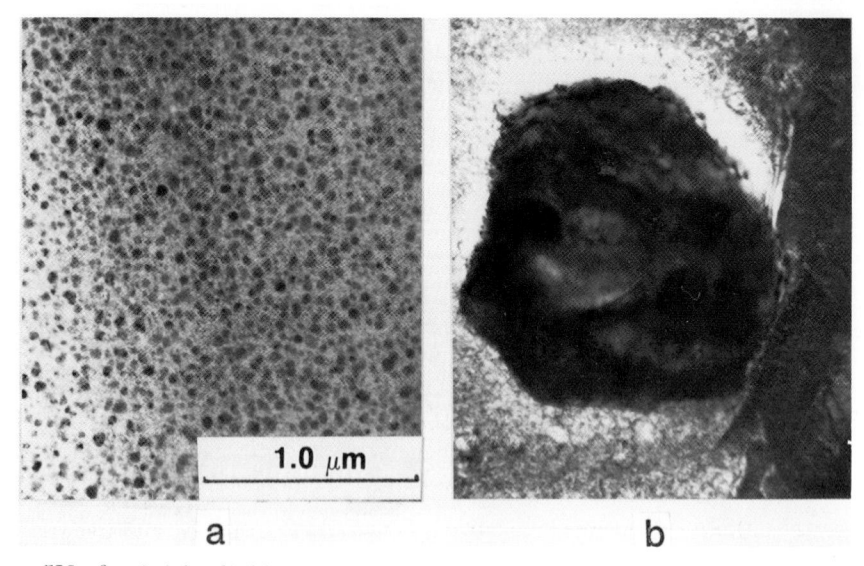

FIG. 2. Bright field TEM showing carbides after (a) one hour at 400°C and (b) one hour at 500°C.

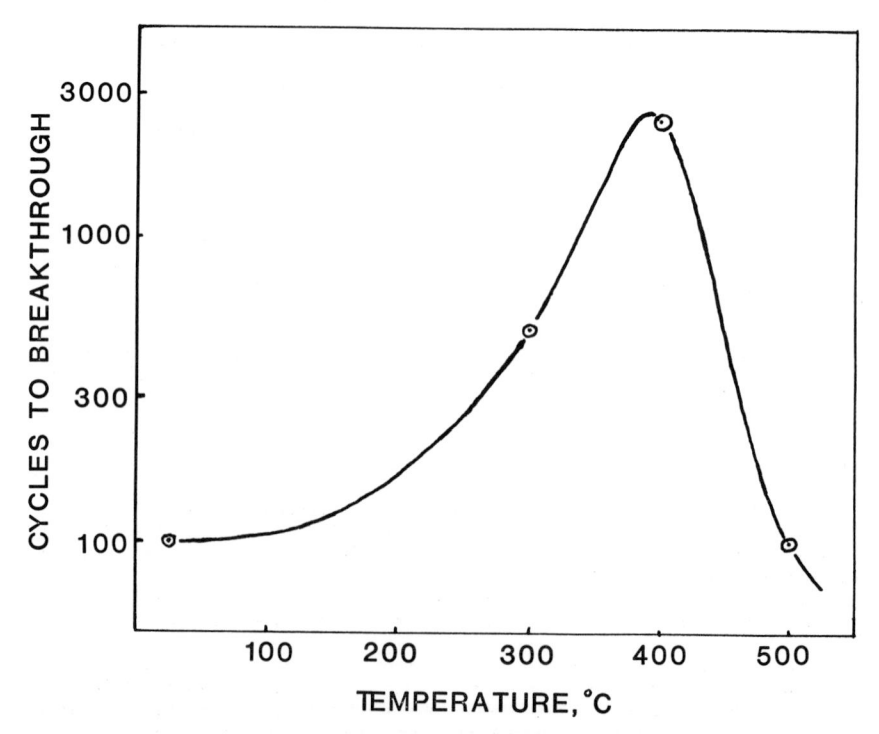

FIG. 3. Cycles to breakthrough of implanted layer vs temperature of heat treatment.

high as 0.35. These higher friction areas were found to be associated with small grooves whose presence did not appear to affect the wear behavior. Fig. 4 shows examples of the various types of friction behavior which were found in this study.

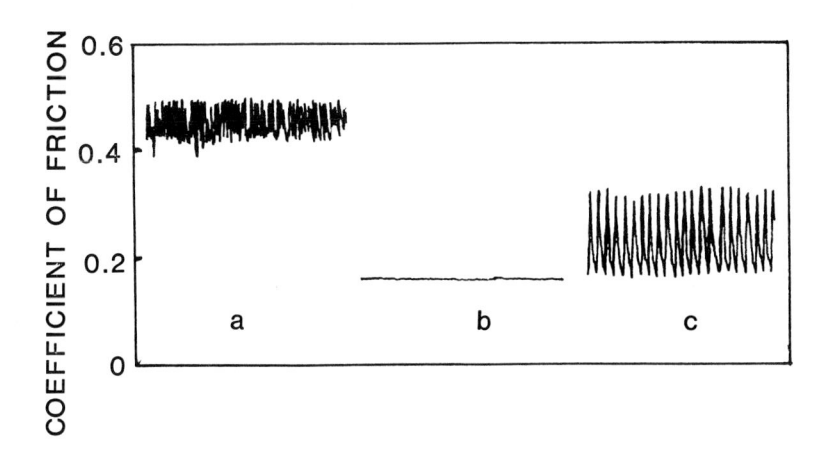

FIG. 4. Coefficient of friction for various conditions: (a) unimplanted surface; (b) and (c) different friction behavior found after carbon implantation plus one hour at 400°C.

By implanting at more than one energy, the implant species concentration can be made nearly constant for extended depths. In order to see the effect of a thicker implanted layer on the wear behavior, one disc was implanted at two energies, as described above, then heat treated one hour at 400°C to optimize wear resistance. Normal loads of 100 and 300 g were used, and even with the 300 g load, no transition occurred after 20,000 cycles, when the test was stopped. No wear was measurable, but the track showed considerable discoloration indicating that the surface had oxidized there.

The nitrogen implanted specimen showed wear characteristics identical to unimplanted material. Tests run with mineral oil was well as hexadecane showed no difference.

DISCUSSION AND CONCLUSIONS

The dual energy implant produces a high concentration carbon layer three to four times thicker than the 75 keV implant. The number of cycles to break through this layer is obviously much greater than the thickness ratio would suggest. The strengthening of a greater depth of material probably contributes strongly to this extra wear resistance. The formation of an oxide layer in the wear track also should decrease the wear rate [5].

Neither nitrogen or carbon improved wear resistance in the as implanted state under the conditions of this experiment. Hutchings and Oliver [5] found that higher concentrations of nitrogen were required to substantially improve wear resistance in Ti-6Al-4V, although Shepard and Suh [4] found wear improvement at the dose and energy used here. The finding here that precipitate growth is needed to improve wear resistance with carbon implantation agrees with the lack of improvement by nitrogen implantation. Heat treatment would not be expected to affect the wear of the nitrogen implanted surface, as temperatures up to 500°C have been found to have little effect on nitride size [6].

In summary, the friction and wear behavior of Ti-6Al-4V is found to change sharply when implanted with carbon and heat treated to increase carbide size. Optimum size and density of precipitates to resist wear is found after heat treatment at 400°C. Even more extensive improvement is achieved by using more than one implantation energy to increase the thickness of the carbide layer.

ACKNOWLEDGMENTS

I wish to thank Dr. I.L. Singer for many stimulating discussions, and the NRL ion implantation group for doing the implantations.

REFERENCES

1. N.J. Finch and J.E. Bowers, B.N.F.M.R.A. Research Report Al536 (1965).
2. S.R. Nutt and A.W. Ruff in Wear of Materials 1983, ed. K.E. Ludema (ASME, New York, 1983) pp. 426-433.
3. N.E.W. Hartley, Thin Solid Films 64, 177-190 (1979).
4. S.R. Shepard and N.P. Suh, J. Lubr. Technol. 104, 29-38 (1982).
5. R. Hutchings and W.C. Oliver, Wear, in press.
6. R.G. Vardiman and R.A. Kant, J. Appl. Phys. 53, 690-694 (1982).

ION IMPLANTED Ti-6Al-4V

W. C. OLIVER[1], R. HUTCHINGS[2], J. B. PETHICA[2], E. L. PARADIS[1], A. J. SHUSKUS[1]
[1]United Technologies Research Center, E. Hartford, CT 06108
[2]Brown Boveri Research Center, CH-5405 Baden, Switzerland

ABSTRACT

Titanium and many of its alloys show very poor wear resistance considering their hardness. This together with high thermodynamic driving forces to form very hard compounds between titanium and nitrogen or carbon made titanium based alloys obvious candidates for ion implantation. In this paper the effects of similar implanted concentration profiles of nitrogen and carbon in two titanium alloys are compared. The wear behavior of pin on disk wear tests are reported along with the ultramicrohardness of the four samples.

INTRODUCTION

Titanium alloys are one group of materials whose wear properties can be extensively improved through ion implantation [1-3]. It has also been shown that their fatigue properties can be improved through implantation [4]. The most effective ion for improving wear properties seems to be N; however, implanting with C yields better fatigue results. Preliminary wear experiments indicated a difference in the effectiveness of C versus N implantations and differences in how these two species affected Ti-6Al-4V with different prior heat treatments. The reasons for such effects are not obvious. TiN and TiC are both hard phases and have been found in implanted specimens [4,5]; hence, one would expect similar hardening effects for similar implantations. To investigate these questions two materials, an α alloy and an α-β alloy, were implanted with N or C such that similar atomic profiles were achieved. By studying the hardness and wear properties of these samples a better understanding of these surfaces should be obtained.

EXPERIMENTAL

Implantation

The two materials studied here are grade 2 commercial purity Ti and Ti-6Al-4V. They were both bar stock used in the as-received (mill annealed) condition. Samples approximately 8 mm thick were cut perpendicular to the bar axis and prepared for implantation by mechanically grinding and polishing to a one µm diamond polish. The implantation was done on a Varion-Extrion 200 CF5 implanter. This implanter is equipped with cryo pumps that maintain a vacuum of 2×10^{-7} torr during implantation, thus minimizing gettering from the ambient atmosphere onto the sample. The implantation condition and the resulting calculated value of the range, range straggling and maximum atomic concentrations are shown in Table I.

Implantation	R (μm)	ΔR (μm)	Peak concentrations atomic %
200 KeV N_2^+ 3.84×10^{21} ions/m^2	0.1769	0.0716	43.1
83 KeV C^+ 5.92×10^{21} ions/m^2	0.1727	0.0732	36.3

TABLE I. Ti implanted with N and C such that similar atomic profiles are achieved.

RESULTS AND DISCUSSION

Hardness

First let us consider the effects of N and C implanatation on the properties of commercial purity Ti. Figure 1 shows the hardness versus depth profile for similar atomic profiles of C and N. The two curves virtually overlay one another. This is the expected result as outlined in the introduction.

The surprising data is shown in Fig. 2. The same implants were performed on the same specimens in Fig. 2 except the substrate was mill annealed Ti-6Al-4V. The C is not nearly as effective at hardening the surface of the Ti-6Al-4V. The hardness increase due to N is ~200% at 50 nm. This has been found in other N implanted specimens with different implant conditions [3]. For C the increase is barely 50%.

It seems unlikely that the Al and V in Ti-6Al-4V could cause a significant difference in the driving force to form TiC. Hence the most probable cause for this difference is the β phase present in the Ti-6Al-4V but absent in the pure Ti. Clearly more experimental work is required to understand this phenomenon.

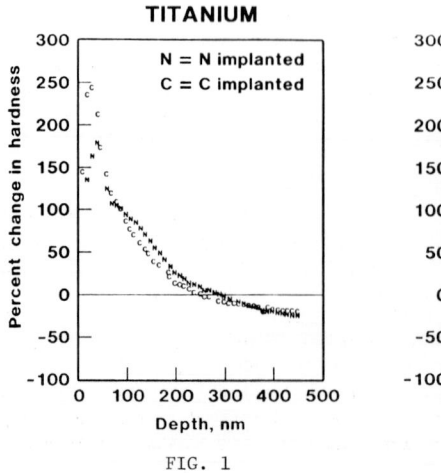

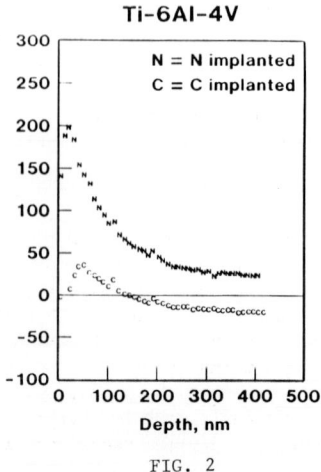

FIG. 1

FIG. 2

WEAR

The data in Figs. 3, 4 and 5 was taken with pin on disk wear tests from the same specimens from which the hardness data was taken. Figure 3 shows the wear of N implanted Ti-6Al-4V compared to the unimplanted case. The single data point plotted for 200 KeV, N_2^+ implanted specimen was taken on a different wear tester, under the same experimental conditions. The other data has been presented previously [3]. Data at lower numbers of turns for the higher energy N implanted specimen was not easily measured with profilometery. It seems that this specimen wears somewhat slower than the other N implanted specimen. This could be due to the deeper penetration and slightly higher dose level. This data simply confirms the low wear rate of N implanted Ti-6Al-4V.

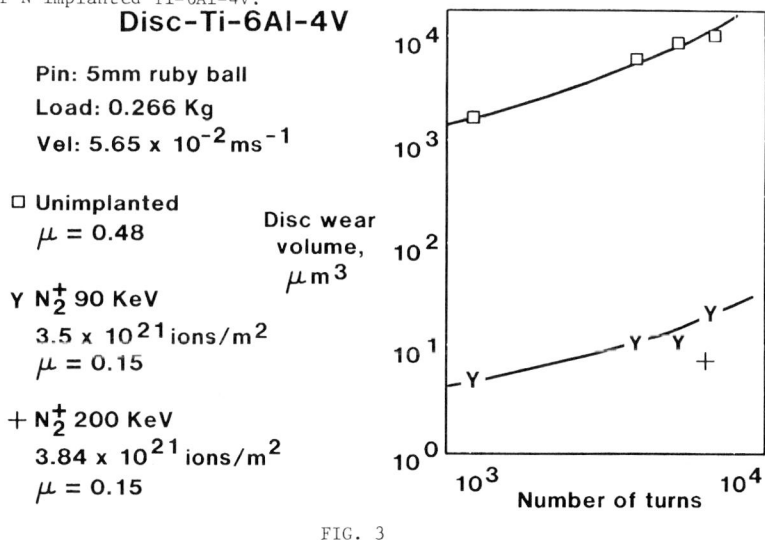

Disc-Ti-6Al-4V

Pin: 5mm ruby ball
Load: 0.266 Kg
Vel: 5.65 x 10^{-2} ms^{-1}

□ Unimplanted
$\mu = 0.48$

Y N_2^+ 90 KeV
3.5 x 10^{21} ions/m^2
$\mu = 0.15$

+ N_2^+ 200 KeV
3.84 x 10^{21} ions/m^2
$\mu = 0.15$

Disc wear volume, μm^3

Number of turns

FIG. 3

Under the test conditions used in Fig. 3 all of the other specimens experienced breakthrough in less than 50 turns. The breakthrough effect is demonstrated in Fig. 4. The friction coefficient starts at the low value (0.1-0.17) characteristic of all of the implanted materials, but after a few turns breakthrough occurs, after which the friction coefficient is higher (0.4-0.5) and there is more stick-slip behavior. This is consistent with previous characterization of failure of the implanted layer [3].

The similarity of the hardness curves for N implanted material presented in Figs. 1 and 2, led to the hypothesis that the breakthrough of the N implanted Ti must be due to the lower substrate strength. To check this, wear tests were performed with various normal loads. It was found that for all the specimens but the C implanted Ti-6Al-4V, there existed a clearly defined breakthrough stress above which failure occurred in 50 turns or less. Below the breakthrough load the low friction and wear rates lasted more than 500 turns. The C implanted Ti-6Al-4V also had a breakthrough stress; however, even at lower load the sample failed between 400 and 500 turns. This is expected considering its low surface hardness. The breakthrough data is presented in Fig. 5. The two Ti specimens have similar hardness curves and similar breakthrough loads. The same hardness curve is once again produced for N implanted Ti-6Al-4V; however, the breakthrough load is 17-21 times as large. This demonstrates that substrate properties are also important.

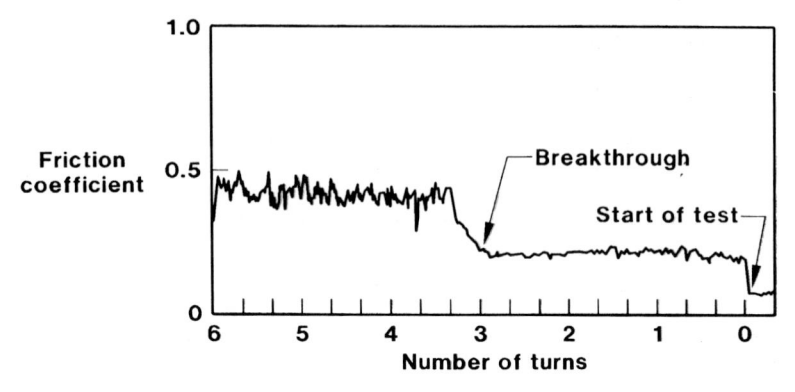

FIG. 4. Breakthrough of implanted layer on Ti, pin-5 mm ruby ball, load-0.266 kg, vel.-5.65 cm s^{-1}

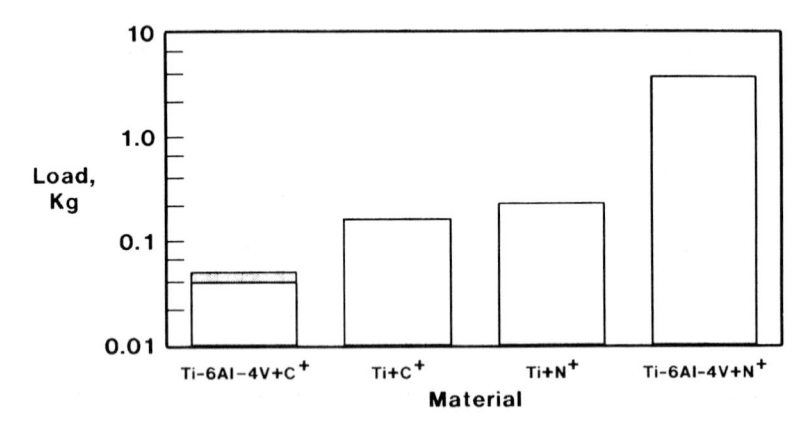

FIG. 5. Breakthrough load for four substrate-ion combinations; pin-5 mm ruby ball; vel.-5.66 cm s^{-1}.

Let us consider at what load breakthrough should occur. It will certainly occur when the implanted layer has worn through. It will also occur when the surface layer is fractured. Assuming the surface layer is fairly brittle, this would occur when the substrate just below the implanted layer yields plastically in a significant way. To determine at what load this should occur one must consider Hertzian [6] loading. Tabor [7] pointed out that subsurface plastic flow will first occur at a depth of,

$$\delta = 0.55 \left\{ \frac{Lgr}{2} \left(\frac{1}{E_1} + \frac{1}{E_2} \right) \right\}^{1/3} \quad (1)$$

where δ = depth below surface
  L = load
  r = radius of the sphere
  E_1 and E_2 = elastic moduli of the sphere and semi-infinite material, respectively
  g = the acceleration due to gravity.

Deformation will occur when the load reaches a value of,

$$W_L = 17.4 \ Y^3 r^2 \left(\frac{1}{E_1} + \frac{1}{E_2} \right)^2 \quad (2)$$

where Y is the yield strength of the substrate. This is the point at which yielding starts. A fully plastic zone will occur at loads 100-200 times greater. Clearly the breakthrough load should lie between these extremes. Table II shows this to be the case for all of the samples on which the implantations result in significant hardening. Equation 2 shows that W_L is a function of Y^3. The yield strength of Ti-6Al-4V is 3 times that of Ti. Thus, the ratio of W_L for these two materials will be 27. This is one possible explanation for the factor of 17-21 difference in the breakthrough loads.

	E (GPa)	Y (MPa)	W_c (Kg)	W_p^{min} (Kg)
Grade 2 Ti	110	275	0.033	3.3
Ti-6Al-4V (annealed)	110	827.2	0.967	96.7
Al_2O_3	384			

TABLE II.

CONCLUSIONS

1. N and C are equally effective in improving the surface hardness and wear properties of grade 2 Ti.

2. N is much more effective than C at hardening and improving the wear properties of Ti-6Al-4V.

3. Ti-6Al-4V implanted with N^+ at 200 keV to a dose of 3.84×10^{21} ions/m^2 has a very wear resistant durable surface capable of sustaining high point loads (3.5 kg on a 5 mm ball).

4. Hard layers on soft substrates are only effective wear barriers below certain point loads. Thus it may be beneficial to combine ion implantation with deeper hardening treatments like shot peening or diffusion coatings.

REFERENCES

1. J. B. Pethica, R. Hutchings, and W. O. Oliver, Nuc. Inst. and Methods, 209/210 (1983) 995-1000.
2. R. Hutchings, W. C. Oliver, and J. Pethica, Proc. of NATO/ASI on Surface Eng., Les Arcs, France, 1983.
3. R. Hutchings and W. C. Oliver, Wear, Vol. 92 (1983) 143-153.
4. R. G. Vardiman and R. A. Kant, J. of App. Phy., Vol. 53 (1982) 690.
5. R. Hutchings, Mat. Letters, Vol. 1, no. 5, 6 (1983) 137.
6. H. Hertz, J. Reine Angew. Math. 92 (1881) 156.
7. Tabor, "The Hardness of Metals", Oxford at the Clarendon Press (1951) 45.

WEAR BEHAVIOR AND STRUCTURAL CHARACTERIZATION OF A NITROGEN IMPLANTED
Ti6Al4V ALLOY AT DIFFERENT TEMPERATURES

R. MARTINELLA*, G. CHEVALLARD*, AND C. TOSELLO**

*C.I.S.E., P.O.B. 12081, Milan Italy; **I.R.S.T., Trento, Italy

ABSTRACT

Mechanically polished Ti6Al4V samples were implanted with 100 keV nitrogen ions to a fluence of $5 \cdot 10^{17}$ ions/cm^2 at two different bulk temperatures: 370 °C and 470 °C. Wear tests were carried out with a reciprocating sliding tribotester. Structural modifications and wear morphologies were studied by TEM and SEM. 370 °C implanted sample showed the same wear behavior as unimplanted ones, while 470 °C implanted sample showed better wear resistance because of a TiN hardened layer. Correlations between microstructural modifications, wear behavior and mechanisms are reported: results agree with the delamination theory. Comparison with ion- and gas-nitrided samples are presented.

1. INTRODUCTION

Titanium alloys, which are widely used because of their high corrosion resistance, mechanical properties and low density, show a very poor wear resistance. Among different anti-wear surface treatments (carburizing, nitriding, hard coating, etc.) ion implantation is interesting because no mechanical, dimensional and strong surface finishing modification is produced [1]. Works regarding ion implantation of Ti and its alloys [2-5] give preferential attention to increased hardness and chemical and/or structural modifications; only few wear tests are reported, but without correlation between microstructural modifications, wear behavior and mechanisms, where there is still a lack of understanding. Shepard et al.[5] reported a wear mechanism, based on the delamination theory, capable of a comprehensive interpretation of wear and structural results, but without any detailed study of microstructural modifications.

Preliminary results on wear behavior and structural modifications of Ti6Al4V samples, nitrogen implanted at two temperatures are reported and are interpreted in terms of the delamination theory. Some results obtained on gas- and ion-nitrided samples are also reported for comparison.

2. EXPERIMENTAL

2.1. Materials

Ti6Al4V bars (HV$_{50}$ ∿ 420 kg/mm^2) and 10% Co H.S. tool steel (HV$_{50}$ ∿ 850 kg/mm^2) sliders were used. Surface for wear tests were optically lapped (R$_a$ on Ti6Al4V ∿ 0.01 µm).

2.2. Surface treatments

At IRST in Trento 100 keV nitrogen ions were implanted on cleaned Ti6Al4V bars at doses of $5 \cdot 10^{17}$ ions/cm^2 with ∿ 80 µA/cm^2 current density in

$\sim 10^{-4}$ Pa vacuum. Different thermal insulation allowed us to reach two different bulk regime temperatures: ~ 370 °C and ~ 470 °C, measured at 1.5 mm under the implanted surface with a Cr-Al thermocouple.

For comparison some Ti6Al4V samples were ion- and N_2 gas-nitrided at 850 °C for 10 and 20 hours, respectively.

2.3. Wear tests

A reciprocating slider on bar tribotester equipped with a proximity transducer was used (fig. 1). Tests were carried out in organic solvent or in air at velocities of 135÷500 cycles/min and loads of 1.5÷10 Kg. Each sample was ultrasonically cleaned, dried and weighed (10^{-5} g accuracy) before and after the test.

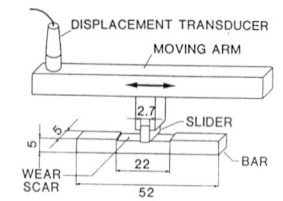

FIG. 1. Reciprocating slider on bar tribotester equipped with a proximity sensor.

2.4. Microstructural observations

TEM samples were prepared by a backthinning technique, masking the implanted surface with lacomit varnish, using a double-jet electrothinner with a solution of 10% perchloric acid in an ethyl-butyl alcohol mixture at temperatures lower than 10 °C.

SEM observations were carried out on wear surfaces and on sections parallel to the sliding direction. No edge damage was produced during polishing. Talysurf measurements were carried out on unimplanted and treated samples with a 2.5 μm pick-up.

3. RESULTS

3.1. Wear tests

FIG. 2. shows untreated and implanted samples weight loss trend obtained by scaling the proximity transducer output with the weight loss at the end of the test (no weight loss was noticed on the slider). The 370 °C implanted sample, just as the unimplanted one, showed linear weight loss trend

FIG. 2. Wear test weight loss trends:
1) unimplanted and 370 °C implanted samples;
2) 470 °C implanted sample;
3) as predicted by the delamination theory:
  a) running-in, b) "no-wear region",
  c) delamination.

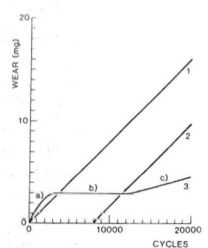

from the beginning of the test. The 470 °C implanted sample showed a delay time of $\sim$8,000 cycles, with no detectable weight loss, before the start of severe wear with the same rate as that found on the untreated ones; at the same time the transducer output changes from a smooth shape to a rough one and a metallic luster groove appears on the brown surface.

A remarkable wear resistance improvement was observed on ion- and gas-nitrided samples, even in more severe test conditions (Table I). No sudden transition to a severe wear regime, similar to that found on the 470 °C implanted sample, was observed.

TABLE I. Wear results.

Sample	Load [kg]	Velocity [cycles/min]	Environment[a]	Test length [kcycles]	Weight loss [10^{-4} g]	Wear rate [10^{-7} g/cycle]
untreated	1.5	135	(1)	20	165.7	8.25
	10	500	(2)	10	621.3	62.1
370 °C implanted	1.5	135	(1)	20	165.0	8.25
470 °C implanted	1.5	135	(1)	20	99.5	--
N_2 gas-nitrided	1.5	135	(1)	56.3	2.4	0.043
	1.5	500	(1)	56.3	unappreciable	
	5	500	(2)	60	unappreciable	
ion-nitrided	1.5	135	(1)	20	unappreciable	
	5	135	(2)	20	1.9	0.095
	10	500	(2)	60	unappreciable	

a) (1) = organic solvent; (2) = air

3.2. Microstructural observations

The microstructure of unimplanted samples is characterized by polygonal α grains (d $\sim$4 µm), with relatively few dislocations, dispersed in a $\alpha + \beta$ Widmanstätten structure, with elongated laths (FIG.3); no evidence of β phase was found on electron diffraction patterns (EDP).

Nitrogen implanted surfaces appear bronzed (while ion- and gas-nitrided ones are dark grey). On implanted samples a thin amorphous oxide layer was found. In the 470 °C implanted sample the most peculiar microstructural modification is the presence of a consistent amount of small crystalline particles (d $\sim$5 nm) with FCC structure, corresponding to TiN (a $\sim$0.425 nm). TiN precipitation, even in the same sample, appears either as total transformation of the primitive structure (as demonstrated by the relative intensities of spots in EDP) or as homogeneous precipitation in α grains, imaged both by the matrix strain field (butterfly-shaped contrast) and by dark field (FIGS.4-5). Probably these different structures correspond to different depths. These observations confirm that TiN particles precipitate at least partially coherently with Tiα matrix.

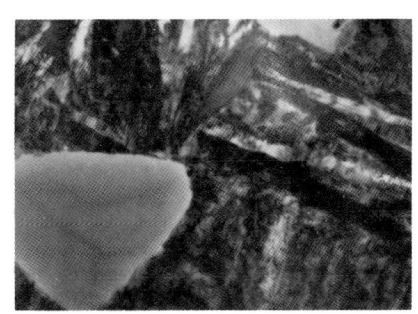

FIG. 3. TEM micrograph of unimplanted Ti6Al4V (10,000 X).

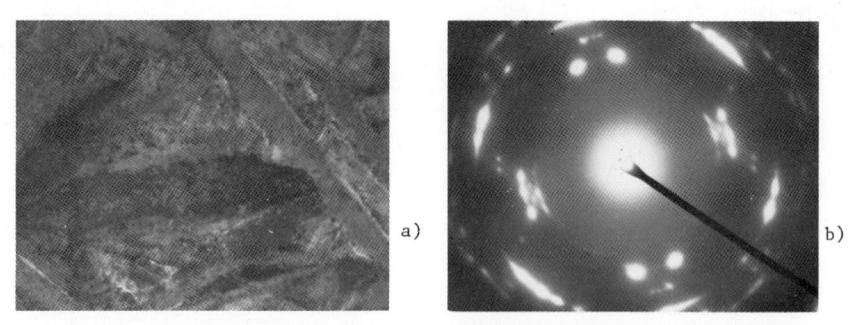

FIG. 4. TEM micrographs of 470 °C implanted sample: a) defected structure (15,000 X); b) EDP (2λL ∿6.1 mm nm).

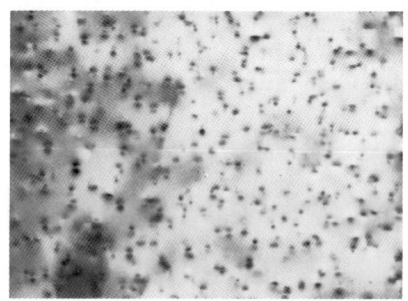

FIG. 5. TEM micrograph of 470 °C implanted sample: TiN particles imaged by matrix strain field (100,000 X).

In the 370 °C implanted sample only increased dislocation density and Tiα rearrangement, with small Tiα grains (50÷100 nm), were found; no evidence of TiN appears in images or in EDP (FIGS.6-7).

The observed TiN precipitation could agree with the Ti-N phase diagram taking into account that the surface temperature is higher than the recorded temperature.

Gas- and ion-nitrided samples show a R_a between 5 and 10 times larger than untreated and nitrogen implanted ones (R_a ∿0.01 μm). On untreated samples the subsurface morphology shows strong plastic deformation, voids and void coalescence that produce cracks, sometimes emerging up to the surface (FIG.8). Grooves were observed on bar and slider surfaces (FIG.9). Moreover, prominent carbides are partially lost on the slider and no Ti particles adhered to the slider were observed; conversely, strong adhesion effects are present in the untreated samples (FIG.10).

FIG. 6. TEM micrograph of 370 °C implanted sample (15,000 X).

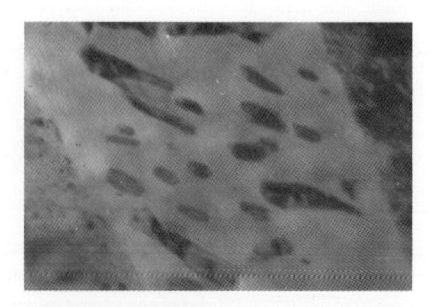

FIG. 7. TEM micrograph of 370 °C implanted sample: Tiα grain rearrangement (100,000 X).

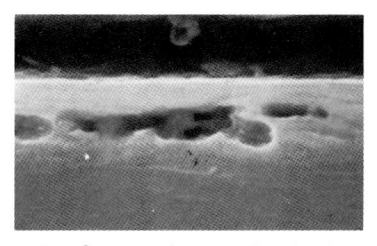

FIG. 8. SEM micrograph of subsurface morphology of untreated wear tested samples (7,000 X).

FIG. 10. SEM micrograph of a slider coupled to an untreated bar (400 X).

a)   b)

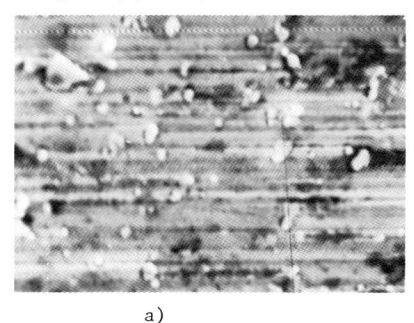

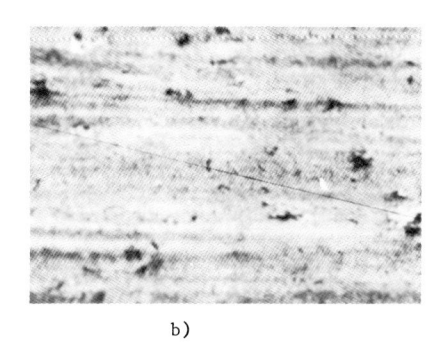

FIG. 9. SEM micrographs of wear grooves on slider (a) and 470 °C implanted bar (b) (1,800 X).

4. DISCUSSION

The TiN hardened layer is responsible for the improved wear resistance in the 470 °C implanted sample: the hardening is caused by increased dislocation density and smaller grain size. As suggested in a previous work [5], nitrogen lattice distortion does not increase the 370 °C implanted sample wear resistance.

Present results are in good agreement with the delamination theory. This theory predicts three stages in weight loss trend (FIG.2): the first stage corresponds to the surface asperity smoothing and fracture. The active mechanism in the second stage is the subsurface deformation, crack nucleation and propagation. The lasting of this "no-wear region" depends on material and test conditions; the third stage (severe wear) starts when cracks emerge thus producing delamination wear sheets. In the present results the first stage is absent because of the high quality finishing of surfaces. The second stage is unappreciable in untreated and 370 °C implanted samples, while it is definitely present in the 470 °C implanted sample. Most probably the TiN layer, while changing the surface chemistry, reduces adhesion (FIG.9) and, while increasing the surface hardness, reduces plowing; in this way also friction turns out to be reduced (the contribution of asperity deformation being negligible) [5]. Subsurface deformation is also reduced and severe wear delayed.

Plowing due to prominent asperities (WC in slider steel: FIG.9) is probably responsible for residual friction (which induces subsurface deformation deeper than TiN layer) and delamination mechanism. The sudden transition to the third stage is probably due to the following catastrophic break away of the hardened layer; after that, the wear mechanism and wear rate are the same as those observed on the untreated material (FIG.2). The better wear resistance of ion- and gas-nitrided samples is probably due to the thicker layer (several tens of μm) of hardened material (up to $HV_{15} \sim 800$ kg/mm^2) which re-

duces subsurface deformation even in more severe test conditions.
Microstructural observations could agree with published results taking into account higher temperatures reached in thin film bombardment (which produces randomly oriented polycrystalline TiN [6-7]) and surface finishing [3] . In fact, TEM observations show that ion-implantation on samples with worse mechanical surface finishing (very high dislocation density) destroys the original Widmanstätten structure and produces a structure with sub-micron cells.

Anyway the ideal surface treatment depends on the actual application and active mechanism: ion implantation could be more useful when mild wear is expected (lubrication, low loads, etc.), when conservation of high surface finishing is required and when bulk mechanical properties could be negatively affected by higher temperature treatments. An example of application is artificial hip joint prosthesis. Conversely ion- and gas-nitriding could be more useful for higher loads and dry conditions.

REFERENCES

1. N.E.W. Hartley, Rad. Effects 44, 19 (1979).
2. J.B.Pethica et al., Nucl.Instr.and Meth. 209/210, 995 (1983).
3. R.Hutchings, Mat. Lett. 1, 5/6 137 (1983).
4. G.Dearnaley et al., Nucl.Instr.and Meth. 189, 117 (1981).
5. S.R. Shepard et al.,J.Lub.Tech. 104, 29 (1982).
6. I.M. Belii et al., Phys.Stat.Sol. 45a, 343 (1978).
7. R.G. Vardiman, NRL Memorandum Report No. 4341(1980).

ACKNOWLEDGEMENTS

Thanks are due to ENEL/DSR for financial support and permission for publishing; to C.A.Be, A.Benvenuti, L.Borghi, V.Cubuzio for microstructure observations; to E.Caruso, S.Giovanardi, A.Gray for useful comments and to R. Anzani,G. D'Antonio,S. Girardi, E. Voltolini for technical assistance.

PART VII

APPLICATIONS: CHEMICAL

ION BEAMS AND CATALYSIS

J.A. CAIRNS
Chemistry Division, AERE Harwell, Didcot, Oxfordshire, OX11 ORA, England

ABSTRACT

Ion beam techniques have found increasing application in recent years to a wide variety of disciplines. It is the purpose of this paper to explore their potential in the field of catalysis, concentrating exclusively on solid heterogeneous catalysts. An initial description of the internal structure and external physical form of some typical catalysts is followed by an assessment of the use of ion beams for the preparation and interrogation of both "real" and model catalysts. These techniques are then compared to some of the modern tools used in current catalysis research. It emerges that ion beams can indeed be used to advantage in certain applications, such as detecting light elements selectively; measuring interaction effects between metals and supports in model systems; highlighting surface relaxation effects in metal single crystals; and even, in special circumstances, in synthesising catalysts.

INTRODUCTION

The purpose of this paper is to explore the potential for interaction between ion beam techniques and the large and apparently distinct discipline of solid heterogeneous catalysis. In order to achieve this objective it is considered necessary to describe, admittedly in a simplified manner, the basic structural features of such catalysts, and to outline some of the techniques used in their preparation and characterisation. We will then be in a position to assess more objectively whether ion beam technology can be applied usefully, either to prepare new types of catalysts, or to interrogate existing ones to yield new information.

SOME BASIC FEATURES OF HETEROGENEOUS CATALYSTS [1]

A catalyst may be designed to perform several functions, but the most basic one is usually to lower the activation energy for a particular reaction, so that it can proceed at a faster rate or at a lower temperature. The catalyst achieves this by adsorbing one or more of the reactants on to its surface, thereby converting them into a more active state for subsequent reaction. Hence typically the first objective in fabricating a catalyst is to prepare it in highly dispersed form - that is to increase its surface area. This is achieved by dispersing the active phase of the catalyst (which usually is a metal) over a high surface area carrier material, known as the support, thereby producing a so-called supported metal catalyst. The support is usually an oxide, such as silica or alumina, although for some reactions a high surface area carbon can be used. The method of preparation may involve impregnating the support with a solution of the appropriate metal salt, followed by drying, calcination, and perhaps reduction at an appropriate temperature, the final result being very finely divided metallic particles distributed over the support, and indeed perhaps within its pores. Fig. 1 shows a transmission electron micrograph of a Pt/Al_2O_3 catalyst [2], where the highly dispersed metal is seen to be distributed throughout the support. Incidentally, the surface area of the metal and of the support can be

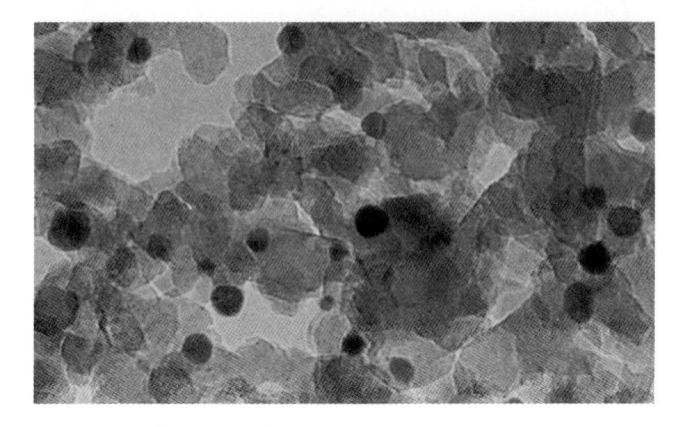

50 nm

FIG. 1. Transmission electron micrograph of Pt/Al$_2$O$_3$ catalyst [2]

measured by gas adsorption techniques. The precise nature of the porosity of the support - that is, the size of the cavities within its structure - can play a vital role in determining one of the most important parameters of the catalyst, namely, its specificity: the extent to which it will yield certain reaction products at the expense of other possible but less desirable ones. This arises because of the support's ability to accommodate preferentially reaction or product molecules of specific dimensions. The ultimate specificity based on pore size is exhibited by zeolites - also known for this reason as molecular sieves. In addition, the support can play an important role in shielding the catalyst against deactivation by poisons, such as trace amounts of Pb in the reactant gas stream, as in vehicle exhaust catalysis. Apart from poisoning, there are other reasons why a catalyst may lose its activity. Two of them are shown in Fig. 2: either the dispersed metallic particles may migrate over the support to form larger globules of lower surface area and hence lower specific activity; or the support may collapse, shielding the metal from the reactants. Both of these effects can result from having subjected the catalyst to relatively high temperatures.

Having prepared the catalyst, which may at this stage be in the form of a powder, it may be necessary to convert it into a physical form which will ensure its efficient interaction with the gaseous reactants. This can be achieved satisfactorily by converting it into pellets. However, for more demanding operations, such as vehicle exhaust emission control, where the

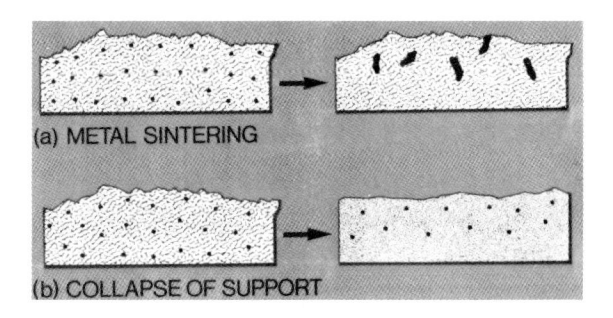

FIG. 2. Schematic illustration of two possible effects
leading to catalyst deactivation

catalyst must interact with exhaust gases with maximum efficiency and minimum physical degradation over long periods of time in an unsupervised situation, a more sophisticated solution is required. This is illustrated in Fig. 3, where the catalyst is mounted on to a ceramic monolith or converted into spheres which are packed into a suitable container.

THE USE OF ION BEAMS TO PREPARE CATALYSTS

(a) Ion Implantation

It would appear, in view of the description given above of the structure of typical heterogeneous catalysts that there would be little point in attempting to prepare such catalysts by implanting metal into porous supports. The apparent limitations are fairly obvious: much of the metal could become buried in the support, and therefore unavailable for surface interactions; and only the outermost regions of the support would be amenable to treatment, due to limitations in the penetrating power of the ion beam [3]. These considerations probably explain why workers in this area have concentrated on producing catalysts which are not readily amenable to conventional preparative methods, or have sought to elucidate some specific aspects of catalyst behaviour.

Thus it has been shown [4] that platinum implanted into solid tungsten can exhibit electrocatalytic activity for cathodic hydrogen evolution approaching that of platinum itself. However this performance was not exhibited by freshly implanted samples but only after various "conditioning" treatments, which, it was proposed, were consistent with the gradual migration of platinum to the surface. Similarly, Pt implanted into single crystal α-Al_2O_3 has been used to study the hydrogenation of ethylene [5]. It was found that the initial activity exhibited by the samples could be lost by

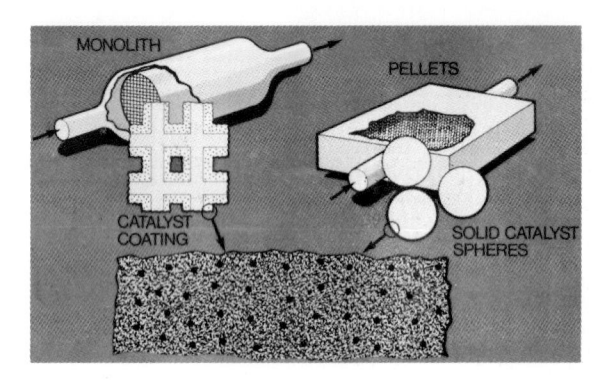

FIG. 3. Two methods of presenting a catalyst to
a flowing gas stream

annealing in argon up to 900K, a behaviour ascribed to surface diffusion of
ions from the support causing pore closure, thereby isolating the implanted
platinum from the reactants. On the other hand, annealing at 900-1300K
caused a recovery of catalytic activity which was correlated with an observed
diffusion of platinum to the surface of the alumina.

Platinum implanted into carbon has been used in electrocatalysis [6]
but the activity for enhancement of the hydrogen redox reaction was smaller
than that exhibited by smooth metallic platinum. This was due to the
unavailability of Pt on the carbon surface: unlike Al_2O_3, heating the
substrate to 1200°C caused no outdiffusion of Pt.

A particularly interesting effect of ion bombardment, by Kr^+ on to a
RuO_2 substrate covered with a film of Pt, is shown in Fig. 4 [7]. This
reveals that the activity of the Pt film for the oxidation of formic acid is
unaffected by the bombardment, even up to high doses (having corrected for
physical losses of Pt during the irradiation). On the other hand, the
efficiency for the oxidation of methanol decreases dramatically. This is
ascribed to a gradual reduction in the size of the Pt clusters during
bombardment: below a critical particle size, formic acid adsorption is still
possible, but not methanol adsorption, which obviously requires larger Pt
clusters. Hence here we have an example of ion bombardment conferring the
highly desirable attribute of selectivity on a catalyst.

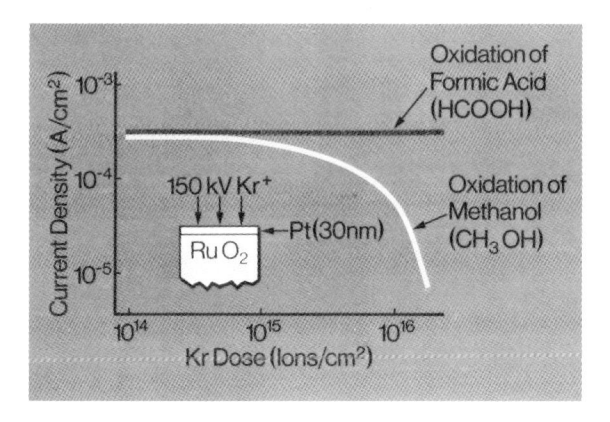

FIG. 4 [7] The comparative effects of Kr^+ irradiation on a Pt/RuO_2 catalyst for the oxidations of formic acid and methanol

(b) Sputtering

Deposition of finely divided (i.e. potentially catalytic) material by sputtering has some advantages over ion implantation: the metal does not penetrate into the support, and therefore is immediately accessible to the reactant gases; and the technique is generally less expensive and easier to scale up. On the other hand, since it is a 'line of sight' deposition procedure, it cannot be used to impregnate a porous support uniformly. For this reason most of the catalysts described in the literature tend to be sputter-deposited on to solid substrates [8, 9]. However, sputtering can be used to prepare very active catalysts of a more conventional type. For example, it has been shown [10], that Pt sputtered on to a porous fibrous alumina of surface area 80 m^2/g produces a catalyst which can out-perform, for the hydrogenation of butadiene, higher loaded catalysts prepared by conventional impregnation. It should be recognised, however, that care must be exercised in comparing catalysts in this way: for example, this particular reaction may have benefitted from an optimum metal cluster size produced by the sputtering. In another approach, platinum was sputtered on to alumina spheres (1-5 μm) in a vibrating cup, in an attempt to achieve a more uniform coating. Although the external surface area of the spheres available to the platinum was rather modest (few m^2/g), the catalytic performance was extremely good [11]. This illustrates a fascinating aspect of sputtering:

unlike conventional catalyst preparation, it does not rely on the support to produce the dispersion. Hence it can be used to separate out metal and support effects in fundamental catalyst studies.

THE USE OF ION BEAMS TO INTERROGATE CATALYSTS

This subject has been considered in more detail elsewhere [12, 13]. Hence for the present purpose only the most important techniques will be highlighted. These include Proton-Induced X-Ray Emission, a multi-element detection technique for elements of atomic numbers from 5-11 (depending on the detector used) upwards. On the other hand, Prompt Nuclear Activation Analysis is specific for individual light elements, including even isotopes. Multiple light elements, including hydrogen, can be detected as a function of depth by the technique of Elastic Recoil Detection Analysis [14]. This is less destructive than secondary ion mass spectrometry, and much more amenable to quantification.

Special mention should be made of Nuclear (Rutherford) Backscattering Spectrometry, a technique well suited to tne quantitative, non-destructive depth profiling of targets consisting of films of relatively heavy elements on low atomic number matrices: hence its widespread use in the electronics industry [15]. In fact it has been used recently [16] in a study of an important phenomenon in current catalysis research known as Strong Metal-Support Interactions.

Turning now to single crystal targets, it emerges that Nuclear Backscattering Spectrometry can be combined with another interesting property of ion beams: their ability to penetrate single crystals along the channels defined by close-packed rows or planes of atoms. This phenomenon arises as a result of the ion beam experiencing a series of Coulombic repulsion forces with the atoms of the rows, and greatly reduces the probability of their participating in close nuclear interactions with the atoms of the target. The channelling phenomenon [17] can be demonstrated by mounting the crystal target on a goniometer so that its alignment with the incoming beam can be adjusted in a controlled manner. There will then be a dramatic fall in the backscattered yield whenever the beam is incident along a well-defined crystal channel direction. Furthermore, when this reduced backscattered yield from the aligned incident beam is examined closely, it is seen to exhibit a peak at its high energy end [18]. The reason for the appearance of this peak may be appreciated by reference to Fig. 5 [19] where it is seen to arise as a result of direct collisions between the incident beam and surface atoms. Fig. 5 shows also that those components of the incoming ion beam which do not make sufficiently close encounters with the surface atoms to be backscattered are deflected away from the atoms of the second and subsequent layers to leave an unreacted region, known as a shadow cone. Hence atoms in the second row remain shielded from close collisions with the incoming ion beam, unless they have a sufficient thermal vibrational amplitude to penetrate the shadow cone. However, there is a situation when the second row atoms can interact with the primary ion beam: this arises when the outermost atomic rows are displaced from their expected positions by, for example, surface reconstruction or surface relaxation phenomena. The latter is illustrated in Fig. 6 [20] for a (111) planar single crystal in which the outer row of atoms has relaxed outwards by a distance of Δd from their expected positions. It is then apparent that the size of the backscattered surface peak will be enhanced when the ion beam is directed along the <110> direction to the surface, particularly at higher energies, which cause the shadow cone to be smaller. On the other hand, when the ion beam is incident normal to the crystal then only the outermost atoms are available for interaction.

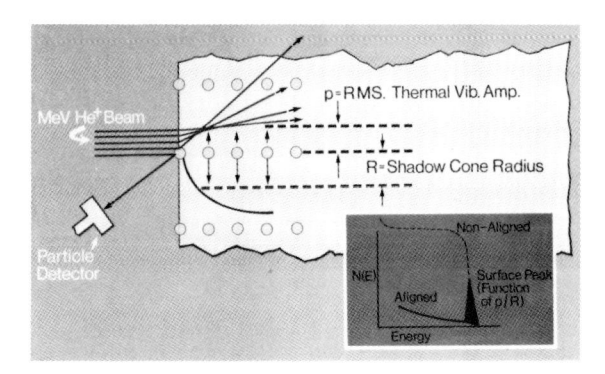

FIG. 5. [18] To illustrate the geometric conditions under which
the backscattered surface peak arises

The use of the surface peak as a highly sensitive surface probe has now
become established [21]. For example, it has been used to measure
crystallinity on a monolayer scale for Au films on a Ag(111) surface [22].
In this case the surface peak from the Ag surface was hardly altered after
the deposition of one monolayer of Au, but rapidly diminished to be replaced
by the Au surface peak, after 3 monolayers. In another study, it was shown
[23] that the clean W(001) surface, which exhibits reconstruction, i.e. a
departure from the periodicity expected from its bulk structure, can become
re-ordered as a consequence of surface adsorption of hydrogen to saturation.
Similarly it was found [24] that the outermost atomic layer of single crystal
Ni(110) is contracted by 4%, but expands by 1% after the adsorption of $\frac{1}{3}$
monolayer of oxygen.

DOES CATALYSIS NEED ION BEAMS?

We have considered above how ion beams can be used to analyse real and
model catalysts, and even to synthesise new ones. It is necessary now to put
this into perspective. The ion beam expert must recognise that catalysis is
an enormous subject, with access to a huge array of preparative and diagnostic
techniques. Thus it should be appreciated that the catalyst specialist,
having measured the basic physical properties of the catalyst, seeks to
understand the mechanism by which it operates. This involves accumulating a
mass of data in an attempt to unravel the crucial steps in the reaction [25]
and thereby to optimise the catalyst formulation. Such an approach is
applicable to "real" - i.e. working heterogeneous catalysts and gas mixtures

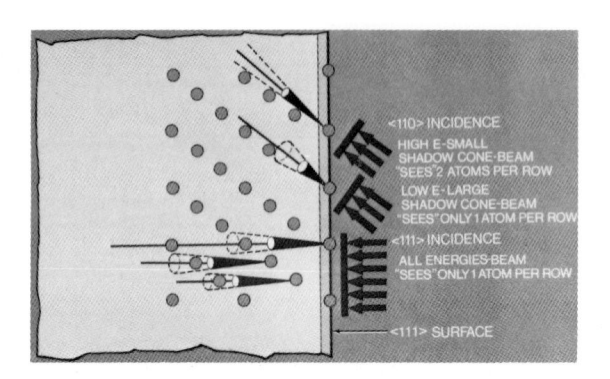

FIG. 6. [19] The use of ion channelling to measure surface relaxation in a single crystal

at appropriate temperatures and pressures. The arising information allows deductions to be made of the events occurring on the atomic scale on the catalyst's surface. This may be done also under more directly controlled conditions, by the use of molecular beams [26]. Another approach involves examining catalyst particles with high spatial resolution in the presence of reactive gas by Controlled Atmosphere Electron Microscopy [27]. However, in order to get closer to the fundamental events, it is necessary to move to single crystals. Gases adsorbed on to such well-defined surfaces can be examined by a variety of techniques, including Infra-Red Spectroscopy and Electron Energy Loss Spectroscopy [28], thereby obtaining direct information on the nature of the chemical bonds between the adsorbed molecules and the substrate. Finally, it is appropriate that we should end by referring to another ion beam technique, now well established in catalysis, namely, Secondary Ion Mass Spectrometry. In a recent study of CO adsorption on Pd(111) it has been used to yield information on the geometric positions occupied by the adsorbed molecules on the surface [29].

CONCLUSION

It has emerged that certain ion beam techniques have the potential for playing a useful role in catalysis research. This is most apparent in their abilities to interrogate for elemental and structural information. However it should be recognised that the catalyst designer has at his disposal a large range of sophisticated techniques, only some of which have been referred to here. For this reason it is clear that significant advances can be only by interaction between specialists in both fields.

ACKNOWLEDGEMENTS

Financial support from the Materials, Chemicals and Vehicles Requirements Board of the U.K. Department of Industry is gratefully acknowledged. Thanks is expressed also to the many colleagues whose data is quoted in this review.

REFERENCES

1. J.R. Anderson, Structure of Metallic Catalysts, Academic Press (1975).
2. P.J.F. Harris, E.D. Boyes and J.A. Cairns, J. Catal. 82, 127 (1983).
3. J.F. Ziegler, Helium, Stopping Powers and Ranges in All Elements, Pergamon, New York (1977).
4. M. Grenness, M.W. Thompson and R.W. Cahn, J. Appl. Electrochem., 4, 211, (1974).
5. I.H.B. Haining, P. Rabette, M. Che, A.M. Deane and A.J. Tench, in Proc. VII Int. Cong. on Catalysis, ed. by T. Seiyama and K. Tanabe, Elsevier, Amsterdam (1981) p317.
6. G.K. Wolf, Radiation Effects 65, 107 (1982).
7. G.K. Wolf, K. Zucholl, H. Folger and W.E. O'Grady, Nucl. Instr. and Meth., 209/210, 835 (1983).
8. G.N. Hays, C.J. Tracy and H.J. Oskam, J. Chem. Phys., 60, 2027 (1974).
9. H. Kasten and G.K. Wolf, Electrochimica Acta, 25, 1581 (1980).
10. J.A. Cairns, IEEE Transactions on Nuclear Science NS-28, 1804 (1981).
11. J.A. Cairns, R.S. Nelson and R.W. Barnfield, U.S. Patent No. 4,046,712 (1977).
12. J.A. Cairns, in Characterisation of Catalysts, ed. by J.M. Thomas and R.M. Lambert, John Wiley and Sons (1980) p185.
13. K.R. Padmanabhan and Y.F. Hsieh, Applications of Surface Science, 15, 205 (1983).
14. P.M. Read, C.J. Sofield, M.C. Franks, G.B. Scott and M.J. Thwaites, Thin Solid Films, 2, 10 (1983).
15. J.F. Ziegler, in New Uses of Ion Accelerators, ed. by J.F. Ziegler, Plenum Press, New York (1975) p75.
16. J.A. Cairns, J.E.E. Baglin, G.J. Clark and J.F. Ziegler, J. Catal. 83, 301 (1983).
17. D.S. Gemmel, Rev. Mod. Phys. 46, 129 (1974).
18. E. Bøgh, in Channelling, ed. by D.V. Morgan, John Wiley and Sons (1973) p435.
19. L.C. Feldman and J.M. Poate, Ann. Rev. Mater. Sci. 12, 149 (1982).
20. J.A. Davies, in Materials Characterisation using Ion Beams, ed. by J.P. Thomas and A. Cachard, Plenum Press, New York (1978) p405.
21. L.C. Feldman, Critical Reviews in Solid State and Materials Science, 10, 143 (1981).
22. R.J. Culbertson, L.C. Feldman, P.J. Silverman and H. Boehm, Phys. Rev. Lett. 47, 657 (1981).
23. I. Stensgaard, L.C. Feldman and P.J. Silverman, Phys. Rev. Lett. 42, 247 (1979).
24. J.F. Van der Veen, R.G. Smeenk, R.M. Tromp and F.W. Saris, Surf. Sci. 79, 212 (1979).
25. D.A. King and D.P. Woodruff (eds.), The Chemical Physics of Solid Surfaces and Heterogeneous Catalysis, Vol. 4, Elsevier Scientific Publishing Co. (1982).
26. D. Löffler, G.L. Haller and J.B. Fenn, J. Catal. 57, 96 (1979).
27. R.T.K. Baker, D.J.C. Yates and J.A. Dumesic in Coke Formation on Metal Surfaces, ed. by L.F. Albright and R.T.K. Baker, Amer. Chem. Soc., Washington (1982).

28. H. Ibach and D.L. Mills, *Electron Energy Loss Spectroscopy and Surface Vibrations*, Academic Press (1982).
29. A. Brown and J.C. Vickerman, Surface Science, **124**, 267 (1983).

CORROSION BEHAVIOR OF Ni$^+$-ION IRRADIATED NiTi ALLOYS

R. WANG AND J. L. BRIMHALL
Pacific Northwest Laboratory, P.O. Box 999, Richland, Washington 99352.

ABSTRACT

Corrosion behavior of Ni$^+$-irradiated NiTi alloys was studied in chloride solutions, together with unirradiated NiTi material with different surface conditions. Ion irradiation with either 2.5 or 5 MeV Ni$^+$ ions transformed the NiTi surface into an amorphous layer up to 1.5 micrometers thick. Studies of corrosion potential vs. time and polarization behavior indicated a small enhancement of the passivation for the Ni$^+$-irradiated NiTi over the unirradiated NiTi. The unirradiated NiTi with a mechanically polished, coarse surface was susceptible to pitting and crevice corrosion attack in 1 N HCl solution. The homogeneous amorphous structure in the irradiated alloy retarded this type of localized corrosion.

INTRODUCTION

Aqueous corrosion behavior of metal and alloy surfaces can be altered by high energy ion-implantation. Depending on the type and the energy of the ions, the surfaces can be chemically and/or structurally modified. Extensive studies have been made of surface alloying with ion-implantation. Typically, improved behavior in both general and pitting corrosion has been observed, such as in Fe and Fe-based alloys implanted with ions of passive metals, e.g., Cr, Ta and Ti. This is true especially if the structure of the metal surface was modified into an amorphous phase [1]. On the other hand, implantation of inert gas ions into pure Fe without the effect of surface alloying resulted in little change in the corrosion behavior [2]. In the latter case, degradation of the corrosion behavior was sometimes observed due to ion-induced defect structures [3]. So far there is little reported work on the effects of amorphization of a crystalline, single-phase alloy by ion-implantation without chemical change of the surface.

This paper concerns the study of the effects of ion induced amorphization on the corrosion behavior of a single-phase NiTi alloy. The fine-grained NiTi structure can be transformed into an amorphous phase by irradiation with one of its own constituents, i.e., Ni ions [4]. Therefore, it is possible to study the effects of this structural modification with minimal effects from the compositional variation.

MATERIALS

The as-cast NiTi alloy, supplied by Raychem Corporation, had a nominal composition of 50% Ni-50% Ti. The microstructure was single-phase NiTi with a few isolated particles of Ti$_2$Ni. Specimen discs of 3 mm in diameter and 0.1 mm thick were extracted from the cast alloy for ion irradiation and corrosion measurements. The specimens were polished to a smooth finish, then surfaces representing three different conditions were prepared: electropolished in perchloric/alcohol solution, abraded on 600 grit paper and abraded on 240 grit paper. A sputter deposited alloy was also prepared using the Raychem material as the sputtering target.

ION IRRADIATION

The specimen discs were irradiated with either 2.5 or 5 MeV Ni ions up to doses of 2×10^{15} ions/cm^2 at ambient temperature. The ion current was kept below 0.7 μamp/cm^2 to avoid excessive beam heating. The damage range extended for $\sim$ 0.8 and 1.5 μm for the two ion energies used [4]. The Ni ion came to rest far from the surface; therefore, there was essentially no influence of the deposited Ni ions on the surface region. The amorphous surface structure after irradiation was confirmed by TEM analysis.

EXPERIMENTAL

Corrosion solutions of 0.1 M NaCl and 1 N HCl were prepared using 13 megohm/cm water and reagent-grade chemicals. The corrosion cell was made of a glass flask containing two graphite counter electrodes, a bubbler and a saturated calomel electrode (SCE) bridged to the solution. The measurement was made with a PAR-350 corrosion measurement apparatus at room temperature. Corrosion specimens were masked with thin polystyrene tape to expose a round area of about 0.03 cm^2 to the solution. Care was taken to assure that no solution penetrated into the masked area or created a crevice under the masked area. Both irradiated and unirradiated specimens were cleaned ultrasonically with acetone followed with methyl alcohol before corrosion measurements were made.

RESULTS

Surface Reactivity

To compare the relative surface reactivities of the irradiated and unirradiated NiTi alloys, the corrosion potential variation with respect to time in the corrosion solutions for the first 1000 seconds was recorded. Figure 1 shows the variation of the corrosion potential, E_{corr} vs. time in 0.1 M NaCl solution. The E_{corr} values of irradiated NiTi, between -0.5 and -0.4 V(SCE), were about 0.1 V higher than those of the unirradiated NiTi, which is near -0.6 V(SCE). Both E_{corr} values of the irradiated NiTi and unirradiated NiTi increase at a similar rate indicating the formation of surface oxide/hydrides with a passive nature.

Similarly, the initial corrosion potential change in 1 N HCl solution, Figure 2, shows that the irradiated NiTi has a higher E_{corr} value initially than the unirradiated NiTi. In this case, the E_{corr} variation of the unirradiated NiTi follows the behavior of pure Ti near -0.2 V(SCE). Pure Ni shows active dissolution compared to NiTi and Ti.

The sputter deposited amorphous NiTi alloy was also examined in both solutions for comparison. The sputter deposited amorphous NiTi has a higher potential than the unirradiated NiTi but follows the lower E_{corr} values of the irradiated NiTi.

Polarization and Pitting

Corrosion behavior was further studied by a potentiodynamic polarization technique for comparisons of surface passivation and the pitting susceptibility. After 1000 seconds in the solution, the surface of the specimen was subjected to an anodic potential scan at a rate of 1 mV/s up to 1.6 V(SCE) for the NiTi alloys. Figure 3 shows the polarization scan in 0.1 M NaCl solution. All the irradiated NiTi data show up to an order of magnitude lower corrosion current density than that of the unirradiated NiTi. Although the irradiated NiTi has a higher E_{corr} to start with, the transpassive potentials for all the NiTi alloys are the same at about

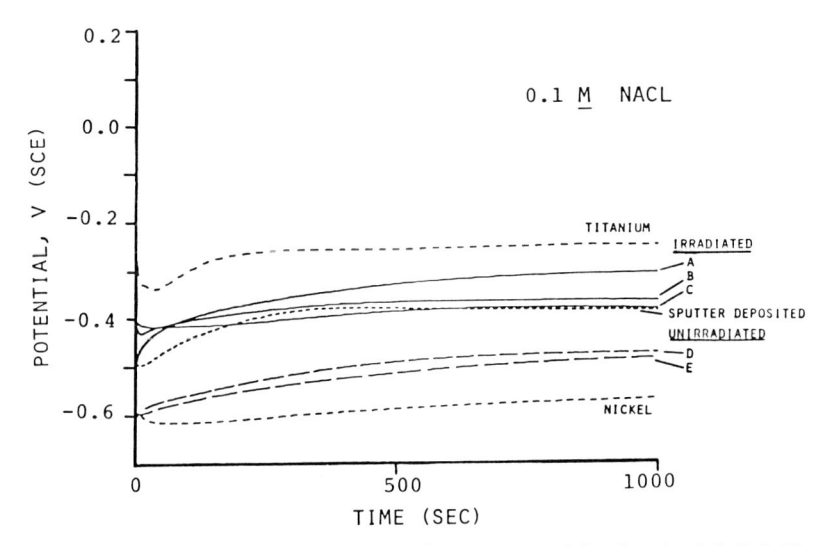

Fig. 1 Variation of Corrosion Potential, E_{corr}, With Time in 0.1 M NaCl Solution. Irradiated NiTi: Curve A, 600 Grit Polish, 5 MeV Ni^{++}; Curve B, Electropolish, 5 MeV Ni^{++}, Then 0.25 μm Diamond Polish; Curve C, 240 Grit Polish, 2.5 MeV Ni$^+$. Unirradiated NiTi: Curve D, 240 Grit Polish; Curve E, 600 Grit Polish. Pure Ni, Ti and Sputter Deposited NiTi Were Polished on 600 Grit Paper.

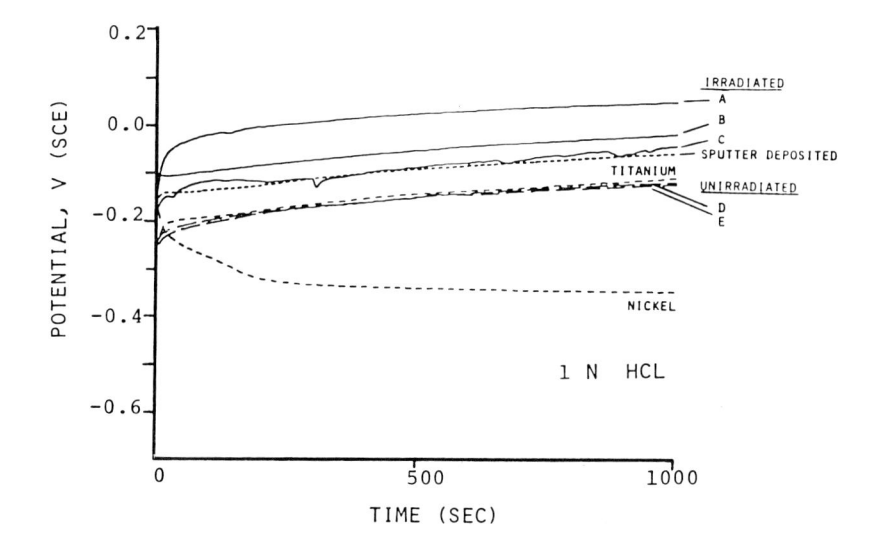

Fig. 2 Variation of Corrosion Potential, E_{corr}, With Time in 1 N HCl Solution. Irradiated NiTi: Curve A, 600 Grit Polish, 2.5 MeV Ni$^+$; Curve B, Electropolish, 5 MeV Ni^{++}, Then 0.25 μm Diamond Polish; Curve C, 240 Grit Polish, 5 MeV Ni^{++}. Unirradiated NiTi: Curve D, 600 Grit Polish; Curve E, 240 Grit Polish. Pure Ni, Ti and Sputter Deposited NiTi Were Polished on 600 Grit Paper.

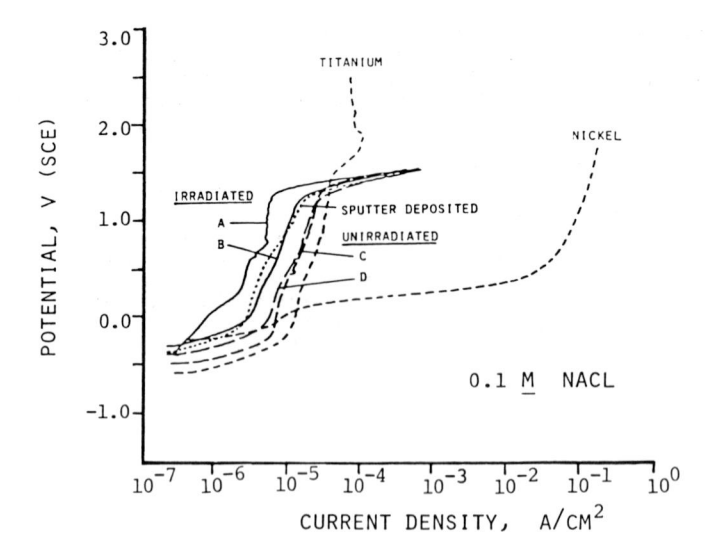

Fig. 3 Potentiostatic Polarization Curves in 0.1 M NaCl Solution. Irradiated NiTi: Curve A, 240 Grit Polish, 2.5 MeV Ni$^+$; Curve B, 600 Grit, 5 MeV Ni^{++}. Unirradiated NiTi: Curve C, 240 Grit Polish; Curve D, 600 Grit Polish.

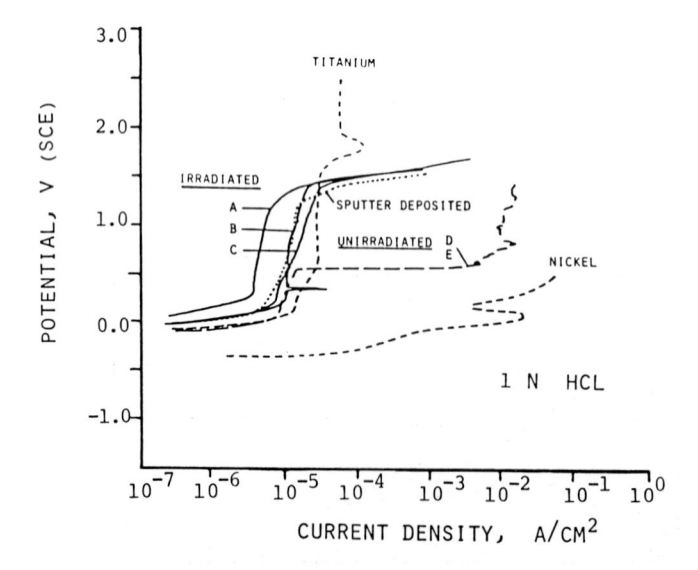

Fig. 4 Potentiostatic Polarization Curves in 1 N HCl Solution. Irradiated NiTi: Curve A, 600 Grit Polish, 2.5 MeV Ni$^+$; Curve B, 240 Grit Polish, 5 MeV Ni^{++}; Curve C, Electropolish, 5 MeV Ni^{++}, Then 0.25 μm Diamond Polish. Unirradiated NiTi: Curve D and E, 240 and 600 Grit Polish.

1.3 V(SCE). No pitting was found in either the irradiated or unirradiated NiTi. The initial polarization behavior of the unirradiated NiTi is very similar to that of the pure Ti, except the NiTi passivity breaks down at 1.3 V(SCE). The irradiated NiTi shows less active dissolution at about -0.2 V(SCE) compared to the unirradiated NiTi.

In 1 N HCl solution, there is a distinct difference between the irradiated NiTi and the unirradiated NiTi alloys, Figure 4. Large pits were initiated at near 0.6 V(SCE) for both 600 and 240 grit polished unirradiated NiTi while the Ni-irradiated NiTi of the same surface condition did not show any pits and proceeded to a transpassive potential at 1.4 V(SCE). Again, the irradiated NiTi shows nearly an order of magnitude less corrosion current density at the passive region than that of the unirradiated NiTi, Figure 4. Both irradiated and unirradiated NiTi alloys show small active dissolution at 0.1 to 0.3 V(SCE). Figure 5 shows the pits formed in the unirradiated NiTi at 0.6 V(SCE) for a surface prepared by 240 grit polishing. Similar pits were seen with a surface prepared by 600 grit polishing.

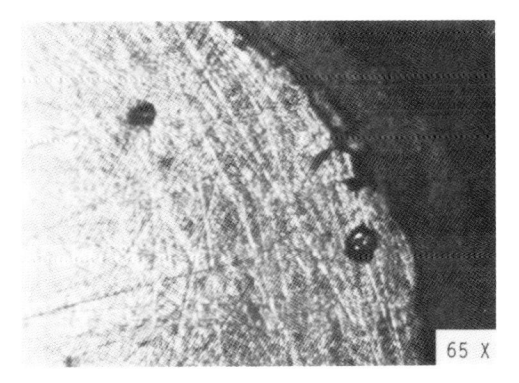

Fig. 5   Pits Developed in 1 N HCl at 0.6 V(SCE) for Unirradiated NiTi With 240 Grit Polish.

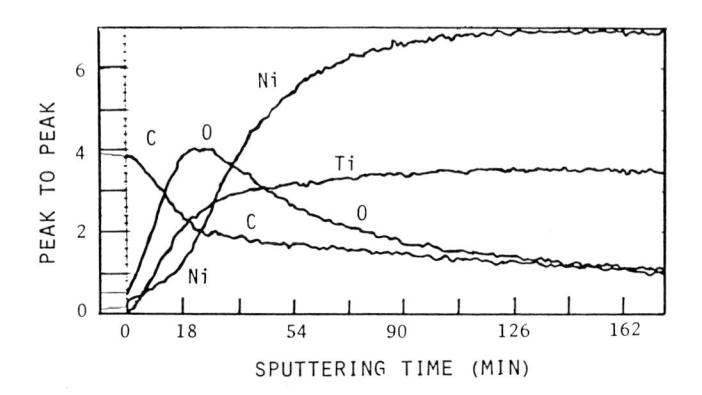

Fig. 6   Auger Sputter Profile for 600 Grit Polished NiTi Indicating Surface Contamination of Oxygen and Carbon.

DISCUSSION

In all cases, the NiTi specimens irradiated by Ni ions were slightly more passive than the single-phase, crystalline NiTi alloys. The improved passivity of the Ni$^+$ irradiated NiTi can be related to the highly disordered amorphous structure caused by irradiation. On the other hand, the observed increase in the passivity did not represent a large improvement in the corrosion behavior over the crystalline NiTi alloy. This is due to the fact that crystalline NiTi has a relatively high general corrosion resistance similar to Ti although it is susceptible to pitting and crevice corrosion attacks in HCl solution, as related to the surface condition. The homogeneous amorphous structure of the irradiated NiTi reduced the susceptibility to this type of corrosion.

Results from irradiation of an electropolished smooth NiTi surface and a sputter deposited 600 grit polished amorphous NiTi surface show that the improved surface passivation for the amorphous phase is not due to the surface contamination. One of the major concerns in these experiments was whether the ion irradiated specimen surface had been contaminated with impurities during the irradiation [2,5], thereby altering the subsequent corrosive behavior. Typically the surface of NiTi would be contaminated with carbon or oxygen as shown by the Auger sputtering profile in Figure 6. To evaluate the effects of possible surface contamination on corrosion behavior, an electropolished NiTi surface was subsequently irradiated with 5 meV Ni^{++} ions to create a 1.5 μm thick amorphous layer. This irradiated surface was then lightly polished with 0.25 μm diamond paste to remove any contamination layer, but not so deep as to penetrate the irradiated layer. This cleaned surface demonstrated more passive behavior than the unirradiated NiTi. Moreover, the sputter deposited amorphous NiTi, which had been polished with 600 grit paper, also showed more passive behavior than the crystalline NiTi. This is further indication that the amorphous NiTi is, in general, more passive than the crystalline phase and the observed effects are a result of the structural changes and not due to surface contamination.

CONCLUSION

- Ni-ion induced amorphization produced a small improvement in passivity in a NiTi alloy.
- The improvement in passivity leads to a decreased susceptibility to pitting and crevice corrosion.
- In general, however, amorphization of the NiTi structure with no chemical change did not drastically change the corrosion behavior.

ACKNOWLEDGMENT

This work supported by Division of Materials Sciences, Office of Basic Energy Sciences, U.S. Department of Energy under Contract DE-AC06-76RL0-1830.

REFERENCES

1. G. K. Hubler, P. P. Trzaskoma, E. McCafferty and I. L. Singer, Proceedings of the Conference on "Modification of Surface Properties of Metals by Ion Implantation," Manchester, England, June 1981.
2. V. Ashworth, W. A. Grant, R.P.M. Procter and T. C. Wellington, Corrosion Science 16, 393 (1976).
3. N. R. Sorensen and S. T. Picraux, Corrosion, 1983, in press.
4. J. L. Brimhall, H. E. Kissinger and L. A. Charlot, Radiation Effects, 77, 237 (1983).
5. C. R. Clayton, Nucl. Instrum. Meth., 182/183, 865 (1981).

EFFECT OF N-IMPLANTATION ON THE CORROSIVE-WEAR PROPERTIES OF SURGICAL
Ti-6Al-4V ALLOY†

J. M. WILLIAMS,* G.M. BEARDSLEY,* R. A. BUCHANAN,** AND R. K. BACON**
*Solid State Division, Oak Ridge National Laboratory, P. O. Box X,
Oak Ridge, Tennessee 37830; **The University of Alabama, Birmingham,
Alabama 35294

ABSTRACT

The effects of N-ion implantation on the corrosive-wear properties of Ti-6Al-4V, an alloy used for construction of the femoral component of artificial hip joints in humans, were tested. In corrosive-wear tests designed to simulate pertinent hip-joint parameters, electrochemical corrosion currents were measured for cylindrical samples in saline electrolyte in an arrangement which allowed the samples to be rotated between loaded polyethylene pads simultaneously with the current measurement. To further quantify material removal, Zr markers were ion-implanted into some samples so that, by use of Rutherford backscattering, material removal could be detected by changes in position of the marker relative to the surface. Corrosion currents were greatly reduced by implantation of approximately 20 at. % N, but even implantation of the Zr markers also reduced corrosion currents. The marker experiments confirmed the low rate of material removal for the implanted samples.

INTRODUCTION

The femoral component of most modern, surgically implanted, total hip-joint systems consists of an integral stem and ball, constructed of a suitable alloy. As the joint moves the ball works in an acetabular cup constructed of ultra-high molecular-weight polyethylene (UHMWPE). The alloy, Ti-6Al-4V, is receiving increased use for the femoral component because of its excellent tensile strength, fatigue strength, low density, high corrosion resistance and substantial ductility [1]. Also, the low elastic modulus of the alloy contributes to good adhesion with living bone. Additionally, the alloy is considered to be more biocompatible than the Cr-containing alloys [2,3]. There is nevertheless a continuing interest in reducing the amount of debris and corrosion products introduced into the body from prosthetic components, even those of Ti-6Al-4V, partly because dispersal of sufficient quantities of substances that are well-tolerated in bulk form, e.g. polyethylene, can cause adverse tissue responses [2]. General concerns are with effects such as inflammatory responses [2], increased susceptibility to infection [4], and sensitizing reactions which can cause loosening of prosthetic components and bone loss [5]. On the long term, systemic effects such as effects on blood-filtering organs cannot be ruled out, and at still longer times carcinogenicity might manifest itself [3]. It is suspected that wear of both Ti-alloy components and their UHMWPE wear partners may be much higher than for other alloys working against UHMWPE [1]. It was considered desirable to investigate, and possibly, ameliorate such effects.

---

†Research sponsored in part by the Division of Materials Sciences, U. S. Department of Energy, under contract W-7405-eng-26 with the Union Carbide Corporation.

Titanium and its alloys are known to exhibit excellent static corrosion resistance but the passivant, TiO_2, while strong in compression, is weak and brittle in tension and has weak adherence to the Ti substrate. Removal of the oxide during rubbing wear and entrapment of oxide particles between wear partners can contribute to abrasion and continuing re-oxidation of the bare metal surface. Thus, Ti and its alloys are not known to perform especially well under wear or corrosive-wear conditions. Conventional nitriding and carburizing are effective in wear prevention, but there are difficulties with adherency of reaction products, and bulk properties can be compromised by the high temperatures required [6].

The present experiments were intended to test the effect of nitrogen ion-implantation on the corrosive-wear of the alloy under conditions similar to service conditions in hip prostheses. The effects of Cr implantation on one sample only were also tested. The test apparatus, described below, was an electrochemical corrosion cell with special features designed to produce approximately the service conditions of the alloy in respect to key parameters that affect corrosion and wear. The rate of material removal was determined from the corrosion current. A further measure of the wear (or corrosion) rate was obtained by use of heavy-ion (Zr) markers, which were ion-implanted to low concentrations in some samples. To detect removal of material from the surfaces, depth distributions of the markers were measured by Rutherford backscattering (RBS) after periods of testing in the corrosive-wear apparatus.

EXPERIMENTAL PROCEDURE

The apparatus (illustrated schematically in Fig. 1) used in the corrosive-wear tests has been described previously [7]. The design incorporates recommendations of Dumbleton [5] for wear testing of orthopedic materials. The electrolyte was O_2 - saturated, 0.9%-NaCl isotonic saline at a pH of 7.00 ± 0.05 and a temperature of 37.0 ± 0.5 °C. The cylindrical alloy sample, 1.27 cm long x 1.27 cm diam, was subjected to a mean surface pressure of 2.30 MPa as it rotated between the closely conforming UHMWPE pads. The load is applied by means of pusher rods attached to the pads. Rotation of the sample produced an interfacial sliding velocity of 5 cm/sec between the sample and the pads. The electrochemical potential, -100 mV with respect to the saturated calomel reference electrode, was chosen on the basis of previously determined polarization curves to place the alloy in the anodic regime of corrosion. Bearing surfaces of pads were finished by reaming and new pads were used for each new alloy specimen. Finish level of the pads was 7 μm RMS.

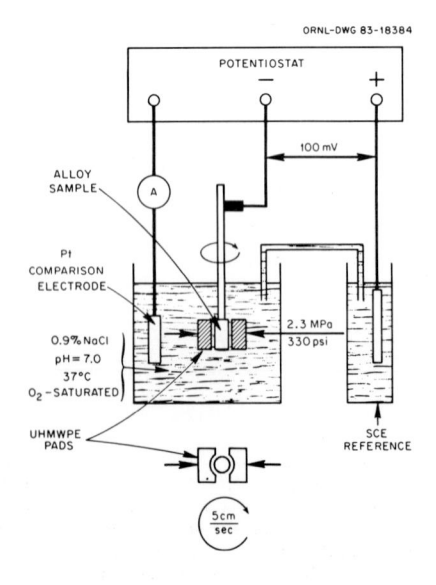

FIG. 1. Schematic representation of the corrosive-wear system.

Samples were made from annealed extra-low interstitial grade Ti-6Al-4V (ASTM F136) rod material, which was supplied to us as a courtesy by Zimmer-USA, Warsaw, Indiana. Sample surfaces were mechanically polished by wet polishing with 240 and 600 grit SiC papers, followed by polishing with 1 μm-sized Al_2O_3 particles in an abrasive slurry. Final cleaning was by use of boiling benzene followed by rinsing in acetone and H_2O.

Samples were implanted with N at dose-energy combinations, indicated in Fig. 2, to produce the approximate calculated individual profiles shown. These individual profiles sum to an approximately flat profile of 20 at. % N to a depth of 100 nm. During implantation the samples were rotated and the entire cylinder walls were exposed to the beam. The calculational method used for Fig. 2 takes account of all angles of incidence. For the calculation, range distributions were assumed Gaussian with identical longitudinal and lateral straggling parameters, so that spherical contours of constant concentration would obtain for a pencil beam at normal incidence ($\bar{R}_p$ and ΔR_p values used were 31.7 and 17.7 nm for 17.5 keV, 63.3 and 29.6 nm for 35 KeV and 109.5 and 44.3 nm for 60 KeV). Assymetry in the profiles is due to cylindrical geometry only; no sputtering was assumed. Assuming a sputtering coefficient of 0.4 produces no practical difference in the characterization of the implant from that shown in Fig. 2, but a thickness of 12.7 nm of material is removed. By strategies similar to those described above implantations with 10 at. % N and with Cr were done (Table I).

Zirconium was chosen as the RBS marker because it is heavy enough to be resolved from the substrate by RBS and, it was thought, would have less structural influence than the inert gas, Xe. Zr-markers were implanted to doses of $5 \times 10^{15}/cm^2$ at energies of 280 KeV. The samples were rotated during implantation, and the beam shape was defined by a slit positioned just ahead of the sample to limit grazing incidence. This procedure produced an implantation profile characteristic of normal incidence. Table I lists treatments and the number of samples at each treatment that were used.

TABLE I. Implantation treatments for Ti-6Al-4V samples.

Treatment	Number of Samples
20 at.% N	3
20 at.% N + Zr Marker	1
10 at.% N + Zr Marker	1
15 at.% Cr + Zr Marker	1
Zr Marker Only	1

RESULTS

Figure 3 shows corrosion currents versus time for the three 20 at. % N samples, and for four control samples which had no implantation treatment of any kind. While the scatter in the data is quite large (>10x), implantation has reduced the corrosion currents by, minimally, a factor of 100. The reduction of corrosion current with time for

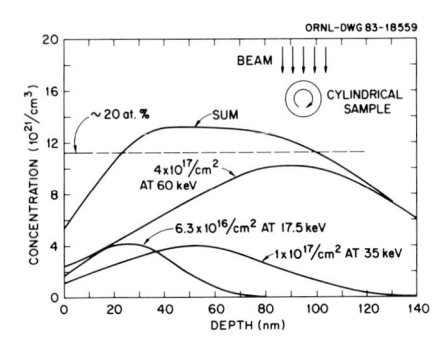

FIG. 2. Calculated individiual profiles and resultant sum for N-implantations in Ti-6Al-4V cylindrical samples. Fluence values pertain to flat area. Actual fluences on rotating cylindrical surface are $(1/\pi)$X fluences given above. Zero sputtering was assumed. Containment fraction is 0.942 assuming zero sputtering and 0.899 assuming sputtering coefficient of 0.4 for all energies and actual order of implantations.

any given sample is attributed to some type of wearing-in effect, but mechanisms are not understood. If tests are interrupted, as was the case for the samples with markers, and then resumed with new pads, corrosion currents revert to higher values, approximately characteristic of zero time. The reason for the large scatter is not known, but it may be related to details of the conformance between pads and sample when the test is initiated, e.g., factors such as slight amounts of taper on pads or sample, axial misalignment, etc.

All four of the samples with markers (Table I) also exhibited considerable reduction in corrosion currents due to the respective implantations. The total time for these samples, 14 h, was divided into segments because of interruptions for RBS analysis. Initial currents were ∿10 µA. Thus the currents tended to range slightly higher than the lower band of values in Fig. 3. Variations with time were more marked, however, than for the samples of Fig. 3. For example, in one time segment the sample with 15% Cr started at 10^3 µA, then dropped to zero or negative corrosion current after 2 h. The "marker only" sample was intended for comparison of RBS results with the electrochemical results characteristic of unimplanted material (upper band, Fig. 3) but, since even implantation of the marker caused a dramatic reduction in corrosion current, this comparison is not possible. Integrated values of material removal for the 14h tests (see Fig. 3 caption) were: <10 Å for (20% N + marker), 25 Å for both (10% N + marker) and marker only, and 120 Å for (Cr + marker).

Figure 4 shows entire RBS histograms for the (20% N + marker) sample, as-implanted and after a 4-h test. Tilting experiments revealed that peaks A and C represent surface contaminants, some of which are apparently deposited on the surface because of impurities in the electrolyte. The 4-h test results of Fig. 4 are quite typical of all marker samples for all testing periods out to 14h. Distortion of peak B, the Zr peak, is entirely due to the growth of peak A. Careful stripping of the background restores peak shape and reveals that there has been no detectable motion of the Zr marker relative to the surface. This statement applies for all four marker samples for all tests out to 14h. This is illustrated by Fig. 5 which shows the pure Zr peak for the "marker only" sample, as-implanted and after 14h of testing. Formal analysis was in terms of the depth of the centroid of the distribution. This value, 0.101 µm, was reproducible to ±0.001 µm among the four as-implanted samples. This resolution apparently decreases with testing time because the increasing amount of surface contamination elevates the background, such that much better RBS statistics are required for precise representation of the peak. Some centroid depths larger than that of the as implanted were recorded. It is estimated that the mean-square error for the entire set of 12 measurements is 0.007 µm.

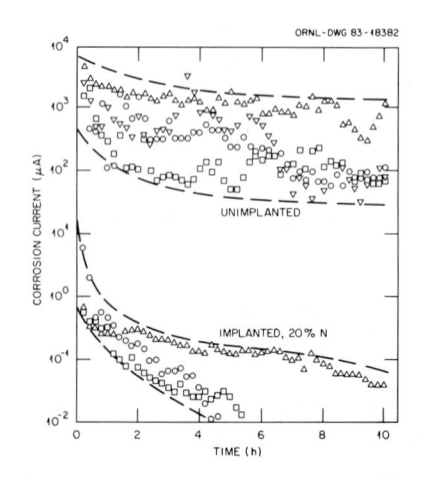

FIG. 3. Corrosion currents versus time for N-implanted and unimplanted Ti-6Al-4V samples. 5 µA-h corresponds to 10 Å of material removal for a Ti valence of 4 and a sample area of 5.07 cm^2.

Thus the shift of 0.01 μm of the centroid toward the surface for the data of Fig. 5 may or may not represent true wear, but in any case, no shifts greater than this were observed for any sample. Circumferential black stripes are visible to the eye after wear testing on some samples, particularly the Cr- implanted sample. No position dependency of the RBS histograms was detected, however, in axial surveys.

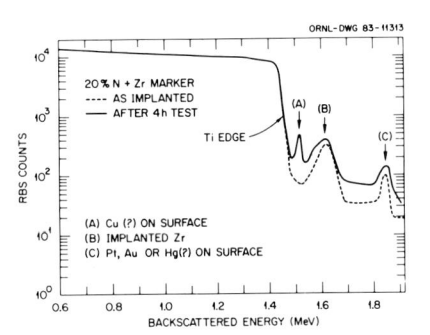

FIG. 4. Rutherford backscattering spectra for the (20% at. N + marker) - sample using 2.0 MeV He ions.

DISCUSSION

The most striking finding of the present study is that, while N-implantation is quite effective in reducing corrosive wear of the Ti-6Al-4V, other implantations including a relatively small dose of Zr are perhaps almost as effective. This fact suggests that we seek explanations or partial explanations in terms of factors that might be common to all implantations. One such factor is the extreme cleanliness and concomitant wetability of ion-implanted surfaces. Conceivably this improved wetability gives rise to much improved lubrication in the test. An improved film of electrolyte between pads and sample could diffuse stresses and reduce the rate at which TiO_2 is removed from the alloy surface. Another possibility is beam carburization. Titanium has a high affinity for C, and ion-implantations of Ti into steels are accompanied by in-situ formation of TiC [8]. Residual hydrocarbons in the vacuum are decomposed by the beam in the vicinity of the sample surface. The resulting C is available for ion-beam mixing or chemical reaction with the substrate. Analysis for C has not yet been performed in the present studies. This will be done by use of the elastic scattering resonance of 1.7 MeV protons on C. The very large benefit produced suggests either that some type of threshold effect or some synergistic set of factors associated with ion-implantation are responsible for the improvements. Surface hardening alone, for example, could hardly account for the effects observed, since N-implantation produces only about a two-fold increase in hardness [9]. In assessing rates of material removal for ion-implanted samples it should be born in mind that tests have little sensitivity, percentage wise, at nearly zero wear rate.

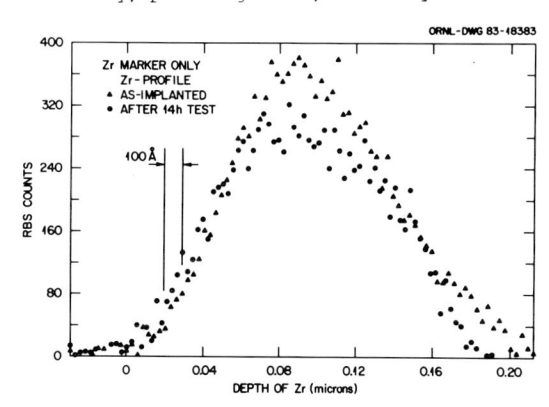

FIG. 5. Depth distributions of the ion-implanted Zr marker for the "marker only" sample before and after corrosive-wear testing. Background counts have been substracted from RBS data in the Zr peak region and the RBS energies have been converted to depths of Zr scatterers.

The marker results confirm the very low rates of material removal for the ion-implanted samples. The high rate of removal indicated by the corrosion currents for the unimplanted samples has no independent confirmation. However, there are large differences in visual appearances of the samples during testing. Implanted samples have only a few black stripes after 14 h of testing, but large areas of unimplanted samples are blackened early in the test. We speculate that the black stripes may consist of TiO_2 particles embedded in films of polyethylene wiped from the pads.

Wear of UHMWPE has generally been thought to be a more important limitation to mechanical functioning of prostheses than wear of the alloy. Yet a corrosion current of 10^3 µA (see Fig. 3) for a duty time (278 h [5]) appropriate for 1 yr of hip service would remove 0.06 mm of material. This is of the same order as the wear rate of the UHMWPE itself [5]. This finding tends to confirm fears of high wear rates for Ti-6Al-4V and UHMWPE as wear partners [1]. Thus, not only biocompatibility but also longevity and mechanical integrity of prostheses might be improved by ion-implantation of the alloy, especially if TiO_2 debris contributes to wear of the UHMWPE.

Large improvements produced by ion implantation on the wear properties of Ti alloys for surgical applications have also been reported by Dearnaley [10].

Clearly ion-beam processing is a promising technique for improving surface interactions between surgical implant materials.

ACKNOWLEDGEMENT

It is a pleasure to thank B. R. Appleton for important suggestions that affected the course of these experiments.

REFERENCES

1. W. Rostoker and J. O. Galante, J. Biomed. Eng. (Trans. ASME) 101, 2(1976).
2. Felix Escalas, Jorge Galante and William Rostoker, J. Biomed. Mater. Res. 10, 175 (1976).
3. D. F. Williams, Phys. Med. Biol. 25, 611 (1980).
4. Katharine Merritt, Judith W. Shafer, and Stanley A. Brown, J. Biomed. Mater. Res. 13, 101 (1979).
5. John H. Dumbleton, Wear, 49, 297 (1978).
6. P. D. Miller and J. W. Holladay, Wear 2, 133 (1958/9).
7. R. A. Buchanan, G. D. Turner, P. D. Gray, J. G. Melendez, T. F. Talbot, and J. L. McDonald, Corrosion 39, 377 (1983).
8. I. L. Singer, C. A. Carosella, and J. R. Reed, Nucl. Inst. Methods 182/183, 923 (1981).
9. J. B. Pethica and W. C. Oliver, Mat. Res. Soc. Symp. Proc. 7, 373 (1982). (Eds. S. Thomas Picraux and W. J. Choyke.)
10. G. Dearnaley, to be published in Surface Engineering, NATO Advanced Study Inst., Les Arcs, France, July 1983, R. Kossowsky and Subhash C. Singhal, eds.

EFFECT OF Cr+ IMPLANTATION ON THE THERMAL OXIDATION OF Ta

K.S. GRABOWSKI AND C.R. GOSSETT
Naval Research Laboratory, Washington, D.C. 20375

ABSTRACT

Cr+ implantation of Ta was undertaken in an attempt to improve oxidation resistance at temperatures between 500 and 750°C, and for oxidation times up to 100h. Samples were implanted with 1.5×10^{17} Cr+/cm^2 at 150 keV, and compared to samples implanted with 1×10^{16} or 1×10^{17} Ta+/cm^2 at 145 keV to evaluate the role of physical effects from ion implantation. Following oxidation, samples were examined using helium and proton backscattering, electron and optical imaging techniques, and auger electron spectroscopy. Improved resistance to oxidation was observed in Cr+-implanted samples oxidized at 500°C for up to 100h, and at 600°C for about 1h. However, some local breakdown occurred in these samples and no protection at all was observed at 750°C. Reasons for this breakdown are discussed and alternate approaches for improving oxidation resistance using ion implantation are proposed.

INTRODUCTION

Ta is one of the refractory metals which suffers from poor resistance to thermal oxidation. Coatings and alloys have been developed to improve its oxidation resistance, but these approaches degrade bulk mechanical properties, are expensive, and provide poor resistance to thermal shock, as reviews have discussed [1,2]. Recent work by Kaufmann et al. [3] attempted to overcome these difficulties by implanting various ions into Ta, thereby maintaining bulk mechanical properties while producing, hopefully, protective well-integrated coatings. This approach was unsuccessful for 1000°C oxidation, but Cr+ implantation was shown to improve the oxidation resistance of Ta at 500°C for 1h. Analysis suggested that a thin Cr-rich oxide at the surface provided the protection. This work expands that earlier investigation of Cr+ implantation to longer times at 500°C and to the intermediate temperatures of 600 and 750°C.

The oxidation of pure Ta between 500 and 750°C has been well studied and reviewed and yet is not completely understood [1,2,4]. The oxidation mechanisms are complex and time dependent. Sub-oxide platelet formation and their incursion into the metal has been observed, and considerable dissolution of O into the metal occurs. In the temperature regime of interest, diffusion controlled oxide growth undergoes a transition to a more rapid oxide growth process. It was hoped that Cr+ implantation might prevent, or at least delay, this transition.

EXPERIMENTAL

Specimens were prepared from 99.9% pure Ta sheet material having a grain size of 20 μm. Large coupons 375 μm thick were cleaned and implanted in the central region with 1.5×10^{17} Cr+/cm^2 at 150 keV, or with 1×10^{16} or 1×10^{17} Ta+/cm^2 at 145 keV. In each case a current density of about 4 μA/cm^2 was used, and the target-chamber pressure was about 1.5x10^{-4} Pa during the implantation. Twelve smaller coupons (6.4x12.7 mm) were cut from each of these larger sheets, producing samples that were each implanted over one-half of their surface area.

The smaller coupons were oxidized in a flowing (1 l/min) 80-20 mixture of research purity Ar and O_2 to represent pure dry air. Oxidations were conducted for 1 to 100h at 500°C, and for 1h at 600 and 750°C.

Helium backscattering at 2.0 MeV was used to investigate the kinetics of oxidation at 500°C. A scattering angle of 135° and a grazing exit angle of 20° were used. Proton backscattering of 1.5-MeV ions was used to investigate the thicker oxides formed at 600 and 750°C. A scattering angle of 135° and symmetric entrance and exit angles were used.

All samples were examined by optical microscopy. Selected samples were also investigated by scanning electron microscopy (SEM), auger electron spectroscopy (AES), or Read camera x-ray diffraction with Cu-Kα radiation and a 10° glancing-incidence angle.

RESULTS AND DISCUSSION

Unimplanted, 500°C Oxidations

The development of oxide film on unimplanted Ta can be understood with the help of the Helium backscattering spectra shown in Fig. 1. These profiles effectively show the Ta concentration in a region close to the sample's surface, which appears at about 1.85 MeV. The lowest energy shown corresponds to a depth of 280 nm in Ta or 390 nm in Ta_2O_5 (equivalent to 160 nm of Ta completely oxidized to Ta_2O_5). The intensity decreases with increasing oxidation time because O diffuses into the sample and causes dilution of the Ta.

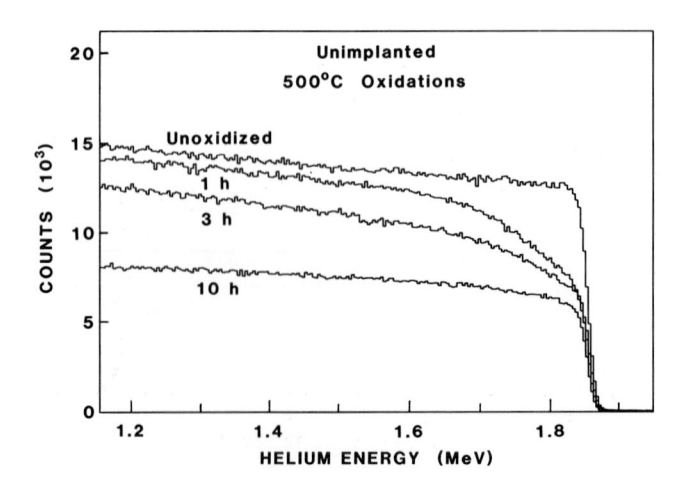

Fig. 1 Helium backscattering spectra for unimplanted Ta, oxidized for various times at 500°C.

For samples oxidized up to about 3h, the Ta concentration increases with depth from about that of Ta$_2$O$_5$ on the surface to a higher plateau level some 100 nm or so below the surface. This plateau level of Ta corresponds to about 16 at.% O in the 1-h sample and about 37 at.% O in the 3-h sample, both well in excess of the 1 at.% solubility limit for O in Ta at 500°C [5]. This large O concentration, and the broad transition region from near Ta$_2$O$_5$ on the surface to the plateau level beneath is probably caused by incursions of sub-oxide platelets into the base metal. Such incursions along {100}$_{Ta}$ planes have been reported for Ta oxidized at 500°C [6-8].

For oxidation times of 10h or more, only compositions near Ta$_2$O$_5$ could be seen by the backscattering measurements. Read-camera x-ray diffraction of a 30-h oxidized sample indicated that the oxide was the low-temperature phase of Ta$_2$O$_5$. After 100h of oxidation, the stresses generated were so large that a complete sheet of oxide film, 70 μm thick, separated from the sample.

These results are consistent with the theory that after an incubation time of 150-250 min at 500°C, Ta$_2$O$_5$ nuclei form on the surface and rapid oxidation commences [8].

Implanted, 500°C Oxidations

Cr-implanted samples showed a remarkable improvement in oxidation resistance for oxidation times up to about 100h, as is indicated by Fig. 2. After 100h of oxidation, regions of both thin and thick oxide were present within the Cr-implanted portion of the sample. The spectrum for the thin region is virtually the same as was found for samples oxidized from 1 to 30h. This indicates that nearly all the oxidation of Cr-implanted Ta occurs in the first 1h or so of oxidation, and that after 100h some type of localized breakdown occurs. Even without oxidation, the Ta backscattering yield was somewhat decreased by an 18 at.% Cr concentration present within the first 50 nm or so of the surface. This dilution effect persisted in the oxidized samples as well.

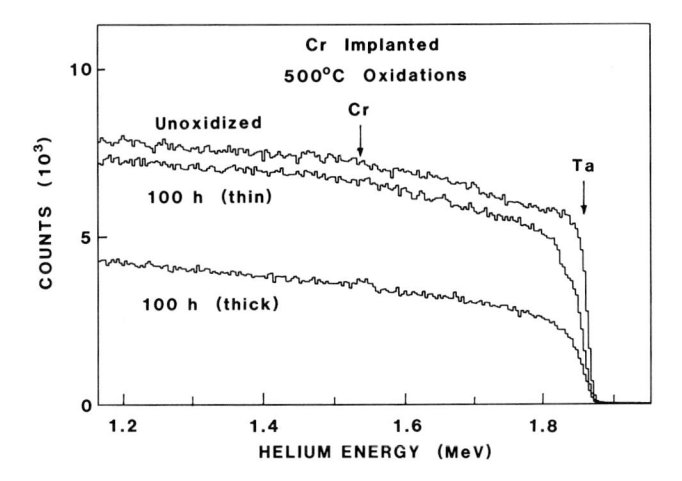

Fig. 2 Helium backscattering spectra for Cr-implanted Ta, unoxidized and oxidized 100h at 500°C. Different regions of the oxidized sample are shown. Same analysis conditions as Fig. 1. Arrows indicate energy for backscattering from the surface of Ta and Cr.

Considerable detail can be gleaned from the spectrum of the thin oxide region shown in Fig. 2. Within the first 1h of oxidation, oxide platelets apparently started to intrude into the base metal, as indicated by the plateau in Ta concentration, but their growth was then checked. The O concentration was lower near the surface compared to that in unimplanted samples oxidized for 1h (see Fig. 1), and the slight shift of the Ta edge to lower energies following oxidation indicates that a Cr-rich oxide layer formed on the surface, consistent with earlier work of Kaufmann et al. [3]. These results suggest that Cr implantation prevented Ta oxidation by forming a thin barrier layer on the surface which blocked the inward diffusion of O.

The consequence of physical effects from ion implantation (e.g., sputtering, radiation damage, vacuum carburization, etc.) were checked by performing Ta-ion implantations followed by oxidations for up to 27h. While the rate of incursion of oxide platelets was slowed slightly for oxidations beyond 1h, no significant long term benefits to oxidation resistance were provided.

The suggestion from helium backscattering that localized breakdown of the barrier layer was able to occur in the 100-h oxidized sample was confirmed using an SEM, as Fig. 3 shows. Sporadic eruptions about 20 µm tall were observed in the implanted region. They were arranged in groups located near the boundary of the implanted region, randomly within the implanted region, and along scratches in the surface. This should be compared to the 100 µm-thick oxide scale which formed on the unimplanted region.

Fig. 3 SEM image of sample oxidized 100h at 500°C. Cr-implanted region is in foreground, unimplanted region (minus 70 µm of scale) is in background. 25° angle of incidence.

Examination of an area adjacent to one of these large eruptions provided evidence that the eruptions nucleate over some type of extended defect in the surface, as shown by Fig. 4. Perhaps after a long enough oxidation period, Ta oxide nuclei form locally beneath the Cr-rich oxide layer, fracture the barrier layer, and cause rapid local oxidation to ensue.

Fig. 4 SEM image of area of incipient breakdown in Cr-implanted sample oxidized 100h at 500°C. 45° angle of incidence.

Implanted, 1-h Oxidations

The durability of oxidation protection provided by Cr implantation was further evaluated through 1 h oxidations conducted at 600 and 750°C. As Fig. 5 shows, while some protection was provided at 600°C, little was provided at 750°C. The lowest energy shown in this figure corresponds to a depth of 4.4 μm in Ta and 6.2 μm in Ta_2O_5 (equivalent to 2.5 μm of Ta). Virtually no difference was observed between spectra from unoxidized and 500°C 1 h oxidized samples.

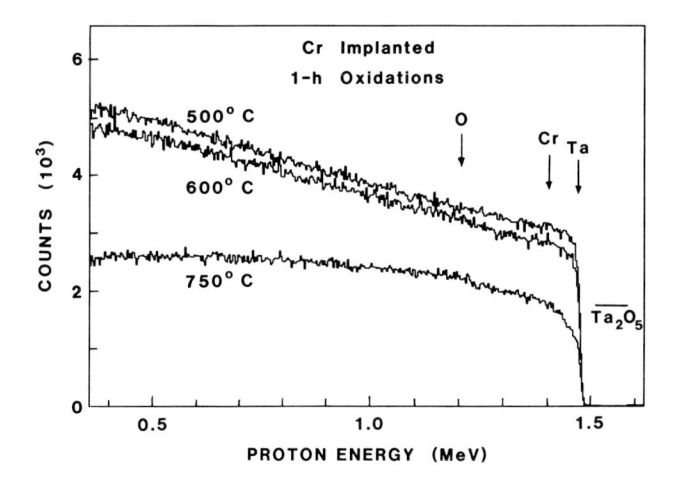

Fig. 5 Proton backscattering spectra for Cr-implanted Ta, oxidized for 1h at 500, 600 or 750°C. The yield for Ta_2O_5 at the surface is shown, as is the energy for backscattering from the surface of O, Cr, and Ta.

The 600°C sample showed evidence that incipient breakdown was occurring. Different regions of the same sample produced different levels of Ta concentration in the plateau region of the spectra. Furthermore, optical microscopy indicated that small eruptions were present on some regions of the surface. AES analysis indicated that, as in 500°C 1h oxidations, a resevoir of Cr was still present in the metal beneath the oxide film. By comparison, the unimplanted area of this sample entirely lost its oxide when the sample was cooled to room temperature. The oxide literally exploded from the surface in small flakes.

Since no reduction in oxide thickness from Cr implantation was observed in the 750°C sample (by either proton backscattering or optical microscopy), an additional oxidation sequence was performed. To ensure sufficient time for a Cr-rich oxide barrier film to form, a sample was first oxidized at 500°C for 1h before further oxidation at 750°C for 1h. The same result as shown in Fig. 5 was obtained. Apparently, at 750°C the barrier layer can no longer sufficiently slow O inward diffusion, or it is no longer stable.

CONCLUSIONS

While Cr implantation has been shown to provide significant improvement to the oxidation resistance of Ta under relatively mild conditions (500°C, t<100h), this protection cannot endure longer times or higher temperatures of oxidation. Improved protection may be possible with a thicker Cr-rich barrier layer, although the AES results indicate that excess Cr was already present in the metal beneath the oxide film. Ultimately such protection will probably fail since Cr_2O_3 is thermodynamically less stable than Ta_2O_5. Ta_2O_5 should nucleate beneath the barrier layer, grow, produce stress, and cause breakdown or dissolution of the protective film. Formation of a more stable diffusion barrier might be able to provide more durable protection. Candidate elements for such protection include Al, Si, and Zr, and further implantation work with these elements is continuing.

ACKNOWLEDGEMENTS

The authors gratefully acknowledge the assistance of Jim Sprague for the SEM work and Dave Baldwin for the AES analysis.

REFERENCES

1. F.E. Bacon and P.M. Moanfeldt in: Columbium and Tantalum, F.T. Sisco and E. Epremian, eds. (Wiley, New York 1963) ch. 9, p. 347.
2. J. Stringer, Reviews on High-Temperature Materials 1, No. 3, 256 (1973).
3. E.N. Kaufmann, R.G. Musket, J.J. Truhan, K.S. Grabowski, C.R. Gossett, and I.L. Singer, Nucl. Instrum. Methods 209/210, 953 (1983).
4. P. Kofstad, High-Temperature Oxidation of Metals (Wiley, New York 1966) pp. 190-227.
5. G. Horz, Acta Metall. 27, 1893 (1979).
6. J. Stringer, J. Less-Common Met. 12, 301 (1967).
7. J.V. Cathcart, R. Bakish, and D.R. Norton, J. Electrochem. Soc. 107, 668 (1960).
8. P. Kofstad and O.J. Krudtaa, J. Less-Common Met. 5, 477 (1963).

ENHANCEMENT OF URANIUM OXIDATION RESISTANCE BY MOLYBDENUM IMPLANTATION

E. N. KAUFMANN,* R. G. MUSKET,* C.A. COLMENARES,* AND B. R. APPLETON**
*Lawrence Livermore National Laboratory, Livermore, CA 94550;
**Oak Ridge National Laboratory, Oak Ridge, TN 37830

ABSTRACT

Initial studies of the oxidation resistance imparted to uranium metal through the implantation of Mo ions have shown that a significantly increased resistance can be achieved for doses of $5 \times 10^{16}/cm^2$. No enhancement of oxidation resistance was found for doses below $1 \times 10^{16}/cm^2$. We report results of weight-gain measurements, ion-beam analysis, and x-ray diffraction, as well as observations of the way in which the protective implanted layer fails at long exposure times.

INTRODUCTION

Pure α-phase uranium is known to be highly susceptible to low-temperature oxidation, particularly in humid atmospheres [1]. Alloys of uranium with additives such as Ti, Nb, or Mo display much-improved oxidation resistance. Bulk alloys, however, also differ from the pure α-phase in crystal structure, density, and mechanical properties, and may for these reasons be less suitable in specific applications. A surface-protection method that prevents oxidation while maintaining bulk properties could prove useful. Of the many surface-modification techniques available, ion implantation is one that does not result in dimensional changes in a finished part, that suffers from no film-adherence problems (since the modified layer is integral with the surface), and that can overcome diffusion barriers and metallurgical incompatibilities.

A preliminary indication that 350-keV Mo implanted in α-uranium to a fluence of $3 \times 10^{16}/cm^2$ provides oxidation resistance was reported by Anthony [2]. This observation is consistent with the behavior of the known alloy U-10wt%Mo, which can be retained metastably in the cubic γ-phase and which is quite corrosion-resistant. Anthony's test conditions [80°C in air with 266-torr H_2O vapor (75% relative humidity)] were considerably less severe than those routinely applied at Lawrence Livermore National Laboratory in evaluating coatings [80°C in N_2 with 100-torr H_2O vapor (27% relative humidity)], primarily because the presence of gaseous O_2 (in air) at greater than 50 ppm inhibits the corrosive action of H_2O vapor [3]. Oxygen content of the N_2 gas used in our experiments was less than 5 ppm. Our intent was to determine if Mo-implanted uranium showed enhanced resistance under the more severe conditions and to reveal the responsible underlying mechanisms.

EXPERIMENTAL PROCEDURES AND RESULTS

Pure depleted-uranium metal disks of 0.565-in. diam and 0.050-in. thickness, with a small hole near the periphery, were machined, mechanically lapped, and finally electropolished for use in this study. The disks were initially implanted, using a rastered beam, on one face with 180-keV $^{96}Mo^+$ ions to a maximum fluence of $5 \times 10^{16}/cm^2$ (projected range $\sim$300 Å). After implantation, the unimplanted face, the edge, and a portion of the implanted face were coated with a polymer

(Kel-F) protective layer on order to slow the corrosion rate of unimplanted uranium and were then exposed to the above quoted wet N_2 atmosphere while weight gain was monitored periodically. Kel-F provides a degree of protection under these test conditions, but has a finite permeability to moisture [4]. Three additional unimplanted control specimens were simultaneously exposed; one uncoated, one fully coated, and one partially coated in a fashion identical to the implanted specimen. An oversized lower-purity sample with a 600-grit finish was also exposed. The data collected are displayed in Fig. 1 in terms of absolute weight gain versus time. Because the polymer coating does not provide an absolute barrier to moisture and actual oxidation beneath the coating would depend sensitively on coating thickness variations, it is not possible to state weight gain per unit area for the exposed-metal regions alone. Thus, the data of Fig. 1 should be viewed as semi-quantitative. Since the implanted and partially coated sample oxidized to the same extent as the fully coated sample, molybdenum implantation provides oxidation resistance comparable to that of the Kel-F coating.

Because the sensitivity of the weight-gain technique for the small areas employed is poor at short times, a second set of samples (without coating) were alternately exposed to the oxidizing environment and analyzed using a combination of MeV-ion backscattering and ion-induced x-ray spectroscopy. These data, converted to equivalent weight gain, are tabulated in the lower left of Fig. 1. Both the ion-beam analysis (2-mm-diam beam spot) and the weight-gain measurements give area-averaged values.

Ion beam analysis:
Equivalent weight gain (μg)

Time (h)	600 grit	Electro-polished	Implanted
1	68	42	2.3
4	232	200	2.0
20	1844		1.9

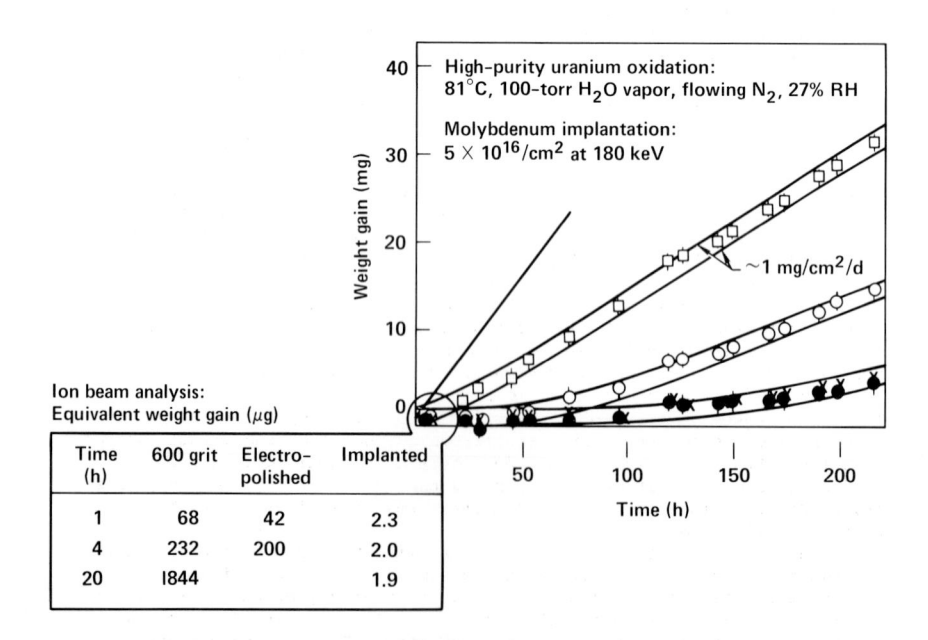

FIG. 1. The plot shows the relative weight gain versus time of a low-purity 600-grit-polished control sample (solid line), a bare electro-polished disk (□), a partially polymer-coated disk (○), a fully polymer-coated disk (●), and a partially coated Mo-implanted disk (X). The table in the lower left of the figure gives equivalent weight-gain results at early times derived from ion-beam analysis.

Details of the simultaneous ion backscattering and ion-induced x-ray analysis approach have been reported previously [5]. In the present study, analyses of oxides thinner than about 1 μm were performed using 2-MeV helium ions; oxides between 1 and ∼5 μm thick were probed with 2.5-MeV protons. Useful ion-induced oxygen x-ray analysis was limited to oxide thicknesses equivalent to 6-1000 Å of UO$_2$; backscattering analysis gave results for oxides greater than ∼300 Å of UO$_2$. The agreement between the backscattering and x-ray results was typically within 15% for UO$_2$ thicknesses between 300 and 1000 Å.

Visual observations of the oxidized sample confirmed that at short times no discernible oxidation occurred in implanted regions. After 220 hours, however, the exposed implanted areas appeared severely oxidized, as did the unimplanted region beneath the polymer coating. Regions that were both implanted and coated were largely free of oxide except for small points of localized attack. It was also noted that although the exposed implanted area was severely attacked after 220 hours to the point where the original surface had sloughed off, a few small plateau-like regions remained where only minor localized oxide clusters were seen. This observation may indicate that the Mo-implant layer is capable of providing greater protection than the area-averaged weight-gain data show, but is defeated by oxidation, which spreads laterally from points of local attack.

An additional set of samples was prepared with the intent of verifying and clarifying the above observations. Some of these were first Ni-plated over the entire surface and then polished on front and back faces, leaving only the edge protected by Ni. Of these, some were electropolished as a final preparatory step and some were only mechanically polished to a 0.25-μm-grit size. Samples intended for ion-beam analysis were Mo-implanted over only one half of one face. Others were implanted over the entire area of both faces. It is known that the electropolishing process provides some degree of passivation in the early stage of oxidation, compared with a purely mechanical polish. A 14% reduction in the oxidation rate of electropolished uranium from that of mechanically polished uranium has been recently measured by Colmenares [6]. Similar reduction in rates has been observed for thorium [7]. For this reason, precautions taken to prevent oxidation before implantation (including storage in vacuum and only limited air exposures) were inadequate for the mechanically polished samples, which developed a blue hue indicative of relatively thick oxide (later determined by ion-beam analysis to be ∼500 Å of UO$_2$). On the basis of the projected range of 180-keV ^{96}Mo ions, we conclude that a significant fraction of the Mo stopped in this oxide for these samples.

Both the electropolished and the mechanically polished samples were alternately exposed to the oxidizing atmosphere and subjected to ion-beam analysis and optical and/or SEM microscopy. Figure 2 shows an SEM micrograph taken in the implanted region of an electropolished sample after 20 hours exposure. This illustrates a general characteristic of the localized attack observed. Individual crystallographic grains showed different susceptibilities to attack and the oxide outcroppings in a given grain appear to form along parallel lines, the directions of which varied from one grain to another.

Figure 3 shows SEM micrographs taken after 52 hours exposure, where implantation has inhibited oxidation to some extent. It is interesting to note that in the electropolished sample, the boundary between unimplanted and implanted areas has been obliterated by a lateral spreading process apparently originating in the unimplanted area. But, in the mechanically polished case where the Mo is believed to be partially trapped in original air-formed oxide, this boundary remained well delineated. Aside from the lateral spreading phenomenon, the implanted regions in both cases show only progressive localized attack. The microscopy performed to date has not revealed the special nature of those points where local attack

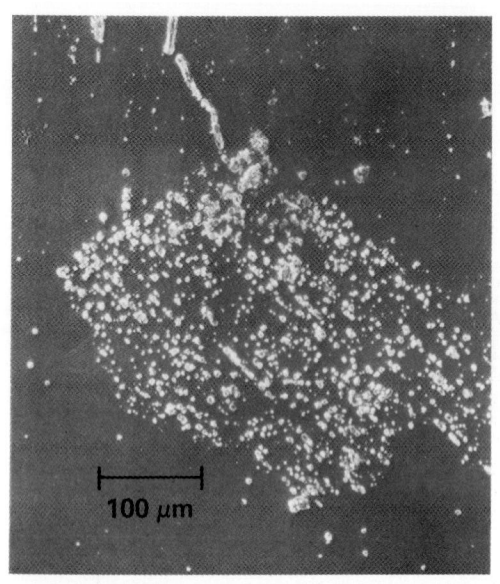

FIG. 2. SEM micrograph of regions of electropolished and implanted sample after 20 hours of oxidation. This shows a correlation of the density and distribution of oxide nuclei with crystalline grain orientation.

nucleates (i.e., no systematic correlation with inclusions, polishing scratches, etc.).

Ion-beam analysis of these samples showed that for 52 hours exposure, the equivalent UO_2 thickness on the electropolished sample grew from ~320 Å to ~750 Å in the implanted region. Without the implant, the oxide grew to over 2200 Å in one hour and subsequently reached a thickness beyond the helium-ion-beam analysis range. The mechanically polished sample behaved similarly to the electropolished case in the unimplanted region, whereas in the implanted area, the oxide thickness, initially at 500 Å, remained below 800 Å during 52 hours of exposure.

A sample of the electropolished and implanted type was studied with glancing-angle x-ray diffraction using a thin-film focusing Guinier camera. These data clearly reveal that the modified surface is no longer single-phase α-uranium. Several additional diffraction lines could be indexed as the more intense lines expected from bcc γ-uranium and the tetragonal intermetallic phase MoU_2 (also denoted γ' or δ in the literature). Lines corresponding to α-phase may have originated from the modified layer and/or from the underlying bulk.

In a parallel fluence-dependence study, it was found that molybdenum fluences of 2, 5, and 10×10^{15} ions/cm^2 provided no observable retardation of oxidation for exposures of more than one hour. In fact, the oxidation rate for all three low-dose cases during the first hour of exposure exceeded that of the electropolished region by more than a factor of two.

DISCUSSION

The data described above clearly show that Mo implantation does enhance the resistance of uranium metal to oxidation in a humid environment and is roughly comparable in protective value to the polymer

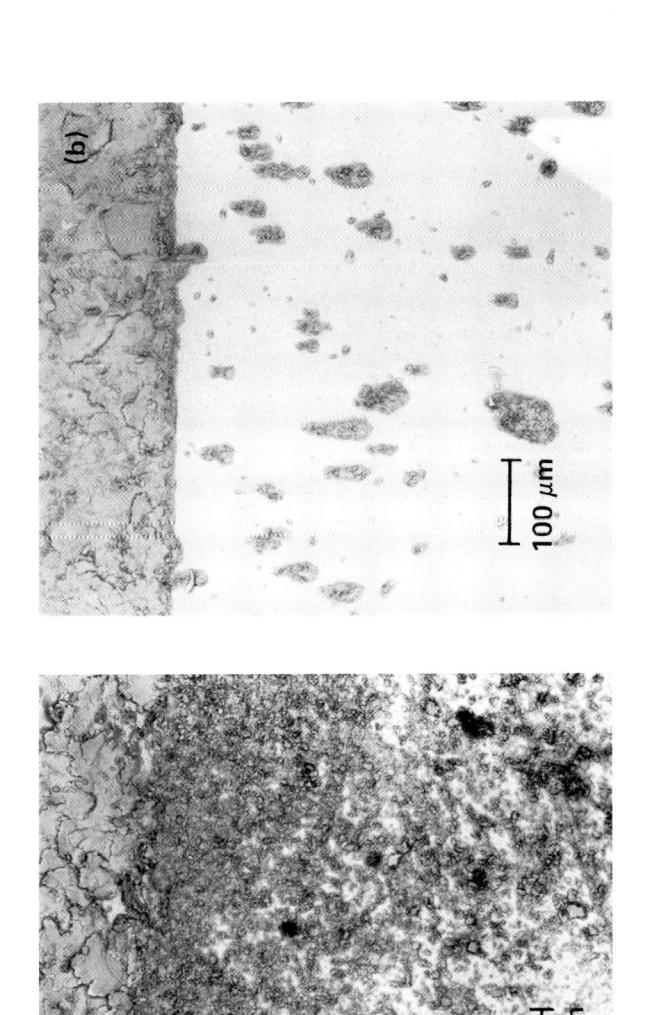

FIG. 3. SEM (backscattered electron) micrographs of the boundary region between implanted and unimplanted areas on (a) an electropolished sample, and (b) a lapped sample (with significant native oxide), after 52-hour exposure to wet N_2.

coating used here. It is also likely that the protective mechanism includes the formation of the cubic γ-phase and MoU$_2$ within the surface layer. The protective action breaks down through localized attack at specific points, followed by continued oxide formation emanating from these points. Several reasons for this can be considered. Native surface imperfections (such as inclusions) may stimulate reactivity through preventing Mo alloy formation or through stress concentrations. The resistant Mo-U phases may not form a continuous layer in the as-implanted state. The micrograph in Fig. 2 indicates that oxide may nucleate preferentially along specific crystallographic directions, possibly because of a favorable density of atomic steps, or slip-induced surface discontinuities. The mechanism of oxidation by H$_2$O may involve factors also relevant to the hydriding reaction because the nature of the corrosive attack appears similar to that observed for the reaction of hydrogen with uranium [8].

Further studies of this system are under way. In particular, the annealing of the implant to achieve more uniform γ-phase formation will be attempted and weight-gain data will be acquired without the added complexity of a polymer coating. An additional aspect needing further study is the different behavior of the sample in which a thick native oxide was Mo-implanted. Here questions dealing with oxide modification by introducing stress, by forming a mixed oxide, or by creating an amorphous phase need resolution.

ACKNOWLEDGMENTS

We would like to thank R. G. Patterson, R. D. Smith, T. G. McCreary, S. Holmes, C. L. Farrell, and J. L. Moore for technical assistance. This work was performed under the auspices of the U.S. Department of Energy by the Lawrence Livermore National Laboratory under Contract W-7405-Eng-48. Aspects of the work were presented at the 11th Annual DOE Compatibility Conference, October 18-20, 1983.

REFERENCES

1. C.A. Colmenares, Progress in Solid State Chemistry 9 (1975)139.
2. S.R. Anthony, Ion Implantation as a Method of Improving the Oxidation Resistance of Uranium, AWRE, Aldermaston, UK, Report 010/79.
3. M. McD. Baker, L.N. Less, and S. Orman, Trans. Faraday Soc. 62 2525 (1966).
4. C. Colmenares, T. McCreary, S. Monaco, C. Walkup, G. Gleeson, J. Kervin, R.L. Smith, and C. McCaffrey, Aluminum and Polymeric Coatings for Protection of Uranium, Lawrence Livermore National Laboratory, Report UCID-19970 (1983).
5. R.G. Musket, Nucl. Inst. Meth. (in press, 1984).
6. C. Colmenares, Lawrence Livermore National Labratory, private communication, 1984.
7. F.T. Zurey, S.H. Gelles, D.I. Phalon, D.A. Vaughan, W.E. Berry, W.K. Boyd, and P.J. Grisphover, The Effect of Surface Treatment on the Corrosion Behavior of Thorium, Battelle Memorial Institute, Columbus, OH, Report BMI-X-572 (1969).
8. R.G. Musket, G. Robinson Weis, and R.G. Patterson, this volume.

MODIFICATION OF THE HYDRIDING OF URANIUM USING ION IMPLANTATION

R. G. MUSKET, G. ROBINSON-WEIS, AND R. G. PATTERSON
Lawrence Livermore National Laboratory
Livermore, California 94550

ABSTRACT

The hydriding of depleted uranium at 76 Torr hydrogen and 130°C has been significantly reduced by implantation of oxygen ions. The high-dose implanted specimens had incubation times for the initiation of the reaction after exposure to hydrogen that exceeded those of the non-implanted specimens by more than a factor of eight. Furthermore, the non-implanted specimens consumed enough hydrogen to cause macroscopic flaking of essentially the entire surface in times much less than the incubation time for the high-dose implanted specimens. In contrast, the ion-implanted specimens reacted only at isolated spots with the major fraction of the surface area unaffected by the hydrogen exposure.

INTRODUCTION

The results presented here represent the first known use of ion implantation for the modification of the hydriding properties of a material. In particular, oxygen ions were implanted in uranium up to concentrations calculated to form surface oxides that should be stable under the hydriding conditions (76 Torr hydrogen and 130°C). The inhibiting effect of surface oxides on the hydrogenation of uranium has been recognized for many years. In addition, uranium hydride ignites in oxygen to form U_3O_8 or, when insufficient oxygen is present, UO_2 and reacts at elevated temperature with water to form UO_2 [1,2]. Since these oxides are thermodynamically more stable than the hydride and inhibit the hydrogenation reaction, it is quite logical to consider using oxide films for protection of uranium from hydrogen. The oxide formed under ambient atmospheric conditions does not form a useful, protective barrier, presumably because of its porosity, defects, impurities, and tendency to spall off the surface. Oxidation at higher temperatures with high-purity oxidizing gas would probably result in an improved barrier. However, heating the uranium may be inconsistent with other constraints for the bulk material (eg, grain size, precipation of undesirable phases, dimensional tolerances). Implantation of oxygen into uranium surfaces under high energy, low current conditions (ie, low power conditions) insures that the bulk properties are retained while the oxygen concentration in the layer increases.

EXPERIMENTAL CONSIDERATIONS

The specimens used all came from one piece of high-purity depleted uranium. Spectrochemical analysis revealed the following impurities or detection limits (in ppm by weight): Tm<60; Ba<40; Zr<10; Fe=7; Zn, Al, Si, Ca, Nb, Sb<4; Mg=2; Cu, Mn, Ni, Bi, Co, Cr, Sn, Sr<1; B, Mo, Ti, Cd, Tl, Pb, V<0.4; Be<0.1. Following EDM machining of the disk specimens from the source piece, standard lapping procedures were used to achieve the nominal

---

*Work performed under the auspices of the U.S. Department of Energy by the Lawrence Livermore National Laboratory under contract number W-7405-ENG-48.

dimensions of 27mm dia by 1.25mm thick. Inadvertent air exposure of the lapped specimens necessitated a repolishing procedure that consisted of an electropolish in perchloric acid to remove the gross oxide, a 600 grit polish to remove pits, and a final mechanical polish with 1μm diamond paste. The specimens were not annealed but were stored in kerosene and/or vacuum until just prior to shipment (in all-metal UHV containers) to the implanters. The last surface treatment immediately before shipment was an electropolishing procedure in a solution of 450 ml ethanol (95%), 275 ml ethylene glycol, and 275 ml phosphoric acid (85%) at room temperature at 13V for 3 seconds. Finally the specimens were rinsed in de-ionized water and dried in flowing argon. These preparation procedures were applied to both the control and the to-be-implanted specimens.

Molecular oxygen ions (O_2^+) were implanted, using an electronically rastered beam, with energies of 180 keV to various doses up to 8×10^{17} O_2^+/cm^2 over the central 25 mm-diameter area of each specimen. The range and standard deviation for 90 keV O^+ ions in pure uranium were calculated using the TRIM program [3] to be 830 Å and 440 Å, respectively. The ion flux during implantation was approximately 1μA/cm^2 (ie, 0.18 W/cm^2). The maximum temperature reached during implantation was estimated to be 50°C. Electropolished control specimens accompanied the specimens for implantation in each UHV shipping container; thus, the control and implanted specimens experienced the same atmospheric exposures. After return from the implanters, all specimens were held under vacuum until they were mounted in the hydriding chamber.

The hydriding studies were performed using a vacuum chamber for edgeless studies of gas-metal reactions which has been detailed previously [4]. In essence, the chamber consisted of two separate vacuum regions with most of the front (ie, implanted) surface of the specimen exposed to the hydrogen while the other surfaces remained under vacuum (<10^{-5} Torr). The extent of the reaction was determined by measuring the quantity of hydrogen taken up by the exposed area (3.0 cm^2) under isobaric (76 Torr) and isothermal (130°C) conditions. To insure uniform temperature across the specimen, the entire chamber was heated to the desired temperature as measured by a sheathed thermocouple in contact with the back of the specimen. Visual observation of the front surface was made through a sapphire window. Hydrogen gas was used directly from tanks of 99.999 percent pure hydrogen. The gas was expanded several times into evacuated, baked volumes of the gas-handling system before each hydriding experiment. The relatively short induction times for the non-implanted specimens that were given a 600-grit polish immediately prior to mounting in the hydriding chamber attest to the cleanliness of the hydrogen over the reacting surface. The reaction was stopped by simply evacuating the hydrogen and removing the heating power.

The reaction conditions were chosen to fulfill several criteria. A relatively slow reaction rate was desired to permit observation of unambiguous distinctions for the various implantation conditions; however, the rate needed to be sufficient to insure fairly prompt reaction with the 600-grit polished specimens. The temperature of 130°C meets the above conditions and, in addition, provides a reaction that follows a linear law [1] and a hydride attack typical of lower temperatures (ie, formation of funnel-like hydride pits which penetrate into the reacting sample) [5]. The gas handling system for hydriding was limited by the available pressure transducers to 1000 Torr in the supply reservoir and 100 Torr in the reaction chamber. We chose to use 800 and 76 Torr, respectively. For our reaction conditions the linear reaction rate was estimated from published data [6] to be approximately 0.8 μm of pure uranium per minute. The reaction was considered initiated when the amount of hydrogen consumed clearly exceeded the drift in the barometric sensors (ie, after consumption of 0.025 Torr-liters). For uniform hydrogen up-take and reaction 0.025 Torr-liters of gas would consume about 0.04 μm of pure uranium (ie, 1.6μm/Torr-liter).

RESULTS AND DISCUSSION

The hydrogen consumption results are displayed in Fig. 1. Reactions of

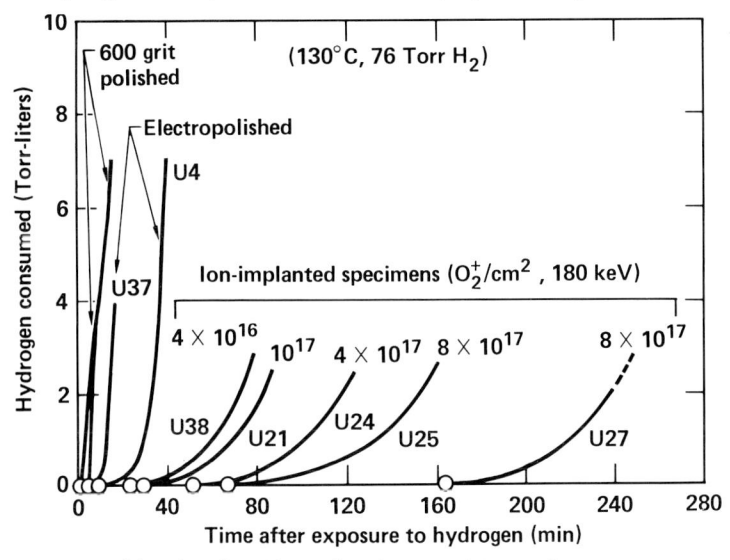

Fig. 1. Reaction of hydrogen with uranium.

the two 600-grit polished specimens and one electropolished specimen were terminated after consumption of about 7 Torr-liters of hydrogen because macroscopic flaking of essentially the entire surface was observed. The reactions with the other specimens were terminated before 4 Torr-liters had been consumed to minimize the surface damage and, hopefully, permit reuse after repolishing. The reaction rates at the end of the exposures were 0.7 and 0.9 μm/min for the two 600-grit polished specimens. These rates are consistent with the anticipated result of 0.8 μm/min for the linear reaction rate [6]. In contrast, the final rates for the two electropolished specimens were 1.3 and 1.5 μm/min. Presumably, the different rates observed for 600-grit polished and electropolished specimens is a consequence of differences in the surface conditions (eg, surface strain). The open circles on the abscissa of Fig. 1 represent the induction times (ie, the times for initiation of the reaction after exposure to hydrogen). Clearly, oxygen-ion implantation can increase the induction time by a considerable amount. In fact, for the oxygen-ion doses employed the induction time increased monotonically with implanted dose. Comparison of the induction times t_i for the two electropolished specimens (t_i = 4 and 8 mins) with the high-dose implanted specimens (t_i = 69 and 163 mins) shows that the induction time can be increased by factors of 8 to 40. The small induction time difference between the electropolished specimens was most probably due to slight differences in the hydrogen gas purity or to differences in the surface films present (these two measurements were made approximately three months apart). For the high dose implanted pair the large discrepancy in t_i is presumably a consequence of the nature of the hydriding process for implanted specimens.

Fig. 2a is an optical micrograph of electropolished uranium (U4) after 40 minutes of exposure to hydrogen and removal of most of the loose hydride powder. The high density of pits that are remanents of the macroscopic flaking process covers essentially the entire surface. Figs. 2b-d are scanning electron micrographs of a spot where the density of pits was lower than

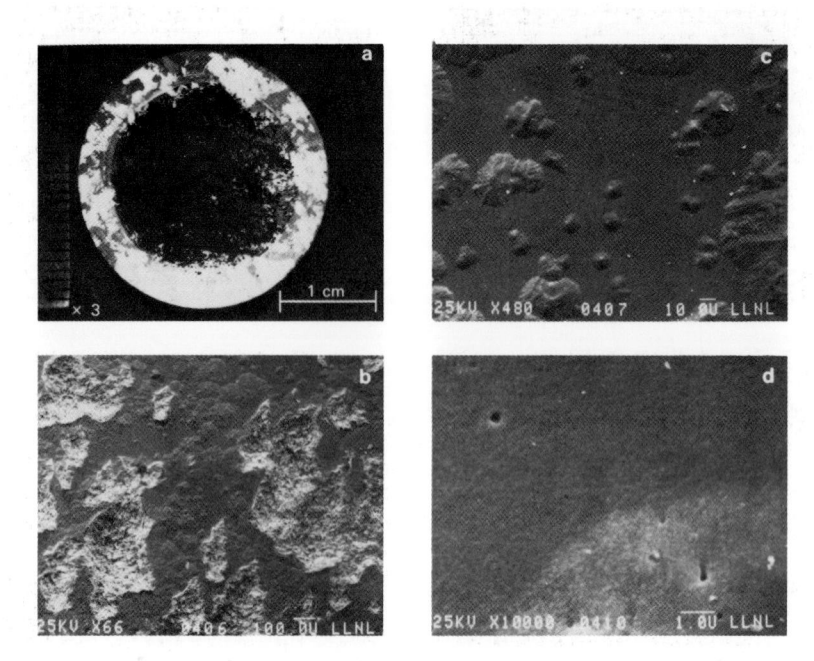

Fig. 2. Micrographs of electropolished uranium (U4) after exposure to 76 Torr H_2 at 130°C for 40 minutes: (a) optical micrograph and (b-d) scanning electron micrographs.

typical. The field of view in Fig. 2b includes all three of the major features of the surface: (1) irregularly shaped pits, (2) pre-flaking hydride mounds, and (3) apparently virgin uranium. Fig. 2c provides a magnified view of the center of Fig. 2b and shows hydride mounds in various stages of development. The small bright spots are residual hydride powder particles. A highly magnified view of the center of Fig. 2c is given in Fig. 2d, which has in its lower right hand corner the edge of a mound. The upper left region of Fig. 2d is typical of the virgin uranium.

Fig. 3 is a display of the surface morphology typical of the implanted specimens after exposure to hydrogen. The optical micrograph of Fig. 3a shows that even after 87 minutes exposure, a surface implanted with 10^{17} $O_2^{\frac{1}{2}}/cm^2$ was attacked only in a few clearly distinguishable spots. Since the attack occurred only over a small fraction of the total surface, the apparent rate of hydrogen consumption (Torr-liters/min) should be less than that for the 600-grit polished and electropolished specimens, which were attacked more uniformly. Thus, for a given amount of hydrogen consumption, the lower rates observed for the implanted specimens (Fig. 1) is a direct measure of the fraction of the surface participating in the reaction. The reaction area decreased with increasing implanted dose. Figs. 3b-d are scanning electron micrographs with magnifications directly comparable to those of Figs. 2b-d. The center of Figs. 3b-d is an apparently undisturbed or virgin uranium area near one of the pits. Removal of the hydride powder debris was more difficult for this specimen as evidenced by the higher density of bright spots in Figs. 3b-d relative to that in Figs. 2b-d. In contrast to the electropolished specimens, there was no evidence of hydride mounds on the implanted samples, except for the lowest dose (4×10^{16} $O_2^{\frac{1}{2}}/cm^2$) specimen. This result was consistent with the visual observations made during the hydriding reactions. The electropolished samples reacted by a

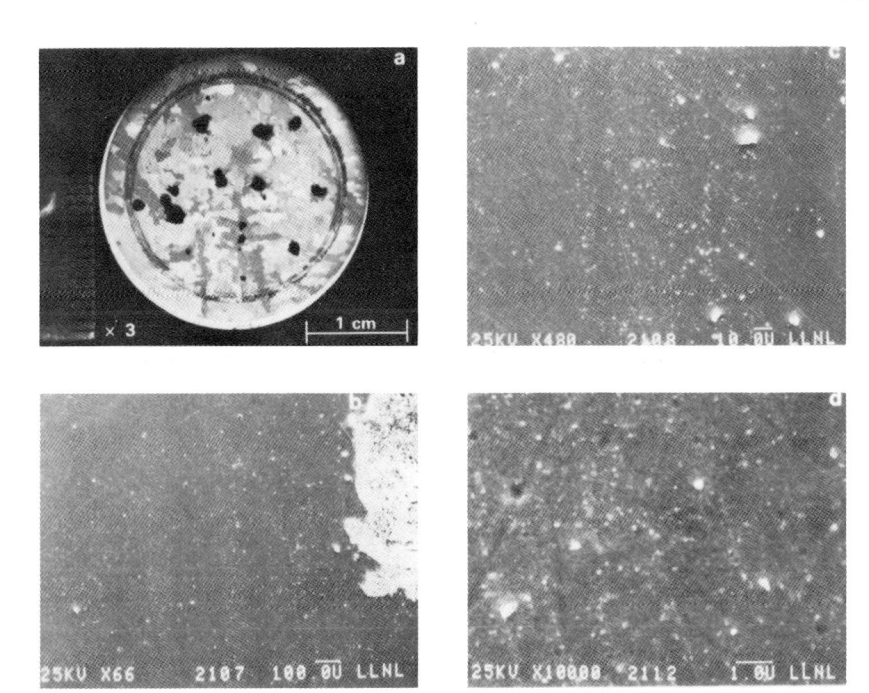

Fig. 3. Micrographs of uranium (U21) implanted with 10^{17} O$\frac{1}{2}$/cm^2 at 180 keV after exposure to 76 Torr H$_2$ at 130°C for 87 minutes: (a) optical micrograph and (b-d) scanning electron micrographs.

combination of (a) creation of small black spots (ie, pits) over the entire surface and (b) growth of the previously created pits by repeated ejection of hydride particles and reaction with the freshly exposed metallic uranium until essentially the entire surface was covered by pits. For doses greater than 4x10^{16} O$\frac{1}{2}$/cm^2, the implanted specimens reacted by creation of a few small, isolated pits and growth of these pits. The relative rate of creation of new pits was very low for these higher dose specimens.

The reasons for the hydriding being restricted to a few isolated spots on the implanted specimens are not known at present. However, assuming that all high-dose implanted uranium surface areas were protected, the hydriding sites must correspond to small areas that were not modified the same as the vast majority of the surface. This could result from implantation into impurity inclusions and from the absence of implantation in surface areas inaccessible to the ions (eg, sides of some defects or areas masked by dust particles). The virgin uranium areas displayed in Figs. 2d and 3d are typical of these areas for exposed specimens and of all non-exposed surfaces. The dark spots shown at this magnification are presumed to be defects in the surface resulting from the preparation procedures and/or voids in the bulk material. Some of these defects may have had nearly vertical or undercut sides, and consequently, the ion implantation process, which is inherently a line-of-sight process, would have been ineffective on such surfaces.

Combined Rutherford backscattering and ion-induced x-ray analysis [7] using 2 MeV helium ions of the as-implanted specimens showed that only the specimens that received 4x10^{17} O$\frac{1}{2}$/cm^2 had a thick, uniform surface oxide. However, this was inconsistent with the highest dose (8x10^{17} O$\frac{1}{2}$/cm^2) specimens, which had a high concentration of oxygen only at a depth approximately equal to the projected range R_p for 90 keV O$^+$ ions by the TRIM program [3]. The specimens that received 4x10^{17} O$\frac{1}{2}$/cm^2 had a blue color in the implanted region, which would be consistent with the formation of a uniform oxide layer by either implantation or by excessive air-exposure. Scanning electron micrographs of the specimens implanted with the highest dose showed evidence of blister formation, which would be consistent with an implantation flux that greatly exceeded the flux of oxygen diffusing away from the implanted depth. The specimens implanted with the lower doses had oxygen concentrations near the expected R_p that were considerably below that necessary for uniform oxide formation.

Even with the microscopic differences revealed by RBS and SEM, all the implanted specimens produced similar optical micrographs (eg, Fig. 3a) after consumption of the same amount of hydrogen. In general, the specimens implanted with higher doses had fewer hydride pits. This suggests that reduction of the hydrogenation of uranium may be achieved without formation of a uniform oxide layer. Two other possibilities may be responsible for the observations: (1) ion-induced stress in the surface and (2) reduction of the solubility of hydrogen in uranium by implanted oxygen. Ion implantation generally induces stress in surfaces. The ion-induced stress may retard the formation of the hydride mounds and, hence, quench the growth of any hydride nuclei that may have formed. Bloch and Mintz [5] noted that strain effects can prevent the development of bulk precipitation even when the concentration of the dissolved hydrogen inside the specimen reaches saturation, which is approximately 2 ppm for α-uranium at temperatures below 668°C [1]. The importance of oxygen inside the uranium during hydrogenation has been discussed by Condon [8, 9]. Hydrogen, occupying interstitial positions, diffuses rapidly in the metal before nucleation and growth of the β-phase UH_3. Oxygen (or other anions) compete for these interstitial sites and considerably modify the kinetics. Thus, the oxygen can limit the hydrogen solubility and reduce the probability of hydride formation.

Although the reason for the reduced hydriding is not clear at this time, the data show conclusively that ion implantation can have a marked influence on the reaction of hydrogen with uranium. Our work is continuing in an effort to understand and refine the processes involved.

ACKNOWLEDGEMENTS

It is a pleasure to acknowledge C. A. Colmenares for useful discussions, for providing the uranium material, and for use of the gas-handling portion of his hydriding system; T. G. McCreary for help with the initial checkout and operation of that system; H. Hayden's group at the University of Connecticut for the expert ion implantations; A. W. Casey for the scanning electron microscopy; and J. B. Condon for a critical review of the manuscript.

REFERENCES

1. W.M. Mueller, J.P. Blackledge, and G.G. Libowitz (editors), Metal Hydrides (Academic Press, NY, 1968).
2. K.M. Mackay, Hydrogen Compounds of the Metallic Elements (E.&F.N. Spon Ltd., London, 1966).
3. J.P. Biersack and L.G. Haggmark, Nucl. Instr. and Meth. 174, 257 (1980).
4. R.G. Patterson and R.G. Musket, J. Vac. Sci. Technol. (accepted for publication).
5. J. Bloch and M.H. Mintz, J. Nucl. Mater. 110, 251 (1982).
6. J. Bloch and M.H. Mintz, J. Less-Common Met. 81, 301 (1981).
7. R.G. Musket, Nucl. Inst. Meth. Phys. Res. (accepted for publication).
8. J.B. Condon, J. Phys. Chem. 79, 42 (1975).
9. J.B. Condon, J. Less Common Metals 73, 105 (1980).

APPLICATIONS OF ION-IMPLANTED COVALENT POLYMERS

DAVID C. WEBER, MARIANNE K. BERNETT, AND HAROLD RAVNER
Naval Research Laboratory, 4555 Overlook Avenue,
Washington, DC 20375

ABSTRACT

Materials presently used for packaging precision instrument bearings contain minor amounts of antistatic agents, shown earlier to be potential contaminants to both bearing surfaces and lubricants. One method to eliminate the antistatic additives is to use electroactive polymers. The materials with the most promise for this application are the F ion implanted polyacetylene and polyparaphenylene. The characteristics and behavior of the polymers will be presented.

INTRODUCTION

Delicate instrument components (i.e., precision gyrobearings) are extremely sensitive to atmospheric particulate contamination. To avoid such effects during storage or transport, these components are often packaged in antistatic polymeric containers, which prevent the accumulation of electrostatic charges. The currently used antistatic agents incorporated into polyethylene or nylon packaging films have shown adverse effects after extended contact with metal surfaces.(1,2) By selective ion implantation of electroactive polymers, remarkable surface resistivity lowering as well as increased oxidation stability were accomplished. Processes for improving the physical properties, such as flexibility, were also studied. After long-term contact with lubricated and unlubricated bearing steels, the modified conducting polymers and commercial antistatic materials were evaluated for adverse effects.

EXPERIMENTAL

Films

Polymers examined were polyacetylene, $(CH)_x$, prepared at the Naval Research Laboratory, polyparaphenylene sulfide (PPS), and a copolymer of ethylene and methacrylic acid (COPOL); the latter two were obtained from commercial suppliers. Since unmodified $(CH)_x$ was susceptible to oxidation when exposed to air, samples were either chemically doped with 2 percent I^- or PF_6^-, or ion implanted with F^+ to improve their stability. To maintain flexibility during air exposure, $(CH)_x$ was deposited on polypropylene (PP) film prior to F^+ implantation. PPS and COPOL, although flexible and stable, were implanted with F^+ to decrease their surface resistivities.

Three flexible polyethylene-based films, and a more rigid gold-coated polyester film, all commercially available antistatic or conductive materials, were also examined. Composition and selected properties of the films are summarized in Table I. All properties of the commercial films were supplied by the vendors.

TABLE I. Physical properties of film materials.

Composition	Code	Flexibility	Thickness (mil)	Surface Resistivity (Ω/sq)
Polyacetylene	(CH)$_x$	flexible	~4	3×10^{11}
(CH)$_x$ I$^-$ doped	(CH)$_x$ + I	fairly flex.	~4	—
(CH)$_x$PF$_6^-$ doped	(CH)$_x$ + PF$_6$	stiff	~4	—
(CH)$_x$F$^+$ implanted	(CH)$_x$ + F	fairly flex.	~5	2×10^7
(CH)$_x$F$^+$ implanted, on polypropylene	(CH)$_x$ + F(PP)	flexible	3	5×10^5
Polyparaphenylene sulfide F$^+$ implanted	(PPS) + F	flexible	3	1×10^8
Copolymer ethylene/ methacrylic acid, F$^+$ implanted	COPOL + F	flexible	5	7×10^7
Polyethylene with slip agent	PE	flexible	4	3×10^9
Polyethylene with antistat + mylar + Al-Ti + antiabrasive	PE-Al-Ti	flexible	3	$<10^{11}$
Polyethylene with antistat + Al + urethane	PE-Al	flexible	3	$<10^{12}$ (int) $<10^6$ (ext)
Mylar + Au + ceramic	MYL + Au	bendable	5	1.6×10

Substrates and Lubricants

Substrates were flats of 440C steels, representative of precision miniature bearing steels (3), metallurgically polished and passivated with sodium dichromate and nitric acid. To ensure surface-chemically clean surfaces, they were ultrasonically cleaned with acetone and fluorinated solvent immediately prior to use.

Four oils were employed in this study. They included: a base stock nonpolar synthetic hydrocarbon derived from the polymerization of α-olefins; a formulated version of the above synthetic hydrocarbon; a formulated synthetic hydrocarbon from a different base stock; and a formulated mixture of a polyolester and a diester (Military Specification MIL-L-81846). The formulations are designed for use with precision miniature bearings. All formulations contained small amounts of antioxidants and other additives.

Preparation and Modification of Conductive Polymers

Polyacetylene was synthesized by a modified Shirakawa technique (4). Polyacetylene on oriented polypropylene was prepared by coating the substrate film on one side with an appropriate catalyst and then adding the acetylene.

Chemical doping with iodine was performed by contacting I$_2$ vapor with the (CH)$_x$ for about 15 hours. For PF$_6^-$ doping, NOPF$_6$ was dissolved in nitromethane and the resulting solution was introduced to the (CH)$_x$. All reactions were carried out under vacuum.

The implantations were performed with a modified ion implanter (5) capable of generating 25-200 keV ions. The ions used were 25 keV F$^+$ generated by fragmentation of BF$_3$. To reduce effects caused by local heating, the ion beam was maintained at 1-10μA while simultaneously cooling the samples. Rastering the beam across the sample ensured a homogeneous distribution of ions laterally across the surface. The samples were implanted to a total fluence of 1x10^{17} F$^+$/cm^2. The energy and fluences used for these implants resulted in a band of implanted ions ~400 Å wide centered at ~300 Å below the surface, achieving a ~20-50 atomic percent F$^+$ content in the implanted region.

TABLE II. Effect on bearing steels and lubricants after extended contact with films.

Film	Lubr.	Visual Film	Visual Lubricant	Visual Steel	FTIR
(CH)$_x$	None	brittle	—	adherent film	unaffected
	SHC-B	brittle	disappeared	adherent film	—
	SHC-F	brittle	disappeared	adherent film	—
	SHC-S	brittle	diminished	oily droplets	lubr. unaffected
	SPDE	brittle	disappeared	dry	clean steel
(CH)$_x$ + I	None	brown ppte	—	brown ppte	unaffected
	SHC-B	iridescent	unaffected	brown ppte + lubr.	lubr. unaffected
	SHC-F	unchanged	unaffected	brown + lubr.	lubr. unaffected
	SHC-S	iridescent	unaffected	brown + lubr.	lubr. unaffected
	SPDE	iridescent	disappeared	brown dry	clean steel
(CH)$_x$ + PF$_6$	None	iridescent	—	opaque adherent	interaction
	SHC-B	unchanged	thickened	opaque adherent	interaction
	SHC-F	unchanged	thickened	opaque adherent	interaction
	SHC-S	unchanged	thickened	opaque adherent	interaction
	SPDE	unchanged	thickened	opaque + lubr.	interaction
(CH)$_x$ + F	None	unchanged	—	unaffected	unaffected
	SHC-B	lubr. seepage	unaffected	unaffected	lubr. unaffected
	SHC-F	lubr. seepage	unaffected	unaffected	lubr. unaffected
	SHC-S	lubr. seepage	unaffected	unaffected	lubr. unaffected
	SPDE	lubr. seepage	unaffected	unaffected	lubr. unaffected
(CH)$_x$ + F (PP)	None	slightly darkened	—	unaffected	—
	SHC-F	darkened	unaffected	unaffected	lubr. unaffected
	SPDE	darkened	unaffected	unaffected	lubr. unaffected
(PPS) + F	None	unaffected	—	unaffected	—
	SHC-F	unaffected	unaffected	unaffected	lubr. unaffected
	SPDE	unaffected	unaffected	unaffected	lubr. unaffected
COPOL + F	None	unaffected	—	unaffected	—
	SHC-F	unaffected	unaffected	unaffected	—
	SPDE	unaffected	unaffected	unaffected	—
PE	None	unaffected	—	smudge	HC contamin.
	SHC-B	unaffected	unaffected	unaffected	lubr. unaffected
	SHC-F	unaffected	unaffected	unaffected	lubr. unaffected
	SPDE	unaffected	unaffected	unaffected	lubr. unaffected
PE-Al-Ti	None	unaffected	—	unaffected	HC contamin.
PE-Al	None	unaffected	—	unaffected	some HC contamination
MYL + Au	None	unaffected	—	unaffected	unaffected
	SHC-B	unaffected	unaffected	unaffected	lubr. unaffected
	SPDE	unaffected	unaffected	unaffected	lubr. unaffected

RESULTS

The acceptable upper limit of surface resistivity for anti-static packaging is 10^{12} Ω/sq. While all films examined in this study have surface resistivities in the acceptable range (Table I), the values for the F^+ implanted specimens are excellent; they are equivalent to- or lower than- the commercial metal-coated laminates (with the exception of MYL).

Table II summarizes the effects of extended contact of the conducting polymers and polymer films with the steel surfaces. The undoped $(CH)_x$ film becomes brittle by air oxidation and adhered strongly to the steel surfaces. Although the $(CH)_x + PF_6^-$ film appeared unchanged, the steel substrate was covered with a spongy opaque adherent residue. Implantation with F^+ ions greatly minimize adverse interactions, both for the sample and the PP-backed $(CH)_x$ films. Equally promising results were obtained with F^+ ion implanted PPS and COPOL. None of the commercial films showed any visually perceptible interactions.

The steel surfaces were analyzed by FTIR for any residues resulting from film contact. Each spectrum was compared to that of a clean steel surface to determine whether additional absorption bands were present. Features in the region of C = O bonds indicated transfer of oxidized hydrocarbons from $(CH)_x$. The $(CH)_x + PF_6^-$ film produced fairly heavy transfers, shown by the presence of several additional absorption bands, including one at 800 cm^{-1}, for the PF bond whereas the spectrum after contact with $(CH)_x$ + F showed no residual matter. The commercial films exhibited the same features in the hydrocarbon region. Such analyses for each system provided the data for the interactions listed in Table II under the FTIR heading.

The steel surfaces were also examined by XPS for contamination from the contacting films. Predominant elements were the steel constituents Fe and O. A binding energy (BE) for O at 530 eV was indicative of iron oxide and at 532-533 eV of organic C-bonded O. A definite shift of the predominant O peak to a higher BE was observed after contact with the unmodified $(CH)_x$ film, which was not too surprising, given the ready oxidizability of $(CH)_x$. A BE of 285eV for C indicated an aliphatic C-C linkage and is therefore characteristic of an almost ubiquitous atmospheric hydrocarbon contamination overlayer, whereas a BE of about 289 eV is characteristic of C bonded to O. Such shifts toward higher BE for C in conjunction with the shifted O signals suggest the presence of organic contamination other than atmospheric. In addition to Fe, O, and C, large signals of F and traces of P were observed in the steel surface after contact with $(CH)_x + PF_6^-$. With the exception of the $(CH)_x + PF_6^-$ specimen, the essentially unvarying C intensities indicate only traces of contaminants from the films.

CONCLUSIONS

Two electroactive polymers, $(CH)_x$ and PPS, when implanted with F^+ ions, have acceptably low surface resistivity for antistatic packaging materials. Furthermore, in the case of $(CH)_x$, implantation also improved oxidation stability. Flexibility of $(CH)_x$ was maintained when backed with polypropylene film. The properties of the F-implanted commercial COPOL were comparable to those of the implanted conducting polymers. A commercial film with a vaporized gold film was the most conductive material examined, but was only moderately flexible. Other commercial polyethylene-based antistatic films were flexible and displayed acceptable resistivity.

The concept of employing conducting polymers as antistatic materials is attractive because, assuming they themselves cause no contamination problems on sensitive surfaces, there are no additional agents that can migrate from the film surface. In the present study, neither the F⁻-implanted conducting polymers nor the F⁻-implanted COPOL evidenced material transfer after long-term contact with 440C steel surfaces. $(CH)_x$ doped with PF_6^- left the surface seriously contaminated. Of the remaining commercial films, only that with the vaporized gold surface caused no contamination of the steel.

Ion-implanted conducting polymers display good conductivity, are flexible and heat-sealable (on backing if necessary), oxidation stable and non-contaminating to contacting metal surfaces. These qualities suggest them to be good packaging materials for sensitive devices. Further research, both with regard to preparation of the conducting polymers and their implantation would probably result in even greater improvements. It must be pointed out, however, that present implant technology and facilites are not yet adequate to produce commercial quantities of film, nor to compete with regard to cost with available antistatic film. Their potential, however, as packaging for contamination-sensitive materials is such as to justify their further study.

ACKNOWLEDGEMENTS

The authors want to thank our collegue, Robert Mowery, for obtaining the FTIR spectra, and Gary O. Head of Lear Siegler, Inc. for performing the surface resistivity measurements on the electroactive polymers and copolymers.

REFERENCES

1. B. J. Kinzig and H. Ravner, 'Problems Encountered with Antistatic Packaging for Miniature Bearings', Lubr. Eng. 36, 219 (1980).

2. M. K. Bernett and H. Ravner, 'Antistatic Agents, Lubricants, and Precision Bearings', Lubr. Eng., 38, 481 (1982).

3. U.S. Department of Defense, 'Bearing, Ball, Precision for Instruments and Rotating Components', Mil. Spec. MIL-B-81793 (1973).

4. T. Ito, H. Shirakawa, and S. Ikeda, 'Simultaneous Polymerization and Formation of Polyacetylene Film on the Surface of Concentrated Soluble Ziegler-Type Catalyst Solution', J. Polym. Sci., Polym. Chem. Ed. 12, 11 (1974).

5. F. A. Smidt, J. K. Hirvonen and S. Ramalingam, 'Preliminary Evaluation of Ion Implantation as a Surface Treatment to Reduce Wear of Tool Bits', NRL Memo Report #4616, (1981).

RECOIL IMPLANTATION OF ITO THIN FILMS ON GLASS SUBSTRATES*

B. H. RABIN, B. B. HARBISON AND S. R. SHATYNSKI
Materials Engineering Department
Rensselaer Polytechnic Institute
Troy, NY 12181

ABSTRACT

Indium-Tin Oxide (ITO) heat mirror films implanted into window glass were obtained by post annealing of argon irradiated coatings of In-5w/o Sn produced by reactive evaporation in oxygen. Characterization of coatings has been carried out using TEM and AES. Optical properties have also been evaluated. The production of acceptable thin films requires low energy deposition rates during ion bombardment. This places a limit on the extent of film-substrate mixing, which is required if increased film lifetimes are to be realized.

INTRODUCTION

The development of a successful solar window relies upon the ability to produce a suitable wavelength selective surface. Heat mirror applications require thin films on window glass that exhibit high visible transmission and high infrared reflection characteristics. The window must also be environmentally stable. Recent work has confirmed the effectiveness of intrinsic and doped semiconducting oxides for heat mirror applications [1]. Thin films are typically produced by a variety of deposition techniques including chemical vapor deposition, sputtering or reactive sputtering, and evaporation or reactive evaporation. Our laboratory has focused on ITO films prepared by a reactive evaporation technique. Suitable optical properties have previously been obtained by reactive evaporation onto substrates heated up to 300°C, or by depositing onto room temperature substrates and then annealing in air. The particular ITO structure produced depends strongly upon the deposition technique and conditions [2]. Sn may substitute for In in In_2O_3 in the form of $In_{2-x}Sn_xO_{3-2x}$, or form small neutral tin clusters [3]. Optical properties and surface analysis [4] have shown films produced in our laboratory consist mostly of bcc In_2O_3 with substitutional Sn. A small amount of unoxidized alloy is also present.

Such ITO films on glass substrates are intended to withstand abrasion and corrosion under prolonged atmospheric exposure. It has been demonstrated that the adhesion of metallic films to glass substrates can be greatly enhanced by ion bombardment [5]. To the best of the authors' knowledge, increased adhesion of oxide films to glass substrates by ion bombardment has not been reported. Oxide adhesion to glass, initially greater than metallic adhesion, can also be improved by ion bombardment based upon structural continuity arguments [6]. The stability of the thin film structure during irradiation must also be considered [7]. This study presents initial

---

*The authors are thankful to Professor Stephen R. Shatynski who unfortunately was killed on September 23, 1983. Without his guidance and energies this research would not have been possible.
Barry Rabin is a General Electric Company Fellow. The authors also acknowledge the support of Consolidated Edison of New York under Grant No.1-04900.

results concerning the structural and compositional changes in metallic and oxide films induced by argon bombardment. Optical properties obtainable by this technique are demonstrated.

EXPERIMENTAL

In-5w/o Sn alloy films were produced by vacuum and reactive evaporation. The details of the reactive evaporation technique can be found in the literature [8]. Films were deposited onto clean soda lime glass substrates and 200Å SiO TEM grids at room temperature. Atmospheres were either 5×10^{-7} torr vacuum or 5×10^{-3} torr total oxygen pressure. Argon implantation was carried out at 100, 120 and 200 keV. Doses of 5×10^{14} and 5×10^{15} ions/cm^2 were obtained with beam currents of $0.2\mu A/cm^2$ and $1.2\mu A/cm^2$ respectively. Transmission microscopy and Auger sputter profiling were carried out before and after ion bombardment. Visible and near infrared transmission were determined relative to uncoated soda lime glass before and after implantation and after annealing in air at 300°C for one hour.

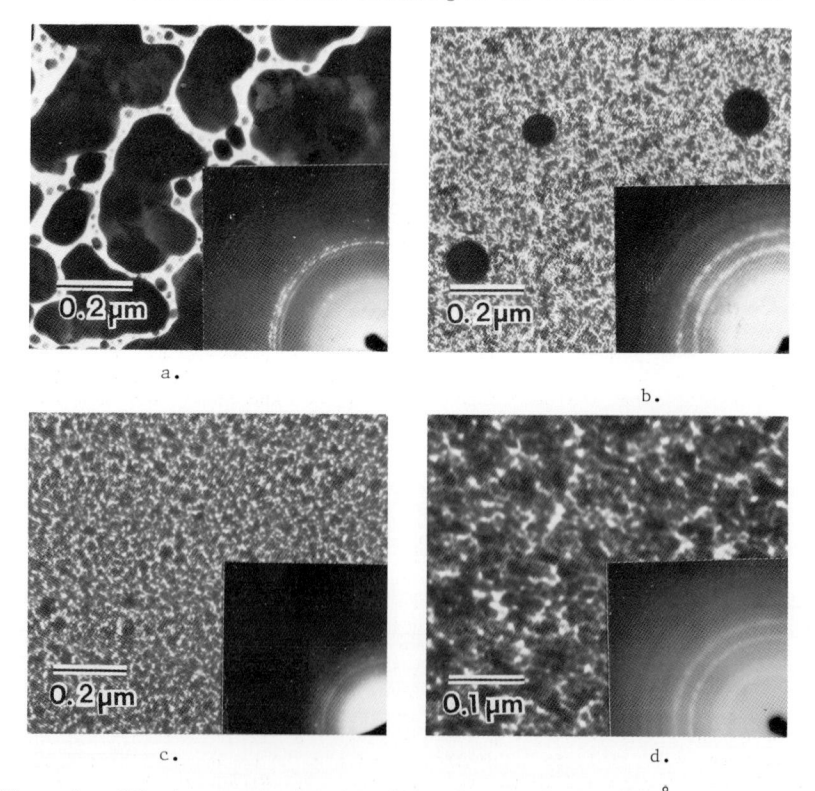

Figure 1: TEM micrographs for thin films deposited onto 200 Å SiO grids a. vacuum evaporated In-5w/o Sn, 1100 Å b. sample a. after implantation with 120 keV Ar$^+$ to dose of 5×10^{15} ions/cm^2 c. reactively evaporated In-5w/o Sn d. sample c. implanted with 120 keV Ar$^+$ to dose of 5×10^{14} ions/cm^2.

RESULTS AND DISCUSSION

The effects of argon ion bombardment on the microstructure of the thin films studied are demonstrated by transmission electron microscopy and selected area diffraction in Figure 1. Prior to irradiation the vacuum evaporated samples consist of islands of a tetragonal solid solution of Sn in In. After 200 keV bombardment to a dose of 5×10^{15} Ar^+/cm^2, electron diffraction confirms the presence of the bcc $In_{2-x} Sn_x O_{3-2x}$ structure. In addition, considerable structural refinement has increased the continuity of the films. Damage created by inert gas bombardment of thin metal films typically results in the formation of point defect clusters, small dislocation loops and gas bubbles [9,10]. Evidence for the presence of such defects in these films has not been found indicating damage effects have been eliminated largely during irradiation or spontaneously at room temperature. The availability of oxygen for the formation of oxides is attributed mainly to backsputtering from the SiO TEM grid and the silica glass substrate. This mechanism has been established by Stroud et. al. [11] who studied recoiling of oxygen atoms from glass substrates into Al films by argon bombardment. The globular regions in Figure 1b indicate that localized melting of the thin film has occurred. The solidus temperature for In-5w/o Sn alloys may be reached at as low as 130°C. It is thus demonstrated that beam heating effects can be detrimental and must be controlled. Large scale processing of low melting temperature films requires low energy deposition rates because substrate cooling is not practical.

Reactive evaporated samples are initially composed mostly of $In_{2-x} Sn_x O_{3-2x}$ with some metal present. The films exhibit a fine granular morphology. Irradiation with excessive dose rates (>0.5μA/cm² at 200 keV) also resulted in partial melting of these films. This condition produces metallic agglomerates in the oxide structure and severely degrades optical properties. Lower dose bombardment does not significantly alter the structure of these films [12]. Radiation damage in the oxide films is not particularly evident at high magnification. Electron diffraction does indicate that some damage persists in the structure, however, most of the defects created during bombardment have been eliminated. This result is not surprising since the implanted species is not chemically active and defect mobility is appreciable at moderate temperatures [13]. These findings indicate that ITO coatings are relatively stable under moderate dose inert gas bombardment, but, implantation effects on optical properties are expected.

Auger depth profiling was carried out on samples before and after recoil implantation to examine compositional changes and determine the extent of film-substrate mixing. Figure 2 illustrates results obtained from a 700Å metallic film and the corresponding reactive evaporated sample. Both metallic and oxide films exhibit interfacial broadening, however, the degree of mixing is greater in the case of the metal. This indicates that the range of argon in ITO was over estimated. Higher beam voltages or thinner oxide films could be used to obtain more efficient recoil mixing. The diffuse nature of the interface suggests this region exhibits the structure and properties of a mixture of film and substrate materials. This is a necessary condition to be met for increasing the adhesion of a thin film to its substrate [6]. Figure 2b shows that ITO films may exhibit a large variation in stoichiometry in the as deposited condition. This is important since the conductivity of ITO is largely determined by the defect structure, particularly the concentration of anion vacancies [3]. Preferential sputtering of oxygen from the surface of ITO samples is observed and this will also affect the conductivity and optical properties. Extensive depletion of oxygen at the sample surface

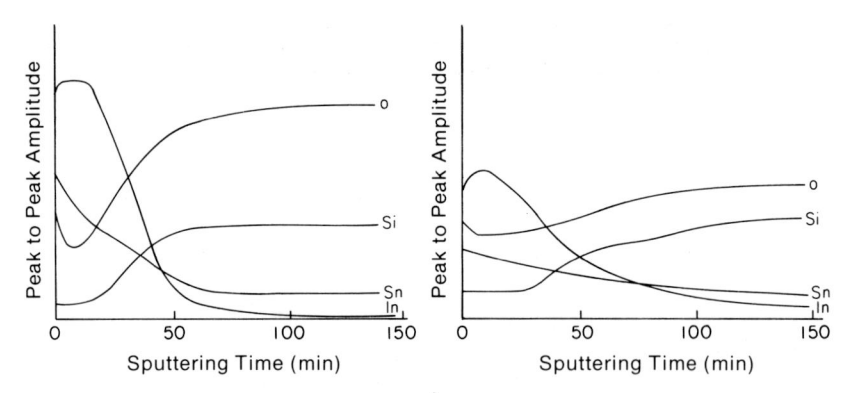

Figure 2a: Auger depth profiles for 700Å In - 5ω/o Sn metal films before and after 200 keV recoil implantation to dose of 5x10^{15}Ar$^+$/cm^2

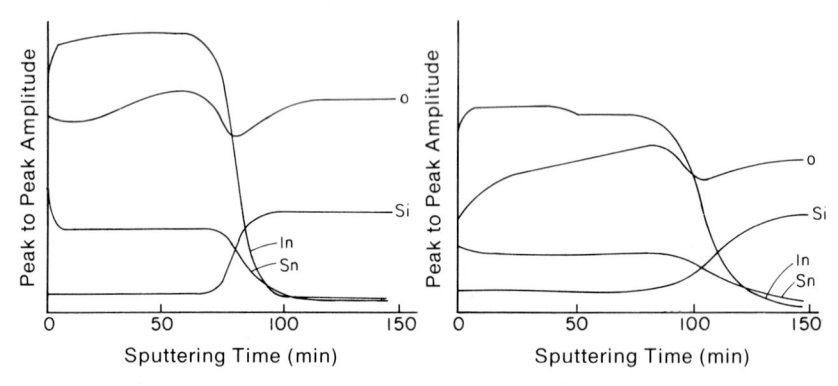

Figure 2b: Auger depth profiles for In - 5ω/o Sn oxide films before and after 200 keV recoil implantation to dose of 5x10^{15}Ar$^+$/cm^2

results in partial or complete reduction of oxides rendering the surface region of the films more metallic in nature. This change is particularly evident in optical properties as will be discussed. Although little research has been conducted on implantation of oxide films on glass substrates these results are consistent with those reported for inert gas bombardment of oxides [14]. Sn surface segregation is initially present in both metallic and oxide films. Ion bombardment is shown to partially eliminate such segregation and gives rise to a more uniform Sn distribution. This effect is probably a combined result of sample heating and enhanced diffusion due to defect migration.

Optical transmission spectra confirm the results described above. Metallic films show an increase in visible transmission confirming the formation of oxide during irradiation. The observed spectra show the majority of the film is still metallic and consequently does not give rise to acceptable heat mirror properties even after annealing. Oxide films show a decrease in visible transmission indicating surface reduction has occurred. Only a small amount of reduction occurs at these dose levels and the metallic region exists very near the surface. Thus, this effect may be offset by a final anneal. Figure 3 summarizes implantation effects on suitable ITO films relative to an unimplanted control film.

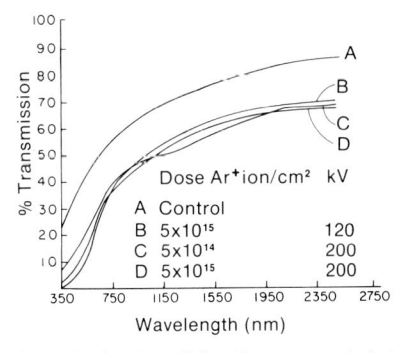

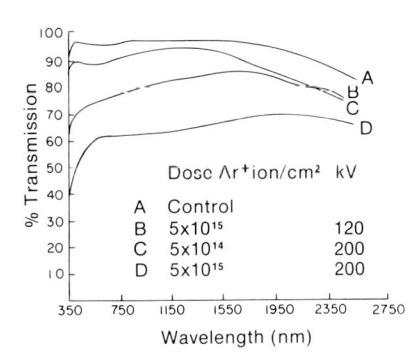

Figure 3a: In - 5 wt. % Sn alloy evaporated at 5x10⁻³ torr before annealing.

Figure 3b: In - 5 wt. % Sn alloy evaporated at 5x10⁻³ torr and annealed at 300° C for 1 hour in air.

Percent transmission is decreased uniformly and is relatively insensitive to implantation parameters. Annealing implanted films at 300°C for one hour in air develops acceptable heat mirror characteristics and this process is strongly affected by irradiation parameters. During annealing the ITO films undergo further oxidation. Metal initially present in the mixture or that formed by surface reduction may be converted to oxide. This gives rise to the observed increase in visible transmission and a corresponding reduced near-IR transmission. It might be expected that higher doses would give rise to more residual damage in the films and hence result in increased oxidation kinetics during the final anneal. However, damage levels are relatively low and this effect is overshadowed by surface reduction and localized melting phenomena. Therefore, it is observed that the most favorable properties are obtained at the lowest energy deposition rates where more complete oxidation is possible.

The need for low energy deposition rates during recoil bombardment of ITO films places constraints on processing conditions. The implantation energy is determined by the need to have the implanted ion range comparable to the film thickness in order to obtain sufficient interfacial mixing. It is therefore necessary to use low beam currents during implantation. Total doses achievable may thus be limited to values that are lower than those typically desired in recoil experiments. This compromise must be made in order to insure that the required optical characteristics can be obtained.

SUMMARY

Argon recoil bombardment of thin reactively evaporated In-5w/o Sn films on window glass is shown to be capable of producing coatings suitable for heat mirror applications. It is expected that these ITO films will exhibit increased lifetime stability under atmospheric exposure. The production of acceptable thin films is sensitive to irradiation parameters and is favored by low energy deposition rates. Future research should focus on evaluating actual film lifetimes and the possibility of implanting with oxygen to remove the need for final annealing.

REFERENCES

1. K. L. Chopra, S. Major, and D. K. Pandya, Thin Solid Films 102, 1 (1983).
2. J. L. Vossen, Physics of Thin Films, Vol. 9 (Academic Press, NY 1977).
3. J. C. C. Fan and J. B. Goodenough, Journal of Applied Physics 48, 3524 (1977).
4. B. B. Harbison, K. D. J. Christian, and S. R. Shatynski, Proceedings of NATO Advanced Study Institute on Surface Engineering (1983).
5. L. E. Collins, J. G. Perkins, and P. T. Stroud, Thin Solid Films 4, 21 (1969).
6. P. T. Stroud, Thin Solid Films 11, 21 (1972).
7. H. M. Naguib and R. Kelly, Radiation Effects 25, 1 (1975).
8. K. D. J. Christian and S. R. Shatynski, Applications of Surface Science 15, 178 (1983).
9. E. Ruedl and R. Kelly, Modern Developments in Powder Metallurgy 2, 145 (1966).
10. G. Dearnally, J. H. Freeman, R. S. Nelson and J. Stephan, Ion Implantation (North-Holland, Amsterdam 1973).
11. P. T. Stroud, L. E. Collins, J. G. Perkins and K. G. Stephens, Proceedings of European Conference on Ion Implantation (Peter Peregrins Limited, England 1970) 166.
12. B. B. Harbison, B. H. Rabin and S. R. Shatynski, Proceedings of NATO Advanced Study Institute on Surface Engineering (1983).
13. M. Mizuhashi, Thin Solid Films 70, 91 (1980).
14. K. G. Stephens and I. H. Wilson, Thin Solid Films 50, 325 (1978).

HIGH ENERGY ION BEAM MIXING IN Al_2O_3*

M. B. LEWIS AND C. J. McHARGUE
Metals and Ceramics Division
Oak Ridge National Laboratory
Oak Ridge, TN 37831 (USA)

ABSTRACT

The ion beam mixing technique has been employed to mix metal atoms into the surface layers of Al_2O_3. Ion beams of Fe^+ and Zr^+ in the 1 to 4 MeV energy range were used to irradiate Al_2O_3 specimens on the surfaces of which films of chromium or zirconium had been evaporated. Some specimens were irradiated at elevated temperatures of 873 or 1173 K. Rutherford backscattering (RBS) and channeling methods were used to measure the metal atom depth profiles near the surface. Analyses of the backscattering data included binary collision calculations using the codes TRIM and MARLOWE. The significance and limitations of high energy ($\geq$1 MeV) beams for ion beam mixing experiments is discussed. Evidence was found for radiation enhanced diffusion and/or solubility of zirconium and chromium in Al_2O_3 at 873 K.

INTRODUCTION

In most ion beam mixing (IBM) experiments, the primary ions are produced by ion implantation accelerators with an energy of about 200 keV [1]. These ions are usually much different atomically (e.g., inert gas ions) than the secondaries or target atoms. Although the mean range of the primaries is usually beyond the interfacial planes of interest, multiple scattering of the primary ions can lead to a fraction of them being implanted at short range near those planes thereby affecting the outcome of the IBM experiment. For example, in the case of 200 keV Ar^+ on Al_2O_3, multiple scattering calculations, discussed below, give a near-surface ($\leq$ 200 Å) argon density that is about 4% of the peak density at 1000 Å. However, it is known that gaseous impurities of only about 100 parts per million in metals can have a dramatic effect on the nucleation of radiation-induced cavities [2] as well as other thermal and mechanical properties. These facts have caused us to investigate both the use of more energetic ($\geq$1 MeV) primary ions and the use of non-gaseous or metallic ions (Fe^+, Zr^+).

This work is a continuation of measurements of the changes in properties of Al_2O_3 due to ion beam irradiation [3-6]. Previous measurements [3] of 1 MeV Fe^{+3} ion beam mixed chromium and zirconium films on Al_2O_3 showed that most of the film mixed atoms occupied substitutional sites and that surface disorder in Al_2O_3 appeared minimal even for a beam fluence of 2 × 10^{17} Fe^+/cm^2. This is believed to be the result of (a) minimal deposition of the ion beam atoms near the region of metal-substrate interface and (b) the self annealing ionization accompanying the high energy primary ion beam.

In this work we have extended the results reported in ref. 3 by continuing to investigate ion beam mixing of metallic films evaporated on

---

*Research sponsored by the Division of Materials Sciences, U.S. Department of Energy, under contract W-7405-eng-26 with Union Carbide Corporation.

Al$_2$O$_3$ crystals but with (a) primary ion beams up to 4 MeV (b) zirconium (self) ion irradiation (c) elevated target temperatures up to 1173 K. Of particular interest has been (a) the chemical stability and (b) the mechanical changes such as hardness of the new ion beam mixed, metal enriched, surface.

MATERIALS AND METHODS

High purity, low-dislocation density crystals of Al$_2$O$_3$ were polished and then annealed for five days at 1500 K in air to remove any mechanical damage introduced during sample preparation. Chromium or zirconium films of thickness 160 Å were evaporated on the (0001) surface of the crystals. Irradiations were carried out with 1 or 4 MeV Fe$^+$ ions or 4 MeV Zr$^+$ ions in a cryogenically pumped target chamber maintained at a pressure of 10^{-8} torr. The temperature of the target surface was measured by securing a chromel-alumel thermocouple between the target surface and a face mask in front of the surface. Due to the low conductivity of Al$_2$O$_3$, the actual surface temperature may be higher than that indicated by the thermocouple. Typical beam heating of the sample raised the thermocouple reading to 573 K from ambient temperature (300 K). Some samples were also heated from the reverse side during irradiation so that the thermocouple reading was 873 or 1173 K; in these cases the beam heating was only about 50 K and the uncertainty in the actual target surface temperature should not have been greater than about 20 K.

Following the irradiations, the remaining chromium film was removed (etched) from the surface by placing the sample in concentrated HCl containing zinc pellets for about two minutes. The zirconium was removed by dipping into a solution of 25 parts HNO$_3$ + 1 part HF for two minutes.

Rutherford backscattering and channeling (0001 axial) techniques were used to characterize the atomic migration and radiation damage. The He$^+$ beam energy was 3.0, 2.0, or 0.5 MeV. Other experimental details can be found in ref. 3. The depth profiles were calculated from the backscattering spectra by direct deconvolution [7]. The data were normalized by the aluminum component in Al$_2$O$_3$ substrate backscattering.

RESULTS AND DISCUSSION

Concentration-Depth Profiles

Rutherford backscattering spectra from Al$_2$O$_3$ following a room temperature, 1 MeV Fe implantation of 2×10^{17} Fe/cm^2 are shown in Fig. 1. The aluminum edge, oxygen edge, and iron peak are labeled. The iron edge is labeled "surface (Fe)" near channel 460. Comparing the random and aligned spectra shows that oxygen, aluminum, and iron all exhibit low radiation damage or disorder near the substrate surface (i.e., near the respective edges in the figure).

Figure 2 shows a concentration versus depth profile in Al$_2$O$_3$ for a specimen irradiated with a 1 MeV iron beam to a fluence of 2×10^{17} Fe/cm^2 for the case in which a 160 Å chromium film had been evaporated onto the surface. The ion mixed chromium distribution and the iron implantation profile overlap near 0.08 μm with minimum Fe + Cr metal concentration of about 2000 appm of which about 500 appm is due to Fe$^+$ (based on data without the Cr film).

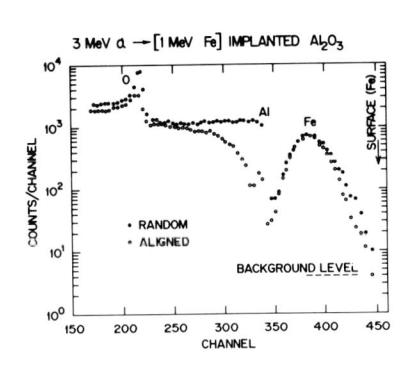

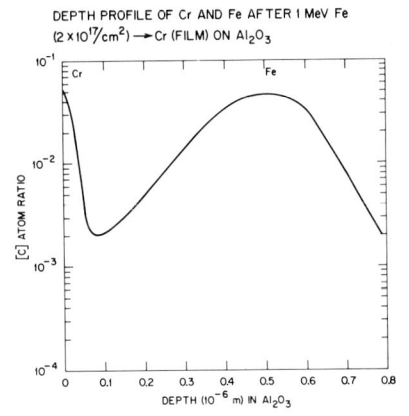

FIG. 1. Pulse height spectra for 1 MeV Fe⁺ ions implanted in Al₂O₃. Aligned (0001) crystal orientation is compared to the random case. Data were taken by backscattering 3 MeV alpha particles. The peak near channel 220 is due to the $O(\alpha,\alpha)$, 3 MeV resonance.

FIG. 2. Concentration versus depth profile of Cr and Fe in Al₂O₃ (after etching) following a 2×10^{17} Fe/cm² irradiation of a 160 Å Cr film on an Al₂O₃ substrate. The surface peak is due to ion beam mixing of the original Cr film while the deeper and broader peak is due to the primary Fe⁺ ion beam.

A similar concentration versus depth profile of zirconium in Al₂O₃ is shown in Fig. 3. In this case, 4 MeV zirconium ions were used to irradiate a 160 Å film of zirconium evaporated on Al₂O₃. The physical separation between the ion mixed film and the deposition of the primary ions is larger than for the previous example and the overlap density is reduced to about 100 appm for the room temperature irradiation (broken curve in Fig. 3).

The shape of the primary ion profile is important for estimating the "contamination" of the ion mixed region by the primary ion beam. In both examples of Fig. 2 and Fig. 3, the primary atom profile exhibits a pronounced asymmetry which we have parameterized by calculating the skewness of the deposition. Table I shows the statistical analysis of primary ions in Al₂O₃ from data and calculations in this work.

Three types of calculations were carried out to compare with the measured values. The first is based on the code E-DEP-1 [8] which assumes a Gaussian distribution for the range profile; skewness and Kurtosis of 0 and 3, respectively, are implied. The model predicts the range and standard deviation slightly lower than measured. The calculation using the codes MARLOWE [9] and TRIM [10] are based on a large series of binary collisions of both primary and secondary ions. While these types of calculations take extensive computer time and have limited statistical accuracy, they calculate multiple scattering explicitly. We find that the skewness exhibited in the measured values can be accounted for by multiple scattering so that it is not necessary to invoke diffusion processes to explain the asymmetry in the data.

TABLE I. Statistical analysis of primary ion implantation profiles. Units are MeV (energy) and micron (range).

Primary ion	Ion energy	Projected range	Standard deviation	Skewness	Kurtosis
Fe (measured)	1.0	0.44	0.107	−0.36	2.7
Fe (E-DEP-1)	1.0	0.40	0.080	0[a]	3.0[a]
Fe (MARLOWE)	1.0	0.45	0.078	−0.46	3.2
Fe (TRIM)	1.0	0.43	0.077	−0.56	3.2
Zr (measured)	4.0	1.17	0.14	−0.18	2.3
Zr (E-DEP-1)	4.0	1.05	0.14	0[a]	3.0[a]
Zr (MARLOWE)	4.0	1.06	0.15	−1.1	4.5
Zr (TRIM)	4.0	1.03	0.14	−0.78	3.6

[a] Calculations based on Gaussian distribution in which the skewness is 0.0 and the Kurtosis is 3.0.

Ion Mixing and Thermal Effects

One measure of the degree of mixing between the surface film and substrate would be the resistance of the metal component to removal by the usual techniques of chemical etching. Table II shows the equivalent film thicknesses before and after etching removal for various primary beam and temperature conditions explored in this work. The thickness measurement is based on the area under the surface peaks such as in Fig. 2 (chromium) and Fig. 3 (zirconium). Thus the measurements reflect short range processes such as cascade mixing, but not long range ballistic recoils which are much less frequent and cannot be distinguished experimentally from the implantation peak.

TABLE II. Summary of metal film or metal enriched surface layer thicknesses on Al_2O_3. Some measurements were made before and after chemical etching. Original films consisted of a 160 Å evaporated metal layer. Final films are listed as fractions of the original.

Film temperature (K)	Primary ion fluence	Ion energy (MeV)	Before etch %	After etch %
Cr (573)	1×10^{17} Fe/cm^2	4.0	57[a]	33
Cr (873)	Thermal condition only		100	16
Cr (873)	1×10^{17} Fe/cm^2	4.0	53[a]	51
Cr (1173)	Thermal condition only		30[b]	9
Cr (1173)	2×10^{17} Fe/cm^2	1.0	13[a,b]	<1[c]
Zr (573)	1×10^{17} Zr/cm^2	4.0	34[a]	2
Zr (873)	Thermal condition only		100	2
Zr (873)	1×10^{17} Zr/cm^2	4.0	32[a]	30

[a] Values less than 100 due to reflective sputtering.
[b] Unusually low value believed to be due to vaporization.
[c] No peak observed; evidence for dissolution at surface.

In contrast to long range recoils, the cascade mixing is expected to give rise to a diffusion like movement of the metal film into the substrate with a temperature independent diffusion constant of approximately

$$D \approx \frac{1}{6} \lambda^2 P \qquad (1)$$

where the jump distance λ is taken as 20 Å and P is given by the E-DEP-1 code [8]. Reflection sputtering of the metal surface film by the primary ions is also significant and we have estimated its yield using the binary collision code MARLOWE [9].

An example of calculated results is given in Fig. 4 for a case of an Ar^+ beam on a chromium film and show the relative importance of these processes at different energies. In the case of the film thickness measurements before etching (see Table II), the reduced film thickness can be explained as reflective sputtering (ordinary Fickian diffusion is expected to be negligible at our experimental temperatures). The interface mixing length, $(Dt)^{1/2}$, where D is given by Eq. (1) and t is the irradiation time, is seen to be less than 100 Å, i.e., near the resolution limit in the RBS method; this suggests that chemical stability tests would be more sensitive to the degree of mixing than RBS profile width measurements. The low values of $(Dt)^{1/2}$ may become a limitation to the high energy ion beam mixing technique.

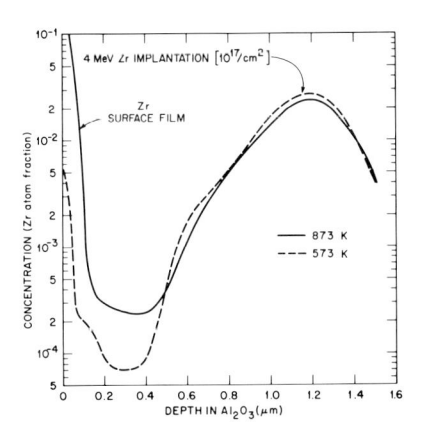

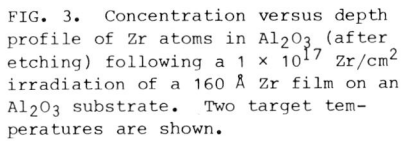

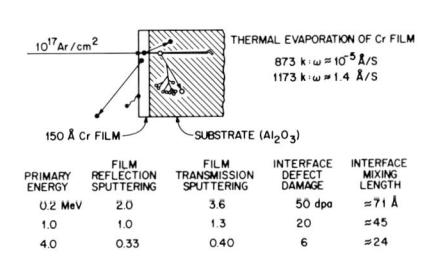

FIG. 3. Concentration versus depth profile of Zr atoms in Al_2O_3 (after etching) following a 1×10^{17} Zr/cm^2 irradiation of a 160 Å Zr film on an Al_2O_3 substrate. Two target temperatures are shown.

FIG. 4. Properties of ion beam mixing for three energy regions. Long range recoils have been ignored. The sputtering coefficients were calculated as explained in the text. The interface mixing is based on the defect damage and Eq. 1 in the text.

The values of the film thicknesses after etching are important because they reflect the chemically stable fraction of film which is mixed with the Al_2O_3 substrate. However, these results of etching can only be considered tentative since the time dependence of the etching of the irradiated films has not yet been measured. In the case of the 873 K irradiation, both for chromium and zirconium, essentially all of the residual film was immuned to chemical etching. In particular, for the zirconium case neither heat nor ion irradiation itself had a significant effect on the zirconium film resistance to etching. However, the combination of irradiation and elevated temperature completely stabilized the film (see also Fig. 3). One explanation for such film stabilization is the possibility that radiation enhanced diffusion, a synergistic effect of temperature and defect production, mixes the unsputtered film completely with the Al_2O_3 surface at 873 K.

However, the data for the Cr film at 1173 K requires a different explanation; in this case, where a more complete mixing is expected, we find that the chemical resistance of the film suddenly decreases as if a phase change has taken place. Again, this property was seen only for the case of simultaneous heat and ion irradiation.

SUMMARY

We have irradiated chromium and zirconium films evaporated on Al_2O_3 substrates with 1 and 4 MeV Fe^+ or Zr^+ ions to fluences up to 2×10^{17} ions/cm^2, for target temperatures up to 1173 K. The concentration profiles of both the primary and secondary ions were measured by Rutherford backscattering. The primary ion profile exhibits an asymmetry (negative skewness) which is believed to be due to multiple scattering. The secondary or film atoms were found to be mixed with the substrate such that at 873 K the films were completely resistant to chemical etching. However, at 1173 K the chemical properties of the chromium film were radically different exhibiting almost no resistance to the etching process. These properties could be a manifestation of radiation enhanced diffusion and/or solubility.

REFERENCES

1. S. Matteson and M.A. Nicolet in: Metastable Materials Formation by Ion Implantation, S.T. Picraux and W.J. Choyke, eds. (Elsevier, Amsterdam 1982) pp. 3—16.
2. K. Farrell, Radiation Effects 53, 175 (1980).
3. M.B. Lewis and C.J. McHargue in: Metastable Materials Formation by Ion Implantation, S.T. Picraux and W.J. Choyke, eds. (Elsevier, Amsterdam 1982) pp. 85—91.
4. C.J. McHargue, M.B. Lewis, B.R. Appleton, H. Naramoto, C.W. White, and J.M. Williams, International Conference on the Science of Hard Materials, Jackson, WY, 1983, J. D. Rowcliffe, R. K. Viswanadham, and J. Gurlon, eds. (Plenum Publishing Co., New York) pp. 451—465.
5. C.J. McHargue, H. Naramoto, B.R. Appleton, C.W. White, and J.M. Williams in: Metastable Materials Formation by Ion Implantation, S.T. Picraux and W.J. Choyke, eds. (Elsevier, Amsterdam 1982) pp. 147—153.
6. H. Naramoto, C.W. White, J.M. Williams, C.J. McHargue, O.W. Holland, M.M. Abraham, and B.R. Appleton, J. Appl. Phys. 54, 683 (1983).
7. M.B. Lewis, Nucl. Inst. Meth. 190, 605 (1981).
8. I. Manning and G.P. Mueller, Comp. Phys. Comm. 7, 85 (1974).
9. M.T. Robinson and I.M. Torrens, Phys. Rev. B9, 5008 (1974).
10. J.B. Biersack and L.G. Haggmark, Nucl. Inst. Meth. 174, 257 (1980).

AUTHOR INDEX

Ahmed, M., 117
Alberts, H.W., 335
Allara, D., 439
Alves, E.J., 187
Amirtharaj, P.M., 305
Angelini, P., 407
Anjum, M., 317
Appleton, B.R., 73,195,385,395,559, 747
Arnold, G.W., 61
Arrowsmith, R.P., 281
Aspnes, D.E., 305
Au, J.J., 679
Averback, R.S., 25

Bacon, R.K., 735
Badwal, S.P.S, 565
Baglin, J.E.E., 55
Baldo, P., 169
Banwell, T.C., 3,109
Battaglin, G., 49
Bauer, C.L., 139
Beardsley, G.M., 735
Benson, Jr., R.B., 151
Bentley, J., 151
Bernett, M.K., 759
Blanchard, B., 175
Boerma, D.O., 259
Bolster, R.N., 637
Bond, P.D., 565
Braunstein, G., 423,461,467,475
Brimhall, J.L., 163,729
Bruel, M., 223
Buchanan, R.A., 735
Budnick, J.I., 341,347
Burkova, R., 323
Burnett, P.J., 401
Butcher, J.B., 281
Byrne, P.F., 253

Cairns, J.A., 719
Calcagno, L., 549
Calliari, L., 85
Campana, J.E., 429
Campisano, S.U., 97
Cannavó, S., 97
Carlson, J.D., 455
Carnera, A., 49
Celotti, G., 49
Chen, L.J., 353
Cheung, N.W., 253
Chevallard, G., 711
Chieu, T.C., 487
Chivers, D.J., 205,247
Choyke, W.J., 359
Clark, G.J., 55,577
Colligen, J.S., 513

Colmenares, C.A., 747
Corazzi, R.J., 241
Cuomo, J.J., 519

d'Heurle, F.M., 55
da Silva, M.F., 187
Delafond, J., 67,133
Della Mea, G., 49
Devenyi, J., 175
Dietrich, H.B., 241
Dillich, S.A., 637
Donovan, E.P., 211
Doyle, N.J., 359,597
Dresshaus, G., 461,467,475,487,493
Dresselhaus, M.S., 413,423,461,467, 475,481,493,553
Dupuy, M., 175
Dynes, R.C., 449

Elliman, R.G., 205,229,247
Elman, B.S., 461,467,475,481,487, 493,553
Endo, M., 487

Farkas, D., 609
Farlow, G.C., 55,385,395
Fasihuddin, A., 347
Fathy, D., 287
Feng, M., 365
Follstaedt, D.M., 655,661
Foti, G., 439,549
Fredrickson, J.E., 217
Freitag, K., 187

Gamo, K., 531
Gerard, P. 175
Gibson, J.M., 449,461
Gill, S.S., 275,317
Glaccum, A.C., 281
Gossett, C.R., 615,741
Grabowski, K.S., 615,741
Gratton, L.M., 85
Gressett, J.D., 287
Grilhe, J., 67,133
Guzman, A.M., 85,139

Haar, T.J., 377
Haberland, D., 371
Hall, B.O., 359,597
Hamdi, A.H., 287
Hamm, R., 449
Harbison, B.B., 765
Harde, P., 371
Harper, J.M.E., 519
Hartley, N.E.W., 615
Hayden, H.C., 341,347
Hemment, P.L.F., 281

Hentzell, H.T.G., 519
Hewett, C.A., 145
Hill, A.E., 513
Hirvonen, J.K., 151,621
Hoff, H.A., 79
Holland, O.W., 73,235,293
Hu, C., 253
Hubler, G.K., 217,603
Hung, L.S., 145
Hutchings, R., 603,705

Ingram, D.C., 455,559
Irwin, R.B., 359
Iverson, R.B., 543

Jacobson, D.C., 211
Janoff, D., 157
Jaouen, C., 67,133
Jata, K.V., 157
Jaussaud, C., 223
Jayram, R., 597
Jeffries, R.A., 667, 673
Jenkinson, H.A., 377
Johnson, S.T., 205,229
Johnson, W.L., 127

Kanber, H., 365
Kant, R.A., 525,649
Kaufmann, E.N., 747
Keenan, J.A., 43
Kelner, G., 329
Kennedy, T.A., 217
Kenny, M.J., 691
Kheyrandish, H., 513
Kilner, J.A., 281
King, B.V., 103
Kissinger, H.E., 163
Knapp, J.A., 661
Koon, N.C., 429,445
Krolikowski, W.F., 287
Kryder, M.H., 139
Kulkarni, V.N., 49
Kumar, K., 649
Kunoff, E.M., 553
Kustas, F.M. 685

Lam, N.Q., 37,79
Lamond, S., 117
Larson, D.C., 377
Lau, S.S., 145
Lawson, E.M., 205
Leavitt, J.A., 103
Lewis, M.B., 407,771
Liu, J., 247
Lo Russo, S., 49
Loh, I.-H., 435
Love, R.P., 537

Madakson, P.B., 643
Maillot, B., 223

Manning, I., 91,615
Marsh, J.H., 317
Martin, P., 175
Martinella, R., 711
Matsui, T., 531
Matteson, S., 43
Maydell-Ondrusz, E.A., 281
Mazzoldi, P., 49
McCormick, A.W., 559
McDaniel, F.D., 287
McGruer, J.N., 359
McHargue, C.J., 385,395,407,771
McMarr, P.J., 299
McNeil, L.E., 493
Melo, A.A., 187
Miller, M., 597
Misra, M.S., 685
Mitchell, I.V., 205
Mogro-Campero, A., 537
Molnar, B., 329
Moriizumi, K., 531
Morrison, G.H., 329
Musket, R.G., 747,753

Namavar, F., 341,347
Namba, S., 531
Narayan, J., 55,235,293,299
Nelkowski, H., 371
Newborn, H., 649
Nicolet, M.-A., 3,31,109,127
Nieh, C.W., 353

Oblas, D.W., 631
Ochmann, F., 571
Oliver, W.C., 603,705
Orrman-Rossiter, K.G., 205
O'Tooni, M., 377
Otter, F.A., 341,347

Page, T.F., 401
Paine, B.M., 3,31
Paradis, E.L., 705
Parikh, N.R., 323
Parrish, P.A., 151
Patarini,V., 341,347
Patterson, P.J.K., 691
Patterson, R.G., 753
Paulson, W.M., 287
Pease, D.A., 347
Pehrsson, P.E., 429,445
Pelton, A.R., 163
Pennycook, S.J., 293
Pethica, J.B., 603,705
Picraux, S.T., 661
Pinizzotto, R.F., 43,265
Poate, J.M., 211
Pogany, A.P., 205,247
Poker, D.B., 73,559
Pollak, F.H., 305
Pollock, J.T.A., 691

Pope, L.E., 655,661
Potter, D.I., 117
Principi, C., 85
Pronko, P.P. 455,559

Rabin, B.H., 765
Raghunathan, V.S., 323
Rai, A.K., 559
Raider, S.I., 577
Ramseyer, G.O., 329
Rangaswamy, M., 609
Ravner, H., 759
Rehn, L.E., 25,37,79
Reif, R., 543
Rimini, E., 97
Riviere, J.P., 67,133
Robinson-Weis, G., 753
Rose, S.L., 429

Sadana, D.K., 311
Salamanca-Riba, L., 481,487
Sands, T., 311
Sartwell, B.D., 525
Schindler, A.I., 445
Schlaak, W., 371
Scott, D.M., 145
Sealy, B.J., 317
Seshan, K., 169
Shahid, M.A., 317
Shatas, S.C., 329
Shatynski, S.R., 765
Sheng, K.L., 549
Shih, Y.C., 253
Short, K.T., 205,229,247
Shreter, U., 31,127
Shuskus, A.J., 705
Singer, I.L., 585,603,609,637,667, 673
Sioshansi, P., 679,685
Sklad, P.S., 407
Smulders, P.J.M., 259
So, F.C.T., 31
Soares, J.C., 187
Sood, D.K., 565
Spaepen, F., 211
Spitzer, W.G., 217
Spitznagel, J.A., 359,597
Starke, Jr., E.A., 157
Stephens, K.G., 281
Stephenson, L.D., 151
Strathman, M., 253
Stritzker, B., 571
Suni, I., 145
Suran, G., 175

Takagi, T., 501
Tam, S., 253
Taylor, G.N., 439
Techang, C., 97
Templier, C., 67

Thompson, D.A., 323
Thompson, L.J., 25
Tiong, K.K., 305
Tonn, D.G., 103
Tosello, C. 85,711
Townsend, J.R., 359,597
Tseng, W.F., 241
Tsong, I.S.T., 103
Turnbull, D., 211

Van Rossum, M., 127
Vardiman, R.G., 699
Vedam, K., 299
Venkatesan, T., 439,449,461,467,475, 481,493
Vianden, R., 187
Vozzo, F.R., 181

Waddell, C.N., 217
Wagenfeld, H.K., 247
Wallace, R.W., 597
Wang, R., 729
Washburn, J., 253,311
Wasserman, B., 413,423,435
Weber, D.C., 429,445,759
Whelan, J.M., 365
White, A.E., 449
White, C.W., 55,385,395
Wiedersich, H., 13,37,79,169
Wierenga, T.S., 259
Wilkens, B., 439,449,461
Williams, J.M., 385,735
Williams, J.S., 205,229,247
Wilson, I.H., 275
Wilson, S.R., 287
Wnek, G.E., 413,423,435
Wolf, T., 439
Wood, S., 359
Woollam, J.A., 559

Yoshiie, T., 139
Yost, F.G., 655,661

Zavada, J.M. 377

SUBJECT INDEX

Adhesion, 55,559,565
Alloy production, 235
Alloys
  Ag-Au implanted with Sb,Ar,As, 181
  Co alloy implanted with Ti,N, 637
  Cu-Au, 13
  Cu-Ni, 13,37,79
  mild steel implanted with N, 691
  M2 steel implanted with Ti and N, 667
  M50 steel implanted with Cr,Ta, 615
  Ni-Si, 13
  NiTi implanted with Ni, 163,729
  Nitronic 60 steel implanted with Ti and C, 655
  Ti-6Al-4V implanted with C, 699,705
  Ti-6Al-4V implanted with N, 699,705,711,735
  WC implanted with N, 631
  1018 steel implanted with Ti and N, 667
  15-5 PH steel implanted with Ti and C, 655
  304 SS implanted with N, 603
  304 SS implanted with Ti and C, 655
  304 SS implanted with Ti and N, 667
  316 SS implanted with He,N, 597
  440 C steel implanted with N,Ti, 685
  440 C steel implanted with Ti and C, 655,661
  52100 steel implanted with Ar,Ta, 615
  52100 steel implanted with Ti, 603,609,615,673,679
Amorphous phase
  buried regions, 235,253,299,311
  in Al_2O_3, 385
  in $CoSi_2$, 145
  in $CrSi_2$, 145
  in garnet, 139
  in MgO, 401
  in Nb_3Si, 55
  in NiHf, 127
  in NiTi, 163,729
  in $PdSi_2$, 145
  in soft magnetic material, 175
  in steels, 655
  melting point in Si, 211
  production, 235,247
  states in Si, 217
Antistatic materials, 759
Auger electron spectroscopy, 37,85,117,341,347,371,429,
  525,549,637,649,685,691,765

Benzene, 549
Boltzman transport, 91
Buried oxide layers, 265,275,281

Carbon films, 449
Cascade effects, 181
Catalysis, 719

Ceramics
  Al_2O_3, 385,395,565
  MgO, 401
  SiC, 385,439
  TiB_2, 407
  yttria stabilized zirconia, 565
Cermet, 519
Chemical driving forces, 3
Complex impedance spectroscopy, 565
Compositional changes, 13,37,61,79,341,347
Corrosive-wear, 735

Damage and damage annealing
  in Al_2O_3, 385,395
  in CdS, 323
  in GaAs, 305,365,377
  in graphite, 461,467,475,487,493
  in InP, 329
  in InSb, 335
  in MgO, 401
  in Ni, 117,169
  in NiTi, 163
  in Si, 287
  in TiB_2, 407
Defect migration, 3
Defect production, 3,13
Differential scanning calorimetry, 211
Diffusion, 127,293,609
Dual ion-beam deposition, 519
Dynamic recoil mixing, 513

Epitaxial growth
  in Ge, 205,211
  in Si, 211,235,247,293,353
  in graphite, 467
  in silicides, 145
  influence of doping, 229
Elastic recoil detection analysis, 455
Electron paramagnetic resonance, 217,413,435
Electrical measurements
  capacitance-voltage, 287,377
  carrier lifetime, 537
  hall mobility, 317
  resistance, 67,133,241,287,413,423,429,449,487
Electrochemical behavior, 565,729,735
Ellipsometry, 299,559

Fast atom bombardment mass spectrometry, 429
Ferromagnetic resonance, 175
Field ion microscopy, 597
Focussed ion beam, 531
Frozen organic molecules, 549

Garnet, 139
Gibbsian adsorption, 13
Grain size modification, 543
Graphite, 429,445,461,467,475,493
Graphite fibers, 487
Graphite intercalation compounds, 481

Heat mirrors, 765
High energy implantation, 163,253,359,559,729,771
Hydriding of uranium, 753

Industrial applications, 621,673
Infrared reflectance, 217,377
Infrared spectrometry, 275,439
Internal oxidation, 169
Ion-beam enhanced deposition, 501,513,519,525
Ion-beam induced annealing, 229,235,241,247
Ion-beam synthesis, 439,513,519,549
Ion channeling analysis, 187,385,395,461,467,475,493,543,771
Ion irradiation smoothing, 559
Ion mixing
  anisotropic transport, 43
  As beams, 43
  ballistic mixing, 37
  bilayer systems, 3,25,97
  cascade mixing and effects, 3,13,25,31,91,103,109,139,163
  dose dependence, 3,73,97
  high temperature regime, 37
  in industrial applications, 621
  in ion-beam assisted deposition, 525,621
  ion-mass dependence, 25,73
  kinetics, 97,133
  low temperature regime, 3
  models, 3,13,31,43,91,97,133,609
  multilayer systems, 67,133
  phase formation, 49,55,73,85,97,127,145,157
  radiation enhanced diffusion, 13,37,49,61,79,97,103,609,771
  radiation induced segregation, 13,37,181
  recoil effects, 3,25,109
  systems
    Ag/Mo, 559
    Ag/Si, 103
    Al/Ag, 67
    Al/Si, 103
    Au/Al, 97
    Au/Fe, 49
    Au/Ge, 73
    Au/Mo, 559
    Au/Si, 3,73
    Au/SiO$_2$, 3
    Cr/Al$_2$O$_3$, 771
    Cr/Si, 31
    Cu/Al, 97
    Cu/Mo, 559
    Cu/SiO$_2$, 55
    Fe/Al, 67,133
    FeNi/Si, 175
    Ge/Sn/Si, 43
    Ge/Sn/Si, 43
    Mo/Si, 103
    Nb/Si, 31
    Nb/SiO$_2$, 55
    Ni/Si, 31
    Ni/SiO$_2$, 3,109
    NiHf, 127
    Pd/SiO$_2$, 55
    Pt/Al$_2$O$_3$, 565

        Pt/Fe, 49
        Pt/Si, 3,25
        Pt/SiO$_2$, 3
        Pt/YSZ, 565
        Si/Sn/Ge, 43
        Si/Sn/Si, 43
        Sn/Ni, 85
        Ti/Si, 103
        V/SiO$_2$, 55
        W/Si, 3
        W/SiO$_2$, 3
        Zr/Al$_2$O$_3$, 771
     thin markers, 3,43,103
     temperature dependence, 3,25,31,109

Ion implantation
  Al implanted with:
     Mo, 151
     Ni, 117
     Pb, 643
     Si, 643
  alkali silicate glasses implanted with Ne,Xe, 61
  Al$_2$O$_3$ implanted with various ions, 385,395
  Be implanted with B, 649
  Be implanted with H,D, 571
  Benzene implanted with H,Ar, 549
  Bi implanted with As,Bi,Sb, 553
  carbon films implanted with various ions, 449
  Cd implanted with H,D, 571
  Cu implanted with Ta, 525
  Fe implanted with Al, 157
  GaAs implanted with:
     Ar, 205,365
     As, 305,365
     P, 377
     Pb, 643
     Se, 205,311
     Sb, 205
     Te, 205
  GaInAs implanted with Se, 317
  Ge implanted with Ar, 211
  Ge implanted with Sb, 205
  graphite fibers implanted with various ions, 487
  graphite implanted with various ions, 429,445,461,467,475,493
  Hf implanted with Be, 187
  indium-tin oxide implanted with Ar, 765
  In implanted with H,D, 571
  InP implanted with Be, 329
  InP implanted with Sn, 371
  InSb implanted with Mg, 335
  MgO implanted with Ti,Cr, 401
  Mo implanted with Mo, 559
  Nb/NbO/PbAuIn implanted with B, 577
  Ni implanted with Li, 169
  Pb implanted with H,D, 571
  polymers implanted with various ions, 413,423,429,435,439,445,449,455,759
  Sb implanted with Cr, 531
  Si implanted with:
     Al, 347
     Ar, 211,247,537

        As,   205,241,253,353
        B,   241,253,353
        BF_2, 241,353
        C,   217
        Cl,   259
        Cr,   341
        H,   241,287,359,537
        He,   229,247
        I,   259
        In,   235
        O,   265,275,281
        P,   217,241,353,543
        Sb,   205,235,247,293
        Si,   205,217,299
        Sn,   217
        Xe,   211
    Sn implanted with H,D, 571
    Ta implanted with Cr,   741
    TiB_2 implanted with Ni, 407
    Tl implanted with H,D, 571
    U implanted with Mo, 747
    U implanted with O_2, 753
    Zn implanted with H,D, 571
Implantation at elevated temperatures,   385,475,493,711
Ionized cluster-beam technique, 501

Josephson tunnel junctions, 577

Laser mirror, modification of, 559
Lateral transport during ion irradiation, 55
Lattice site location, 187,259
Low energy ion scattering spectrometry, 79
Low temperature crystal growth, 501

Magnetic properties, 429,445
Magnetorefection, 493,553
Magnetic alloys, 175
Maskless patterning using ion beams, 531
Mechanical properties
  fracture, 401
  friction, 513,585,637,643,649,655,661,667,699
  microhardness, 385,395,401,513,585,603,643,649,705
  plasticity, 401
  rolling contact fatigue, 685
  stress-strain response, 597
  wear, 385,401,513,637,643,649,655,661,667,679,691,699,705,711,735
  wear mechanisms, 585,603
  work hardening, 585
Mossbauer spectroscopy, 85

Nuclear reaction analysis, 439

Optical properties, 413,559,765
Oxidation, 643,741,747

Phase formation, 117,151,157,187,699
Plasma etching, 531
Polycrystalline silicon, 543
Polymer materials, 413,423,429,435,439,445,449,455,501,549,759
Preferential sputtering, 13,181,371
Pyrolysis, 449

Raman spectroscopy, 305,385,461,475,493
Reactive ionized cluster beam technique, 501
Reactive ion mixing, 513
Recoil implantation, 765
Reflection high energy electron diffraction, 175
Retention of implanted ions, 615,673

Scanning electron microscopy, 55,85,223
Secondary ion mass spectroscopy, 103,175,371,753
Solid state reaction, 127,151
Sputtering, 13,181,371,525,609,615
Stoichiometric determination, 481
Structural modifications
  in semiconductors, 195,205,543
  layered samples, 223
Superconductivity, 571
Surface mobility, 501,525,559
Surgical alloys, 735

Tarnishing resistance, 85
Thermoelectric power, 413,423,435
Thermal desorption mass spectrometry, 631
Thermal spike effects, 31
Time differential perturbed angular correlation method, 187
Transmission electron microscopy, 55,117,139,151,157,163,169,235,253,287, 293,299,311,317,323,377,385,407,449,461,487,543,699,711
Tungsten carbide tools, 621,631

X-ray diffraction, 49,55,85,481,519,747
X-ray emission, 459
X-ray photoelectron spectroscopy, 429